Lecture Notes in Computer Science   7237

Commenced Publication in 1973
Founding and Former Series Editors:
Gerhard Goos, Juris Hartmanis, and Jan van Leeuwen

David Pointcheval   Thomas Johansson (Eds.)

Advances in Cryptology – EUROCRYPT 2012

31st Annual International Conference
on the Theory and Applications of Cryptographic Techniques
Cambridge, UK, April 15-19, 2012
Proceedings

 Springer

Volume Editors

David Pointcheval
École Normale Supérieure
45 rue d'Ulm, 75005 Paris, France
E-mail: david.pointcheval@ens.fr

Thomas Johansson
Lund University
Department of Electrical and Information Technology
P.O. Box 118, 221 00, Lund, Sweden
E-mail: thomas.johansson@eit.lth.se

ISSN 0302-9743   e-ISSN 1611-3349
ISBN 978-3-642-29010-7   ISBN 978-3-642-29011-4 (eBook)
DOI 10.1007/978-3-642-29011-4
Springer Heidelberg Dordrecht London New York

Library of Congress Control Number: 2012933758

CR Subject Classification (1998): E.3, F.2.1-2, G.2.1, D.4.6, K.6.5, C.2, J.1

LNCS Sublibrary: SL 4 – Security and Cryptology

Typesetting: Camera-ready by author, data conversion by Scientific Publishing Services, Chennai, India

Printed on acid-free paper

Springer is part of Springer Science+Business Media (www.springer.com)

Preface

These are the proceedings of Eurocrypt 2012, the 31st Annual IACR Eurocrypt Conference. The conference, sponsored by the International Association for Cryptologic Research, was held April 15–19, 2012, in Cambridge, UK, within the celebrations of Alan Turing Year. The General Chair was Nigel Smart, from University of Bristol.

The Eurocrypt 2012 Program Committee (PC) consisted of 32 members. There were 195 papers submitted to the conference. Each paper was assigned to at least three PC members, while submissions co-authored by PC members were reviewed by at least four PC members. Papers were refereed anonymously. Due to the large number of high-quality submissions, the review process was challenging: the PC, aided by reports from 177 external reviewers, produced a total of 604 reviews in all. After the reviews were submitted, the committee deliberated online for several weeks, exchanging 738 discussion messages. All of our deliberations were aided by the iChair Web submission and review software written by Thomas Baignères and Matthieu Finiasz. We are indebted to them for letting us use their software and for providing us with some help.

The PC eventually selected 41 submissions for presentation during the conference and these are the articles that are included in this volume. Note that these proceedings contain the revised versions of the selected papers. Since the revisions were not checked again before publication, the authors (and not the committee) bear full responsibility of the contents of their papers.

The PC decided to give the Best Paper Award to Antoine Joux and Vanessa Vitse for their paper "Cover and Decomposition Index Calculus on Elliptic Curves made practical. Application to a previously unreachable curve over F_{p^6}." The conference program also included two invited lectures, and short abstracts are provided in the proceedings: one by Antoine Joux entitled "A Tutorial on High-Performance Computing Applied to Cryptanalysis," and the other by Alfred Menezes on "Another Look at Provable Security." We would like to thank them for accepting our invitation and for contributing to the success of Eurocrypt 2012.

We wish to warmly thank the authors who submitted their papers. The hard task of reading, commenting, debating and finally selecting the papers for the conference fell on the PC members. We are very grateful to the committee members and their sub-reviewers for their hard and conscientious work. We would like to thank Jacques Beigbeder for setting up and maintaining the submission and review server at ENS, and Nigel Smart for his great help.

Finally, we would like to say it has been a great honor to be PC Chairs for Eurocrypt 2012!

April 2012

David Pointcheval
Thomas Johansson

Organization

General Chair

Nigel Smart   University of Bristol, UK

Program Chairs

David Pointcheval   ENS, CNRS, and INRIA, Paris, France
Thomas Johansson   Lund University, Sweden

Program Committee

Masayuki Abe   NTT, Japan
John Black   University of Colorado at Boulder and UC
                               Santa Barbara, USA
David Cash   IBM Research, USA
Dario Catalano   Università di Catania, Italy
Jean-Sébastien Coron   University of Luxembourg
Orr Dunkelman   University of Haifa and Weizmann Institute,
                               Israel
Marc Fischlin   TU Darmstadt, Germany
Pierre-Alain Fouque   ENS, France
Steven Galbraith   University of Auckland, New Zealand
Henri Gilbert   ANSSI, France
Louis Goubin   University of Versailles, France
Jens Groth   University College London, UK
Dennis Hofheinz   Karlsruher Institut für Technologie, Germany
Tetsu Iwata   Nagoya University, Japan
John Kelsey   NIST, USA
Aggelos Kiayias   University of Athens, Greece
Arjen Lenstra   EPFL, Switzerland
Benoit Libert   UC Louvain, Belgium
Yehuda Lindell   Bar-Ilan University, Israel
Kaisa Nyberg   Aalto University and Nokia, Finland
Thomas Peyrin   Nanyang Technological University, Singapore
Krzysztof Pietrzak   CWI, The Netherlands
Vincent Rijmen   KU Leuven and TU Graz, Belgium/Austria
Thomas Ristenpart   University of Wisconsin, USA
Kazue Sako   NEC, Japan
Palash Sarkar   Indian Statistical Institute, India
Igor Shparlinski   Macquarie University, Australia

Martijn Stam	University of Bristol, UK	
Vinod Vaikuntanathan	Microsoft Research and University of Toronto, Canada	
Ivan Visconti	University of Salerno, Italy	
Xiaoyun Wang	Tsinghua University, China	
Duncan Wong	City University of Hong Kong, SAR China	

External Reviewers

Michel Abdalla	Mario Di Raimondo	Antoine Joux
Adi Akavia	Yevgeniy Dodis	Pascal Junod
Joël Alwen	Nico Döttling	Bhavana Kanukurthi
Elena Andreeva	Pooya Farshim	Eike Kiltz
Giuseppe Ateniese	Jean-Charles Faugère	Thorsten Kleinjung
Nuttapong Attrapadung	Sebastian Faust	David Kohel
Man Ho Au	Serge Fehr	Yuichi Komano
Paul Baecher	Dario Fiore	Takeshi Koshiba
Thomas Baignères	David Mandell Freeman	Daniel Kraschewski
Foteini Baldimtsi	Georg Fuchsbauer	Kaoru Kurosawa
Paulo Barreto	Thomas Fuhr	Fabien Laguillaumie
Aurélie Bauer	Eichiro Fujisaki	Mario Larangeira
Stephanie Bayer	Jun Furukawa	Dong Hoon Lee
David Bernhard	David Galindo	Jooyoung Lee
Daniel J. Bernstein	Nicolas Gama	Kwangsu Lee
Sanjay Bhattacherjee	Sanjam Garg	Kaitai Liang
Joppe Bos	Essam Ghadafi	Dongdai Lin
Christoph Bösch	Benedikt Gierlichs	Zhen Liu
Zvika Brakerski	Domingo Gomez	Victor Lomné
Billy Brumley	Sergey Gorbunov	Adriana Lopez-Alt
Christina Brzuska	Dov Gordon	Stefan Lucks
Jesper Buus Nielsen	Robert Granger	Anna Lysyanskaya
Ran Canetti	Adam Groce	Vadim Lyubashevsky
Debrup Chakraborty	Jian Guo	Hemanta Maji
Nishanth Chandran	Carmit Hazay	Avradip Mandal
Donghoon Chang	Javier Herranz	Joana Marim
Lidong Chen	Shoichi Hirose	Damian Markham
Jung Hee Cheon	Susan Hohenberger	Alexander May
Céline Chevalier	Qiong Huang	Florian Mendel
Seung Geol Choi	Toshiyuki Isshiki	Rachel Miller
Ashish Choudhury	Tibor Jager	Kazuhiko Minematsu
Özgür Dagdelen	Abhishek Jain	Payman Mohassel
Bernardo David	Kimmo Järvinen	Michael Naehrig
Emiliano De Cristofaro	Dimitar Jetchev	Koh-ichi Nagao
Jean Paul Degabriele	Shaoquan Jiang	Svetla Nikova
Claus Diem	Stephen Jordan	Takashi Nishide

Ryo Nishimaki
Ryo Nojima
Satoshi Obana
Miyako Ohkubo
Adam O'Neill
Cristina Onete
Claudio Orlandi
Alina Ostafe
Jong Hwan Park
Kenneth Paterson
Alain Patey
Souradyuti Paul
Chris Peikert
Rene Peralta
Olivier Pereira
Ray Perlner
Ludovic Perret
Edoardo Persichetti
Marcel Pfaffhauser
Benny Pinkas
Axel Poschmann
Carla Ràfols
Ananth Raghunathan
Somindu C. Ramanna

Oded Regev
Leonid Reyzin
Yannis Rouselakis
Subhabrata Samajder
Bagus Santoso
Santanu Sarkar
Alessandra Scafuro
Christian Schaffner
Sven Schäge
Werner Schindler
Martin Schläffer
Yannick Seurin
Barhum Kfir Shlomo
Thomas Shrimpton
Shashank Singh
Daniel Smith
Damien Stehlé
John Steinberger
Ron Steinfeld
Fatih Sulak
Koutarou Suzuki
Xiao Tan
Isamu Teranishi
Stefano Tessaro

Nicolas Theriault
Mehdi Tibouchi
Elmar Tischhauser
Tomas Toft
Deniz Toz
Meltem Sonmez Turan
Dominique Unruh
Kerem Varıcı
Muthu
    Venkitasubramaniam
Akshay Wadia
Bogdan Warinschi
Brent Waters
Daniel Wichs
Keita Xagawa
Dongsheng Xing
Guomin Yang
Kan Yasuda
Bingsheng Zhang
Yunlei Zhao
Hong-Sheng Zhou
Vassilis Zikas

Table of Contents

Protocols

Lossy Trapdoor Functions

Tools

Symmetric Constructions II

Public-Key Schemes

Security Models

Lattices

A Tutorial
on High Performance Computing
Applied to Cryptanalysis

(Invited Talk Abstract)

Antoine Joux

DGA and
Université de Versailles Saint-Quentin-en-Yvelines, Laboratoire PRISM,
45 avenue des États-Unis, F-78035 Versailles Cedex, France
antoine.joux@m4x.org

Abstract. Cryptology and computers have a long common history; in fact, some of the early computers were created as cryptanalytic tools. The development of faster and widely deployed computers also had a great impact on cryptology, allowing modern cryptography to become a practical tool. Today, both computers and cryptology are not only practical, but they have became ubiquitous tools. In truth, computing devices incorporating cryptography features range from very small low-end devices to supercomputer, going through all possible intermediate sizes; these devices include both general purpose computing devices and specific, often embedded, processors which enable computing and security features in hundreds of technological objects.

In this invited talk, we mostly consider the cryptanalytic side of things, where it is fair to use very large amounts of computing power to break cryptographic primitives or protocols. As a consequence, demonstrating the feasibility of new cryptanalytic methods often requires large scale computations. Most articles describing such cryptanalyses usually focus on the mathematical or algorithmic advances and gloss over the implementation details, giving only sufficient data to show that the computations are feasible. The goal of the present abstract is to give an idea of the difficulty facing implementers of large scale cryptanalytic attacks.

Computers and cryptanalysis have a long common history. This is well-emphasized by the location of this Eurocrypt conference located near Bletchley Park, the home of the UK code-breaking during World War II. In particular, the park features a working replica of the first digital computer, the Colossus and of the Turing-Welchman Bombe, which was initially developed for cryptanalytic purposes. The organization of the park itself reflects the duality of computers and cryptanalysis. Indeed, the park hosts two museums, the "National Codes and Ciphers Centre" and the "National Museum of Computing".

Even if computers and other computing devices have become general purpose tools in the present days, they still have a lot in common with cryptography. Today, almost all computing devices, from credit cards to high-end computers

D. Pointcheval and T. Johansson (Eds.): EUROCRYPT 2012, LNCS 7237, pp. 1–7, 2012.

implement some cryptographic functionality and cryptography is an essential tool for securing the digital world. Moreover, most of the recent cryptographic advances rely on the enhanced performances of modern computing devices.

On the cryptanalytic side, we encounter a similar situation. Having faster and bigger computers allows cryptanalysts to run huge computations which would not have been possible in their wildest dreams a few decades ago.

This article discuss this cryptanalytic application of supercomputers. In Section 1, we classify the typical cryptanalytic applications. In Section 2, we describe the hardware context of the last decade and discuss some possible evolutions. Section 3 explains the practical issues that can be encountered while managing the necessary computations to set new cryptanalysis records. Finally, Section 4 describes some algorithm challenges that need to be solved to efficiently use the potential power of forthcoming computers.

1   Typical Cryptanalytic Applications

The applications of high-performance computing to cryptanalysis are numerous and varied. They range from attacks which are "embarrassingly-parallel" and can trivially use a large distributed computing power to algorithms which are essentially sequential by nature and are very difficult to adapt to take advantage of the power of supercomputers.

The easiest case of embarrassingly parallel computations contains brute-force attacks and their variants. In this case, each task can run completely independently of the others, it only needs to receive a small amount of input data (such as a plaintext/ciphertext pair) and a description of the part of the key space it should work on. Note that this description is not enough necessary and, especially when the control loop is loose, simply letting each task try a random subset of the key might even be preferable. Some other attacks, such as differential collision searches on hash functions are also of an embarrassingly parallel nature [4].

A slightly harder class of computations which can be parallelized in a reasonable straightforward way, but require communications to send back some partial results in a centralized place. This centralized place then redistributes the values in order to conclude the computation. This is typically the case of parallelized collision-finding algorithms [14, 21].

The next important class of problem contains the sieving-based index calculus algorithm for factoring [1, 5, 15] and discrete logarithms [11–13]. In this class, the largest phase of the computation (the sieving phase) is embarrassingly parallel, however, it produces a large amount of data which needs to be collected in a centralized place. Note that this amount of data is small compared to the magnitude of the computation but it is still a difficult task to centralize this data without introducing errors. The next phase consists in transforming this data into a linear system of equations and then in solving this system. This offers much more difficulty than the initial computation. Currently, this task is achieve by first reducing the size of the system using ad'hoc heuristics called, structured Gaussian Elimination. This is usually done on a small number of processor, but the computational cost required here is low enough and this is not a problem. The reduced system is then solved using an iterative linear solver such as the Lanczos or Wiedemann algorithms. The main problem is that these algorithms can be distributed but require a large amount of communications between the individual tasks. As a consequence, even when using the block Wiedemann variant [6, 20] which lowers the amount of communications, this is usually the computational bottleneck.

Finally, some cryptanalyses rely on algorithmic tasks for which no satisfactory parallel descriptions are known. This is the case for many advanced algorithms used in cryptanalysis (see Section 4). Note that even in the best cases, writing record-breaking codes is a very specific programming activity, which rarely follows the tenants of modern software engineering. The reason for this discrepancy is that the use of modern programming features has a cost in terms of performance, which is rarely acceptable in this specific context.

2   Hardware Context

During the last decades of the twentiest century, the speed of processor increased at the steady rhythm. More precisely, clock rates were at the MHz level in the 80s and raised to the GHz level in the 2000s. This increased the performance of individual processors and permitted to do bigger computations while using at most a small amount of parallelism. However, the clock rates of processors are no longer increasing and the additional computing power of recent processors come from their ability to perform more computations in parallel. This capability is obtained either by allowing the machines to work on larger data types, by giving CPUs the ability to parallelize micro-instructions or by building multi-core processors. As a consequence, despite the stopped growth of clock rate, the raw computing power of processors is still increasing steadily. However, taking advantage of this power for cryptanalytic tasks requires much more effort on the programmer's part.

At the same time, the amount of memory available in modern machines has increased considerably. In the 80s, 64 Kbytes of memory for a personal computer was above standard, in the present days, the equivalent would be around 8 Gbytes. However, on modern processors, accessing memory is proportionally more expensive. To palliate this problem, designers have added several levels of memory-cache that greatly increase the memory accesses as long as they remain reasonably localized. This is also an important constraint since in this model some algorithmic techniques such as sieving are considerably slowed down.

Where personal computers are concerned, the main processor(s) is no longer the only available computing ressource. Indeed, with the development of 3D-games, graphics cards have progressively been transformed into massively parallel computing ressources, capable of performing quite general computations. As a consequence, it has become worthwhile to consider their potential as computing devices in massive computations.

Above the personal computer scale, the development of cloud computing is offering a new opportunity to run medium-scale computation at a moderate cost.

At the present time, these infrastructures seem are more suited to embarrassingly parallel tasks than to communication bounded computations. One advantage of using cloud-computing, emphasized in [16], it that it gives a simple metric to compare computations: their monetary cost.

Finally, turning to supercomputers, it is interesting to see that, even at this large scale, many computers among the most powerful are built by assembling many high-end "personal computers" tied together by a high-performance network. As a consequence, running embarrassingly-parallel task on such computers does not require much programming beyond the initial work of writing the program for a general purpose computer. It also means that the previous considerations about parallelism and memory accesses remain true. Of course, thanks to the high-performance network, it is possible to perform tasks that require a fairly high amount of communications. However, despite this improve performance, communications often remain the bottleneck point for algorithms which are not straightforward to parallelize.

Another possibility to perform very large computations is to consider the use of specific hardware. However, the cost of building such hardware is high. As a consequence, many papers [10, 18, 19] dealing with specific hardware remain theoretical and aim at finding the limit of feasible computations. A notable exception is the development of the DES-Cracker [9].

3   Running Record Computations

Once a new cryptanalytic algorithm has been discovered or improved and implemented, running the algorithm to set some record computation is a nice way to demonstrate the potential of the algorithm. This being said, one could easily imagine that this final step of performing the computation is just a routine task. Unfortunately, this is not the case and running record computations is a difficult and tedious task.

The first step is to obtain the necessary computing resources. This can be easy if the computation only requires dozens of desktop computers for a few weeks or become a real nightmare for people trying to run a large scale computation by recruiting tens of thousands of computers on the Internet. The easiest approach for computations that requires significantly more power than a dozen of desktop computers is to apply for computing time one or several supercomputers. Throughout the world, there exists several supercomputing organizations that let researchers apply for computing power.

The next step, once computing power has been granted, is to port the computing code to the computers that have been made available. Even when great care has been taken to write portable code in the first place, there are always specific "features" that call for modifications. This is especially true for complex computations that have successive computing phases. Indeed, in that case, one often discovers that one of the "negligible" phases of the computation does not scale well and needs a complete rewriting to run correctly for the record being considered.

Once all this preparation has been settled and, contrary to what might be expected, the really hard part of the computation starts. Indeed, while large computations may become routine when series of similar computations are performed[1], record computations never are. A first problem is that by going to larger sizes, one often triggers unexpected bugs, with rare probability of occurrences. This may lead the program to make several passes over the same search space, which not only wastes computing power but may trigger other bugs when unexpected data collisions are encountered later on. Another possible consequence is that some intermediate computational data may contain corrupted information. While benign in some application such as brute force, incorrect data may cause major failures in other cases. For example, if a single incorrect equation is added to a large linear system, then any hope of recovering the solution is lost. Note that corrupted data is not always a consequence of coding bugs, supercomputers are often experimental machines which may suffer from occasional hardware problem and the sheer scale of the computation also makes physical corruption of data, whether in memory or on disk, possible.

As a consequence, when programming with record computations in mind, it is essential to add extra robustness in the processing. A good practice is to check intermediate computational results whenever this can be done at a reasonable cost. Such checking should use independently written code and should do the verification at the highest achievable level of mathematical abstraction. For example, before performing the iterative linear algebra step of a sieving algorithm it is very good practice to pull back the equation to the mathematical group being considered and check them on this group.

Also note that closely monitoring the computation is a must: processes may get stuck, they may fail to restart after maintenance. To make it short, when running record computations, one should always expect the unexpected.

4   Algorithmic Challenges

As computations grow bigger, the relevant metric to measure the computation is shifting. We can no longer focus on running time and ignore other parameters. Of course, running time has never been a perfect metric but it still gave a good approximation of the efficiency of algorithms. With modern supercomputers, the pictures is much more complicated. First, the gap between the cost of time and memory is growing bigger. Second, another, very important parameter should be taken into account: the cost of communication between the parts of the supercomputer.

Combining all the relevant parameters is not easy because there the parameters are not independent. Indeed, programs which require a lot of memory cannot store their data locally in a single node and are going to use larger amount of communications.

---

[1] A typical example is the weather forecast computations which despite their large scale become routine once the production code becomes stable enough.

As a consequence, in order to use supercomputers to their full power, new algorithms are becoming necessary. These new algorithms should be designed with new metrics in mind. Basically, processes should be as independent of each others as possible and memory use should be limited to fit within local memory (or even better within cache memory).

Of course, embarrassingly parallel tasks are not going to be a problem. However, there are many more algorithms which need to be adapted or improved to become more efficient on supercomputer. To give some example, let us mention iterative linear algebra [17, 20], structured Gaussian elimination, computation of Gröbner bases [8], SAT solvers [7], collision-search techniques [14, 21], large-scale lattice reduction [2], generalized birthday algorithms [3], ...

5   Conclusion

Performing cryptanalytic records computation is a very efficient tool to understand the concrete security level of cryptographic primitive and this should remain true in the future. In particular, such computations can be used to benchmark lower security levels. Indeed, on the one hand, many low-end cryptographic devices rely on a 80-bit security level. On the other hand, the current fastest computer can perform more than 2^{73} floating-point instructions per year. As a consequence, since the figures are getting close, studying the evolution of record computations is essential in order to decide when to phase out such low-end systems before they become insufficiently secure.

References

1. Aoki, K., Franke, J., Kleinjung, T., Lenstra, A.K., Osvik, D.A.: A Kilobit Special Number Field Sieve Factorization. In: Kurosawa, K. (ed.) ASIACRYPT 2007. LNCS, vol. 4833, pp. 1–12. Springer, Heidelberg (2007)
2. Backes, W., Wetzel, S.: Parallel lattice basis reduction - the road to many-core. In: Thulasiraman, P., Yang, L.T., Pan, Q., Liu, X., Chen, Y.-C., Huang, Y.-P., Chang, L.H., Hung, C.-L., Lee, C.-R., Shi, J.Y., Zhang, Y. (eds.) 13th IEEE International Conference on High Performance Computing & Communication, pp. 417–424. IEEE (2011)
3. Bernstein, D.J.: Better price-performance ratios for generalized birthday attacks (2007), http://cr.yp.to/rumba20/genbday-20070904.pdf
4. Biham, E., Chen, R., Joux, A., Carribault, P., Lemuet, C., Jalby, W.: Collisions of SHA-0 and Reduced SHA-1. In: Cramer, R. (ed.) EUROCRYPT 2005. LNCS, vol. 3494, pp. 36–57. Springer, Heidelberg (2005)
5. Brent, R.P.: Recent Progress and Prospects for Integer Factorisation Algorithms. In: Du, D.-Z., Eades, P., Sharma, A.K., Lin, X., Estivill-Castro, V. (eds.) COCOON 2000. LNCS, vol. 1858, pp. 3–22. Springer, Heidelberg (2000)
6. Coppersmith, D.: Solving linear equations over GF(2) via block Wiedemann algorithm. Mathematics of Computation 62, 333–350 (1994)
7. Hamadi, Y. (ed.). Special issue on parallel SAT solving. Journal on Satisfiability, Boolean Modeling and Computation 6, 203–262 (2009)

8. Faugère, J.-C., Lachartre, S.: Parallel Gaussian elimination for Gröbner bases computations in finite fields. In: Maza, M.M., Roch, J.-L. (eds.) Proceedings of the 4th International Workshop on Parallel Symbolic Computation, pp. 89–97. ACM (2010)
9. Electronic Frontier Foundation. Cracking DES: Secrets of Encryption Research, Wiretap Politics and Chip Design. O'Reilly & Associates, Inc. (1998)
10. Franke, J., Kleinjung, T., Paar, C., Pelzl, J., Priplata, C., Stahlke, C.: SHARK: A Realizable Special Hardware Sieving Device for Factoring 1024-Bit Integers. In: Rao, J.R., Sunar, B. (eds.) CHES 2005. LNCS, vol. 3659, pp. 119–130. Springer, Heidelberg (2005)
11. Hayashi, T., Shinohara, N., Wang, L., Matsuo, S., Shirase, M., Takagi, T.: Solving a 676-Bit Discrete Logarithm Problem in $GF(3^{6n})$. In: Nguyen, P.Q., Pointcheval, D. (eds.) PKC 2010. LNCS, vol. 6056, pp. 351–367. Springer, Heidelberg (2010)
12. Joux, A., Lercier, R.: The Function Field Sieve in the Medium Prime Case. In: Vaudenay, S. (ed.) EUROCRYPT 2006. LNCS, vol. 4004, pp. 254–270. Springer, Heidelberg (2006)
13. Joux, A., Lercier, R., Smart, N.P., Vercauteren, F.: The Number Field Sieve in the Medium Prime Case. In: Dwork, C. (ed.) CRYPTO 2006. LNCS, vol. 4117, pp. 326–344. Springer, Heidelberg (2006)
14. Joux, A., Lucks, S.: Improved Generic Algorithms for 3-Collisions. In: Matsui, M. (ed.) ASIACRYPT 2009. LNCS, vol. 5912, pp. 347–363. Springer, Heidelberg (2009)
15. Kleinjung, T., Aoki, K., Franke, J., Lenstra, A.K., Thomé, E., Bos, J.W., Gaudry, P., Kruppa, A., Montgomery, P.L., Osvik, D.A., te Riele, H., Timofeev, A., Zimmermann, P.: Factorization of a 768-Bit RSA Modulus. In: Rabin, T. (ed.) CRYPTO 2010. LNCS, vol. 6223, pp. 333–350. Springer, Heidelberg (2010)
16. Kleinjung, T., Lenstra, A.K., Page, D., Smart, N.P.: Using the cloud to determine key strengths. IACR Cryptology ePrint Archive, p. 254 (2011)
17. Kleinjung, T., Nussbaum, L., Thomé, E.: Using a grid platform for solving large sparse linear systems over gf(2). In: Proceedings of the 2010 11th IEEE/ACM International Conference on Grid Computing, pp. 161–168. IEEE (2010)
18. Lenstra, A.K., Shamir, A.: Analysis and Optimization of the TWINKLE Factoring Device. In: Preneel, B. (ed.) EUROCRYPT 2000. LNCS, vol. 1807, pp. 35–52. Springer, Heidelberg (2000)
19. Shamir, A., Tromer, E.: Factoring Large Numbers with the TWIRL Device. In: Boneh, D. (ed.) CRYPTO 2003. LNCS, vol. 2729, pp. 1–26. Springer, Heidelberg (2003)
20. Thomé, E.: Subquadratic computation of vector generating polynomials and improvement of the block wiedemann algorithm. J. Symb. Comput. 33(5), 757–775 (2002)
21. van Oorschot, P.C., Wiener, M.J.: Parallel collision search with cryptanalytic applications. Journal of Cryptology 12(1), 1–28 (1999)

Another Look at Provable Security

Alfred Menezes

Department of Combinatorics & Optimization
University of Waterloo
ajmeneze@uwaterloo.ca

Abstract. Many cryptographers believe that the only way to have confidence in the security of a cryptographic protocol is to have a mathematically rigorous proof that the protocol meets its stated goals under certain assumptions. However, it is often difficult to assess what such proofs really mean in practice especially if the proof is non-tight, the underlying assumptions are contrived, or the security definition is in the single-user setting. We will present some examples that illustrate this difficulty and highlight the important role that old-fashioned cryptanalysis and sound engineering practices continue to play in establishing and maintaining confidence in the security of a cryptographic protocol.

This talk is based on joint work with Neal Koblitz [2,3] and with Sanjit Chatterjee and Palash Sarkar [1].

References

1. Chatterjee, S., Menezes, A., Sarkar, P.: Another Look at Tightness. In: Vaudenay, S. (ed.) SAC 2011. LNCS, vol. 7118, pp. 293–319. Springer, Heidelberg (2011)
2. Koblitz, N., Menezes, A.: Another look at provable security, http://anotherlook.ca
3. Koblitz, N., Menezes, A.: Another look at security definitions, Cryptology ePrint Archive: Report 2011/343

D. Pointcheval and T. Johansson (Eds.): EUROCRYPT 2012, LNCS 7237, p. 8, 2012.

Cover and Decomposition Index Calculus on Elliptic Curves Made Practical*
Application to a Previously Unreachable Curve over $\mathbb{F}_{p^6}$

Antoine Joux[1] and Vanessa Vitse[2]

[1] DGA and Université de Versailles Saint-Quentin, Laboratoire PRISM, 45 avenue des États-Unis, F-78035 Versailles cedex, France
`antoine.joux@m4x.org`
[2] Université de Versailles Saint-Quentin, Laboratoire PRISM, 45 avenue des États-Unis, F-78035 Versailles cedex, France
`vanessa.vitse@prism.uvsq.fr`

Abstract. We present a new "cover and decomposition" attack on the elliptic curve discrete logarithm problem, that combines Weil descent and decomposition-based index calculus into a single discrete logarithm algorithm. This attack applies, at least theoretically, to all composite degree extension fields, and is particularly well-suited for curves defined over $\mathbb{F}_{p^6}$. We give a real-size example of discrete logarithm computations on a curve over a 151-bit degree 6 extension field, which would not have been practically attackable using previously known algorithms.

Keywords: elliptic curve, discrete logarithm, index calculus, Weil descent, decomposition attack.

1 Introduction

Elliptic curves are used in cryptography to provide groups where the discrete logarithm problem is thought to be difficult. We recall that given a finite group G (written additively) and two elements $P, Q \in G$, the discrete logarithm problem (DLP) consists in computing, when it exists, an integer x such that $Q = xP$. When elliptic curves are used in cryptographic applications, the DLP is usually considered to be as difficult as in a generic group of the same size [31]. As a consequence, for a given security level, the key size is much smaller than for other popular cryptosystems based on factorization or discrete logarithms in finite fields. The first elliptic curves considered in cryptography were defined over either binary or prime fields [20,24]. But to speed up arithmetic computations, it has been proposed to use various forms of extension fields. In particular, Optimal Extension Fields have been proposed in [4] to offer high performance in hardware implementations. They are of the form $\mathbb{F}_{p^d}$ where p is a pseudo-Mersenne prime and d is such that there exists an irreducible polynomial of

---

* This work was granted access to the HPC resources of CCRT under the allocation 2010-t201006445 made by GENCI (Grand Equipement National de Calcul Intensif).

D. Pointcheval and T. Johansson (Eds.): EUROCRYPT 2012, LNCS 7237, pp. 9–26, 2012.

the form $X^d - \omega \in \mathbb{F}_p[X]$. In most examples, the degree d of the extension is rather small. However, when curves defined over extension fields are considered, some non-generic attacks, such as the Weil descent or decomposition attacks, can be applied. The first one aims at transferring the DLP from $E(\mathbb{F}_{q^n})$ to the Jacobian of a curve $\mathcal{C}$ defined over $\mathbb{F}_q$ and then uses index calculus on this Jacobian [2,12,15] to compute the logarithm; it works well when the genus of the curve $\mathcal{C}$ is small, ideally equal to n, but this occurs quite infrequently in practice. Many articles have studied the scope of this technique (cf. [7,10,11,14,16]), but even on vulnerable curves, the Weil descent approach is often just a little more efficient than generic attacks on the DLP. Decomposition-based index calculus, or decomposition attack, is a more recent algorithm (see [9,13,18,26]), which applies equally well to all (hyper-)elliptic curves defined over an extension field. Its asymptotic complexity is promising, but in practice, due to large hidden constants in the complexity, it becomes better than generic attacks for group sizes too large to be threatened anyway.

In this article, we combine both techniques into a cover and decomposition attack, which applies as soon as the extension degree is composite. The idea is to first transfer the DLP to the Jacobian of a curve defined on an intermediate field, then use the decomposition method on this sub-extension instead of the classical index calculus. This new attack is not a mere theoretical possibility: we give concrete examples of curves defined over $\mathbb{F}_{p^6}$ that are practically secure against all other attacks, but for which our method allows to solve the DLP in a reasonable time. In particular, we have been able to compute logarithms on a 149-bit elliptic curve group defined over a degree 6 extension field in about a month real-time, using approximately 110 000 CPU.hours.

The paper is organized as follows: first we briefly recall the principles of Weil descent and of the decomposition method. We then give an explicit description of our attack in Section 3, introducing a useful variant of the decomposition step that can be of independent interest. In particular, we study the case of elliptic curves defined over $\mathbb{F}_{p^6}$, list all the potentially vulnerable curves and give a complexity analysis and a comparison with previously known attacks. Finally, in Section 5, we describe in details the computations on our 149-bit example.

2 Survey of Previous Work

2.1 Weil Descent and Cover Attacks

Weil descent has been first introduced in cryptography by Frey [10]; the idea is to view an abelian variety A of dimension d defined over an extension field K/k as an abelian variety $W_{K/k}$ of dimension $d \cdot [K:k]$ over k. If $W_{K/k}$ turns out to be the Jacobian of a curve $\mathcal{C}_{|k}$ or can be mapped into such a Jacobian, then the discrete logarithm in $A(K)$ can be transferred to $\mathrm{Jac}_\mathcal{C}(k)$, where it may become much weaker due to the existence of efficient index calculus algorithms. When the genus of $\mathcal{C}$ is small relative to the cardinality p of k, the complexity is in $O((g^2 \log^3 p)g! \, p + (g^2 \log p)p^2)$ as p grows to infinity [12]; the first term comes from the relation search and the second from the sparse linear algebra.

Following [15], it is possible to rebalance these two terms by using a double large prime variation. In this variant, only a small number p^α of *prime divisors*[1] are considered as genuine, while the rest of the prime divisors are viewed as "large primes". The optimal value of α depends of the cost of the two phases; asymptotically the choice that minimizes the total running time is $1 - 1/g$, yielding a complexity in $\tilde{O}(p^{2-2/g})$ for fixed g as p goes to infinity.

The main difficulty of this Weil descent method is to find the curve $\mathcal{C}$. This problem was first addressed for binary fields by Gaudry, Hess and Smart (GHS [14]) and further generalized by Diem [7] in odd characteristic. To attack an elliptic curve E defined over $\mathbb{F}_{p^n}$ (p a prime power), the GHS algorithm builds a curve $\mathcal{C}$ defined over $\mathbb{F}_p$ such that there exists a cover map $\pi : \mathcal{C} \to E$ defined over $\mathbb{F}_{p^n}$. The construction is more easily explained in terms of function fields: the Frobenius automorphism $\sigma_{\mathbb{F}_{p^n}/\mathbb{F}_p}$ can be extended to the composite field $F' = \prod_{i=0}^{n-1} \mathbb{F}_{p^n}(E^{\sigma^i})$, and the function field $F = \mathbb{F}_p(\mathcal{C})$ is defined as the subfield of F' fixed by σ. The GHS algorithm then uses the so-called conorm-norm map $N_{F'/F} \circ Con_{F'/\mathbb{F}_{p^n}(E)}$ to transfer the discrete logarithm from $E(\mathbb{F}_{p^n})$ to $\mathrm{Jac}_{\mathcal{C}}(\mathbb{F}_p)$. An important condition is that the kernel of this map must not intersect the subgroup in which the discrete logarithm takes place, but as remarked in [7,16], this is not a problem in most cryptographically relevant situations. This technique is efficient when the genus g of $\mathcal{C}$ is close to n. In particular, for some specific finite fields most elliptic curves are "weak" in the sense that Weil descent algorithms are better, if only by a small margin, than generic attacks [23]. Indeed, when the GHS method does not provide any low genus cover for E, it may be possible to find a sequence of low degree isogenies (a.k.a. an *isogeny walk*) from E to another, more vulnerable elliptic curve E' [11]. Nevertheless, we emphasize that the security loss is quite small for a random curve, and for most curves on most fields $\mathbb{F}_{p^n}$, g is of the order of 2^n which means that index calculus in the Jacobian of $\mathcal{C}$ is slower than generic attacks on $E(\mathbb{F}_{p^n})$.

2.2 Decomposition Attack

Index calculus has become ubiquitous in the last decades for the DLP resolution. However its direct application to elliptic curves faces two major challenges: contrarily to finite fields or hyperelliptic curves, there is no natural choice of factor base and no equivalent of the notion of factorization of group elements. The first main breakthrough was achieved in 2004 by Semaev [30] when he suggested to replace factorization by decomposition into a fixed number of points; for that purpose, he introduced the summation polynomials which give an algebraic expression of the fact that a given point decomposes into a sum of factor base elements. But for a lack of an adequate factor base, this approach fails in the general case. Then Gaudry and Diem [9,13] independently proposed to use Semaev's idea to attack all curves defined over small degree extension fields $\mathbb{F}_{p^n}/\mathbb{F}_p$. Their method shares the basic outline of index calculus, but to distinguish it from what has been presented in the previous subsection, we follow [26]

---

[1] The term *prime divisor* is an abuse of language that denotes the linear irreducible polynomials that are used in the index calculus algorithm on $\mathrm{Jac}_{\mathcal{C}}(k)$.

and call it the decomposition attack. On $E(\mathbb{F}_{p^n})$, a convenient choice of factor base is the set of rational points of the curve with abscissae in $\mathbb{F}_p$. By combining Semaev's summation polynomials and restriction of scalars, the relation search then becomes a resolution of a multivariate polynomial system over $\mathbb{F}_p$. The complexity of this approach can be estimated using double large prime variation by $\tilde{O}\left(p^{2-2/n}\right)$ for fixed n as p grows to infinity. Unfortunately, the hidden constants in this complexity become very large as n grows, and the resolution of the systems is intractable as soon as $n \geq 4$ (or $n \geq 5$ with the variant of [18]).

The decomposition attacks can also be applied to higher genus curves. However, Semaev's polynomials are no longer available in this case and the algebraic expression of the group law is more complicated. In [26], Nagao proposes an elegant way to circumvent this problem, using divisors and Riemann-Roch spaces. For hyperelliptic curves, the decomposition search then amounts to solving a quadratic multivariate polynomial system. This approach is less efficient than Semaev's in the elliptic case, but is the simplest otherwise. For fixed extension degree n and genus g, the complexity of a decomposition attack is in $\tilde{O}\left(p^{2-2/ng}\right)$ with a double large prime variation. Again, the resolution of the polynomial system is the main technical difficulty, and is easily feasible for only very few couples (n, g), namely $(2, 2)$, $(2, 3)$ and $(3, 2)$.

3   Cover and Decomposition Attack

Let $\mathbb{F}_{q^d}/\mathbb{F}_p$ be an extension of finite fields, where q is a power of p (in most applications p denotes a large prime but in general, it can be any prime power), and let E be an elliptic curve defined over $\mathbb{F}_{q^d}$ of cryptographic interest, i.e. containing a subgroup G of large prime order. As E is defined over an extension field, it is subject to the attacks presented above. But if the degree $[\mathbb{F}_{q^d} : \mathbb{F}_p]$ of the extension is larger than 5, then we have seen that E is practically immune to decomposition attacks. In the following, we assume that the potential reduction provided by the GHS attack or its variants is not significant enough to threaten the security of the DLP on the chosen curve E.

When q is a strict power of p, we have a tower of extensions given by $\mathbb{F}_{q^d}/\mathbb{F}_q$ and $\mathbb{F}_q/\mathbb{F}_p$. In this context, it becomes possible to combine both cover and decomposition methods and obtain an efficient attack of the DLP on E. The idea is to use Weil descent on the first extension $\mathbb{F}_{q^d}/\mathbb{F}_q$ to get a cover defined over $\mathbb{F}_q$, with small enough[2] genus g. Then we can apply a decomposition attack on the curve thus obtained, making use of the second extension $\mathbb{F}_q/\mathbb{F}_p$. As this *cover and decomposition* attack is more efficient when Weil descent provides a hyperelliptic cover over the intermediate field, we focus on this case in the following.

3.1   Description of the Attack

We now explicitly detail this cover and decomposition approach. We suppose first that there exists an imaginary hyperelliptic curve $\mathcal{H}$ of small genus g with

---

[2] Meaning that g should be small relatively to the genus that could be obtained by direct Weil descent, using the extension $\mathbb{F}_{q^d}/\mathbb{F}_p$.

equation $y^2 = h(x)$, defined over $\mathbb{F}_q$, together with a covering map $\pi : \mathcal{H} \to E$ defined over $\mathbb{F}_{q^d}$. This can be obtained by the GHS attack or its variants, possibly preceded by an isogeny walk. This cover classically allows to transfer the DLP from G to a subgroup $G' \subset \mathrm{Jac}_{\mathcal{H}}(\mathbb{F}_q)$ via the conorm-norm map $N_{\mathbb{F}_{q^d}/\mathbb{F}_q} \circ \pi^* :$ $E(\mathbb{F}_{q^d}) \simeq \mathrm{Jac}_E(\mathbb{F}_{q^d}) \to \mathrm{Jac}_{\mathcal{H}}(\mathbb{F}_q)$, assuming that $\ker(N_{\mathbb{F}_{q^d}/\mathbb{F}_q} \circ \pi^*) \cap G = \{\mathcal{O}_E\}$.

The decomposition part of the attack is adapted from Gaudry and Nagao; since it is quite recent, we detail the method. As in all index calculus approaches, there are two time-consuming steps: first we have to collect relations between factor base elements, then we compute discrete logarithms by using linear algebra on the matrix of relations. We consider the same factor base as [13,26]

$$\mathcal{F} = \{D_Q \in \mathrm{Jac}_{\mathcal{H}}(\mathbb{F}_q) \ : \ D_Q \sim (Q) - (\mathcal{O}_{\mathcal{H}}), Q \in \mathcal{H}(\mathbb{F}_q), x(Q) \in \mathbb{F}_p\},$$

which contains about p elements. As usual, we can use the hyperelliptic involution to reduce the size of $\mathcal{F}$ by a factor 2, so that only $p/2$ relations are needed.

Let n be the extension degree $[\mathbb{F}_q : \mathbb{F}_p]$. In Nagao's decomposition method, one tries to decompose an arbitrary divisor D (typically obtained by considering a large multiple of some element in $\mathcal{F}$) into a sum of ng divisors of $\mathcal{F}$

$$D \sim \sum_{i=1}^{ng} ((Q_i) - (\mathcal{O}_{\mathcal{H}})). \tag{1}$$

Heuristically, there exist approximately $p^{ng}/(ng)!$ distinct sums of ng elements of $\mathcal{F}$, so the probability that a given divisor D is decomposable can be estimated by $1/(ng)!$. To check if D can be decomposed, one considers the Riemann-Roch $\mathbb{F}_q$-vector space

$$\mathcal{L}\left(ng(\mathcal{O}_{\mathcal{H}}) - D\right) = \{f \in \mathbb{F}_q(\mathcal{H})^* : \mathrm{div}(f) \geq D - ng(\mathcal{O}_{\mathcal{H}})\} \cup \{0\}.$$

We can assume that the divisor D is reduced and has Mumford representation $(u(x), v(x))$ with $\deg u = g$, so that this $\mathbb{F}_q$-vector space is spanned by $u(x)x^i, (y - v(x))x^j, 1 \leq i \leq m_1, 1 \leq j \leq m_2$, where $m_1 = \lfloor (n-1)g/2 \rfloor$ and $m_2 = \lfloor ((n-1)g - 1)/2 \rfloor$. A function $f = \lambda_0 u(x) + \lambda_1 u(x)x + \ldots + \lambda_{m_1} u(x)x^{m_1} + \mu_0(y - v(x)) + \mu_1 x(y - v(x)) + \ldots + \mu_{m_2} x^{m_2}(y - v(x))$ vanishes on the support of D and exactly ng other points (counted with multiplicity and possibly defined on the algebraic closure of $\mathbb{F}_q$) if its top-degree coefficient is not zero. We are looking for a condition on $\lambda_0, \ldots, \lambda_{m_1}, \mu_0, \ldots, \mu_{m_2} \in \mathbb{F}_q$ such that the zeroes $Q_1 \ldots, Q_{ng}$ of f disjoint from $Supp(D)$ have x-coordinate in $\mathbb{F}_p$; this event yields a relation as in (1). Therefore, we consider the polynomial $F(x) = f(x,y)f(x,-y)/u(x)$ where y^2 has been replaced by $h(x)$. Without loss of generality, we can fix either $\lambda_{m_1} = 1$ or $\mu_{m_2} = 1$ in order to have F monic of degree ng. The roots of F are exactly the x-coordinates of the zeroes of f distinct from $Supp(D)$, thus we are looking for the values of λ and μ for which F splits in linear factors over $\mathbb{F}_p$. A first necessary condition is that all of its coefficients, which are quadratic polynomials in λ and μ, belong to $\mathbb{F}_p$; a scalar restriction on these coefficients then yields a quadratic polynomial system of $(n-1)ng$ equations and variables coming from the components of the variables λ and μ. The corresponding ideal is generically of dimension 0, and the solutions of the system can be found using e.g. a Gröbner basis computation. Since the number of systems to solve is huge (on average $(ng)! \cdot p/2$, or more if a large prime variation is applied), techniques such as the F4 variant of [19] should be preferred. Once the solutions are obtained, it remains to check if the resulting polynomial F splits in $\mathbb{F}_p[x]$, and if it is the case, to compute the corresponding decomposition of D.

In this article, we also consider a somewhat different approach to the relation search that offers some similarity with the method used in the number field and function field sieves [1,22]. More precisely, we no longer have a divisor D to decompose, but instead search for sums of factor base elements equal to 0:

$$\sum_{i=1}^{m}((Q_i) - (\mathcal{O}_{\mathcal{H}})) \sim 0. \tag{2}$$

Heuristically, the expected number of relations of the form (2) involving m points of the factor base is approximately $p^{m-ng}/m!$. Since we need to collect at least about $p/2$ relations, we look for sums of $m = ng + 2$ points, assuming that $p \geq (ng+2)!/2$. As before, we work with the $\mathbb{F}_q$-vector space $\mathcal{L}(m(\mathcal{O}_{\mathcal{H}}))$, which is spanned by $1, x, \ldots, x^{m_1}, y, xy, \ldots, x^{m_2}y$, where $m_1 = \lfloor m/2 \rfloor$ and $m_2 = \lfloor (m+1)/2 \rfloor - g$. We consider the function $f = \lambda_0 + \lambda_1 x + \ldots + \lambda_{m_1} x^{m_1} + \mu_0 y + \mu_1 xy + \ldots + \mu_{m_2} x^{m_2}y$: it vanishes in exactly m points if its top-degree coefficient is not zero, and the abscissae of its zeroes are the roots of

$$F(x) = f(x,y)f(x,-y) = (\lambda_0 + \lambda_1 x + \ldots + \lambda_{m_1} x^{m_1})^2 - h(x)(\mu_0 + \mu_1 x + \ldots + \mu_{m_2} x^{m_2})^2.$$

Again, we fix $\lambda_{m_1} = 1$ if m is even or $\mu_{m_2} = 1$ otherwise, so that F is monic. In order to obtain a relation of the form (2), we look for values of λ and μ for which F splits over $\mathbb{F}_p$. The first condition is that F belongs to $\mathbb{F}_p[x]$; after a scalar restriction on its coefficients, this translates as a quadratic polynomial system of $(n-1)m$ equations and $n(m-g)$ variables. With our choice of $m = ng + 2$, this corresponds to an underdetermined system of $n(n-1)g + 2n - 2$ equations in $n(n-1)g + 2n$ variables. When the parameters n and g are not too large, we remark that it is possible to compute once for all the corresponding lexicographic Gröbner basis. Each specialization of the last two variables should then provide an easy to solve system, namely triangular with low degrees. It remains to check whether the corresponding expression of F is indeed split and to deduce the corresponding relations between the points of $\mathcal{F}$.

Once enough relations of the form (2) have been collected, and possibly after a structured Gaussian elimination or a large prime variation, we can deduce with linear algebra the logarithms of all elements in $\mathcal{F}$ (up to a multiplicative constant, since we have not specified the logarithm base). To compute the discrete logarithm of an arbitrary divisor D, we proceed to a descent phase: we need to decompose D as a sum of factor base elements. This decomposition search can be done using the first method described above. Note that, if D does not decompose as a sum, it suffices to try small multiples $2D, 3D \ldots$ until we find one correct decomposition. Thanks to this descent step, it is possible to compute many discrete logarithms in the same group for negligible additional cost.

When the cover of E is not hyperelliptic, one can still use the Riemann-Roch based approach. It is not difficult to compute a basis of the vector spaces $\mathcal{L}(ng(\mathcal{O}_\mathcal{H}) - D)$ or $\mathcal{L}(m(\mathcal{O}_\mathcal{H}))$ and to consider a function $f(x, y)$ (depending of parameters λ and μ) in these spaces. Getting rid of the y-variable can be done easily by computing the resultant in y of f and the equation of the curve (or multiresultant if the curve is not planar); however, the resulting polynomial $F(x)$ no longer depends quadratically of the parameters λ and μ. The system obtained by scalar restriction still has the same number of equations and variables but its degree is greater than 2, so that the resolution is more complicated.

3.2   Sieving for Quadratic Extensions

This new decomposition technique is already faster than Nagao's when the lex Gröbner basis of the system coming from (2) is efficiently computable, but can still be further improved. Indeed, checking that F is split has a non-negligible cost, since we need to factor a polynomial of degree $ng + 2$ into linear terms. To avoid this, it is possible to modify the search for relations of the form (2) using a sieving technique when $[\mathbb{F}_q : \mathbb{F}_p] = 2$ in the odd characteristic case. Let t be an element such that $\mathbb{F}_{p^2} = \mathbb{F}_p(t)$; we assume wlog that $t^2 = \omega \in \mathbb{F}_p$. In this case, $f = \lambda_0 + \cdots + \lambda_g x^g + \mu y$ and the polynomial F is of the form

$$F(x) = (\lambda_0 x + \cdots + \lambda_g x^g + x^{g+1})^2 - \mu^2 h(x).$$

In particular, when the parameter μ equals 0, f is independent of the y variable; the corresponding relation of type (2) is thus necessarily of the form $(P_1) + (\iota(P_1)) + \ldots + (P_{g+1}) + (\iota(P_{g+1})) - (2g+2)\mathcal{O}_\mathcal{H} \sim 0$, where $\iota(P)$ is the image of P by the hyperelliptic involution. To avoid these trivial relations, we look only for solutions $(\lambda_{0,0}, \ldots, \lambda_{g,0}, \lambda_{0,1}, \ldots, \lambda_{g,1}, \mu_0, \mu_1) \in \mathbb{V}_{\mathbb{F}_p}(I : (\mu_0, \mu_1)^\infty)$, where I is the ideal corresponding to the $2(g + 1)$ quadratic polynomials in $2(g + 2)$ variables arising from the scalar restriction on $F \in \mathbb{F}_{p^2}[x]$, setting $\lambda_i = \lambda_{i,0} + t\lambda_{i,1}$ and $\mu^2 = \mu_0 + t\mu_1$. More precisely, with the type of extension considered here, the ideal I is given by the equations corresponding to the vanishing of the coefficients of the $\mathbb{F}_p[x]$-polynomial

$$2(1 \cdot x^{g+1} + \lambda_{g,0} x^g + \cdots + \lambda_{0,0})(\lambda_{g,1} x^g + \cdots + \lambda_{0,1}) - \mu_0 h_1(x) - \mu_1 h_0(x),$$

where $h(x) = h_0(x) + t h_1(x)$, $h_0, h_1 \in \mathbb{F}_p[x]$. This ideal I is not 2-dimensional, but its saturation is. An easy but crucial remark is that the ideal is multi-homogeneous, generated by polynomials of bi-degree $(1, 1)$ in the variables $(1 : \lambda_{0,0} : \ldots : \lambda_{g,0}), (\lambda_{0,1} : \ldots : \lambda_{g,1} : \mu_0 : \mu_1)$. This additional structure has two major consequences. First, the lexicographic Gröbner basis computation is much faster than for a generic quadratic system of the same size. Second, if we denote by π_1 the projection on the first block of variables $(\lambda_{0,0}, \ldots, \lambda_{g,0})$, then the image $\pi_1(\mathbb{V}_{\mathbb{F}_p}(I : (\mu_0, \mu_1)^\infty)) = \pi_1(\overline{\mathbb{V}_{\mathbb{F}_p}(I) \setminus \mathbb{V}_{\mathbb{F}_p}(\mu_0, \mu_1)})$ is a dimension 1 variety (whose equations are easily deduced from the lex Gröbner basis of I), and each fiber is a 1-dimensional vector space.

From this, we can simplify the relation search. Rather than evaluating a first variable, we choose a point $(\lambda_{0,0}, \ldots, \lambda_{g,0}) \in \pi_1(\mathbb{V}_{\mathbb{F}_p}(I : (\mu_0, \mu_1)^\infty))$ and express the remaining variables linearly in terms of $\lambda_{0,1}$, so that now F belongs to $\mathbb{F}_p[x, \lambda_{0,1}]$ and has degree $2g + 2$ in x and 2 in $\lambda_{0,1}$. Instead of trying to factor F for many values of $\lambda_{0,1}$, the key idea is to compute for each $x \in \mathbb{F}_p$ the values of $\lambda_{0,1}$ such that $F(x, \lambda_{0,1}) = 0$. Since F has degree 2 in $\lambda_{0,1}$, this can be done very efficiently by computing the square root of the discriminant. In fact, we can speed the process even more by tabulating the square roots of $\mathbb{F}_p$. Our sieving process consists, for each root $\lambda_{0,1}$, to increment a counter corresponding to this value of $\lambda_{0,1}$; when one of these counters reaches $2g + 2$, then the polynomial F evaluated at the corresponding value of $\lambda_{0,1}$ splits into $2g + 2$ distinct linear terms, yielding a relation. This technique not only allows to skip the factorization of a degree $2g + 2$ polynomial, but is also well-suited to the double large prime variation, as explained in next section.

3.3   Complexity Analysis

Constructing the cover $\mathcal{H}_{|\mathbb{F}_q}$ of an elliptic curve $E_{|\mathbb{F}_{q^d}}$ with the GHS method and transferring the DLP from $G \subset E(\mathbb{F}_{q^d})$ to $G' \subset \mathrm{Jac}_{\mathcal{H}}(\mathbb{F}_q)$ has essentially a unit cost, which is negligible compared to the rest of the attack. The complexity of the decomposition phase is divided between the relation search and the linear algebra steps. In order to collect about $p/2$ relations using Nagao's decomposition method, we need to solve on average $(ng)! \cdot p/2$ quadratic polynomial systems. The resolution cost of this kind of system using e.g. Gröbner bases is hard to estimate precisely, but is at least polynomial in the degree $2^{(n-1)ng}$ of the corresponding zero-dimensional ideal. The linear algebra step then costs $O(ngp^2)$ operations modulo $\#G$, using sparse linear algebra techniques. With the second decomposition method, we need to compute first the lexicographic Gröbner basis of an ideal generated by $n(n-1)g + 2n - 2$ quadratic equations in $n(n-1)g + 2n$ variables. This cost is also at least exponential in n^2g, but the Gröbner basis computation has to be done only once. Afterwards, we have to solve on average $(ng + 2)! \cdot p/2$ "easy" systems. The complexity of the linear algebra step is the same (the cost of the descent is negligible compared to the sieving phase).

When p is large relatively to n and g, the linear algebra becomes the dominating phase. It is nevertheless possible to rebalance the cost of the two steps. Indeed, collecting extra relations can speed up the logarithm computations; this is the idea behind structured Gaussian elimination [21] and double large prime variation. The analysis of [15] shows that with the latter, the asymptotic complexity of our cover and decomposition attack becomes either $\tilde{O}(p^{2-2/ng})$ or $\tilde{O}(p^{2-2/(ng+2)})$ with the decomposition variant, as p grows to infinity for fixed n and g. Although the complexity of the variant is asymptotically higher, the much smaller hidden constant means that it is actually faster for accessible values of p. Note that it is straightforward to parallelize the relation search phase; this is also possible, but much less efficiently, for the linear algebra step. In particular, the optimal choice of the balance depends not only of the implementation but also of the computing power available.

When $n = 2$, it is possible to improve the double large prime variation by sieving only among the values of x corresponding to the abscissae of points of the "small primes" factor base. As soon as $2g$ values of x are associated to one value of $\lambda_{0,1}$, we obtain a relation involving at most 2 large primes (if the remaining degree 2 factor is split, which occurs with probability close to $1/2$). This speeds up the relation search and decreases the overall complexity from $\tilde{O}(p^{2-2/(2g+2)})$ to $\tilde{O}(p^{2-2/(2g+1)})$ as p grows to infinity, thus reducing the asymptotic gap between the two decomposition methods without degrading the practical performances.

Obviously, our approach outperforms generic algorithms only if the genus of the intermediate cover is not too large. Otherwise, it may be possible to transfer the DLP from E to a more vulnerable isogenous curve E'. There exist two "isogeny walk" strategies to find E' (if it exists) [17]: one can sample the isogeny class of E via low-degree isogenies until a weak curve is found, or one can try all the weak curves until a curve isogenous to E is found. The best strategy to use depends on the size of the isogeny class, on the number of weak curves and on the availability of an efficient algorithm for constructing these weak curves. For the cases we have considered, this isogeny walk can become the dominating part in the overall complexity (see below for details).

4   Application to Elliptic Curves Defined over $\mathbb{F}_{p^6}$

For an elliptic curve E defined over an extension field $\mathbb{F}_{p^6}$, we can apply our cover and decomposition attack either with the tower $\mathbb{F}_{p^6} - \mathbb{F}_{p^2} - \mathbb{F}_p$ or with the tower $\mathbb{F}_{p^6} - \mathbb{F}_{p^3} - \mathbb{F}_p$. We have seen in Section 2.2 that in practice, we can compute decompositions only for a very limited number of values of (n, g). In particular, our attack is feasible only if E admits a genus 3 (resp. 2) cover; we give examples of such curves below. Of course, this attack needs to be compared with the classic cover attacks or decomposition attacks using the base field $\mathbb{F}_{p^3}, \mathbb{F}_{p^2}$ or $\mathbb{F}_p$, as recalled in Section 2.

4.1   Using a Genus 3 Cover

In the present subsection, we apply our cover and decomposition attack using the first tower $\mathbb{F}_{p^6} - \mathbb{F}_{p^2} - \mathbb{F}_p$. Thanks to the results of [7,25,32], in odd characteristic, we know that the only elliptic curves defined over $\mathbb{F}_{q^3}$ (in our case, $q = p^2$) for which the GHS attack yields a cover by a hyperelliptic curve $\mathcal{H}$ of genus 3 defined over $\mathbb{F}_q$, are of the form

$$y^2 = h(x)(x - \alpha)(x - \sigma(\alpha)) \tag{3}$$

where σ is the Frobenius automorphism of $\mathbb{F}_{q^3}/\mathbb{F}_q$, $\alpha \in \mathbb{F}_{q^3} \setminus \mathbb{F}_q$ and $h \in \mathbb{F}_q[x]$ is of degree 1 or 2. Similar results are also available in characteristic 2 (see [27]), thus our attack is also applicable in characteristic 2; we give details of the construction of the cover in both cases in Appendix A. The number of curves admitting an equation of the form (3) is $\Theta(q^2)$, thus only a small proportion of curves is directly vulnerable to the cover and decomposition attack using this extension tower. However, since this number of weak curves is much larger than the number of isogeny classes (which is about $q^{3/2}$), a rough reasoning would conclude that essentially all curves should be insecure using an isogeny walk strategy. Assuming that the probability for a curve to be weak is independent from its isogeny class, we obtain that the average number of steps before reaching a weak isogenous curve should be about $q = p^2$ steps. It is thus the dominating phase of the algorithm, but is still better than the $\tilde{O}(p^3)$-generic attacks. Nevertheless, all the curves of the form (3) have a cardinality divisible by 4, so obviously not all curves are vulnerable to this isogeny walk. Still, we conjecture that all curves with cardinality divisible by 4 are vulnerable to this cover and decomposition attack using an isogeny walk.

We can also consider non-hyperelliptic genus 3 covers. In this case, weak curves have equation $y^2 = c(x - \alpha)(x - \sigma(\alpha))(x - \beta)(x - \sigma(\beta))$, where $c \in \mathbb{F}_{q^3}$ and either $\alpha, \beta \in \mathbb{F}_{q^3} \setminus \mathbb{F}_q$ or $\alpha \in \mathbb{F}_{q^6} \setminus (\mathbb{F}_{q^2} \cup \mathbb{F}_{q^3})$ and $\beta = \sigma^3(\alpha)$. This targets much more curves: actually, about half of the curves having their full 2-torsion defined over $\mathbb{F}_{q^3}$ admit an equation of this form [25]. For a genus 3 hyperelliptic cover over $\mathbb{F}_{p^2}$, the quadratic polynomial systems to solve over $\mathbb{F}_p$ are composed of 6 variables and 6 equations, or 8 equations and 10 variables with our variant. Such systems can be solved very quickly by any computational algebra system. Unfortunately, with non-hyperelliptic covers, the systems of equations are much more complicated, and we have not been able to compute decompositions with available Gröbner basis implementations.

4.2   Using a Genus 2 Cover

We now consider the tower $\mathbb{F}_{p^6} - \mathbb{F}_{p^3} - \mathbb{F}_p$. The existence of genus 2 covers (which are necessarily hyperelliptic) defined over $\mathbb{F}_q$, where $q = p^3$, has been studied in [3,29]. In odd characteristic, vulnerable curves admit an equation in so-called Scholten form

$$y^2 = ax^3 + bx^2 + \sigma(b)x + \sigma(a) \qquad (4)$$

where $a, b \in \mathbb{F}_{q^2}$ and σ is the Frobenius automorphism of $\mathbb{F}_{q^2}/\mathbb{F}_q$. An elliptic curve E can be transformed into Scholten form as soon as its full 2-torsion is defined over $\mathbb{F}_{q^2}$ [29] or its cardinality is odd and $j(E) \notin \mathbb{F}_q$ [3]. Consequently, a large proportion of curves are vulnerable to our cover and decomposition attack. Moreover, any curve without full 2-torsion but still with a cardinality divisible by 4, is 2-isogenous to a curve with full 2-torsion. In this setting, the quadratic polynomial systems to solve over $\mathbb{F}_p$ are composed of 12 variables and 12 equations, or 16 equations and 18 variables with the decomposition variant. Solving such systems is still feasible on current personal computers, but is much lower than in the case of hyperelliptic genus 3 cover defined over $\mathbb{F}_{p^2}$.

4.3   Complexity and Comparison with Other Attacks

Apart from the generic algorithms, the existing ECDLP attacks over sextic extensions are either Gaudry's decomposition method [13] or the GHS attack followed by Gaudry's or Diem's index calculus [8,15], with base field $\mathbb{F}_p$ or $\mathbb{F}_{p^2}$ (using $\mathbb{F}_{p^3}$ as base field does not provide any advantage in this context). When the base field is $\mathbb{F}_{p^2}$, the asymptotic complexity is in $\tilde{O}\left(p^{8/3}\right)$ for both decomposition and GHS (assuming a genus 3 cover), or even in $\tilde{O}\left(p^2\right)$ with a degree 4 planar cover. But in all cases, the memory requirement is then very large, in $\tilde{O}(p^2)$. When the base field is $\mathbb{F}_p$, computing direct decompositions is completely out of reach, and the GHS attack very rarely provides low genus covers: the smallest possible genus is actually 9 (with corresponding degree 10 plane model), which occurs for at most p^3 curves; the resulting genus is much higher for most curves [5], implying that this attack is rarely practical.

We give below a summary of the performances of the presented approaches. In order to obtain actual (and not just asymptotic) comparisons, we also consider the cryptographically significant example of a curve $E_{|\mathbb{F}_{p^6}}$ where p is a prime close to 2^{27}, whose cardinality is divisible by a 160-bit prime number. The values given are obviously just estimates relying on extrapolations of relation searches done on Magma V2-17-5 with an Intel Core 2 Duo processor (see details in Appendix B); in particular, the two last estimates are greater than what could be expected from the results obtained with optimized implementation presented in Section 5.

Attack	Asymptotic complexity	Memory complexity	Time estimates (years)
Pollard on $E(\mathbb{F}_{p^6})$ [28]	$\tilde{O}(p^3)$	$\tilde{O}(1)$	5×10^{13}
Ind. calc. on $\mathrm{Jac}_{\mathcal{H}}(\mathbb{F}_{p^2})$, $g = 3$ [15]	$\tilde{O}(p^{8/3})$	$\tilde{O}(p^2)$	6×10^{10} †
Ind. calc. on $\mathrm{Jac}_{\mathcal{C}}(\mathbb{F}_{p^2})$, $d = 4$ [8]	$\tilde{O}(p^2)$	$\tilde{O}(p^2)$	700 000
Decompositions on $E((\mathbb{F}_{p^2})^3)$ [13]	$\tilde{O}(p^{8/3})$	$\tilde{O}(p^2)$	10^{12}
Ind. calc. on $\mathrm{Jac}_{\mathcal{C}}(\mathbb{F}_p)$, $d = 10$ [8]	$\tilde{O}(p^{7/4})$	$\tilde{O}(p)$	$1\,500^{(*)}$
Decomp. on $\mathrm{Jac}_{\mathcal{H}}(\mathbb{F}_{p^3})$, $g = 2$ [this work]	$\tilde{O}(p^{5/3})$	$\tilde{O}(p)$	4×10^6
Decomp. on $\mathrm{Jac}_{\mathcal{H}}(\mathbb{F}_{p^2})$, $g = 3$ [this work]	$\tilde{O}(p^{5/3})$	$\tilde{O}(p)$	$750^{\dagger}$
Sieving on $\mathrm{Jac}_{\mathcal{H}}(\mathbb{F}_{p^2})$, $g = 3$ [this work]	$\tilde{O}(p^{12/7})$	$\tilde{O}(p)$	$300^{\dagger}$

†: only for $\Theta(p^4)$ curves before isogeny walk   (*): only for $O(p^3)$ curves

5 A 149-Bit Example

In this section, we give a practical example of the cover and decomposition attack for an elliptic curve defined over the Optimal Extension Field $\mathbb{F}_{p^6}$, where $p = 2^{25} + 35$ is the smallest 26-bit prime. We define $\mathbb{F}_{p^2}$ as $\mathbb{F}_p[t]$ where $t^2 = 2$ and $\mathbb{F}_{p^6}$ as $\mathbb{F}_{p^2}[\theta]$ where $\theta^3 = t$. The elliptic curve E is given by the equation

$$y^2 = x(x - \alpha)(x - \sigma(\alpha)),$$

where $\sigma : x \mapsto x^{p^2}$ and $\alpha = 9\,819\,275 + 31\,072\,607\,\theta + 17\,686\,237\,\theta^2 + 31\,518\,659\,\theta^3 + 22\,546\,098\,\theta^4 + 17\,001\,125\,\theta^5$. It has a genus 3 cover by the hyperelliptic curve $\mathcal{H}$ defined over $\mathbb{F}_{p^2}$

which is of the form $y^2 = \left(x + \phi(x) + \phi^\sigma(x) + \phi^{\sigma^2}(x)\right) N(x)^2$, with $N(x)$ the minimal polynomial of α over $\mathbb{F}_{p^2}$ and $\phi : x \mapsto \dfrac{\left(\alpha - \sigma^2(\alpha)\right)\left(\sigma(\alpha) - \sigma^2(\alpha)\right)}{x - \sigma^2(\alpha)} + \sigma^2(\alpha)$. The cover map π is given by:

$$\pi(x, y) = \left(\frac{x + \phi(x) + \phi^\sigma(x) + \phi^{\sigma^2}(x)}{4}, \; \frac{y(x - \phi^\sigma(x))(x - \phi^{\sigma^2}(x))}{8N(x)(x - \sigma^2(\alpha))} \right).$$

The common cardinality of E over $\mathbb{F}_{p^6}$ and of the Jacobian of $\mathcal{H}$ over $\mathbb{F}_{p^2}$ is four times the 149-bit prime $\ell = 356814156285346166966901450449051336101786213$, and the number of elements in the factor base is $16\,775\,441$.

For best performances, we use the sieving approach described in Section 3.2. As a first step, we compute a lex Gröbner basis of the system composed of 10 quadratic equations in 8 variables in about $3\,\mathrm{s}$ on a 2.6 GHz Intel Core 2 Duo processor with Magma V2.16-12 [6]. Instead of the double large prime variation, we execute a structured Gaussian elimination. During the sieving phase, we used $1\,024$ cores of quadri-core Intel Xeon 5570 processors at 2.93 GHz. After 62 h, we had collected about $1.4 \times 10^{10} \simeq p^2/(2 \cdot 8!)$ relations, that is all the possible relations of the form (2). For comparison, we also tested Nagao-style decompositions on the same type of processors. Such a test takes about 22 ms on a single core, showing that our decomposition variant is about 960 times faster.

Thanks to the large number of extra relations, structured Gaussian elimination performed quite well and, after $25.5\,\mathrm{h}$ on 32 cores, it reduces the number of unknowns to $3\,092\,914$ (a fivefold reduction). For safety, the output system contains $15\,000$ more equations than this number of unknowns, and each equation involves between 8 and 182 basis elements. The total number of non-zero entries in all the equations is $191\,098\,665$ and all these entries are equal to ± 1. The most time-consuming step is the iterative linear algebra, which is done with a MPI implementation of the Lanczos algorithm. It took about 28.5 days on 64 cores of the same Intel Xeon processors. A large fraction of this time was taken by the MPI communications, since at each round 200 MB of data had to be broadcast between the 2 involved machines (32 cores/machine). This linear algebra phase produced discrete logarithms for all the basis elements that remained after structured Gaussian elimination. Substituting these values back in the initial linear system, we recovered, in less than $12\,\mathrm{h}$ using 32 cores, the discrete logarithms modulo ℓ of all elements in the basis (given by their coordinates on $\mathcal{H}$):

$$\log(5, 1\,646\,475 + 19\,046\,912\,t) = 32409023361661844778889944628386231778304600\,6$$
$$\log(6, 2\,062\,691 + 792\,228\,t) = 1344249878422626954866114769897690521528324\,41$$
$$\vdots$$
$$\log(33\,554\,465, 4\,471\,075 + 14\,598\,628\,t) = 34071346746090041947316793372263165411114515\,1$$

With the results of the above precomputation, computing logarithm of arbitrary points on the elliptic curve becomes easy. To demonstrate this, we constructed points on E with the following process and computed their logarithms. First, we let $X_0 = \sum_{j=0}^5 \left(\lceil \pi \cdot p^{j+1} \rceil \bmod p \right) \theta^j = 4\,751\,066 + 748\,974\,\theta + 8\,367\,234\,\theta^2 + 24\,696\,290\,\theta^3 + 1\,372\,315\,\theta^4 + 7\,397\,713\,\theta^5$. We then constructed points on E with abscissa $X_0 + \delta$ for small offsets δ. Let P_1, P_2, P_3, P_4, and P_5 be the points corresponding to offsets

3, 4, 11, 14 and 15. We lift each of these points to the Jacobian of $\mathcal{H}$ using the conorm-norm map, which takes negligible time in Magma. After that, we apply the descent method of Section 3.1 to small multiples of the lifted element, until we find a multiple that decomposes as a sum of elements from the smoothness basis. Looking up the corresponding logarithms (and dividing back by the small multiples that have been included) yields the logarithm of each point. On average, we expect to try $6! = 720$ multiples before finding a decomposition. To actually decompose the five considered points, we needed $61.3\,\mathrm{s}$. As a consequence, each individual logarithm on E can be performed in less than one minute. We give details below: the points involved in the decomposition are described by their abscissa together with a $+$ or $-$ sign that indicates whether the "real" part of the ordinate has a positive or negative representative in $(-p/2, p/2)$. Similarly, we indicate the choice of the points on E (as produced by Magma) with a $+$ or a $-$ depending on the representative of the constant term in the ordinate.

Points	Mult. Nagao	Points in decomposition					
$(X_0 + 3)^+$	97	2844007^+	3819744^-	5618276^-	8396644^-	11841629^-	23771773^-
$(X_0 + 4)^-$	36	4673075^-	11272201^+	12937918^-	13869464^-	14428213^+	21399158^-
$(X_0 + 11)^+$	742	4884810^-	6230068^-	8411592^-	12188294^+	20118618^+	20945232^-
$(X_0 + 14)^-$	956	3660673^-	4314732^-	20180301^+	22563519^+	26157093^-	27107773^-
$(X_0 + 15)^-$	682	780652^+	8444164^+	10116987^+	11070139^-	14566563^-	32232816^+

The group structure of E is $\mathbb{Z}/2\mathbb{Z} \times \mathbb{Z}/(2\ell)\mathbb{Z}$ and all the logarithms are computed mod ℓ. Thus, in order to obtain points of order ℓ, we multiply each of the points P_j by 2. To obtain the discrete logarithms in base P_1, we simply divide the results by the logarithm of P_1. Finally, we obtain:

$$2\cdot P_2 = 448534294563063755208836857225511427502049\,29\cdot 2\cdot P_1$$
$$2\cdot P_3 = 2458287441772026421671868666551887048603090\,93\cdot 2\cdot P_1$$
$$2\cdot P_4 = 2417736985739928971972614543487604064993258\,84\cdot 2\cdot P_1$$
$$2\cdot P_5 = 479144347310864978609802733270378337321097\,67\cdot 2\cdot P_1$$

6 Conclusion and Perspectives

In this paper, we have proposed a new index calculus algorithm to compute discrete logarithms on elliptic curves defined over extension fields of composite degree. In particular, sextic extensions are very well-suited to this method, as we have practically demonstrated on a 149-bit example. This combination of cover and decomposition techniques raises many questions. For example, it would be interesting to know if elliptic curves of prime cardinality defined over a degree 6 extension field can be efficiently attacked. A related problem is how to target more curves easily: this requires either an improvement of the isogeny walk, or an efficient use of non-hyperelliptic covers. Finally, whether our method applies to different extension degrees is an important issue; as will be explained in the extended version of this article, degree 4 extensions are also susceptible, but the advantage over generic methods is then less significant.

Acknowledgements. We acknowledge that the results in this paper have been achieved using the PRACE Research Infrastructure resource Curie based in France at TGCC, Bruyères-le-Chatel.

References

1. Adleman, L.M.: The Function Field Sieve. In: Huang, M.-D.A., Adleman, L.M. (eds.) ANTS 1994. LNCS, vol. 877, pp. 108–121. Springer, Heidelberg (1994)
2. Adleman, L.M., DeMarrais, J., Huang, M.-D.: A Subexponential Algorithm for Discrete Logarithms over the Rational Subgroup of the Jacobians of Large Genus Hyperelliptic Curves over Finite Fields. In: Huang, M.-D.A., Adleman, L.M. (eds.) ANTS 1994. LNCS, vol. 877, pp. 28–40. Springer, Heidelberg (1994)
3. Arita, S., Matsuo, K., Nagao, K.-I., Shimura, M.: A Weil descent attack against elliptic curve cryptosystems over quartic extension fields. IEICE Trans. Fundam. Electron. Commun. Comput. Sci. E89-A, 1246–1254 (2006)
4. Bailey, D.V., Paar, C.: Efficient arithmetic in finite field extensions with application in elliptic curve cryptography. J. Cryptology 14(3), 153–176 (2001)
5. Blake, I.F., Seroussi, G., Smart, N.P. (eds.): Advances in elliptic curve cryptography. London Mathematical Society Lecture Note Series, vol. 317. Cambridge University Press, Cambridge (2005)
6. Bosma, W., Cannon, J., Playoust, C.: The Magma algebra system. I. The user language. J. Symbolic Comput. 24(3-4), 235–265 (1997)
7. Diem, C.: The GHS attack in odd characteristic. J. Ramanujan Math. Soc. 18(1), 1–32 (2003)
8. Diem, C.: An Index Calculus Algorithm for Plane Curves of Small Degree. In: Hess, F., Pauli, S., Pohst, M. (eds.) ANTS 2006. LNCS, vol. 4076, pp. 543–557. Springer, Heidelberg (2006)
9. Diem, C.: On the discrete logarithm problem in elliptic curves. Compos. Math. 147(1), 75–104 (2011)
10. Frey, G.: How to disguise an elliptic curve (Weil descent). Talk at the 2nd Elliptic Curve Cryptography Workshop (ECC) (1998)
11. Galbraith, S.D., Hess, F., Smart, N.P.: Extending the GHS Weil Descent Attack. In: Knudsen, L.R. (ed.) EUROCRYPT 2002. LNCS, vol. 2332, pp. 29–44. Springer, Heidelberg (2002)
12. Gaudry, P.: An Algorithm for Solving the Discrete Log Problem on Hyperelliptic Curves. In: Preneel, B. (ed.) EUROCRYPT 2000. LNCS, vol. 1807, pp. 19–34. Springer, Heidelberg (2000)
13. Gaudry, P.: Index calculus for abelian varieties of small dimension and the elliptic curve discrete logarithm problem. J. Symbolic Comput. 44(12), 1690–1702 (2008)
14. Gaudry, P., Hess, F., Smart, N.P.: Constructive and destructive facets of Weil descent on elliptic curves. J. Cryptology 15(1), 19–46 (2002)
15. Gaudry, P., Thomé, E., Thériault, N., Diem, C.: A double large prime variation for small genus hyperelliptic index calculus. Math. Comp. 76, 475–492 (2007)
16. Hess, F.: Generalising the GHS attack on the elliptic curve discrete logarithm problem. LMS J. Comput. Math. 7, 167–192 (2004) (electronic)
17. Hess, F.: Weil descent attacks. In: Advances in Elliptic Curve Cryptography. London Math. Soc. Lecture Note Ser, vol. 317, pp. 151–180. Cambridge Univ. Press, Cambridge (2005)

18. Joux, A., Vitse, V.: Elliptic curve discrete logarithm problem over small degree extension fields. J. Cryptology, 1–25 (2011), doi:10.1007/s00145-011-9116-z
19. Joux, A., Vitse, V.: A Variant of the F4 Algorithm. In: Kiayias, A. (ed.) CT-RSA 2011. LNCS, vol. 6558, pp. 356–375. Springer, Heidelberg (2011)
20. Koblitz, N.: Elliptic curve cryptosystems. Math. Comp. 48(177), 203–209 (1987)
21. LaMacchia, B.A., Odlyzko, A.M.: Computation of discrete logarithms in prime fields. Des. Codes Cryptogr. 1(1), 47–62 (1991)
22. Lenstra, A.K., Lenstra Jr., H.W. (eds.): The development of the number field sieve. Lecture Notes in Math., vol. 1554. Springer, Berlin (1993)
23. Menezes, A., Teske, E., Weng, A.: Weak Fields for ECC. In: Okamoto, T. (ed.) CT-RSA 2004. LNCS, vol. 2964, pp. 366–386. Springer, Heidelberg (2004)
24. Miller, V.S.: Use of Elliptic Curves in Cryptography. In: Williams, H.C. (ed.) CRYPTO 1985. LNCS, vol. 218, pp. 417–426. Springer, Heidelberg (1986)
25. Momose, F., Chao, J.: Scholten forms and elliptic/hyperelliptic curves with weak Weil restrictions. Cryptology ePrint Archive, Report 2005/277 (2005)
26. Nagao, K.-i.: Decomposition Attack for the Jacobian of a Hyperelliptic Curve over an Extension Field. In: Hanrot, G., Morain, F., Thomé, E. (eds.) ANTS-IX. LNCS, vol. 6197, pp. 285–300. Springer, Heidelberg (2010)
27. Nart, E., Ritzenthaler, C.: Genus 3 curves with many involutions and application to maximal curves in characteristic 2. In: Arithmetic, Geometry, Cryptography and Coding Theory 2009. Contemp. Math., vol. 521, pp. 71–85. Amer. Math. Soc., Providence (2010)
28. Pollard, J.M.: Monte Carlo methods for index computation (mod p). Math. Comp. 32(143), 918–924 (1978)
29. Scholten, J.: Weil restriction of an elliptic curve over a quadratic extension, http://homes.esat.kuleuven.be/~jscholte/weilres.pdf
30. Semaev, I.A.: Summation polynomials and the discrete logarithm problem on elliptic curves. Cryptology ePrint Archive, Report 2004/031 (2004)
31. Shoup, V.: Lower Bounds for Discrete Logarithms and Related Problems. In: Fumy, W. (ed.) EUROCRYPT 1997. LNCS, vol. 1233, pp. 256–266. Springer, Heidelberg (1997)
32. Thériault, N.: Weil descent attack for Kummer extensions. J. Ramanujan Math. Soc. 18(3), 281–312 (2003)

A   Genus 3 Cover

A.1   Odd Characteristic

We consider elliptic curves defined over $\mathbb{F}_{q^3}$ of the form

$$y^2 = h(x)(x - \alpha)(x - \sigma(\alpha)) \tag{5}$$

where σ is the Frobenius automorphism of $\mathbb{F}_{q^3}/\mathbb{F}_q$, $\alpha \in \mathbb{F}_{q^3} \setminus \mathbb{F}_q$ and $h \in \mathbb{F}_q[x]$ is of degree 1 or 2. Such elliptic curves were studied in [7,32]; they are the only elliptic curves for which the GHS attack yields a cover by a hyperelliptic curve $\mathcal{H}$ of genus 3 defined over $\mathbb{F}_q$. We give an explicit description of the cover $\pi : \mathcal{H} \to E$; following [25], we express this cover as a quotient by a bi-elliptic involution, instead of using the GHS approach. For simplicity, we will assume that $h(x) = x$ (this can always be achieved by an appropriate change of coordinates if h has a root in $\mathbb{F}_q$). Let $\phi : x \mapsto \frac{D}{x - \sigma^2(\alpha)} + \sigma^2(\alpha)$ be the unique involution of $\mathbb{P}^1(\overline{\mathbb{F}_q})$ sending $\sigma^2(\alpha)$ to ∞ and α to $\sigma(\alpha)$, so that $D = (\alpha - \sigma^2(\alpha))(\sigma(\alpha) - \sigma^2(\alpha))$. If ϕ lifts to an involution of a hyperelliptic curve $\mathcal{H}_{|\mathbb{F}_q}$, then necessarily ϕ^σ and ϕ^{σ^2} will be also involutions of $\mathcal{H}$. Observing that $\{Id, \phi, \phi^\sigma, \phi^{\sigma^2}\}$ forms a group, this leads us to consider the curve of equation $y^2 = x + \phi(x) + \phi^\sigma(x) + \phi^{\sigma^2}(x)$; a more usual form for this equation is

$$\mathcal{H} : \quad y^2 = F(x)N(x) \tag{6}$$

where $N(x) = (x - \alpha)(x - \sigma(\alpha))(x - \sigma^2(\alpha))$ is the minimal polynomial of α over $\mathbb{F}_q$ and $F(x) = N(x)(x + \phi(x) + \phi^\sigma(x) + \phi^{\sigma^2}(x)) \in \mathbb{F}_q[x]$. It is clear that ϕ gives an involution of $\mathcal{H}$, still denoted by $\phi : (x, y) \mapsto \left(\frac{D}{x - \sigma^2(\alpha)} + \sigma^2(\alpha), \frac{yD^2}{(x - \sigma^2(\alpha))^4}\right)$. The quotient of this genus 3 hyperelliptic curve $\mathcal{H}$ by ϕ is the elliptic curve

$$E' : \quad y^2 = (x - \alpha - \sigma(\alpha))(x^2 - 4\alpha\sigma(\alpha))$$

and the quotient map $\pi' : \mathcal{H} \to E'$ satisfies $\pi'(x, y) = (x + \phi(x), y/(x - \sigma^2(\alpha))^2)$. The curve E' is 2-isogenous to the original curve $E : y^2 = x(x - \alpha)(x - \sigma(\alpha))$ via the map $(x, y) \mapsto \left(\frac{x^2 - 4\alpha\sigma(\alpha)}{4(x - \alpha - \sigma(\alpha))}, y\frac{(x - 2\alpha)(x - 2\sigma(\alpha))}{8(x - \alpha - \sigma(\alpha))^2}\right)$. Finally, the cover map $\pi : \mathcal{H} \to E$ has the expression

$$\pi(x, y) = \left(\frac{F(x)}{4N(x)}, \frac{y(x - \phi^\sigma(x))(x - \phi^{\sigma^2}(x))}{8N(x)(x - \sigma^2(\alpha))}\right). \tag{7}$$

In the general case, when E has equation (5), the cover (7) remains the same and the corresponding hyperelliptic curve $\mathcal{H}$ of genus 3 defined over $\mathbb{F}_q$ has the following equation:

$$\mathcal{H} : \quad y^2 = 4N(x)^2 h\left(\frac{F(x)}{4N(x)}\right).$$

A.2   Characteristic 2

Let $E \ y^2 + xy = x^3 + ax^2 + b$ be an ordinary curve defined over a binary field $\mathbb{F}_{q^3}$, where $b = 1/j(E)$. As already apparent in [14], the GHS attack produces a genus 3 hyperelliptic cover of E when $\mathrm{Tr}_{\mathbb{F}_{q^3}/\mathbb{F}_q}(b) = 0$, so that $\Theta(q^2)$ curves are directly vulnerable. To describe this cover, we slightly adapt the description of [25,27], already used in the previous subsection. Let $\sigma : x \mapsto x^q$ be the Frobenius automorphism and let $v = \sqrt[4]{b}$; by assumption its trace over $\mathbb{F}_q$ is zero. As in the case of odd characteristic, we consider the involution $\phi : x \mapsto \sigma(v)\sigma^2(v)/(x+v) + v$ of $\mathbb{P}^1(\overline{\mathbb{F}_q})$ sending v to infinity and $\sigma(v)$ to $\sigma^2(v)$. We denote by N the minimal polynomial of v over $\mathbb{F}_q$ and by F the product $N(x)(x + \phi(x) + \phi^\sigma(x) + \phi^{\sigma^2}(x)) \in \mathbb{F}_q[x]$. Then, ϕ lifts to a bi-elliptic involution of the hyperelliptic curve $\mathcal{H}_{|\mathbb{F}_q}$ defined by

$$\mathcal{H} : \quad y^2 + N(x)y = F(x)N(x) + aN(x)^2. \tag{8}$$

The curve E is up to a change of variable the quotient of $\mathcal{H}$ by ϕ and the cover map from $\mathcal{H}$ to E is given by:

$$\pi : (x,y) \mapsto \left(x + \phi(x) + v, \frac{y(x+\phi(x)+v)}{N(x)} + v^2 \right). \tag{9}$$

B   Complexity Comparisons of Different Attacks on $E(\mathbb{F}_{p^6})$ with $\log_2 p \approx 27$

The basis of comparison for all attacks on the ECDLP comes from generic algorithms such as Pollard's Rho [28]. Using Floyd's cycle-finding algorithm, the expected number of iterations is approximately $0.94\sqrt{\ell} \approx 1.14 \times 10^{24}$ where ℓ is the 160-bit prime dividing the cardinality of $E(\mathbb{F}_{p^6})$. With Magma V2-17-5 on Intel Core 2 Duo 2.6 GHz, it takes 13.91 s to compute 10 000 iterations, corresponding to 5×10^{13} years for the complete DLP resolution.

The main difficulty with the index calculus methods is the estimation of the linear algebra cost, which is needed to find the optimal balance in the large primes variation. We base our extrapolations on the experiment of Section 5, where the resolution of a sparse system of size 3×10^6 took about 44 000 h·CPU. Thus we assume that for a factor base of size n, the linear algebra costs $(n/3\,000\,000)^2 \cdot 44\,000 \cdot 160/148$ h, or $n^2 \cdot 2 \times 10^{-5}$ s. On the other hand, all the relation timings are obtained with Magma as we did not implement optimized versions of all the different attacks.

We first consider index calculus methods for which the size of the factor base is in $p^2/2$. The corresponding memory complexity is clearly problematic for any real implementation, since the sole storage of the factor base elements requires about 2^{60} bits. When E admits a hyperelliptic genus 3 cover $\mathcal{H}_{|\mathbb{F}_{p^2}}$, we can apply index calculus after transfer to its Jacobian. Our experiment takes 13.27 s to complete 10 000 tests, yielding 1 689 relations; the complete relation search thus requires 2×10^6 years. With our assumption, the linear algebra step (memory issues notwithstanding) takes 5×10^{19} years, a much more larger time. To rebalance the two phases using double large primes, we divide the size of factor base by about 40 000; the total computation time then becomes 6×10^{10} years. If E admits a non-hyperelliptic genus 3 cover $\mathcal{C}_{|\mathbb{F}_{p^2}}$, this cover admits a degree 4 plane model on which we can apply Diem's index calculus [8]. It then takes 11.74 s to complete 10 000 tests, yielding 4 972 relations. This means that 700 000 years are necessary to collect $p^2/2$ relations. With the adapted double large prime variation, the optimal small factor base contains about p elements, and the linear algebra cost becomes negligible compared to the relation search. We can finally apply directly Gaudry's attack [13] to E with base field $\mathbb{F}_{p^2}$. Our experiment needs 22.35 s for 100 tests, yielding 36 relations. This is 80 times slower than with the genus 3 hyperelliptic cover, and so the optimal balances are different. The size of the factor base should then be divided by 9 000, for an overall computation time of 10^{12} years.

Now, we consider the index calculus method for which the size of the factor base is in $p/2$. We recall that it is not possible to use Gaudry's decomposition attack with base field $\mathbb{F}_p$. In the very rare case where E admits a non-hyperelliptic genus 9 cover, it is possible to use the attack of [8] on a degree 10 plane model and obtain after 200 000 tests 5 relations in 123 s, implying a time of 50 years for the relation step. With our assumption, the linear algebra costs 3 000 years. The rebalanced optimal size of the factor base corresponds to a twofold reduction, for an overall computation time of 1 500 years. If E admits a genus 3 hyperelliptic cover $\mathcal{H}_{|\mathbb{F}_{p^2}}$, we can apply the techniques presented in this article and search for decompositions in $\mathrm{Jac}_{\mathcal{H}}(\mathbb{F}_{p^2})$ either with Nagao's method or our sieving variant. In the first case, it takes 126 s to run 5 000 tests yielding 9 relations. This means that the relation search would need 30 years, the linear algebra still lasting 3 000 years. The optimal balance corresponds to a reduction by a factor 2.7 of the size of the factor base, for a total computation time of 750 years. In the second case, using the sieving technique we obtained 3 300 relations in 1 800 s, which is 25 times faster than with Nagao's technique (in practice, we have seen in Section 5 that with optimized implementation, the ratio is rather of the order of 900). With the adapted large prime variation, the optimal size of the factor base corresponds to a factor 4.4 reduction, for an overall computation time of 300 years. Note that for this sieving method, we have more accurate experimental data obtained with an optimized implementation in C instead of Magma. We detail below the timings obtained for curves defined over OEF of sizes 138, 144 and 150 bits; the sieving times are given for the collection of all $p^2/(2 \cdot 8!)$ relations, and the linear algebra is done after a structured Gaussian elimination. Based on these figures, we estimate more accurately that breaking the DLP over a 160-bit elliptic curve group would take about 200 years on a single core.

Size of p	Sieving (CPU.hours)	Sieving (real time)	Lanczos (CPU.hours)	Lanczos (real time)
$\log_2 p \approx 23$	3 600	3.5 hours	4 900	77 hours
$\log_2 p \approx 24$	15 400	15 hours	16 000	250 hours
$\log_2 p \approx 25$	63 500	62 hours	43 800	28.5 days

Eventually, it is possible to apply our cover and decomposition technique on a hyperelliptic genus 2 cover defined over $\mathbb{F}_{p^3}$, but without the sieving improvement. On this curve, our experiment takes 3 780 s for a single decomposition test, which is 150 000 times slower than with the same method on a genus 3 cover defined over $\mathbb{F}_{p^2}$. In particular, no rebalance is needed since the relation search dominates the computation time of about 4×10^6 years.

Improving the Complexity of Index Calculus Algorithms in Elliptic Curves over Binary Fields

Jean-Charles Faugère[1,**], Ludovic Perret[1,**], Christophe Petit[2,*], and Guénaël Renault[1,**]

[1] UPMC, Université Paris 06, LIP6
INRIA, Centre Paris-Rocquencourt, PolSys Project-team
CNRS, UMR 7606, LIP6
4 place Jussieu, 75252 Paris, Cedex 5, France
{Jean-Charles.Faugere,Ludovic.Perret,Guenael.Renault}@lip6.fr
[2] UCL Crypto Group,
Université catholique de Louvain
Place du levant 3, 1348 Louvain-la-Neuve, Belgium
christophe.petit@uclouvain.be

Abstract. The goal of this paper is to further study the index calculus method that was first introduced by Semaev for solving the ECDLP and later developed by Gaudry and Diem. In particular, we focus on the step which consists in decomposing points of the curve with respect to an appropriately chosen factor basis. This part can be nicely reformulated as a purely algebraic problem consisting in finding solutions to a multivariate polynomial $\mathbf{f}(\mathbf{x_1}, \ldots, \mathbf{x_m}) = \mathbf{0}$ such that $\mathbf{x_1}, \ldots, \mathbf{x_m}$ all belong to some vector subspace of $\mathbb{F}_{2^n}/\mathbb{F}_2$. Our main contribution is the identification of particular structures inherent to such polynomial systems and a dedicated method for tackling this problem. We solve it by means of Gröbner basis techniques and analyze its complexity using the multi-homogeneous structure of the equations. A direct consequence of our results is an index calculus algorithm solving ECDLP over any binary field $\mathbb{F}_{2^n}$ in time $O(2^{\omega\,t})$, with $t \approx n/2$ (provided that a certain heuristic assumption holds). This has to be compared with Diem's [14] index calculus based approach for solving ECDLP over $\mathbb{F}_{q^n}$ which has complexity $\exp\big(O(n \log(n)^{1/2})\big)$ for $q = 2$ and n a prime (but this holds without any heuristic assumption). We emphasize that the complexity obtained here is very conservative in comparison to experimental results. We hope the new ideas provided here may lead to efficient index calculus based methods for solving ECDLP in theory and practice.

Keywords: Elliptic Curve Cryptography, Index Calculus, Polynomial System Solving.

* F.R.S.-FNRS postdoctoral research fellow at Université Catholique de Louvain, Louvain-la-Neuve. This work was partly supported by the Belgian State's IAP program P6/26 BCRYPT.
** This work was partly supported by the Commission of the European Communities through the ICT program under contract ICT-2007-216676 (ECRYPT-II) and by the French ANR under the Computer Algebra and Cryptography (CAC) project (ANR-09-JCJCJ-0064-01).

D. Pointcheval and T. Johansson (Eds.): EUROCRYPT 2012, LNCS 7237, pp. 27–44, 2012.

1   Introduction

Elliptic curves were independently introduced to cryptography by Miller and Koblitz in 1985 [37,32]. The security of curve-based cryptosystems often relies on the difficulty of solving the well-known Discrete Logarithm Problem on Elliptic Curves (ECDLP). Given a finite cyclic group $G = \langle P \rangle$ and given an element Q in G, the discrete logarithm problem asks for an integer k such that $Q = kP$. For an elliptic curve E defined over a finite field K, the group G can chosen to be the set $E(K)$ of rational points on E.

One of the main method for solving (EC)DLP is *Index Calculus*. This approach, which was first introduced by Kraitchik [33] and later optimized by Adleman [1], can be seen as a general framework [15]. To summarize, Index Calculus algorithms are composed of the following three steps:

1. **Factor Basis definition.** Identify a subset $\mathcal{F} = \{\pi_1, \dots, \pi_s\} \subset G$.
2. **Sieving step.** Collect more than $\#\mathcal{F}$ relations of the form $aP + bQ = \sum_{i=1}^{s} e_i \pi_i$ where a, b are random integers.
3. **Linear Algebra step.** From these relations, construct a matrix with a non-trivial kernel. Then, find a non-trivial vector in this kernel and deduce k (a discrete logarithm) from such vector.

While the last step is independent of the choice of G, the efficiency of the first two steps relies on specific properties of the group. Depending on the ability to find a factor basis together with an algorithm for computing relations during the sieving step, the index calculus method may achieve a *subexponential* complexity. For instance, subexponential algorithms have been obtained for multiplicative groups of finite fields [3,2,4,30] and for Jacobian groups of hyperelleptic curves with large genus [3,27,26].

Related Works and Contributions. Our results are related to the sieving step of the ECDLP index calculus method proposed by Semaev [40] and later developed by Gaudry [28] and Diem [13,14]. The main idea introduced by Semaev is the use of so-called *summation polynomials.* As soon as a factor basis is fixed, summation polynomials can be used for sieving elements of an elliptic curve E and thus an index calculus method follows. The main problem to get an efficient index calculus is to find a factor basis together with an efficient algorithm for solving summation polynomials. During the ECC conference following publication of Semaev's result, Gaudry and Diem independently proposed such solutions. More precisely, when E is defined over an extension $\mathbb{F}_{q^n}$ with $n > 1$ a composite integer, Gaudry [28] proposed to use the following set $\mathcal{F} = \{(\mathbf{x}, \mathbf{y}) \in E(\mathbb{F}_{q^n}) \mid \mathbf{x} \in \mathbb{F}_q\}$ as a factor basis and provides an algorithm for solving the ECDLP with a complexity better than generic methods. Next, the problem of finding $P_1, \dots, P_m$ in $\mathcal{F}$ such that $R = P_1 + \cdots + P_m$ for a given $R \in E$ is reduced to solve the equation $\mathbf{S}_{m+1}((P_1)_x, \dots, (P_m)_x, R_x) = 0$, where $\mathbf{S}_r$ is the r-th Semaev's summation polynomial [40] and $(P)_x$ stands for the x-coordinate of P. Diem proposed a generalization of this approach by considering (in a simpler form here) the factor basis $\mathcal{F}_V := \{(\mathbf{x}, \mathbf{y}) \in E(K) \mid \mathbf{x} \in V\}$, with V a vector subspace of $\mathbb{F}_{q^n}/\mathbb{F}_q$. Thus, the main computational tool for sieving here is an algorithm solving efficiently a specific polynomial system.

Recently, Diem presented in [14] new complexity results for this generalization. He succeeds to prove – without any heuristic assumption – some subexponential complexity results for solving ECDLP. But, for $q = 2$ and n is a prime – the setting considered here – he has an index calculus algorithm of exponential complexity $e^{O\left(n \log(n)^{1/2}\right)}$. The polynomial systems occurring in [14] are solved with a geometrical algorithm proposed by Rojas [39]. Whilst such algorithm has a good complexity estimate, it is well known that its hard to implement it in practice.

In this work, we focus on the specific case $q = 2$ and n prime. We show that the polynomial systems occurring have a very specific structure. We provide a new (heuristic) algorithm taking advantage of the structure for the sieving step. We focus our study on the following point decomposition problem related to some vector space:

Problem 1 (Point Decomposition Problem associated to a vector space V). Let V be a vector space of $\mathbb{F}_{2^n}/\mathbb{F}_2$. Given a point $R \in E(\mathbb{F}_{2^n})$, the problem is to find – if any – m points $P_1, \ldots, P_m$ such that $R = P_1 + \cdots + P_m$ with the additional constraint that $(P_i)_x \in V$ for all $i \in \{1, \ldots, m\}$.

This problem can be reduced to a polynomial system solving problem using Semaev's summation polynomials (or using any other polynomial system modeling). Here, we show that the *multi-homogeneous* structure of the system constructed from Semaev's summation polynomials can be used to design a Gröbner based algorithm. To be more precise, we design an efficient algorithm for:

Problem 2. Let $t \geq 1$, V be a vector space of $\mathbb{F}_{2^n}/\mathbb{F}_2$ and $\mathbf{f} \in \mathbb{F}_{2^n}[\mathbf{x_1}, \ldots, \mathbf{x_m}]$ be any multivariate polynomial of degree bounded by $2^t - 1$ in each variable. The problem is to find $(\mathbf{z_1}, \ldots, \mathbf{z_m}) \in V^m$ such that $\mathbf{f}(\mathbf{z_1}, \ldots, \mathbf{z_m}) = \mathbf{0}$.

Since $\mathbb{F}_{2^n}$ is a vector space over $\mathbb{F}_2$, $\mathbf{f}$ can be rewritten (or deployed) as a polynomial system of m equations over $\mathbb{F}_2$ and then can be solved using Gröbner bases algorithms. The prominent observation is to remark that this system is (affine) *multi-homogeneous*. While the complexity of solving bi-linear systems using Gröbner bases – that is to say polynomials of bi-degree $(1, 1)$ – is now well understood [24], the general case is not known. Consequently, we propose a simple *ad-hoc* algorithm to take advantage of the multihomogeneous structure. This is of independent interest in the more general context of computer algebra.

The main idea is to show that starting from the unique equation $\mathbf{f} = \mathbf{0}$, we can generate many low-degree equations by deploying the equations $\mathbf{mf} = \mathbf{0}$ over $\mathbb{F}_2$ for a large number of appropriately chosen monomials $\mathbf{m}$. Another main difference with unstructured polynomials is that – in some degree d – the number of monomials occurring in the polynomials coming from $\mathbf{mf}$ is much smaller than the number of all possible monomials in degree d. Indeed, due to the choice of $\mathbf{m}$, the monomials occurring in the equations constructed from $\mathbf{mf}$ have still a multi-homogeneous structure and their degrees are well controlled.

As usual, to estimate the maximum degree D reached during the computation we study the number of equations minus the number of monomials. When this number is > 0, and under a reasonable linear independence assumption confirmed by our experimental results, the computation is finished. Using the structure of the polynomials, we prove that this degree D is much smaller than expected; assuming that a reasonable heuristic is true. It is worth noticing that although we describe our algorithm as a linearization method [35], Gröbner basis algorithms like F_4 or F_5 [16,17] can be advantageously used in practice to solve the corresponding polynomial systems. The algorithm presented in this paper, together with its complexity analysis, can thus be understood as a method to (heuristically) bound the complexity of computing the corresponding Gröbner basis similarly to the Macaulay's bound obtain by Lazard [34] to bound the complexity of Gröbner bases in the general case. More precisely, we obtain:

Theorem 2. *Assuming some linear independence assumption (see Assumption 1, p. 36), Problem 2 can asymptotically be solved in time $O(2^{\omega \tau})$ and memory $O(2^{2\tau})$, where ω is the linear algebra constant and $\tau \approx n/2$. Under the same hypothesis, there exists an index calculus based algorithm solving ECDLP over $\mathbb{F}_{2^n}$ in the same time complexity.*

The index calculus algorithm presented here has a better complexity than the one proposed recently by Diem [14]. Moreover, we propose a novel approach for solving the sieving step. We consider that it is a major open challenge to further exploit the intrinsic algebraic structure of Problem 2 using Gröbner bases algorithms. The complexity obtained here for solving ECDLP is still exponential. We hope that the structures identified here can be further used to get a complexity better than generic algorithms (see preliminary experiments in Section 5.2) or to get a subexponential algorithm in a near future. Finally, we emphasize that the complexity analysis of our algorithm relies on a *heuristic* assumption on the rank of a linearized system. The validity of this assumption was experimentally checked (see Section 5.1). Whilst we pushed the experiments as far as possible, we pointed out that – due to the size of the systems involved – it is very difficult to verify experimentally the assumption for large parameters. Consequently, it is an open issue to prove Assumption 1.

Outline. The remaining of this paper is organized as follows. In Section 2, we detail our notations and we provide some background on Gröbner bases. In Section 3, we introduce and analyze our algorithm for solving Problem 2. In Section 4, we apply our new result to the ECDLP over binary fields. In Section 5, we present experimental results to give first evidences for Assumption 1. Section 6 concludes the paper and introduces future extensions of our work.

2   Preliminaries

In this section, we introduce definitions, notations and recall well known results concerning polynomial system solving.

2.1 Definition and Notation

Let $\mathbb{F}_2$ be the finite field of cardinality 2. We will consider a degree n extension $\mathbb{F}_{2^n}$ of $\mathbb{F}_2$. We will often see $\mathbb{F}_{2^n}$ as an n dimensional vector space over $\mathbb{F}_2$. Let $\{\theta_1, \ldots, \theta_n\}$ be a basis of $\mathbb{F}_{2^n}$ over $\mathbb{F}_2$. We will use bold letters for elements, variables and polynomials over $\mathbb{F}_{2^n}$ and normal letters for elements, variables and polynomials over $\mathbb{F}_2$. If $x_1, \ldots, x_m$ are algebraic independent variables over a finite field $\mathbb{K}$, we write $R = \mathbb{K}[x_1, \ldots, x_m]$ for the polynomial ring in those variables. Given a set of polynomials $\{f_1, \ldots, f_\ell\} \in R$, the *ideal* generated by this set will be denoted by $\langle f_1, \ldots, f_\ell \rangle \subset R$. We write $\mathrm{Res}_{x_i}(f_1, f_2)$ for the *resultant* of $f_1 \in R$ and $f_2 \in R$ with respect to the variable x_i. A power product, i.e. $\prod_{i=1}^{k} x_i^{e_i}$ where $e_i \in \mathbb{N}$, is called a *monomial*. Finally, we introduce a structure which will be very useful in this paper.

Definition 1 ([24]). *Let $X_1 \cup X_2 \cup \cdots \cup X_t = \{x_1, \ldots, x_m\}$ be a partition of the variables set. We shall say that a polynomial $f \in \mathbb{K}[x_1, \ldots, x_m] = \mathbb{K}[X_1, \ldots, X_t]$ is* multi-homogeneous *of multi-degree $(d_1, d_2, \ldots, d_t)$ if:*

$$\forall(\alpha_1, \ldots, \alpha_t) \in \mathbb{K}^t, \ f(\alpha_1 X_1, \ldots, \alpha_t X_t) = \alpha_1^{d_1} \cdots \alpha_t^{d_t} f(X_1, \ldots, X_t).$$

For all $i, 1 \leq i \leq t$, let $X_i = \{x_{i,1}, \ldots, x_{i,n_i}\}$. We shall say that f is affine multi-homogeneous *if there exists $f^h \in \mathbb{K}[X_1, \ldots, X_t]$ a multi-homogeneous polynomial of same degree such that when one replaces (homogenization) variables x_{i,n_i} by 1 we obtains f, i.e.: $f(x_{1,1}, \ldots, x_{1,n_1-1}, \ldots, x_{t,1}, \ldots, x_{t,n_t-1})$ is equal to $f_i^h(x_{1,1}, \ldots, x_{1,n_1-1}, 1, \ldots, x_{t,1}, \ldots, x_{t,n_t-1}, 1)$. Finally, we shall say that f has a* multi-homogeneous structure *if it is multi-homogeneous or affine multi-homogeneous. A system of equation has a* multi-homogeneous structure *if each equation has a multi-homogeneous structure (the equations can have different multi-degrees).*

Given a number $e = \sum_{i=0}^{\infty} e_i 2^i$ with $e_i \in \{0, 1\}$, we define its *Hamming weight* as the number of non-zero elements in its binary expansion, i.e. $W(e) := \sum_{i=0}^{\infty} e_i$. We write $\binom{n}{k}$ for the number of choices of k elements among a set of n elements without repetition. We write O for the "big O" notation: given two functions f and g of n, we say that $f = O(g)$ if there exist $N, c \in \mathbb{Z}^+$ such that $n > N \Rightarrow f(n) \leq cg(n)$. The notation log stands for the binary logarithm. Finally, we write ω for the *linear algebra constant*. Depending on the algorithm used for linear algebra, we have $2.376 \leq \omega \leq 3$.

2.2 Gröbner Bases [10]

Recent methods such as Faugère's F_4 and F_5 [16,17] algorithms reduce Gröbner basis computation to Gaussian eliminations on several matrices. The link between linear algebra and Gröbner bases has been established by Lazard [34]. He showed that computing a Gröbner basis is equivalent to perform Gaussian elimination on the so-called *Macaulay matrices* as defined below:

Definition 2 (Macaulay Matrix [35,36]).

$$t_{i,j}f_i \begin{pmatrix} & m_1 > m_2 > \dots & \\ & \vdots & \\ & c_{i,j}^1 \quad c_{i,j}^2 \quad \dots & \end{pmatrix}$$

Let $F = \{f_1, \dots, f_\ell\} \subset R$ be a set of polynomials of degree $\leq d$. Let $\mathcal{B} = \{m_1 > m_2 > \dots\} \subset R$ be the sorted set (w.r.t. a fixed monomial ordering) of degree $\leq d$ monomials. The set $\mathcal{B}$ is a basis of the vector space of degree $\leq d$ polynomials in R. The Macaulay matrix $\mathcal{M}_d(F)$ of degree d is defined as follows. We consider all the polynomials $t_{i,j}f_i$ of degree $\leq d$ with $t_{i,j} \in \mathcal{B}$ and $f_i \in F$. Rows of $\mathcal{M}_d(F)$ correspond to the coefficients vectors $(c_{i,j}^1, c_{i,j}^2, \dots)$ of these polynomials $t_{i,j}f_i = \sum_k c_{i,j}^k m_k$ with respect to the basis $\mathcal{B}$.

Precisely, Lazard [34] proved the following fundamental result:

Theorem 1. *Let $F = \{f_1, \dots, f_\ell\} \subset R$. There exists a positive integer D for which Gaussian elimination on all matrices $\mathcal{M}_1(F), \mathcal{M}_2(F), \dots, \mathcal{M}_D(F)$ computes a Gröbner basis of $\langle f_1, \dots, f_\ell \rangle$.*

F_4 [16] can be seen as another way to use linear algebra without knowing an a priori bound. It successively constructs and reduces matrices until a Gröbner basis is found. The same is true for F_5 when considered in "F_4-style" [25].

It is clear that an important parameter in Gröbner basis computation is the maximal degree D reached during the computation. This maximal degree is called the *degree of regularity*. However, it is a difficult problem to estimate *a priori* the degree of regularity. This degree is known and well mastered for specific families of systems called regular and semi-regular [5,6,7]. It is classical to assume that the regularity of regular/semi-regular systems provides an extremely tight upper bound on the regularity of random system of equations (most of the times, we have equality). For example, the regularity degree of a regular sequence $f_1, \dots, f_\ell \in R$ (with $\ell \leq n$) is $D = 1 + \sum_{i=1}^{\ell}(\deg(f_i) - 1)$. Ideals with special structures – typically arising in cryptography – may have a much lower regularity degree, hence a much better time complexity. This has permitted Gröbner bases techniques to successfully attack many cryptosystems (e.g. the *Hidden Field Equation* cryptosystem (HFE) [38,31,18,29], Multi-HFE [9]). In many cases, the algebraic systems appearing in these applications were not generic and could be solved more efficiently than generic systems, sometimes using *dedicated* Gröbner basis algorithms.

In our case, the algebraic system considered in Problem 2 has a *multi-homogeneous* structure (more precisely *multi-linear*). Interestingly, the formal study of computing Gröbner basis of multi-homogeneous systems has been initiated by Faugère, Safey El Din and Spaenlehauer for bilinear systems [24] leading already to new cryptanalytic results [19,23,21]. However, the general case (more blocks, larger degrees) remains to be investigated.

As a consequence, we design in this paper a simple *ad hoc* algorithm to take advantage of the multi-homogeneous structure arising in Problem 2. Basically, the idea is to generate a submatrix of the Macaulay matrix at a degree D_{Lin} which allows to linearize the system derived from Problem 2. To do so, we strongly exploit the multi-homogeneous structure of the equations. The fundamental remark is that many low-degree relations exist and can be explicitly predicted. Moreover, the number of columns in the Macaulay matrix is less than usual. This is due to the fact that all monomials occurring have still the multi-homogeneous structure. Roughly, this allows to predict many zero columns for a suitably chosen subset of the rows. In section 3.2, we derive a bound on the degree D_{Lin} which needs to be considered.

It is worth noticing that although we describe our algorithm as a linearization method, computing a Gröbner basis with F_4 or F_5 [16,17] can be advantageously used in practice to solve the corresponding polynomial systems. The D_{Lin} obtained has to be understood as an upper bound on the real regularity degree of the system. Any new result on multi-homogeneous systems would allow to improve the complexity of solving Problem 2 (partial results are presented in Section 5.2).

3 Solving Multivariate Polynomials with Linear Constraints

In this section, we describe the main result of this paper: an algorithm for solving Problem 2. Let V be a $\mathbb{F}_2$-vector subspace $\subset \mathbb{F}_{2^n}$ of dimension n' and let $\mathbf{f} \in \mathbb{F}_{2^n}[\mathbf{x_1}, \ldots, \mathbf{x_m}]$ be a multivariate polynomial with degree $< 2^t$ in each variable. We want to solve $\mathbf{f}(\mathbf{x_1}, \ldots, \mathbf{x_m}) = \mathbf{0}$ under the *linear constraints* $\mathbf{x_1}, \ldots, \mathbf{x_m} \in V$. To tackle this problem, we generalize the algorithm and analysis of [22] from the multi-linear case ($t = 1$) to arbitrary values of t. From now on, we assume that $m \cdot n' \approx n$ so that the problem has about one solution on average.

3.1 Modeling the Linear Constraints

Let $\{\nu_1, \ldots, \nu_{n'}\} \subset \mathbb{F}_{2^n}$ be a basis of V as a $\mathbb{F}_2$-vector space. Let $y_{i,j}$ be $m \cdot n'$ variables defined by $x_i = \nu_1 y_{i,1} + \nu_2 y_{i,2} + \cdots + \nu_{n'} y_{i,n'}$. We apply a *Weil descent* to Problem 2 (see e.g. [11, Chapter 7]). By replacing the variables x_i in the polynomial $\mathbf{f}$, we get a new polynomial $\mathbf{f}_V \in \mathbb{F}_{2^n}[y_{1,1}, \ldots, y_{m,n'}]$ with $m \cdot n'$ variables. The linear constraints on $\mathbf{f}$ are translated to *Galoisian constraints* by constraining the solutions of $\mathbf{f}_V$ to $\mathbb{F}_2$. Using the *field equations*, $\mathbf{f}_V$ is viewed more precisely as a a polynomial in the affine algebra $\mathcal{A}(\mathbb{F}_{2^n}) := \mathbb{F}_{2^n}[y_{1,1}, \ldots, y_{m,n'}]/\langle \mathcal{S}_{\mathrm{fe}} \rangle$, where $\langle \mathcal{S}_{\mathrm{fe}} \rangle$ is the 0-dimensional ideal generated by the field equations:

$$\mathcal{S}_{\mathrm{fe}} = \{y_{i,j}^2 - y_{i,j}\}_{1 \leq i \leq m}^{1 \leq j \leq n'}.$$

Problem 2 is then equivalent to compute a Gröbner basis of the ideal $\langle \mathbf{f}_V, \mathcal{S}_{\mathrm{fe}} \rangle \subset \mathbb{F}_{2^n}[y_{1,1}, \ldots, y_{m,n'}]$. It is generally more efficient to consider its resolution over $\mathbb{F}_2$. To do so, we consider $\mathcal{A}(\mathbb{F}_{2^n})$ as a module over $\mathcal{A}(\mathbb{F}_2) = \mathbb{F}_2[y_{1,1}, \ldots, y_{m,n'}]/\langle \mathcal{S}_{\mathrm{fe}} \rangle$ whose basis is $\{\theta_1, \ldots, \theta_n\}$. We consider $\mathbf{f}_V$ as an element of $\mathcal{A}(\mathbb{F}_{2^n})$ and we *deploy* it as a $\mathcal{A}(\mathbb{F}_2)$-linear combination of the basis $\{\theta_1, \ldots, \theta_n\}$. Namely:

$$\mathbf{f_V} = [\mathbf{f_V}]_1^{\downarrow} \theta_1 + [\mathbf{f_V}]_2^{\downarrow} \theta_2 + \cdots + [\mathbf{f_V}]_n^{\downarrow} \theta_n \tag{1}$$

for some $[\mathbf{f_V}]_1^\downarrow, \ldots, [\mathbf{f_v}]_n^\downarrow \in \mathcal{A}(\mathbb{F}_2)$ that depend on $\mathbf{f}$ and the vector subspace V. Due to the linear independence of the $\theta_1, \ldots, \theta_n$, Problem 2 is equivalent to solve:

$$\mathcal{S}_{\text{alg}}: \quad [\mathbf{f_v}]_1^\downarrow = [\mathbf{f_v}]_2^\downarrow = \cdots = [\mathbf{f_v}]_n^\downarrow = 0. \tag{2}$$

In order to solve $\mathcal{S}_{\text{alg}}$, we will generate many new equations by deploying multiples of $\mathbf{f} \in \mathbb{F}_{2^n}[\mathbf{x_1}, \ldots, \mathbf{x_m}]$ over the vector subspace V. The key point of this strategy is the existence of abnormally high number of low-degree equations arising as algebraic combinations of the equations in (2).

From now on, we represent the classes of polynomials $\mathbf{g_V} \in \mathcal{A}(\mathbb{F}_{2^n})$ and $[\mathbf{g_V}]_i^\downarrow \in \mathcal{A}(\mathbb{F}_2)$ corresponding to $\mathbf{g} \in \mathbb{F}_{2^n}[x_1, \ldots, x_m]$ by their minimal elements, in other words by their normal forms modulo $\mathcal{I}_{\text{fe}}$ (whose generators form a Gröbner basis). By abuse of notation, we use the same symbol for a class and its minimal representative in the underlying polynomial ring. When the context is not clear, we precise the algebra where the element is lying.

3.2   Low-Degree Equations

Let $e_1, \ldots, e_m \in \mathbb{N}$ and let $\mathbf{m} = \prod_{i=1}^m \mathbf{x_i}^{e_i}$ be a monomial of $\mathbb{F}_{2^n}[\mathbf{x_1}, \ldots, \mathbf{x_m}]$. Following the descent described in Section 3.1, we have

$$\mathcal{A}(\mathbb{F}_{2^n}) \ni (\mathbf{mf})_V = \sum_{k=1}^n [(\mathbf{mf})_\mathbf{v}]_k^\downarrow \, \theta_k, \text{ with } [(\mathbf{mf})_\mathbf{v}]_k^\downarrow \in \mathcal{A}(\mathbb{F}_{2^n}).$$

The equation $\mathbf{f} = \mathbf{0}$ clearly implies $\mathbf{mf} = \mathbf{0}$, hence $[(\mathbf{mf})_\mathbf{V}]_k^\downarrow = 0$ for $k = 1, \ldots, n$. We can then add these new equations to the polynomial system (2). The equations obtained in this way all share the same structure. More precisely, their minimal representatives, due to the normal form computation modulo $\mathcal{I}_{\text{fe}}$, are all *affine multi-linear* in $\mathbb{F}_{2^n}[y_{1,1}, \ldots, y_{m,n'}]$. Moreover, thanks to the evaluations done during the deployments, each *block of variables* $X_i = \{y_{i,1}, \ldots y_{i,n'}\}$ naturally corresponds to the variable x_i. From these structures, we deduce the following result.

Lemma 1. *Let* $\mathbf{f} \in \mathbb{F}_{2^n}[\mathbf{x_1}, \ldots, \mathbf{x_m}]$ *be a multivariate polynomial with degree* $< 2^t$ *in each variable. Let* $e_1, \ldots, e_m \in \mathbb{N}$ *and* $\mathbf{m} = \prod_{i=1}^m \mathbf{x_i}^{e_i}$ *be a monomial of* $\mathbb{F}_{2^n}[\mathbf{x_1}, \ldots, \mathbf{x_m}]$. *There exist polynomials* $p_{j,k} \in \mathbb{F}_2[y_{1,1}, \ldots, y_{m,n'}]$ *such that* $[(\mathbf{mf})_\mathbf{v}]_k^\downarrow = \sum_{j=1}^n p_{j,k} [\mathbf{f_v}]_j^\downarrow$. *Each polynomial* $p_{j,k}$ *has degree* $\leq W(e_i)$ *with respect to every block of variables* $X_i = \{y_{i,1}, \ldots y_{i,n'}\}$, $1 \leq i \leq m$. *Moreover, each minimal representative of* $[(\mathbf{mf})_\mathbf{v}]_k^\downarrow$ *has degree* $\leq \max_{0 \leq e_i' < 2^t} W(e_i + e_i')$ *w.r.t. each block of variables* $X_i, 1 \leq i \leq m$.

This lemma implies that the new equations (obtained from $\mathbf{mf}$) are algebraic combinations of the original ones (obtained from $\mathbf{f}$). In particular, they can *a priori* be recovered "in a hidden form" with any Gröbner basis algorithm at degree $D_{apriori} = mt + \sum_{j=1}^m W(e_j)$. The value $D_{apriori}$ is the degree that the equations $[(\mathbf{mf})_\mathbf{V}]_k^\downarrow$ should have *a priori* from the algebraic dependencies of Lemma 1. It is the sum of the degree of the deployments of $\mathbf{f}$ (at most mt) and the degree of each polynomial $p_{j,k}$ $\left(\text{at most } \sum_{j=1}^{m} W(e_j)\right)$. However, Lemma 1 also implies that the $[(\mathbf{mf})_{\mathbf{V}}]_k^{\downarrow}$ only have degree $D_{actual} = \sum_{j=1}^{m}(\max_{0 \le e'_j < 2^t} W(e_j + e'_j))$. Thus:

$$D_{apriori} - mt \le D_{actual} \le D_{apriori}.$$

Therefore, $[(\mathbf{mf})_{\mathbf{V}}]_k^{\downarrow}$ may have a *degree drop* as large as mt depending on the monomial $\mathbf{m}$ chosen. The existence such low-degree relations compared generic systems makes Gröbner basis algorithms faster in practice and allows a linearization strategy.

Following the general method of Macaulay [35], we will linearize the polynomial system $\mathcal{S}_{\mathrm{alg}} \cup \{\ldots, \ldots, [(\mathbf{mf})_{\mathbf{V}}]_1^{\downarrow}, [(\mathbf{mf})_{\mathbf{V}}]_2^{\downarrow}, \ldots, [(\mathbf{mf})_{\mathbf{V}}]_n^{\downarrow}, \ldots, \ldots\}$ using the low-degree equations identified in this section. The choice of the monomials $\mathbf{m}$ used to generate the equations are particularly important for the efficiency of the linearization strategy. In particular, the equations with the lowest degrees are the most interesting ones since they involve less monomial terms. Of course, this strategy requires that a substantial subset of all low-degree relations are linearly independent.

3.3   Linear Dependencies

In the previous section, we explained how low-degree relations can be produced. To be used in a linearization strategy, these equations must be linearly independent. In this section, we describe two sources of linear dependencies.

Frobenius Transforms. The first source of linear dependencies is due to the Frobenius endomorphism (as identified in [22]). Let $\{\theta_1, \ldots, \theta_n\}$ be a basis of $\mathbb{F}_{2^n}/\mathbb{F}_2$. The set $\{\theta_1^2, \ldots, \theta_n^2\}$ is another basis of $\mathbb{F}_{2^n}/\mathbb{F}_2$. Let $a_{ij} \in \mathbb{F}_2$ be such that $\theta_j^2 = \sum_i a_{ij}\theta_i$. We have $\mathbf{f_V}^2 = \sum_{j=1}^n \left[\mathbf{f_V}^2\right]_j^{\downarrow} \theta_j$. However, $\mathbf{f_V}^2 = \left(\sum_{j=1}^n [\mathbf{f_V}]_j^{\downarrow} \theta_j\right)^2 = \sum_{j=1}^n [\mathbf{f_V}]_j^{\downarrow} \theta_j^2 = \sum_{i=1}^n \left(\sum_{j=1}^n a_{ij} [\mathbf{f_V}]_j^{\downarrow}\right) \theta_i$. Thus, we obtain $\left[\mathbf{f_V}^2\right]_i^{\downarrow} = \sum_{j=1}^n a_{ij} [\mathbf{f_V}]_j^{\downarrow}$. In other words, the polynomials $\left[\mathbf{f_V}^2\right]_1^{\downarrow}, \ldots, \left[\mathbf{f_V}^2\right]_n^{\downarrow}$ are linear combinations of $[\mathbf{f_V}]_1^{\downarrow}, \ldots, [\mathbf{f_V}]_n^{\downarrow}$. Decomposing $\mathbf{f_V}$ as a sum of monomials, we deduce that $\left[\mathbf{f_V}^2\right]_i^{\downarrow} = \sum_{\mathbf{m}\in\mathrm{Mon}(\mathbf{f_V})} [(\mathbf{mf})_{\mathbf{V}}]_i^{\downarrow} = \sum_{j=1}^n a_{ij} [\mathbf{f_V}]_j^{\downarrow}$. This provides a non trivial linear relation between some low-degree equations. More generally, we obtain similar relations if we replace $\mathbf{f}$ by $\mathbf{m'f}$ in the above equation (for any monomial $\mathbf{m'}$). These linear dependencies can be easily detected and prevented. Indeed, we can simply avoid every monomial $\mathbf{m}$ that is the leading term of $(\mathbf{m'f})$ for some $\mathbf{m'}$.

Vector Dependencies. Another source of dependency, that we call *vector dependencies*, is induced by the vector space V. To illustrate the phenomena, consider the simplest polynomial $\mathbf{g} = \mathbf{x_1} \in \mathbb{F}_{2^n}[\mathbf{x_1}, \ldots, \mathbf{x_m}]$. We have $\mathbf{g_V} = \nu_1 y_{1,1} + \nu_2 y_{1,2} + \cdots + \nu_{n'} y_{1,n'} \in \mathcal{A}(\mathbb{F}_{2^n})$. More generally, $(\mathbf{g}^{2^i})_{\mathbf{V}} = \nu_1^{2^i} y_{1,1} + \nu_2^{2^i} y_{1,2} + \cdots + \nu_{n'}^{2^i} y_{1,n'} \in \mathcal{A}(\mathbb{F}_{2^n})$. Since $y_{1,1}, \ldots, y_{1,n'}$ are linearly independent, we obtain for any $k > n'$ a non trivial linear dependency by considering different g^{2^i}'s, i.e. $\exists \beta_1, \ldots, \beta_k \in \mathbb{F}_{2^n} \setminus \{(0, \ldots, 0)\}$ such that $\beta_1 (\mathbf{g}^{2^{i_1}})\mathbf{v} + \beta_2 (\mathbf{g}^{2^{i_2}})\mathbf{v} + \cdots + \beta_k (\mathbf{g}^{2^{i_k}})\mathbf{v} = 0$. This simple example can be easily generalized to $\mathbf{g} = \mathbf{mf}$ with $\mathbf{m}$ a monomial. Such linear dependency can be clearly prevented during the generation of equations $[(\mathbf{mf})_V]_i^\downarrow$'s by considering monomials $\mathbf{m} = \prod_{i=1}^m \mathbf{x_i}^{e_i}$, with $0 \leq e_i < 2^{n'} < 2^n$.

3.4   Description of the Linearization Algorithm

For any positive integer d, let MLinMonB(d) be the set of multi-linear monomials in $\mathbb{F}_2[y_{1,1}, \ldots, y_{m,n'}]$ of degrees $\leq d$ with respect to each block $X_i = \{y_{i,1}, \ldots, y_{i,n'}\}$. The image of MLinMonB($d$) in $\mathcal{A}(\mathbb{F}_2)$ is a basis of the vector subspace $\mathcal{A}(\mathbb{F}_2)_d$ composed of elements in $\mathcal{A}(\mathbb{F}_2)$ with a minimal representative having degrees $\leq d$ with respect to each block X_i. Let also Mon(d) be the set of monomials $\mathbf{m} = \prod_{i=1}^m \mathbf{x_i}^{e_i} \in \mathbb{F}_{2^n}[\mathbf{x_1}, \ldots, \mathbf{x_n}]$ with $1 \leq e_i < 2^{n'}$, such that all $[(\mathbf{mf})_\mathbf{V}]_k^\downarrow$ ($1 \leq k \leq n$) are in $\mathcal{A}(\mathbb{F}_2)_d$. Finally, let $E(d) := n \cdot \#\text{Mon}(d)$ and $M(d) := \#\text{MLinMonB}(d)$.

We are now ready to describe Algorithm 1, a simple linearization algorithm for solving Problem 2. The algorithm constructs a sub-matrix of the Macaulay (see Definition 2) matrix $\mathcal{M}_d$ for system 2. We first gather $M(d)$ equations $[(\mathbf{mf})_\mathbf{V}]_i^\downarrow$ with $\mathbf{m} \in \text{Mon}(d)$. By definition, all these equations are in $\mathcal{A}(\mathbb{F}_2)_d$. Hence, they can be decomposed with respect to the basis MLinMonB(d). We then form the corresponding linear system $\mathcal{S}_{lin}$ over $\mathbb{F}_2$, where each row corresponds to the coefficients involves in the equations and each column corresponds to an element in MLinMonB(d). We finally solve the linear system. This simple algorithm, that we call *Sub-Macaulay*, is not aimed to be optimal in practice but to derive complexity bounds.

The general linearization strategy and our analysis below rely on a heuristic assumption formalized below:

Assumption 1. *With a probability exponentially close to one, the equations generated by Algorithm 1 are linearly independent.*

Particularly, the assumption states that the solutions of $\mathcal{S}_{lin}$ are in one-to-one correspondence with the solutions of Problem 2.

3.5   Complexity Bounds for Solving Problem 2

We now derive an upper bound on the complexity of Algorithm 1. The main task is to estimate the values of $M(d)$ (number of columns in $\mathcal{S}_{lin}$) and $E(d)$ (number of equations in $\mathcal{S}_{lin}$). Due to the field equations, we only have multi-linear monomials, i.e. variables can only have exponents 0 or 1. Therefore, the number of monomials of total degree $d', 0 \leq d' \leq d$ involving variables of the block X_i is $\binom{n'}{d'}$. For the m blocks, we get:

$$M(d) = \left(\sum_{d'=0}^d \binom{n'}{d'} \right)^m. \tag{3}$$

Input: $\mathbf{f} \in \mathbb{F}_{2^n}[\mathbf{x_1}, \ldots, \mathbf{x_m}]$ of degree in each variable is bounded by $2^t - 1$, and V a $\mathbb{F}_2$-vector subspace $\subset \mathbb{F}_{2^n}$ of dimension n'.
Result: If not empty, the finite set $\{s_1, \ldots\} \subset V^m$ such that $\mathbf{f}(s_i) = \mathbf{0}$.

```
1 begin
2    Let d be the smallest integer such that E(d) ≥ M(d);
3    Mon ← [];  S_alg ← []   // Empty lists of monomials and equations
4    for k = 1, ..., ⌈E(d)/n⌉  do
5        Randomly pick a monomial m ∈ Mon(d) \ Mon;
6        Let [(mf)v]₁↓, ..., [(mf)v]ₙ↓ ∈ F₂[y₁₁, ..., y_{m nₘ}] be such that
             (mf)v = Σⁿ_{k=1} [(mf)v]_k↓ θ_k;
7        S_alg ← S_alg ∪ ([(mf)v]₁↓, ..., [(mf)v]ₙ↓);  Mon ← Mon ∪ [m];
8    end
9    Construct S_lin the linear system over F₂ obtained by linearizing S_alg;
10   If S_lin has solutions then  solve S_lin and return {s₁, ..., } else  return no
     solution end;
11 end
```

Algorithm 1. Sub-Macaulay

By definition, the degree of each variable $\mathbf{x_i}$ occurring in the monomials of $\mathbf{f}$ may have any degree between 0 and $2^t - 1$. By Lemma 1, the degree of $[(\mathbf{mf})\mathbf{v}]_k^{\downarrow}$ with respect to X_i is $\max_{0 \leq e_i'' < 2^t} W(e_i' + e_i'') = W\left(\left\lfloor \frac{e_i'}{2^t} \right\rfloor\right) + t$. Therefore, the number of exponents e_i', $1 \leq e_i' \leq 2^{n'}$ leading to degree d' with respect to the block X_i is $2^t \binom{n'-t}{d'-t}$. As a consequence[1]

$$E(d) := n \, 2^{tm} \left(\sum_{d'=t}^{d} \binom{n'-t}{d'-t} \right)^m. \tag{4}$$

We derive the following asymptotic bound (proven in Appendix A) on the minimal value d that allows linearization.

Lemma 2. *Let α be such that $1 - \alpha < 1/2 < \alpha < 1$. Assuming that $n' = n^{\alpha}, m = n^{1-\alpha}$ and $t = m - 1$, then $E(d) \geq M(d)$ for $d \approx \frac{n^{\alpha}}{2}$ when n is large enough.*

In the next table, we have computed the smallest d_{real} such that $E(d_{\text{real}}) \geq M(d_{\text{real}})$ for different values of n and α. Then, we compute $\beta_{\text{real}} = \log_n(d_{\text{real}})$. According to Lemma 2, the theoretical value predicted for $\beta = \log_n(d)$ is $\beta_{\text{theo}} = \alpha - \frac{1}{\log(n)}$. The last column columns shows that β_{theo} is extremely close to β_{real}.

Finally, we obtain an estimate on the complexity:

Theorem 2. *Let α, n', m, t be as in Lemma 2. Under Assumption 1, Problem 2 can asymptotically be solved in time $O(2^{\omega \tau})$ and memory $O(2^{2\tau})$, where ω is the linear algebra constant and $\tau \approx \frac{n}{2}$.*

---

[1] Note that we assume that $d \geq t$.

n	d_{real}	α	β_{real}	$\left\|\left(\alpha - \frac{1}{\log(n)}\right) - \beta_{\text{real}}\right\|$
100000	1114	2/3	0.6093	-0.0029
1000000	5102	2/3	0.617956	-0.0014
10000000	23466	2/3	0.62435	-0.00069
100000000	108353	2/3	0.62936	-0.00032
1000000	1285	0.55	0.51815	-0.018
10000000	4331	0.55	0.51951	-0.012
100000000	14738	0.55	0.52105	-0.0089
1000000000	50577	0.55	0.52266	-0.0061
10000000000	174773	0.55	0.52425	-0.0043

Proof. The above linearization algorithm reduces the problem to linear algebra on a matrix of rank $M(d)$. According to Lemma 4 (Appendix A), we get that

$$\log(M(d)) \approx m \log\binom{n^{\alpha}}{n^{\beta}} \approx n^{(1-\alpha)} n^{\beta}(\alpha - \beta) \log(n) = n^{\left(1 - \frac{1}{\log(n)}\right)} = \frac{n}{2}. \qquad \square$$

4 Application to ECDLP over Binary Fields

4.1 Diem's Variant of Index Calculus

Let E be an elliptic curve defined over $\mathbb{F}_{2^n}$ by the equation

$$E : y^2 + xy = x^3 + x^2 + \mathbf{a_6}, \text{ for some } \mathbf{a_6} \in \mathbb{F}_{2^n}. \tag{5}$$

Semaev' summation polynomials $\mathbf{S_r}$ [40] are multivariate polynomials with the following property: $\mathbf{S_r}(\mathbf{x_1}, \ldots, \mathbf{x_r}) = \mathbf{0}$ for some $\mathbf{x_1}, \ldots, \mathbf{x_r} \in \overline{\mathbb{F}_{2^n}}$ if and only if there exist $\mathbf{y_1}, \ldots, \mathbf{y_r} \in \overline{\mathbb{F}_{2^n}}$ such that $(\mathbf{x_i}, \mathbf{y_i}) \in E(\overline{\mathbb{F}_{2^n}})$ and $(\mathbf{x_1}, \mathbf{y_1}) + \cdots + (\mathbf{x_r}, \mathbf{y_r}) = P_{\infty}$.

Proposition 1 ([40]). *The summation polynomials of the elliptic curve* (5) *are recursively given by:* $\mathbf{S_2}(\mathbf{x_1}, \mathbf{x_2}) := \mathbf{x_2} + \mathbf{x_1}$, $\mathbf{S_3}(\mathbf{x_1}, \mathbf{x_2}, \mathbf{x_3}) := \mathbf{x_1}^2 \mathbf{x_2}^2 + \mathbf{x_1}^2 \mathbf{x_3}^2 + \mathbf{x_1} \mathbf{x_2} \mathbf{x_3} + \mathbf{x_2}^2 \mathbf{x_3}^2 + \mathbf{a_6}$ *and for* $r \geq 4$ *and any* $k, 1 \leq k \leq r-3$, *the* r-th *summation polynomial is*

$$\mathbf{S_r}(\mathbf{x_1}, \ldots, \mathbf{x_r}) := Res_{\mathbf{X}}\left(\mathbf{S_{r-k}}(\mathbf{x_1}, \ldots, \mathbf{x_{m-k-1}}, \mathbf{X}), \mathbf{S_{k+2}}(\mathbf{x_{r-k}}, \ldots, \mathbf{x_r}, \mathbf{X})\right).$$

Moreover, the polynomial $\mathbf{S_r}$ *is symmetric and has degree* 2^{r-2} *in each variable* $\mathbf{x_i}$ *as soon as* $r \geq 2$. *The cost of constructing this polynomial is bounded by* $2^{O((r-1)^2)}$.

Following Diem [14], we use summation polynomials in the sieving stage of an index calculus algorithm. Let V be a vector subspace of $\mathbb{F}_{2^n}/\mathbb{F}_2$ with a dimension n' to be fixed later. We define the *factor basis* $\mathcal{F}_V$ as

$$\mathcal{F}_V := \{(\mathbf{x}, \mathbf{y}) \in E(\mathbb{F}_{2^n}) | \mathbf{x} \in V\}.$$

Since the abscissas of points $\in E$ are uniformly distributed in $\mathbb{F}_{q^n}$ [13,14], we can assume that the set $\mathcal{F}_V$ has size about $2^{n'}$. During the sieving stage, we compute about $2^{n'}$ relations $P_{\infty} = a_i P + b_i Q + \sum_{P_j \in \mathcal{F}_V} e_j^i P_j$ with $P_j \in \mathcal{F}_V$ for randomly chosen integer couples (a_i, b_i). Each relation is obtained by solving an instance of the following problem, for some integer parameter m to be fixed later.

Problem 3. Let a_i, b_i be fixed random integers and $R = a_i P + b_i Q$. Find - if any - $(\mathbf{x_1}, \ldots, \mathbf{x_m}) \in V^m$ such that $\mathbf{S_{m+1}}(\mathbf{x_1}, \ldots, \mathbf{x_m}, (R)_x) = \mathbf{0}$.

Clearly, this problem is a particular instance of Problem 2.

4.2   A Linearization Strategy for Solving ECDLP over $\mathbb{F}_{2^n}$

We now apply the analysis of Section 3 to Problem 3. Let $\alpha, 1/2 < \alpha < 1$. be a parameter that will be optimized later. We set $n' := n^\alpha$ and $m := n^{1-\alpha}$ as in Lemma 2. According to Proposition 1, the $(m+1)$th Semaev's polynomial $\mathbf{S_{m+1}}$ can be computed in time $O(2^{t_1})$, where

$$t_1 \approx m^2 \approx n^{2(1-\alpha)}.$$

For each relation computed in the sieving stage, we generate and solve an instance of Problem 2 where $\mathbf{f}$ has degree 2^{m-1} with respect to each of the m variables. According to Theorem 2, each instance can be solved in time $O(2^{\omega\tau})$ and memory $O(2^{2\tau})$, where $\tau \approx n/2$. The probability that a point R can be written as a sum of m factor basis elements is $\frac{1}{m!}$ [14]. Hence, we need $m!2^{n'}$ trials on average to obtain $2^{n'}$ valid relations. Since $\log(m!) \approx n^{(1-\alpha)} \log n^{(1-\alpha)}$, the total cost of the sieving stage is bounded by $O(2^{t_2})$ where

$$t_2 \approx n^{(1-\alpha)} \log n^{(1-\alpha)} + n^\alpha + \omega \frac{n}{2}.$$

Finally, the last step of our algorithm consists in (sparse) linear algebra on a matrix of rank about $2^{n'}$ with elements of size about n bits. The computation time of this part can be approximated by $O(2^{t_3})$, where $t_3 \approx \omega' n^\alpha$, ω' being the sparse linear algebra constant. Finally, we obtain the following theorem.

Theorem 3. *Under Assumption 1, ECDLP over $\mathbb{F}_{2^n}$ can asymptotically be solved in time $O(2^{\omega t})$, where $t \approx n/2$.*

This has to be compared with the algorithm presented [14] which is was so far the best algebraic index calculus based approach. For $q = 2$ and n prime, the algorithm of [14] has complexity $e^{O(n \log(n)^{1/2})}$ (but such complexity holds without any heuristic assumption). The complexity of our approach is also very close to $O(2^{n/2})$; the complexity of generic algorithms (e.g. Pollard's rho algorithm).

5   Experimental Results

5.1   Validation of the Heuristic Assumption

We have experimentally checked the validity of Assumption 1 using the computer algebra system MAGMA. During the sieving stage, most of polynomial systems generated have no solution (only $1/m!$ polynomial systems will produce a solution). Thus we mainly check the assumption for polynomial systems with *no solution* in the vector space V. In order to validate this assumption we proceeded in the following way. We chose many random polynomials

f with degree $2^t < 2^{m-1}$ in each of its $m < 5$ variables. The coefficients are in $\mathbb{F}_{2^n}$ with n a prime less than 40. Then, for each of these polynomials, we construct a large binary matrix $\mathfrak{M}$ of size $N \times M$. This matrix represents the deployed polynomials $[(\mathbf{mf})_V]_1^{\downarrow}, \ldots, [(\mathbf{mf})_V]_n^{\downarrow}$ for all monomials in $\mathrm{Mon}(d)$ (with d is the smallest integer with $M(d) < E(d)$). We avoid the ones corresponding to possible linear dependencies as identified in Section 3.3. We want to demonstrate that a random square submatrix of size $M \times M$ of $\mathfrak{M}$ is full rank. We recall that the probability that a random $N \times M$ boolean matrix has rank r is $P(N, M, r) = 2^{-NM} \prod_{j=0}^{r-1}(2^N - 2^j) \prod_{j=0}^{r-1}(2^M - 2^j) / \prod_{j=0}^{r-1}(2^r - 2^j)$. Hence $P(100, 100, 100)$ is only 28.8% but $P(105, 100, 100) = 96.9\%$. For this reason, we consider submatrices $\mathfrak{M}'$ of $\mathfrak{M}$ of size $(M+5) \times M$. We check that the rank is M or $M-1$ or $M-2$; for a random Boolean matrix the probability is 99.999982%. We repeated the test 100 times and deduced an approximation of the success probability. In all our experiments we always obtain $\approx 100\%$ of success. In particular, we validated the assumption for Problem 3 using Semaev's summation polynomials with $m+1$ variables ($m = 2, \ldots, 4$). These validations represent a huge computational effort since some of the matrices $\mathfrak{M}$ encountered during the experiments had more than 10000 columns.

5.2   Gröbner Basis Computations

We performed actual Gröbner basis computation using the FGb software [20] to compare the theoretical number of operations with a more realistic value.

n	m	Number of Operations (GB)	Theoretical bound bound (Algo 1)
41	2	$2^{23.5}$	$M(d)^2 \approx 2^{60}$
67	2	$2^{37.1}$	$M(d)^2 \approx 2^{90}$
97	2	$2^{51.1}$	$M(d)^2 \approx 2^{125}$
131	2	$2^{74.5}$	$M(d)^2 \approx 2^{160}$

For instance, for $n = 131$, $m = 2$ we can solve the point decomposition problem in $2^{74.5}$ operations using a variant of the hybrid method [8]. We compare the cost of solving Semaev's equations using Algorithm 1 (complexity of solving a linear system of rank $M(d)$) and a Gröbner basis computation. This is the dominating part in the complexity of our index calculus based algorithm. Remark that the number of operations reported in this table are much below the theoretical estimate for Algorithm 1. We have a gain of (at least) a factor 2 in the exponent in favor of Gröbner computations. Further experiments suggest that a more advanced approach (i.e. Gröbner basis instead of linearization) can lead to an algorithm of better complexity. One can hope a gain of a factor m in the exponent. However, we are, for the moment, not able to provide theoretical evidences supporting such an improvement.

6   Conclusion and Perspectives

As a conclusion, we emphasize that the algorithm of Section 3 is of independent interest in the more general context of polynomial system solving. It shows that algebraic systems arising by deploying a multivariate polynomial equation from

$\mathbb{F}_{2^n}$ to $\mathbb{F}_2$ are easier to solve than generic systems. We have underlined the intrinsic structures which help for solving such systems. This is an open problem how to use such structure in an F_4/F_5 based algorithm. We hope that such improvements may lead to a subexponential algorithm. Finally, the approach generalizes quite easily to other composite fields with "small" characteristics, resulting in similar algorithms with comparable asymptotic complexities. All these extensions will be discussed in an extended version. Finally, although the paper is mainly theoretical, we hope that it could be the building block towards the development of more efficient methods to solve the ECDLP problem.

Acknowledgements. We thank the reviewers of Eurocrypt 2012 for their very careful evaluation of our paper and their meaningful comments. We are also indebted to Sylvie Baudine for her comments on a preliminary version of this paper. The third author would like to thank the PolSys team for hosting him in the beginning of 2011, as well as Kristin Lauter and Daniele Micciancio since this paper was partially written in San Diego.

References

1. Adleman, L.M.: A Subexponential Algorithm for the Discrete Logarithm Problem with Applications to Cryptography. In: Proceedings of the 20th Annual Symposium on Foundations of Computer Science, SFCS 1979, pp. 55–60. IEEE Computer Society, Washington, DC, USA (1979)
2. Adleman, L.M.: The Function Field Sieve. In: Huang, M.-D.A., Adleman, L.M. (eds.) ANTS 1994. LNCS, vol. 877, pp. 108–121. Springer, Heidelberg (1994)
3. Adleman, L.M., DeMarrais, J., Huang, M.: A Subexponential Algorithm for Discrete Logarithms over the Rational Subgroup of the Jacobians of Large Genus Hyperelliptic Curves over Finite Fields. In: Huang, M.-D.A., Adleman, L.M. (eds.) ANTS 1994. LNCS, vol. 877, pp. 28–40. Springer, Heidelberg (1994)
4. Adleman, L.M., Huang, M.: Function Field Sieve Method for Discrete Logarithms over Finite Fields. Inform. and Comput. 151(1-2), 5–16 (1999)
5. Bardet, M.: Étude des Systèmes Algébriques Surdéterminés. Applications aux codes Correcteurs et à la Cryptographie. PhD thesis, Université Paris VI (2004)
6. Bardet, M., Faugère, J.-C., Salvy, B.: Complexity of Gröbner Basis Computation for Semi-Regular Overdetermined Sequences over F_2 with Solutions in F_2. Technical Report 5049, INRIA (December 2003), http://www.inria.fr/rrrt/rr-5049.html
7. Bardet, M., Faugère, J.-C., Salvy, B., Yang, B.-Y.: Asymptotic Expansion of the Degree of Regularity for Semi-Regular Systems of Equations. In: Gianni, P. (ed.) The Effective Methods in Algebraic Geometry Conference, Mega 2005, pp. 1–14 (May 2005)
8. Bettale, L., Faugère, J.-C., Perret, L.: Hybrid Approach for Solving Multivariate Systems over Finite Fields. Journal of Math. Cryptology 3(3), 177–197 (2010)
9. Bettale, L., Faugère, J.-C., Perret, L.: Cryptanalysis of HFE, multi-HFE and Variants for Odd and Even Characteristic. Des. Codes Cryptography, 1–46 (2012)
10. Buchberger, B.: Ein Algorithmus zum Auffinden der Basiselemente des Restklassenringes nach einem nulldimensionalen Polynomideal. PhD thesis, Universität Innsbruck (1965)

11. Cohen, H., Frey, G. (eds.): Handbook of Elliptic and Hyperelliptic Curve Cryptography. Discrete Mathematics and its Applications. Chapman & Hall/CRC (2005)
12. Coppersmith, D.: Fast Evaluation of Logarithms in Fields of Characteristic Two. IEEE Transactions on Information Theory 30(4), 587–593 (1984)
13. Diem, C.: On the Discrete Logarithm Problem in Elliptic Curves. Compositio Mathematica 147, 75–104 (2011)
14. Diem, C.: On the Discrete Logarithm Problem in Elliptic Curves II. Presented at ECC 2011 (2011), http://www.math.uni-leipzig.de/diem/preprints/dlp-ell-curves-II.pdf
15. Enge, A., Gaudry, P.: A General Framework for Subexponential Discrete Logarithm Algorithms. Acta Arith. 102(1), 83–103 (2002)
16. Faugère, J.-C.: A New Efficient Algorithm for Computing Gröbner Basis (F4). Journal of Pure and Applied Algebra 139(1-3), 61–88 (1999)
17. Faugère, J.-C.: A New Efficient Algorithm for Computing Gröbner Bases without Reduction to Zero (F5). In: Proceedings of the 2002 International Symposium on Symbolic and Algebraic Computation, ISSAC 2002, pp. 75–83. ACM, New York (2002)
18. Faugère, J.-C., Joux, A.: Algebraic Cryptanalysis of Hidden Field Equation (HFE) Cryptosystems Using Gröbner Bases. In: Boneh, D. (ed.) CRYPTO 2003. LNCS, vol. 2729, pp. 44–60. Springer, Heidelberg (2003)
19. Faugère, J.-C., Otmani, A., Perret, L., Tillich, J.-P.: Algebraic Cryptanalysis of McEliece Variants with Compact Keys. In: Gilbert, H. (ed.) EUROCRYPT 2010. LNCS, vol. 6110, pp. 279–298. Springer, Heidelberg (2010)
20. Faugère, J.-C.: FGb: A Library for Computing Gröbner Bases. In: Fukuda, K., van der Hoeven, J., Joswig, M., Takayama, N. (eds.) ICMS 2010. LNCS, vol. 6327, pp. 84–87. Springer, Heidelberg (2010)
21. Faugère, J.-C., Levy-dit-Vehel, F., Perret, L.: Cryptanalysis of MinRank. In: Wagner, D. (ed.) CRYPTO 2008. LNCS, vol. 5157, pp. 280–296. Springer, Heidelberg (2008)
22. Faugère, J.-C., Perret, L., Petit, C., Renault, G.: New Subexponential Algorithms for Factoring in $SL(2, \mathbb{F}_2^n)$. Preprint (2011)
23. Faugère, J.-C., Safey El Din, M., Spaenlehauer, P.-J.: Computing Loci of Rank Defects of Linear Matrices using Gröbner Bases and Applications to Cryptology. In: ISSAC 2010: Proceedings of the 2010 International Symposium on Symbolic and Algebraic Computation, ISSAC 2010, pp. 257–264. ACM, New York (2010)
24. Faugère, J.-C., Safey El Din, M., Spaenlehauer, P.-J.: Gröbner Bases of Bihomogeneous Ideals Generated by Polynomials of Bidegree (1,1): Algorithms and Complexity. Journal of Symbolic Computation 46(4), 406–437 (2011)
25. Faugère, J.-C., Rahmany, S.: Solving Systems of Polynomial Equations with Symmetries using SAGBI-Gröbner bases. In: ISSAC 2009: Proceedings of the 2009 International Symposium on Symbolic and Algebraic Computation, ISSAC 2009, pp. 151–158. ACM, New York (2009)
26. Gaudry, P., Thomé, E., Thériault, N., Diem, C.: A Double Large Prime Variation for Small Genus Hyperelliptic Index Calculus. Math. Comp. 76(257), 475–492 (electronic) (2007)
27. Gaudry, P.: An Algorithm for Solving the Discrete Log Problem on Hyperelliptic Curves. In: Preneel, B. (ed.) EUROCRYPT 2000. LNCS, vol. 1807, pp. 19–34. Springer, Heidelberg (2000)
28. Gaudry, P.: Index Calculus for Abelian Varieties of Small Simension and the Elliptic Curve Discrete Logarithm Problem. J. Symb. Comput. 44(12), 1690–1702 (2009)

29. Granboulan, L., Joux, A., Stern, J.: Inverting HFE Is Quasipolynomial. In: Dwork, C. (ed.) CRYPTO 2006. LNCS, vol. 4117, pp. 345–356. Springer, Heidelberg (2006)
30. Joux, A., Lercier, R.: The Function Field Sieve in the Medium Prime Case. In: Vaudenay, S. (ed.) EUROCRYPT 2006. LNCS, vol. 4004, pp. 254–270. Springer, Heidelberg (2006)
31. Kipnis, A., Shamir, A.: Cryptanalysis of the HFE Public Key Cryptosystem by Re-linearization. In: Wiener, M.J. (ed.) CRYPTO 1999. LNCS, vol. 1666, pp. 19–30. Springer, Heidelberg (1999)
32. Koblitz, N.: Elliptic Curve Cryptosystems. Mathematics of Computation 48, 203–209 (1987)
33. Kraitchik, M.: Théorie des Nombres. Gauthier–Villards (1922)
34. Lazard, D.: Gröbner-Bases, Gaussian Elimination and Resolution of Systems of Algebraic Equations. In: van Hulzen, J.A. (ed.) EUROCAL 1983. LNCS, vol. 162, pp. 146–156. Springer, Heidelberg (1983)
35. Macaulay, F.S.: The Algebraic Theory of Modular Systems. Cambridge Mathematical Library, vol. xxxi. Cambridge University Press (1916)
36. Macaulay, F.S.: Some Properties of Enumeration in the Theory of Modular Systems. Proc. London Math. Soc. 26, 531–555 (1927)
37. Miller, V.S.: Use of Elliptic Curves in Cryptography. In: Williams, H.C. (ed.) CRYPTO 1985. LNCS, vol. 218, pp. 417–426. Springer, Heidelberg (1986)
38. Patarin, J.: Hidden Fields Equations (HFE) and Isomorphisms of Polynomials (IP): Two New Families of Asymmetric Algorithms. In: Maurer, U.M. (ed.) EUROCRYPT 1996. LNCS, vol. 1070, pp. 33–48. Springer, Heidelberg (1996)
39. Rojas, J.M.: Solving Degenerate Sparse Polynomial Systems Faster. J. Symbolic Computation 28, 155–186 (1999)
40. Semaev, I.: Summation Polynomials and the Discrete Logarithm Problem on Elliptic Curves (2004), http://eprint.iacr.org/2004/031.pdf

A   Proof of Lemma 2

In order to show Lemma 2, We use the following well-known technical result (a proof is given in [22] for instance).

Lemma 3. *Let n be an integer and let $\delta, 0 < \delta < 1/2$ be a number such that $\delta n \in \mathbb{N}$. Finally, let $\nu := \frac{\delta}{1-\delta}$. Then, the following inequalities hold:*

$$\binom{n}{\delta n} < \sum_{i=0}^{\delta n} \binom{n}{i} < \frac{1}{1-\nu} \binom{n}{\delta n}.$$

We can now come back to the proof of Lemma 2.

Proof (of Lemma 2). We recall that $1 - \alpha < \frac{1}{2} < \alpha < 1$, and $n' = n^\alpha, m = n^{(1-\alpha)}, t = m-1$. The number of equations in the system generated in Algorithm 1 is $E(d) = n \, 2^{tm} \left(\sum_{d'=t}^{d} \binom{n'-t}{d'-t} \right)^m$ and the number of columns is $M(d) = \left(\sum_{d'=0}^{d} \binom{n'}{d'} \right)^m$. We try to find $\beta, 1/2 < \beta < \alpha < 1$ such that $E(d) \geq M(d)$, where $d = n^\beta$ for n big enough.

A sum of binomials can be bounded from above by the last binomial occurring in the sum. We then get:

$$E(d)^{\frac{1}{m}} = n^{\frac{1}{m}} 2^t \sum_{d'=t}^{d} \binom{n'-t}{d'-t} \geq n^{\frac{1}{m}} 2^t \binom{n'-t}{d-t}.$$

Now, let $\delta := \frac{d}{n^\alpha}$ and $\nu := \frac{\delta}{1-\delta}$. When n is large enough, we have $\delta < \frac{1}{3}$. This leads to $\nu \leq \frac{1}{2}$ and $\frac{1}{1-\nu} \leq 2$. We are now in position to apply Lemma 3.

$$M(d)^{\frac{1}{m}} \leq 2\binom{n'}{d}. \tag{6}$$

Hence, we want to find d such that $n^{\frac{1}{m}} 2^t \binom{n'-d}{d-t} \geq 2\binom{n'}{d}$. We search now for the equality and consider the logarithm of each side of (6):

$$\log(n^{\frac{1}{m}}) + t + \log\binom{n'-t}{d-t} = 1 + \log\binom{n'}{d}. \tag{7}$$

The following result is useful to derive an asymptotical equivalent of this equality:

Lemma 4. *Let* $\gamma < 1/2 < \beta < \alpha < 1$. *For* n *large enough, we have:* $\log\binom{n^\alpha-n^\gamma}{n^\beta-n^\gamma} \approx (n^\beta - n^\gamma)(\alpha - \beta)\log(n)$.

Proof. We have $\log\binom{n^\alpha-n^\gamma}{n^\beta-n^\gamma} \approx (n^\alpha - n^\gamma)\log(n^\alpha - n^\gamma) - (n^\beta - n^\gamma)\log(n^\beta - n^\gamma) - (n^\alpha - n^\beta)\log(n^\alpha - n^\beta)$ and thus
$\log\binom{n^\alpha-n^\gamma}{n^\beta-n^\gamma} \approx (n^\alpha - n^\gamma)\alpha\log(n) - (n^\beta - n^\gamma)\beta\log(n) - (n^\alpha - n^\beta)\alpha\log(n)$
$= -n^\gamma\alpha\log(n) - (n^\beta - n^\gamma)\beta\log(n) + n^\beta\alpha\log(n) = (n^\beta - n^\gamma)(\alpha - \beta)\log(n)$

Taking $\gamma = 0$, we deduce $\log\binom{n^\alpha}{n^\beta} \approx n^\beta(\alpha - \beta)\log(n)$. Then, using $\gamma = 1 - \alpha$ we get $\log\binom{n^\alpha-n^{1-\alpha}}{n^\beta-n^{1-\alpha}} \approx (n^\beta - n^{1-\alpha})(\alpha - \beta)\log(n)$. As a consequence (7) yields:

$$n^{\alpha-1}\log n + n^{1-\alpha} + (n^\beta - n^{1-\alpha})(\alpha - \beta)\log(n) = \log(2) + n^\beta(\alpha - \beta)\log(n)$$
$$n^{\alpha-1}\log n + n^{1-\alpha} - n^{1-\alpha}(\alpha - \beta)\log(n) = \log(2) \text{ and } n^{1-\alpha} \approx n^{1-\alpha}(\alpha - \beta)\log(n)$$

So that $1 \approx (\alpha - \beta)\log(n)$. Thus, $\beta \approx \alpha - 1/\log(n)$ and $d^\beta \approx n^{\left(\alpha - \frac{1}{\log(n)}\right)} = n^\alpha/2$. This concludes the proof of Lemma 2. $\qquad\qquad\square$

Key-Alternating Ciphers in a Provable Setting: Encryption Using a Small Number of Public Permutations
(Extended Abstract[*])

Andrey Bogdanov[1], Lars R. Knudsen[2], Gregor Leander[2],
Francois-Xavier Standaert[3], John Steinberger[4], and Elmar Tischhauser[1]

[1] KU Leuven and IBBT
{Andrey.Bogdanov,Elmar.Tischhauser}@esat.kuleuven.be
[2] Technical University of Denmark
{G.Leander,Knudsen}@mat.dtu.dk
[3] Université catholique de Louvain, UCL Crypto Group
fstandae@uclouvain.be
[4] Tsinghua University
jpsteinb@gmail.com

Abstract. This paper considers—for the first time—the concept of key-alternating ciphers in a provable security setting. Key-alternating ciphers can be seen as a generalization of a construction proposed by Even and Mansour in 1991. This construction builds a block cipher PX from an n-bit permutation P and two n-bit keys k_0 and k_1, setting $PX_{k_0,k_1}(x) = k_1 \oplus P(x \oplus k_0)$. Here we consider a (natural) extension of the Even-Mansour construction with t permutations $P_1, \ldots, P_t$ and $t+1$ keys, $k_0, \ldots, k_t$. We demonstrate in a formal model that such a cipher is secure in the sense that an attacker needs to make at least $2^{2n/3}$ queries to the underlying permutations to be able to distinguish the construction from random. We argue further that the bound is tight for $t = 2$ but there is a gap in the bounds for $t > 2$, which is left as an open and interesting problem. Additionally, in terms of statistical attacks, we show that the distribution of Fourier coefficients for the cipher over all keys is close to ideal. Lastly, we define a practical instance of the construction with $t = 2$ using AES referred to as AES^2. Any attack on AES^2 with complexity below 2^{85} will have to make use of AES with a fixed known key in a non-black box manner. However, we conjecture its security is 2^{128}.

Keywords: Block ciphers, provable security, Even-Mansour construction, AES.

1 Introduction

Block ciphers are one of the fundamental primitives in symmetric cryptography. Often called the work horses of cryptography, they form the backbone of today's

---

[*] Due to page limitations, several proofs are omitted in this proceedings version. A full version is available at [9].

D. Pointcheval and T. Johansson (Eds.): EUROCRYPT 2012, LNCS 7237, pp. 45–62, 2012.

secure communication. Therefore, their design has been an important research focus over the last 20 years, giving rise to different well-established strategies to prevent large classes of attacks. As typical examples, one can mention the practical security approach against linear and differential cryptanalysis [23], and the wide-trail strategy [15] that lead to the design of the AES Rijndael [14]. Another line of research is the so-called provable security approach against statistical attacks, that served as foundation for the block cipher MISTY [27, 28]. One can also mention the decorrelation theory [33] and the design of the ciphers C [1] and KFC [2]. At a high level, the three main design paradigms for block ciphers are Feistel structures such as DES, Lai-Massey ciphers such as IDEA [24], and key-alternating ciphers [12,14,15] for which the AES Rijndael is a prominent representative. State-of-the-art block ciphers are quite well understood and provide security against all known attacks. Though there has recently been remarkable progress in the cryptanalysis of AES [7], these results are far from being any threat for the use of AES in practice. Thus, from a practical point of view, block ciphers in general and key-alternating ciphers in particular can be seen as a success story.

Given the degree of confidence in properly designed key-alternating ciphers on the practical side (e.g. with AES approved for the encryption of secret and top secret data in the USA), it is even more surprising that there has been no provable setting developed so far for the design of key-alternating ciphers on the theoretical side. Nobody seems to have even formulated the problem of whether the key-alternating cipher makes sense from this point of view. Clearly, given the state of the art, proving AES secure in any strict sense is out of reach. However, by modeling the round functions as fixed public randomly chosen permutations, we are able to precisely formulate and—as we shall see—prove the soundness of the key-alternating cipher design. The cipher we are dealing with is depicted in Figure 2 and detailed in Section 2.

We note the difference of our setting to that of an idealized Feistel cipher, often called the Luby-Rackoff construction [26], or to that of similar results obtained for the Lai-Massey schemes [34]. In these former works, for each key it is assumed that the function used in the Feistel (resp. Lai-Massey) construction is chosen at random. Directly adopting this model to the case of a key-alternating cipher immediately results in an ideal cipher (even for one round). At the same time, in most key-alternating ciphers including AES, the key is the only part of the design to define the cipher permutation and all round permutations are fixed for the entire cipher, not varying from key to key. In other words, working along the lines of [26] does not elucidate how to mix the key into the state. It is exactly this point we deal with in the present paper, both at a high-level, i.e. in a provable setting, as well as at lower-levels, i.e. considering statistical attacks and as a guideline for actually designing ciphers.

Interestingly, another look at the construction and its properties arises from the question of how to design the key schedule of a block cipher. This has been an open problem in symmetric-key cryptography for decades. While some ciphers are based upon simple linear or nearly linear key schedules [8, 18], a number of others opt for heavier and often highly nonlinear key schedules, sometimes as complex as the round functions [3] or the cipher itself [31]. In the prominent case of AES, for instance, the key schedule is iterative, mainly linear, and provides relatively slow diffusion in the backward direction. It is precisely these properties that facilitated the related-key cryptanalysis of the full AES-192 and AES-256, e.g. [5,6] as well as the recent biclique cryptanalysis of all three full AES versions in the classical single-key model [7]. In general, these examples emphasize a relatively weak understanding of key scheduling algorithms, compared to the design of block cipher rounds. In this context, the results of this paper can be seen as a case for simple key schedules (or even no key scheduling at all). Hence, they provide new insights into the design of block ciphers.

1.1 Related Work

An exception from the above-mentioned lack of theoretical studies of key-alternating block ciphers is the Even-Mansour construction [16] depicted in Figure 1. This construction can be seen as a one-round variant of a key-alternating

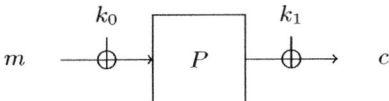

Fig. 1. The Even-Mansour construction

cipher. Informally, Even and Mansour proved that in order to have a reasonable success probability in decrypting an (unqueried) message, an attacker has to make roughly $2^{n/2}$ queries to the permutation P. In this setting, the attacker is given oracle access to P, its inverse, and to an encryption and decryption oracle. Later, Daemen [11] showed that this bound is actually tight. He presented a differential attack on the Even-Mansour scheme that allows to successfully recover the key with a good probability, after $2^{n/2}$ evaluations of both the permutation P and the encryption oracle.

1.2 Our Contribution

Our contributions in this paper are twofold.

On the theoretical side (cf. Section 3), we provide the first treatment of the concept of key-alternating ciphers in a provable security setting. We prove below that, for any t-round version of the cipher with randomly drawn and fixed underlying permutations, $t \geq 2$, depicted in Figure 2, an attacker needs to make at least $2^{2n/3}$ queries before being able to distinguish the encryption oracle from a random permutation. Here n is the block size of the cipher. Furthermore, we provide a simple attack that shows that an attacker, by making $2^{\frac{t}{t+1}n}$ queries, is able to recover the secret key used in the decryption oracle. We do conjecture that this lower bound — being tight only for $t = 2$ — is the actual bound. We leave proving this as an important open question (see also Section 7). Note that in this setup, we necessarily only consider the query complexity of an attacker, ignoring the computational complexity. It seems unlikely that an attack with a comparable computational complexity exists. Such an attack would in particular imply an attack on e.g. AES-256 with a complexity of around 2^{120} operations.

On the practical side, we propose to actually use the construction of Figure 2. Given our theoretical results, the merit of this approach is the following: Any attack on a key-alternating cipher with complexity below $2^{2n/3}$ will have to make use of the round functions in a non-black box manner.

However, and we feel that it is important to make this point explicit even though it might be obvious, the theoretical result does not carry over to any efficient instance, as one must consider the round functions as black-boxes— i.e. objects which the adversary must query to evaluate—in order to meaningfully discuss the distinguishability of the cipher from a random permutation by an information-theoretic adversary.

This fact and the fact that, as mentioned above, the theoretical bounds are likely to be lower than the computational complexity of any attack, motivates us to study the security of our proposal with respect to such statistical attacks as linear cryptanalysis (see Section 5).

To capture the difference between the single-round Even-Mansour cipher and the multiple-round key-alternating construction with respect to linear cryptanalysis, we study the Fourier spectrum of the ciphers. We prove that once the fixed underlying permutations are close to average (which is the case for randomly drawn permutations with high probability), the distribution of Fourier coefficients for the key-alternating cipher over all keys for $t \geq 2$ gets close to that over all permutations — the natural reference point for any block cipher. At the same time, we demonstrate that this is not the case for the original Even-Mansour construction with $t = 1$ where the Fourier coefficients almost do not change from key to key. It seems therefore unlikely that linear attacks are able to break the multiple-round key-alternating cipher with $t \geq 2$.

Finally, as the crypto community likes targets and we anticipate that having a concrete proposal is a valuable stimulation for further research, we propose an actual cipher called AES^2 following the 2-round version of the general construction (see Section 6). Here we replace the random permutations by two instantiations of AES-128 with fixed known keys. Given the new AES instructions on recent Intel processors, AES^2 performs very competitively on those platforms, with as few as 2.65 cycles per byte required in the counter mode.

We conclude with a section dedicated to open questions and further work (Section 7), discussing how to possibly improve and extend the research we consider in the paper.

2   The Construction

The cipher we consider is an idealized model of a key-alternating cipher — the notion introduced under this name in [14,15] in connection with the design of AES and used without being explicitly named even before that [12] in similar contexts. Such a cipher consists of round functions interleaved with xoring round keys to the current state. In our idealized model, the round functions are the public, randomly chosen permutations P_i and the key consists of $t + 1$ independent round-keys are k_i. More precisely, let $P_1, \ldots, P_t$ be permutations from $\{0,1\}^n$ to $\{0,1\}^n$, $t \geq 1$. Let $k_0, \ldots, k_t \in \{0,1\}^n$ be keys. The block cipher $E = E_{k_0,\ldots,k_t} : \{0,1\}^n \to \{0,1\}^n$ we consider is defined by

$$E(x) = E_{k_0 \cdots k_t}(x) = P_t(\ldots P_2(P_1(x \oplus k_0) \oplus k_1) \ldots) \oplus k_t \qquad (1)$$

for $x \in \{0,1\}^n$. The cipher is shown in Figure 2.

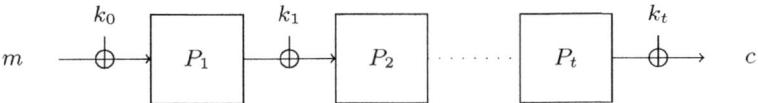

Fig. 2. A key-alternating cipher

3   Indistinguishability Analysis

Putting $N = 2^n$, we define the PRP security of E against an adversary A expecting a $(t + 1)$-tuple of oracles as

$$\mathbf{Adv}_{E,N,t}^{\mathrm{PRP}}(A) = \Pr[k_0 \cdots k_t \leftarrow \{0,1\}^n; A^{E_{k_0 \cdots k_t}, P_1, \ldots, P_t} = 1] - \Pr[A^{Q, P_1, \ldots, P_t} = 1]$$

where in each experiment $Q, P_1, \ldots, P_t$ are independent and uniformly sampled random permutations. Here A can make inverse queries to each of its oracles. Thus, an attacker has to tell apart two worlds, depicted below.

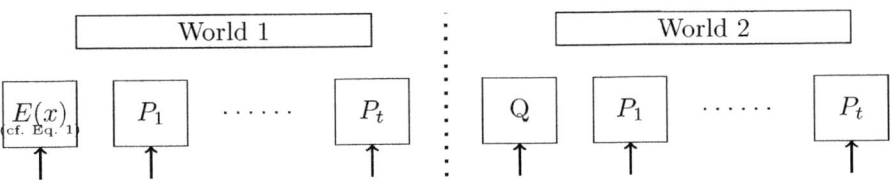

We note that one *must* consider the permutations $P_1, \ldots, P_t$ as random (or pseudorandom) black-boxes—i.e. objects which the adversary must query to evaluate—in order to meaningfully discuss the distinguishability of $E_{k_0,\ldots,k_t}$ from a random permutation by an information-theoretic adversary.

We define

$$\mathbf{Adv}_{E,N,t}^{\mathrm{PRP}}(q) = \max_{A} \mathbf{Adv}_{E}^{\mathrm{PRP}}(A)$$

where the maximum is taken over all adversaries A making at most q queries. (We note the parameters n and t are elided from both of the notations $\mathbf{Adv}_{E}^{\mathrm{PRP}}(A)$ and $\mathbf{Adv}_{E}^{\mathrm{PRP}}(q)$; but it should be understood that $\mathbf{Adv}_{E}^{\mathrm{PRP}}(q)$ is a function n and t as well as of q.)

Our main security result is the following:

Theorem 1. *Let $N = 2^n$ and let $q = N^{\frac{t}{t+1}}/Z$ for some $Z \geq 1$. Then, for any $t \geq 1$, and assuming $q < N/100$, we have*

$$\mathbf{Adv}_{E,N,t}^{\mathrm{PRP}}(q) \leq \frac{4.3q^3 t}{N^2} + \frac{t+1}{Z^t}.$$

For $t \geq 2$ the limiting term in the above bound is $4q^3 t/N^2$, which caps q at around $N^{2/3}$. The following corollary is more telling.

Corollary 1. *Assume $t \geq 2$. Let $q = N^{\frac{2}{3}}/\lambda \sqrt[3]{t}$ for some $\lambda \geq 1$. Then, assuming $q < N/100$,*

$$\mathbf{Adv}_{E,N,t}^{\mathrm{PRP}}(q) \leq \frac{4.3}{\lambda^3} + \frac{t+1}{(\sqrt[3]{t}\lambda)^t}.$$

We also note that $q < N/100$ as long as $n \geq 20$; this condition is therefore compatible with practical parameters. We note that Corollary 1's security of $q \approx N^{\frac{2}{3}}$ is optimal for $t = 2$ (cf. Section 3.1) and suboptimal for $t > 2$, in which case we conjecture a security of $q \approx N^{\frac{t}{t+1}}$. Closing this gap might be obtained by a tightening of Proposition 2 below.

Theorem 1 is proved by a hybrid argument involving an intermediate game. In order to outline this hybrid argument we start by developing some new notation.

Note firstly that if E is defined as in (1) then, putting $P_0 = E^{-1}$, we have

$$P_0(P_t(\cdots P_1(\cdot \oplus k_0) \cdots) \oplus k_t) = id.$$

Applying P_0^{-1} to both sides and then substituting $P_0(\cdot)$ for the input, we find

$$P_t(\cdots P_2(P_1(P_0(\cdot) \oplus k_0) \oplus k_1) \cdots) \oplus k_t = id. \tag{2}$$

It is easy to see that, for fixed $k_0, \ldots, k_t$, randomly sampling $P_1, \ldots, P_t$, defining E as in (1) and giving an adversary access to the tuple of oracles $(E, P_1, \ldots, P_t)$ (and their inverses) is equivalent to sampling $P_0, \ldots, P_t$ uniformly at random from all $(t+1)$-tuples of permutations satisfying (2) and giving the adversary access to $(P_0^{-1}, P_1, \ldots, P_t)$ (and their inverses). Moreover, it is just a notational change to give the adversary access to $(P_0, P_1, \ldots, P_t)$, since the adversary is allowed inverse queries anyway (of course, the adversary is alerted to the fact that its first oracle is now P_0 and not P_0^{-1}).

We now formally implement the interface $(P_0, \ldots, P_t)$ via an oracle $O(N, t)$ taking $k_0, \ldots, k_t$ as implicit parameters. Rather than sampling $P_0, \ldots, P_t$ uniformly at random from those sequences satisfying (2) at the start of the experiment, $O(N, t)$ implements the permutations $P_0, \ldots, P_t$ by lazy sampling. More precisely, $P_0, \ldots, P_t$ are initially set to be undefined everywhere. When the adversary makes a query $P_i(x)$ or $P_i^{-1}(y)$, the adversary defines P_i at the relevant point using the following procedure, illustrated for the case of a forward query $P_i(x)$ (the case of a backward query is analogous):

• Let $\mathcal{P} = \mathcal{P}(P_0, \ldots, P_t)$ be the set of all $(t+1)$-tuples of permutations $(\overline{P}_0, \ldots, \overline{P}_t)$ such that $\overline{P}_i$ extends the currently defined portion of P_i, and such that

$$\overline{P}_t(\cdots \overline{P}_2(\overline{P}_1(\overline{P}_0(\cdot) \oplus k_0) \oplus k_1) \cdots \oplus k_{t-1}) \oplus k_t = id. \tag{3}$$

Then $O(N,t)$ samples uniformly at random an element $(\overline{P}_0, \ldots, \overline{P}_t)$ from $\mathcal{P}$. The adversary sets $P_i(x) = \overline{P}_i(x)$ and returns this value.

After the above, the adversary "forgets" about $\overline{P}_0, \ldots, \overline{P}_t$, and samples these afresh at the next query. It is clear that this lazy sampling process gives the same distribution as sampling the tuple $(P_0, \ldots, P_t)$ at the start of the game. Thus, giving the adversary oracle access to $O(N,t)$ is equivalent to giving the adversary oracle access to $(E, P_1, \ldots, P_t)$, up to the cosmetic change that E is replaced by E^{-1}. We therefore have:

Proposition 1. *With $O(N,t)$ defined as above, we have:*

$$\mathbf{Adv}_{E,N,t}^{PRP}(A) = \Pr[k_0 \cdots k_t \leftarrow \{0,1\}^n; A^{O(N,t)} = 1] - \Pr[A^{Q_0, Q_1, \ldots, Q_t} = 1]$$

where $Q_0, \ldots, Q_t$ are independent random permutations.

(We emphasize that $k_0, \ldots, k_t$ are implicit arguments to $O(N,t)$.)

Our hybrid will be an oracle $\tilde{O}(N,t)$ (also taking $k_0, \ldots, k_t$ as implicit inputs) that uses a slightly different lazy sampling procedure to define the permutations $P_0, \ldots, P_t$. Say that a sequence of partially defined permutations is *consistent* if $\mathcal{P}(P_0, \ldots, P_t) \neq \emptyset$, with $\mathcal{P}(\cdot)$ defined as in the description of $O(N,t)$ above. Initially, $\tilde{O}(N,t)$ also sets the permutations $P_0, \ldots, P_t$ to be undefined everywhere. Upon receiving (say) a forward query $P_i(x)$, $\tilde{O}(N,t)$ uses the following lazy sampling procedure to answer:

• Let $U \subseteq \{0,1\}^n$ be the set of values y such that defining $P_i(x) = y$ maintains the consistency of $P_0, \ldots, P_t$, besides maintaining the fact that P_i is a permutation. Then $\tilde{O}(N,t)$ samples a value y uniformly from U, sets $P_i(x) = y$, and returns y.

Inverse queries are lazy sampled the same way. While not immediately apparent, the above lazy sampling procedure produces a slightly *different* distribution of outputs than the first lazy sampling procedure.

Theorem 1 is an direct consequence of Proposition 1 and of the following two propositions.

Proposition 2. *Let $q < N/100$. With $O(N,t)$ and $\tilde{O}(N,t)$ defined as above,*

$$\Pr[k_0, \ldots, k_t \leftarrow \{0,1\}^n; A^{O(N,t)} = 1] - \Pr[k_0, \ldots, k_t \leftarrow \{0,1\}^n; A^{\tilde{O}(N,t)} = 1] \leq \frac{4.3q^3 t}{N^2}$$

for every adversary A making at most q queries.

Proposition 3. *Let* $q = N^{\frac{t}{t+1}}/Z$ *for some* $Z \geq 1$ *be such that* $q < N/3$. *With* $\tilde{O}(N,t)$ *defined as above,*

$$\Pr[k_0, \ldots, k_t \leftarrow \{0,1\}^n; A^{\tilde{O}(N,t)} = 1] - \Pr[A^{Q_0, \ldots, Q_t} = 1] \leq \frac{t+1}{Z^{t+1}}.$$

for every adversary A *making at most* q *queries, where* $Q_0, \ldots, Q_t$ *are independent random permutations.*

Proposition 2 is the main technical hurdle in our proof. Its proof, however, is entirely combinatorial, given that we actually show this bound holds even when A sees the keys $k_0, \ldots, k_t$. The presence of keys is therefore actually irrelevant for this proposition[1]. We refer to the full version for more details and a proof of Proposition 2.

The proof of Proposition 3, on the other hand, is fairly accessible, and also contains those ingredients that have the most "cryptographic interest".

Proof (of Proposition 3.). We make the standard assumption that the adversary never makes a redundant query (querying $P_i^{\pm 1}(x)$ twice or querying, e.g., $P_i(x)$ after obtaining x as an answer to a query $P_i^{-1}(y)$).

We modify $\tilde{O}(N,t)$ to use a slightly different lazy sampling method, equivalent to $\tilde{O}(N,t)$'s original sampling method. In this new method, we also maintain a flag bad which is originally set to false.

$\tilde{O}(N,t)$'s new sampling method is as follows: when faced with a query $P_i(x)$, $\tilde{O}(N,t)$ samples a value y uniformly at random from the remaining range of $P_i(x)$, that is, uniformly at random from

$$\{0,1\}^n \backslash \{P_i(x') : x' \in \{0,1\}^n, P_i(x') \text{ is defined}\}.$$

$\tilde{O}(N,t)$ then checks if setting $P_i(x) = y$ would make $P_0, \ldots, P_t$ inconsistent; if so, it sets bad = true, and resumes its original sampling method for the rest of the game (including to answer the last query); otherwise, it sets $P_i(x) = y$, and returns y. Inverse queries are treated the same.

We can also define a value for the bad flag when the adversary has oracle access to the random permutations $(Q_0, Q_1, \ldots, Q_t)$. Originally, set bad = false and select random values $k_0, \ldots, k_t$. Set $Q_0, \ldots, Q_t$ to be undefined at all points, and use lazy sampling to define them by simulating the lazy sampling process for $P_0, \ldots, P_t$ up until bad = true; after bad = true, simply keep lazy sampling each permutation Q_i while ignoring bad as well as $k_0, \ldots, k_t$.

Obviously, the probability bad is set to true is equal in both worlds, and the two worlds behave identically up until bad = true. Thus (a standard argument shows that) the adversary's advantage is upper bounded by the probability that bad is set to true.

For simplicity, we upper bound the probability that bad becomes true when the adversary has oracle access to $Q_0, \ldots, Q_t$. In this case, note that it is equivalent

---

[1] We note that the bound of Proposition 2 is the bottleneck of Theorem 1. A potential improvement of Proposition 2 might exploit the fact that $k_0, \ldots, k_t$ aren't known to the adversary.

to set the bad flag by sampling the values $k_0, \ldots, k_t$ randomly at the end of the game, and then checking whether these values are inconsistent with the partially defined permutations $Q_0, \ldots, Q_t$. (To recall, $k_0, \ldots, k_t$ are inconsistent with $Q_0, \ldots, Q_t$ if there exist no permutations $\overline{Q}_0, \ldots, \overline{Q}_t$ such that

$$\overline{Q}_t(\cdots \overline{Q}_2(\overline{Q}_1(\overline{Q}_0(\cdot) \oplus k_0) \oplus k_1) \cdots \oplus k_{t-1}) \oplus k_t = id.)$$

Given the partially defined permutations $Q_0, \ldots, Q_t$ and values $k_0, \ldots, k_t$ a *contradictory path* is a sequence of values $(x_0, y_0), \ldots, (x_t, y_t)$ such that (i) $Q_i(x_i) = y_i$ for all i and (ii) $|\{i : y_i \oplus x_{i+1} = k_i, 0 \leq i \leq t\}| = t$, where we put $x_{t+1} = x_0$. Because $q < N/3$, one can show that $Q_0, \ldots, Q_t$ is consistent with $k_0, \ldots, k_t$ if and only if there exists no contradictory path (again, we have to refer to the full versions for details). Since each Q_i contains at most q defined input-output pairs (x_i, y_i) at the end of the game, there are at most q^{t+1} possible different sequences $((x_0, y_0), \ldots, (x_t, y_t))$ such that $Q(x_i) = y_i$ for $0 \leq i \leq t$. For each of these sequences, the probability that the random selection of $k_0, \ldots, k_t$ creates a contradictory path is upper bounded by $(t + 1)N^{-t}$, since the condition $k_i = y_i \oplus x_{i+1}$ must be satisfied for all but one value of i, $0 \leq i \leq t$, and we can union bound over this value of i. Hence, by a union bound over the (at most) q^{t+1} possible different sequences, the probability that bad is set to true is at most $\frac{(t+1)q^{t+1}}{N^t} = \frac{t+1}{Z^t}$ as desired.

3.1 An Upper Bound

For any number of rounds t, there is an (non-adaptive) attack with a query complexity of roughly $t2^{\frac{t}{t+1}n}$, thus meeting the bound on the query complexity for $t = 2$. Note that this is not an attack in the practical sense, as the computational cost is higher than brute force. The idea of this attack is to actually construct (with high probability) a contradictory path for each possible key.

1. Make $2^{\frac{t}{t+1}n}$ queries to E and each of the oracles P_1 to P_t. Denote the set of queries to P_i by $\mathcal{P}_i$ and queries to E_k by $\mathcal{M}$.
2. For each key candidate $(k_0, k_1, \ldots, k_t)$ do:
   (a) Find all sequences of values $(x_1, \ldots, x_{t-1})$ such that $x_1 \in \mathcal{M}$ and $x_i \oplus k_{i-1} \in \mathcal{P}_i$, $\forall 1 \leq i \leq t$ and $P_i(x_i \oplus k_{i-1}) = x_{i+1}$, $\forall 1 \leq i \leq t - 1$.
   (b) Check if $P_t(x_t \oplus k_{t-1}) \oplus k_t = E(x_1)$ for all these sequences.
   (c) If so, assume $(k_0, k_1, \ldots, k_t)$ is the correct value of the key;
   (d) otherwise, it is certainly the wrong value of the key.

To get a better reduction on key-candidates, a bit more than $t2^{\frac{t}{t+1}n}$ queries are sufficient.

4 Attacks

The bounds proved earlier are information-theoretic bounds which take into account only the number of queries of the random permutations made by an adversary. Of equal interest are attacks which take the computational complexity into account. In this section we consider only attacks in the single key-model. Note that, in the case where all round-keys are independent, related-key attacks exist trivially. However, the situation might be very different in the case where all round-keys are identical, see Section 7 for further discussion on this point.

4.1   Daemen's Attack for $t = 1$

For the original Even-Mansour construction (in our setting, this corresponds to $t = 1$), a differential attack has been published by Daemen [11] meeting the lower bound of $2^{n/2}$ evaluations of P proven by Even and Mansour. It can be described as follows:

1. Choose s plaintext pairs (m_i, m_i^*), $1 \le i \le s$, with $m_i \oplus m_i^* = \Delta$ for any nonzero constant Δ.
2. Get the encryptions (c_i, c_i^*) of the s pairs.
3. For $2^n/s$ values v:
   (a) Compute $w' := P(v) \oplus P(v \oplus \Delta)$.
   (b) If $w' = c_i \oplus c_i^*$ for some i: Output $k_0 := v \oplus m_1$ and $k_1 := c_1 \oplus P(m_1 \oplus k_0)$ and stop.

For a random permutation P, only very few values of v are expected to satisfy $P(v) + P(v + \Delta) = c_i \oplus c_i^*$. The wrong candidates can be easily filtered in step (3b) by testing them on a few additional encryptions. After encrypting s plaintext pairs, one has to perform about $2 \cdot 2^n/s$ evaluations of P. The expression $2(s + 2^n/s)$ is minimal for $s = 2^{n/2}$. In this case, the time complexity is $2^{n/2}$ with a storage requirement of $2^{n/2}$ plaintext pairs.

4.2   A Meet in the Middle Attack

There is a meet in the middle attack on the t-permutation construction which finds the keys in time and space $2^{tn/2}$ for $t > 1$. This is a straight-forward attack given here for the case $t = 2$:

1. From a pair of messages (m_1, m_2), compute and save in a sorted table, T, the values $P(m_1 \oplus k) \oplus P(m_2 \oplus k)$ for all possible 2^n values of k.
2. Get the encryptions c_1 and c_2 of m_1 respectively m_2.
3. For all 2^n possible values of k' compute $Q^{-1}(c_1 \oplus k') \oplus Q^{-1}(c_2 \oplus k')$ and look for a match in T.
4. Each match gives candidate values for the three keys, which are tested against additional encryptions.

5   Statistical Properties

A fundamental cryptographic property of a block cipher is its Fourier spectrum that completely defines the cipher via the Fourier transform and whose distribution is closely related to the resistance against linear cryptanalysis [10].

To support security claims, block cipher designs usually come with arguments why these Fourier coefficients cannot take values exploitable by an attacker. In most cases, however, formal proofs of these properties appear technically infeasible and designers limit themselves to demonstrating upper bounds on trail probabilities, that can be seen as summands to obtain the actual Fourier coefficients. This solution is usually denoted as the practical security approach for statistical cryptanalysis. Such an approach does not allow an accurate estimation of the data complexity of statistical attacks, that typically depends on numerous trails [25, 29].

As opposed to that, we analyze the construction of key alternating cipher following a provable security approach, by directly investigating its Fourier coefficients. In addition, we provide a more informative analysis than for standard block ciphers, as we study the distribution of the Fourier coefficients for the cipher over all keys, rather than bounding the mean value of this distribution. This is made possible by the use of fixed public permutations in our construction. More precisely, in a key-alternating cipher using $t \geq 2$ fixed public permutations, we study the distribution of the Fourier coefficients over all cipher keys. If these permutations are close to the average over all permutations, we show that this distribution turns out to be very close to that over all permutations, suggesting that the t-round key-alternating construction is theoretically sound from this perspective. This implies that it behaves well with respect to linear cryptanalysis.

On the contrary, the distribution of Fourier coefficients for a fixed point in the Fourier spectrum is nearly degenerated for the key-alternating cipher with $t = 1$ (the Even-Mansour cipher). This emphasizes the constructive effect of having 2 and more rounds in the key-alternating cipher.

5.1   Fourier Coefficients over All Permutations

Here we recall the definitions of Fourier coefficients and Fourier spectrum as well as the distribution of Fourier coefficients over all permutations. We also introduce some notations we will be using throughout the section.

Notations. The canonical scalar product of two vectors $a, b \in \{0, 1\}^n$ is denoted by $a^T b$. We denote the normal distribution with mean μ and variance σ^2 as $\mathcal{N}(\mu, \sigma^2)$. By $X \sim_v \mathcal{D}$, we denote a random variable X following a distribution $\mathcal{D}$ taken over all values of v. The expectation of X with respect to v is denoted by $\mathbf{E}_v[X]$, its variance (with respect to v) by $\mathbf{Var}_v[X]$.

Fourier Coefficients and Fourier Spectrum. For a permutation $P : \{0, 1\}^n \to \{0, 1\}^n$, its *Fourier coefficient* at point (α, β) is defined as

$$W_{\alpha,\beta}^P \stackrel{\text{def}}{=} \sum_{x \in \{0,1\}^n} (-1)^{\alpha^T x + \beta^T P(x)}.$$

The collection of Fourier coefficients at all points $(\alpha, \beta) \in \{0, 1\}^n \times \{0, 1\}^n$ is called the *Fourier spectrum* of P. For a block cipher F, we denote the Fourier coefficient at point (α, β) as $W^F_{\alpha,\beta}[K]$ to emphasize its dependency on key K. If F is the t-round key-alternating cipher, this is denoted by $W^{P_1,\ldots,P_t}_{\alpha,\beta}[K]$.

The following characterisation for the distribution of Fourier coefficients in a Boolean permutation has been proven.

Fact 1 ([13, Corollary 4.3, Lemma 4.6]). *When $n \geq 5$, the distribution of the Fourier coefficient $W^P_{\alpha_0,\beta_0}$ with $\alpha_0, \beta_0 \neq 0$ over all n-bit permutations can be approximated by the following distribution up to continuity correction:*

$$W^P_{\alpha_0,\beta_0} \sim_P \mathcal{N}(0, 2^n). \tag{4}$$

The distribution of Fact 1 is the reference point throughout the section: A block cipher cannot have a better distribution of Fourier coefficients than that close to Fact 1.

5.2 Fourier Coefficients in the Single-Round Even-Mansour Cipher

Let F be the basic single-round Even-Mansour cipher, that is, a fixed public permutation P surrounded by two additions with keys k_0 and k_1, respectively (see Figure 1). If $W^P_{\beta_0,\beta_1}$ is the Fourier coefficient for the underlying permutation P at point (β_0, β_1), then the Fourier coefficient for the cipher at this point is

$$W^F_{\beta_0,\beta_1} = (-1)^{\beta_0^T k_0 \oplus \beta_1^T k_1} W^P_{\beta_0,\beta_1}.$$

Now consider the distribution of $W^F_{\beta_0,\beta_1}$ with $\beta_0 \neq 0$, $\beta_1 \neq 0$ taken over all keys (k_0, k_1). Its support contains exactly two points: $W^P_{\beta_0,\beta_1}$ and $-W^P_{\beta_0,\beta_1}$. Thus, the value of $W^F_{\beta_0,\beta_1}$ almost does not vary from key to key. This is crucially different from the reference point – the distribution over all permutations of Fact 1.

5.3 Fourier Coefficients in the t-Round Key-Alternating Cipher

Now we state the main result of this section. The proof is given omitted in this extended abstract and we refer to the full version.

Theorem 2. *Fix a point (β_0, β_t) with $\beta_0, \beta_t \neq 0$ in the Fourier spectrum of the t-round key-alternating n-bit block cipher with round permutations $P_1, \ldots, P_t$ for $t \geq 2$ and sufficiently high n. Then the distribution of the Fourier coefficient $W^{P_1,\ldots,P_t}_{\beta_0,\beta_t}$ at this point over all keys K is approximated by:*

$$W^{P_1,\ldots,P_t}_{\beta_0,\beta_t}[K] \sim_K \mathcal{N}\left(0, (1+\varepsilon)\left(\frac{2^n - 1}{2^n}\right)^{t-1} 2^n\right), \tag{5}$$

assuming that the distributions over points of the Fourier spectra of the permutations P_i, $1 \leq i \leq t$, have variances satisfying

$$\mathbf{Var}_{(\beta_{i-1},\beta_i)}\left[W^{P_i}_{\beta_{i-1},\beta_i}\right] \geq 2^{n/2}, \tag{6}$$

and that for any given key K, the signs of the Fourier coefficients behave independently for different points. The deviation of the permutations P_i from the mean over all permutations Q_i is quantified by factor $(1 + \varepsilon)$:

$$\sum_{(\beta_1, \ldots, \beta_{t-1})} \left(W^{P_1}_{\beta_0, \beta_1} \cdots W^{P_t}_{\beta_{t-1}, \beta_t} \right)^2$$
$$= (1 + \varepsilon) \cdot \mathbf{E}_{Q_1, \ldots, Q_t} \left[\sum_{(\beta_1, \ldots, \beta_{t-1})} \left(W^{Q_1}_{\beta_0, \beta_1} \cdots W^{Q_t}_{\beta_{t-1}, \beta_t} \right)^2 \right]. \tag{7}$$

Interestingly, the latter deviation ε from the mean in (7) is small for most choices of the P_i. For instance, in case $t = 2$, it can be shown that over all permutations, mean and variance of each summand in (7) are 2^{2n} and 2^{4n+2}, respectively. The whole sum then approximately follows a normal distribution $\mathcal{N}(2^{3n} - 2^{2n}, 2^{5n+2} - 2^{4n+2})$. This means that for *randomly drawn permutations* P_1, P_2, the sum $\sum_{\beta_1} \left(W^{P_1}_{\beta_0, \beta_1} W^{P_2}_{\beta_1, \beta_2} \right)^2$ will be within d standard deviations from its mean with probability $\operatorname{erf}\left(d/\sqrt{2}\right)$. Notably, this implies $\Pr(|\varepsilon| \leq 2^{-n/2+3}) \approx 0.9999$, i.e. $|\varepsilon|$ only very rarely exceeds $2^{-n/2+3}$.

Theorem 2 gives the distribution over all keys of the Fourier coefficient $W^{P_1, \ldots, P_t}_{\beta_0, \beta_t}$ individually for each nontrivial point (β_0, β_t). Appropriate choices for the P_i should have distributions close to $\mathcal{N}(0, 2^n)$ for each nontrivial point, not only for some of them. Conversely, the distribution of the Fourier coefficient at the (trivial) point $(\beta_0, 0)$ differs from (5) for any choice of the P_i, since it is constant over the keys.

Note also that the result of Theorem 2 does not require the underlying permutations to be different. Moreover, it does not require the permutations P_i to be randomly drawn from the set of all permutations, but holds for any fixed choice of permutations satisfying (6). To obtain a distribution close to ideal, however, the set of underlying permutations has to ensure a small deviation ε in (7). As argued above, drawing the underlying permutations at random from the set of all permutations is highly likely to result in a very small deviation ε from the average.

Summarising, the results of Theorem 2 suggest that once the small number of $t \geq 2$ underlying permutations are carefully chosen and fixed, the t-round key-alternating cipher for each secret key is likely to be statistically sound which rules out some crucial cryptanalytic distinguishers. More precisely, the distributions of the Fourier coefficients for the t-round key-alternating cipher over all keys become close to those over all permutations.

Note that, in contrast to the reference point, it is possible to identify large but efficiently representable subsets of keys where the distribution is again degenerated, as in the case for $t = 1$. Examples of such sets are sets of keys where one fixes all keys k_1 up to k_{t-1}. For any point (β_0, β_1) the value of $W^{P_1, \ldots, P_t}_{\beta_0, \beta_t}$ takes on only two possible values - over all possible sub-keys k_0, k_t. However, it seems unlikely that this can be used in an attack.

6   Practical Constructions

In this section, we discuss possible practical realisations of the t-round key-alternating cipher.

A natural approach to building a practical cipher following the t-permutation construction is to base the t fixed permutations on a block cipher by fixing some keys. With $t = 1$, this corresponds to the original Even-Mansour construction, so the security level is limited to $2^{n/2}$ operations with n denoting the cipher's block length. With a 128-bit block cipher such as the AES, we therefore only obtain a security level of 2^{64} in terms of computational complexity, so it is advisable to choose $t > 1$.

In the following we describe a sample construction with $t = 2$, that is, we consider the 2-round key alternating construction with permutations P_1 and P_2 and the keys k_0, k_1, k_2.

6.1   AES²: A Block Cipher Proposal Based on AES

The construction is defined by fixing two randomly chosen 128-bit AES-128 keys, which specifies the permutations P_1 and P_2. The key is comprised by three independently chosen 128-bit secret keys k_0, k_1, k_2.

Let AES$[k]$ denote the (10-round) AES-128 algorithm with the 128-bit key k and the 128-bit quantities π_1, π_2 be defined based on the first 256 bits of the binary digit expansion of $\pi = 3.1415\ldots$:

$$\pi_1 := 0\mathtt{x}243f6a8885a308d313198a2e03707344 \quad \text{and}$$
$$\pi_2 := 0\mathtt{x}a4093822299f31d0082efa98ec4e6c89.$$

Then we denote the resulting 2-permutation construction by AES$^2[k_0, k_1, k_2]$. Its action on the 128-bit plaintext m is defined as:

$$\text{AES}^2[k_0, k_1, k_2](m) := \text{AES}[\pi_2](\text{AES}[\pi_1](m \oplus k_0) \oplus k_1) \oplus k_2. \qquad (8)$$

Security. Any attack on AES^2 in the single secret-key model with complexity below 2^{85} will have to make use of AES with a fixed known key in a non-black box manner. On the other hand, we are aware of no attack with a computational complexity of less than 2^{128}. Moreover, if the distribution of Fourier coefficients for AES$[\pi_1]$ and AES$[\pi_2]$ meets the assumption of average behaviour, Theorem 2 suggests that the Fourier coefficients for AES^2 are distributed close to ideal which implies resistance against basic linear cryptanalysis and some of its variants. Intuitively, this construction can be seen to arguably transfer the security properties for AES with a single randomly fixed key to the entire cipher as a set of permutations. For AES^2, we explicitly do not claim any related-, known- or chosen-key security.

Performance. AES^2 can be implemented very efficiently in software on general-purpose processors. The two AES keys π_1 and π_2 are fixed and, therefore, the round keys for the two AES transformations can be precomputed, so there is no need to implement the key scheduling algorithm of AES. This ensures high key agility of AES^2.

On the Westmere architecture generation of Intel general-purpose processors, AES^2 can be implemented using the AES-NI instruction set [19]. As the AES round instructions are pipelined, we fully utilise the pipeline by processing four independent plaintext blocks in parallel implementing the basic electronic code-book mode (ECB) and counter mode (CTR). The performance of these implementations on recent processors is demonstrated and compared to two conventional implementations of AES-128 (i.e. without AES-NI instructions) – the bitsliced implementation of [21] and the OpenSSL 1.0.0e implementation based on lookup tables. All numbers are given in cycles per byte (cpb).

	Intel Xeon X5670	Intel Core i7 640M
	2.93 GHz, 12 MB L3 cache	2.8 GHz, 4 MB L3 cache
AES^2, AES-NI, ECB	2.54 cpb	2.69 cpb
AES^2, AES-NI, CTR	2.65 cpb	2.76 cpb
AES-128, AES-NI, ECB	1.18 cpb	1.25 cpb
AES-128, AES-NI, CTR	1.32 cpb	1.36 cpb
AES-128, bitsliced, CTR	7.08 cpb	7.84 cpb
AES-128, OpenSSL, CTR	15.73 cpb	16.76 cpb

It turns out that on both platforms, the performance of AES^2 is almost equal to half that of AES, indicating that the overhead is very low. Compared to the best implementations of the AES which are in widespread use now on standard platforms, AES^2 provides a performance improvement of almost factor three and higher with the AES-NI instruction set.

7 Conclusion, Open Problems and Future Work

In this paper we gave the first formal treatment of the key-alternating cipher in a provable setting. For two or more rounds an attacker needs to query the oracles at least $2^{2n/3}$ times for having a reasonable success probability. Furthermore, we studied the security of the construction with respect to statistical attacks, arguing that even for $t = 2$ linear attacks do not seem to be applicable. Finally we gave a concrete proposal mimicking the construction for $t = 2$. There are several lines of future work and open problems we like to mention.

On the theoretical side, it seems unlikely that the bounds given here are tight. Thus, improving them is an important open problem. We actually conjecture that the correct bound on the query complexity is roughly $2^{t/(t+1)n}$. As a first step, deriving bounds that increase with the number of rounds is a goal worth aiming for. Secondly, for now, we have to assume that all round keys are independent. For aesthetical reasons, but also from a practical point of view (see below) it would be nice to prove bounds for the case that all round keys are identical.

On the practical side, mainly for efficiency reasons but also due to resistance against related-key attacks, several variants for $t = 2$ are worth studying. First of all, since the security level is at most 2^n, due to the meet in the middle attack, one could be tempted to derive three n-bit keys k_0, k_1, and k_2 from one n-bit word. The simplest case here is to have all three keys identical. Taking P and Q different, we are not aware of any attack with computational complexity below 2^n. Furthermore, it seems reasonable to assume that such a construction provides some security against certain types of related-key attacks as well. The best attacks we are aware of in such a setting has birthday complexity $2^{n/2}$. See the full version for details.

Eventually, it is an interesting open problem to determine whether the results in this work can be used as directions for alternative block cipher designs, e.g. with minimum key scheduling algorithms. As a typical example, one could consider the possibility to generate public permutations from a variant of the AES, where the round keys would be replaced with simple constants. In general, such an approach could lead to efficient lightweight designs. Interestingly, it is also the direction taken, to a certain extent, by the recently proposed block cipher LED [20]. In its 64-bit version, this cipher just iterates blocks made of 4 rounds and the addition of the master key.

Another tempting way, in order to increase efficiency, is to choose $Q = P$. Similarly, it may be advantageous to have $Q = P^{-1}$, which has the further advantage that the decryption and encryption operations are similar, except for using the keys in reverse order. However, with $Q = P^{-1}$ there is an attack which finds the value of $k_0 \oplus k_2$ using $2^{n/2}$ queries and similar time. After $k_0 \oplus k_2$ is known the cipher is easily distinguishable from a random permutation. Also, with $Q = P$ but now assuming that $k_0 \oplus k_2$ is known, one finds the secret keys using $2^{n/2}$ queries and similar time.

Acknowledgements. Andrey Bogdanov is a postdoctoral fellow of the Fund for Scientific Research - Flanders (FWO). Francois-Xavier Standaert is associate researcher of the Belgian fund for scientific research (FNRS-F.R.S.). This work has been funded in parts by the ERC project 280141 (acronym CRASH). John Steinberger is supported by the National Basic Research Program of China Grant 2011CBA00300, 2011CBA00301, the National Natural Science Foundation of China Grant 61033001, 61061130540, 61073174, and by NSF grant 0994380. Elmar Tischhauser is a doctoral fellow of the Fund for Scientific Research - Flanders (FWO). This work is supported in part by the IAP Programme P6/26 BCRYPT of the Belgian State, by the European Commission under contract number ICT-2007-216676 ECRYPT NoE phase II, by KU Leuven-BOF (OT/08/027), and by the Research Council KU Leuven (GOA TENSE).

References

1. Baignères, T., Finiasz, M.: Dial C for Cipher. In: Biham, E., Youssef, A.M. (eds.) SAC 2006. LNCS, vol. 4356, pp. 76–95. Springer, Heidelberg (2007)
2. Baignères, T., Finiasz, M.: KFC - The Krazy Feistel Cipher. In: Lai, X., Chen, K. (eds.) ASIACRYPT 2006. LNCS, vol. 4284, pp. 380–395. Springer, Heidelberg (2006)
3. Barreto, P.S.L.M., Rijmen, V.: The KHAZAD Legacy-Level Block Cipher. In: First open NESSIE Workshop, Leuven, Belgium, 15 pages (November 2000)
4. Bertoni, G., Daemen, J., Peeters, M., Van Assche, G.: Keccak sponge function family main document. Submission to NIST (Round 2) (2009)
5. Biryukov, A., Dunkelman, O., Keller, N., Khovratovich, D., Shamir, A.: Key Recovery Attacks of Practical Complexity on AES-256 Variants with up to 10 Rounds. In: Gilbert, H. (ed.) EUROCRYPT 2010. LNCS, vol. 6110, pp. 299–319. Springer, Heidelberg (2010)
6. Biryukov, A., Khovratovich, D.: Related-Key Cryptanalysis of the Full AES-192 and AES-256. In: Matsui, M. (ed.) ASIACRYPT 2009. LNCS, vol. 5912, pp. 1–18. Springer, Heidelberg (2009)
7. Bogdanov, A., Khovratovich, D., Rechberger, C.: Biclique Cryptanalysis of the Full AES. In: Lee, D.H., Wang, X. (eds.) ASIACRYPT 2011. LNCS, vol. 7073, pp. 344–371. Springer, Heidelberg (2011)
8. Bogdanov, A.A., Knudsen, L.R., Leander, G., Paar, C., Poschmann, A., Robshaw, M.J.B., Seurin, Y., Vikkelsoe, C.: PRESENT: An Ultra-Lightweight Block Cipher. In: Paillier, P., Verbauwhede, I. (eds.) CHES 2007. LNCS, vol. 4727, pp. 450–466. Springer, Heidelberg (2007)
9. Bogdanov, A., Knudsen, L.R., Leander, G., Standaert, F.-X., Steinberger, J., Tischhauser, E.: Key-Alternating Ciphers in a Provable Setting: Encryption Using A Small Number of Public Permutations. IACR Eprint Report 2012/035
10. Chabaud, F., Vaudenay, S.: Links between Differential and Linear Cryptanalysis. In: De Santis, A. (ed.) EUROCRYPT 1994. LNCS, vol. 950, pp. 356–365. Springer, Heidelberg (1995)
11. Daemen, J.: Limitations of the Even-Mansour Construction. In: Matsumoto, T., Imai, H., Rivest, R.L. (eds.) ASIACRYPT 1991. LNCS, vol. 739, pp. 495–498. Springer, Heidelberg (1993)
12. Daemen, J., Govaerts, R., Vandewalle, J.: Correlation Matrices. In: Preneel, B. (ed.) FSE 1994. LNCS, vol. 1008, pp. 275–285. Springer, Heidelberg (1995)
13. Daemen, J., Rijmen, V.: Probability distributions of correlations and differentials in block ciphers. Journal on Mathematical Cryptology 1(3), 221–242 (2007)
14. Daemen, J., Rijmen, V.: The Design of Rijndael. Springer, Heidelberg (2002)
15. Daemen, J., Rijmen, V.: The Wide Trail Design Strategy. In: Honary, B. (ed.) Cryptography and Coding 2001. LNCS, vol. 2260, pp. 222–238. Springer, Heidelberg (2001)
16. Even, S., Mansour, Y.: A Construction of a Cipher from a Single Pseudorandom Permutation. J. Cryptology 10(3), 151–162 (1997)
17. Even, S., Mansour, Y.: A Construction of a Cipher From a Single Pseudorandom Permutation. In: Matsumoto, T., Imai, H., Rivest, R.L. (eds.) ASIACRYPT 1991. LNCS, vol. 739, pp. 210–224. Springer, Heidelberg (1993)
18. FIPS PUB 46-3: Data Encryption Standard (DES) (1999)
19. Gueron, S.: Intel Mobility Group, Israel Development Center, Israel: Intel Advanced Encryption Standard (AES) Instructions Set (2010),
http://software.intel.com/file/24917

20. Guo, J., Peyrin, T., Poschmann, A., Robshaw, M.: The LED Block Cipher. In: Preneel, B., Takagi, T. (eds.) CHES 2011. LNCS, vol. 6917, pp. 326–341. Springer, Heidelberg (2011)

21. Käsper, E., Schwabe, P.: Faster and Timing-Attack Resistant AES-GCM. In: Clavier, C., Gaj, K. (eds.) CHES 2009. LNCS, vol. 5747, pp. 1–17. Springer, Heidelberg (2009)

22. Keliher, L., Meijer, H., Tavares, S.: Improving the Upper Bound on the Maximum Average Linear Hull Probability for Rijndael. In: Vaudenay, S., Youssef, A.M. (eds.) SAC 2001. LNCS, vol. 2259, pp. 112–128. Springer, Heidelberg (2001)

23. Knudsen, L.R.: Practically Secure Feistel Ciphers. In: Anderson, R. (ed.) FSE 1993. LNCS, vol. 809, pp. 211–221. Springer, Heidelberg (1994)

24. Lai, X., Massey, J.L.: A Proposal for a New Block Encryption Standard. In: Damgård, I.B. (ed.) EUROCRYPT 1990. LNCS, vol. 473, pp. 389–404. Springer, Heidelberg (1991)

25. Lai, X., Massey, J.L.: Markov Ciphers and Differential Cryptanalysis. In: Davies, D.W. (ed.) EUROCRYPT 1991. LNCS, vol. 547, pp. 17–38. Springer, Heidelberg (1991)

26. Luby, M., Rackoff, C.: How to Construct Pseudorandom Permutations from Pseudorandom Functions. SIAM J. Comput. 17(2), 373–386 (1988)

27. Matsui, M.: New Block Encryption Algorithm MISTY. In: Biham, E. (ed.) FSE 1997. LNCS, vol. 1267, pp. 54–68. Springer, Heidelberg (1997)

28. Matsui, M.: New Structure of Block Ciphers with Provable Security against Differential and Linear Cryptanalysis. In: Gollmann, D. (ed.) FSE 1996. LNCS, vol. 1039, pp. 205–218. Springer, Heidelberg (1996)

29. Nyberg, K.: Linear Approximation of Block Ciphers. In: De Santis, A. (ed.) EUROCRYPT 1994. LNCS, vol. 950, pp. 439–444. Springer, Heidelberg (1995)

30. O'Connor, L.: Properties of Linear Approximation Tables. In: Preneel, B. (ed.) FSE 1994. LNCS, vol. 1008, pp. 131–136. Springer, Heidelberg (1995)

31. Rijmen, V., Daemen, J., Preneel, B.: Antoon Bosselaers and Erik De Win. The Cipher SHARK. In: Gollmann, D. (ed.) FSE 1996. LNCS, vol. 1039, pp. 99–111. Springer, Heidelberg (1996)

32. Spanos, A.: Probability Theory and Statistical Inference: Econometric Modeling with Observational Data. Cambridge University Press (1999)

33. Vaudenay, S.: Decorrelation: A Theory for Block Cipher Security. J. Cryptology 16(14), 249–286 (2003)

34. Vaudenay, S.: On the Lai-Massey Scheme. In: Lam, K.-Y., Okamoto, E., Xing, C. (eds.) ASIACRYPT 1999. LNCS, vol. 1716, pp. 8–19. Springer, Heidelberg (1999)

Efficient and Optimally Secure Key-Length Extension for Block Ciphers via Randomized Cascading

Peter Gaži[1,2] and Stefano Tessaro[3,4]

[1] Department of Computer Science, Comenius University, Bratislava, Slovakia
[2] Department of Computer Science, ETH Zurich, Switzerland
peter.gazi@inf.ethz.ch
[3] Department of Computer Science and Engineering
University of California San Diego, La Jolla CA, USA
[4] MIT, Cambridge MA, USA
stessaro@cs.ucsd.edu

Abstract. We consider the question of efficiently extending the key length of block ciphers. To date, the approach providing highest security is triple encryption (used e.g. in Triple-DES), which was proved to have roughly $\kappa + \min\{n/2, \kappa/2\}$ bits of security when instantiated with ideal block ciphers with key length κ and block length n, at the cost of three block-cipher calls per message block.

This paper presents a new practical key-length extension scheme exhibiting $\kappa + n/2$ bits of security – hence improving upon the security of triple encryption – solely at the cost of *two* block cipher calls and a key of length $\kappa + n$. We also provide matching generic attacks showing the optimality of the security level achieved by our approach with respect to a general class of two-query constructions.

Keywords: Block ciphers, Cascade encryption, Provable security.

1 Introduction

1.1 Key-Length Extension for Block Ciphers

Several practical block cipher designs have been proposed over the last decades and have been the object of extensive cryptanalytic efforts. Examples include DES [1], IDEA [19], BLOWFISH [28], and the currently in-use AES [4]. Within applications, we typically demand that these block ciphers are a good *pseudorandom permutation* (PRP), i.e., in the eyes of a computationally bounded attacker, they behave as a randomly chosen permutation under a random secret key. For instance, PRP security of the underlying block cipher is necessary to infer security of all modes of operations for message encryption (such as counter-mode and CBC encryption [8]) as well as of message authentication codes like CBC-MAC [9] and PMAC [12].

In practice, we define the PRP security level of a block cipher as the complexity required to distinguish it from a random permutation with non-negligible

D. Pointcheval and T. Johansson (Eds.): EUROCRYPT 2012, LNCS 7237, pp. 63–80, 2012.

advantage. The *key length* κ of a block cipher crucially limits the achievable security level, since the secret key K can be recovered given black-box access to $E(K, \cdot)$ evaluating $E(\cdot, \cdot)$ approximately 2^κ times; obviously, this also yields a PRP distinguishing adversary with equal complexity. Such weakness is *generic*, in the sense that it only depends on κ, and even an *ideal block cipher* suffers from the same attack.[1] In contrast, no real dependency exists between security and the block length n of a block cipher: No generic attack faster than 2^κ exists even if $n = 1$. In the following, let us refer to a block cipher with key and block lengths κ and n, respectively, as a (κ, n)-block cipher.

KEY LENGTH EXTENSION. With a continuous increase of the availability of computing resources, the role of the key length has hence never been more important. Key lengths of say fewer than 64 bits are no longer sufficient to ensure security, making key recovery a matter of a few hours even on modest architectures. This is a serious problem for legacy designs such as DES which have very short keys of length 56 bits, but which otherwise do not seem to present significant non-generic security weaknesses. Constructions based on DES also remain very attractive because of its short block length $n = 64$ which allows enciphering short inputs. This is for example crucial in current applications in the financial industry, such as the EMV standard [6], where the block cipher is applied to PIN numbers, which are very short.

The above described situation motivates the problem of *key-length extension*, which is the main object of this paper: We seek for very efficient constructions provably transforming any (κ, n)-block cipher E into a (κ', n)-block cipher E' with both $\kappa' > \kappa$ and higher PRP security, i.e., the PRP security of E' should be higher than 2^κ whenever E does not exhibit any non-generic weaknesses. We aim both at providing very efficient approaches to key length extension and at understanding the optimal security achievable by such constructions. Our main contribution will be a new and very efficient two-call key-length extension method outperforming the efficiency of existing solutions by a large margin, and achieving security levels which we prove optimal, and which are comparable (and even better) than those of earlier, less efficient, designs.

IDEAL CIPHER MODEL. In our proofs, we model the absence of generic weaknesses of the underlying block cipher by analyzing our constructions when instantiated with an ideal block cipher **E**. In this model, complexity is measured in terms of the number of queries to **E** (so-called *ideal block cipher queries*) and to E' or the given random permutation (we refer to these as *construction* queries). It should be noted that proving security of key-length extension in the ideal cipher model implies absence of generic attacks, treating the underlying cipher as a black-box, and as we will explain in the next section, all attacks on existing schemes are indeed generic.

---

[1] As usual, an *ideal block cipher* $\mathbf{E} : \{0,1\}^\kappa \times \{0,1\}^n \times \{+, -\} \to \{0,1\}^n$ is the system associating with each key $k \in \{0,1\}^\kappa$ an independent randomly chosen permutation $E(k, \cdot)$ and allowing the adversary to learn $E(k, x)$ and $E^{-1}(k, y)$ for k, x, y of her choice.

1.2 Existing Approaches to Key-Length Extension

The short key length $\kappa = 56$ of DES has constituted the main motivation behind previous work on key-length extension. However, we stress that all previous constructions are generic, and can be applied to *any* block cipher with short keys, hence extending the applicability of these results (as well as the results of this paper) way beyond the specific case of DES.

A first proposal called DESX (due to Rivest) stretches the key length of DES employing a technique called *key whitening* (this approach was later used by Even and Mansour [15]): It is defined such that

$$\text{DESX}_{k_i, k_o, k}(m) = k_o \oplus \text{DES}_k(k_i \oplus m)$$

for all $m, k_i, k_o \in \{0,1\}^{64}$ and $k \in \{0,1\}^{56}$. DESX can be generalized to a generic transformation from a (κ, n)-block cipher to a $(\kappa + 2n, n)$-block cipher whose security was studied by Kilian and Rogaway [18]: They proved that any successful PRP distinguishing attack requires $2^{\frac{\kappa+n}{2}}$ queries.[2] They also observe that the same key may be used in both whitening steps (i.e., $k_i = k_o$) and provide an attack using $2^{\max\{\kappa, n\}}$ queries.

An alternative to whitening is *cascading* (or cascade encryption), i.e., sequentially composing ℓ block-cipher calls with usually different keys. (This is referred to as a cascade of length ℓ.) It is well known that a cascade of length two does not substantially increase security due to the meet-in-the-middle attack [13]. (Even though a security increase in terms of distinguishing advantage is achieved for low attack complexities [7].) The security properties of a cascade of different ciphers was studied by Even and Goldreich [14] showing that a cascade is at least as strong as the strongest of the ciphers used; and by Maurer and Massey [23] proving that it is at least as secure as the *first* cipher of the cascade, however in a more general attack model.

The meet-in-the-middle attack makes triple encryption the shortest cascade with a potential for significant security gain and indeed it has found widespread usage as *Triple-DES* (3DES) [2,3,5], where given keys $k_1, k_2, k_3 \in \{0,1\}^{56}$, a 64-bit message m is mapped to

$$\text{3DES}_{k_1, k_2, k_3}(m) = \text{DES}_{k_1}(\text{DES}_{k_2}(\text{DES}_{k_3}(m))) \ .$$

(A variant with shorter keys $\text{3DES}'_{k_1, k_2}(m) = \text{DES}_{k_1}(\text{DES}_{k_2}^{-1}(\text{DES}_{k_1}(m)))$ is also sometimes used.) For 3DES (and a variant of 3DES' with independent keys), Bellare and Rogaway [11] and subsequently Gaži and Maurer [16] have shown security up to roughly $2^{\kappa + \min\{n, \kappa\}/2}$ queries when DES is replaced by an ideal block cipher. For the case of DES parameters, their result gives concretely security up to 2^{78} queries, whereas the best known attack due to Lucks [21] shows

---

[2] Their result is in fact more fine-grained, as they show that 2^ρ construction and $2^{\kappa+n-\rho}$ ideal block cipher queries, respectively, are necessary for all integers ρ; while different bounds for both query types are sometimes justified, we adopt a (by now more standard) worst-case approach only bounding the *sum* of both query numbers.

that no security better than 2^{90} can be expected. (It should also be noted that the proof of [16] extends to prove that longer cascades can achieve better security for short keys.)

We emphasize that despite the availability of modern designs with built-in larger keys (e.g., $\kappa \in \{128, 192, 256\}$ for AES), Triple-DES remains nowadays popular, not only because of backwards compatibility, but also because its short block size ($n = 64$ vs. $n \geq 128$ for AES) is well suited to applications enciphering short inputs such as personal-identification numbers (PINs). For example, it is the basis of the EMV standard for PIN-based authentication of debit and credit card transactions [6]. However, the use of three calls per processed message block is widely considered a drawback within applications which we address and solve in this paper.

OTHER RELATED WORK. It is worth mentioning that several works have studied cascading-based security amplification of block ciphers only assumed to satisfy weaker forms of PRP security, both in the information-theoretic [32,24,25,17] as well as in the computational settings [20,26,31]. These results however consider an orthogonal model to ours and are hence incomparable.

1.3   Our Results

None of the above efficient constructions provably achieves security beyond $2^{\kappa+\min\{\kappa,n\}/2}$, and such security is achieved only at the fairly expensive cost of at least three block-cipher calls per message block. This paper aims at improving the efficiency-security trade-off in key-length extension. We ask the following question: Suppose that we only consider constructions making at most *two* calls to the underlying cipher. *What is the best security level we are expected to achieve?*

BETTER SECURITY AND BETTER EFFICIENCY. Quite surprisingly, our main result (presented in Section 4) exposes a "win-win" situation: We devise a *two*-call construction of a $(\kappa + n, n)$-block cipher from any (κ, n)-block cipher with security $2^{\kappa+n/2}$ in the ideal block cipher model, i.e., the obtained security is *higher* than that of existing three-call designs studied in [11,16].[3] Our construction – which we refer to as the *double XOR-cascade* (2XOR) – is obtained by careful insertion of two randomization steps (with the *same* key value) to a related-key version of double encryption. Concretely, we map each n-bit message m to

$$2\mathrm{XOR}_{k,z}(m) = E_{\widetilde{k}}(E_k(m \oplus z) \oplus z)$$

for all key values $k \in \{0,1\}^{\kappa}$ and $z \in \{0,1\}^{n}$, and where $\widetilde{k}$ is, for example, obtained from k by flipping one individual bit.

We note that the key length is comparable to the one of the two-key variant of 3DES (assuming $\kappa \approx n$). Intuitively, our construction requires some mild

---

[3] In fact, our construction tolerates arbitrarily many construction queries (i.e., up to 2^n) *and* $2^{\kappa+n/2}$ ideal block cipher queries. However, we stress that in all practically relevant cases $\kappa \geq n/2$, hence we tacitly assume this property throughout the paper.

Table 1. Required number of block-cipher queries, key lengths, security lower bounds and best known attacks for various key-length extension schemes. The bounds are parameterized by the key length of the underlying block cipher (denoted by κ) and its block size (denoted by n), and are for the usual case where $\kappa \geq n/2$.

construction	# of queries	key length	log of the number of queries security lower bound	log of the number of queries best known attack
(κ, n)-block cipher	1	κ	κ	κ
DESX [15,18]	1	$\kappa + n$	$(\kappa + n)/2$	$\max\{\kappa, n\}$
double encryption [13]	2	2κ	κ	κ
triple encryption [11,16,21]	3	3κ	$\kappa + \min\{\kappa, n\}/2$	90 (for 3DES)
double XOR-cascade [here]	2	$\kappa + n$	$\kappa + n/2$ (Thm. 3)	$\kappa + n/2$ (Thm. 2)

form of related-key security [10] which we obtain for free when the underlying block cipher is ideal, but may be a concern in practice. However, it should be noted that an alternative version of the construction where $\widetilde{k}$ is replaced by an independent and unrelated key value k' achieves the same security level at the cost of a longer $(2\kappa + n)$-bit key, which is for instance still shorter than in DESX with independent whitening keys (for DES parameters).

The core of our security proof (cf. Theorem 3) is a technical argument of independent interest: Namely, we prove that it is hard to distinguish two random, independent, permutations π_1, π_2 on the n-bit strings from two randomly chosen permutations π_1, π_2 with the property that $\pi_2(\pi_1(x \oplus Z) \oplus Z) = x$ for all x and a random secret value Z even if we are allowed arbitrary queries to each of π_1, π_2, π_1^{-1}, and π_2^{-1}. This fact yields our main theorem by a careful adaptation of the techniques from [11,16] to take into account both randomization and the use of related keys.

GENERIC ATTACKS AND OPTIMALITY. With the above result at hand, it is legitimate to ask whether we should expect two-call constructions with even better security: In Section 3, we answer this in the negative, at least for a class of natural constructions.

As a warm up of independent interest, we confirm that only much weaker security can be achieved by a one-call construction: Regardless of the amount of key material employed in the construction, an attack with query complexity $2^{\max\{\kappa,n\}}$ always exists (using memory $2^{\max\{\kappa,n\}}$),[4] showing the optimality of DESX-like constructions in the case $\kappa = n$. We then turn to two-call constructions, which are necessary to achieve higher security: Here, we prove that any construction for which distinct inputs map to distinct first queries and distinct answers from the first call imply distinct inputs to the second call admits a distinguishing attack making $2^{\kappa+n/2}$ ideal block cipher queries and 2^n construction queries. This class contains as a special case all constructions obtained by randomizing the cascade of length two using arbitrarily many key bits, including ours.

---

[4] More precisely, our attack requires roughly 2^κ ideal block cipher queries and 2^n construction queries.

In addition, we also show that simpler randomization methods for length-two cascades admit distinguishing attacks with even lower complexity. For example, randomizing the cascade of length two as $E_{k_2}(E_{k_1}(m \oplus z_1)) \oplus z_2$ instead of using our approach yields a simple $2^{\max\{\kappa,n\}}$ meet-in-the middle attack. This shows an interesting feature of our constructions, namely that while targeting CCA security (i.e., we allow for forward and backward queries to the construction), our design requires *asymmetry*, a fact which seems to contradict common wisdom.

Finally, note that all generic attacks presented in this paper (both against one-query and two-query constructions) can be mounted even if the distinguisher is only allowed to ask forward construction queries (i.e., in the CPA setting). In contrast, the construction we propose is proven to be secure even with respect to an adversary allowed to ask inverse construction queries (CCA adversary).

FINAL REMARKS. Table 1 summarizes the results of this paper in the context of previously known results. To serve as an overview, some bounds are presented in a simplified form. Note that the security of *any* key-length extension construction in our model can be upper-bounded by $2^{\kappa+n}$ which corresponds to the trivial attack asking all possible block cipher and construction queries.

Our results and proofs are presented using Maurer's random systems framework [22], which we review in Section 2 in a self-contained way sufficient to follow the contents of the paper.

2   Preliminaries

2.1   Basic Notation

We denote sets by calligraphic letters $\mathcal{X}, \mathcal{Y}, \ldots$, and by $|\mathcal{X}|, |\mathcal{Y}|, \ldots$ their cardinalities. We also let $\mathcal{X}^k$ be the set of k-tuples $x^k = (x_1, \ldots, x_k)$ of elements of $\mathcal{X}$. Strings are elements of $\{0,1\}^k$ and are usually denoted as $s = s_1 s_2 \ldots s_k$, with $\|$ denoting the usual string concatenation. Additionally, we let $\mathrm{Func}(m, \ell)$ be the set of all functions from $\{0,1\}^m$ to $\{0,1\}^\ell$ and $\mathrm{Perm}(n)$ be the set of all permutations of $\{0,1\}^n$. In particular, $id \in \mathrm{Perm}(n)$ represents the identity mapping when n is understood from the context. Throughout this paper logarithms will always be to the base 2.

We denote random variables and concrete values they can take by upper-case letters $X, Y, \ldots$ and lower-case letters $x, y, \ldots$, respectively. For events A and B and random variables U and V with ranges $\mathcal{U}$ and $\mathcal{V}$, respectively, we let $\mathsf{P}_{UA|VB}$ be the corresponding conditional probability distribution, seen as a (partial) function $\mathcal{U} \times \mathcal{V} \to [0,1]$. Here the value $\mathsf{P}_{UA|VB}(u,v) = \mathsf{P}[U = u \wedge A | V = v \wedge B]$ is well defined for all $u \in \mathcal{U}$ and $v \in \mathcal{V}$ such that $\mathsf{P}_{VB}(v) > 0$ and undefined otherwise. Two probability distributions P_U and $\mathsf{P}_{U'}$ on the same set $\mathcal{U}$ are equal, denoted $\mathsf{P}_U = \mathsf{P}_{U'}$, if $\mathsf{P}_U(u) = \mathsf{P}_{U'}(u)$ for all $u \in \mathcal{U}$. Conditional probability distributions are equal if the equality holds for all arguments for which both of them are defined. To emphasize the random experiment $\mathcal{E}$ in consideration, we sometimes write it in the superscript, e.g. $\mathsf{P}_{U|V}^{\mathcal{E}}(u,v)$. Finally, the complement of an event A is denoted by $\overline{A}$.

2.2 Random Systems

The presentation of this paper relies on Maurer's random systems framework [22]. However, we stress that most of the paper remains understandable at a very high level, even without the need of a deeper understanding of the techniques behind the framework; we provide a self-contained introduction.

The starting point of the random-system framework is the basic observation that the input-output behavior of any kind of discrete system with inputs in $\mathcal{X}$ and outputs in $\mathcal{Y}$ can be described by an infinite family of functions describing, for each $i \geq 1$, the probability distribution of the i-th output $Y_i \in \mathcal{Y}$ given the values of the first i inputs $X^i \in \mathcal{X}^i$ and the previous $i-1$ outputs $Y^{i-1} \in \mathcal{Y}^{i-1}$. Formally, hence, an $(\mathcal{X}, \mathcal{Y})$-*(random) system* $\mathbf{F}$ is an infinite sequence of functions $\mathsf{p}^{\mathbf{F}}_{Y_i|X^iY^{i-1}} : \mathcal{Y} \times \mathcal{X}^i \times \mathcal{Y}^{i-1} \to [0,1]$ such that, $\sum_{y_i} \mathsf{p}^{\mathbf{F}}_{Y_i|X^iY^{i-1}}(y_i, x^i, y^{i-1}) = 1$ for all $i \geq 1$, $x^i \in \mathcal{X}^i$ and $y^{i-1} \in \mathcal{Y}^{i-1}$. We stress that the notation $\mathsf{p}^{\mathbf{F}}_{Y_i|X^iY^{i-1}}$, by itself, involves some abuse, as we are not considering any particular random experiment with well-defined random variables Y_i, X^i, Y^{i-1} until the system will be interacting with a distinguisher (see below), in which case the random variables will exist and take the role of the transcript. In general, we shall also typically define discrete systems by a high level description, as long as the resulting conditional probability distributions can be derived uniquely from this description.

We say that a system $\mathbf{F}$ is *deterministic* if the range of $\mathsf{p}^{\mathbf{F}}_{Y_i|X^iY^{i-1}}$ is $\{0,1\}$ for all $i \geq 1$. Moreover, it is *stateless* if the probability distribution of each output depends only on the current input, i.e., if there exists a distribution $\mathsf{p}_{Y|X} : \mathcal{Y} \times \mathcal{X} \to [0,1]$ such that $\mathsf{p}^{\mathbf{F}}_{Y_i|X^iY^{i-1}}(y_i, x^i, y^{i-1}) = \mathsf{p}_{Y|X}(y_i, x_i)$ for all y_i, x^i and y^{i-1}.

We also consider systems $\mathbf{C}^{\mathbf{F}}$ that arise from *constructions* $\mathbf{C}^{(\cdot)}$ accessing a sub-system $\mathbf{F}$. Note that while a construction $\mathbf{C}^{(\cdot)}$ does not define a random system by itself, $\mathbf{C}^{\mathbf{F}}$ does define a random system. The notions of being deterministic and of being stateless naturally extend to constructions.[5] We also consider the *parallel composition* of two (possibly dependent) discrete systems $\mathbf{F}$ and $\mathbf{G}$, denoted $(\mathbf{F}, \mathbf{G})$, which is the system that allows queries to both systems $\mathbf{F}$ and $\mathbf{G}$.

EXAMPLES. A *random function* $\mathbf{F} : \{0,1\}^m \to \{0,1\}^n$ is a system which implements a function f initially chosen according to some distribution on $\mathrm{Func}(m, n)$.[6] In particular, the *uniform random function (URF)* $\mathbf{R} : \{0,1\}^m \to \{0,1\}^\ell$ realizes a uniformly chosen function $f \in \mathrm{Func}(m, \ell)$, whereas the *uniform random permutation (URP)* on $\{0,1\}^n$, denoted $\mathbf{P} : \{0,1\}^n \times \{+, -\} \to \{0,1\}^n$, realizes a uniformly chosen permutation $P \in \mathrm{Perm}(n)$ allowing both forward queries of the form $(x, +)$ returning $P(x)$ as well as backward queries $(y, -)$ returning $P^{-1}(y)$. More generally, we meet the convention (for the purpose of this paper) that any

---

[5] We dispense with a formal definition. However, we point out that we allow a stateless construction to keep a state during invocations of its subsystem.

[6] As is the case with the notion of a random variable, the word "random" does not imply uniformity of the distribution.

system realizing a random function (possibly by means of a construction) which is a permutation will *always* allow both forward and backward queries.

Another important example of a random function is the *ideal block cipher* $\mathbf{E} : \{0,1\}^{\kappa} \times \{0,1\}^{n} \times \{+,-\} \to \{0,1\}^{n}$ which realizes an independent uniform random permutation $\mathbf{E}_k \in \mathrm{Perm}(n)$ for each key $k \in \{0,1\}^{\kappa}$; in particular, the system allows both forward and backward queries to each $\mathbf{E}_k$.

Finally, note that with some abuse of notation, we often write $\mathbf{E}_k$ or $\mathbf{P}$ to refer to the randomly chosen permutation P implemented by the system $\mathbf{E}_k$ or $\mathbf{P}$, respectively.

DISTINGUISHERS AND INDISTINGUISHABILITY. A *distinguisher* $\mathbf{D}$ for an $(\mathcal{X},\mathcal{Y})$-random system asking q queries is a $(\mathcal{Y},\mathcal{X})$-random system which is "one query ahead:" its input-output behavior is defined by the conditional probability distributions of its queries $\mathsf{p}^{\mathbf{D}}_{X_i|X^{i-1}Y^{i-1}}$ for all $1 \leq i \leq q$. (The first query of $\mathbf{D}$ is determined by $\mathsf{p}^{\mathbf{D}}_{X_1}$.) After the distinguisher asks all q queries, it outputs a bit W_q depending on the transcript (X^q, Y^q). For a random system $\mathbf{F}$ and a distinguisher $\mathbf{D}$, let $\mathbf{DF}$ be the random experiment where $\mathbf{D}$ interacts with $\mathbf{F}$, with the distributions of the transcript (X^q, Y^q) and of the bit W_q being uniquely defined by their conditional probability distributions. Then, for two $(\mathcal{X},\mathcal{Y})$-random systems $\mathbf{F}$ and $\mathbf{G}$, the *distinguishing advantage* of $\mathbf{D}$ in distinguishing systems $\mathbf{F}$ and $\mathbf{G}$ by q queries is the quantity $\Delta^{\mathbf{D}}(\mathbf{F},\mathbf{G}) = |\mathsf{P}^{\mathbf{DF}}(W_q = 1) - \mathsf{P}^{\mathbf{DG}}(W_q = 1)|$. We are usually interested in the maximal distinguishing advantage over all distinguishers asking q queries, which we denote by $\Delta_q(\mathbf{F},\mathbf{G}) = \max_{\mathbf{D}} \Delta^{\mathbf{D}}(\mathbf{F},\mathbf{G})$ (with $\mathbf{D}$ ranging over all such distinguishers).

For a random system $\mathbf{F}$, we often consider an internal *monotone condition* defined on it. Such a condition is initially satisfied (true), but once it gets violated, it cannot become true again (hence the name monotone). Typically, the condition captures whether the behavior of the system meets some additional requirement (e.g. distinct outputs, consistent outputs) or this was already violated during the interaction. We formalize such a condition by a sequence of events $\mathcal{A} = A_0, A_1, \ldots$ such that A_0 always holds, and A_i holds if the condition holds after query i. The probability that a distinguisher $\mathbf{D}$ issuing q queries to $\mathbf{F}$ makes a monotone condition $\mathcal{A}$ fail in the random experiment $\mathbf{DF}$ is denoted by $\nu^{\mathbf{D}}(\mathbf{F}, \overline{A_q}) = \mathsf{P}^{\mathbf{DF}}(\overline{A_q})$ and we are again interested in the maximum over all such distinguishers, denoted by $\nu(\mathbf{F}, \overline{A_q}) = \max_{\mathbf{D}} \nu^{\mathbf{D}}(\mathbf{F}, \overline{A_q})$. For a random system $\mathbf{F}$ with a monotone condition $\mathcal{A} = A_0, A_1, \ldots$ and a random system $\mathbf{G}$, we say that $\mathbf{F}$ *conditioned on* $\mathcal{A}$ *is equivalent to* $\mathbf{G}$, denoted $\mathbf{F}|\mathcal{A} \equiv \mathbf{G}$, if $\mathsf{p}^{\mathbf{F}}_{Y_i|X^iY^{i-1}A_i} = \mathsf{p}^{\mathbf{G}}_{Y_i|X^iY^{i-1}}$ for $i \geq 1$, for all arguments for which $\mathsf{p}^{\mathbf{F}}_{Y_i|X^iY^{i-1}A_i}$ is defined. Intuitively, this captures the fact that as long as the condition $\mathcal{A}$ holds in $\mathbf{F}$, it behaves the same as $\mathbf{G}$.

Let $\mathbf{F}$ be a random system with a monotone condition $\mathcal{A}$. Following [25], we define $\mathbf{F}$ *blocked by* $\mathcal{A}$ to be a new random system that behaves exactly like $\mathbf{F}$ while the condition $\mathcal{A}$ is satisfied. Once $\mathcal{A}$ is violated, it only outputs a special blocking symbol $\bot$ not contained in the output alphabet of $\mathbf{F}$.

We make use of the following helpful claims proven in previous papers. Below, we also present an informal explanation of their merits.

Lemma 1. *Let* $\mathbf{C}^{(\cdot)}$ *and* $\mathbf{C}'^{(\cdot)}$ *be two constructions invoking a subsystem, and let* $\mathbf{F}$ *and* $\mathbf{G}$ *be random systems. Let* $\mathcal{A}$ *and* $\mathcal{B}$ *be two monotone conditions defined on* $\mathbf{F}$ *and* $\mathbf{G}$, *respectively.*

(i) [22, Theorem 1] If $\mathbf{F}|\mathcal{A} \equiv \mathbf{G}$ *then* $\Delta_q(\mathbf{F}, \mathbf{G}) \leq \nu(\mathbf{F}, \overline{\mathcal{A}_q})$.

(ii) [16, Lemma 2] Let $\mathbf{F}^{\perp}$ *denote the random system* $\mathbf{F}$ *blocked by* $\mathcal{A}$ *and let* $\mathbf{G}^{\perp}$ *denote* $\mathbf{G}$ *blocked by* $\mathcal{B}$. *Then for every distinguisher* $\mathbf{D}$ *asking* q *queries we have* $\Delta^{\mathbf{D}}(\mathbf{F}, \mathbf{G}) \leq \Delta_q(\mathbf{F}^{\perp}, \mathbf{G}^{\perp}) + \nu^{\mathbf{D}}(\mathbf{F}, \overline{\mathcal{A}_q})$.

(iii) [22, Lemma 5] $\Delta_q(\mathbf{C}^{\mathbf{F}}, \mathbf{C}^{\mathbf{G}}) \leq \Delta_{q'}(\mathbf{F}, \mathbf{G})$, *where* q' *is the maximum number of invocations of any internal system* $\mathbf{H}$ *for any sequence of* q *queries to* $\mathbf{C}^{\mathbf{H}}$, *if such a value is defined.*

(iv) [16, Lemma 3] There exists a fixed permutation $S \in \mathrm{Perm}(n)$ *(represented by a deterministic stateless system) such that* $\Delta_q(\mathbf{C}^{\mathbf{P}}, \mathbf{C}'^{\mathbf{P}}) \leq \Delta_q(\mathbf{C}^S, \mathbf{C}'^S)$.

The first claim can be seen as a generalized version of the Fundamental Lemma of Game-Playing for the context of random systems, stating that if two systems are equivalent as long as some condition is satisfied, then the advantage in distinguishing these systems can be upper-bounded by the probability of violating this condition. The second claim is even more general, analyzing the situation where the systems are not equivalent even if the conditions defined on them are satisfied, but their behavior is similar (which is captured by the term $\Delta_q(\mathbf{F}^{\perp}, \mathbf{G}^{\perp})$). The third claim states the intuitive fact that interacting with the distinguished systems through an additional enveloping construction $\mathbf{C}$ cannot improve the distinguishing advantage and the last claim is just an averaging argument over all the possible values taken by $\mathbf{P}$.

3 Generic Attacks against Efficient Key-Length Extension Schemes

We start by addressing the following question: *What is the maximum achievable security level for very efficient key-length extension schemes?* To this end, this section presents generic distinguishing attacks against one- and two-call block-cipher constructions in Sections 3.1 and 3.2, respectively. These attacks are in the same spirit as the recent line of work on generic attacks on hash functions (cf. e.g. [27,29,30]). Along the same lines, here attack complexity will be measured in terms of query- rather than time-complexity. This allows us to consider arbitrary constructions, while being fully sufficient to assess security in the ideal cipher model, where distinguishers are computationally unrestricted.

More formally, we consider stateless and deterministic (keyed) constructions $\mathbf{C}^{(\cdot)}$ invoking an ideal cipher $\mathbf{E} : \{0,1\}^{\kappa} \times \{0,1\}^n \times \{+, -\} \to \{0,1\}^n$ to implement a function $\mathbf{C}^{\mathbf{E}} : \{0,1\}^{\kappa'} \times \{0,1\}^n \times \{+, -\} \to \{0,1\}^n$ to serve as a block cipher with key length κ'. We assume that the construction $\mathbf{C}^{\mathbf{E}}$ realizes a permutation for each $k' \in \{0,1\}^{\kappa'}$ and hence it also provides the interface for inverse queries as indicated. Consequently, for a random (secret) κ'-bit string K', we let $\mathbf{C}^{\mathbf{E}}_{K'}$ denote the system which only gives access to the permutation

$\mathbf{C^E}(K', \cdot)$ and its inverse (i.e., takes inputs from $\{0,1\}^n \times \{+,-\}$). (In fact, none of the attacks in this section will require backward queries.)

Since the goal of this section is mainly to serve as a supporting argument for the optimality of our construction presented in Section 4, due to space restrictions we omit the proofs of our claims and only provide some intuition. All statements are proved in a more general setting in the full version of this paper.

3.1   One-Query Constructions

Throughout this section, we assume that $\mathbf{C}^{(\cdot)}$, to evaluate input $(x,+)$ for $x \in \{0,1\}^n$ under a key $k' \in \{0,1\}^{\kappa'}$, makes exactly one query to the underlying subsystem, and we denote this query as $q(k',x)$. We consider two different cases, depending on the structure of $q(\cdot,\cdot)$, before deriving the final attack.

THE INJECTIVE CASE. We first consider the case where the mapping $x \mapsto q(k',x)$ is injective for each k'. We shall denote this as a *one-injective-query construction*. In this case, distinct queries to $\mathbf{C}^E_{k'}$ lead to distinct internal queries to $\mathbf{E}$ and hence if the distinguisher queries both $\mathbf{C}^E_{K'}$ and $\mathbf{E}$ at sufficiently many random positions, one can expect that during the evaluation of the outer queries, $\mathbf{C}^{(\cdot)}_{K'}$ asks $\mathbf{E}$ for a value that was also asked by the distinguisher. If this occurs, the distinguisher can, while trying all possible keys k', evaluate $\mathbf{C}^{(\cdot)}_{k'}$ on its own by simulating $\mathbf{C}^{(\cdot)}$ and using the response from $\mathbf{E}$; and by comparing the outcomes it can distinguish the construction from a truly random permutation. This is the main idea behind the following lemma.

Lemma 2. *Let* $\mathbf{E}: \{0,1\}^\kappa \times \{0,1\}^n \times \{+,-\} \to \{0,1\}^n$ *be an ideal block cipher, let* $\mathbf{C}^{(\cdot)}: \{0,1\}^{\kappa'} \times \{0,1\}^n \times \{+,-\} \to \{0,1\}^n$ *be a one-injective-query construction and let* $\mathbf{P}$ *be a URP on* $\{0,1\}^n$. *Then, for a random key* $K' \in \{0,1\}^{\kappa'}$ *and every parameter* $0 < t < 2^{\min\{n,k\}-1}$,[7] *there exists a distinguisher* $\mathbf{D}$ *such that*

$$\Delta^{\mathbf{D}}((\mathbf{E},\mathbf{C}^E_{K'}),(\mathbf{E},\mathbf{P})) \geq 1 - 2/t - 2^{\kappa'-t\cdot(n-1)},$$

and which makes at most $4t \cdot 2^{\max\{(\kappa+n)/2,\kappa\}}$ *queries to the block cipher* $\mathbf{E}$, *as well as at most* $2 \cdot 2^{\min\{(\kappa+n)/2,n\}}$ *forward queries to either of* $\mathbf{C}^E_{K'}$ *and* $\mathbf{P}$.

The above lemma covers most of the natural one-query constructions, since these typically satisfy the injectivity requirement (e.g. the DESX construction). In the following we see that constructions asking non-injective queries do not achieve any improvement in security.

NON-INJECTIVE QUERIES. We now permit that the construction $\mathbf{C}^{(\cdot)}$ might, for some key k', invoke the underlying ideal cipher in a *non-injective* way, i.e., $q(k',\cdot)$ is not an injective map. We prove that, roughly speaking, such a construction $\mathbf{C}^E_{K'}$ might be distinguishable from a URP $\mathbf{P}$ based solely on an entropy argument. The intuitive reasoning is that if $\mathbf{C}^{(\cdot)}$ allows on average (over the choice of

---

[7] Roughly speaking, higher t increases the advantage but also the required number of queries; we obtain the desired bound using a constant t. For a first impression, consider e.g. $t = 4$ and $\kappa' \approx 2n$.

the key k') that too many queries x map to the same $q(k', x)$, then it also does not manage to obtain sufficient amount of randomness from the underlying random function to simulate $\mathbf{P}$ convincingly, opening the door to a distinguishing attack. In the following, let $q(k') = |\{q(k', x) : x \in \{0, 1\}^n\}|$ for all $k' \in \{0, 1\}^\kappa$.

Lemma 3. *Let $\mathbf{C}^{(\cdot)} \colon \{0, 1\}^{\kappa'} \times \{0, 1\}^n \times \{+, -\} \to \{0, 1\}^n$ be a one-query construction, let $\mathbf{P}$ be a URP on $\{0, 1\}^n$ and let $\mathbf{E} \colon \{0, 1\}^\kappa \times \{0, 1\}^n \times \{+, -\} \to \{0, 1\}^n$ be an ideal block cipher. Also, let $K' \in \{0, 1\}^{\kappa'}$ be a random key, and assume that there exists q^* such that $q(K') \leq q^*$ with probability at least $\frac{1}{2}$. Then, there exists a distinguisher $\mathbf{D}$ asking 2^n forward queries such that*

$$\Delta^{\mathbf{D}}\left(\mathbf{C}_{K'}^{\mathbf{E}}, \mathbf{P}\right) \geq \tfrac{1}{2} - 2^{\kappa' + n \cdot q^* - \log(2^n!)} \ .$$

PUTTING THE PIECES TOGETHER. We can combine the techniques used to prove Lemma 2 (somewhat relaxing the injectivity requirement) and Lemma 3, to obtain the following final theorem yielding an attack for arbitrary one-query block-cipher constructions.

Theorem 1. *Let $n \geq 6$ and $\kappa' \leq 2^n - 1$, let $\mathbf{E} \colon \{0, 1\}^\kappa \times \{0, 1\}^n \times \{+, -\} \to \{0, 1\}^n$ be an ideal block cipher, let $\mathbf{C}^{(\cdot)} \colon \{0, 1\}^{\kappa'} \times \{0, 1\}^n \times \{+, -\} \to \{0, 1\}^n$ be a one-query construction, and let $\mathbf{P}$ be a URP on $\{0, 1\}^n$. Then, for a random key $K' \in \{0, 1\}^{\kappa'}$ and for all parameters $0 < t < 2^{n-2}$, there exists a distinguisher $\mathbf{D}$ such that*

$$\Delta^{\mathbf{D}}\left((\mathbf{E}, \mathbf{C}_{K'}^{\mathbf{E}}), (\mathbf{E}, \mathbf{P})\right) \geq \min\left\{\tfrac{1}{4}, \tfrac{1}{2} - \tfrac{2}{t} - 2^{\kappa' - t \cdot (n-1)}\right\} \ ,$$

and which asks at most $8t \cdot 2^\kappa$ queries to $\mathbf{E}$ and 2^n forward queries to either of $\mathbf{C}_{K'}^{\mathbf{E}}$ and $\mathbf{P}$.

Theorem 1 shows that no one-query construction can achieve security beyond $2^{\max\{\kappa, n\}}$ queries, hence in our search for efficient key-length extension schemes we have to we turn our attention towards constructions issuing at least two queries.

3.2   Two-Query Constructions

We now consider an arbitrary deterministic stateless construction $\mathbf{C}^{(\cdot)} \colon \{0, 1\}^{\kappa'} \times \{0, 1\}^n \times \{+, -\} \to \{0, 1\}^n$ that makes exactly two queries to an ideal block cipher $\mathbf{E} \colon \{0, 1\}^k \times \{0, 1\}^n \times \{+, -\} \to \{0, 1\}^n$ to evaluate each query. In the following, these constructions shall be referred to as *two-query constructions*. We denote by $q_1(k', x) \in \{0, 1\}^\kappa \times \{0, 1\}^n \times \{+, -\}$ the first query $\mathbf{C}^{(\cdot)}$ asks its subsystem when it is itself being asked a forward query $(k', x, +)$. Moreover, we denote by $q_2(k', x, s) \in \{0, 1\}^\kappa \times \{0, 1\}^n \times \{+, -\}$ the second query it asks when it is itself being asked a forward query $(k', x, +)$ and the answer to the first query $q_1(k', x)$ was $s \in \{0, 1\}^n$. Since $\mathbf{C}^{(\cdot)}$ is deterministic and stateless, both q_1 and q_2 are well-defined mappings.

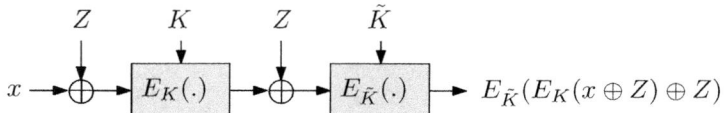

Fig. 1. The double XOR-cascade construction analyzed in Theorem 3

Theorem 2. *Let* $\mathbf{C}^{(\cdot)}: \{0,1\}^{\kappa'} \times \{0,1\}^n \times \{+,-\} \to \{0,1\}^n$ *be a two-query construction satisfying the following two conditions:*

1. *for every* $k' \in \{0,1\}^{\kappa'}$ *the mapping* $q_1(k', \cdot)$ *is injective,*
2. *distinct answers to the first query imply distinct second queries, i.e., for every* $k' \in \{0,1\}^{\kappa'}$ *and every* $x, x' \in \{0,1\}^n$ *if* $s \neq s'$ *then* $q_2(k', x, s) \neq q_2(k', x', s')$.

Then for a random key $K' \in \{0,1\}^{\kappa'}$, *for a URP* $\mathbf{P}$ *on* $\{0,1\}^n$ *and for every parameter* $0 < t < 2^{n/2-1}$, *there exists a distinguisher* $\mathbf{D}$ *such that*

$$\Delta^{\mathbf{D}}((\mathbf{E}, \mathbf{C}^{\mathbf{E}}_{K'}), (\mathbf{E}, \mathbf{P})) \geq 1 - 2/t - 13 \cdot 2^{-\frac{n}{2}} - 2^{\kappa'-t\cdot(n-1)},$$

where $\mathbf{D}$ *makes at most* $2(t+4) \cdot 2^{\kappa+n/2}$ *queries to* $\mathbf{E}$ *as well as* 2^n *forward queries to either of* $\mathbf{C}^{\mathbf{E}}_{K'}$ *and* $\mathbf{P}$.

Hence, no two-query construction from a large class described in the above theorem can achieve security beyond $2^{\kappa+n/2}$ queries. In the following section we present a simple and efficient construction from this class that achieves the above limit.

4   The Double XOR-Cascade Construction

We present a two-query construction matching the upper bound $2^{\kappa+n/2}$ on security proved in the previous section. The construction, which we call the *double XOR-cascade construction* (2XOR), consists of two applications of the block-cipher interleaved with two XOR operations: Given a (κ, n)-block cipher E, we define the $(\kappa + n, n)$-block cipher $2\mathrm{XOR}^E$ such that

$$2\mathrm{XOR}^E_{k,z}(m) = E_{\widetilde{k}}(E_k(m \oplus z) \oplus z)$$

for all $k \in \{0,1\}^{\kappa}$, $z, m \in \{0,1\}^n$, and where $\widetilde{k} = \pi(k)$ for some understood fixpoint-free permutation $\pi \in \mathrm{Perm}(\kappa)$ (e.g., $\pi(k) = k \oplus 0^{\kappa-1}1$, i.e., π flips the last bit). The construction is depicted in Figure 1. Note that both XOR transformations use the same value z and the two block-cipher calls use two distinct keys such that one can be deterministically derived from the other one. We also consider a construction $2\mathrm{XOR}'$ of a $(2\kappa + n)$-block cipher where $\widetilde{k}$ is replaced by an (independent) κ-bit key.

SECURITY OF 2XOR. We now discuss the security of the double XOR-cascade construction in the ideal cipher model. To this end, let $\mathbf{X}^{(\cdot)} \colon \{0,1\}^\kappa \times \{0,1\}^n \times \{0,1\}^n \times \{+,-\} \to \{0,1\}^n$ denote a (deterministic stateless) construction which expects a subsystem $\mathbf{E} \colon \{0,1\}^\kappa \times \{0,1\}^n \times \{+,-\} \to \{0,1\}^n$ realizing a block cipher. $\mathbf{X}^{\mathbf{E}}$ then answers each query $(k, z, x, +)$ by $\mathbf{E}_{\widetilde{k}}\left(\mathbf{E}_k\left(x \oplus z\right) \oplus z\right)$ and each query $(k, z, y, -)$ by $\mathbf{E}_k^{-1}(\mathbf{E}_{\widetilde{k}}^{-1}(y) \oplus z) \oplus z$. As before, for randomly chosen (secret) keys $(K, Z) \in \{0,1\}^\kappa \times \{0,1\}^n$, we let $\mathbf{X}_{K,Z}^{\mathbf{E}}$ be the system which gives access to the permutation $\mathbf{X}^{\mathbf{E}}(K, Z, \cdot)$ in both directions (i.e., takes inputs from $\{0,1\}^n \times \{+,-\}$).

Theorem 3. *Let* $\mathbf{P}$ *and* $\mathbf{E}$ *denote a URP on* $\{0,1\}^n$ *and an ideal* (κ, n)-*block cipher, respectively; let* $(K, Z) \in \{0,1\}^\kappa \times \{0,1\}^n$ *be uniformly chosen keys. For the construction* $\mathbf{X}_{K,Z}^{(\cdot)}$ *defined as above, and for every distinguisher* $\mathbf{D}$ *making* q *queries to* $\mathbf{E}$,

$$\Delta^{\mathbf{D}}\left(\left(\mathbf{E}, \mathbf{X}_{K,Z}^{\mathbf{E}}\right), (\mathbf{E}, \mathbf{P})\right) \leq 4 \cdot \left(\frac{q}{2^{\kappa + n/2}}\right)^{2/3} .$$

In particular, $\mathbf{D}$ *can make arbitrarily many queries to either of* $\mathbf{X}_{K,Z}^{\mathbf{E}}$ *and* $\mathbf{P}$.

We also note that an analogous statement for the construction 2XOR′ can be easily derived from the presented claim.

PROOF INTUITION. The proof, given below, follows a two-step approach. In the first part, we prove that for any parameter $h \leq 2^{n/2}$, the above advantage is upper bounded by $\varepsilon(h) + \frac{q}{h 2^{\kappa-1}}$, where $\varepsilon(h)$ is an upper bound on the advantage of a h-query distinguisher in telling apart the following two settings, in both of which it issues both forward and backward queries to two permutations $\pi_1, \pi_2 \in \mathrm{Perm}(n)$:

1. In the first case, π_1, π_2 are chosen uniformly and independently.
2. In the second setting, a uniform n-bit string Z is chosen, and π_1 and π_2 are chosen uniformly at random such that $\pi_2(\pi_1(\cdot \oplus Z) \oplus Z) = id$.

This step follows a pattern similar to the one used in [11,16] to analyze the security of plain cascades, but with obvious modifications and extra care to take into account randomization as well as key dependency.

Then, the main technical part of the proof consists of proving a bound $3h^2/2^{n+1}$ on $\varepsilon(h)$, which is a new result of independent interest. The intuition here is that without knowing Z, it is hard to come up with two queries, one to π_1 and one to π_2, which result in input-output pairs $\pi_1(x) = y$ and $\pi_2(x') = y'$ satisfying $x = y' \oplus Z$ and $x' = y \oplus Z$ simultaneously. However, as long as this does not happen, both permutations appear independent and random.

We stress that our double-randomization is crucial here: omitting one of the randomization steps, as well as adding a third randomization step for the same Z, would all result in invalidating the argument. The full version of this paper also provides some useful extra intuition as for why other simpler randomization methods for the cascade fail to provide the required security level.

Proof. We start by noting that the system $(\mathbf{E}, \mathbf{X}^{\mathbf{E}}_{K,Z})$ simply can be seen as providing an interface to query $2^\kappa + 1$ (dependent) permutations

$$\mathbf{E}_{k_1}, \mathbf{E}_{k_2}, \ldots, \mathbf{E}_{k_{2^\kappa}}, \mathbf{E}_{\tilde{K}}\left(\mathbf{E}_K\left(\cdot \oplus Z\right) \oplus Z\right) \,,$$

each both in forward and backward direction, where $k_1, k_2, \ldots, k_{2^\kappa}$ is an enumeration of the κ-bit strings. By the group structure of $\mathrm{Perm}(n)$ under composition, the joint distribution of these permutations does not change if we start by choosing the last permutation uniformly at random, i.e., we replace it by $\mathbf{P}$, then choose K and Z and finally choose the permutations of the block cipher independently and randomly except for the one corresponding to the key $\tilde{K}$, which we set to $x \mapsto \mathbf{P}\left(\mathbf{E}_K^{-1}\left(x \oplus Z\right) \oplus Z\right)$. Hence, let $\mathbf{G}^{(\cdot)}$ be a system that expects a single permutation as its subsystem (let us denote it P) and itself provides an interface to a block cipher (let us denote it G). It answers queries to G in the following way: in advance, it chooses random keys (K, Z) and then generates random independent permutations for G used with any key except $\tilde{K}$. For $\tilde{K}$, $\mathbf{G}$ realizes the permutation $x \mapsto P\left(G_K^{-1}\left(x \oplus Z\right) \oplus Z\right)$, querying P for any necessary values. Then the above argument shows that $(\mathbf{E}, \mathbf{X}^{\mathbf{E}}_{K,Z}) = (\mathbf{G}^{\mathbf{P}}, \mathbf{P})$ and hence we obtain

$$\Delta_q\left((\mathbf{E}, \mathbf{X}^{\mathbf{E}}_{K,Z}), (\mathbf{E}, \mathbf{P})\right) = \Delta_q\left((\mathbf{G}^{\mathbf{P}}, \mathbf{P}), (\mathbf{E}, \mathbf{P})\right) \leq \Delta_q\left((\mathbf{G}^S, S), (\mathbf{E}, S)\right),$$

where the last inequality follows from claim (iv) in Lemma 1 and S denotes the fixed permutation whose existence is guaranteed by this claim. Since S is fixed and hence known to the distinguisher, it makes no sense to query it and thus it remains to bound $\Delta_q\left(\mathbf{G}^S, \mathbf{E}\right)$ for any permutation S. From now on, we denote the system $\mathbf{G}^S$ by $\mathbf{G}$ to simplify the notation.

We shall refer to all forward or backwards queries to G involving the permutations indexed by K or $\tilde{K}$ as *relevant*. Similarly, the system $\mathbf{E}$ can be seen as also choosing some random key K (and hence $\tilde{K}$), this just does not affect its behavior, and we can hence define relevant queries for $\mathbf{E}$ in an analogous way. Let $\mathcal{A}^h$ and $\mathcal{B}^h$ denote monotone conditions defined on systems $\mathbf{E}$ and $\mathbf{G}$ respectively, such that each of these conditions remains satisfied as long as at most h of the queries asked so far were relevant. The parameter h will be chosen optimally at the end of the proof. We require $h < 2^{n/2}$.

It is easy to upper-bound the probability of asking more than h relevant queries in $\mathbf{E}$: since the key K does not affect the responses of the system (and therefore the behavior is also independent of the associated monotone condition), we only have to consider non-adaptive strategies. Hence, for any distinguisher $\mathbf{D}$ asking q queries, the expected number of relevant queries among them is $q \cdot 2^{1-\kappa}$ and using Markov inequality, we obtain $\nu(\mathbf{E}, \overline{\mathcal{A}}^h_q) \leq q/h2^{\kappa-1}$. Let $\mathbf{E}^\perp$ and $\mathbf{G}^\perp$ denote the systems $\mathbf{E}$ and $\mathbf{G}$ blocked by $\mathcal{A}^h$ and $\mathcal{B}^h$, respectively. Then we can apply claim (ii) of Lemma 1 to obtain

$$\Delta_q(\mathbf{G}, \mathbf{E}) \leq \Delta_q(\mathbf{G}^\perp, \mathbf{E}^\perp) + \nu(\mathbf{E}, \overline{\mathcal{A}}^h_q) \leq \Delta_q(\mathbf{G}^\perp, \mathbf{E}^\perp) + q/h2^{\kappa-1} \,.$$

Now, one can observe that the systems $\mathbf{G}^\perp$ and $\mathbf{E}^\perp$ only differ in a small part. More specifically, we have $\mathbf{G}^\perp = \mathbf{C}^{\mathbf{S}}$ and $\mathbf{E}^\perp = \mathbf{C}^{\mathbf{T}}$, where:

- **S** is a system that chooses $Z \in \{0,1\}^n$ at random and provides access (by means of both forward and backward queries) to two randomly chosen permutations π_1, π_2 on $\{0,1\}^n$ such that they satisfy the equation $\pi_2(\pi_1(\cdot \oplus Z) \oplus Z) = id$;
- **T** is a system providing access (by means of both forward and backward queries) to two independent uniformly random permutations $\pi_1, \pi_2 \in$ Perm(n);
- $\mathbf{C}^{(\cdot)}$ is a (randomized) construction that expects a subsystem which provides two permutations π_1 and π_2. $\mathbf{C}^{(\cdot)}$ itself provides access to a block cipher C as follows: it chooses a uniformly random key K and sets $C_K := \pi_1$ and $C_{\bar{K}} := \pi_2 \circ S$. (**C** only queries its subsystem once it is asked a relevant query). The permutations for all other keys are chosen independently at random. Moreover, **C** only allows h relevant queries, after that it returns $\perp$.

By Lemma 1(iii), the above observation gives us $\Delta_q(\mathbf{G}^\perp, \mathbf{E}^\perp) \le \Delta_h(\mathbf{S}, \mathbf{T})$ and hence it remains to bound $\Delta_h(\mathbf{S}, \mathbf{T})$. We start by taking a different view of the internal workings of the system **S**. Once the values Z, π_1, π_2 are chosen, the internal state of **S** can be represented by a set $\mathcal{T}$ of 2^n 4-tuples (x_1, y_1, x_2, y_2) such that $\pi_1(x_1) = y_1$ and $\pi_2(x_2) = y_2$, and $x_2 = y_1 \oplus Z$ and $x_1 = y_2 \oplus Z$. For any $\mathcal{I} \subseteq \{1, \ldots, 4\}$, let $\mathcal{T}_\mathcal{I}$ be the projection of $\mathcal{T}$ on the components in $\mathcal{I}$. Then note that for any two distinct tuples $(x_1, y_1, x_2, y_2), (x'_1, y'_1, x'_2, y'_2) \in \mathcal{T}$ we have $x_1 \ne x'_1$, $y_1 \ne y'_1$, $x_2 \ne x'_2$, and $y_2 \ne y'_2$, in other words $\mathcal{T}_{\{i\}} = \{0,1\}^n$ for every $i \in \{1, \ldots, 4\}$.

Equivalently, it is not hard to verify that **S** can be implemented using lazy-sampling to set up $\mathcal{T}$: Initially, $\mathcal{T} = \emptyset$ and Z is a uniform n-bit string. Then, **S** answers queries as follows:

- Upon a query $\pi_1(x)$, it returns y if $(x, y) \in \mathcal{T}_{\{1,2\}}$ for some y. Otherwise, it returns a random $y \in \{0,1\}^n \setminus \mathcal{T}_{\{2\}}$ and adds $(x, y, y \oplus Z, x \oplus Z)$ to $\mathcal{T}$.
- Upon a query $\pi_1^{-1}(y)$, it returns x if $(x, y) \in \mathcal{T}_{\{1,2\}}$ for some x. Otherwise, it returns a random $x \in \{0,1\}^n \setminus \mathcal{T}_{\{1\}}$ and adds $(x, y, y \oplus Z, x \oplus Z)$ to $\mathcal{T}$.
- Upon a query $\pi_2(x)$, it returns y if $(x, y) \in \mathcal{T}_{\{3,4\}}$ for some y. Otherwise, it returns a random $y \in \{0,1\}^n \setminus \mathcal{T}_{\{4\}}$ and adds $(y \oplus Z, x \oplus Z, x, y)$ to $\mathcal{T}$.
- Upon a query $\pi_2^{-1}(y)$, it returns x if $(x, y) \in \mathcal{T}_{\{3,4\}}$ for some x. Otherwise, it returns a random $x \in \{0,1\}^n \setminus \mathcal{T}_{\{3\}}$ and adds $(y \oplus Z, x \oplus Z, x, y)$ to $\mathcal{T}$.

We consider an intermediate system $\mathbf{S}'$ obtained from **S**: In addition to $\mathcal{T}$, it also keeps track of sets $\mathcal{P}_1$ and $\mathcal{P}_2$, both consisting of ordered pairs of n-bit strings. (Again $\mathcal{P}_{i,1}$ and $\mathcal{P}_{i,2}$ denote the strings appearing as first and second component in $\mathcal{P}_i$, respectively.) Initially each $\mathcal{P}_i$ is empty and during the experiment, $\mathcal{P}_i$ keeps track of input-output pairs for π_i which were already defined by directly answering a π_i query in either direction (as opposed to those that were defined internally by $\mathbf{S}'$ when answering a π_{3-i} query). Concretely, $\mathbf{S}'$ answers a query $\pi_1(x)$ by y if $(x, y) \in \mathcal{T}_{\{1,2\}} \cup \mathcal{P}_1$ for some y. Otherwise, it returns a uniformly chosen $y \in \{0,1\}^n \setminus \mathcal{P}_{1,2}$ and adds (x, y) to $\mathcal{P}_1$. Moreover, if $y \notin \mathcal{T}_{\{2\}}$, it also adds the tuple $(x, y, y \oplus Z, x \oplus Z)$ to $\mathcal{T}$. Queries $\pi_1^{-1}(y)$, $\pi_2(x)$, and $\pi_2^{-1}(y)$ are answered in a symmetric fashion. Having this description of $\mathbf{S}'$, note that we obtain the system $\mathbf{T}$ if a query $\pi_1(x)$ is answered by some given y only if $(x, y) \in \mathcal{P}_1$, and otherwise a fresh random output is generated (but the 4-tuples are still added to $\mathcal{T}$ as above).

We now define two monotone conditions $\mathcal{A}$ and $\mathcal{B}$ on $\mathbf{S}'$:

- $\mathcal{A} = A_0, A_1, \ldots$ fails at the first query $\pi_i(x)$ answered by a random y which satisfies $y \in \mathcal{T}_{\{2(i-1)+2\}}$, or $\pi_i^{-1}(y)$ answered by a random x such that $x \in \mathcal{T}_{\{2(i-1)+1\}}$.
- $\mathcal{B} = B_0, B_1, \ldots$ fails at the first query $\pi_i(x)$ such that there exists y satisfying $(x, y) \in \mathcal{T}_{\{2(i-1)+1, 2(i-1)+2\}} \setminus \mathcal{P}_i$, or $\pi_i^{-1}(y)$ such that there exists x satisfying $(x, y) \in \mathcal{T}_{\{2(i-1)+1, 2(i-1)+2\}} \setminus \mathcal{P}_i$.

By the above representations of $\mathbf{S}$ and $\mathbf{T}$, one can easily verify that $\mathbf{S}'|\mathcal{A} \equiv \mathbf{S}$ and $\mathbf{S}'|\mathcal{B} \equiv \mathbf{T}$. Therefore, by the triangle inequality and by claim (i) from Lemma 1,

$$\Delta_h(\mathbf{S}, \mathbf{T}) \leq \Delta_h(\mathbf{S}, \mathbf{S}') + \Delta_h(\mathbf{S}', \mathbf{T}) \leq \nu(\mathbf{S}', \overline{A}_h) + \nu(\mathbf{S}', \overline{B}_h).$$

To upper bound $\nu(\mathbf{S}', \overline{A}_h)$, note that each time a fresh random value is chosen from $\{0, 1\}^n \setminus \mathcal{P}_{i,j}$ when answering the ith query, it is in $\mathcal{T}_{2(i-1)+j}$ with probability at most $\frac{i-1}{2^n - i} \leq 2\frac{i-1}{2^n}$, hence the union bound gives us $\nu(\mathbf{S}', \overline{A}_h) \leq \frac{h^2}{2^n}$.

In order to bound $\nu(\mathbf{S}', \overline{B}_h)$, let us introduce a monotone condition $\mathcal{C} = C_0, C_1, \ldots$ on $\mathbf{T}$ which fails under the same circumstances as $\mathcal{B}$ in $\mathbf{S}'$ (note that this can be done since $\mathbf{T}$ also keeps track of the sets $\mathcal{T}$ and $\mathcal{P}_i$). As a consequence of these equivalent definitions and the fact that the behaviors of $\mathbf{S}'$ and $\mathbf{T}$ are the same as long as the respective associated conditions are satisfied, we have $\nu(\mathbf{S}', \overline{B}_h) = \nu(\mathbf{T}, \overline{C}_h)$. However, the input-output behavior of $\mathbf{T}$ is independent of Z (and $\mathcal{C}$ failing), and hence we can equivalently postpone the sampling of Z to the end of the interaction, go through the generated transcript to construct $\mathcal{T}$, and upper bound the probability that $\mathcal{C}$ has failed at some query. This implies that for the choice of Z, one query must have been *bad* in the following sense:

- query $\pi_1(x)$ is preceded by a π_2-query resulting in an input-output pair (x', y') such that $y' \oplus Z = x$;
- query $\pi_1^{-1}(y)$ preceded by a π_2-query resulting in pair (x', y') s.t. $x' \oplus Z = y$;
- query $\pi_2(x')$ preceded by a π_1-query resulting in pair (x, y) s.t. $y \oplus Z = x'$;
- query $\pi_2^{-1}(y')$ is preceded by a π_1-query resulting in pair (x, y) s.t. $x \oplus Z = y'$.

Given the transcript, and for randomly chosen Z, the ith query is bad with probability at most $(i-1)/2^n$, and the probability that at least one query is bad is thus at most $\frac{h^2}{2^{n+1}}$ by the union bound.

Putting all the obtained terms together, the part of the distinguisher's advantage that depends on h is $f(h) = q/h2^{\kappa-1} + 3h^2/2^{n+1}$. This term is minimal for $h^* = (\frac{1}{3}q2^{n-\kappa+1})^{1/3}$ which gives us $f(h^*) < 4 \cdot \left(\frac{q}{2^{\kappa+n/2}}\right)^{2/3}$ as desired. $\qquad\square$

Acknowledgements. We thank Mihir Bellare and Ueli Maurer for insightful feedback. Peter Gaži was partially supported by grants SNF 200020-132794, VEGA 1/0266/09 and UK/95/2011. Stefano Tessaro's work was done while at UC San Diego, partially supported by NSF grant CNS-0716790 and by Calit2. He is currently supported by DARPA, contract number FA8750-11-2-0225.

References

1. FIPS PUB 46: Data Encryption Standard (DES). National Institute of Standards and Technology (1977)
2. ANSI X9.52: Triple Data Encryption Algorithm Modes of Operation (1998)
3. FIPS PUB 46-3: Data Encryption Standard (DES). National Institute of Standards and Technology (1999)
4. FIPS PUB 197: Advanced Encryption Standard (AES). National Institute of Standards and Technology (2001)
5. NIST SP 800-67: Recommendation for the Triple Data Encryption Algorithm (TDEA) Block Cipher. National Institute of Standards and Technology (2004)
6. EMV Integrated Circuit Card Specifications for Payment Systems. Book 2: Security and Key Management, v.4.2. EMVCo (June 2008)
7. Aiello, W., Bellare, M., Di Crescenzo, G., Venkatesan, R.: Security Amplification by Composition: The Case of Doubly-Iterated, Ideal Ciphers. In: Krawczyk, H. (ed.) CRYPTO 1998. LNCS, vol. 1462, pp. 390–407. Springer, Heidelberg (1998)
8. Bellare, M., Desai, A., Jokipii, E., Rogaway, P.: A concrete security treatment of symmetric encryption. In: FOCS 1997: Proceedings of the 38th IEEE Annual Symposium on Foundations of Computer Science, pp. 394–403 (1997)
9. Bellare, M., Kilian, J., Rogaway, P.: The Security of Cipher Block Chaining Message Authentication Code. In: Desmedt, Y.G. (ed.) CRYPTO 1994. LNCS, vol. 839, pp. 341–358. Springer, Heidelberg (1994)
10. Bellare, M., Kohno, T.: A Theoretical Treatment of Related-key Attacks: RKA-PRPs, RKA-PRFs, and Applications. In: Biham, E. (ed.) EUROCRYPT 2003. LNCS, vol. 2656, pp. 491–506. Springer, Heidelberg (2003)
11. Bellare, M., Rogaway, P.: The Security of Triple Encryption and a Framework for Code-Based Game-Playing Proofs. In: Vaudenay, S. (ed.) EUROCRYPT 2006. LNCS, vol. 4004, pp. 409–426. Springer, Heidelberg (2006), http://eprint.iacr.org/2004/331
12. Black, J., Rogaway, P.: A Block-Cipher Mode of Operation for Parallelizable Message Authentication. In: Knudsen, L.R. (ed.) EUROCRYPT 2002. LNCS, vol. 2332, pp. 384–397. Springer, Heidelberg (2002)
13. Diffie, W., Hellman, M.E.: Exhaustive Cryptanalysis of the NBS Data Encryption Standard. Computer 10(6), 74–84 (1977)
14. Even, S., Goldreich, O.: On the power of cascade ciphers. ACM Trans. Comput. Syst. 3(2), 108–116 (1985)
15. Even, S., Mansour, Y.: A construction of a cipher from a single pseudorandom permutation. In: Journal of Cryptology, pp. 151–161. Springer, Heidelberg (1991)
16. Gaži, P., Maurer, U.: Cascade Encryption Revisited. In: Matsui, M. (ed.) ASIACRYPT 2009. LNCS, vol. 5912, pp. 37–51. Springer, Heidelberg (2009)
17. Gaži, P., Maurer, U.: Free-Start Distinguishing: Combining Two Types of Indistinguishability Amplification. In: Kurosawa, K. (ed.) ICITS 2009. LNCS, vol. 5973, pp. 28–44. Springer, Heidelberg (2010)
18. Kilian, J., Rogaway, P.: How to Protect DES Against Exhaustive Key Search (an Analysis of DESX). Journal of Cryptology 14, 17–35 (2001)
19. Lai, X., Massey, J.L.: A Proposal for a New Block Encryption Standard. In: Damgård, I.B. (ed.) EUROCRYPT 1990. LNCS, vol. 473, pp. 389–404. Springer, Heidelberg (1991)
20. Luby, M., Rackoff, C.: Pseudo-random permutation generators and cryptographic composition. In: STOC 1986: Proceedings of the Eighteenth Annual ACM Symposium on Theory of Computing, pp. 356–363 (1986)

21. Lucks, S.: Attacking Triple Encryption. In: Vaudenay, S. (ed.) FSE 1998. LNCS, vol. 1372, pp. 239–253. Springer, Heidelberg (1998)
22. Maurer, U.: Indistinguishability of Random Systems. In: Knudsen, L.R. (ed.) EUROCRYPT 2002. LNCS, vol. 2332, pp. 110–132. Springer, Heidelberg (2002)
23. Maurer, U., Massey, J.L.: Cascade ciphers: The importance of being first. Journal of Cryptology 6(1), 55–61 (1993)
24. Maurer, U., Pietrzak, K.: Composition of Random Systems: When Two Weak Make One Strong. In: Naor, M. (ed.) TCC 2004. LNCS, vol. 2951, pp. 410–427. Springer, Heidelberg (2004)
25. Maurer, U., Pietrzak, K., Renner, R.: Indistinguishability Amplification. In: Menezes, A. (ed.) CRYPTO 2007. LNCS, vol. 4622, pp. 130–149. Springer, Heidelberg (2007)
26. Maurer, U., Tessaro, S.: Computational Indistinguishability Amplification: Tight Product Theorems for System Composition. In: Halevi, S. (ed.) CRYPTO 2009. LNCS, vol. 5677, pp. 355–373. Springer, Heidelberg (2009)
27. Rogaway, P., Steinberger, J.P.: Security/Efficiency Tradeoffs for Permutation-Based Hashing. In: Smart, N.P. (ed.) EUROCRYPT 2008. LNCS, vol. 4965, pp. 220–236. Springer, Heidelberg (2008)
28. Schneier, B.: Description of a New Variable-Length Key, 64-bit Block Cipher (Blowfish). In: Anderson, R. (ed.) FSE 1993. LNCS, vol. 809, pp. 191–204. Springer, Heidelberg (1994)
29. Stam, M.: Beyond Uniformity: Better Security/Efficiency Tradeoffs for Compression Functions. In: Wagner, D. (ed.) CRYPTO 2008. LNCS, vol. 5157, pp. 397–412. Springer, Heidelberg (2008)
30. Steinberger, J.P.: Stam's Collision Resistance Conjecture. In: Gilbert, H. (ed.) EUROCRYPT 2010. LNCS, vol. 6110, pp. 597–615. Springer, Heidelberg (2010)
31. Tessaro, S.: Security Amplification for the Cascade of Arbitrarily Weak PRPs: Tight Bounds via the Interactive Hardcore Lemma. In: Ishai, Y. (ed.) TCC 2011. LNCS, vol. 6597, pp. 37–54. Springer, Heidelberg (2011)
32. Vaudenay, S.: Decorrelation: a theory for block cipher security. Journal of Cryptology 16(4), 249–286 (2003)

Fair Computation with Rational Players

Adam Groce and Jonathan Katz*

Department of Computer Science, University of Maryland
{agroce,jkatz}@cs.umd.edu

Abstract. We consider the problem of *fairness* in two-party computation, where this means (informally) that both parties should learn the correct output. A seminal result of Cleve (STOC 1986) shows that fairness is, in general, impossible to achieve for *malicious* parties. Here, we treat the parties as *rational* and seek to understand what can be done.

Asharov et al. (Eurocrypt 2011) recently considered this problem and showed impossibility of rational fair computation for a particular function and a particular set of utilities. We observe, however, that in their setting the parties have no incentive to compute the function *even in an ideal world where fairness is guaranteed*. Revisiting the problem, we show that rational fair computation *is* possible, for arbitrary functions and utilities, as long as at least one of the parties has a strict incentive to compute the function in the ideal world. This gives a novel setting in which game-theoretic considerations can be used to circumvent an impossibility result in cryptography.

1 Introduction

Cryptography and game theory are both concerned with understanding interactions between mutually distrusting parties with potentially conflicting interests. Cryptography typically adopts a "worst case" viewpoint; that is, cryptographic protocols are designed to protect the interests of each party against *arbitrary* (i.e., malicious) behavior of the other parties. The game-theoretic perspective, however, views parties as being *rational*; game-theoretic protocols, therefore, only need to protect against rational deviations by other parties.

Significant effort has recently been devoted to bridging cryptography and game theory; see [9,21] for surveys. This work has tended to focus on two general sets of questions:

"Using cryptographic protocols to implement games" (e.g., [7,10,4,8,25,1,19]). Given a game played by parties relying an external trusted entity (a *mediator*), when can the externa; trusted party be replaced by a cryptographic protocol executed by the parties themselves?

"Applying game-theoretic analysis to cryptographic protocols" (e.g., [17,21,13,26,1,15,16]). What game-theoretic definitions are appropriate for computationally bounded, rational players executing a network protocol? Can

---

* Research supported by NSF awards #0830464 and #1111599.

D. Pointcheval and T. Johansson (Eds.): EUROCRYPT 2012, LNCS 7237, pp. 81–98, 2012.

impossibility results in the cryptographic setting be circumvented if we are willing to take a game-theoretic approach?

Here, we turn our attention to the question of *fair two-party computation* in a rational setting, where fairness means that both parties should learn the value of some function f evaluated on the two parties' inputs. Following recent work of Asharov et al. [2] (see further below), our goal is to understand when fairness is achievable by rational parties running some cryptographic protocol, without the aid of any external trusted entity. Our work touches on both the issues outlined above. Our motivation was to circumvent the strong impossibility result of Cleve [6] for fair two-party computation in a malicious setting. In this sense, our work is a generalization of results on *rational secret sharing* [17,13,26,1,22,23,29,27,3,11], which can be seen as a special case of fair computation for a specific function with parties' inputs provided by a trusted dealer. It is also possible to view the problem of rational fair computation from a different perspective. Specifically, one could define a natural "fairness game" involving a trusted mediator who computes a function f on behalf of the parties (and gives both parties the result), and where parties can choose whether or not to participate and, if so, what input to send to the mediator. One can then ask whether there exists a real-world protocol (replacing the mediator) that preserves equilibria of the original mediated game. Our work demonstrates a close connection between these two complementary viewpoints; see further below.

1.1   Our Results

Our setting is the same as that studied by Asharov, Canetti, and Hazay [2]. Informally, there are two parties P_0 and P_1 who wish to compute a function f of their respective inputs x_0 and x_1, where the joint distribution of x_0 and x_1 is common knowledge. (In [2] independent uniform distributions were assumed but we consider arbitrary joint distributions.) Following work on rational secret sharing, the parties' utilities are such that each party prefers to learn the correct answer $f(x_0, x_1)$ and otherwise prefers that the other party outputs an *incorrect* answer. Informally, a cryptographic protocol computing f is *fair* if having both parties run the protocol is a (computational) Nash equilibrium with respect to fail-stop deviations,[1] i.e., it is assumed that parties may abort the protocol early, but cannot otherwise deviate.

Asharov et al. show a strong negative result in this context: they give a function f, a pair of distributions on the inputs, and a set of utilities for which they prove there is *no* fair protocol computing f with correctness better than $1/2$. They also show that correctness $1/2$ *can* be achieved for that function and utilities, but their work seemed to suggest that the power of rational fair computation is relatively limited.

Looking more closely at the impossibility result of Asharov et al., we observe that for their specific choices of f, the input distributions, and the utility

---

[1] We will consider Byzantine deviations as well, but stick to fail-stop deviations here for simplicity.

functions, *the parties have no incentive to run a protocol at all!* Namely, the utility each party obtains by running *any* protocol that correctly (and fairly) computes f is equal to the expected utility that each party obtains if it simply guesses the input of the other party and computes the function on its own (without any interaction).[2] A cleaner way of stating this is that *even in an ideal world* with a trusted entity computing f with complete fairness, the parties would be indifferent between using the trusted entity or not. In game-theoretic terms, computing f in this ideal world is not a *strict* Nash equilibrium for the specific setting considered in [2]. If running a (real-world) protocol incurs any cost at all, there is thus little hope that parties will prefer to run *any* protocol for computing f.

Asharov et al. rule out rational fair computation for a *specific* function, *specific* input distributions, and a *specific* set of utilities. Are there any settings where rational fair computation (with complete correctness) *is* possible? Assuming the existence of (standard) secure computation, we show a strong, general result for when this is the case:

Main Theorem. (Informal)   *Fix f, a distribution on inputs, and utility functions such that computing f in the ideal world (with complete fairness) is a* **strict** *Nash equilibrium for at least one party. Then, for the same input distributions and utility functions, there is a protocol Π for computing f (where correctness holds with all but negligible probability) such that following Π is a computational Nash equilibrium. This holds in both the fail-stop and Byzantine settings.*

In addition to the fact that we show a positive result, our work goes beyond the setting considered in [2] in several respects: we handle (deterministic) functions over arbitrary domains where parties receive possibly different outputs, and treat arbitrary distributions over the parties' inputs. (In [2], only single-output functions and independent, uniform input distributions were considered.) Moreover, we also treat the *Byzantine* setting where, in particular, parties have the option of changing their inputs; Asharov et al. [2] only treat the fail-stop case.

1.2   Other Related Work

The most relevant prior work is that of Asharov et al. [2], already discussed extensively above. Here we merely add that an additional contribution of their work was to develop formal definitions of various cryptographic goals (with fairness being only one of these) in a game-theoretic context. Their paper takes an important step toward that worthy goal.

As observed earlier, work on rational secret sharing [17,13,26,1,22,23,29,27,3,11] can be viewed as a special case of fair secure computation, where the function

---

[2] Specifically, using the input distributions and utility functions from [2], P_0's utility if both parties (run some protocol and) output the correct answer is 0, whereas if both parties guess then each party is (independently) correct with probability $1/2$ and so the expected utility of P_0 is $\frac{1}{4} \cdot 1 + \frac{1}{4} \cdot (-1) + \frac{1}{2} \cdot 0 = 0$. A similar calculation holds for P_1.

being computed is the reconstruction function of the secret-sharing scheme being used, and the parties' inputs are generated by a dealer. Depending on the utilities and input distributions, our results would give rational secret-sharing protocols where following the protocol is a (computational) Nash equilibrium. In contrast, most of the work on rational secret sharing has focused on achieving stronger equilibrium notions, in part because constructing Nash protocols for rational secret sharing is trivial in the multi-party setting. (We stress that in the two-party setting we consider here, constructing Nash protocols for rational secret sharing is *not* trivial.) We leave for future work consideration of stronger equilibrium notions for rational fair computation of general functions.

An analogue of our results is given by Izmalkov et al. [24,25,20,18] who, essentially, also show protocols for rational fair computation whenever parties would prefer to compute the function in the ideal world. The main difference is that we work in the cryptographic setting where parties communicate using standard channels, whereas the protocols of Izmalkov et al. require strong *physical* assumptions such as secure envelopes and ballot boxes.

There has recently been a significant amount of work on fairness in the cryptographic setting, showing functions that can be computed with complete fairness [12] and exploring various notions of partial fairness (see [14] and references therein). The relationship between complete fairness and rational fairness is not clear. In particular, completely fair protocols are not necessarily rationally fair: if the distributions and utilities are such that aborting is preferable even in the ideal world then no protocol can be rationally fair (with respect to the same distributions and utilities). In any case, relatively few functions are known that can be computed with complete fairness. Partial fairness is similarly incomparable to rational fairness.

2   Model and Definitions

Given a deterministic function $f : X \times Y \to \{0,1\}^* \times \{0,1\}^*$, we let f_0 (resp., f_1) denote the first (resp., second) output of f, so that $f(x,y) = (f_0(x,y), f_1(x,y))$. We consider two settings where parties P_0 and P_1 wish to compute f on their respective inputs x_0 and x_1, with P_0 receiving $f_0(x_0, x_1)$ and P_1 receiving $f_1(x_0, x_1)$: an *ideal world* computation of f using a trusted third party, and a *real-world* computation of f using some protocol Π. In each setting, x_0 and x_1 are chosen according to some joint probability distribution D, and in each setting we consider both fail-stop and Byzantine strategies.

The output of P_0 is correct if it is equal to $f_0(x_0, x_1)$ and incorrect otherwise; this is defined analogously for P_1. The parties' utilities are given by the following table, where the first value in each ordered pair is the utility of P_0 on the specified outcome, and the second is the utility of P_1:

		P_1's output	
		correct	incorrect
P_0's output	correct	(a_0, a_1)	(b_0, c_1)
	incorrect	(c_0, b_1)	(d_0, d_1)

We assume that parties prefer to output the correct answer, and otherwise prefer that the other party outputs an incorrect answer; i.e., we assume $b_0 > a_0 \geq d_0 \geq c_0$ (and analogously for P_1's utilities). Asharov et al. [2] assume that the parties' utilities are symmetric, with $b_0 = b_1 = 1, a_0 = a_1 = d_0 = d_1 = 0$, and $c_0 = c_1 = -1$; we consider more general utilities here.

2.1   Execution in the Ideal World

Our ideal world includes a trusted third party who computes f with complete fairness. This defines a natural game that proceeds as follows:

1. Inputs x_0 and x_1 are sampled according to a joint probability distribution D over input pairs. Then x_0 (resp., x_1) is given to P_0 (resp., P_1).
2. Each player sends a value to the trusted third party. We also allow parties to send a special value $\perp$ denoting an abort. Let x_0' (resp., x_1') denote the value sent by P_0 (resp., P_1).
3. If $x_0' = \perp$ or $x_1' = \perp$, the trusted party sends $\perp$ to both parties. Otherwise, the trusted party sends $f_0(x_0', x_1')$ to P_0, and $f_1(x_0', x_1')$ to P_1.
4. Each party outputs some value, and obtains a utility that depends on whether the parties' outputs are correct or not. (We stress that correctness is defined with respect to the "real" inputs x_0, x_1, not the effective inputs x_0', x_1'.)

In the *fail-stop* setting, we restrict $x_0' \in \{x_0, \perp\}$ and $x_1' \in \{x_1, \perp\}$. In the *Byzantine* setting we allow x_0', x_1' to be arbitrary.

The "desired" play in this game is for each party to send its input to the trusted third party, and then output the value returned by the trusted party. To fully define an honest strategy, however, we must specify what each party does for every possible value (including $\perp$) it receives from the trusted third party. We formally define strategy (cooperate, W_0) for P_0 as follows:

> P_0 sends its input x_0 to the trusted party. If the trusted party returns anything other than $\perp$, then P_0 outputs that value. If instead $\perp$ is returned, then P_0 generates output according to the distribution $W_0(x_0)$.

The strategy (cooperate, W_1) for P_1 is defined analogously. The situation in which P_0 plays (cooperate, W_0) and P_1 plays (cooperate, W_1) is a *(Bayesian) strict Nash equilibrium* if every (allowed[3]) deviation that has $x_0' \neq x_0$ with nonzero probability results in a strictly lower expected utility for P_0, and analogously for P_1. (The expectation is computed over the distribution of the other party's input.) Since the utility obtained by P_0 when the honest strategies are followed is exactly a_0, this means that the honest strategies form a Bayesian strict Nash equilibrium if, for every possible input x_0 of P_0, the expected utility of P_0 is strictly less than a_0 if it sends any (allowed) $x_0' \neq x_0$ to the third party, and similarly for P_1.

---

[3] I.e., in the fail-stop case the only allowed deviation is aborting, whereas in the Byzantine case parties are allowed to send arbitrary inputs. In either case a deviating party may determine its output any way it likes.

As long as both parties follow honest strategies, W_0 and W_1 are irrelevant (as they are never used). They are important, however, insofar as they serve as "empty threats" in case of an abort by the other party: namely, P_0 knows that if he aborts then P_1 will determine its own output according to $W_1(x_1)$, and so P_0 must take this into account when deciding whether to abort or not. We now define what it means for the parties to have an incentive to compute f.

Definition 1. *Fix f, a distribution D, and utilities for the parties. We say these are* incentive compatible in the fail-stop *(resp.,* Byzantine) *setting* if there exist W_0, W_1 such that the strategy profile $\big((\text{cooperate}, W_0), (\text{cooperate}, W_1)\big)$ is a Bayesian strict Nash equilibrium in the game above.

As discussed in the Introduction, we require the Nash equilibrium to be *strict* to ensure that the parties have *some* incentive to use the trusted party to compute the function. If carrying out the computation with the trusted party is a Nash (but not strict Nash) equilibrium, then the parties are indifferent between using the trusted party and just guessing the output on their own.

We remark that for our positive results it is sufficient for the ideal-world equilibrium to be strict Nash for only *one* of the parties; this follows directly from the proofs and we omit further discussion.

The Setting of Asharov et al. [2]. For completeness, we show that the setting considered by Asharov et al. is *not* incentive compatible. Recall that they fix the utilities such that (1) getting the correct answer while the other party outputs an incorrect answer gives utility 1; (2) getting an incorrect answer while the other party outputs the correct answer gives utility -1; and (3) any other outcome gives utility 0. Furthermore (cf. [2, Definition 4.6]), f can be taken to be boolean XOR, with inputs for each party chosen uniformly and independently. We claim that there is no choice of W_0, W_1 for which $\big((\text{cooperate}, W_0), (\text{cooperate}, W_1)\big)$ is a Bayesian strict Nash equilibrium. To see this, fix W_0, W_1 and note that playing $\big((\text{cooperate}, W_0), (\text{cooperate}, W_1)\big)$ gives utility 0 to both parties. On the other hand, if P_0 aborts and outputs a random bit, then regardless of the guessing strategy W_1 employed by P_1, we see that P_0 and P_1 are each correct with independent probability $1/2$ and so the expected utility of P_0 remains 0. This is an allowed deviation that results in no change to the expected utility.

In contrast, if the utilities are modified so that when both parties get the correct answer they each obtain utility $1/2$ (and everything else is unchanged), the setup *is* incentive compatible in the fail-stop setting. To see this, let W_0, W_1 be the uniform distribution. Playing $\big((\text{cooperate}, W_0), (\text{cooperate}, W_1)\big)$ gives utility $1/2$ to both parties. If, on the other hand, P_0 ever aborts then — no matter how P_0 determines its output — P_0 and P_1 are each correct with independent probability $1/2$. (Recall that P_1 is assumed to guess according to W_1 when we consider possible deviations by P_0.) The expected utility of deviating is $1/8$, which is strictly smaller than $1/2$; thus, we have a Bayesian *strict* Nash equilibrium. Our results imply that a rational fair protocol *can* be constructed for this setting.

Incentive Compatibility in the Byzantine Setting. The preceding discussion gives an example of an incentive-compatible setup in the fail-stop setting. In Appendix A we show an example of a simple function, a distribution on inputs, and utilities that are incentive compatible in the Byzantine setting.

2.2 Execution in the Real World

In the real world there is no trusted party, and the players instead must communicate in order to compute f. We thus have a real-world game in which inputs x_0 and x_1 are jointly sampled according to D, then x_0 (resp., x_1) is given to P_0 (resp., P_1), and the parties finally execute some strategy (i.e., protocol) and decide on their respective outputs. The goal is to construct a protocol Π such that running the protocol is a (computational) Nash equilibrium. The running times of the parties, as well as the protocol itself, are parameterized in terms of a security parameter n; however, the function f as well as the parties' utilities are fixed and independent of n. We only consider protocols where correctness holds with all but negligible probability.

We again consider two types of deviations. In the *fail-stop* setting, each party follows the protocol as directed except that it may choose to abort at any point. Upon aborting, a party may output whatever value it likes (and not necessarily the value prescribed by the protocol). We stress that in the fail-stop setting a party is assumed not to change its input when running the protocol. In the *Byzantine* setting, parties may behave arbitrarily (and, in particular, may run the protocol using a different input). In either setting, we will only be interested in players whose strategies can be implemented in probabilistic polynomial-time.

We now define what it means for Π to induce a game-theoretic equilibrium. We consider computational Nash equilibria, rather than computational *strict* Nash equilibria, since the latter are notoriously difficult to define [11]; also, the goal of our work is only to construct real-world protocols that induce a Nash equilibrium. (We define strict Nash equilibria in the ideal world because we use it for our results.) The following definition is equivalent to (a generalized version of) the definition used by Asharov et al. [2, Definition 4.6].

Definition 2. *Fix f, a distribution D, utilities for the parties, and a protocol Π computing f. We say Π is a* rational fair protocol *(with respect to these parameters) in the fail-stop (resp., Byzantine) setting if running the protocol is a Bayesian computational Nash equilibrium in the game defined above.*

For example, if we let Π_1 denote the algorithm that honestly implements P_1's role in Π, then Π is a rational fair protocol in the fail-stop setting if for all (efficient) fail-stop algorithms P_0' there is a negligible function μ such that the expected utility of $P_0'(1^n)$ (when running against P_1 playing $\Pi_1(1^n)$) is at most $a_0 + \mu(n)$ (and analogously for deviations by P_1). We stress that if P_0' aborts here, then P_1 determines its output as directed by Π_1.

Note that it only makes sense to speak of Π being a rational fair protocol with regard to some input distribution and utilities. In particular, it is possible for Π to be rational for one set of utilities but not another.

3   Positive Results for Rational Fair Computation

We show broad positive results for rational fair computation in both the fail-stop and Byzantine settings. Specifically, we show that whenever computing the function honestly is a Bayesian strict Nash equilibrium in the ideal world, then there exists a protocol Π computing f such that running Π is a Bayesian computational Nash equilibrium in the real world.

Our protocols all share a common structure. As in prior work on fairness (in the cryptographic setting) [12,28,14], our protocols have two stages. The first stage is a "pre-processing" step that uses any protocol for (standard) secure two-party computation, and the second stage takes place in a sequence of n iterations. In our work, the stages have the following form:

First stage:

1. A value $i^* \in \{1, \ldots\}$ is chosen according to a geometric distribution. This represents the iteration (unknown to the parties) in which both parties will learn the correct output.
2. Values $r_1^0, r_1^1, \ldots, r_n^0, r_n^1$ are chosen, with the $\{r_i^0\}_{i=1}^n$ intended for P_0 and the $\{r_i^1\}_{i=1}^n$ intended for P_1. For $i \geq i^*$ we have $r_i^0 = f_0(x_0, x_1)$ and $r_i^1 = f_1(x_0, x_1)$, while for $i < i^*$ the $\{r_i^0\}$ (resp., $\{r_i^1\}$) values depend on P_0's (resp., P_1's) input only.
3. Each r_i^b value is randomly shared as s_i^b and t_i^b (with $r_i^b = s_i^b \oplus t_i^b$), where s_i^b is given to P_0 and t_i^b is given to P_1.

Second stage: For n iterations, each consisting of two rounds, the parties alternate sending shares to each other. In the ith iteration, P_1 sends t_i^0 to P_0, enabling P_0 to learn r_i^0; then P_0 sends s_i^1 to P_1, enabling P_1 to learn r_i^1. When the protocol ends (either through successful termination or an abort by the other party) a party outputs the most-recently-learned r_i.

The key difference with respect to prior work is how we set the distribution of the $\{r_i^b\}$ for $i < i^*$. Here we use the assumption that f, D, and the utilities are incentive compatible, and thus there are "guessing strategies" $W_0(x_0)$ and $W_1(x_1)$ for the parties (in case the other party aborts) that are in equilibrium (see Section 2.1). We use exactly these distributions in our protocol.

3.1   The Fail-Stop Setting

We first analyze the fail-stop setting. We let W_0 (resp., W_1) denote the distribution that P_0 (resp., P_1) uses to determine its output in the ideal world in case P_1 (resp., P_0) aborts, where this distribution may depend on P_0's input x_0 (resp., P_1's input x_1). We say W_0 has *full support* if for every x_0 the distribution $W_0(x_0)$ puts non-zero probability on every element in the range of f; we define this notion analogously for W_1. We begin with a technical claim.

Lemma 1. *Fix a function f, a distribution D, and utilities for the parties that are incentive compatible in the fail-stop (resp., Byzantine) setting. Then there* *exist W_0, W_1 **with full support** such that $\big((\mathsf{cooperate}, W_0), (\mathsf{cooperate}, W_1)\big)$ is a Bayesian strict Nash equilibrium in the fail-stop (resp., Byzantine) setting.*

Proof. We focus on the fail-stop setting, though the proof follows along the same lines for the Byzantine case. Incentive compatibility implies that there exist W_0', W_1' such that the strategy vector $\big((\mathsf{cooperate}, W_0'), (\mathsf{cooperate}, W_1')\big)$ is a Bayesian strict Nash equilibrium. Distributions W_0', W_1' may not have full support, but we show that they can be modified so that they do. Specifically, we simply modify each distribution so that with some sufficiently small probability it outputs a uniform element from the range of f. Details follow.

When P_0 and P_1 cooperate and both output the correct answer, P_0 obtains utility a_0. Consider now some input x_0 for P_0, and let $u_0^*(x_0)$ denote the maximum utility P_0 can obtain if it aborts on input x_0. (Recall that when P_0 aborts, P_1 chooses its output according to $W_1'(x_1)$. Here, P_0 knows W_1' as well as the marginal distribution of x_1 conditioned on P_0's input x_0.) Because cooperating is a Bayesian strict Nash equilibrium, we must have $u_0^*(x_0) < a_0$. Define

$$u_0^* \stackrel{\text{def}}{=} \max_x \{u_0^*(x)\} < a_0$$

(the maximum is taken over all x that have non-zero probability as input to P_0); i.e., u_0^* denotes the highest expected utility P_0 can hope to obtain when aborting on some input. Define u_1^* analogously with respect to deviations by P_1. Set

$$\lambda \stackrel{\text{def}}{=} \frac{1}{2} \cdot \min \left\{ \frac{a_0 - u_0^*}{b_0 - c_0}, \frac{a_1 - u_1^*}{b_1 - c_1} \right\} > 0.$$

We define a distribution $W_0(x_0)$ as follows: with probability λ output a uniform element from the range of f, and with probability $(1 - \lambda)$ choose an output according to $W_0'(x_0)$; define W_1 similarly. Note that W_0 and W_1 have full support. We claim that $\big((\mathsf{cooperate}, W_0), (\mathsf{cooperate}, W_1)\big)$ is a Bayesian strict Nash equilibrium. Assume the contrary; then, without loss of generality, there is an input x_0 such that P_0 obtains expected utility at least a_0 by aborting on input x_0. (Note that now P_1 chooses its output according to $W_1(x_1)$ when P_0 aborts.) But then, by following the same strategy, P_0 can obtain utility at least $a_0 - \lambda \cdot (b_0 - c_0)$ when playing against a P_1 who chooses his output according to $W_1'(x_1)$. Since $a_0 - \lambda \cdot (b_0 - c_0) > u_0^*$, this is a contradiction to the way u_0^* was defined. □

Theorem 1. *Fix a function f, a distribution D, and utilities for the parties. If these are incentive compatible in the fail-stop setting, then (assuming the existence of general secure two-party computation for semi-honest adversaries) there exists a protocol Π computing f such that Π is a rational fair protocol (with respect to the same distribution and utilities) in the fail-stop setting.*

Proof. By definition of incentive compatibility, there exist distributions W_0, W_1 (that can depend on x_0 and x_1, respectively) for which the strategy profile

Functionality ShareGen

Inputs: ShareGen takes as input a value x_0 from P_0 and a value x_1 from P_1.
  If either input is invalid, then ShareGen simply outputs $\perp$ to both parties.
Computation: Proceed as follows:
  1. Choose i^* according to a geometric distribution with parameter p.
  2. Set the values of r_i^0 and r_i^1 for $i \in \{1, \ldots, n\}$ as follows:
     – If $i < i^*$, choose $r_i^0 \leftarrow W_0(x_0)$ and $r_i^1 \leftarrow W_1(x_1)$.
     – If $i \geq i^*$, set $r_i^0 = f_0(x_0, x_1)$ and $r_i^1 = f_1(x_0, x_1)$.
  3. For each r_i^b, choose two values s_i^b and t_i^b as random secret shares of r_i^b.
     (I.e., s_i^b is random and $s_i^b \oplus t_i^b = r_i^b$.)
Output: Send $s_1^0, s_1^1, \ldots, s_n^0, s_n^1$ to P_0, and $t_1^0, t_1^1, \ldots, t_n^0, t_n^1$ to P_1.

Fig. 1. Functionality ShareGen. The security parameter is n. This functionality is parameterized by a real number $p > 0$.

$\big((\text{cooperate}, W_0), (\text{cooperate}, W_1)\big)$ is a Bayesian strict Nash equilibrium. By Lemma 1, we may assume that W_0 and W_1 both have full support. We define a functionality ShareGen (cf. Figure 1) based on these distributions; this functionality is parameterized by a real number $p > 0$ that we will set later. We define our protocol Π, that uses ShareGen as a building block, in Figure 2.

Since p is a constant (independent of n), we have $i^* \leq n$ with all but negligible probability and hence both parties obtain the correct answer with all but negligible probability. In the analysis it is easiest to simply assume that $i^* \leq n$; this does not affect the analysis since computational Nash equilibria are robust to negligible changes in the utility. Alternately, one could simply modify ShareGen to enforce that $i^* \leq n$ always (namely, by setting $i^* = n$ in case $i^* > n$).

We will analyze Π in a hybrid world where there is a trusted entity computing ShareGen. One can show (following [5]) that if Π is a computational Nash equilibrium in this hybrid world, then so is Π when executed in the real world

Protocol Π

Stage one: Both players use their inputs to execute a secure protocol for
  computing ShareGen. This results in P_0 obtaining output $s_1^0, s_1^1, \ldots, s_n^0, s_n^1$,
  and P_1 obtaining output $t_1^0, t_1^1, \ldots, t_n^0, t_n^1$.
Stage two: There are n iterations. In each iteration $i \in \{1, \ldots, n\}$ do:
  1. P_1 sends t_i^0 to P_0, and P_0 computes $r_i^0 := t_i^0 \oplus s_i^0$.
  2. P_0 sends s_i^1 to P_1, and P_1 computes $r_i^1 := t_i^1 \oplus s_i^1$.
Output: Players determine their outputs as follows:
  – If P_{1-i} aborts before P_i has computed any r_i value, then P_i chooses
    its output according to $W_i(x_i)$.
  – If P_{1-i} aborts at any other point, or the protocol completes success-
    fully, then P_i outputs the last r_i value it received.

Fig. 2. Formal definition of our protocol

(with a secure protocol implementing ShareGen). Once we have moved to this hybrid world, we may in fact take the parties to be computationally unbounded.

Our goal is to show that there exists a $p > 0$ for which Π (in the hybrid world described above) is a rational fair protocol. We first observe that there are no profitable deviations for a fail-stop P_1; this is because P_0 always "gets the output first" in every iteration. (More formally, say P_1 aborts after receiving its iteration-i message. If $i \geq i^*$ then P_0 will output the correct answer and so P_1 cannot possibly get utility greater than a_1. If $i < i^*$ then P_0 has no information beyond what it could compute from its input x_1, and P_0 will generate output according to $W_0(x_0)$; by incentive compatibility, P_1 will obtain utility strictly lower than a_1 regardless of how it determines its output.) We are thus left with the more difficult task of analyzing deviations by P_0.

Before continuing, it is helpful to introduce two modifications to the protocol that can only increase P_0's utility. First, in each iteration i we tell P_0 whether $i^* < i$. (One can easily see that P_0 cannot increase its utility by aborting when $i^* < i$, and so the interesting question is whether P_0 can improve its utility by aborting when $i^* \geq i$.) Second, if P_0 ever aborts the protocol in some iteration i with $i^* \geq i$, then we tell P_0 whether $i^* = i$ before P_0 generates its output. (P_0 is not, however, allowed to change its decision to abort.)

So, fix some input x_0 for P_0, and consider some iteration $i < n$. Say P_0 has just learned that $r_i = y$ (for some y in the range of f) and is told that $i^* \geq i$. If P_0 does not abort, but instead runs the protocol honestly to the end, then it obtains utility a_0. If P_0 aborts, then with some probability α it learns that $i^* = i$; in that case, P_0 may possibly get utility b_0. Otherwise, with probability $1 - \alpha$ it learns that $i^* > i$. In this latter case, P_0 has no information beyond what it could compute from its input, and P_1 will output a value distributed according to $W_1(x_1)$; hence, incentive compatibility implies that the maximum expected utility of P_0 is $u_0^* < a_0$. (This u_0^* is the same as defined in the proof of Lemma 1; for our purposes all we need is that u_0^* is strictly less than a_0.) That is, the expected utility of aborting is at most $\alpha \cdot b_0 + (1 - \alpha) \cdot u_0^*$. If

$$\alpha < \frac{a_0 - u_0^*}{b_0 - u_0^*} \tag{1}$$

then $\alpha \cdot b_0 + (1 - \alpha) \cdot u_0^* < a_0$, implying that P_0 has no incentive to deviate. We show that p can be set such that (1) holds.

We have

$$\alpha \stackrel{\text{def}}{=} \Pr[i^* = i \mid r_i = y \wedge i^* \geq i] = \frac{\Pr[i^* = i \wedge r_i = y \mid i^* \geq i]}{\Pr[r_i = y \mid i^* \geq i]}$$

$$= \frac{\Pr[i^* = i \mid i^* \geq i] \cdot \Pr[r_i = y \mid i^* = i]}{\Pr[i^* = i \mid i^* \geq i] \cdot \Pr[r_i = y \mid i^* = i] + \Pr[i^* > i \mid i^* \geq i] \cdot \Pr[r_i = y \mid i^* > i]}$$

$$= \frac{p \cdot \Pr[r_i = y \mid i^* = i]}{p \cdot \Pr[r_i = y \mid i^* = i] + (1 - p) \cdot \Pr[r_i = y \mid i^* > i]}.$$

Let $q \stackrel{\text{def}}{=} \min_{x_0,y} \{ \Pr[W_0(x_0) = y] \}$, where the minimum is taken over all inputs x_0 for P_0 and all y in the range of f. Since W_0 has full support, we have $q > 0$. Thus,

$$
\begin{aligned}
\alpha &= \frac{p \cdot \Pr[r_i = y \mid i^* = i]}{p \cdot \Pr[r_i = y \mid i^* = i] + (1 - p) \cdot \Pr[r_i = y \mid i^* > i]} \\
&\leq \frac{p}{p + (1 - p) \cdot q} \\
&= \frac{p}{p \cdot (1 - q) + q} \leq \frac{p}{q},
\end{aligned}
$$

and by setting $p < q \cdot (a_0 - u_0^*)/(b_0 - u_0^*)$ we ensure that (1) holds.

Assuming p is set as just discussed, the above analysis shows that in any iteration $i < n$ and for any value $r_i = y$ received by P_0 in that iteration, P_0 has no incentive to abort. (In fact, P_0 has strict incentive *not* to abort.) The only remaining case to analyze is when $i = n$. In this case it would indeed be advantageous for P_0 to abort when $i^* \geq i$; however, this occurs with only negligible probability and so does not impact the fact that we have a computational Nash equilibrium (which is insensitive to negligible changes in the utility). □

Although security notions other than fairness are not the focus of our work, we note that the protocol Π presented in the proof of the previous theorem is private in addition to being rationally fair. That is, the parties learn the function output only, but nothing else regarding the other party's input. We omit formal definitions and the straightforward proof.

3.2   The Byzantine Setting

We next consider the Byzantine setting, where in the ideal world a deviating party can change the input it sends to the trusted third party (or may choose to abort, as before), and in the real world a deviating party may behave arbitrarily.

The protocol and proof of fairness in the Byzantine setting are similar to those of the fail-stop setting, however we must modify our protocol to ensure that it will work in the Byzantine setting. In particular, we require ShareGen to now apply a message-authentication code (MAC) to each s_i^b and t_i^b value so that parties can detect if these values have been modified. The remaining issue to deal with is the effect of changing inputs; however, we show that if incentive compatibility holds — so parties have disincentive to change their inputs in the ideal world — then parties have no incentive to change their inputs in the real world either.

Theorem 2. *Fix a function f, a distribution D, and utilities for the parties. If these are incentive compatible in the Byzantine setting, then (assuming the existence of general secure two-party computation for malicious adversaries) there exists a protocol Π computing f such that Π is a rational fair protocol (with respect to the same distribution and utilities) in the Byzantine setting.*

Functionality ShareGen

Inputs: ShareGen takes as input a value x_0 from P_0 and a value x_1 from P_1.
If either input is invalid, then ShareGen simply outputs $\perp$ to both parties.
Computation: Proceed as follows:
1. Choose i^* according to a geometric distribution with parameter p.
2. Choose MAC keys $k^0, k^1 \leftarrow \{0,1\}^n$.
3. Set the values of r_i^0 and r_i^1 for $i \in \{1, \ldots, n\}$ as follows:
   - If $i < i^*$, choose $r_i^0 \leftarrow W_0(x_0)$ and $r_i^1 \leftarrow W_1(x_1)$.
   - If $i \geq i^*$, set $r_i^0 = f_0(x_0, x_1)$ and $r_i^1 = f_1(x_0, x_1)$.
4. For each r_i^b, choose two values s_i^b and t_i^b as random secret shares of r_i^b.
   (I.e., s_i^b is random and $s_i^b \oplus t_i^b = r_i^b$.)
5. For $i = 1, \ldots, n$, set $\text{tag}_i^1 \leftarrow \text{Mac}_{k^1}(i\|s_i^1)$ and $\text{tag}_i^0 \leftarrow \text{Mac}_{k^0}(i\|t_i^0)$.
Output: Send k^0, s_1^0, s_1^1, tag_1^1, $\ldots$, s_n^0, s_n^1, tag_n^1 to P_0, and k^1, t_1^0, tag_1^0, t_1^1, $\ldots$, t_n^0, tag_n^0, t_n^1 to P_1.

Fig. 3. Functionality ShareGen. The security parameter is n. This functionality is parameterized by a real number $p > 0$.

Proof. By definition of incentive compatibility, there exist distributions W_0, W_1 (that can depend on x_0 and x_1, respectively) for which the strategy profile $\big((\text{cooperate}, W_0), (\text{cooperate}, W_1)\big)$ is a Bayesian strict Nash equilibrium. By Lemma 1, we may assume that W_0 and W_1 both have full support. We define a protocol Π based on a functionality ShareGen (cf. Figures 3 and 4), where the latter is parameterized by a real number $p > 0$. These are largely identical to the protocols used in the proof of Theorem 1, with the exception that the secret shares exchanged by the parties are authenticated by a message-authentication code (MAC) as part of the computation of ShareGen, and the resulting tags are verified by the parties (as part of Π). For our proof, we assume the MAC being used is an information-theoretically secure, n-time MAC; a computationally secure MAC would also be fine.

Since p is a constant (independent of n), it is again easy to check that correctness holds with all but negligible probability. As in the proof of Theorem 1, in our analysis we assume that $i^* \leq n$ always and this does not affect our results.

The proof that Π is rationally fair in the Byzantine setting is similar to the proof of Theorem 1, and we assume familiarity with that proof here. Once again, we analyze Π in a hybrid world where there is a trusted entity computing ShareGen on behalf of the parties. We also ignore the possibility of a MAC forgery, and treat a party who sends a different share/tag from the one it received from ShareGen as if that party had simply aborted. This is justified by the fact that a successful forgery occurs with only negligible probability.

Our goal, as in the proof of Theorem 1, is to show that there exists a $p > 0$ for which Π (in the hybrid world described above, and ignoring the possibility of a MAC forgery) is a rational fair protocol. As in the preceding proof, there are no profitable deviations for P_1. Note that here, P_1 may either abort early or change its input to ShareGen. The former does not help because P_0 always "gets the

Protocol Π

Stage one: Both players use their inputs to execute a secure protocol for computing ShareGen. This results in P_0 obtaining output $k^0, s_1^0, s_1^1, \mathsf{tag}_1^1, \ldots, s_n^0, s_n^1, \mathsf{tag}_n^1$, and P_1 obtaining output $k^1, t_1^0, \mathsf{tag}_1^0, t_1^1, \ldots, t_n^0, \mathsf{tag}_n^0, t_n^1$.

Stage two: There are n iterations. In each iteration $i \in \{1, \ldots, n\}$ do:
1. P_1 sends t_i^0 and tag_i^0 to P_0. If $\mathsf{Vrfy}_{k^0}(i\|t_i^0, \mathsf{tag}_i^0) = 1$, then P_0 computes $r_i^0 := t_i^0 \oplus s_i^0$. Otherwise, this is treated as if P_1 had aborted.
2. P_0 sends s_i^1 and tag_i^1 to P_1. If $\mathsf{Vrfy}_{k^1}(i\|s_i^1, \mathsf{tag}_i^1) = 1$, then P_1 computes $r_i^1 := t_i^1 \oplus s_i^1$. Otherwise, this is treated as if P_0 had aborted.

Output: Players determine their outputs as follows:
- If P_{1-i} aborts before P_i has computed any r_i value, then P_i chooses its output according to $W_i(x_i)$.
- If P_{1-i} aborts at any other point, or the protocol completes successfully, then P_i outputs the last r_i value it received.

Fig. 4. Formal definition of our protocol

output first" in every iteration; incentive compatibility in the Byzantine setting implies that the latter — whether in combination with aborting early or not — cannot help, either.

We are thus left with analyzing deviations by P_0. We again introduce two modifications that can only increase P_0's utility. First, in each iteration i we tell P_0 whether $i^* < i$. It follows immediately from incentive compatibility that P_0 cannot increase its utility by aborting when $i^* < i$ (regardless of what input it sends to ShareGen), and so we assume that P_0 never does so. Second, if P_0 ever aborts the protocol in some iteration i, then we tell P_0 whether $i^* = i$ before P_0 generates its output. (P_0 may not, however, change its decision to abort.)

There are two cases to analyze: either P_0 sends its actual input x_0 to ShareGen, or P_0 sends a different input $x_0' \neq \perp$. In the former case, the analysis is exactly the same as in the proof of Theorem 1, and one can show that p can be set such that P_0 has no incentive to abort the protocol at any point. It remains to show that, in the second case, P_0 cannot increase its expected utility beyond a_0, the utility it obtains by running the protocol honestly using its actual input x_0.

If P_0 substitutes its input and then runs the protocol to the end, then (by incentive compatibility) P_0's expected utility is strictly less than a_0. Can P_0 do better by aborting early? Fix some input x_0 for P_0, let $x_0' \neq \perp$ be the input that P_0 sends to the trusted entity computing ShareGen, and consider some iteration $i < n$. (The case of $i = n$ is handled as in the proof of Theorem 1.) Say P_0 has just learned that $r_i = y$ (for some y in the range of f) and is told that $i^* \geq i$. If P_0 aborts, then with some probability α it learns that $i^* = i$; in that case, P_0 may possibly get utility b_0. Otherwise, with probability $1 - \alpha$ it learns that $i^* > i$. In this latter case, P_0 has no information beyond what it could compute from its input, and P_1 will output a value distributed according to $W_1(x_1)$; hence, incentive compatibility implies that the maximum expected utility of P_0 is $u_0^* < a_0$. (This u_0^* is the same as in the proof of Theorem 1.) That is, the expected utility of aborting is at most $\alpha \cdot b_0 + (1 - \alpha) \cdot u_0^*$. If

$$\alpha < \frac{a_0 - u_0^*}{b_0 - u_0^*}$$

then $\alpha \cdot b_0 + (1 - \alpha) \cdot u_0^* < a_0$, implying that P_0 does not gain anything beyond what it could have obtained by running the protocol honestly with its actual input.[4] A calculation exactly as in the proof of Theorem 1 shows that p can be set in such a way that this condition holds.

4 Conclusions and Future Work

Given the stark impossibility results for fairness in a purely *malicious* context [6], it is natural to try to understand whether, or to what extent, fairness is achievable in a *rational* setting. Recent work of Asharov et al. [2] seemed to give a pessimistic answer to this question, as they show a specific case where rational fairness *cannot* be achieved (if correctness better than $1/2$ is desired). Our work, in contrast, shows broad feasibility results for rational fairness: roughly, we show that whenever computing the function is a *strict* Nash equilibrium in the ideal world, then it is possible to construct a rational fair protocol computing the function in the real world.

Within the broader context of research at the intersection of game theory and cryptography, our result can be interpreted in two ways:

- Given a "fairness game" defined in an ideal world where there is a trusted entity (i.e., a "mediator") computing some function on behalf of the parties, a natural question to ask is when a game-theoretic equilibrium in the ideal world can be implemented via a real-world protocol. We do not provide a complete answer to this question, but we do show a partial characterization: roughly, whenever there is a *strict* Nash equilibrium in the ideal world, there is a protocol that induces a computational Nash equilibrium in the real world.
- We show a new setting in which cryptographic impossibility results can be circumvented by assuming rational behavior. Viewed in this light, our results can be seen as a generalization of work on rational secret sharing.

Our work suggests several interesting directions for future research. First, it would be interesting to prove a converse of our result. Fix some f, a distribution on the inputs, and utility functions for the parties. We conjecture that if there exists a rational fair protocol Π for computing f (with respect to this distribution and utilities), then incentive compatibility holds.[5]

---

[4] It is possible that, conditioned on the fact that P_0 sent input $x_0' \neq x_0$ to ShareGen, party P_0 can obtain better expected utility by aborting early than by running the protocol to the end. What we claim here, though, is that P_0 will never obtain better expected utility than it would have obtained by *using its actual input x_0* and then running the protocol to the end.

[5] As noted earlier, for our positive results it is sufficient for the ideal-world equilibrium to be strict for only one of the parties. With regard to this conjecture, then, the definition of incentive compatibility should be modified appropriately.

It will also be interesting to explore stronger game-theoretic solution concepts in the real world. We construct real-world protocols that induce a computational Nash equilibrium, but one could also aim to construct protocols satisfying some of the stronger equilibrium notions proposed, e.g., in [17,22,23,11].

Another natural extension is the multi-party case. Our protocols can easily be extended to handle *single-player deviations* in that setting; we leave the general case (where parties may collude) as an interesting open question.

Finally, one could consider even more complex settings of the players' utilities, e.g., where the utilities depend on the true output and the actual output of the parties and not just on whether the outputs are correct or incorrect. This would model situations where being "closer" to the right answer is better, or where some answers are more important to get right than others.

References

1. Abraham, I., Dolev, D., Gonen, R., Halpern, J.: Distributed computing meets game theory: robust mechanisms for rational secret sharing and multiparty computation. In: 25th Annual ACM Symposium on Principles of Distributed Computing (PODC), pp. 53–62. ACM Press (2006)
2. Asharov, G., Canetti, R., Hazay, C.: Towards a Game Theoretic View of Secure Computation. In: Paterson, K.G. (ed.) EUROCRYPT 2011. LNCS, vol. 6632, pp. 426–445. Springer, Heidelberg (2011), Full version available at, http://eprint.iacr.org/2011/137
3. Asharov, G., Lindell, Y.: Utility dependence in correct and fair rational secret sharing. Journal of Cryptology 24(1), 157–202 (2011)
4. Barany, I.: Fair distribution protocols, or how the players replace fortune. Mathematics of Operations Research 17, 327–340 (1992)
5. Canetti, R.: Security and composition of multiparty cryptographic protocols. Journal of Cryptology 13(1), 143–202 (2000)
6. Cleve, R.: Limits on the security of coin flips when half the processors are faulty. In: 18th Annual ACM Symposium on Theory of Computing (STOC), pp. 364–369. ACM Press (1986)
7. Crawford, V., Sobel, J.: Strategic information transmission. Econometrica 50, 1431–1451 (1982)
8. Dodis, Y., Halevi, S., Rabin, T.: A Cryptographic Solution to a Game Theoretic Problem. In: Bellare, M. (ed.) CRYPTO 2000. LNCS, vol. 1880, pp. 112–130. Springer, Heidelberg (2000)
9. Dodis, Y., Rabin, T.: Cryptography and game theory. In: Nisan, N., Roughgarden, T., Tardos, E., Vazirani, V. (eds.) Algorithmic Game Theory, pp. 181–207. Cambridge University Press (2007)
10. Forges, F.: Universal mechanisms. Econometrica 58, 1342–1364 (1990)
11. Fuchsbauer, G., Katz, J., Naccache, D.: Efficient Rational Secret Sharing in Standard Communication Networks. In: Micciancio, D. (ed.) TCC 2010. LNCS, vol. 5978, pp. 419–436. Springer, Heidelberg (2010)
12. Gordon, S.D., Hazay, C., Katz, J., Lindell, Y.: Complete fairness in secure two-party computation. J. ACM 58(6) (2011)
13. Gordon, S.D., Katz, J.: Rational Secret Sharing, Revisited. In: De Prisco, R., Yung, M. (eds.) SCN 2006. LNCS, vol. 4116, pp. 229–241. Springer, Heidelberg (2006)

14. Gordon, S.D., Katz, J.: Partial fairness in secure two-party computation. Journal of Cryptology 25(1), 14–40 (2012)
15. Gradwohl, R.: Rationality in the Full-Information Model. In: Micciancio, D. (ed.) TCC 2010. LNCS, vol. 5978, pp. 401–418. Springer, Heidelberg (2010)
16. Gradwohl, R., Livne, N., Rosen, A.: Sequential rationality in cryptographic protocols. In: 51st FOCS Annual Symposium on Foundations of Computer Science (FOCS), pp. 623–632. IEEE (2010)
17. Halpern, J., Teague, V.: Rational secret sharing and multiparty computation. In: 36th Annual ACM Symposium on Theory of Computing (STOC), pp. 623–632. ACM Press (2004)
18. Izmalkov, S., Lepinski, M., Micali, S.: Verifiably Secure Devices. In: Canetti, R. (ed.) TCC 2008. LNCS, vol. 4948, pp. 273–301. Springer, Heidelberg (2008)
19. Izmalkov, S., Lepinski, M., Micali, S.: Perfect implementation. Games and Economic Behavior 71(1), 121–140 (2011), http://hdl.handle.net/1721.1/50634
20. Izmalkov, S., Micali, S., Lepinski, M.: Rational secure computation and ideal mechanism design. In: 46th FOCSAnnual Symposium on Foundations of Computer Science (FOCS), pp. 585–595. IEEE (2005), Full version available at, http://dspace.mit.edu/handle/1721.1/38208
21. Katz, J.: Bridging Game Theory and Cryptography: Recent Results and Future Directions. In: Canetti, R. (ed.) TCC 2008. LNCS, vol. 4948, pp. 251–272. Springer, Heidelberg (2008)
22. Kol, G., Naor, M.: Cryptography and Game Theory: Designing Protocols for Exchanging Information. In: Canetti, R. (ed.) TCC 2008. LNCS, vol. 4948, pp. 320–339. Springer, Heidelberg (2008)
23. Kol, G., Naor, M.: Games for exchanging information. In: 40th Annual ACM Symposium on Theory of Computing (STOC), pp. 423–432. ACM Press (2008)
24. Lepinski, M., Micali, S., Peikert, C., Shelat, A.: Completely fair SFE and coalition-safe cheap talk. In: 23rd ACM PODCAnnual ACM Symposium on Principles of Distributed Computing (PODC), pp. 1–10. ACM Press (2004)
25. Lepinski, M., Micali, S., Shelat, A.: Collusion-free protocols. In: 37th Annual ACM Symposium on Theory of Computing (STOC), pp. 543–552. ACM Press (2005)
26. Lysyanskaya, A., Triandopoulos, N.: Rationality and Adversarial Behavior in Multi-party Computation. In: Dwork, C. (ed.) CRYPTO 2006. LNCS, vol. 4117, pp. 180–197. Springer, Heidelberg (2006)
27. Micali, S., Shelat, A.: Purely Rational Secret Sharing (Extended Abstract). In: Reingold, O. (ed.) TCC 2009. LNCS, vol. 5444, pp. 54–71. Springer, Heidelberg (2009)
28. Moran, T., Naor, M., Segev, G.: An Optimally Fair Coin Toss. In: Reingold, O. (ed.) TCC 2009. LNCS, vol. 5444, pp. 1–18. Springer, Heidelberg (2009)
29. Ong, S.J., Parkes, D.C., Rosen, A., Vadhan, S.: Fairness with an Honest Minority and a Rational Majority. In: Reingold, O. (ed.) TCC 2009. LNCS, vol. 5444, pp. 36–53. Springer, Heidelberg (2009)

A   Incentive Compatibility in the Byzantine Setting

Here, for the sake of illustration, we present a function f, input distributions, and a set of utilities that are incentive compatible in the Byzantine setting. Let f be the equality predicate over domain $\{0,1,2\}$, and assume the parties' inputs are independently and uniformly distributed over this set. The utilities of the parties are symmetric, and correspond to each player receiving utility 3 from getting the correct answer and (independently) utility 1 if the other player outputs the wrong answer. I.e.,

$$a_0 = a_1 = 3; \quad b_0 = b_1 = 4; \quad c_0 = c_1 = 0; \quad d_0 = d_1 = 1.$$

Let W_0, W_1 be point distributions on the output '0'. (These do not have full support, though Lemma 1 shows that they can be modified so that they do.) We want to show that $\big((\text{cooperate}, W_0), (\text{cooperate}, W_1)\big)$ is a strict Nash equilibrium in the Byzantine setting.

By playing honestly, each party gets utility 3. Consider deviations by P_0 (the case of a deviating P_1 is analogous), and assume without loss of generality that the true input x_0 is equal to 0.

P_0 could send $\perp$ to the trusted third party. In that case, P_0 does best by outputting '0'; its expected utility (recalling that P_1 also outputs '0') is then

$$3 \cdot \Pr[x_1 \neq 0] + 1 \cdot \Pr[x_1 = 0] = \frac{2}{3} * 3 + \frac{1}{3} * 1 = 7/3 < 3.$$

Alternately, P_0 may change its input; without loss of generality, say it uses input $x_0' = 1$. Here there are two sub-cases, depending on the output returned by the trusted party. If this output is '0', then $x_1 \neq 1$ but P_1 will output '0'. Regardless of what P_0 outputs, it is correct (and gets utility 3) with probability 1/2 and P_1 is incorrect (and so P_0 gets utility 1) with probability 1/2; the expected utility of P_0 is thus 2. If the trusted party returns '1' then P_1 outputs '1' (recall that P_1 outputs what it is given by the trusted party) which is wrong, whereas P_0 learns P_1's input and so can output the correct answer. Thus, in this case, P_0 obtains utility 4. Overall, then, the expected utility P_0 obtains by changing its input is

$$\frac{2}{3} \cdot 2 + \frac{1}{3} \cdot 4 = 8/3 < 3.$$

We conclude that the given scenario is a (Bayesian) strict Nash equilibrium.

Concurrently Secure Computation
in Constant Rounds

Sanjam Garg[1], Vipul Goyal[2], Abhishek Jain[1], and Amit Sahai[1]

[1] UCLA
[2] MSR India

Abstract. We study the problem of constructing concurrently secure computation protocols in the plain model, where no trust is required in any party or setup. While the well established UC framework for concurrent security is impossible to achieve in this setting, meaningful relaxed notions of concurrent security have been achieved.

The main contribution of our work is a new technique useful for designing protocols in the concurrent setting (in the plain model). The core of our technique is a new rewinding-based extraction procedure which only requires the protocol to have a constant number of rounds. We show two main applications of our technique.

We obtain the *first* concurrently secure computation protocol in the plain model with super-polynomial simulation (SPS) security that uses only a constant number of rounds and requires only standard assumptions. In contrast, the only previously known result (Canetti et al., FOCS'10) achieving SPS security based on standard assumptions requires polynomial number of rounds. Our second contribution is a new definition of input indistinguishable computation (IIC) and a constant round protocols satisfying that definition. Our definition of input indistinguishable computation is a simplification and strengthening of the definition of Micali et al. (FOCS'06) in various directions. Most notably, our definition provides meaningful security guarantees even for randomized functionalities.

1   Introduction

The notion of *secure computation* is central to cryptography. Introduced in the seminal works of [49,19], secure multi-party computation allows a group of (mutually) distrustful parties $P_1, \ldots, P_n$, with private inputs $x_1, \ldots, x_n$, to jointly compute any functionality f in such a manner that the honest parties obtain correct outputs and no group of malicious parties learn anything beyond their inputs and prescribed outputs. The original definition of secure computation, although very useful and fundamental to cryptography, is only relevant to the *stand-alone* setting where security holds only if a single protocol session is executed in isolation. As it has become increasingly evident over the last two decades, stand-alone security does not suffice in real-world scenarios where several protocol sessions may be executed *concurrently* – a typical example being protocols executed over modern networked environments such as the Internet.

D. Pointcheval and T. Johansson (Eds.): EUROCRYPT 2012, LNCS 7237, pp. 99–116, 2012.

Concurrent Security. Towards that end, the last decade has seen a push towards obtaining protocols that have strong concurrent *composability* properties. For example, we could require concurrent self-composability: the protocol should remain secure even when there are multiple copies executing concurrently. The framework of *universal composability* (UC) was introduced by Canetti [10] to capture the more general security requirements when a protocol may be executed concurrently with not only several copies of itself but also with other protocols in an arbitrary manner.

Unfortunately, strong impossibility results have been shown ruling out the existence of secure protocols in the concurrent setting. UC secure protocols for most functionalities of interest have been ruled out in [11,8]. These results were further generalized [35] to rule out the existence of protocols providing even concurrent self-composability. Protocols in even less demanding settings (where all honest party inputs are fixed in advance) were ruled out in [4]. All these impossibility results refer to the "plain model," where parties do not trust any external entity or setup. We stress that, in fact, some of these impossibility results provide an *explicit attack* in the concurrent setting using which the adversary may even fully recover the input of an honest party (see, e.g., the chosen protocol attack in [4]). Hence, designing secure protocols in the concurrent setting is a question of great theoretical as well as practical interest. Unfortunately, the only known positive results for concurrent composition in the plain model are for the zero-knowledge functionality [46,30,44].

To overcome these impossibility results, UC secure protocols were proposed based on various "trusted setup assumptions" such as a common random string that is published by a *trusted party* [11,9,1,14,28,15]. Nevertheless, a driving goal in cryptographic research is to eliminate the need to trust other parties. In the context of UC secure protocols based on setup assumptions, while there has been some recent effort [26,24,18] towards reducing the extent of trust in any single party (or entity), obviously this approach cannot completely eliminate trust in other parties (since that is the very premise of a trusted setup assumption). Ideally, we would like to obtain concurrently-secure protocols in the *plain model* (which is the main focus of this paper).

Relaxing the Security Notion. To address the problem of concurrent security for secure computation in the plain model, a few candidate definitions have been proposed, including input-indistinguishable security [37] and super-polynomial simulation [40,45,5]. We discuss the each of these notions (and the state of the art) separately.

Super-Polynomial Simulation. The notion of security with *super-polynomial simulators* (SPS) is one where the adversary in the ideal world is allowed to run in (fixed) super-polynomial time. Very informally, SPS security guarantees that any polynomial-time attack in the real execution can also be mounted in the ideal world execution, albeit in super-polynomial time. This is directly applicable and meaningful in settings where ideal world security is guaranteed statistically or information-theoretically (which would be the case in most "end-user" functionalities that have been considered, from privacy-preserving data mining to electronic voting). SPS security for concurrently composable zero knowledge proofs was first studied by [40], and SPS security for concurrently composable secure computation protocols was first studied by [45,5]. The SPS definition guarantees security with respect to concurrent self-composition of the secure computation protocol being studied, and guarantees security with respect to general concurrent composition with arbitrary other protocols in the context of super-polynomial adversaries.

In recent years, the design of secure computation protocols in the plain model with SPS security has been the subject of several works [45,5,34,13]. Very recently, Canetti, Lin, and Pass [13] obtained the first secure computation protocol that achieves SPS security based on *standard assumptions*[1].

Unfortunately, however, the improvement in terms of assumptions comes at the cost of the round complexity of the protocol. Specifically, the protocol of [13] incurs *polynomial-round complexity*. The latency of sending messages back and forth has been shown to often be the dominating factor in the running time of cryptographic protocols [36,6]. Indeed, round complexity has been the subject of a great deal of research in cryptography. For example, in the context of concurrent zero knowledge (ZK) proofs, round complexity was improved in a sequence of works [46,30,44] from polynomial to slightly super-logarithmic (that nearly matches the lower bound w.r.t. black-box simulation [12]). The round complexity of non-malleable commitments in the stand-alone and concurrent settings has also been studied in several works [17,2,43,42,31,48,22,32], improving the round complexity from logarithmic rounds to constant rounds under minimal assumptions. We observe that for the setting of concurrently secure computation protocols with SPS security, the situation is much worse since the only known protocol that achieves SPS security based on standard assumptions incurs polynomial-round complexity [13].

Input-Indistinguishable Computation. The notion of input indistinguishable computation [37] is a relaxation of the standard notion of secure computation akin to how witness indistinguishability is a relaxation of the notion of zero-knowledge. In input indistinguishable computation (IIC), very roughly, given the output vector (consisting of outputs in all concurrent sessions), consider any two honest party input vectors x_1 and x_2 "consistent" with the output vector. The security guarantee requires the adversary to have only a negligible advantage in distinguishing which of these is the actual input vector. While SPS security definition is based on the ideal/real world paradigm, the security definition of IIC is a game based one where various required properties (such as input independence) are formalized separately. In IIC, no guarantees are provided for any two input vectors which don't lead to the identical output (e.g., the functionality may be randomized; furthermore, the outputs may only be computationally indistinguishable as opposed to coming from identical or statistically close distributions).

---

[1] In fact, the work of [13], together with [45,5], considers the stronger "angel-based security model" of [45]. In this work, we focus only on SPS security.

1.1   Our Contributions

The main contribution of our work is a new technique useful for designing protocols in the concurrent setting (in the plain model). The core of our technique
is a new rewinding-based extraction procedure which only requires the protocol
to have a constant number of rounds. Overall, our technique allows us to improve upon the previous works in terms of round complexity, the security notion
being achieved as well the assumptions. We show two main applications of our
technique in this work.

Super Polynomial Simulation. We construct the first *constant-round* concurrently composable secure computation protocol that achieves SPS security based
on only *standard assumptions*. In addition, our construction only uses black-box
simulation techniques.

In contrast to prior works where several powerful tools were employed to
obtain positive results, e.g., CCA-secure commitments [13], our new proof technique allows us to only use relatively less powerful primitives, such as standard
non-malleable commitments. Our positive result relies on the nearly minimal assumptions that constant-round (semi-honest) oblivious transfer (OT) exists and
collision-resistant hash functions (CRHFs) exist.[2]

Input Indistinguishable Computation. We introduce a new definition of
input indistinguishable computation and prove that, in fact, the same protocol
(as for constant round super-polynomial simulation) satisfies this notion as well.
Our definition of input indistinguishable computation is a simplification and
strengthening of the definition in [37] in various directions. In particular, our
definition provides meaningful security guarantees even for randomized functionalities. Furthermore, the security guarantees hold even when the output distributions resulting from the two honest party inputs (among which the adversary
is trying to distinguish) are computationally indistinguishable (as opposed to
coming from identical distributions)[3]. We follow the real/ideal world paradigm
for formalizing the security guarantees which leads to an arguably simpler definition. Additionally, we show that our definition *implies* the definition of [37].

The essence of our new definition can be understood as follows. Consider a real
world adversary. For any two input vectors x_1 and x_2, we require the existence
of a (PPT) ideal world simulator such that the output distribution in the ideal
and the real world are indistinguishable. Hence, the only relaxation compared to
the standard ideal/real world definition is now the ideal world simulator could be
different for different pairs (x_1, x_2). The key intuition behind such a guarantee is
that for any two honest party input vectors (x_1, x_2) leading to the same output
vector (on the input vector chosen by the adversary), the simulator in the ideal
world has no advantage in distinguishing which of the two was used. This implies

---

[2] We believe that our assumption of CRHFs can be removed by employing techniques
from the recent work of [33], leaving only the minimal assumption that constant-
round OT exists. We leave this for the full version of this paper.

[3] This is comparable to the relationship between witness indistinguishability and
strong witness indistinguishability.

that even to the real world adversary should only have a negligible distinguishing advantage. We stress that in our definition, this holds even if the functionality is randomized and the outputs are computationally indistinguishable (as opposed to being identical). In addition, as opposed to [37], our ideal world simulator is required to extract the input being used by the adversary (in PPT) and send it to the trusted party. This provides a form of "input-awareness" guarantee.

While the above simple definition already provides meaningful security guarantees, the guarantees are unsatisfactory if there exists a "splitting input" which the ideal world simulator uses even when the real world adversary is such that it does not use a splitting input. A more detailed discussion of such issues can be found in [37]. Towards that end, we propose an extension of our definition and finally show that it implies the definition in [37]. To see an example of a functionality for which our definition provides meaningful security guarantees which neither the definition in [37] nor the SPS definition provide, please refer to the full version.

1.2 The Main Technique

A ubiquitous technique for simulation-based proofs in cryptography is that of *rewinding* the adversary. In the concurrent setting (which is the setting we consider in this paper), where an adversary can interleave messages from different protocols in any arbitrary manner, rewinding an adversary (to correctly simulate each session) is often problematic. The rewinding becomes recursive because of which the protocols typically requires a large number of rounds (in a single protocol). For example, in the context of concurrent zero knowledge, the best known result [44] requires super-logarithmic round complexity, which nearly matches the lower bound w.r.t. black-box simulation [12].

To deal with the problem of concurrent rewinding, we develop a novel proof technique using which we can limit the depth of such recursion to at most 2. Such a significant relaxation of the properties we need from our rewinding technique allows us to obtain our result. In the following discussion, we give a more detailed intuition behind our techniques, where we assume somewhat greater familiarity with recent work in this area. The discussion is primarily for obtaining constant round providing with SPS security although similar intuition applies for IIC as well.

We first note that all prior works on obtaining secure computation protocols with SPS security crucially use the super-polynomial time simulator to "break" some cryptographic scheme and extract some "secret information". Then, to avoid any complexity-leveraging type technique (which would lead to non-standard assumptions), and yet argue security, the technique used in [13] was to replace the super-polynomial time simulator with a polynomial-time rewinding "hybrid experiment" via a hybrid argument in the security proof. Indeed, this is why their protocol incurs large round complexity (so as to facilitate concurrent-rewinding). We also make use of rewinding, but crucially, in a weaker way. The main insights behind our rewinding technique are explained as follows:

– We first note that (like other works) we will restrict our usage of rewinding only to the creation of "look-ahead threads". Very roughly, this means that a rewinding simulator never changes its actions on the "main thread" of execution; and as such, the rewinding is employed only to extract some information from the adversary. Here, we again stress that our final simulator does not perform any rewinding, and that we only perform rewindings in hybrid experiments to bridge the gap between the real and ideal world executions.

– Now that we use rewindings only to extract some information from the adversary, and only in hybrid experiments, we make the critical observation that, in fact, we can make use of the secret inputs of the honest parties in the look-ahead threads. Indeed, in *all* our intermediate hybrid experiments, we perform rewindings to create look-ahead threads where we make "judicious" use of the honest party's inputs. In this manner, we eventually end up with a rewinding (hybrid) simulator that simulates the *main thread* without the honest party's inputs, but still uses them in the look-ahead threads (in a manner that guarantees extraction). This is our main conceptual deviation from prior work, where, to the best of our knowledge, honest party's inputs were only used in *some* intermediary hybrids, with the main goal being to eventually remove their usage even from the look-ahead threads. We show that this is in fact unnecessary, since our final simulator does not perform any rewindings, but instead runs in super-polynomial time to extract the same information that was being earlier extracted via rewinding in the hybrid experiments. We only need to argue that the main thread output by the rewinding (hybrid) experiment and the main thread output by the final simulator be indistinguishable. Indeed, we are able to argue that there is only a small statistical distance between our final simulator (that corresponds to the ideal execution) and the previous rewinding-based hybrid experiment. This statistical distance corresponds to the probability that the rewinding-based extraction is unsuccessful, since the SPS extraction is always successful.

– We further note that since we use the honest party's inputs in the look-ahead threads, we can bypass complex *recursive* rewinding schedules used in previous works and simply use "local rewindings" that only require constant rounds (in fact, only "one slot").

– Finally, we observe that since we perform rewindings only in hybrid experiments, we do not need the rewinding to succeed with probability negligibly close to 1, as is needed for concurrent ZK. Instead, we only require rewinding to succeed with probability $1 - \epsilon$, where ϵ is related to the success probability of the distinguisher that is assumed to exist for the sake of contradiction. This observation, yet again, allows us to use a simpler rewinding strategy.

– Our overall proof strategy only makes use of relatively well understood primitives like standard non-malleable commitments. This is a departure from [13] which introduces a new primitive called CCA-secure commitment schemes.

At this point, an informed reader may question the feasibility of a "sound implementation" of the above approach. Indeed, a-priori it is not immediately clear whether it is even possible for the simulator to "cheat" on the main thread, yet behave honestly in look-ahead threads *at the same time*. In a bit more detail, recall that any given look-ahead thread shares a prefix with the main thread of execution. Now consider any session i on a look-ahead thread. Note that since some part of session i may already be executed on the *shared prefix*, it is not clear how the simulator can continue simulating session i on the look-ahead thread *without ever performing any recursive rewindings* if it was already cheating in session i on the shared prefix.

We address the above issues by a careful protocol design that guarantees that a rewinding simulator can always extract some "trapdoor" information *before* it "commits" to cheating in any session. As a result, during the simulation, whenever a look-ahead thread is forked at any point from the main thread, the simulator can either always continue cheating, or simply behave honestly (without any conflict with the main thread) in any session.

In our overall proof, SPS is used only at the very last step to stop the look-ahead threads (which required knowledge of honest party inputs to execute). A modification of this step is required to prove that the protocols satisfies our new notion of IIC as well. Instead of stopping the look-ahead threads (which used honest party inputs), we will now run "two-sets" of look-ahead threads one for each input vector given to the ideal world simulator. Since of these two is the real honest party input vector, at least one of the sets of look-ahead threads is guaranteed to be successful.

1.3   Other Related Work

Here we discuss some additional prior work related to the work in this paper. We note that while the focus of this work is on the notions of SPS security and IIC as means to obtain concurrently-secure protocols in the plain model, some recent works have investigated alternative security models for the same. Very recently, [25,23] considered a model where the ideal world adversary is allowed to make additional queries (as compared to a single query, as per the standard definition) to the ideal functionality per session. While our protocol bears much similarity to the construction in [23], our rewinding technique (and the overall proof) is quite different.

Independent of our work, a constant round protocol providing SPS security was recently obtained by Lin, Pass and Venkitasubramaniam [41]. Their technique are quite different from ours and make use of a non-uniform argument. An advantage of our work over that of Lin et. al. is that we provide a *uniform* reduction to the underlying hardness assumptions. Hence, our construction guarantees security against uniform adversaries assuming that the underlying primitives are only secure against uniform adversaries. Lin et. al. crucially require the underlying primitives to be secure against non-uniform adversaries to provide any meaningful security guarantees.

We note that their techniques seem not to apply to get a construction satisfying our IIC security notion. Since the IIC simulator has to extract the adversarial inputs in PPT, a rewinding technique in the concurrent setting is crucially required.

2   Our Definitions

2.1   UC Security and SPS

In this section we briefly review UC security. For full details see [10]. Following [21,20], a protocol is represented as an interactive Turing machine (ITM), which represents the program to be run within each participant.

Security of Protocols. Protocols that securely carry out a given task (or, protocol problem) are defined in three steps, as follows. First, the process of executing a protocol in an adversarial environment is formalized. Next, an "ideal process" for carrying out the task at hand is formalized. In the ideal process the parties do not communicate with each other. Instead they have access to an "ideal functionality," which is essentially an incorruptible "trusted party" that is programmed to capture the desired functionality of the task at hand. A protocol is said to securely realize an ideal functionality if the process of running the protocol amounts to "emulating" the ideal process for that ideal functionality.

Securely Realizing an Ideal Functionality. We say that a protocol Π *emulates* protocol ϕ if for any adversary $\mathcal{A}$ there exists an adversary $\mathcal{S}$ such that no environment $\mathcal{Z}$, on any input, can tell with non-negligible probability whether it is interacting with $\mathcal{A}$ and parties running Π, or it is interacting with $\mathcal{S}$ and parties running ϕ. This means that, from the point of view of the environment, running protocol Π is 'just as good' as interacting with ϕ. We say that Π *securely realizes* an ideal functionality $\mathcal{F}$ if it emulates the ideal protocol $\Pi(\mathcal{F})$. More precise definitions follow. A distribution ensemble is called *binary* if it consists of distributions over $\{0, 1\}$.

Definition 1. *Let Π and ϕ be protocols. We say that Π UC-emulates ϕ if for any adversary $\mathcal{A}$ there exists an adversary $\mathcal{S}$ such that for any environment $\mathcal{Z}$ that obeys the rules of interaction for UC security we have $\mathrm{EXEC}_{\phi,\mathcal{S},\mathcal{Z}} \approx \mathrm{EXEC}_{\pi,\mathcal{A},\mathcal{Z}}$.*

Definition 2. *Let $\mathcal{F}$ be an ideal functionality and let Π be a protocol. We say that Π UC-realizes $\mathcal{F}$ if Π UC-emulates the ideal process $\Pi(\mathcal{F})$.*

UC Security with Super-Polynomial Simulation. We next provide a relaxed notion of UC security by giving the simulator access to super-poly computational resources. The universal composition theorem generalizes naturally to the case of UC-SPS, the details of which we skip.

Definition 3. *Let Π and ϕ be protocols. We say that Π UC-SPS-emulates ϕ if for any adversary $\mathcal{A}$ there exists a super-polynomial time adversary $\mathcal{S}$ such that for any environment $\mathcal{Z}$ that obeys the rules of interaction for UC security we have $\mathrm{EXEC}_{\phi,\mathcal{S},\mathcal{Z}} \approx \mathrm{EXEC}_{\pi,\mathcal{A},\mathcal{Z}}$.*

Definition 4. *Let $\mathcal{F}$ be an ideal functionality and let Π be a protocol. We say that Π UC-SPS-realizes $\mathcal{F}$ if Π UC-SPS-emulates the ideal process $\Pi(\mathcal{F})$.*

For simplicity of exposition, in the rest of this paper we assume authenticated communication; that is, the adversary may deliver only messages that were actually sent. (This is however not essential as shown previously [3].)

2.2   Input Indistinguishable Computation

Under our notion, very roughly, an adversaries' goal is to guess the input, among two pre-specified inputs, used by the honest party. We say that a protocol is *input indistinguishable* if an adversary can not guess the honest parties input in the protocol execution any better than what it could have done in the ideal scenario. We formalize this by saying that the adversary learns nothing more than the two pre-specified inputs (which it already knows) and the output it learns in the ideal world. This naturally implies that if the adversary can not guess the honest parties input in the ideal scenario then it can not do so in the protocol execution as well.

CONCURRENT EXECUTION IN THE IDEAL MODEL. In the ideal model, there is a trusted party $\mathcal{F}$ that computes the functionality f (described above) based on the inputs handed to it by the two parties – P_1, P_2 which are involved in $m = m(n)$ sessions (polynomial in the security parameter, n). An execution in the ideal model with an adversary that controls P_1 or P_2 proceeds as follows:

Inputs: The honest party and adversary each obtain a vector of m inputs each of length n; denote this vector by $\boldsymbol{w}$ (i.e., $\boldsymbol{w} = \boldsymbol{x}$ or $\boldsymbol{w} = \boldsymbol{y}$).

Honest Parties Send Inputs to Trusted Party: The honest party sends its entire input vector $\boldsymbol{w}$ to the trusted party $\mathcal{F}$.

Adversary Interacts with Trusted Party: For every $i = 1, \ldots, m$, the adversary can send (i, w_i') to the trusted party, for any $w_i' \in \{0,1\}^*$ of its choice. Upon sending this pair, it receives back its output based on w_i' and the input sent by the honest party. (That is, if P_1 is corrupted, then the adversary receives $f_1(w_i', y_i)$ and if P_2 is corrupted then it receives $f_2(x_i, w_i')$.) The adversary can send the (i, w_i') pairs in any order it wishes and can also send them *adaptively* (i.e., choosing inputs based on previous outputs). The only limitation is that for any i, *at most one pair* indexed by i can be sent to the trusted party.

Adversary Answers Honest Party: Having received all of its own outputs, the adversary specifies which outputs the honest party receives. That is, the adversary sends the trusted party a set $I \subseteq \{1, \ldots, m\}$. Then, the trusted party supplies the honest party with a vector $\boldsymbol{v}$ of length m such that for every $i \notin I$, $v_i = \bot$ and for every $i \in I$, v_i is the party's output from the ith execution. (That is, if P_1 is honest, then for every $i \in I$, $v_i = f_1(x_i, w_i')$ and if P_2 is honest, then $v_i = f_2(w_i', y_i)$.)

Outputs: The honest party always outputs the vector $\boldsymbol{v}$ that it obtained from the trusted party. The adversary may output an arbitrary (probabilistic polynomial-time computable) function of its initial-input and the messages obtained from the trusted party.

Let $\mathcal{S}$ be a non-uniform probabilistic polynomial-time ideal-model machine (representing the ideal-model adversary). Then, the ideal execution of f (on input vectors $(\boldsymbol{x}, \boldsymbol{y})$ of length m and auxiliary input z to $\mathcal{S}$) denoted by $\mathrm{IDEAL}_{\mathcal{F},\mathcal{S}}(\boldsymbol{x}, \boldsymbol{y}, z)$, is defined as the output pair of the honest party and $\mathcal{S}$ from the above ideal execution.

EXECUTION IN THE REAL MODEL. We next consider the real model in which a real two-party protocol is executed (and there exists no trusted third party). Let $m = m(n)$ be a polynomial, let f be as above and let Π be a two-party protocol for computing f. Furthermore, let $\mathcal{A}$ be a non-uniform probabilistic polynomial-time machine that controls either P_1 or P_2. Then, the real concurrent execution of Π (on input vectors $(\boldsymbol{x}, \boldsymbol{y})$ of length $m(n)$ and auxiliary input z to $\mathcal{A}$), denoted $\mathrm{REAL}_{\Pi,\mathcal{A}}(\boldsymbol{x}, \boldsymbol{y}, z)$, is defined as the output pair of the honest party and $\mathcal{A}$, resulting from $m(n)$ executions of the protocol interaction, where the honest party always inputs its ith input into the ith execution. The scheduling of all messages throughout the executions is controlled by the adversary. That is, the execution proceeds as follows. The adversary sends a message of the form (i, α) to the honest party. The honest party then adds α to the view of its ith execution of Π and replies according to the instructions of Π and this view. The adversary continues by sending another message (j, β), and so on. Adversary can schedule these the messages in any way it likes. (Formally, view the schedule as the ordered series of messages of the form $(index, message)$ that are sent by the adversary.)

Definition 5 (Input Indistinguishable Computation (IIC)). *Let $\mathcal{F}$ and Π be the ideal trusted parted and the protocol realizing functionality f, as defined above. Protocol Π is said to* input indistinguishably compute *(or, IIC) f for P_1 under concurrent composition if for every polynomial $m = m(n)$, for every inputs $\boldsymbol{x}_0, \boldsymbol{x}_1 \in (\{0,1\}^n)^m$ of the honest party P_1, for every real-model non-uniform probabilistic polynomial-time adversary $\mathcal{A}$ controlling party P_2, there exists an ideal-model non-uniform probabilistic polynomial-time adversary $\mathcal{S}$ controlling P_2 such that $\forall \boldsymbol{x} \in \{\boldsymbol{x}_0, \boldsymbol{x}_1\}$,*

$$\left\{\mathrm{IDEAL}_{\mathcal{F},\mathcal{S}}(\boldsymbol{x}, \boldsymbol{y}, z)\right\}_{n \in \mathbb{N}; z \in \{0,1\}^*} \stackrel{c}{\equiv} \left\{\mathrm{REAL}_{\Pi,\mathcal{A}}(\boldsymbol{x}, \boldsymbol{y}, z)\right\}_{n \in \mathbb{N}; z \in \{0,1\}^*}$$

Protocol Π is said to input indistinguishably compute *(or, IIC) f if it input indistinguishably computes f both for P_1 and P_2.*

The above definition has various shortcomings and can be seen as only a stepping stone to our final definition (which implies the one in [37]). We refer the reader to the full version for our extended definition and for the relationship between various notions.

3   Building Blocks

We now discuss the main cryptographic primitives that we use in our construction.

Statistically Binding String Commitments. In our protocol, we will use a (2-round) statistically binding string commitment scheme, e.g., a parallel version of Naor's bit commitment scheme [38] based on one-way functions. For simplicity of exposition, in the presentation of our results in this manuscript, we will actually use a non-interactive perfectly binding string commitment.[4] Such a scheme can be easily constructed based on a 1-to-1 one way function. Let COM(·) denote the commitment function of the string commitment scheme. For simplicity of exposition, in the sequel, we will assume that random coins are an implicit input to the commitment function.

Extractable Commitment Scheme. We will also use a simple challenge-response based extractable statistically-binding string commitment scheme $\langle C, R \rangle$ that has been used in several prior works, most notably [44,47]. We note that in contrast to [44] where a multi-slot protocol was used, here (similar to [47]), we only need a one-slot protocol.

Protocol $\langle C, R \rangle$. Let COM(·) denote the commitment function of a non-interactive perfectly binding string commitment scheme (as described in Section 3). Let n denote the security parameter. The commitment scheme $\langle C, R \rangle$ is described as follows.

COMMIT PHASE:

1. To commit to a string str, C chooses $k = \omega(\log(n))$ independent random pairs $\{\alpha_i^0, \alpha_i^1\}_{i=1}^k$ of strings such that $\forall i \in [k]$, $\alpha_i^0 \oplus \alpha_i^1 = $ str; and commits to all of them to R using COM. Let $B \leftarrow \text{COM}(\text{str})$, and $A_i^0 \leftarrow \text{COM}(\alpha_i^0)$, $A_i^1 \leftarrow \text{COM}(\alpha_i^1)$ for every $i \in [k]$.
2. R sends k uniformly random bits $v_1, \ldots, v_n$.
3. For every $i \in [k]$, if $v_i = 0$, C opens A_i^0, otherwise it opens A_i^1 to R by sending the appropriate decommitment information.

OPEN PHASE: C opens all the commitments by sending the decommitment information for each one of them.

This completes the description of $\langle C, R \rangle$.

Modified Commitment Scheme. Due to technical reasons, we will also use a minor variant, denoted $\langle C', R' \rangle$, of the above commitment scheme. Protocol $\langle C', R' \rangle$ is the same as $\langle C, R \rangle$, except that for a given receiver challenge string, the committer does not "open" the commitments, but instead simply reveals the appropriate committed values (without revealing the randomness used to create the corresponding commitments). More specifically, in protocol $\langle C', R' \rangle$, on receiving a challenge string $v_1, \ldots, v_n$ from the receiver, the committer uses the following strategy: for every $i \in [k]$, if $v_i = 0$, C' sends α_i^0, otherwise it sends α_i^1 to R'. Note that C' does not reveal the decommitment values associated with the revealed shares.

---

[4] It is easy to see that the construction given in Section 4 does not necessarily require the commitment scheme to be non-interactive, and that a standard 2-round scheme works as well. As noted above, we choose to work with non-interactive schemes only for simplicity of exposition.

When we use $\langle C', R' \rangle$ in our main construction, we will require the committer C' to prove the "correctness" of the values (i.e., the secret shares) it reveals in the last step of the commitment protocol. In fact, due to technical reasons, we will also require the the committer to prove that the commitments that it sent in the first step are "well-formed". Below we formalize both these properties in the form of a *validity* condition for the commit phase.

Proving Validity of the Commit Phase. We say that commit phase between C' and R' is *valid* with respect to a value str if there exist values $\{\hat{\alpha}_i^0, \hat{\alpha}_i^1\}_{i=1}^k$ such that:

1. For all $i \in [k]$, $\hat{\alpha}_i^0 \oplus \hat{\alpha}_i^1 = \mathsf{str}$, and
2. Commitments B, $\{A_i^0, A_i^1\}_{i=1}^k$ can be decommitted to str, $\{\hat{\alpha}_i^0, \hat{\alpha}_i^1\}_{i=1}^k$ respectively.
3. Let $\bar{\alpha}_1^{v_1}, \ldots, \bar{\alpha}_k^{v_k}$ denote the secret shares revealed by C in the commit phase. Then, for all $i \in [k]$, $\bar{\alpha}_i^{v_i} = \hat{\alpha}_i^{v_i}$.

We can define *validity* condition for the commitment protocol $\langle C, R \rangle$ in a similar manner.

Constant-Round Non-Malleable Zero Knowledge Argument. In our main construction, we will use a constant-round non-malleable zero knowledge (NMZK) argument for every language in **NP** with perfect completeness and negligible soundness error. In particular, we will use a specific (stand-alone) NMZK protocol, denoted $\langle P, V \rangle$, based on the concurrent-NMZK protocol of Barak et al [4]. Specifically, we make the following two changes to Barak et al's protocol: (a) Instead of using an $\omega(\log n)$-round PRS preamble [44], we simply use the one-slot commitment scheme $\langle C, R \rangle$ (described above). (b) Further, we require that the non-malleable commitment scheme being used in the protocol be constant-round and public-coin w.r.t. receiver. We note that such commitment schemes are known due to Pass, Rosen [43]. Further, in full version, we show how to adapt the scheme of Goyal [22] to incorporate the public-coin property.[5] We now describe the protocol $\langle P, V \rangle$.

Protocol $\langle P, V \rangle$. Let P and V denote the prover and the verifier respectively. Let L be an NP language with a witness relation R. The common input to P and V is a statement $\pi \in L$. P additionally has a private input w (witness for π). Protocol $\langle P, V \rangle$ consists of two main phases: (a) the *preamble phase*, where the verifier commits to a random secret (say) σ via an execution of $\langle C, R \rangle$ with the prover, and (b) the *post-preamble phase*, where the prover proves an NP statement. In more detail, protocol $\langle P, V \rangle$ proceeds as follows.

PREAMBLE PHASE.

1. P and V engage in the execution of $\langle C, R \rangle$ where V commits to a random string σ.

---

[5] We note that while the commitment scheme of [43] admits a non black-box security proof, the security proof of Goyal's scheme is black-box. As such, the resultant NMZK protocol has a black-box security proof as well.

POST-PREAMBLE PHASE.

2. P commits to 0 using a statistically-hiding commitment scheme. Let c be the commitment string. Additionally, P proves the knowledge of a valid decommitment to c using a statistical zero-knowledge argument of knowledge (SZKAOK).

3. V now reveals σ and sends the decommitment information relevant to $\langle C, R \rangle$ that was executed in step 1.

4. P commits to the witness w using a constant-round public-coin extractable non-malleable commitment scheme.

5. P now proves the following statement to V using SZKAOK:

   (a) *either* the value committed to in step 4 is a valid witness to π (i.e., $R(\pi, w) = 1$, where w is the committed value), *or*

   (b) the value committed to in step 2 is the trapdoor secret σ.

   P uses the witness corresponding to the first part of the statement.

Decoupling the Preamble Phase from the Protocol. Note that the preamble phase in $\langle P, V \rangle$ is independent of the proof statement and can therefore be executed by P and V *before* the proof statement is fixed. Indeed, this is the case when we use $\langle P, V \rangle$ in our main construction in Section 4. Specifically, in our main construction, the parties first engage in multiple executions of $\langle C, R \rangle$ at the beginning of the protocol. Later, when a party (say) P_i wishes to prove the validity of a statement π to (say) P_j, then P_i and P_j engage in an execution of the post-preamble phase of $\langle P, V \rangle$ for statement π. The protocol specification fixes a particular instance of $\langle C, R \rangle$ that was executed earlier as the preamble phase of this instance of $\langle P, V \rangle$. In the description of our main construction, we will abuse notation and sometimes refer to the post-preamble phase as $\langle P, V \rangle$.

Straight-line Simulation of $\langle P, V \rangle$. A nice property of protocol $\langle P, V \rangle$ is that it allows *straight-line* simulation of the prover if the trapdoor secret σ is available to the simulator S. (Note that S can rewind V during the execution of $\langle C, R \rangle$ in order to extract σ.) See the full version for a description of the simulation strategy.

Constant-Round Statistically Witness Indistinguishable Arguments. In our construction, we shall use a constant-round statistically witness indistinguishable (SWI) argument $\langle P_{\mathsf{swi}}, V_{\mathsf{swi}} \rangle$ for proving membership in any **NP** language with perfect completeness and negligible soundness error. Such a protocol can be constructed by using $\omega(\log n)$ copies of Blum's Hamiltonicity protocol [7] in parallel, with the modification that the prover's commitments in the Hamiltonicity protocol are made using a constant-round statistically hiding commitment scheme [39,27,16].

Semi-Honest Two Party Computation. We will also use a constant-round semi-honest two party computation protocol $\langle P_1^{\mathsf{sh}}, P_2^{\mathsf{sh}} \rangle$ for any functionality $\mathcal{F}$ in the stand-alone setting. The existence of such a protocol follows from the existence of constant-round semi-honest 1-out-of-2 oblivious transfer [49,19,29].

4   Our Construction

Let $\mathcal{F}$ be any well-formed functionality[6] that admits a constant round two-party computation protocol in the semi-honest setting. In particular, $\mathcal{F}$ can be a universal functionality. In this section we will give a protocol $\varPi$ that UC-SPS-realizes $\mathcal{F}$. Note that in the UC framework any two parties (say P_i and P_j) might interact as per the protocol $\varPi$ on initiation by the environment for some session corresponding to a SID sid. For simplicity of notation, we will describe the protocol in terms of two parties P_1 and P_2, where these roles could be taken by any two parties in the system. Further we will skip mentioning the SID to keep the protocol specification simple.

In order to describe our construction, we first recall the notation associated with the primitives that we use in our protocol. Let $\mathrm{COM}(\cdot)$ denote the commitment function of a non-interactive perfectly binding commitment scheme, and let $\langle C, R \rangle$ denote the one-slot extractable commitment scheme, and $\langle C', R' \rangle$ be its modified version (see Section 3). Further, we will use a constant-round NMZK protocol $\langle P, V \rangle$ (see Section 3), a constant-round SWI argument $\langle P_{\mathsf{swi}}, V_{\mathsf{swi}} \rangle$, and a constant-round *semi-honest* two party computation protocol $\langle P_1^{\mathsf{sh}}, P_2^{\mathsf{sh}} \rangle$ that securely computes $\mathcal{F}$ as per the standard simulation-based definition of secure computation.

Let P_1 and P_2 be two parties with inputs x_1 and x_2 provided to them by the environment $\mathcal{Z}$. Let n be the security parameter. Protocol $\varPi = \langle P_1, P_2 \rangle$ proceeds as follows.

I. Trapdoor Creation Phase.

1. $P_1 \Rightarrow P_2$: P_1 samples a random string σ_1 (of appropriate length; see below) and engages in an execution of $\langle C, R \rangle$ with P_2, where P_1 commits to σ_1. We will denote this commitment protocol by $\langle C, R \rangle_{1 \to 2}$.
2. $P_2 \Rightarrow P_1$: P_2 now acts symmetrically. That is, P_2 samples a random string σ_2 and commits it via an execution of $\langle C, R \rangle$ (denoted as $\langle C, R \rangle_{2 \to 1}$) with P_1.
3. $P_1 \Rightarrow P_2$: P_1 creates a commitment $com_1 = \mathrm{COM}(0)$ to bit 0 and sends com_1 to P_2. P_1 and P_2 now engage in an execution of (the post-preamble phase of) $\langle P, V \rangle$, where P_1 proves that com_1 is a commitment to bit 0. The commitment protocol $\langle C, R \rangle_{2 \to 1}$ (executed earlier in step 2) is fixed as the preamble phase for this instance of $\langle P, V \rangle$ (see Section 3).
4. $P_2 \Rightarrow P_1$: P_2 now acts symmetrically.

Informally speaking, the purpose of this phase is to aid the simulator in obtaining a "trapdoor" to be used during the simulation of the protocol. As discussed earlier in Section 1.2, in order to bypass the need of recursive rewindings (even though we consider concurrent security), we want to ensure that a "hybrid" simulator (that performs rewindings) can always extract a "trapdoor" *before*

---

[6] See [9] for a definition of well-formed functionalities.

it begins cheating in any protocol session. Here, we achieve this effect by decoupling the preamble phase of $\langle P, V \rangle$ from the post-preamble phase (see Section 3) and executing the preamble phase at the very beginning of our protocol.

II. Input Commitment Phase. In this phase, the parties commit to their inputs and random coins (to be used in the next phase) via the commitment protocol $\langle C', R' \rangle$.

1. $P_1 \Rightarrow P_2$: P_1 first samples a random string r_1 (of appropriate length, to be used as P_1's randomness in the execution of $\langle P_1^{\mathsf{sh}}, P_2^{\mathsf{sh}} \rangle$ in phase III) and engages in an execution of $\langle C', R' \rangle$ (denoted as $\langle C', R' \rangle_{1 \rightarrow 2}$) with P_2, where P_1 commits to $x_1 \| r_1$. Next, P_1 and P_2 engage in an execution of $\langle P_{\mathsf{swi}}, V_{\mathsf{swi}} \rangle$ where P_1 proves the following statement to P_2: (a) *either* there exist values $\hat{x}_1$, $\hat{r}_1$ such that the commitment protocol $\langle C', R' \rangle_{1 \rightarrow 2}$ is *valid* with respect to the value $\hat{x}_1 \| \hat{r}_1$ (see Section 3), *or* (b) com_1 is a commitment to bit 1.
2. $P_2 \Rightarrow P_1$: P_2 now acts symmetrically. Let r_2 (analogous to r_1 chosen by P_1) be the random string chosen by P_2 (to be used in the next phase).

Informally speaking, the purpose of this phase is aid the simulator in extracting the adversary's input and randomness.

III. Secure Computation Phase. In this phase, P_1 and P_2 engage in an execution of $\langle P_1^{\mathsf{sh}}, P_2^{\mathsf{sh}} \rangle$ where P_1 plays the role of P_1^{sh}, while P_2 plays the role of P_2^{sh}. Since $\langle P_1^{\mathsf{sh}}, P_2^{\mathsf{sh}} \rangle$ is secure only against semi-honest adversaries, we first enforce that the coins of each party are truly random, and then execute $\langle P_1^{\mathsf{sh}}, P_2^{\mathsf{sh}} \rangle$, where with every protocol message, a party gives a proof using $\langle P_{\mathsf{swi}}, V_{\mathsf{swi}} \rangle$ of its honest behavior "so far" in the protocol. We now describe the steps in this phase.

1. $P_1 \leftrightarrow P_2$: P_1 samples a random string r_2' (of appropriate length) and sends it to P_2. Similarly, P_2 samples a random string r_1' and sends it to P_1. Let $r_1'' = r_1 \oplus r_1'$ and $r_2'' = r_2 \oplus r_2'$. Now, r_1'' and r_2'' are the random coins that P_1 and P_2 will use during the execution of $\langle P_1^{\mathsf{sh}}, P_2^{\mathsf{sh}} \rangle$.
2. Let t be the number of rounds in $\langle P_1^{\mathsf{sh}}, P_2^{\mathsf{sh}} \rangle$, where one round consists of a message from P_1^{sh} followed by a reply from P_2^{sh}. Let transcript $T_{1,j}$ (resp., $T_{2,j}$) be defined to contain all the messages exchanged between P_1^{sh} and P_2^{sh} before the point P_1^{sh} (resp., P_2^{sh}) is supposed to send a message in round j. For $j = 1, \ldots, t$:
   (a) $P_1 \Rightarrow P_2$: Compute $\beta_{1,j} = P_1^{\mathsf{sh}}(T_{1,j}, x_1, r_1'')$ and send it to P_2. P_1 and P_2 now engage in an execution of $\langle P_{\mathsf{swi}}, V_{\mathsf{swi}} \rangle$, where P_1 proves the following statement:
      i. *either* there exist values $\hat{x}_1$, $\hat{r}_1$ such that (a) the commitment protocol $\langle C', R' \rangle_{1 \rightarrow 2}$ is *valid* with respect to the value $\hat{x}_1 \| \hat{r}_1$ (see Section 3), and (b) $\beta_{1,j} = P_1^{\mathsf{sh}}(T_{1,j}, \hat{x}_1, \hat{r}_1 \oplus r_1')$
      ii. *or,* com_1 is a commitment to bit 1.
   (b) $P_2 \Rightarrow P_1$: P_2 now acts symmetrically.

This completes the description of protocol $\varPi$. We now claim the following.

Theorem 1. *Assume the existence of constant round semi-honest OT and collision resistant hash functions. Then for every well-formed functionality $\mathcal{F}$, there exists a constant-round protocol that UC-SPS-realizes $\mathcal{F}$.*

We prove the above claim by arguing that the protocol $\Pi = \langle P_1, P_2 \rangle$ described earlier UC-SPS-realizes $\mathcal{F}$. Note that our simulator will run in sub-exponential time, where the desired parameters can be obtained by using a "scaled-down" security parameter of the commitment scheme COM. See the full version for the proof.

Acknowledgements. Research supported in part from a DARPA/ONR PRO-CEED award, NSF grants 1136174, 1118096, 1065276, 0916574 and 0830803, a Xerox Faculty Research Award, a Google Faculty Research Award, an equipment grant from Intel, and an Okawa Foundation Research Grant. This material is based upon work supported by the Defense Advanced Research Projects Agency through the U.S. Office of Naval Research under Contract N00014-11-1-0389. The views expressed are those of the author and do not reflect the official policy or position of the Department of Defense or the U.S. Government.

References

1. Barak, B., Canetti, R., Nielsen, J., Pass, R.: Universally composable protocols with relaxed set-up assumptions. In: FOCS, pp. 186–195 (2004)
2. Barak, B.: Constant-round coin-tossing with a man in the middle or realizing the shared random string model. In: FOCS, pp. 345–355 (2002)
3. Barak, B., Canetti, R., Lindell, Y., Pass, R., Rabin, T.: Secure Computation Without Authentication. In: Shoup, V. (ed.) CRYPTO 2005. LNCS, vol. 3621, pp. 361–377. Springer, Heidelberg (2005)
4. Barak, B., Prabhakaran, M., Sahai, A.: Concurrent non-malleable zero knowledge. In: FOCS, pp. 345–354 (2006)
5. Barak, B., Sahai, A.: How to play almost any mental game over the net - concurrent composition via super-polynomial simulation. In: FOCS, pp. 543–552. IEEE Computer Society (2005)
6. Ben-David, A., Nisan, N., Pinkas, B.: Fairplaymp: a system for secure multi-party computation. In: ACM Conference on Computer and Communications Security, pp. 257–266 (2008)
7. Blum, M.: How to prove a theorem so no one else can claim it. In: International Congress of Mathematicians, pp. 1444–1451 (1987)
8. Canetti, R., Kushilevitz, E., Lindell, Y.: On the limitations of universally composable two-party computation without set-up assumptions. J. Cryptology 19(2), 135–167 (2006)
9. Canetti, R., Lindell, Y., Ostrovsky, R., Sahai, A.: Universally composable two-party and multi-party secure computation. In: STOC, pp. 494–503 (2002)
10. Canetti, R.: Universally composable security: A new paradigm for cryptographic protocols. In: FOCS, pp. 136–145 (2001)
11. Canetti, R., Fischlin, M.: Universally Composable Commitments. In: Kilian, J. (ed.) CRYPTO 2001. LNCS, vol. 2139, pp. 19–40. Springer, Heidelberg (2001)

12. Canetti, R., Kilian, J., Petrank, E., Rosen, A.: Black-box concurrent zero-knowledge requires $\tilde{\Omega}(\log n)$ rounds. In: STOC, pp. 570–579 (2001)
13. Canetti, R., Lin, H., Pass, R.: Adaptive hardness and composable security in the plain model from standard assumptions. In: FOCS, pp. 541–550 (2010)
14. Canetti, R., Pass, R., Shelat, A.: Cryptography from sunspots: How to use an imperfect reference string. In: FOCS, pp. 249–259 (2007)
15. Chandran, N., Goyal, V., Sahai, A.: New Constructions for UC Secure Computation Using Tamper-Proof Hardware. In: Smart, N.P. (ed.) EUROCRYPT 2008. LNCS, vol. 4965, pp. 545–562. Springer, Heidelberg (2008)
16. Damgård, I., Pedersen, T.P., Pfitzmann, B.: On the existence of statistically hiding bit commitment schemes and fail-stop signatures. J. Cryptology 10(3), 163–194 (1997)
17. Dolev, D., Dwork, C., Naor, M.: Nonmalleable cryptography. SIAM J. Comput. 30(2), 391–437 (2000)
18. Garg, S., Goyal, V., Jain, A., Sahai, A.: Bringing People of Different Beliefs Together to Do UC. In: Ishai, Y. (ed.) TCC 2011. LNCS, vol. 6597, pp. 311–328. Springer, Heidelberg (2011)
19. Goldreich, O., Micali, S., Wigderson, A.: How to play any mental game. In: STOC 1987: Proceedings of the 19th Annual ACM Conference on Theory of Computing, pp. 218–229. ACM Press, New York (1987)
20. Goldreich, O.: Foundation of Cryptography - Basic Tools. Cambridge University Press (2001)
21. Goldwasser, S., Micali, S., Rackoff, C.: The knowledge complexity of interactive proof systems. SIAM J. Comput. 18(1), 186–208 (1989)
22. Goyal, V.: Constant round non-malleable protocols using one-way functions. In: STOC (2011)
23. Goyal, V., Jain, A., Ostrovsky, R.: Password-Authenticated Session-Key Generation on the Internet in the Plain Model. In: Rabin, T. (ed.) CRYPTO 2010. LNCS, vol. 6223, pp. 277–294. Springer, Heidelberg (2010)
24. Goyal, V., Katz, J.: Universally Composable Multi-party Computation with an Unreliable Common Reference String. In: Canetti, R. (ed.) TCC 2008. LNCS, vol. 4948, pp. 142–154. Springer, Heidelberg (2008)
25. Goyal, V., Sahai, A.: Resettably Secure Computation. In: Joux, A. (ed.) EUROCRYPT 2009. LNCS, vol. 5479, pp. 54–71. Springer, Heidelberg (2009)
26. Groth, J., Ostrovsky, R.: Cryptography in the Multi-string Model. In: Menezes, A. (ed.) CRYPTO 2007. LNCS, vol. 4622, pp. 323–341. Springer, Heidelberg (2007)
27. Halevi, S., Micali, S.: Practical and Provably-Secure Commitment Schemes from Collision-Free Hashing. In: Koblitz, N. (ed.) CRYPTO 1996. LNCS, vol. 1109, pp. 201–215. Springer, Heidelberg (1996)
28. Katz, J.: Universally Composable Multi-party Computation Using Tamper-Proof Hardware. In: Naor, M. (ed.) EUROCRYPT 2007. LNCS, vol. 4515, pp. 115–128. Springer, Heidelberg (2007)
29. Kilian, J.: Founding cryptography on oblivious transfer. In: STOC, pp. 20–31 (1988)
30. Kilian, J., Petrank, E.: Concurrent and resettable zero-knowledge in polyloalgorithm rounds. In: STOC, pp. 560–569 (2001)
31. Lin, H., Pass, R.: Non-malleability amplification. In: STOC, pp. 189–198 (2009)
32. Lin, H., Pass, R.: Constant-round non-malleable commitments from any one-way function. In: STOC (2011)

33. Lin, H., Pass, R., Tseng, W.-L.D., Venkitasubramaniam, M.: Concurrent Non-Malleable Zero Knowledge Proofs. In: Rabin, T. (ed.) CRYPTO 2010. LNCS, vol. 6223, pp. 429–446. Springer, Heidelberg (2010)

34. Lin, H., Pass, R., Venkitasubramaniam, M.: A unified framework for concurrent security: universal composability from stand-alone non-malleability. In: STOC, pp. 179–188 (2009)

35. Lindell, Y.: Lower Bounds for Concurrent Self Composition. In: Naor, M. (ed.) TCC 2004. LNCS, vol. 2951, pp. 203–222. Springer, Heidelberg (2004)

36. Malkhi, D., Nisan, N., Pinkas, B., Sella, Y.: Fairplay - secure two-party computation system. In: USENIX Security Symposium, pp. 287–302 (2004)

37. Micali, S., Pass, R., Rosen, A.: Input-indistinguishable computation. In: FOCS, pp. 367–378. IEEE Computer Society (2006)

38. Naor, M.: Bit commitment using pseudorandomness. J. Cryptology 4(2), 151–158 (1991)

39. Naor, M., Yung, M.: Universal one-way hash functions and their cryptographic applications. In: STOC, pp. 33–43 (1989)

40. Pass, R.: Simulation in Quasi-Polynomial Time, and its Application to Protocol Composition. In: Biham, E. (ed.) EUROCRYPT 2003. LNCS, vol. 2656, pp. 160–176. Springer, Heidelberg (2003)

41. Pass, R.: Personal Communication (2011)

42. Pass, R., Rosen, A.: Concurrent non-malleable commitments. In: FOCS, pp. 563–572 (2005)

43. Pass, R., Rosen, A.: New and improved constructions of non-malleable cryptographic protocols. In: STOC, pp. 533–542 (2005)

44. Prabhakaran, M., Rosen, A., Sahai, A.: Concurrent zero knowledge with logarithmic round-complexity. In: FOCS, pp. 366–375 (2002)

45. Prabhakaran, M., Sahai, A.: New notions of security: achieving universal composability without trusted setup. In: STOC, pp. 242–251 (2004)

46. Richardson, R., Kilian, J.: On the Concurrent Composition of Zero-Knowledge Proofs. In: Stern, J. (ed.) EUROCRYPT 1999. LNCS, vol. 1592, pp. 415–431. Springer, Heidelberg (1999)

47. Rosen, A.: A Note on Constant-Round Zero-Knowledge Proofs for NP. In: Naor, M. (ed.) TCC 2004. LNCS, vol. 2951, pp. 191–202. Springer, Heidelberg (2004)

48. Wee, H.: Black-box, round-efficient secure computation via non-malleability amplification. In: FOCS, pp. 531–540 (2010)

49. Yao, A.C.C.: How to generate and exchange secrets (extended abstract). In: FOCS, pp. 162–167. IEEE (1986)

Identity-Based Encryption
Resilient to Continual Auxiliary Leakage

Tsz Hon Yuen[1], Sherman S.M. Chow[2], Ye Zhang[3,*], and Siu Ming Yiu[1]

[1] University of Hong Kong, Hong Kong
{thyuen,smyiu}@cs.hku.hk
[2] University of Waterloo, Canada
smchow@math.uwaterloo.ca
[3] Pennsylvania State University, USA
yxz169@cse.psu.edu

Abstract. We devise the first identity-based encryption (IBE) that remains secure even when the adversary is equipped with *auxiliary input* (STOC '09) – any computationally uninvertible function of the master secret key and the identity-based secret key. In particular, this is more general than the tolerance of Chow *et al.*'s IBE schemes (CCS '10) and Lewko *et al.*'s IBE schemes (TCC '11), in which the leakage is bounded by a pre-defined number of bits; yet our construction is also fully secure in the standard model based on only static assumptions, and can be easily extended to give the first hierarchical IBE with auxiliary input.

Furthermore, we propose the model of *continual auxiliary leakage* (CAL) that can capture both memory leakage and continual leakage. The CAL model is particularly appealing since it not only gives a clean definition when there are multiple secret keys (the master secret key, the identity-based secret keys, and their refreshed versions), but also gives a generalized definition that does not assume secure erasure of secret keys after each key update. This is different from previous definitions of continual leakage (FOCS '10, TCC '11) in which the length-bounded leakage is only the secret key in the current time period. Finally, we devise an IBE scheme which is secure in this model. A major tool we use is the modified Goldreich-Levin theorem (TCC '10), which until now has only been applied in traditional public-key encryption with a single private key.

1   Introduction

In cryptography, security guarantees are usually proven under the assumption that the secret key must be kept safely and other internal (random) state is not leaked to the adversary. Even if a single bit of these secrets is leaked, the protection guaranteed by the proof is lost. In practice, however, it is difficult to avoid all possible kinds of leakage, such as side-channel attacks that exploit the physical nature of cryptographic operations (e.g., timing, power, radiation, cold

---

* Part of this work was done while the third author was at University of Hong Kong.

D. Pointcheval and T. Johansson (Eds.): EUROCRYPT 2012, LNCS 7237, pp. 117–134, 2012.

boot attacks, etc.) or the reuse of the secret key and/or the randomness in a number of applications.

Leakage-resilient cryptography was introduced to provide formal security guarantees even when the leakage of the secret keying material is allowed. In this paper, we focus on public key encryption that is secure against memory leakage. More precisely, the adversary is allowed to specify an efficiently computable leakage function f and obtain the output of f applied to the secret key and other internal state. This function f aims to model the possible leakage that the adversary can learn in practice.

In recent years, we have seen a number of leakage models that impose different restriction on f. Recall the major (open) problem in leakage-resilient cryptography [10]:

> *"allowing for continual (overall unbounded) leakage, without additionally restricting its type."*

An ongoing line of research is on loosening the restriction of leakage types. The *relative leakage* model [1] restricts the function f to output at most l bits, where l is smaller than the secret key size. This restriction is later relaxed by allowing f to lower the entropy of the secret key by at most l bits [15]. To sum up these two models, l is defined as a fraction of the key (either in terms of the bit size or the entropy). The *bounded retrieval model* (e.g., see [2,8]), on the other hand, treats the leakage l as a system parameter. The size of the secret key can be increased to allow l bits of leakage, without affecting the public key size, communication and computation efficiency. To further relax the restriction, Dodis *et al.* [9] considered *auxiliary inputs*, which allow any f that no polynomial time adversary can invert (i.e., to output the secret key being leaked) with non-negligible probability. For example, any (computationally) one-way permutation can be used by the adversary as its auxiliary input, but is not allowed in the relative (leakage/entropy) model (as a permutation information-theoretically reveals the entire key). Therefore, auxiliary inputs allow us to consider a larger class of leakage functions.

The above line of research bounds the leakage *throughout the entire lifetime* of the secret key. Another paradigm considers a key update algorithm that continually refreshes the secret key, while bounding the leakage between updates. It addresses the first part of the aforementioned major problem. This model is known as the *continual leakage* model. There are signature, identification [10] and public key encryption schemes [6] secure in this model. Lewko *et al.* [12] recently proposed signature and encryption schemes that allow a constant fraction leakage of the secret key and the randomness during updates. In these papers, the leakage bound between updates is either based on the relative leakage or the bounded retrieval model. Therefore, the number of bits leaked between updates is still restricted.

IBE with Auxiliary Inputs. Dodis *et al.* [9] only considered public key encryption and there is no known identity-based encryption (IBE) scheme that is secure with auxiliary input, even though there are a number of (bounded) leakage-resilient IBE schemes [1,2,8,6,13]. A distinctive feature of IBE is that any string can be used as a public key and potentially an exponential number of identities can be supported. This feature has many applications (e.g., see [2,3,5,8]), and furthermore makes the auxiliary input model appealing to IBE for various reasons. First, the auxiliary input model is useful in the context of composition. One may use the encryption public key (and corresponding secret key) in other applications (e.g., digital signatures and identification), and their composition remains secure as long as these other schemes were proved secure in the standard (no leakage) sense [9]. With the versatility of identity-based (ID-based) keys this guarantee appears to be more desirable. Second, the auxiliary input model not only tolerates a wider class of leakage, but also gives a "clean" definition of the necessary restriction on leakage. The model is free from numeric bounds (e.g., number of bits leaked from the master secret key, or number of bits leaked from the ID-based secret key of the target user) which are necessary in bounded retrieval model.

Continual Auxiliary Leakages. Recall that the key idea to achieve continual memory leakage (CML) resilience is to refresh the secret key in each time period. Previous CML models for IBE [6,13] only consider leakage of the current secret key for a given time. In other words, after a user has computed a new secret key for the next period, the old secret key should be *securely erased* from memory (so the leakage via the key-update query is the "last chance" for the adversary)[1]. This greatly diminishes the benefits offered by the formal leakage-resilience guarantee since with frequent secure erasures it is less disastrous to have memory leakage.

Combining the concept of auxiliary inputs with CML brings the possibility of new leakage-resilience guarantees. Our continual auxiliary leakage (CAL) model allows continual leakage, and the leakage between updates has minimal restriction: no polynomial time algorithm can use the leaked information to output a valid secret key. The CAL model still inherits the simplicity of the standard (non-continual) auxiliary input/leakage model. In particular, we do not need to keep track of the "version number" of keys.

Our Contributions. We tackle the problem of allowing continual (overall unbounded) leakage, without additionally restricting its type, for IBE. Brakerski *et al.*'s CML-resilient IBE [6] does not tolerate leakage from the master secret key and is only selective-secure. Lewko *et al.* [13] proposed a fully-secure CML-resilient IBE, but the leakage size during updates is limited to logarithmic. It is fair to say there is no complete solution for this major problem in leakage-resilient cryptography. To achieve our goal, we propose the continual auxiliary leakage (CAL) model and construct an IBE scheme secure in this strong model.

---

[1] We do not claim to have discovered any attack against the schemes in the respective papers exploiting this more general form of leakage. We are merely pointing out that the stronger attack is not covered by the current proofs.

We begin by constructing the *first* IBE scheme that is secure in the presence of *auxiliary inputs*. Our construction in §3.3 preserves the nice features of recent leakage-resilient IBE schemes [8,13]: adaptive security in the standard model and based on static assumptions, and moderate increases the size of the ciphertext and computational complexity.

Our work combines a number of techniques in the literature. We use the dual system encryption [14] for both adaptive security regarding the ID-based keys, and for the leakage-resilience via the proof technique [13] which allows the leakage of the master secret key and the ID-based secret keys. For leakage in the form of auxiliary input (or "auxiliary leakage"), we use the modified Goldreich-Levin theorem [9]. We overcome a number of technical difficulties when combining these techniques. Firstly, we cannot directly use the modified Goldreich-Levin theorem as it restricts the blinding factor of the semi-functional key (an artifact in the security reduction for allowing bounded leakage, which is created using a blinding factor from $\mathbb{Z}_{p_2}^m$ where $m \in \mathbb{N}$ and p_2 is a large prime of size 2^λ; see [13, Lemmata 6.2, 6.7]) to be a λ-bit number. Therefore, we need to construct the semi-functional key subject to this design constraint. It is also interesting to note that the λ-bit number is used as a real secret key of the public key encryption with auxiliary input [9], but in our case it just appears in the "imaginary" semi-functional secret key in the simulation. Secondly, Lewko *et al.*'s IBE [13] does not allow any leakage during setup. We twist their idea of using multiple tags (instead of a single tag in [8]) and do the "replication" in another level. (Thus we retain the same order of complexity for performance.) Although this technique by itself (see §3.3) does not allow leakage during setup, this structure leads us to construct an IBE scheme (in §4.4) which can be proven secure in the CAL model (i.e., leakage is allowed during setup). Furthermore, our scheme in §3.3 can be extended to give the *first* hierarchical IBE with auxiliary inputs.

In §4, we present the CAL model and propose the *first* IBE scheme secure in this strengthened model. There are a few problems that arise when we tried to borrow the construction technique from the (Diffie-Hellman-based) BHHO encryption in [9] to extend our IBE scheme in §3.3. Firstly, the modified Goldreich-Levin theorem [9] states that if the master secret keys α_i belongs to a subgroup H of $\mathbb{Z}_{p_1}$, then there exists an inverter with running time $\text{poly}(|H|)$. If $H = \mathbb{Z}_{p_1}$ and p_1 is a λ-bit prime, the running time of the inverter is $\text{poly}(2^\lambda)$, which is undesirable. Secondly, if we simply change the scheme such that α_i belongs to a subgroup $H = \{0, 1\}$, then the master public key becomes $y_i = \hat{e}(g_1, v_i)^0$ or $\hat{e}(g_1, v_i)^1$. Then any adversary can determine α_i by brute-force. The same attack applies if $|H| \ll p_1$. We then try to construct the public key as in the BHHO encryption in [9], and the master public key becomes $y = \prod_{i=1}^n \hat{e}(g_1, v_i)^{\alpha_i}$. The master secret key msk becomes $v_i^{\alpha_i}$ for $i \in [1, n]$. (It seems one may set msk be $\prod_{i=1}^n v_i^{\alpha_i}$, but we want to split it into n pieces in order to apply the modified Goldreich-Levin theorem.) That means leakage will be in the form of $f(v_1^{\alpha_1}, \ldots, v_n^{\alpha_n})$. Intuitively, to simulate all possible uninvertible function f, the knowledge $v_i^{\alpha_i}$ is needed, which implies the knowledge of α_i since the brute-force attack is easy on α_i. This leads to a contradiction since the α_i's is the solution of the underlying intractability problem. To resolve these issues, we change the structure of msk to $\prod_{j=1}^{m} v_{i,j}^{\alpha_j}$. We use a random $(n \times m)$ matrix $\boldsymbol{V} = \{v_{i,j}\}_{i \in [1,n], j \in [1,m]}$ and m random numbers $(\alpha_1, \ldots, \alpha_m)$ to obtain n master secret keys and public keys. Similar to (the semi-functional key in) our IBE in §3.3, this $\boldsymbol{V}$ is a conceptual building block for leakage-resilience and the knowledge of these group elements is not required anywhere else including key-update, encryption and decryption. This new msk does not reveal α_i under some intractability assumptions. Interestingly, these assumptions are required for a variant of Gentry-Peikert-Vaikuntanathan's (GPV) encryption scheme [11] based on learning with error (LWE) [9], which gives some evidence that our construction is not a trivial extension from the IBE in §3.3. Using lattice-based assumptions to help constructing pairing-based cryptosystems seems to be interesting on its own right. We leave it as an open problem to build a CAL-resilient IBE scheme without these assumptions.

2 Background

Composite Order Bilinear Groups [4]. Let $\mathcal{G}$ be a group generator, that takes a security parameter 1^λ as input where $\lambda \in \mathbb{N}$, outputs a description of bilinear group $(N = p_1 p_2 p_3, \mathbb{G}, \mathbb{G}_T, \hat{e})$, where p_1, p_2, p_3 are distinct λ-bit primes, $\mathbb{G}$ and $\mathbb{G}_T$ are cyclic groups of order N, and $\hat{e} : \mathbb{G} \times \mathbb{G} \to \mathbb{G}_T$ is a bilinear map such that $\forall g, h \in \mathbb{G}$ and $a, b \in \mathbb{Z}_N$, $\hat{e}(g^a, h^b) = \hat{e}(g, h)^{ab}$; $\hat{e}(g, g)$ generates $\mathbb{G}_T$ if g is a generator of $\mathbb{G}$. We denote $\mathbb{G}_{p_i}$ as the subgroup of order p_i in $\mathbb{G}$ ($i = 1, 2, 3$). Let g_i be the generator of the subgroup $\mathbb{G}_{p_i}$. For all $h_i \in \mathbb{G}_{p_i}$ and $h_j \in \mathbb{G}_{p_j}$, if $i \neq j$, $\hat{e}(h_i, h_j) = 1$. We denote $\mathbb{G}_{p_1 p_2}$ as the subgroup of order $p_1 p_2$ in $\mathbb{G}$. For all $T \in \mathbb{G}_{p_1 p_2}$, T can be written uniquely as the product of an element of $\mathbb{G}_{p_1}$ and an element of $\mathbb{G}_{p_2}$. We refer to these elements as the "$\mathbb{G}_{p_1}$ part of T" and the "$\mathbb{G}_{p_2}$ part of T" respectively. We also define $\mathbb{G}_{p_1 p_3}$ and $\mathbb{G} = \mathbb{G}_{p_1 p_2 p_3}$ similarly.

Decisional Problems [14]. For a group generator $\mathcal{G}$, the following experiments define subgroup decision problem for $\mathbb{G}_{p_1}$ and $\mathbb{G}_{p_1 p_2}$, subgroup decision problem for $\mathbb{G}_{p_1 p_3}$ and $\mathbb{G}$, and subgroup decisional bilinear Diffie-Hellman problem.

Experiment $\mathbf{Exp}_{\mathcal{G}, \mathcal{A}_1, \beta}^{(1)}(1^\lambda)$

   $(N = p_1 p_2 p_3, \mathbb{G}, \mathbb{G}_T, \hat{e}) \xleftarrow{R} \mathcal{G}(1^\lambda), \quad g, X_1 \xleftarrow{R} \mathbb{G}_{p_1}, \quad X_2 \xleftarrow{R} \mathbb{G}_{p_2}, \quad X_3 \xleftarrow{R} \mathbb{G}_{p_3}$

   $T_0 \xleftarrow{R} \mathbb{G}_{p_1 p_2}, \quad T_1 \xleftarrow{R} \mathbb{G}_{p_1}$.

   Return $\beta' \leftarrow \mathcal{A}_1(N, \mathbb{G}, \mathbb{G}_T, \hat{e}, g, X_1 X_2, X_3, T_\beta)$.

Experiment $\mathbf{Exp}_{\mathcal{G}, \mathcal{A}_2, \beta}^{(2)}(1^\lambda)$

   $(N = p_1 p_2 p_3, \mathbb{G}, \mathbb{G}_T, \hat{e}) \xleftarrow{R} \mathcal{G}(1^\lambda), g, X_1, Z_1 \xleftarrow{R} \mathbb{G}_{p_1}, X_i, Y_i, Z_i \xleftarrow{R} \mathbb{G}_{p_i} (i = 2, 3)$,

   $T_0 = Z_1 Z_3, \quad T_1 = Z_1 Z_2 Z_3$.

   Return $\beta' \leftarrow \mathcal{A}_2(N, \mathbb{G}, \mathbb{G}_T, \hat{e}, g, X_1 X_2, X_3, Y_2 Y_3, T_\beta)$.

Experiment $\mathbf{Exp}^{(3)}_{\mathcal{G},\mathcal{A}_3,\beta}(1^\lambda)$

$(N = p_1 p_2 p_3, \mathbb{G}, \mathbb{G}_T, \hat{e}) \xleftarrow{R} \mathcal{G}(1^\lambda), \quad g \xleftarrow{R} \mathbb{G}_{p_1}, X_2, Y_2, Z_2 \xleftarrow{R} \mathbb{G}_{p_2}, X_3 \xleftarrow{R} \mathbb{G}_{p_3},$

$\alpha, s \xleftarrow{R} \mathbb{Z}_N, \quad T_0 = \hat{e}(g,g)^{\alpha s}, \quad T_1 \xleftarrow{R} \mathbb{G}_T.$

Return $\beta' \leftarrow \mathcal{A}_3(N, \mathbb{G}, \mathbb{G}_T, \hat{e}, g, g^\alpha X_2, g^s Y_2, Z_2, X_3, T_\beta).$

For $i = 1, 2, 3$, we define the advantage of an algorithm $\mathcal{A}_i$ in breaking Assumption i to be $Adv^{(i)}_{\mathcal{G},\mathcal{A}_i}(\lambda) :=$

$$\left| \Pr[\mathbf{Exp}^{(i)}_{\mathcal{G},\mathcal{A}_i,1}(1^\lambda) = 1] - \Pr[\mathbf{Exp}^{(i)}_{\mathcal{G},\mathcal{A}_i,0}(1^\lambda) = 1] \right|.$$

Definition 1. *For $i = 1, 2, 3$, we say that $\mathcal{G}$ satisfies Assumption i if $Adv^{(i)}_{\mathcal{G},\mathcal{A}_i}(\lambda)$ is a negligible function of λ for any polynomial time algorithm $\mathcal{A}_i$.*

Goldreich-Levin Theorem for Large Fields. Dodis *et al.* [9] proved the following theorem – Goldreich-Levin theorem over any field $GF(q)$ for prime q.

Theorem 2 ([9]). *Let q be a prime, and let H be an arbitrary subset of $GF(q)$. Let $f : H^m \to \{0,1\}^*$ be any function. $s \leftarrow H^m, y \leftarrow f(s), r \leftarrow GF(q)^m$. If there is a distinguisher $\mathcal{D}$ that runs in time t such that*

$$\left| \Pr[\mathcal{D}(y, r, \langle r, s \rangle) = 1] - \Pr[u \leftarrow GF(q) : \mathcal{D}(y, r, u) = 1] \right| = \epsilon,$$

then there is an inverter $\mathcal{A}$ that runs in time $t' = t \cdot \mathsf{poly}(m, |H|, 1/\epsilon)$ such that

$$\Pr[s \leftarrow H^m, y \leftarrow f(s) : \mathcal{A}(y) = s] \geq \frac{\epsilon^3}{512 \cdot m \cdot q^2}.$$

Modular Lattices. Here we review some theorems for modular lattices. The first is a lemma on additive groups simplified from the lemma in [16].

Lemma 3 ([11]). *Let q be prime and let $m \geq 2n \log q$. For all but an at most q^n fraction of $\mathbf{A} \in \mathbb{Z}_q^{n \times m}$, the subset-sums of the columns of $\mathbf{A}$ generate $\mathbb{Z}_q^n$; i.e., for every $\mathbf{u} \in \mathbb{Z}_q^n$ there exists $\mathbf{e} \in \{0,1\}^m$ such that $\mathbf{A}\mathbf{e} = \mathbf{u}$.*

We give the definition of the average-case problem of inhomogeneous small integer solution problem (ISIS), which is related to the shortest independent vectors problem and decision shortest vector problem [11].

Definition 4 (ISIS$_{q,m,\beta}$). *Given an integer q, a uniformly random matrix $\mathbf{A} \in \mathbb{Z}_q^{n \times m}$, a random $\mathbf{u} \in \mathbb{Z}_n^q$, and a real β, find an integer vector $\mathbf{e} \in \mathbb{Z}^m$ such that $\mathbf{A}\mathbf{e} = \mathbf{u}$ mod q and $||\mathbf{e}||_2 \leq \beta$, where $||\mathbf{e}||_2$ is the Euclidean ℓ_2 norm. We say that the ISIS$_{q,m,\beta}$ assumption holds if no polynomial time algorithm can output $\mathbf{e}$ with a non-negligible probability.*

3   Identity-Based Encryption with Auxiliary Inputs

An IBE scheme consists of four probabilistic polynomial-time (PPT) algorithms:

1. Setup: On input a security parameter 1^λ, it generates a master public key mpk and a master secret key msk.
2. Ext: On input msk and an identity ID from an identity space $\mathcal{I}$, it outputs an identity-based secret key sk_{ID}.
3. Enc: On input mpk, ID and a message M from a message space $\mathcal{M}$, it outputs a ciphertext $\mathfrak{C}$.
4. Dec: On input sk_{ID}, and $\mathfrak{C}$, it outputs a message M or $\perp$ symbolizing the failure of decryption.

$\forall M \in \mathcal{M}$ and $\forall \mathsf{ID} \in \mathcal{I}$, $M \leftarrow \mathsf{Dec}(sk_{\mathsf{ID}}, \mathsf{Enc}(\mathsf{mpk}, \mathsf{ID}, M))$, where $(\mathsf{mpk}, \mathsf{msk}) \leftarrow \mathsf{Setup}(1^\lambda)$, $sk_{\mathsf{ID}} \leftarrow \mathsf{Ext}(\mathsf{msk}, \mathsf{ID})$.

We denote the space of the master secret key and that of ID-based secret keys by $\mathcal{MK}$ and $\mathcal{SK}$ respectively.

3.1   Auxiliary Input Model for Confidentiality

We consider the following indistinguishability based game against adaptive chosen identity and chosen plaintext attacks (IND-ID-CPA) for semantic security with leakage in the form of auxiliary inputs. Denote a polynomial-time (in λ) computable function family $\mathcal{F}$. We define the attack game as follows.

1. Setup. The challenger runs $(\mathsf{mpk}, \mathsf{msk}) \leftarrow \mathsf{Setup}(1^\lambda)$ and gives mpk to the adversary $\mathcal{A}$. The challenger also constructs an initially empty list $\mathcal{L}_{\mathsf{ID}}$.
2. Query 1. The following oracles can be queried by $\mathcal{A}$:
   - Extraction Oracle $\mathcal{KEO}(\mathsf{ID}, i)$: On input $\mathsf{ID} \in \mathcal{I}, i \in \mathbb{N}^+$, it first checks the list $\mathcal{L}_{\mathsf{ID}}$ for the tuple in the form of $(sk_{\mathsf{ID}}, \mathsf{ID}, j)$. If there is no such tuple, $\bar{j}$ is set to 1, then it runs $sk_{\mathsf{ID}} \leftarrow \mathsf{Ext}(\mathsf{msk}, \mathsf{ID})$ and puts $(sk_{\mathsf{ID}}, \mathsf{ID}, \bar{j})$ in the list $\mathcal{L}_{\mathsf{ID}}$. Otherwise, the maximum j is retrieved. If $i \leq j$, then sk_{ID} from the tuple $(sk_{\mathsf{ID}}, \mathsf{ID}, i)$ in the list $\mathcal{L}_{\mathsf{ID}}$ is returned.
   - Leakage Oracle $\mathcal{LO}(f, \mathsf{ID})$: On input $f \in \mathcal{F}$, it returns $f(\mathsf{msk}, \mathcal{L}_{\mathsf{ID}}, \mathsf{mpk}, \mathsf{ID})$.
   - UpdateUSK Oracle $\mathcal{USO}(\mathsf{ID})$: This oracle is useful for schemes with probabilistic ID-based secret key generation, where a user of identity ID may request for another ID-based secret key after obtained the first copy. It first checks the list $\mathcal{L}_{\mathsf{ID}}$ for the tuple in the form of $(sk_{\mathsf{ID}}, \mathsf{ID}, j)$ where j is a positive integer. If there is no such tuple, $\bar{j}$ is set to 1. Otherwise, the maximum j is retrieved and $\bar{j}$ is set to $(j + 1)$. Then, it runs $\bar{sk}_{\mathsf{ID}} \leftarrow \mathsf{Ext}(\mathsf{msk}, \mathsf{ID})$. It puts $(\bar{sk}_{\mathsf{ID}}, \mathsf{ID}, \bar{j})$ in the list $\mathcal{L}_{\mathsf{ID}}$ and returns $\bar{j}$.
   $\mathcal{KEO}, \mathcal{USO}$ and $\mathcal{LO}$ can be queried for at most q_e, q_u and q_ℓ times throughout this game respectively.
3. Challenge. $\mathcal{A}$ sends two messages $M_0, M_1 \in \mathcal{M}$ and an identity $\mathsf{ID}^* \in \mathcal{I}$ to the challenger. The challenger picks a random bit b' and computes $\mathfrak{C}^* \leftarrow \mathsf{Enc}(\mathsf{mpk}, \mathsf{ID}^*, M_{b'})$. The challenger sends $\mathfrak{C}^*$ to $\mathcal{A}$.
4. Query 2. $\mathcal{A}$ is allowed to query the Extraction Oracle adaptively.

5. Output. $\mathcal{A}$ returns a guess b^* of b'.

$\mathcal{A}$ wins the game if $b' = b^*$ and there was no query in the form of $\mathcal{KEO}(\mathsf{ID}^*, \cdot)$. The advantage of $\mathcal{A}$ is $\left| \Pr[\mathcal{A} \text{ wins}] - \frac{1}{2} \right|$. An IBE scheme is IND-ID-CPA secure w.r.t. auxiliary inputs from $\mathcal{F}$ if there is no PPT $\mathcal{A}$ with non-negligible advantage in the game above.

It remains to define a class of function families $\mathcal{F}$. For convenience, we will parametrize these families by the min-entropy k_u of the ID-based secret key respectively, as opposed to the security parameter 1^λ. (In our schemes the secret keys will be random, so k_u is simply the length of the ID-based secret key.)

Let $\mathcal{S}^*$ denote a set of all possible valid identity-based secret keys with respect to ID^{*2}. Let $\mathcal{S}$ denote a set of q_e identity-based secret keys such that $\mathcal{S}^* \cap \mathcal{S} = \emptyset$. Finally, let $\mathcal{F}_{id-ow}(g^u(k_u))$ be the class of all polynomial-time computable functions f; such that for all $i \in [1, q_\ell]$, given

$$\mathsf{mpk}, \quad \mathsf{ID}^*, \quad \mathcal{S}, \text{ and } \quad \{f_i(\mathsf{msk}, \mathcal{L}_{\mathsf{ID}}, \mathsf{mpk}, \mathsf{ID}_i)\}_{i \in [1, q_\ell]},$$

(for $(\mathsf{msk}, \mathsf{mpk}, \{sk_{\mathsf{ID}_i}\}_{i \in [1, q_\ell]}, \mathcal{S}, \mathcal{L}_{\mathsf{ID}})$ that is randomly generated, and $\{\mathsf{ID}^*\} \cup \{\mathsf{ID}_i\}_{i \in [1, q_\ell]} \subseteq \mathcal{I}$), no PPT algorithm can find a valid secret key sk_{ID^*} of ID^* with probability greater than $g^u(k_u)$, where $g^u(k_u) \geq 2^{-k_u}$ is the hardness parameter. Our goal is to make $g^u(k_u)$ as large (i.e., as close to $\mathrm{negl}(k_u)$) as possible.

Definition 5. *An identity-based encryption is said to be $(g^u(k_u))$-AI-CPA (auxiliary input CPA) secure if it is IND-ID-CPA secure w.r.t. family $\mathcal{F}_{id-ow}(g^u(k_u))$.*

Discussions. Our model for IBE with auxiliary inputs bears some differences from the existing model of *public key encryption (PKE) with auxiliary input* [9] and *IBE with length-bounded leakage* [13].

- For PKE, the public key itself leaks some information about the secret key. Therefore, in [9], they define the family $\mathcal{F}_{pk-ow}$ such that, given $f(\mathsf{msk}, \mathsf{mpk})$ and mpk (where $f \in \mathcal{F}_{pk-ow}$), it is difficult to output msk. For IBE, the master public key leaks some information about the master secret key, which may be exploited to compromise the security since ID-based secret key can be computed from the master secret. Therefore, we define the family $\mathcal{F}_{id-ow}$ such that given the above information, it is difficult to output sk_{ID^*}.
- The CPA security for PKE only has one single oracle which is for leakage, so adaptive leakage can be modeled by a single oracle query [9]. On the other hand, for IBE (e.g., the bounded retrieval model in [13]), the adversary can either query the leakage of msk or sk_{ID}, or unlock all bits of sk_{ID}. Thus we need to model the leakage via multiple queries for adaptive adversary.
- We combine the two separate leakage oracles in [13], for modeling the case that *the adversary may obtain leakage from msk and sk_{ID} at the same time*, and they may share the same internal randomness.

---

[2] A valid ID-based secret key for ID^* can decrypt all ciphertext encrypted to ID^*, hence every ID-based secret key for ID^* from Ext are in the set $\mathcal{S}^*$.

– Same as [13], multiple keys are allowed for each identity. For leakage oracle, we allow leakage of all these keys. Moreover, we do not need to store the amount of leakage for the master secret key and the identity-based secret keys, therefore we do not need to create a set which stores the handles of secret keys. For extraction, we could just return a new key upon each query (and store them for later leakage), but we chose to allow the adversary to supply a handle to specify which particular key to be extracted. The generality here may be useful for other higher applications of IBE, for example, those assigning keys for the same identity to different users.

3.2   Intuition

The Scheme. This construction is a parallel repetition of the Lewko-Waters IBE [14]. The key generation centre (KGC) splits the master secret key msk into m pieces $\{\alpha_i\}$. This idea can be found in many leakage-resilient schemes, e.g. [9]. The ID-based secret keys contain n components, each of which is created from a share of msk. The ciphertext also contains n components. Pairing the secret key and the ciphertext component-wise recovers a encapsulated key corresponding to α_i. Their product is the padding for hiding the actual message.

Like Lewko-Waters IBE [14] (and the underlying IBE [3]), an identity $\mathsf{ID} \in \mathbb{Z}_N$ is mapped to a group element $u^{\mathsf{ID}}h$, where $u, h \in \mathbb{G}_{p_1}$. To allow leakage of the master secret key, instead of keeping $\alpha_i \in \mathbb{Z}_N$, we store msk in a form similar to the structure of an ID-based secret key [14]. Recall that a "basic" [3] ID-based secret key contains $(g_1^{\alpha_i} \cdot (u^{\mathsf{ID}}h)^{r_i}, v_i^{r_i})$ (the second term embeds r_i for cancelling the $(u^{\mathsf{ID}}h)^{r_i}$ part in decryption), one can store $(g_1^{\alpha_i} \cdot h^{r_i}, u^{r_i}, v_i^{r_i})$ as the master secret for "undetermined" ID.

The Proof. Our proof uses the dual-system encryption technique [17,14,13]. The keys and the ciphertexts are masked by random group elements in $\mathbb{G}_{p_3}$ for adaptive security, and in the proof all these will be turned into their semi-functional (SF) version by introducing random factors from $\mathbb{G}_{p_2}$ [17]. The basic technique [14] ensures that the real key is indistinguishable to an SF key. For leakage, an SF key is further classified into two types: truly SF and nominally SF. The latter can still decrypt an SF ciphertext, but the former will make the decryption fails. Thus, a truly SF key is used to simulate the leakage oracle, which does not help the adversary.

If the adversary can distinguish between these two types of SF keys, we hope to leverage this to break the underlying assumption. In our case, we want to invert the leakage function which is supposed to be uninvertible. For Lewko-Waters IBE, it was done by applying their Lemma 6.2 for bounded leakage [14]. However, we cannot directly replace it with Theorem 2 for auxiliary input as it restricts the blinding factor of an SF key to be a λ-bit number. Therefore, these SF structures have to be changed accordingly. Since they only appear in the proof, the actual scheme is not affected.

3.3   Concrete Construction

Setup(1^λ): The KGC runs the bilinear group generator $\mathcal{G}(1^\lambda)$ to get ($N = p_1 p_2 p_3, \mathbb{G}, \mathbb{G}_T, \hat{e}$) as defined in §2. Suppose $\mathcal{G}$ also gives g_1 and X_3 which are generators of the subgroups $\mathbb{G}_{p_1}$ and $\mathbb{G}_{p_3}$ respectively. Let $0 < \epsilon < 1$ and $m = (3\lambda)^{1/\epsilon}$. The KGC randomly picks $\alpha_1, \ldots, \alpha_m \in \mathbb{Z}_N$, $u, h, v_1, \ldots, v_m \in \mathbb{G}_{p_1}$. The master public key is

$$\left\{ N, \mathbb{G}, \mathbb{G}_T, \hat{e}, g_1, u, h, X_3, \{v_i, y_i = \hat{e}(g_1, v_i)^{\alpha_i}\}_{i\in[1,m]} \right\}.$$

The KGC also randomly picks $t_i \in \mathbb{Z}_N$, and $T_{1,i}, T_{2,i}, T_{3,i} \in \mathbb{G}_{p_3}$ for $i \in [1, m]$. The master secret key is $\{K_{1,i}, K_{2,i}, K_{3,i}\}_{i\in[1,m]}$ where

$$K_{1,i} = g_1^{\alpha_i} \cdot h^{t_i} \cdot T_{1,i}, \qquad K_{2,i} = u^{t_i} \cdot T_{2,i}, \qquad K_{3,i} = v_i^{t_i} \cdot T_{3,i}.$$

The message space $\mathcal{M}$ is $\mathbb{G}_T$ and the identity space $\mathcal{I}$ is $\mathbb{Z}_N$.

Ext(msk, ID): For $i \in [1, m]$, randomly picks $r_i \in \mathbb{Z}_N$ and $R_{1,i}, R_{2,i} \in \mathbb{G}_{p_3}$, it outputs the identity-based secret key $sk_{\text{ID}} = \{D_1, E_1, \ldots, D_m, E_m\}$ where

$$D_i = K_{1,i} \cdot K_{2,i}^{\text{ID}} \cdot (u^{\text{ID}} h)^{r_i} \cdot R_{1,i}, \qquad E_i = K_{3,i} \cdot v_i^{r_i} \cdot R_{2,i}.$$

Enc(ID, M): For $i \in [1, m]$, it randomly picks $s_i \in \mathbb{Z}_N$ and outputs the ciphertext $\mathfrak{C} = \{A, \{B_i\}_{i\in[1,m]}, \{C_i\}_{i\in[1,m]}\}$ where

$$A = M \cdot \prod_{i=1}^{m} y_i^{s_i}, \qquad B_i = v_i^{s_i}, \qquad C_i = (u^{\text{ID}} h)^{s_i}.$$

Dec($sk_{\text{ID}}, \mathfrak{C}$): Given a ciphertext $\mathfrak{C} = \{A, \{B_i\}_{i\in[1,m]}, \{C_i\}_{i\in[1,m]}\}$, and a secret key $sk_{\text{ID}} = \{D_1, E_1, \ldots, D_m, E_m\}$ for an identity ID, it outputs

$$M = A \cdot \frac{\prod_{i=1}^{m} \hat{e}(C_i, E_i)}{\prod_{i=1}^{m} \hat{e}(B_i, D_i)}.$$

Hierarchical Extension. Similar to existing HIBE schemes [3,14], one can extend the above IBE to n-level HIBE by extending the master public key. Due to page limitation, we just outline the modifications and omit its security proof. Specifically, mpk contains $u_1, \ldots, u_n \in \mathbb{G}_{p_1}$ instead of a single u. Accordingly, to extract a key or encrypt a message for a vector of identity ($\text{ID}_1, \ldots, \text{ID}_n$) instead of just ID, $(u^{\text{ID}} h)$ in the computation of D_i in Ext and in the computation of C_i in Enc is replaced by $(\prod_{j=1}^{n} u_j^{\text{ID}_j} h)$. The Dec algorithm remains the same.

3.4   Security

Under the dual system encryption paradigm [17], we define the following three semi-functional (SF) structures which are used in the security proofs only. These

SF structures just like their normal version in the actual scheme, but "perturbed" by a $\mathbb{G}_{p_2}$ generator, denoted by either $\bar{g}_2$ or $\hat{g}_2$ below.

An *SF master-key* $\{K'_{1,i}, K'_{2,i}, K'_{3,i}\}_{i \in [1,m]}$ is given by:

$$K'_{1,i} = K_{1,i} \cdot \bar{g}_2^{\theta_i}, \qquad K'_{2,i} = K_{2,i} \cdot \bar{g}_2^{\tau \theta_i}, \qquad K'_{3,i} = K_{3,i} \cdot \bar{g}_2^{w_i},$$

where $\theta_1, \ldots, \theta_m \in [0, \lambda]$, $\tau, w_1, \ldots, w_m \in \mathbb{Z}_N$ and $\{K_{1,i}, K_{2,i}, K_{3,i}\}$ is a normal master key.

An *SF ID-based key* (or just *SF key*) is in the form of

$$\{D'_i = D_i \cdot \bar{g}_2^{\gamma_i}, \qquad E'_i = E_i \cdot \bar{g}_2^{z_i}\}_{i \in [1,m]},$$

where $z_1, \ldots, z_m, \gamma_1, \ldots, \gamma_m \in \mathbb{Z}_N$ and $\{D_i, E_i\}_{i \in [1,m]}$ is a normal ID-based key.

An *SF ciphertext* is in the form of

$$\left\{A' = A, \qquad \{B'_i = B_i \cdot \hat{g}_2^{\delta_i}, \qquad C'_i = C_i \cdot \hat{g}_2^{x_i}\}_{i \in [1,m]}\right\},$$

where $\delta_1, x_1, \ldots, \delta_m, x_m \in \mathbb{Z}_N$ and $\{A, \{B_i, C_i\}_{i \in [1,m]}\}$ is a normal ciphertext.

Decryption will succeed if an SF key is used to decrypt a normal ciphertext, or a normal key is used to decrypt an SF ciphertext. However, decrypting an SF ciphertext using an SF key will result in a message "blinded" by

$$\hat{e}(\bar{g}_2, \hat{g}_2)^{\sum_{i=1}^m z_i x_i - \sum_{i=1}^m \gamma_i \delta_i}.$$

Furthermore, the ID-based secret key generated by applying Ext with an SF master key is also semi-functional. If we use it to decrypt an SF ciphertext, result will be shifted by a factor

$$\hat{e}(\bar{g}_2, \hat{g}_2)^{\sum_{i=1}^m w_i x_i - \sum_{i=1}^m (1 + \tau \mathsf{ID}) \theta_i \delta_i}.$$

In case that the exponents in these extra blinding factors are zeros, decryption still works and this leads us to the notion of *nominally semi-functional (NSF) keys*. An NSF ID-based key is a special kind of SF key which can be used to decrypt SF ciphertext, that means $\sum_{i=1}^m \gamma_i \delta_i = \sum_{i=1}^m z_i x_i$. Similarly, an NSF master-key is a special kind of SF master-key which can be used to decrypt SF ciphertext, that means $\sum_{i=1}^m (1 + \tau \mathsf{ID}) \theta_i \delta_i = \sum_{i=1}^m w_i x_i$. If an SF identity-based / master key is not nominally semi-functional, then it is *truly semi-functional*.

Theorem 6. *Our IBE scheme is (2^{-m^ϵ})-AI-ID-CPA secure under Assumptions 1, 2 and 3.*

Proof. We prove by a hybrid argument using a sequence of games. The first game Game_{real} is the real AI-ID-CPA game and we denote the challenge identity as ID^*. The second game $\mathrm{Game}_{restricted}$ is the same as Game_{real} except that the adversary cannot ask for the secret key of identity ID where $\mathsf{ID} \equiv \mathsf{ID}^* \mod p_2$. This restriction will be retained throughout the subsequent games. After that, we denote $q := q_e + q_u + q_\ell$ as the number of extraction oracle, UpdateUSK oracle and leakage oracle queries. For $k = 0$ to q, we define Game_k as follows.

Game_k: It is the same as Game_{real}, except that both the challenge ciphertext and the keys used to answer first k-th *distinct* oracle queries[3] are semi-functional. The keys for the rest of the queries are normal. So, for the first k-th queries:

1. If it is for extraction oracle, it returns the semi-functional key sk'_{ID}.
2. If it is for leakage oracle, it returns $f(\text{msk}', \mathcal{L}_{\text{ID}}, \text{mpk}, \text{ID})$ where msk' is semi-functional and for the last entry $(sk'_{\text{ID}}, \text{ID}, \cdot) \in \mathcal{L}_{ID}$, sk'_{ID} is semi-functional.
3. If it is for UpdateUSK oracle, it puts a semi-functional key sk'_{ID} into $\mathcal{L}_{ID}$.

As a result, all keys are normal and the challenge ciphertext is semi-functional in Game_0. In Game_q, all keys and the challenge ciphertext are semi-functional.

The last game is Game_{final}, which is the same as Game_q except that the challenge ciphertext is a semi-functional encryption of a random message, instead of one of the two challenge messages.

We will prove the indistinguishability between these games.

Lemma 7. *If there exists an adversary $\mathcal{A}$ such that $Adv_\mathcal{A}(\text{Game}_{real})$ - $Adv_\mathcal{A}(\text{Game}_{restricted}) = \epsilon$, then we can construct an algorithm $\mathcal{B}$ with non-negligible advantage in breaking Assumption 2.*

Lemma 8. *If there exists an adversary $\mathcal{A}$ such that $Adv_\mathcal{A}(\text{Game}_{restricted})$ - $Adv_\mathcal{A}(\text{Game}_0) = \epsilon$, then we can construct an algorithm $\mathcal{B}$ with advantage ϵ in breaking Assumption 1.*

Lemma 9. *If there exists an adversary $\mathcal{A}$ such that $Adv_\mathcal{A}(\text{Game}_{\ell-1})$ - $Adv_\mathcal{A}(\text{Game}_\ell) = \epsilon$, then we can construct an algorithm $\mathcal{B}$ with advantage ϵ in breaking Assumption 2.*

Lemma 10. *If there exists an adversary $\mathcal{A}$ such that $Adv_\mathcal{A}(\text{Game}_q)$ - $Adv_\mathcal{A}(\text{Game}_{final}) = \epsilon$, then we can construct an algorithm $\mathcal{B}$ with advantage ϵ in breaking Assumption 3.*

The proofs of lemma 7, 8, 9 and 10 are given in the the full version.

Finally in Game_{final}, the value of b is information theoretically hidden from $\mathcal{A}$. Hence $\mathcal{A}$ has no advantage in winning Game_{final}. If Assumptions 1, 2 and 3 hold, Game_{real} is indistinguishable from Game_{final}. Hence the attacker has negligible advantage in winning Game_{real}. Therefore, our scheme is (2^{-m^ϵ})-AI-ID-CPA secure.   □

---

[3] We consider the following parameters to determine if two queries are the same, i.e., *not distinct*. For leakage queries, we consider the function f and its argument. In particular, when they are the same, the same version of the secret key for the same ID is leaked in the same way. For extraction, we consider ID and the counter i.

4   IBE with Continual Auxiliary Inputs

4.1   Continual Auxiliary Leakage Model

We propose the continual auxiliary leakage model for IBE. First, we separate the setup algorithm into two, one for common reference string (CRS) generation and another for master key pair generation. This separation has been done previously [7] for specific security goals. It is necessary in our case since leakage is only allowed from the later part. We also introduce two additional update algorithms.

- CRSGen: On input a security parameter 1^λ, it generates a CRS param.
- Setup: On input a CRS param, it generates a master public key mpk and a master secret key msk. Denote the randomness used (msk, mpk) as r_s.
- UpdateMSK: On input a master key pair (msk, mpk), it outputs a re-randomized master secret key $\bar{\text{msk}}$. Denote the randomness used as r_m.
- UpdateUSK: On input an identity-based secret key sk_{ID} for the identity ID and mpk, it outputs a re-randomized identity-based secret key $\bar{sk}_{\text{ID}}$. Denote the randomness used as r_u.

After running both UpdateMSK and UpdateUSK algorithms, the corresponding public keys remain unchanged after the re-randomization; and the size of the secret keys also remain unchanged.

Denote the master secret key's, identity-based secret keys', messages' and identities' spaces as $\mathcal{MK}$, $\mathcal{SK}$, $\mathcal{M}$ and $\mathcal{I}$ respectively. Denote a polynomial-time computable function family $\mathcal{F}$. The security of IBE in the continual auxiliary leakage model is defined via the following game.

1. **Setup.** The challenger firstly runs param $\leftarrow$ CRSGen(1^λ) and (mpk, msk) $\leftarrow$ Setup(param). Denote the randomness used in Setup as r_s. The adversary specifies a function $f_0 \in \mathcal{F}$. Denote ϵ as an empty string. The challenger gives param, mpk and $f_0(r_s, \epsilon, \epsilon, \epsilon, \epsilon, \epsilon, \epsilon)$ to the adversary $\mathcal{A}$. The challenger constructs the list $\mathcal{L}_{\text{msk}}$, which stores the tuples $(\text{msk}, \cdot)^4$, and the lists $\mathcal{L}_e$ and $\mathcal{L}_{\text{ID}}$, which are initially empty.

2. **Phase 1.** The following oracles can be queried by $\mathcal{A}$ adaptively:
   - Extraction Oracle $\mathcal{KEO}(\text{ID})$: On input an identity ID $\in \mathcal{I}$, it looks for the last $(\text{ID}, sk_{\text{ID}})$ entry from the list $\mathcal{L}_e$. If such entry does not exist, it runs $sk_{\text{ID}} \leftarrow \text{Ext}(\text{msk}, \text{ID})$ and stores $(\text{ID}, sk_{\text{ID}})$ in the list $\mathcal{L}_e$. Finally, it returns the identity-based secret key sk_{ID}.
   - Leakage Oracle $\mathcal{LO}(f)$: On input a polynomial-time computable function $f \in \mathcal{F}$, it returns $f(\epsilon, \mathcal{L}_{\text{msk}}, \epsilon, \text{msk}, \epsilon, \text{mpk}, \epsilon)$.
   - UpdateMSK Oracle $\mathcal{UMO}$: It runs $\bar{\text{msk}} \leftarrow \text{UpdateMSK}(\text{msk})$. Denote the randomness used as r_m. It puts (msk, r_m) in the list $\mathcal{L}_{\text{msk}}$. After that, it sets msk $\leftarrow \bar{\text{msk}}$ and outputs msk.

   Denote q_e, q_ℓ, q_m as the number of oracle queries to the $\mathcal{KEO}$, $\mathcal{LO}$ and $\mathcal{UMO}$ respectively in this game.

---

[4] An alternative definition is to include r_s in the list $\mathcal{L}_{\text{msk}}$, which is part of the input for the leakage queries in the later phase.

3. **Challenge Identity.** $\mathcal{A}$ sends a challenge identity $\mathsf{ID}^* \in \mathcal{I}$ to the challenger. The challenger runs $sk_{\mathsf{ID}^*} \leftarrow \mathtt{Ext}(\mathsf{msk}, \mathsf{ID}^*)$.
4. **Phase 2.** The following oracles can be queried by $\mathcal{A}$ adaptively:
   - Extraction Oracle $\mathcal{KEO}(\mathsf{ID})$: Same as that in Phase 1.
   - Leakage Oracle $\mathcal{LO}(f)$: On input a polynomial-time computable function $f \in \mathcal{F}$, it returns $f(\epsilon, \mathcal{L}_{\mathsf{msk}}, \mathcal{L}_{\mathsf{ID}}, \mathsf{msk}, sk_{\mathsf{ID}^*}, \mathsf{mpk}, \mathsf{ID}^*)$.
   - UpdateMSK Oracle $\mathcal{UMO}$: Same as that in Phase 1.
   - UpdateUSK Oracle $\mathcal{USO}$: It runs $\bar{sk}_{\mathsf{ID}^*} \leftarrow \mathtt{UpdateUSK}(sk_{\mathsf{ID}^*})$. Denote the randomness used as r_u. It puts $(sk_{\mathsf{ID}^*}, r_u)$ in $\mathcal{L}_{\mathsf{ID}}$ and sets $sk_{\mathsf{ID}^*} = \bar{sk}_{\mathsf{ID}^*}$.
   Denote q_u as the number of oracle queries to the $\mathcal{USO}$ in this game.
5. **Challenge.** $\mathcal{A}$ sends two messages $M_0, M_1 \in \mathcal{M}$ to the challenger. The challenger picks a random bit b' and computes $\mathfrak{C}^* \leftarrow \mathtt{Enc}(\mathsf{mpk}, \mathsf{ID}^*, M_{b'})$.
6. **Phase 3.** The challenger sends $\mathfrak{C}^*$ to $\mathcal{A}$. $\mathcal{A}$ is allowed to query $\mathcal{KEO}$ adaptively.
7. **Output.** $\mathcal{A}$ returns a guess b^* of b'. $\mathcal{A}$ wins the game if $b' = b^*$ and there was no $\mathcal{KEO}(\mathsf{ID}^*)$ query.

The advantage of $\mathcal{A}$ is $\left| \Pr[\mathcal{A} \text{ wins}] - \frac{1}{2} \right|$. An IBE scheme is IND-ID-CPA secure in the continual auxiliary leakage model for $\mathcal{F}$ (or CAL-CPA secure) if there is no PPT $\mathcal{A}$ with non-negligible advantage in the game above.

Class of Auxiliary Functions. Let $\mathcal{S}^*$ denote a set of all possible valid identity-based secret keys with respect to ID^*. Let $\mathcal{S}$ denote a set of q_e identity-based secret keys such that $\mathcal{S}^* \cap \mathcal{S} = \emptyset$. Let $\mathcal{F}_{id-ow}(g^{\mathrm{u}}(k_u))$ be the class of all polynomial-time computable functions f; such that given

mpk, ID^*, $\mathcal{S}$, $f_0(r_s, \cdots)$ and $\{f_i(\epsilon, \mathcal{L}_{\mathsf{msk}}, \mathcal{L}_{\mathsf{ID}}, \mathsf{msk}, sk_{\mathsf{ID}^*}, \mathsf{mpk}, \mathsf{ID}^*)\}_{i \in [1, q_\ell]}$,

(for a randomly generated $(\mathsf{msk}, \mathsf{mpk}, r_s, sk_{\mathsf{ID}^*}, \mathcal{S}, \mathcal{L}_{\mathsf{msk}}, \mathcal{L}_{\mathsf{ID}})$, and $\mathsf{ID}^* \subseteq \mathcal{I})$[5], no PPT algorithm can find a $sk_{\mathsf{ID}^*} \in \mathcal{S}^*$ with probability greater than $g^{\mathrm{u}}(k_u)$, where $g^{\mathrm{u}}(k_u) \geq 2^{-k_u}$ is the hardness parameter.

Definition 11. *An IBE scheme is said to be $(g^{\mathrm{u}}(k_u))$-CAL-CPA secure if it is IND-ID-CPA secure w.r.t. family $\mathcal{F}^{\mathrm{u}}_{id-ow}(g^{\mathrm{u}}(k_u))$.*

4.2   Construction in the Continual Auxiliary Leakage Model

We can extend our basic IBE in §3.3 to give an IBE scheme secure in the continual leakage model. The advantage is that it does not have much difference from our basic IBE. However, the extended scheme does not allow leakage during the setup phase. It implies that in the security model, the leakage f_0 is not allowed. Firstly, the common reference string is $(N = p_1p_2p_3, \mathbb{G}, \mathbb{G}_T, \hat{e})$ as generated by the bilinear group generator, i.e. $\mathtt{CRSGen}(\cdot) = \mathcal{G}()$. Then, the rest of $\mathtt{Setup}$ in §3.3 constitutes our new $\mathtt{Setup}$. Finally, we introduce the two algorithms below.

---

[5] The msk here is the current value of msk when the leakage oracle query for f is made. It may be changed by the UpdateMSK oracle, hence it is not a fixed value. The same applies for other variables such as $\mathcal{L}_{\mathsf{msk}}$ and $\mathcal{L}_{\mathsf{ID}}$.

UpdateMSK: Given $\{K_{1,i}, K_{2,i}, K_{3,i}\}$, the KGC randomly picks $t_i' \in \mathbb{Z}_N$ and $T_{1,i}', T_{2,i}', T_{3,i}' \in \mathbb{G}_{p_3}$, for $i \in [1, m]$. The new master secret key is defined by:

$$K_{1,i}' = K_{1,i} \cdot h^{t_i'} \cdot T_{1,i}', \qquad K_{2,i}' = K_{2,i} \cdot u^{t_i'} \cdot T_{2,i}', \qquad K_{3,i}' = K_{3,i} \cdot v^{t_i'} \cdot T_{3,i}'.$$

UpdateUSK: Given $sk_{\mathsf{ID}} = \{D_1, E_1, \ldots, D_m, E_m\}$ for the identity ID, it randomly picks $r_i' \in \mathbb{Z}_N$ and $R_{1,i}', R_{2,i}' \in \mathbb{G}_{p_3}$ for $i \in [1, m]$, then the new key is given by:

$$D_i' = D_i \cdot (u^{\mathsf{ID}}h)^{r_i'} \cdot R_{1,i}', \qquad E_i' = E_i \cdot v^{r_i'} \cdot R_{2,i}'.$$

Theorem 12. *Our IBE scheme is (2^{-m^ϵ})-CAL-CPA secure if Assumption 1, Assumption 2 and Assumption 3 hold.*

Compared with our basic IBE, we have to additionally simulate the oracles for updating and leak the randomness used. These updates all used random elements in $\mathbb{G}_{p_3}$, which which has no impact to the previous proof. So the proof is similar to that of our basic IBE and hence is omitted.

4.3   Further Discussions on the Continual Auxiliary Input Model

Our continual auxiliary input model extends the traditional continual memory leakage model in two dimensions. Previous definitions consider only length-bounded leakage with the requirement of secure erasure. For length-bounded leakage, continual leakage is obviously stronger than non-continual leakage since it allows more bits to be leaked, or, there cannot be arbitrarily large leakage on the same copy of the (static) secret key in the non-continual model.

Here, we consider "continual leakage without erasure" for an even stronger attack model – for example, an adversary may decide to leak more bits of the old secret key after seeing some bits of its refreshed version. But that seems to bring us back to the original non-continual scenario. If the old keys are not erased, the adversary can always choose to keep leaking the old keys even in the later "epochs" when the secret keys are refreshed. One may consider a model which allows "fine-grained" leakage, say, only allowing a particular query to leak a certain number of bits of an old key and keeping track of the number of bits leaked from the refreshed key via the same leakage function. But it might be difficult to have a clean definition for that. On the other hand, in the auxiliary input model, we can capture this stronger attack model by a simple uninvertibility condition.

Indeed, the basic auxiliary input model seems to be so "powerful" that any scheme that is secure in the basic model for a certain function family $\mathcal{F}$ is also secure in the continual auxiliary leakage model for another family $\mathcal{F}'$. However, it does not mean that the additional key-refreshing algorithms we just introduced have no significance. Instead, a careful design of these algorithms can enlarge the size of the allowed function family, which means a more general form of leakage is allowed. To see, consider an artificial scheme which "keeps state" across epochs and puts one bit of the same identity-based secret key in each version of a certain secret key. Eventually, this secret key can be recovered by leaking a single bit of each of these keys, so any set of queries containing these leakages is ruled out by the uninvertibility condition, i.e., they are excluded from the function family. On the other hand, our scheme could allow this set of leakage queries.

4.4 Construction Supporting Leakage-Resilient Setup

Intuition. In the IBE scheme in [13], the KGC first picks an $\alpha \in \mathbb{Z}_N$ and uses n random tags (and other elements) to blind it. The master secret key contains multiple elements but α is enough for decryption of every ciphertexts. It seems leaking a part of the randomness is sufficient to break the scheme. To allow auxiliary leakage in the setup, we resort to Theorem 2 again. Our scheme picks a random $n \times m$ matrix $\boldsymbol{V}$ and multiplies it with α_i's to obtain m master public keys $\boldsymbol{Y}$, where $\alpha_i \in \{0, 1\}$ for $i \in [1, n]$. Similar method is used in (LWE-based) GPV encryption in [9]. Denote the randomness $\boldsymbol{\alpha} = (\alpha_1, \ldots, \alpha_n)$ and the generator $\boldsymbol{g} = (g_1, \ldots, g_n)$. Roughly speaking, we set $\boldsymbol{Y} = \hat{e}(\boldsymbol{g}, \boldsymbol{V}^{\boldsymbol{\alpha}})$, where the pairing operation is taken entry-wise. The master secret key is $\boldsymbol{V}^{\boldsymbol{\alpha}} = \{\prod_{j=1}^{m} v_{i,j}^{\alpha_j}\}_{i \in [1,n]}$. In the security proof, the simulator can use the uninvertible function $F(\alpha_1, \ldots, \alpha_n)$ to output $v_{i,j}^{\alpha_j}$. It is difficult to invert if the ISIS assumption holds.

Construction.
CRSGen: On input the security parameter 1^λ, the setup algorithm runs $(N = p_1 p_2 p_3, \mathbb{G}, \mathbb{G}_T, \hat{e}) \leftarrow \mathcal{G}(1^\lambda)$. We suppose the group generator $\mathcal{G}$ also gives the generators (u, h) and X_3 of the subgroups $\mathbb{G}_{p_1}$ and $\mathbb{G}_{p_3}$ respectively.

Setup: Let $0 < \epsilon < 1$, $n = O(\lambda)$ and $m = ((n+4)\lambda)^{1/\epsilon}$. The KGC randomly picks $\alpha_1, \ldots, \alpha_m \in \{0, 1\}$, a random matrix $\mathbf{V} \in \mathbb{G}_{p_1}^{n \times m}$ and a random $\mathbf{G} \in \mathbb{G}_{p_1}^n$:

$$\mathbf{V} = \begin{bmatrix} v_{1,1} & v_{1,2} & \cdots & v_{1,m} \\ v_{2,1} & v_{2,2} & \cdots & v_{2,m} \\ \vdots & \vdots & \ddots & \vdots \\ v_{n,1} & v_{n,2} & \cdots & v_{n,m} \end{bmatrix}, \quad \mathbf{G} = \begin{bmatrix} g_1 \\ g_2 \\ \vdots \\ g_n \end{bmatrix}.$$

Then we define $q_i = \prod_{j=1}^{m} v_{i,j}^{\alpha_j}$ for $i \in [1, n]$, and

$$\boldsymbol{\alpha} = \begin{bmatrix} \alpha_1 \\ \alpha_2 \\ \vdots \\ \alpha_m \end{bmatrix}, \quad \mathbf{Y} = \begin{bmatrix} y_1 = \hat{e}(g_1, q_1) \\ y_2 = \hat{e}(g_2, q_2) \\ \vdots \\ y_n = \hat{e}(g_n, q_n) \end{bmatrix}.$$

The system parameter param is $\{N, \mathbb{G}, \mathbb{G}_T, \hat{e}, u, h, X_3, \mathbf{G}\}$. The master public key mpk is $\mathbf{Y}$. The KGC randomly picks $t_i \in \mathbb{Z}_N$ and $T_{1,i}, T_{2,i}, T_{3,i} \in \mathbb{G}_{p_3}$ for $i \in [1, n]$. The master secret key msk is $\{\{K_{1,i}, K_{2,i}, K_{3,i}\}_{i \in [1,n]}\}$ where

$$K_{1,i} = q_i \cdot h^{t_i} \cdot T_{1,i}, \qquad K_{2,i} = u^{t_i} \cdot T_{2,i}, \qquad K_{3,i} = g_i^{t_i} \cdot T_{3,i}.$$

The randomness used to generate msk are $\{\alpha_i, t_i, T_{1,i}, T_{2,i}, T_{3,i}\}$ for $i \in [1, n]$.

Define the message space $\mathcal{M}$ as $\mathbb{G}_T$ and the identity space $\mathcal{I}$ as a λ-bit integer.

Ext: Given the master secret key $\left\{\{K_{1,i}, K_{2,i}, K_{3,i}\}_{i\in[1,n]}\right\}$, and an identity ID, it picks a random $r_i \in \mathbb{Z}_N$ and some random $R_{1,i}, R_{2,i} \in \mathbb{G}_{p_3}$, then it calculates:

$$D_i = K_{1,i} \cdot K_{2,i}^{\mathsf{ID}} \cdot (u^{\mathsf{ID}}h)^{r_i} R_{1,i}, \qquad E_i = K_{3,i} \cdot g_i^{r_i} R_{2,i},$$

for $i \in [1,n]$. It is equivalent to $D_i = \prod_{j=1}^{m} v_{i,j}^{\alpha_j} \cdot (u^{\mathsf{ID}}h)^{\bar{r}_i} \cdot \bar{R}_{1,i}$ and $E_i = g_i^{\bar{r}_i} \cdot \bar{R}_{2,i}$, for random $\bar{r}_i \in \mathbb{Z}_N$, $\bar{R}_{1,i}, \bar{R}_{2,i} \in \mathbb{G}_{p_3}$. The output is $sk_{\mathsf{ID}} = \{D_1, E_1, \ldots, D_n, E_n\}$.

Enc: To encrypt a message $M \in \mathbb{G}_T$ for a user ID, for $i \in [1,n]$, it randomly picks $s_i \in \mathbb{Z}_N$ and calculates the ciphertext is $\{A, B_1, C_1, \ldots, B_n, C_n\}$ as:

$$A = M \cdot \prod_{i=1}^{n} y_i^{s_i}, \qquad B_i = g_i^{s_i}, \qquad C_i = (u^{\mathsf{ID}}h)^{s_i}.$$

Dec: Given $\mathfrak{C} = \{A, B_1, C_1, \ldots, B_n, C_n\}$, and $sk_{\mathsf{ID}} = \{D_1, E_1, \ldots, D_n, E_n\}$, the message is recovered from the ciphertext $\mathfrak{C}$ by:

$$M = A \cdot \frac{\prod_{i=1}^{n} \hat{e}(C_i, E_i)}{\prod_{i=1}^{n} \hat{e}(B_i, D_i)}.$$

UpdateMSK: Given $\left\{\{K_{1,i}, K_{2,i}, K_{3,i}\}_{i\in[1,n]}\right\}$, the KGC randomly picks $t'_i \in \mathbb{Z}_N$ and $T'_{1,i}, T'_{2,i}, T'_{3,i} \in \mathbb{G}_{p_3}$, for $i \in [1,n]$, it sets the new master secret key as:

$$K'_{1,i} = K_{1,i} \cdot h^{t'_i} \cdot T'_{1,i}, \qquad K'_{2,i} = K_{2,i} \cdot u^{t'_i} \cdot T'_{2,j}, \qquad K'_{3,i} = K_{3,i} \cdot g_i^{t'_i} \cdot T'_{3,i}.$$

UpdateUSK: Given $sk_{\mathsf{ID}} = \{D_1, E_1, \ldots, D_n, E_n\}$ for the identity ID, it picks some random $r'_i \in \mathbb{Z}_N$ and some random $R'_{1,i}, R'_{2,i} \in \mathbb{G}_{p_3}$ for $i \in [1,n]$, then it calculates the new identity-based secret key by:

$$D'_i = D_i \cdot (u^{\mathsf{ID}}h)^{r'_i} \cdot R'_{1,i}, \qquad E'_i = E_i \cdot v^{r'_i} \cdot R'_{2,i}.$$

Theorem 13. *Our IBE scheme is (2^{-m^e})-CAL-CPA secure if Assumptions 1, 2, 3 and the $\mathsf{ISIS}_{p_1,m,\sqrt{m}}$ assumption hold.*

The structure of the proof is similar to the proof of Theorem 6. We prove by a hybrid argument using a sequence of games. The $\mathsf{Game}_{restricted}$ and Game_k are almost the same as the proof of Theorem 6. After that, a new Game_{leak_i} is defined as the same as Game_q, except that $q_1, \ldots, q_i$ in the mpk are replaced by random elements in $\mathbb{G}_{p_1}$. The Game_{final} is also defined as the previous proof: the challenge ciphertext is changed to a semi-functional ciphertext encrypting a random message. The details of the proof are given in the full version.

Acknowledgement. Sherman Chow would like to thank Alfred Menezes and Dale Brydon for discussions, especially to Alfred for commenting on an earlier draft, and Jonathan Katz for pointing out the relation between the basic and continual auxiliary leakages.

References

1. Akavia, A., Goldwasser, S., Vaikuntanathan, V.: Simultaneous Hardcore Bits and Cryptography against Memory Attacks. In: Reingold, O. (ed.) TCC 2009. LNCS, vol. 5444, pp. 474–495. Springer, Heidelberg (2009)
2. Alwen, J., Dodis, Y., Naor, M., Segev, G., Walfish, S., Wichs, D.: Public-Key Encryption in the Bounded-Retrieval Model. In: Gilbert, H. (ed.) EUROCRYPT 2010. LNCS, vol. 6110, pp. 113–134. Springer, Heidelberg (2010)
3. Boneh, D., Boyen, X., Goh, E.-J.: Hierarchical Identity Based Encryption with Constant Size Ciphertext. In: Cramer, R. (ed.) EUROCRYPT 2005. LNCS, vol. 3494, pp. 440–456. Springer, Heidelberg (2005)
4. Boneh, D., Goh, E.-J., Nissim, K.: Evaluating 2-DNF Formulas on Ciphertexts. In: Kilian, J. (ed.) TCC 2005. LNCS, vol. 3378, pp. 325–341. Springer, Heidelberg (2005)
5. Boneh, D., Katz, J.: Improved Efficiency for CCA-Secure Cryptosystems Built Using Identity-Based Encryption. In: Menezes, A. (ed.) CT-RSA 2005. LNCS, vol. 3376, pp. 87–103. Springer, Heidelberg (2005)
6. Brakerski, Z., Kalai, Y.T., Katz, J., Vaikuntanathan, V.: Overcoming the hole in the bucket: Public-key cryptography resilient to continual memory leakage. In: FOCS 2010. IEEE Computer Society (2010)
7. Chow, S.S.M.: Removing Escrow from Identity-Based Encryption. In: Jarecki, S., Tsudik, G. (eds.) PKC 2009. LNCS, vol. 5443, pp. 256–276. Springer, Heidelberg (2009)
8. Chow, S.S.M., Dodis, Y., Rouselakis, Y., Waters, B.: Practical leakage-resilient identity-based encryption from simple assumptions. In: Al-Shaer, E., Keromytis, A.D., Shmatikov, V. (eds.) CCS 2010, pp. 152–161. ACM (2010)
9. Dodis, Y., Goldwasser, S., Kalai, Y.T., Peikert, C., Vaikuntanathan, V.: Public-Key Encryption Schemes with Auxiliary Inputs. In: Micciancio, D. (ed.) TCC 2010. LNCS, vol. 5978, pp. 361–381. Springer, Heidelberg (2010)
10. Dodis, Y., Haralambiev, K., López-Alt, A., Wichs, D.: Cryptography against continuous memory attacks. In: FOCS 2010, pp. 511–520. IEEE Computer Society (2010)
11. Gentry, C., Peikert, C., Vaikuntanathan, V.: Trapdoors for hard lattices and new cryptographic constructions. In: Dwork, C. (ed.) STOC 2008, pp. 197–206. ACM (2008)
12. Lewko, A.B., Lewko, M., Waters, B.: How to leak on key updates. In: Fortnow, L., Vadhan, S.P. (eds.) STOC 2011, pp. 725–734. ACM (2011)
13. Lewko, A.B., Rouselakis, Y., Waters, B.: Achieving Leakage Resilience through Dual System Encryption. In: Ishai, Y. (ed.) TCC 2011. LNCS, vol. 6597, pp. 70–88. Springer, Heidelberg (2011)
14. Lewko, A., Waters, B.: New Techniques for Dual System Encryption and Fully Secure HIBE with Short Ciphertexts. In: Micciancio, D. (ed.) TCC 2010. LNCS, vol. 5978, pp. 455–479. Springer, Heidelberg (2010)
15. Naor, M., Segev, G.: Public-Key Cryptosystems Resilient to Key Leakage. In: Halevi, S. (ed.) CRYPTO 2009. LNCS, vol. 5677, pp. 18–35. Springer, Heidelberg (2009)
16. Regev, O.: On lattices, learning with errors, random linear codes, and cryptography. In: Gabow, H.N., Fagin, R. (eds.) STOC 2005, pp. 84–93. ACM (2005)
17. Waters, B.: Dual System Encryption: Realizing Fully Secure IBE and HIBE under Simple Assumptions. In: Halevi, S. (ed.) CRYPTO 2009. LNCS, vol. 5677, pp. 619–636. Springer, Heidelberg (2009)

Quantum Proofs of Knowledge

Dominique Unruh

University of Tartu, Estonia

Abstract. We motivate, define and construct quantum proofs of knowledge, proofs of knowledge secure against quantum adversaries. Our constructions are based on a new quantum rewinding technique that allows us to extract witnesses in many classical proofs of knowledge. We give criteria under which a classical proof of knowledge is a quantum proof of knowledge. Combining our results with Watrous' results on quantum zero-knowledge, we show that there are zero-knowledge quantum proofs of knowledge for all languages in NP (assuming quantum 1-1 one-way functions).

1 Introduction

Cryptographic protocols, with few exceptions, are based on the assumption that certain problems are computationally hard. Typical examples include specific number-theoretic problems such as the difficulty of finding discrete logarithms, and general problems such as inverting one-way functions. It is well-known, however, that many such problems would become easy in the advent of quantum computers. Shor's algorithm [16], e.g., efficiently solves the discrete logarithm problem and allows to factor large integers. While quantum computers do not exist today, it is not unreasonable to expect quantum computers to be available in the future. To meet this threat, we need cryptographic protocols that are secure even in the presence of an adversary with a quantum computer. We stress that this does not necessarily imply that the protocol itself should make use of quantum technology; instead, it is preferable that the protocol itself can be easily implemented on today's readily-available classical computers.

Finding such quantum-secure protocols, however, is not trivial. Even when we have found suitable complexity-theoretic assumptions such as the hardness of certain lattice problems, a classical protocol based on these assumptions may fail to be secure against quantum computers. The reason for this is that many cryptographic proofs use a technique called rewinding. This technique requires that it is possible, when simulating some machine, to make snapshots of the state of that machine and then later to go back to that snapshot. As first observed by van de Graaf [9], classical rewinding-based proofs do not carry over to the quantum case. Two features unique to the quantum setting prohibit (naive) rewinding: The no-cloning theorem [21] states that quantum-information cannot be copied, so we cannot make snapshots. Furthermore, measurements destroy information, so interacting with a simulated machine may destroy information that would be needed later.

D. Pointcheval and T. Johansson (Eds.): EUROCRYPT 2012, LNCS 7237, pp. 135–152, 2012.

This leads to the following observation: Even if a classical protocol is proven secure based on the hardness of some problem, and even if that problem is hard even for quantum computers, we have no guarantee that the protocol is secure against quantum computers. The reduction of the protocol's security to the problem's hardness may be based on inherently classical features such as the possibility of rewinding.

An example of a protocol construction that suffers from this difficulty is zero-knowledge proofs. Zero-knowledge proofs are interactive proofs with the special property that the verifier does not learn anything except the validity of the proven statement. Zero-knowledge proofs are inherently based on rewinding (at least as long as we do not assume additional trusted setup such as so-called common-reference strings). Yet, zero-knowledge proofs are one of the most powerful tools available to the cryptographer; a multitude of protocol constructions use zero-knowledge proofs. These protocol constructions cannot be proven secure without using rewinding. To resolve this issue, Watrous [19] introduced a quantum rewinding technique. This technique allows to prove the quantum security of many common zero-knowledge proofs. One should note, however, that Watrous' technique is restricted to a specific type of rewinding: If we use Watrous' technique, whenever some machine rewinds another machine to an earlier point, the rewinding machine forgets everything it learned after that point (we call this oblivious rewinding). That is, we can only use Watrous' technique to backtrack if the rewinding machine made a mistake that should be corrected, but it cannot be used to collect and combine information from different branches of an execution.

Constructing quantum zero-knowledge proofs solves, however, only half of the problem. In many, if not most, applications of zero-knowledge proofs one needs zero-knowledge *proofs of knowledge*. A proof of knowledge [7,3] is a proof system which does not only show the truth of a certain statement, but also that the prover knows a witness for that statement. This is made clearer by an example: Assume that Alice wishes to convince Bob that she (the prover) is in possession of a signature issued by some certification authority. For privacy reasons, Alice does not wish to reveal the signature itself. If Alice uses a zero-knowledge *proof*, she can only show the statement "there exists a signature with respect to the CA's public key". This does not, however, achieve anything: A signature always exists in a mathematical sense, even if it has never been computed. What Alice wishes to say is: "I *know* a signature with respect to the CA's public key." To prove such a statement, Alice needs a zero-knowledge *proof of knowledge*; a proof of knowledge would convince Bob that Alice indeed knows a witness, i.e., a signature. Very roughly, the definition of a proof of knowledge is the following: Whenever the prover can convince the verifier, one can extract the witness from the prover given oracle access to the prover. Here oracle access means that one can interact with the prover and *rewind* him. Thus, we have the same problem as in the case of quantum zero-knowledge proofs: To get proofs of knowledge that are secure against quantum adversaries, we need to use quantum rewinding. Unfortunately, Watrous' *oblivious* rewinding does not work here; proofs of knowledge use rewinding to produce two (or more) different protocol traces and compute the witness by combining the information from both traces. Thus, we are back to where we started: to make classical cryptographic protocols work in a quantum setting, we need (in many cases) quantum zero-knowledge *proofs of knowledge*, but we only have constructions for quantum zero-knowledge *proofs*.

Our Contribution. We define and construct quantum proofs of knowledge. Our protocols are classical (i.e., honest parties do not use quantum computation or communication) but secure against quantum adversaries. Our constructions are based on a new quantum rewinding technique (different from Watrous' technique) that allows us to extract witnesses in many classical proofs of knowledge. We give criteria under which a classical proof of knowledge is a quantum proof of knowledge. Combining our results with Watrous' results on zero-knowledge, we can show that there are zero-knowledge quantum proofs of knowledge for all languages in NP (assuming quantum 1-1 one-way functions). (We leave it as an open question whether unconditionally secure protocols exist for more restricted languages related, e.g., to lattice-problems.)

Also, we believe that the use of our rewinding technique is not limited to QPoKs. For example, we encourage the reader to try to prove the following without using our technique: Given a quantum computationally binding commitment scheme, first let the adversary commit, and then give a random value v to the adversary. Then the probability that the adversary opens the commitment to v is negligible.[1]

Follow-up Work. In subsequent work, Lunemann and Nielsen [14] and Hallgren, Smith, and Song [12] developed zero-knowledge QPoKs with the additional advantage of allowing to simultaneously simulate an interaction with the malicious prover and extract the witness; this property is necessary in some multi-party computations. (In contrast, in our setting the initial state of the prover could be lost after extracting.) We stress, however, that this powerful feature comes at a cost: They need considerably stronger assumptions, namely quantum mixed commitments (while we only need quantum 1-1 one-way functions). Both their zero-knowledge property and their extractability hold only against polynomial-time adversaries. In contrast, we get unconditional extractability and computational zero-knowledge; and by adapting our construction to unconditionally hiding commitments, we could instead make the zero-knowledge property unconditional – this would be necessary, e.g., for constructions that achieve everlasting security. Finally, note that the protocols from [14,12] are much more involved than their classical counterparts while we only slightly modify existing classical protocols. Thus, [14,12] give valuable alternatives to our protocols but do not supersede them.

---

[1] The definition of a computationally binding commitment only guarantees that the adversary cannot simultaneously produce opening information for two different values. Thus, to get a contradiction, we need to rewind the adversary to extract two values. If the commitment is strictly binding (Definition 9), our rewinding technique can be used.

Organization. In Section 1.1, we give an overview over the techniques underlying our results. In Section 2 we present and discuss the definition of quantum proofs of knowledge (QPoKs). In Section 3, we give criteria under which a proof system is a QPoK. In Section 4, we show that zero-knowledge QPoKs exist for all languages in NP. Omitted proofs and definitions are presented in the full version [18].

1.1 Our Techniques

Defining Proofs of Knowledge. In the classical setting, proofs of knowledge are defined as follows:[2] A proof system consisting of a prover P and a verifier V is a proof of knowledge (PoK) with knowledge error κ if there is a polynomial-time machine K (the extractor) such that the following holds: For any prover P*, if P* convinces V with probability $\mathrm{Pr}_V \geq \kappa$, then K^{P^*} (the extractor K with rewinding black-box access to P*) outputs a witness with probability $\mathrm{Pr}_K \geq \frac{1}{p}(\mathrm{Pr}_V - \kappa)^d$ for some polynomial p and constant $d > 0$. In order to transfer this definition to the quantum setting, we need to specify what it means that K has quantum rewinding black-box access to P*. We choose the following definition: Let U denote the unitary transformation describing one activation of P* (if P* is not unitary, this needs to work for *all* purifications of P*). K may invoke U (this corresponds to running P*), he may invoke the inverse $U^\dagger$ of U (this corresponds to rewinding P* by one activation), and he may read/write a shared register N for exchanging messages with P*. But K may not make snapshots of the state of P*. Allowing K to invoke $U^\dagger$ is justified by the fact that all quantum circuits are reversible; given a circuit for U, we can efficiently apply $U^\dagger$. Note that previous black-box constructions such as Watrous' rewinding technique and Grover's algorithm [10] make use of this fact. We can now define quantum proofs of knowledge: (P, V) is a quantum proof of knowledge (QPoK) with knowledge error κ iff there is a polynomial-time quantum algorithm K such that for all malicious provers P*, K^{P^*} (the extractor K with quantum rewinding black-box access to P*) outputs a witness with probability $\mathrm{Pr}_K \geq \frac{1}{p}(\mathrm{Pr}_V - \kappa)^d$ for some polynomial p and constant $d > 0$.

We illustrate that QPoKs according to this definition are indeed useful for analyzing cryptographic protocols. Assume the following toy protocol: In phase 1, a certification authority (CA) signs the pair (Alice, a) where a is Alice's age. In phase 2, Alice uses a zero-knowledge QPoK with negligible knowledge error κ to prove to Bob that she possesses a signature σ on (Alice, a') for some $a' \geq 21$. That is, a witness in this QPoK would consist of an integer $a' \geq 21$ and a signature σ on (Alice, a') with respect to the CA's public key. We can now show that, if Alice is underage, i.e., if $a < 21$, Bob accepts the QPoK only with negligible probability: Assume that Bob accepts with non-negligible probability ν. Then, by the definition of QPoKs, K^{Alice} will, with probability $\frac{1}{p}(\nu - \kappa)^d$,

---

[2] This is one of different possible definitions, loosely following [11]. It permits us to avoid the use of expected polynomial-time. We discuss alternatives in Section 2.2 "On the success probability of the extractor".

output an integer $a' \geq 21$ and a (forged) signature σ on $(\texttt{Alice}, a')$ with respect to the CA's public key (given the information learned in phase 1 as auxiliary input). Notice that $\frac{1}{p}(\nu - \kappa)^d$ is non-negligible. However, the CA only signed $(\texttt{Alice}, a)$ with $a < 21$. This implies that $\mathsf{K}^{\texttt{Alice}}$ can produce with non-negligible probability a valid signature of a message that has never been signed by the CA. This contradicts the security of the signature scheme (assuming, e.g., existential unforgeability [8]). This shows the security of our toy protocol.

Relation to Classical Proofs of Knowledge. Notice that a quantum proof of knowledge according to our definition is not necessarily a classical PoK because the quantum extractor might have more computational power. (E.g., in a proof system where the witness is a factorization, a quantum extractor could just compute this witness himself.) We stress that this "paradox" is not particular to our definition, it occurs with all simulation-based definitions (e.g., zero-knowledge [19], universal composability [17]). If needed, one can avoid this "paradox" by requiring the extractor/simulator to be classical if the malicious prover/verifier is. (This would actually be equivalent to requiring that the scheme is both a classical ZK PoK and a quantum one.)

Amplification. Our toy example shows that QPoKs with negligible knowledge error can be used to show the security of protocols. But what about QPoKs with non-negligible knowledge error? In the classical case, we know that the knowledge error of a PoK can be made exponentially small by sequential repetition. Fortunately, this result carries over to the quantum case; its proof follows the same lines.

Elementary Constructions. In order to understand our constructions of QPoKs, let us first revisit a common method for constructing classical PoKs. Assume a protocol that consists of three messages: the commitment (sent by the prover), the challenge (picked from a set C and sent by the verifier), and the response (sent by prover). Assume that there is an efficient algorithm $\mathsf{K_0}$ that computes a witness given two conversations with the same commitment but different challenges; this property is called special soundness. Then we can construct the following (classical) extractor K: $\mathsf{K}^{\mathsf{P^*}}$ runs $\mathsf{P^*}$ using a random challenge ch. Then $\mathsf{K}^{\mathsf{P^*}}$ rewinds $\mathsf{P^*}$ to the point after it produced the commitment, and then $\mathsf{K}^{\mathsf{P^*}}$ runs $\mathsf{P^*}$ with a random challenge ch'. If both executions lead to an accepting conversation, and $ch \neq ch'$, $\mathsf{K_0}$ can compute a witness. The probability of getting two accepting conversations can be shown to be Pr_V^2, where Pr_V is the probability of the verifier accepting $\mathsf{P^*}$'s proof. From this, a simple calculation shows that the knowledge error of the protocol is $1/\#C$.

If we directly translate this approach to the quantum setting, we end up with the following extractor: K runs one step of $\mathsf{P^*}$, measures the commitment com, provides a random challenge ch, runs the second step of $\mathsf{P^*}$, measures the response, runs the inverse of the second step of $\mathsf{P^*}$, provides a random challenge ch', runs the second step of $\mathsf{P^*}$, and measures the response $resp'$. If $ch \neq ch'$, and both $(com, ch, resp)$ and $(com, ch', resp')$ are accepting conversations, then we get a witness using $\mathsf{K_0}$. We call this extractor the canonical extractor. The problem is to bound the probability F of getting two accepting conversations. In the classical setting, one uses that the two conversations are essentially independent (given a fixed commitment), and each of them is, from the point of view of P^*, the same as an interaction with the honest verifier V. In the quantum setting, this is not the case. Measuring $resp$ disturbs the state of P^*; we hence cannot make any statement about the probability that the second conversation is accepting.

How can we solve this problem? Note that we cannot use Watrous' oblivious rewinding since we need to remember both responses $resp$ and $resp'$ from two different execution paths of P^*. Instead, we observe that, the more information we measure in the first conversation (i.e., the longer $resp$ is), the more we destroy the state of P^* used in the second conversation. Conversely, if would measure only one bit, the disturbance of P^*'s state would be small enough to still get a sufficiently high success probability. But if $resp$ would contain only one bit, it would clearly be too short to be of any use for K_0. Yet, it turns out that this conflict can be resolved: In order not to disturb P^*'s state, we only need that the $resp$ information-theoretically contains little information. For K_0, however, even an information-theoretically determined $resp$ is still useful; it might, for example, reveal a value which P^* was already committed to. To make use of this observation, we introduce an additional condition on our proof systems, strict soundness. A proof system has strict soundness if for any commitment and challenge, there is at most one response that makes the conversation accepting. Given a proof system with special and strict soundness, we can show that measuring $resp$ does not disturb P^*'s state too much; the canonical extractor is successful with probability approximately $\Pr_V^3$. A precise calculation shows that a proof system with special and strict soundness has knowledge error $1/\sqrt{\#C}$.

QPoKs for All Languages in NP. Blum [4] presents a classical zero-knowledge PoK for showing the knowledge of a Hamiltonian cycle. Using a suitable commitment scheme (it should have the property that the opening information is uniquely determined by the commitment), the proof system is easily seen to have special and strict soundness, thus it is a QPoK. By sequential repetition, we get a QPoK for Hamiltonian cycles. Using the Watrous' results, we get that the QPoK is also zero-knowledge. Using the fact that the Hamiltonian cycle problem is NP-complete, we get zero-knowledge QPoKs for all languages in NP (assuming quantum 1-1 one-way functions).

1.2   Preliminaries

General. A non-negative function μ is called negligible if for all $c > 0$ and all sufficiently large k, $\mu(k) < k^{-c}$. $\oplus$ denotes the XOR operation on bitstrings. $\#C$ is the cardinality of the set C.

Quantum Systems. We can only give a terse overview over the formalism used in quantum computing. For a thorough introduction, we recommend the textbook by Nielsen and Chuang [15, Chap. 1–2]. A (pure) state in a quantum system is described by a unit vector $|\Phi\rangle$ in some Hilbert space $\mathcal{H}$. We always assume a designated orthonormal basis for each Hilbert space, called the computational basis. The tensor product of several states (describing a joint system) is written $|\Phi\rangle \otimes |\Psi\rangle$. We write $\langle\Psi|$ for the linear transformation mapping $|\Phi\rangle$ to the scalar product $\langle\Psi|\Phi\rangle$. The norm $\||\Phi\rangle\|$ is defined as $\sqrt{\langle\Phi|\Phi\rangle}$. A unit vector is a vector with $\||\Phi\rangle\| = 1$. The Hermitean transpose of a linear operator A is written $A^\dagger$.

2   Quantum Proofs of Knowledge

2.1   Definitions

Interactive Machines. A *quantum interactive machine* M (machine, for short) is a machine that gets two inputs, a classical input x and a quantum input $|\Phi\rangle$. M operates on two quantum registers; a network register N and a register S_M for the state. S_M is initialized with $|\Phi\rangle$. The operation of M is described by a unitary transformation M_x (depending on the classical input x). In each activation of M, M_x is applied to N, S_M. We write $\langle M(x, |\Phi\rangle), M'(x', |\Phi'\rangle)\rangle$ for the classical output of M' in an interaction where M is activated first (and where M and M' share the register N). Often, we will omit the quantum input $|\Phi\rangle$ or $|\Phi'\rangle$. In this case, we assume the input $|0\rangle$.

Oracles Algorithms with Rewinding. A *quantum oracle algorithm* A is an algorithm that has oracle access to a machine M. In an execution $A^{M(x', |\Phi\rangle)}(x)$, two registers N, S_M are used for the communication with and the state of M. A's behavior is described by a quantum circuit; A has access to two special gates $\square$ and $\square^\dagger$ that invoke the unitary transformations $M_{x'}$ and $M_{x'}^\dagger$, respectively. This corresponds to running and rewinding M. A is not allowed to access S_M directly, and he is allowed to apply $\square$ and $\square^\dagger$ only to N, S_M. (I.e., A has no access to the internal state and the quantum input of the prover. Any access to this information is done by communicating with M.) Details on the definitions of interactive quantum machines and quantum oracle algorithms are given in the full version [18].

Proof Systems. A *quantum proof system* for a relation R is a pair of two machines (P, V). We call P the prover and V the verifier. The prover expects a classical input (x, w) with $(x, w) \in R$, the verifier expects only the input x. We call (P, V) *complete* if there is a negligible function μ such that for all $(x, w) \in R$, we have that $\Pr[\langle P(x, w), V(x)\rangle = 1] \geq 1 - \mu(|x|)$. (Remember that, if we do not explicitly specify a quantum input, we assume the quantum input $|0\rangle$.) Although we allow P and V to be quantum machines, and in particular to send and receive quantum messages, we will not need this property in the following; all protocols constructed in this paper will consist of classical machines. We call a (P, V) *sound* with soundness error s iff for all malicious prover P^*, all auxiliary inputs $|\Phi\rangle$, and all x with $\not\exists w : (x, w) \in R$, we have $\Pr[\langle P^*(x, |\Phi\rangle), V(x)\rangle = 1] \leq s(|x|)$. A proof system is computational zero-knowledge iff for all polynomial-time verifiers V^* there is a polynomial-time machine S (the simulator) such that for all auxiliary inputs $|\Phi\rangle$, and all $(x, w) \in R$, we have that the quantum state of V^* after an interaction $\langle \mathsf{P}(x, w), \mathsf{V}^*(x, |\Phi\rangle)\rangle$ is computationally indistinguishable from the output of $\mathsf{S}(x, |\Phi\rangle)$; we refer to [19] for details.

Quantum Proofs of Knowledge. We can now define quantum proofs of knowledge (QPoKs). Roughly, a quantum proof system (P, V) is a QPoK if there is a quantum oracle algorithm K (the extractor) that achieves the following: Whenever some malicious prover P^* convinces V that a certain statement holds, the extractor $\mathsf{K}^{\mathsf{P}^*}$ with oracle access to P^* is able to return a witness. Here, we allow a certain knowledge error κ; if P^* convinces V with a probability smaller than κ, we do not require anything. Furthermore, we also do not require that the success probability of $\mathsf{K}^{\mathsf{P}^*}$ is as high as the success probability of P^*; instead, we only require that it is polynomially related. Finally, to facilitate the use of QPoKs as subprotocols, we give the malicious prover an auxiliary input $|\Phi\rangle$. We get the following definition:

Definition 1 (Quantum Proofs of Knowledge). *We call a proof system* (P, V) *for a relation R quantum extractable with knowledge error κ if there exists a constant $d > 0$, a polynomially-bounded function $p > 0$, and a polynomial-time quantum oracle machine K such that for any interactive quantum machine P^*, any state $|\psi\rangle$, and any $x \in \{0,1\}^*$, we have that*

$$\Pr[\langle \mathsf{P}^*(x, |\psi\rangle), \mathsf{V}(x)\rangle = 1] \geq \kappa(|x|) \implies$$

$$\Pr[(x, w) \in R : w \leftarrow \mathsf{K}^{\mathsf{P}^*(x, |\psi\rangle)}(x)] \geq \tfrac{1}{p(|x|)}\Big(\Pr\big[\langle \mathsf{P}^*(x, |\psi\rangle), \mathsf{V}(x)\rangle = 1\big] - \kappa(|x|)\Big)^d.$$

A quantum proof of knowledge for R with knowledge error κ (QPoK, for short) is a complete[3] quantum extractable proof system for R with knowledge error κ.

Note that by quantifying over all unitary provers P^*, we implicitly quantify over *all* purifications of *all* possible non-unitary provers. Note that extractability with knowledge error κ implies soundness with soundness error κ. We thus do not need to explicitly require soundness in Definition 1. The knowledge error κ can be made exponentially small by sequential repetition:

Theorem 2. *Let n be a polynomially bounded and efficiently computable function. Let (P, V) be extractable with knowledge error κ. Let $(\mathsf{P}', \mathsf{V}')$ be the proof system consisting of n-sequential executions of (P, V). Then $(\mathsf{P}', \mathsf{V}')$ is extractable with knowledge error κ^n.*

2.2 Discussion

In this section, we motivate various design choices made in the definition of QPoKs.

Access to the Black-Box Prover's State and Input. The extractor has no access to the prover's state nor to its quantum input. (This is modeled by the fact

---

[3] I.e., for honest prover and verifier, the proof succeeds with overwhelming probability.

that an oracle algorithm may not apply any gates except for $\square, \square^{\dagger}$ to the register containing the oracle's state and quantum input.) In this, we follow [3] who argue in Section 4.3 that a proof of knowledge is supposed to "capture the knowledge of the prover *demonstrated by the interaction*" and that thus the extractor is not supposed to see the internal state of the prover. We stress, however, that our results are independent of this issue; they also hold if we allow the extractor to access the prover's state directly.

Unitary and Invertible Provers – Technical View. Probably the most important design choice in our definition is to require the prover to be a unitary operation, and to allow the extractor to also execute the inverse of this operation. We begin with a discussion of this design choice from a technical point of view. First, we stress that seems that these assumptions are necessary: Since in a quantum world, making a snapshot/copy of a state is not possible or even well-defined, we have to allow the extractor to run the prover "backwards". But the inverse of a non-unitary quantum operation does not, in general, exist. Thus rewinding seems only possible with respect to unitary provers. Second, the probably most important question is: Does the definition, from an operational point of view, make sense? That is, does our definition behave well in cryptographic, reduction-based proofs? A final answer to this question can only be given when more protocols using QPoKs have been analyzed. However, the toy protocol discussed on page 138 gives a first indication that our definition can be used in a similar fashion to classical proofs of knowledge. Third, we would like to remind the reader that any non-unitary prover can be transformed into a unitary one by purification before applying the definition of QPoKs. Thus allowing only unitary malicious provers does not seem to be a restriction in practice.

Unitary and Invertible Provers – Philosophical View. Intuitively, a QPoK should guarantee that a prover that convinces the verifier "knows" the witness.[4] The basic idea is that if an extractor can extract the witness using only what is available to the prover, then the prover "knew" the witness (or could have computed it). In particular, we may allow the extractor to run a purified (unitary) version of the prover because the prover himself could have done so. Similarly for the inverse of that operation. Of course, this leaves the question why we give these two capabilities to the extractor but not others (e.g., access to the circuit of the prover)? We would like to stress that analogous questions are still open (from a philosophical point) even in the classical case: Why is it natural to allow an extractor to rewind the prover? Why is it natural to give a trapdoor for a common reference string to the extractor? We would like to point out one justification for the assumption that the prover is unitary, though: [3] suggests that we "capture the knowledge of the prover *demonstrated by the interaction*". A prover that performs non-unitary operations is identical in terms of its interaction to one that is purified. Thus, by restricting to unitary provers, we come closer to only capturing the interaction but not the inner workings of the prover.

---

[4] We believe, though, that this issue is secondary to the technical suitability; it is much more important that a QPoK is useful as a cryptographic subprotocol.

On the Success Probability of the Extractor. We require the extractor to run in polynomial-time and to succeed with probability $\frac{1}{p}(\mathrm{Pr}_V - \kappa)^d$ where Pr_V is the probability that the prover convinces the verifier. (We call this an A-style definition.) In classical PoKs, a more common definition is to require the extractor to have expected runtime $\frac{p}{\mathrm{Pr}_V - \kappa}$ and to succeed with probability 1. (We call this a B-style definition.) This definition is known to be equivalent to the definition in which the extractor runs in expected polynomial-time and succeeds with probability $\frac{1}{p}(\mathrm{Pr}_V - \kappa)$. (We call this a C-style definition.) The advantage of an A-style definition (which follows [11]) is that we can consider polynomial-time extractors (instead of expected polynomial-time extractors). To get extractors for B-style and C-style definitions, one has to increase the success probability of an extractor by repeatedly invoking it until it outputs a correct witness. In the quantum case, however, this does not work directly: If the invoked extractor fails once, the auxiliary input of the prover is destroyed. The oblivious rewinding technique by Watrous' would seem to help here, but when trying to apply that technique one gets the requirement that the invoked extractors' success probability must be independent of the auxiliary input. This condition is not necessarily fulfilled. To summarize, all three styles of definitions have their advantages, but it is not clear how one could fulfil B- and C-style definitions in the quantum case. This is why we chose an A-style definition. There are, however, applications that would benefit from a proof system fulfilling a C-style definition. For example, general multi-party computation protocols such as [5] use extractors as part of the construction of the simulator for the multi-party computation; these extractors must then succeed with probability close to 1. We leave the construction of C-style QPoKs as an open problem.

3   Elementary Constructions

In this section, we show that under certain conditions, a classical PoK is also a QPoK (i.e., secure against malicious quantum provers). The first condition refers to the outer form of the protocol; we require that the proof systems is a protocol with three messages (commitment, challenge, and response) with a public-coin verifier. Such protocols are called Σ-protocols. Furthermore, we require that the proof system has special soundness. This means that given two accepting conversations between prover and verifier that have the same commitment but different challenges, we can efficiently compute a witness. Σ-protocols with special soundness are well-studied in the classical case; many efficient classical protocols with these properties exist. The third condition (strict soundness) is non-standard. We require that given the commitment and the challenge of a conversation, there is at most one response that would make the verifier accept. We require strict soundness to ensure that the response given by the prover does not contain too much information; measuring it will then not disturb the state of the prover too much. Not all known protocols have strict soundness (the proof for graph isomorphism [6] is an example). Fortunately, many protocols do satisfy strict soundness; a slight variation of the proof for Hamiltonian cycles [4] is an example (see Section 4).

Definition 3 (Σ-protocol). *A proof system* (P, V) *is called a Σ-protocol if* P *and* V *are classical, the interaction consists of three messages com, ch, resp (sent by* P, V, *and* P, *respectively, and called* commitment, challenge, *and* response*), and ch is uniformly chosen from some set C_x (the* challenge space*) that may only depend on the statement x. Furthermore, the verifier decides whether to accept or not by a deterministic polynomial-time computation on $x, com, ch, resp$. (We call (com, ch, resp) an* accepting conversation *for x if the verifier would accept it.) We also require that it is possible in polynomial time to sample uniformly from C_x, and that membership in C_x should be decidable in polynomial time.*

Definition 4 (Special soundness). *We say a Σ-protocol* (P, V) *for a relation R has* special soundness *if there is a deterministic polynomial-time algorithm K_0 (the* special extractor*) such that the following holds: For any two accepting conversations (com, ch, resp) and (com, ch', resp') for x such that $ch \neq ch'$ and $ch, ch' \in C_x$, we have that $w := K_0(x, com, ch, resp, ch', resp')$ satisfies $(x, w) \in R$.*

Definition 5 (Strict soundness). *We say a Σ-protocol* (P, V) *has* strict soundness *if for any two accepting conversations (com, ch, resp) and (com, ch, resp') for x, we have that $resp = resp'$.*

Canonical Extractor. Let (P, V) be a Σ-protocol with special soundness and strict soundness. Let K_0 be the special extractor for that protocol. We define the *canonical extractor* K for (P, V). K will use measurements, even though our definition of quantum oracle algorithms only allows for unitary operations. This is only for the sake of presentation; by purifying K one can derive a unitary algorithm with the same properties. Given a malicious prover P*, $K^{P^*(x,|\Phi\rangle)}(x)$ operates on two quantum registers N, S_{P^*}. N is used for communication with P*, and S_{P^*} is used for the state of P*. The registers N, S_{P^*} are initialized with $|0\rangle, |\Phi\rangle$. Let P_x^* denote the unitary transformation describing a single activation of P. First, K applies P_x^* to N, S_{P^*}. (This can be done using the special gate $\square$.) This corresponds to running the first step of P*; in particular, N should now contain the commitment. Then K measures N in the computational basis; call the result *com*. Then K initializes N with $|0\rangle$. Then K chooses uniformly random values $ch, ch' \in C_x$. Let U_{ch} denote the unitary transformation operating on N such that $U_{ch}|x\rangle = |x \oplus ch\rangle$. Then K applies $P_x^* U_{ch}$. (Now N is expected to contain the response for challenge *ch*.) Then K measures N in the computational basis; call the result *resp*. Then K applies $(P_x^* U_{ch})^\dagger$ (we rewind the prover). Then $P_x^* U_{ch'}$ is applied. (Now N is expected to contain the response for challenge *ch'*.) Then N is measured in the computational basis; call the result *resp'*. Then $(P_x^* U_{ch'})^\dagger$ is applied. Finally, K outputs $w := K_0(x, com, ch, resp, ch', resp')$.

Analysis of the Canonical Extractor. In order to analyze the canonical extractor (Theorem 8 below), we first need a lemma that bounds the probability that two consecutive binary measurements P_{ch} and $P_{ch'}$ with random $ch \neq ch'$ succeed in terms of the probability that a single such measurement succeeds. In a classical setting (or in the case of commuting measurements), the answer is simple: the outcomes of the measurements are independent; thus the probability that two measurements succeed is the square of the probability that a single measurement succeeds. In the quantum case, however, the first measurement may disturb the state; this makes the analysis considerably more involved. We first prove some inequalities needed in the proof:

Lemma 6. *Let C be a set with $\#C = c$. Let $(P_i)_{i \in C}$ be orthogonal projectors on a Hilbert space $\mathcal{H}$. Let $|\Phi\rangle \in \mathcal{H}$ be a unit vector. Let $V := \sum_{i \in C} \frac{1}{c} \||P_i|\Phi\rangle\|^2$ and $F := \sum_{i,j \in C} \frac{1}{c^2} \||P_i P_j|\Phi\rangle\|^2$. Then $F \geq V^3$.*

Proof. To prove the lemma, we first show two simple facts:

Claim. For any positive operator A on $\mathcal{H}$ and any unit vector $|\Phi\rangle \in \mathcal{H}$, we have that $(\langle\Phi|A|\Phi\rangle)^3 \leq \langle\Phi|A^3|\Phi\rangle$.

Since A is positive, it is diagonalizable. Thus we can assume without loss of generality that A is diagonal (by applying a suitable basis transform to A and $|\Phi\rangle$). Let a_i be the i-th diagonal element of A, and let f_i be the i-th component of $|\Phi\rangle$. Then

$$(\langle\Phi|A|\Phi\rangle)^3 = \left(\sum_i |f_i|^2 a_i\right)^3 \overset{(*)}{\leq} \sum_i |f_i|^2 a_i^3 = \langle\Phi|A^3|\Phi\rangle.$$

Here $(*)$ uses Jensen's inequality [13] and the facts that $a_i \geq 0$, that $a_i \mapsto a_i^3$ is a convex function on nonnegative numbers, and that $\sum_i |f_i|^2 = 1$. This concludes the proof of Lemma 3.

Claim. For vectors $|\Psi_1\rangle, \ldots, |\Psi_c\rangle \in \mathcal{H}$, it holds that $\|\frac{1}{c}\sum_i |\Psi_i\rangle\|^2 \leq \frac{1}{c}\sum_i \||\Psi_i\rangle\|^2$.

To show the claim, let $|\bar\Psi\rangle := \sum_i \frac{1}{c}|\Psi_i\rangle$. Then

$$\sum_i \left(\||\Psi_i\rangle\|^2 - \||\bar\Psi\rangle\|^2\right) = \sum_i \left(\||\Psi_i\rangle\| - \||\bar\Psi\rangle\|\right)\left(\||\Psi_i\rangle\| - \||\bar\Psi\rangle\| + 2\||\bar\Psi\rangle\|\right)$$

$$= \sum_i \left(\||\Psi_i\rangle\| - \||\bar\Psi\rangle\|\right)^2 + 2\||\bar\Psi\rangle\| \sum_i \left(\||\Psi_i\rangle\| - \||\bar\Psi\rangle\|\right)$$

$$\geq 2\||\bar\Psi\rangle\| \sum_i \left(\||\Psi_i\rangle\| - \||\bar\Psi\rangle\|\right) = 2\||\bar\Psi\rangle\| \left(\sum_i \||\Psi_i\rangle\| - \|n|\bar\Psi\rangle\|\right) \tag{1}$$

$$= 2\||\bar\Psi\rangle\| \left(\sum_i \||\Psi_i\rangle\| - \left\|\sum_i |\Psi_i\rangle\right\|\right) \tag{2}$$

From the triangle inequality, it follows that $\sum_i \||\Psi_i\rangle\| \geq \|\sum_i |\Psi_i\rangle\|$, hence with (2), we have $\sum_i \left(\||\Psi_i\rangle\|^2 - \||\bar\Psi\rangle\|^2\right) \geq 0$. Since $\frac{1}{c}\sum_i \||\Psi_i\rangle\|^2 - \|\frac{1}{c}\sum_i |\Psi_i\rangle\|^2 = \frac{1}{c}\sum_i \left(\||\Psi_i\rangle\|^2 - \||\bar\Psi\rangle\|^2\right) \geq 0$, Lemma 3 follows.

We proceed to prove Lemma 6. Let $A := \sum_i \frac{1}{c} P_i$, let $|\Psi_{ij}\rangle := P_j P_i |\Phi\rangle$. Then A is positive. Furthermore,

$$V^3 = \left(\sum_i \frac{1}{c}\langle\Phi|P_i|\Phi\rangle\right)^3 = \left(\langle\Phi|A|\Phi\rangle\right)^3 \overset{(*)}{\leq} \langle\Phi|A^3|\Phi\rangle = \sum_{i,j,k} \frac{1}{c^3}\langle\Phi|P_iP_jP_k|\Phi\rangle$$

$$= \sum_{i,j,k} \frac{1}{c^3}\langle\Psi_{ij}|\Psi_{kj}\rangle = \sum_j \frac{1}{c}\left(\sum_i \frac{1}{c}\langle\Psi_{ij}|\right)\left(\sum_k \frac{1}{c}|\Psi_{kj}\rangle\right) = \sum_j \frac{1}{c}\left\|\sum_i \frac{1}{c}|\Psi_{ij}\rangle\right\|^2$$

$$\overset{(**)}{\leq} \sum_j \frac{1}{c}\sum_i \frac{1}{c}\||\Psi_{ij}\rangle\|^2 = F.$$

Here $(*)$ uses Lemma 3 and $(**)$ uses Lemma 3. Thus we have $F \geq V^3$ and Lemma 6 follows.

Lemma 7. *Let C be a set with $\#C = c$. Let $(P_i)_{i \in C}$ be orthogonal projectors on a Hilbert space $\mathcal{H}$. Let $|\Phi\rangle \in \mathcal{H}$ be a unit vector. Let $V := \sum_{i \in C} \frac{1}{c}\|P_i|\Phi\rangle\|^2$ and $E := \sum_{i,j \in C, i \neq j} \frac{1}{c^2}\|P_iP_j|\Phi\rangle\|^2$. Then, if $V \geq \frac{1}{\sqrt{c}}$, $E \geq V(V^2 - \frac{1}{c})$.*

Proof. Let F be as in Lemma 6. Then

$$E = \sum_{\substack{i,j \in C \\ i \neq j}} \frac{1}{c^2}\|P_iP_j|\Phi\rangle\|^2 = \sum_{i,j \in C} \frac{1}{c^2}\|P_iP_j|\Phi\rangle\|^2 - \sum_{i \in C} \frac{1}{c^2}\|P_iP_i|\Phi\rangle\|^2$$

$$\overset{(*)}{=} \sum_{i,j \in C} \frac{1}{c^2}\|P_iP_j|\Phi\rangle\|^2 - \sum_{i \in C} \frac{1}{c^2}\|P_i|\Phi\rangle\|^2 = F - \frac{V}{c} \overset{(**)}{\geq} V^3 - \frac{V}{c} = V(V^2 - \frac{1}{c})$$

Here $(*)$ uses that $P_i = P_iP_i$ since P_i is a projection, and $(**)$ uses Lemma 6.   □

Theorem 8. *A Σ-protocol (P,V) for a relation R with special and strict soundness and challenge space C_x is extractable with knowledge error $\frac{1}{\sqrt{\#C_x}}$.*

Proof. To show that (P,V) is extractable, we will use the canonical extractor K. Fix a malicious prover P^*, a statement x, and an auxiliary input $|\Phi\rangle$. Let $\mathrm{Pr_V}$ denote the probability that the verifier accepts when interacting with P^*. Let $\mathrm{Pr_K}$ denote the probability that $\mathsf{K}^{\mathsf{P}^*(x,|\Phi\rangle)}(x)$ outputs some w with $(x,w) \in R$. We will show that $\mathrm{Pr_K} \geq \mathrm{Pr_V} \cdot (\mathrm{Pr_V^2} - \frac{1}{\#C_x})$. For $\mathrm{Pr_V} \geq \frac{1}{\sqrt{\#C_x}}$, we have that $\mathrm{Pr_V}(\mathrm{Pr_V^2} - \frac{1}{\#C_x}) \geq (\mathrm{Pr_V} - \frac{1}{\sqrt{\#C_x}})^3$. Since furthermore K is polynomial-time, this implies that (P,V) is extractable with knowledge error $\frac{1}{\sqrt{\#C_x}}$.

In order to show $\mathrm{Pr_K} \geq \mathrm{Pr_V} \cdot (\mathrm{Pr_V^2} - \frac{1}{\#C_x})$, we will use a short sequence of games. Each game will contain an event Succ, and in the first game, we will have $\Pr[\mathsf{Succ} : \mathsf{Game\ 1}] = \mathrm{Pr_K}$. For any two consecutive games, we will have $\Pr[\mathsf{Succ} : \mathsf{Game}\ i] \geq \Pr[\mathsf{Succ} : \mathsf{Game}\ i+1]$, and for the final game, we will have $\Pr[\mathsf{Succ} : \mathsf{Game\ 7}] \geq \mathrm{Pr_V} \cdot (\mathrm{Pr_V^2} - \frac{1}{\#C_x})$. This will then conclude the proof. The description of each game will only contain the changes with respect to the preceding game.

Game 1. An execution of $\mathsf{K}^{\mathsf{P}^*(x,|\Phi\rangle)}(x)$. Succ denotes the event that K outputs a witness for x. By definition, $\Pr_{\mathsf{K}} = \Pr[\mathsf{Succ} : \text{Game 1}]$.

Game 2. Succ denotes the event that $(com, ch, resp)$ and $(com, ch', resp')$ are accepting conversations for x and $ch \neq ch'$. (The variables $(com, ch, resp)$ and $(com, ch', resp')$ are as in the definition of the canonical extractor.) Since (P, V) has special soundness, if Succ occurs, K outputs a witness. Thus $\Pr[\mathsf{Succ} : \text{Game 1}] \geq \Pr[\mathsf{Succ} : \text{Game 2}]$.

Game 3. Before K measures $resp$, it first measures whether measuring $resp$ would yield an accepting conversation. More precisely, it measures N with the orthogonal projector P_{ch} projecting onto $V_{ch} := \text{span}\{|resp\rangle :$ $(com, ch, resp)$ is accepting$\}$. Analogously for the measurement of $resp'$ (using the projector $P_{ch'}$.) Since a complete measurement (of $resp$ and $resp'$, respectively) is performed on N after applying the measurement P_{ch} and $P_{ch'}$, introducing the additional measurements does not change the outcomes $resp$ and $resp'$ of these complete measurements, nor their post-measurement state. Thus $\Pr[\mathsf{Succ} : \text{Game 2}] = \Pr[\mathsf{Succ} : \text{Game 3}]$.

Game 4. Succ denotes the event that $ch \neq ch'$ and both measurements P_{ch} and $P_{ch'}$ succeed. By definition of these measurements, this happens iff $(com, ch, resp)$ and $(com, ch', resp')$ are accepting conversations. Thus $\Pr[\mathsf{Succ} : \text{Game 3}] = \Pr[\mathsf{Succ} : \text{Game 4}]$.

Game 5. We do not execute K_0, i.e., we stop after applying $(\mathsf{P}_x^* U_{ch'})^\dagger$. Since at that point, Succ has already been determined, $\Pr[\mathsf{Succ} : \text{Game 4}] = \Pr[\mathsf{Succ} : \text{Game 5}]$.

Game 6. We remove the measurements of $resp$ and $resp'$. Note that the outcomes of these measurements are not used any more. Since (P, V) has strict soundness, $V_{ch} = \text{span}\{|resp_0\rangle\}$ for a single value $resp_0$ (depending on com and ch, of course). Thus if the measurement P_{ch} succeeds, the post-measurement state in N is $|resp_0\rangle$. That is, the state in N is classical at this point. Thus, measuring N in the computational basis does not change the state. Hence, the measurement of $resp$ does not change the state. Analogously for the measurement of $resp'$. It follows that $\Pr[\mathsf{Succ} : \text{Game 5}] = \Pr[\mathsf{Succ} : \text{Game 6}]$.

Game 7. First, N and $S_{\mathsf{P}*}$ are initialized with $|0\rangle$ and $|\Phi\rangle$. Then the unitary transformation P_x^* is applied. Then com is measured (complete measurement on N), and N is initialized to $|0\rangle$. Random $ch, ch' \in C_x$ are chosen. Then $\mathsf{P}_x^* U_{ch}$ is applied. Then the measurement P_{ch} is performed. Then $(\mathsf{P}_x^* U_{ch})^\dagger$ is applied. Then $\mathsf{P}_x^* U_{ch'}$ is applied. Then the measurement $P_{ch'}$ is performed. Then $(\mathsf{P}_x^* U_{ch'})^\dagger$ is applied. The event Succ holds if $ch \neq ch'$ and both measurements succeed. Games 6 and 7 are identical; we have just recapitulated the game for clarity. Thus, $\Pr[\mathsf{Succ} : \text{Game 6}] = \Pr[\mathsf{Succ} : \text{Game 7}]$.

In Game 7, for some value d, let p_d denote the probability that $com = d$ is measured. Let $|\Phi_d\rangle$ denote the state of $N, S_{\mathsf{P}*}$ after measuring $com = d$ and initializing N with $|0\rangle$. (I.e., the state directly before applying $\mathsf{P}_x^* U_{ch}$.) Let K_d denote the probability that starting from state $|\Phi_d\rangle$, both measurements P_{ch} and

$P_{ch'}$ succeed. Let $c := \#C_x$. Then we have that $\Pr[\text{Succ} : \text{Game } 7] = \sum_d p_d K_d$ and

$$K_d = \sum_{\substack{ch, ch' \in C_x \\ ch \neq ch'}} \frac{1}{c^2} \|(\mathsf{P}_x^* U_{ch'})^\dagger P_{ch'} (\mathsf{P}_x^* U_{ch'})(\mathsf{P}_x^* U_{ch})^\dagger P_{ch} (\mathsf{P}_x^* U_{ch}) |\Phi_d\rangle\|^2$$

$$= \sum_{\substack{ch, ch' \in C_x \\ ch \neq ch'}} \frac{1}{c^2} \| P_{ch'}^* P_{ch}^* |\Phi_d\rangle \|^2$$

where $P_{ch}^* := (\mathsf{P}_x^* U_{ch})^\dagger P_{ch} (\mathsf{P}_x^* U_{ch})$. Since P_{ch} is an orthogonal projector and $\mathsf{P}_x^* U_{ch}$ is unitary, P_{ch}^* is an orthogonal projector. Let $\varphi(v) := v(v^2 - \frac{1}{c})$ for $v \in [\frac{1}{\sqrt{c}}, 1]$ and $\varphi(v) := 0$ for $v \in [0, \frac{1}{\sqrt{c}}]$. Then, by Lemma 7, $K_d \geq \varphi(V_d)$ for $V_d := \sum_{ch \in C_x} \frac{1}{c} \| P_{ch}^* |\Phi_d\rangle \|^2$.

Furthermore, by construction of the honest verifier V, we have that

$$\Pr_\mathsf{V} = \sum_d p_d \sum_{ch \in C_x} \frac{1}{c} \| P_{ch} \mathsf{P}_x^* U_{ch} |\Phi_d\rangle \|^2$$

$$\overset{(*)}{=} \sum_d p_d \sum_{ch \in C_x} \frac{1}{c} \|(\mathsf{P}_x^* U_{ch})^\dagger P_{ch} (\mathsf{P}_x^* U_{ch}) |\Phi_d\rangle\|^2 = \sum_d p_d V_d$$

where $(*)$ uses that $(\mathsf{P}_x^* U_{ch})^\dagger$ is unitary. Finally, we have

$$\Pr_\mathsf{K} = \Pr[\text{Succ} : \text{Game } 1] \geq \Pr[\text{Succ} : \text{Game } 7]$$

$$= \sum_d p_d K_d \geq \sum_d p_d \varphi(V_d) \overset{(*)}{\geq} \varphi(\Pr_\mathsf{V}).$$

Here $(*)$ uses Jensen's inequality [13] and the fact that φ is convex on $[0, 1]$. As discussed in the beginning of the proof, $\Pr_\mathsf{K} \geq \varphi(\Pr_\mathsf{V}) = \Pr_\mathsf{V} \cdot (\Pr_\mathsf{V}^2 - \frac{1}{c})$ for $\Pr_\mathsf{V} \geq \frac{1}{\sqrt{c}}$ implies that (P, V) is a QPoK with knowledge error $1/\sqrt{\#C_x}$.

4 QPoKs for All Languages in NP

In the preceding section, we have seen that complete proof systems with strict and special soundness are QPoKs. The question that remains to be asked is: do such proof systems, with the additional property of being zero-knowledge, exist for interesting languages? In this section, we will show that for any language in NP (more precisely, for any NP-relation), there is a zero-knowledge QPoK. (Assuming the existence of quantum 1-1 one-way functions.) Here and in the following, by zero-knowledge we mean quantum computational zero-knowledge.

The starting point for our construction will be the Blum's zero-knowledge PoK for Hamiltonian cycles [4]. In this Σ-protocol, the prover's commits to the vertices of a graph using a perfectly binding commitment scheme. In the prover's response, some of these commitments are opened. That is, the response contains the opening information for some of the commitments. The problem is that standard definitions of commitment schemes do not guarantee that the opening information is unique; only the actual content of the commitment has to be determined by the commitment. This means that the prover's response is not unique. Thus, with a standard commitment scheme we do not get strict soundness. Instead we need a commitment scheme such that the sender of the commitment scheme is committed not only to the actual content of the commitment, but also to the opening information.

Definition 9 (Strict binding). *A commitment scheme* COM *is a deterministic polynomial-time function taking two arguments* a, y, *the opening information* a *and the message* y. *We say* COM *is* strictly binding *if for all* a, y, a', y' *with* $(a, y) \neq (a', y')$, *we have that* $\mathrm{COM}(a, y) \neq \mathrm{COM}(a', y')$.

Furthermore, in order to get the zero-knowledge property, we will need that our commitment schemes are quantum computationally concealing. We refer to [19] for a precise definition of this property. In [2], an unconditionally binding, quantum computationally concealing commitment scheme based on quantum 1-1 one-way function is presented.[5] Unfortunately, to the best of our knowledge, no candidates for quantum 1-1 functions are known. Their definitions differ somewhat from those of [19], but as mentioned in [19], their proof carries over to the definitions from [19]. Furthermore, in the scheme from [2], the commitment contains the image of the opening information under a quantum 1-1 one-way function. Thus the strict binding property is trivially fulfilled. Thus strictly binding, quantum computationally concealing commitment schemes exist under the assumption that quantum 1-1 one-way functions exist.

Given such a commitment scheme COM, we can construct the proof system (P, V). This proof system differs from the original proof system for Hamiltonian cycles [4] only in the following aspect: The prover does not only commit to the vertices in the graph $\pi(x)$, but also to the permutation π and the cycle H. Without these additional commitments, we would not get strict soundness; there might be several permutations leading to the same graph, or the graph might contain several Hamiltonian cycles. The full description of the protocol is given in Figure 1.

Theorem 10. *Let* $(x, w) \in R$ *iff* w *is a Hamiltonian cycle of the graph* x. *Assume that* COM *is a strictly binding, quantum computationally concealing commitment scheme. Then the proof system* (P, V) *is a zero-knowledge QPoK for* R *with knowledge error* $\frac{1}{\sqrt{2}}$.

The zero-knowledge property is proven using the techniques from [19]. Extractability is shown by proving special and strict soundness. The strict soundness follows from the fact that the prover is committed to all the information sent in his response using a strictly binding commitment.

---

[5] In [2], the result is stated for quantum one-way permutations $f : \{0, 1\}^n \to \{0, 1\}^n$. (To the best of our knowledge, no candidates for quantum one-way permutations are known.) Inspection of their proof reveals, however, that the result also holds for families of quantum 1-1 one-way functions $f_i : \{0, 1\}^n \to D$ for arbitrary domain D and efficiently samplable indices i, assuming that given an index i, it can be efficiently verified that f_i is injective.

Inputs: A directed graph x (the statement) with vertices W, and a Hamiltonian cycle w in x (the witness).
Protocol:
1. P picks a random permutation π on W. Let A be the adjacency matrix of the graph $\pi(x)$. Let $H := \{(\pi(i), \pi(j)) : (i,j) \in w\}$. Using COM, P commits to π, H, and to each entry A_{ij} of A. P sends the resulting commitments to V.
2. V picks $ch \in \{0,1\}$ and sends ch to P.
3. If $ch = 0$, P opens the commitments to π and A. If $ch = 1$, P opens the commitments to H and to all A_{ij} with $(i,j) \in H$.
4. If ch $= 0$, V checks that the commitments are opened correctly, that π is a permutation, and that A is the adjacency matrix of $\pi(x)$. If $ch = 1$, V checks that the commitments are opened correctly, that H is a cycle, that exactly the A_{ij} with $(i,j) \in H$ are opened, and that $A_{ij} = 1$ for all $(i,j) \in H$. If all checks succeed, V outputs 1.

Fig. 1. A QPoK (P, V) for Hamiltonian cycles

Corollary 11 (QPoKs for all languages in NP). *Let R be an NP-relation.*[6] *Then there is a zero-knowledge QPoK for R with negligible knowledge error.*

Acknowledgements. We thank the anonymous referees, Märt Põldvere, and Rainis Haller for suggestions on how to significantly simplify the proof of Lemma 7. We thank Dennis Hofheinz, Chris Peikert, and Vinod Vaikuntanathan for discussions on candidates for quantum 1-1 one-way functions. We thank Claude Crépeau and Louis Salvail for inspiring discussions on the difficulties of quantum proofs of knowledge. This research was supported by the Cluster of Excellence "Multimodal Computing and Interaction", by the European Social Fund's Doctoral Studies and Internationalisation Programme DoRa, by the European Regional Development Fund through the Estonian Center of Excellence in Computer Science, EXCS, and by European Social Fund through the Estonian Doctoral School in Information and Communication Technology.

References

1. Aaronson, S.: Limitations of quantum advice and one-way communication. Theory of Computing 1(1), 1–28 (2005), http://www.theoryofcomputing.org/articles/v001a001
2. Adcock, M., Cleve, R.: A Quantum Goldreich-Levin Theorem with Cryptographic Applications. In: Alt, H., Ferreira, A. (eds.) STACS 2002. LNCS, vol. 2285, pp. 323–334. Springer, Heidelberg (2002)
3. Bellare, M., Goldreich, O.: On Defining Proofs of Knowledge. In: Brickell, E.F. (ed.) CRYPTO 1992. LNCS, vol. 740, pp. 390–420. Springer, Heidelberg (1993), http://www-cse.ucsd.edu/users/mihir/papers/pok.ps

---

[6] An NP-relation is a relation R such that $(x, w) \in R$ is decidable in deterministic polynomial time, and there is a polynomial p such that for all $(x, w) \in R$, $|w| \leq p(|x|)$.

4. Blum, M.: How to prove a theorem so no one else can claim it. In: Proceedings of the International Congress of Mathematicians, Berkeley, pp. 1444–1451 (1986)
5. Goldreich, O., Micali, S., Wigderson, A.: How to play any mental game – or – a completeness theorem for protocols with honest majority. In: STOC 1987, pp. 218–229 (1987)
6. Goldreich, O., Micali, S., Wigderson, A.: Proofs that yield nothing but their validity or all languages in NP have zero-knowledge proof systems. Journal of the ACM 38(3), 690–728 (1991), http://www.wisdom.weizmann.ac.il/~oded/X/gmw1j.pdf
7. Goldwasser, S., Micali, S., Rackoff, C.: The knowledge complexity of interactive proof-systems. In: Proceedings of the Seventeenth Annual ACM Symposium on Theory of Computing, pp. 291–304. ACM Press (1985)
8. Goldwasser, S., Micali, S., Rivest, R.L.: A digital signature scheme secure against adaptive chosen-message attacks. SIAM Journal on Computing 17(2), 281–308 (1988), http://theory.lcs.mit.edu/ rivest/GoldwasserMicaliRivest-ADigitalSignatureSchemeSecureAgainstAdaptiveChosenMessageAttacks.ps
9. van de Graaf, J.: Towards a formal definition of security for quantum protocols. Ph.D. thesis, Départment d'informatique et de r.o., Université de Montréal (1998), http://www.cs.mcgill.ca/~crepeau/PS/these-jeroen.ps
10. Grover, L.K.: A fast quantum mechanical algorithm for database search. In: STOC, pp. 212–219 (1996)
11. Halevi, S., Micali, S.: More on proofs of knowledge. IACR ePrint 1998/015 (1998)
12. Hallgren, S., Smith, A., Song, F.: Classical Cryptographic Protocols in a Quantum World. In: Rogaway, P. (ed.) CRYPTO 2011. LNCS, vol. 6841, pp. 411–428. Springer, Heidelberg (2011)
13. Jensen, J.L.W.V.: Sur les fonctions convexes et les inégalités entre les valeurs moyennes. Acta Mathematica 30(1), 175–193 (1906) (in French)
14. Lunemann, C., Nielsen, J.B.: Fully Simulatable Quantum-Secure Coin-Flipping and Applications. In: Nitaj, A., Pointcheval, D. (eds.) AFRICACRYPT 2011. LNCS, vol. 6737, pp. 21–40. Springer, Heidelberg (2011)
15. Nielsen, M.A., Chuang, I.L.: Quantum Computation and Quantum Information. Cambridge University Press (2000)
16. Shor, P.W.: Algorithms for quantum computation: Discrete logarithms and factoring. In: Proceedings of 35th Annual Symposium on Foundations of Computer Science, FOCS 1994, pp. 124–134. IEEE Computer Society (1994)
17. Unruh, D.: Universally Composable Quantum Multi-party Computation. In: Gilbert, H. (ed.) EUROCRYPT 2010. LNCS, vol. 6110, pp. 486–505. Springer, Heidelberg (2010) preprint on arXiv:0910.2912 [quant-ph]
18. Unruh, D.: Quantum proofs of knowledge. IACR ePrint 2010/212 (2012), full version
19. Watrous, J.: Zero-knowledge against quantum attacks. SIAM J. Comput. 39(1), 25–58 (2009)
20. Winter, A.: Coding Theorems of Quantum Information Theory, Ph.D. thesis, Universität Bielefeld (1999), arXiv:quant-ph/9907077v1
21. Wootters, W.K., Zurek, W.H.: A single quantum cannot be cloned. Nature 299, 802–803 (1982)

On Round-Optimal Zero Knowledge
in the Bare Public-Key Model

Alessandra Scafuro and Ivan Visconti

Dipartimento di Informatica, University of Salerno, Italy
{scafuro,visconti}@dia.unisa.it

Abstract. In this paper we revisit previous work in the BPK model and point out subtle problems concerning security proofs of concurrent and resettable zero knowledge (c$\mathcal{ZK}$ and r$\mathcal{ZK}$, for short). Our analysis shows that the c$\mathcal{ZK}$ and r$\mathcal{ZK}$ simulations proposed for previous (in particular *all* round-optimal) protocols are distinguishable from real executions. Therefore some of the questions about achieving round optimal c$\mathcal{ZK}$ and r$\mathcal{ZK}$ in the BPK model are still open. We then show our main protocol, $\Pi_{c\mathcal{ZK}}$, that is a round-optimal concurrently sound c$\mathcal{ZK}$ argument of knowledge (AoK, for short) for **NP** under standard complexity-theoretic assumptions. Next, using complexity leveraging arguments, we show a protocol $\Pi_{r\mathcal{ZK}}$ that is round-optimal and concurrently sound r$\mathcal{ZK}$ for **NP**. Finally we show that $\Pi_{c\mathcal{ZK}}$ and $\Pi_{r\mathcal{ZK}}$ can be instantiated efficiently through transformations based on number-theoretic assumptions. Indeed, starting from any language admitting a perfect Σ-protocol, they produce concurrently sound protocols $\bar{\Pi}_{c\mathcal{ZK}}$ and $\bar{\Pi}_{r\mathcal{ZK}}$, where $\bar{\Pi}_{c\mathcal{ZK}}$ is a round-optimal c$\mathcal{ZK}$AoK, and $\bar{\Pi}_{r\mathcal{ZK}}$ is a 5-round r$\mathcal{ZK}$ argument. The r$\mathcal{ZK}$ protocols are mainly inherited from the ones of Yung and Zhao [31].

1   Introduction

The notion of concurrent zero knowledge (c$\mathcal{ZK}$, for short) introduced in [11] deals with proofs given in asynchronous networks controlled by the adversary.

In [3] Canetti et al. studied the case of an adversary that can reset the prover, forcing it to re-use the same randomness in different executions. They defined as resettable zero knowledge (r$\mathcal{ZK}$, for short) the security of a proof system against such attacks. Very interestingly, r$\mathcal{ZK}$ is proved to be stronger than c$\mathcal{ZK}$.

Motivated by the need of achieving round-efficient r$\mathcal{ZK}$, in [3] the Bare Public-Key (BPK, for short) model has been introduced, with the goal of relying on a setup assumption that is as close as possible to the standard model. Indeed, round-efficient c$\mathcal{ZK}$ and r$\mathcal{ZK}$ are often easy to achieve in other models (e.g., with trusted parameters) that unfortunately are hard to justify in practice.

The BPK *Model.* The sole assumption of the BPK model is that when proofs are played, identities of (polynomially many) verifiers interacting with honest provers are fixed. Identities have to be posted to a public directory before proofs start. This registration phase is non-interactive, does not involve trusted parties

D. Pointcheval and T. Johansson (Eds.): EUROCRYPT 2012, LNCS 7237, pp. 153–171, 2012.

or other assumptions, and can be fully controlled by an adversarial verifier. When proofs starts, it is assumed that honest provers interact with registered verifiers only. The BPK model is very close to the standard model, indeed the proof phase does not have any requirement beyond the availability of the directory to all provers, and for verifiers, of a secret key associated to their identities. Moreover, in both phases the adversary has full control of the communication network, and of corrupted players.

The first constant-round $r\mathcal{ZK}$ argument for **NP** in the BPK model has been given in [3]. Then in [18] it is pointed out the subtle separations among soundness notions in the BPK model. Indeed, in contrast to the standard model, the notions of one-time, sequential and concurrent soundness, are distinct in the BPK model. In [18] it is then proved that the protocol of [3] is actually sequentially sound only. Moreover in [18] it is proven that 4 rounds are necessary for concurrent soundness and finally, they showed a 4-round $r\mathcal{ZK}$ argument with sequential soundness. In light of the impossibility proved by [1] (i.e., there exists no 3 round sequentially sound $c\mathcal{ZK}$ conversation-based argument in the BPK model for non-trivial languages) the above 4-round $r\mathcal{ZK}$ argument is round optimal. Concurrent soundness along with $r\mathcal{ZK}$ was achieved in [7], requiring 4 rounds. Further improvements on the required complexity assumptions have been showed in [31] where a 4-round protocol under generic assumptions and an efficient 5-round protocol under number-theoretic assumptions are shown. All previously discussed results on constant-round $r\mathcal{ZK}$ in the BPK model relied on the assumptions that some cryptographic primitives are secure against subexponential time adversaries (i.e., complexity leveraging) and obtained black-box simulation.

The question of achieving a constant-round black-box $c\mathcal{ZK}$ argument of knowledge (AoK, for short) in the BPK model without relying on complexity leveraging has been first addressed in [32] and then in [9]. The protocol of [32] needs 4 rounds and enjoys sequential soundness. The protocol given in [9] needs only 4 rounds and enjoys concurrent soundness. A follow up result of [27] showed an efficient transformation that starting from a language admitting a Σ-protocol produces a $c\mathcal{ZK}$ AoK with concurrent soundness needing only 4 rounds and adding only a constant number of modular exponentiations. A more recent result [6] obtains both round optimality and optimal complexity assumptions (i.e., OWFs) in a concurrently sound $c\mathcal{ZK}$ AoK. More sophisticated notions of arguments of knowledge have been given in [10] and in [29,28]. Indeed these papers focus on concurrent knowledge extraction (under different formulations). All above results achieving $c\mathcal{ZK}$ are based on hardness assumptions with respect to polynomial-time adversaries.

1.1   Our Results and Techniques

In this paper we show subtle problems concerning security proofs of various $c\mathcal{ZK}$ and $r\mathcal{ZK}$ arguments in the BPK model [18,32,7,9,27,31,6,29], including *all* round-optimal constructions published so far.

The Source of the Problem: Parallel Execution of Different Sub-Protocols. In order to achieve round efficiency, various known protocols, including all round-optimal protocols, consist in parallel executions of sub-protocols that are useful in different ways in the proofs of soundness and $c\mathcal{ZK}/r\mathcal{ZK}$. Roughly speaking, there is always a sub-protocol π_0 where in 3 rounds the verifier is required to use a secret related to its identity. Then there is a 3-round sub-protocol π_1 in which the prover convinces the verifier about the validity of the statement and the simulator can do the same by using knowledge of a secret information obtained by rewinding π_0 (in this session or in other sessions corresponding to the same identity). To obtain a 4-round protocol[1], π_1 starts during the second round of π_0. Such round combination yields the following two cases.

First we consider the case in which the simulator needs the extracted secret in order to play the first message of π_1 so that such a message can appear in the final transcript of the simulation. In this case when the simulator needs to run π_1 for the first time with a given identity, it needs first to obtain some secret information by the verifier in π_0. The use of look-ahead threads (i.e., trying to go ahead with a virtual simulation with the purpose of obtaining the required information needed in the main thread of the simulation) would not help here since only a limited polynomial amount of work can be invested for them, and there is always a non-negligible probability that look-ahead threads fail, while in the main thread the verifier plays the next message. Given the above difficulty, the simulator needs to play a *bad* first round in π_1 so that later, when the needed secret information is obtained, the simulator can play again the second round of the protocol, this time playing a *good* first round in π_1. However, this approach suffers of a problem too. Indeed, stopping the main thread and trying to start and complete a new thread leads to a detectable deviation in the final transcript that the simulator will output. Indeed, the fact that the simulator gives up with a thread each time it is stuck, and then starts a new one, as we shall see later, modifies the distribution of the output of the simulator, since the output will then include with higher probability threads that are "easier" to complete (e.g., where the simulator does not get stuck because new sessions for new identities do not appear). Notice that this issue motivates the simulation strategies adopted in previous work on $c\mathcal{ZK}$ (e.g., [25,23]) where the main thread corresponds to the construction of the view that will be given in output, while other threads are started with the sole purpose of extracting secrets useful to go ahead in the main thread. Similar issues concerning the use of a main thread during the whole simulation have been recently considered in [21] for analyzing previous work on selective decommitments.

We now consider the second case where the simulator does not need any secret to compute the first round of π_1. We observe that this approach could hurt the proof of concurrent soundness, when the latter is proved by means of witness extraction. Indeed, a malicious concurrent prover can exploit the execution of π_0 in a session j, for completing the execution of π_1 in another concurrent session $j' \neq j$ by playing a man-in-the-middle attack such that, when (in the proof

---

[1] Similar discussions hold for some 5-round protocols when π_0 requires 4 rounds.

of concurrent soundness) one tries to reach a contradiction by extracting the witness from the proof π_1 given in session j', it instead obtains the secret used to simulate π_0 in session j. Instead, if the secret to be extracted from π_1 is fixed from the very first round of π_1, then one can show that it is either independent from the one used in session j (this happens when the secret is used in π_0 of session j after the first round of π_1 in session j' is played), or is dependent but not affected by the rewind of the extraction of session j' (this happens when the secret is used in π_0 of session j before the first round of π_1 in session j' is played).

The use of the secret in the last round of π_1 only, could instead be helpful in the following three cases: I) when one is interested in $r\mathcal{ZK}$ since in this case soundness is proved through a reduction based on complexity leveraging; II) when $c\mathcal{ZK}$ with sequential soundness only is desired; III) when the secret needed by the simulator when running π_1 in a session j' is different from the witness used by the verifier in the execution of π_0 in the other sessions. Indeed, in those cases the above discussion does not necessarily apply, and indeed some proposed round-optimal protocols might be secure (see discussion in Section 2), even though their security proofs seem to ignore at least in part the problems that we are pointing out.

Because of the above case I, we believe that achieving 4-round $c\mathcal{ZK}$ with concurrent soundness in the BPK model under standard assumptions is definitively harder[2] than obtaining 4-round $r\mathcal{ZK}$ with concurrent soundness in the BPK model through complexity leveraging. This is the reason why we mainly concentrate on achieving $\Pi_{c\mathcal{ZK}}$ and this will require a new technique. Instead, to obtain $\Pi_{r\mathcal{ZK}}$, we will just rely on a previous protocol given in [31] and make some minor variations in order to recycle part of the analysis given for $\Pi_{c\mathcal{ZK}}$.

We stress that in all previous constructions, one could obtain a different protocol that satisfies the desired soundness and zero-knowledge properties by simply running π_0 and π_1 sequentially. Indeed, in this case the simulator can complete π_0 in the main thread, then can run the extractor in another thread, and finally can continue the main thread running π_1 having the secret information. We also stress that all papers that we revisit in this work, achieved also other results that are not affected by our analysis.

We finally note that we did not investigate other round-efficient results in variations of the BPK model [17,33,8], and other results in the BPK model that do not focus on (almost) optimal round complexity [20,30,4].

New Techniques for Round-Optimal $c\mathcal{ZK}$ *and* $r\mathcal{ZK}$ *in the* BPK *Model.* In the main contribution of this paper we show a protocol and a security proof that close the gap in between lower and upper bounds for the round complexity of concurrently sound $c\mathcal{ZK}AoK$ in the BPK model. The result is achieved by using a new technique where in addition to the secret of the verifier corresponding to her identity, there is a temporary secret per session that enables the simulator to

---

[2] However, when we will then focus on efficient instantiations, we will obtain a 4-round protocol for $c\mathcal{ZK}$ while $r\mathcal{ZK}$ will require 5 rounds.

proceed in two modes. Indeed, knowledge of the permanent secret of the verifier allows the simulator to proceed in straight-line in the main thread in sessions started after the extraction of the permanent secret. We show that temporary secrets allow the simulator to proceed with the main thread even for sessions started before the extraction of such secrets.

We implement this technique by means of trapdoor commitments. The proof of $c\mathcal{ZK}$ will be tricky since it requires the synergy of the two above simulation modes. Each time an extraction procedure is started, the simulator is straight-line in the new thread, and aborts in case an unknown secret key is needed to proceed. Essentially, we can show that the number of extraction procedures of temporary and permanent secret keys correspond to the number of sessions[3]. The proof of concurrent soundness also requires special attention. Indeed while the interplay of temporary and permanent secrets helps the simulator, it could also be exploited by the malicious prover.

Our specifically designed protocol $\Pi_{c\mathcal{ZK}}$ is a round-optimal concurrently sound black-box perfect $c\mathcal{ZK}$AoK for **NP**, under standard complexity-theoretic assumptions. Then, we show that by using complexity leveraging (and thus assuming the existence of complexity-theoretic primitives secure against sub-exponential time adversaries) a variation of a previous protocol due to Yung and Zhao [31] produces a protocol $\Pi_{r\mathcal{ZK}}$ that is black-box $r\mathcal{ZK}$, round-optimal and concurrently sound for **NP**. The variations with respect to the work of [31] allow us to recycle part of the analysis used for $\Pi_{c\mathcal{ZK}}$. Indeed, as we show in Section 2.2, although fixable, the security proof provided in [31] relies on a simulator that outputs a transcript that is distinguishable from the real execution.

We then show that $\Pi_{c\mathcal{ZK}}$ and $\Pi_{r\mathcal{ZK}}$ admit efficient transformations that starting from any language admitting a perfect Σ-protocol, produce concurrently-sound protocols $\bar{\Pi}_{c\mathcal{ZK}}$ and $\bar{\Pi}_{r\mathcal{ZK}}$, where $\bar{\Pi}_{c\mathcal{ZK}}$ is a round-optimal black-box perfect $c\mathcal{ZK}$AoK, while $\bar{\Pi}_{r\mathcal{ZK}}$ is a 5-round black-box $r\mathcal{ZK}$ argument. Both transformations only require a constant number of modular exponentiations, and $\bar{\Pi}_{c\mathcal{ZK}}$ is secure under standard number-theoretic assumptions, while $\bar{\Pi}_{r\mathcal{ZK}}$ also needs number-theoretic assumptions w.r.t. sub-exponential time adversaries. $\bar{\Pi}_{r\mathcal{ZK}}$ will again correspond to a variation of a protocol presented in [31].

It is plausible that motivated by different purposes one can get more general constructions or constructions with better efficiency, assumptions or security, but this is out of the scope of this work.

Notation and Tools. We denote by $n \in \mathbb{N}$ the security parameter and by PPT the property of an algorithm of running in probabilistic polynomial-time. We assume confidence with the concepts of witness indistinguishability (WI) and of proof of knowledge. A Σ-protocol $(\mathtt{pok_1}, \mathtt{pok_2}, \mathtt{pok_3})$ is a 3-round public-coin WI proof of knowledge enjoying the honest-verifier zero knowledge property (HVZK), that is, there exists a PPT simulator that on input the theorem to be proved and the message $\mathtt{pok_2}$, outputs a transcript that is indistinguishable from the transcript

---

[3] This contrasts with the main technique used in the past in the BPK model, where the extraction procedure were applied only for the identities registered in the directory.

output by the prover. If the output is perfectly indistinguishable the Σ-protocol is called *perfect*. We call *special* a Σ-protocol in which the prover can compute the message pok_1 without knowing the theorem to be proved. We refer to [5] for details on Σ-protocols and to [16] for details on *special* Σ-protocols.

2 Issues in Security Proofs of Previous Results

We now show issues in the proofs of $\mathsf{c}\mathcal{ZK}$ and $\mathsf{r}\mathcal{ZK}$ of known protocols.

2.1 The Case of Π_{MR} [18]

Description of Π_{MR}. In [18], it is shown a 4-round $\mathsf{r}\mathcal{ZK}$ argument with sequential soundness, Π_{MR}, in the BPK model. The identity of the verifier $\mathcal{V}$ is a public-key pk for a semantically secure encryption scheme, and the secret key sk is the corresponding private key. In the 1st round, $\mathcal{V}$ sends an encryption c under pk of a random string $\sigma_{\mathcal{V}}$. The prover $\mathcal{P}$ sends in the 2nd round a random string $\sigma_{\mathcal{P}}$. In the 3rd round $\mathcal{V}$ sends $\sigma_{\mathcal{V}}$ and the randomness used to compute c. Moreover in these first 3 rounds, $\mathcal{V}$ proves to $\mathcal{P}$ knowledge of sk using Blum's protocol for Hamiltonicity [2]. In the 4th round $\mathcal{P}$ sends a non-interactive zero knowledge (NIZK, for short) proof [12] on string $\sigma = \sigma_{\mathcal{V}} \oplus \sigma_{\mathcal{P}}$ proving that $x \in L$.

The Proof of $\mathsf{r}\mathcal{ZK}$ for Π_{MR}. The simulator S discussed in [18] (see also [24]) for Π_{MR} goes as follows. It runs the extractor associated to the proof of knowledge, therefore obtaining sk. Then, it can run in straight-line since the encryption c of $\sigma_{\mathcal{V}}$ can be decrypted using sk, and thus S can choose $\sigma_{\mathcal{P}}$ so that the resulting σ corresponds to the fake random string generated by the NIZK simulator. Then S can complete the protocol running in the 4th round the NIZK simulator.

The above simulation produces a transcript that is distinguishable from the one generated by an honest prover. Indeed, we can give two interpretations to the above simulation and in both cases there exists a successful adversary.

Case 1. The first interpretation is to assume that the extraction of sk is performed in a look-ahead thread that is played before the main thread (where the simulator computes the actual messages to be given in output). In this case, notice that the proof of knowledge of sk could be completed by V^* with some probability p unknown to S (S is black-box, and there can be different adversarial verifiers using different values for p, and some of them can be negligible). Therefore, since the attempt of S to extract sk can not be (in order to have an expected polynomial time simulation) unlimited in time, S must give up if after some polynomial effort sk has not been extracted. When such a look-ahead thread is aborted, then S continues the main thread and it can happen with non-negligible probability p (since S stopped after a polynomial number of attempts) that V^* completes the proof of knowledge of sk. Since in this case S has already played the second round σ_p, the outcome σ of the coin flipping does not allow S to complete the protocol; if one gives in output such a failure, then the transcript of the simulation would be easily distinguishable. S will therefore need to abort this main thread and start a new one, having now sk as input. The problem in this case corresponds to Case 2 below.

Case 2. The second interpretation consists in assuming that once the verifier completes the proof of knowledge, then S solves the identity by running the extractor of the proof of knowledge, therefore obtaining the secret in time roughly $\texttt{poly}(n)/p$, where p is the probability that $\mathcal{V}$ completes the proof of knowledge. Once the secret key is obtained, S can rewind the verifier and start the proof phase of the simulation from scratch, without changing the key generation phase. S now using knowledge of the secret key can complete in straight-line all sessions that correspond to that solved identity.

The above approach is often used in literature in the BPK model and consists therefore in dividing the simulation in phases. Each time the current phase is not completed in straight-line, an extraction is performed, one more identity is solved and then a new phase is started. Since the number of identities is polynomial, at some point there will be a phase that can be executed in straight-line by the simulator. We show now that this approach is affected by a subtle problem. Indeed the approach of S in this case follows the standard procedure of [14] for the case of stand-alone zero knowledge. Here however, a concurrent malicious verifier V^* can nest polynomially many other sessions each one corresponding to a different identity, and each one using a different abort probability.

Consider the simple case of V^* that only runs two nested sessions, corresponding to two different identities and such that in each session the 3rd round is played with probability $1/2$, adaptively to the transcript so far (i.e., this can be easily done by assuming that the coins used for such a probability are taken from the output of a PRF on input the transcript so far and a seed hardwired in V^*). The nesting is performed by including the whole execution of the 2nd session in between the 2nd and 3rd round of the first session. The view of V^* in the real game with probability $1/4$ includes the two sessions both aborted.

Instead, the output of S will be computed as follows. First of all, it can happen that the simulation is straight-line when V^* aborts in both sessions, and this event happens with probability $1/4$. Then, it can happen that the second session is aborted (probability $1/2$) and the first one is not aborted (probability $1/2$). In this case S performs the extraction of the secret from the first session, and once this is done, since it can not continue the previous execution (in the previous execution σ has been already computed and does not allow S to finish the protocol), it will have to start a new phase, this time having the secret key of the first identity as input. However notice that in this new phase (that happens with probability $\frac{1}{4}$), it can happen that both executions abort since the messages sent by S are different, and therefore the coins used by V^* to decide whether to abort or not, will be computationally independent. Since when an execution starts the case of getting two aborts happens with probability $1/4$, and since this new phase of S starts with probability $1/4$, we have that this produces in the output of S both sessions aborted with probability $\frac{1}{16} = \frac{1}{4} \cdot \frac{1}{4}$. Therefore we have that with probability at least $\frac{5}{16} = \frac{1}{4} + \frac{1}{16}$ the simulator outputs a transcript where both sessions are aborted. Given that in the real game this probability is only 1/4, we have that the output of the simulator is trivially distinguishable. For simplicity in the above analysis we have ignored the fact that V^* uses a PRF instead of independent coins.

Given the above explicit attack, one might wonder if the protocol is anyway valid and a different simulator or a different interpretation of S can be used to prove the same theorem. Indeed, the above attack is certainly addressable with a slightly more sophisticated ad-hoc simulator. However other more sophisticated attacks can easily hurt the new simulation strategy, as in a cat and mouse game where given an adversary one can find a valid simulator for it; but given the valid simulator for that adversary one can find another adversary that requires another simulator. It is not clear at all whether one can finally design a simulator that works against any adversary, as required by the definition of black-box zero knowledge.

The above difficulties[4] are not an issue when considering the simulators for concurrent zero knowledge [25,23] that indeed use the following strategy: the simulator starts a main thread that is updated with new messages exchanged with V^*; other threads are started only to allow the main thread to proceed successfully, but no thread ever replaces the main thread. This is a well understood strategy that we will also use in our constructions. It however will require a new technique to design a protocol where new threads can help the execution of the main thread (this is precisely the problem of some of previous constructions). The strategy of [25] is actually based on starting look-ahead threads, and the large round complexity tolerates failures of look-ahead threads.

The Same Attack Can Be Replicated to All Other Simulators. We have given details to explain the problem with the simulation of Π_{MR}, and under minor variations, all other results [18,32,7,9,27,31,6,29] suffer of similar problems. We will now focus on protocols that however could have a different simulation.

2.2 Replacing Simulation in Phases by Threads

We now discuss 4 previous protocols that besides the issues in the proposed simulation strategies discussed above, seem (in some cases with some fixes) still to be able to admit a simulation strategy based on maintaining a main thread. We stress that later we will show a new technique based on the use of temporary keys along with permanent keys so that the simulator works in two modes that allow it to stick with the main thread. Our technique was never used in previous papers. Protocols below when using a different simulation strategy (in some cases, our new simulation strategy) can potentially achieve some of the 4 results that we will achieve in the next sections. We did not go through details of the proofs of such (in some cases, fixed) 4 protocols. Summing up, we do not claim their security and here we only explain how such protocols and their (distinguishable) simulations in phases could potentially be adjusted in light of our results and techniques.

---

[4] We stress that such difficulties disappears if round optimality is not needed.

The Case of Π_Z [32]. A 4-round conversation-based c$\mathcal{ZK}$ argument enjoying sequential soundness only is shown in [32]. While the security proof still relies on the use of a simulator that works in phases, we notice that a different simulator based on keeping a main thread could be used instead. The reason, is that the secret information is needed by the simulator only in the 3rd round of π_1 (see our discussion in Section 1) and, since the achieved result is only sequentially sound, there is no concurrent attack to soundness to take care of.

The Case of Π_{YZ} [31]. In [31], Yung and Zhao showed protocols Π_{YZ} and $\bar{\Pi}_{YZ}$ that are respectively a 4-round concurrently sound r$\mathcal{ZK}$ argument in the BPK model under general complexity-theoretic assumptions and an efficient 5-round concurrently sound r$\mathcal{ZK}$ argument under number theoretic assumptions. Both protocols use complexity leveraging and we will now concentrate on Π_{YZ} since the analysis extends also to $\bar{\Pi}_{YZ}$ with one more round.

Π_{YZ} consists of 3 sub-protocols played in parallel. In the first three rounds the verifier, using a special Σ-protocol Σ^{fls}, gives a proof of knowledge of its secret key sk or of a solution of a puzzle. The puzzle was sent by the prover during the second round, and Σ^{fls} is such that knowledge of the theorem (and therefore of the witness) is not required in the first round. The prover gives a resettable WI proof (i.e., the verifier commits to the challenge in the first round) in rounds 2, 3 and 4 where it proves that $x \in L$ or it knows sk. Since black-box extraction of the witness (necessary for the proof of concurrent soundness) is not allowed in the resetting verifier setting, they enforce the extraction using complexity leveraging as follows. The challenge is committed through a trapdoor commitment scheme with a 2-round decommitment phase, where the trapdoor, that is needed only in the opening, corresponds to the solution of the puzzle sent by the prover. Therefore there exists a sub-exponential time extractor that can find the solution of the puzzle, open the commitment in multiple ways and thus extract the actual witness of the prover. This proof of concurrent soundness falls down when one would like to use standard hardness assumptions only (e.g., to prove c$\mathcal{ZK}$ under standard assumptions). The technical difficulty of implementing efficiently Σ^{fls} is solved by requiring the prover to send the puzzle in a first round, so that an OR composition of Σ-protocols can be used, therefore obtaining a 5-round protocol $\bar{\Pi}_{YZ}$.

As discussed in [31] (see page 136), the simulator runs in different phases, trying in each phase to complete the simulation, but in case it can not, it obtains a new secret key and starts a new phase, with new randomness. This approach as previously discussed makes the output of the simulator distinguishable when playing with some specific adversarial concurrent verifiers. However, we notice that in this case an alternative simulation strategy could be possible. Indeed, when the simulator starts the main thread and gets stuck, it does not actually need to abort it, but instead can start a new thread just to get the secret information to complete the main thread. The reason why this can be possible here (in contrast to previous protocols), is that the simulator needs the secret of the verifier only when it plays the last message of the protocol, therefore it can always perform the extraction (in a new thread) before being stuck. However, as discussed in the introduction, playing the extracted secret only in the last round exposes the protocol to concurrent soundness attacks. In the very specific case of r$\mathcal{ZK}$, since soundness is proved through complexity leveraging, the proof of soundness could go through.

The Case of Π_{YYZ} [29,28]. In the concurrently sound c$\mathcal{ZK}$ protocol presented in [29,28], the simulator is required to commit in the second round to one of the two secret keys of the verifier. This must be done before the verifier completes its proof of knowledge of one of her secret keys. It is immediate to see that precisely as we discussed above, this requires the simulator to try to complete the simulation using new phases (see page 24 of [28]). Therefore the same attacks showed before can be mounted against this simulator too.

In Section 6.2 of [28] an update of the protocol yielding round optimality is suggested. The update consists in replacing a strong WI proof with a 4-round zero-knowledge AoK due to Feige and Shamir [13] (FSZK, for short) such that this protocol can share the statistical WI proof of knowledge given by the verifier. However, in the same section it is then observed that such update hurts the concurrent soundness of their scheme.

Here we observe that since their first subprotocol is a statistical WI argument of knowledge given by the verifier, it can be instantiated under general complexity-theoretic assumptions only requiring a first round from prover to verifier. Indeed this message is needed to establish the parameters for a statistically hiding commitment scheme to be used in the statistical WI argument. Therefore, the resulting construction can be round optimal only when using number-theoretic assumptions, that can be used to implement the statistical WI proof in 3 rounds [26,5].

Our technique based on temporary keys and simulation in two modes can potentially be applied when using FSZK differently, so that concurrent soundness could be preserved. This could be possible when FSZK is played independently of the public keys of the verifier, therefore including some session keys (which would have a role similar to the temporary keys of our technique). Then our new simulation technique could be used to maintain a main thread working in two modes (in one mode using the extracted permanent keys, in the other mode using the simulator of FSZK that use the extracted session keys).

The Case of Π_D [6]. A 4-round concurrently sound c$\mathcal{ZK}$ argument Π_D in the BPK model under the existence of one-way functions only is showed in [6]. In the first round, the verifier sends a message m_v. Then in the second round the prover sends a statistically binding commitments of potential signatures of messages (under the public-key of the verifier) and a message m_p. In the third round the verifier sends a signature of $(m_v|m_p)$ (instead of the usual proof of knowledge of a secret). In the last 3 rounds $\mathcal{P}$ proves that $x \in L$ or the commitment sent in the second round corresponds to messages $(m'|m_0')$ and $(m'|m_1')$ and their signatures, where $m_0' \neq m_1'$.

Of course since the concurrent adversarial prover can not rewind the verifier, the above argument is sufficient to prove concurrent soundness. Indeed, signatures received in concurrent proofs always correspond to messages with a different prefix selected by the verifier. The $c\mathcal{ZK}$ property of the protocol however is problematic again for the very same reasons discussed above. Indeed, the simulator does not have any signature at all when it plays the second round, and thus later on, in order to be able to complete proofs it will have to start new phases where knowledge of the signatures accumulated during previous executions is sufficient to run in straight-line. Indeed, the simulator presented in [6] rewinds the verifier when it is stuck, and produces a new transcript committing to the extracted signatures. As already explained, this makes distinguishable its output w.r.t. real executions.

We finally argue that the protocol could be adjusted to then admit a simulator that keeps a main thread. In contrast to previously discussed protocols, the main advantage of Π_D is that the verifier uses his secret keys to generate signatures of messages with different formats in different sessions. This makes problematic the attack of concurrent soundness, since the execution of concurrent sessions does not provide useful messages to cheat in a specific session. Therefore one could tweak the protocol so that the simulator needs to use the obtained signatures only at the last round. In this way, the simulator could obtain through rewinds two signatures for messages with the same structure, and could use them in the main thread and in all future sessions that correspond to that verifier.

3   Round-Optimal $c\mathcal{ZK}$ and $r\mathcal{ZK}$ in the BPK Model

We show under standard complexity-theoretic assumptions round-optimal concurrently sound $c\mathcal{ZK}$AoK and a $r\mathcal{ZK}$ argument with complexity leveraging.

3.1   Concurrent Zero Knowledge in the BPK Model

Overview, Techniques and Proof Intuition. In light of the attacks shown in the previous section, we construct a protocol that allows a simulation strategy in which the transcript generated in main thread is kept unchanged. In the following we describe the protocol incrementally.

The public identity of the verifier $\mathcal{V}$ corresponds to a pair of public keys $\mathrm{pk}_0 = f(\mathrm{sk}_0), \mathrm{pk}_1 = f(\mathrm{sk}_1)$, where f is a one-way function, for which $\mathcal{V}$ knows one of the pre-images sk_b, that we call the secret key. Following the usual paradigm used in the BPK model, we require that $\mathcal{V}$ provides a proof of knowledge, using a Σ-protocol, that we denote by Σ^{pk_j}, of the secret key associated to the public identity $\mathrm{pk}_j = (\mathrm{pk}_0, \mathrm{pk}_1)$. In the second round, the prover $\mathcal{P}$ first commits to a bit (representing the selection of one of $\mathcal{V}$'s public keys), then it provides a proof that either $x \in L$ or it knows the secret corresponding to the public key selected in the commitment, using a Σ-protocol as well, which we denote by Σ^{L_j}. Requiring that $\mathcal{P}$ selects the key already in the first message allows to use the witness-indistinguishability property of Σ-protocols and the binding property of the commitment scheme to prove the concurrent-soundness property.

A simulator for this protocol would extract the secret (by exploiting the proof of knowledge property of Σ^{pk_j}) and would complete the protocol Σ^{L_j} using the extracted secret key as witness. However, this step is done without changing the commitment sent in the first message only if S has committed to the bit corresponding to the extracted secret key. Instead, if this is not the case, S has to rewind the verifier and change the commitment, therefore changing the transcript of the main thread, that is precisely the problem of previous works.

We overcome this problem by tweaking the protocol in two ways. First, we require that upon each new execution, $\mathcal{V}$ freshly generates a pair of public parameters and the respective trapdoors for a two-round trapdoor commitment scheme. Such parameters can be seen as temporary keys. $\mathcal{V}$ then sends the public parameters to $\mathcal{P}$ and proves knowledge of one of the trapdoors running an additional Σ-protocol that we denote by Σ^{trap}. This will allow the simulator to extract the trapdoor. Second, we require that $\mathcal{P}$, instead of sending the first message of Σ^{L_j} in clear, it sends a trapdoor commitment of it, using both public parameters received from $\mathcal{V}$, i.e., $\mathcal{P}$ computes two commitments, each one with a distinct parameter, of two shares of the first message. The shares are revealed only in the third round of Σ^{L_j}, precisely only after $\mathcal{P}$ has seen the challenge for Σ^{L_j} sent by $\mathcal{V}$. Intuitively, due to the binding of the commitment scheme $\mathcal{P}$ is not able to take advantage of the knowledge of the challenge. Furthermore, since the parameters of the trapdoor commitment are freshly generated by $\mathcal{V}$ upon each execution, due to the witness indistinguishability property of Σ^{trap}, $\mathcal{P}$ cannot take advantage of concurrent executions with many verifiers, thus concurrent soundness still holds[5]. Indeed, we are able to prove concurrent soundness by showing a concurrent extractor, that extracts the witness from any accepting transcript obtained by any malicious prover. The guarantee of the witness extraction is necessary for the proof of soundness to go through.

Instead the simulator can use its rewinding capabilities to extract the trapdoor by exploiting the proof of knowledge property Σ^{trap}, so that in the main thread it can open the commitments of the first round of Σ^{L_j}, according to the challenge received from $\mathcal{V}$. Here, the simulator uses the HVZK property of Σ^{L_j}. We stress that the simulator does not change messages previously played in the main thread, i.e., the commitments of the first round of Σ^{L_j}, but it cheats only in the third round by equivocating one of the commitments by using the trapdoor extracted in the rewinding thread. Since commitments computed by the prover are perfectly hiding and Σ^{L_j} is a perfect Σ-protocol, the simulation will be perfectly indistinguishable from a real execution.

Note that, in order to prevent the blow-up of the running time, it is crucial that the simulator extracts the trapdoor only for sessions for which it has not extracted the secret key yet. Once a secret corresponding to an identity has been extracted, all sessions played by the malicious verifier with such identity are simulated in straight-line.

---

[5] If instead parameters of the trapdoor commitments were fixed for all executions, then in the proof of soundness one can not derive a contradiction in case $\mathcal{P}$ equivocates the commitment associated to the same trapdoor used as witness in Σ^{trap}.

Formal construction. In the following we provide the formal specification of the c$\mathcal{ZK}$AoK protocol that we denote by $\Pi_{c\mathcal{ZK}}$.

The Public File. Let f be a given one-way function $f : \{0,1\}^{\texttt{poly}(n)} \rightarrow \{0,1\}^*$. The jth identity of the public file F is $\text{pk}_j := (\text{pk}_j^0 = f(\text{sk}_j^0), \text{pk}_j^1 = f(\text{sk}_j^1))$ for some values $\text{sk}_j^0, \text{sk}_j^1 \in \{0,1\}^n$.

Sub-Protocols. Let $\texttt{TCom} = (\texttt{TSen}, \texttt{TRec}, \texttt{TDec})$ be a two-round perfectly-hiding trapdoor commitment scheme and $\texttt{PHCom} = (\texttt{PHSen}, \texttt{PHRec})$ a two-round perfectly-hiding commitment scheme. For simplicity we assume that the public parameter pk for the scheme $\texttt{PHCom}$ is included in the identity of $\mathcal{V}$. The parameters' generation procedure is denoted by $pk \leftarrow \texttt{PHRec}(1^n, r)$ (resp. $(pk, trap) \leftarrow \texttt{TRec}(1^n, r)$) where r is a random string and n is the security parameter. The commitment procedure $(\texttt{C}, \texttt{D}) \leftarrow \texttt{PHSen}(pk, m)$ (resp. $\texttt{TSen}$) takes as input the public parameter pk and a message m and outputs the commitment $\texttt{C}$ and the decommitment $\texttt{D}$. The verification procedure $\texttt{PHRec}(pk, \texttt{C}, \texttt{D}, m)$ (resp. $\texttt{TRec}$) outputs 1 if $\texttt{D}$ is a valid decommitment of $\texttt{C}$ for the message m, under pk. Finally, $(m', \texttt{D}) \leftarrow \texttt{TDec}(trap, \texttt{C}, m', z)$ is the procedure that allows to open $\texttt{C}$ as any message using the trapdoor $trap$ and some auxiliary information z inherited from the commitment phase.

Auxiliary Languages. We use the following **NP** relations and in turn the respective **NP**-languages $L_{\text{pk}_j}, L_{\text{trap}}, L_{\text{sk}_j}, L_j$:

- $\mathcal{R}_{\text{pk}_j} = \{(\text{pk}_j^0, \text{pk}_j^1), \text{sk}) \text{ s.t. } \text{pk}_j^0 = f(\text{sk}) \text{ OR } \text{pk}_j^1 = f(\text{sk})\}$;
- $\mathcal{R}_{\text{trap}} = \{((k_0, k_1), (t, r)) \text{ s.t. } (k_0, t) \leftarrow \texttt{TRec}(1^n, r) \text{ OR } (k_1, t) \leftarrow \texttt{TRec}(1^n, r)\}$;
- $\mathcal{R}_{\text{sk}_j} = \{((\texttt{C}, \text{pk}_j^0, \text{pk}_j^1), (d, \texttt{D}, \text{sk})) \text{ s.t. } \texttt{PHRec}(pk, \texttt{C}, \texttt{D}, d) = 1 \wedge \text{pk}_d^j = f(\text{sk})\}$;
- $\mathcal{R}_{\text{L}_j} := \mathcal{R}_L \vee \mathcal{R}_{\text{sk}_j} = \{(x, \texttt{C}, \text{pk}_j^0, \text{pk}_j^1), (w, d, \texttt{D}, \text{sk})) \text{ s.t. } (x, w) \in R_L \vee ((\texttt{C}, \text{pk}_j^0, \text{pk}_j^1), (d, \texttt{D}, \text{sk})) \in \mathcal{R}_{\text{sk}_j}\}$.

Σ-Protocols. The languages showed above are proved by means of Σ-protocols. We denote by $\Sigma^{\text{pk}_j} = (\text{pok}_1^{\text{pk}_j}, \text{pok}_2^{\text{pk}_j}, \text{pok}_3^{\text{pk}_j})$, $\Sigma^{\text{trap}} = (\text{pok}_1^{\text{trap}}, \text{pok}_2^{\text{trap}}, \text{pok}_3^{\text{trap}})$ the Σ-protocols run by $\mathcal{V}$ with identity $\text{pk}_j = (\text{pk}_j^0, \text{pk}_j^1)$ to prove instances of relations $\mathcal{R}_{\text{pk}_j}$ and $\mathcal{R}_{\text{trap}}$ respectively. We denote by $\Sigma^{\text{L}_j} = (\text{pok}_1^{\text{L}_j}, \text{pok}_2^{\text{L}_j}, \text{pok}_3^{\text{L}_j})$ the perfect Σ-protocol run by $\mathcal{P}$ for instances of $\mathcal{R}_{\text{L}_j}$ when interacting with the verifier with identity pk_j.

The Protocol. The protocol is depicted in Fig. 1. By noticing that Blum's protocol [2] is a perfect HVZK Σ-protocol (when the first message is computed with a perfectly-hiding commitment scheme) for **NP** languages, we conclude that Protocol $\Pi_{c\mathcal{ZK}}$ is a black-box perfect c$\mathcal{ZK}$AoK for all **NP**.

Theorem 1. *If $\Sigma^{\text{pk}_j}, \Sigma^{\text{trap}}$ are Σ-protocols, Σ^{L_j} is a perfect Σ-protocol, $\texttt{PHCom}$ is a two-round perfectly-hiding commitment scheme and $\texttt{TCom}$ is a two-round perfectly-hiding trapdoor commitment scheme then $\Pi_{c\mathcal{ZK}}$ is a 4-round concurrently sound black-box perfect c$\mathcal{ZK}$AoK in the BPK model for all* **NP**.

Common input: the public file F, n-bit string $x \in L$ and index j specifying the jth entry of F, i.e. $(\mathrm{pk}_j^0 = f(\mathrm{sk}_j^0), \mathrm{pk}_j^1 = f(\mathrm{sk}_j^1))$. $\mathcal{P}$'s **private input:** a witness w for $x \in L$. $\mathcal{V}$'s **private input:** a randomly chosen secret sk_j^b between sk_j^0 and sk_j^1.

$\mathcal{V}$-round-1:

- $r_0, r_1 \xleftarrow{\$} \{0,1\}^n$, $(k_0, t_0) \leftarrow \mathrm{TRec}(1^n, r_0)$; $(k_1, t_1) \leftarrow \mathrm{TRec}(1^n, r_1)$;
- compute $\mathrm{pok}_1^{\mathrm{pk}_j}$ and $\mathrm{pok}_1^{\mathrm{trap}}$;
- send $k_0, k_1, \mathrm{pok}_1^{\mathrm{pk}_j}, \mathrm{pok}_1^{\mathrm{trap}}$ to $\mathcal{P}$.

$\mathcal{P}$-round-2:

- $(\mathtt{C}, \mathtt{D}) \leftarrow \mathrm{PHSen}(pk, d)$ for a randomly chosen bit d; compute $\mathrm{pok}_2^{\mathrm{pk}_j}$, $\mathrm{pok}_2^{\mathrm{trap}}$;
- compute $\mathrm{pok}_1^{L_j}$ and compute shares s_0, s_1 s.t. $s_0 \oplus s_1 = \mathrm{pok}_1^{L_j}$;
- $(\mathtt{tcom}_0, \mathtt{tdec}_0) \leftarrow \mathrm{TSen}(k_0, s_0)$, $(\mathtt{tcom}_1, \mathtt{tdec}_1) \leftarrow \mathrm{TSen}(k_1, s_1)$;
- send $\mathrm{pok}_2^{\mathrm{pk}_j}, \mathrm{pok}_2^{\mathrm{trap}}, \mathtt{C}, \mathtt{tcom}_0, \mathtt{tcom}_1$ to $\mathcal{V}$.

$\mathcal{V}$-round-3:

- compute $\mathrm{pok}_3^{\mathrm{pk}_j}$ using as witness sk_j^b;
- compute $\mathrm{pok}_3^{\mathrm{trap}}$ using as witness t_e, r_e for a randomly selected bit e;
- compute $\mathrm{pok}_2^{L_j}$;
- send $\mathrm{pok}_3^{\mathrm{pk}_j}, \mathrm{pok}_3^{\mathrm{trap}}, \mathrm{pok}_2^{L_j}$ to $\mathcal{P}$.

$\mathcal{P}$-round-4:

- verify that $(\mathrm{pok}_1^{\mathrm{pk}_j}, \mathrm{pok}_2^{\mathrm{pk}_j}, \mathrm{pok}_3^{\mathrm{pk}_j})$ is an accepting transcript of Σ^{pk_j} for the statement $(\mathrm{pk}_j^0, \mathrm{pk}_j^1) \in L_{\mathrm{pk}_j}$, if not abort;
- verify that $(\mathrm{pok}_1^{\mathrm{trap}}, \mathrm{pok}_2^{\mathrm{trap}}, \mathrm{pok}_3^{\mathrm{trap}})$ is an accepting transcript of Σ^{trap} for the statement $(k_0, k_1) \in L_{\mathrm{trap}}$, if not abort;
- compute $\mathrm{pok}_3^{L_j}$ using the witness w;
- send $\mathrm{pok}_3^{L_j}, \mathtt{tdec}_0, \mathtt{tdec}_1, s_0, s_1$ to $\mathcal{V}$.

$\mathcal{V}$-**decision:** if $\mathrm{TRec}(k_0, \mathtt{tcom}_0, \mathtt{tdec}_0, s_0) = 1$ AND $\mathrm{TRec}(k_1, \mathtt{tcom}_1, \mathtt{tdec}_1, s_1) = 1$ then $\mathrm{pok}_1^{L_j} \leftarrow s_0 \oplus s_1$ and accept iff $(\mathrm{pok}_1^{L_j}, \mathrm{pok}_2^{L_j}, \mathrm{pok}_3^{L_j})$ is an accepting transcript of Σ^{L_j} for the statement $(x, \mathtt{C}, \mathrm{pk}_j^0, \mathrm{pk}_j^1) \in L'_j$; else, abort.

Fig. 1. $\Pi_{\mathrm{c}\mathcal{ZK}}$: 4–round concurrently-sound c$\mathcal{ZK}$AoK in the BPK model for all **NP**

3.2 Resettable Zero Knowledge in the BPK Model

In this section we discuss the updates to (a simpler version of) $\Pi_{\mathrm{c}\mathcal{ZK}}$ to deal with the resetting power of the adversarial verifier V^*.

To suppress the resetting power of V^* we add the commitment of the challenge $\mathrm{pok}_2^{L_j}$ in the first round, and we require that the randomness of $\mathcal{P}$ is computed by applying a PRF on the transcript obtained so far. This ensures that on the same prefix of interaction the verifier will always get the same response from the prover. $\mathcal{V}$ will then provide the opening of the commitment in the third round.

Unfortunately, this approach prevents the (black-box) extraction of the witness that we need to prove concurrent soundness.

Thus, to allow extraction we need to resort to complexity leveraging arguments. As mentioned in Section 1.1, the use of such techniques allows one to design round-optimal protocols in which the use of the secret extracted from the malicious verifier can be postponed to the last round, ruling out the issues about the indistinguishability of the transcript pointed out in this work (of course only if the simulation strategy does not work in phases).

Therefore, designing a $\mathsf{r}\mathcal{ZK}$ protocol using complexity leveraging is a much simpler task that does not require the two-mode simulation that we used for $\Pi_{\mathsf{c}\mathcal{ZK}}$. Thus, in $\Pi_{\mathsf{r}\mathcal{ZK}}$ we do not need the use of temporary keys along with protocol $\Sigma^{\mathtt{trap}}$, and the prover sends the first round of Σ^{L_j} in clear (we assume that the witness is used only in the third round of Σ^{L_j}), instead of sending a trapdoor commitment of it. Moreover in $\Pi_{\mathsf{r}\mathcal{ZK}}$, we do not need that $\mathcal{P}$ commits to one secret already in the second round, and thus the theorem proved with Σ^{L_j} is that either $\mathcal{P}$ knows the witness for $x \in L$ or it knows one of the secret keys (instead of proving that the opening of the commitment points to one of the secrets keys). Instead we need that $\mathcal{P}$, in the second round, computes and sends a puzzle that is solvable in sub-exponential time. Then, we require that instead of the opening of the commitment, in the third round $\mathcal{V}$ sends only the message $\mathsf{pok}_2^{\mathsf{L}_j}$ and it proves, using again a Σ-protocol that we denote by $\mathsf{FLS}^{\mathtt{com}}$, that either message $\mathsf{pok}_2^{\mathsf{L}_j}$ is the valid opening or it knows the solution of the puzzle. Moreover, while $\Sigma^{\mathtt{trap}}$ disappear, in $\Sigma^{\mathtt{pk}_j}$ the verifier proves knowledge of one of the secret keys or of the solution of the puzzle (this update is necessary for the proof of concurrent soundness giving that the prover does not commit in the second round). Note that to preserve round-optimality, $\mathsf{FLS}^{\mathtt{com}}$ and $\Sigma^{\mathtt{pk}_j}$ must be *special* Σ-protocols [16] since the puzzle, that is part of the theorem, is sent by $\mathcal{P}$ only in the second round.

Obviously any malicious verifier, running in polynomial time is not able to solve the puzzle, and is bound on the challenge committed in the first round (thus the zero-knowledge property is preserved). Instead, the extraction of the witness is possible by running in sub-exponential time and solving the puzzle. When the theorem proved is instead false, the extraction will produce a contradiction, breaking the WI of $\mathsf{FLS}^{\mathtt{com}}$ or $\Sigma^{\mathtt{pk}_j}$, or inverting the one-way function used to produce the public keys. All these primitives are setup with ad-hoc security parameters so that they are secure against adversaries that by exhaustive search can solve the puzzle and check membership of the common instances (i.e., the size of such instances must be known before the experiment starts) of the $\mathsf{r}\mathcal{ZK}$ protocol in the language. The final protocol is a variation of the one proposed in [31].

Theorem 2. *If 2-round perfectly hiding commitments and OWPs secure against sub-exponential time adversaries exist then protocol $\Pi_{\mathsf{r}\mathcal{ZK}}$ is a 4-round $\mathsf{r}\mathcal{ZK}$ argument in the* BPK *model with concurrent soundness for all* **NP**.

4   Efficient Instantiations

Here we show efficient transformations that starting from any language L admitting a perfect Σ-protocol, and adding a constant number of modular exponentiations, produce: 1) a 4-round concurrently sound c$\mathcal{ZK}$AoK in the BPK model $\bar{\Pi}_{c\mathcal{ZK}}$ based on the Discrete Logarithm (DL) assumption; 2) a 5-round concurrently sound r$\mathcal{ZK}$ argument in the BPK model $\bar{\Pi}_{r\mathcal{ZK}}$ based on the hardness of the $\mathcal{DDH}$ assumption w.r.t. sub-exponential time adversaries.

Interestingly, both protocols are obtained essentially for free, by properly instantiating the sub-protocols in the constructions of $\Pi_{c\mathcal{ZK}}$ and $\Pi_{r\mathcal{ZK}}$. All Σ-protocols used in the following transformations are perfect.

$\bar{\Pi}_{c\mathcal{ZK}}$. Let (G, p, q, g) such that p, q are primes, $p = 2q+1$ and g is a generator of the only subgroup G of order q of $\mathbb{Z}_p^\star$. The one-way function f used to compute the identities of the public file is instantiated with the DL function. Therefore, $\mathrm{pk}_0 = g^{\mathrm{sk}_0} \pmod{p}$ and $\mathrm{pk}_1 = g^{\mathrm{sk}_1} \pmod{p}$, where $\mathrm{sk}_0, \mathrm{sk}_1 \xleftarrow{\$} \mathbb{Z}_q$. An identity also contains a pairs (g, h) of generators of G as parameters for a perfectly-hiding commitment scheme. To prove knowledge of one of the secret keys associated to identity $\mathrm{pk}_j = (\mathrm{pk}_0, \mathrm{pk}_1)$ is sufficient to prove the knowledge of the DL of either pk_0 or pk_1, that can be instantiated with Schnorr's [26] Σ-protocol under OR composition, as discussed in [5]. This is the efficient implementation of Σ^{pk_j}.

The trapdoor commitment scheme is instantiated with the scheme proposed by Pedersen [22]. Thus, temporary keys consist of the parameters for Pedersen commitment, i.e., $k_b = (g_b, h_b)$, where $h_b = g_b^{t_b} \pmod{p}$, and g_b is a generator of G, for $b = 0, 1$ and the corresponding trapdoors are t_0, t_1. We stress that t_0, t_1 are generated on the fly and are not contained in the public file. Thus, Σ^{trap} is instantiated again with Schnorr's protocol under OR composition.

The most interesting part consists in the implementation of protocol Σ^{L_j}, more specifically the implementation of the Σ-protocol for the relation $\mathcal{R}_{\mathrm{sk}_j}$. The perfectly-hiding commitment of a bit b (i.e., the commitment C of Fig. 1) is replaced by the Pedersen commitment of sk_b computed as $\mathsf{C} = h^r \mathrm{pk}_b$ for some random string r and bit b. Then, to prove that C corresponds to a commitment of sk_0 or sk_1, $\mathcal{P}$ is required to prove the AND of the following statements: 1) knowledge of the DL of $(\mathsf{C}/\mathrm{pk}_b)$, for some bit b, 2) knowledge of the decommitment of C. Both statements can be proved by efficient Σ-protocols based on DL assumption. Since we have a Σ-protocol for L too, putting everything together, Σ^{L_j} is obtained as the composition of these Σ-protocols by means of AND and OR logic operators. All the above computations require a constant number of modular exponentiations. $\bar{\Pi}_{c\mathcal{ZK}}$ is secure under the DL assumption.

$\bar{\Pi}_{r\mathcal{ZK}}$. The PRF is implemented by the efficient Naor-Reingold PRF [19] based on $\mathcal{DDH}$ assumption. The commitment $\mathrm{pok}_2^{L_j}$ is implemented with the El Gamal encryption scheme (based on the $\mathcal{DDH}$ assumption), i.e., the commitment of a string m corresponds to the pair $\mathsf{com} = (u = g_c^r \pmod{p}, v = h_c^r m \pmod{p})$, for a randomly chosen r, where g_c is a generator of G and $h_c = g_c^\beta \pmod{p}$

for some $\beta \leftarrow \mathbb{Z}_q$. Proving knowledge of the decommitment of com corresponds to prove that $(G, g, h, u, v/m)$ is a $\mathcal{DDH}$ tuple. The puzzle can be implemented by using again the DL assumption (obviously the use of complexity leveraging requires to work with groups of appropriate size). Having a Σ-protocol (i.e., Schnorr's protocol) to prove knowledge of a solution for the puzzle, and having a Σ-protocol for $\mathcal{DDH}$ problem [15], $\Sigma^{\texttt{fls}}$ is implemented as the OR composition of these two Σ-protocols. In $\bar{\Pi}_{r\mathcal{ZK}}$ the protocol $\Sigma^{\texttt{pk}_j}$ is used for an augmented theorem in which $\mathcal{V}$ proves also knowledge of the solution of the puzzle. Thus in $\bar{\Pi}_{r\mathcal{ZK}}$, the protocol $\Sigma^{\texttt{pk}_j}$ implemented above is used in OR composition with Schnorr's protocol for DL. However, there is a technicality here. When instantiating $\Sigma^{\texttt{fls}}$, $\Sigma^{\texttt{pk}_j}$ with the OR composition protocol as shown in [5], we have that $\mathcal{V}$ needs to know the theorem already when computing the first round. Therefore, as already discussed in Section 2 for the case of $\bar{\Pi}_{YZ}$ the puzzle must be sent in the first round and thus $\bar{\Pi}_{r\mathcal{ZK}}$ is a 5 round protocol. All the above computations require a constant number of modular exponentiations. The resulting protocol is secure under the $\mathcal{DDH}$ assumption w.r.t. sub-exponential time adversaries. The perfect Σ-protocol has to require the use of the witness in the last round only.

Acknowledgments. We thank the anonymous referees for their comments. Research supported in part by the European Commission through the FP7 programme under contract 216676 ECRYPT II, and done in part at UCLA.

References

1. Alwen, J., Persiano, G., Visconti, I.: Impossibility and Feasibility Results for Zero Knowledge with Public Keys. In: Shoup, V. (ed.) CRYPTO 2005. LNCS, vol. 3621, pp. 135–151. Springer, Heidelberg (2005)
2. Blum, M.: How to Prove a Theorem So No One Else Can Claim It. In: Proceedings of the International Congress of Mathematicians, pp. 1444–1451 (1986)
3. Canetti, R., Goldreich, O., Goldwasser, S., Micali, S.: Resettable Zero-Knowledge (Extended Abstract). In: STOC 2000, pp. 235–244. ACM (2000)
4. Cho, C., Ostrovsky, R., Scafuro, A., Visconti, I.: Simultaneously Resettable Arguments of Knowledge. In: TCC 2012. LNCS. Springer, Heidelberg (2012)
5. Cramer, R., Damgård, I., Schoenmakers, B.: Proof of Partial Knowledge and Simplified Design of Witness Hiding Protocols. In: Desmedt, Y.G. (ed.) CRYPTO 1994. LNCS, vol. 839, pp. 174–187. Springer, Heidelberg (1994)
6. Di Crescenzo, G.: Minimal Assumptions and Round Complexity for Concurrent Zero-Knowledge in the Bare Public-Key Model. In: Ngo, H.Q. (ed.) COCOON 2009. LNCS, vol. 5609, pp. 127–137. Springer, Heidelberg (2009)
7. Di Crescenzo, G., Persiano, G., Visconti, I.: Constant-Round Resettable Zero Knowledge with Concurrent Soundness in the Bare Public-Key Model. In: Franklin, M. (ed.) CRYPTO 2004. LNCS, vol. 3152, pp. 237–253. Springer, Heidelberg (2004)
8. Di Crescenzo, G., Persiano, G., Visconti, I.: Improved Setup Assumptions for 3-Round Resettable Zero Knowledge. In: Lee, P.J. (ed.) ASIACRYPT 2004. LNCS, vol. 3329, pp. 530–544. Springer, Heidelberg (2004)

9. Di Crescenzo, G., Visconti, I.: Concurrent Zero Knowledge in the Public-Key Model. In: Caires, L., Italiano, G.F., Monteiro, L., Palamidessi, C., Yung, M. (eds.) ICALP 2005. LNCS, vol. 3580, pp. 816–827. Springer, Heidelberg (2005)
10. Di Crescenzo, G., Visconti, I.: On Defining Proofs of Knowledge in the Bare Public Key Model. In: ICTCS 2007, pp. 187–198. World Scientific (2007)
11. Dwork, C., Naor, M., Sahai, A.: Concurrent Zero-Knowledge. In: STOC 1998, pp. 409–418. ACM (1998)
12. Feige, U., Lapidot, D., Shamir, A.: Multiple Non-Interactive Zero Knowledge Proofs Based on a Single Random String. In: FOCS 1990, pp. 308–317. IEEE (1990)
13. Feige, U., Shamir, A.: Zero Knowledge Proofs of Knowledge in Two Rounds. In: Brassard, G. (ed.) CRYPTO 1989. LNCS, vol. 435, pp. 526–544. Springer, Heidelberg (1990)
14. Goldreich, O., Kahan, A.: How to Construct Constant-Round Zero-Knowledge Proof Systems for NP. J. Cryptology 9(3), 167–190 (1996)
15. Hazay, C., Lindell, Y.: Efficient Secure Two-Party Protocols Techniques and Constructions. Springer (2010)
16. Lapidot, D., Shamir, A.: Publicly Verifiable Non-Interactive Zero-Knowledge Proofs. In: Menezes, A., Vanstone, S.A. (eds.) CRYPTO 1990. LNCS, vol. 537, pp. 353–365. Springer, Heidelberg (1991)
17. Micali, S., Reyzin, L.: Min-Round Resettable Zero-Knowledge in the Public-Key Model. In: Pfitzmann, B. (ed.) EUROCRYPT 2001. LNCS, vol. 2045, pp. 373–393. Springer, Heidelberg (2001)
18. Micali, S., Reyzin, L.: Soundness in the Public-Key Model. In: Kilian, J. (ed.) CRYPTO 2001. LNCS, vol. 2139, pp. 542–565. Springer, Heidelberg (2001)
19. Naor, M., Reingold, O.: Number-Theoretic Constructions of Efficient Pseudo-Random Functions. J. ACM 51(2), 231–262 (2004)
20. Ostrovsky, R., Persiano, G., Visconti, I.: Constant-Round Concurrent Non-Malleable Zero Knowledge in the Bare Public-Key Model. In: Aceto, L., Damgård, I., Goldberg, L.A., Halldórsson, M.M., Ingólfsdóttir, A., Walukiewicz, I. (eds.) ICALP 2008, Part II. LNCS, vol. 5126, pp. 548–559. Springer, Heidelberg (2008)
21. Ostrovsky, R., Rao, V., Scafuro, A., Visconti, I.: Revisiting Lower and Upper Bounds for Selective Decommitments. In: Cryptology ePrint Archive, Report 2011/536 (2011)
22. Pedersen, T.P.: Non-Interactive and Information-Theoretic Secure Verifiable Secret Sharing. In: Feigenbaum, J. (ed.) CRYPTO 1991. LNCS, vol. 576, pp. 129–140. Springer, Heidelberg (1992)
23. Prabhakaran, M., Rosen, A., Sahai, A.: Concurrent Zero Knowledge with Logarithmic Round-Complexity. In: FOCS 2002, pp. 366–375 (2002)
24. Reyzin, L.: Zero-Knowledge with Public Keys, Ph.D. Thesis. MIT (2001)
25. Richardson, R., Kilian, J.: On the Concurrent Composition of Zero-Knowledge Proofs. In: Stern, J. (ed.) EUROCRYPT 1999. LNCS, vol. 1592, pp. 415–431. Springer, Heidelberg (1999)
26. Schnorr, C.P.: Efficient Signature Generation for Smart Cards. Journal of Cryptology 4(3), 239–252 (1991)
27. Visconti, I.: Efficient Zero Knowledge on the Internet. In: Bugliesi, M., Preneel, B., Sassone, V., Wegener, I. (eds.) ICALP 2006, Part II. LNCS, vol. 4052, pp. 22–33. Springer, Heidelberg (2006)
28. Yao, A.C.C., Yung, M., Zhao, Y.: Concurrent Knowledge-Extraction in the Public-Key model. ECCC 14(002) (2007)

29. Yao, A.C.C., Yung, M., Zhao, Y.: Concurrent Knowledge Extraction in the Public-Key Model. In: Abramsky, S., Gavoille, C., Kirchner, C., Meyer auf der Heide, F., Spirakis, P.G. (eds.) ICALP 2010, Part I. LNCS, vol. 6198, pp. 702–714. Springer, Heidelberg (2010)
30. Deng, Y., Feng, D., Goyal, V., Lin, D., Sahai, A., Yung, M.: Resettable Cryptography in Constant Rounds – The Case of Zero Knowledge. In: Lee, D.H., Wang, X. (eds.) ASIACRYPT 2011. LNCS, vol. 7073, pp. 390–406. Springer, Heidelberg (2011)
31. Yung, M., Zhao, Y.: Generic and Practical Resettable Zero-Knowledge in the Bare Public-Key Model. In: Naor, M. (ed.) EUROCRYPT 2007. LNCS, vol. 4515, pp. 129–147. Springer, Heidelberg (2007)
32. Zhao, Y.: Concurrent/Resettable Zero-Knowledge with Concurrent Soundness in the Bare Public-Key Model and its Applications. In: Cryptology ePrint Archive, Report 2003/265 (2003)
33. Zhao, Y., Deng, X., Lee, C.H., Zhu, H.: Resettable Zero-Knowledge in the Weak Public-Key Model. In: Biham, E. (ed.) EUROCRYPT 2003. LNCS, vol. 2656, pp. 123–139. Springer, Heidelberg (2003)

Robust Coin Flipping

Gene S. Kopp and John D. Wiltshire-Gordon*

University of Michigan

Abstract. Alice seeks an information-theoretically secure source of private random data. Unfortunately, she lacks a personal source and must use remote sources controlled by other parties. Alice wants to simulate a coin flip of specified bias α, as a function of data she receives from p sources; she seeks privacy from any coalition of r of them. We show: If $p/2 \leq r < p$, the bias can be any rational number and nothing else; if $0 < r < p/2$, the bias can be any algebraic number and nothing else. The proof uses projective varieties, convex geometry, and the probabilistic method. Our results improve on those laid out by Yao, who asserts one direction of the $r = 1$ case in his seminal paper [Yao82]. We also provide an application to secure multiparty computation.

Keywords: multiparty computation, outsourcing randomness, biased coin flip, algebraic number, projective duality, hyperdeterminant.

1 Introduction

Alice has a perfectly fair penny—one that lands heads exactly 50% of the time. Unfortunately, the penny is mixed in with a jar of ordinary, imperfect pennies. The truly fair penny can never be distinguished from the other pennies, since no amount of experimentation can identify it with certainty. Still, Alice has discovered a workable solution. Whenever she needs a fair coin flip, she flips all the pennies and counts the Lincolns; an even number means heads, and an odd number means tails.

Alice's technique is an example of "robust coin flipping." She samples many random sources, some specified number of which are unreliable, and still manages to simulate a desired coin flip. Indeed, Alice's technique works even if the unreliable coin flips somehow fail to be independent.

Bob faces a sort of converse problem. He's marooned on an island, and the nearest coin is over three hundred miles away. Whenever *he* needs a fair coin flip, he calls up two trustworthy friends who don't know each other, asking for random equivalence classes modulo two. Since the sum of the classes is completely

* It is our pleasure to thank Tom Church, for helping simplify our original proof of algebraicity of mystery-values; László Babai, for providing guidance with respect to publication; László Csirmaz, for discussing secret-sharing with us; Victor Protsak, for pointing us to Lind's article [Lin84]; and Matthew Woolf, Nic Ford, Vipul Naik, and Steven J. Miller for reading drafts and providing helpful comments. We are also grateful to several anonymous referees for their suggestions.

mysterious to either of the friends, Bob may safely use the sum to make private decisions.

Bob's technique seems similar to Alice's, and indeed we shall see that the two predicaments are essentially the same. We shall also see that the story for biased coin flips is much more complex.

1.1 Preliminaries and Definitions

Informally, we think of a random source as a (possibly remote) machine capable of sampling from certain probability spaces. Formally, a **random source** is a collection $\mathcal{C}$ of probability spaces that is closed under quotients. That is, if $X \in \mathcal{C}$ and there is a measure-preserving map[1] $X \to Y$, then $Y \in \mathcal{C}$. Random sources are partially ordered by inclusion: We say that $\mathcal{C}$ is **stronger than** $\mathcal{D}$ iff $\mathcal{C} \supset \mathcal{D}$.

The quotients of a probability space X are precisely the spaces a person can model with X. For example, one can model a fair coin with a fair die: Label three of the die's faces "heads" and the other three "tails." Similarly, one can model the uniform rectangle $[0,1]^2$ with the uniform interval $[0,1]$: Take a decimal expansion of each point in $[0,1]$, and build two new decimals, one from the odd-numbered digits and one from the even-numbered digits.[2] Thus, forcing $\mathcal{C}$ to be closed under quotients is not a real restriction; it allows us to capture the notion that "a fair die is more powerful that a fair coin."[3]

We define an **infinite random source** to be one that contains an infinite space.[4] A **finite random source**, on the other hand, contains only finite probability spaces. Further, for any set of numbers $\mathbb{S}$, we define an $\mathbb{S}$**-random source** to be one which is forced to take probabilities in $\mathbb{S}$. That is, all the measurable sets in its probability spaces have measures in $\mathbb{S}$.

Sometimes we will find it useful to talk about the strongest random source in some collection of sources. We call such a random source **full-strength** for that collection. For instance, a full-strength finite random source can model any finite probability space, and a full-strength $\mathbb{S}$-random source can model any $\mathbb{S}$-random source.

In practice, when p people simulate a private random source for someone else, they may want to make sure that privacy is preserved even if a few people blab about the data from their random sources or try to game the system. Define an r-**robust** function of p independent random variables to be one whose distribution

---

[1] A measure-preserving map (morphism in the category of probability spaces) is a function for which the inverse image of every measurable set is measurable and has the same measure. Any measure-preserving map may be thought of as a quotient "up to measure zero."

[2] In fact, this defines an isomorphism of probability spaces between the rectangle and the interval.

[3] It would also be natural (albeit unnecessary) to require that $\mathcal{C}$ is closed under finite products.

[4] An infinite space is one that is not isomorphic to any finite space. A space with exactly 2012 measurable sets will always be isomorphic to a finite space, no matter how large it is as a set.

does not change when the joint distribution of any r of the random variables is altered. Saying that p people simulate a random source r-robustly is equivalent to asserting that the privacy of that source is preserved unless someone learns the data of more than r participants. Similarly, to simulate a random source using p sources, at least q of which are working properly, Alice must run a $(p - q)$-robust simulation.

By a **robust** function or simulation, we mean a 1-robust one.

We use J to denote the all-ones tensor of appropriate dimensions. When we apply J to a vector or hypermatrix, we always mean "add up the entries."

1.2   Results

This paper answers the question "When can a function sampling from p independent random sources be protected against miscalibration or dependency among $p - q$ of them?" (Alice's predicament), or equivalently, "When can p people with random sources simulate a *private* random source for someone else[5] in a way that protects against gossip among any $p - q$ of them?" (Bob's predicament). In the first question, we assume that at least q of the sources are still functioning correctly, but we don't know which. In the second question, we assume that at least q of the people keep their mouths shut, but we don't know who. In the terminology just introduced, we seek a $(p - q)$-robust simulation.

Consider the case of p full-strength finite random sources. We prove: If $1 \leq q \leq p/2$, the people may simulate any finite $\mathbb{Q}$-random source and nothing better; if $p/2 < q < p$, they may simulate any finite $\overline{\mathbb{Q}}$-random source and nothing better. The proof uses projective varieties, convex geometry, and the probabilistic method. We also deal briefly with the case of infinite random sources, in which full-strength simulation is possible, indeed easy (see Appendix C).

1.3   Yao's Robust Coin Flipping

Our work fits in the context of secure multiparty computation, a field with roots in A. C. Yao's influential paper [Yao82]. In the last section of his paper, entitled "What cannot be done?", Yao presents (a claim equivalent to) the following theorem:

Theorem 1 (A. C. Yao). *Alice has several finite random sources, and she wants to generate a random bit with bias α. Unfortunately, she knows that one of them may be miscalibrated, and she doesn't know which one. This annoyance actually makes her task impossible if α is a transcendental number.*

It does not not suffice for Alice to just program the distribution $(\alpha \quad 1 - \alpha)$ into one of the random sources and record the result; this fails because she might use the miscalibrated one! We require—as in our jar of pennies example—that

---

Alice's algorithm be robust enough to handle unpredictable results from any single source.

Unfortunately, Yao provides no proof of the theorem, and we are not aware of any in the literature. Yao's theorem is a special case of the results we described in the previous section.

2   Simulating Finite Random Sources

The following result is classical.

Proposition 2. *If p players are equipped with private d-sided dice, they may $(p-1)$-robustly simulate a d-sided die.*

Proof. We provide a direct construction. Fix a group G of order d (such as the cyclic group $\mathbb{Z}/d\mathbb{Z}$). The ith player uses the uniform measure to pick $g_i \in G$ at random. The roll of the simulated die will be the product $g_1 g_2 \cdots g_p$.

It follows from the G-invariance of the uniform measure that any p-subset of

$$\{g_1, g_2, \ldots, g_p, g_1 g_2 \cdots g_p\} \tag{1}$$

is independent! Thus, this is a $(p-1)$-robust simulation.

For an example of this construction, consider how Alice and Bob may robustly flip a coin with bias $2/5$. Alice picks an element $a \in \mathbb{Z}/5\mathbb{Z}$, and Bob picks an element $b \in \mathbb{Z}/5\mathbb{Z}$; both do so using the uniform distribution. Then, a, b, and $a+b$ are pairwise independent! We say that the coin came up heads if $a + b \in \{0, 1\}$ and tails if $a + b \in \{2, 3, 4\}$.

This construction exploits the fact that several random variables may be pairwise (or $(p-1)$-setwise) independent but still dependent overall. In cryptology, this approach goes back to the one-time pad. Shamir [Sha79] uses it to develop secret-sharing protocols, and these are exploited in multiparty computation to such ends as playing poker without cards [GM82, GMW87].

Corollary 3. *If p players are equipped with private, full-strength finite $\mathbb{Q}$-random sources, they may $(p-1)$-robustly simulate a private, full-strength finite $\mathbb{Q}$-random source for some other player.*

Proof. Follows from Proposition 2 because any finite rational probability space is a quotient of some finite uniform distribution.

2.1   Cooperative Numbers

We define a useful class of numbers.

Definition 4. *If p people with private full-strength finite random sources can robustly simulate a coin flip with bias α, we say α is **p-cooperative**. We denote the set of p-cooperative numbers by $\mathfrak{C}(p)$.*

The ability to robustly simulate coin flips of certain bias is enough to robustly simulate any finite spaces with points having those biases, assuming some hypotheses about $\mathfrak{C}(p)$ which we will later see to be true.

Lemma 5. *Suppose that, if $\alpha, \alpha' \in \mathfrak{C}(p)$ and $\alpha < \alpha'$, then $\alpha/\alpha' \in \mathfrak{C}(p)$. If p people have full-strength finite random sources, they can robustly simulate precisely finite $\mathfrak{C}(p)$-random sources.*

Proof. Clearly, any random source they simulate must take p-cooperative probabilities, because any space with a subset of mass α has the space $(\alpha \quad 1 - \alpha)$ as a quotient.

In the other direction, consider a finite probability space with point masses

$$(\alpha_1 \ \alpha_2 \ \cdots \ \alpha_n) \tag{2}$$

in $\mathfrak{C}(p)$. Robustly flip a coin of bias α_1. In the heads case, we pick the first point. In the tails case, we apply induction to robustly simulate

$$(\alpha_2/(1 - \alpha_1) \ \cdots \ \alpha_n/(1 - \alpha_1)). \tag{3}$$

This is possible because $1 - \alpha_1 \in \mathfrak{C}(p)$ by symmetry, and so the ratios $\alpha_i/(1 - \alpha_1) \in \mathfrak{C}(p)$ by assumption.

2.2 Restatement Using Multilinear Algebra

Consider a {heads, tails}-valued function of several independent finite probability spaces that produces an α-biased coin flip when random sources sample the spaces. If we model each probability space as a stochastic vector—that is, a nonnegative vector whose coordinates sum to one—we may view the product probability space as the Kronecker product of these vectors. Each entry in the resulting tensor represents the probability of a certain combination of outputs from the random sources. Since the sources together determine the flip, some of these entries should be marked "heads," and the rest "tails."

For instance, if we have a fair die and a fair coin at our disposal, we may cook up some rule to assign "heads" or "tails" to each combination of results:

$$\begin{pmatrix} \frac{1}{6} \\ \frac{1}{6} \\ \frac{1}{6} \\ \frac{1}{6} \\ \frac{1}{6} \\ \frac{1}{6} \end{pmatrix} \otimes \begin{pmatrix} \frac{1}{2} & \frac{1}{2} \end{pmatrix} = \begin{pmatrix} \frac{1}{12} & \frac{1}{12} \\ \frac{1}{12} & \frac{1}{12} \\ \frac{1}{12} & \frac{1}{12} \\ \frac{1}{12} & \frac{1}{12} \\ \frac{1}{12} & \frac{1}{12} \\ \frac{1}{12} & \frac{1}{12} \end{pmatrix} \longrightarrow \begin{pmatrix} H & T \\ H & T \\ T & H \\ H & T \\ T & H \\ T & H \end{pmatrix} \tag{4}$$

If we want to calculate the probability of heads, we can substitute 1 for H and 0 for T in the last matrix and evaluate

$$
\left(\begin{smallmatrix} \frac{1}{6} & \frac{1}{6} & \frac{1}{6} & \frac{1}{6} & \frac{1}{6} & \frac{1}{6} \end{smallmatrix} \right)
\begin{pmatrix} 1 & 0 \\ 1 & 0 \\ 0 & 1 \\ 1 & 0 \\ 0 & 1 \\ 0 & 1 \end{pmatrix}
\begin{pmatrix} \frac{1}{2} \\ \frac{1}{2} \end{pmatrix} = \frac{1}{2}.
\tag{5}
$$

This framework gives an easy way to check if the algorithm is robust in the sense of Yao. If one of the random sources is miscalibrated (maybe the die is a little uneven), we may see what happens to the probability of heads:

$$
\left(\begin{smallmatrix} \frac{1}{12} & \frac{1}{10} & \frac{1}{6} & \frac{1}{4} & \frac{1}{15} & \frac{1}{3} \end{smallmatrix} \right)
\begin{pmatrix} 1 & 0 \\ 1 & 0 \\ 0 & 1 \\ 1 & 0 \\ 0 & 1 \\ 0 & 1 \end{pmatrix}
\begin{pmatrix} \frac{1}{2} \\ \frac{1}{2} \end{pmatrix} = \frac{1}{2}.
\tag{6}
$$

It's unaffected! In fact, defining

$$
A\left(x^{(1)}, x^{(2)}\right) = x^{(1)}
\begin{pmatrix} 1 & 0 \\ 1 & 0 \\ 0 & 1 \\ 1 & 0 \\ 0 & 1 \\ 0 & 1 \end{pmatrix}
{x^{(2)}}^{\top},
\tag{7}
$$

we see that letting $\beta^{(1)} = \left(\begin{smallmatrix} \frac{1}{6} & \frac{1}{6} & \frac{1}{6} & \frac{1}{6} & \frac{1}{6} & \frac{1}{6} \end{smallmatrix} \right)$ and $\beta^{(2)} = \left(\begin{smallmatrix} \frac{1}{2} & \frac{1}{2} \end{smallmatrix} \right)$ gives us

$$
A\left(x^{(1)}, \beta^{(2)}\right) = \frac{1}{2}
$$
$$
A\left(\beta^{(1)}, x^{(2)}\right) = \frac{1}{2}
\tag{8}
$$

for all $x^{(1)}$ and $x^{(2)}$ of mass one. These relations express Yao's notion of robustness; indeed, changing at most one of the distributions to some other distribution leaves the result unaltered. As long as no two of the sources are miscalibrated, the bit is generated with probability $1/2$.

If α denotes the bias of the bit, we may write the robustness condition as

$$
A\left(x^{(1)}, \beta^{(2)}\right) = \alpha J\left(x^{(1)}, \beta^{(2)}\right)
$$
$$
A\left(\beta^{(1)}, x^{(2)}\right) = \alpha J\left(\beta^{(1)}, x^{(2)}\right)
\tag{9}
$$

since the $\beta^{(i)}$ both have mass one. (Here as always, J stands for the all-ones tensor of appropriate dimensions.) These new equations hold for all $x^{(i)}$, by linearity. Subtracting, we obtain

$$0 = (\alpha J - A)\left(x^{(1)}, \beta^{(2)}\right)$$

$$0 = (\alpha J - A)\left(\beta^{(1)}, x^{(2)}\right) \tag{10}$$

which says exactly that the bilinear form $(\alpha J - A)$ is degenerate, i.e., that

$$\mathrm{Det}(\alpha J - A) = 0.^{6} \tag{11}$$

These conditions seem familiar: Changing the all-ones matrix J to the identity matrix I would make α an eigenvalue for the left and right eigenvectors $\beta^{(i)}$. By analogy, we call α a *mystery-value* of the matrix A and the vectors $\beta^{(i)}$ *mystery-vectors*. Here's the full definition:

Definition 6. *A p-linear form A is said to have **mystery-value** α and corresponding **mystery-vectors** $\beta^{(i)}$ when, for any $1 \leq j \leq p$,*

$$0 = (\alpha J - A)\left(\beta^{(1)}, \ldots, \beta^{(j-1)}, x^{(j)}, \beta^{(j+1)}, \ldots, \beta^{(p)}\right) \text{ for all vectors } x^{(j)}. \tag{12}$$

We further require that $J(\beta^{(i)}) \neq 0$.

We will see later that these conditions on $(\alpha J - A)$ extend the notion of degeneracy to multilinear forms in general. This extension is captured by a generalization of the determinant—the hyperdeterminant.[7] Hyperdeterminants will give meaning to the statement $\mathrm{Det}(\alpha J - A) = 0$, even when A is not bilinear.

   This organizational theorem summarizes our efforts to restate the problem using multilinear algebra.

Theorem 7. *A function from the product of several finite probability spaces to the set $\{H, T\}$ generates an α-biased bit robustly iff the corresponding multilinear form has mystery-value α with the probability spaces as the accompanying mystery-vectors.*

We may now show the equivalence of robustness and privacy more formally. Privacy requires that $(\alpha J - A)\left(\otimes \beta^{(i)}\right)$ remains zero, even if one of the distributions in the tensor product collapses to some point mass, that is, to some basis vector.[8] This condition must hold for all basis vectors, so it extends by linearity to Yao's robustness.

---

[6] If the matrix $(\alpha J - A)$ is not square, this equality should assert that all determinants of maximal square submatrices vanish.

[7] Hyperdeterminants were first introduced in the $2 \times 2 \times 2$ case by Cayley [Cay45], and were defined in full generality and studied by Gelfand, Kapranov, and Zelevinsky [GKZ94, Chapter 14].

[8] That is, the simulated bit remains a "mystery" to each player, even though she can see the output of her own random source.

2.3 Two Players

The case $p = 2$ leaves us in the familiar setting of bilinear forms.

Proposition 8 (Uniqueness). *Every bilinear form has at most one mystery-value.*

Proof. Suppose α and α' are both mystery-values for the matrix A with mystery-vectors $\beta^{(i)}$ and $\beta^{(i)'}$, respectively. We have four equations at our disposal, but we will only use two:

$$A\left(x^{(1)}, \beta^{(2)}\right) = \alpha$$
$$A\left(\beta^{(1)'}, x^{(2)}\right) = \alpha' \tag{13}$$

We observe that a compromise simplifies both ways:

$$\alpha = A\left(\beta^{(1)'}, \beta^{(2)}\right) = \alpha', \tag{14}$$

so any two mystery-values are equal.

Corollary 9. *Two players may not simulate an irrationally-biased coin.*

Proof. Say the $\{0,1\}$-matrix A has mystery-value α. Any field automorphism $\sigma \in \mathrm{Gal}(\mathbb{C}/\mathbb{Q})$ respects all operations of linear algebra, so $\sigma(\alpha)$ is a mystery-value of the matrix $\sigma(A)$. But the entries of A are all rational, so $\sigma(A) = A$. Indeed, $\sigma(\alpha)$ must also be a mystery-value of A itself. By the uniqueness proposition, $\sigma(\alpha) = \alpha$. Thus, α is in the fixed field of every automorphism over $\mathbb{Q}$ and cannot be irrational.

Theorem 10. $\mathfrak{C}(2) = \mathbb{Q} \cap [0, 1]$. *Two people with finite random sources can robustly simulate only $\mathbb{Q}$-random sources; indeed, they can already simulate a full-strength finite $\mathbb{Q}$-random source if they have full-strength finite $\mathbb{Q}$-random sources.*

Proof. The previous corollary shows that no probability generated by the source can be irrational, since it could be used to simulate an irrationally-biased coin. The other direction has already been shown in Corollary 3.

Proposition 11. *If p people have full-strength finite $\mathbb{Q}$-random sources, they may $(p-1)$-robustly simulate any finite $\mathbb{Q}$-random source.*

Proof. Follows from Proposition 2 just as the constructive direction of Theorem 10 does.

2.4   Three or More Players: What Can't Be Done

Even if three or more players have private finite random sources, it remains impossible to robustly simulate a transcendentally-biased coin. The proof makes use of algebraic geometry, especially the concept of the dual of a complex projective variety. We describe these ideas briefly in Appendix A. For a more thorough introduction, see [Har92, Lec. 14, 15, 16] or [GKZ94, Ch. 1].

Let A be a rational multilinear functional of format $n_1 \times \cdots \times n_p$ (see Section A.2), and let X be the Segre variety of the same format. Set $n := n_1 \cdots n_p - 1$, the dimension of the ambient projective space where X lives. In what follows, we prove that A has algebraic mystery-values. This is trivial when A is a multiple of J, and for convenience we exclude that case.

Proposition 12. *Let A have mystery-value α with corresponding mystery-vectors $\beta^{(i)}$. Define $\beta = \otimes \beta^{(i)}$, and let $\mathfrak{B}$ denote the hyperplane of elements of $(\mathbb{P}^n)^*$ that yield zero when applied to β. Now $(\mathfrak{B}, (\alpha J - A))$ is in the incidence variety $W_{X^\vee}$ (see Section A.1).*

Proof. By the biduality theorem 31, the result would follow from the statement,

$$\text{"The hyperplane } \{\, x : (\alpha J - A)(x) = 0 \,\} \text{ is tangent to } X \text{ at } \beta.\text{"} \qquad (15)$$

But this statement is true by the partial derivatives formulation (Definition 32) of the degeneracy of $(\alpha J - A)$.

It is a standard fact (see *e.g.* [Mum95, p. 6]) that any variety has a stratification into locally closed smooth sets. The first stratum of $X^\vee$ is the Zariski-open set of smooth points of the variety. This leaves a subvariety of strictly smaller dimension, and the procedure continues inductively. Equations for the next stratum may be found by taking derivatives and determinants.

Since $X^\vee$ itself is defined over $\mathbb{Q}$, it follows that each of its strata is as well. We conclude that there must be some subvariety $S \subseteq X^\vee$, defined over $\mathbb{Q}$, that contains $(\alpha J - A)$ as a smooth point.

Theorem 13. *Any mystery-value of A must be an algebraic number.*

Proof. Let $A' = \alpha J - A$, and let ℓ be the unique projective line through A and J. Let $\mathbb{A}$ be some open affine in $(\mathbb{P}^n)^*$ containing A' and J. The hyperplane $\mathfrak{B} \cap \mathbb{A}$ is the zero locus of some degree one regular function f on $\mathbb{A}$. On $\ell \cap \mathbb{A}$, this function will be nonzero at J (since $J(\beta) \neq 0$), so f is linear and not identically zero. It follows that $f(A) = 0$ is the unique zero of f on ℓ, occurring with multiplicity one. Thus, the restriction of f to the local ring of ℓ at A' is in the maximal ideal but not its square:

$$f \neq 0 \in \mathfrak{m}_\ell / \mathfrak{m}_\ell^2 = T_{A'}^*(\ell) \qquad \text{where } \mathfrak{m}_\ell \text{ denotes the maximal ideal in } \mathcal{O}_{\ell, A'}. \quad (16)$$

On the other hand, Proposition 12 shows that $(\mathfrak{B}, A') \in W_{X^\vee}$. Consequently, $\mathfrak{B}$ must be tangent to S, that is, f restricted to S *is* in the square of the maximal ideal of the local ring of S at A':

$f = 0 \in \mathfrak{m}_S/\mathfrak{m}_S^2 = T_{A'}^*(S)$ where $\mathfrak{m}_S$ denotes the maximal ideal in $\mathcal{O}_{S,A'}$. (17)

The function f must be zero in the cotangent space of the intersection $S \cap \ell$ since the inclusion $S \cap \ell \hookrightarrow S$ induces a surjection

$$T_{A'}^*(S) \twoheadrightarrow T_{A'}^*(S \cap \ell), \tag{18}$$

so the corresponding surjection

$$T_{A'}^*(\ell) \twoheadrightarrow T_{A'}^*(S \cap \ell) \tag{19}$$

must kill f. This first space is the cotangent space of a line, hence one dimensional. But f is nonzero in the first space, so the second space must be zero. It follows that $S \cap \ell$ is a zero dimensional variety.

Of course, $[\alpha : 1]$ lies in $S \cap \ell$, which is defined over $\mathbb{Q}$! The number α must be algebraic.

Therefore, the set of p-cooperative numbers is contained in $\overline{\mathbb{Q}} \cap [0, 1]$, and we have established the following proposition:

Proposition 14. *If several people with finite random sources simulate a private random source for someone else, that source must take probabilities in $\overline{\mathbb{Q}}$.*

2.5   Three Players: What Can Be Done

We prove that three players with private full-strength finite random sources are enough to simulate any private finite $\overline{\mathbb{Q}}$-random source. First, we give a construction for a hypermatrix with stochastic mystery-vectors for a given algebraic number α, but whose entries may be negative. Next, we use it to find a nonnegative hypermatrix with mystery-value $(\alpha+r)/s$ for some suitable natural numbers r and s. Then, after a bit of convex geometry to "even out" this hypermatrix, we scale and shift it back, completing the construction.

Remark 15. *Our construction may easily be made algorithmic, but in practice it gives hypermatrices that are far larger than optimal. An optimal algorithm would need to be radically different to take full advantage of the third person. The heart of our construction (see Proposition 18) utilizes $2 \times (n+1) \times (n+1)$ hypermatrices, but the degree of the hyperdeterminant polynomial grows much more quickly for (near-)diagonal formats [GKZ94, Ch. 14]. We would be excited to see a method of producing (say) small cubic hypermatrices with particular mystery-values.*

Hypermatrices with Cooperative Entries   Recall that a {heads, tails}-function of several finite probability spaces may be represented by a {1,0}-hypermatrix. The condition that the entries of the matrix are either 1 or 0 is inconvenient when we want to build simulations for a given algebraic bias. Fortunately, constructing a matrix with cooperative entries will suffice.

Lemma 16. *Suppose that A is a p-dimensional hypermatrix with p-cooperative entries in $[0,1]$ and stochastic mystery-vectors $\beta^{(1)}, \ldots, \beta^{(p)}$ for the mystery-value α. Then, α is p-cooperative.*

Proof. Let the hypermatrix A have entries $w_1, w_2, \ldots, w_n$. Each entry w_k is p-cooperative, so it is the mystery-value of some p-dimensional $\{0, 1\}$-hypermatrix A_k with associated stochastic mystery-vectors $\beta_k^{(1)}, \beta_k^{(2)}, \ldots, \beta_k^{(p)}$. We now build a $\{0, 1\}$-hypermatrix A' with α as a mystery-value. The hypermatrix A' has blocks corresponding to the entries of A. We replace each entry w_i of A with a Kronecker product:

$$w_i \text{ becomes } J_1 \otimes J_2 \otimes \cdots \otimes J_{i-1} \otimes A_i \otimes J_{i+1} \otimes \cdots \otimes J_n. \tag{20}$$

It is easy to check that the resulting tensor A' has α as a mystery-value with corresponding mystery-vectors $\beta^{(i)} \otimes \beta_1^{(i)} \otimes \beta_2^{(i)} \otimes \cdots \otimes \beta_n^{(i)}$.

Because rational numbers are 2-cooperative, this lemma applies in particular to rational p-dimensional hypermatrices, for $p \geq 2$. In this case and in others, the construction can be modified to give an A' of smaller format.

Readers who have been following the analogy between mystery-values and eigenvalues will see that Lemma 16 corresponds to an analogous result for eigenvalues of matrices. Nonetheless, there are striking differences between the theories of mystery-values and eigenvalues. For instance, we are in the midst of showing that it is always possible to construct a nonnegative rational hypermatrix with a given nonnegative algebraic mystery-value and stochastic mystery-vectors. The analogous statement for matrix eigenvalues is false, by the Perron-Frobenius theorem: any such algebraic number must be greater than or equal to all of its Galois conjugates (which will also occur as eigenvalues). Encouragingly, the inverse problem for eigenvalues has been solved: Every "Perron number" may be realized as a "Perron eigenvalue" [Lin84]. Our solution to the corresponding inverse problem for mystery-values uses different techniques. It would be nice to see if either proof sheds light on the other.

Constructing Hypermatrices from Matrices

Proposition 17. *If λ is a real algebraic number of degree n, then there is some $M \in \mathrm{M}_n(\mathbb{Q})$ having λ as an eigenvalue with non-perpendicular positive left and right eigenvectors.*

Proof. Let $f \in \mathbb{Q}[x]$ be the minimal polynomial for λ over $\mathbb{Q}$, and let L be the companion matrix for f. That is, if

$$f(x) = x^n + \sum_{k=0}^{n-1} a_k x^k \text{ for } a_k \in \mathbb{Q}, \tag{21}$$

then

$$L = \begin{pmatrix} 0 & 0 & \cdots & 0 & -a_0 \\ 1 & 0 & \cdots & 0 & -a_1 \\ 0 & 1 & \cdots & 0 & -a_2 \\ \vdots & \vdots & \ddots & \vdots & \vdots \\ 0 & 0 & \cdots & 1 & -a_{n-1} \end{pmatrix}. \tag{22}$$

The polynomial f is irreducible over $\mathbb{Q}$, so it has no repeated roots in $\mathbb{C}$. The matrix L is therefore diagonalizable, with diagonal entries the roots of f. Fix a basis for which L is diagonal, with λ in the upper-left entry. In this basis, the right and left eigenvectors, v_0 and w_0, corresponding to λ are zero except in the first coordinate. It follows that $v_0(w_0) \neq 0$.

The right and left eigenvectors may now be visualized as two geometric objects: a real hyperplane and a real vector not contained in it. It's clear that $\mathrm{GL}_n(\mathbb{R})$ acts transitively on the space $\mathcal{S} := \{(v, w) \in (\mathbb{R}^n)^* \times \mathbb{R}^n : v(w) = v_0(w_0)\}$. Moreover, $\mathrm{GL}_n(\mathbb{Q})$ is dense in $\mathrm{GL}_n(\mathbb{R})$, so the orbit of (v_0, w_0) under the action of $\mathrm{GL}_n(\mathbb{Q})$ is dense in $\mathcal{S}$. The set of positive pairs in $\mathcal{S}$ is non-empty and open, so we may rationally conjugate L to a basis which makes v_0 and w_0 positive.

Proposition 18. *If λ is real algebraic, then there exist integers $r \geq 0$, $s > 0$ such that $(\lambda + r)/s \in \mathfrak{C}(3)$.*

Proof. By Proposition 17, there is a rational $n \times n$ matrix M with non-perpendicular positive right and left eigenvectors v, w for the eigenvalue λ. Rescale w so that $v(w) = 1$, and choose an integer $q \geq \max\{J(v), J(w)\}$. Define the block $2 \times (n+1) \times (n+1)$ hypermatrix

$$A := \left(\begin{array}{ccccc|ccccc} 0 & 0 & \cdots & 0 & 1 & 1 & \cdots & 1 \\ \hline 0 & & & & 1 & & & \\ \vdots & & q^2 M & & \vdots & & q^2(M-I)+J & \\ 0 & & & & 1 & & & \end{array} \right), \tag{23}$$

where I and J are the $n \times n$ identity and all-ones matrices, respectively. Consider A as a trilinear form, where the metacolumns correspond to the coordinates of the first vector, the rows the second, and the columns the third. Define the block vectors

$$\beta^{(1)} = (\, 1 - \lambda \;\; \lambda \,),$$
$$\beta^{(2)} = (\, 1 - J(v)/q \mid v_1/q \;\; v_2/q \;\; \cdots \;\; v_n/q \,), \text{ and} \tag{24}$$
$$\beta^{(3)} = (\, 1 - J(w)/q \mid w_1/q \;\; w_2/q \;\; \cdots \;\; w_n/q \,).$$

Clearly, these are all probability vectors. It's easy to verify that

$$A\left(x^{(1)}, \beta^{(2)}, \beta^{(3)}\right) = \lambda J\left(x^{(1)}\right),$$
$$A\left(\beta^{(1)}, x^{(2)}, \beta^{(3)}\right) = \lambda J\left(x^{(2)}\right), \text{ and}$$
$$A\left(\beta^{(1)}, \beta^{(2)}, x^{(3)}\right) = \lambda J\left(x^{(3)}\right). \tag{25}$$

Choose a nonnegative integer r large enough so that all the entries of $A + rJ$ are positive, and then a positive integer s so that all the entries of $A' := (A + rJ)/s$ are between 0 and 1.

$$A' \left(x^{(1)}, \beta^{(2)}, \beta^{(3)} \right) = \frac{\lambda + r}{s} J \left(x^{(1)} \right),$$

$$A' \left(\beta^{(1)}, x^{(2)}, \beta^{(3)} \right) = \frac{\lambda + r}{s} J \left(x^{(2)} \right), \text{ and}$$

$$A' \left(\beta^{(1)}, \beta^{(2)}, x^{(3)} \right) = \frac{\lambda + r}{s} J \left(x^{(3)} \right). \tag{26}$$

By Lemma 16, it follows that $(\lambda + r)/s$ is 3-cooperative.

Finishing the Proof The following lemma, which we we prove later, enables us to complete the goal of this section: to classify which private random sources three or more people can simulate.

Lemma 19 (Approximation lemma). *Let α be a p-cooperative number. Now for any $\varepsilon > 0$ there exists a p-dimensional rational hypermatrix whose entries are all within ε of α, having α as a mystery-value with stochastic mystery-vectors.*

Theorem 20. $\mathfrak{C}(p) = \overline{\mathbb{Q}} \cap [0, 1]$ *for each $p \geq 3$.*

Proof. Certainly 0 and 1 are 3-cooperative. Let α be an algebraic number in $(0, 1)$. By Proposition 18, there are integers $r \geq 0$, $s > 0$ so that $(\alpha + r)/s$ is 3-cooperative. Let $\varepsilon := (\min\{\alpha, 1 - \alpha\})/s$.

By Proposition 19, there is some three-dimensional rational hypermatrix A whose entries are all within ε of $(\alpha + r)/s$, having $(\alpha + r)/s$ as a mystery-value with stochastic mystery-vectors. Then, $sA - rJ$ is a three-dimensional rational hypermatrix with entries between 0 and 1, having α as a mystery-value with stochastic mystery-vectors. By Lemma 16, α is 3-cooperative.

We already showed that all cooperative numbers are algebraic. Thus, for $p \geq 3$,

$$\overline{\mathbb{Q}} \cap [0, 1] \subseteq \mathfrak{C}(3) \subseteq \mathfrak{C}(p) \subseteq \overline{\mathbb{Q}} \cap [0, 1], \tag{27}$$

so $\mathfrak{C}(p) = \overline{\mathbb{Q}} \cap [0, 1]$.

In conclusion, we have the following theorem.

Theorem 21. *Three or more people with finite random sources can robustly simulate only $\overline{\mathbb{Q}}$-random sources. Indeed, if they have full-strength finite $\overline{\mathbb{Q}}$-random sources, they can already robustly simulate a full-strength finite $\overline{\mathbb{Q}}$-random source.*

Proof of the Approximation Lemma. The proof that follows is a somewhat lengthy "delta-epsilon" argument broken down into several smaller steps. As we believe our construction of a hypermatrix with mystery-value α to be far from optimal, we strive for ease of exposition rather than focusing on achieving tight bounds at each step along the way.

Recall that a finite probability space may be usefully modeled by a positive[9] vector of mass one. Let β be such a vector. We denote by $\#\beta$ the number of coordinates of β . We say β' is a *refinement* of β when β is the image of a measure-preserving map from β'; that is, when the coordinates of β' may be obtained by splitting up the coordinates of β.

The following easy lemma states that any positive vector of unit mass can be refined in such a way that all the coordinates are about the same size.

Lemma 22 (Refinement lemma). *Let β be a positive vector of total mass 1. For any $\delta > 0$ there exists a refinement β' of β with the property that*

$$\min_j \beta'_j \geq \frac{1 - \delta}{\#\beta'}. \tag{28}$$

Proof. Without loss of generality, assume that β_1 is the smallest coordinate of β. Let $\gamma = \beta_1\delta$, and let $k = \#\beta$. The vector β is in the standard open k-simplex

$$\Delta^k = \{\text{positive vectors of mass 1 and dimension } k\}. \tag{29}$$

The rational points in Δ^k are dense (as in any rational polytope), and

$$U := \{x \in \Delta^k : (\forall i)\, |\beta_i - x_i| < \gamma \text{ and } \beta_1 < x_1\} \tag{30}$$

is an open subset of the simplex. So U contain a rational point $\left(\frac{n_1}{n}, \ldots, \frac{n_k}{n}\right)$, with $n = \sum n_i$. Thus, $\left|\beta_i - \frac{n_i}{n}\right| < \gamma$ and $\beta_1 < \frac{n_1}{n}$, so

$$\left|\frac{\beta_i}{n_i} - \frac{1}{n}\right| < \frac{\gamma}{n_i} \leq \frac{\gamma}{n_1} < \frac{\gamma}{\beta_j n} = \frac{\delta}{n}. \tag{31}$$

Let β' be the refinement of β obtained by splitting up β_i into n_i equal-sized pieces. We have $\#\beta' = n$, and the claim follows from this last inequality.

Remark 23. *The best general bounds on the smallest possible $\#\beta'$ given β and δ are not generally known, but fairly good bounds may be obtained from the multidimensional version of Dirichlet's theorem on rational approximation, which is classical and elementary [Dav54]. Actually calculating good simultaneous rational approximations is a difficult problem, and one wishing to make an algorithmic version of our construction should consult the literature on multidimensional continued fractions and Farey partitions, for example, [Lag82, NS06].*

The next proposition is rather geometrical. It concerns the $n \times n$ matrix $S_\delta := (1 - \delta)(J/n) + \delta I$, which is a convex combination of two maps on the standard simplex: the averaging map and the identity map. Each vertex gets mapped almost to the center, so the action of S_δ can be visualized as shrinking the standard simplex around its center point. The proposition picks up where the refinement lemma left off:

---

[9] We may leave out points of mass zero.

Proposition 24. *If a stochastic vector β satisfies*

$$\min_i \beta_i \geq \frac{1 - \delta}{\#\beta} \tag{32}$$

then its image under the map S_δ^{-1} is still stochastic.

Proof. First note that $[(1 - \delta)\,(J/\#\beta) + \delta I]\,[(1 - 1/\delta)\,(J/\#\beta) + (1/\delta)I] = I$, so we have an explicit form for S_δ^{-1}. We know that $\min_i \beta_i \geq (1 - \delta)/\#\beta$, so the vector

$$E = \frac{1}{\delta}\left[\beta - \left(\frac{1 - \delta}{\#\beta}\right)J\right] \tag{33}$$

is still positive. Now $\beta = (1 - \delta)\,(J/\#\beta) + \delta E$, a convex combination of two positive vectors. The vector β has mass 1, and $(J/\#\beta)$ as well, so E also has mass 1.

Now compute:

$$\begin{aligned}
S_\delta^{-1}\beta &= \left[(1 - 1/\delta)\,(J/\#\beta) + (1/\delta)I\right]\left[(1 - \delta)\,(J/\#\beta) + \delta E\right]\\
&= \left[(1 - 1/\delta)(1 - \delta) + (1/\delta)(1 - \delta) + (1 - 1/\delta)\delta\right](J/\#\beta) + E\\
&= E.
\end{aligned} \tag{34}$$

This completes the proof.

The following proposition shows that applying the matrix S_δ in all arguments of some multilinear functional forces the outputs to be close to each other.

Proposition 25. *Let A be a hypermatrix of format $n_1 \times n_2 \times \cdots \times n_p$ with entries in $[0, 1]$, and take $\delta := \varepsilon/(2p)$. Now the matrix A' defined by*

$$A'\left(\otimes x^{(i)}\right) := A\left(\otimes S_\delta x^{(i)}\right) \tag{35}$$

satisfies $|A'(x) - A'(x')| \leq \varepsilon$ for any two stochastic tensors x and x'.

Proof. Let $m := A\left(\otimes(J/n_i)\right)$, the mean of the entries of A. We show that for any stochastic vectors $x^{(i)}$,

$$\left|A'\left(\otimes x^{(i)}\right) - m\right| \leq \varepsilon/2. \tag{36}$$

Since any other stochastic tensor is a convex combination of stochastic pure tensors, it will follow that $|A'(x) - m| \leq \varepsilon/2$. Then the triangle inequality will yield the result.

It remains to show that A' applied to a stochastic pure tensor gives a value within $\varepsilon/2$ of m.

$$\begin{aligned}
A'\left(\otimes x^{(i)}\right) &= A\left(\otimes S_\delta x^{(i)}\right)\\
&= A\left(\otimes\left[(1 - \delta)(J/n_i) + \delta I\right]x^{(i)}\right)\\
&= A\left(\otimes\left[(1 - \delta)(J/n_i) + \delta x^{(i)}\right]\right).
\end{aligned} \tag{37}$$

Each argument of A—that is, factor in the tensor product—is a convex combination of two stochastic vectors. Expanding out by multilinearity, we get convex combination with 2^p points. Each point—let's call the kth one y_k—is an element of $[0,1]$ since it is some weighted average of the entries of A. This convex combination has positive μ_k such that $\sum \mu_k = 1$ and

$$A'\left(\otimes x^{(i)}\right) = \sum_{k=1}^{2^p} \mu_k y_k. \tag{38}$$

Taking the first vector in each argument of A in (37), we see that $y_1 = A\left(\otimes(J/n_i)\right)$ $= m$, the average entry of A. Thus, the first term in the convex combination is $\mu_1 y_1 = (1-\delta)^p m$.

The inequality $(1-\varepsilon/2) \leq (1-\delta)^p$ allows us to split up the first term. Let $\mu_0 := 1 - \varepsilon/2$ and $\mu_1' := \mu_1 - \mu_0 \geq 0$. We have $\mu_1 y_1 = (\mu_0 + \mu_1')y_1 = (1-\varepsilon/2)m + \mu_1' m$. After splitting this term, the original convex combination becomes

$$A'\left(\otimes x^{(i)}\right) = (1 - \varepsilon/2)m + \mu_1' m + \sum_{k=2}^{2^p} \mu_k y_k. \tag{39}$$

Let e denote the weighted average of the terms after the first. We may rewrite the convex combination

$$A'\left(\otimes x^{(i)}\right) = (1 - \varepsilon/2)m + (\varepsilon/2)e. \tag{40}$$

Since $m, e \in [0,1]$,

$$m - \varepsilon/2 \leq (1 - \varepsilon/2)m \leq A'\left(\otimes x^{(i)}\right) \leq (1 - \varepsilon/2)m + \varepsilon/2 \leq m + \varepsilon/2, \tag{41}$$

and

$$\left| A'\left(\otimes x^{(i)}\right) - m \right| \leq \varepsilon/2, \tag{42}$$

so we are done.

These results are now strong enough to prove the approximation lemma 19.

Proof. The number α is p-cooperative, so it comes with some p-dimensional nonnegative rational hypermatrix A and positive vectors $\beta^{(1)}, \beta^{(2)}, \ldots, \beta^{(p)}$ of mass one, satisfying (in particular) $A\left(\otimes\beta^{(i)}\right) = \alpha$. The refinement lemma allows us to assume that each $\beta^{(i)}$ satisfies

$$\min_j \beta_j^{(i)} \geq \frac{1-\delta}{\#\beta^{(i)}}. \tag{43}$$

If one of the $\beta^{(i)}$ fails to satisfy this hypothesis, we may replace it with the refinement given by the lemma, and duplicate the corresponding slices in A to match.

Now, by Proposition 24, each $S_\delta^{-1}\beta^{(i)}$ is a stochastic vector.

Let A' be as in Proposition 25. It will still be a rational hypermatrix if we pick ε to be rational. We know

$$A'\left(\otimes S_\delta^{-1}\beta^{(i)}\right) = \alpha. \tag{44}$$

On the other hand, any entry of the matrix A' is given by evaluation at a tensor product of basis vectors. Both α and any entry of A' can be found by evaluating A' at a stochastic tensor. Thus, by Proposition 25, each entry of A' is within ε of α.

2.6   Higher-Order Robustness

We complete the proof of our main theorem.

Proposition 26. *If $r \geq p/2$, then p people with finite random sources may r-robustly simulate only finite $\mathbb{Q}$-random sources.*

Proof. Consider an r-robust simulation. Imagine that Alice has access to half of the random sources (say, rounded up), and Bob has access to the remaining sources. Because Alice and Bob have access to no more than r random sources, neither knows anything about the source being simulated. But this is precisely the two-player case of ordinary 1-robustness, so the source being simulated is restricted to rational probabilities.

In the constructive direction, we show the following:

Proposition 27. *If $r < p/2$, then p people with full-strength finite $\overline{\mathbb{Q}}$-random sources may r-robustly simulate a full-strength finite $\overline{\mathbb{Q}}$-random source.*

The proof is to simulate simulations (and simulate simulations of simulations, etc.). We treat the $p = 3$ case of our 1-robust simulation protocol as a black box. If a majority of the random sources put into it are reliable, the one that comes out (the simulated random source) will also be reliable. This viewpoint leads us into a discussion of majority gates.

Definition 28. *A **p-ary majority gate** is a logic gate that computes a boolean function returning 1 if a majority of its inputs are 1 and 0 if a majority of its inputs are 0. (The output doesn't matter when there are ties.)*

Lemma 29 (Bureaucracy). *A p-ary majority gate may be built by wiring together ternary majority gates.*

The proof of the bureaucracy lemma is a straightforward application of the probabilistic method, and is covered in detail in Appendix B. Now, by iterating simulations of simulations according to the wiring provided by the bureaucracy lemma, we can overcome any minority of malfunctioning sources. So the bureaucracy lemma, together with the "black box" of our three-player construction, implies Proposition 27.

Now we're finally ready to prove our main result. The statement here is equivalent to the ones in the abstract and in Section 1.2 but uses the language of robustness.

Theorem 30. *Say p people have full-strength finite random sources. If $p/2 \leq r < p$, the people may r-robustly simulate any finite $\mathbb{Q}$-random source and nothing better; if $1 \leq r < p/2$, they may r-robustly simulate any finite $\overline{\mathbb{Q}}$-random source and nothing better.*

Proof. The claim simply combines Proposition 11, Proposition 26, Theorem 30, and Proposition 27.

3   Application to Secure Multiparty Computation and Mental Poker

We begin with the classical case: Three gentlemen wish to play poker, but they live far away from each other, so playing with actual cards is out of the question. They could play online poker, in which another party (the remotely hosted poker program) acts as a dealer and moderator, keeping track of the cards in each player's hand, in the deck, etc., and giving each player exactly the information he would receive in a physical game. But this solution require our gentlemen to trust the moderator! If they fear the moderator may favor one of them, or if they wish to keep their game and its outcome private, they need another system.

A better solution is to use secure multiparty computation. Our gentlemen work to *simulate* a moderator in a way that keeps the outcomes of the moderator's computations completely hidden from each of them. An unconditionally-secure method of playing poker (and running other games/computations) "over the phone" has been described in [GM82].

In the classical case, the players may perform finite computations, communicate along private channels, and query full-strength finitary private random sources. The simulated moderator has the almost same abilities as the players, except that its private random source is limited to rational probabilities. The work of this paper expands this to all algebraic probabilities, and shows that one can do no better.

To see how this may be useful, think back to our poker players. They may be preparing for a poker tournament, and they may want to simulate opponents who employ certain betting strategies. But poker is a complicated multiplayer game (in the sense of economic game theory), and Nash equilibria will occur at mixed strategies with algebraic coefficients.[10]

A   Relevant Constructions in Algebraic Geometry

Comprehensive introductions to these constructions may be found in [Har92, Lec. 14, 15, 16] and [GKZ94, Ch. 1].

---

[10] The appearance of algebraic (but not transcendental) coefficients in mixed strategies is explained by R. J. Lipton and E. Markakis in [LM04].

A.1 Tangency and Projective Duality

Let k be an algebraically closed field of characteristic zero. (For our purposes, it would suffice to take $k = \mathbb{C}$, but the methods are completely general.) Let $X \subseteq \mathbb{P}^n$ be a projective variety over k. A hyperplane $H \in (\mathbb{P}^n)^*$ is *(algebraically) tangent* to X at a point z if every regular function on an affine neighborhood of z vanishing on H lies in the square of the maximal ideal of the local ring $\mathcal{O}_{X,z}$.

This notion of tangency agrees with geometric intuition on the set of smooth points X^{sm} of X. To get a more complete geometric picture, we define an incidence variety:

$$W_X := \overline{\{(z, H) : z \in X^{\mathrm{sm}}, H \text{ is tangent to } X \text{ at } z\}} \subseteq \mathbb{P}^n \times (\mathbb{P}^n)^*. \qquad (45)$$

The bar denotes Zariski closure. Membership in W_X may be thought of as extending the notion of tangency at a smooth point to include singular points "by continuity."

The image of a projective variety under a regular map is Zariski closed, so the projection of W_X onto the second coordinate is a variety, called the dual variety and denoted $X^\vee$.

The following theorem explains why projective duality is called "duality." We omit the proof; see [Har92, p. 208–209] or [GKZ94, p. 27–30].

Theorem 31 (Biduality theorem). *Let X be a variety in $\mathbb{P}^n$. For $z \in \mathbb{P}^n$, let z^{**} be the image under the natural isomorphism to $(\mathbb{P}^n)^{**}$. Then, $(z, H) \mapsto (H, z^{**})$ defines an isomorphism $W_X \cong W_{X^\vee}$. (Specializing to the case when (z, H) and (H, z^{**}) are smooth points X and $X^\vee$, respectively, this says that H is tangent to X at z if and only if z is tangent to $X^\vee$ at H.) Moreover, $z \mapsto z^{**}$ defines an isomorphism $X \cong (X^\vee)^\vee$.*

A.2 Segre Embeddings and Their Duals

Consider the natural map $k^{n_1} \times \cdots \times k^{n_p} \to k^{n_1} \otimes \cdots \otimes k^{n_p} = k^{n_1 \cdots n_p}$ given by the tensor product. Under this map, the fiber of a line through the origin is a tuple of lines through the origin. Thus, this map induces an embedding $\mathbb{P}^{n_1-1} \times \cdots \times \mathbb{P}^{n_p-1} \hookrightarrow \mathbb{P}^{n_1 \cdots n_p-1}$. The map is known as the Segre embedding, and the image is known as the Segre variety X of *format* $n_1 \times \cdots \times n_p$. It is, in other words, the pure tensors considered as a subvariety of all tensors, up to constant multiples. This variety is cut out by the determinants of the 2×2 subblocks. Also, it is smooth because it is isomorphic as a variety to $\mathbb{P}^{n_1-1} \times \cdots \times \mathbb{P}^{n_p-1}$.

When a projective variety is defined over the rational numbers,[11] its dual is also defined over the rationals, by construction [GKZ94, p. 14]. In particular, the dual $X^\vee$ of the Segre embedding is defined over $\mathbb{Q}$.

When the dimensions n_i satisfy the "p-gon inequality"

$$(n_j - 1) \leq \sum_{i \neq j}(n_i - 1), \qquad (46)$$

---

[11] That is, it is the zero set of a system of homogeneous rational polynomials.

Gelfand, Kapranov, and Zelevinsky [GKZ94, p. 446] show that the dual of the Segre variety is a hypersurface. The polynomial for this hypersurface is irreducible, has integer coefficients, and is known as the *hyperdeterminant* of format $n_1 \times \cdots \times n_p$. It is denoted by Det. When $p = 2$ and $n_1 = n_2$, the hyperdeterminant is the same as the determinant of a square matrix [GKZ94, p. 36].

Gelfand, Kapranov, and Zelevinsky provide us with two equivalent definitions of degeneracy.

Definition 32. *A p-linear form T is said to be **degenerate** if either of the following equivalent conditions holds:*

- *there exist nonzero vectors $\beta^{(i)}$ so that, for any $0 \leq j \leq p$,*

$$T\left(\beta^{(1)}, \ldots, \beta^{(j-1)}, x^{(j)}, \beta^{(j+1)}, \ldots, \beta^{(p)}\right) = 0 \text{ for all } x^{(j)}; \qquad (47)$$

- *there exist nonzero vectors $\beta^{(i)}$ so that T vanishes at $\otimes \beta^{(i)}$ along with every partial derivative with respect to an entry of some $\beta^{(i)}$:*

$$T \text{ and } \frac{\partial T}{\partial \beta_j^{(i)}} \text{ vanish at } \otimes \beta^{(i)}. \qquad (48)$$

The dual of the Segre variety is useful to us because it can tell whether a multilinear form is degenerate.

Theorem 33 (Gelfand, Kapranov, and Zelevinsky). *For any format, the dual $X^\vee$ of the Segre embedding is defined over $\mathbb{Q}$ and satisfies, for every multilinear form T of that format,*

$$T \in X^\vee \quad \Longleftrightarrow \quad T \text{ is degenerate.} \qquad (49)$$

When the format satisfies the "p-gon inequality," $X^\vee$ is defined by a polynomial in the entries of T with coefficients in $\mathbb{Z}$, called the hyperdeterminant:

$$\operatorname{Det}(T) = 0 \quad \Longleftrightarrow \quad T \text{ is degenerate.} \qquad (50)$$

B   Proof of the Bureaucracy Lemma

Here, we show that a p-ary majority gate may be built out of ternary majority gates.

Proof. We prove the existence of the majority gate by showing that a random gate built in a certain way has a positive probability of being a majority gate. For simplicity, we assume p is odd. The even case follows from the odd case: A $(2k - 1)$-ary majority gate functions as a $(2k)$-ary majority gate if we simply ignore one of the inputs.

Make a balanced ternary tree of depth n out of $3^0 + 3^1 + \cdots + 3^{n-1}$ ternary majority gates, where n is to be specified later. Let S be the set of possible assignments of p colors (one for each input slot) to the 3^n leaves of the tree. Each $s \in S$ defines a p-ary gate; we prove that, for n large enough, a positive fraction of these are majority gates. Let T be the set of p-tuples of input values with exactly $\frac{p+1}{2}$ coordinates equal to 1. For $(s, t) \in S \times T$, let $\chi(s, t)$ be the bit returned by the gate defined by s on input t.

If each input of a 3-ary majority gate is chosen to be 1 with probability x, and 0 with probability $1 - x$, we may compute the probability $f(x)$ that the resulting bit is 1:

$$f(x) = \binom{3}{2} x^2 (1 - x) + \binom{3}{3} x^3 = x^2 (3 - 2x). \tag{51}$$

Fixing the choice of $t \in T$ and letting s vary uniformly, it's as if we're assigning 1 or 0 to each leaf with probabilities $\frac{p+1}{2}$ and $\frac{p-1}{2}$, respectively. We have

$$\frac{1}{|S|} \sum_{s \in S} \chi(s, t) = f^n \left(\frac{p+1}{2} \right), \tag{52}$$

where f^n denotes iterated composition. Whenever $\frac{1}{2} < \xi \le 1$, it's easy to see that $f^n(\xi)$ approaches 1 as n becomes large.[12] Choose n so that $f^n \left(\frac{p+1}{2} \right) > 1 - \frac{1}{|T|}$. Now,

$$\frac{1}{|S|} \sum_{s \in S} \sum_{t \in T} \chi(s, t) = \sum_{t \in T} \frac{1}{|S|} \sum_{s \in S} \chi(s, t)$$

$$= \sum_{t \in T} f^n \left(\frac{p+1}{2} \right)$$

$$= |T| f^n \left(\frac{p+1}{2} \right)$$

$$> |T| \left(1 - \frac{1}{|T|} \right) = |T| - 1. \tag{53}$$

This is an average over S, and it follows that there must be some particular $s_0 \in S$ so that the inner sum $\sum_{t \in T} \chi(s_0, t)$ is greater than $|T| - 1$. But that sum clearly takes an integer value between 0 and $|T|$, so it must take the value $|T|$, and we have $\chi(s_0, t) = 1$ for every $t \in T$. That is, the gate specified by s_0 returns 1 whenever exactly $\frac{p+1}{2}$ of the inputs are 1. By construction, setting more inputs to 1 will not alter this outcome, so the gate returns 1 whenever a majority of the inputs are 1. By the symmetry between 1 and 0 in each ternary component, the gate returns 0 whenever a majority of the inputs are 0. Thus, s_0 defines a p-ary majority gate.

We illustrate a 5-ary majority gate of the type obtained in the bureaucracy lemma:

---

[12] In fact, the convergence is very fast. While we're ignoring computational complexity questions in this paper, more careful bookkeeping shows that this proof gives a polynomial bound (in p) on the size of the tree.

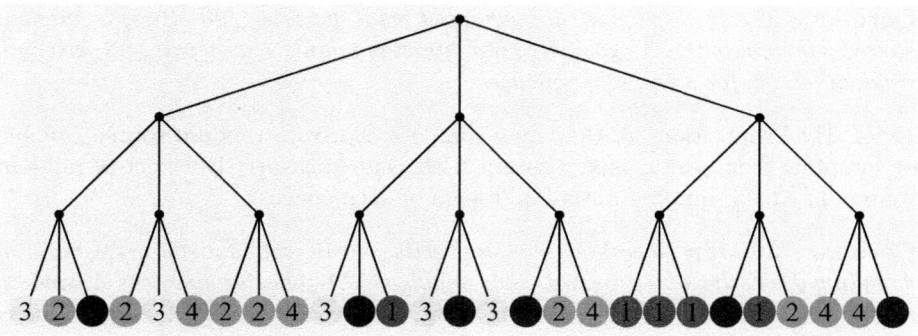

C   Simulating Infinite Random Sources

Say Alice and Bob are both equipped with private, full-strength random sources; they wish to simulate a private, full-strength random source for some other player.

For technical reasons, we will take "full-strength random source" to mean "a random source capable of sampling from any Haar measure." This restriction is mostly to avoid venturing into the wilds of set theory. After all, the pathologies available to probability spaces closely reflect the chosen set-theoretic axioms. We call these restricted spaces "Haar spaces."

Definition 34. *A probability space P is a **Haar space** if there exists some compact topological group G, equipped with its normalized Haar measure, admitting a measure-preserving map to P.*

Remark 35. *The following probability spaces are all Haar spaces: any continuous distribution on the real line; any standard probability space in the sense of Rokhlin [Rok49]; any Borel space or Borel measure on a Polish space; any finite probability space; arbitrary products of the above.*

The following construction is an easy generalization of the classical construction given in Proposition 2.

Proposition 36. *Let G be a compact group with normalized Haar measure. Now, p players equipped with private sources that sample from G may $(p-1)$-robustly simulate an source that samples from G.*

Proof. We provide a direct construction. The ith player uses the Haar measure to pick $g_i \in G$ at random. The output of the simulated source will be the product $g_1 g_2 \cdots g_p$.

It follows from the invariance of the Haar measure that any p-subset of

$$\{g_1, g_2, ..., g_p, g_1 g_2 \cdots g_p\} \tag{54}$$

is independent! Thus, this is a $(p-1)$-robust simulation.

Corollary 37. *If* p *players are equipped with private, full-strength random sources, they may* $(p-1)$*-robustly simulate may simulate a private, full-strength random source for some other player.*

Proof. By Proposition 36, they may simulate a private random source capable of sampling from any compact group with Haar measure. But such a random source may also sample from all quotients of such spaces.

Corollary 38. *If* p *players are equipped with private random sources capable of sampling from the unit interval, they may* $(p-1)$*-robustly simulate a random source capable of sampling from any standard probability space—in particular, any finite probability space.*

Proof. Immediate from Proposition 36.

References

[Cay45]   Cayley, A.: On the theory of linear transformations. Cambridge Math. J. 4 (1845)
[Dav54]   Davenport, H.: Simultaneous Diophantine approximation. In: Proc. of ICM, vol. 3, pp. 9–12 (1954)
[GKZ94]   Gelfand, I.M., Kapranov, M.M., Zelevinsky, A.V.: Discriminants, Resultants and Multidimensional Determinants. Birkhäuser (1994)
[GM82]   Goldwasser, S., Micali, S.: Probabilistic encryption & how to play mental poker keeping secret all partial information. In: Proc. of ACM Symp. on TOC (1982)
[GMW87]   Goldreich, O., Micali, S., Wigderson, A.: How to play ANY mental game. In: Proc. of ACM Symp. on TOC, pp. 218–229. ACM Press (1987)
[Har92]   Harris, J.: Algebraic Geomerty: A First Course. Springer (1992)
[Lag82]   Lagarias, J.C.: Best simultaneously Diophantine approximations. I. growth rates of best approximation denominators. Trans. of AMS 272(2), 545–554 (1982)
[Lin84]   Lind, D.A.: The entropies of topological Markov shifts and a related class of algebraic integers. Ergodic Theory Dynam. Systems 4(2), 283–300 (1984)
[LM04]   Lipton, R.J., Markakis, E.: Nash equilibria Via polynomial equations. In: Farach-Colton, M. (ed.) LATIN 2004. LNCS, vol. 2976, pp. 413–422. Springer, Heidelberg (2004)
[Mum95]   Mumford, D.: Algebraic Geometry I: Complex Projective Varieties. Springer (1995)
[NS06]   Nogueira, A., Sevennec, B.: Multidimensional farey partitions. Indag. Mathem. 17(3), 437–456 (2006)
[Rok49]   Rokhlin, V.A.: On the fundamental ideas of measure theory. Mat. Sbornik N.S. 25(67), 107–150 (1949)
[Sha79]   Shamir, A.: How to share a secret. CACM 22(11), 612–613 (1979)
[Yao82]   Yao, A.C.: Protocols for secure computations (extended abstract). In: Proc. of FOCS, pp. 160–164 (November 1982)

Unconditionally-Secure Robust Secret Sharing with Compact Shares

Alfonso Cevallos[1], Serge Fehr[2,⋆], Rafail Ostrovsky[3], and Yuval Rabani[4,⋆]

[1] Mathematical Institute, Leiden University, The Netherlands
alfonsoc@gmail.com
[2] Centrum Wiskunde & Informatica (CWI), Amsterdam, The Netherlands
serge.fehr@cwi.nl
[3] Department of Computer Science, Department of Mathematics, UCLA
rafail@cs.ucla.edu
[4] The Rachel and Selim Benin School of Computer Science and Engineering,
The Hebrew University of Jerusalem, Jerusalem 91904, Israel
yrabani@cs.huji.ac.il

Abstract. We consider the problem of reconstructing a shared secret in the presence of faulty shares, with unconditional security. We require that any t shares give no information on the shared secret, and reconstruction is possible even if up to t out of the n shares are incorrect. The interesting setting is $n/3 \le t < n/2$, where reconstruction of a shared secret in the presence of faulty shares is possible, but only with an increase in the share size, and only if one admits a small failure probability. The goal of this work is to minimize this overhead in the share size. Known schemes either have a $\Omega(\kappa n)$-overhead in share size, where κ is the security parameter, or they have a close-to-optimal overhead of order $O(\kappa + n)$ but have an exponential running time (in n).

In this paper, we propose a new scheme that has a close-to-optimal overhead in the share size of order $\tilde{O}(\kappa + n)$, *and* a polynomial running time. Interestingly, the shares in our new scheme are prepared in the very same way as in the well-known scheme by Rabin and Ben-Or, which relies on message authentication, but we use a message authentication code with *short* tags and keys and with correspondingly *weak* security. The short tags and keys give us the required saving in the share size. Surprisingly, we can compensate for the weakened security of the authentication and achieve an exponentially small (in κ) failure probability by means of a more sophisticated reconstruction procedure.

1 Introduction

BACKGROUND. *Secret sharing*, invented independently by Shamir [18] and Blakley [2] in 1979, is a fundamental cryptographic primitive that has found numerous applications. In its basic form, it permits a dealer to share a secret s among a set of n players in such a way that: (1) up to t of the players learn no information

---

⋆ Part of this research was done while visiting UCLA.

D. Pointcheval and T. Johansson (Eds.): EUROCRYPT 2012, LNCS 7237, pp. 195–208, 2012.

on s by means of their shares, and (2) any $t + 1$ of the players can (efficiently) recover s from their shares. The most famous example, Shamir's secret sharing scheme, works by choosing a random polynomial $f(X) \in \mathbb{F}[X]$ of degree at most t with s as constant coefficient (assuming that s comes from a finite field $\mathbb{F}$), and the n shares are computed as $s_1 = f(x_1), \ldots, s_n = f(x_n)$ for publicly known pairwise-distinct non-vanishing interpolation points $x_1, \ldots, x_n$. Properties (1) and (2) follow easily from Lagrange's interpolation theorem.

In its basic form, secret sharing assumes the players to be honest and to provide the correct shares when asked. However, in cryptographic scenarios we often want/need to protect against malicious behavior of the participants. Therefore, strengthened versions of secret sharing have been proposed and studied over the years. One natural strengthening is to require that the shared secret can be recovered even if some players hand in incorrect shares. This is sometimes referred to as *robust* secret sharing. Formally, it is required that if all the n players pool together their shares, but up to t of them are incorrect (and it is not known which ones), then the shared secret can still be reconstructed (except maybe with small probability). Robust secret sharing has direct applications to *secure storage* and *unconditionally secure message transmission*. The goal of secure storage is to outsource the storing of sensitive data to a group of servers, in such a way that any coalition of up to t dishonest servers does not compromise the privacy nor the retrievability of the data. In unconditionally secure message transmission, as introduced in [8], (for follow-up works, see [9,10]) a sender wants to send some message to a receiver via a communication network that consists of n wires of which up to t may be under the control of an adversary, and privacy and receipt of the message should be guaranteed. It is immediate that "good" robust secret sharing schemes lead to "good" secure storage and "good" secure message transmission schemes. Furthermore, robust secret sharing schemes may act as stepping stone towards secret sharing schemes with yet stronger security guarantees. For instance, a *verifiable* secret sharing (VSS) scheme, as introduced in [4], additionally protects against a possibly malicious dealer who hands out inconsistent shares.

It follows immediately from the theory of Reed-Solomon error correcting codes that Shamir's secret sharing scheme is robust if (and only if) $t < n/3$. On the other hand, it is easy to see that robust secret sharing is impossible if $t \geq n/2$, and alternative definitions are needed [12]. Therefore, in this paper, we consider the range $n/3 \leq t < n/2$. In this range, robust secret sharing is possible, but only if one admits a small but positive failure probability. What makes robust secret sharing in the range $n/3 \leq t < n/2$ tricky is the fact that, say, Shamir shares alone do not carry enough redundancy to recover the correct secret in the presence of faulty shares, not even in principle. Indeed, if $n = 2t + 1$, and $s_1, \ldots, s_n$ are Shamir shares of a secret s but t of the shares are incorrect, then *any* $t + 1$ of the shares lie on a degree t polynomial and thus could actually be the $t + 1$ correct shares. There is no way to reconstruct the correct secret s from the list of partly modified Shamir shares. Additional redundancy needs to be added to the shares to permit reconstruction in the presence of incorrect shares. In the computational setting, this can be done by means of *commitments* (as e.g. in [16]); however, we aim for *unconditional security*, i.e., we do not put any computational restrictions on the adversary.

In this paper, we address the question of *how much* redundancy needs to be added to the shares in order to obtain robustness, for the maximal possible value of t, i.e., when $n = 2t + 1$. In other words, how small can the shares be in robust secret sharing?

KNOWN SCHEMES.   Interestingly, rather little is known about robust secret sharing with unconditional security for the range $n/3 \leq t < n/2$. To the best of our knowledge, up to small modifications, there exist two known (classes of) robust secret sharing schemes for this setting; we briefly discuss them here.

The first one is due to Rabin and BenOr [17] (called *"secret sharing when the dealer is a knight"* there). The Rabin-BenOr scheme consists of standard Shamir secret sharing, but enhanced my means of an (unconditionally secure) message authentication code. Specifically, for every pair of players P_i and P_j, P_i's Shamir share s_i is authenticated with an authentication tag τ_{ij}, where the corresponding authentication key key_{ji} is given to player P_j. During reconstruction, every player P_j can then verify the correctness of all the shares with the help of his authentication keys (and he detects every incorrect share except with small probability). If the reconstruction is performed by an outside reconstructor R that has no authentication keys, the reconstruction is slightly trickier. Every share s_i is then declared to be correct and used for reconstructing the secret by means of Lagrange interpolation if and only if it is accepted by the authentication keys of at least $t + 1$ players.

In order for this scheme to have a failure probability (in reconstructing the correct secret) of $2^{-\kappa}$, the message authentication code must have a failure probability smaller than $2^{-\kappa}$, which means that keys and tags must be of bitsize at least κ. As a consequence, beyond the actual Shamir share, every player gets another $\Omega(n\kappa)$ bits of redundancy as part of his share.[1]

The other scheme was first pointed out by Cramer, Damgård and Fehr [5], based on an idea by [3]. This scheme works as follows. Using standard Shamir secret sharing, the dealer shares independently the actual secret $s \in \mathbb{F}$, a randomly chosen field element $r \in \mathbb{F}$, and its product $p = s \cdot r$. To reconstruct the secret, the reconstructor does the following *for every subset* of $t + 1$ players. He reconstructs s', r' and p', supposed to be s, r and p, respectively, using the (possibly partly incorrect) shares of these $t + 1$ players, checks if $s' \cdot r' = p'$, and halts and outputs s' if it is the case. One can show that for any subset of $t + 1$ players: if $s' \neq s$ then $s' \cdot r' \neq p'$ except with probability $1/|\mathbb{F}|$. Thus, taking into account union bound over all subsets of size $t + 1$, choosing $\mathbb{F}$ to be of cardinality $2^{\kappa+n}$ gives a robust secret sharing scheme with failure probability $2^{-\kappa}$ and shares of size $O(\kappa + n)$.

---

[1] There are some additional log terms that we ignore, for instance due to applying union bound over the players.

Hence, much less redundancy is added to the actual share than in the Rabin-BenOr scheme.[2] Furthermore, it is not too hard to see that an increase in share size of κ bits is *necessary* for robust reconstruction (with $t < n/2$); thus, this scheme has close-to-optimal share size (at least if n is of order κ). The obvious downside of the scheme is that the reconstruction has exponential (in n) running time, as it loops over all possible subsets of size $t + 1$. Up to now, it is not known if there is an *efficient* reconstruction procedure for this robust secret sharing scheme. Another drawback of this scheme is that it is insecure in case the dishonest players get to see the shares (of r) of the honest players *before* they have to submit their own shares. Thus, it cannot be used, say, if reconstruction is performed by the (partly corrupted) players, and the adversary is *rushing*, meaning that the corrupt players wait with announcing their shares and then rush to announce their (correspondingly modified) shares.[3]

The latter scheme can be understood as being obtained, in a generic way, from a secret sharing scheme that allows error *detection*, i.e., that detects if a set of $t + 1$ shares contains some incorrect ones (but can not necessarily tell which ones). Indeed, as rigorously analyzed in [14], any secret sharing scheme with error detection (as in [20,3,15]) can be transformed into a robust secret sharing scheme by looping over all sets of size $t + 1$; but of course, any such scheme will suffer from the same exponential running time. For dishonest majority, a notion of *identifiable* secret sharing is explored in [12].

OUR CONTRIBUTION.   We propose a new robust secret sharing scheme that combines the advantages of both the above schemes. Our new scheme has a similar overhead in share size as the scheme by Cramer *et al.*, i.e., of the order $\tilde{O}(\kappa + n)$ rather than $\Omega(\kappa n)$, yet it is *computationally efficient*, meaning that sharing and reconstruction run in polynomial time in n, κ and the bitsize of the secret. Furthermore, security is preserved when reconstruction takes place among the partly corrupted players and the adversary is rushing.

Maybe somewhat surprisingly, the sharing procedure of our new robust secret sharing scheme is identical to that of the Rabin-BenOr scheme, except that we use a message authentication code with *short* keys and tags and with correspondingly *weak* security. In order to compensate for that, we need a more sophisticated reconstruction procedure, which inspects the *acceptance graph*, which describes which share is accepted by which player's key, more carefully. Essentially, the idea is to accept a share as being correct not as soon as it is correctly verified by $t+1$ players (as is the case in Rabin-BenOr), but only if it is correctly verified by $t + 1$ players *that hold accepted shares*. In other words, once a share is declared incorrect, then this player's vote is not counted anymore, making it harder for

---

[2] If the bitsize of the secret is much bigger than κ, then one can employ an adaptation of the scheme for which r and p can still live in a field of order $2^{\kappa+n}$, and thus the redundancy added to the actual share of s remains $O(\kappa + n)$ (see [6]).

[3] We stress that the Rabin-BenOr scheme does not suffer from this when the reconstruction is done in *two* rounds, where the players first announce their shares and tags, and only once everyone has revealed their shares and tags, then the keys are revealed. Looking ahead, the same will hold for our new scheme.

other incorrect shares to be accepted. In order to take care of a few incorrect shares that might survive, Reed-Solomon error correction is applied to the set of accepted shares. As will be seen later, although the basic idea of the new scheme is rather simple, its analysis is not. What makes the analysis tricky is that the probability of a bad share being detected now depends on how many other bad shares are detected. Thus, we cannot analyze the bad shares independently.

Interestingly, in [5] Cramer *et al.* prove a lower bound of $\Omega(\kappa n)$ on the necessary redundancy in the shares necessary to reconstruct a shared secret in the presence of up to a minority of incorrect shares. The discrepancy to our positive result stems from the fact that they consider a slightly stronger notion of robust secret sharing (which they called *"single-round honest-dealer VSS"* there): the reconstruction procedure must produce the correct secret except with probability $2^{-\kappa}$, but if it fails then it must output "failure". Thus, in their definition, reconstructing an incorrect secret is strictly prohibited, whereas we allow reconstruction of an incorrect secret with negligible probability. Also, they assume that reconstruction is done by the players with *one* round of communication and then each player deciding locally (possibly based on some part of his share he did not announce) on the reconstructed secret. Our new scheme does not seem to fit into this model since its security crucially relies on the fact that players release their shares in two rounds.

2   Preliminaries

2.1   Robust Secret Sharing

In order to define the robustness property of a secret sharing scheme, we formalize the latter by means of two interactive protocols, Share and Rec, where Share involves a *dealer D* and n players $P_1, \ldots, P_n$, and Rec involves the n players and a *reconstructor R*. More formally, an *n-player secret sharing scheme* for a message space $\mathcal{S}$ consists of two phases, the *sharing* and the *reconstruction* phase, specified by two protocols Share and Rec. During the sharing phase, the dealer D takes as input a secret $s \in \mathcal{S}$, locally computes shares $\sigma_1, \ldots, \sigma_n$, and sends the i-th share σ_i to player P_i for every $i \in [n]$. During reconstruction, player P_i (for every $i \in [n]$) communicates, possibly by means of several synchronous communication rounds, σ_i to the reconstructor R. Based on the received shares, R then produces an output s', which is supposed to be the original secret s.

Before we formalize the security requirements, we specify the capabilities (and limitations) of the *adversary* that tries to break the scheme. During the sharing phase, the adversary remains inactive, and he does not get to learn any information at all. In particular, he does not get to see the shares that D sends to the players. After the sharing phase, the adversary can adaptively corrupt up to t of the players P_i (but not D), where t is some parameter.[4] Once a player P_i is corrupted, the adversary learns P_i's share σ_i, and from now on, the

---

[4] Since the sharing phase only involves one round of communication from D to the players, it does not help the adversary to corrupt players *during* the sharing phase.

adversary has full control over P_i. The corruptions being adaptive means that after each corruption, the adversary can decide on who to corrupt next depending on the shares he has seen so far. During the reconstruction phase, the adversary gets to see the communication between *all* players P_i and the reconstructor R.[5] Furthermore, he controls the information that the dishonest players send to R. Namely, in every communication round, he can decide for every dishonest player on what this player should send to R, depending on what he has seen so far and depending on what the honest players have sent to R in the current round. The latter means that the adversary is *rushing*. Finally, if he has not yet corrupted t players, he can between each round of communication adaptively corrupt additional players, as long as the total number of corrupt players does not exceed t. We stress that the adversary cannot corrupt D or R.

Definition 2.1. *An n-player secret sharing scheme* (Share, Rec) *is* (t, δ)*-robust if the following properties hold for any distribution of* $s \in S$ *and for any adversary as specified above.*

   Privacy: *Before* Rec *is started, the adversary has no more information on the shared secret s than he had before the execution of* Share.

   Reconstructability: *At the end of* Rec, *the reconstructor R outputs* $s' = s$ *except with probability at most* δ.

It is known that in any (not necessarily robust but perfectly private) secret sharing scheme, the bit-size of every share σ_i is at least the bit-size $\log |S|$ of the secret. In this paper, we are interested in how much redundancy needs to be added to this minimal share size in order to achieve robustness, i.e., in the quantity $\max_i(\log |\Sigma_i|) - \log |S|$, where Σ_i denotes the set of all possible shares σ_i for player i. We call this quantity the *overhead* of a scheme.

2.2   Message Authentication Codes

A message authentication code (MAC) is a tool that enables to verify the integrity of a message. Unconditionally secure MACs were initially invented by Carter and Wegman [21,22]. We give here a definition that suits our needs.

Definition 2.2. *A message authentication code (or* MAC*) for a finite message space* $\mathcal{M}$ *consists of a function* $MAC : \mathcal{M} \times \mathcal{K} \to \mathcal{T}$ *for finite sets* $\mathcal{K}$ *and* $\mathcal{T}$. *It is called* ε*-secure if for all* $m, \hat{m} \in \mathcal{M}$ *with* $m \neq \hat{m}$ *and for all* $\tau, \hat{\tau} \in \mathcal{T}$:

$$P[MAC(\hat{m}, K) = \hat{\tau} \mid MAC(m, K) = \tau] \leq \delta \,,$$

where the random variable K *is uniformly distributed over* $\mathcal{K}$.

---

[5] It may look unnatural at first glance that the adversary does not get to see the communication between D and the players, but he does get to see the communication between the players and R. The reason why we want to allow him to observe the communication with R is that in certain applications, it is actually the set of all players that wants/needs to reconstruct the secret. In this case, whenever the reconstruction procedure dictates player P_i to send some information to R, it has to send that information to all the players. But this of course then means that if at least one of the players is corrupt, then the adversary gets to see all the communication intended for R.

It is well known that if $\mathcal{M}$ is a finite field $\mathbb{F}$, then $MAC : \mathbb{F} \times \mathbb{F}^2 \to \mathbb{F}$ with $(m, (\alpha, \beta)) \mapsto \alpha \cdot m + \beta$ is a ε-secure MAC with $\varepsilon = 1/|\mathbb{F}|$. More generally, as first shown in [7,13,19],

$$MAC : \mathbb{F}^d \times \mathbb{F}^2 \to \mathbb{F}, \ ((m_1 \ldots, m_d), (\alpha, \beta)) \mapsto \sum_{k=1}^{d} \alpha^i \cdot m_i + \beta$$

is a ε-secure MAC with $\varepsilon = d/|\mathbb{F}|$.

2.3 Reed-Solomon Error Correction

Let $\mathbb{F}$ be a finite field, let n' be a positive integer, and let $x_1, \ldots, x_{n'}$ be pairwise distinct interpolation points in $\mathbb{F}$. We consider the problem of recovering a polynomial $f(X) \in \mathbb{F}[X]$ of degree at most t, when given a perturbed version of its evaluations $(f(x_1), \ldots, f(x_{n'}))$, i.e., when given a vector $(y_1, \ldots, y_{n'})$ for which it is promised that $y_i = f(x_i)$ for all but e of the indexes $i \in \{1, \ldots, n'\}$, where e is some parameter, but it is not known for *which* indices. This is known as Reed-Solomon error correction. It is not hard to see, using Lagrange interpolation, that $f(X)$ is uniquely determined from $(y_1, \ldots, y_{n'})$ if (and only if) $n' \geq t + 1 + 2e$. Indeed, if there are two such polynomials, then they must coincide in at least $t + 1$ points, and hence are identical. Furthermore, there exist algorithms that permit to *efficiently compute* $f(X)$ from $(y_1, \ldots, y_{n'})$ in case $n' \geq t + 1 + 2e$, for instance the Berlekamp-Welch algorithm [1]. A simplified version of the original Berlekamp-Welch algorithm, provided by Gemmell and Sudan, can be found in [11].

3   The New Scheme and Its Analysis

Let t be an arbitrary positive integer, and $n = 2t + 1$. Consider Shamir's secret sharing scheme over a field $\mathbb{F}$ with $|\mathbb{F}| > n$, with pairwise-distinct non-vanishing interpolation points $x_1, \ldots, x_n \in \mathbb{F}$. Furthermore, let $MAC : \mathbb{F} \times \mathcal{K} \to \mathcal{T}$ be an ε-secure MAC with message space $\mathbb{F}$. The sharing procedure Share of our new scheme is presented in Figure 1.

Local computation: On input $s \in \mathbb{F}$, the dealer D chooses a random sharing polynomial $f(X) \in \mathbb{F}[X]$ with degree at most t and $f(0) = s$, and he computes the Shamir shares $s_1 = f(x_1), \ldots, s_n = f(x_n)$. Furthermore, for every pair $i, j \in [n]$, he chooses a random $key_{ij} \in \mathcal{K}$ and computes $\tau_{ij} = MAC(key_{ji}, s_i)$.

Share distribution: For every $i \in [n]$, the dealer D sends to player P_i the share $\sigma_i = (s_i, \tau_{i1}, \ldots, \tau_{in}, key_{i1}, \ldots, key_{in})$.

Fig. 1. Sharing procedure Share

The new sharing procedure is identical to the sharing procedure of the Rabin-BenOr robust secret sharing scheme, except that we describe it by means of an arbitrary MAC. However, in the end we will use a MAC with *short* keys and tags and a correspondingly weak security, which would render the original Rabin-BenOr scheme insecure. The reader may think of ε being $1/n$; indeed, as we will see later, this will give us δ-robustness with δ approximately $2^{-n/4}$.

In order to deal with a non-negligible ε for the security of the MAC, we need a sophisticated reconstruction procedure. The idea is to inspect the *acceptance graph*, which describes which share s_i is consistent (together with the corresponding tag) with which authentication key key_{ji}, more carefully. Instead of accepting a share s_i as being correct as soon as it is consistent with (the keys of) at least $t + 1$ players (which means that a dishonest player only needs to fool *one* honest player to get his share accepted), we will require, for a share to be accepted, that it is consistent with at least $t + 1$ players *that hold accepted shares*. In other words, once a share is declared incorrect, then this player's vote is not counted anymore, making it harder for other incorrect shares to be accepted. Some might still survive, though; to take care of that, Reed-Solomon error correction is then applied to the set of accepted shares. The procedure is described in Figure 2. It is easy to see that the set $\mathcal{I}$ in step 2 is well defined and can efficiently be computed by starting with the set of all $i \in [n]$ and inductively eliminating "bad" players.

First round: Every player P_i sends s_i and $\tau_{i1}, \ldots, \tau_{in}$ to the reconstructor R.

Second round: Every player P_i sends $key_{i1}, \ldots, key_{in}$ to R.

Local computation:

1. For every $i, j \in [n]$, R sets v_{ij} to be 1 if the share s_i is accepted by (the key of) player P_j, i.e., if $\tau_{ij} = MAC(key_{ji}, s_i)$, and else to 0.

2. R computes the largest set $\mathcal{I} \subseteq [n]$ with the property that

$$\forall\, i \in \mathcal{I} : \left| \{ j \in \mathcal{I} \mid v_{ij} = 1 \} \right| = \sum_{j \in \mathcal{I}} v_{ij} \geq t + 1\,;$$

in other words, such that every share of a player in $\mathcal{I}$ is accepted by at least $t + 1$ players in $\mathcal{I}$.

Clearly, $\mathcal{I}$ contains all honest players. Let $c = |\mathcal{I}| - (t+1)$ be the maximum number of corrupt players in $\mathcal{I}$.

3. Using Berlekamp-Welch, R computes a polynomial $f(X) \in \mathbb{F}[X]$ of degree at most t such that $f(x_i) = s_i$ for at least $(t + 1) + \frac{c}{2}$ players i in $\mathcal{I}$. If no such polynomial exists then R outputs $\perp$; otherwise, he outputs $s = f(0)$.

Fig. 2. Reconstruction procedure Rec

The intuition behind the security is the following. If the corrupt players hand in only a few incorrect shares and many correct shares, then the incorrect shares have a good chance of surviving since they only need to be consistent with the keys of a few honest players. However, since there are only a few incorrect shares, the Reed-Solomon decoding will take care of them. On the other hand, if the corrupt players hand in many incorrect shares, then, because there are many of them, some will probably be detected as being incorrect, which will make it harder for the remaining incorrect shares because they now need to be consistent with more honest players, which means that some more incorrect shares will probably be detected, which will make it even harder for the remaining ones, etc., so that in the end, hopefully only a few survive so that again Reed-Solomon error correction takes care. The following theorem shows that the above intuition is indeed correct. However, the formal reasoning is quite involved, as we will see later.

Theorem 3.1. *For any positive integer t, any finite field $\mathbb{F}$ with $|\mathbb{F}| > n = 2t+1$, and any ε-secure MAC $: \mathbb{F} \times \mathcal{K} \to \mathcal{T}$ with $\varepsilon \leq 1/(t+1)$, the pair $(\mathsf{Share}, \mathsf{Rec})$ forms an n-player (t, δ)-robust secret sharing scheme for message space $\mathbb{F}$ with*

$$\delta \leq \mathrm{e} \cdot \big((t+1)\varepsilon\big)^{(t+1)/2},$$

where $\mathrm{e} = \exp(1)$.

The crucial property on δ is that it is not of order ε, as in the Rabin-BenOr scheme, but of order $\varepsilon^{\Omega(n)}$. This allows us to reduce the authentication key and tag sizes by a factor (linear in) n.

Specifically, we can get the following instantiation. Let λ be an arbitrary parameter, and let $GF(2^m)$ be the binary field with $2^m > n$ elements. By Section 2.2, there exists an ε-secure MAC $: GF(2^m) \times \mathcal{K} \to \mathcal{T}$ with $\mathcal{K} = GF(2^\lambda)^2$ and $\mathcal{T} = GF(2^\lambda)$ and $\varepsilon \leq m/2^\lambda$. By Theorem 3.1, the resulting secret sharing scheme is δ-robust for $\delta \leq \mathrm{e} \cdot ((t+1)m/2^\lambda)^{(t+1)/2}$. Therefore, for a given security parameter κ, setting $\lambda = \lceil \log(t+1) + \log(m) + \frac{2}{t+1}(\kappa + \log(e)) \rceil$, we obtain $\delta \leq 2^{-\kappa}$, and every share consists of the ordinary m-bit Shamir share plus an overhead of

$$3n\lambda \leq 12\kappa + 3n(\log(t+1) + \log(m) + 3)$$

bits.

Corollary 3.2. *For any positive integers t, m, κ, and for $n = 2t+1$, there exists an n-player (t, δ)-robust secret sharing scheme for message space $\mathcal{S} = \{0, 1\}^m$, with $\delta = 2^{-\kappa}$ and an overhead of $O\big(\kappa + n(\log n + \log m)\big)$.*

We will now prove Theorem 3.1. Although the idea for the new scheme is rather simple and natural, the security analysis is non-trivial. One reason is that it is not clear what the optimal strategy for the adversary is. In comparison, in the Rabin-BenOr scheme, it is obvious that the best the adversary can do is to have every corrupt player hand in an incorrect share and hope that at least one gets accepted. In our new scheme, however, it might be advantages to have some corrupt players hand in *correct* shares; the reason being that such players could support incorrect shares of other corrupt players, making it easier for them to survive the elimination round. On the other hand, having too many corrupt players handing in correct shares will facilitate Reed-Solomon decoding. Another reason is that there seems to be some circularity: in order to argue that many incorrect shares get eliminated, we want to argue that incorrect shares need to be accepted by many honest players in order to survive, but this is only true once many incorrect shares got eliminated.

Our proof below is pretty much "brute force". We work out a bound on the failure probability (for an arbitrary strategy) by essentially listing all possible scenarios of which incorrect shares might be accepted by which honest players, and then we simplify the resulting unhandy expression.

Proof (of Theorem 3.1). Privacy is obvious. It remains to prove the reconstructability property. Consider the state of the reconstruction phase right before the second round of communication, i.e., after R has received the shares and tags, but before the keys are communicated. We may assume that at this stage, the adversary has corrupted t players. We define the following sets. $\mathcal{A} \subset [n]$ is the set of corrupt players i that have handed in a *modified* Shamir share s_i, and $\mathcal{P} \subset [n]$ is the set of corrupt players i that have handed in the *correct* Shamir share s_i. It holds that $|\mathcal{A}| + |\mathcal{P}| = t$. The remaining set $\mathcal{H} = [n] \setminus (\mathcal{A} \cup \mathcal{P})$ is the set of uncorrupt players.[6]

We consider the probability space specified by the random choices of the authentication keys held by the uncorrupt players, conditioned on the shares and tags handed out by the dealer D during the sharing procedure, plus the choices of the (possibly modified) authentication keys claimed in the second round of the reconstruction procedure by the corrupt players. For every pair $i, j \in [n]$, we can define the binary random variable V_{ij} that specifies if player P_i's (possibly incorrect) share with the corresponding tag is accepted by player P_j's key. Since the authentication keys of uncorrupt players have been chosen independently, all the V_{ij} with $i \in [n]$ and $j \in \mathcal{H}$ are independent. Also, $V_{ij} = 1$ with probability 1 for every pair $i, j \in \mathcal{H}$, i.e., honest players accept each others shares. Furthermore, by the security of the MAC (Definition 2.2), $P[V_{ij} = 1] \leq \varepsilon$ for all $i \in \mathcal{A}$ and $j \in \mathcal{H}$. Finally, it is not too hard to see that it does not help the corrupt players to hand in correct Shamir shares but incorrect authentication tags: a player in $\mathcal{P}$ that is eliminated is of no use for the adversary; thus, we may assume that $V_{ij} = 1$ for every pair $i \in \mathcal{P}$, $j \in [n]$.

It follows that the set $\mathcal{I}$ computed during Rec (which depends on the V_{ij}'s and thus we treat it as a random variable here) contains $\mathcal{H}$ and $\mathcal{P}$ with certainty. Thus, the reconstruction procedure is guaranteed to output the correct secret if at most p players $i \in \mathcal{A}$ end up in $\mathcal{I}$, where $p = |\mathcal{P}|$. Indeed, if $|\mathcal{A} \cap \mathcal{I}| \leq p$, then the requirement for Reed-Solomon decoding is satisfied (see Section 2.3 with $n' = |\mathcal{I}| = t + 1 + c = t + 1 + p + e$ where $e = |\mathcal{A} \cap \mathcal{I}| \leq p$), and the polynomial

---

[6] The mnemonic is: $\mathcal{A}$ for *actively* corrupt, $\mathcal{P}$ for *passively* corrupt, and $\mathcal{H}$ for *honest*, but we stress that the players in $\mathcal{P}$ are merely passive with respect to their respective Shamir shares s_i; they may very well lie about their authentication keys and tags.

$f(X)$ computed during Rec is guaranteed to satisfy $f(x_i) = s_i$ for at least $t + 1$ correct shares, and thus it is the correct sharing polynomial and $f(0)$ the correct secret.

It thus remains to analyze the probability $P[|\mathcal{A} \cap \mathcal{I}| > p]$. For this, it is sufficient to consider the case $p \leq (t - 1)/2$; indeed, if $p > (t - 1)/2$ and thus $p \geq t/2$ then obviously $|\mathcal{A}| \leq p$ and hence $P[|\mathcal{A} \cap \mathcal{I}| \leq p]$ with certainty. Actually, we will now show that $P[|\mathcal{A} \cap \mathcal{I}| > 0]$ is small if $p \leq (t - 1)/2$.

We can write $P[|\mathcal{A} \cap \mathcal{I}| > 0] = \sum_\ell P[|\mathcal{A} \cap \mathcal{I}| = \ell]$ where the sum ranges from $\ell = 1$ to $t - p$. In order to bound the probability $P[|\mathcal{A} \cap \mathcal{I}| = \ell]$, it is convenient to introduce for every $i \in \mathcal{A}$ the random variable

$$N_i = \sum_{j \in \mathcal{H}} V_{ij} = |\{j \in \mathcal{H} \mid V_{ij} = 1\}|,$$

i.e., the number of honest players that accept P_i's incorrect share. Note that since the V_{ij}'s are independent for all $i \in [n]$ and $j \in \mathcal{H}$, so are all the N_i's. We can now bound $P[|\mathcal{A} \cap \mathcal{I}| = \ell]$ for an arbitrary ℓ in the range $1 \leq \ell \leq t - p$ as follows.

$$P[|\mathcal{A} \cap \mathcal{I}| = \ell] \leq P\big[\exists \mathcal{A}_\circ \subseteq \mathcal{A} : (|\mathcal{A}_\circ| = \ell) \wedge (\forall i \in \mathcal{A}_\circ : N_i \geq t + 1 - p - \ell)\big]$$

$$\leq \sum_{\substack{\mathcal{A}_\circ \subseteq \mathcal{A} \\ |\mathcal{A}_\circ| = \ell}} P\big[\forall i \in \mathcal{A}_\circ : N_i \geq t + 1 - p - \ell\big]$$

$$= \sum_{\substack{\mathcal{A}_\circ \subseteq \mathcal{A} \\ |\mathcal{A}_\circ| = \ell}} \prod_{i \in \mathcal{A}_\circ} P\big[N_i \geq t + 1 - p - \ell\big]$$

$$\leq \sum_{\substack{\mathcal{A}_\circ \subseteq \mathcal{A} \\ |\mathcal{A}_\circ| = \ell}} \prod_{i \in \mathcal{A}_\circ} P\big[\exists \mathcal{H}_\circ \subseteq \mathcal{H} : (|\mathcal{H}_\circ| = t + 1 - p - \ell) \wedge (\forall j \in \mathcal{H}_\circ : V_{ij} = 1)\big]$$

$$\leq \sum_{\substack{\mathcal{A}_\circ \subseteq \mathcal{A} \\ |\mathcal{A}_\circ| = \ell}} \prod_{i \in \mathcal{A}_\circ} \sum_{\substack{\mathcal{H}_\circ \subseteq \mathcal{H} \\ |\mathcal{H}_\circ| = t + 1 - p - \ell}} P\big[\forall j \in \mathcal{H}_\circ : V_{ij} = 1\big]$$

Now, since $P\big[\forall j \in \mathcal{H}_\circ : V_{ij} = 1\big] = \prod_{j \in \mathcal{H}_\circ} P[V_{ij} = 1]$ and $P[V_{ij} = 1] \leq \varepsilon$ for all $i \in \mathcal{A}$ and $j \in \mathcal{H}$, we can proceed as follows, where we write $a = |\mathcal{A}| = t - p \geq (t + 1)/2$ and $\tilde{\varepsilon} = (t + 1)\varepsilon$.

$$P[|\mathcal{A} \cap \mathcal{I}| = \ell] \leq \binom{a}{\ell} \cdot \left(\binom{t + 1}{a - \ell + 1} \cdot \varepsilon^{a - \ell + 1}\right)^\ell$$

$$\leq \binom{a}{\ell} \cdot \left(\frac{(t + 1)^{a - \ell + 1}}{(a - \ell + 1)!} \cdot \varepsilon^{a - \ell + 1}\right)^\ell = \frac{a!}{\ell!(a - \ell)!} \cdot \left(\frac{\tilde{\varepsilon}^{a - \ell + 1}}{(a - \ell + 1)!}\right)^\ell$$

$$= \frac{\tilde{\varepsilon}^{\ell(a - \ell + 1)}}{((a - \ell)!)^\ell} \cdot \frac{a!/(a - \ell)!}{\ell!(a - \ell + 1)^\ell} = \frac{\tilde{\varepsilon}^{\ell(a - \ell + 1)}}{((a - \ell)!)^\ell} \cdot \underbrace{\prod_{k=1}^{\ell} \frac{a - \ell + k}{k(a - \ell + 1)}}_{\leq 1}$$

$$\leq \frac{\tilde{\varepsilon}^{\ell(a - \ell + 1)}}{((a - \ell)!)^\ell} \leq \frac{\tilde{\varepsilon}^{\ell(a - \ell + 1)}}{(a - \ell)!} \leq \frac{\tilde{\varepsilon}^a}{(a - \ell)!}$$

where the very last inequality follows from $\tilde{\varepsilon} = (t+1)\varepsilon \leq 1$ (by assumption on ε) and the fact that $\min\{\ell(a - \ell + 1) \mid 1 \leq \ell \leq t - p = a\} = a$, which can easily be verified.[7] We can now conclude that

$$P[|\mathcal{A} \cap \mathcal{I}| > 0] = \sum_{\ell=1}^{t-p} P[|\mathcal{A} \cap \mathcal{I}| = \ell] \leq \sum_{\ell=1}^{t-p} \frac{\tilde{\varepsilon}^a}{(a-\ell)!}$$

$$\leq \tilde{\varepsilon}^a \sum_{k=0}^{a-1} \frac{1}{k!} \leq \tilde{\varepsilon}^a \sum_{k=0}^{\infty} \frac{1}{k!} \leq \tilde{\varepsilon}^{(t+1)/2} \, e$$

which proves the claim.   □

4   Conclusion and Open Questions

We have shown and analyzed a new robust secret sharing scheme, which combines the computational efficiency of the Rabin-BenOr scheme [17] with a close-to-optimal overhead of $\tilde{O}(\kappa + n)$ in the share size, as featured by the (computationally inefficient) scheme of Cramer et al. [5].

It is interesting to see that our new scheme is based on a completely different approach than the scheme of Cramer et al., but displays the same order-n gap to the known lower bound of $\Omega(\kappa)$ for the share size in robust secret sharing. This raises the question of the true optimal share size in robust secret sharing: is the linear term in n inherent, or is it an artifact of current constructions?

Acknowledgments. We thank Brett Hemenway for multiple helpful discussions. R.O. was supported in part by NSF grants 0830803, 09165174, 1065276, 1118126 and 1136174, US-Israel BSF grant 2008411, B. John Garrick Foundation, Teradata award, Intel equipment grant, NSF Cybertrust grant, OKAWA Award, Xerox Innovation Group Award, IBM Faculty Award, Lockheed-Martin Corporation and the Defense Advanced Research Projects Agency through the U.S. Office of Naval Research under Contract N00014-11-1-0392. The views expressed are those of the author and do not reflect the official policy or position of the Department of Defense or the U.S. Government. Y.R was supported in part by the I-CORE program 4/11, by ISF grant 856-11, and by BSF grant 2008059. The research was initiated while S.F. and Y.R. were visiting UCLA, and S.F. and Y.R. would like to thank UCLA for the hospitality.

References

1. Berlekamp, E.R., Welch, L.R.: Error correction of algebraic block codes. U.S. Patent Number 4.633.470 (1986)
2. Blakley, G.R.: Safeguarding cryptographic keys. In: National Computer Conference, vol. 48, pp. 313–317. AFIPS Press (1979)

---

[7] Indeed, the function $\ell \mapsto \ell(a - \ell + 1)$ is concave and thus reaches its minimum at one of the boundaries.

3. Cabello, S., Padró, C., Sáez, G.: Secret Sharing Schemes with Detection of Cheaters for a General Access Structure. In: Ciobanu, G., Păun, G. (eds.) FCT 1999. LNCS, vol. 1684, pp. 185–194. Springer, Heidelberg (1999)
4. Chor, B., Goldwasser, S., Micali, S., Awerbuch, B.: Verifiable secret sharing and achieving simultaneity in the presence of faults (extended abstract). In: 26th Annual IEEE Symposium on Foundations of Computer Science (FOCS), pp. 383–395 (1985)
5. Cramer, R., Damgård, I., Fehr, S.: On the Cost of Reconstructing a Secret, or VSS with Optimal Reconstruction Phase. In: Kilian, J. (ed.) CRYPTO 2001. LNCS, vol. 2139, pp. 503–523. Springer, Heidelberg (2001)
6. Cramer, R., Dodis, Y., Fehr, S., Padró, C., Wichs, D.: Detection of Algebraic Manipulation with Applications to Robust Secret Sharing and Fuzzy Extractors. In: Smart, N.P. (ed.) EUROCRYPT 2008. LNCS, vol. 4965, pp. 471–488. Springer, Heidelberg (2008)
7. den Boer, B.: A simple and key-economical unconditional authentication scheme. Journal of Computer Security 2, 65–72 (1993)
8. Dolev, D., Dwork, C., Waarts, O., Yung, M.: Perfectly secure message transmission. In: 31st Annual IEEE Symposium on Foundations of Computer Science (FOCS), vol. I, pp. 36–45 (1990)
9. Garay, J., Givens, C., Ostrovsky, R.: Secure Message Transmission with Small Public Discussion. In: Gilbert, H. (ed.) EUROCRYPT 2010. LNCS, vol. 6110, pp. 177–196. Springer, Heidelberg (2010)
10. Garay, J., Givens, C., Ostrovsky, R.: Secure Message Transmission by Public Discussion: A Brief Survey. In: Chee, Y.M., Guo, Z., Ling, S., Shao, F., Tang, Y., Wang, H., Xing, C. (eds.) IWCC 2011. LNCS, vol. 6639, pp. 126–141. Springer, Heidelberg (2011)
11. Gemmell, P., Sudan, M.: Highly resilient correctors for polynomials. Information Processing Letters 43(4), 169–174 (1992)
12. Ishai, Y., Ostrovsky, R., Seyalioglu, H.: Identifying Cheaters without an Honest Majority. In: Cramer, R. (ed.) TCC 2012. LNCS, vol. 7194, pp. 21–38. Springer, Heidelberg (2012)
13. Johansson, T., Kabatianskii, G., Smeets, B.: On the Relation between A-Codes and Codes Correcting Independent Errors. In: Helleseth, T. (ed.) EUROCRYPT 1993. LNCS, vol. 765, pp. 1–11. Springer, Heidelberg (1994)
14. Kurosawa, K., Suzuki, K.: Almost secure (1-round, n -channel) message transmission scheme. IEICE Transactions 92-A(1) (2009)
15. Ogata, W., Kurosawa, K., Stinson, D.R.: Optimum secret sharing scheme secure against cheating. SIAM Journal on Discrete Mathematics 20(1), 79–95 (2006)
16. Pedersen, T.P.: Non-interactive and Information-Theoretic Secure Verifiable Secret Sharing. In: Feigenbaum, J. (ed.) CRYPTO 1991. LNCS, vol. 576, pp. 129–140. Springer, Heidelberg (1992)
17. Rabin, T., Ben-Or, M.: Verifiable secret sharing and multiparty protocols with honest majority. In: 21st Annual ACM Symposium on Theory of Computing (STOC), pp. 73–85 (1989)
18. Shamir, A.: How to share a secret. Communications of the ACM 22(11), 612–613 (1979)
19. Taylor, R.: An Integrity Check Value Algorithm for Stream Ciphers. In: Stinson, D.R. (ed.) CRYPTO 1993. LNCS, vol. 773, pp. 40–48. Springer, Heidelberg (1994)

20. Tompa, M., Woll, H.: How to Share a Secret with Cheaters. In: Odlyzko, A.M. (ed.) CRYPTO 1986. LNCS, vol. 263, pp. 261–265. Springer, Heidelberg (1987)
21. Wegman, M.N., Lawrence Carter, J.: New classes and applications of hash functions. In: 20th Annual IEEE Symposium on Foundations of Computer Science (FOCS), pp. 175–182 (1979)
22. Wegman, M.N., Lawrence Carter, J.: New hash functions and their use in authentication and set equality. Journal of Computer and System Science 22(3), 265–279 (1981)

All-But-Many Lossy Trapdoor Functions

Dennis Hofheinz

Karlsruhe Institute of Technology, Karlsruhe, Germany

Abstract. We put forward a generalization of lossy trapdoor functions (LTFs). Namely, all-but-many lossy trapdoor functions (ABM-LTFs) are LTFs that are parametrized with tags. Each tag can either be injective or lossy, which leads to an invertible or a lossy function. The interesting property of ABM-LTFs is that it is possible to generate an arbitrary number of lossy tags by means of a special trapdoor, while it is not feasible to produce lossy tags without this trapdoor.

Our definition and construction can be seen as generalizations of all-but-one LTFs (due to Peikert and Waters) and all-but-N LTFs (due to Hemenway et al.). However, to achieve ABM-LTFs (and thus a number of lossy tags which is not bounded by any polynomial), we have to employ some new tricks. Concretely, we give two constructions that use "disguised" variants of the Waters, resp. Boneh-Boyen signature schemes to make the generation of lossy tags hard without trapdoor. In a nutshell, lossy tags simply correspond to valid signatures. At the same time, tags are disguised (i.e., suitably blinded) to keep lossy tags indistinguishable from injective tags.

ABM-LTFs are useful in settings in which there are a polynomial number of adversarial challenges (e.g., challenge ciphertexts). Specifically, building on work by Hemenway et al., we show that ABM-LTFs can be used to achieve selective opening security against chosen-ciphertext attacks. One of our ABM-LTF constructions thus yields the first SO-CCA secure encryption scheme with compact ciphertexts ($\mathbf{O}(1)$ group elements) whose efficiency does not depend on the number of challenges. Our second ABM-LTF construction yields an IND-CCA (and in fact SO-CCA) secure encryption scheme whose security reduction is independent of the number of challenges and decryption queries.

Keywords: lossy trapdoor functions, public-key encryption, selective opening attacks.

1 Introduction

Lossy Trapdoor Functions. Lossy trapdoor functions (LTFs) have been formalized by Peikert and Waters [30], in particular as a means to construct chosen-ciphertext (CCA) secure public-key encryption (PKE) schemes from lattice assumptions. In a nutshell, LTFs are functions that may be operated with an injective key (in which case a trapdoor allows to efficiently invert the function), or with a lossy key (in which case the function is highly non-injective, i.e., loses information). The key point is that injective and lossy keys are computationally indistinguishable. Hence, in a security proof (say, for a PKE scheme), injective keys can be replaced with lossy keys without an adversary noticing. But once all keys are lossy, a ciphertext does not contain any (significant) information anymore about the encrypted message. There exist quite efficient constructions of LTFs

D. Pointcheval and T. Johansson (Eds.): EUROCRYPT 2012, LNCS 7237, pp. 209–227, 2012.

based on a variety of assumptions (e.g., [30, 7, 10, 20]). Besides, LTFs have found various applications in public-key encryption [22, 7, 6, 5, 23, 19] and beyond [16, 30, 27] (where [16] implicitly uses LTFs to build commitment schemes).

LTFs with Tags and All-but-one LTFs. In the context of CCA-secure PKE schemes, it is useful to have LTFs which are parametrized with a tag[1]. In all-but-one LTFs (ABO-LTFs), all tags are injective (i.e., lead to an injective function), except for one single lossy tag. During a proof of CCA security, this lossy tag will correspond to the (single) challenge ciphertext handed to the adversary. All decryption queries an adversary may make then correspond to injective tags, and so can be handled successfully. ABO-LTFs have been defined, constructed, and used as described by Peikert and Waters [30].

Note that ABO-LTFs are not immediately useful in settings in which there is more than one challenge ciphertext. One such setting is the selective opening (SO) security of PKE schemes ([6], see also [11, 18]). Here, an adversary A is presented with a vector of ciphertexts (which correspond to eavesdropped ciphertexts), and gets to choose a subset of these ciphertexts. This subset is then opened for A; intuitively, this corresponds to a number of corruptions performed by A. A's goal then is to find out any nontrivial information about the *unopened* ciphertexts. It is currently not known how to reduce this multi-challenge setting to a single-challenge setting (such as IND-CCA security). In particular, ABO-LTFs are not immediately useful to achieve SO-CCA security. Namely, if we follow the described route to achieve security, we would have to replace all challenge ciphertexts (and only those) with lossy ones. However, an ABO-LTF has only one lossy tag, while there are many challenge ciphertexts.

All-but-N LTFs and their Limitations. A natural solution has been given by Hemenway et al. [23], who define and construct all-but-N LTFs (ABN-LTFs). ABN-LTFs have exactly N lossy tags; all other tags are injective. This can be used to equip exactly the challenge ciphertexts with the lossy tags; all other ciphertexts then correspond to injective tags, and can thus be decrypted. Observe that ABN-LTFs encode the set of lossy tags in their key. (That is, a computationally unbounded adversary could always brute-force search which tags lead to a lossy function.) For instance, the construction of [23] embeds a polynomial in the key (hidden in the exponent of group elements) such that lossy tags are precisely the zeros of that polynomial.

Hence, ABN-LTFs have a severe drawback: namely, the space complexity of the keys is at least linear in N. In particular, this affects the SO secure PKE schemes derived in [23]: there is no single scheme that would work in arbitrary protocols (i.e., for arbitrary N). Besides, their schemes quickly become inefficient as N gets larger, since each encryption requires to evaluate a polynomial of degree N in the exponent.

Our Contribution: LTFs with Many Lossy Tags. In this work, we define and construct all-but-many LTFs (ABM-LTFs). An ABM-LTF has superpolynomially many lossy tags, which however require a special trapdoor to be found. This is the most crucial difference to ABN-LTFs: with ABN-LTFs, the set of lossy tags is specified initially, at construction time. Our ABM-LTFs have a trapdoor that allows to sample on the fly

---

[1] What we call "tag" is usually called "branch." We use "tag" in view of our later construction, in which tags have a specific structure, and cannot be viewed as branches of a (binary) tree.

from a superpolynomially large pool of lossy tags. (Of course, without that trapdoor, and even given arbitrarily many lossy tags, another lossy tag is still hard to find.) This in particular allows for ABM-LTF instantiations with compact keys and images whose size is independent of the number of lossy tags.

Our constructions can be viewed as "disguised" variants of the Waters, resp. Boneh-Boyen (BB) signature schemes [33, 8]. Specifically, lossy tags correspond to valid signatures. However, to make lossy and injective tags appear indistinguishable, we have to blind signatures by encrypting them, or by multiplying them with a random subgroup element. We give more details on our constructions below.

A DCR-Based Construction. Our first construction operates in $\mathbb{Z}_{N^{s+1}}$. (Larger s yield lossier functions. For our applications, $s = 2$ will be sufficient.) A tag consists of two Paillier/Damgård-Jurik encryptions $\mathsf{E}(x) \in \mathbb{Z}_{N^{s+1}}$. At the core of our construction is a variant of Waters signatures over $\mathbb{Z}_{N^{s+1}}$ whose security can be reduced to the problem of computing $\mathsf{E}(ab)$ from $\mathsf{E}(a)$ and $\mathsf{E}(b)$, i.e., of multiplying Paillier/DJ-encrypted messages. This "multiplication problem" may be interesting in its own right. If it is easy, then Paillier/DJ is *fully* homomorphic; if it is infeasible, then we can use it as a "poor man's CDH assumption" in the plaintext domain of Paillier/DJ.

We stress that our construction does not yield a signature scheme; verification of Waters signatures requires a pairing operation, to which we have no equivalent in $\mathbb{Z}_{N^{s+1}}$. However, we will be able to construct a matrix $M \in \mathbb{Z}_{N^{s+1}}^{3\times 3}$ out of a tag, such that the "decrypted matrix" $\widetilde{M} = \mathsf{D}(M) \in \mathbb{Z}_{N^s}^{3\times 3}$ has low rank iff the signature embedded in the tag is valid. Essentially, this observation uses products of plaintexts occurring in the determinant $\det(\widetilde{M})$ to implicitly implement a "pairing over $\mathbb{Z}_{N^{s+1}}$" and verify the signature. Similar techniques to encode arithmetic formulas in the determinant of a matrix have been used, e.g., by [25, 2] in the context of secure computation.

Our function evaluation is now a suitable multiplication of the encrypted matrix M with a plaintext vector $X \in \mathbb{Z}_{N^s}^3$, similar to the one from Peikert and Waters [30]. Concretely, on input X, our function outputs an encryption of the ordinary matrix-vector product $\widetilde{M} \cdot X$. If $\widetilde{M}$ is non-singular, then we can invert this function using the decryption key. If $\widetilde{M}$ has low rank, however, the function becomes lossy. This construction has compact tags and function images; both consist only of a (small) constant number of group elements, and only the public key has $\mathbf{O}(k)$ group elements, where k is the security parameter. Thus, our construction does not scale in the number N of lossy tags.

A Pairing-Based Construction. Our second uses a product group $\mathbb{G}_1 = \langle g_1 \rangle \times \langle h_1 \rangle$ that allows for a pairing. We will implement BB signatures in $\langle h_1 \rangle$, while we blind with elements from $\langle g_1 \rangle$. Consequently, our security proof requires both the Strong Diffie-Hellman assumption (SDH, [8]) in $\langle h_1 \rangle$ and a subgroup indistinguishability assumption.

Tags are essentially matrices $(W_{i,j})_{i,j}$ for $W_{i,j} \in \mathbb{G}_1 = \langle g_1 \rangle \times \langle h_1 \rangle$. Upon evaluation, this matrix is first suitably paired entry-wise to obtain a matrix $(M_{i,j})_{i,j}$ over $\mathbb{G}_T = \langle g_T \rangle \times \langle h_T \rangle$, the pairing's target group. This operation will ensure that (a) $M_{i,j}$ (for $i \neq j$) always lies in $\langle g_T \rangle$, and (b) $M_{i,i}$ lies in $\langle g_T \rangle$ iff the h_1-factor of $W_{i,i}$ constitutes a valid BB signature for the whole tag. With these ideas in mind, we revisit the original matrix-based LTF construction from [30] to obtain a function with trapdoors.

Unfortunately, using the matrix-based construction from [30] results in rather large tags (of size $\mathbf{O}(n^2)$ group elements for a function with domain $\{0,1\}^n$). On the bright side, a number of random self-reducibility properties allow for a security proof whose reduction quality does *not* degrade with the number N of lossy tags (i.e., challenge ciphertexts) around. Specifically, neither construction nor reduction scale in N.

Applications. Given the work of [23], a straightforward application of our results is the construction of an SO-CCA secure PKE scheme. (However, a slight tweak is required compared to the construction from [23] — see Section 5.3 for details.) Unlike the PKE schemes from [23], both of our ABM-LTFs give an SO-CCA construction that is independent of N, the number of challenge ciphertexts. Moreover, unlike the SO-CCA secure PKE scheme from [19], our DCR-based SO-CCA scheme has compact ciphertexts of $\mathbf{O}(1)$ group elements. Finally, unlike both [23] and [19], our pairing-based scheme has a reduction that does not depend on N and the number of decryption queries (see the full version for details).

As a side effect, our pairing-based scheme can be interpreted as a new kind of CCA secure PKE scheme with a security proof that is tight in the number of challenges and decryption queries. This solves an open problem of Bellare et al. [4], although the scheme should be seen as a (relatively inefficient) proof of concept rather than a practical system. Also, to be fair, we should mention that the SDH assumption we use in our pairing-based ABM-LTF already has a flavor of accommodating multiple challenges: an SDH instance contains polynomially many group elements.

Open Problems. An interesting open problem is to find different, and in particular efficient *and* tightly secure ABM-LTFs under reasonable assumptions. This would imply efficient *and* tightly (SO-)CCA-secure encryption schemes. (With our constructions, one basically has to choose between efficiency and a tight reduction.) Also, our pairing-based PKE scheme achieves only indistinguishability-based, but not (in any obvious way) simulation-based SO security [6]. (To achieve simulation-based SO security, a simulator must essentially be able to efficiently explain lossy ciphertexts as encryptions of any given message, see [6, 19].) However, as we demonstrate in case of our DCR-based scheme, in some cases ABM-LTFs can be equipped with an additional "explainability" property that leads to simulation-based SO security (see the full version for details). It would be interesting to find other applications of ABM-LTFs. One reviewer suggested that ABM-LTFs can be used instead of ABO-LTFs in the commitment scheme from Nishimaki et al. [27], with the goal of attaining *reusable* commitments.

Organization. After fixing some notation in Section 2, we proceed to our definition of ABM-LTFs in Section 3. We define and analyze our DCR-based ABM-LTF in Section 4. We then show how ABM-LTFs imply CCA-secure (indistinguishability-based) selective-opening security in Section 5. Due to lack of space, we postpone a detailed description and analysis of our pairing-based ABM-LTF to the full version.

2   Preliminaries

Notation. For $n \in \mathbb{N}$, let $[n] := \{1, \ldots, n\}$. Throughout the paper, $k \in \mathbb{N}$ denotes the security parameter. For a finite set S, we denote by $s \leftarrow S$ the process of sampling s uniformly from $\mathcal{S}$. For a probabilistic algorithm A, we denote $y \leftarrow A(x; R)$ the process of running A on input x and with randomness R, and assigning y the result. We let $\mathcal{R}_A$ denote the randomness space of A; we require $\mathcal{R}_A$ to be of the form $\mathcal{R}_A = \{0,1\}^r$. We write $y \leftarrow A(x)$ for $y \leftarrow A(x; R)$ with uniformly chosen $R \in \mathcal{R}_A$, and we write $y_1, \ldots, y_m \leftarrow A(x)$ for $y_1 \leftarrow A(x), \ldots, y_m \leftarrow A(x)$ with fresh randomness in each execution. If A's running time is polynomial in k, then A is called probabilistic polynomial-time (PPT). The statistical distance of two random variables X and Y over some countable domain S is defined as $\mathsf{SD}(X\,;\,Y) := \frac{1}{2} \sum_{s \in S} |\Pr[X = s] - \Pr[Y = s]|$.

Chameleon Hashing. A chameleon hash function (CHF, see [26]) is collision-resistant when only the public key of the function is known. However, this collision-resistance can be broken (in a very strong sense) with a suitable trapdoor. We will assume an input domain of $\{0,1\}^*$. We do not lose (much) on generality here, since one can always first apply a collision-resistant hash function on the input to get a fixed-size input.

Definition 1 (Chameleon Hash Function). *A chameleon hash function* CH *consists of the following PPT algorithms:*
Key Generation. CH.Gen(1^k) *outputs a key* pk_{CH} *along with a trapdoor* td_{CH}.
Evaluation. CH.Eval($pk_{\mathsf{CH}}, X; R_{\mathsf{CH}}$) *maps an input* $X \in \{0,1\}^*$ *to an image* Y. *By* R_{CH}, *we denote the randomness used in the process. We require that if* R_{CH} *is uniformly distributed, then so is* Y *(over its respective domain).*
Equivocation. CH.Equiv($td_{\mathsf{CH}}, X, R_{\mathsf{CH}}, X'$) *outputs randomness* R'_{CH} *with*

$$\mathsf{CH.Eval}(pk_{\mathsf{CH}}, X; R_{\mathsf{CH}}) = \mathsf{CH.Eval}(pk_{\mathsf{CH}}, X'; R'_{\mathsf{CH}}) \tag{1}$$

for the corresponding key pk_{CH}. *We require that for any* X, X', *if* R_{CH} *is uniformly distributed, then so is* R'_{CH}.
We require that CH *is **collision-resistant** in the sense that given* pk_{CH}, *it is infeasible to find* $X, R_{\mathsf{CH}}, X', R'_{\mathsf{CH}}$ *with* $X \neq X'$ *that meet (1). Formally, for every PPT* B,

$$\mathsf{Adv}^{\mathsf{cr}}_{\mathsf{CH}, B}(k) := \Pr\left[X \neq X' \text{ and (1) holds} \mid (X, R_{\mathsf{CH}}, X', R'_{\mathsf{CH}}) \leftarrow B(1^k, pk_{\mathsf{CH}}) \right]$$

is negligible, where $(pk_{\mathsf{CH}}, td_{\mathsf{CH}}) \leftarrow \mathsf{CH.Gen}(1^k)$.

Lossy Trapdoor Functions. Lossy trapdoor functions (see [30]) are a variant of trapdoor one-way functions. They may be operated in an "injective mode" (which allows to invert the function) and a "lossy mode" in which the function is non-injective. For simplicity, we restrict to an input domain $\{0,1\}^n$ for polynomially bounded $n = n(k) > 0$.

Definition 2 (Lossy Trapdoor Function). *A lossy trapdoor function (LTF)* LTF *with domain* Dom *consists of the following algorithms:*
Key generation. LTF.IGen(1^k) *yields an evaluation key* ek *and an inversion key* ik.
Evaluation. LTF.Eval(ek, X) *(with* $X \in$ Dom*) yields an image* Y. *Write* $Y = f_{ek}(X)$.
Inversion. LTF.Invert(ik, Y) *outputs a preimage* X. *Write* $X = f_{ik}^{-1}(Y)$.
Lossy key generation. LTF.LGen(1^k) *outputs an evaluation key* ek'.
We require the following:
Correctness. *For all* $(ek, ik) \leftarrow$ LTF.IGen(1^k), $X \in$ Dom, *it is* $f_{ik}^{-1}(f_{ek}(X)) = X$.

Indistinguishability. *The first output of* LTF.IGen(1^k) *is indistinguishable from the output of* LTF.LGen(1^k), *i.e.,*

$$\text{Adv}_{\text{LTF},A}^{\text{ind}}(k) := \Pr\left[A(1^k, ek) = 1\right] - \Pr\left[A(1^k, ek') = 1\right]$$

is negligible for all PPT A, for $(ek, ik) \leftarrow$ LTF.IGen(1^k), $ek' \leftarrow$ LTF.LGen(1^k).
Lossiness. *We say that* LTF *is ℓ-lossy if for all possible* $ek' \leftarrow$ LTF.LGen(1^k), *the image set* $f_{ek'}(\text{Dom})$ *is of size at most* $|\text{Dom}|/2^\ell$.

3 Definition of ABM-LTFs

We are now ready to define ABM-LTFs. As already discussed in Section 1, ABM-LTFs generalize ABO-LTFs and ABN-LTFs in the sense that there is a superpolynomially large pool of lossy tags from which we can sample. We require that even given oracle access to such a sampler of lossy tags, it is not feasible to produce a (fresh) non-injective tag. Furthermore, it should be hard to distinguish lossy from injective tags.

Definition 3 (ABM-LTF). *An* all-but-many lossy trapdoor function *(ABM-LTF) ABM with domain* Dom *consists of the following PPT algorithms:*
Key generation. ABM.Gen(1^k) *yields an evaluation key* ek, *an inversion key* ik, *and a tag key* tk. *The evaluation key* ek *defines a set* $\mathcal{T} = \mathcal{T}_p \times \{0,1\}^*$ *that contains the disjoint sets of* lossy tags $\mathcal{T}_{\text{loss}} \subseteq \mathcal{T}$ *and* injective tags $\mathcal{T}_{\text{inj}} \subseteq \mathcal{T}$. *Tags are of the form* $t = (t_p, t_a)$, *where* $t_p \in \mathcal{T}_p$ *is the* core part *of the tag, and* $t_a \in \{0,1\}^*$ *is the* auxiliary part *of the tag.*
Evaluation. ABM.Eval(ek, t, X) *(for* $t \in \mathcal{T}, X \in$ Dom*) produces* $Y =: f_{ek,t}(X)$.
Inversion. ABM.Invert(ik, t, Y) *(with* $t \in \mathcal{T}_{\text{inj}}$*) outputs a preimage* $X =: f_{ik,t}^{-1}(Y)$.
Lossy tag generation. ABM.LTag(tk, t_a) *takes as input an auxiliary part* $t_a \in \{0,1\}^*$ *and outputs a core tag* t_p *such that* $t = (t_p, t_a)$ *is lossy.*
We require the following:
Correctness. *For all possible* $(ek, ik, tk) \leftarrow$ ABM.Gen(1^k), $t \in \mathcal{T}_{\text{inj}}$, *and* $X \in$ Dom, *it is always* $f_{ik,t}^{-1}(f_{ek,t}(X)) = X$.
Lossiness. *We say that* ABM *is ℓ-lossy if for all possible* $(ek, ik, tk) \leftarrow$ ABM.Gen(1^k), *and all lossy tags* $t \in \mathcal{T}_{\text{loss}}$, *the image set* $f_{ek,t}(\text{Dom})$ *is of size at most* $|\text{Dom}|/2^\ell$.
Indistinguishability. *Even multiple lossy tags are indistinguishable from random tags:*

$$\text{Adv}_{\text{ABM},A}^{\text{ind}}(k) := \Pr\left[A(1^k, ek)^{\text{ABM.LTag}(tk, \cdot)} = 1\right] - \Pr\left[A(1^k, ek)^{\mathcal{O}_{\mathcal{T}_p}(\cdot)} = 1\right]$$

is negligible for all PPT A, where $(ek, ik, tk) \leftarrow$ ABM.Gen(1^k), *and* $\mathcal{O}_{\mathcal{T}}(\cdot)$ *ignores its input and returns a uniform and independent core tag* $t_p \leftarrow \mathcal{T}_p$.
Evasiveness. *Non-injective tags are hard to find, even given multiple lossy tags:*

$$\text{Adv}_{\text{ABM},A}^{\text{eva}}(k) := \Pr\left[A(1^k, ek)^{\text{ABM.LTag}(tk, \cdot)} \in \mathcal{T} \setminus \mathcal{T}_{\text{inj}}\right]$$

is negligible with $(ek, ik, tk) \leftarrow$ ABM.Gen(1^k), *and for any PPT algorithm A that never outputs tags obtained through oracle queries (i.e., A never outputs tags* $t = (t_p, t_a)$, *where* t_p *has been obtained by an oracle query* t_a).

On Our Tagging Mechanism. Our tagging mechanism is different from the mechanism from ABO-, resp. ABN-LTFs. In particular, our tag selection involves an auxiliary and a core tag part; lossy tags can be produced for arbitrary auxiliary tags. (Conceptually, this resembles the two-stage tag selection process from Abe et al. [1] in the context of hybrid encryption.) On the other hand, ABO- and ABN-LTFs simply have fully arbitrary (user-selected) bitstrings as tags.

The reason for our more complicated tagging mechanism is that during a security proof, tags are usually context-dependent and not simply random. For instance, a common trick in the public-key encryption context is the following: upon encryption, choose a one-time signature keypair (v, s), set the tag to the verification key v, and then finally sign the whole ciphertext using the signing key s. This trick has been used numerous times (e.g., [17, 12, 30, 31]) and ensures that a tag cannot be re-used by an adversary in a decryption query. (To re-use that tag, an adversary would essentially have to forge a signature under v.)

However, in our constructions, in particular lossy tags cannot be freely chosen. (This is different from ABO- and ABN-LTFs and stems from the fact that there are superpolynomially many lossy tags.) But as outlined, during a security proof, we would like to embed auxiliary information in a tag, while being able to force the tag to be lossy. We thus divide the tag into an auxiliary part (which can be used to embed, e.g., a verification key for a one-time signature), and a core part (which will be used to enforce lossiness).

4 A DCR-Based ABM-LTF

We now construct an ABM-LTF ABMD in rings $\mathbb{Z}_{N^{s+1}}$ for composite N. Domain and codomain of our function will be $\mathbb{Z}_{N^s}^3$ and $(\mathbb{Z}_{N^{s+1}}^*)^3$, respectively. One should have in mind a value of $s \geq 2$ here, since we will prove that ABMD is $((s-1)\log_2(N))$-lossy.

4.1 Setting and Assumptions

In the following, let $N = PQ$ for primes P and Q, and fix a positive integer s. Write $\varphi(N) := (P-1)(Q-1)$. We will silently assume that P and Q are chosen from a distribution that depends on the security parameter. Unless indicated otherwise, all computations will take place in $\mathbb{Z}_{N^{s+1}}$, i.e., modulo N^{s+1}. It will be useful to establish the notation $\mathfrak{h} := 1 + N \in \mathbb{Z}_{N^{s+1}}$. We also define algorithms E and D by $\mathsf{E}(x) = r^{N^s}\mathfrak{h}^x$ for $x \in \mathbb{Z}_{N^s}$ and a uniformly and independently chosen $r \in \mathbb{Z}_{N^{s+1}}^*$, and $\mathsf{D}(c) = ((c^{\varphi(N)})^{1/\varphi(N) \bmod N^s} - 1)/N \in \mathbb{Z}_{N^s}$ for $c \in \mathbb{Z}_{N^{s+1}}$. That is, E and D are Paillier/Damgård-Jurik encryption and decryption operations as in [28, 14], so that $\mathsf{D}(r^{N^s}\mathfrak{h}^x) = x$ and $\mathsf{D}(\mathsf{E}(x)) = x$. Moreover, D can be efficiently computed using the factorization of N. We will also apply D to vectors or matrices over $\mathbb{Z}_{N^{s+1}}$, by which we mean component-wise application. We make the following assumptions:

Assumption 1. *The s-Decisional Composite Residuosity (short: s-DCR) assumption holds iff*

$$\mathsf{Adv}_D^{\text{s-dcr}}(k) := \Pr\left[D(1^k, N, r^{N^s}) = 1\right] - \Pr\left[D(1^k, N, r^{N^s}\mathfrak{h}) = 1\right]$$

is negligible for all PPT D, where $r \leftarrow \mathbb{Z}_{N^{s+1}}^*$ is chosen uniformly.

Assumption 1 is rather common and equivalent to the semantic security of the Paillier [28] and Damgård-Jurik (DJ) [14] encryption schemes. In fact, it turns out that all s-DCR assumptions are (tightly) equivalent to 1-DCR [14]. Nonetheless, we make s explicit here to allow for a simpler exposition. Also note that Assumption 1 supports a form of random self-reducibility. Namely, given one challenge element $c \in \mathbb{Z}^*_{N^{s+1}}$, it is possible to generate many fresh challenges c_i with the same decryption $D(c_i) = D(c)$ by re-randomizing the r^{N^s} part.

Assumption 2. *The No-Multiplication (short: No-Mult) assumption holds iff*

$$\mathsf{Adv}^{\mathsf{mult}}_A(k) := \Pr\left[A(1^k, N, c_1, c_2) = c_* \in \mathbb{Z}^*_{N^2} \text{ for } D(c_*) = D(c_1) \cdot D(c_2) \bmod N^s\right]$$

is negligible for all PPT A, where $c_1, c_2 \leftarrow \mathbb{Z}^*_{N^2}$ are chosen uniformly.

The No-Mult assumption stipulates that it is infeasible to *multiply* Paillier-encrypted messages. If No-Mult (along with s-DCR and a somewhat annoying technical assumption explained below) hold, then our upcoming construction will be secure. But if the No-Mult problem is easy, then Paillier encryption is fully homomorphic.[2]

The following technical lemma will be useful later on, because it shows how to lift $\mathbb{Z}_{N^2}$-encryptions to $\mathbb{Z}_{N^{s+1}}$-encryptions.

Lemma 1 (Lifting, Implicit in [14]). *Let $s \geq 1$ and $\tau : \mathbb{Z}_{N^2} \to \mathbb{Z}_{N^{s+1}}$ be the canonical embedding with $\tau(c \bmod N^2) = c \bmod N^{s+1}$ for $c \in \mathbb{Z}_{N^2}$ interpreted as an integer from $\{0, \ldots, N^2 - 1\}$. Then, for any $c \in \mathbb{Z}^*_{N^2}$, and $X := D(\tau(c)) \in \mathbb{Z}_{N^s}$ and $x := D(c) \in \mathbb{Z}_N$, we have $X = x \bmod N$.*

Proof. Consider the canonical homomorphism $\pi : \mathbb{Z}_{N^{s+1}} \to \mathbb{Z}_{N^2}$. Write $\mathbb{Z}^*_{N^{s+1}} = \langle \mathfrak{g}_s \rangle \times \langle \mathfrak{h}_s \rangle$ for some $\mathfrak{g}_s \in \mathbb{Z}^*_{N^{s+1}}$ of order $\varphi(N)$ and $\mathfrak{h}_s := 1 + N \bmod N^{s+1}$. We have $\pi(\langle \mathfrak{g}_s \rangle) = \langle \mathfrak{g}_1 \rangle$ and $\pi(\mathfrak{h}_s^x) = \mathfrak{h}_1^{x \bmod N}$. Since $\pi \circ \hat{\pi} = \mathrm{id}_{\mathbb{Z}_{N^2}}$, this gives $\hat{\pi}(\mathfrak{g}_1^u \mathfrak{h}_1^x) = \mathfrak{g}_s^{u'} \mathfrak{h}_s^{x + x'N}$ for suitable u', x'.

Unfortunately, we need another assumption to exclude certain corner cases:

Assumption 3. *We require that the following function is negligible for all PPT A:*

$$\mathsf{Adv}^{\mathsf{noninv}}_A(k) := \Pr\left[A(1^k, N) = c \in \mathbb{Z}_{N^2} \text{ such that } 1 < \gcd(D(c), N) < N\right].$$

Intuitively, Assumption 3 stipulates that it is infeasible to generate Paillier encryptions of "funny messages." Note that actually *knowing* any such message allows to factor N.

4.2 Our Construction

Overall Idea. The first idea in our construction will be to use the No-Mult assumption as a "poor man's CDH assumption" in order to implement Waters signatures [33] over $\mathbb{Z}_{N^{s+1}}$. Recall that the verification of Waters signatures requires a pairing operation, which corresponds to the multiplication of two Paillier/DJ-encrypted messages

---

[2] Of course, there is a third, less enjoyable possibility. It is always conceivable that an algorithm breaks No-Mult with low but non-negligible probability. Such an algorithm may not be useful for constructive purposes. Besides, if either s-DCR or the annoying technical assumption do not hold, then our construction may not be secure.

in our setting. We do not have such a multiplication operation available; however, for our purposes, signatures will never actually have to be verified, so this will not pose a problem. We note that the original Waters signatures from [33] are re-randomizable and thus not *strongly* unforgeable. To achieve the evasiveness property from Definition 3, we will thus combine Waters signatures with a chameleon hash function, much like Boneh et al. [9] did to make Waters signatures strongly unforgeable.

Secondly, we will construct 3×3-matrices $M = (M_{i,j})_{i,j}$ over $\mathbb{Z}_{N^{s+1}}$, in which we carefully embed our variant of Waters signatures. Valid signatures will correspond to singular "plaintext matrices" $\widetilde{M} := (\mathsf{D}(M_{i,j}))_{i,j}$; invalid signatures correspond to full-rank matrices $\widetilde{M}$. We will define our ABM-LTF f as a suitable matrix-vector multiplication of M with an input vector $X \in \mathbb{Z}_{N^s}^3$. For a suitable choice of s, the resulting f will be lossy if $\det(\widetilde{M}) = 0$.

Key Generation. ABM.Gen(1^k) first chooses $N = PQ$, and a key pk_{CH} along with trapdoor td_{CH} for a chameleon hash function CH. Finally, ABM.Gen chooses $a, b \leftarrow \mathbb{Z}_{N^s}$, as well as $k + 1$ values $h_i \leftarrow \mathbb{Z}_{N^s}$ for $0 \leq i \leq k$, and sets

$$A \leftarrow \mathsf{E}(a) \qquad\qquad B \leftarrow \mathsf{E}(b) \qquad H_i \leftarrow \mathsf{E}(h_i) \quad (\text{for } 0 \leq i \leq k)$$

$$ek = (N, A, B, (H_i)_{i=0}^k, pk_{\mathsf{CH}}) \quad ik = (ek, P, Q) \quad tk = (ek, a, b, (h_i)_{i=0}^k, td_{\mathsf{CH}}).$$

Tags. Recall that a tag $t = (t_\mathsf{p}, t_\mathsf{a})$ consists of a core part t_p and an auxiliary part $t_\mathsf{a} \in \{0,1\}^*$. Core parts are of the form $t_\mathsf{p} = (R, Z, R_{\mathsf{CH}})$ with $R, Z \in \mathbb{Z}_{N^{s+1}}^*$ and randomness R_{CH} for CH. (Thus, random core parts are simply uniform values $R, Z \in \mathbb{Z}_{N^{s+1}}^*$ and uniform CH-randomness.) With t, we associate the chameleon hash value $T := \mathsf{CH.Eval}(pk_{\mathsf{CH}}, (R, Z, t_\mathsf{a}))$, and a group hash value $H := H_0 \prod_{i \in T} H_i$, where $i \in T$ means that the i-th bit of T is 1. Let $h := \mathsf{D}(H) = h_0 + \sum_{i \in T} h_i$. Also, we associate with t the matrices

$$M = \begin{pmatrix} Z & A & R \\ B & \mathfrak{h} & 1 \\ H & 1 & \mathfrak{h} \end{pmatrix} \in \mathbb{Z}_{N^{s+1}}^{3 \times 3} \qquad\qquad \widetilde{M} = \begin{pmatrix} z & a & r \\ b & 1 & 0 \\ h & 0 & 1 \end{pmatrix} \in \mathbb{Z}_{N^s}^{3 \times 3}, \qquad (2)$$

where $\widetilde{M} = \mathsf{D}(M)$ is the component-wise decryption of M, and $r = \mathsf{D}(R)$ and $z = \mathsf{D}(Z)$. It will be useful to note that $\det(\widetilde{M}) = z - (ab + rh)$. We will call t *lossy* if $\det(\widetilde{M}) = 0$, i.e., if $z = ab + rh$; we say that t is *injective* if $\widetilde{M}$ is invertible.

Lossy Tag Generation. ABM.LTag(tk, t_a), given $tk = ((N, A, B, (H_i)_i), a, b, (h_i)_{i=0}^k, td_{\mathsf{CH}})$ and an auxiliary tag part $t_\mathsf{a} \in \{0,1\}^*$, picks an image T of CH that can later be explained (using td_{CH}) as the image of an arbitrary preimage (R, Z, t_a). Let $h := h_0 + \sum_{i \in T} h_i$ and $R \leftarrow \mathsf{E}(r)$ for uniform $r \leftarrow \mathbb{Z}_{N^s}$, and set $Z \leftarrow \mathsf{E}(z)$ for $z = ab + rh$. Finally, let R_{CH} be CH-randomness for which $T = \mathsf{CH.Eval}(pk_{\mathsf{CH}}, (R, Z, t_\mathsf{a}))$. Obviously, this yields uniformly distributed lossy tag parts (R, Z, R_{CH}).

Evaluation. ABM.Eval(ek, t, X), for $ek = (N, A, B, (H_i)_i, pk_{\mathsf{CH}})$, $t = ((R, Z, R_{\mathsf{CH}}), t_\mathsf{a})$, and a preimage $X = (X_i)_{i=1}^3 \in \mathbb{Z}_{N^s}^3$, first computes the matrix $M = (M_{i,j})_{i,j}$ as in (2). Then, ABM.Eval computes and outputs

$$Y := M \circ X := \left(\prod_{j=1}^{3} M_{i,j}^{X_j} \right)_{i=1}^{3} .$$

Note that the decryption $\mathsf{D}(Y)$ is simply the ordinary matrix-vector product $\mathsf{D}(M) \cdot X$.

Inversion and Correctness. ABM.Invert(ik, t, Y), given an inversion key ik, a tag t, and an image $Y = (Y_i)_{i=1}^{3}$, determines $X = (Y_i)_{i=1}^{3}$ as follows. First, ABM.Invert computes the matrices M and $\widetilde{M} = \mathsf{D}(M)$ as in (2), using P, Q. For correctness, we can assume that the tag t is injective, so $\widetilde{M}$ is invertible; let $\widetilde{M}^{-1}$ be its inverse. Since $\mathsf{D}(Y) = \widetilde{M} \cdot X$, ABM.Invert can retrieve X as $\widetilde{M}^{-1} \cdot \mathsf{D}(Y) = \widetilde{M}^{-1} \cdot \widetilde{M} \cdot X$.

4.3 Security Analysis

Theorem 1 (Security of ABMD**).** *Assume that Assumption 1, Assumption 2, and Assumption 3 hold, that* CH *is a chameleon hash function, and that* $s \geq 2$. *Then the algorithms described in Section 4.2 form an ABM-LTF* ABMD *as per Definition 3.*

We have yet to prove lossiness, indistinguishability, and evasiveness.

Lossiness. Our proof of lossiness loosely follows Peikert and Waters [30]:

Lemma 2 (Lossiness of ABMD**).** ABMD *is* $((s-1)\log_2(N))$*-lossy.*

Proof. Assume an evaluation key $ek = (N, A, B, (H_i)_i, pk_{\mathsf{CH}})$, and a lossy tag t, so that the matrix $\widetilde{M}$ from (2) is of rank ≤ 2. Hence, any fixed decrypted image

$$\mathsf{D}(f_{ek,t}(X)) = \mathsf{D}(M \circ X) = \widetilde{M} \cdot X$$

leaves at least one inner product $\langle C, X \rangle \in \mathbb{Z}_{N^s}$ (for $C \in \mathbb{Z}_{N^s}^3$ that only depends on $\widetilde{M}$) completely undetermined. The additional information contained in the encryption randomness of an image $Y = f_{ek,t}(X)$ fixes the components of X and thus $\langle C, X \rangle$ only modulo $\varphi(N) < N$. Thus, for any given image Y, there are at least $\lfloor N^s/\varphi(N) \rfloor \geq N^{s-1}$ possible values for $\langle C, X \rangle$ and thus possible preimages. The claim follows.

Indistinguishability. Observe that lossy tags can be produced without knowledge of the factorization of N. Hence, even while producing lossy tags, we can use the indistinguishability of Paillier/DJ encryptions $\mathsf{E}(x)$. This allows to substitute the encryptions $R = \mathsf{E}(r), Z = \mathsf{E}(z)$ in lossy tags by independently uniform encryptions. This step also makes the CH-randomness independently uniform, and we end up with random tags. We omit the straightforward formal proof and state:

Lemma 3 (Indistinguishability of ABMD**).** *Given the assumptions from Theorem 1,* ABMD *is indistinguishable. Concretely, for any PPT adversary* A, *there exists an* s-DCR *distinguisher* D *of roughly the same complexity as* A, *such that*

$$\mathsf{Adv}_{\mathsf{ABMD},A}^{\mathsf{ind}}(k) = \mathsf{Adv}_{D}^{\text{s-dcr}}(k). \tag{3}$$

The tightness of the reduction in (3) stems from the random self-reducibility of s-DCR.

Evasiveness. It remains to prove evasiveness.

Lemma 4 (Evasiveness of ABMD**).** *Given the assumptions from Theorem 1, ABMD is evasive. Concretely, for any PPT adversary A that makes at most $Q = Q(k)$ oracle queries, there exist adversaries B, D, and F of roughly the same complexity as A, with*

$$\mathsf{Adv}_{\mathsf{ABMD},A}^{\mathsf{eva}}(k) \leq \mathbf{O}(kQ(k)) \cdot \mathsf{Adv}_F^{\mathsf{mult}}(k) + \mathsf{Adv}_E^{\mathsf{noninv}}(k) + \left|\mathsf{Adv}_D^{\mathsf{s\text{-}dcr}}(k)\right| + \mathsf{Adv}_{\mathsf{CH},B}^{\mathsf{cr}}(k).$$

At its core, the proof of Lemma 4 adapts the security proof of Waters signatures to $\mathbb{Z}_{N^{s+1}}$. That is, we will create a setup in which we can prepare $Q(k)$ lossy tags (which correspond to valid signatures), and the tag the adversary finally outputs will be interpreted as a forged signature. Crucial to this argument will be a suitable setup of the group hash function $(H_i)_{i=0}^k$. Depending on the (group) hash value, we will either be able to create a lossy tag with that hash, or use any lossy tag with that hash to solve a underlying No-Mult challenge. With a suitable setup, we can hope that with probability $\mathbf{O}(1/(kQ(k)))$, $Q(k)$ lossy tags can be created, and the adversary's output can be used to solve an No-Mult challenge. The proof of Lemma 4 is somewhat complicated by the fact that in order to use the collision-resistance of the employed CHF, we have to first work our way towards a setting in which the CHF trapdoor is not used. This leads to a somewhat tedious "deferred analysis" (see [21]) and the s-DCR term in the lemma.

Proof. We turn to the full proof of Lemma 4. Fix an adversary A. We proceed in games. In **Game** 1, $A(ek)$ interacts with an ABM.LTag$(tk, \cdot)$ oracle that produces core tag parts for lossy tags $t_\mathsf{p} = ((R, Z, R_{\mathsf{CH}}), t_\mathsf{a})$ that satisfy $z = ab + rh$ for $r = \mathsf{D}(R)$, $z = \mathsf{D}(Z)$, and $h = \mathsf{D}(H)$ with $H = H_0 \prod_{i \in T} H_i$ and $T = \mathsf{CH.Eval}(pk_{\mathsf{CH}}, (R, Z, t_\mathsf{a}))$. Without loss of generality, we assume that A makes exactly Q oracle queries, where $Q = Q(k)$ is a suitable polynomial. Let bad_i denote the event that the output of A in Game i is a lossy tag, i.e., lies in $\mathcal{T}_{\mathsf{loss}}$. By definition,

$$\Pr\left[\mathsf{bad}_1\right] = \Pr\left[A^{\mathsf{ABM.LTag}(tk, \cdot)}(ek) \in \mathcal{T}_{\mathsf{loss}}\right], \tag{4}$$

where the keys ek and tk are generated via $(ek, ik, tk) \leftarrow \mathsf{ABM.Gen}(1^k)$.

Getting Rid of (Chameleon) Hash Collisions. To describe **Game** 2, let $\mathsf{bad}_{\mathsf{hash}}$ be the event that A finally outputs a tag $t = ((R, Z, R_{\mathsf{CH}}), t_\mathsf{a})$ with a CHF hash $T = \mathsf{CH.Eval}((R, Z, t_\mathsf{a}); R_{\mathsf{CH}})$ that has already appeared as the CHF hash of an ABM.LTag output (with the corresponding auxiliary tag part input). Now Game 2 is the same as Game 1, except that we abort (and do not raise event bad_2) if $\mathsf{bad}_{\mathsf{hash}}$ occurs. Obviously,

$$\Pr\left[\mathsf{bad}_1\right] - \Pr\left[\mathsf{bad}_2\right] \leq \Pr\left[\mathsf{bad}_{\mathsf{hash}}\right]. \tag{5}$$

It would seem intuitive to try to use CH's collision resistance to bound $\Pr\left[\mathsf{bad}_{\mathsf{hash}}\right]$. Unfortunately, we cannot rely on CH's collision resistance in Game 2 yet, since we use CH's trapdoor in the process of generating lossy tags. So instead, we use a technique called "deferred analysis" [21] to bound $\Pr\left[\mathsf{bad}_{\mathsf{hash}}\right]$. The idea is to forget about the storyline of our evasiveness proof for the moment and develop Game 2 further up to a point at which we can use CH's collision resistance to bound $\Pr\left[\mathsf{bad}_{\mathsf{hash}}\right]$.

This part of the proof largely follows the argument from Lemma 3. Concretely, we can substitute the lossy core tag parts output by ABM.LTag by uniformly random core tag parts. At this point, CH's trapdoor is no longer required to implement the oracle A interacts with. Hence we can apply CH's collision resistance to bound $\Pr\left[\mathsf{bad_{hash}}\right]$ in this modified game. This also implies a bound on $\Pr\left[\mathsf{bad_{hash}}\right]$ in Game 2: since the occurrence of $\mathsf{bad_{hash}}$ is obvious from the interaction between A and the experiment, $\Pr\left[\mathsf{bad_{hash}}\right]$ must be preserved across these transformations. We omit the details, and state the result of this deferred analysis:

$$\Pr\left[\mathsf{bad_{hash}}\right] \leq \left|\mathsf{Adv}_D^{\text{s-dcr}}(k)\right| + \mathsf{Adv}_{\mathsf{CH},B}^{\text{cr}}(k) \tag{6}$$

for suitable adversaries D, E, and B. This ends the deferred analysis step, and we are back on track in our evasiveness proof.

Preparing the Setup for Our Reduction. In **Game** 3, we set up the group hash function given by $(H_i)_{i=0}^{k}$ differently. Namely, for $0 \leq i \leq k$, we choose independent $\gamma_i \leftarrow \mathbb{Z}_{N^s}$, and set

$$H_i := A^{\alpha_i}\mathsf{E}(\gamma_i), \qquad \text{so that} \quad h_i := \mathsf{D}(H_i) = \alpha_i a + \gamma_i \bmod N^s \tag{7}$$

for independent $\alpha_i \in \mathbb{Z}$ yet to be determined. Note that this yields an identical distribution of the H_i no matter how concretely we choose the α_i. For convenience, we write $\alpha = \alpha_0 + \sum_{i \in T} \alpha_i$ and $\gamma = \gamma_0 + \sum_{i \in T} \gamma_i$ for a given tag t with associated CH-image T. This in particular implies $h := \mathsf{D}(H) = \alpha a + \gamma$ for the corresponding group hash $H = H_0 \prod_{i \in T} H_i$. Our changes in Game 3 are purely conceptual, and so

$$\Pr\left[\mathsf{bad_3}\right] = \Pr\left[\mathsf{bad_2}\right]. \tag{8}$$

To describe **Game** 4, let $t^{(i)}$ denote the i-th lossy core tag part output by ABM.LTag (including the corresponding auxiliary part $t_{\mathsf{a}}^{(i)}$), and let t^* be the tag finally output by A. Similarly, we denote with $T^{(i)}$, α^*, etc. the intermediate values for the tags output by ABM.LTag and A. Now let $\mathsf{good_{setup}}$ be the event that $\gcd(\alpha^{(i)}, N) = 1$ for all i, and that $\alpha^* = 0$. In Game 4, we abort (and do not raise event $\mathsf{bad_4}$) if $\neg\mathsf{good_{setup}}$ occurs. (In other words, we only continue if each $H^{(i)}$ has an invertible A-component, and if H^* has no A-component.)

Waters [33] implicitly shows that for a suitable distribution of α_i, the probability $\Pr\left[\mathsf{good_{setup}}\right]$ can be kept reasonably high:

Lemma 5 (Waters [33], Claim 2, Adapted to Our Setting). *In the situation of Game 4, there exist efficiently computable distributions α_i, such that for every possible view view that A could experience in Game 4, we have*

$$\Pr\left[\mathsf{good_{setup}} \mid \text{view}\right] \geq \mathbf{O}(1/(kQ(k))). \tag{9}$$

This directly implies

$$\Pr\left[\mathsf{bad_4}\right] \geq \Pr\left[\mathsf{good_{setup}}\right] \cdot \Pr\left[\mathsf{bad_3}\right]. \tag{10}$$

An Annoying Corner Case. Let $\Pr[\mathsf{bad_{tag}}]$ be the event that A outputs a tag t^* for which $\det(\widetilde{M}^*) = z^* - (ab + r^*h^*)$ is neither invertible nor 0 modulo N. This in particular means that t^* is neither injective nor lossy. A straightforward reduction to Assumption 3 shows that

$$\Pr[\mathsf{bad_{tag}}] \leq \mathsf{Adv}_E^{\mathsf{noninv}}(k) \tag{11}$$

for an adversary E that simulates Game 4 and outputs $Z^*/(\mathfrak{h}^{ab} \cdot (R^*)^{h^*}) \bmod N^2$.

The Final Reduction. We now claim that

$$\Pr[\mathsf{bad_4}] \leq \mathsf{Adv}_F^{\mathsf{mult}}(k) + \Pr[\mathsf{bad_{tag}}] \tag{12}$$

for the following adversary F on the No-Mult assumption. Our argument follows in the footsteps of the security proof of Waters' signature scheme [33]. Our No-Mult adversary F obtains as input $c_1, c_2 \in \mathbb{Z}_{N^2}$, and is supposed to output $c_* \in \mathbb{Z}_{N^2}$ with $\mathsf{D}(c_1) \cdot \mathsf{D}(c_2) = \mathsf{D}(c_*) \in \mathbb{Z}_N$. In order to do so, F simulates Game 4. F incorporates its own challenge as $A := \tau(c_1)\mathsf{E}(a'N)$ and $B := \tau(c_2)\mathsf{E}(b'N)$ for the embedding τ from Lemma 1 and uniform $a', b' \in \mathbb{Z}_{N^{s-1}}$. This gives uniformly distributed $A, B \in \mathbb{Z}_{N^{s+1}}^*$. Furthermore, by Lemma 1, we have $a = \mathsf{D}(c_1) \bmod N$ and $b = \mathsf{D}(c_2) \bmod N$ for $a := \mathsf{D}(A)$ and $b := \mathsf{D}(B)$. Note that F can still compute all H_i and thus ek efficiently using (7).

We now describe how F constructs lossy tags, as required to implement oracle $\mathcal{T}_{\mathsf{loss}}$ of Game 4. Since the CH trapdoor td_{CH} is under F's control, we can assume a given CH-value T to which we can later map our tag $((R, Z), t_a)$. By our changes from Game 4, we can also assume that the corresponding $\alpha = \alpha_0 + \sum_{i \in T} \alpha_i$ is invertible modulo N^s and known. We pick $\delta \leftarrow \mathbb{Z}_{N^s}$, and set

$$R := B^{-1/\alpha \bmod N^s}\mathsf{E}(\delta) \qquad\qquad Z := A^{\alpha\delta}B^{\gamma/\alpha}\mathsf{E}(\gamma\delta).$$

With the corresponding CH-randomness R_{CH}, this yields perfectly distributed lossy tags satisfying

$$\mathsf{D}(A)\cdot\mathsf{D}(B)+\mathsf{D}(R)\cdot\mathsf{D}(H) = ab+(-b/\alpha+\delta)(\alpha a+\gamma) = \alpha\delta a-(\gamma/\alpha)b+\gamma\delta = \mathsf{D}(Z).$$

Note that this generation of lossy tags is *not* possible when $\alpha = 0$.

So far, we have argued that F can simulate Game 4 perfectly for A. It remains to show how F can extract an No-Mult solution out of a tag t^* output by A. Unless $\mathsf{bad_{tag}}$ occurs or t^* is injective, we have

$$z^* = \mathsf{D}(Z^*) = \mathsf{D}(A) \cdot \mathsf{D}(B) + \mathsf{D}(R^*) \cdot \mathsf{D}(H^*) = ab + r^*h^* \bmod N.$$

Since we abort otherwise, we may assume that $\alpha^* = 0$, so that $z^* = ab + r^*h^* = ab + \gamma^*r^* \bmod N$ for known γ^*. This implies $ab = z^* - \gamma^*r^* \bmod N$, so F can derive and output a $\mathbb{Z}_{N^2}$-encryption of ab as $Z^*/(R^*)^{\gamma^*} \bmod N^2$. This shows (12).

Taking (4)-(12) together shows Lemma 4.

5   Application: Selective Opening Security

5.1   ABM-LTFs with Explainable Tags

For the application of SOA-CCA security, we need a slight variant of ABM-LTFs. Concretely, we require that values that are revealed during a ciphertext opening can be explained as uniformly chosen "without ulterior motive," if only their distribution is uniform. (This is called "invertible sampling" by Damgård and Nielsen [15].)

Definition 4 (Efficiently samplable and explainable). *A finite set S is efficiently samplable and explainable if any element of S can be explained as the result of a uniform sampling. Formally, there are PPT algorithms* Samp_S, Expl_S, *such that*
1. $\mathsf{Samp}_S(1^k)$ *uniformly samples from S, and*
2. *for any $s \in S$, $\mathsf{Expl}_S(s)$ outputs random coins for Samp that are uniformly distributed among all random coins R with $\mathsf{Samp}_S(1^k; R) = s$.*

Definition 5 (ABM-LTF with explainable tags). *An ABM-LTF has* explainable tags *if the core part of tags is efficiently samplable and explainable. Formally, if we write* $\mathcal{T} = \mathcal{T}_\mathsf{p} \times \mathcal{T}_\mathsf{aux}$, *where $\mathcal{T}_\mathsf{p}$ and $\mathcal{T}_\mathsf{aux}$ denote the core and auxiliary parts of tags, then $\mathcal{T}_\mathsf{p}$ is efficiently samplable and explainable.*

Explainable tags and our ABM-LTFs. Our DCR-based ABM-LTF ABMD has explainable tags, as $\mathbb{Z}_{N^{s+1}}^*$ is efficiently explainable. Concretely, $\mathsf{Samp}_{\mathbb{Z}_{N^{s+1}}^*}$ can choose a uniform $s \leftarrow \mathbb{Z}_{N^{s+1}}$ and test s for invertibility. If s is invertible, we are done; if not, we can factor N and choose a uniform $s' \leftarrow \mathbb{Z}_{N^{s+1}}^*$ directly, using the group order of $\mathbb{Z}_{N^{s+1}}^*$. Similarly, our pairing-based ABM-LTF ABMP has explainable tags as soon as the employed group $\mathbb{G}_1$ is efficiently samplable and explainable. We will also have to explain the CHF randomness R_{CH} in both of our constructions. Fortunately, the CHF randomness of many known constructions [29, 26, 16, 3, 13, 24] consists of uniform values (over an explainable domain), which are efficiently samplable and explainable.

5.2   Selective Opening Security

PKE Schemes. A public-key encryption (PKE) scheme consists of three PPT algorithms $(\mathsf{PKE.Gen}, \mathsf{PKE.Enc}, \mathsf{PKE.Dec})$. Key generation $\mathsf{PKE.Gen}(1^k)$ outputs a public key pk and a secret key sk. Encryption $\mathsf{PKE.Enc}(pk, msg)$ takes a public key pk and a message msg, and outputs a ciphertext C. Decryption $\mathsf{PKE.Dec}(sk, C)$ takes a secret key sk and a ciphertext C, and outputs a message msg. For correctness, we want $\mathsf{PKE.Dec}(sk, C) = msg$ for all msg, all $(pk, sk) \leftarrow \mathsf{PKE.Gen}(1^k)$, and all $C \leftarrow (pk, msg)$. For simplicity, we only consider message spaces $\{0,1\}^k$.

Definition of Selective Opening Security. Following [18, 6, 23], we present a definition for security under selective openings that captures security under adaptive attacks. The definition is indistinguishability-based; it demands that even an adversary that gets to see a vector of ciphertexts cannot distinguish the true contents of the ciphertexts from independently sampled plaintexts.[3] To model adaptive corruptions, our notion also allows the adversary to request "openings" of adaptively selected ciphertexts.

Definition 6 (Efficiently Re-samplable). *Let* $N = N(k) > 0$, *and let* dist *be a joint distribution over* $(\{0,1\}^k)^N$. *We say that* dist *is efficiently re-samplable if there is a PPT algorithm* $\mathsf{ReSamp}_{\mathsf{dist}}$ *such that for any* $\mathcal{I} \subseteq [N]$ *and any partial vector* $\mathbf{msg}'_{\mathcal{I}} := (msg'^{(i)})_{i \in \mathcal{I}} \in (\{0,1\}^k)^{|\mathcal{I}|}$, $\mathsf{ReSamp}_{\mathsf{dist}}(\mathbf{msg}'_{\mathcal{I}})$ *samples from the distribution* dist, *conditioned on* $msg^{(i)} = msg'^{(i)}$ *for all* $i \in \mathcal{I}$.

Definition 7 (IND-SO-CCA Security). *A PKE scheme* $\mathsf{PKE} = (\mathsf{PKE.Gen}, \mathsf{PKE.Enc}, \mathsf{PKE.Dec})$ *is IND-SO-CCA secure iff for every polynomially bounded function* $N = N(k) > 0$, *and every stateful PPT adversary* A, *the function*

$$\mathsf{Adv}^{\mathsf{cca\text{-}so}}_{\mathsf{PKE},A}(k) := \Pr\left[\mathsf{Exp}^{\mathsf{ind\text{-}so\text{-}cca\text{-}b}}_{\mathsf{PKE},A}(k) = 1\right] - \frac{1}{2}$$

is negligible. Here, the experiment $\mathsf{Exp}^{\mathsf{ind\text{-}so\text{-}cca\text{-}b}}_{\mathsf{PKE},A}(k)$ *is defined as follows:*

Experiment $\mathsf{Exp}^{\mathsf{ind\text{-}so\text{-}cca\text{-}b}}_{\mathsf{PKE},A}$

$b \leftarrow \{0,1\}$
$(pk, sk) \leftarrow \mathsf{PKE.Gen}(1^k)$
$(\mathsf{dist}, \mathsf{ReSamp}_{\mathsf{dist}}) \leftarrow A^{\mathsf{PKE.Dec}(sk,\cdot)}(pk)$
$\mathbf{msg}_0 := (msg^{(i)})_{i \in [n]} \leftarrow \mathsf{dist}$
$\mathbf{R} := (R^{(i)})_{i \in [n]} \leftarrow (\mathcal{R}_{\mathsf{PKE.Enc}})^N$
$\mathbf{C} := (C^{(i)})_{i \in [n]} := (\mathsf{PKE.Enc}(pk, msg^{(i)}; R^{(i)}))_{i \in [n]}$
$\mathcal{I} \leftarrow A^{\mathsf{PKE.Dec}(sk,\cdot)}(\texttt{select}, \mathbf{C})$
$\mathbf{msg}_1 := \mathsf{ReSamp}_{\mathsf{dist}}(\mathbf{msg}_{\mathcal{I}})$
$out_A \leftarrow A^{\mathsf{PKE.Dec}(sk,\cdot)}(\texttt{output}, (msg^{(i)}, R^{(i)})_{i \in \mathcal{I}}, \mathbf{msg}_b)$
return $(out_A = b)$

We only allow A *that (a) always output efficiently re-samplable distributions* dist *over* $(\{0,1\}^k)^N$ *with corresponding efficient re-sampling algorithms* $\mathsf{ReSamp}_{\mathsf{dist}}$, *(b) never submit a received challenge ciphertext* $C^{(i)}$ *to their decryption oracle* $\mathsf{PKE.Dec}(sk, \cdot)$, *and (c) always produce binary final output* out_A.

This definition can be generalized in many ways, e.g., to more opening phases, or more encryption keys. We focus on the one-phase, one-key case for ease of presentation; our techniques apply equally to a suitably generalized security definitions.

5.3   IND-SO-CCA Security from ABM-LTFs

The Construction. To construct our IND-SO-CCA secure PKE scheme, we require the following ingredients:

---

[3] Like previous works, we restrict ourselves to message distributions that allow for an efficient re-sampling. We explain in the full version how to achieve simulation-based selective opening security for *arbitrary* message spaces.

- an LTF LTF = (LTF.IGen, LTF.Eval, LTF.Invert, LTF.LGen) with domain $\{0,1\}^n$ (as in Definition 2) that is ℓ'-lossy,
- an efficiently explainable ABM-LTF ABM = (ABM.Gen, ABM.Eval, ABM.Invert, ABM.LTag) with domain[4] $\{0,1\}^n$ and tag set $\mathcal{T} = \mathcal{T}_\mathsf{p} \times \mathcal{T}_\mathsf{aux}$ (as in Definition 5) that is ℓ-lossy, and
- a family $\mathcal{UH}$ of universal hash functions $h : \{0,1\}^n \to \{0,1\}^k$, so that for any $f : \{0,1\}^n \to \{0,1\}^{\ell'+\ell}$, it is SD $((h, f(X), h(X)) ; (h, f(X), U)) = \mathbf{O}(2^{-2k})$, where $h \leftarrow \mathcal{UH}$, $X \leftarrow \{0,1\}^n$, and $U \leftarrow \{0,1\}^k$.

Then, consider the following PKE scheme PKE = (PKE.Gen, PKE.Enc, PKE.Dec):

Alg. PKE.Gen(1^k)	**Alg.** PKE.Enc(pk, msg)	**Alg.** PKE.Dec(sk, C)
$(ek', ik') \leftarrow$ LTF.IGen(1^k)	parse $pk =: (ek', ek)$	parse $sk =: (ik', ek)$,
$(ek, ik, tk) \leftarrow$ ABM.Gen(1^k)	$X \leftarrow \{0,1\}^n$	$C =: (\rho, Y', t_\mathsf{p}, Y)$
$pk := (ek', ek)$	$\rho := h(X) \oplus msg$	$X \leftarrow f_{ik'}^{-1}(Y')$
$sk := (ik', ek)$	$Y' := f_{ek'}(X)$	if $Y \neq f_{ek,(t_\mathsf{p},(\rho,Y'))}(X)$
return (pk, sk)	$t_\mathsf{p} := \mathsf{Samp}_{\mathcal{T}_\mathsf{p}}(1^k; R_{t_\mathsf{p}})$	return $\perp$
	$Y := f_{ek,(t_\mathsf{p},(\rho,Y'))}(X)$	$msg := h(X) \oplus \rho$
	$C := (\rho, Y', t_\mathsf{p}, Y)$	return msg
	return C	

The core of this scheme is a (deterministic) double encryption as in [30, 23]. One encryption (namely, Y') is generated using an LTF, and the other (namely, Y) is generated using an ABM-LTF. In the security proof, the LTF will be switched to lossy mode, and the ABM-LTF will be used with lossy tags precisely for the (IND-SO-CCA) challenge ciphertexts. This will guarantee that all challenge ciphertexts will be lossy. At the same time, the evasiveness property of our ABM-LTF will guarantee that no adversary can come up with a decryption query that corresponds to a lossy ABM-LTF tag. As a consequence, we will be able to answer all decryption queries during the security proof.

Relation to the Construction of Hemenway et al.. Our construction is almost identical to the one of Hemenway et al. [23], which in turn builds upon the construction of an IND-CCA secure encryption scheme from an all-but-one lossy trapdoor function [30]. However, while we employ ABM-LTFs, [23] employ "all-but-N lossy trapdoor functions" (ABN-LTFs), which are defined similarly to ABM-LTFs, only with the number of lossy tags fixed in advance (to a polynomial value N). Thus, unlike in our schemes, the number of challenge ciphertexts N has to be fixed in advance with [23]. Furthermore, the complexity of the schemes from [23] grows (linearly) in the number N of challenge ciphertexts. On the other hand, ABN-LTFs also allow to explicitly determine all lossy tags in advance, upon key generation. (For instance, all lossy tags can be chosen as suitable signature verification keys or chameleon hash values.) With ABM-LTFs, lossy tags are generated on the fly, through ABM.LTag. This difference is the reason for the auxiliary tag parts in the ABM-LTF definition, cf. Section 3.

---

[4] In case of our DCR-based ABM-LTF ABMD, the desired domain $\{0,1\}^n$ must be suitably mapped to ABMD's "native domain" $\mathbb{Z}_{N^s}^3$.

Theorem 2. *If* LTF *is an LTF,* ABM *is an efficiently explainable ABM-LTF, and* $\mathcal{UH}$ *is an UHF family as described, then* PKE *is IND-SO-CCA secure. In particular, for every IND-SO-CCA adversary A on* PKE *that makes at most* $q = q(k)$ *decryption queries, there exist adversaries* $B, C,$ *and* D *of roughly same complexity as* A, *and such that*

$$\left|\mathsf{Adv}^{\mathsf{cca\text{-}so}}_{\mathsf{PKE},A}(k)\right| \leq \left|\mathsf{Adv}^{\mathsf{ind}}_{\mathsf{ABM},B}(k)\right| + q(k) \cdot \mathsf{Adv}^{\mathsf{eva}}_{\mathsf{ABM},C}(k) + \left|\mathsf{Adv}^{\mathsf{ind}}_{\mathsf{LTF},D}(k)\right| + \mathbf{O}(2^{-k}).$$

Note that the reduction does not depend on N, the number of challenge ciphertexts. On the other hand, the number of an adversary's decryption queries goes linearly into the reduction factor. We can get rid of this factor of $q(k)$ in case of our pairing-based ABM-LTF ABMP; see the full version. We also prove Theorem 2 in the full version.

Acknowledgements. The author would like to thank Florian Böhl, Serge Fehr, Eike Kiltz, and Hoeteck Wee for helpful discussions concerning SO-CCA security. The anonymous Crypto 2011 and Eurocrypt 2012 referees, and in particular one Eurocrypt referee have given very useful comments that helped to improve the paper. Jorge Villar pointed me to [32], a result that improves the reduction of our pairing-based ABM-LTF.

References

[1] Abe, M., Gennaro, R., Kurosawa, K.: Tag-KEM/DEM: A new framework for hybrid encryption. Journal of Cryptology 21(1), 97–130 (2008)

[2] Applebaum, B., Ishai, Y., Kushilevitz, E.: How to garble arithmetic circuits. In: 52nd FOCS. IEEE Computer Society Press (2011)

[3] Ateniese, G., de Medeiros, B.: Identity-Based Chameleon Hash and Applications. In: Juels, A. (ed.) FC 2004. LNCS, vol. 3110, pp. 164–180. Springer, Heidelberg (2004)

[4] Bellare, M., Boldyreva, A., Micali, S.: Public-Key Encryption in a Multi-user Setting: Security Proofs and Improvements. In: Preneel, B. (ed.) EUROCRYPT 2000. LNCS, vol. 1807, pp. 259–274. Springer, Heidelberg (2000)

[5] Bellare, M., Brakerski, Z., Naor, M., Ristenpart, T., Segev, G., Shacham, H., Yilek, S.: Hedged Public-Key Encryption: How to Protect against Bad Randomness. In: Matsui, M. (ed.) ASIACRYPT 2009. LNCS, vol. 5912, pp. 232–249. Springer, Heidelberg (2009)

[6] Bellare, M., Hofheinz, D., Yilek, S.: Possibility and Impossibility Results for Encryption and Commitment Secure under Selective Opening. In: Joux, A. (ed.) EUROCRYPT 2009. LNCS, vol. 5479, pp. 1–35. Springer, Heidelberg (2009)

[7] Boldyreva, A., Fehr, S., O'Neill, A.: On Notions of Security for Deterministic Encryption, and Efficient Constructions without Random Oracles. In: Wagner, D. (ed.) CRYPTO 2008. LNCS, vol. 5157, pp. 335–359. Springer, Heidelberg (2008)

[8] Boneh, D., Boyen, X.: Short signatures without random oracles and the SDH assumption in bilinear groups. Journal of Cryptology 21(2), 149–177 (2008)

[9] Boneh, D., Shen, E., Waters, B.: Strongly Unforgeable Signatures Based on Computational Diffie-Hellman. In: Yung, M., Dodis, Y., Kiayias, A., Malkin, T. (eds.) PKC 2006. LNCS, vol. 3958, pp. 229–240. Springer, Heidelberg (2006)

[10] Boyen, X., Waters, B.: Shrinking the Keys of Discrete-Log-Type Lossy Trapdoor Functions. In: Zhou, J., Yung, M. (eds.) ACNS 2010. LNCS, vol. 6123, pp. 35–52. Springer, Heidelberg (2010)

[11] Canetti, R., Feige, U., Goldreich, O., Naor, M.: Adaptively secure multi-party computation. In: 28th ACM STOC, pp. 639–648. ACM Press (May 1996)

[12] Canetti, R., Halevi, S., Katz, J.: Chosen-Ciphertext Security from Identity-Based Encryption. In: Cachin, C., Camenisch, J.L. (eds.) EUROCRYPT 2004. LNCS, vol. 3027, pp. 207–222. Springer, Heidelberg (2004)

[13] Chevallier-Mames, B., Joye, M.: A Practical and Tightly Secure Signature Scheme Without Hash Function. In: Abe, M. (ed.) CT-RSA 2007. LNCS, vol. 4377, pp. 339–356. Springer, Heidelberg (2006)

[14] Damgård, I., Jurik, M.: A Generalisation, a Simplification and Some Applications of Paillier's Probabilistic Public-Key System. In: Kim, K. (ed.) PKC 2001. LNCS, vol. 1992, pp. 119–136. Springer, Heidelberg (2001)

[15] Damgård, I., Nielsen, J.B.: Improved Non-committing Encryption Schemes Based on a General Complexity Assumption. In: Bellare, M. (ed.) CRYPTO 2000. LNCS, vol. 1880, pp. 432–450. Springer, Heidelberg (2000)

[16] Damgård, I., Nielsen, J.B.: Perfect Hiding and Perfect Binding Universally Composable Commitment Schemes with Constant Expansion Factor. In: Yung, M. (ed.) CRYPTO 2002. LNCS, vol. 2442, pp. 581–596. Springer, Heidelberg (2002)

[17] Dolev, D., Dwork, C., Naor, M.: Non-malleable cryptography. In: 23rd ACM STOC, pp. 542–552. ACM Press (May 1991)

[18] Dwork, C., Naor, M., Reingold, O., Stockmeyer, L.J.: Magic functions. In: 40th FOCS, pp. 523–534. IEEE Computer Society Press (October 1999)

[19] Fehr, S., Hofheinz, D., Kiltz, E., Wee, H.: Encryption Schemes Secure against Chosen-Ciphertext Selective Opening Attacks. In: Gilbert, H. (ed.) EUROCRYPT 2010. LNCS, vol. 6110, pp. 381–402. Springer, Heidelberg (2010)

[20] Freeman, D.M., Goldreich, O., Kiltz, E., Rosen, A., Segev, G.: More Constructions of Lossy and Correlation-Secure Trapdoor Functions. In: Nguyen, P.Q., Pointcheval, D. (eds.) PKC 2010. LNCS, vol. 6056, pp. 279–295. Springer, Heidelberg (2010)

[21] Gennaro, R., Shoup, V.: A note on an encryption scheme of Kurosawa and Desmedt. Cryptology ePrint Archive, Report 2004/194 (2004), http://eprint.iacr.org/

[22] Goldwasser, S., Micali, S.: Probabilistic encryption. Journal of Computer and System Sciences 28(2), 270–299 (1984)

[23] Hemenway, B., Libert, B., Ostrovsky, R., Vergnaud, D.: Lossy Encryption: Constructions from General Assumptions and Efficient Selective Opening Chosen Ciphertext Security. In: Lee, D.H., Wang, X. (eds.) ASIACRYPT 2011. LNCS, vol. 7073, pp. 70–88. Springer, Heidelberg (2011)

[24] Hohenberger, S., Waters, B.: Realizing Hash-and-Sign Signatures under Standard Assumptions. In: Joux, A. (ed.) EUROCRYPT 2009. LNCS, vol. 5479, pp. 333–350. Springer, Heidelberg (2009)

[25] Ishai, Y., Kushilevitz, E.: Perfect Constant-Round Secure Computation via Perfect Randomizing Polynomials. In: Widmayer, P., Triguero, F., Morales, R., Hennessy, M., Eidenbenz, S., Conejo, R. (eds.) ICALP 2002. LNCS, vol. 2380, pp. 244–256. Springer, Heidelberg (2002)

[26] Krawczyk, H., Rabin, T.: Chameleon signatures. In: NDSS 2000. The Internet Society (February 2000)

[27] Nishimaki, R., Fujisaki, E., Tanaka, K.: Efficient Non-interactive Universally Composable String-Commitment Schemes. In: Pieprzyk, J., Zhang, F. (eds.) ProvSec 2009. LNCS, vol. 5848, pp. 3–18. Springer, Heidelberg (2009)

[28] Paillier, P.: Public-Key Cryptosystems Based on Composite Degree Residuosity Classes. In: Stern, J. (ed.) EUROCRYPT 1999. LNCS, vol. 1592, pp. 223–238. Springer, Heidelberg (1999)

[29] Pedersen, T.P.: Non-interactive and Information-Theoretic Secure Verifiable Secret Sharing. In: Feigenbaum, J. (ed.) CRYPTO 1991. LNCS, vol. 576, pp. 129–140. Springer, Heidelberg (1992)

[30] Peikert, C., Waters, B.: Lossy trapdoor functions and their applications. In: Ladner, R.E., Dwork, C. (eds.) 40th ACM STOC, pp. 187–196. ACM Press (May 2008)

[31] Rosen, A., Segev, G.: Chosen-Ciphertext Security via Correlated Products. In: Reingold, O. (ed.) TCC 2009. LNCS, vol. 5444, pp. 419–436. Springer, Heidelberg (2009)

[32] Villar, J.: An efficient reduction from DDH to the rank problem (2011)

[33] Waters, B.: Efficient Identity-Based Encryption Without Random Oracles. In: Cramer, R. (ed.) EUROCRYPT 2005. LNCS, vol. 3494, pp. 114–127. Springer, Heidelberg (2005)

Identity-Based (Lossy) Trapdoor Functions and Applications

Mihir Bellare[1], Eike Kiltz[2], Chris Peikert[3], and Brent Waters[4]

[1] Department of Computer Science & Engineering,
University of California San Diego, USA
http://cseweb.ucsd.edu/~mihir/
[2] Horst Görtz Institut für IT-Sicherheit, Ruhr-Universität Bochum, Germany
http://www.cits.rub.de/personen/kiltz.html
[3] School of Computer Science, College of Computing,
Georgia Institute of Technology, USA
http://www.cc.gatech.edu/~cpeikert/
[4] Department of Computer Science, University of Texas at Austin, USA
http://userweb.cs.utexas.edu/~bwaters/

Abstract. We provide the first constructions of identity-based (injective) trapdoor functions. Furthermore, they are lossy. Constructions are given both with pairings (DLIN) and lattices (LWE). Our lossy identity-based trapdoor functions provide an automatic way to realize, in the identity-based setting, many functionalities previously known only in the public-key setting. In particular we obtain the first deterministic and efficiently searchable IBE schemes and the first hedged IBE schemes, which achieve best possible security in the face of bad randomness. Underlying our constructs is a new definition, namely *partial* lossiness, that may be of broader interest.

1 Introduction

A trapdoor function F specifies, for each public key pk, an injective, *deterministic* map F_{pk} that can be inverted given an associated secret key (trapdoor). The most basic measure of security is one-wayness. The canonical example is RSA [49].

Suppose there is an algorithm that generates a "fake" public key pk^* such that F_{pk^*} is no longer injective but has image much smaller than its domain and, moreover, given a public key, you can't tell whether it is real or fake. Peikert and Waters [47] call such a TDF lossy. Intuitively, F_{pk} is close to a function F_{pk^*} that provides information-theoretic security. Lossiness implies one-wayness [47].

Lossy TDFs have quickly proven to be a powerful tool. Applications include IND-CCA [47], deterministic [16], hedged [7] and selective-opening secure public-key encryption [9]. Lossy TDFs can be constructed from DDH [47], QR [33], DLIN [33], DBDH [23], LWE [47] and HPS (hash proof systems) [38]. RSA was shown in [41] to be lossy under the Φ-hiding assumption of [25], leading to the first proof of security of RSA-OAEP [13] without random oracles.

D. Pointcheval and T. Johansson (Eds.): EUROCRYPT 2012, LNCS 7237, pp. 228–245, 2012.

Lossy TDFs and their benefits belong, so far, to the realm of public-key cryptography. The purpose of this paper is to bring them to identity-based cryptography, defining and constructing identity-based TDFs (IB-TDFs), both one-way and lossy. We see this as having two motivations, one more theoretical, the other more applied, yet admittedly both foundational, as we discuss before moving further.

THEORETICAL ANGLE. Trapdoor functions are the primitive that began public key cryptography [30,49]. Public-key encryption was built from TDFs. (Via hardcore bits.) Lossy TDFs enabled the first DDH and lattice (LWE) based TDFs [47].

It is striking that identity-based cryptography developed entirely differently. The first realizations of IBE [21,29,52] directly used randomization and were neither underlain by, nor gave rise to, any IB-TDFs.

We ask whether this asymmetry between the public-key and identity-based worlds (TDFs in one but not the other) is inherent. This seems to us a basic question about the nature of identity-based cryptography that is worth asking and answering.

APPLICATION ANGLE. Is there anything here but idle curiosity? IBE has already been achieved *without* IB-TDFs, so why go backwards to define and construct the latter? The answer is that *losssy* IB-TDFs enable new applications that we do not know how to get in other ways.

Stepping back, identity-based cryptography [53] offers several advantages over its public-key counterpart. Key management is simplified because an entity's identity functions as their public key. Key revocation issues that plague PKI can be handled in alternative ways, for example by using identity+date as the key under which to encrypt to identity [21]. There is thus good motivation to go beyond basics like IBE [21,29,52,17,18,55,34] and identity-based signatures [11,31] to provide identity-based counterparts of other public-key primitives.

Furthermore we would like to do this in a systematic rather than ad hoc way, leading us to seek tools that enable the transfer of multiple functionalities in relatively blackbox ways. The applications of lossiness in the public-key realm suggest that lossy IBTDFs will be such a tool also in the identity-based realm. As evidence we apply them to achieve identity-based deterministic encryption and identity-based hedged encryption. The first, the counterpart of deterministic public-key encryption [6,16], allows efficiently searchable identity-based encryption of database entries while maintaining the maximal possible privacy, bringing the key-management benefits of the identity-based setting to this application. The second, counterpart of hedged symmetric and public-key encryption [50,7], makes IBE as resistant as possible in the face of low-quality randomness, which is important given the widespread deployment of IBE and the real danger of bad-randomness based attacks evidenced by the ones on the Sony Playstation and Debian Linux. We hope that our framework will facilitate further such transfers.

We clarify that the solutions we obtain are not practical but they show that the security goals can be achieved in principle, which was not at all clear prior to our work. Allowed random oracles, we can give solutions that are much more efficient and even practical.

CONTRIBUTIONS IN BRIEF. We define IB-TDFs and two associated security notions, one-wayness and lossiness, showing that the second implies the first.

The first wave of IBE schemes was from pairings [21,52,17,18,55,54] but another is now emerging from lattices [34,28,2,3]. We aim accordingly to reach our ends with either route and do so successfully. We provide lossy IB-TDFs from a standard pairings assumption, namely the Decision Linear (DLIN) assumption of [19]. We also provide IB-TDFs based on Learning with Errors (LWE) [48], whose hardness follows from the worst-case hardness of certain lattice-related problems [48,46]. (The same assumption underlies lattice-based IBE [34,28,2,3] and public-key lossy TDFs [47].) None of these results relies on random oracles.

Existing work brought us closer to the door with lattices, where one-way IB-TDFs can be built by combining ideas from [34,28,2]. Based on techniques from [46,42] we show how to make them lossy. With pairings, however it was unclear how to even get a one-way IB-TDF, let alone one that is lossy. We adapt the matrix-based framework of [47] so that by populating matrix entries with ciphertexts of a very special kind of *anonymous* IBE scheme it becomes possible to implicitly specify per-identity matrices defining the function. No existing anonymous IBE has the properties we need but we build one that does based on methods of [22]. Our results with pairings are stronger because the lossy branches are universal hash functions which is important for applications.

Public-key lossy TDFs exist aplenty and IBE schemes do as well. It is natural to think one could easily combine them to get IB-TDFs. We have found no simple way to do this. Ultimately we do draw from both sources for techniques but our approaches are intrusive. Let us now look at our contributions in more detail.

NEW PRIMITIVES AND DEFINITIONS. Public parameters *pars* and an associated master secret key having been chosen, an IB-TDF F associates to any identity a map $F_{pars,id}$, again injective and deterministic, inversion being possible given a secret key derivable from id via the master secret key. One-wayness means F_{pars,id^*} is hard to invert on random inputs for an adversary-specified challenge identity id^*. Importantly, as in IBE, this must hold even when the adversary may obtain, via a key-derivation oracle, a decryption key for any non-challenge identity of its choice [21]. This key-derivation capability contributes significantly to the difficulty of realizing the primitive. As with IBE, security may be selective (the adversary must specify id^* before seeing *pars*) [27] or adaptive (no such restriction) [21].

The most direct analog of the definition of lossiness from the public-key setting would ask that there be a way to generate "fake" parameters $pars^*$, indistinguishable from the real ones, such that F_{pars^*,id^*} is lossy (has image smaller than domain). In the selective setting, the fake parameter generation algorithm Pg^* can take id^* as input, making the goal achievable at least in principle, but in the adaptive setting it is impossible to achieve, since, with id^* not known in advance, Pg^* is forced to make $F_{pars^*,id}$ lossy for all id, something the adversary can immediately detect using its key-derivation oracle.

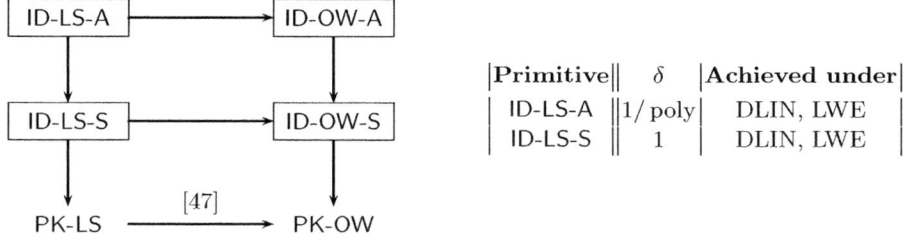

Primitive	δ	**Achieved under**
ID-LS-A	$1/\text{poly}$	DLIN, LWE
ID-LS-S	1	DLIN, LWE

Fig. 1. Types of TDFs based on setting (PK=Public-key, ID=identity-based), security (OW=one-way, LS=loss) and whether the latter is selective (S) or adaptive (A). An arrow A → B in the diagram on the left means that TDF of type B is implied by (can be constructed from) TDF of type A. Boxed TDFs are the ones we define and construct. The table on the right shows the δ for which we prove δ-lossiness and the assumptions used. In both the S and A settings the δ we achieve is best possible and suffices for applications.

We ask whether there is an adaptation of the definition of lossiness that is achievable in the adaptive case while sufficing for applications. Our answer is a definition of δ-*lossiness*, a metric of partial lossiness parameterized by the probability δ that F_{pars^*,id^*} is lossy. The definition is unusual, involving an adversary advantage that is the difference, not of two probabilities as is common in cryptographic metrics, but of two differently weighted ones. We will achieve selective lossiness with degree $\delta = 1$, but in the adaptive case the best possible is degree $1/\text{poly}$ with the polynomial depending on the number of key-derivation queries of the adversary, and this what we will achieve. We show that lossiness with degree δ implies one-wayness, in both the selective and adaptive settings, as long as δ is at least $1/\text{poly}$.

In summary, in the identity-based setting (ID) there are two notions of security, one-wayness (OW) and lossiness (LS), each of which could be selective (S) or adaptive (A), giving rise to four kinds of IB-TDFs. The left side of Fig. 1 shows how they relate to each other and to the two kinds of TDFs —OW and LS— in the public-key setting (PK). The un-annotated implications are trivial, ID-LS-A → ID-LS-S meaning that δ-lossiness of the first type implies δ-lossiness of the other for all δ. It is not however via this implication that we achieve ID-LS-S, for, as the table shows, we achieve it with degree higher than ID-LS-A.

CLOSER LOOK. One's first attempt may be to build an IB-TDF from an IBE scheme. In the random oracle (RO) model, this can be done by a method of [8], namely specify the coins for the IBE scheme by hashing the message with the RO. It is entirely unclear how to turn this into a standard model construct and it is also unclear how to make it lossy.

To build ID-TDFs from lattices we consider starting from the public-key TDF of [47] (which is already lossy) and trying to make it identity-based, but it is unclear how to do this. However, Gentry, Peikert and Vaikuntanathan (GPV) [34] showed that the function $g_{\mathbf{A}} : B_\alpha^{n+m} \rightarrow \mathbb{Z}_q^n$ defined by $g_{\mathbf{A}}(\mathbf{x}, \mathbf{e}) = \mathbf{A}^T \cdot \mathbf{x} + \mathbf{e}$ is a TDF for appropriate choices of the domain and parameters, where matrix $\mathbf{A} \in \mathbb{Z}_q^{n \times m}$ is a uniformly random public key which is constructed together with a trapdoor as for example in [4,5,43]. We make this function identity-based using the trapdoor extension and delegation methods introduced by Cash, Hofheinz, Kiltz and Peikert [28], and improved in efficiency by Agrawal, Boneh and Boyen [2] and Micciancio and Peikert [43]. Finally, we obtain a lossy IB-TDF by showing that this construction is already lossy.

With pairings there is no immediate way to get an IB-TDF that is even one-way, let alone lossy. We aim for the latter, there being no obviously simpler way to get the former. In the selective case we need to ensure that the function is lossy on the challenge identity id^* yet injective on others, this setup being indistinguishable from the one where the function is always injective. Whereas the matrix diagonals in the construction of [47] consisted of ElGamal ciphertexts, in ours they are ciphertexts for identity id^* under an anonymous IBE scheme, the salient property being that the "anonymity" property should hide whether the underlying ciphertext is to id^* or is a random group element. Existing anonymous IBE schemes, in particular that of Boyen and Waters (BW) [22], are not conducive and we create a new one. A side benefit is a new anonymous IBE scheme with ciphertexts and private keys having one less group element than BW but still proven secure under DLIN.

A method of Boneh and Boyen [17] can be applied to turn selective into adaptive security but the reduction incurs a factor that is equal to the size of the identity space and thus ultimately exponential in the security parameter, so that adaptive security according to the standard asymptotic convention would not have been achieved. To achieve it, we want to be able to "program" the public parameters so that they will be lossy on about a $1/Q$ fraction of "random-ish" identities, where Q is the number of key-derivation queries made by the attacker. Ideally, with probability around $1/Q$ all of (a successful) attacker's queries will land outside the lossy identity-space, but the challenge identity will land inside it so that we achieve δ-lossiness with δ around $1/Q$.

This sounds similar to the approach of Waters [55] for achieving adaptively secure IBE but there are some important distinctions, most notably that the technique of Waters is information-theoretic while ours is of necessity computational, relying on the DLIN assumption. In the reduction used by Waters the partitioning of the identities into two classes was based solely on the reduction algorithm's internal view of the public parameters; the parameters themselves were distributed independently of this partitioning and thus the adversary view was the same as in a normal setup. In contrast, the partitioning in our scheme will actually directly affect the parameters and how the system behaves. This is why we must rely on a computational assumption to show that the partitioning in undetectable. A key novel feature of our construction is the introduction of a system that will produce lossy public parameters for about a $1/Q$ fraction of the identities.

APPLICATIONS. Deterministic PKE is a TDF providing the best possible privacy subject to being deterministic, a notion called PRIV that is much stronger than one-wayness [6]. An application is encryption of database records in a way that permits logarithmic-time search, improving upon the linear-time search of PEKS [20]. Boldyreva, Fehr and O'Neill [16] show that lossy TDFs whose lossy branch is a universal hash (called universal lossy TDFs) achieve (via the LHL [15,37]) PRIV-security for message sequences which are blocksources, meaning each message has some min-entropy even given the previous ones, which remains the best result without ROs. Deterministic IBE and the resulting efficiently-searchable IBE are attractive due to the key-management benefits. We can achieve them because our DLIN-based lossy IB-TDFs are also universal lossy. (This is not true, so far, for our LWE based IB-TDFs.)

To provide IND-CPA security in practice, IBE relies crucially on the availability of fresh, high-quality randomness. This is fine in theory but in practice RNGs (random number generators) fail due to poor entropy gathering or bugs, leading to prominent security breaches [35,36,24,45,44,1,56,32]. Expecting systems to do a better job is unrealistic. Hedged encryption [7] takes poor randomness as a fact of life and aims to deliver best possible security in the face of it, providing privacy as long as the message together with the "randomness" have some min-entropy. Hedged PKE was achieved in [7] by combining IND-CPA PKE with universal lossy TDFs. We can adapt this to IBE and combine existing (randomized) IBE schemes with our DLIN-based universal lossy IB-TDFs to achieved hedged IBE. This is attractive given the widespread use of IBE in practice and the real danger of randomness failures.

Both applications are for the case of selective security. It remains open to achieve them in the adaptive case.

RELATED WORK. A number of papers have studied security notions of trapdoor functions beyond traditional one-wayness. Besides lossiness [47] there is Rosen and Segev's notion of correlated-product security [51], and Canetti and Dakdouk's extractable trapdoor functions [26]. The notion of adaptive one-wayness for tag-based trapdoor functions from Kiltz, Mohassel and O'Neill [40] can be seen as the special case of our selective IB-TDF in which the adversary is denied key-derivation queries. Security in the face of these queries was one of the main difficulties we faced in realizing IB-TDFs.

ORGANIZATION. We define IB-TDFs, one-wayness and δ-lossiness in Section 2. We also define extended IB-TDFs, an abstraction that will allow us to unify and shorten the analyses for the selective and adaptive security cases. In [10] we show that δ-lossiness implies one-wayness as long as δ is at least $1/\operatorname{poly}$. This allows us to focus on achieving δ-lossiness. In Section 3 we provide our pairing-based schemes and in [10] our lattice-based schemes. In [10] we sketch how to apply δ-lossy IB-TDFs to achieve deterministic and hedged IBE.

2 Definitions

NOTATION AND CONVENTIONS. If $\mathbf{x}$ is a vector then $|\mathbf{x}|$ denotes the number of its coordiates and $\mathbf{x}[i]$ denotes its i-th coordinate. Coordinates may be numbered

proc Initialize(id)   // $\mathrm{OW_F, Real_F}$
$\overline{\qquad\qquad\qquad\qquad\qquad\qquad}$
$(pars, msk) \xleftarrow{\$} \mathsf{F.Pg}$; $IS \leftarrow \emptyset$; $id^* \leftarrow id$
Return $pars$

proc GetDK(id)   // $\mathrm{OW_F, Real_F}$
$\overline{\qquad\qquad\qquad\qquad}$
$IS \leftarrow IS \cup \{id\}$
$dk \leftarrow \mathsf{F.Kg}(pars, msk, id)$
Return dk

proc Ch(id)   // $\mathrm{OW_F}$
$\overline{\qquad\qquad\qquad}$
$id^* \leftarrow id$; $x \xleftarrow{\$} \mathsf{InSp}$
$y \leftarrow \mathsf{F.Ev}(pars, id^*, x)$
Return y

proc Finalize(x')   // $\mathrm{OW_F}$
$\overline{\qquad\qquad\qquad\qquad}$
Return $((x' = x)$ and $(id^* \notin IS))$

proc Initialize(id)   // $\mathrm{Lossy_{F,LF,\ell}}$
$\overline{\qquad\qquad\qquad\qquad\qquad\qquad}$
$(pars, msk) \xleftarrow{\$} \mathsf{LF.Pg}(id)$; $IS \leftarrow \emptyset$; $id^* \leftarrow id$
Return $pars$

proc GetDK(id)   // $\mathrm{Lossy_{F,LF,\ell}}$
$\overline{\qquad\qquad\qquad\qquad}$
$IS \leftarrow IS \cup \{id\}$
$dk \leftarrow \mathsf{LF.Kg}(pars, msk, id)$
Return dk

proc Ch(id)   // $\mathrm{Real_F, Lossy_{F,LF,\ell}}$
$\overline{\qquad\qquad\qquad}$
$id^* \leftarrow id$

proc Finalize(d')   // $\mathrm{Real_F}$
$\overline{\qquad\qquad\qquad\qquad}$
Return $((d' = 1)$ and $(id^* \notin IS))$

proc Finalize(d')   // $\mathrm{Lossy_{F,LF,\ell}}$
$\overline{\qquad\qquad\qquad\qquad}$
If $(\lambda(\mathsf{F.Ev}(pars, id^*, \cdot)) < \ell)$ then return false
Return $((d' = 1)$ and $(id^* \notin IS))$

Fig. 2. Games defining one-wayness and δ-lossiness of IBTDF F with sibling LF

$1, \dots, |\mathbf{x}|$ or $0, \dots, |\mathbf{x}| - 1$ as convenient. A string x is identified with a vector over $\{0, 1\}$ so that $|x|$ denotes its length and $x[i]$ its i-th bit. The empty string is denoted ε. If S is a set then $|S|$ denotes its size, S^a denotes the set of a-vectors over S, $S^{a \times b}$ denotes the set of a by b matrices with entries in S, and so on. The (i, j)-th entry of a 2 dimensional matrix $\mathbf{M}$ is denoted $\mathbf{M}[i, j]$ and the (i, j, k)-th entry of a 3 dimensional matrix $\mathbf{M}$ is denoted $\mathbf{M}[i, j, k]$. If $\mathbf{M}$ is a n by μ matrix then $\mathbf{M}[j, \cdot]$ denotes the vector $(\mathbf{M}[j, 1], \dots, \mathbf{M}[j, \mu])$. If $a = (a_1, \dots, a_n)$ then $(a_1, \dots, a_n) \leftarrow a$ means we parse a as shown. Unless otherwise indicated, an algorithm may be randomized. By $y \xleftarrow{\$} A(x_1, x_2, \dots)$ we denote the operation of running A on inputs $x_1, x_2, \dots$ and fresh coins and letting y denote the output. We denote by $[A(x_1, x_2, \dots)]$ the set of all possible outputs of A on inputs $x_1, x_2, \dots$. The (Kronecker) delta function Δ is defined by $\Delta(a, b) = 1$ if $a = b$ and 0 otherwise. If a, b are equal-length vectors of reals then $\langle a, b \rangle = a[1]b[1] + \dots + a[|a|]b[|b|]$ denotes their inner product.

GAMES. A game —look at Fig. 2 for an example— has an **Initialize** procedure, procedures to respond to adversary oracle queries, and a **Finalize** procedure. To execute a game G is executed with an adversary A means to run the adversary and answer its oracle queries by the corresponding procedures of G. The adversary must make exactly one query to **Initialize**, this being its first oracle query. (This means the adversary can give **Initialize** an input, an extension of the usual convention [14].) It must make exactly one query to **Finalize**, this being its last oracle query. The reply to this query, denoted G^A, is called the output of the game, and we let "G^A" denote the event that this game output takes value true. Boolean flags are assumed initialized to false.

IBTDFs. An *identity-based trapdoor function* (IBTDF) is a tuple $\mathsf{F} = (\mathsf{F.Pg}, \mathsf{F.Kg}, \mathsf{F.Ev}, \mathsf{F.Ev}^{-1})$ of algorithms with associated input space InSp and identity space IDSp. The parameter generation algorithm F.Pg takes no input and returns common parameters *pars* and a master secret key *msk*. On input *pars*, *msk*, *id*, the key generation algorithm F.Kg produces a decryption key *dk* for identity *id*. For any *pars* and *id* ∈ IDSp, the *deterministic* evaluation algorithm F.Ev defines a function F.Ev(*pars*, *id*, ·) with domain InSp. We require *correct inversion*: For any *pars*, any *id* ∈ IDSp and any *dk* ∈ [F.Kg(*pars*, *id*)], the deterministic inversion algorithm F.Ev^{-1} defines a function that is the inverse of F.Ev(*pars*, *id*, ·), meaning F.Ev^{-1}(*pars*, *id*, *dk*, F.Ev(*pars*, *id*, *x*)) = *x* for all *x* ∈ InSp.

E-IBTDF. To unify and shorten the selective and adaptive cases of our analyses it is useful to define and specify a more general primitive. An extended IBTDF (E-IBTDF) E = (E.Pg, E.Kg, E.Ev, E.Ev^{-1}) consists of four algorithms that are just like the ones for an IBTDF except that F.Pg takes an additional *auxiliary* input from an auxiliary input space AxSp. Fixing a particular auxiliary input *aux* ∈ AxSp for F.Pg results in an IBTDF scheme that we denote E(*aux*) and call the IBTDF induced by *aux*. Not all these induced schemes need, however, satisfy the correct inversion requirement. If the one induced by *aux* does, we say that *aux* grants invertibility. Looking ahead we will build an E-IBTDF and then obtain our IBTDF as the one induced by a particular auxiliary input, the other induced schemes being the basis of the siblings and being used in the proof.

ONE-WAYNESS. One-wayness of IBTDF F = (F.Pg, F.Kg, F.Ev, F.Ev^{-1}) is defined via game OW$_F$ of Fig. 2. The adversary is allowed only one query to its challenge oracle **Ch**. The advantage of such an adversary *I* is $\mathbf{Adv}_F^{ow}(I) = \Pr\left[\,OW_F^I\,\right]$.

SELECTIVE VERSUS ADAPTIVE ID. We are interested in both these variants for all the notions we consider. To avoid a proliferation of similar definitions, we capture the variants instead via different adversary classes relative to the same game. To exemplify, consider game OW$_F$ of Fig. 2. Say that an adversary *A* is *selective-id* if the identity *id* in its queries to **Initialize** and **Ch** is always the same, and say it is *adaptive-id* if this is not necessarily true. Selective-id security for one-wayness is thus captured by restricting attention to selective-id adversaries and full (adaptive-id) security by allowing adaptive-id adversaries. Now, adopt the same definitions of selective and adaptive adversaries relative to *any* game that provides procedures called **Initialize** and **Ch**, regardless of how these procedures operate. In this way, other notions we will introduce, including partial lossiness defined via games also in Fig. 2, will automatically have selective-id and adaptive-id security versions.

PARTIAL LOSSINESS. We first provide the formal definitions and later explain them and their relation to standard definitions. If *f* is a function with domain a (non-empty) set Dom(*f*) then its image is Im(*f*) = { *f*(*x*) : *x* ∈ Dom(*f*) }. We define the *lossiness* $\lambda(f)$ of *f* via $\lambda(f) = \lg(|Dom(f)|/|Im(f)|)$ or equivalently $|Im(f)| = |Dom(f)| \cdot 2^{-\lambda(f)}$. We say that *f* is *ℓ-lossy* if $\lambda(f) \geq \ell$. Let IBTDF F = (F.Pg, F.Kg, F.Ev, F.Ev^{-1}) be an IBTDF with associated input space InSp and identity space IDSp. A *sibling* for F is an E-IBTDF LF = (LF.Pg, LF.Kg, F.Ev, F.Ev^{-1}) whose evaluation and inversion algorithms, as the notation indicates, are those of F and whose auxiliary input space is IDSp. Algorithm LF.Pg will use this input in the selective-id case and ignore it in the adaptive-id case. Consider games Real_F and $\text{Lossy}_{F,LF,\ell}$ of Fig. 2. The first uses the real parameter and key-generation algorithms while the second uses the sibling ones. A los-adversary A is allowed just one **Ch** query, and the games do no more than record the challenge identity id^*. The advantage $\mathbf{Adv}^{\delta\text{-los}}_{F,LF,\ell}(A) = \delta \cdot \Pr[\text{Real}^A_F] - \Pr[\text{Lossy}^A_{F,LF,\ell}]$ of the adversary is *not*, as usual, the difference in the probabilities that the games return true, but is instead parameterized by a probability $\delta \in [0,1]$.

DISCUSSION. The PW [47] notion of lossy TDFs in the public-key setting asks for an alternative "sibling" key-generation algorithm, producing a public key but no secret key, such that two conditions hold. The first, which is combinatorial, asks that the functions defined by sibling keys are lossy. The second, which is computational, asks that real and sibling keys are indistinguishable. The first change for the IB setting is that one needs an alternative parameter generation algorithm which produces not only *pars* but a master secret key *msk*, and an alternative key-generation algorithm that, based on *msk*, can issue decryption keys to users. Now we would like to ask that the function $F.Ev(pars, id^*, \cdot)$ be lossy on the challenge identity id^* when *pars* is generated via LF.Pg, but, in the adaptive-id case, we do not know id^* in advance. Thus the requirement is made via the games.

We would like to define the advantage normally, meaning with $\delta = 1$, but the resulting notion is not achievable in the adaptive-id case. (This can be shown via attack.) With the relaxation, a low (close to zero) advantage means that the probability that the adversary finds a lossy identity id^* and then outputs 1 is less than the probability that it merely outputs 1 by a factor not much less than δ. Roughly, it means that a δ fraction of identities are lossy. The advantage represents the computational loss while δ represents a necessary information-theoretic loss.

IBE. Recall that an IBE scheme $\text{IBE} = (\text{IBE.Pg}, \text{IBE.Kg}, \text{IBE.Enc}, \text{IBE.Dec})$ is a tuple of algorithms with associated message space InSp and identity space IDSp. The parameter generation algorithm IBE.Pg takes no input and returns common parameters *pars* and a master secret key *msk*. On input *pars, msk, id*, the key generation algorithm IBE.Kg produces a decryption key *dk* for identity *id*. On input *pars*, $id \in \text{IDSp}$ and a message $M \in \text{InSp}$ the encryption algorithm IBE.Enc returns a ciphertext. The decryption algorithm IBE.Dec is deterministic. The scheme has decryption error ϵ if $\Pr[\text{IBE.Dec}(pars, id, dk, \text{IBE.Enc}(pars, id, M)) \neq M] \leq \epsilon$ for all *pars*, all $id \in \text{IDSp}$, all $dk \in [F.Kg(pars, id)]$ and all $M \in \text{InSp}$. We say that IBE is deterministic if IBE.Enc is deterministic. A deterministic IBE scheme is identical to an IBTDF.

3   IB-TDFs from Pairings

In [10] we show that δ-lossiness implies one-wayness in both the selective and adaptive cases. We now show how to achieve δ-lossiness using pairings.

SETUP. Throughout we fix a bilinear map $\mathbf{e}\colon \mathbb{G} \times \mathbb{G} \to \mathbb{G}_T$ where $\mathbb{G}, \mathbb{G}_T$ are groups of prime order p. By $\mathbf{1}, \mathbf{1}_T$ we denote the identity elements of $\mathbb{G}, \mathbb{G}_T$, respectively. By $\mathbb{G}^* = \mathbb{G} - \{\mathbf{1}\}$ we denote the set of generators of $\mathbb{G}$. The advantage of a dlin-adversary B is $\mathbf{Adv}^{\mathrm{dlin}}(B) = 2\Pr[\mathrm{DLIN}^B] - 1$, where game DLIN is as follows. The **Initialize** procedure picks $g, \hat{g}$ at random from $\mathbb{G}^*$, s at random from $\mathbb{Z}_p^*$, $\hat{s}$ at random from $\mathbb{Z}_p$ and X at random from $\mathbb{G}$. It picks a random bit b. If $b = 1$ it lets $T \leftarrow X^{s+\hat{s}}$ and otherwise picks T at random from $\mathbb{G}$. It returns $(g, \hat{g}, g^s, \hat{g}^{\hat{s}}, X, T)$ to the adversary B. The adversary outputs a bit b' and **Finalize**, given b' returns true if $b = b'$ and false otherwise. For integer $\mu \geq 1$, vectors $\mathbf{U} \in \mathbb{G}^{\mu+1}$ and $\mathbf{y} \in \mathbb{Z}_p^{\mu+1}$, and vector $id \in \mathbb{Z}_p^\mu$ we let $\overline{id} = (1, id[1], \ldots, id[\mu]) \in \mathbb{Z}_p^{\mu+1}$ and $\mathcal{H}(\mathbf{U}, id) = \prod_{k=0}^{\mu} \mathbf{U}[k]^{\overline{id}[k]}$. $\mathcal{H}$ is the BB hash function [17] when $\mu = 1$, and the Waters' one [22] when $\mathsf{IDSp} = \{0,1\}^\mu$ and an $id \in \mathsf{IDSp}$ is viewed as a μ-vector over $\mathbb{Z}_p$. We also let $f(\mathbf{y}, id) = \sum_{k=0}^{\mu} \mathbf{y}[k]\overline{id}[k]$ and $\overline{f}(\mathbf{y}, id) = f(\mathbf{y}, id) \bmod p$.

OVERVIEW. In the Peikert-Waters [47] design, the matrix entries are ciphertexts of an underlying homomorphic encryption scheme, and the function output is a vector of ciphertexts of the same scheme. We begin by presenting an IBE scheme, that we call the basic IBE scheme, such that the function outputs of our eventual IB-TDF will be a vector of ciphertexts of this IBE scheme. Towards building the IB-TDF, the first difficulty we run into in setting up the matrix is that ciphertexts depend on the identity and we cannot have a different matrix for every identity. Thus, our approach is more intrusive. We will have many matrices which contain certain "atoms" from which, given an identity, one can reconstruct ciphertexts of the IBE scheme. The result of this intrusive approach is that security of the IB-TDF relies on more than security of the base IBE scheme. Our ciphertext pseudorandomness lemma (Lemma 1) shows something stronger, namely that even the atoms from which the ciphertexts are created look random under DLIN. This will be used to establish Lemma 2, which moves from the real to the lossy setup. The heart of the argument is the proofs of the lemmas, which are in the appendices.

We introduce a general framework that allows us to treat both the selective-id and adaptive-id cases in as unified a way as possible. We will first specify an E-IBTDF. The selective-id and adaptive-id IB-TDFs are obtained via different auxiliary inputs. Furthermore, the siblings used to prove lossiness also emanate from this E-IBTDF. With this approach, the main lemmas become usable in both the selective-id and adaptive-id cases with only minor adjustments for the latter due to artifical aborts. This saves us from repeating similar arguments and significantly compacts the proof.

OUR BASIC IBE SCHEME. We associate to any integer $\mu \geq 1$ and any identity space $\mathsf{IDSp} \subseteq \mathbb{Z}_p^\mu$ an IBE scheme $\mathsf{IBE}[\mu, \mathsf{IDSp}]$ that has message space $\{0, 1\}$ and algorithms as follows:

1. <u>Parameters:</u> Algorithm $\mathsf{IBE}[\mu, \mathsf{IDSp}].\mathsf{Pg}$ lets $g \xleftarrow{\$} \mathbb{G}^*$; $t \xleftarrow{\$} \mathbb{Z}_p^*$; $\hat{g} \leftarrow g^t$. It then lets $H, \hat{H} \xleftarrow{\$} \mathbb{G}$; $\mathbf{U}, \hat{\mathbf{U}} \xleftarrow{\$} \mathbb{G}^{\mu+1}$. It returns $pars = (g, \hat{g}, H, \hat{H}, \mathbf{U}, \hat{\mathbf{U}})$ as the public parameters and $msk = t$ as the master secret key.

2. Key generation: Given parameters $(g, \hat{g}, H, \hat{H}, \mathbf{U}, \hat{\mathbf{U}})$, master secret t and identity $id \in \mathsf{IDSp}$, algorithm $\mathsf{IBE}[\mu, \mathsf{IDSp}].\mathsf{Kg}$ returns decryption key (D_1, D_2, D_3, D_4) computed by letting $r, \hat{r} \xleftarrow{\$} \mathbb{Z}_p$ and setting

$$D_1 \leftarrow \mathcal{H}(\mathbf{U}, id)^{tr} \cdot H^{t\hat{r}} \; ; \; D_2 \leftarrow \mathcal{H}(\hat{\mathbf{U}}, id)^r \cdot \hat{H}^{\hat{r}} \; ; \; D_3 \leftarrow g^{-tr} \; ; \; D_4 \leftarrow g^{-t\hat{r}} \; .$$

3. Encryption: Given parameters $(g, \hat{g}, H, \hat{H}, \mathbf{U}, \hat{\mathbf{U}})$, identity $id \in \mathsf{IDSp}$ and message $M \in \{0,1\}$, algorithm $\mathsf{IBE}[\mu, \mathsf{IDSp}].\mathsf{Enc}$ returns ciphertext (C_1, C_2, C_3, C_4) computed as follows. If $M = 0$ then it lets $s, \hat{s} \xleftarrow{\$} \mathbb{Z}_p$ and $C_1 \leftarrow g^s \; ; \; C_2 \leftarrow \hat{g}^{\hat{s}} \; ; \; C_3 \leftarrow \mathcal{H}(\mathbf{U}, id)^s \cdot \mathcal{H}(\hat{\mathbf{U}}, id)^{\hat{s}} \; ; \; C_4 \leftarrow H^s \hat{H}^{\hat{s}}$. If $M = 1$ it lets $C_1, C_2, C_3, C_4 \xleftarrow{\$} \mathbb{G}$.

4. Decryption: Given parameters $(g, \hat{g}, H, \hat{H}, \mathbf{U}, \hat{\mathbf{U}})$, identity $id \in \mathsf{IDSp}$, decryption key (D_1, D_2, D_4, D_4) for id and ciphertext (C_1, C_2, C_3, C_4), algorithm $\mathsf{IBE}[\mu, \mathsf{IDSp}].\mathsf{Dec}$ returns 0 if $\mathbf{e}(C_1, D_1)\mathbf{e}(C_2, D_2)\mathbf{e}(C_3, D_3)\mathbf{e}(C_4, D_4) = \mathbf{1}_T$ and 1 otherwise.

This scheme has non-zero decryption error (at most $2/p$) yet our IBTDF will have zero inversion error. This scheme turns out to be IND-CPA+ANON-CPA although we will not need this in what follows. Instead we will have to consider a distinguishing game related to this IBE scheme and our IBTDF. In [10] we give a (more natural) variant of $\mathsf{IBE}[\mu, \mathsf{IDSp}]$ that is more efficient and encrypts strings rather than bits. The improved IBE scheme can still be proved IND-CPA+ANON-CPA but it cannot be used for our purpose of building IB-TDFs.

Our E-IBTDF and IB-TDF. Our E-IBTDF $\overline{\mathsf{E}}[n, \mu, \mathsf{IDSp}]$ is associated to any integers $n, \mu \geq 1$ and any identity space $\mathsf{IDSp} \subseteq \mathbb{Z}_p^\mu$. It has message space $\{0,1\}^n$ and auxiliary input space $\mathbb{Z}_p^{\mu+1}$, and the algorithms are as follows:

1. Parameters: Given auxiliary input $\mathbf{y}$, algorithm $\overline{\mathsf{E}}[n, \mu, \mathsf{IDSp}].\mathsf{Pg}$ lets $g \xleftarrow{\$} \mathbb{G}^*$; $t \xleftarrow{\$} \mathbb{Z}_p^*$; $\hat{g} \leftarrow g^t$; $U \xleftarrow{\$} \mathbb{G}^*$. It then lets $\mathbf{H}, \hat{\mathbf{H}} \xleftarrow{\$} \mathbb{G}^n$; $\mathbf{V}, \hat{\mathbf{V}} \xleftarrow{\$} \mathbb{G}^{n \times (\mu+1)}$ and $\mathbf{s} \xleftarrow{\$} (\mathbb{Z}_p^*)^n$; $\hat{\mathbf{s}} \xleftarrow{\$} \mathbb{Z}_p^n$. It returns $pars = (g, \hat{g}, \mathbf{G}, \hat{\mathbf{G}}, \mathbf{J}, \mathbf{W}, \mathbf{H}, \hat{\mathbf{H}}, \mathbf{V}, \hat{\mathbf{V}}, U)$ as the public parameters and $msk = t$ as the master secret key where for $1 \leq i, j \leq n$ and $0 \leq k \leq \mu$:

$$\mathbf{G}[i] \leftarrow g^{\mathbf{s}[i]} \; ; \; \hat{\mathbf{G}}[i] \leftarrow \hat{g}^{\hat{\mathbf{s}}[i]} \; ; \; \mathbf{J}[i,j] \leftarrow \mathbf{H}[j]^{\mathbf{s}[i]} \hat{\mathbf{H}}[j]^{\hat{\mathbf{s}}[i]}$$
$$\mathbf{W}[i,j,k] \leftarrow \mathbf{V}[j,k]^{\mathbf{s}[i]} \hat{\mathbf{V}}[j,k]^{\hat{\mathbf{s}}[i]} U^{\mathbf{s}[i]\mathbf{y}[k]\Delta(i,j)} \; ,$$

where we recall that $\Delta(i,j) = 1$ if $i = j$ and 0 otherwise is the Kronecker Delta function.

2. Key generation: Given parameters $(g, \hat{g}, \mathbf{G}, \hat{\mathbf{G}}, \mathbf{J}, \mathbf{W}, \mathbf{H}, \hat{\mathbf{H}}, \mathbf{V}, \hat{\mathbf{V}}, U)$, master secret t and identity $id \in \mathsf{IDSp}$, algorithm $\overline{\mathsf{E}}[n, \mu, \mathsf{IDSp}].\mathsf{Kg}$ returns decryption key $(\mathbf{D}_1, \mathbf{D}_2, \mathbf{D}_3, \mathbf{D}_4)$ where $\mathbf{r} \xleftarrow{\$} (\mathbb{Z}_p^*)^n$; $\hat{\mathbf{r}} \xleftarrow{\$} \mathbb{Z}_p^n$ and for $1 \leq i \leq n$

$$\mathbf{D}_1[i] \leftarrow \mathcal{H}(\mathbf{V}[i,\cdot], id)^{t\mathbf{r}[i]} \cdot \mathbf{H}[i]^{t\hat{\mathbf{r}}[i]} \; ; \; \mathbf{D}_2[i] \leftarrow \mathcal{H}(\hat{\mathbf{V}}[i,\cdot], id)^{\mathbf{r}[i]} \cdot \hat{H}[i]^{\hat{\mathbf{r}}[i]}$$
$$\mathbf{D}_3[i] \leftarrow g^{-t\mathbf{r}[i]} \; ; \; \mathbf{D}_4[i] \leftarrow g^{-t\hat{\mathbf{r}}[i]} \; .$$

3. <u>Evaluate:</u> Given parameters $(g, \hat{g}, \mathbf{G}, \hat{\mathbf{G}}, \mathbf{J}, \mathbf{W}, \mathbf{H}, \hat{\mathbf{H}}, \mathbf{V}, \hat{\mathbf{V}}, U)$, identity $id \in$ IDSp and input $x \in \{0,1\}^n$, algorithm $\overline{\mathsf{E}}[n, \mu, \mathsf{IDSp}].\mathsf{Ev}$ returns $(C_1, C_2, \mathbf{C}_3, \mathbf{C}_4)$ where for $1 \leq j \leq n$

$$C_1 \leftarrow \prod_{i=1}^n \mathbf{G}[i]^{x[i]} \; ; \; C_2 \leftarrow \prod_{i=1}^n \hat{\mathbf{G}}[i]^{x[i]}$$
$$\mathbf{C}_3[j] \leftarrow \prod_{i=1}^n \prod_{k=0}^\mu \mathbf{W}[i,j,k]^{x[i]\overline{id}[k]} \; ; \; \mathbf{C}_4[j] \leftarrow \prod_{i=1}^n \mathbf{J}[i,j]^{x[i]}$$

4. <u>Invert:</u> Given parameters $(g, \hat{g}, \mathbf{G}, \hat{\mathbf{G}}, \mathbf{J}, \mathbf{W}, \mathbf{H}, \hat{\mathbf{H}}, \mathbf{V}, \hat{\mathbf{V}}, U)$, identity $id \in$ IDSp, decryption key $(\mathbf{D}_1, \mathbf{D}_2, \mathbf{D}_3, \mathbf{D}_4)$ for id and output (ciphertext) $(C_1, C_2, \mathbf{C}_3, \mathbf{C}_4)$, algorithm $\overline{\mathsf{E}}[n, \mu, \mathsf{IDSp}].\mathsf{Ev}^{-1}$ returns $x \in \{0,1\}^n$ where for $1 \leq j \leq n$ it sets $x[j] = 0$ if $\mathbf{e}(C_1, \mathbf{D}_1[j])\mathbf{e}(C_2, \mathbf{D}_2[j])\mathbf{e}(\mathbf{C}_3[j], \mathbf{D}_3[j])\mathbf{e}(\mathbf{C}_4[j], \mathbf{D}_4[j]) = \mathbf{1}_T$ and 1 otherwise.

INVERTIBILITY. We observe that if parameters $(g, \hat{g}, \mathbf{G}, \hat{\mathbf{G}}, \mathbf{J}, \mathbf{W}, \mathbf{H}, \hat{\mathbf{H}}, \mathbf{V}, \hat{\mathbf{V}}, U)$ were generated with auxiliary input $\mathbf{y}$ and $(C_1, C_2, \mathbf{C}_3, \mathbf{C}_4) = \overline{\mathsf{E}}[n, \mu, \mathsf{IDSp}].\mathsf{Ev}((g, \hat{g}, \mathbf{G}, \hat{\mathbf{G}}, \mathbf{J}, \mathbf{W}), id, x)$ then for $1 \leq j \leq n$

$$C_1 = \prod_{i=1}^n g^{\mathbf{s}[i]x[i]} = g^{\langle \mathbf{s}, x \rangle} \tag{1}$$

$$C_2 = \prod_{i=1}^n \hat{g}^{\hat{\mathbf{s}}[i]x[i]} = \hat{g}^{\langle \hat{\mathbf{s}}, x \rangle} \tag{2}$$

$$\mathbf{C}_3[j] = \prod_{i=1}^n \prod_{k=0}^\mu \mathbf{V}[j,k]^{\mathbf{s}[i]x[i]\overline{id}[k]} \hat{\mathbf{V}}[j,k]^{\hat{\mathbf{s}}[i]x[i]\overline{id}[k]} U^{\mathbf{s}[i]x[i]\mathbf{y}[k]\overline{id}[k]\Delta(i,j)}$$
$$= \prod_{i=1}^n \mathcal{H}(\mathbf{V}[j,\cdot], id)^{\mathbf{s}[i]x[i]} \mathcal{H}(\hat{\mathbf{V}}[j,\cdot], id)^{\hat{\mathbf{s}}[i]x[i]} U^{\mathbf{s}[i]x[i]f(\mathbf{y},id)\Delta(i,j)}$$
$$= \mathcal{H}(\mathbf{V}[j,\cdot], id)^{\langle \mathbf{s}, x \rangle} \mathcal{H}(\hat{\mathbf{V}}[j,\cdot], id)^{\langle \hat{\mathbf{s}}, x \rangle} U^{\mathbf{s}[j]x[j]f(\mathbf{y},id)} \tag{3}$$

$$\mathbf{C}_4[j] = \prod_{i=1}^n \mathbf{H}[j]^{\mathbf{s}[i]x[i]} \hat{\mathbf{H}}[j]^{\hat{\mathbf{s}}[i]x[i]} = \mathbf{H}[j]^{\langle \mathbf{s}, x \rangle} \hat{\mathbf{H}}[j]^{\langle \hat{\mathbf{s}}, x \rangle} \; . \tag{4}$$

Thus if $x[j] = 0$ then $(C_1, C_2, \mathbf{C}_3[j], \mathbf{C}_4[j])$ is an encryption, under our base IBE scheme, of the message 0, with coins $\langle \mathbf{s}, x \rangle \bmod p$, $\langle \hat{\mathbf{s}}, x \rangle \bmod p$, parameters $(g, \hat{g}, \mathbf{H}[j], \hat{\mathbf{H}}[j], \mathbf{V}[j, \cdot], \hat{\mathbf{V}}[j, \cdot])$ and identity id. The inversion algorithm will thus correctly recover $x[j] = 0$. On the other hand suppose $x[j] = 1$. Then $\mathbf{e}(C_1, \mathbf{D}_1[j])\mathbf{e}(C_2, \mathbf{D}_2[j])\mathbf{e}(\mathbf{C}_3[j], \mathbf{D}_3[j])\mathbf{e}(\mathbf{C}_4[j], \mathbf{D}_4[j]) = \mathbf{e}(U^{\mathbf{s}[j]x[j]f(\mathbf{y},id)}, \mathbf{D}_3[j])$. Now suppose $f(\mathbf{y}, id) \bmod p \neq 0$. Then $U^{\mathbf{s}[j]x[j]f(\mathbf{y},id)} \neq \mathbf{1}$ because we chose $\mathbf{s}[j]$ to be non-zero modulo p and $\mathbf{D}_3[j] \neq \mathbf{1}$ because we chose $\mathbf{r}[j]$ to be non-zero modulo p. So the result of the pairing is never $\mathbf{1}_T$, meaning the inversion algorithm will again correctly recover $x[j] = 1$. We have established that auxiliary input $\mathbf{y}$ grants invertibility, meaning induced IBTDF $\overline{\mathsf{E}}[n, \mu, \mathsf{IDSp}](\mathbf{y})$ satisfies the correct inversion condition, if $f(\mathbf{y}, id) \bmod p \neq 0$ for all $id \in$ IDSp.

OUR IBTDF. We associate to any integers $n, \mu \geq 1$ and any identity space IDSp $\subseteq \mathbb{Z}_p^\mu$ the IBTDF scheme induced by our E-IBTDF $\overline{\mathsf{E}}[n, \mu, \mathsf{IDSp}]$ via auxiliary input $\mathbf{y} = (1, 0, \ldots, 0) \in \mathbb{Z}_p^{\mu+1}$, and denote this IBTDF scheme by $\overline{\mathsf{F}}[n, \mu, \mathsf{IDSp}]$. This IBTDF satisfies the correct inversion requirement because $f(\mathbf{y}, id) = \overline{id}[0] = 1 \neq 0 \pmod{p}$ for all id. We will show that this IBTDF is selective-id secure when $\mu = 1$ and IDSp $= \mathbb{Z}_p$, and adaptive-id secure when IDSp $= \{0, 1\}^\mu$. In the first case, it is fully lossy (i.e. 1-lossy) and in the second it is δ-lossy for appropriate δ. First we prove two technical lemmas that we will use in both cases.

proc **Initialize**(y)   // ReC, RaC

$(pars, msk) \xleftarrow{\$} \mathsf{IBE}[\mu, \mathsf{IDSp}].\mathsf{Pg}$
$(g, \hat{g}, H, \hat{H}, \mathbf{U}, \hat{\mathbf{U}}) \leftarrow pars$
$U \xleftarrow{\$} \mathbb{G}^*$
Return $(g, \hat{g}, H, \hat{H}, \mathbf{U}, \hat{\mathbf{U}}, U)$

proc **GetDK**(id)   // ReC, RaC

If $f(\mathbf{y}, id) = 0$ then $dk \leftarrow \perp$
Else $dk \leftarrow \mathsf{IBE}[\mu, \mathsf{IDSp}].\mathsf{Kg}(pars, msk, id)$
Return dk

proc **Ch**()   // ReC

$s \xleftarrow{\$} \mathbb{Z}_p^*$; $\hat{s} \xleftarrow{\$} \mathbb{Z}_p$
$G \leftarrow g^s$; $\hat{G} \leftarrow \hat{g}^{\hat{s}}$; $S \leftarrow H^s \hat{H}^{\hat{s}}$
For $k = 0, \ldots, \mu$ do
   $\mathbf{Z}[k] \leftarrow (U^{\mathbf{y}[k]} \mathbf{U}[k])^s \hat{\mathbf{U}}[k]^{\hat{s}}$
Return $(G, \hat{G}, S, \mathbf{Z})$

proc **Ch**()   // RaC

$G, \hat{G}, S \xleftarrow{\$} \mathbb{G}$; $\mathbf{Z} \xleftarrow{\$} \mathbb{G}^{\mu+1}$
Return $(G, \hat{G}, S, \mathbf{Z})$

proc **Finalize**(d')   // ReC, RaC

Return $(d' = 1)$

Fig. 3. Games ReC ("Real Ciphertexts") and RaC ("Random Ciphertexts") associated to $\mathsf{IDSp} \subseteq \mathbb{Z}_p^\mu$

CIPHERTEXT PSEUDORANDOMNESS LEMMA. Consider games ReC, RaC of Fig. 3 associated to some choice of $\mathsf{IDSp} \subseteq \mathbb{Z}_p^\mu$. The adversary provides the **Initialize** procedure with an auxiliary input $\mathbf{y} \in \mathbb{Z}_p^{\mu+1}$. Parameters are generated as per our base IBE scheme with the addition of U. The decryption key for id is computed as per our base IBE scheme except that the games refuse to provide it when $f(\mathbf{y}, id) = 0$. The challenge oracle, however, does not return ciphertexts of our IBE scheme. In game ReC, it returns group elements that resemble diagonal entries of the matrices in the parameters of our E-IBTDF, and in game RaC it returns random group elements. Notice that the challenge oracle does not take an identity as input. (Indeed, it has no input.) As usual it must be invoked exactly once. The following lemma says the games are indistinguishable under DLIN. The proof is in [10].

Lemma 1. *Let $\mu \geq 1$ be an integer and $\mathsf{IDSp} \subseteq \mathbb{Z}_p^\mu$. Let P be an adversary. Then there is an adversary B such that* $\Pr\left[\,\mathrm{ReC}^P\,\right] - \Pr\left[\,\mathrm{RaC}^P\,\right] \leq (\mu + 2) \cdot \mathbf{Adv}^{\mathrm{dlin}}(B)$. *The running time of B is that of P plus some overhead.*

REAL-TO-LOSSY LEMMA. Consider games $\mathrm{RL}_0, \mathrm{RL}_n$ of Fig. 4 associated to some choice of $n, \mu, \mathsf{IDSp} \subseteq \mathbb{Z}_p^\mu$ and auxiliary input generator Aux for $\overline{\mathsf{E}}[n, \mu, \mathsf{IDSp}]$. The latter is an algorithm that takes input an identity in IDSp and returns an auxiliary input in $\mathbb{Z}_p^{\mu+1}$. Game RL_0 obtains an auxiliary input $\mathbf{y}_0$ via Aux but generates parameters exactly as $\overline{\mathsf{E}}[n, \mu, \mathsf{IDSp}].\mathsf{Pg}$ with the real auxiliary input $\mathbf{y}_1$. The game will return true under the same condition as game Real but additionally requiring that $f(\mathbf{y}_0, id) \neq 0$ for all **GetDK**(id) queries and $f(\mathbf{y}_0, id) = 0$ for the **Ch**(id) query. Game RL_n generates parameters with the auxiliary input provided by Aux but is otherwise identical to game RL_0. The following lemma says it is hard to distinguish these games. We will apply this by defining Aux in such a way that its output $\mathbf{y}_0$ results in a lossy setup. The proof of the following is in [10].

proc **Initialize**(*id*)   // RL$_0$

$\mathbf{y}_0 \xleftarrow{\$} \mathsf{Aux}(id)$; $\mathbf{y}_1 \leftarrow (1, 0, \ldots, 0)$
$(pars, msk) \xleftarrow{\$} \overline{\mathsf{E}}[n, \mu, \mathsf{IDSp}].\mathsf{Pg}(\mathbf{y}_1)$
$IS \leftarrow \emptyset$; $id^* \leftarrow id$; W$_\mathrm{IN}$ $\leftarrow$ true
Return *pars*

proc **Initialize**(*id*)   // RL$_n$

$\mathbf{y}_0 \xleftarrow{\$} \mathsf{Aux}(id)$; $\mathbf{y}_1 \leftarrow (1, 0, \ldots, 0)$
$(pars, msk) \xleftarrow{\$} \overline{\mathsf{E}}[n, \mu, \mathsf{IDSp}].\mathsf{Pg}(\mathbf{y}_0)$
$IS \leftarrow \emptyset$; $id^* \leftarrow id$; W$_\mathrm{IN}$ $\leftarrow$ true
Return *pars*

proc **GetDK**(*id*)   // RL$_0$, RL$_n$

$IS \leftarrow IS \cup \{id\}$
If $f(\mathbf{y}_0, id) = 0$ then W$_\mathrm{IN}$ $\leftarrow$ false ; $dk \leftarrow \perp$
Else $dk \leftarrow \overline{\mathsf{E}}[n, \mu, \mathsf{IDSp}].\mathsf{Kg}(pars, msk, id)$
Return dk

proc **Ch**(*id*)   // RL$_0$, RL$_n$

$id^* \leftarrow id$
If $f(\mathbf{y}_0, id) \neq 0$ then W$_\mathrm{IN}$ $\leftarrow$ false

proc **Finalize**(*d'*)   // RL$_0$, RL$_n$

Return $((d' = 1)$ and $(id^* \notin IS)$ and W$_\mathrm{IN})$

Fig. 4. Games RL$_0$, RL$_n$ ("Real-to-Losssy") associated to $n, \mu, \mathsf{IDSp} \subseteq \mathbb{Z}_p^\mu$ and auxiliary input generator algorithm Aux

Lemma 2. *Let* $n, \mu \geq 1$ *be integers and* $\mathsf{IDSp} \subseteq \mathbb{Z}_p^\mu$. *Let* Aux *be an auxiliary input generator for* $\overline{\mathsf{E}}[n, \mu, \mathsf{IDSp}]$ *and A an adversary. Then there is an adversary P such that* $\Pr[\mathrm{RL}_0^A] - \Pr[\mathrm{RL}_n^A] \leq 2n \cdot (\Pr[\mathrm{ReC}^P] - \Pr[\mathrm{RaC}^P])$. *The running time of P is that of A plus some overhead. If A is selective-id then so is P.*

The last statement allows us to use the lemma in both the selective-id and adaptive-id cases.

SELECTIVE-ID SECURITY. We show that IBTDF $\overline{\mathsf{F}}[n, 1, \mathbb{Z}_p]$ is selective-id δ-lossy for $\delta = 1$, meaning fully selective-id lossy, and hence selective-id one-way. To do this we define a sibling $\overline{\mathsf{LF}}[n, 1, \mathbb{Z}_p]$. It preserves the key-generation, evaluation and inversion algorithms of $\overline{\mathsf{F}}[n, 1, \mathbb{Z}_p]$ and alters parameter generation to

    Algorithm $\overline{\mathsf{LF}}[n, 1, \mathbb{Z}_p].\mathsf{Pg}(id)$

$\mathbf{y} \leftarrow (-id, 1)$; $(pars, msk) \xleftarrow{\$} \overline{\mathsf{E}}[n, 1, \mathbb{Z}_p].\mathsf{Pg}(\mathbf{y})$; Return $(pars, msk)$

The following says that our IBTDF is 1-lossy under the DLIN assumption with lossiness $\ell = n - 2 \lg(p)$. The proof is in [10].

Theorem 3. *Let* $n > 2 \lg(p)$ *and let* $\ell = n - 2 \lg(p)$. *Let* $\mathsf{F} = \overline{\mathsf{F}}[n, 1, \mathbb{Z}_p]$ *be the IBTDF associated by our construction to parameters* n, $\mu = 1$ *and* $\mathsf{IDSp} = \mathbb{Z}_p$. *Let* $\mathsf{LF} = \overline{\mathsf{LF}}[n, 1, \mathbb{Z}_p]$ *be the sibling associated to it as above. Let* $\delta = 1$ *and let be A a selective-id adversary. Then there is an adversary B such that* $\mathbf{Adv}_{\mathsf{F}, \mathsf{LF}, \ell}^{\delta\text{-los}}(A) \leq 2n(\mu + 2) \cdot \mathbf{Adv}^{\mathrm{dlin}}(B)$. *The running time of B is that of A plus overhead.*

ADAPTIVE-ID SECURITY. We show that IBTDF $\overline{\mathsf{F}}[n, \mu, \{0, 1\}^\mu]$ is adaptive-id δ-lossy for $\delta = (4(\mu+1)Q)^{-1}$ where Q is the number of key-derivation queries of the adversary. By [10] this means $\overline{\mathsf{F}}[n, \mu, \{0, 1\}^\mu]$ is adaptive-id one-way. To do this we define a sibling $\overline{\mathsf{LF}}_Q[n, \mu, \{0, 1\}^\mu]$. It preserves the key-generation, evaluation and inversion algorithms of $\overline{\mathsf{F}}[n, \mu, \{0, 1\}^\mu]$ and alters parameter generation to $\overline{\mathsf{LF}}[n, \mu, \{0, 1\}^\mu].\mathsf{Pg}(id)$ defined via

$\mathbf{y} \leftarrow \mathsf{Aux}$; $(pars, msk) \xleftarrow{\$} \overline{\mathsf{E}}[n, \mu, \{0, 1\}^\mu].\mathsf{Pg}(\mathbf{y})$; Return $(pars, msk)$.

where algorithm Aux is defined via

$$\mathbf{y}'[0] \xleftarrow{\$} \{0, \ldots, 2Q - 1\} \, ; \, \ell \xleftarrow{\$} \{0, \ldots, \mu + 1\} \, ; \, \mathbf{y}[0] \leftarrow \mathbf{y}'[0] - 2\ell Q$$
For $i = 1$ to μ do $\mathbf{y}[i] \xleftarrow{\$} \{0, \ldots, 2Q - 1\}$
Return $\mathbf{y} \in \mathbb{Z}_p^{\mu+1}$

The following says that our IBTDF is δ-lossy under the DLIN assumption with lossiness $\ell = n - 2\lg(p)$. The proof is in [10].

Theorem 4. *Let $n > 2\lg(p)$ and let $\ell = n - 2\lg(p)$. Let $\mathsf{F} = \overline{\mathsf{F}}[n, \mu, \{0, 1\}^\mu]$ be the IBTDF associated by our construction to parameters n, μ and $\mathsf{IDSp} = \{0, 1\}^\mu$. Let A be an adaptive-id adversary that makes a maximal number of $Q < p/(3m)$ queries and let $\delta = (4(\mu + 1)Q)^{-1}$. Let $\mathsf{LF} = \overline{\mathsf{LF}}_Q[n, \mu, \{0, 1\}^\mu]$ be the sibling associated to F, A as above. Then there is an adversary B such that $\mathbf{Adv}_{\mathsf{F},\mathsf{LF},\ell}^{\delta\text{-los}}(A) \leq 2n(\mu + 2) \cdot \mathbf{Adv}^{\mathrm{dlin}}(B)$. The running time of B is that of A plus $O(\mu^2 \rho^{-1}((\mu Q \rho)^{-1}))$ overhead, where $\rho = \frac{1}{2} \cdot \mathbf{Adv}_{\mathsf{F},\mathsf{LF},\ell}^{\delta\text{-los}}(A)$.*

We remark that we could use the proof technique of [12] which avoids the artificial abort but this increases the value of δ, making it dependent on the adversary advantage. The proof technique of [39] could be used to strengthen δ in Theorem 4 to $O(\sqrt{m}Q)^{-1}$ which is close to the optimal value Q^{-1}.

Acknowledgments. We thank Xiang Xie and the (anonymous) reviewers of Eurocrypt 2012 for their careful reading and valuable comments.

Bellare was supported in part by NSF CNS-0627779 and CCF-0915675. Kiltz was supported by a Sofja Kovalevskaja Award of the Alexander von Humboldt Foundation, funded by the German Federal Ministry for Education and Research. Waters was supported in part by NSF CNS-0915361 and CNS-0952692, AFOSR FA9550-08-1-0352, DARPA PROCEED, DARPA N11AP20006, a Google Faculty Research award, the Alfred P. Sloan Fellowship, a Microsoft Faculty Fellowship, and a Packard Foundation Fellowship.

References

1. Abeni, P., Bello, L., Bertacchini, M.: Exploiting DSA-1571: How to break PFS in SSL with EDH (July 2008),
   http://www.lucianobello.com.ar/exploiting_DSA-1571/index.html
2. Agrawal, S., Boneh, D., Boyen, X.: Efficient Lattice (H)IBE in the Standard Model. In: Gilbert, H. (ed.) EUROCRYPT 2010. LNCS, vol. 6110, pp. 553–572. Springer, Heidelberg (2010)
3. Agrawal, S., Boneh, D., Boyen, X.: Lattice Basis Delegation in Fixed Dimension and Shorter-Ciphertext Hierarchical IBE. In: Rabin, T. (ed.) CRYPTO 2010. LNCS, vol. 6223, pp. 98–115. Springer, Heidelberg (2010)
4. Ajtai, M.: Generating Hard Instances of the Short Basis Problem. In: Wiedermann, J., Van Emde Boas, P., Nielsen, M. (eds.) ICALP 1999. LNCS, vol. 1644, pp. 1–9. Springer, Heidelberg (1999)
5. Alwen, J., Peikert, C.: Generating shorter bases for hard random lattices. Theory of Computing Systems 48(3), 535–553 (2009); Preliminary version in STACS 2009

6. Bellare, M., Boldyreva, A., O'Neill, A.: Deterministic and Efficiently Searchable Encryption. In: Menezes, A. (ed.) CRYPTO 2007. LNCS, vol. 4622, pp. 535–552. Springer, Heidelberg (2007)
7. Bellare, M., Brakerski, Z., Naor, M., Ristenpart, T., Segev, G., Shacham, H., Yilek, S.: Hedged Public-Key Encryption: How to Protect against Bad Randomness. In: Matsui, M. (ed.) ASIACRYPT 2009. LNCS, vol. 5912, pp. 232–249. Springer, Heidelberg (2009)
8. Bellare, M., Halevi, S., Sahai, A., Vadhan, S.P.: Many-to-One Trapdoor Functions and Their Relation to Public-Key Cryptosystems. In: Krawczyk, H. (ed.) CRYPTO 1998. LNCS, vol. 1462, pp. 283–298. Springer, Heidelberg (1998)
9. Bellare, M., Hofheinz, D., Yilek, S.: Possibility and Impossibility Results for Encryption and Commitment Secure under Selective Opening. In: Joux, A. (ed.) EUROCRYPT 2009. LNCS, vol. 5479, pp. 1–35. Springer, Heidelberg (2009)
10. Bellare, M., Kiltz, E., Peikert, C., Waters, B.: Identity-based (lossy) trapdoor functions and applications. IACR ePrint Archive, Report 2011/479, Full version of this abstract (2011), http://eprint.iacr.org/
11. Bellare, M., Namprempre, C., Neven, G.: Security proofs for identity-based identification and signature schemes. Journal of Cryptology 22(1), 1–61 (2009)
12. Bellare, M., Ristenpart, T.: Simulation without the Artificial Abort: Simplified Proof and Improved Concrete Security for Waters' IBE Scheme. In: Joux, A. (ed.) EUROCRYPT 2009. LNCS, vol. 5479, pp. 407–424. Springer, Heidelberg (2009)
13. Bellare, M., Rogaway, P.: Optimal Asymmetric Encryption. In: De Santis, A. (ed.) EUROCRYPT 1994. LNCS, vol. 950, pp. 92–111. Springer, Heidelberg (1995)
14. Bellare, M., Rogaway, P.: The Security of Triple Encryption and a Framework for Code-Based Game-Playing Proofs. In: Vaudenay, S. (ed.) EUROCRYPT 2006. LNCS, vol. 4004, pp. 409–426. Springer, Heidelberg (2006)
15. Bennet, C., Brassard, G., Crépeau, C., Maurer, U.: Generalized privacy amplification. IEEE Transactions on Information Theory 41(6) (1995)
16. Boldyreva, A., Fehr, S., O'Neill, A.: On Notions of Security for Deterministic Encryption, and Efficient Constructions without Random Oracles. In: Wagner, D. (ed.) CRYPTO 2008. LNCS, vol. 5157, pp. 335–359. Springer, Heidelberg (2008)
17. Boneh, D., Boyen, X.: Efficient Selective-ID Secure Identity-Based Encryption Without Random Oracles. In: Cachin, C., Camenisch, J.L. (eds.) EUROCRYPT 2004. LNCS, vol. 3027, pp. 223–238. Springer, Heidelberg (2004)
18. Boneh, D., Boyen, X.: Secure Identity Based Encryption Without Random Oracles. In: Franklin, M. (ed.) CRYPTO 2004. LNCS, vol. 3152, pp. 443–459. Springer, Heidelberg (2004)
19. Boneh, D., Boyen, X., Shacham, H.: Short Group Signatures. In: Franklin, M. (ed.) CRYPTO 2004. LNCS, vol. 3152, pp. 41–55. Springer, Heidelberg (2004)
20. Boneh, D., Di Crescenzo, G., Ostrovsky, R., Persiano, G.: Public Key Encryption with Keyword Search. In: Cachin, C., Camenisch, J.L. (eds.) EUROCRYPT 2004. LNCS, vol. 3027, pp. 506–522. Springer, Heidelberg (2004)
21. Boneh, D., Franklin, M.K.: Identity based encryption from the Weil pairing. SIAM Journal on Computing 32(3), 586–615 (2003)
22. Boyen, X., Waters, B.: Anonymous Hierarchical Identity-Based Encryption (Without Random Oracles). In: Dwork, C. (ed.) CRYPTO 2006. LNCS, vol. 4117, pp. 290–307. Springer, Heidelberg (2006)
23. Boyen, X., Waters, B.: Shrinking the Keys of Discrete-Log-Type Lossy Trapdoor Functions. In: Zhou, J., Yung, M. (eds.) ACNS 2010. LNCS, vol. 6123, pp. 35–52. Springer, Heidelberg (2010)

24. Brown, D.R.: A weak randomizer attack on RSA-OAEP with e=3. IACR ePrint Archive, Report 2005/189 (2005), http://eprint.iacr.org/
25. Cachin, C., Micali, S., Stadler, M.A.: Computationally Private Information Retrieval with Polylogarithmic Communication. In: Stern, J. (ed.) EUROCRYPT 1999. LNCS, vol. 1592, pp. 402–414. Springer, Heidelberg (1999)
26. Canetti, R., Dakdouk, R.R.: Towards a Theory of Extractable Functions. In: Reingold, O. (ed.) TCC 2009. LNCS, vol. 5444, pp. 595–613. Springer, Heidelberg (2009)
27. Canetti, R., Halevi, S., Katz, J.: A Forward-Secure Public-Key Encryption Scheme. In: Biham, E. (ed.) EUROCRYPT 2003. LNCS, vol. 2656, pp. 255–271. Springer, Heidelberg (2003)
28. Cash, D., Hofheinz, D., Kiltz, E., Peikert, C.: Bonsai Trees, or How to Delegate a Lattice Basis. In: Gilbert, H. (ed.) EUROCRYPT 2010. LNCS, vol. 6110, pp. 523–552. Springer, Heidelberg (2010)
29. Cocks, C.: An Identity Based Encryption Scheme Based on Quadratic Residues. In: Honary, B. (ed.) Cryptography and Coding 2001. LNCS, vol. 2260, pp. 360–363. Springer, Heidelberg (2001)
30. Diffie, W., Hellman, M.E.: New directions in cryptography. IEEE Transactions on Information Theory 22(6), 644–654 (1976)
31. Dodis, Y., Katz, J., Xu, S., Yung, M.: Strong Key-Insulated Signature Schemes. In: Desmedt, Y.G. (ed.) PKC 2003. LNCS, vol. 2567, pp. 130–144. Springer, Heidelberg (2002)
32. Dorrendorf, L., Gutterman, Z., Pinkas, B.: Cryptanalysis of the windows random number generator. In: Ning, P., di Vimercati, S.D.C., Syverson, P.F. (eds.) ACM CCS 2007, pp. 476–485. ACM Press (October 2007)
33. Freeman, D.M., Goldreich, O., Kiltz, E., Rosen, A., Segev, G.: More Constructions of Lossy and Correlation-Secure Trapdoor Functions. In: Nguyen, P.Q., Pointcheval, D. (eds.) PKC 2010. LNCS, vol. 6056, pp. 279–295. Springer, Heidelberg (2010)
34. Gentry, C., Peikert, C., Vaikuntanathan, V.: Trapdoors for hard lattices and new cryptographic constructions. In: Ladner, R.E., Dwork, C. (eds.) 40th ACM STOC, pp. 197–206. ACM Press (May 2008)
35. Goldberg, I., Wagner, D.: Randomness in the Netscape browser. Dr. Dobb's Journal (January 1996)
36. Gutterman, Z., Malkhi, D.: Hold Your Sessions: An Attack on Java Session-Id Generation. In: Menezes, A. (ed.) CT-RSA 2005. LNCS, vol. 3376, pp. 44–57. Springer, Heidelberg (2005)
37. Håstad, J., Impagliazzo, R., Levin, L.A., Luby, M.: A pseudorandom generator from any one-way function. SIAM Journal on Computing 28(4), 1364–1396 (1999)
38. Hemenway, B., Ostrovsky, R.: Lossy trapdoor functions from smooth homomorphic hash proof systems. Electronic Colloquium on Computational Complexity TR09-127 (2009)
39. Hofheinz, D., Kiltz, E.: Programmable Hash Functions and Their Applications. In: Wagner, D. (ed.) CRYPTO 2008. LNCS, vol. 5157, pp. 21–38. Springer, Heidelberg (2008)
40. Kiltz, E., Mohassel, P., O'Neill, A.: Adaptive Trapdoor Functions and Chosen-Ciphertext Security. In: Gilbert, H. (ed.) EUROCRYPT 2010. LNCS, vol. 6110, pp. 673–692. Springer, Heidelberg (2010)
41. Kiltz, E., O'Neill, A., Smith, A.: Instantiability of RSA-OAEP under Chosen-Plaintext Attack. In: Rabin, T. (ed.) CRYPTO 2010. LNCS, vol. 6223, pp. 295–313. Springer, Heidelberg (2010)

42. Lyubashevsky, V., Micciancio, D.: On Bounded Distance Decoding, Unique Shortest Vectors, and the Minimum Distance Problem. In: Halevi, S. (ed.) CRYPTO 2009. LNCS, vol. 5677, pp. 577–594. Springer, Heidelberg (2009)
43. Micciancio, D., Peikert, C.: Trapdoors for Lattices: Simpler, Tighter, Faster, Smaller. In: Pointcheval, D., Johansson, T. (eds.) EUROCRYPT 2012. LNCS, vol. 7237, pp. 700–718. Springer, Heidelberg (2012)
44. Mueller, M.: Debian OpenSSL predictable PRNG bruteforce SSH exploit (May 2008), http://milw0rm.com/exploits/5622
45. Ouafi, K., Vaudenay, S.: Smashing SQUASH-0. In: Joux, A. (ed.) EUROCRYPT 2009. LNCS, vol. 5479, pp. 300–312. Springer, Heidelberg (2009)
46. Peikert, C.: Public-key cryptosystems from the worst-case shortest vector problem: extended abstract. In: Mitzenmacher, M. (ed.) 41st ACM STOC, pp. 333–342. ACM Press (May/June 2009)
47. Peikert, C., Waters, B.: Lossy trapdoor functions and their applications. In: Ladner, R.E., Dwork, C. (eds.) 40th ACM STOC, pp. 187–196. ACM Press (May 2008)
48. Regev, O.: On lattices, learning with errors, random linear codes, and cryptography. In: Gabow, H.N., Fagin, R. (eds.) 37th ACM STOC, pp. 84–93. ACM Press (May 2005)
49. Rivest, R.L., Shamir, A., Adleman, L.M.: A method for obtaining digital signature and public-key cryptosystems. Communications of the Association for Computing Machinery 21(2), 120–126 (1978)
50. Rogaway, P., Shrimpton, T.: A Provable-Security Treatment of the Key-Wrap Problem. In: Vaudenay, S. (ed.) EUROCRYPT 2006. LNCS, vol. 4004, pp. 373–390. Springer, Heidelberg (2006)
51. Rosen, A., Segev, G.: Chosen-Ciphertext Security via Correlated Products. In: Reingold, O. (ed.) TCC 2009. LNCS, vol. 5444, pp. 419–436. Springer, Heidelberg (2009)
52. Sakai, R., Ohgishi, K., Kasahara, M.: Cryptosystems based on pairing. In: SCIS 2000, Okinawa, Japan (January 2000)
53. Shamir, A.: Identity-Based Cryptosystems and Signature Schemes. In: Blakely, G.R., Chaum, D. (eds.) CRYPTO 1984. LNCS, vol. 196, pp. 47–53. Springer, Heidelberg (1985)
54. Waters, B.: Dual System Encryption: Realizing Fully Secure IBE and HIBE under Simple Assumptions. In: Halevi, S. (ed.) CRYPTO 2009. LNCS, vol. 5677, pp. 619–636. Springer, Heidelberg (2009)
55. Waters, B.: Efficient Identity-Based Encryption Without Random Oracles. In: Cramer, R. (ed.) EUROCRYPT 2005. LNCS, vol. 3494, pp. 114–127. Springer, Heidelberg (2005)
56. Yilek, S., Rescorla, E., Shacham, H., Enright, B., Savage, S.: When private keys are public: Results from the 2008 Debian OpenSSL vulnerability. In: IMC 2009. ACM (2009)

Dual Projective Hashing and Its Applications — Lossy Trapdoor Functions and More

Hoeteck Wee*

George Washington University
hoeteck@gwu.edu

Abstract. We introduce the notion of dual projective hashing. This is similar to Cramer-Shoup projective hashing, except that instead of smoothness, which stipulates that the output of the hash function looks random on NO instances, we require invertibility, which stipulates that the output of the hash function on NO instances uniquely determine the hashing key, and moreover, that there is a trapdoor which allows us to efficiently recover the hashing key.

- We show a simple construction of lossy trapdoor functions via dual projective hashing. Our construction encompasses almost all known constructions of lossy trapdoor functions, as given in the works of Peikert and Waters (STOC '08) and Freeman et al. (PKC '10).

- We also provide a simple construction of deterministic encryption schemes secure with respect to hard-to-invert auxiliary input, under an additional assumption about the projection map. Our construction clarifies and encompasses all of the constructions given in the recent work of Brakerski and Segev (Crypto '11). In addition, we obtain a new deterministic encryption scheme based on LWE.

1   Introduction

In [14], Cramer and Shoup introduced a new primitive called smooth projective hashing as an abstraction of their earlier chosen-ciphertext (CCA) secure encryption scheme [13]. This primitive has since found numerous applications beyond CCA security, notably password-authenticated key exchange, two-message oblivious transfer, and leakage-resilient encryption [20, 22, 31]. In each of these cases, the connection to smooth projective hashing provided two important benefits: first, a more intuitive description and analysis of previous (sometimes seemingly ad-hoc) schemes given respectively in [27], [30, 1] and [2]; second, new instantiations based on different cryptographic assumptions, such as quadratic residuocity (QR) and decisional composite residuocity (DCR) [33].

Informally, smooth projective hashing refers to a family of hash functions $\{H_k\}$ indexed by a hashing key k and whose input u comes from some "hard" language. The *projective* property stipulates that there is a projective map α defined on hashing keys such that for all YES instances u, the hash value $H_k(u)$ is completely determined by u

---

* Supported by NSF CAREER Award CNS-0953626.

D. Pointcheval and T. Johansson (Eds.): EUROCRYPT 2012, LNCS 7237, pp. 246–262, 2012.

and $\alpha(k)$. In contrast, the *smoothness* property stipulates that on NO instances, $H_k(\cdot)$ should be completely undetermined. Typically in applications, the hash value $H_k(\cdot)$ is used to "mask" and hide an input (e.g. the plaintext in encryption, or sender's input in oblivious transfer).

1.1 Our Contributions

We introduce the notion of *dual projective hashing*. As with smooth projective hashing, we consider a family of projective hash functions $\{H_k\}$ indexed by a hashing key k and whose input u comes from some "hard" language. As before, we require that on YES instances u, the hash value $H_k(u)$ is completely determined by u and $\alpha(k)$. On the other hand, for NO instances u, we require *invertibility* — that $\alpha(k)$ and $H_k(u)$ jointly determine k; moreover, there is some inversion trapdoor that allows us to efficiently recover k given $(\alpha(k), H_k(u))$ along with u. We proceed to describe two applications of dual projective hashing. In both of these applications, we will think of u as an index and k as an input to some hash function. As such, we will henceforth use $\Lambda_u^*(k)$ to denote $H_k(u)$ whenever we refer to dual projective hashing.

Lossy Trapdoor Functions. Lossy trapdoor functions (TDF) [34] is a strengthened variant of the classical notion of trapdoor functions and were introduced with the main goal of enabling simple and black-box constructions of CCA-secure encryption. A collection of lossy trapdoor functions consists of two families of functions. Functions in one family are injective and can be efficiently inverted using a trapdoor. Functions in the other family are "lossy," which means that the size of their image is signicantly smaller than the size of their domain. The only computational requirement is that a description of a randomly chosen function from the family of injective functions is computationally indistinguishable from a description of a randomly chosen function from the family of lossy functions.

Lossy trapdoor functions were introduced by Peikert and Waters [34], who showed that they imply fundamental cryptographic primitives such as trapdoor functions, collision-resistant hash functions, oblivious transfer, and CCA-secure public-key encryption. In addition, lossy trapdoor functions have already found various other applications, including deterministic public-key encryption [7], OAEP-based public-key encryption [28], "hedged" public-key encryption for protecting against bad randomness [5], security against selective opening attacks [6], and efficient non-interactive string commitments [32].

Starting from dual projective hashing, we may derive a family of lossy trapdoor functions indexed by u and given by:

$$F_u : x \mapsto \alpha(x) \| \Lambda_u^*(x)$$

The injective mode is given by uniformly sampling u from NO instances, and the lossy mode is given by uniformly sampling u from YES instances. The *injective* property guarantees that if u is a NO instance, then we can efficiently recover x from the output of the function. On the other hand, the *projective* property guarantees that if u is a YES instance, then the output is fully determined by $\alpha(x)$, and therefore reveals at most $\log |\alpha(x)|$ bits of information about x.

Deterministic Encryption. Deterministic public-key encryption (where the encryption algorihtm is deterministic) was introduced by Bellare, Boldyreva and O'Neil [3], with additional constructions given in [7, 4, 11] and in concurrent works [19, 26]. Deterministic encryption has a number of practical applications, such as efficient search on encrypted data and securing legacy protocols. Our framework further clarify the constructions of deterministic encryption schemes of Boldyreva, Fehr and O'Neill [7] for high-entropy inputs and of Brakerski and Segev [11] for hard-to-invert auxiliary input (which in particular, generalize high-entropy inputs). The former combines lossy trapdoor functions and extractors, whereas the latter rely on algebraic properties of specific instantiations of lossy trapdoor functions. Specifically, the latter presented two seemingly different schemes, one based on DDH/DLIN and the other based on QR and DCR.

We consider the deterministic encryption scheme that follows from our lossy trapdoor function (following the approach used in [7]):

- the public key is a random NO instance u and the secret key is the inversion trapdoor;
- to encrypt a message M, we output $\alpha(M) \| \Lambda_u^*(M)$.

We show that:

- if $\alpha(\cdot)$ is a strong extractor (where the seed is provided by the public parameter) for high min-entropy sources, then we obtain a deterministic encryption for high min-entropy message distributions;
- if $\alpha(\cdot)$ is a "reconstructive" extractor (which is similar to a hard-core function), then we obtain a deterministic encryption secure with respect to hard-to-invert auxiliary input.

In particular, a reconstructive extractor is also a strong extractor [35], and random linear functions are both good strong extractors and good reconstructive extractors (via the left-over hash lemma [23], the Goldreich-Levin theorem [21] and generalizations there-of). These results provide a unifying framework for deterministic encryption and clarify the relation between the previous schemes and the connections to the literature on pseudorandomness. It is also interesting to contrast this construction with leakage-resilient public-key encryption derived from smooth projective hashing [31], where the extractor comes from the hash function instead of the projection map (and the seed comes from the instance instead of the public parameter).

Instantiations. We present instantiations of dual projective hashing from all three major classes of cryptographic assumptions: (1) Diffie-Hellman assumptions like DDH and DLIN, (2) number-theoretic assumptions like QR and DCR, and (3) lattice-based assumptions like LWE. Most of these instantiations are already implicit in recent works. In fact, the early constructions of hash proof systems based on DDH and QR in [14] already satisfy the *invertibility* property, albeit inefficiently. However, since the hashing key k is encoded in the exponent, efficiently recovering the key seems as hard as computing discrete log. Instead, we will rely on hashing keys that are vectors and/or matrices over $\{0, 1\}$. (Similar constructions have been used in KDM-security [9, 10] for different technical issues.) On the other hand, the DCR-based hash proof system in [14, 12] does yield a dual projective hash function, since we can efficiently solve discrete log for base $1 + N$ over $\mathbb{Z}_{N^2}^*$, given the factorization of N.

Combining these instantiations with our generic transformations, we obtain:

- a unified treatment of almost all known constructions of lossy trapdoor functions, as given in [34, 18] (the exceptions being the QR-based constructions in [18, 29] and the one based on the Φ-hiding assumption in [28]);
- a unified treatment of both of the deterministic encryption schemes secure with respect to hard-to-invert auxiliary input given in [11];
- the first lattice-based deterministic encryption scheme that is secure with respect to hard-to-invert auxiliary input.

Relation to Smooth Projective Hashing. Having presented the applications, we would like to highlight several conceptual differences between smooth projective hashing and dual projective hashing. In smooth projective hashing, we are interested in quantifying what the projected key and the instance tells us about the hash value, whereas in dual projective hashing, we want to know what the projected key and hash value tells us about the hashing key. Moreover, in essentially all applications of smooth projective hashing, YES instances are used for functionality/correctness and NO are used to establish security; it is the other way around for dual projective hashing. Finally, in smooth projective hashing, we use the hash value to hide information; in dual projective hashing, we publish the hash value as part of the output.

1.2   Previous Work

Comparison with Previous Constructions. Peikert and Waters constructed lossy trapdoor functions from the DDH and LWE assumptions, and more generally, from any homomorphic encryption schemes with reusable randomness. The description of the trapdoor functions in their constructions are a matrix of ciphertexts, and evaluation corresponds to matrix multiplication. Hemenway and Ostrovsky [24] constructed lossy trapdoor functions from smooth projective hashing where the hash function is homomorphic, which may in turn be instantiated from the QR, DDH and DCR assumptions. The construction is syntactically very similar to the matrix-based construction in [34] (although the analysis is somewhat different): the description of the trapdoor functions are a matrix of hash values, and evaluation corresponds to matrix multiplication.

Freeman et al. [18] gave direct constructions of lossy trapdoor functions from the QR, DCR and DLIN assumptions. Each of these constructions are fairly different and there is no evident unifying theme to these constructions. Specifically, the DLIN construction is a variant of the matrix-based scheme in [34]. Mol and Yilek [29] and Bellare et al. [4] independently constructed lossy trapdoor functions from the QR and the DCR assumptions respectively. We note that the QR-based schemes in these two papers only handle bounded lossiness.

In contrast to the Hemenway-Ostrovsky construction, our construction does not rely on smoothness nor any algebraic structure on the underlying hash proof system; we also have a more direct transformation from the hash function to the lossy trapdoor function, which are syntactically and conceptually quite different from that in [24]. (For instance, we use NO instances for injective functions and YES instances for lossy functions, and it is the other way around in [24].) On the other hand, in order to instantiate the hash functions, we do rely on a vector/matrix of values, similar to the constructions developed in the different context of key dependent message security and leakage resilience [9, 31, 10].

Additional Related Work. A lossy encryption scheme is a standard public-key encryption scheme where the public key may be generated in one of two modes: in the *injective* mode, the ciphertext uniquely determines the plaintext and there is an associated secret key which allows correct decryption, and in the *lossy* mode, the ciphertext reveals no information about the plaintext [6]. Given a lossy trapdoor function, it is easy to construct a lossy encryption scheme [6]. Hemenway et al. [25] gave a direct construction of a lossy encryption scheme from any hash proof system. In the construction, the public key is also an instance from the language. However, the usage is reversed: for lossy encryption, the injective mode uses a YES instance, and the lossy mode uses a NO instance.

Organization. We formalize dual projective hashing in Section 2. We present both the definition and our results on lossy trapdoor functions in Section 3, and those for deterministic encryption in Section 4. We present the instantiations of dual projective hashing in Sections 5 through 8.

Notation. We denote by $s \leftarrow_R S$ the fact that s is picked uniformly at random from a finite set S and by $x, y, z \leftarrow_R S$ that all x, y, z are picked independently and uniformly at random from S. We denote by $\mathrm{negl}(\cdot)$ a negligible function. By PPT, we denote a probabilistic polynomial-time algorithm. Throughout, we use 1^λ as the security parameter. We use $\cdot$ to denote multiplication (or group operation) as well as component-wise multiplication. We use boldface to denote vectors (always column vectors) and matrices.

2 Dual Projective Hashing

In this section, we describe dual projective hashing more formally. We warn the reader that we use slightly different notation from the outline given in the introduction (in particular, we denote the input to $\Lambda_u^*(\cdot)$ by x instead of k).

Setup. There is a setup algorithm Setup that given the security parameter 1^λ, outputs the public parameters HP for the hash function. All algorithms are given HP as part of its input; we omit HP henceforth whenever the context is clear. Associated with each HP are a pair of disjoint sets Π_Y and Π_N corresponding to YES and NO instances respectively. We require that the uniform distributions over each of Π_Y and Π_N be efficiently samplable. Specifically, there exist a pair of sampling algorithms: SampYes(HP) outputs a random pair of values (u, w) where the first output u is uniformly distributed over Π_Y and w is the corresponding witness; SampNo(HP) outputs a random pair of values (u, τ) where the first output u is uniformly distributed over Π_N and τ is the corresponding trapdoor. We discuss the roles of the witness and the trapdoor below.

Subset Membership Assumption. The *subset membership assumption* states that the uniform distributions over Π_Y and Π_N are computationally indistinguishable, even given HP. More formally, for an adversary $\mathcal{A}$, we consider the advantage function AdvSubset$^{\mathcal{A}}(\lambda)$ given by

$$\Pr\left[\mathcal{A}(\text{HP}, u) = 1 : \text{HP} \leftarrow \text{Setup}(1^\lambda), u \leftarrow_R \Pi_Y\right] - \Pr\left[\mathcal{A}(\text{HP}, u) = 1 : \text{HP} \leftarrow \text{Setup}(1^\lambda), u \leftarrow_R \Pi_N\right]$$

The subset membership assumption states that for all PPT $\mathcal{A}$, the advantage AdvSubset$^{\mathcal{A}}(\lambda)$ is a negligible function in λ.

Projective Hashing. Fix a public parameter HP. We consider a family of hash functions $\{\Lambda_u^*(\cdot)\}$ indexed by an instance $u \in \Pi_Y \cup \Pi_N$. We also require that the hash function be efficiently computable; we call the algorithm for computing $\Lambda_u^*(\cdot)$ the private evaluation algorithm. We say that $\Lambda_u^*(\cdot)$ is *projective* if there exists a projection map $\alpha(\cdot)$ such that for all $u \in \Pi_Y$ and for all inputs x, $\alpha(x)$ completely determines $\Lambda_u^*(x)$. Specifically, we require that there exists an efficient public evaluation algorithm Pub that on input $\alpha(x)$ and for all $(u, w) \leftarrow \text{SampYes}(\text{HP})$, outputs $\Lambda_u^*(x)$. That is,

$$\text{Pub}(\alpha(x), u, w) = \Lambda_u^*(x)$$

Invertibility. We say that $\Lambda_u^*(\cdot)$ is *invertible* if there is an efficient trapdoor inversion algorithm TdInv that for all $(u, \tau) \leftarrow \text{SampNo}(\text{HP})$ and for all x, recovers x given $(\alpha(x), \Lambda_u^*(x))$ and the trapdoor τ. That is,

$$\text{TdInv}(\tau, \alpha(x), \Lambda_u^*(x)) = x$$

We note here that for two of our factoring-related instantiations, SampNo(HP) also requires as input the coin tosses used to sample HP in order to compute the inversion trapdoor (there, HP is a public RSA modulus N and τ is the factorization of N). For these instantiations, we cannot treat HP as a global system parameter; instead, it will be part of the public key in the case of deterministic encryption and part of the function index in the case of lossy trapdoor functions. We suppress this subtlety in our main constructions since SampNo is only used for functionality and not in the proof of security.

3   Lossy Trapdoor Functions

In this section, we present our results on lossy trapdoor functions. We first describe the definition of lossy TDFs given in [34].

Definition 1 (Lossy Trapdoor Functions). A collection of (m, k)-*lossy trapdoor functions* is a 4-tuple of probabilistic polynomial-time algorithms (G_0, G_1, F, F^{-1}) such that:

1. (SAMPLING A LOSSY FUNCTION.) $G_0(1^\lambda)$ outputs a function index u.
2. (SAMPLING AN INJECTIVE FUNCTION.) $G_1(1^\lambda)$ outputs a pair (u, τ) where u is a function index and τ is a trapdoor.

3. (EVALUATION OF LOSSY FUNCTIONS.) For every function index u produced by G_0, the algorithm $F(u, \cdot)$ computes a function $f_u : \{0,1\}^m \to \{0,1\}^*$, whose image is of size at most 2^{m-k}.

4. (EVALUATION OF INJECTIVE FUNCTIONS.) For every pair (u, τ) produced by G_1, the algorithm $F(u, \cdot)$ computes a injective function $f_u : \{0,1\}^m \to \{0,1\}^*$.

5. (INVERSION OF INJECTIVE FUNCTIONS.) For every pair (u, τ) produced by G_1 and every $x \in \{0,1\}^m$, we have $F^{-1}(\tau, F(\sigma, x)) = x$.

6. (SECURITY.) The first outputs of $G_0(1^\lambda)$ and $G_1(1^\lambda)$ are computationally indistinguishable.

Here λ is the security parameter, and the value k is called the *lossiness*.

Our Construction. Given a dual projective hash function, we may construct a family of lossy trapdoor functions, as shown in Fig 1.

Lossy TDF

(SAMPLING A LOSSY FUNCTION.) $G_0(1^\lambda)$: Run $(u, w) \leftarrow \mathsf{SampYes}(\mathrm{HP})$. Output $\mathrm{HP}\|u$.

(SAMPLING AN INJECTIVE FUNCTION.) $G_1(1^\lambda)$: Run $(u, \tau) \leftarrow \mathsf{SampNo}(\mathrm{HP})$. Output $(\mathrm{HP}\|u, \tau)$.

(EVALUATION.) $F(u, x)$: Output $\alpha(x)\|\Lambda_u^*(x)$.

(INVERSION.) $F^{-1}(\tau, y_0\|y_1)$: Output $\mathsf{TdInv}(\tau, y_0, y_1)$.

Note: We assume here all algorithms receive as input $\mathrm{HP} \leftarrow \mathsf{Setup}(1^\lambda)$.

Fig. 1. Lossy TDF from dual projective hashing

Theorem 1. *Under the subset membership assumption, the above construction yields a collection of $(m, m - \log | \mathrm{Im}\, \alpha|)$-lossy trapdoor functions.*

Proof. Correctness for injective functions follows readily from the invertibility property. Lossiness for lossy functions follows readily from the projective property, which implies that for $u \in \Pi_Y$, $|\mathrm{Im}\, f_u| \le |\mathrm{Im}\, \alpha|$. Security is equivalent to the subset membership assumption. □

4   Deterministic Encryption

In this section, we present our results for deterministic encryption. We begin with the definition, then some results about extractors, and finally our construction.

4.1 Deterministic Encryption

A deterministic encryption scheme is a triplet of algorithms (Gen, Enc, Dec) where Gen is randomized and Enc, Dec are deterministic. Via $(\text{PK}, \text{SK}) \leftarrow \text{Gen}(1^\lambda)$, the randomized key-generation algorithm, produces public/secret keys for security parameter 1^λ. Enc on input a public key PK and a message M, produces a ciphertext. Dec(SK, ψ) on input a secret key SK and a ciphertext ψ, outputs a message. We require correctness, namely that with overwhelming probability over (PK, SK), for all M, $\text{Dec}(\text{Enc}(M)) = M$.

Hard-to-Invert Auxiliary Inputs. Following [11, 16, 17], we consider auxiliary input $f(x)$ from which it is hard to recover x. The source of hardness may be any combination of information-theoretic hardness (where the function is many-to-one) and computational hardness (e.g. if f is a one-way permutation). An efficiently computable function $\mathcal{F} = \{f_\lambda\}$ is δ-*hard-to-invert w.r.t. an efficiently samplable distribution* $\mathcal{D}$ if for every PPT algorithm $\mathcal{A}$, it holds that $\Pr[\mathcal{A}(1^\lambda, f_\lambda(x)) = x] \leq \delta$ where the probability is taken over $x \leftarrow_\text{R} \mathcal{D}$ and over the internal coin tosses of $\mathcal{A}$.

Security with Auxiliary Input. We follow the definition of security for deterministic encryption with auxiliary input from [11, 3, 4, 7].[1] For simplicity, we will only consider security while encrypting a single message, although our proofs of security extend to multiple messages and block-wise hard-to-invert auxiliary inputs. For an adversary $\mathcal{A}$, auxiliary input function $\mathcal{F}$ and message distribution $\mathcal{M}$ over $\{0,1\}^m$, we define the advantage function

$$\mathsf{AdvPrivSInd}^{\mathcal{A},\mathcal{F},\mathcal{M}}(\lambda) := \Pr \left[b = b' : \begin{array}{l} (M_0, M_1) \leftarrow \mathcal{M}; \\ (\text{PK}, \text{SK}) \leftarrow \text{Gen}(1^\lambda); \\ b \leftarrow_\text{R} \{0,1\}; \\ \psi \leftarrow \text{Enc}(\text{PK}, M_b); \\ b' \leftarrow \mathcal{A}(\text{PK}, \psi, f(M_0), f(M_1)) \end{array} \right] - \frac{1}{2}$$

A deterministic encryption scheme is $(\mathcal{F}, \mathcal{M})$-PrivSInd secure if for all PPT $\mathcal{A}$, the advantage $\mathsf{AdvPrivSInd}^{\mathcal{A},\mathcal{F},\mathcal{M}}(\lambda)$ is a negligible function in λ.

4.2 Extractors

Reconstructive Extractors. A (ϵ, δ)-*reconstructive extractor* is a pair of functions (Ext, Rec):

- an extractor $\text{Ext} : \{0,1\}^n \times \{0,1\}^d \to \Sigma$
- a (uniform) oracle machine Rec that on input $(1^n, 1/\epsilon)$ runs in time $\text{poly}(n, 1/\epsilon, \log|\Sigma|)$.

---

[1] Specifically, we use the notion of *strong indistinguishability* (PRIV-sIND) [11, Definition 4.4] restricted to single messages.

that satisfy the following property: for every $x \in \{0,1\}^n$ and every function D such that

$$\left| \Pr_{r \leftarrow_R \{0,1\}^d} [D(r, \mathsf{Ext}(x,r)) = 1] - \Pr_{r \leftarrow_R \{0,1\}^d, \sigma \leftarrow_R \Sigma} [D(r, \sigma) = 1] \right| \geq \epsilon$$

we have:

$$\Pr[\mathsf{Rec}^D(1^n, 1/\epsilon) = x] \geq \delta$$

where the probability is over the coin tosses of Rec.

It was shown in [35] that any (ϵ, δ)-reconstructive extractor is a (strong) extractor for sources of min-entropy roughly $\log 1/\delta$. It is also easy to show that the output of any (ϵ, δ)-reconstructive extractor is pseudorandom for $\delta \cdot \mathsf{negl}(\cdot)$-hard-to-invert auxiliary inputs.

Extractors from Linear Functions. It turns out that random linear functions are not only good randomness extractors (a fact commonly referred to as the left-over hash lemma), but also good reconstructive extractors.

Lemma 1 ([21, 17, 10]). *Let q be a prime. Then, the function* $\mathsf{Ext} : \{0,1\}^n \times \mathbb{Z}_q^n \to \mathbb{Z}_q$ *given by* $(\mathbf{x}, \mathbf{a}) \mapsto \mathbf{x}^\top \mathbf{a}$ *is a* $(\epsilon, \frac{\epsilon^3}{512 \cdot n \cdot q^2})$-*reconstructive extractor.*

That is, Ext maps $(x_1, \ldots, x_n), (a_1, \ldots, a_n)$ to $a_1 x_1 + \cdots a_n x_n \pmod{q}$. Moreover, the lemma extends to the following settings:

– q is a random RSA modulus, assuming that factoring is hard on average.
– $\mathbb{G}$ is a group of prime order q with generator g, and we consider the extractor $\mathsf{Ext} : \{0,1\}^n \times \mathbb{G}^n \to \mathbb{G}$ given by $(\mathbf{x}, g^{\mathbf{a}}) \mapsto g^{\mathbf{x}^\top \mathbf{a}}$.

4.3   Our Construction

Given a dual projective hash function, we may construct a deterministic encryption scheme, as shown in Fig 2. For this construction, it is important that we state explicitly that the projection map $\alpha(\cdot)$ takes the public parameter HP as its first input.

Theorem 2. *If* $(x, \mathsf{HP}) \mapsto \alpha(\mathsf{HP}, x)$ *is a* (ϵ, δ)-*reconstructive extractor and the subset membership assumption holds, then the encryption scheme as shown above is* PrivSInd-*secure with respect to hard-to-invert auxiliary input.*

Correctness of the encryption scheme follows readily the invertibility property of dual projective hashing. IND-PRIV security follows from the next technical claim.

Lemma 2. *Let $\mathcal{A}$ be an adversary against $(\mathcal{F}, \mathcal{M})$-PrivSInd security of the above encryption scheme* $(\mathsf{Gen}, \mathsf{Enc}, \mathsf{Dec})$. *Then, we can construct adversaries $\mathcal{A}_0$ and $\mathcal{A}_1$ such that for any ϵ:*

$$\textit{either} \qquad \mathsf{AdvPrivSInd}^{\mathcal{A}, \mathcal{F}, \mathcal{M}}(\lambda) \leq \mathsf{AdvSubset}^{\mathcal{A}_0}(\lambda) + 2\epsilon$$
$$\textit{or} \qquad \Pr_{M \leftarrow \mathcal{M}} [\mathcal{A}_1(f(M)) = M] \geq \delta\epsilon$$

The running time of $\mathcal{A}_0$ is roughly that of $\mathcal{A}$ and the running time of $\mathcal{A}_1$ is $\mathsf{poly}(n, 1/\epsilon, \log |\Sigma|)$ *times that of $\mathcal{A}$.*

Deterministic Encryption Scheme

(KEY GENERATION.) $\mathsf{Gen}(1^\lambda)$: Run HP $\leftarrow$ $\mathsf{Setup}(1^\lambda)$ and $(u, \tau) \leftarrow \mathsf{SampNo}(\text{HP})$.
Output
$$\text{PK} := \text{HP}\|u \quad \text{and} \quad \text{SK} := \tau$$

(ENCRYPTION.) $\mathsf{Enc}(\text{PK}, M)$: On input PK $= \text{HP}\|u$ and message M, output the ciphertext
$$\alpha(\text{HP}, M)\|\Lambda_u^*(M)$$

(DECRYPTION.) $\mathsf{Dec}(\text{SK}, \psi)$: On input SK $= \tau$ and ciphertext $\psi = y_0\|y_1$, output
$$\mathsf{TdInv}(\tau, y_0, y_1)$$

Fig. 2. Deterministic encryption scheme from dual projective hashing

Proof. We proceed via a sequence of games. We start with Game 0 as in the PrivSInd experiment and end up with a game where the view of $\mathcal{A}$ is statistically independent of the challenge bit b. We write $u \in \Pi_N$ to denote the public key PK in Game 0. This means that the view of the adversary $\mathcal{A}$ is given by:
$$\left\langle \text{HP}\|u, \alpha(\text{HP}, M_b) \| \Lambda_u^*(M_b), f(M_0), f(M_1) \right\rangle$$

GAME 1: SWITCHING TO $u \leftarrow_{\text{R}} \Pi_{\text{Y}}$. We replace $u \leftarrow_{\text{R}} \Pi_{\text{N}}$ with sampling $(u, w) \leftarrow \mathsf{SampYes}(\text{HP})$. Clearly, Game 0 and 1 are computationally indistinguishable by hardness of subset membership, and the advantage of the adversary changes by at most $\mathsf{AdvSubset}(\lambda)$.

GAME 2: ENCRYPTING USING Pub. In the challenge ciphertext, we replace $\Lambda_u^*(M_b)$ with $\mathsf{Pub}(\alpha(\text{HP}, M_b), u, w)$. By the projective property, Games 1 and 2 are identically distributed.

GAME 3: SWITCHING THE OUTPUT OF $\alpha(\cdot)$ TO RANDOM. We replace $\alpha(\text{HP}, M_b)$ in the challenge ciphertext with a random $\sigma \leftarrow_{\text{R}} \Sigma$. That is, we change the ciphertext from
$$\alpha(\text{HP}, M_b)\|\mathsf{Pub}(\alpha(\text{HP}, M_b), u, w) \quad \text{to} \quad \sigma\|\mathsf{Pub}(\sigma, u, w)$$

If the advantage of the adversary from Game 2 to Game 3 changes by at most 2ϵ, then we are done. Otherwise, we may use $\mathcal{A}$ to construct a distinguisher D such that
$$\left| \Pr\left[D\big(\text{HP}, \alpha(\text{HP}, m), f(M)\big) = 1 \right] - \Pr\left[D\big(\text{HP}, \sigma, f(M)\big) = 1 \right] \right| > 2\epsilon$$

where HP $\leftarrow \mathsf{Setup}(1^\lambda)$, $M \leftarrow \mathcal{M}, \sigma \leftarrow_{\text{R}} \Sigma$. ($D$ simply chooses $b \leftarrow_{\text{R}} \{0, 1\}$, uses its input as M_b, chooses $M_{1-b} \leftarrow_{\text{R}} \mathcal{M}$, simulates the view of $\mathcal{A}$ using $\mathsf{Pub}(\cdot, u, w)$ to obtain an output b' and outputs 1 if $b' = b$.) By an averaging argument, with probability ϵ over $M \leftarrow \mathcal{M}$, D achieves distinguishing probability ϵ, upon which we can use Rec^D to compute M from $f(M)$ with probability δ. This means that we can invert f on the distribution $\mathcal{M}$ with probability $\epsilon \cdot \delta$.

We conclude by observing that in Game 3, the view of the adversary is statistically independent of the challenge bit b. Hence, the probability that $b' = b$ is exactly $1/2$.

$\square$

Remark 1. It follows fairly readily from the analysis that if $(x, \mathrm{HP}) \mapsto \alpha(\mathrm{HP}, x)$ is a strong extractor (which is a weaker guarantee than a reconstructive extractor), then the above encryption scheme is PrivSInd-secure with respective to high min-entropy inputs. We defer the details and a more precise statement to the full version of this paper. We also point out here that the distribution for HP must be independent of the message distribution $\mathcal{M}$ (for the same reason the seed to an extractor must be chosen independently of the weaker random source). For this reason, all known constructions of deterministic encryption only achieve security for message distributions that do not depend on the public key.

5   Instantiations from DDH and DLIN

Let $\mathbb{G}$ be a group of prime order q specified using a generator g. The DDH assumption asserts that g^{ab} is pseudorandom given g, g^a, g^b where $g \leftarrow_R \mathbb{G}; a, b \leftarrow_R \mathbb{Z}_q$. The d-LIN assumption asserts that $g_{d+1}^{r_1 + \cdots + r_d}$ is pseudorandom given $g_1, \ldots, g_{d+1}, g_1^{r_1}, \ldots, g_d^{r_d}$ where $g_1, \ldots, g_{d+1} \leftarrow_R \mathbb{G}; r_1, \ldots, r_d \leftarrow_R \mathbb{Z}_q$. DDH is equivalent to 1-LIN. We present the DLIN-based hash proof system in [9, 31], also used in [18, 11]. When instantiated with our generic transformations, this yields the DLIN-based $(m, m - d\log q)$-lossy trapdoor functions given in [18] and the DLIN-based deterministic encryption scheme in [11].

Setup. $\mathrm{HP} := (\mathbb{G}, g^{\mathbf{P}}), \mathbf{P} \leftarrow_R \mathbb{Z}_q^{d \times m}$. The language is given by

$$\Pi_Y := \left\{ g^{\mathbf{WP}} : \mathbf{W} \in \mathbb{Z}_q^{m \times d} \right\} \quad \text{and} \quad \Pi_N := \left\{ g^{\mathbf{A}} : \mathbf{A} \in \mathbb{Z}_q^{m \times m} \text{ with full rank} \right\}$$

A uniformly chosen matrix $\mathbf{A} \leftarrow_R \mathbb{Z}_q^{m \times m}$ has full rank with overwhelming probability, so Π_N is efficiently samplable via rejection sampling. The uniform distributions over Π_Y and Π_N are computationally distinguishable under the d-LIN assumption as shown in [31, 9].

Hashing. The hashing input is given by $\mathbf{x} \in \{0, 1\}^m$, with

$$\alpha(g^{\mathbf{P}}, \mathbf{x}) := g^{\mathbf{P}\mathbf{x}}$$

Private and public evaluation are given by:

$$\Lambda_{\mathbf{U}}^*(\mathbf{x}) := \mathbf{U}^{\mathbf{x}} \in \mathbb{G}^m \quad \text{and} \quad \mathsf{Pub}(g^{\mathbf{P}\mathbf{x}}, \mathbf{U}, \mathbf{W}) := g^{\mathbf{W} \cdot \mathbf{P}\mathbf{x}}$$

where $(\mathbf{U}^{\mathbf{x}})_i := \sum_{j=1}^m \mathbf{U}_{ij}^{\mathbf{x}_j}$. Observe that for $\mathbf{U} = g^{\mathbf{WP}} \in \Pi_Y$, we have

$$\Lambda_{\mathbf{U}}^*(\mathbf{x}) = g^{\mathbf{WPx}} = \mathsf{Pub}(g^{\mathbf{P}\mathbf{x}}, \mathbf{U}, \mathbf{W})$$

Inversion. The inversion trapdoor is $\mathbf{A}^{-1}$. Observe that for $\mathbf{U} = g^{\mathbf{A}} \in \Pi_{\mathsf{N}}$, we have

$$\Lambda_{\mathbf{U}}^*(\mathbf{x}) = g^{\mathbf{A}\mathbf{x}}$$

Given the inversion trapdoor $\mathbf{A}^{-1}$ and $\Lambda_{\mathbf{U}}^*(\mathbf{x})$, we can compute $g^{\mathbf{x}}$ and thus $\mathbf{x}$.

6 Instantiations from QR

Fix a Blum integer $N = PQ$ for safe primes $P, Q \equiv 3 \pmod 4$ (such that $P = 2p + 1$ and $Q = 2q + 1$ for primes p, q). Let $\mathbb{J}_N$ denote the subgroup of $\mathbb{Z}_N^*$ with Jacobi symbol $+1$, and let $\mathbb{QR}_N$ denote the cyclic subgroup of quadratic residues. Observe that $|\mathbb{J}_N| = 2pq = 2|\mathbb{QR}_N|$. The QR assumption states that the uniform distributions over $\mathbb{QR}_N$ and $\mathbb{J}_N \setminus \mathbb{QR}_N$ are computationally indistinguishable.

First Construction. We present a QR-based hash proof system based on the IBE scheme of Boneh et. al [8]. When instantiated with our generic transformations, this yields a new family of QR-based $(\log \phi(N) - 1, 1)$-lossy trapdoor functions; however, it is less efficient than that given in [18].

Setup. HP $:= (N)$. The language is given by

$$\Pi_{\mathsf{Y}} := \mathbb{QR}_N \qquad \text{and} \qquad \Pi_{\mathsf{N}} := \mathbb{J}_N \setminus \mathbb{QR}_N$$

The uniform distributions over Π_{Y} and Π_{N} are computationally indistinguishable under the QR assumption.

Hashing. The hashing input is given by $x \in \mathbb{Z}_N^*/\{\pm 1\}$, with

$$\alpha(x) := x^2$$

Private and public evaluation are given by:

$$\Lambda_u^*(x) := f(x) \qquad \text{and} \qquad \mathsf{Pub}(N, u, w) := g(w)$$

where f, g are the polynomials obtained by running the "IBE compatible algorithm" [8, Section 4] on inputs x^2, u. For $u = w^2 \in \Pi_{\mathsf{Y}}$, we have $f(x) = g(w)$ by correctness of the IBE compatible algorithm.

Inversion. The inversion trapdoor (which depends on HP) is the factorization of N. For $u = -w^2 \in \Pi_{\mathsf{N}}$, we have $J(f(x))$ is equally likely to be 1 and -1 given x^2. Given the inversion trapdoor (i.e. the factorization of N), we can compute all four square roots $\pm x_0, \pm x_1$ of x^2 along with both $J(f(x_0))$ and $J(f(x_1))$; we can then recover x.

Second Construction. We present a QR-based hash proof system implicit in [11, 24], which is a matrix analogue of original Cramer-Shoup construction [14]. When instantiated with our generic transformations, this yields the QR-based $(m, m - \log|\phi(N)|)$-lossy trapdoor functions in [24]. and the QR-based deterministic encryption scheme in [11]

Setup. HP $:= (N, g^\mathbf{P}), \mathbf{p} \leftarrow_\mathrm{R} \mathbb{Z}_{N/2}^m, g \leftarrow_\mathrm{R} \mathbb{QR}_N$. The language is given by

$$\Pi_\mathrm{Y} := \left\{ g^{\mathbf{w}\mathbf{P}^\top} : \mathbf{w} \in \mathbb{Z}_{N/2}^m \right\} \qquad \text{and} \qquad \Pi_\mathrm{N} := \left\{ (-1)^{\mathbf{I}_m} \cdot g^{\mathbf{w}\mathbf{P}^\top} : \mathbf{w} \in \mathbb{Z}_{N/2}^m \right\}$$

where in the expression for Π_N, the matrix dot product refers to element-wise multiplication. The uniform distributions over Π_Y and Π_N are computationally indistinguishable under the QR assumption as shown in [11, 24, 10].

Hashing. The hashing input is given by $\mathbf{x} \in \{0, 1\}^m$, with

$$\alpha(g^\mathbf{P}, \mathbf{x}) := g^{\mathbf{P}^\top \mathbf{x}} \in \mathbb{Z}_N^*$$

Here, $\Lambda_\mathbf{U}^* : \{0, 1\}^m \to (\mathbb{Z}_N^*)^m$, with private and public evaluation given by:

$$\Lambda_\mathbf{U}^*(\mathbf{x}) := \mathbf{U}^\mathbf{x} \qquad \text{and} \qquad \mathsf{Pub}(\mathrm{PK}, \mathbf{U}, \mathbf{w}) := \mathrm{PK}^\mathbf{w}$$

where $(\mathbf{U}^\mathbf{x})_i := \sum_{j=1}^m \mathbf{U}_{ij}^{\mathbf{x}_j}$. Observe that for $\mathbf{U} = g^{\mathbf{w}\mathbf{P}^\top} \in \Pi_\mathrm{Y}$, we have

$$\Lambda_\mathbf{U}^*(\mathbf{x}) = g^{\mathbf{w}\mathbf{P}^\top \mathbf{x}} = (g^{\mathbf{P}^\top \mathbf{x}})^\mathbf{w} = \mathsf{Pub}(\mathrm{PK}, \mathbf{U}, \mathbf{w})$$

Inversion. The inversion trapdoor is the vector $\mathbf{w}$. Observe that for $\mathbf{U} = (-1)^{\mathbf{I}_m} \cdot g^{\mathbf{w}\mathbf{P}^\top} \in \Pi_\mathrm{N}$, we have

$$\Lambda_\mathbf{U}^*(\mathbf{x}) = (-1)^\mathbf{x} \cdot g^{\mathbf{w}\mathbf{P}^\top \mathbf{x}} = (-1)^\mathbf{x} \cdot \mathrm{PK}^\mathbf{w}$$

Given the inversion trapdoor $\mathbf{w}$ and $\Lambda_\mathbf{U}^*(\mathbf{x})$, we can compute $(-1)^\mathbf{x}$ and thus $\mathbf{x}$.

7   Instantiations from DCR

Fix a Blum integer $N = PQ$ for safe primes $P, Q \equiv 3 \pmod 4$ (such that $P = 2p + 1$ and $Q = 2q + 1$ for primes p, q). Let $m \in \mathbb{Z}^+$ be a parameter. The group $\mathbb{Z}_{N^{m+1}}^*$ is isomorphic to $\mathbb{Z}_{\phi(N)} \times \mathbb{Z}_{N^m}$.

First Construction. We present the Cramer-Shoup DCR-based hash proof system [14], extended to the Damgård-Jurik scheme [15]. When instantiated with our generic transformation, this yields the DCR-based $(m \log N, m \log N - \log |\phi(N)|)$-lossy trapdoor functions given in [18].

Setup. HP $:= (N, g^{N^m}), g \leftarrow_\mathrm{R} \mathbb{Z}_{N^{m+1}}^*$. The language is given by

$$\Pi_\mathrm{Y} := \left\{ g^{N^m w} : w \in \mathbb{Z}_{N^m} \right\} \qquad \text{and} \qquad \Pi_\mathrm{N} := \left\{ g^{N^m w}(1 + N) : w \in \mathbb{Z}_{N^m} \right\}$$

The uniform distributions over Π_Y and Π_N are computationally indistinguishable under the DCR assumption, as shown in [15].

Hashing. The hashing input is given by $x \in \mathbb{Z}_{N^m}$, with

$$\alpha(g^{N^m}, x) := g^{N^m x}$$

Private and public evaluation are given by:

$$\Lambda_u^*(x) := u^x \quad \text{and} \quad \mathsf{Pub}(\mathsf{PK}, u, w) := \mathsf{PK}^w$$

Observe that for $u = g^{N^m w} \in \Pi_Y$, we have

$$\Lambda_u^*(x) = g^{N^m w x} = (g^{N^m x})^w = \mathsf{Pub}(\mathsf{PK}, u, w)$$

Inversion. The inversion trapdoor (which depends on HP) is the factorization of N. For $u = g^{N^m w}(1 + N) \in \Pi_N$, we have

$$\Lambda_u^*(x) = g^{N^m w x}(1 + N)^x$$

Given the inversion trapdoor (i.e. factorization of N), we can efficiently compute x from $g^{N^m w x}(1 + N)^x$, c.f. [15].

Second Construction. There is a second DCR-based hash proof system implicit in [11], which is a matrix analogue of original Cramer-Shoup construction [14]. It is similar to the second QR-based construction, except we replace (-1) with $1 + N$. When instantiated with our generic transformations, this yields the DCR-based deterministic encryption scheme in [11].

8 Instantiations from LWE

We present the LWE-based construction, which is based on the lossy trapdoor functions in [34, Section 6.3]. For a real parameter $\beta \in (0, 1)$, we denote by Ψ_β the distribution over $\mathbb{R}/\mathbb{Z}$ of a normal variable with mean 0 and standard deviation $\beta/\sqrt{2\pi}$ then reduced modulo 1. Denote by $\bar{\Psi}_\beta$ the discrete distribution over $\mathbb{Z}_q$ of the random variable $\lfloor q X \rceil$ mod q where the random variable X has distribution Ψ_β.

In the following, we consider the standard LWE parameters m, n, q as well as additional parameters $\tilde{n}, p$ such that

$$m = O(n \log q) \quad \text{and} \quad \alpha = \Theta(1/q) \quad \text{and} \quad p \leq q/4n \quad \text{and} \quad \tilde{n} = m/\log p$$

In particular, fix $\gamma < 1$ to be a constant. Then, we will set

$$q = \Theta(n^{1+1/\gamma}) \quad \text{and} \quad p = \Theta(n^{1/\gamma})$$

When instantiated with our generic transformations, this yields the LWE-based lossy trapdoor functions in [34] and a new LWE-based deterministic encryption scheme.

Setup. $\mathsf{HP} := \mathbf{A} \leftarrow_{\mathsf{R}} \mathbb{Z}_q^{n \times m}$. The language is given by

$$\Pi_Y \leftarrow_{\mathsf{R}} \mathbf{A}^\top \mathbf{S} + \mathbf{E} \quad \text{and} \quad \Pi_N \leftarrow_{\mathsf{R}} \mathbf{A}^\top \mathbf{S} + \mathbf{E} + \mathbf{G}$$

where $\mathbf{S} \leftarrow_{\mathsf{R}} \mathbb{Z}_q^{n \times \tilde{n}}, \mathbf{E} \leftarrow_{\mathsf{R}} (\bar{\Psi}_\beta)^{m \times \tilde{n}}$. Here, $\mathbf{G} \in \mathbb{Z}_q^{m \times \tilde{n}}$ is a fixed public matrix with special structure for which the bounded error-decoding problem is easy (see [34, Section 6.3.2]). These distributions are computationally distinguishable under LWE.

Hashing. The hashing input is given by a column vector $\mathbf{x} \leftarrow_{\mathrm{R}} \{0,1\}^m$, with

$$\alpha(\mathbf{A}, \mathbf{x}) := \mathbf{A}\mathbf{x} \in \mathbb{Z}_q^n$$

Here, $\Lambda_{\mathbf{U}}^* : \{0,1\}^m \to \mathbb{Z}_q^{\tilde{n}}$, with private and public evaluation given by:

$$\Lambda_{\mathbf{U}}^*(\mathbf{x}) := \mathbf{x}^\top \mathbf{U} \qquad \text{and} \qquad \mathrm{Pub}(\mathbf{p}, \mathbf{U}, \mathbf{S}) := \mathbf{p}^\top \mathbf{S}$$

The projective property is approximate, that is,

$$\mathbf{x}^\top (\mathbf{A}^\top \mathbf{S} + \mathbf{E}) \approx (\mathbf{A}\mathbf{x})^\top \mathbf{S}$$

In fact, for all $\mathbf{x}$, with overwhelming probability over $\mathbf{E}$, we have $\mathbf{x}^\top \mathbf{E} \subseteq [q/p]^{\tilde{n}}$. That is, the projective property holds up to an additive error term in $[q/p]^{\tilde{n}}$.

Inversion. The inversion trapdoor is the matrix $\mathbf{S}$. For $\mathbf{U} \leftarrow \Pi_{\mathbf{N}}$, we have

$$(\alpha(\mathbf{A}, \mathbf{x}), \Lambda_{\mathbf{U}}^*(\mathbf{x})) = (\mathbf{A}\mathbf{x}, (\mathbf{A}\mathbf{x})^\top \mathbf{S} + \mathbf{x}^\top \mathbf{E} + \mathbf{x}^\top \mathbf{G})$$

Given $\mathbf{S}$, we can recover $\mathbf{x}^\top \mathbf{E} + \mathbf{x}^\top \mathbf{G}$. The quantity $\mathbf{x}^\top \mathbf{E}$ has small norm, so we can do bounded-error decoding to recover $\mathbf{x}^\top \mathbf{G}$ and thus $\mathbf{x}$.

Lossy TDF. For lossy TDF, in the lossy mode, we can bound the size of the image by $|\operatorname{Im}\alpha| \cdot (q/p)^{\tilde{n}}$, where the latter term accounts for the error incurred by the approximate projective property. That is, the lossiness is given by

$$m - \left(n \log q + \frac{m}{\log p} \log \frac{q}{p} \right) = (1 - \gamma)m - n \log q$$

Deterministic Encryption. For deterministic encryption, the adversary $\mathcal{A}_1$ will guess the error term $\mathbf{x}^\top \mathbf{E}$, which incurs a multiplicative loss of $(p/q)^{\tilde{n}} = 1/2^{\gamma m}$. The rest of the security loss is $q^{2n} \cdot \mathrm{poly}(m, \lambda)$. This means that for every constant $\gamma < 1$, we have a deterministic encryption scheme for m-bit messages, secure with respect to $2^{-\gamma m}$-hard-to-invert auxiliary input, based on the hardness of solving certain lattice problems with approximation factor better than $\tilde{O}(n^{2+1/\gamma})$.

Acknowledgments. I would like to thank Gil Segev and the anonymous referees for helpful and detailed comments.

References

[1] Aiello, W., Ishai, Y., Reingold, O.: Priced Oblivious Transfer: How to Sell Digital Goods. In: Pfitzmann, B. (ed.) EUROCRYPT 2001. LNCS, vol. 2045, pp. 119–135. Springer, Heidelberg (2001)

[2] Akavia, A., Goldwasser, S., Vaikuntanathan, V.: Simultaneous Hardcore Bits and Cryptography against Memory Attacks. In: Reingold, O. (ed.) TCC 2009. LNCS, vol. 5444, pp. 474–495. Springer, Heidelberg (2009)

[3] Bellare, M., Boldyreva, A., O'Neill, A.: Deterministic and Efficiently Searchable Encryption. In: Menezes, A. (ed.) CRYPTO 2007. LNCS, vol. 4622, pp. 535–552. Springer, Heidelberg (2007)

[4] Bellare, M., Fischlin, M., O'Neill, A., Ristenpart, T.: Deterministic Encryption: Definitional Equivalences and Constructions without Random Oracles. In: Wagner, D. (ed.) CRYPTO 2008. LNCS, vol. 5157, pp. 360–378. Springer, Heidelberg (2008)

[5] Bellare, M., Brakerski, Z., Naor, M., Ristenpart, T., Segev, G., Shacham, H., Yilek, S.: Hedged Public-Key Encryption: How to Protect against Bad Randomness. In: Matsui, M. (ed.) ASIACRYPT 2009. LNCS, vol. 5912, pp. 232–249. Springer, Heidelberg (2009)

[6] Bellare, M., Hofheinz, D., Yilek, S.: Possibility and Impossibility Results for Encryption and Commitment Secure under Selective Opening. In: Joux, A. (ed.) EUROCRYPT 2009. LNCS, vol. 5479, pp. 1–35. Springer, Heidelberg (2009)

[7] Boldyreva, A., Fehr, S., O'Neill, A.: On Notions of Security for Deterministic Encryption, and Efficient Constructions without Random Oracles. In: Wagner, D. (ed.) CRYPTO 2008. LNCS, vol. 5157, pp. 335–359. Springer, Heidelberg (2008)

[8] Boneh, D., Gentry, C., Hamburg, M.: Space-efficient identity based encryption without pairings. In: FOCS, pp. 647–657 (2007)

[9] Boneh, D., Halevi, S., Hamburg, M., Ostrovsky, R.: Circular-Secure Encryption from Decision Diffie-Hellman. In: Wagner, D. (ed.) CRYPTO 2008. LNCS, vol. 5157, pp. 108–125. Springer, Heidelberg (2008)

[10] Brakerski, Z., Goldwasser, S.: Circular and Leakage Resilient Public-Key Encryption under Subgroup Indistinguishability- (or: Quadratic Residuosity Strikes Back). In: Rabin, T. (ed.) CRYPTO 2010. LNCS, vol. 6223, pp. 1–20. Springer, Heidelberg (2010); Also, Cryptology ePrint Archive, Report 2010/522

[11] Brakerski, Z., Segev, G.: Better Security for Deterministic Public-Key Encryption: The Auxiliary-Input Setting. In: Rogaway, P. (ed.) CRYPTO 2011. LNCS, vol. 6841, pp. 543–560. Springer, Heidelberg (2011)

[12] Camenisch, J.L., Shoup, V.: Practical Verifiable Encryption and Decryption of Discrete Logarithms. In: Boneh, D. (ed.) CRYPTO 2003. LNCS, vol. 2729, pp. 126–144. Springer, Heidelberg (2003)

[13] Cramer, R., Shoup, V.: A Practical Public Key Cryptosystem Provably Secure against Adaptive Chosen Ciphertext Attack. In: Krawczyk, H. (ed.) CRYPTO 1998. LNCS, vol. 1462, pp. 13–25. Springer, Heidelberg (1998)

[14] Cramer, R., Shoup, V.: Universal Hash Proofs and a Paradigm for Adaptive Chosen Ciphertext Secure Public-Key Encryption. In: Knudsen, L.R. (ed.) EUROCRYPT 2002. LNCS, vol. 2332, pp. 45–64. Springer, Heidelberg (2002); Also, Cryptology ePrint Archive, Report 2001/085

[15] Damgård, I., Jurik, M.: A Generalisation, a Simplification and Some Applications of Paillier's Probabilistic Public-Key System. In: Kim, K.-c. (ed.) PKC 2001. LNCS, vol. 1992, pp. 119–136. Springer, Heidelberg (2001)

[16] Dodis, Y., Kalai, Y.T., Lovett, S.: On cryptography with auxiliary input. In: STOC, pp. 621–630 (2009)

[17] Dodis, Y., Goldwasser, S., Kalai, Y.T., Peikert, C., Vaikuntanathan, V.: Public-Key Encryption Schemes with Auxiliary Inputs. In: Micciancio, D. (ed.) TCC 2010. LNCS, vol. 5978, pp. 361–381. Springer, Heidelberg (2010)

[18] Freeman, D.M., Goldreich, O., Kiltz, E., Rosen, A., Segev, G.: More Constructions of Lossy and Correlation-Secure Trapdoor Functions. In: Nguyen, P.Q., Pointcheval, D. (eds.) PKC 2010. LNCS, vol. 6056, pp. 279–295. Springer, Heidelberg (2010); Also, Cryptology ePrint Archive, Report 2009/590

[19] Fuller, B., O'Neill, A., Reyzin, L.: A unified approach to deterministic encryption: New constructions and a connection to computational entropy. In: TCC 2012 (2012); To appear, also Cryptology ePrint Archive, Report 2012/005

[20] Gennaro, R., Lindell, Y.: A framework for password-based authenticated key exchange. ACM Trans. Inf. Syst. Secur. 9(2), 181–234 (2006)

[21] Goldreich, O., Levin, L.A.: A hard-core predicate for all one-way functions. In: STOC, pp. 25–32 (1989)

[22] Halevi, S., Kalai, Y.T.: Smooth Projective Hashing and Two-Message Oblivious Transfer. In: Cramer, R. (ed.) EUROCRYPT 2005. LNCS, vol. 3494, pp. 78–95. Springer, Heidelberg (2005)

[23] Håstad, J., Impagliazzo, R., Levin, L.A., Luby, M.: A pseudorandom generator from any one-way function. SIAM J. Comput. 28(4), 1364–1396 (1999)

[24] Hemenway, B., Ostrovsky, R.: Lossy trapdoor functions from smooth homomorphic hash proof systems. In: Electronic Colloquium on Computational Complexity, ECCC (2009)

[25] Hemenway, B., Libert, B., Ostrovsky, R., Vergnaud, D.: Lossy Encryption: Constructions from General Assumptions and Efficient Selective Opening Chosen Ciphertext Security. In: Lee, D.H., Wang, X. (eds.) ASIACRYPT 2011. LNCS, vol. 7073, pp. 70–88. Springer, Heidelberg (2011); also Cryptology ePrint Archive, Report 2009/088

[26] Ilya Mironov, O.R., Pandey, O., Segev, G.: Incremental Deterministic Public-key Encryption. In: Pointcheval, D., Johansson, T. (eds.) EUROCRYPT 2012. LNCS, vol. 7237, pp. 628–644. Springer, Heidelberg (2012)

[27] Katz, J., Ostrovsky, R., Yung, M.: Efficient and secure authenticated key exchange using weak passwords. J. ACM 57(1) (2009)

[28] Kiltz, E., O'Neill, A., Smith, A.: Instantiability of RSA-OAEP under Chosen-Plaintext Attack. In: Rabin, T. (ed.) CRYPTO 2010. LNCS, vol. 6223, pp. 295–313. Springer, Heidelberg (2010)

[29] Mol, P., Yilek, S.: Chosen-Ciphertext Security from Slightly Lossy Trapdoor Functions. In: Nguyen, P.Q., Pointcheval, D. (eds.) PKC 2010. LNCS, vol. 6056, pp. 296–311. Springer, Heidelberg (2010)

[30] Naor, M., Pinkas, B.: Efficient oblivious transfer protocols. In: SODA, pp. 448–457 (2001)

[31] Naor, M., Segev, G.: Public-Key Cryptosystems Resilient to Key Leakage. In: Halevi, S. (ed.) CRYPTO 2009. LNCS, vol. 5677, pp. 18–35. Springer, Heidelberg (2009)

[32] Nishimaki, R., Fujisaki, E., Tanaka, K.: Efficient Non-interactive Universally Composable String-Commitment Schemes. In: Pieprzyk, J., Zhang, F. (eds.) ProvSec 2009. LNCS, vol. 5848, pp. 3–18. Springer, Heidelberg (2009)

[33] Paillier, P.: Public-Key Cryptosystems Based on Composite Degree Residuosity Classes. In: Stern, J. (ed.) EUROCRYPT 1999. LNCS, vol. 1592, pp. 223–238. Springer, Heidelberg (1999)

[34] Peikert, C., Waters, B.: Lossy trapdoor functions and their applications. In: STOC, pp. 187–196 (2008)

[35] Trevisan, L.: Extractors and pseudorandom generators. JACM 48(4), 860–879 (2001)

Efficient Zero-Knowledge Argument
for Correctness of a Shuffle

Stephanie Bayer and Jens Groth*

University College London
{s.bayer,j.groth}@cs.ucl.ac.uk

Abstract. Mix-nets are used in e-voting schemes and other applications that require anonymity. Shuffles of homomorphic encryptions are often used in the construction of mix-nets. A shuffle permutes and re-encrypts a set of ciphertexts, but as the plaintexts are encrypted it is not possible to verify directly whether the shuffle operation was done correctly or not. Therefore, to prove the correctness of a shuffle it is often necessary to use zero-knowledge arguments.

We propose an honest verifier zero-knowledge argument for the correctness of a shuffle of homomorphic encryptions. The suggested argument has sublinear communication complexity that is much smaller than the size of the shuffle itself. In addition the suggested argument matches the lowest computation cost for the verifier compared to previous work and also has an efficient prover. As a result our scheme is significantly more efficient than previous zero-knowledge schemes in literature.

We give performance measures from an implementation where the correctness of a shuffle of 100,000 ElGamal ciphertexts is proved and verified in around 2 minutes.

Keywords: Shuffle, zero-knowledge, ElGamal encryption, mix-net, voting, anonymous broadcast.

1 Introduction

A mix-net [4] is a multi-party protocol which is used in e-voting or other applications which require anonymity. It allows a group of senders to input a number of encrypted messages to the mix-net, which then outputs the messages in random order. It is common to construct mix-nets from shuffles.

Informally, a shuffle of ciphertexts $C_1, \ldots, C_N$ is a set of ciphertexts $C'_1, \ldots, C'_N$ with the same plaintexts in permuted order. In our work we will examine shuffle protocols constructed from homomorphic encryption schemes. That means for a given public key pk, messages M_1, M_2, and randomness ρ_1, ρ_2 the encryption function satisfies $\mathcal{E}_{pk}(M_1 M_2; \rho_1 + \rho_2) = \mathcal{E}_{pk}(M_1; \rho_1)\mathcal{E}_{pk}(M_2; \rho_2)$. Thus, we may construct a shuffle of $C_1, \ldots, C_N$ by selecting a permutation $\pi \in \Sigma_N$ and randomizers $\rho_1, \ldots \rho_N$, and calculating $C'_1 = C_{\pi(1)}\mathcal{E}_{pk}(1; \rho_1), \ldots, C'_N = C_{\pi(N)}\mathcal{E}_{pk}(1; \rho_N)$.

A common construction of mix-nets is to let the mix-servers take turns in shuffling the ciphertexts. If the encryption scheme is semantically secure the shuffle $C'_1, \ldots, C'_N$

---

* Both authors are supported by EPSRC grant number EP/G013829/1.

D. Pointcheval and T. Johansson (Eds.): EUROCRYPT 2012, LNCS 7237, pp. 263–280, 2012.

output by a mix-server does not reveal the permutation or the messages. But this also means that a malicious mix-server in the mix-net could substitute some of the cipher-texts without being detected. In a voting protocol, it could for instance replace all ciphertexts with encrypted votes for candidate X. Therefore, our goal is to construct an interactive zero-knowledge argument that makes it possible to verify that the shuffle was done correctly (soundness), but reveals nothing about the permutation or the randomizers used (zero-knowledge).

Efficiency is a major concern in arguments for the correctness of a shuffle. In large elections it is realistic to end up shuffling millions of votes. This places considerable strain on the performance of the zero-knowledge argument both in terms of communication and computation. We will construct an honest verifier zero-knowledge argument for correctness of a shuffle that is highly efficient both in terms of communication and computation.

1.1   Related Work

The idea of a shuffle was introduced by Chaum [4] but he didn't give any method to guarantee the correctness. Many suggestions had been made how to build mix-nets or prove the correctness of a shuffle since then, but many of these approaches have been partially or fully broken, and the remaining schemes sometimes suffer from other drawbacks. None of these drawbacks are suffered by the shuffle scheme of Wikström [27] and approaches based on zero-knowledge arguments. Since zero-knowledge arguments achieve better efficiency they will be the focus of our paper.

Early contributions using zero-knowledge arguments were made by Sako and Killian [23], and Abe [1–3]. Furukawa and Sako [10] and Neff [20, 21] proposed the first shuffle arguments for ElGamal encryption with a complexity that depends linearly on the number of ciphertexts.

Furukawa and Sako's approach is based on permutation matrices and has been refined further [7, 16]. Furukawa, Miyachi, Mori, Obana, and Sako [8] presented an implementation of a shuffle argument based on permutation matrices and tested it on mix-nets handling 100,000 ElGamal ciphertexts. Recently, Furukawa and Sako [9] have reported on another implementation based on elliptic curve groups.

Wikström [28] also used the idea of permutation matrices and suggested a shuffle argument which splits in an offline and online phase. Furthermore, Terelius and Wikström [25] constructed conceptually simple shuffle arguments that allowed the restriction of the shuffles to certain classes of permutations. Both protocols are implemented in the Verificatum mix-net library [29].

Neff's approach [20] is based on the invariance of polynomials under permutation of the roots. This idea was picked up by Groth who suggested a perfect honest verifier zero-knowledge protocol [14]. Later Groth and Ishai [15] proposed the first shuffle argument where the communication complexity is sublinear in the number of ciphertexts.

1.2   Our Contribution

Results. We propose a practical efficient honest verifier zero-knowledge argument for the correctness of a shuffle. Our argument is very efficient; in particular we drastically decrease the communication complexity compared to previous shuffle arguments. We cover the case of shuffles of ElGamal ciphertexts but it is possible to adapt our argument to other homomorphic cryptosystems as well.

Our argument has sublinear communication complexity. When shuffling N ciphertexts, arranged in an $m \times n$ matrix, our argument transmits $O(m + n)$ group elements giving a minimal communication complexity of $O(\sqrt{N})$ if we choose $m = n$. In comparison, Groth and Ishai's argument [15] communicates $\Theta(m^2 + n)$ group elements and all other state of the art shuffle arguments communicate $\Theta(N)$ elements.

The disadvantage of Groth and Ishai's argument compared to the schemes with linear communication was that the prover's computational complexity was on the order of $O(Nm)$ exponentiations. It was therefore only possible to choose small m. In comparison, our prover's computational complexity is $O(N \log m)$ exponentiations for constant round arguments and $O(N)$ exponentiations if we allow a logarithmic number of rounds. In practice, we do not need to increase the round complexity until m gets quite large, so the speedup in the prover's computation is significant compared to Groth and Ishai's work and is comparable to the complexity seen in arguments with linear communication. Moreover, the verifier is fast in our argument making the entire process very light from the verifier's point of view.

In Sect. 6 we report on an implementation of our shuffle argument using shuffles of 100,000 ElGamal ciphertexts. We compare this implementation on the parameter setting for ElGamal encryption used in [8] and find significant improvements in both communication and computation. We also compare our implementation to the shuffle argument in the Verificatum mix-net [29] and find significant improvements in communication and moderate improvements in computation.

New Techniques. Groth [13] proposed efficient sublinear size arguments to be used in connection with linear algebra over a finite field. We combine these techniques with Groth and Ishai's sublinear size shuffle argument. The main problem in applying Groth's techniques to shuffling is that they were designed for use in finite fields and not for use with group elements or ciphertexts. It turns out though that the operations are mostly linear and therefore it is possible to carry them out "in the exponent"; somewhat similar to what is often done in threshold cryptography. Using this adaptation we are able to construct an efficient multi-exponentiation argument that a ciphertext C is the product of a set of known ciphertexts $C_1, \ldots, C_N$ raised to a set of hidden committed values $a_1, \ldots, a_N$. This is the main bottleneck in our shuffle argument and therefore gives us a significant performance improvement.

Groth's sublinear size zero-knowledge arguments also suffered from a performance bottleneck in the prover's computation. At some juncture it is necessary to compute the sums of the diagonal strips in a product of two matrices. This problem is made even worse in our setting because when working with group elements we have to compute these sums in the exponents. By adapting techniques for polynomial multiplication such as Toom-Cook [5, 26] and the Fast Fourier Transform [6] we are able to reduce this computation. Moreover, we generalize the interactive technique of Groth [13] to further reduce the prover's computation.

2   Preliminaries

We use vector notation in the paper, and we write $\boldsymbol{xy} = (x_1y_1, \ldots, x_ny_n)$ for the entry-wise product and correspondingly $\boldsymbol{x}^z = (x_1^z, \ldots, x_n^z)$ for vectors of group elements. Similar, we write $\boldsymbol{x}_\pi$ if the entries of vector $\boldsymbol{x}$ are permuted by the permutation π, i.e., $\boldsymbol{x}_\pi = (x_{\pi(1)}, \ldots, x_{\pi(n)})$. We use the standard inner product $\boldsymbol{x} \cdot \boldsymbol{y} = \sum_{i=1}^{n} x_iy_i$ for vectors of field elements .

Our shuffle argument is constructed with homomorphic encryption. An encryption scheme is homomorphic if for a public key pk, messages M_1, M_2, and randomness ρ_1, ρ_2 the encryption function satisfies $\mathcal{E}_{pk}(M_1M_2; \rho_1 + \rho_2) = \mathcal{E}_{pk}(M_1; \rho_1)\mathcal{E}_{pk}(M_2; \rho_2)$. We will focus on ElGamal encryption, but our construction works with many different homomorphic encryption schemes where the message space has large prime order q. To simplify the presentation, we will use notation from linear algebra. We define $\boldsymbol{C}^{\boldsymbol{a}} = \prod_{i=1}^{n} C_i^{a_i}$ for vectors $(C_1, \ldots, C_n) \in \mathbb{H}^n$ and $(a_1, \ldots, a_n)^T \in \mathbb{Z}_q^n$, where $\mathbb{H}$ is the ciphertext space.

Likewise, we need a homomorphic commitment scheme in our protocol. Again informally, a commitment scheme is homomorphic if for a commitment key ck, messages a, b, and randomizers r, s it holds that $\text{com}_{ck}(a + b; r + s) = \text{com}_{ck}(a; r)\text{com}_{ck}(b; s)$. We also demand that it is possible to commit to n elements in $\mathbb{Z}_q$, where q is a large prime, at the same time. I.e., given a vector $(a_1, \ldots, a_n)^T \in \mathbb{Z}_q^n$ we can compute a single short commitment $c = \text{com}_{ck}(\boldsymbol{a}; r) \in \mathbb{G}$, where $\mathbb{G}$ is the commitment space. The length-reducing property of the commitment scheme mapping n elements to a single commitment is what allows us to get sublinear communication complexity. Many homomorphic commitment schemes with this property can be used, but for convenience we just focus on a generalization of the Pedersen commitment scheme [22]. To simplify notation, we write $\boldsymbol{c}_A = \text{com}_{ck}(A; \boldsymbol{r})$ for the vector $(c_{A_1}, \ldots, c_{A_m}) = (\text{com}_{ck}(\boldsymbol{a}_1; r_1), \ldots, \text{com}_{ck}(\boldsymbol{a}_m; r_m))$ when A is a matrix with column vectors $\boldsymbol{a}_1, \ldots, \boldsymbol{a}_m$.

2.1   Special Honest Verifier Zero-Knowledge Argument of Knowledge

In the shuffle arguments we consider a prover $\mathcal{P}$ and a verifier $\mathcal{V}$ both of which are probabilistic polynomial time interactive algorithms. We assume the existence of a probabilistic polynomial time setup algorithm $\mathcal{G}$ that when given a security parameter λ returns a common reference string σ.

The common reference string will be $\sigma = (pk, ck)$, where pk and ck are public keys for the ElGamal encryption scheme and the generalized Pedersen commitment scheme. The encryption scheme and the commitment scheme may use different underlying groups, but we require that they have the same prime order q. We will write $\mathbb{G}$ for the group used by the commitment scheme and write $\mathbb{H}$ for the ciphertext space.

The setup algorithm can also return some side-information that may be used by an adversary; however, we require that even with this side-information the commitment scheme should remain computationally binding. The side-information models that the keys may be set up using some multi-party computation protocol that leaks some information, the adversary may see some decryptions or even learn the decryption key, etc. Our protocol for verifying the correctness of a shuffle is secure in the presence of such leaks as long as the commitment scheme is computationally binding.

Let R be a polynomial time decidable ternary relation, we call w a witness for a statement x if $(\sigma, x, w) \in R$. We define the language

$$L_\sigma := \{x \mid \exists w : (\sigma, x, w) \in R\}$$

as the set of statements x that have a witness w for the relation R.

The public transcript produced by $\mathcal{P}$ and $\mathcal{V}$ when interacting on inputs s and t is denoted by $tr \leftarrow \langle \mathcal{P}(s), \mathcal{V}(t) \rangle$. The last part of the transcript is either accept or reject from the verifier. We write $\langle \mathcal{P}(s), \mathcal{V}(t) \rangle = b$, $b \in \{0, 1\}$ for rejection or acceptance.

Definition 1 (Argument). *The triple $(\mathcal{G}, \mathcal{P}, \mathcal{V})$ is called an* argument *for a relation R with perfect completeness if for all non-uniform polynomial time interactive adversaries $\mathcal{A}$ we have:*
Perfect completeness:

$$\Pr[(\sigma, \text{hist}) \leftarrow \mathcal{G}(1^\lambda); (x, w) \leftarrow \mathcal{A}(\sigma, \text{hist}) : (\sigma, x, w) \notin R \text{ or } \langle \mathcal{P}(\sigma, x, w), \mathcal{V}(\sigma, x) \rangle = 1] = 1$$

Computational soundness:

$$\Pr[(\sigma, \text{hist}) \leftarrow \mathcal{G}(1^\lambda); x \leftarrow \mathcal{A}(\sigma, \text{hist}) : x \notin L_\sigma \text{ and } \langle \mathcal{A}, \mathcal{V}(\sigma, x) \rangle = 1] \approx 0$$

Definition 2 (Public coin). *An argument $(\mathcal{G}, \mathcal{P}, \mathcal{V})$ is called* public coin *if the verifier chooses his messages uniformly at random and independently of the messages sent by the prover, i.e., the challenges correspond to the verifier's randomness ρ.*

Definition 3 (Special honest verifier zero-knowledge). *A public coin argument $(\mathcal{G}, \mathcal{P}, \mathcal{V})$ is called a* perfect special honest verifier zero knowledge (SHVZK) *argument for R with common reference string generator $\mathcal{G}$ if there exists a probabilistic polynomial time simulator $\mathcal{S}$ such that for all non-uniform polynomial time interactive adversaries $\mathcal{A}$ we have*

$$\Pr[(\sigma, \text{hist}) \leftarrow \mathcal{G}(1^\lambda); (x, w, \rho) \leftarrow \mathcal{A}(\sigma, \text{hist});$$
$$tr \leftarrow \langle \mathcal{P}(\sigma, x, w), \mathcal{V}(\sigma, x; \rho) \rangle : (\sigma, x, w) \in R \text{ and } \mathcal{A}(tr) = 1]$$
$$= \Pr[(\sigma, \text{hist}) \leftarrow \mathcal{G}(1^\lambda); (x, w, \rho) \leftarrow \mathcal{A}(\sigma, \text{hist});$$
$$tr \leftarrow \mathcal{S}(\sigma, x, \rho) : (\sigma, x, w) \in R \text{ and } \mathcal{A}(tr) = 1]$$

To construct a fully zero-knowledge argument secure against *arbitrary* verifiers in the common reference string model one can first construct a SHVZK argument and then convert it into a fully zero-knowledge argument [11, 12]. This conversion has constant additive overhead, so it is very efficient and allows us to focus on the simpler problem of getting SHVZK against honest verifiers.

To define an argument of knowledge we follow the approach of Groth and Ishai [15] and do it through witness-extended emulation first introduced by Lindell [19]. This definition informally says that given an adversary that produces an acceptable argument with some probability, there exist an emulator that produces a similar argument with the same probability and at the same time provides a witness w.

Definition 4 (Witness-extended emulation). *A public coin argument* $(\mathcal{G}, \mathcal{P}, \mathcal{V})$ *has witness extended emulation if for all deterministic polynomial time* $\mathcal{P}^*$ *there exists an expected polynomial time emulator* $\mathcal{X}$ *such that for all non-uniform polynomial time interactive adversaries* $\mathcal{A}$ *we have*

$$\Pr[(\sigma, \text{hist}) \leftarrow \mathcal{G}(1^\lambda); (x, s) \leftarrow \mathcal{A}(\sigma, \text{hist}); tr \leftarrow \langle \mathcal{P}^*(\sigma, x, s), \mathcal{V}(\sigma, x) \rangle : \mathcal{A}(tr) = 1]$$
$$\approx \Pr[(\sigma, \text{hist}) \leftarrow \mathcal{G}(1^\lambda); (x, s) \leftarrow \mathcal{A}(\sigma, \text{hist}); (tr, w) \leftarrow \mathcal{X}^{\langle \mathcal{P}^*(\sigma, x, s), \mathcal{V}(\sigma, x) \rangle}(\sigma, x, \rho) :$$
$$\mathcal{A}(tr) = 1 \text{ and if } tr \text{ is accepting then } (\sigma, x, w) \in R].$$

In the definition, s can be interpreted as the state of $\mathcal{P}^*$, *including the randomness. So whenever* $\mathcal{P}^*$ *is able to make a convincing argument when in state s, the emulator can extract a witness at the same time giving us an argument of knowledge. This definition automatically implies soundness.*

3   Shuffle Argument

We will give an argument of knowledge of a permutation $\pi \in \Sigma_N$ and randomness $\{\rho_i\}_{i=1}^N$ such that for given ciphertexts $\{C_i\}_{i=1}^N$, $\{C_i'\}_{i=1}^N$ we have $C_i' = C_{\pi(i)} \mathcal{E}_{pk}(1; \rho_i)$. The shuffle argument combines a multi-exponentiation argument, which allows us to prove that the product of a set of ciphertexts raised to a set of committed exponents yields a particular ciphertext, and a product argument, which allows us to prove that a set of committed values has a particular product. The multi-exponentiation argument is given in Sect. 4 and the product argument is given in Sect. 5. In this section, we will give an overview of the protocol and explain how a multi-exponentiation argument can be combined with a product argument to yield an argument for the correctness of a shuffle.

The first step for the prover is to commit to the permutation. This is done by committing to $\pi(1), \ldots, \pi(N)$. The prover will now receive a challenge x and commit to $x^{\pi(1)}, \ldots, x^{\pi(N)}$. The prover will give an argument of knowledge of openings of the commitments to permutations of $1, \ldots, N$ and $x^1, \ldots, x^N$ and demonstrate that the same permutation has been used in both cases. This means the prover has a commitment to $x^1, \ldots, x^N$ permuted in an order that was fixed before the prover saw x.

To check that the same permutation has been used in both commitments the verifier sends random challenges y and z. By using the homomorphic properties of the commitment scheme the prover can in a verifiable manner compute commitments to $d_1 - z = y\pi(1) + x^{\pi(1)} - z, \ldots, d_N - z = y\pi(N) + x^{\pi(N)} - z$. Using the product argument from Sect. 5 the prover shows that $\prod_{i=1}^N (d_i - z) = \prod_{i=1}^N (yi + x^i - z)$. Observe that we have two identical degree N polynomials in z since the only difference is that the roots have been permuted. The verifier does not know a priori that the two polynomials are identical but can by the Schwartz-Zippel lemma deduce that the prover has negligible chance over the choice of z of making a convincing argument unless indeed there is a permutation π such that $d_1 = y\pi(1) + x^{\pi(1)}, \ldots, d_N = y\pi(N) + x^{\pi(N)}$. Furthermore, there is negligible probability over the choice of y of this being true unless the first commitment contains $\pi(1), \ldots, \pi(N)$ and the second commitment contains $x^{\pi(1)}, \ldots, x^{\pi(N)}$.

The prover has commitments to $x^{\pi(1)}, \ldots, x^{\pi(N)}$ and uses the multi-exponentiation argument from Sect. 4 to demonstrate that there exists a ρ such that $\prod_{i=1}^{N} C_i^{x^i} = \mathcal{E}_{pk}(1;\rho) \prod_{i=1}^{N} (C_i')^{x^{\pi(i)}}$. The verifier does not see the committed values and thus does not learn what the permutation is. However, from the homomorphic properties of the encryption scheme the verifier can deduce $\prod_{i=1}^{N} M_i^{x^i} = \prod_{i=1}^{N} (M_i')^{x^{\pi(i)}}$ for some permutation π that was chosen before the challenge x was sent to the prover. Taking discrete logarithms we have the polynomial identity $\sum_{i=1}^{N} \log(M_i)x^i = \sum_{i=1}^{N} \log(M_{\pi^{-1}(i)}')x^i$. There is negligible probability over the choice of x of this equality holding true unless $M_1' = M_{\pi(1)}, \ldots, M_N' = M_{\pi(N)}$. This shows that we have a correct shuffle.

Common reference string: pk, ck.
Statement: $C, C' \in \mathbb{H}^N$ with $N = mn$.
Prover's witness: $\pi \in \Sigma_N$ and $\rho \in \mathbb{Z}_q^N$ such that $C' = \mathcal{E}_{pk}(1;\rho)C_\pi$.
Initial message: Pick $r \leftarrow \mathbb{Z}_q^m$, set $a = \{\pi(i)\}_{i=1}^N$ and compute $c_A = \mathrm{com}_{ck}(a;r)$.
    Send: c_A
Challenge: $x \leftarrow \mathbb{Z}_q^*$.
Answer Pick $s \in \mathbb{Z}_q^m$, set $b = \{x^{\pi(i)}\}_{i=1}^N$ and compute $c_B = \mathrm{com}_{ck}(b;s)$.
    Send: c_B
Challenge: $y, z \leftarrow \mathbb{Z}_q^*$.
Answer: Define $c_{-z} = \mathrm{com}_{ck}(-z, \ldots, -z; 0)$ and $c_D = c_A^y c_B$. Compute $d = ya + b$, and $t = yr + s$. Engage in a product argument as described in Sect. 5 of openings $d_1 - z, \ldots, d_N - z$ and t such that

$$c_D c_{-z} = \mathrm{com}_{ck}(d - z; t) \qquad \text{and} \qquad \prod_{i=1}^{N}(d_i - z) = \prod_{i=1}^{N}(yi + x^i - z) \, .$$

Compute $\rho = -\rho \cdot b$ and set $x = (x, x^2, \ldots, x^N)^T$. Engage in a multi-exponentiation argument as described in Sect. 4 of b, s and ρ such that

$$C^x = \mathcal{E}_{pk}(1;\rho)C'^b \qquad \text{and} \qquad c_B = \mathrm{com}_{ck}(b;s)$$

The two arguments can be run in parallel. Furthermore, the multi-exponentiation argument can be started in round 3 after the computation of the commitments c_B.
Verification: The verifier checks $c_A, c_B \in \mathbb{G}^m$ and computes c_{-z}, c_D as described above and computes $\prod_{i=1}^{N}(yi + x^i - z)$ and C^x. The verifier accepts if the product and multi-exponentiation arguments both are valid.

Theorem 5 (Full paper). *The protocol is a public coin perfect SHVZK argument of knowledge of $\pi \in \Sigma_N$ and $\rho \in \mathbb{Z}_q^N$ such that $C' = \mathcal{E}_{pk}(1;\rho)C_\pi$.*

4   Multi-exponentiation Argument

Given ciphertexts $C_{11}, \ldots, C_{mn}$, and C we will in this section give an argument of knowledge of openings of commitments c_A to $A = \{a_{ij}\}_{i,j=1}^{n,m}$ such that

$$C = \mathcal{E}_{pk}(1;\rho) \prod_{i=1}^{m} C_i^{a_i} \qquad \text{and} \qquad c_A = \mathrm{com}_{ck}(A;r) \, ,$$

where $C_i = (C_{i1}, \ldots, C_{in})$ and $a_j = (a_{1j}, \ldots, a_{nj})^T$.

To explain the idea in the protocol let us for simplicity assume $\rho = 0$ and the prover knows the openings of c_A, and leave the question of SHVZK for later. In other words, we will for now just explain how to convince the verifier in a communication-efficient manner that $C = \prod_{i=1}^{m} C_i^{a_i}$. The prover can calculate the ciphertexts

$$E_k = \prod_{\substack{1 \le i,j \le m \\ j = (k-m)+i}} C_i^{a_j} \, ,$$

where $E_m = C$. To visualize this consider the following matrix

$$
\begin{pmatrix} a_1 & \cdots & a_m \end{pmatrix}
$$

$$
\begin{pmatrix} C_1 \\ C_2 \\ \vdots \\ C_m \end{pmatrix}
\begin{pmatrix}
C_1^{a_1} & & \ddots & & C_1^{a_m} \\
C_2^{a_1} & & \ddots & & C_2^{a_m} \\
& \ddots & \ddots & \ddots & \ddots \\
C_m^{a_1} & & \ddots & & C_m^{a_m}
\end{pmatrix}
\begin{matrix} \\ E_{2m-1} \\ \vdots \\ E_{m+1} \end{matrix}
$$

$$E_1 \ \cdots \ E_{m-1} \quad E_m$$

The prover sends the ciphertexts $E_1, \ldots, E_{2m-1}$ to the verifier. The ciphertext $C = E_m$ is the product of the main diagonal and the other E_k's are the products of the other diagonals. The prover will use a batch-proof to simultaneously convince the verifier that all the diagonal products give their corresponding E_k.

The verifier selects a challenge $x \leftarrow \mathbb{Z}_q^*$. The prover sets $x = (x, x^2, \ldots, x^m)^T$, opens c_A^x to $a = \sum_{j=1}^{m} x^j a_j$, and the verifier checks

$$C^{x^m} \prod_{\substack{k=1 \\ k \ne m}}^{2m-1} E_k^{x^k} = \prod_{i=1}^{m} C_i^{(x^{m-i} a)} \, .$$

Since x is chosen at random, the prover has negligible probability of convincing the verifier unless the x^k-related terms match on each side of the equality for all k. In particular, since $a = \sum_{j=1}^{m} x^j a_j$ the x^m-related terms give us

$$C^{x^m} = \prod_{i=1}^{m} C_i^{x^{m-i} \sum_{\substack{1 \le j \le m \\ m = m-i+j}} x^j a_j} = \left(\prod_{i=1}^{m} C_i^{a_i} \right)^{x^m}$$

and allow the verifier to conclude $C = \prod_{i=1}^{m} C_i^{a_i}$.

Finally, to make the argument honest verifier zero-knowledge we have to avoid leaking information about the exponent vectors $a_1, \ldots, a_m$. The prover therefore commits to a random vector $a_0 \leftarrow \mathbb{Z}_q^n$ and after she sees the challenge x she reveals $a = a_0 + \sum_{j=1}^{m} x^j a_j$. Since a_0 is chosen at random this vector does not leak any information about the exponents.

Another possible source of leakage is the products of the diagonals. The prover will therefore randomize each E_k by multiplying it with a random ciphertext $\mathcal{E}_{pk}(G^{b_k}; \tau_k)$. Now each E_k is a uniformly random group element in $\mathbb{H}$ and will therefore not leak information about the exponents. Of course, this would make it possible to encrypt anything in the E_k and allow cheating. To get around this problem the prover has to commit to the b_k's used in the random encryptions and the verifier will check that the prover uses $b_m = 0$. The full argument that also covers the case $\rho \neq 0$ can be found below.

Common reference string: pk, ck.
Statement: $C_1, \ldots, C_m \in \mathbb{H}^n$, $C \in \mathbb{H}$, and $c_A \in \mathbb{G}^m$
Prover's witness: $A = \{a_j\}_{j=1}^m \in \mathbb{Z}_q^{n \times m}$, $r \in \mathbb{Z}_q^m$, and $\rho \in \mathbb{Z}_q$ such that

$$C = \mathcal{E}_{pk}(1; \rho) \prod_{i=1}^m C_i^{a_i} \qquad \text{and} \qquad c_A = \mathrm{com}_{ck}(A; r)$$

Initial message: Pick $a_0 \leftarrow \mathbb{Z}_q^n, r_0 \leftarrow \mathbb{Z}_q$, and $b_0, s_0, \tau_0 \ldots, b_{2m-1}, s_{2m-1}, \tau_{2m-1} \leftarrow \mathbb{Z}_q$ and set $b_m = 0, s_m = 0, \tau_m = \rho$. Compute for $k = 0, \ldots, 2m - 1$

$$c_{A_0} = \mathrm{com}_{ck}(a_0; r_0), \quad c_{B_k} = \mathrm{com}_{ck}(b_k; s_k), \quad E_k = \mathcal{E}_{pk}(G^{b_k}; \tau_k) \prod_{\substack{i=1, j=0 \\ j=(k-m)+i}}^{m,m} C_i^{a_j}$$

Send: $c_{A_0}, \{c_{B_k}\}_{k=0}^{2m-1}, \{E_k\}_{k=0}^{2m-1}$.
Challenge: $x \leftarrow \mathbb{Z}_q^*$.
Answer: Set $x = (x, x^2, \ldots, x^m)^T$ and compute

$$a = a_0 + Ax \qquad r = r_0 + r \cdot x \qquad b = b_0 + \sum_{k=1}^{2m-1} b_k x^k$$

$$s = s_0 + \sum_{k=1}^{2m-1} s_k x^k \qquad \tau = \tau_0 + \sum_{k=1}^{2m-1} \tau_k x^k .$$

Send: a, r, b, s, τ.
Verification: Check $c_{A_0}, c_{B_0}, \ldots, c_{B_{2m-1}} \in \mathbb{G}$, and $E_0, \ldots, E_{2m-1} \in \mathbb{H}$, and $a \in \mathbb{Z}_q^n$, and $r, b, s, \tau \in \mathbb{Z}_q$, and accept if $c_{B_m} = \mathrm{com}_{ck}(0; 0)$ and $E_m = C$, and

$$c_{A_0} c_A^x = \mathrm{com}_{ck}(a; r) \qquad c_{B_0} \prod_{k=1}^{2m-1} c_{B_k}^{x^k} = \mathrm{com}_{ck}(b; s)$$

$$E_0 \prod_{k=1}^{2m-1} E_k^{x^k} = \mathcal{E}_{pk}(G^b; \tau) \prod_{i=1}^m C_i^{x^{m-i} a} .$$

Theorem 6 (Full paper). *The protocol above is a public coin perfect SHVZK argument of knowledge of openings* $a_1, \ldots, a_m, r$ *and randomness* ρ *such that* $C = \mathcal{E}_{pk}(1; \rho) \prod_{i=1}^{m} C_i^{a_i}$.

4.1   The Prover's Computation

The argument we just described has efficient verification and very low communication complexity, but the prover has to compute

$$E_0, \ldots, E_{2m-1} .$$

In this section we will for clarity ignore the randomization needed to get honest verifier zero-knowledge, which can be added in a straightforward manner at little extra computational cost. So let us say we need to compute for $k = 1, \ldots, 2m - 1$ the elements

$$E_k = \prod_{\substack{i=1, j=1 \\ j=(k-m)+i}}^{m,m} C_i^{a_j} .$$

This can be done by first computing the m^2 products $C_i^{a_j}$ and then computing the E_k's as suitable products of some of these values. Since each product $C_i^{a_j}$ is of the form $\prod_{\ell=1}^{n} C_{i\ell}^{a_{j\ell}}$ this gives a total of $m^2 n$ exponentiations in $\mathbb{H}$. For large m this cost is prohibitive.

It turns out that we can do much better by using techniques inspired by multiplication of integers and polynomials, such as Karatsuba [17], Toom-Cook [5, 26] and using the Fast Fourier Transform [6]. A common theme in these techniques is to compute the coefficients of the product $p(x)q(x)$ of two degree $m - 1$ polynomials $p(x)$ and $q(x)$ by evaluating $p(x)q(x)$ in $2m - 1$ points $\omega_0, \ldots, \omega_{2m-2}$ and using polynomial interpolation to recover the coefficients of $p(x)q(x)$ from $p(\omega_0)q(\omega_0), \ldots, p(\omega_{2m-2})q(\omega_{2m-2})$.

If we pick $\omega \in \mathbb{Z}_q$ we can evaluate the vectors

$$\prod_{i=1}^{m} C_i^{\omega^{m-i}} \quad \text{and} \quad \sum_{j=1}^{m} \omega^{j-1} a_j .$$

This gives us

$$\left(\prod_{i=1}^{m} C_i^{\omega^{m-i}} \right)^{\sum_{j=1}^{m} \omega^{j-1} a_j} = \prod_{k=1}^{2m-1} \left(\prod_{\substack{i=1, j=1 \\ j=(k-m)+i}}^{m,m} C_i^{a_j} \right)^{\omega^{k-1}} = \prod_{k=1}^{2m-1} E_k^{\omega^{k-1}} .$$

Picking $2m - 1$ different $\omega_0, \ldots, \omega_{2m-2} \in \mathbb{Z}_q$ we get the $2m - 1$ ciphertexts

$$\prod_{k=1}^{2m-1} E_k^{\omega_0^{k-1}}, \ldots, \prod_{k=1}^{2m-1} E_k^{\omega_{2m-2}^{k-1}} .$$

The $\omega_0, \ldots, \omega_{2m-2}$ are different and therefore the transposed Vandermonde matrix

$$\begin{pmatrix} 1 & \cdots & 1 \\ \vdots & & \vdots \\ \omega_0^{2m-2} & \cdots & \omega_{2m-2}^{2m-2} \end{pmatrix}$$

is invertible. Let $\boldsymbol{y}_i = (y_0, \ldots, y_{2m-2})^T$ be the ith column of the inverse matrix. We can now compute E_i as

$$E_i = \prod_{\ell=0}^{2m-2} \left(\prod_{k=1}^{2m-1} E_k^{\omega_\ell^{k-1}} \right)^{y_\ell} = \prod_{\ell=0}^{2m-2} \left(\left(\prod_{i=1}^{m} C_i^{\omega_\ell^{m-i}} \right)^{\sum_{j=1}^{m} \omega_\ell^{j-1} \boldsymbol{a}_j} \right)^{y_\ell}.$$

This means the prover can compute $E_1, \ldots, E_{2m-1}$ as linear combinations of

$$\left(\prod_{i=1}^{m} C_i^{\omega_0^{m-i}} \right)^{\sum_{j=1}^{m} \omega_0^{j-1} \boldsymbol{a}_j} \qquad \cdots \qquad \left(\prod_{i=1}^{m} C_i^{\omega_{2m-2}^{m-i}} \right)^{\sum_{j=1}^{m} \omega_{2m-2}^{j-1} \boldsymbol{a}_j}.$$

The expensive step in this computation is to compute $\prod_{i=1}^{m} C_i^{\omega_0^{m-i}}, \ldots, \prod_{i=1}^{m} C_i^{\omega_{2m-2}^{m-i}}$.

If $2m - 2$ is a power of 2 and $2m - 2 | q - 1$ we can pick $\omega_1, \ldots, \omega_{2m-2}$ as roots of unity, i.e., $\omega_k^{2m-2} = 1$. This allows us to use the Fast Fourier Transformation "in the exponent" to simultaneously calculate $\prod_{i=1}^{m} C_i^{\omega_k^{m-i}}$ in all of the roots of unity using only $O(mn \log m)$ exponentiations. This is asymptotically the fastest technique we know for computing $E_0, \ldots, E_{2m-2}$.

Unfortunately, the FFT is not well suited for being used in combination with multi-exponentiation techniques and in practice it therefore takes a while before the asymptotic behavior kicks in. For small m it is therefore useful to consider other strategies. Inspired by the Toom-Cook method for integer multiplication, we may for instance choose $\omega_0, \omega_1, \ldots, \omega_{2m-2}$ to be small integers. When m is small even the largest exponent ω_k^{2m-2} will remain small. For instance, if $m = 4$ we may choose $\omega_k \in \{0, -1, 1, -2, 2, -3, 3\}$, which makes the largest exponent $\omega_k^{m-1} = 3^3 = 27$. This makes it cheap to compute each $\prod_{i=1}^{m} C_i^{\omega_k^{m-i}}$ because the exponents are very small.

The basic step of Toom-Cook sketched above can be optimized by choosing the evaluation points carefully. However, the performance degrades quickly as m grows. Using recursion it is possible to get subquadratic complexity also for large m, however, the cost still grows relatively fast. In the next section we will therefore describe an interactive technique for reducing the prover's computation. In our implementation, see Sect. 6, we have used a combination of the interactive technique and Toom-Cook as the two techniques work well together.

4.2 Trading Computation for Interaction

We will present an interactive technique that can be used to reduce the prover's computation. The prover wants to show that C has the same plaintext as the product of the main diagonal of following matrix (here illustrated for $m = 16$).

$$
\begin{pmatrix}
C_1^{a_1} \; C_1^{a_2} \; C_1^{a_3} \; C_1^{a_4} & & \ddots \\
C_2^{a_1} \; C_2^{a_2} \; C_2^{a_3} \; C_2^{a_4} & & \\
C_3^{a_1} \; C_3^{a_2} \; C_3^{a_3} \; C_3^{a_4} & & \\
C_4^{a_1} \; C_4^{a_2} \; C_4^{a_3} \; C_4^{a_4} & & \\
& \ddots & \\
& & C_{13}^{a_{13}} \; C_{13}^{a_{14}} \; C_{13}^{a_{15}} \; C_{13}^{a_{16}} \\
\ddots & & C_{14}^{a_{13}} \; C_{14}^{a_{14}} \; C_{14}^{a_{15}} \; C_{14}^{a_{16}} \\
& & C_{15}^{a_{13}} \; C_{15}^{a_{14}} \; C_{15}^{a_{15}} \; C_{15}^{a_{16}} \\
& & C_{16}^{a_{13}} \; C_{16}^{a_{14}} \; C_{16}^{a_{15}} \; C_{16}^{a_{16}}
\end{pmatrix}
$$

In the previous section the prover calculated all m^2 entries of the matrix. But we are only interested in the product along the diagonal so we can save computation by just focusing on the blocks close to the main diagonal.

Let us explain the idea in the case of $m = 16$. We can divide the matrix into 4×4 blocks and only use the four blocks that are on the main diagonal. Suppose the prover wants to demonstrate $C = \prod_{i=1}^{16} C_i^{a_i}$. Let us for now just focus on soundness and return to the question of honest verifier zero-knowledge later. The prover starts by sending $E_0, E_1, E_2, E_3, E_4, E_5, E_6$ that are the products along the diagonals of the elements in the blocks that we are interested in. I.e., $E_0 = \prod_{i=1}^{4} C_{4i}^{a_{4i-3}}, \ldots, E_6 = \prod_{i=1}^{4} C_{4i-3}^{a_{4i}}$ and $E_3 = C$. The verifier sends a random challenge x and using the homomorphic properties of the encryption scheme and of the commitment scheme both the prover and the verifier can compute $C_1', \ldots, C_4'$ and $c_{A_1'}, \ldots, c_{A_4'}$ as

$$
C_i' = C_{4i-3}^{x^3} C_{4i-2}^{x^2} C_{4i-1}^{x} C_{4i} \qquad c_{A_j'} = c_{A_{4j-3}}^{x} c_{A_{4j-2}}^{x} c_{A_{4j-1}}^{x^2} c_{A_{4j}}^{x^3} .
$$

They can also both compute $C' = \prod_{k=0}^{6} E_k^{x^k}$. The prover and the verifier now engage in an SHVZK argument for the smaller statement $C' = \prod_{i=1}^{4} C_i'^{a_i'}$. The prover can compute a witness for this statement with $a_i' = a_{4i-3} + x a_{4i-2} + x^2 a_{4i-1} + x^3 a_{4i}$. This shows

$$
C^{x^3} \prod_{\substack{k=0 \\ k \neq 3}}^{6} E_k^{x^k} = \prod_{i=1}^{4} (C_{4i-3}^{x^3} C_{4i-2}^{x^2} C_{4i-1}^{x} C_{4i})^{(a_{4i-3} + x a_{4i-2} + x^2 a_{4i-1} + x^3 a_{4i})} .
$$

Looking at the x^3-related terms, we see this has negligible chance of holding for a random x unless $C = \prod_{i=1}^{16} C_i^{a_i}$, which is what the prover wanted to demonstrate.

We will generalize the technique to reduce a statement $C_1, \ldots, C_m, C, c_{A_1}, \ldots, c_{A_m}$ with a factor μ to a statement $C_1', \ldots, C_{m'}', C', c_{A_1'}, \ldots, c_{A_{m'}'}$, where $m = \mu m'$. To add honest verifier zero-knowledge to the protocol, we have to prevent the E_k's from leaking information about $a_1, \ldots, a_m$. We do this by randomizing each E_k with a random ciphertext $\mathcal{E}_{pk}(G^{b_k}; t_k)$. To prevent the prover to use the randomization to cheat she will have to commit the b_k's before seeing the challenge x.

Common Reference string: pk, ck.

Statement: $C_1, \ldots, C_m \in \mathbb{H}^n$ and $C \in \mathbb{H}$ and $c_{A_1}, \ldots, c_{A_m} \in \mathbb{G}$ where $m = \mu m'$.

Prover's witness: $A \in \mathbb{Z}_q^{n \times m}, r \in \mathbb{Z}_q^m$ and $\rho \in \mathbb{Z}_q$ such that

$$C = \mathcal{E}_{pk}(1; \rho) \prod_{i=1}^{m} C_i^{a_i} \quad \text{and} \quad c_A = \text{com}_{ck}(A; r) .$$

Initial message: Pick $b = (b_0, \ldots, b_{2\mu-2}), s, \tau \leftarrow \mathbb{Z}_q^{2\mu-1}$ and set $b_{\mu-1} = 0, s_{\mu-1} = 0, \tau_{\mu-1} = \rho$. Compute for $k = 0, \ldots, 2\mu - 2$

$$c_{b_k} = \text{com}_{ck}(b_k; s_k) \qquad E_k = \mathcal{E}_{pk}(G^{b_k}; \tau_k) \prod_{\ell=0}^{m'-1} \prod_{\substack{i=1,j=1 \\ j=(k+1-\mu)+i}}^{\mu,\mu} C_{\mu\ell+i}^{a_{\mu\ell+j}} .$$

Send: $c_b = (c_{b_0}, \ldots, c_{b_{2\mu-2}})$ and $E = (E_0, \ldots, E_{2\mu-2})$.

Challenge: $x \leftarrow \mathbb{Z}_q^*$.

Answer: Set $x = (1, x, \ldots, x^{2\mu-2})^T$ and send $b = b \cdot x$ and $s = s \cdot x$ to the verifier.
Compute for $\ell = 1, \ldots, m'$

$$a'_\ell = \sum_{j=1}^{\mu} x^{j-1} a_{\mu(\ell-1)+j} \qquad r'_\ell = \sum_{j=1}^{\mu} x^{j-1} r_{\mu(\ell-1)+j} \qquad \rho' = \tau \cdot x .$$

Define $C'_1, \ldots, C'_{m'}$ and $c_{A'_1}, \ldots, c_{A'_{m'}}$ and C' by

$$C'_\ell = \prod_{i=1}^{\mu} C_{\mu(\ell-1)+i}^{x^{\mu-i}} \qquad c_{A'_\ell} = \prod_{j=1}^{\mu} c_{A_{\mu(\ell-1)+j}}^{x^{j-1}} \qquad C' = \mathcal{E}_{pk}(G^{-b}; 0) E^x .$$

Engage in an SHVZK argument of openings $a'_1, \ldots, a'_{m'}, r'$, and ρ' such that $C' = \mathcal{E}_{pk}(1; \rho') \prod_{\ell=1}^{m'} C'_\ell{}^{a'_\ell}$.

Verification: Check $c_b \in \mathbb{G}^{2\mu-1}$ and $E_0, \ldots, E_{2\mu-2} \in \mathbb{H}$ and $b, s \in \mathbb{Z}_q$. Accept if

$$c_{b_{\mu-1}} = \text{com}_{ck}(0; 0) \qquad E_{\mu-1} = C \qquad c_b^x = \text{com}_{ck}(b; s)$$

and if the SHVZK argument for $C'_1, \ldots, C'_{m'}, C', c_{A'_1}, \ldots, c_{A'_{m'}}$ is valid.

Theorem 7 (Full paper). *The protocol above is a public coin perfect SHVZK argument of knowledge of $a_1, \ldots, a_m, r$ such that $C = \mathcal{E}_{pk}(1; \rho) \prod_{i=1}^{m} C_i^{a_i}$*

5   Product Argument

We will sketch an argument that a set of committed values have a particular product. More precisely, given commitments $A_1, \ldots, A_m$ to $a_{11}, \ldots, a_{mn}$ and a value b we want to give an argument of knowledge for $\prod_{i=1}^{m} \prod_{j=1}^{n} a_{ij} = b$. Our strategy is to compute a commitment

$$B = \text{com}_{ck}(\prod_{i=1}^{m} a_{i1}, \ldots, \prod_{i=1}^{m} a_{in}; s) .$$

We give an argument of knowledge that B is true, i.e., it contains $\prod_{i=1}^{m} a_{i1}, \ldots,$ $\prod_{i=1}^{m} a_{in}$. Groth [13] described how to do this efficiently. Next, we give an argument of knowledge that b is the product of the values inside B. This can be done using an argument given in [14]. Here, we just give an overview of the protocol.

Common reference string: pk, ck.
Statement: $A_1, \ldots, A_m \in \mathbb{G}$ and $b \in \mathbb{Z}_q$.
Prover's witness: $a_{11}, \ldots, a_{mn}, r_1, \ldots, r_m \in \mathbb{Z}_q$ such that

$$A_1 = \mathrm{com}_{ck}(a_{11}, \ldots, a_{1n}; r_1)$$
$$\vdots \qquad \qquad \vdots$$
$$A_m = \mathrm{com}_{ck}(a_{m1}, \ldots, a_{mn}; r_m),$$

$$\text{and} \qquad \prod_{i=1}^{m} \prod_{j=1}^{n} a_{ij} = b.$$

Initial message: Pick $s \leftarrow \mathbb{Z}_q$ and compute $B = \mathrm{com}_{ck}(\prod_{i=1}^{m} a_{i1}, \ldots, \prod_{i=1}^{m} a_{in}; s)$. Send B to the verifier. Engage in an SHVZK argument of knowledge as described in [13] of $B = \mathrm{com}_{ck}(\prod_{i=1}^{m} a_{i1}, \ldots, \prod_{i=1}^{m} a_{in}; s)$, where $a_{11}, \ldots, a_{mn}$ are the committed values in $A_1, \ldots, A_m$. Engage (in parallel) in an SHVZK argument of knowledge as described in [14] of b being the product of the values in B.
Verification: The verifier accepts if $B \in \mathbb{G}$ and both SHVZK arguments are valid.

Theorem 8. *The protocol is a public coin perfect SHVZK argument of knowledge of openings* $a_{11}, \ldots, a_{mn}, r_1, \ldots, r_m \in \mathbb{Z}_q$ *such that* $b = \prod_{i=1}^{m} \prod_{i=1}^{n} a_{ij}$.

The proof along with details of the underlying arguments can be found in the full paper.

6   Implementation and Comparison

We will now compare our protocol with the most efficient shuffle arguments for ElGamal encryption. First, we compare the theoretical performance of the schemes without any optimization. Second, we compare an implementation of our protocol with the implementation by Furukawa et al. [8] and with the implementation in the Verificatum mix-net library [29].

Theoretical Comparison. Previous work in the literature mainly investigated the case where we use ElGamal encryption and commitments over the same group $\mathbb{G}$, i.e., $\mathbb{H} = \mathbb{G} \times \mathbb{G}$. Table 1 gives the asymptotic behavior of these protocols compared to our protocol for $N = mn$ as m and n grows.

In our protocol, we may as detailed in Sect. 4.1 use FFT techniques to reduce the prover's computation to $O(N \log m)$ exponentiations as listed in Table 1. Furthermore, by increasing the round complexity as in Sect. 4.2 we could even get a linear complexity of $O(N)$ exponentiations. These techniques do not apply to the other shuffle arguments; in particular it is not possible to use FFT techniques to reduce the factor m in the shuffle by Groth and Ishai [15].

Table 1. Comparison of the protocols with ElGamal encryption

SHVZK argument	Rounds	Time $\mathcal{P}$ Expos	Time $\mathcal{V}$ Expos	Size Elements
[10]	3	$8N$	$10N$	$5N\ \mathbb{G} + N\ \mathbb{Z}_q$
[8]	5	$9N$	$10N$	$5N\ \mathbb{G} + N\ \mathbb{Z}_q$
[14]	7	$6N$	$6N$	$3N\ \mathbb{Z}_q$
[7]	3	$7N$	$8N$	$N\ \mathbb{G} + 2N\ \mathbb{Z}_q$
[25]	5	$9N$	$11N$	$3N\ \mathbb{G} + 4N\ \mathbb{Z}_q$
[15]	7	$3mN$	$4N$	$3m^2\ \mathbb{G} + 3n\ \mathbb{Z}_q$
This paper	9	$2\log(m)N$	$4N$	$11m\ \mathbb{G} + 5n\ \mathbb{Z}_q$

As the multi-exponentiation argument, which is the most expensive step, already starts in round 3 we can insert two rounds of interactive reduction as described in Sect. 4.2 without increasing the round complexity above 9 rounds. For practical parameters this would give us enough of a reduction to make the prover's computation comparable to the schemes with linear $O(N)$ computation.

The figures in Table 1 are for non-optimized versions of the schemes. All of the schemes may for instance benefit from the use of multi-exponentiation techniques, see e.g. Lim [18] for how to compute a product of n exponentiations using only $O(\frac{n}{\log n})$ multiplications. The schemes may also benefit from randomization techniques, where the verifier does a batch verification of all the equations it has to check.

Experimental Results. We implemented our shuffle argument in C++ using the NTL library by Shoup [24] for the underlying modular arithmetic. We experimented with five different implementations to compare their relative merit:

1. Without any optimizations at all.
2. Using multi-exponentiation techniques.
3. Using multi-exponentiation and the Fast Fourier transform.
4. Using multi-exponentiation and a round of the interactive technique with $\mu = 4$ and Toom-Cook for $m' = 4$ giving $m = \mu m' = 16$.
5. Using multi-exponentiation and two rounds of the interactive technique first with $\mu = 4$ and Toom-Cook for $m' = 4$ giving $m = \mu^2 m' = 64$.

In our experiments we used ElGamal encryption and commitments over the same group $\mathbb{G}$, which was chosen as an order q subgroup of $\mathbb{Z}_p^*$, where $|q| = 160$ and $|p| = 1024$. These security parameters are on the low end for present day use but facilitate comparison with earlier work. The results can be found in Table 2 for $N = 100,000$, $m = 8, 16, 64$ on our machine. We see that the plain multi-exponentiation techniques yield better results than the FFT method for small m; the better asymptotic behavior of the FFT only kicks in for $m > 16$. As expected the Toom-Cook inspired version with added interaction has the best running time and communication cost.

Table 2. Run time of the shuffle arguments in seconds on a Core2Duo 2.53 GHz, 3 MB L2-Cache, 4 GB Ram machine for $N = 100,000$ and $m = 8, 16, 64$

	Optimization	Total time	Time $\mathcal{P}$	Time $\mathcal{V}$	Size
$m = 8$	Unoptimized	570	462	108	4.3 MB
	Multi-expo	162	125	37	
	FFT	228	190	38	
$m = 16$	Unoptimized	900	803	97	2.2 MB
	Multi-expo	193	169	24	
	FFT	245	221	24	
	Toom-Cook	139	101	38	
$m = 64$	Multi-expo	615	594	21	0.7MB
	FFT	328	307	20	
	Toom-Cook	128	91	18	

Comparison with Other Implementations. Furukawa, Miyauchi, Mori, Obana, and Sako [8] gave performance results for a mix-net using a version of the Furukawa-Sako [10] shuffle arguments. They optimized the mix-net by combining the shuffling and decryption operations into one. They used three shuffle centers communicating with each other and their results included both the process of shuffling and the cost of the arguments. So, to compare the values we multiply our shuffle argument times with 3 and add the cost of our shuffling operation on top of that. The comparison can be found in Table 3.

Table 3. Runtime comparison of [8] (CPU: 1 GHz, 256 MB) to our shuffle argument (Toom-Cook with $m = 64$, CPU: 1.4 GHz, 256 MB)

$N = 100,000$	[8]	This paper
Single argument	51 min	15 min
Argument size	66 MB	0.7 MB
Total mix-net time	3 hrs 44 min	53 min

We expected to get better performance than they did and indeed we see that our argument is faster and the communication is a factor 100 smaller. When adding the cost of shuffling and decryption to our argument we still have a speedup of a factor 3 in Table 3 when comparing the two mix-net implementations and taking the difference in the machines into account.

Recently, Furukawa et al. [9] announced a new implementation based on elliptic curve groups. Due to the speed of using elliptic curves this gave them a speedup of a factor 3. A similar speedup can be expected for our shuffle argument if we switch to using elliptic curves in our implementation.

Recently Wikström made a complete implementation of a mix-net in Java in [29] called Verificatum, which is based on the shuffle argument in [25]. To produce compa-rable data, we ran the demo file with only one mix party in the non-interactive mode

Table 4. Runtime comparison of [25] to our shuffle argument on our machine (CPU: 2.53 GHz, 4 GB)

$N = 100,000$	[25]	This paper Toom-Cook
Single argument	5 min	2 min
Argument size	37.7 MB	0.7 MB

using the same modular group as in our protocol. Verificatum is a full mix-net implementation; for fairness in the comparison we only counted the time of the relevant parts for the shuffle argument. As described in Table 1 the theoretical performance of Verificatum's shuffle argument is $20N$ exponentiations, while our prover with Toom-Cook and 2 extra rounds of interaction uses $12N$ exponentiations and our verifier $4N$, so in total $16N$ exponentiations. So we expect a similar running time for the Verificatum mix-net. As shown in Table 4 we perform better, but due to the different programming languages used and different levels of optimization in the code we will not draw any conclusion except that both protocols are efficient and usable in current applications. In terms of size it is clear that our arguments leave a much smaller footprint than Verificatum; we save a factor 50 in the communication.

Acknowledgment. We would like to thank Douglas Wikström for discussions and help regarding our comparison with the shuffle argument used in Verificatum [29].

References

1. Abe, M.: Universally Verifiable Mix-Net with Verification Work Independent of the Number of Mix-Servers. In: Nyberg, K. (ed.) EUROCRYPT 1998. LNCS, vol. 1403, pp. 437–447. Springer, Heidelberg (1998)
2. Abe, M.: Mix-Networks on Permutation Networks. In: Lam, K.-Y., Okamoto, E., Xing, C. (eds.) ASIACRYPT 1999. LNCS, vol. 1716, pp. 258–273. Springer, Heidelberg (1999)
3. Abe, M., Hoshino, F.: Remarks on Mix-Network Based on Permutation Networks. In: Kim, K.-c. (ed.) PKC 2001. LNCS, vol. 1992, pp. 317–324. Springer, Heidelberg (2001)
4. Chaum, D.: Untraceable electronic mail, return addresses, and digital pseudonyms. Commun. ACM 24(2), 84–88 (1981)
5. Cook, S.: On the minimum computation time of functions. PhD thesis, Department of Mathematics, Harvard University (1966), http://cr.yp.to/bib/1966/cook.html
6. Cooley, J.W., Tukey, J.W.: An algorithm for the machine calculation of complex fourier series. Math. Comp. 19, 297–301 (1965)
7. Furukawa, J.: Efficient and verifiable shuffling and shuffle-decryption. IEICE Transactions 88-A(1), 172–188 (2005)
8. Furukawa, J., Miyauchi, H., Mori, K., Obana, S., Sako, K.: An Implementation of a Universally Verifiable Electronic Voting Scheme Based on Shuffling. In: Blaze, M. (ed.) FC 2002. LNCS, vol. 2357, pp. 16–30. Springer, Heidelberg (2003)
9. Furukawa, J., Mori, K., Sako, K.: An Implementation of a Mix-Net Based Network Voting Scheme and Its Use in a Private Organization. In: Chaum, D., Jakobsson, M., Rivest, R.L., Ryan, P.Y.A., Benaloh, J., Kutylowski, M., Adida, B. (eds.) Towards Trustworthy Elections. LNCS, vol. 6000, pp. 141–154. Springer, Heidelberg (2010)
10. Furukawa, J., Sako, K.: An Efficient Scheme for Proving a Shuffle. In: Kilian, J. (ed.) CRYPTO 2001. LNCS, vol. 2139, pp. 368–387. Springer, Heidelberg (2001)

11. Garay, J., MacKenzie, P., Yang, K.: Strengthening zero-knowledge protocols using signatures. J. Cryptology 19(2), 169–209 (2006)
12. Groth, J.: Honest verifier zero-knowledge arguments applied. Dissertation Series DS-04-3, BRICS, 2004. PhD thesis. xii+119 (2004)
13. Groth, J.: Linear Algebra with Sub-linear Zero-Knowledge Arguments. In: Halevi, S. (ed.) CRYPTO 2009. LNCS, vol. 5677, pp. 192–208. Springer, Heidelberg (2009)
14. Groth, J.: A verifiable secret shuffle of homomorphic encryptions. J. Cryptology 23(4), 546–579 (2010)
15. Groth, J., Ishai, Y.: Sub-linear Zero-Knowledge Argument for Correctness of a Shuffle. In: Smart, N.P. (ed.) EUROCRYPT 2008. LNCS, vol. 4965, pp. 379–396. Springer, Heidelberg (2008)
16. Groth, J., Lu, S.: Verifiable Shuffle of Large Size Ciphertexts. In: Okamoto, T., Wang, X. (eds.) PKC 2007. LNCS, vol. 4450, pp. 377–392. Springer, Heidelberg (2007)
17. Karatsuba, A., Ofman, Y.: Multiplication of multidigit numbers on automata. Soviet Physics Dokl. 7, 595–596 (1963)
18. Lim, C.: Efficient multi-exponentiation and application to batch verification of digital signatures (2000), http://dasan.sejong.ac.kr/~chlim/pub/multiexp.ps
19. Lindell, Y.: Parallel coin-tossing and constant-round secure two-party computation. J. Cryptology 16(3), 143–184 (2003)
20. Neff, C.A.: A verifiable secret shuffle and its application to e-voting. In: ACM CCS, pp. 116–125 (2001)
21. Neff, C.A.: Verifiable mixing (shuffling) of elgamal pairs (2003), http://people.csail.mit.edu/rivest/voting/papers/Neff-2004-04-21-ElGamalShuffles.pdf
22. Pedersen, T.P.: Non-interactive and Information-Theoretic Secure Verifiable Secret Sharing. In: Feigenbaum, J. (ed.) CRYPTO 1991. LNCS, vol. 576, pp. 129–140. Springer, Heidelberg (1992)
23. Sako, K., Kilian, J.: Receipt-Free Mix-Type Voting Scheme - A Practical Solution to the Implementation of a Voting Booth. In: Guillou, L.C., Quisquater, J.-J. (eds.) EUROCRYPT 1995. LNCS, vol. 921, pp. 393–403. Springer, Heidelberg (1995)
24. Shoup, V.: Ntl library (2009), http://www.shoup.net/ntl/
25. Terelius, B., Wikström, D.: Proofs of Restricted Shuffles. In: Bernstein, D.J., Lange, T. (eds.) AFRICACRYPT 2010. LNCS, vol. 6055, pp. 100–113. Springer, Heidelberg (2010)
26. Toom, A.: The complexity of a scheme of functional elements realizing the multiplication of integers (2000), http://www.de.ufpe.br/~toom/my_articles/engmat/MULT-E.PDF
27. Wikström, D.: The Security of a Mix-Center Based on a Semantically Secure Cryptosystem. In: Menezes, A., Sarkar, P. (eds.) INDOCRYPT 2002. LNCS, vol. 2551, pp. 368–381. Springer, Heidelberg (2002)
28. Wikström, D.: A Commitment-Consistent Proof of a Shuffle. In: Boyd, C., González Nieto, J. (eds.) ACISP 2009. LNCS, vol. 5594, pp. 407–421. Springer, Heidelberg (2009)
29. Wikström, D.: Verificatum (2010), http://www.verificatum.com/

Malleable Proof Systems and Applications

Melissa Chase[1], Markulf Kohlweiss[2],
Anna Lysyanskaya[3], and Sarah Meiklejohn[4,*]

[1] Microsoft Research Redmond
melissac@microsoft.com
[2] Microsoft Research Cambridge
markulf@microsoft.com
[3] Brown University
anna@cs.brown.edu
[4] UC San Diego
smeiklej@cs.ucsd.edu

Abstract. Malleability for cryptography is not necessarily an opportunity for attack; in many cases it is a potentially useful feature that can be exploited. In this work, we examine notions of malleability for non-interactive zero-knowledge (NIZK) proofs. We start by defining a malleable proof system, and then consider ways to meaningfully *control* the malleability of the proof system, as in many settings we would like to guarantee that only certain types of transformations can be performed.

As our motivating application, we consider a shorter proof for verifiable shuffles. Our controlled-malleable proofs allow us for the first time to use one compact proof to prove the correctness of an entire multi-step shuffle. Each authority takes as input a set of encrypted votes and a controlled-malleable NIZK proof that these are a shuffle of the original encrypted votes submitted by the voters; it then permutes and re-randomizes these votes and updates the proof by exploiting its controlled malleability. As another application, we generically use controlled-malleable proofs to realize a strong notion of encryption security.

Finally, we examine malleability in existing proof systems and observe that Groth-Sahai proofs are malleable. We then go beyond this observation by characterizing all the ways in which they are malleable, and use them to efficiently instantiate our generic constructions from above; this means we can instantiate our proofs and all their applications using only the Decision Linear (DLIN) assumption.

1 Introduction

Let L be a language in NP. For concreteness, consider the language of Diffie-Hellman tuples: $(G, g, X, Y, Z) \in L_{DH}$ if there exist (x, y) such that g, X, Y, Z are elements of the group G, $X = g^x$, $Y = g^y$, and $Z = g^{xy}$. Suppose that we have a polynomial time prover P, and a verifier V, and P wants to convince V that $(G, g, X, Y, Z) \in L_{DH}$. Does the efficient prover need to know the values

---

* Work done as an intern at Microsoft Research Redmond.

D. Pointcheval and T. Johansson (Eds.): EUROCRYPT 2012, LNCS 7237, pp. 281–300, 2012.

(x, y) in order to convince the verifier? Not necessarily. Suppose that P is in possession of a non-interactive zero-knowledge (NIZK) proof π' that another tuple, $(G, g, X', Y', Z') \in L_{DH}$; suppose in addition that P happens to know (a, b) such that $X = (X')^a$, $Y = (Y')^b$, and $Z = (Z')^{ab}$. Can he, using the fact that he knows (a, b), transform π' into a NIZK π for the related instance (G, g, X, Y, Z)? In the sequel, we say that a proof system is *malleable* if it allows a prover to derive proofs of statements (such as $(G, g, X, Y, Z) \in L_{DH}$) not just from witnesses for their truth, but also from proofs of related statements (such as the proof π' that $(G, g, X', Y', Z') \in L_{DH}$).

In this paper, we consider malleability for non-interactive zero-knowledge proof systems. Our contributions are threefold: (1) definitions; (2) constructions; and (3) applications.

Motivating Application. Why is malleability for non-interactive zero-knowledge proof systems an interesting feature? Let us present, as a motivating application, a verifiable vote shuffling scheme that becomes much more efficient if constructed using malleable proofs.

In a vote shuffling scheme, we have a set of encrypted votes $(v_1, \ldots, v_n)$ submitted by n voters; each vote v_i is an encryption of the voter's ballot under some trusted public key pk. The set of encrypted votes is then re-randomized[1] and shuffled, in turn, by several shuffling authorities. More precisely, let $(v_1^{(0)}, \ldots, v_n^{(0)})$ $= (v_1, \ldots, v_n)$; then each authority A_j takes as input $(v_1^{(j-1)}, \ldots, v_n^{(j-1)})$, picks a random permutation ρ and outputs $(v_1^{(j)}, \ldots, v_n^{(j)}) = (\tilde{v}_{\rho(1)}^{(j)}, \ldots, \tilde{v}_{\rho(n)}^{(j)})$, where $\tilde{v}_i^{(j)}$ is a randomization of $v_i^{(j-1)}$. At the end, the final set of encrypted votes $(v_1^{(\ell)}, \ldots, v_n^{(\ell)})$ is decrypted (for example, by a trustee who knows the decryption key corresponding to pk, or via a threshold decryption protocol) and the election can be tallied.

It is easy to see that, if we are dealing with an honest-but-curious adversary, this scheme guarantees both correctness and privacy as long as at least one of the authorities is honest. To make it withstand an active adversary, however, it is necessary for all participants (both the voters and the shuffling authorities) to prove (using a proof system with appropriate, technically subtle soundness and zero-knowledge guarantees) that they are correctly following the protocol. If these proofs are non-interactive, then the protocol gets the added benefit of being universally verifiable: anyone with access to the original encrypted votes and the output and proofs of each authority can verify that the votes were shuffled correctly. Thus, any party wishing to verify an election with n voters and ℓ shuffling authorities (and ℓ can potentially be quite large, for example a large polynomial in n for cases where a small group is voting on a very sensitive issue) will have to access $\Omega(n\ell)$ data just to read all the proofs.

Can the proof that the verifier needs to read be shorter than that? The statement that needs to be verified is that the ciphertexts $(v_1^{(\ell)}, \ldots, v_n^{(\ell)})$ can be

---

[1] It is therefore important that the encryption scheme used is randomizable, so that on input a ciphertext $c = \mathsf{Enc}_{pk}(m; r)$ and randomness r' one can compute $c' = \mathsf{Enc}_{pk}(m; r * r')$, where $*$ is some group operation.

obtained by randomizing and permuting the original votes $(v_1, \ldots, v_n)$. The witness for this statement is just some permutation (that is obtained by composing the permutations applied by individual authorities) and randomness that went into randomizing each ciphertext (that can be obtained by applying the group operation repeatedly to the randomness used by each authority); thus, ignoring the security parameter, the length of the witness can potentially be only $O(n)$.[2]

Of course, no individual authority knows this witness. But each authority A_j is given a proof π_{j-1} that, up until now, everything was permuted and randomized correctly. Using controlled malleable proofs, from this π_{j-1} and its own secret permutation ρ_j and vector of random values $(r_1^{(j)}, \ldots, r_n^{(j)})$, A_j should be able to compute the proof π that his output is a permutation and randomization of the original votes.

In this paper, we give a construction that roughly corresponds to this outline, and prove its security. We must stress that even though this construction is a more or less direct consequence of the new notion of controllable malleability, and therefore may seem obvious in hindsight, it is actually a significant breakthrough as far as the literature on efficient shuffles is concerned: for the first time, we obtain a non-interactive construction in which the complexity of verifying the tally with ℓ authorities is not ℓ times the complexity of verifying the tally with one authority!

Our Definitions. Care needs to be taken when defining malleable NIZKs suitable for the above application. We first need malleability itself: from an instance x' and a proof π' that $x' \in L$, we want to have an efficient algorithm ZKEval that computes another instance $x = T(x')$ and a proof π that $x \in L$, where T is some transformation (in the above example, x' is a set of ciphertexts, and T is a re-randomization and permutation of these ciphertexts). We want the resulting proof to be *derivation private*, so that, from x and π, it is impossible to tell from which T and x' they were derived. (In the above example, it should be impossible to tell how the ciphertexts were shuffled.) Finally, we want to ensure that the proof system is sound, even in the presence of a zero-knowledge simulator that provides proofs of adversarially chosen statements (so that we can relate the real-world experiment where the adversary participates in shuffling the ciphertexts to an ideal-world process that only has access to the final tally). To this end, we define *controlled malleability* (as opposed to malleability that is out of control!) that guarantees that, from proofs computed by an adversary, an extractor (with a special extracting trapdoor) can compute either a witness to the truth of the statement, or the transformation T and some statement for which the simulator had earlier provided a proof.

---

[2] In our concrete construction we use a very simple approach to proving a shuffle in which we represent the permutation as a matrix, thus the length of a single shuffle proof is $O(n^2)$. This could potentially be improved using more sophisticated verifiable shuffle techniques as we will mention later. Additionally, because we want to be able to verify the fact that each authority participated in the shuffle, we will include a public key for each authority involved and the size will actually grow to $O(n^2 + \ell)$.

Our definitional approach to derivation privacy is inspired by circuit privacy for fully homomorphic encryption [27,40,39,13], also called function privacy or unlinkability. Our definitional approach to controlled malleability is inspired by the definitions of HCCA (homomorphic-CCA) secure encryption due to Prabhakaran and Rosulek [36]; it is also related to the recently proposed notion of targeted malleability due to Boneh, Segev, and Waters [12]. (See the full version of our paper [15] for more detailed comparison with these notions.)

Our Construction. Our construction of controlled-malleable and derivation-private NIZK proof systems consists of two steps. First, in Section 3, we show how to construct a controlled-malleable derivation-private NIZK from any derivation-private non-interactive witness-indistinguishable (NIWI) proof system and secure signature scheme. Then, in Section 4.1 we show how to instantiate the appropriate NIWI proof system and signature scheme using the Groth-Sahai proof system [33] and a recent structure-preserving signature due to Chase and Kohlweiss [14]; this combination means we can instantiate our proofs (and in fact all of the constructions in our paper) using the Decision Linear (DLIN) assumption [11]. The size of the resulting proof is linear in the size of the statement, although the size of the structure-preserving signature does make it admittedly much less efficient than Groth-Sahai proofs alone.

At the heart of our construction is the observation that the Groth-Sahai (GS) proof system is malleable in ways that can be very useful. This feature of GS proofs has been used in prior work in a wide variety of applications: Belenkiy et al. [8] use the fact that the GS proof system can be randomized in order to construct delegatable anonymous credentials; Dodis et al. [20] uses homomorphic properties of GS proofs in order to create a signature scheme resilient to continuous leakage; Acar and Nguyen [7] use malleability to delegate and update non-membership proofs for a cryptographic accumulator in their implementation of a revocation mechanism for delegatable anonymous credentials; and Fuchsbauer [24] uses malleability to transform a proof about the contents of a commitment into a proof of knowledge of a signature on the committed message in his construction of commuting signatures.

Compact Verifiable Shuffles. Armed with a construction of a controlled-malleable and derivation-private NIZK, we proceed, in Section 6, to consider the problem of obtaining a verifiable shuffle with compact proofs. We formally define this concept, describe a generic construction from a semantically-secure encryption scheme and a controlled-malleable and derivation-private NIZK following the outline above, and finally argue that we can in fact construct such a proof system for the appropriate set of transformations based on the instantiation described in Section 4.1.

An Application to Encryption. Can controlled malleability of NIZKs give us controlled malleability for encryption? That is to say, can we achieve a meaningful notion of adaptively secure encryption, even while allowing computations on encrypted data? Similarly to controlled malleability for proofs, we define in

Section 5 controlled malleability for encryption (directly inspired by the notion of HCCA security; in this, our work can be considered closely related to that of Prabhakaran and Rosulek), and show a *general* method for realizing it for broad classes of unary transformations, using a semantically secure encryption scheme with appropriate homomorphic properties and a controlled-malleable and derivation-private NIZK for an appropriate language as building blocks. Our construction follows easily from these properties, resulting in a much simpler proof of security than was possible in previous works. (We note that our methods do not extend to n-ary transformations for $n > 1$, because the same limitations that apply for HCCA security, pointed out by Prabhakaran and Rosulek, also apply here. The work of Boneh et al. overcomes this and allows for binary transformations as well, with the sacrifice that, unlike both our scheme and the Prabhakaran-Rosulek scheme, the encryption scheme can no longer satisfy function privacy.)

Related Work on Shuffling Ciphertexts. Shuffles and mixing in general were introduced by Chaum in 1981 [16], and the problem of verifiable shuffles was introduced by Sako and Kilian in 1995 [38]; the work on verifiable shuffles in the ensuing sixteen years has been extensive and varied [2,26,6,34,29,25,41,31]. In 1998, Abe [1] considered the problem of compact proofs of shuffles. Unlike our non-interactive solution, his solution is based on an interactive protocol[3] wherein all mixing authorities must jointly generate a proof with size independent of ℓ; in comparison, our solution allows each authority to be offline before and after it performs its shuffling of the ciphertexts. In terms of approaches most similar to our own, Furukawa and Sako [26] use a permutation matrix to shuffle the ciphertexts; they then prove that the matrix used was in fact a permutation matrix, and that it was applied properly. Most recently, Groth and Lu [31] give a verifiable shuffle that is non-interactive (the only one to do so without use of the Fiat-Shamir heuristic [23]), uses pairing-based verifiability, and obtains $O(n)$ proof size for a single shuffle. The advantage, as outlined above, that our construction has over all of these is that one proof suffices to show the security of the entire shuffle; we do not require a separate proof from each mix server. An interesting open problem is to see if there is some way to combine some of these techniques with an appropriate controlled-malleable proof system to obtain a multi-step shuffle with optimal proof size $O(n + \ell)$.

2   Definitions and Notation

Our definitional goal is to formulate what it means to construct a proof of a particular statement using proofs of related statements. Let $R(\cdot, \cdot)$ be some relation that is polynomial-time computable in the size of its first input; in the sequel we call such a relation an *efficient* relation. Associated with R, there is an NP language $L_R = \{x \mid \exists\, w \text{ such that } R(x, w) = TRUE\}$.[4] For example, let

---

[3] The protocol could in fact be made non-interactive, but only using the Fiat-Shamir heuristic [23] and thus the random oracle model.

[4] Without the restriction that R is efficient in its first input, the resulting language won't necessarily be in NP.

$R(x, w)$ be a relation that holds if the witness $w = (a, b)$ demonstrates that the instance $x = (G, g, A, B, C)$ is a Diffie-Hellman tuple; i.e. it holds if $g, A, B, C \in G$ and $A = g^a$, $B = g^b$, $C = g^{ab}$. Then the language associated with R is L_{DH} defined in the introduction. We often write $(x, w) \in R$ to denote that $R(x, w) = TRUE$.

Let $T = (T_x, T_w)$ be a pair of efficiently computable n-ary functions, where $T_x : \{\{0,1\}^*\}^n \to \{0,1\}^*$, $T_w : \{\{0,1\}^*\}^n \to \{0,1\}^*$. In what follows, we refer to such a tuple T as an n-ary *transformation*.

Definition 2.1. *An efficient relation R is* closed *under an n-ary transformation $T = (T_x, T_w)$ if for any n-tuple $\{(x_1, w_1), \ldots, (x_n, w_n)\} \in R^n$, the pair $(T_x(x_1, \ldots, x_n), T_w(w_1, \ldots, w_n)) \in R$. If R is closed under T, then we say that T is* admissible *for R. Let $\mathcal{T}$ be some set of transformations; if for every $T \in \mathcal{T}$, T is admissible for R, then $\mathcal{T}$ is an* allowable *set of transformations.*

For example, for the DH relation R described above, consider $T = (T_x, T_w)$ where for some (a', b'), $T_x(G, g, A, B, C) = (G, g, A^{a'}, B^{b'}, C^{a'b'})$ and $T_w(a, b) = (aa', bb')$; then the Diffie-Hellman relation R is closed under transformation T, and additionally the set $\mathcal{T}$ of transformations of this form (i.e., where there is a transformation T corresponding to any pair (a', b')) is an allowable set of transformations.

Our goal is to define non-interactive zero-knowledge and witness-indistinguishable proof systems for efficient relations R that are (1) malleable with respect to an allowable set of transformations $\mathcal{T}$; that is to say, for any $T \in \mathcal{T}$, given proofs for $x_1, \ldots x_n \in L_R$, they can be transformed into a proof that $T_x(x_1, \ldots, x_n) \in L_R$; and (2) derivation-private; that is to say, the resulting proof cannot be distinguished from one freshly computed by a prover on input $(T_x(x_1, \ldots, x_n), T_w(w_1, \ldots, w_n))$. Before we can proceed, however, we need to recall the definition of a non-interactive zero-knowledge proof system.

A proof system for an efficient relation R allows a prover to prove that a value x is in the associated language L_R. A non-interactive (NI) proof system with efficient provers [10,21] consists of three PPT algorithms: the algorithm CRSSetup(1^k) that generates a common reference string (CRS) σ_{crs}, the algorithm $\mathcal{P}(\sigma_{\mathrm{crs}}, x, w)$ that outputs a proof π that $x \in L_R$, and the algorithm $\mathcal{V}(\sigma_{\mathrm{crs}}, x, \pi)$ that verifies the proof; such a proof system must be complete (meaning the verifier will always accept an honestly generated proof) and sound (meaning that a verifier cannot be fooled into accepting a proof for a false statement). A NI zero-knowledge proof (NIZK) [28,10], additionally requires the existence of a simulator S that can generate proofs without access to a witness, while a NI witness-indistinguishable proof system [22] has the requirement that proofs generated using two different witnesses for the same x are indistinguishable from each other. A NI proof of knowledge [28,9] additionally requires an efficient extractor algorithm E that, on input a proof that $x \in L_R$, finds a witness for the instance x.

We use the original definitions for completeness and soundness of NI proof systems in the common-reference-string model [10]. The version of the definition of zero-knowledge for NIZK we give is originally due to Feige, Lapidot and

Shamir (FLS) [21]; they call it "adaptive multi-theorem NIZK." We also use the FLS definition of witness indistinguishability. The version of knowledge extraction we use is a generalization of the definition of knowledge extraction given by Groth, Ostrovsky and Sahai (GOS) [32]: they defined the notion of *perfect* knowledge extraction, while here we find it useful to generalize their definition, in the straightforward way, to the case when extraction is not perfect. Due to space constraints, a formal definition for non-interactive zero-knowledge proofs of knowledge (NIZKPoK) systems and non-interactive witness-indistinguishable proofs of knowledge (NIWIPoK) systems combining all of these concepts can be found in the full version of our paper [15].

Next, we define a malleable proof system; i.e., one in which, from proofs $(\pi_1, \ldots, \pi_n)$ that $(x_1, \ldots, x_n) \in L$, one can compute a proof π that $T_x(x_1, \ldots, x_n) \in L$, for an admissible transformation $T = (T_x, T_w)$:

Definition 2.2 (Malleable non-interactive proof system). *Let* (CRSSetup, $\mathcal{P}, \mathcal{V}$) *be a non-interactive proof system for a relation R. Let $\mathcal{T}$ be an allowable set of transformations for R. Then this proof system is* malleable with respect to $\mathcal{T}$ *if there exists an efficient algorithm* ZKEval *that on input* $(\sigma_{crs}, T, \{x_i, \pi_i\})$, *where $T \in \mathcal{T}$ is an n-ary transformation and $\mathcal{V}(\sigma_{crs}, x_i, \pi_i) = 1$ for all i, $1 \leq i \leq n$, outputs a valid proof π for the statement $x = T_x(\{x_i\})$ (i.e., a proof π such that $\mathcal{V}(\sigma_{crs}, x, \pi) = 1$).*

Going back to our above example, the algorithm ZKEval will take as input the transformation T (which is equivalent to taking as input the values a' and b'), and a proof π_1 that $x_1 = (G, g, A, B, C)$ is a DH tuple, and output a proof π that $x = T_x(x_1) = (G, g, A^{a'}, B^{b'}, C^{a'b'})$ is a DH tuple.

2.1 Derivation Privacy for Proofs

In addition to malleability, we must also consider a definition of derivation privacy analogous to the notion of function privacy for encryption. (In the encryption setting this is also called unlinkability [36]; for a formal definition see the full version [15].) We have the following definition:

Definition 2.3 (Derivation privacy). *For a NI proof system* (CRSSetup, $\mathcal{P}, \mathcal{V}$, ZKEval) *for an efficient relation R malleable with respect to $\mathcal{T}$, an adversary $\mathcal{A}$, and a bit b, let $p_b^{\mathcal{A}}(k)$ be the probability of the event that $b' = 0$ in the following game:*

- *Step 1.* $\sigma_{crs} \xleftarrow{\$} $ CRSSetup(1^k).
- *Step 2.* (state, $x_1, w_1, \pi_1, \ldots, x_q, w_q, \pi_q, T$) $\xleftarrow{\$} \mathcal{A}(\sigma_{crs})$.
- *Step 3. If $\mathcal{V}(\sigma_{crs}, x_i, \pi_i) = 0$ for some i, $(x_i, w_i) \notin R$ for some i, or $T \notin \mathcal{T}$, abort and output $\perp$. Otherwise, form*

$$\pi \xleftarrow{\$} \begin{cases} \mathcal{P}(\sigma_{crs}, T_x(x_1, \ldots, x_q), T_w(w_1, \ldots, w_q)) & \text{if } b = 0 \\ \text{ZKEval}(\sigma_{crs}, T, \{x_i, \pi_i\}) & \text{if } b = 1. \end{cases}$$

- *Step 4.* $b' \xleftarrow{\$} \mathcal{A}(\text{state}, \pi)$.

We say that the proof system is derivation private *if for all PPT algorithms $\mathcal{A}$ there exists a negligible function $\nu(\cdot)$ such that $|p_0^{\mathcal{A}}(k) - p_1^{\mathcal{A}}(k)| < \nu(k)$.*

In some cases, we would like to work with a stronger definition that applies only for NIZKs. In this case, the adversary will not be asked to provide witnesses or distinguish between the outputs of the prover and ZKEval, but instead between the zero-knowledge simulator and ZKEval. It will also be given the simulation trapdoor so that it can generate its own simulated proofs.

Definition 2.4 (Strong derivation privacy). *For a malleable NIZK proof system* $(\mathsf{CRSSetup}, \mathcal{P}, \mathcal{V}, \mathsf{ZKEval})$ *with an associated simulator* (S_1, S_2)*, a given adversary $\mathcal{A}$, and a bit b, let $p_b^{\mathcal{A}}(k)$ be the probability of the event that $b' = 0$ in the following game:*

- *Step 1.* $(\sigma_{sim}, \tau_s) \xleftarrow{\$} S_1(1^k)$.
- *Step 2.* $(\text{state}, x_1, \pi_1, \ldots, x_q, \pi_q, T) \xleftarrow{\$} \mathcal{A}(\sigma_{sim}, \tau_s)$.
- *Step 3. If* $\mathcal{V}(\sigma_{sim}, x_i, \pi_i) = 0$ *for some i, $(x_1, \ldots, x_q)$ is not in the domain of T_x, or $T \notin \mathcal{T}$, abort and output $\perp$. Otherwise, form*

$$\pi \xleftarrow{\$} \begin{cases} S_2(\sigma_{sim}, \tau_s, T_x(x_1, \ldots, x_q)) & \text{if } b = 0 \\ \mathsf{ZKEval}(\sigma_{sim}, T, \{x_i, \pi_i\}) & \text{if } b = 1. \end{cases}$$

- *Step 4.* $b' \xleftarrow{\$} \mathcal{A}(\text{state}, \pi)$.

We say that the proof system is strongly derivation private *if for all PPT algorithms $\mathcal{A}$ there exists a negligible function $\nu(\cdot)$ such that $|p_0^{\mathcal{A}}(k) - p_1^{\mathcal{A}}(k)| < \nu(k)$.*

As we will see in Section 3, schemes that satisfy the weaker notion of derivation privacy can in fact be generically "boosted" to obtain schemes that satisfy the stronger notion. We can also show a generic way to obtain derivation privacy using malleability and the notion of *randomizability* for proofs, defined by Belenkiy et al. [8]; this can be found in the full version of the paper [15].

3 Controlled Malleability for NIZKs

Is the notion of malleability compatible with the notion of a proof of knowledge or with strong notions like simulation soundness? Recall that to achieve simulation soundness, as defined by Sahai and de Santis et al. [37,19], we intuitively want an adversary $\mathcal{A}$ to be unable to produce a proof of a new false statement even if it can request many such proofs from the simulator; for the even stronger notion of simulation-extractability as defined by de Santis et al. and Groth [19,30], a proof system must admit an efficient extractor that finds witnesses to all statements proved by an adversary, again even when the adversary has access to a simulator.

Malleability, in contrast, explicitly allows an adversary to take as input the values x', π', apply some admissible transformation T to x' to obtain $x = T_x(x')$, and compute a proof π that $x \in L_R$; importantly, the adversary can do all this without knowing the original witness w'. Suppose, for a malleable proof system, that the adversary is given as input a *simulated* proof π' that was generated without access to the witness w' for x', and for concreteness let T be the identity transformation. Then requiring that, on input (x, π), the extractor should output w, implies that membership in L_R can be tested for a given x by computing a simulated proof, mauling it, and then extracting the witness from the resulting proof (formally, this would mean that $L_R \in \mathbf{RP}$). Thus, seemingly, one cannot reconcile the notion of malleability with that of a simulation-extractable proof of knowledge.

Surprisingly, however, under a relaxed but still meaningful extractability requirement, we can have a proof system that is both malleable and simulation-extractable to a satisfactory extent; we call this notion *controlled malleability*. Essentially this definition will require that the extractor can extract either a valid witness, or a previously proved statement x' and a transformation T in our allowed set $\mathcal{T}$ that could be used to transform x' into the new statement x. To demonstrate that our definition is useful, we will show in Section 5 that it can be used to realize a strong notion of encryption security, and in Section 6 that it can also be used to reduce the overall size of proofs for verifiable shuffles.

Definition 3.1 (Controlled-malleable simulation sound extractability).
Let $(\mathsf{CRSSetup}, \mathcal{P}, \mathcal{V})$ be a NIZKPoK system for an efficient relation R, with a simulator (S_1, S_2) and an extractor (E_1, E_2). Let $\mathcal{T}$ be an allowable set of unary transformations for the relation R such that membership in $\mathcal{T}$ is efficiently testable. Let SE_1 be an algorithm that, on input 1^k outputs $(\sigma_{crs}, \tau_s, \tau_e)$ such that (σ_{crs}, τ_s) is distributed identically to the output of S_1. Let $\mathcal{A}$ be given, and consider the following game:

- *Step 1.* $(\sigma_{crs}, \tau_s, \tau_e) \xleftarrow{\$} SE_1(1^k)$.
- *Step 2.* $(x, \pi) \xleftarrow{\$} \mathcal{A}^{S_2(\sigma_{crs}, \tau_s, \cdot)}(\sigma_{crs}, \tau_e)$.
- *Step 3.* $(w, x', T) \leftarrow E_2(\sigma_{crs}, \tau_e, x, \pi)$.

We say that the NIZKPoK satisfies controlled-malleable simulation-sound extractability (CM-SSE, for short) if for all PPT algorithms $\mathcal{A}$ there exists a negligible function $\nu(\cdot)$ such that the probability (over the choices of SE_1, $\mathcal{A}$, and S_2) that $\mathcal{V}(\sigma_{crs}, x, \pi) = 1$ and $(x, \pi) \notin Q$ (where Q is the set of queried statements and their responses) but either (1) $w \neq \perp$ and $(x, w) \notin R$; (2) $(x', T) \neq (\perp, \perp)$ and either $x' \notin Q_x$ (the set of queried instances), $x \neq T_x(x')$, or $T \notin \mathcal{T}$; or (3) $(w, x', T) = (\perp, \perp, \perp)$ is at most $\nu(k)$.

This definition is actually closely related to simulation-extractability; in fact, if we restrict our set of transformations to be $\mathcal{T} = \emptyset$, we obtain exactly Groth's notion of simulation-sound extractability. Note also that this definition does not require that a proof system actually be malleable, it only requires that, should it happen to be malleable, this malleability be limited in a controlled way. Thus, a simulation-sound extractable proof system would also satisfy our definition, for any set $\mathcal{T}$, even though it is not malleable. We refer to a proof system that is both strongly derivation private and controlled-malleable simulation-sound extractable as a *controlled-malleable NIZK* (cm-NIZK).

Finally, note that our definition applies only to unary transformations. This is because our requirement that we can extract the transformation T means we cannot hope to construct cm-NIZKs for n-ary transformations where $n > 1$, as this would seem to necessarily expand the size of the proof (similarly to what Prabhakaran and Rosulek show for HCCA encryption [36]). We therefore achieve cm-NIZKs for classes of unary transformations that are closed under composition (i.e., $T' \circ T \in \mathcal{T}$ for all $T, T' \in \mathcal{T}$). In addition, our simulation strategy depends on the identity transformation being a member of $\mathcal{T}$, so we can achieve cm-NIZKs only for classes of transformations that include the identity transformation.

A Generic Construction

Let R be an efficient relation, and suppose $\mathcal{T}$ is an allowable set of transformations for R that contains the identity transformation; suppose further that membership in $\mathcal{T}$ is efficiently testable. Let $(\mathsf{KeyGen}, \mathsf{Sign}, \mathsf{Verify})$ be a secure signature scheme. Let $(\mathsf{CRSSetup}_{\mathsf{WI}}, \mathcal{P}_{\mathsf{WI}}, \mathcal{V}_{\mathsf{WI}})$ be a NIWIPoK for the following relation R_{WI}: $((x, vk), (w, x', T, \sigma)) \in R_{\mathsf{WI}}$ if $(x, w) \in R$ or $\mathsf{Verify}(vk, \sigma, x') = 1$, $x = T_x(x')$, and $T \in \mathcal{T}$. Consider the proof system $(\mathsf{CRSSetup}, \mathcal{P}, \mathcal{V})$ defined as follows:

- $\mathsf{CRSSetup}(1^k)$: First generate $\sigma_{\mathsf{WI}crs} \xleftarrow{\$} \mathsf{CRSSetup}_{\mathsf{WI}}(1^k)$ and $(vk, sk) \xleftarrow{\$} \mathsf{KeyGen}(1^k)$; then output $\sigma_{\mathrm{crs}} := (\sigma_{\mathsf{WI}crs}, vk)$.
- $\mathcal{P}(\sigma_{\mathrm{crs}}, x, w)$: Output $\pi \xleftarrow{\$} \mathcal{P}_{\mathsf{WI}}(\sigma_{\mathsf{WI}crs}, x_{\mathsf{WI}}, w_{\mathsf{WI}})$, where $x_{\mathsf{WI}} = (x, vk)$ and $w_{\mathsf{WI}} = (w, \perp, \perp, \perp)$.
- $\mathcal{V}(\sigma_{\mathrm{crs}}, x, \pi)$: Output $\mathcal{V}_{\mathsf{WI}}(\sigma_{\mathsf{WI}crs}, x_{\mathsf{WI}}, \pi)$ where $x_{\mathsf{WI}} = (x, vk)$.

To obtain strong derivation privacy with respect to R and $\mathcal{T}$ we also require the NIWIPoK to be derivation private with respect to R_{WI} and a set of transformations $\mathcal{T}_{\mathsf{WI}}$ such that for every $T' = (T'_x, T'_w) \in \mathcal{T}$ there exists a $T_{\mathsf{WI}}(T') \in \mathcal{T}_{\mathsf{WI}}$. For $T_{\mathsf{WI}}(T') = (T_{\mathsf{WI},x}, T_{\mathsf{WI},w})$ we require that $T_{\mathsf{WI},x}(x, vk) = (T'_x(x), vk)$, and $T_{\mathsf{WI},w}(w, x', T, \sigma) = (T'_w(w), x', T' \circ T, \sigma)$. Assuming our underlying NIWI is malleable, we can define ZKEval in terms of $\mathsf{ZKEval}_{\mathsf{WI}}$:

- $\mathsf{ZKEval}(\sigma_{\mathrm{crs}}, T, x, \pi)$: Output $\mathsf{ZKEval}_{\mathsf{WI}}(\sigma_{\mathsf{WI}crs}, T_{\mathsf{WI}}(T), x_{\mathsf{WI}}, \pi)$ where $x_{\mathsf{WI}} = (x, vk)$.

To see that this construction gives us the desired properties, we have the following three theorems; due to space constraints, the proofs can be found in the full version of our paper [15]:

Theorem 3.1. *If the underlying non-interactive proof system is witness indistinguishable, the scheme described above is zero knowledge.*

Theorem 3.2. *If the underlying signature scheme is EUF-CMA secure and the underlying NIWIPoK is extractable, the scheme described above satisfies controlled-malleable simulation-sound extractability.*

Theorem 3.3. *If the underlying NIWIPoK is derivation private for $\mathcal{T}_{\mathsf{WI}}$ (as defined in Definition 2.3), then the scheme described above is strongly derivation private for $\mathcal{T}$ (as defined in Definition 2.4).*

In addition, we would like to ensure that this construction can in fact be instantiated efficiently for many useful sets $\mathcal{T}$ with a derivation-private NIWIPoK; it turns out that this can be done by combining Groth-Sahai proofs [33] with a special type of signature called a *structure-preserving signature*. For more details, we defer to Section 4.2.

4   Instantiating cm-NIZKs Using Groth-Sahai Proofs

In this section, we explore the malleability of Groth-Sahai (GS) proofs [33]. This will allow us to efficiently instantiate controlled-malleable proofs for a large class of transformations.

4.1   Malleability for Groth-Sahai Proofs

We aim to fully characterize the class of transformations with respect to which GS proofs can be made malleable. First, we recall that GS proofs allow a prover to prove knowledge of a satisfying assignment to a list of (homogeneous) *pairing product equations* eq of the form $\prod_{i,j\in[1..n]} e(x_i, x_j)^{\gamma_{ij}} = 1$ concerning the set of variables $x_1, \ldots, x_n \in \mathbb{G}$. Furthermore, some of the variables in these equations may be fixed to be specific constant values (for example, the public group generator g). In what follows we will use $a, b, c, \ldots$ to denote fixed constants, and $\mathbf{x}, \mathbf{y}, \mathbf{z}, \ldots$ to denote unconstrained variables. An instance x of such a *pairing product statement* consists of the list of equations $\mathsf{eq}_1, \ldots, \mathsf{eq}_\ell$ (fully described by their exponents $\{\gamma_{ij}^{(1)}\}, \ldots, \{\gamma_{ij}^{(\ell)}\}$) and the values of the constrained variables (fully described by the list $a_1, \ldots, a_{n'} \in \mathbb{G}$ for $n' \leq n$).

In the existing literature, there are already various examples [20,7,24] of ways in which pairing product statements and the accompanying Groth-Sahai proofs can be mauled. Here, we attempt to generalize these previous works by providing a characterization of *all* the ways in which GS proofs of pairing product statements can be mauled; we then show, in the full version of our paper [15], how these previous examples can be obtained as special cases of our general characterization.

To start, we describe transformations on pairing product instances in terms of a few basic operations. We will say that any transformation that can be described as a composition of these operations is a *valid transformation*. For each valid transformation we show, in the full version, that there is a corresponding ZKEval procedure that updates the GS proof to prove the new statement. Finally, we present in the full version some other convenient operations that can be derived from our minimal set.

To help illustrate the usage of our basic transformations, we consider their effect on the pairing product instance $(\mathsf{eq}_1, \mathsf{eq}_2, a, b)$, where $\mathsf{eq}_1 := e(\mathbf{x}, b)e(a, b) = 1$

and $\mathsf{eq}_2 := e(a, \mathbf{y}) = 1$. Note that here we will describe the transformations in terms of their effect on the instances, but in all of these operations the corresponding witness transformations T_w are easily derived from the instance transformations T_x.

Definition 4.1. *(Informal) A* valid transformation *is one that can be expressed as some combination of (a polynomial number of) the following six operations:*

1. *Merge equations:* $\mathsf{MergeEq}(\mathsf{eq}_i, \mathsf{eq}_j)$ *adds the product of* eq_i *and* eq_j *as a new equation.*
   Ex. $\mathsf{MergeEq}(\mathsf{eq}_1, \mathsf{eq}_2)$ *adds the equation* $e(\mathbf{x}, b)e(a, b)e(a, \mathbf{y}) = 1$
2. *Merge variables:* $\mathsf{MergeVar}(x, y, z, S)$ *generates a new variable* z. *If* x *and* y *are both constants,* z *will have value* xy. *Otherwise* $\mathbf{z}$ *will be unconstrained. For every variable* w *in the set* S, *we add the equation* $e(xy, w)^{-1}e(z, w) = 1$.[5]
   Ex. $\mathsf{MergeVar}(x, a, z, \{x, b, z\})$ *adds the variable* z *and the equations* $e(\mathbf{x}a, \mathbf{x})^{-1}e(\mathbf{z}, \mathbf{x}) = 1$, $e(\mathbf{x}a, b)^{-1}e(\mathbf{z}, b) = 1$, *and* $e(\mathbf{x}a, \mathbf{z})^{-1}e(\mathbf{z}, \mathbf{z}) = 1$.
3. *Exponentiate variable:* $\mathsf{ExpVar}(x, \delta, z, S)$ *generates a new variable* z. *If* x *is a constant,* $z = x^\delta$, *otherwise* $\mathbf{z}$ *will be unconstrained. For every variable* $w \in S$, *we add the equation* $e(x, w)^{-\delta}e(z, w) = 1$.
   Ex. $\mathsf{ExpVar}(x, \delta, z, \{x, b, z\})$ *adds the variable* z *and the equations* $e(\mathbf{x}, \mathbf{x})^{-\delta}e(\mathbf{z}, \mathbf{x}) = 1$, $e(\mathbf{x}, b)^{-\delta}e(\mathbf{z}, b) = 1$, *and* $e(\mathbf{x}, \mathbf{z})^{-\delta}e(\mathbf{z}, \mathbf{z}) = 1$.
4. *Add constant equation:* $\mathsf{Add}(\{a_i\}, \{b_j\}, \{\gamma_{ij}\})$ *takes a set of constants* a_i, b_i, *satisfying a pairing product equation* $\prod e(a_i, b_j)^{\gamma_{ij}} = 1$ *and adds these variables and the new equation to the statement.*
   Ex. $\mathsf{Add}(\{g\}, \{1\}, \{1\})$ *adds the variables* $g, 1$ *and equation* $\mathsf{eq}_3 := e(g, 1) = 1$. *We often write as a shorthand* $\mathsf{Add}(\mathsf{eq}_3 := e(g, 1) = 1)$.
5. *Remove equation:* $\mathsf{RemoveEq}(\mathsf{eq}_i)$ *simply removes equation* eq_i *from the list.*
   Ex. $\mathsf{RemoveEq}(\mathsf{eq}_2)$ *removes the equation* $e(a, \mathbf{y}) = 1$ *from the equation list.*
6. *Remove variable:* $\mathsf{RemoveVar}(x)$ *removes the variable* x *from the variable set iff* x *does not appear in any of the listed equations.*
   Ex. We cannot remove any of the variables from the example statement. However, we could do $\mathsf{RemoveEq}(\mathsf{eq}_2)$ *and then* $\mathsf{RemoveVar}(y)$, *which would remove the equation* $e(a, \mathbf{y}) = 1$ *from the equation list and the variable* y *from the set of variables.*

A proof of the following lemma appears in the full version:

Lemma 4.1. *There exists an efficient procedure* ZKEval *such that given any pairing product instance* x, *any valid transformation* T, *and any accepting Groth-Sahai proof* π *for* x, $\mathsf{ZKEval}(x, \pi, T)$ *produces an accepting proof for* $T(x)$.

4.2   An Efficient Instantiation of Controlled Malleable NIZKs

Looking back at Section 3 we see that there are two main components needed to efficiently instantiate a controlled-malleable NIZK proof system: appropriately malleable proofs and signatures that can be used in conjunction with these proofs.

---

[5] This is shorthand for $e(x, w)^{-1}e(y, w)^{-1}e(z, w) = 1$.

First we consider the set of relations and tranformations for which we can use Groth-Sahai proofs to construct the necessary malleable NIWIPoKs.

Definition 4.2. *For a relation R and a class of transformations $\mathcal{T}$, we say $(R, \mathcal{T})$ is* CM-friendly *if the following six properties hold: (1) representable statements: any instance and witness of R can be represented as a set of group elements; (2) representable transformations: any transformation in $\mathcal{T}$ can be represented as a set of group elements; (3) provable statements: we can prove the statement $(x, w) \in R$ using pairing product equations; (4) provable transformations: we can prove the statement "$T_x(x') = x$ for $T \in \mathcal{T}$" using pairing product equations; (5) transformable statements: for any $T \in \mathcal{T}$ there is a valid transformation from the statement "$(x, w) \in R$" to the statement "$(T_x(x), T_w(w)) \in R$"; and (6) transformable transformations: for any $T, T' \in \mathcal{T}$ there is a valid transformation from the statement "$T_x(x') = x$ for $T = (T_x, T_w) \in \mathcal{T}$" to the statement "$T'_x \circ T_x(x') = T'_x(x)$ for $T' \circ T \in \mathcal{T}$."*

In order for the signatures to be used within our construction, we know that they need to have pairing-based verifiability (i.e., we can represent the Verify algorithm in terms of a set of GS equations), and that the values being signed must be group elements so that they can be efficiently extracted from the proof (as GS proofs are extractable for group elements only, not exponents). These requirements seem to imply the need for *structure-preserving signatures* [3], which we can define for the symmetric setting as follows:

Definition 4.3. *A signature scheme* (KeyGen, Sign, Verify) *over a bilinear group (p, G, G_T, g, e) is said to be* structure preserving *if the verification key, messages, and signatures all consist of group elements in G, and the verification algorithm evaluates membership in G and pairing product equations.*

Since their introduction, three structure-preserving signature schemes have emerged that would be suitable for our purposes; all three have advantages and disadvantages. The first, due to Abe, Haralambiev, and Ohkubo [5,3] is quite efficient but uses a slightly strong q-type assumption. The second, due to Abe et al. [4], is optimally efficient but provably secure only in the generic group model. The third and most recent, due to Chase and Kohlweiss [14], is significantly less efficient than the previous two, but relies for its security on Decision Linear (DLIN) [11], which is already a relatively well-established assumption.

Because we can also instantiate GS proofs using DLIN, we focus on this last structure-preserving signature, keeping in mind that others may be substituted in for the sake of efficiency (but at the cost of adding an assumption). Putting these signatures and GS proofs together, we can show our main result of this section: given any CM-friendly relation and set of transformations $(R, \mathcal{T})$, we can combine structure-preserving signatures and malleable proofs to obtain a cm-NIZK. This can be stated as the following theorem (a proof of which can be found in the full version of our paper):

Theorem 4.1. *Given a derivation private NIWIPoK for pairing product statements that is malleable for the set of all valid transformations, and a structure preserving signature scheme, we can construct a cm-NIZK for any CM-friendly relation and transformation set $(R, \mathcal{T})$.*

In the full version of our paper, we show that Groth-Sahai proofs are malleable for the set of all valid transformations (as outlined in Definition 4.1). As Groth-Sahai proofs and structure-preserving signatures can both be constructed based on DLIN, we obtain the following theorem:

Theorem 4.2. *If DLIN holds, then we can construct a cm-NIZK that satisfies strong derivation privacy for any CM-friendly relation and transformation set $(R, \mathcal{T})$.*

5   Controlled Malleability for Encryption

As we mentioned earlier, malleability can also be an attractive feature for a cryptosystem: it allows computation on encrypted data. On the other hand, it seems to be in conflict with security: if a ciphertext can be transformed into a ciphertext for a related message, then the encryption scheme is clearly not secure under an adaptive chosen ciphertext attack, which is the standard notion of security for encryption.

Prabhakaran and Rosulek [35,36] were the first to define and realize a meaningful notion of security in this context. They introduced re-randomizable CCA security (RCCA) [35] and homomorphic CCA security (HCCA) [36]. In a nutshell, their definition of security is given as a game between a challenger and an adversary; the adversary receives a public key and a challenge ciphertext and can query the challenger for decryptions of ciphertexts. The challenger's ciphertext c^* is either a valid encryption of some message, or a dummy ciphertext; in the former case, the challenger answers the decryption queries honestly; in the latter case, the challenger may decide that a decryption query is a "derivative" ciphertext computed from c^* using some transformation T; if this is an allowed transformation, the challenger responds with $T(m)$, else it rejects the query. The adversary wins if it correctly guesses whether its challenge ciphertext was meaningful.[6] Prabhakaran and Rosulek achieve their notion of security under the decisional Diffie-Hellman assumption using ad-hoc techniques reminiscent of the Cramer-Shoup [17] cryptosystem.

In this section, we show that controlled-malleable NIZKs can be used as a general tool for achieving RCCA and HCCA security. Our construction is more modular than that of Prabhakaran and Rosulek: we construct a controlled-malleable-CCA-secure encryption scheme generically from a semantically secure one and a cm-NIZK for an appropriate language; where controlled-malleable-CCA security is our own notion of security that is, in some sense, a generalization

---

[6] A formal definition and more detailed explanation of their notion of homomorphic-CCA (HCCA) security can be found in the full version of our paper [15].

of RCCA security and also captures the security goals of HCCA security. We then show how our construction can be instantiated using Groth-Sahai proofs, under the DLIN assumption in groups with bilinear maps.

5.1   Definition of Controlled-Malleable CCA Security

Our definitional goals here are (1) to give a definition of controlled malleability for encryption that closely mirrors our definition of controlled malleability for proofs, and (2) to give a definition that can be easily related to previous notions such as CCA, RCCA, and HCCA. We call this notion of security *controlled-malleable CCA* (CM-CCA) security.

Following Prabhakaran and Rosulek [36], CM-CCA requires the existence of two algorithms, SimEnc and SimExt. SimEnc creates ciphertexts that are distributed indistinguishably from regular ciphertexts (those generated using the encryption algorithm Enc), but contain no information about the queried message; this is modeled by having SimEnc not take any message as input. SimExt allows the challenger to track "derivative" ciphertexts. That is to say, on input a ciphertext c, SimExt determines if it was obtained by transforming some ciphertext c' previously generated using SimEnc; if so, SimExt outputs the corresponding transformation T.

The game between the challenger and the adversary in the definition of security is somewhat different from that in the definition by Prabhakaran and Rosulek. Specifically, we do not have a single challenge ciphertext c^*; instead, the adversary has access to encryption and decryption oracles. Intuitively, for our definition we would like to say that an adversary cannot distinguish between two worlds: the real world in which it is given access to honest encryption and decryption oracles, and an ideal world in which it is given access to an ideal encryption oracle (which outputs ciphertexts containing no information about the queried message) and a decryption oracle that outputs a special answer for ciphertexts derived from the ideal ciphertexts (by using SimExt to track such ciphertexts) and honestly decrypts otherwise.

Let us consider transformations more closely. Recall that, for proofs of language membership, a transformation $T \in \mathcal{T}$ consists of a pair of transformations (T_x, T_w), where T_x acts on the instances, and T_w on the witnesses. What is the analogue for ciphertexts? A legal transformation T_x on a ciphertext implies some legal transformation T_m on an underlying message and a corresponding transformation T_r on the underlying randomness. Thus, here we view transformations as tuples $T = (T_x, (T_m, T_r))$, where T_x acts on the ciphertexts, T_m acts on the plaintexts, and T_r acts on the randomness.

In the full version of our paper [15], we relate CM-CCA security to CCA, RCCA and HCCA security. Specifically, we show that (1) when the class of allowed transformation $\mathcal{T}$ is the empty set, CM-CCA implies regular CCA security; (2) when the class of allowed transformations is as follows: $T \in \mathcal{T}$ if $T = (T_x, (T_m, T_r))$ where T_m is the identity transformation, then CM-CCA security implies RCCA security; (3) in more general cases we show that it implies the notion of targeted malleability introduced by Boneh et al. [12]; in addition, we show that our notion satisfies the UC definition given by Prabhakaran and Rosulek, so that it captures the desired HCCA security goals, even if it does not satisfy their definition of HCCA security (which is in fact a stronger notion).

Finally, because our cm-NIZK is malleable only with respect to unary transformations, we inherit the limitation that our encryption scheme is malleable only with respect to unary transformations as well; as our security definition is closely related to HCCA security and Prabharakan and Rosulek in fact prove HCCA security (combined with unlinkability) is impossible with respect to binary transformations, this is perhaps not surprising.

Definition 5.1. *For an encryption scheme* (KeyGen, Enc, Dec), *a class of transformations* $\mathcal{T}$, *an adversary* $\mathcal{A}$, *and a bit* b, *let* $p_b^{\mathcal{A}}(k)$ *be the probability of the event* $b' = 0$ *in the following game: first* $(pk, sk) \stackrel{\$}{\leftarrow} K(1^k)$, *and next* $b' \stackrel{\$}{\leftarrow} \mathcal{A}^{E_{pk}(\cdot), D_{sk}(\cdot)}(pk)$, *where* (K, E, D) *are defined as* (KeyGen, Enc, Dec) *if* $b = 0$, *and the following algorithms (defined for a state set* $Q = Q_m \times Q_c = \{(m_i, c_i)\}$*) if* $b = 1$:

Procedure $K(1^k)$	Procedure $E(pk, m)$	Procedure $D(sk, c)$
(pk, sk, τ_1, τ_2)	$c \stackrel{\$}{\leftarrow} \mathsf{SimEnc}(pk, \tau_1)$	$(c', T) \leftarrow \mathsf{SimExt}(sk, \tau_2, c)$
$\quad \stackrel{\$}{\leftarrow} \mathsf{SimKeyGen}(1^k)$	add (m, c) to Q	if $\exists i$ s.t. $c' = c_i \in Q_c$ and $T \neq \perp$
return pk	return c	$\quad$ return $T_m(m_i)$
		else
		$\quad$ return $\mathsf{Dec}(sk, c)$

We say that the encryption scheme is controlled-malleable-CCA secure *(or CM-CCA secure for short) if there exist PPT algorithms* SimKeyGen, SimEnc, *and* SimExt *as used above such that for all PPT algorithms* $\mathcal{A}$ *there exists a negligible function* $\nu(\cdot)$ *such that* $|p_0^{\mathcal{A}}(k) - p_1^{\mathcal{A}}(k)| < \nu(k)$.

As mentioned earlier, we can obtain an encryption scheme that achieves this notion of security; we do this by combining a cm-NIZK (CRSSetup, $\mathcal{P}, \mathcal{V}$) for the relation R such that $((pk, c), (m, r)) \in R$ iff $c := \mathsf{Enc}'(pk, m; r)$ and an IND-CPA-secure encryption scheme (KeyGen', Enc', Dec'). Due to space constraints, our construction and efficient instantiation of this scheme can be found in the full version of our paper [15].

6   Compactly Proving Correctness of a Shuffle

As described in the introduction, we achieve a notion of verifiability for shuffles that does not require each mix server to output its own proof of correctness; instead, using the malleability of our proofs, each mix server can maul the proof of the previous one. One point that is important to keep in mind with this approach is that the soundness of the scheme does not follow directly from the soundness of each of the individual proofs anymore; instead, one proof must somehow suffice to prove the validity of the entire series of shuffles, yet still remain compact. To capture this requirement, we define a new notion for the security of a shuffle, that we call *compact verifiability*.

To define our notion, we assume that a verifiable shuffle consists of three algorithms: a Setup algorithm that outputs the parameters for the shuffle and the identifying public keys for the honest mix servers, a Shuffle algorithm that takes in a set of ciphertexts and outputs both a shuffle of these ciphertexts and a proof that the shuffle was done properly, and finally a Verify algorithm that checks the validity of the proofs.

In our definition, the adversary is given the public keys of all the honest shuffling authorities, as well as an honestly generated public key for the encryption scheme. It can then provide a list of ciphertexts and ask that they be shuffled by one of the honest authorities (we call this an initial shuffle), or provide a set of input ciphertexts, a set of shuffled ciphertexts, and a proof, and ask one of the honest authorities to shuffle the ciphertexts again and update the proof. Finally, the adversary produces challenge values consisting of a set of input ciphertexts, a set of shuffled ciphertexts and a proof that includes the public key of at least one of the honest authorities. If this proof verifies, it receives either the decryption of the shuffled ciphertexts, or a random permutation of the decryptions of the initial ciphertexts. Our definition requires that it should be hard for the adversary to distinguish which of the two it is given.

We also require that the input ciphertexts are always accompanied by a proof that they are well-formed; i.e., a proof of knowledge of a valid message and the randomness used in encryption. This is usually necessary in many applications (for example in voting when each voter must prove that he has encrypted a valid vote), and in our construction it means that we can easily handle an adversary who produces the input ciphertexts in invalid ways; e.g., by mauling ciphertexts from a previous shuffle, or by submitting malformed ciphertexts.

Definition 6.1. *For a verifiable shuffle* (Setup, Shuffle, Verify) *with respect to an encryption scheme* (KeyGen, Enc, Dec), *a given adversary $\mathcal{A}$ and a bit $b \in \{0,1\}$, let $p_b^{\mathcal{A}}(k)$ be the probability that $b' = 0$ in the following experiment:*

- *Step 1.* $(params, sk, S = \{pk_i\}, \{sk_i\}) \xleftarrow{\$} \mathsf{Setup}(1^k)$.
- *Step 2.* $\mathcal{A}$ *gets params, S, and access to the following two oracles: an initial shuffle oracle that, on input* $(\{c_i, \pi_i\}, pk_\ell)$ *for $pk_\ell \in S$, outputs* $(\{c_i'\}, \pi, \{pk_\ell\})$ *(if all the proofs of knowledge π_i verify), where π is a proof that the $\{c_i'\}$ constitute a valid shuffle of the $\{c_i\}$ performed by the user corresponding to pk_ℓ (i.e., the user who knows sk_ℓ), and a shuffle oracle that, on input* $(\{c_i, \pi_i\}, \{c_i'\}, \pi, \{pk_j\}, pk_m)$ *for $pk_m \in S$, outputs* $(\{c_i''\}, \pi', \{pk_j\} \cup pk_m)$.
- *Step 3. Eventually, $\mathcal{A}$ outputs a tuple* $(\{c_i, \pi_i\}, \{c_i'\}, \pi, S' = \{pk_j\})$.
- *Step 4. If* $\mathsf{Verify}(params, (\{c_i, \pi_i\}, \{c_i'\}, \pi, \{pk_j\})) = 1$ *and $S \cap S' \neq \emptyset$ then continue; otherwise simply abort and output $\perp$. If $b = 0$ give $\mathcal{A}$ $\{\mathsf{Dec}(sk, c_i')\}$, and if $b = 1$ then give $\mathcal{A}$ $\varphi(\{\mathsf{Dec}(sk, c_i)\})$, where φ is a random permutation* $\varphi \xleftarrow{\$} S_n$.
- *Step 5. $\mathcal{A}$ outputs a guess bit b'.*

We say that the shuffle is compactly verifiable *if for all PPT algorithms $\mathcal{A}$ there exists a negligible function $\nu(\cdot)$ such that $|p_0^{\mathcal{A}}(k) - p_1^{\mathcal{A}}(k)| < \nu(k)$.*

Our compactly-verifiable shuffle construction will utilize four building blocks: a *hard relation* R_{pk} (as defined by Damgård [18, Definition 3]), a re-randomizable IND-CPA-secure encryption scheme (KeyGen, ReRand, Enc, Dec), a proof of knowledge (CRSSetup, $\mathcal{P}$, $\mathcal{V}$), and a cm-NIZK (CRSSetup$'$, $\mathcal{P}'$, $\mathcal{V}'$). The hard relation will be used to ensure that the secret key sk_j known to the j-th mix server cannot be derived from its public key pk_j,[7] the proof of knowledge will be created by the users performing the initial encryptions to prove knowledge of their votes, and the cm-NIZK will be used to prove that a given collection $\{c_i'\}$ is a valid shuffle of a collection $\{c_i\}$, performed by the mix servers corresponding to a set of public keys $\{pk_j\}$. This means that the instances are of the form $x = (pk, \{c_i\}, \{c_i'\}, \{pk_j\})$, witnesses are of the form $w = (\varphi, \{r_i\}, \{sk_j\})$ (where φ is the permutation used, $\{r_i\}$ is the randomness used to re-randomize the ciphertexts, and $\{sk_j\}$ are the secret keys corresponding to $\{pk_j\}$), and the relation R is $((pk, \{c_i\}, \{c_i'\}, \{pk_j\}_{i=1}^{\ell'}), (\varphi, \{r_i\}, \{sk_j\})) \in R$ iff $\{c_i'\} = \{\mathsf{ReRand}(pk, \varphi(c_i); r_i)\} \wedge (pk_j, sk_j) \in R_{pk} \; \forall j \in [1..\ell']$. The valid transformations are then $T_{(\varphi', \{r_i'\}, \{sk_j^+, pk_j^+\}, \{pk_j^-\})} = (T_x, T_w)$, where $T_x(pk, \{c_i\}, \{c_i'\}, \{pk_j\}) := (pk, \{c_i\}, \{\mathsf{ReRand}(pk, \varphi'(c_i); r_i')\}, \{pk_j\} \cup (\{pk_j^+\} \setminus \{pk_j^-\}))$ and T_w transforms the witness accordingly. Due to space constraints, a formal outline of how these primitives are combined can be found in the full version of our paper, along with a proof of the following theorem:

Theorem 6.1. *If the encryption scheme is re-randomizable and IND-CPA secure, R_{pk} is a hard relation, the proofs π_i are NIZKPoKs, and the proof π is a cm-NIZK, then the above construction gives a compactly verifiable shuffle.*

Acknowledgments. Anna Lysyanskaya was supported by NSF grants 1012060, 0964379, 0831293, and by a Sloan Foundation fellowship, and Sarah Meiklejohn was supported in part by a MURI grant administered by the Air Force Office of Scientific Research and in part by a graduate fellowship from the Charles Lee Powell Foundation.

References

1. Abe, M.: Universally Verifiable Mix-Net with Verification Work Independent of the Number of Mix-Servers. In: Nyberg, K. (ed.) EUROCRYPT 1998. LNCS, vol. 1403, pp. 437–447. Springer, Heidelberg (1998)
2. Abe, M.: Mix-Networks on Permutation Networks. In: Lam, K.-Y., Okamoto, E., Xing, C. (eds.) ASIACRYPT 1999. LNCS, vol. 1716, pp. 258–273. Springer, Heidelberg (1999)
3. Abe, M., Fuchsbauer, G., Groth, J., Haralambiev, K., Ohkubo, M.: Structure-Preserving Signatures and Commitments to Group Elements. In: Rabin, T. (ed.) CRYPTO 2010. LNCS, vol. 6223, pp. 209–236. Springer, Heidelberg (2010)

---

[7] It is worth mentioning that generically we can use a one-way function to obtain this property, but that we cannot efficiently instantiate this in our setting and so use instead a hard relation (for more on this see the full version of our paper).

4. Abe, M., Groth, J., Haralambiev, K., Ohkubo, M.: Optimal Structure-Preserving Signatures in Asymmetric Bilinear Groups. In: Rogaway, P. (ed.) CRYPTO 2011. LNCS, vol. 6841, pp. 649–666. Springer, Heidelberg (2011)

5. Abe, M., Haralambiev, K., Ohkubo, M.: Signing on elements in bilinear groups for modular protocol design. Cryptology ePrint Archive, Report 2010/133 (2010), http://eprint.iacr.org/2010/133

6. Abe, M., Hoshino, F.: Remarks on Mix-Networks Based on Permutation Networks. In: Kim, K.-c. (ed.) PKC 2001. LNCS, vol. 1992, pp. 317–324. Springer, Heidelberg (2001)

7. Acar, T., Nguyen, L.: Revocation for Delegatable Anonymous Credentials. In: Catalano, D., Fazio, N., Gennaro, R., Nicolosi, A. (eds.) PKC 2011. LNCS, vol. 6571, pp. 423–440. Springer, Heidelberg (2011)

8. Belenkiy, M., Camenisch, J., Chase, M., Kohlweiss, M., Lysyanskaya, A., Shacham, H.: Randomizable Proofs and Delegatable Anonymous Credentials. In: Halevi, S. (ed.) CRYPTO 2009. LNCS, vol. 5677, pp. 108–125. Springer, Heidelberg (2009)

9. Bellare, M., Goldreich, O.: On Defining Proofs of Knowledge. In: Brickell, E.F. (ed.) CRYPTO 1992. LNCS, vol. 740, pp. 390–420. Springer, Heidelberg (1993)

10. Blum, M., de Santis, A., Micali, S., Persiano, G.: Non-interactive zero-knowledge. SIAM Journal of Computing 20(6), 1084–1118 (1991)

11. Boneh, D., Boyen, X., Shacham, H.: Short Group Signatures. In: Franklin, M. (ed.) CRYPTO 2004. LNCS, vol. 3152, pp. 41–55. Springer, Heidelberg (2004)

12. Boneh, D., Segev, G., Waters, B.: Targeted malleability: homomorphic encryption for restricted computations. In: Proceedings of ITCS 2012 (2012)

13. Brakerski, Z., Vaikuntanathan, V.: Fully Homomorphic Encryption from Ring-LWE and Security for Key Dependent Messages. In: Rogaway, P. (ed.) CRYPTO 2011. LNCS, vol. 6841, pp. 505–524. Springer, Heidelberg (2011)

14. Chase, M., Kohlweiss, M.: A domain transformations for structure-preserving signatures on group elements. Cryptology ePrint Archive, Report 2011/342 (2011), http://eprint.iacr.org/2011/342

15. Chase, M., Kohlweiss, M., Lysyanskaya, A., Meiklejohn, S.: Malleable proof systems and applications. Cryptology ePrint Archive, Report 2012/012 (2012), http://eprint.iacr.org/2012/012

16. Chaum, D.: Untraceable electronic mail, return addresses, and digital pseudonyms. Communications of the ACM 24(2), 84–88 (1981)

17. Cramer, R., Shoup, V.: A Practical Public Key Cryptosystem Provably Secure against Adaptive Chosen Ciphertext Attack. In: Krawczyk, H. (ed.) CRYPTO 1998. LNCS, vol. 1462, pp. 13–25. Springer, Heidelberg (1998)

18. Damgård, I.: On sigma protocols, http://www.daimi.au.dk/~ivan/Sigma.pdf

19. De Santis, A., Di Crescenzo, G., Ostrovsky, R., Persiano, G., Sahai, A.: Robust Non-interactive Zero Knowledge. In: Kilian, J. (ed.) CRYPTO 2001. LNCS, vol. 2139, pp. 566–598. Springer, Heidelberg (2001)

20. Dodis, Y., Haralambiev, K., López-Alt, A., Wichs, D.: Cryptography against continuous memory attacks. In: Proceedings of FOCS 2010, pp. 511–520 (2010)

21. Feige, U., Lapidot, D., Shamir, A.: Multiple non-interactive zero knowledge proofs under general assumptions. SIAM Journal of Computing 29, 1–28 (1999)

22. Feige, U., Shamir, A.: Witness indistinguishable and witness hiding protocols. In: Proceedings of STOC 1990, pp. 416–426 (1990)

23. Fiat, A., Shamir, A.: How to Prove Yourself: Practical Solutions to Identification and Signature Problems. In: Odlyzko, A.M. (ed.) CRYPTO 1986. LNCS, vol. 263, pp. 186–194. Springer, Heidelberg (1987)

24. Fuchsbauer, G.: Commuting Signatures and Verifiable Encryption. In: Paterson, K.G. (ed.) EUROCRYPT 2011. LNCS, vol. 6632, pp. 224–245. Springer, Heidelberg (2011)
25. Furukawa, J.: Efficient and verifiable shuffling and shuffle-decryption. IEICE Transactions on Fundamentals of Electronic, Communications and Computer Science 88(1), 172–188 (2005)
26. Furukawa, J., Sako, K.: An Efficient Scheme for Proving a Shuffle. In: Kilian, J. (ed.) CRYPTO 2001. LNCS, vol. 2139, pp. 368–387. Springer, Heidelberg (2001)
27. Gentry, C.: Fully homomorphic encryption using ideal lattices. In: Proceedings of STOC 2009, pp. 169–178 (2009)
28. Goldwasser, S., Micali, S., Rackoff, C.: The knowledge complexity of interactive proof systems. In: Proceedings of STOC 1985, pp. 186–208 (1985)
29. Groth, J.: A Verifiable Secret Shuffle of Homomorphic Encryptions. In: Desmedt, Y.G. (ed.) PKC 2003. LNCS, vol. 2567, pp. 145–160. Springer, Heidelberg (2002)
30. Groth, J.: Simulation-Sound NIZK Proofs for a Practical Language and Constant Size Group Signatures. In: Lai, X., Chen, K. (eds.) ASIACRYPT 2006. LNCS, vol. 4284, pp. 444–459. Springer, Heidelberg (2006)
31. Groth, J., Lu, S.: A Non-interactive Shuffle with Pairing Based Verifiability. In: Kurosawa, K. (ed.) ASIACRYPT 2007. LNCS, vol. 4833, pp. 51–67. Springer, Heidelberg (2007)
32. Groth, J., Ostrovsky, R., Sahai, A.: Perfect Non-interactive Zero Knowledge for NP. In: Vaudenay, S. (ed.) EUROCRYPT 2006. LNCS, vol. 4004, pp. 339–358. Springer, Heidelberg (2006)
33. Groth, J., Sahai, A.: Efficient Non-interactive Proof Systems for Bilinear Groups. In: Smart, N.P. (ed.) EUROCRYPT 2008. LNCS, vol. 4965, pp. 415–432. Springer, Heidelberg (2008)
34. Neff, A.: A verifiable secret shuffle and its applications to e-voting. In: Proceedings of CCS 2001, pp. 116–125 (2001)
35. Prabhakaran, M., Rosulek, M.: Rerandomizable RCCA Encryption. In: Menezes, A. (ed.) CRYPTO 2007. LNCS, vol. 4622, pp. 517–534. Springer, Heidelberg (2007)
36. Prabhakaran, M., Rosulek, M.: Homomorphic Encryption with CCA Security. In: Aceto, L., Damgård, I., Goldberg, L.A., Halldórsson, M.M., Ingólfsdóttir, A., Walukiewicz, I. (eds.) ICALP 2008, Part II. LNCS, vol. 5126, pp. 667–678. Springer, Heidelberg (2008)
37. Sahai, A.: Non-malleable non-interactive zero knowledge and adaptive chosen-ciphertext security. In: Proceedings of FOCS 1999, pp. 543–553 (1999)
38. Sako, K., Kilian, J.: Receipt-Free Mix-Type Voting Scheme. In: Guillou, L.C., Quisquater, J.-J. (eds.) EUROCRYPT 1995. LNCS, vol. 921, pp. 393–403. Springer, Heidelberg (1995)
39. Smart, N., Vercauteren, F.: Fully Homomorphic Encryption with Relatively Small Key and Ciphertext Sizes. In: Nguyen, P.Q., Pointcheval, D. (eds.) PKC 2010. LNCS, vol. 6056, pp. 420–443. Springer, Heidelberg (2010)
40. van Dijk, M., Gentry, C., Halevi, S., Vaikuntanathan, V.: Fully Homomorphic Encryption over the Integers. In: Gilbert, H. (ed.) EUROCRYPT 2010. LNCS, vol. 6110, pp. 24–43. Springer, Heidelberg (2010)
41. Wikström, D.: A Sender Verifiable Mix-Net and a New Proof of a Shuffle. In: Roy, B. (ed.) ASIACRYPT 2005. LNCS, vol. 3788, pp. 273–292. Springer, Heidelberg (2005)

Group to Group Commitments Do Not Shrink

Masayuki Abe, Kristiyan Haralambiev, and Miyako Ohkubo

NTT Information Sharing Platform Laboratories
abe.masayuki@lab.ntt.co.jp
New York University
kkh@cs.nyu.edu
Security Architecture Laboratory, NSRI, NICT
m.ohkubo@nict.go.jp

Abstract. We investigate commitment schemes whose messages, keys, commitments, and decommitments are elements of bilinear groups, and whose openings are verified by pairing product equations. Such commitments facilitate efficient zero-knowledge proofs of knowledge of a correct opening. We show two lower bounds on such schemes: a commitment cannot be shorter than the message and verifying the opening in a symmetric bilinear group setting requires evaluating at least two independent pairing product equations. We also present optimal constructions that match the lower bounds in symmetric and asymmetric bilinear group settings.

Keywords: Structure-Preserving Commitments, Homomorphic Trapdoor Commitments.

1 Introduction

Efficient cryptographic protocols are often hand-crafted and their underlying idea is hardly visible. On the other hand, modular design offers conceptual simplicity in exchange of losing efficiency. Structure-preserving cryptography [1] is a concept that facilitates modular yet reasonably efficient construction of cryptographic protocols. It provides inter-operable cryptographic building blocks whose input/output data consist only of group elements and their computations preserve the group structure. Combined with the Groth-Sahai (GS) proof system [18], such structure-preserving schemes allow proofs of knowledge about privacy-sensitive data present in their inputs and outputs. Commitments [9,1], various signatures [1,10,2], and adaptive chosen-ciphertext secure public-key encryption [8] have been presented in the context of structure-preserving cryptography. They yield a number of applications including various privacy-protecting signatures [1], efficient zero-knowledge arguments [17], and efficient leakage-resilient signatures [13].

We revisit structure preserving commitment schemes. Their keys, messages, commitments, and decommitments are elements of bilinear groups, and the opening is verified by evaluating pairing product equations. Using a bilinear map $\mathbb{G} \times \mathbb{G} \rightarrow \mathbb{G}_T$, messages from the base group are either committed to target

D. Pointcheval and T. Johansson (Eds.): EUROCRYPT 2012, LNCS 7237, pp. 301–317, 2012.

group elements and the commitments are shrinking, or committed to group elements from the same group but commitments are larger than the messages. In other words, there are two types of commitment functions: either "$\mathbb{G} \to \mathbb{G}_T$ and shrinking" or "$\mathbb{G} \to \mathbb{G}$ and expanding". The former type, [1,16], takes multiple elements in the base group $\mathbb{G}$ as input and shrinks them into a constant number of elements in the target group $\mathbb{G}_T$ by exploiting the one-way nature of the mapping from $\mathbb{G}$ to $\mathbb{G}_T$. Involving elements in $\mathbb{G}_T$ in a commitment is acceptable as long as witness-indistinguishability is sufficient for the accompanying GS proofs, but it is problematic if zero-knowledge is necessary. The latter type, [9,3], which we call *strictly* structure-preserving schemes, takes messages in $\mathbb{G}$ and also yields commitments in $\mathbb{G}$. Unfortunately, due to the absence of a one-way structure in the mapping from $\mathbb{G}$ to $\mathbb{G}$, their construction is more involved. Moreover, they are expanding: commitments are 2-3 times larger than messages in the known constructions. Nothing is known about the lower bound, and constructing more efficient commitment schemes of the latter type has been an open problem.

Our Results. This paper presents two lower bounds on strictly structure-preserving commitment schemes. First, we show that for a message of size k the commitment must be at least size k; thus, negatively answering to the above-stated open problem. This lower bound highlights the gap from the known upper bound of $2k$ in [3]. The lower bound is obtained by assuming that key generation and commitment functions are algebraic. By algebraic algorithms we mean any computation conditioned so that, when outputting a group element, the algorithm "knows" its representation with respect to given bases. The class covers a wide range of algorithms including all constructions in the standard model to the best our knowledge. See Section 2.5 for more detailed discussion.

Next, we show that strictly structure-preserving commitment schemes for symmetric bilinear groups require at least two pairing product equations in the verification. The number of equations, as well as the size of commitments, is an important factor in determining efficiency since the size of a zero-knowledge proof of a correct opening grows linearly with the number of verification equations. A scheme described in [3] achieves this bound but verifies k elements from a commitment in one equation and other k elements in the other equation, which requires $2k$ elements for a commitment. Thus it does not match to the first lower bound. Because the lower bounds of a commitment size and the number of equations are independent, we see that a scheme that achieves both bounds is missing.

We close the gap by presenting two optimal constructions (except for small additive constants). The first construction works over asymmetric bilinear groups, yields commitments of size $k + 1$, and verifies with one equation. The second construction works over symmetric bilinear groups, yields commitments of $k + 2$, and verifies with two equations. Both constructions implement trapdoor and homomorphic properties. The schemes are computationally binding based on simple standard computational assumptions. Finally, we assess their efficiency in combination with GS zero-knowledge proofs of correct opening.

2   Preliminaries

2.1   Bilinear Groups

Let $\mathcal{G}$ be a bilinear group generator that takes security parameter 1^λ and outputs a description of bilinear groups $\Lambda := (p, \mathbb{G}_1, \mathbb{G}_2, \mathbb{G}_T, e, G, \tilde{G})$ where $\mathbb{G}_1$, $\mathbb{G}_2$ and $\mathbb{G}_T$ are groups of prime order p, e is an efficient and non-degenerating bilinear map $e : \mathbb{G}_1 \times \mathbb{G}_2 \to \mathbb{G}_T$, and G and $\tilde{G}$ are generators of $\mathbb{G}_1$ and $\mathbb{G}_2$, respectively. By Λ^*, we denote Λ without the generators G and $\tilde{G}$, i.e., $\Lambda^* = (p, \mathbb{G}_1, \mathbb{G}_2, \mathbb{G}_T, e)$.

By Λ_{sym} we denote a special case of Λ where $\mathbb{G}_1 = \mathbb{G}_2$ (and $G = \tilde{G}$), which is also referred to as a symmetric setting. Λ_{sxdh} denotes a case where the decision Diffie-Hellman (DDH) assumption holds in $\mathbb{G}_1$ and $\mathbb{G}_2$. This means that no efficient mapping is available for either direction. Λ_{sxdh} is usually referred to as the symmetric external DDH (SXDH) setting [22,6,15,23]. For practical differences between Λ_{sym} and Λ_{sxdh}, please refer to [14].

2.2   Notations

By $\mathbb{G}$, we denote a base group, $\mathbb{G}_1$ or $\mathbb{G}_2$, when the difference is not important. By $\mathbb{G}^*$ we denote $\mathbb{G} \setminus \{1_{\mathbb{G}}\}$. We use upper case letters to group elements and corresponding lower case letters to represent the discrete-log of the group element with respect to a fixed (but not necessarily explicit) base. For a set or a vector of group elements, $\boldsymbol{X} \in \mathbb{G}^n$, the size of $\boldsymbol{X}$ refers to n and is denoted as $|\boldsymbol{X}|$. We consider $\boldsymbol{X}$ as a row vector. For a vector or an ordered set $\boldsymbol{X}$, the i-th element is denoted as $\boldsymbol{X}[i]$ or X_i.

We use multiplicative notations for group operations and additive notation for vector operations. The transpose of $\boldsymbol{X}$ is denoted as $\boldsymbol{X}^t$. A concatenation of vectors $\boldsymbol{X} \in \mathbb{G}^n$ and $\boldsymbol{Y} \in \mathbb{G}^k$ is denoted as $\boldsymbol{X}||\boldsymbol{Y} \stackrel{\text{def}}{=} (X_1, \ldots, X_n, Y_1, \ldots, Y_k)$. For $\boldsymbol{X} \in \mathbb{G}^n$ and $\boldsymbol{a} \in \mathbb{Z}_p^n$, we define $\boldsymbol{a}\boldsymbol{X}^t \stackrel{\text{def}}{=} \prod_{i=1}^n X_i^{a_i}$. For a matrix $A \in \mathbb{Z}_p^k \times \mathbb{Z}_p^n$ and $\boldsymbol{X} \in \mathbb{G}^n$, $A\,\boldsymbol{X}^t \stackrel{\text{def}}{=} (\prod_{i=1}^n X_i^{a_{1,i}}, \cdots, \prod_{i=1}^n X_i^{a_{k,i}})^t$, where $a_{i,j}$ is entry (i,j) of A. For $\boldsymbol{X}, \boldsymbol{Y} \in \mathbb{G}^n$, $\boldsymbol{X} + \boldsymbol{Y} \stackrel{\text{def}}{=} (X_1 \cdot Y_1, \ldots, X_n \cdot Y_n)$. For $\boldsymbol{X} \in \mathbb{G}_1^n$ and $\boldsymbol{Y} \in \mathbb{G}_2^n$, $\boldsymbol{X} \cdot \boldsymbol{Y}^t$ is defined as $\prod_{i=1}^n e(X_i, Y_i)$. By $\boldsymbol{0} \in \mathbb{G}^n$ we denote additive unity vector $\boldsymbol{0} = \{1_{\mathbb{G}}, \ldots, 1_{\mathbb{G}}\}$.

For $a_{ij} \in \mathbb{Z}_p, T \in \mathbb{G}_T$, $X_i \in \mathbb{G}_1$, and $Y_j \in \mathbb{G}_2$, an equation of the form

$$\prod_i \prod_j e(X_i, Y_j)^{a_{ij}} = T$$

is called a pairing product equation (PPE). With our notation, any pairing product equation for variables $\boldsymbol{X} \in \mathbb{G}_1^k$ and $\boldsymbol{Y} \in \mathbb{G}_2^n$ can be represented as $\boldsymbol{X} A \boldsymbol{Y}^t = T$ where A is a $k \times n$ matrix over $\mathbb{Z}_p$ and T is a constant in $\mathbb{G}_T$. For convenience, we may abuse these notations for vectors that consist of elements from both $\mathbb{G}_1$ and $\mathbb{G}_2$ assuming that relevant entries of a multiplied scaler matrix are zero so that the computation is well defined in either $\mathbb{G}_1$ or $\mathbb{G}_2$.

For a sequence of events, $E_1, \ldots, E_n$ and a statement S, $\Pr[E_1, \ldots, E_n : S]$ denotes the probability that S is satisfied when events $E_1, \ldots, E_n$ occur. The probability is taken over the random coins used in the events.

2.3   Commitment Schemes

We focus on non-interactive commitment schemes and follow a standard syntactical definition with the following setup.

Definition 1 (Commitment Scheme). *A commitment scheme C is a quadruple of efficient algorithms $C = (Setup, Key, Com, Vrf)$ in which;*

- *$gk \leftarrow Setup(1^\lambda)$ is a common parameter generator that takes security parameter λ and outputs a set of common parameters, gk.*
- *$ck \leftarrow Key(gk)$ is a key generator that takes gk as input and outputs commitment-key ck. It may take extra parameters as input if needed. It is assumed that ck determines the message space $\mathcal{M}_{ck}$. A messages is valid if it is in $\mathcal{M}_{ck}$.*
- *$(com, open) \leftarrow Com(ck, msg)$ is a commitment function that takes ck and message, msg, and outputs commitment, com, and opening information, $open$.*
- *$1/0 \leftarrow Vrf(ck, com, msg, open)$ is a verification function that takes ck, com, msg, and $open$ as input, and outputs 1 or 0 representing acceptance or rejection, respectively.*

It is required that $\Pr[gk \leftarrow Setup(1^\lambda),\ ck \leftarrow Key(gk),\ msg \leftarrow \mathcal{M}_{ck},\ (com, open) \leftarrow Com(ck, msg) : 1 \leftarrow Vrf(ck, com, msg, open)] = 1$.

Definition 2 (Binding and Hiding Properties). *A commitment scheme is binding if, for any polynomial-time adversary $\mathcal{A}$, $\Pr[gk \leftarrow Setup(1^\lambda), ck \leftarrow Key(gk), (com, msg, open, msg', open') \leftarrow \mathcal{A}(ck) : 1 \leftarrow Vrf(ck, com, msg, open) \wedge 1 \leftarrow Vrf(ck, com, msg', open')]$ is negligible. It is hiding if, for any polynomial-time adversary $\mathcal{A}$, advantage $\Pr[1 \leftarrow \text{Hide}_\mathcal{A}^{TC}(1)] - \Pr[1 \leftarrow \text{Hide}_\mathcal{A}^{TC}(0)]$ is negligible in λ where $b' \leftarrow \text{Hide}_\mathcal{A}^{TC}(b)$ is the process that $gk \leftarrow Setup(1^\lambda)$, $ck \leftarrow Key(gk)$, $(msg_0, msg_1, \omega) \leftarrow \mathcal{A}(ck)$, $(com, -) \leftarrow Com(ck, msg_b)$, $b' \leftarrow \mathcal{A}(\omega, com)$.*

Definition 3 (Trapdoor Commitment Scheme). *A commitment scheme is called a trapdoor commitment scheme if Key additionally outputs a trapdoor-key tk, and there is an efficient algorithm Equiv called equivocation algorithm that takes $(ck, tk, com, msg, open, msg')$ as input and outputs $open'$ such that, for legitimately generated ck, tk, and any valid messages msg and msg', it holds that $(com, open) \leftarrow Com(ck, msg)$, $open' \leftarrow Equiv(ck, tk, com, msg, open, msg')$, $1 \leftarrow Vrf(ck, com, msg', open')$, and two distributions $(ck, com, msg, open)$ and $(ck, com, msg', open')$ over all choices of msg and msg' are indistinguishable.*

Definition 3 is usually referred to as chameleon hash [20] and, in fact, is a stronger requirement than the common definition of a trapdoor commitment scheme (e.g., see [16]), which allows a different algorithm (taking tk as an input) to compute equivocalable commitments.

Definition 4 (Homomorphic Commitment Scheme). *A commitment scheme is homomorphic if, for any legitimately generated ck, three binary operations, $\cdot$, $\odot$, $\otimes$, are defined, and for any valid messages, msg and msg', it holds that $(com, open) \leftarrow Com(ck, msg)$, $(com, open) \leftarrow Com(ck, msg)$, $1 \leftarrow Vrf(ck, com \cdot com', msg \odot msg', open \otimes open')$ with probability 1.*

2.4   Strictly Structure-Preserving Commitments

Definition 5 (Strictly Structure-Preserving Commitments). *A commitment scheme* C *is strictly structure-preserving with respect to a bilinear group generator* $\mathcal{G}$ *if*

- *Setup*(1^λ) *outputs* gk *that consists of* $\Lambda = (p, \mathbb{G}_1, \mathbb{G}_2, \mathbb{G}_T, e, G, \tilde{G})$ *generated by* $\mathcal{G}(1^\lambda)$,
- *Key outputs* ck *that consists of* Λ^* *and group elements in* $\mathbb{G}_1$ *and* $\mathbb{G}_2$,
- *the messages consist of group elements in* $\mathbb{G}_1$ *and* $\mathbb{G}_2$,
- *Com outputs com and open that consist of elements in* $\mathbb{G}_1$ *and* $\mathbb{G}_2$, *and*
- *Vrf evaluates membership in* $\mathbb{G}_1$ *and* $\mathbb{G}_2$ *and evaluating pairing product equations over* Λ^*.

Function Setup may also determine non-group elements, such as constants in $\mathbb{Z}_p$, which are given implicitly to other functions as system parameters. Note that the size of a message, denoted by k, may be limited by the size of ck. Also note that, in a previous work [1], *com* is allowed to include elements in $\mathbb{G}_T$ while it is limited to $\mathbb{G}$ in the above *strict* case. This results in limiting the pairing product equations in Vrf to have $T = 1_{\mathbb{G}_T}$ since none of ck, *com*, *msg*, *open* could include elements from $\mathbb{G}_T$. Our lower bounds, however, hold even if ck and *open* include $T \neq 1$ used for verification.

2.5   Algebraic Algorithms

Roughly, an algorithm $\mathcal{A}$ is algebraic over Λ if, whenever $\mathcal{A}$ is given elements $(X_1, \ldots, X_n)$ of a group and outputs an element Y in the same group, $\mathcal{A}$ should "know" a representation $(r_1, \ldots, r_n)$ of Y that fulfils $Y = \prod X_i^{r_i}$. We require the property only with respect to the base groups. A formal definition follows.

Definition 6 (Algebraic Algorithm). *Let* $\mathcal{A}$ *be a probabilistic polynomial time algorithm that takes a bilinear group description* Λ, *a string aux* $\in \{0,1\}^*$, *and base group elements* $\boldsymbol{X} \in \mathbb{G}^k$ *for some* k *as input; and outputs a group element in* $\mathbb{G}$ *and a string ext* $\in \{0,1\}^*$. *Algorithm* $\mathcal{A}$ *is called algebraic with respect to* $\mathcal{G}$ *if there exists a probabilistic polynomial-time algorithm,* Ext, *receiving the same input as* $\mathcal{A}$ *including the same random coins such that for any* $\Lambda \leftarrow \mathcal{G}(1^\lambda)$, *all polynomial size* $\boldsymbol{X} \neq (1, \ldots, 1)$, *and aux, the following probability, taken over coin* r, *is negligible in* λ.

$$\Pr\left[\begin{array}{l} (Y_1, \ldots, Y_n, ext) \leftarrow \mathcal{A}(\Lambda, \boldsymbol{X}, aux\,; r), \\ (\boldsymbol{y}_1, \ldots, \boldsymbol{y}_n, ext) \leftarrow \mathsf{Ext}(\Lambda, \boldsymbol{X}, aux\,; r) \end{array} : \exists i \in \{1, \ldots, n\} \; s.t. \; Y_i \neq \prod_{j=1}^{k} X_j^{y_{i,j}} \right]$$

The notion is often used for restricting a class of reduction algorithms for showing impossibility of security proofs for practical cryptographic schemes by black-box reduction, e.g., [7,11]. The notion in this case implies the limitation of current reduction techniques and considered as "not overly restrictive" as it covers all known efficient reductions.

The notion is also used for characterising constructions of cryptographic schemes. In [2], the signing function is assumed computable only with generic operations, which implies that it is algebraic. A closely related concept is known as the knowledge of exponent assumption [12,19,5]. It is applied to adversary $\mathcal{A}$ and considered as a "very strong assumption" since it is hardly falsifiable. It is also generally undesirable to put a limitation on the ability of a malicious party.

Similar to [2], but with slightly more generality, we put a restriction on the key generation and commitment algorithms so that they are algebraic. Though this narrows the coverage of our result, it still covers quite a wide range of approaches. It also suggests a direction to find a new construction that includes non-algebraic operations yet the relation can be efficiently verified by generic operations through pairing product equations.

2.6   Assumptions

Assumption 7 (Double Pairing Assumption (DBP)). Given Λ and $(G_z, G_r) \leftarrow \mathbb{G}_1^{*2}$, it is hard to find $(Z, R) \in \mathbb{G}_2^* \times \mathbb{G}_2^*$ that satisfies

$$1 = e(G_z, Z)\, e(G_r, R). \tag{1}$$

Assumption 8 (Simultaneous Double Pairing Assumption (SDP)). Given Λ and $(G_z, G_r, F_z, F_s) \leftarrow \mathbb{G}_1^{*4}$, it is hard to find $(Z, R, S) \in \mathbb{G}_2^{*3}$ that satisfies

$$1 = e(G_z, Z)\, e(G_r, R) \qquad \text{and} \qquad 1 = e(F_z, Z)\, e(F_s, S). \tag{2}$$

DBP is implied by DDH in $\mathbb{G}_1$. It does not hold for Λ_{sym}. SDP is implied by DLIN [9] for Λ_{sym}. When Λ_{sxdh} is considered, we can assume the dual version of these assumptions that swap $\mathbb{G}_1$ and $\mathbb{G}_2$.

3   Lower Bounds

We show two lower bounds for strictly structure-preserving commitment scheme C over $\mathcal{G}$. Let $\Lambda \leftarrow \mathcal{G}(1^\lambda)$, $ck := (\Lambda^*, \boldsymbol{V})$, $msg := \boldsymbol{M}$, $com := \boldsymbol{C}$, $open := \boldsymbol{D}$, where $\boldsymbol{V}, \boldsymbol{M}, \boldsymbol{C}, \boldsymbol{D}$ are vectors of elements in $\mathbb{G}_1$ and $\mathbb{G}_2$ in Λ. Let ℓ_v, ℓ_m, and ℓ_c denote the size of $\boldsymbol{V}$, $\boldsymbol{M}$, and $\boldsymbol{C}$, respectively.

3.1   Commitment Size

Theorem 9. *If the discrete-logarithm problem in the base groups of Λ is hard, Key and Com are algebraic, and $\ell_c < \ell_m$, then C is not binding.*

Proof. Algorithm Com takes $(\Lambda^*, \boldsymbol{V}, \boldsymbol{M})$ as input and outputs $(\boldsymbol{C}, \boldsymbol{D})$ under the constraint that $\ell_c < \ell_m$. Since Com is algebraic, there exists an associated algorithm $\mathsf{Ext_{Com}}$ that takes the same input as Com does and outputs matrices B_1, B_2, B_3, B_4 over $\mathbb{Z}_p$ for which

$$(\boldsymbol{C})^t = B_1 (\boldsymbol{M})^t + B_2 (\boldsymbol{V})^t \quad \text{and} \quad (\boldsymbol{D})^t = B_3 (\boldsymbol{M})^t + B_4 (\boldsymbol{V})^t \tag{3}$$

hold. Note that B_1 is an $\ell_c \times \ell_m$ rectangular matrix. We first consider the symmetric bilinear setting where $\mathbb{G}_1 = \mathbb{G}_2$ and represent the group by $\mathbb{G}$. We later argue that the same argument holds for asymmetric setting with trivial modifications.

We construct an adversary $\mathcal{A}$ that breaks the binding property of C. First $\mathcal{A}$ selects arbitrary $\boldsymbol{M}$ and computes $(\boldsymbol{C}, \boldsymbol{D}) \leftarrow \mathsf{Com}(\varLambda^*, \boldsymbol{V}, \boldsymbol{M})$. It then runs $\mathsf{Ext}_{\mathsf{Com}}(\varLambda^*, \boldsymbol{V}, \boldsymbol{M})$ and obtains $B_1, \ldots, B_4$. If an i-th column of B_1 is zero, then $\boldsymbol{M}'$ is formed by replacing M_i in $\boldsymbol{M}$ with a fresh arbitrary M'_i. If none of the columns of B_1 are zero, $\mathcal{A}$ finds a non-zero vector $\boldsymbol{R}$ that satisfies $B_1 (\boldsymbol{R})^t = \boldsymbol{0}$. Then it computes $\boldsymbol{M}' = \boldsymbol{M} + \boldsymbol{R}$. In either case, $\mathcal{A}$ then computes $(\boldsymbol{D}')^t := B_3 (\boldsymbol{M}')^t + B_4 (\boldsymbol{V})^t$, and outputs $(\boldsymbol{C}, \boldsymbol{M}, \boldsymbol{D}, \boldsymbol{M}', \boldsymbol{D}')$. This completes the description of $\mathcal{A}$.

We first show that the above $\boldsymbol{R}$ can be efficiently found. By applying standard Gaussian elimination to B_1, one can efficiently find S_1 that is the largest regular sub-matrix of B_1. Let I and J be the set of indexes of rows and columns of B_1, respectively, that form S_1. By $\bar{I}$ and $\bar{J}$, we denote the rest of the indexes in B_1. Note that $|I| = |J|$ and $|J| + |\bar{J}| = \ell_m$. Consider matrix S_2 of size $|I| \times |\bar{J}|$ formed by selecting entries $B_1[i][j]$, $i \in I$, and $j \in \bar{J}$. Such S_2 can be formed since $\bar{J}$ is not empty due to $\ell_c < \ell_m$. Select arbitrary non-zero vector $\boldsymbol{R}_2$ of size $|\bar{J}|$ and compute $(\boldsymbol{R}_1)^t = -S_1^{-1} S_2 (\boldsymbol{R}_2)^t$. Then $\boldsymbol{R}_1$ is a vector of size $|J|$. Then compose $\boldsymbol{R}$ from $\boldsymbol{R}_1$ and $\boldsymbol{R}_2$ in such a way that $\boldsymbol{R}[J[i]] := \boldsymbol{R}_1[i]$ and $\boldsymbol{R}[\bar{J}[i]] := \boldsymbol{R}_2[i]$. Since $\boldsymbol{R}_2$ is not zero, the resulting $\boldsymbol{R}$ is not zero as well. Let S be a matrix consisting of rows of B_1 that belong to I. It then holds that $S \cdot (\boldsymbol{R})^t = S_1 (\boldsymbol{R}_1)^t + S_2 (\boldsymbol{R}_2)^t = \boldsymbol{0}$. Since other rows of B_1 are linearly dependent on S, we have $B_1 (\boldsymbol{R})^t = \boldsymbol{0}$ as expected.

We next show that $\mathcal{A}$ outputs a valid answer. First, $1 \leftarrow \mathsf{Vrf}(\varLambda, \boldsymbol{V}, \boldsymbol{C}, \boldsymbol{M}, \boldsymbol{D})$ holds due to the correctness of C. Recall that Vrf consists of evaluating PPEs. Every PPE in Vrf can be represented by

$$\mathsf{PPE}_i : \quad (\boldsymbol{V} \| \boldsymbol{C} \| \boldsymbol{M} \| \boldsymbol{D}) \, A_i \, (\boldsymbol{V} \| \boldsymbol{C} \| \boldsymbol{M} \| \boldsymbol{D})^t = 1 \tag{4}$$

with some constant matrix A_i over $\mathbb{Z}_p$. Suppose that $\mathsf{Ext}_{\mathsf{Com}}$ is successful and (3) indeed holds. Then (4) can be rewritten by

$$(\boldsymbol{V} \| \boldsymbol{M}) \, E_i \, (\boldsymbol{V} \| \boldsymbol{M})^t = 1 \tag{5}$$

with matrix E_i in which

$$E_i = F \, A_i \, F^t \quad \text{where} \quad F = \begin{pmatrix} \boldsymbol{1}_{\ell_v} & B_2^t & \boldsymbol{0}_{\ell_v} & B_4^t \\ \boldsymbol{0}_{\ell_m} & B_1^t & \boldsymbol{1}_{\ell_m} & B_3^t \end{pmatrix} \tag{6}$$

where $\boldsymbol{1}_n$ and $\boldsymbol{0}_n$ denote $n \times n$ identity and zero matrices over $\mathbb{Z}_p$, respectively. Note that E_i depends on $\boldsymbol{M}$ (through the computation of B_1 to B_4); hence, (5) holds for that $\boldsymbol{M}$. Nevertheless, we claim that any $\boldsymbol{M}'$ that is even unrelated to E_i fulfils (4) as long as (5) is fulfilled and $\boldsymbol{C}$ and $\boldsymbol{D}$ are computed as in (3).

Claim. For valid $\boldsymbol{M}'(\neq \boldsymbol{M})$ that fulfils

$$(\boldsymbol{V} \| \boldsymbol{M}') \, E_i \, (\boldsymbol{V} \| \boldsymbol{M}')^t = 1, \tag{7}$$

for all i, relation

$$(V||C'||M'||D') A_i (V||C'||M'||D')^t = 1 \tag{8}$$

holds for all i with respect to

$$(C')^t := B_1 (M')^t + B_2 (V)^t \quad \text{and} \quad (D')^t := B_3 (M')^t + B_4 (V)^t. \tag{9}$$

Proof is trivial by converting (7) into (8) by using (6) and (9). As a consequence, such (C', M', D') fulfils $1 \leftarrow \mathsf{Vrf}(\Lambda^*, V, C', M', D')$. We next make a strong claim that any M' satisfies (7).

Claim. If the discrete-logarithm problem in $\mathbb{G}$ is hard, the relation (7) holds for any $M' \in \mathbb{G}^{\ell_m}$.

Intuition is that Com and $\mathsf{Ext_{Com}}$ do not know the discrete-log of M in computing B_1 to B_4. Thus the only way to fulfil (5) is to set B_1 to B_4 so that (5) is trivial for M. It then holds for any M' as in (7). To formally reduce to the hardness of the discrete-logarithm problem, we also require $\mathsf{Ext_{Key}}$ to be algebraic so that v is available to our reduction algorithm.

Proof. Consider the relation in the exponents of (7) where V is a constant and M' is a variable. The relation is in a quadratic form, say $Q_i(m') = 0$, whose coefficients can be computed efficiently from E_i. To prove the statement, it suffices to show that Q_i is a constant polynomial for all i.

Suppose, on the contrary, that there exists i where Q_i is a non-trivial polynomial with probability ϵ_Q that is not negligible. The probability is taken over the choice of V, M. (Recall that E_i depends on V and M. It also depends on the randomness of the extractor of Com, but the theorem statement is conditional on the success of the extractor.) We construct algorithm $\mathcal{D}$ that solves the discrete logarithm problem by using Key, Com, and their extractors $\mathsf{Ext_{Key}}$ and $\mathsf{Ext_{Com}}$ as follows. Let (Λ, Y) be an instance of the discrete-logarithm problem where Λ includes base G. The goal is to compute $x := \log_G Y$. Given (Λ, Y), algorithm $\mathcal{D}$ first generates commitment key $(ck, tk) \leftarrow \mathsf{Key}(\Lambda, k)$ where $ck = (\Lambda^*, V)$. By invoking $\mathsf{Ext_{Key}}$, algorithm $\mathcal{D}$ obtains discrete-log v of V with respect to G. ($\mathcal{D}$ halts if negligible extraction error occurs.) It then forms M by setting $M_j := Y^{\gamma_j}$ with random γ_j, and runs $(C, D) \leftarrow \mathsf{Com}(\Lambda^*, V, M)$. By running $\mathsf{Ext_{Com}}$, algorithm $\mathcal{D}$ obtains B_1, B_2, B_3 and B_4. It then computes E_i and further obtains quadratic polynomial Q_i that is non-trivial by hypothesis. By using the relation that $m_j = \gamma_j \cdot x$, $\mathcal{D}$ converts Q_i into quadratic polynomial Q_i' in x, which is also non-trivial except for negligible probability. (The probability is over the choice of every γ_i. Rigorously, the bound is given by Schwartz's lemma [21] since Q_i is a low-degree polynomial in γ_j.) Finally, $\mathcal{D}$ solves $Q_i'(x) = 0$ and outputs x. The running time of $\mathcal{D}$ is polynomial since Key, Com, and their extractors run in polynomial-time and other computations are obviously executable in polynomial-time. The success probability of $\mathcal{D}$ is almost the same as ϵ_Q except for the negligible errors. This contradicts the hardness of the discrete-logarithm problem in $\mathbb{G}$. ∎

Now recall that M' is set to $M + R$ and that $B_1\,R = 0$. Thus,

$$(C')^t = B_1\,(M')^t + B_2\,(V)^t = B_1\,(M)^t + B_2\,(V)^t = (C)^t. \tag{10}$$

Due to the above claims, $1 \leftarrow \mathsf{Vrf}(\Lambda, V, C, M', D')$ holds. Furthermore, $M \neq M'$ since $R \neq 0$. Thus, (C, M, D, M', D') is a valid solution that breaks the binding property of C. This completes the proof in the symmetric group setting.

In the asymmetric setting where M and other vectors consist of elements from both $\mathbb{G}_1$ and $\mathbb{G}_2$, essentially the same argument holds since elements in the gruops do not mix each other. In the following, we only describe the points where the argument has to be adjusted.

– Every vector is split into $\mathbb{G}_1$ vector and $\mathbb{G}_2$ vector, e.g., $M = (M_1, M_2) \in \mathbb{G}_1^{\ell_{m1}} \times \mathbb{G}_2^{\ell_{m2}}$ for $\ell_{m1} + \ell_{m2} = \ell_m$.
– By running $\mathsf{Ext_{Com}}$, we obtain B_j in the form of

$$B_j = \begin{pmatrix} B_{j1} & 0 \\ 0 & B_{j2} \end{pmatrix} \tag{11}$$

so that linear computation such as (3) is well defined.
– Without loss of generality, we assume that $|C_1| < |M_1|$. (Otherwise, $|C_2| < |M_2|$ holds.) Then, we can obtain non-zero vector R_1 from B_{11} in the same way as we obtain R from B_1 in the symmetric case. By setting $R = (R_1, 0)$, we have $B_1 R = 0$ as desired.
– Pairing product equations (4), (5), (7) and (8) are modified so that their left and right vectors consist only of $\mathbb{G}_1$ and $\mathbb{G}_2$, respectively, for computational consistency. Also, matrix E_i in (6) is modified to $E_i = F_1\,A_i\,(F_2)^t$ where F_i is formed by using B_{1i}, B_{2i}, B_{3i}, and B_{4i} in the same manner as in F in (6).
– In the second claim, we require hardness of the discrete-logarithm problem in both $\mathbb{G}_1$ and $\mathbb{G}_2$. Depending on which of M_1 and M_2 polynomial Q_i is non-trivial, we solve the discrete-logarithm problem in $\mathbb{G}_1$ or $\mathbb{G}_2$, respectively.

∎

3.2 Number of Verification Equations

Theorem 10. *If $\Lambda = \Lambda_{\mathsf{sym}}$, $\ell_m \geq 2$, and Vrf evaluates only one PPE, then C is not binding.*

Proof. By focusing on M_1 and M_2 in M, the PPE in the verification can be written as

$$e(M_1, M_1)^{a_1}\,e(M_1, K_1)^{b_1}\,e(M_2, M_2)^{a_2}\,e(M_2, K_2)^{b_2}\,e(M_1, M_2)^c\,P = 1 \tag{12}$$

where $a_1, b_1, a_2, b_2, c \in \mathbb{Z}_p$ are constants determined by the common parameter, and K_1 and K_2 are linear combinations of elements in V, C, D and $M \setminus \{M_1, M_2\}$, and P is a product of pairings that does not involve M_1 and M_2. Let f be the polynomial that represents the relation in the exponent of the leftmost five pairings of (12). Namely,

$$f := a_1 \, m_1^2 + b_1 \, k_1 \, m_1 + c \, m_1 m_2 + b_2 \, k_2 \, m_2 + a_2 \, m_2^2, \tag{13}$$

where m_1, m_2, k_1, and k_2 are the discrete-logs of M_1, M_2, K_1, and K_2 with respect to the generator, say G, in Λ.

Given a commitment-key $(\Lambda^*, \boldsymbol{V})$, we set $\boldsymbol{M} = 1$ and honestly compute $\boldsymbol{C}$ and $\boldsymbol{D}$ by running Com. These $\boldsymbol{C}$ and $\boldsymbol{D}$ define K_1, K_2, and P in (12). Let $f(m_1, m_2)$ be f, as defined in (13), with k_1 and k_2 determined by these K_1 and K_2. We have $f(0,0) = 0$ and look for another pair $(m_1', m_2') \neq (0,0)$ that fulfils $f(m_1', m_2') = 0$. Such (m_1', m_2') yield $(M_1', M_2') = (G^{m_1'}, G^{m_2'}) \neq (1,1)$.

Next, we show how to obtain such (M_1', M_2'):

- If $(a_1, a_2, c) = (0, 0, 0)$, we have $f(m_1, m_2) = b_1 \, k_1 \, m_1 + b_2 \, k_2 \, m_2$. We then proceed with the following sub-cases.
  - If $b_1 \, k_1 \neq 0$ and $b_2 \, k_2 \neq 0$, then $m_1' := k_2$ and $m_2' := (-b_1/b_2) \cdot k_1$ results in $(m_1', m_2') \neq (0,0)$ and $f(m_1', m_2') = 0$. Thus, setting $M_1' := K_2$ and $M_2' := K_1^{-b_1/b_2}$ works.
  - If $b_i k_i = 0$ for $i = 1$ or $i = 2$, or both, $f(m_1, m_2)$ is independent of m_i. Therefore, any non-zero m_i' suffices. Simply select arbitrary non-zero m_i' and compute $M_i' = G^{m_i'}$.
- If $(a_1, a_2, c) \neq (0, 0, 0)$, we do as follows.
  - If $b_1 \, k_1 = 0$ and $b_2 \, k_2 = 0$, we have $f(m_1, m_2) = a_1 \, m_1^2 + c \, m_1 \, m_2 + a_2 \, m_2^2$. By selecting non-zero m_1' and solving m_2' for $f = 0$ (if $f(m_1, m_2) = 0$ is independent of m_2, arbitrary m_2' suffices), we have $(M_1', M_2') = (G^{m_1}, G^{m_2}) \neq (1,1)$.
  - If $b_1 \, k_1 \neq 0$ or $b_2 \, k_2 \neq 0$, we consider setting $m_2 = \delta \, m_1$ for some δ. With this relation, (13) is written as

$$f(m_1, m_2) = m_1 \left\{ (a_1 + a_2 \, \delta^2 + c \, \delta) \, m_1 + (b_1 \, k_1 + b_2 \, k_2 \, \delta) \right\}. \tag{14}$$

We need (14) to have a non-zero solution for m_1. Therefore, we set δ so that $a_1 + a_2 \, \delta^2 + c \, \delta \neq 0$ and $b_1 \, k_1 + b_2 \, k_2 \, \delta \neq 0$ hold. (There are at most two δ for which these inequalities do not hold. For an arbitrary δ, the first inequality can be tested directly, whereas the second is through the relation $K_1^{b_1} K_2^{b_2 \delta} \neq 1$. Thus, by trying at most three non-zero different δ, we have an appropriate δ.) Then

$$m_1' = -\frac{b_1 \, k_1 + b_2 \, k_2 \delta}{a_1 + a_2 \, \delta^2 + c \, \delta} \quad \text{and} \quad m_2' = \delta \, m_1'$$

fulfil $(m_1', m_2') \neq (0,0)$ and $f(m_1', m_2') = 0$. This corresponds to setting

$$M_1' := (K_1^{b_1} K_2^{b_2 \, \delta})^{\frac{1}{a_1 + a_2 \, \delta^2 + c \, \delta}} \quad \text{and} \quad M_2' := (M_1')^{\delta}.$$

By replacing M_1 and M_2 in $\boldsymbol{M}$ with M_1' and M_2' computed as described above, we obtain $\boldsymbol{M}' \neq \boldsymbol{M}$, which is consistent with $\boldsymbol{C}$ and $\boldsymbol{D}$; Hence, the binding property breaks.

4   Optimal Constructions

4.1   In Asymmetric Setting

Let $\mathcal{G}$ be a generator of asymmetric bilinear groups. Scheme 1 in Fig. 1 is for messages $\boldsymbol{M} = (M_1, \ldots, M_k) \in \mathbb{G}_2^k$ for some fixed constant k specified at the time of commitment-key generation. The default generators G and $\tilde{G}$ in Λ can be used as G_0 and H, respectively. One can switch $\mathbb{G}_1$ and $\mathbb{G}_2$ in the description to obtain a dual scheme that accepts messages in $\mathbb{G}_1$. It also implies a scheme for messages from both $\mathbb{G}_1$ and $\mathbb{G}_2$. We show that the scheme is correct, perfectly hiding, and computationally binding as well as trapdoor and homomorphic.

[**Scheme 1**]

> Setup(1^λ): Run $\mathcal{G}(1^\lambda)$ and obtain $\Lambda := (p, \mathbb{G}_1, \mathbb{G}_2, \mathbb{G}_T, e, G, \tilde{G})$. Output Λ.
> Key(Λ, k): Select G_0 and H uniformly from $\mathbb{G}_1^*$ and $\mathbb{G}_2^*$, respectively. For $i = 1, \ldots, k$, compute $G_i := G_0^{\gamma_i}$ for random $\gamma_i \in \mathbb{Z}_p^*$. Output commitment-key $ck = (\Lambda^*, H, G_0, \ldots, G_k)$ and trapdoor $tk = (\gamma_1, \ldots, \gamma_k)$.
> Com($ck, \boldsymbol{M}$): Randomly select $\tau_0, \ldots, \tau_k \in \mathbb{Z}_p$ and compute

$$C_i := M_i \cdot H^{\tau_i} \text{ (for } i = 1, \ldots, k), \quad C_{k+1} := \prod_{j=0}^{k} G_j^{\tau_j}, \text{ and} \quad (15)$$

$$D := H^{\tau_0}. \quad (16)$$

> Then output $\boldsymbol{C} := (C_1, \ldots, C_{k+1})$ and D.
> Vrf($ck, \boldsymbol{C}, \boldsymbol{M}, D$): Output 1 if

$$e(C_{k+1}, H) = e(G_0, D) \prod_{i=1}^{k} e(G_i, C_i/M_i) \quad (17)$$

> holds. Output 0, otherwise.
> Equiv($ck, tk, \boldsymbol{C}, \boldsymbol{M}, D, \boldsymbol{M}'$): Take $(\gamma_1, \ldots, \gamma_k)$ from tk. Then output D' such that

$$D' := D \cdot \prod_{i=1}^{k} (M_i'/M_i)^{\gamma_i}. \quad (18)$$

Fig. 1. Homomorphic trapdoor commitment scheme in asymmetric bilinear group setting

Theorem 11. *Scheme 1 is correct.*

Proof. For any $\boldsymbol{C}$ and D correctly computed for ck and $\boldsymbol{M}$ as in (15), the right-hand of verification equation (17) is

$$e(G_0, D) \prod_{i=1}^{k} e(G_i, C_i/M_i) = e(G_0, H^{\tau_0}) \prod_{i=1}^{k} e(G_i, H^{\tau_i}) = e(C_{k+1}, H). \quad (19)$$

Thus $(ck, \boldsymbol{C}, \boldsymbol{M}, D)$ passes the verification with probability 1.

Theorem 12. *Scheme 1 is perfectly hiding and computationally binding if the DBP assumption holds for Λ.*

Proof. It is perfectly hiding because, for every commitment $\boldsymbol{C} = (C_1, \ldots, C_{k+1})$ $\in \mathbb{G}_1 \times \mathbb{G}_2^k$ and every message $\boldsymbol{M} = (M_1, \ldots, M_k) \in \mathbb{G}_2^k$, there exists a unique $(\tau_0, \ldots, \tau_k) \in \mathbb{Z}_p^{k+1}$ that is consistent with relations (15), (16) and (17).

The binding property is proven by constructing an algorithm $\mathcal{B}$ that breaks DBP using an adversary $\mathcal{A}$ that successfully computes two openings for a commitment. Given an instance (Λ, G_z, G_r) of DBP, algorithm $\mathcal{B}$ works as follows.

- Randomly select ρ_0 from $\mathbb{Z}_p^*$ and compute $G_0 := G_r^{\rho_0}$.
- For $i = 1, \ldots, k$, randomly select $\zeta_i \in \mathbb{Z}_p^*$ and $\rho_i \in \mathbb{Z}_p$ and compute $G_i := G_z^{\zeta_i} G_r^{\rho_i}$. If $G_i = 1$ for any i, $\mathcal{B}$ aborts; since the probability for this is negligible, this does not affect the overall success of $\mathcal{B}$.
- Run $\mathcal{A}$ with input $ck = (\Lambda^*, H, G_0, \ldots, G_k)$.
- Given commitment $\boldsymbol{C}$ and two openings $(\boldsymbol{M}, \boldsymbol{D})$ and $(\boldsymbol{M}', \boldsymbol{D}')$ from $\mathcal{A}$, compute

$$Z^\star = \prod_{i=1}^{k} \left(\frac{M_i'}{M_i} \right)^{\zeta_i} \quad \text{and} \quad R^\star = \left(\frac{D}{D'} \right)^{\rho_0} \prod_{i=1}^{k} \left(\frac{M_i'}{M_i} \right)^{\rho_i}. \tag{20}$$

- Output $(Z^\star, R^\star)$.

Since both $(\boldsymbol{M}, \boldsymbol{D})$ and $(\boldsymbol{M}', \boldsymbol{D}')$ fulfil (17) for the same commitment $\boldsymbol{C}$, dividing the two verification equations yields

$$1 = e\left(G_0, \frac{D}{D'} \right) \prod_{i=1}^{k} e\left(G_i, \frac{M_i'}{M_i} \right) = e\left(G_r^{\rho_0}, \frac{D}{D'} \right) \prod_{i=1}^{k} e\left(G_z^{\zeta_i} G_r^{\rho_i}, \frac{M_i'}{M_i} \right) \tag{21}$$

$$= e\left(G_z, \prod_{i=1}^{k} \left(\frac{M_i'}{M_i} \right)^{\zeta_i} \right) e\left(G_r, \left(\frac{D}{D'} \right)^{\rho_0} \prod_{i=1}^{k} \left(\frac{M_i'}{M_i} \right)^{\rho_i} \right) \tag{22}$$

$$= e(G_z, Z^\star)\, e(G_r, R^\star). \tag{23}$$

Since $\boldsymbol{M} \neq \boldsymbol{M}'$, there exists i such that $M_i'/M_i \neq 1$. Also, ζ_i is independent from the view of the adversary, i.e., for every choice of ζ_i, there exist a corresponding ρ_i that gives the same G_i. Accordingly, $Z^\star = \prod_i (M_i'/M_i)^{\zeta_i} \neq 1$ holds with overwhelming probability, and $(Z^\star, R^\star)$ is a valid answer to the instance of DBP. Therefore, $\mathcal{B}$ breaks DBP with the same probability that $\mathcal{A}$ breaks the binding property of Scheme 1 (minus a negligible difference). ∎

Theorem 13. *Scheme 1 is trapdoor and homomorphic.*

Proof. For the trapdoor property, observe that, for any trapdoor tk generated by Key, and for any valid $\boldsymbol{M}$ and $(\boldsymbol{C}, D)$ generated by Com, and D' generated by Equiv for any valid $\boldsymbol{M}'$, it holds that

$$e(G_0, D') \prod_{i=1}^{k} e(G_i, C_i/M_i') = e\left(G_0, D \cdot \prod_{i=1}^{k} (M_i'/M_i)^{\gamma_i}\right) \prod_{i=1}^{k} e(G_i, C_i/M_i') \quad (24)$$

$$= e(G_0, D) \prod_{i=1}^{k} e(G_0^{\gamma_i}, M_i'/M_i) \prod_{i=1}^{k} e(G_i, C_i/M_i') \quad (25)$$

$$= e(G_0, D) \prod_{i=1}^{k} e(G_i, C_i/M_i) \quad (26)$$

$$= e(C_{k+1}, H). \quad (27)$$

Thus (M', D') is a correct opening of C computed from M. Also observe that (ck, M, C) uniquely determines D and so is (ck, M', C) and D'. Therefore, distributions (ck, M, C, D) and (ck, M', C, D') over all choices of M and M' are identical.

To check the homomorphic property, let (ck, C, M, D) and (ck, C', M', D') satisfy verification equation (17). Also, let $M^{\star} := M + M'$, $C^{\star} := C + C'$, and $D^{\star} := D \cdot D'$. Then it holds that

$$e(G_0, D^{\star}) \prod_{i=1}^{k} e(G_i, C_i^{\star}/M_i^{\star}) \quad (28)$$

$$= e(G_0, D) \, e(G_0, D') \prod_{i=1}^{k} e(G_i, C_i/M_i) \prod_{i=1}^{k} e(G_i, C_i'/M_i') \quad (29)$$

$$= e(C_{k+1}, H) \, e(C_{k+1}', H) \quad (30)$$

$$= e(C_{k+1}^{\star}, H). \quad (31)$$

4.2 In Symmetric Setting

Let $\mathcal{G}$ be a generator of symmetric bilinear groups. Scheme 2 in Fig. 2 is for messages $M = (M_1, \ldots, M_k) \in \mathbb{G}_1^k$ for some fixed constant k specified at the time of commitment-key generation. The default generator G in Λ can be used as H in the key generation.

Theorem 14. *Scheme 2 is correct.*

Proof. For correctly generated/computed (ck, C, M, D), the following holds:

$$e(G_0, D_1) \prod_{i=1}^{k} e(G_i, C_i/M_i) = e(G_0, H^{\tau_0}) \prod_{i=1}^{k} e(G_i, H^{\tau_i}) = e(C_{k+1}, H) \quad (37)$$

$$e(F_0, D_2) \prod_{i=1}^{k} e(F_i, C_i/M_i) = e(F_0, H^{\mu_0}) \prod_{i=1}^{k} e(F_i, H^{\tau_i}) = e(C_{k+2}, H). \quad (38)$$

Thus it passes the verification with probability 1.

[Scheme 2]

Setup(1^λ): Run $\mathcal{G}(1^\lambda)$ and obtain $\Lambda := (p, \mathbb{G}_1, \mathbb{G}_T, e, G)$. Output Λ.

Key(Λ, k): Select H, G_0 and F_0 from $\mathbb{G}_1$ uniformly. For $i = 1, \ldots, k$, compute $G_i := G_0^{\gamma_i}$ and $F_i := F_0^{\delta_i}$ for random $\gamma_i, \delta_i \in \mathbb{Z}_p^*$. Output $ck := (\Lambda^*, H, (G_i, F_i)_{i=0}^k)$ and $tk := ((\gamma_i, \delta_i)_{i=1}^k)$.

Com($ck, \boldsymbol{M}$): Choose $\mu_0, \tau_0, \ldots, \tau_k \in \mathbb{Z}_p$ randomly and compute (for $i = 1, \ldots k$)

$$C_i := M_i \cdot H^{\tau_i}, \quad C_{k+1} := G_0^{\tau_0} \prod_{j=1}^k G_j^{\tau_j}, \quad C_{k+2} := F_0^{\mu_0} \prod_{j=1}^k F_j^{\tau_j}, \quad (32)$$

$$D_1 := H^{\tau_0}, \text{ and } \quad D_2 := H^{\mu_0}. \quad (33)$$

Output $\boldsymbol{C} := (C_1, \ldots, C_{k+2})$ and $\boldsymbol{D} = (D_1, D_2)$.

Vrf($ck, \boldsymbol{C}, \boldsymbol{M}, \boldsymbol{D}$): Output 1 if the following equations hold. Output 0, otherwise.

$$e(C_{k+1}, H) = e(G_0, D_1) \prod_{i=1}^k e(G_i, C_i/M_i) \quad (34)$$

$$e(C_{k+2}, H) = e(F_0, D_2) \prod_{i=1}^k e(F_i, C_i/M_i) \quad (35)$$

Equiv($ck, tk, \boldsymbol{C}, \boldsymbol{M}, \boldsymbol{D}, \boldsymbol{M}'$): Parse tk as $((\gamma_i, \delta_i)_{i=1}^k)$. Output $\boldsymbol{D}' = (D_1', D_2')$ such that

$$D_1' := D_1 \cdot \prod_{i=1}^k (M_i'/M_i)^{\gamma_i}, \quad \text{and} \quad D_2' := D_2 \cdot \prod_{i=1}^k (M_i'/M_i)^{\delta_i}. \quad (36)$$

Fig. 2. Homomorphic trapdoor commitment scheme in symmetric bilinear group setting

Theorem 15. *Scheme 2 is perfectly hiding and computationally binding if the SDP assumption holds for Λ.*

Proof. It is perfectly hiding due to the uniform choice of $(\mu_0, \tau_0, \tau_1, \ldots, \tau_k)$ when committing, and due to the fact that for every commitment $\boldsymbol{C} = (C_1, \ldots, C_{k+2}) \in \mathbb{G}_1^{k+2}$ and for every message $\boldsymbol{M} = (M_1, \ldots, M_k) \in \mathbb{G}_1^k$ there exists a unique pair (D_1, D_2) that satisfies equations (34)-(35).

The binding property is shown by constructing an algorithm $\mathcal{B}$ that breaks SDP using an adversary $\mathcal{A}$ that successfully computes two openings for a commitment. Given an instance $(\Lambda, G_z, G_r, F_z, F_s)$ of SDP, algorithm $\mathcal{B}$ works as follows.

- Pick random ρ_0 and ω_0 from $\mathbb{Z}_p^*$ and compute $G_0 := G_r^{\rho_0}$, and $F_0 := F_s^{\omega_0}$.
- For $i = 1, \ldots, k$, pick random $\zeta_i \in \mathbb{Z}_p^*$ and $\rho_i, \omega_i \in \mathbb{Z}_p$ and compute $G_i := G_z^{\zeta_i} G_r^{\rho_i}$, and $F_i := F_z^{\zeta_i} F_s^{\omega_i}$. If $G_i = 1$ or $F_1 = 1$ for any i, $\mathcal{B}$ aborts; since the probability for this is negligible, we can ignore such cases.

- Run $\mathcal{A}$ with input $ck = (\Lambda^*, H, G_0, F_0, \ldots, G_k, F_k)$.
- Given commitment $\boldsymbol{C}$ and two openings $(\boldsymbol{M}, \boldsymbol{D})$ and $(\boldsymbol{M}', \boldsymbol{D}')$ from $\mathcal{A}$, compute

$$Z^\star = \prod_{i=1}^{k} \left(\frac{M_i'}{M_i} \right)^{\zeta_i}, R^\star = \left(\frac{D_1}{D_2'} \right)^{\rho_0} \prod_{i=1}^{k} \left(\frac{M_i'}{M_i} \right)^{\rho_i}, S^\star = \left(\frac{D_2}{D_2'} \right)^{\omega_0} \prod_{i=1}^{k} \left(\frac{M_i'}{M_i} \right)^{\omega_i}.$$

- Output $(Z^\star, R^\star, S^\star)$.

Since both $(\boldsymbol{M}, D_1)$ and $(\boldsymbol{M}', D_1')$ fulfils (34) with $\boldsymbol{C}$, dividing the two equations yields

$$
\begin{aligned}
1 &= e\left(G_0, \frac{D_1}{D_1'} \right) \prod_{i=1}^{k} e\left(G_i, \frac{M_i'}{M_i} \right) = e\left(G_r^{\rho_0}, \frac{D_1}{D_1'} \right) \prod_{i=1}^{k} e\left(G_z^{\zeta_i} G_r^{\rho_i}, \frac{M_i'}{M_i} \right) \\
&= e\left(G_z, \prod_{i=1}^{k} \left(\frac{M_i'}{M_i} \right)^{\zeta_i} \right) e\left(G_r, \left(\frac{D_1}{D_1'} \right)^{\rho_0} \prod_{i=1}^{k} \left(\frac{M_i'}{M_i} \right)^{\rho_i} \right) \\
&= e(G_z, Z^\star)\, e(G_r, R^\star).
\end{aligned}
$$

Similarly, from $(\boldsymbol{M}, D_2)$ and $(\boldsymbol{M}', D_2')$ fulfilling (35) with $\boldsymbol{C}$, we have

$$1 = e\left(F_0, \frac{D_2}{D_2'} \right) \prod_{i=1}^{k} e\left(F_i, \frac{M_i'}{M_i} \right) = e(F_z, Z^\star)\, e(F_s, S^\star).$$

Since $\boldsymbol{M} \neq \boldsymbol{M}'$, there exists i such that $M_i'/M_i \neq 1$. Observe that ζ_i is independent from the view of the adversary, i.e., for every choice of ζ_i, there exist corresponding ρ_i and ω_i that give the same G_i and F_i, respectively. Thus, $Z^\star = \prod_i (M_i'/M_i)^{\zeta_i} \neq 1$ holds with overwhelming probability, and $(Z^\star, R^\star, S^\star)$ is a valid answer to the instance of SDP. Accordingly, $\mathcal{B}$ breaks SDP if $\mathcal{A}$ can break the binding property with a non-negligible probability. ∎

Theorem 16. *Scheme 2 is trapdoor and homomorphic.*

The proof is analogous to that that of Theorem 13; thus, it is omitted.

4.3 Efficiency

Table 1 compares storage and computation costs to commit to a message consisting of k group elements. Schemes for symmetric setting are above the line and those for asymmetric setting are below the line. In [3], another scheme in an asymmetric setting is discussed without details. The scheme yields a commitment of at least $2k$, which is not optimal.

We also assess the efficiency in combination of GS proofs. A typical proof statement would be "I can open the commitment." It uses $(\boldsymbol{M}, \boldsymbol{D})$ as witness and $(\boldsymbol{V}, \boldsymbol{C})$ as constants in the theorem statement represented by PPEs in the

Table 1. Efficiency comparison. The size indicates the number of elements in a commitment-key V, a commitment C, and a decommitment D for a message M consisting of k group elements. For Scheme 1, (x, y) means x elements in $\mathbb{G}_1$ (or $\mathbb{G}_2$) and y elements in $\mathbb{G}_2$ (or $\mathbb{G}_1$, resp.). #(pairings) and #(PPE) indicate the number of pairings and pairing product equations in the verification predicate, respectively.

Scheme	Setting	$\|V\|$	$\|M\|$	$\|C\|$	$\|D\|$	#(pairings)	#(PPE)	assumption
CLY09 [9]	Λ_{sym}	5	k	$3k$	$3k$	$9k$	$3k$	DLIN
AHO10 [3]	Λ_{sym}	$2k+2$	k	$2k+2$	2	$2k+2$	2	SDP
Scheme 2	Λ_{sym}	$2k+2$	k	$k+2$	2	$2k+4$	2	SDP
Scheme 1	Λ_{sxdh}	$(k, 0)$	$(0, k)$	$(1, k)$	$(0, 1)$	$k+2$	1	DBP

verification predicate. Table 2 shows the size of the witness, theorem, and proof in the example. We also show the total size for a theorem and a proof in bits with a reasonable parameter setting (which is considered as comparable security to an RSA modulus of 2000 bits) where elements in $\mathbb{G}$ are 380 bits in the symmetric setting, and elements in $\mathbb{G}_1$ and $\mathbb{G}_2$ are 224 bits and 448 bits, respectively, assuming the use of point compression [4]. Scheme 1 is optimized by considering the dual scheme taking messages from $\mathbb{G}_1$.

Table 2. Storage costs for proving correct opening in zero-knowledge by GS proofs. Figures for |proof| include commitments of the witness and a proof. Size in bits indicates |theorem| + |proof| in bits.

Scheme	Setting	|witness|	|theorem|	|proof|	Size in Bits $k = 1$	5	10
CLY09 [9]	Λ_{sym}	$4k$	$3k+5$	$39k$	17860	81700	161500
AHO10 [3]	Λ_{sym}	$k+2$	$4k+4$	$15k+24$	17860	46740	82840
Scheme 2	Λ_{sym}	$k+2$	$3k+4$	$12k+21$	15200	38000	66500
Scheme 1	Λ_{sxdh}	$(0, k+1)$	$(k+1, k)$	$(0, 6k+8)$	4256	12320	22400

References

1. Abe, M., Fuchsbauer, G., Groth, J., Haralambiev, K., Ohkubo, M.: Structure-Preserving Signatures and Commitments to Group Elements. In: Rabin, T. (ed.) CRYPTO 2010. LNCS, vol. 6223, pp. 209–236. Springer, Heidelberg (2010)
2. Abe, M., Groth, J., Haralambiev, K., Ohkubo, M.: Optimal Structure-Preserving Signatures in Asymmetric Bilinear Groups. In: Rogaway, P. (ed.) CRYPTO 2011. LNCS, vol. 6841, pp. 649–666. Springer, Heidelberg (2011)
3. Abe, M., Haralambiev, K., Ohkubo, M.: Signing on group elements for modular protocol designs. IACR ePrint Archive, Report 2010/133 (2010)
4. Barreto, P.S.L.M., Naehrig, M.: Pairing-Friendly Elliptic Curves of Prime Order. In: Preneel, B., Tavares, S. (eds.) SAC 2005. LNCS, vol. 3897, pp. 319–331. Springer, Heidelberg (2006)

5. Bellare, M., Palacio, A.: The knowledge-of-exponent assumptions and 3-round zero-knowledge protocols. In: Franklin, M. (ed.) CRYPTO 2004. LNCS, vol. 3152, pp. 273–289. Springer, Heidelberg (2004)
6. Boneh, D., Boyen, X., Shacham, H.: Short Group Signatures. In: Franklin, M. (ed.) CRYPTO 2004. LNCS, vol. 3152, pp. 41–55. Springer, Heidelberg (2004)
7. Boneh, D., Venkatesan, R.: Breaking RSA May Not Be Equivalent to Factoring. In: Nyberg, K. (ed.) EUROCRYPT 1998. LNCS, vol. 1403, pp. 59–71. Springer, Heidelberg (1998)
8. Camenisch, J., Haralambiev, K., Kohlweiss, M., Lapon, J., Naessens, V.: Structure Preserving CCA Secure Encryption and Applications. In: Lee, D.H., Wang, X. (eds.) ASIACRYPT 2011. LNCS, vol. 7073, pp. 89–106. Springer, Heidelberg (2011)
9. Cathalo, J., Libert, B., Yung, M.: Group Encryption: Non-interactive Realization in the Standard Model. In: Matsui, M. (ed.) ASIACRYPT 2009. LNCS, vol. 5912, pp. 179–196. Springer, Heidelberg (2009)
10. Chase, M., Kohlweiss, M.: A domain transformation for structure-preserving signatures on group elements. IACR ePrint Archive, Report 2011/342 (2011)
11. Coron, J.-S.: Optimal Security Proofs for PSS and Other Signature Schemes. In: Knudsen, L.R. (ed.) EUROCRYPT 2002. LNCS, vol. 2332, pp. 272–287. Springer, Heidelberg (2002)
12. Damgaard, I.: Towards Practical Public Key Systems Secure against Chosen Ciphertext Attacks. In: Feigenbaum, J. (ed.) CRYPTO 1991. LNCS, vol. 576, pp. 445–456. Springer, Heidelberg (1992)
13. Dodis, Y., Haralambiev, K., López-Alt, A., Wichs, D.: Efficient Public-Key Cryptography in the Presence of Key Leakage. In: Abe, M. (ed.) ASIACRYPT 2010. LNCS, vol. 6477, pp. 613–631. Springer, Heidelberg (2010)
14. Galbraith, S., Paterson, K., Smart, N.: Pairings for cryptographers. IACR ePrint archive, Report 2006/165 (2006)
15. Galbraith, S.D., Rotger, V.: Easy decision-Diffie-Hellman groups. LMS Journal of Computation and Mathematics 7 (2004)
16. Groth, J.: Homomorphic trapdoor commitments to group elements. IACR ePrint Archive, Report 2009/007 (January 2009)
17. Groth, J.: Efficient Zero-Knowledge Arguments from Two-Tiered Homomorphic Commitments. In: Lee, D.H., Wang, X. (eds.) ASIACRYPT 2011. LNCS, vol. 7073, pp. 431–448. Springer, Heidelberg (2011)
18. Groth, J., Sahai, A.: Efficient Non-interactive Proof Systems for Bilinear Groups. In: Smart, N.P. (ed.) EUROCRYPT 2008. LNCS, vol. 4965, pp. 415–432. Springer, Heidelberg (2008); Full version available: IACR ePrint Archive 2007/155
19. Hada, S., Tanaka, T.: On the Existence of 3-Round Zero-Knowledge Protocols. In: Krawczyk, H. (ed.) CRYPTO 1998. LNCS, vol. 1462, pp. 354–369. Springer, Heidelberg (1998); Full version available from IACR e-print archive 1999/009
20. Krawczyk, H., Rabin, T.: Chameleon hashing and signatures. IACR ePrint archive, Report 1998/010 (1998)
21. Schwartz, J.T.: Fast probabilistic algorithms for verification of polynomial identities. Journal of the ACM 27(4) (1980)
22. Scott, M.: Authenticated id-based key exchange and remote log-in with simple token and pin number. IACR ePrint Archive, Report 2002/164 (2002)
23. Verheul, E.R.: Evidence that xtr is more secure than supersingular elliptic curve cryptosystems. J. Cryptology 17(4), 277–296 (2004)

Tools for Simulating Features
of Composite Order Bilinear Groups
in the Prime Order Setting

Allison Lewko[*]

The University of Texas at Austin
alewko@cs.utexas.edu

Abstract. In this paper, we explore a general methodology for converting composite order pairing-based cryptosystems into the prime order
setting. We employ the dual pairing vector space approach initiated by
Okamoto and Takashima and formulate versatile tools in this framework
that can be used to translate composite order schemes for which the prior
techniques of Freeman were insufficient. Our techniques are typically applicable for composite order schemes relying on the canceling property
and proven secure from variants of the subgroup decision assumption,
and will result in prime order schemes that are proven secure from the
decisional linear assumption. As an instructive example, we obtain a
translation of the Lewko-Waters composite order IBE scheme. This provides a close analog of the Boneh-Boyen IBE scheme that is proven fully
secure from the decisional linear assumption. In the full version of this
paper, we also provide a translation of the Lewko-Waters unbounded
HIBE scheme.

1 Introduction

Recently, several cryptosystems have been constructed in composite order bilinear groups and proven secure from instances (and close variants) of the general subgroup decision assumption defined in [3]. For example, the systems
presented in [27,25,29,28,26] provide diverse and advanced functionalities like
identity-based encryption (IBE), hierarchical identity-based encryption (HIBE),
and attribute-based encryption with strong security guarantees (e.g. full security, leakage-resilience) proven from static assumptions. These works leverage
convenient features of composite order bilinear groups that are not shared by
prime order bilinear groups, most notably the presence of orthogonal subgroups
of coprime orders. Up to isomorphism, a composite order bilinear group has the
structure of a direct product of prime order subgroups, so every group element
can be decomposed as the product of components in the separate subgroups.
However, when the group order is hard to factor, such a decomposition is hard
to compute. The orthogonality of these subgroups means that they can function
as independent spaces, allowing a system designer to use them in different ways

---

[*] Supported by a Microsoft Research Ph.D. Fellowship.

D. Pointcheval and T. Johansson (Eds.): EUROCRYPT 2012, LNCS 7237, pp. 318–335, 2012.

without any cross interactions between them destroying correctness. Security relies on the assumption that these subgroups are essentially inseparable: given a random group element, it should be hard to decide which subgroups contribute non-trivial components to it.

Though composite order bilinear groups have appealing features, it is desirable to obtain the same functionalities and strong guarantees achieved in composite order groups from other assumptions, particularly from the decisional linear assumption (DLIN) in prime order bilinear groups. The ability to work with prime order bilinear groups instead of composite order ones offers several advantages. First, we can obtain security under the more standard decisional linear assumption. Second, we can achieve much more efficient systems for the same security levels. This is because in composite order groups, security typically relies on the hardness of factoring the group order. This requires the use of large group orders, which results in considerably slower pairing operations.

There have been many previous examples of cryptosystems that were first built in composite order groups while later analogs were obtained in prime order groups. These include Groth-Ostrovsky-Sahai proofs [22,21], the Boneh-Sahai-Waters traitor tracing scheme [10,15], and the functional encryption schemes of Lewko-Okamoto-Sahai-Takashima-Waters [25,33]. Waters also notes that the dual system encryption techniques in [38] used to obtain prime order systems were first instantiated in composite order groups. These results already suggest that there are strong parallels between the composite order and prime order settings, but the translation techniques are developed in system-specific ways.

Beyond improving the assumptions and efficiency for particular schemes, our goal in this paper is to expand our general understanding of how tools that are conveniently inherent in the composite order setting can be simulated in the prime order setting. We begin by asking: what are the basic features of composite order bilinear groups that are typically exploited by cryptographic constructions and security proofs? Freeman considers this question in [14] and identifies two such features, called *projecting* and *canceling* (we also refer to canceling as "orthogonality"). Freeman then provides examples of how to construct either of these properties using pairings of vectors of group elements in prime order groups. Notably, Freeman does not provide a way of simultaneously achieving *both* projecting and canceling. There may be good reason for this, since Meiklejohn, Shacham, and Freeman [30] have shown that both properties cannot be simultaneously achieved in prime order groups when one relies on the decisional linear assumption in a "natural way". By instantiating either projecting or canceling in prime order groups, Freeman [14] successfully translates several composite order schemes into prime order schemes: the Boneh-Goh-Nissim encryption scheme [9], the Boneh-Sahai-Waters traitor tracing system [10], and the Katz-Sahai-Waters predicate encryption scheme [24]. These translations use a three step process. The first step is to write the scheme in an abstract framework (replacing subgroups by subspaces of vectors in the exponent), the second step is to translate the assumptions into prime order analogs, and the third step is to transfer the security proof.

There are two aspects of Freeman's approach that can render the results unsatisfying in certain cases. First, the step of translating the assumptions often does not result in standard assumptions like DLIN. A reduction to DLIN is only provided for the most basic variant of the subgroup decision assumption, and does not extend (for example) to the general subgroup decision assumption from [3]. Second, the step of translating the proof fails for many schemes, including all of the recent composite order schemes employing the dual system encryption proof methodology [27,25,29,28,26]. These schemes use only canceling and not projecting, and so this is unrelated to the limitations discussed in [30].

The reason for this failure is instructive to examine. As Freeman points out, "the recent identity-based encryption scheme of Lewko and Waters [27] uses explicitly in its security proof the fact that the group G has two subgroups of relatively prime order". The major obstacle here is not translating the description of the scheme or its assumptions - instead the problem lies in translating a trick in the security proof. The trick works as follows. Suppose we have a group G of order $N = p_1 p_2 \ldots p_m$, where $p_1, \ldots, p_m$ are distinct primes. Then if we take an element $g_1 \in G$ of order p_1 (i.e. an element of the subgroup of G with order p_1) and a random exponent $a \in \mathbb{Z}_N$, the group element g_1^a *reveals no information* about the value of a modulo the other primes. Only $a \bmod p_1$ is revealed. The fact that $a \bmod p_2$, for instance, is uniformly random even conditioned on $a \bmod p_1$ follows from the Chinese Remainder Theorem. In the security proof of the Lewko-Waters scheme, there are elements of the form g_1^a in the public parameters, and the fact that $a \bmod p_2$ remains information-theoretically hidden is later used to argue that all the keys and ciphertext received by the attacker are properly distributed in the midst of a hybrid argument.

Clearly, in a prime order group, we cannot hope to construct subgroups with coprime orders. There are a few possible paths for resolving this difficulty. We could start by reworking proofs in the composite order setting to avoid using this trick and then hope to apply the techniques of [14] without modification. This approach is likely to result in more complicated (though still static) assumptions in the composite order setting, which will translate into more complicated assumptions in the prime order setting. Since we prefer to rely only on the decisional linear assumption, we follow an alternate strategy: finding a version of this trick in prime order groups that does not rely on coprimeness. This is possible because coprimeness here is used a mechanism for achieving "parameter hiding," meaning that some useful information is information-theoretically hidden from the attacker, even after the public parameters are revealed. We can construct an alternate mechanism in prime order groups that similarly enables a form of parameter hiding.

Our Contribution. We present versatile tools that can be used to translate composite order bilinear systems relying on canceling to prime order bilinear systems, particularly those whose security proofs rely on general subgroup decision assumptions and employ the coprime mechanism discussed above. This includes schemes like [27], which could not be handled by Freeman's methods. Our tools are based in the dual pairing vector space framework initiated by Okamoto and

Takashima [31,32]. We observe that dual pairing vector spaces provide a mechanism for parameter hiding that can be used in place of coprimeness. We then formulate an assumption in prime order groups that can be used to mimic the effect of the general subgroup decision assumption in composite order groups. We prove that this assumption is implied by DLIN. Putting these ingredients together, we obtain a flexible toolkit for turning a class of composite order constructions into prime order constructions that can be proven secure from DLIN.

We demonstrate the use of our toolkit by providing a translation of the composite order Lewko-Waters IBE construction [27]. This yields a prime order IBE construction that is proven fully secure from DLIN and also inherits the intuitive structure of the Boneh-Boyen IBE [5]. Compared to the fully secure prime order IBE construction in [38], our scheme achieves comparable efficiency and security with a simpler structure. As a second application, we provide a translation of the Lewko-Waters unbounded HIBE scheme [29] in the full version. This additionally demonstrates how to handle delegation of secret keys with our tools.

We note that some composite order systems employing dual system encryption, such as the attribute-based encryption scheme in [25], already have analogs in prime order groups proven secure from DLIN using dual pairing vector spaces. In [33], Okamoto and Takashima provide a functional encryption scheme in prime order bilinear groups that is proven fully secure under DLIN. Their construction encompasses both attribute-based and inner product encryption, and their proof relies on dual system encryption techniques, similarly to [25]. While they focus on providing a particular construction and proof, our goal is to formulate a more general strategy for translating composite order schemes into prime order schemes with analogous proofs.

Other Related Work. The concept of identity-based encryption was first proposed by Shamir [36] and later constructed by Boneh and Franklin [8] and Cocks [13]. In an identity-based encryption scheme, users are associated with identities and obtain secret keys from a master authority. Encryption to any identity can be done knowing only the identity and some global public parameters. Both of the initial constructions of IBE were proven secure in the random oracle model. The first standard model constructions, by Canetti, Halevi, and Katz [11] and Boneh and Boyen [5] relied on selective security, which is a more restrictive security model requiring the attacker to announce the identity to be attacker prior to viewing the public parameters. Subsequently, Boneh and Boyen [6], Gentry [16], and Waters [37,38] provided constructions proven fully secure in the standard model from various assumptions. Except for the scheme of [13], which relied on the quadratic residuousity assumption, all of the schemes we have cited above rely on bilinear groups. A lattice-based IBE construction was first provided by Gentry, Peikert, and Vaikuntanathan in [18].

Hierarchical identity-based encryption was proposed by Horwitz and Lynn [23] and then constructed by Gentry and Silverberg [19] in the random oracle model. In a HIBE scheme, users are associated with identity vectors that indicate their places in a hierarchy (a user Alice is a superior of the user Bob if her identity vector is a prefix of his). Any user can obtain a secret key for his identity vector either from the master authority or from one of his superiors (i.e. a mechanism for key delegation to subordinates is provided). Selectively secure standard model constructions of HIBE were provided by Boneh and Boyen [5] and Boneh, Boyen, and Goh [7] in the bilinear setting and by Cash, Hofheinz, Kiltz, and Peikert [12] and Agrawal, Boneh, and Boyen [1,2] in the lattice-based setting. Fully secure constructions allowing polynomial depth were given by Gentry and Halevi [17], Waters [38], and Lewko and Waters [27]. The first unbounded construction (meaning that the maximal depth is not bounded by the public parameters) was given by Lewko and Waters in [29].

Attribute-based encryption (ABE) is a more flexible functionality than (H)IBE, first introduced by Sahai and Waters in [35]. In an ABE scheme, keys and ciphertexts are associated with attributes and access policies instead of identities. In a ciphertext-policy ABE scheme, keys are associated with attributes and ciphertexts are associated with access policies. In a key-policy ABE scheme, keys are associated with access policies and ciphertexts are associated with attributes. In both cases, a key can decrypt a ciphertext if and only if the attributes satisfy the formula. There are several constructions of both kinds of ABE schemes, e.g. [35,20,34,4,25,33,39].

The dual system encryption methodology was introduced by Waters in [38] as a tool for proving full security of advanced functionalities such as (H)IBE and ABE. It was further developed in several subsequent works [27,25,33,26,29,28]. Most of these works have used composite order groups as a convenient setting for instantiating the dual system methodology, with the exception of [33]. Here, we extend and generalize the techniques of [33] to demonstrate that this use of composite order groups can be viewed as an intermediary step in the development of prime order systems whose security relies on the DLIN assumption.

2   Background

2.1   Composite Order Bilinear Groups

When G is a bilinear group of composite order $N = p_1 p_2 \ldots p_m$ (where p_1, p_2, $\ldots, p_m$ are distinct primes), we let $e : G \times G \rightarrow G_T$ denote its bilinear map (also referred to as a pairing). We note that both G and G_T are cyclic groups of order N. For each p_i, G has a subgroup of order p_i denoted by G_{p_i}. We let $g_1, \ldots, g_m$ denote generators of G_{p_1} through G_{p_m} respectively. Each element $g \in G$ can be expressed as $g = g_1^{a_1} g_2^{a_2} \cdots g_m^{a_m}$ for some $a_1, \ldots, a_m \in \mathbb{Z}_N$, where each a_i is unique modulo p_i. We will refer to $g_i^{a_i}$ as the "G_{p_i} component" of g. When a_i is congruent to zero modulo p_i, we say that g has no G_{p_i} component. The subgroups $G_{p_1}, \ldots, G_{p_m}$ are "orthogonal" under the bilinear map e, meaning that if $h \in G_{p_i}$ and $u \in G_{p_j}$ for $i \neq j$, then $e(h, u) = 1$, where 1 denotes the identity element in G_T.

General Subgroup Decision Assumption. The general subgroup decision assumption for composite order bilinear groups (formulated in [3]) is a family of static complexity assumptions based on the intuition that it should be hard to determine which components are present in a random group element, except for what can be trivially determined by testing for orthogonality with other given group elements. More precisely, for each non-empty subset $S \subseteq [m]$, there is an associated subgroup of order $\prod_{i \in S} p_i$ in G, which we will denote by G_S. For two distinct, non-empty subsets S_0 and S_1, we assume it is hard to distinguish a random element of G_{S_0} from a random element of G_{S_1}, when one is only given random elements of $G_{S_2}, \ldots, G_{S_k}$ where for each $2 \leq j \leq k$, either $S_j \cap S_0 = \emptyset = S_j \cap S_1$ or $S_j \cap S_0 \neq \emptyset \neq S_j \cap S_1$.

More formally, we let $\mathcal{G}$ denote a group generation algorithm, which takes in m and a security parameter λ and outputs a bilinear group G of order $N = p_1 \cdots p_m$, where $p_1, \ldots, p_m$ are distinct primes. The General Subgroup Decision Assumption with respect to $\mathcal{G}$ is defined as follows.

Definition 1. *General Subgroup Decision Assumption. Let $S_0, S_1, S_2, \ldots, S_k$ be non-empty subsets of $[m]$ such that for each $2 \leq j \leq k$, either $S_j \cap S_0 = \emptyset = S_j \cap S_1$ or $S_j \cap S_0 \neq \emptyset \neq S_j \cap S_1$. Given a group generator $\mathcal{G}$, we define the following distribution:*

$$\mathbb{G} := (N = p_1 \cdots p_m, G, G_T, e) \xleftarrow{R} \mathcal{G},$$

$$Z_0 \xleftarrow{R} G_{S_0}, \ Z_1 \xleftarrow{R} G_{S_1}, \ Z_2 \xleftarrow{R} G_{S_2}, \ldots, Z_k \xleftarrow{R} G_{S_k},$$

$$D := (\mathbb{G}, Z_2, \ldots, Z_k).$$

We assume that for any PPT algorithm $\mathcal{A}$ (with output in $\{0, 1\}$),

$$Adv_{\mathcal{G}, \mathcal{A}} := |\mathbb{P}\left[\mathcal{A}(D, Z_0) = 1\right] - \mathbb{P}\left[\mathcal{A}(D, Z_1) = 1\right]|$$

is negligible in the security parameter λ.

We note that this assumption holds in the generic group model, assuming it is hard to find a non-trivial factor of the group order N.

Restricting to Challenge Sets Differing by One Element. We observe that it suffices to consider challenge sets S_0 and S_1 of the form $S_1 = S_0 \cup \{i\}$ for some $i \in [m]$, $i \notin S_0$. We refer to this restricted class of subgroup decision assumptions as the 1-General Subgroup Decision Assumption. To see that the 1-general subgroup decision assumption implies the general subgroup decision assumption, we show that any instance of the general subgroup decision assumption is implied by a sequence of the more restricted instances. More precisely, for general S_0, S_1, we let U denote the set $S_0 \cup S_1 - S_0$. For any i in U, the 1-general subgroup decision assumption implies that it hard to distinguish a random element of G_{S_0} from a random element of $G_{S_0 \cup \{i\}}$, even given random elements from $G_{S_2}, \ldots, G_{S_k}$. That is because each of the sets $S_2, \ldots, S_k$ either does not intersect S_1 or S_0 and hence does not intersect S_0 or $S_0 \cup \{i\} \subseteq S_1$, or intersects both S_0 and $S_0 \cup \{i\}$. We can now incrementally add the other elements of U using instances of the 1-general subgroup decision assumption, ultimately showing that it is hard to distinguish a random element of G_{S_0} from a random element of $G_{S_0 \cup S_1}$. We can reverse the process and subtract one element at a time from $S_0 \cup S_1$ until we arrive at S_1. Thus, the seemingly more restrictive 1-general subgroup decision assumption implies the general subgroup decision assumption.

2.2   Prime Order Bilinear Groups

We now let G denote a bilinear group of prime order p, with bilinear map $e :$ $G \times G \to G_T$. More generally, one may have a bilinear map $e : G \times H \to G_T$, where G and H are different groups. For simplicity in this paper, we will always consider groups where $G = H$.

In addition to referring to individual elements of G, we will also consider "vectors" of group elements. For $\boldsymbol{v} = (v_1, \ldots, v_n) \in \mathbb{Z}_p^n$ and $g \in G$, we write $g^{\boldsymbol{v}}$ to denote a n-tuple of elements of G:

$$g^{\boldsymbol{v}} := (g^{v_1}, g^{v_2}, \ldots, g^{v_n}).$$

We can also perform scalar multiplication and vector addition in the exponent. For any $a \in \mathbb{Z}_p$ and $\boldsymbol{v}, \boldsymbol{w} \in \mathbb{Z}_p^n$, we have:

$$g^{a\boldsymbol{v}} := (g^{av_1}, \ldots, g^{av_n}), \quad g^{\boldsymbol{v}+\boldsymbol{w}} = (g^{v_1+w_1}, \ldots, g^{v_n+w_n}).$$

We define e_n to denote the product of the componentwise pairings:

$$e_n(g^{\boldsymbol{v}}, g^{\boldsymbol{w}}) := \prod_{i=1}^{n} e(g^{v_i}, g^{w_i}) = e(g,g)^{\boldsymbol{v} \cdot \boldsymbol{w}}.$$

Here, the dot product is taken modulo p.

Dual Pairing Vector Spaces. We will employ the concept of dual pairing vector spaces from [31,32]. For a fixed (constant) dimension n, we will choose two random bases $\mathbb{B} := (\boldsymbol{b}_1, \ldots, \boldsymbol{b}_n)$ and $\mathbb{B}^* := (\boldsymbol{b}_1^*, \ldots, \boldsymbol{b}_n^*)$ of $\mathbb{Z}_p^n$, subject to the constraint that they are "dual orthonormal", meaning that

$$\boldsymbol{b}_i \cdot \boldsymbol{b}_j^* = 0 \ (\mathrm{mod}\, p),$$

whenever $i \neq j$, and

$$\boldsymbol{b}_i \cdot \boldsymbol{b}_i^* = \psi$$

for all i, where ψ is a uniformly random element of $\mathbb{Z}_p$. (This is a slight abuse of the terminology "orthonormal", since ψ is not constrained to be 1.)

For a generator $g \in G$, we note that

$$e_n(g^{\boldsymbol{b}_i}, g^{\boldsymbol{b}_j^*}) = 1$$

whenever $i \neq j$, where 1 here denotes the identity element in G_T.

We note that choosing random dual orthonormal bases $(\mathbb{B}, \mathbb{B}^*)$ can equivalently be thought of as choosing a random basis $\mathbb{B}$, choosing a random vector $\boldsymbol{b}_1^*$ subject to the constraint that it is orthogonal to $\boldsymbol{b}_2, \ldots, \boldsymbol{b}_n$, defining $\psi = \boldsymbol{b}_1 \cdot \boldsymbol{b}_1^*$, and then choosing $\boldsymbol{b}_2^*$ so that it is orthogonal to $\boldsymbol{b}_1, \boldsymbol{b}_3, \ldots, \boldsymbol{b}_n$, and has dot product with $\boldsymbol{b}_2$ equal to ψ, and so on. We will later use the notation $(\mathbb{D}, \mathbb{D}^*)$ and $\boldsymbol{d}_1, \ldots,$ etc. to also denote dual orthonormal bases and their vectors (and even $\mathbb{F}, \mathbb{F}^*$ and $\boldsymbol{f}_1$, etc.). This is because we will sometimes be handling more than one pair of dual orthonormal bases at a time, and we use different notation to avoid confusing them.

Decisional Linear Assumption. The complexity assumption we will rely on in prime order bilinear groups is the Decisional Linear Assumption. To define this formally, we let $\mathcal{G}$ denote a group generation algorithm, which takes in a security parameter λ and outputs a bilinear group G of order p.

Definition 2. *Decisional Linear Assumption. Given a group generator $\mathcal{G}$, we define the following distribution:*

$$\mathbb{G} := (p, G, G_T, e) \xleftarrow{R} \mathcal{G},$$

$$g, f, v, w \xleftarrow{R} G, \; c_1, c_2, w \xleftarrow{R} \mathbb{Z}_p,$$

$$D := (g, f, v, f^{c_1}, v^{c_2}).$$

We assume that for any PPT algorithm $\mathcal{A}$ (with output in $\{0, 1\}$),

$$Adv_{\mathcal{G}, \mathcal{A}} := \left| \mathbb{P}\left[\mathcal{A}(D, g^{c_1 + c_2}) = 1 \right] - \mathbb{P}\left[\mathcal{A}(D, g^{c_1 + c_2 + w}) = 1 \right] \right|$$

is negligible in the security parameter λ.

3   Our Main Tools

There is an additional feature of composite order groups that is often exploited along with canceling/orthogonality in the security proofs for composite order constructions: we call this *parameter hiding*. In composite order groups, parameter hiding takes the following form. Consider a composite order group G of order $N = p_1 p_2$ and an element $g_1 \in G_{p_1}$ (an element of order p_1). Then if we sample a uniformly random exponent $a \in Z_N$ and produce g_1^a, this reveals nothing about the value of a modulo p_2. More precisely, the Chinese Remainder theorem guarantees that the value of a modulo p_2 conditioned on the value of a modulo p_1 is still uniformly random, and g_1^a only depends on the value of a modulo p_1. This allows a party choosing a to publish g_1^a and still *hide* some information about a, namely its value modulo p_2. Note that this party only needs to know N and g_1: it does not need to know the factorization of N.

This is an extremely useful tool in security proofs, enabling a simulator to choose some secret random exponents, publish the public parameters by raising known subgroup elements to these exponents, and still information-theoretically hide the values of these exponents modulo some of the primes. These hidden values can be leveraged later in the security game to argue that something looks well-distributed in the attacker's view, even if this does not hold in the simulator's view. This sort of trick is crucial in proofs employing the dual system encryption methodology.

Replicating this trick in prime order groups seems challenging, since if one is given g and g^a in a prime order group, a is completely revealed modulo p in an information-theoretic sense. To resolve this issue, we use dual pairing vector spaces. We observe that a form of parameter hiding is achieved by using dual orthonormal bases: one can generate a random pair of dual orthonormal bases

$(\mathbb{B}, \mathbb{B}^*)$ for $\mathbb{Z}_p^n$, apply an invertible change of basis matrix A to a subset of these basis vectors, and produce a new pair of dual orthonormal bases which is also randomly distributed, *independently of A*. This allows us to *hide* a random matrix A. We formulate this precisely below.

3.1   Parameter Hiding in Dual Orthonormal Bases

We consider taking dual orthonormal bases and applying a linear change of basis to a subset of their vectors. We do this in such a way that we produce new dual orthonormal bases. In this subsection, we prove that if we start with randomly sampled dual orthonormal bases, then the resulting bases will also be random - in particular, the distribution of the final bases reveals nothing about the change of basis matrix that was employed. This "hidden" matrix can then be leveraged in security proofs as a way of separating the simulator's view from the attacker's.

To describe this formally, we let $m \leq n$ be fixed positive integers and $A \in \mathbb{Z}_p^{m \times m}$ be an invertible matrix. We let $S_m \subseteq [n]$ be a subset of size m ($|S| = m$). For any dual orthonormal bases $\mathbb{B}, \mathbb{B}^*$, we can then define new dual orthonormal bases $\mathbb{B}_A, \mathbb{B}_A^*$ as follows. We let B_m denote the $n \times m$ matrix over $\mathbb{Z}_p$ whose columns are the vectors $\boldsymbol{b}_i \in \mathbb{B}$ such that $i \in S_m$. Then $B_m A$ is also an $n \times m$ matrix. We form $\mathbb{B}_A$ by retaining all of the vectors $\boldsymbol{b}_i \in \mathbb{B}$ for $i \notin S_m$ and exchanging the $\boldsymbol{b}_i$ for $i \in S_m$ with the columns of $B_m A$. To define $\mathbb{B}_A^*$, we similarly let B_m^* denote the $n \times m$ matrix over $\mathbb{Z}_p$ whose columns are the vectors $\boldsymbol{b}_i^* \in \mathbb{B}^*$ such that $i \in S_m$. Then $B_m^*(A^{-1})^t$ is also an $n \times m$ matrix, where $(A^{-1})^t$ denotes the transpose of A^{-1}. We form $\mathbb{B}_A^*$ by retaining all of the vectors $\boldsymbol{b}_i^* \in \mathbb{B}^*$ for $i \notin S_m$ and exchanging the $\boldsymbol{b}_i$ for $i \in S_m$ with the columns of $B_m^*(A^{-1})^t$.

To see that $\mathbb{B}_A$ and $\mathbb{B}_A^*$ are dual orthonormal bases, note that for $i \in S_m$, the corresponding basis vector in $\mathbb{B}_A$ can be expressed as a linear combination of the basis vectors $\boldsymbol{b}_j \in \mathbb{B}$ with $j \in S_m$, and the coefficients of this linear combination correspond to a column of A, say the ℓth column (equivalently, say i is the ℓth element of S_m). When $\ell \neq \ell'$, the ℓth column of A is orthogonal to the $(\ell')^{th}$ column of $(A^{-1})^t$. This means that the ith vector of $\mathbb{B}_A$ will be orthogonal to the $(i')^{th}$ vector of $\mathbb{B}_A^*$ whenever $i \neq i'$. Moreover, the ℓth column of A and the ℓth column of $(A^{-1})^t$ have dot product equal to 1, so the dot product of the ith vector of $\mathbb{B}_A$ and the ith vector of $\mathbb{B}_A^*$ will be equal to the same value ψ as in the original bases $\mathbb{B}$ and $\mathbb{B}^*$.

For a fixed dimension n and prime p, we let $(\mathbb{B}, \mathbb{B}^*) \xleftarrow{R} Dual(\mathbb{Z}_p^d)$ denote choosing random dual orthonormal bases $\mathbb{B}$ and $\mathbb{B}^*$ of $\mathbb{Z}_p^n$. Here, $Dual(\mathbb{Z}_p^n)$ denotes the set of dual orthonormal bases.

Lemma 1. *For any fixed positive integers $m \leq n$, any fixed invertible $A \in \mathbb{Z}_p^{m \times m}$ and set $S_m \subseteq [n]$ of size m, if $(\mathbb{B}, \mathbb{B}^*) \xleftarrow{R} Dual(\mathbb{Z}_p^d)$, then $(\mathbb{B}_A, \mathbb{B}_A^*)$ is also distributed as a random sample from $Dual(\mathbb{Z}_p^d)$. In particular, the distribution of $(\mathbb{B}_A, \mathbb{B}_A^*)$ is independent of A.*

Proof. There is a one-to-one correspondence between $(\mathbb{B}, \mathbb{B}^*)$ and $(\mathbb{B}_A, \mathbb{B}_A^*)$: given $(\mathbb{B}_A, \mathbb{B}_A^*)$, one can recover $(\mathbb{B}, \mathbb{B}^*)$ by applying A^{-1} to the vectors in $\mathbb{B}_A$ whose indices are in S_m, and applying A^t to the corresponding vectors in $\mathbb{B}_A^*$. This shows that every pair of dual orthonormal bases is equally likely to occur as $\mathbb{B}_A, \mathbb{B}_A^*$.

3.2   The Subspace Assumption

We now state a complexity assumption in prime order groups that we will use to simulate the effects of subgroup decision assumptions in composite order groups. We call this the Subspace Assumption. In the full version, we show that the subspace assumption is implied by the decisional linear assumption.

In prime order groups, basis vectors in the exponent take the place of subgroups. Since we are using dual orthonormal bases, our new concept of orthogonality between "subgroups" becomes asymmetric. If we have dual orthonormal bases $\mathbb{B}, \mathbb{B}^*$ and we think of "subgroup 1" in $\mathbb{B}$ as corresponding to the span of $\boldsymbol{b}_1, \ldots, \boldsymbol{b}_4$, then this is not orthogonal to the other vectors in $\mathbb{B}$, but it is orthogonal to vectors $\boldsymbol{b}_5^*, \ldots, \boldsymbol{b}_n^*$ in $\mathbb{B}^*$. Essentially, the notion of a single subgroup has now been split into a pair of "subgroups", one for each side of the pairing, and orthogonality between different subgroups now only holds for elements on opposite sides.

This sort of asymmetry can be quite useful. For example, consider an instance of the general subgroup decision assumption in composite order groups, where the task is to distinguish a random element of G_{p_1} from $G_{p_1 p_2}$. In this case, we cannot give out an element of G_{p_2}, since it can trivially be used to break the assumption by pairing it with the challenge term and seeing if the result is the identity. If we instead use dual orthonormal bases in a prime order group, the situation is a bit different. Suppose that given $g^{\boldsymbol{v}}$, the task is to distinguish whether the exponent vector $\boldsymbol{v}$ is in the span of $\boldsymbol{b}_1^*, \boldsymbol{b}_2^*$ or in the larger span of $\boldsymbol{b}_1^*, \boldsymbol{b}_2^*, \boldsymbol{b}_3^*$. We cannot give out $g^{\boldsymbol{b}_3}$, since one could then break the assumption by testing if $e_n(g^{\boldsymbol{v}}, g^{\boldsymbol{b}_3}) = e(g,g)^{\boldsymbol{v} \cdot \boldsymbol{b}_3}$ is the identity, but *we can give out* $g^{\boldsymbol{b}_3^*}$.

Our definition of the subspace assumption is motivated by this and our observation in Section 2.1 that the general subgroup decision assumption in composite order groups can be restricted to distinguishing between sets that differ by one element. What this means is that to simulate the uses of the general subgroup decision in composite order groups, one can focus merely on creating an analog for expansion into one new "subgroup" at a time. At its core, our subspace assumption says that if one is given $g^{\boldsymbol{v}}$, then it is hard to tell if $\boldsymbol{v}$ is randomly chosen from the span of $\boldsymbol{b}_1^*, \boldsymbol{b}_2^*$ or from the larger span of $\boldsymbol{b}_1^*, \boldsymbol{b}_2^*, \boldsymbol{b}_3^*$, even if one is given scalar multiples of all bases vectors in $\mathbb{B}$ and $\mathbb{B}^*$ in the exponent, *except for* $\boldsymbol{b}_3$. We augment this by also given out a random linear combination of $\boldsymbol{b}_1, \boldsymbol{b}_2, \boldsymbol{b}_3$ in the exponent. We then generalize this by replicating the same structure for k 3-tuples of vectors, with the random linear combinations having the *same* coefficients. (The fact that these coefficients are the same prevents this from following immediately from the assumption for a single 3-tuple applied in hybrid fashion.)

We now give the formal description of the subspace assumption. For a fixed dimension $n \geq 3$ and prime p, we recall that $(\mathbb{B}, \mathbb{B}^*) \xleftarrow{R} Dual(\mathbb{Z}_p^n)$ denotes choosing random dual orthonormal bases $\mathbb{B}$ and $\mathbb{B}^*$ of $\mathbb{Z}_p^n$, and $Dual(\mathbb{Z}_p^n)$ denotes the set of dual orthonormal bases. Our assumption is additionally parameterized by a positive integer $k \leq \frac{n}{3}$.

Definition 3. *(Subspace Assumption) Given a group generator* $\mathcal{G}$, *we define the following distribution:*

$$\mathbb{G} := (p, G, G_T, e) \xleftarrow{R} \mathcal{G}, \ (\mathbb{B}, \mathbb{B}^*) \xleftarrow{R} Dual(\mathbb{Z}_p^n),$$

$$g \xleftarrow{R} G, \ \eta, \beta, \tau_1, \tau_2, \tau_3, \mu_1, \mu_2, \mu_3 \xleftarrow{R} \mathbb{Z}_p,$$

$$U_1 := g^{\mu_1 \boldsymbol{b}_1 + \mu_2 \boldsymbol{b}_{k+1} + \mu_3 \boldsymbol{b}_{2k+1}}, \ U_2 := g^{\mu_1 \boldsymbol{b}_2 + \mu_2 \boldsymbol{b}_{k+2} + \mu_3 \boldsymbol{b}_{2k+2}}, \ \dots,$$

$$U_k := g^{\mu_1 \boldsymbol{b}_k + \mu_2 \boldsymbol{b}_{2k} + \mu_3 \boldsymbol{b}_{3k}}, V_1 := g^{\tau_1 \eta \boldsymbol{b}_1^* + \tau_2 \beta \boldsymbol{b}_{k+1}^*}, \ V_2 := g^{\tau_1 \eta \boldsymbol{b}_2^* + \tau_2 \beta \boldsymbol{b}_{k+2}^*}, \ \dots,$$

$$V_k := g^{\tau_1 \eta \boldsymbol{b}_k^* + \tau_2 \beta \boldsymbol{b}_{2k}^*}, \ W_1 := g^{\tau_1 \eta \boldsymbol{b}_1^* + \tau_2 \beta \boldsymbol{b}_{k+1}^* + \tau_3 \boldsymbol{b}_{2k+1}^*},$$

$$W_2 := g^{\tau_1 \eta \boldsymbol{b}_2^* + \tau_2 \beta \boldsymbol{b}_{k+2}^* + \tau_3 \boldsymbol{b}_{2k+2}^*}, \ \dots, W_k := g^{\tau_1 \eta \boldsymbol{b}_k^* + \tau_2 \beta \boldsymbol{b}_{2k}^* + \tau_3 \boldsymbol{b}_{3k}^*}$$

$$D := \left(g^{\boldsymbol{b}_1}, g^{\boldsymbol{b}_2}, \dots, g^{\boldsymbol{b}_{2k}}, g^{\boldsymbol{b}_{3k+1}}, \dots, g^{\boldsymbol{b}_n}, g^{\eta \boldsymbol{b}_1^*}, \dots, g^{\eta \boldsymbol{b}_k^*}, \right.$$

$$\left. g^{\beta \boldsymbol{b}_{k+1}^*}, \dots, g^{\beta \boldsymbol{b}_{2k}^*}, g^{\boldsymbol{b}_{2k+1}^*}, \dots, g^{\boldsymbol{b}_n^*}, U_1, U_2, \dots, U_k, \mu_3 \right).$$

We assume that for any PPT algorithm $\mathcal{A}$ *(with output in* $\{0,1\}$*),*

$$Adv_{\mathcal{G},\mathcal{A}} := |\mathbb{P}\left[\mathcal{A}(D, V_1, \dots, V_k) = 1\right] - \mathbb{P}\left[\mathcal{A}(D, W_1, \dots, W_k) = 1\right]|$$

is negligible in the security parameter λ.

We have included in D more terms than will be necessary for many applications of this assumption. We will work exclusively with the $k = 1$ and $k = 2$ cases. We present the assumption in the form above to make it more versatile for use in future applications. We additionally note that the form stated above can be further generalized to involve multiple, independently generated dual orthonormal bases $(\mathbb{B}_1, \mathbb{B}_1^*), (\mathbb{B}_2, \mathbb{B}_2^*), \dots, (\mathbb{B}_j, \mathbb{B}_j^*)$, for any fixed j. The terms in the assumption would be duplicated for each pair of bases, with the same values of $\eta, \beta, \tau_1, \tau_2, \tau_3, \mu_1, \mu_2, \mu_3$. We will not need this generalization for the applications we present. To help the reader see the main structure of this assumption through the burdensome notation, we include a heuristic illustration of the $k = 2$ case.

In the diagram, the top rows illustrate the U terms, while the bottom rows illustrate the V, W terms. The solid ovals and rectangles indicate the presence of basis vectors. The crossed rectangles indicate basis elements of $\mathbb{B}$ which are present in U_1, U_2 but are not given out in isolation. The dotted ovals adorned by question marks indicate the basis vectors whose presence depends on whether we consider the V's or the W's.

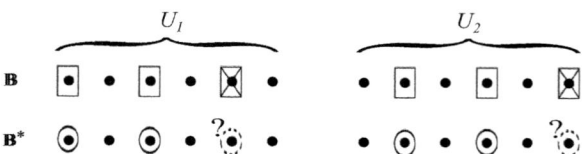

Fig. 1. Subspace Assumption with $k = 2$

4   Analog of the Boneh-Boyen IBE Scheme

In this section, we employ our subspace assumption and our parameter hiding technique for dual orthonormal bases to prove full security for a close analog of the Boneh-Boyen IBE scheme from the decisional linear assumption. This is the same security guarantee achieved for the IBE scheme in [38] and our efficiency is also similar. The advantage of our scheme is that it is a much closer analog to the original Boneh-Boyen IBE, and resultingly has a simpler, more intuitive structure.

Our security proof essentially mirrors the structure of the security proof given in [27], which provides a fully secure variant of the Boneh-Boyen IBE scheme in composite order groups. This serves as an illustrative example of how our techniques can be used to simulate dual system encryption proofs in the prime order setting that were originally presented in composite order groups.

4.1   Our Construction

We will use dual orthonormal bases $(\mathbb{D}, \mathbb{D}^*)$ of $\mathbb{Z}_p^6$, where p is the prime order of our bilinear group G. Public parameters and ciphertexts will have exponents described in terms of the basis vectors in $\mathbb{D}$, while secret keys will have exponents described in terms of $\mathbb{D}^*$. The first four basis vectors of each will constitute the "normal space" (like G_{p_1} in the LW scheme), and the last two basis vectors of each will constitute the "semi-functional space" (like G_{p_2} in the LW scheme).

By using dual pairing vector spaces, we avoid the need to simulate G_{p_3}. In the LW scheme, the purpose of G_{p_3} is to allow the creation of other semi-functional keys while a challenge key is changing from normal to semi-functional. More precisely, it allows the subgroup decision assumption to give out an element of $G_{p_2 p_3}$ that can be used to generate semi-functional keys when the task is to distinguish a random element of $G_{p_1 p_3}$ from a random element of G. We note that if we did not use G_{p_3} here and instead tried to create all of the semi-functional keys from a term in $G_{p_1 p_2}$, then these keys would not be properly randomized in the G_{p_2} subgroup because the structure of the scheme is enforced in the G_{p_1} subgroup. Pairwise independence cannot save us here because there are many keys. However, the asymmetry of dual pairing vector spaces avoids this issue: while we are expanding the challenge key into the "semi-functional space" in $\mathbb{D}^*$, we can still know a basis for the semi-functional space of $\mathbb{D}^*$ in the exponent - it is only the corresponding terms in the semi-functional space of $\mathbb{D}$ that we do not have access to in isolation. This allows us to make the other semi-functional keys without needing to create an analog of the G_{p_3} subgroup.

The core of the Boneh-Boyen scheme is a cancelation between terms in two pairings, one with the identity appearing on the ciphertext side and the other with the identity appearing on the key side. This is combined with a mechanism for preventing multiplication manipulation of the identity. In our scheme, this core cancelation is duplicated: instead of having one cancelation, we have two, each with its own random coefficients. The first cancelation will occur for the $\boldsymbol{d}_1, \boldsymbol{d}_2$ and $\boldsymbol{d}_1^*, \boldsymbol{d}_2^*$ components, and the second will occur for the $\boldsymbol{d}_3, \boldsymbol{d}_4$ and $\boldsymbol{d}_3^*, \boldsymbol{d}_4^*$ components.

This expansion gives us room to use the subspace assumption with parameter $k = 2$ to transition from 4-dimensional exponents for normal keys and ciphertexts to 6-dimensional exponents for semi-functional keys and ciphertexts. Having a 2-dimensional semi-functional space allows us to implement nominal semi-functionality. To prevent multiplicative manipulations of the identities in our scheme is rather easy, since the orthogonality of the dual bases allows us to "tie" all the components of the keys and ciphertexts together without causing cross interactions that interfere with decryption.

We assume that messages M are elements of G_T (the target group of the bilinear map) and that identities ID are elements of $\mathbb{Z}_p$.

$Setup(\lambda) \to \text{MSK}, \text{PP}$. The setup algorithm takes in the security parameter λ and chooses a bilinear group G of sufficiently large prime order p. We let $e : G \times G \to G_T$ denote the bilinear map. We set $n = 6$. The algorithm samples random dual orthonormal bases, $(\mathbb{D}, \mathbb{D}^*) \xleftarrow{R} Dual(\mathbb{Z}_p^n)$. We let $d_1, \ldots, d_6$ denote the elements of $\mathbb{D}$ and $d_1^*, \ldots, d_6^*$ denote the elements of $\mathbb{D}^*$. It also chooses random values $\alpha, \theta, \sigma \in \mathbb{Z}_p$. The public parameters are computed as:

$$\text{PP} := \left\{ G, p, e(g,g)^{\alpha\theta d_1 \cdot d_1^*}, g^{d_1}, \ldots, g^{d_4} \right\}.$$

(We note that $d_1 \cdot d_1^* = \psi$ by definition of $\mathbb{D}, \mathbb{D}^*$, but we write out the dot product when we feel it is more instructive.) The master secret key is:

$$\text{MSK} := \left\{ g^{\theta d_1^*}, g^{\alpha\theta d_1^*}, g^{\theta d_2^*}, g^{\sigma d_3^*}, g^{\sigma d_4^*} \right\}.$$

$KeyGen(\text{MSK}, ID) \to \text{SK}_{ID}$. The key generation algorithm chooses random values $r_1, r_2 \in \mathbb{Z}_p$ and forms the secret key as:

$$\text{SK}_{ID} := g^{(\alpha + r_1 ID)\theta d_1^* - r_1\theta d_2^* + r_2 ID\sigma d_3^* - r_2\sigma d_4^*}.$$

$Encrypt(M, ID, \text{PP}) \to \text{CT}$. The encryption algorithm chooses random values $s_1, s_2 \in \mathbb{Z}_p$ and forms the ciphertext as:

$$\text{CT} := \left\{ C_1 := M \left(e(g,g)^{\alpha\theta d_1 \cdot d_1^*} \right)^{s_1}, \ C_2 := g^{s_1 d_1 + s_1 ID d_2 + s_2 d_3 + s_2 ID d_4} \right\}.$$

$Decrypt(\text{CT}, \text{SK}_{ID}) \to M$. The decryption algorithm computes the message as:

$$M := C_1 / e_n(\text{SK}_{ID}, C_2).$$

Recall that $n = 6$, so this requires six pairings.

4.2 Semi-functional Algorithms

We choose to define our semi-functional objects by providing algorithms that generate them. We note that these algorithms are only provided for definitional purposes, and are not part of the IBE system. In particular, they do not need to be efficiently computable from the public parameters and master secret key alone.

KeyGenSF. The semi-functional key generation algorithm chooses random values $r_1, r_2, t_5, t_6 \in \mathbb{Z}_p$ and forms the secret key as

$$\mathrm{SK}_{ID} := g^{(\alpha + r_1 ID)\theta \boldsymbol{d}_1^* - r_1 \theta \boldsymbol{d}_2^* + r_2 ID\sigma \boldsymbol{d}_3^* - r_2 \sigma \boldsymbol{d}_4^* + t_5 \boldsymbol{d}_5^* + t_6 \boldsymbol{d}_6^*}.$$

This is distributed like a normal key with additional random multiples of $\boldsymbol{d}_5^*$ and $\boldsymbol{d}_6^*$ added in the exponent.

EncryptSF. The semi-functional encryption algorithm chooses random values $s_1, s_2, z_5, z_6 \in \mathbb{Z}_p$ and forms the ciphertext as:

$$\mathrm{CT} := \left\{ C_1 := M \left(e(g,g)^{\alpha \theta \boldsymbol{d}_1 \cdot \boldsymbol{d}_1^*} \right)^{s_1}, C_2 := g^{s_1 \boldsymbol{d}_1 + s_1 ID\boldsymbol{d}_2 + s_2 \boldsymbol{d}_3 + s_2 ID\boldsymbol{d}_4 + z_5 \boldsymbol{d}_5 + z_6 \boldsymbol{d}_6} \right\}.$$

This is distributed like a normal ciphertext with additional random multiples of $\boldsymbol{d}_5$ and $\boldsymbol{d}_6$ added in the exponent.

We observe that if one applies the decryption procedure with a semi-functional key and a normal ciphertext, decryption will succeed because $\boldsymbol{d}_5^*, \boldsymbol{d}_6^*$ are orthogonal to all of the vectors in exponent of C_2, and hence have no effect on decryption. Similarly, decryption of a semi-functional ciphertext by a normal key will also succeed because $\boldsymbol{d}_5, \boldsymbol{d}_6$ are orthogonal to all of the vectors in the exponent of the key. When *both* the ciphertext and key are semi-functional, the result of $e_n(\mathrm{SK}_{ID}, C_2)$ will have an additional term, namely $e(g,g)^{t_5 z_5 \boldsymbol{d}_5 \cdot \boldsymbol{d}_5^* + t_6 z_6 \boldsymbol{d}_6 \cdot \boldsymbol{d}_6^*} = e(g,g)^{(t_5 z_5 + t_6 z_6)\psi}$. Decryption will then fail unless $t_5 z_5 + t_6 z_6 \equiv 0 \bmod p$. If this modular equation holds, we say that the key and ciphertext pair is *nominally semi-functional*. We note that this is possible, even when none of t_5, z_5, t_6, z_6 are congruent to zero modulo p (this is why we have designated a semi-functional space of dimension two).

In the full version, we prove the following theorem. Here, we sketch the outline of the proof.

Theorem 1. *Under the decisional linear assumption, the IBE scheme presented in Section 4.1 is fully secure.*

We prove this using a hybrid argument over a sequence of games, following the LW strategy. We start with the real security game, denoted by Game_{real}. We let q denote the number of keys requested by the attacker. We define the following additional games.

Game_i for $i = 0, 1, \ldots, q$. Game_i is like Game_{real}, except the ciphertext given to the attacker is semi-functional (i.e. generated by a call to EncryptSF instead of Encrypt) and the first i keys given to the attacker are semi-functional (generated by KeyGenSF). The remaining keys are normal. We note that in Game_0, all of the keys are normal, and in Game_q, all of the keys are semi-functional.

Game_{final}. Game_{final} is like Game_q, except that the ciphertext is a semi-functional encryption of a *random* message in G_T, instead of one of the messages supplied by the attacker.

We transition from Game_{real} to Game_0, then to Game_1, and so on, until we arrive at Game_q. We prove that with each transition, the attacker's advantage cannot change by a non-negligible amount. As a last step, we transition to Game_{final}, where it is clear that the attacker's advantage is zero. These transitions are accomplished in the following lemmas, all using the subspace assumption. We let $Adv_{\mathcal{A}}^{real}$ denote the advantage of an algorithm $\mathcal{A}$ in the real game, $Adv_{\mathcal{A}}^{i}$ denote its advantage in Game_i, and $Adv_{\mathcal{A}}^{final}$ denote its advantage in Game_{final}.

We begin with the transition from Game_{real} to Game_0. At the analogous step in the LW proof, a subgroup decision assumption is used to expand the ciphertext from G_{p_1} into $G_{p_1 p_2}$. Here, we use the subspace assumption with $k = 2$ to expand the ciphertext exponent vector from the span of $\boldsymbol{d}_1, \dots, \boldsymbol{d}_4$ into the larger span of $\boldsymbol{d}_1, \dots, \boldsymbol{d}_6$. We use a very basic instance of the parameter hiding technique to argue that the resulting coefficients of $\boldsymbol{d}_5$ and $\boldsymbol{d}_6$ are randomly distributed: this is done by initially embedding a random 2×2 change of basis matrix A into our setting of the basis vectors $\boldsymbol{d}_5, \boldsymbol{d}_6$.

We now handle the transition from Game_{i-1} to Game_i. At this step in the LW proof, a subgroup decision assumption is used to expand the ith secret key from $G_{p_1 p_3}$ into $G = G_{p_1 p_2 p_3}$. Analogously, we will use the subspace assumption to expand the ith secret key exponent vector from the span of $\boldsymbol{d}_1^*, \dots, \boldsymbol{d}_4^*$ into the larger span of $\boldsymbol{d}_1^*, \dots, \boldsymbol{d}_6^*$. We will embed a 2×2 change of basis matrix A and set $\mathbb{D} = \mathbb{B}_A$ and $\mathbb{D}^* = \mathbb{B}_A^*$, where A is applied to $\boldsymbol{b}_5, \boldsymbol{b}_6$ to form $\boldsymbol{d}_5, \boldsymbol{d}_6$. As in the LW proof, we cannot be given an object that resides solely in the semi-functional space of the ciphertext (e.g. we cannot be given $g^{\boldsymbol{d}_5}, g^{\boldsymbol{d}_6}$), but we are given objects that have semi-functional components attached to normal components, and we can use these to create the semi-functional ciphertext. In the LW proof, a term in $G_{p_1 p_2}$ in used. Here, an exponent vector that is a linear combination of $\boldsymbol{b}_1, \boldsymbol{b}_3, \boldsymbol{b}_5$ and another exponent vector that is a linear combination of $\boldsymbol{b}_2, \boldsymbol{b}_4, \boldsymbol{b}_6$ are used. In our case, making the other normal and semi-functional keys is straightforward, since we are given scalar multiples of all of the vectors of $\mathbb{D}^*$ in the exponent. We use the fact that the matrix A is hidden from the attacker in order to argue that the semi-functional parts of the ciphertext and i^{th} key appear well-distributed.

The final step of the LW proof uses an assumption that it is not technically an instance of the general subgroup decision assumption, but is of a similar flavor. In our case, we use a slightly different strategy: we use the subspace assumption with $k = 1$ twice to randomize each appearance of s_1 in the C_2 term of the ciphertext, thereby severing its link with the blinding factor. The end result is the same - we obtain a semi-functional encryption of a random message. This randomization of s_1 is accomplished by first expanding an exponent vector from the span of $\boldsymbol{d}_5, \boldsymbol{d}_6$ into the larger span of $\boldsymbol{d}_5, \boldsymbol{d}_6, \boldsymbol{d}_2$ and then expanding an exponent vector from the span of $\boldsymbol{d}_5, \boldsymbol{d}_6$ into the larger span of $\boldsymbol{d}_5, \boldsymbol{d}_6, \boldsymbol{d}_1$. We note that the knowledge of the μ_3 value in the subspace assumption is used here to ensure that while we are doing the first expansion, for example, we can make the two occurrences of r_1 in the keys match consistently (this is necessary because $g^{\boldsymbol{d}_2^*}$ by itself will not be known during this step).

5   Further Applications

As a second demonstration of our tools, in the full version of this paper we consider a variant of the Lewko-Waters unbounded HIBE construction [29]. The composite order construction we present is simpler than the one presented in [29], at the cost of using more subgroups. Since we will ultimately simulate these subgroups in a prime order group, such a cost is no longer a significant detriment. In designing our prime order translation and proof, we will proceed along a path that is very similar to the path we took to translate the more basic IBE scheme. However, we now must take care to preserve delegation ability throughout our proof. As a result, we employ a different strategy for the final step of the proof. The details of our composite order construction, its prime order translation, and security proofs in both settings can be found in the full version of this paper.

In applying our tools to the both IBE and unbounded HIBE applications, we see that there is some flexibility in how we choose the construction, organize the hybrid games, and embed the subspace assumption in our reductions. All of these considerations interact, allowing us to make tradeoffs. The amount of flexibility available in applying our tools make them suitably versatile to handle a wider variety of applications as well. In particular, they can be applied in the attribute-based encryption setting. We suspect that applying our techniques to the composite order ABE constructions in [25] would result in a system and proof quite similar to the functional encryption schemes presented by Okamoto and Takashima in [33], who obtain security from the decisional linear assumption through dual pairing vector spaces.

References

1. Agrawal, S., Boneh, D., Boyen, X.: Efficient Lattice (H)IBE in the Standard Model. In: Gilbert, H. (ed.) EUROCRYPT 2010. LNCS, vol. 6110, pp. 553–572. Springer, Heidelberg (2010)
2. Agrawal, S., Boneh, D., Boyen, X.: Lattice Basis Delegation in Fixed Dimension and Shorter-Ciphertext Hierarchical IBE. In: Rabin, T. (ed.) CRYPTO 2010. LNCS, vol. 6223, pp. 98–115. Springer, Heidelberg (2010)
3. Bellare, M., Waters, B., Yilek, S.: Identity-Based Encryption Secure against Selective Opening Attack. In: Ishai, Y. (ed.) TCC 2011. LNCS, vol. 6597, pp. 235–252. Springer, Heidelberg (2011)
4. Bethencourt, J., Sahai, A., Waters, B.: Ciphertext-policy attribute-based encryption. In: Proceedings of the IEEE Symposium on Security and Privacy, pp. 321–334.
5. Boneh, D., Boyen, X.: Efficient Selective-ID Secure Identity-Based Encryption Without Random Oracles. In: Cachin, C., Camenisch, J.L. (eds.) EUROCRYPT 2004. LNCS, vol. 3027, pp. 223–238. Springer, Heidelberg (2004)
6. Boneh, D., Boyen, X.: Secure Identity Based Encryption Without Random Oracles. In: Franklin, M. (ed.) CRYPTO 2004. LNCS, vol. 3152, pp. 443–459. Springer, Heidelberg (2004)
7. Boneh, D., Boyen, X., Goh, E.-J.: Hierarchical Identity Based Encryption with Constant Size Ciphertext. In: Cramer, R. (ed.) EUROCRYPT 2005. LNCS, vol. 3494, pp. 440–456. Springer, Heidelberg (2005)

8. Boneh, D., Franklin, M.: Identity-Based Encryption from the Weil Pairing. In: Kilian, J. (ed.) CRYPTO 2001. LNCS, vol. 2139, pp. 213–229. Springer, Heidelberg (2001)

9. Boneh, D., Goh, E.-J., Nissim, K.: Evaluating 2-DNF Formulas on Ciphertexts. In: Kilian, J. (ed.) TCC 2005. LNCS, vol. 3378, pp. 325–341. Springer, Heidelberg (2005)

10. Boneh, D., Sahai, A., Waters, B.: Fully Collusion Resistant Traitor Tracing with Short Ciphertexts and Private Keys. In: Vaudenay, S. (ed.) EUROCRYPT 2006. LNCS, vol. 4004, pp. 573–592. Springer, Heidelberg (2006)

11. Canetti, R., Halevi, S., Katz, J.: A Forward-Secure Public-Key Encryption Scheme. In: Biham, E. (ed.) EUROCRYPT 2003. LNCS, vol. 2656, pp. 255–271. Springer, Heidelberg (2003)

12. Cash, D., Hofheinz, D., Kiltz, E., Peikert, C.: Bonsai Trees, or How to Delegate a Lattice Basis. In: Gilbert, H. (ed.) EUROCRYPT 2010. LNCS, vol. 6110, pp. 523–552. Springer, Heidelberg (2010)

13. Cocks, C.: An Identity Based Encryption Scheme Based on Quadratic Residues. In: Honary, B. (ed.) Cryptography and Coding 2001. LNCS, vol. 2260, pp. 360–363. Springer, Heidelberg (2001)

14. Freeman, D.M.: Converting Pairing-Based Cryptosystems from Composite-Order Groups to Prime-Order Groups. In: Gilbert, H. (ed.) EUROCRYPT 2010. LNCS, vol. 6110, pp. 44–61. Springer, Heidelberg (2010)

15. Garg, S., Kumarasubramanian, A., Sahai, A., Waters, B.: Building efficient fully collusion-resilient traitor tracing and revocation schemes. In: ACM Conference on Computer and Communications Security, pp. 121–130 (2010)

16. Gentry, C.: Practical Identity-Based Encryption Without Random Oracles. In: Vaudenay, S. (ed.) EUROCRYPT 2006. LNCS, vol. 4004, pp. 445–464. Springer, Heidelberg (2006)

17. Gentry, C., Halevi, S.: Hierarchical Identity Based Encryption with Polynomially Many Levels. In: Reingold, O. (ed.) TCC 2009. LNCS, vol. 5444, pp. 437–456. Springer, Heidelberg (2009)

18. Gentry, C., Peikert, C., Vaikuntanathan, V.: Trapdoors for hard lattices and new cryptographic constructions. In: Proceedings of the 40th Annual ACM Symposium on Theory of Computing, pp. 197–206 (2008)

19. Gentry, C., Silverberg, A.: Hierarchical ID-Based Cryptography. In: Zheng, Y. (ed.) ASIACRYPT 2002. LNCS, vol. 2501, pp. 548–566. Springer, Heidelberg (2002)

20. Goyal, V., Pandey, O., Sahai, A., Waters, B.: Attribute based encryption for fine-grained access control of encrypted data. In: ACM Conference on Computer and Communications Security, pp. 89–98 (2006)

21. Groth, J., Ostrovsky, R., Sahai, A.: Non-interactive Zaps and New Techniques for NIZK. In: Dwork, C. (ed.) CRYPTO 2006. LNCS, vol. 4117, pp. 97–111. Springer, Heidelberg (2006)

22. Groth, J., Ostrovsky, R., Sahai, A.: Perfect Non-interactive Zero Knowledge for NP. In: Vaudenay, S. (ed.) EUROCRYPT 2006. LNCS, vol. 4004, pp. 339–358. Springer, Heidelberg (2006)

23. Horwitz, J., Lynn, B.: Toward Hierarchical Identity-Based Encryption. In: Knudsen, L.R. (ed.) EUROCRYPT 2002. LNCS, vol. 2332, pp. 466–481. Springer, Heidelberg (2002)

24. Katz, J., Sahai, A., Waters, B.: Predicate Encryption Supporting Disjunctions, Polynomial Equations, and Inner Products. In: Smart, N.P. (ed.) EUROCRYPT 2008. LNCS, vol. 4965, pp. 146–162. Springer, Heidelberg (2008)

25. Lewko, A., Okamoto, T., Sahai, A., Takashima, K., Waters, B.: Fully Secure Functional Encryption: Attribute-Based Encryption and (Hierarchical) Inner Product Encryption. In: Gilbert, H. (ed.) EUROCRYPT 2010. LNCS, vol. 6110, pp. 62–91. Springer, Heidelberg (2010)

26. Lewko, A., Rouselakis, Y., Waters, B.: Achieving Leakage Resilience through Dual System Encryption. In: Ishai, Y. (ed.) TCC 2011. LNCS, vol. 6597, pp. 70–88. Springer, Heidelberg (2011)

27. Lewko, A., Waters, B.: New Techniques for Dual System Encryption and Fully Secure HIBE with Short Ciphertexts. In: Micciancio, D. (ed.) TCC 2010. LNCS, vol. 5978, pp. 455–479. Springer, Heidelberg (2010)

28. Lewko, A., Waters, B.: Decentralizing Attribute-Based Encryption. In: Paterson, K.G. (ed.) EUROCRYPT 2011. LNCS, vol. 6632, pp. 568–588. Springer, Heidelberg (2011)

29. Lewko, A., Waters, B.: Unbounded HIBE and Attribute-Based Encryption. In: Paterson, K.G. (ed.) EUROCRYPT 2011. LNCS, vol. 6632, pp. 547–567. Springer, Heidelberg (2011)

30. Meiklejohn, S., Shacham, H., Freeman, D.M.: Limitations on Transformations from Composite-Order to Prime-Order Groups: The Case of Round-Optimal Blind Signatures. In: Abe, M. (ed.) ASIACRYPT 2010. LNCS, vol. 6477, pp. 519–538. Springer, Heidelberg (2010)

31. Okamoto, T., Takashima, K.: Homomorphic Encryption and Signatures from Vector Decomposition. In: Galbraith, S.D., Paterson, K.G. (eds.) Pairing 2008. LNCS, vol. 5209, pp. 57–74. Springer, Heidelberg (2008)

32. Okamoto, T., Takashima, K.: Hierarchical Predicate Encryption for Inner-Products. In: Matsui, M. (ed.) ASIACRYPT 2009. LNCS, vol. 5912, pp. 214–231. Springer, Heidelberg (2009)

33. Okamoto, T., Takashima, K.: Fully Secure Functional Encryption with General Relations from the Decisional Linear Assumption. In: Rabin, T. (ed.) CRYPTO 2010. LNCS, vol. 6223, pp. 191–208. Springer, Heidelberg (2010)

34. Ostrovksy, R., Sahai, A., Waters, B.: Attribute based encryption with non-monotonic access structures. In: ACM Conference on Computer and Communications Security, pp. 195–203 (2007)

35. Sahai, A., Waters, B.: Fuzzy Identity-Based Encryption. In: Cramer, R. (ed.) EUROCRYPT 2005. LNCS, vol. 3494, pp. 457–473. Springer, Heidelberg (2005)

36. Shamir, A.: Identity-Based Cryptosystems and Signature Schemes. In: Blakely, G.R., Chaum, D. (eds.) CRYPTO 1984. LNCS, vol. 196, pp. 47–53. Springer, Heidelberg (1985)

37. Waters, B.: Efficient Identity-Based Encryption Without Random Oracles. In: Cramer, R. (ed.) EUROCRYPT 2005. LNCS, vol. 3494, pp. 114–127. Springer, Heidelberg (2005)

38. Waters, B.: Dual System Encryption: Realizing Fully Secure IBE and HIBE under Simple Assumptions. In: Halevi, S. (ed.) CRYPTO 2009. LNCS, vol. 5677, pp. 619–636. Springer, Heidelberg (2009)

39. Waters, B.: Ciphertext-Policy Attribute-Based Encryption: An Expressive, Efficient, and Provably Secure Realization. In: Catalano, D., Fazio, N., Gennaro, R., Nicolosi, A. (eds.) PKC 2011. LNCS, vol. 6571, pp. 53–70. Springer, Heidelberg (2011)

Minimalism in Cryptography: The Even-Mansour Scheme Revisited

Orr Dunkelman[1,2], Nathan Keller[2,3], and Adi Shamir[2]

[1] Computer Science Department
University of Haifa
Haifa 31905, Israel
orrd@cs.haifa.ac.il
[2] Faculty of Mathematics and Computer Science
Weizmann Institute of Science
P.O. Box 26, Rehovot 76100, Israel
{nathan.keller,adi.shamir}@weizmann.ac.il
[3] Department of Mathematics
Bar-Ilan University
Ramat Gan 52900, Israel

Abstract. In this paper we consider the following fundamental problem: What is the simplest possible construction of a block cipher which is provably secure in some formal sense? This problem motivated Even and Mansour to develop their scheme in 1991, but its exact security remained open for more than 20 years in the sense that the lower bound proof considered known plaintexts, whereas the best published attack (which was based on differential cryptanalysis) required chosen plaintexts. In this paper we solve this open problem by describing the new *Slidex attack* which matches the $T = \Omega(2^n/D)$ lower bound on the time T for any number of known plaintexts D. Once we obtain this tight bound, we can show that the original two-key Even-Mansour scheme is not minimal in the sense that it can be simplified into a single key scheme with half as many key bits which provides exactly the same security, and which can be argued to be the simplest conceivable provably secure block cipher. We then show that there can be no comparable lower bound on the memory requirements of such attacks, by developing a new memoryless attack which can be applied with the same time complexity but only in the special case of $D = 2^{n/2}$. In the last part of the paper we analyze the security of several other variants of the Even-Mansour scheme, showing that some of them provide the same level of security while in others the lower bound proof fails for very delicate reasons.

Keywords: Even-Mansour block cipher, whitening keys, minimalism, provable security, tight security bounds, slide attacks, slidex attack.

1 Introduction

A major theme in cryptographic research over the last thirty years was the analysis of minimal constructions. For example, many papers were published on

D. Pointcheval and T. Johansson (Eds.): EUROCRYPT 2012, LNCS 7237, pp. 336–354, 2012.

the minimal cryptographic assumptions which are necessary and sufficient in order to construct various types of secure primitives. Other examples analyzed the smallest number of rounds required to make Feistel structures with truly random functions secure, the smallest possible size of shares in various types of secret sharing schemes, and the simplest way to transform one primitive into another by using an appropriate mode of operation. Since the vague notion of conceptual simplicity only partially orders all the possible schemes, in many cases we have to consider minimal schemes (which are local minima that become insecure when we eliminate any one of their elements) rather than minimum schemes (which are global minima among all the possible constructions).

In the case of stream ciphers, one can argue that the simplest possible secure scheme is the one-time pad, since any encryption algorithm requires a secret key, and XORing is the simplest conceivable way to mix it with the plaintext bits. The question we address in this paper is its dual: What is the simplest possible construction of a block cipher which has a formal proof of security?

This problem was first addressed by Even and Mansour [8,9] in 1991. They were motivated by the DESX construction proposed by Ron Rivest in 1984 [15], in which he proposed to protect DES against exhaustive search attacks by XORing two independent prewhitening and postwhitening keys to the plaintext and ciphertext (respectively). The resultant scheme increased the key size from 56 to 184 bits without changing the definition of DES and with almost no additional complexity. The Even-Mansour scheme used such whitening keys but eliminated the keyed block cipher in the middle, replacing it with a fixed random permutation that everyone can share. The resultant scheme is extremely simple: To encrypt a plaintext, XOR it with one key, apply to it a publicly known permutation, and XOR the result with a second key.

To argue that the Even-Mansour scheme is minimal, its designers noted in [9] that eliminating either one of the two XORed keys makes it easy to invert the known effect of the permutation on the plaintext or ciphertext, and thus to recover the other key from a single known plaintext/ciphertext pair. Eliminating the permutation is also disastrous, since it makes the scheme completely linear. In fact, the two-key EM block cipher is not minimal in the sense that it can be further simplified into a single-key variant with half as many key bits which has exactly the same security.

To compare various variants of the Even-Mansour scheme, we need tight bounds on the exact level of security they provide. Unfortunately, all the bounds published so far are not tight in the sense that the lower bound allows known message attacks whereas the best known upper bounds require either chosen plaintexts or an extremely large number of known plaintexts.

One of the main tools used in previous attacks was the slide attack [3]. Originally, slide attacks were developed in order to break iterated cryptosystems with an arbitrarily large number of rounds by exploiting their self similarity under small shifts. The attack searched the given data for a slid pair of encryptions which have identical values along their common part (see Section 3.2 for formal definitions). For each candidate pair, the attack uses the two known plaintexts and two known ciphertexts to analyze the two short non-common parts in order to verify the assumption that the two encryptions are indeed a slid pair, and if so to derive some key material. A different variant of this attack, called *slide with a twist* [4], tries to find a slid pair consisting of one encryption and one decryption, which have identical values along their common parts (i.e., the attack considers both shifts and reversals of the encryption rounds). In both cases, the existence of slid pairs is a random event which is expected to have a sharp threshold: Regardless of whether we use known or chosen messages, we do not expect to find any slid pairs if we are given fewer than $2^{n/2}$ encryptions where n is the size of the internal state.[1] Consequently, we cannot apply the regular or twisted slide attack unless we are given a sufficiently large number of encryptions, even if we are willing to trade off the lower amount of data with higher time and space complexities.

In this paper we propose the *slidex attack*, which is a new extended version of the slide attack that can efficiently use any amount of given data, even when it is well below the $2^{n/2}$ threshold for the existence of slid pairs. Its main novelty is that we no longer require equality between the values along the common part, but only the existence of some known relationship between these values. By using this new attack, we can finally close the gap between the upper and lower bounds on the security of the Even-Mansour scheme.

To demonstrate the usefulness and versatility of the new slidex attack, we apply it to several additional schemes which are unrelated to Even-Mansour. In particular, we show how to break 20 rounds of GOST using 2^{33} known plaintexts in 2^{77} time. In the extended version of this paper [7] we show several additional attacks, such as how to use the complementation property of DES in order to attack it with a slide attack even when it is surrounded by Vaudenay's decorrelation modules.

The paper is organized as follows. In Section 2 we introduce the Even-Mansour scheme, describe its formal proof of security, and survey all the previously published attacks on the scheme. In Section 3 we describe the known types of slide attacks, and explain why they cannot efficiently exploit a small number of known plaintexts. We then introduce our new Slidex attack, and use it to develop a new upper bound for the security of the Even-Mansour scheme which matches the proven lower bound for any number of known plaintexts. In Section 4 we describe the single-key variant of the Even-Mansour scheme, which is strictly simpler but has the same level of provable security. In Section 5 we analyze the security of several other variants of the Even-Mansour scheme, demonstrating both the generality and the fragility of its formal proof of security. Another limitation of the proof technique is described in Section 6, where we show that no comparable lower bound on the memory complexity of our attacks can exist. Finally, in the Appendix we describe the *mirror slide* attack, which is a generalization of the slidex attack.

---

[1] We note that for specific block cipher structures, e.g., Feistel networks, a dedicated slide attack can require fewer than $2^{n/2}$ plaintexts. However, there is no such method that works for general structures.

2   The Even-Mansour Scheme

In this section we present the Even-Mansour (EM) scheme, review its security proof given in [9] and describe previous attacks on it presented in [5] and [4].

2.1   Definition of the EM Scheme and Its Notation

The Even-Mansour scheme is a block cipher which consists of a single publicly known permutation $\mathcal{F}$ over n-bit strings, preceded and followed by n-bit whitening keys K_1 and K_2, respectively, i.e.,

$$EM_{K_1,K_2}^{\mathcal{F}}(P) = \mathcal{F}(P \oplus K_1) \oplus K_2.$$

It is assumed that the adversary is allowed to perform two types of queries:

- Queries to a full encryption/decryption oracle, called an E-oracle, that computes either $E(P) = EM_{K_1,K_2}^{\mathcal{F}}(P)$ or $D(C) = (EM_{K_1,K_2}^{\mathcal{F}})^{-1}(C)$.
- Queries to an $\mathcal{F}$-oracle, that computes either $\mathcal{F}(x)$ or $\mathcal{F}^{-1}(y)$.

The designers of EM considered two types of attacks. In the first type, called *existential forgery attack*, the adversary tries to find a *new* pair (P, C) such that $E(P) = C$. The second type is the more standard security game, where the adversary tries to decrypt a message C, i.e., to find P for which $E(P) = C$.[2] The data complexity of an attack on the scheme is determined by the number D of queries to the E-oracle and their type (i.e., known/chosen/adaptively chosen etc.), and the time complexity of the attack is lower bounded by the number T of queries to the $\mathcal{F}$-oracle.[3] The success probability of an attack is the probability that the single guess it produces (either a pair (P, C) for the first type of attack, or a plaintext P for the second type) is correct.

2.2   The Lower Bound Security Proof

The main rigorously proven result in [9] was an upper bound of $O(DT/2^n)$ on the success probability of any cryptanalytic attack (of either type) on EM that uses at most D queries to the E-oracle and T queries to the $\mathcal{F}$-oracle. This result implies that in order to attack EM with a constant probability of success, we must have $DT = \Omega(2^n)$. Since this security proof is crucial for some of our results, we briefly describe its main steps.

---

[2] These security notions are significantly different than the indistinguishability notions of [12] which proved similar lower bounds on the inability of the adversary to distinguish the given instance of the cipher from a random permutation. Finding the actual keys not only allows distinguishing the construction from a random permutation, but also allows winning the two security games considered in [9].

[3] In concrete implementations, this oracle is usually replaced by some publicly known program which the attacker can run on its own. In this case the type of query (e.g., whether the inputs are adaptively chosen or not) can determine whether the attack can be parallelized on multiple processors, but we ignore such low level details in our analysis.

The proof requires several definitions. Consider a cryptanalytic attack on EM, and assume that at some stage of the attack, the adversary already performed s queries to the E-oracle and t queries to the $\mathcal{F}$-oracle, and obtained sets $\mathcal{S}$ and $\mathcal{T}$ of E-pairs and $\mathcal{F}$-pairs, respectively, i.e.,

$$\mathcal{D} = \{(P_i, C_i)\}_{i=1,\ldots,d}, \qquad \text{and} \qquad \mathcal{T} = \{(X_j, Y_j)\}_{j=1,\ldots,t}.$$

We say that the key K_1 is *bad* with respect to the sets of queries $\mathcal{D}$ and $\mathcal{T}$, if there exist i, j such that $P_i \oplus K_1 = X_j$. Otherwise, K_1 is *good* with respect to $\mathcal{D}, \mathcal{T}$. Intuitively, a good key is one whose feasibility can not be deduced from the available data, whereas a bad key is one whose feasibility has to be analyzed further (but not necessarily discarded). Similarly, K_2 is bad w.r.t. $\mathcal{D}, \mathcal{T}$ if there exist i, j such that $Y_j \oplus K_2 = C_i$, and K_2 is good otherwise. The key $K = (K_1, K_2)$ is *good* with respect to $\mathcal{D}, \mathcal{T}$ if both K_1 and K_2 are good. It is easy to show that the number of good keys w.r.t. $\mathcal{D}$ and $\mathcal{T}$ is at least $2^{2n} - 2st \cdot 2^n$. A pair $(K = (K_1, K_2), \mathcal{F})$ is *consistent* w.r.t. $\mathcal{D}$ and $\mathcal{T}$ if for any pair $(P_i, C_i) \in \mathcal{D}$ we have $C_i = K_2 \oplus \mathcal{F}(P_i \oplus K_1)$, and for any pair $(X_j, Y_j) \in \mathcal{T}$, we have $\mathcal{F}(X_j) = Y_j$.

The proof consists of two main steps.

1. The first step shows that all good keys are, in some sense, equally likely to be the correct key. Formally, if the probability over the keys and over the permutations is uniform, then for all $\mathcal{D}, \mathcal{T}$, the probability

$$\Pr_{K,\mathcal{F}} \left[K = k \middle| (K, \mathcal{F}) \text{ is consistent with } \mathcal{D}, \mathcal{T} \right]$$

is the same for any key $k \in \{0,1\}^{2n}$ that is good with respect to $\mathcal{D}, \mathcal{T}$. We present the proof of this step, since it will be crucial in the sequel. It follows from Bayes' formula that it suffices to prove that the probability

$$p = \Pr_{K,\mathcal{F}} \left[(K, \mathcal{F}) \text{ is consistent with } \mathcal{D}, \mathcal{T} \middle| K = k \right] \tag{1}$$

is the same for all good keys. Given a good key $k = (k_1, k_2)$, it is possible to transform the set $\mathcal{D}$ of E-pairs to an equivalent set $\mathcal{D}'$ of $\mathcal{F}$-pairs by transforming the E-pair (P_i, C_i) to the $\mathcal{F}$-pair $(P_i \oplus k_1, C_i \oplus k_2)$. Since the key k is good, the pairs in $\mathcal{D}'$ and $\mathcal{T}$ do not overlap, and hence p is simply the probability of consistency of a random permutation $\mathcal{F}$ with $d + t$ given distinct input/output pairs. This probability clearly does not depend on k, which proves the assertion.

2. The second step shows that the success probability of any attack is bounded by the sum of the probability that in some step of the attack, the right key becomes a bad key, and the probability that the adversary can successfully generate a "new" consistent E-pair (P, C) if the right key is still amongst the good keys. The first probability can be bounded by $4DT/(2^n - 2DT)$, and the second probability can be bounded by $1/(2^n - D - T)$. Hence, the total success probability of the attack is bounded by $O(DT/2^n)$. We omit the proof of this step since it is not used in the sequel.

We note that obtaining non-trivial information about the key (e.g., that the least significant bit of the K_1 is zero, or the value of $K_1 \oplus K_2$), is also covered by this proof. Hence, throughout the paper we treat such leakage of information as a "problem" in the security of the construction (even if the exact keys are not found).

Finally, we note that in [12] a slightly different model is considered. The analyzed construction is a one where besides the pre-/post-whitening keys, the internal permutation $\mathcal{F}$ is keyed with a k-bit key. For such a construction, Kilian and Rogaway prove that given D queries to the construction and time T evaluations of $\mathcal{F}$, one cannot succeed in distinguishing the construction from a random permutation with probability higher than $DT/2^{n+k-1}$. Obviously, when $k = 0$, i.e., the internal permutation is fixed, one can view this as a proof that indeed the Even-Mansour is indistinguishable from a random permutation with success rate over $DT/2^{n-1}$. Note that in this paper we consider the stronger notion of attack (namely, finding the actual keys) and the thus the results are not identical.

2.3   Previous Attacks on the Even-Mansour Scheme

The first proposed attack on the Even-Mansour scheme was published by Joan Daemen at Asiacrypt 1991 [5]. Daemen used the framework of differential cryptanalysis [2] to develop a *chosen plaintext* attack which matched the Even-Mansour lower bound for any amount of given data. The approach is to pick D pairs of chosen plaintexts whose XOR difference is some nonzero constant Δ. This plaintext difference is preserved by the XOR with the prewhitening key K_1, and similarly, the ciphertext difference is preserved by the XOR with the postwhitening key K_2. For a known permutation $\mathcal{F}$, most combinations of input and output differences suggest only a small number of possible input and output values, but it is not easy to find them. To carry out the attack, all we have to do is to sample $2^n/D$ pairs of inputs to $\mathcal{F}$ whose difference is Δ, and with constant non-negligible probability we can find an output difference which already exists among the chosen data pairs. This equality suggests actual input and output values to/from $\mathcal{F}$ for that pair, and thus recovers the two keys. We note that a similar chosen-plaintext attack was suggested in [12] for constructions where $\mathcal{F}$ is keyed (where $DT \geq 2^{n+k-1}$ for a k-bit keyed $\mathcal{F}$).

This attack matches the time/data relationship of the lower bound, but it is not tight since it requires chosen plaintexts, whereas the lower bound allows known plaintexts. This discrepancy was handled ten years later by a new attack called *slide with a twist* which was developed by Alex Biryukov and David Wagner, and presented at Eurocrypt 2000 [4]. By taking two Even-Mansour encryptions, sliding one of them and reversing the other, they showed how to attack the scheme with known instead of chosen plaintexts.[4] However, in order to find at least one slid pair, their attack requires at least $\Omega(2^{n/2})$ known plaintext/ciphertext pairs, and thus it could not be applied with a reasonable probability of success given any smaller number of known pairs.

---

[4] The slide with a twist attack on EM is described in detail in Section 3.1.

These two cryptanalytic attacks were thus complementary: One of them matched the full time/data tradeoff curve but required chosen plaintexts, while the other could use known plaintexts but only if at least $\Omega(2^{n/2})$ of them were given. In the next section we present the new slidex technique that closes this gap: it allows to use any number of known plaintexts with the same time/data tradeoff as in the lower bound proof, thus providing an optimal attack on the Even-Mansour scheme.

3   The Slidex Attack and a Tight Bound on the Security of the Even-Mansour Scheme

In this section we present the new Slidex attack and use it to obtain a tight bound on the security of the Even-Mansour scheme. We start with a description of the slide with a twist attack on EM [4] which serves as a basis for our attack, and then we present the slidex technique and apply it to EM. For more information on slide attacks, we refer the reader to [1,3,4].

3.1   The Slide with a Twist Attack

The main idea of the slide with a twist attack on EM is as follows. Assume that two plaintexts P, P^* satisfy

$$P \oplus P^* = K_1.$$

In such a case, we have

$$E(P) = \mathcal{F}(P \oplus K_1) \oplus K_2 = \mathcal{F}(P^*) \oplus K_2,$$

and similarly,

$$E(P^*) = \mathcal{F}(P^* \oplus K_1) \oplus K_2 = \mathcal{F}(P) \oplus K_2$$

(see Figure 1(a)). Hence,

$$E(P) \oplus E(P^*) = \mathcal{F}(P) \oplus \mathcal{F}(P^*),$$

or equivalently,

$$E(P) \oplus \mathcal{F}(P) = E(P^*) \oplus \mathcal{F}(P^*).$$

This relation allows to mount the following attack:

1. Query both the E-oracle and the $\mathcal{F}$-oracle at the same $2^{(n+1)/2}$ known values $P_1, P_2, \ldots$.[5] Store in a hash table the pairs $(E(P_i) \oplus \mathcal{F}(P_i)), i)$, sorted by the first coordinate.
2. For each collision in the table, i.e., $E(P_i) \oplus \mathcal{F}(P_i) = E(P_j) \oplus \mathcal{F}(P_j)$, check the guess $K_1 = P_i \oplus P_j$ and $K_2 = E(P_i) \oplus \mathcal{F}(P_j)$.

---

[5] Formally, the adversary obtains known plaintext/ciphertext pairs $(P_i, E(P_i))$ and queries the $\mathcal{F}$-oracle at the value P_i.

By the birthday paradox, it is expected that the data set contains a slid pair, i.e., a pair satisfying $P_i \oplus P_j = K_1$, with a non-negligible constant probability. For a random pair (P_i, P_j), the probability that $E(P_i) \oplus \mathcal{F}(P_i) = E(P_j) \oplus \mathcal{F}(P_j)$ is 2^{-n}, and thus, only a few collisions are expected in the table. These collisions include the collision induced by the slid pair, which suggests the correct values of K_1 and K_2. The data complexity of the attack is $D = 2^{(n+1)/2}$ known plaintexts, and the number of queries to $\mathcal{F}$ it requires is $T = 2^{(n+1)/2}$. Thus, $DT = 2^{n+1}$, which matches the lower bound up to a constant factor of 2.

3.2   The New Slidex Attack

The *slidex* attack is an enhancement of the slide with a twist technique, which makes it possible to use a smaller number of known plaintexts (i.e., queries to the E-oracle), in exchange for a higher number of queries to the $\mathcal{F}$-oracle. The basic idea of the attack is as follows: Assume that a pair of plaintexts P, P^* satisfies

$$P \oplus P^* = K_1 \oplus \Delta,$$

for some $\Delta \in \{0,1\}^n$. In such a case,

$$E(P) = \mathcal{F}(P \oplus K_1) \oplus K_2 = \mathcal{F}(P^* \oplus \Delta) \oplus K_2,$$

and similarly,

$$E(P^*) = \mathcal{F}(P^* \oplus K_1) \oplus K_2 = \mathcal{F}(P \oplus \Delta) \oplus K_2$$

(see Figure 1(b)). Hence,

$$E(P) \oplus E(P^*) = \mathcal{F}(P^* \oplus \Delta) \oplus \mathcal{F}(P \oplus \Delta),$$

or equivalently,

$$E(P) \oplus \mathcal{F}(P \oplus \Delta) = E(P^*) \oplus \mathcal{F}(P^* \oplus \Delta).$$

This allows to mount the following attack, for any $d \leq n$:

1. Query the E-oracle at $2^{(d+1)/2}$ arbitrary values (i.e., known plaintexts) $P_1, P_2, \ldots$.
2. Choose 2^{n-d} arbitrary values $\Delta_1, \Delta_2, \ldots$ of Δ. For each Δ_ℓ, query the $\mathcal{F}$-oracle at the values $\{P_i \oplus \Delta_\ell\}_{i=1,2,\ldots,2^{(d+1)/2}}$, store in a hash table the pairs $(E(P_i) \oplus \mathcal{F}(P_i \oplus \Delta_\ell)), i)$, sorted by the first coordinate, and search for a collision.
3. For each collision in any of the hash tables, i.e., when P_i, P_j for which $E(P_i) \oplus \mathcal{F}(P_i \oplus \Delta_\ell) = E(P_j) \oplus \mathcal{F}(P_j \oplus \Delta_\ell)$ are detected, check the guess $K_1 = P_i \oplus P_j \oplus \Delta_\ell$ and $K_2 = E(P_i) \oplus \mathcal{F}(P_j \oplus \Delta_\ell)$.

For each triplet (P_i, P_j, Δ_ℓ), the probability that $P_i \oplus P_j \oplus \Delta_\ell = K_1$ is 2^{-n}. Since the data contains $2^d \cdot 2^{n-d} = 2^n$ such triplets, it is expected that with a non-negligible constant probability the data contains at least one *slidex triplet*

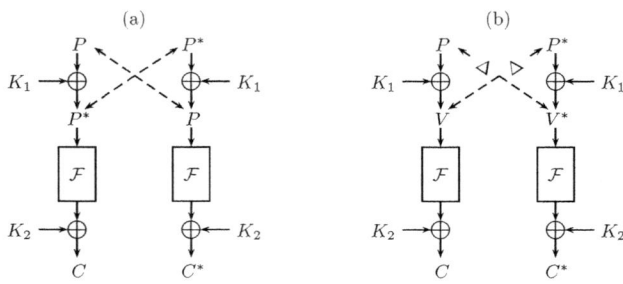

Fig. 1. (a) A twisted-slid pair; (b) A slidex pair

Table 1. Comparison of Results on the Even-Mansour scheme

Known Plaintext Attacks				
Attack	Data	Time	Memory	Tradeoff
Guess and determine [9]	2	2^n	2	—
Slide with a twist [4]	$2^{n/2}$	$2^{n/2}$	$2^{n/2}$	—
Slidex (Sect. 3.2)	D	T	D	$DT = 2^n$
Chosen Plaintext Attacks				
Attack	Data	Time	Memory	Tradeoff
Differential [5]	D	T	D	$DT = 2^n$
Adaptive Chosen Plaintext Attacks				
Attack	Data	Time	Memory	Tradeoff
Slide (Sect. 6)	D	T	1	$DT = 2^n, D \geq 2^{n/2}$

(i.e., a triplet for which $P_i \oplus P_j \oplus \Delta_\ell = K_1$). On the other hand, since the probability of a collision in each hash table is 2^{d-n} and there are 2^{n-d} tables, it is expected that only a few collisions occur, and one of them suggests the correct key guess.

The number of queries to the E-oracle in the attack is $D = 2^{(d+1)/2}$, and the number of queries to the $\mathcal{F}$-oracle is $T = 2^{n-(d-1)/2}$. Thus, $DT = 2^{n+1}$, which matches the lower bound of [9] up to a constant factor of 2.

A summary of the complexities of all the old and new attacks on the Even-Mansour scheme appears in Table 1.

4   The Single-Key Even-Mansour Scheme

In this section we analyze the single-key variant of the Even-Mansour scheme (abbreviated in the sequel as "SEM"), which has the same level of security while using only n secret key bits (compared to $2n$ bits in EM).[6] First, we define the

---

[6] Kurosawa uses such SEMs in his constructions [13], where in each block the pre-/post-whitening keys are changed.

scheme and show that the security proof of [9] can be adapted to yield a similar lower bound on its security. Then, we present a simple attack on the new scheme which matches the lower bound, thus proving its optimality.

4.1 Definition of the Scheme and Its Security Proof

Given a publicly known permutation $\mathcal{F}$ over n-bit strings and an n-bit secret key K, the Single-Key Even-Mansour (SEM) scheme is defined as follows:

$$SEM_K^{\mathcal{F}}(P) = \mathcal{F}(P \oplus K) \oplus K.$$

The attack model is the same as in the EM scheme. That is, the adversary can query an encryption/decryption E-oracle and an $\mathcal{F}$-oracle, and the complexity of an attack is determined by the number D of queries to the E-oracle and their type (known/chosen etc.), and the number T of queries to the $\mathcal{F}$-oracle.

Surprisingly, the security proof of the EM scheme [9] holds almost without a change when we apply it to the single-key SEM variant. The only modification we have to make is to define a key K as *bad* with respect to sets of oracle queries S and T if there exist i, j such that either $P_i \oplus K = X_j$ or $C_i \oplus K = Y_j$, and K as good otherwise. It is easy to see that if $|S| = s$ and $|T| = t$, then at least $2^n - 2st$ keys are still "good" keys. Exactly the same proof as for EM shows that all the good keys are equally likely to be the right key, and the bounds on the success probability of an attack apply without change for SEM.[7]

Therefore, for any successful attack on SEM, we must have $DT = \Omega(2^n)$, which means that SEM provides the same security as EM, using only half as many key bits.

4.2 A Simple Optimal Attack on SEM

The slidex attack presented in Section 3 applies also to SEM, and is optimal since it uses only known plaintexts and matches everywhere the tradeoff curve of the security proof.

However, in the case of SEM, there is an even simpler attack (though, with the same complexity). Consider an encryption of a plaintext P through SEM, and denote the intermediate values in the encryption process by:

$$x = P, \qquad y = P \oplus K, \qquad z = \mathcal{F}(P \oplus K), \qquad w = E(P) = \mathcal{F}(P \oplus K) \oplus K.$$

Note that $x \oplus w = y \oplus z$. This allows to mount the following simple attack, applicable for any $D \leq 2^n$:

1. Query the E-oracle at D arbitrary values $P_1, P_2, \ldots, P_D$ and store in a hash table the values $(P_i \oplus E(P_i), i)$, sorted by the first coordinate.

---

[7] We note that the indistinguishability of this construction was also studied in [12], and it was shown that also the indistinguishability of SEM is the same as regular EM.

2. Query the $\mathcal{F}$-oracle at $2^n/D$ arbitrary values $X_1, X_2, \ldots, X_{2^n/D}$, insert the values $X_j \oplus \mathcal{F}(X_j)$ to the hash table and search for a match.
3. If a match is found, i.e., $P_i \oplus E(P_i) = X_j \oplus \mathcal{F}(X_j)$, check the guess $K = P_i \oplus X_j$.

The analysis of the attack is exactly the same as that of the slide with a twist attack (see Section 3.1).

5   The Security of Other Variants of the Even-Mansour Scheme

In this section we consider two natural variants of the Even-Mansour scheme, and analyze their security.

The first variant replaces the XOR operations with modular additions, which are not involutions and are thus immune to standard slide-type attacks. However, we show that a new *addition slidex* attack can break it with the same complexity as that of the slidex attack on the original EM scheme.

The second variant considers the case in which the mapping $\mathcal{F}$ is chosen as an involution. This is motivated by the fact that in many "real-life" implementations of the EM scheme we would like to instantiate $\mathcal{F}$ by a keyless variant of a block cipher. Since in Feistel structures and many other schemes (e.g., KHAZAD, Anubis, Noekeon) the only difference between the encryption and decryption processes is the key schedule, such schemes become involutions when we make them keyless. In this section we show that this seemingly mild weakness of $\mathcal{F}$ can be used to mount a devastating attack on the EM scheme. In particular, we show that even when $\mathcal{F}$ is chosen uniformly at random among the set of all the possible involutions on n-bit strings, the adversary can recover the value $K_1 \oplus K_2$ with $O(2^{n/2})$ queries to the E-oracle and no queries at all (!) to the $\mathcal{F}$-oracle. This clearly violates the lower bound proof that no significant information about the key can be obtained unless $DT = \Omega(2^n)$ (which was proven for random permutations but seems to be equally applicable to random involutions), and is achieved by a new variant of the slide attack, which we call the *mirror slide* attack.

5.1   Even-Mansour with Addition

Consider the following scheme:

$$AEM_{K_1,K_2}^{\mathcal{F}}(P) = \mathcal{F}(P + K_1) + K_2,$$

where $\mathcal{F}$ is a publicly known permutation over n-bit strings, and '+' denotes modular addition in the additive group Z_{2^n}. In the sequel, we call it "Addition Even-Mansour" (AEM).

It is clear that the lower bound security proof of EM holds without any change for AEM. Similarly, it is easy to see that Daemen's differential attack on EM [5] can be easily adapted to AEM, by replacing XOR differences with modular differences.

It may seem that the new variant has better security with respect to slide-type attacks. As noted in [4], ordinary slide attacks can be applied only for ciphers in which the secret key is inserted through a *symmetric* operation such as XOR, and not through modular addition. In the specific case of EM, the slide with a twist attack relies on the observation that if for two plaintexts P, P^*, we have $P^* = P \oplus K_1$, then surely, $P = P^* \oplus K_1$ as well. This observation fails for AEM: If $P^* = P + K_1$, then $P^* + K_1 = P + 2K_1 \neq P$ (unless $K_1 = 0$ or $K = 2^{n-1}$). The slidex attack presented in Section 3.2 fails against AEM for the same reason. Hence, it seems that none of the previously known attacks can break AEM in the *known plaintext* model.

We present an extension of the slidex attack, which we call *addition slidex*, which can break AEM with data complexity of D known plaintexts and time complexity of T $\mathcal{F}$-oracle queries, for any D, T such that $DT = 2^n$, hence showing that the security of AEM is identical to that of EM.

The basic idea of the attack is as follows: Assume that a pair of plaintexts P, P^* satisfies $P + P^* = -K_1 + \Delta$. (Note that somewhat counter intuitive, we consider the modular sum of the plaintexts rather than their modular difference!). In such a case,

$$E(P) = \mathcal{F}(P + K_1) + K_2 = \mathcal{F}(-P^* + \Delta) + K_2,$$

and similarly,

$$E(P^*) = \mathcal{F}(P^* + K_1) + K_2 = \mathcal{F}(-P + \Delta) + K_2.$$

Hence,

$$E(P) - E(P^*) = \mathcal{F}(-P^* + \Delta) - \mathcal{F}(-P + \Delta),$$

or equivalently,

$$E(P) + \mathcal{F}(-P + \Delta) = E(P^*) + \mathcal{F}(-P^* + \Delta). \tag{2}$$

Equation (2) allows us to mount an attack similar to the slidex attack, with the only change that instead of the values $(E(P_i) \oplus \mathcal{F}(P_i \oplus \Delta)), i)$, the adversary stores in the hash table the values $(E(P_i) + \mathcal{F}(-P_i + \Delta)), i)$.

We note that actually, the slidex attack can be considered as a special case of the addition slidex attack, since the addition slidex attack clearly applies to modular addition in any group, and the XOR operation corresponds to addition in the group Z_2.

5.2 Even-Mansour with a Random Involution as the Permutation

Let Involutional Even-Mansour (IEM) be the following scheme:

$$IEM^{\mathcal{I}}_{K_1, K_2}(P) = \mathcal{I}(P \oplus K_1) \oplus K_2,$$

where $\mathcal{I}$ is chosen uniformly at random amongst the set of involutions on n-bit strings. We present a new technique, which we call *mirror slide*, that allows to recover the value $K_1 \oplus K_2$ using $2^{n/2}$ queries to the E-oracle, and with no queries to the $\mathcal{I}$-oracle.

The idea of the technique is as follows. Consider two input/output pairs $(P, C), (P^*, C^*)$ for IEM. Assume that we have

$$P \oplus C^* = K_1 \oplus K_2. \tag{3}$$

In such case,

$$P \oplus K_1 = C^* \oplus K_2,$$

and hence, since $\mathcal{I}$ is an involution,

$$\mathcal{I}(P \oplus K_1) = \mathcal{I}^{-1}(C^* \oplus K_2).$$

However, by the construction we have

$$C = \mathcal{I}(P \oplus K_1) \oplus K_2, \qquad \text{and} \qquad P^* = \mathcal{I}^{-1}(C^* \oplus K_2) \oplus K_1,$$

and thus,

$$C \oplus K_2 = P^* \oplus K_1,$$

or equivalently,

$$P^* \oplus C = K_1 \oplus K_2 = P \oplus C^*,$$

where the last equality follows from Equation (3). Therefore, assuming that $P \oplus C^* = K_1 \oplus K_2$, we must have:

$$P \oplus C = P^* \oplus C^*.$$

This allows to mount a simple attack, similar to the slide with a twist attack. In the attack, the adversary queries the E-oracle at $2^{(n+1)/2}$ arbitrary values $P_1, P_2, \ldots$, and stores in a hash table the pairs $(E(P_i) \oplus P_i, i)$, sorted by the first coordinate. It is expected that only a few collisions exist, and that with a non-negligible probability, one of them results from a pair (P_i, P_j), for which $P_i \oplus E(P_j) = K_1 \oplus K_2$.

Therefore, the attack supplies the adversary with only a few possible values of $K_1 \oplus K_2$, after performing $2^{(n+1)/2}$ queries to the E-oracle and no queries at all to the $\mathcal{I}$-oracle. As we show later, the adversary cannot obtain K_1 or K_2 themselves (without additional effort or data), but at the same time, the adversary does learn a nontrivial information about the key, which contradicts the security proof of the original EM scheme.

We note that this is an example for the gap between the indistinguishability security notion and the cost of finding a key. Obviously, when $K_1 = K_2$ is known (or when $K_1 \oplus K_2$ is known), one can easily distinguish the single-key involution Even-Mansour (ISEM) from a random permutation using two adaptive queries with extremely high probability. At the same time, the lower bounds of the Even-Mansour security proof assure us that it is impossible to decrypt a ciphertext C encrypted by single-key involution Even-Mansour without first obtaining $DT = O(2^n)$ (similar result holds with respect to the existential forgery attack of producing another valid plaintext/ciphertext pair).

Where the Security Proof Fails. One may wonder, which part of the formal security proof fails when $\mathcal{F}$ is an involution. It turns out that the only part that fails is the argument in the first step of the proof showing that all good keys are equally likely to be the right key. Recall that in order to show this, one has to show that the probability

$$p = \Pr_{K,\mathcal{F}}[(K,\mathcal{F}) \text{ is consistent with } \mathcal{D}, \mathcal{T} | K = k]$$

is the same for all good keys. In the case of EM, p is shown to be the probability of consistence of a random permutation $\mathcal{F}$ with $d+t$ given distinct input/output pairs, which indeed does not depend on k (since such pairs are independent). In the case of IEM, the input/output pairs may be dependent, since it may occur that an encryption query to the E-oracle results in querying $\mathcal{I}$ at some value x, while a decryption query to the E-oracle results in querying $\mathcal{I}^{-1}$ at the same value x. Since $\mathcal{I}$ is an involution, these queries are not independent and thus, the probability p depends on whether such dependency has occurred, and this event does depend on k. An examination of the mirror slide attack shows that this property is exactly the one exploited by the attack.

It is interesting to note that in the single-key case (i.e., for SEM where $\mathcal{F}$ is an involution, which we denote by ISEM), such event cannot occur, as in order to query $\mathcal{I}$ and $\mathcal{I}^{-1}$ at the same value, one must query E and E^{-1} at the same value. Since in the single-key case, the entire construction is an involution, such two queries result in the same answer for any value of the secret key, and hence, do not create dependence on the key. It can be shown, indeed, that the security proof does hold for ISEM and yields the same security bound, thus showing that in the case of involutions, the single-key variant is even stronger than the original two-key variant! Moreover, it can be noticed that in the case of EM, after the adversary recovers the value $K_1 \oplus K_2$, the encryption scheme becomes equivalent to a single-key Even-Mansour scheme with the key K_1, i.e., $E'(P) = \mathcal{I}(P \oplus K_1) \oplus K_1$. Thus, using two different keys in this case is totally obsolete, and also creates a security flaw which can be deployed by an adversary if the keys K_1 and K_2 are used also in other systems.

5.3   Addition Even-Mansour with an Involution as the Permutation

In this subsection we consider a combination of the two variants discussed in the previous subsections, i.e., AEM where $\mathcal{F}$ is a random involution. We abbreviate this variant as AIEM.

It can be easily shown that the mirror slide attack can be adapted to the case of AIEM, by modifying the assumption to $C^* - P = K_1 + K_2$, and the conclusion to $P + C = P^* + C^*$. The attack allows to recover the value $K_1 + K_2$, and then the scheme becomes equivalent to a *conjugation* EM scheme with a single key: $CISEM(P) = \mathcal{I}(P + K_1) - K_1$, and it can be shown that the security proof of EM applies also to CISEM. Thus, the security of AIEM under the assumption that $\mathcal{F}$ is an involution is identical to that of the original EM.

An interesting phenomenon is that in the involution case, the security of single-key AEM (which we denote by AISEM) is much worse than that of AIEM. Indeed, the mirror slide attack allows to recover $K_1 + K_1 = 2K_1$, and hence to find K_1 (up to the value of the MSB), which breaks the scheme completely. This suggests that in the case of addition, the "natural" variant of single-key AEM is the conjugation variant, i.e., $CSEM(P) = \mathcal{F}(P + K_1) - K_1$, for which the security proof of EM indeed applies even if $\mathcal{F}$ is an involution, as mentioned above.

In the extended version of this paper, available at [7], we consider all 12 variants of Even-Mansour (single key/two keys, random permutation/random involution, and whether the keys are XORed, added, or conjugated).

6   Memoryless Attacks on the Even-Mansour Scheme

All previous papers on the Even-Mansour scheme, including the lower bounds proved by the designers [9], Daemen's attack [5], and Biryukov-Wagner's slide attack [4], considered only the data and time complexities of attacks, but not the memory complexity. Analysis of the previously proposed attacks shows that in all of them, the memory complexity is $\min\{D, T\}$, where D is the data complexity (i.e., the number of E-queries) and T is the time complexity (i.e., the number of $\mathcal{F}$-queries). Thus, it is natural to ask whether the memory complexity can also be inserted into the lower bound security proofs, e.g., in the form $M \geq \min(D, T)$.

In this section we show that such a general lower bound can not exist, by constructing an attack with the particular data and time complexities of $O(2^{n/2})$, and with only a constant memory complexity. The attack is a memoryless variant of the slide with a twist attack described in Section 3.1. Recall that the main step of the slide with a twist attack is to find collisions of the form $E(P) \oplus \mathcal{F}(P) = E(P^*) \oplus \mathcal{F}(P^*)$.

We observe that such collisions can be found in a memoryless manner. We treat the function

$$\mathcal{G} : P \to E(P) \oplus \mathcal{F}(P)$$

as a random function, and apply Floyd's cycle finding algorithm [10] (or any of its variants, such as Nivasch's algorithm [14]) to find a collision in $\mathcal{G}$. The attack algorithm is as follows:

1. Query the E-oracle at a sequence of $O(2^{n/2})$ adaptively chosen values $P_1, P_2, \ldots$, such that P_1 is arbitrary and for $k > 1$, $P_k = E(P_{k-1}) \oplus \mathcal{F}(P_{k-1})$. (Here, after each query to the E-oracle, the adversary queries the $\mathcal{F}$-oracle at the same value and uses its answer in choosing the next query to the E-oracle).
2. Use Floyd's cycle finding algorithm to find P_i, P_j such that $E(P_i) \oplus \mathcal{F}(P_i) = E(P_j) \oplus \mathcal{F}(P_j)$.
3. For each colliding pair, check the guess $K_1 = P_i \oplus P_j$ and $K_2 = E(P_i) \oplus \mathcal{F}(P_j)$.

The analysis of the attack is identical to the analysis of the slide with a twist attack. The memory complexity is negligible, and the data and time complexities remain $O(2^{n/2})$. As the attack algorithm succeeds once a pair P_i, P_j satisfying $E(P_i) \oplus \mathcal{F}(P_i) = E(P_j) \oplus \mathcal{F}(P_j)$ is found, the expected number of queries is determined by the random function G's graph. The analysis of graphs induced by random functions such as G shows that the expected number of queries in the tail (the steps until entering the cycle) is $\pi m/8$ and the length of the cycle itself is $\pi m/8$ [14]. We note that these incur a small overhead in terms of query complexity (up to a factor of 5 in the case of Floyd's algorithm or 2 in the case Nivasch's cycle finding algorithm of [14] is used in exchange for a logarithmic memory). The only downside of this algorithm is the fact that the queries to the E-oracle are chosen adaptively, whereas in the slide with a twist attack we could choose arbitrary queries to the E-oracle.

7   Open Problems

If the amount of available E-oracle queries is smaller than $2^{n/2}$, the adversary can still apply the slidex attack described in Section 3.2, but there seems to be no way to convert it into a memoryless attack by using the strategy described above. The main obstacle is that the adversary has to reuse the data many times in order to construct the hash tables for different values of Δ, which can be done only if the data is stored somewhere rather than used in an on-line manner which discards it after computing the next plaintext. This leads to the following open problem:

Problem 1. Does there exist a memoryless attack on the Even-Mansour scheme with D E-oracle queries and $2^n/D$ $\mathcal{F}$-oracle queries, where $D \ll 2^{n/2}$?

A similar question can be asked with respect to the Single-Key Even-Mansour scheme, where in addition to the slidex attack, the simple attack presented in Section 4.2 can also break the scheme when $D \ll 2^{n/2}$. The attack of Section 4.2 can also be transformed to a memoryless attack, by defining a random function:

$$\mathcal{H}(X) = \begin{cases} X \oplus E(X), & LSB(X) = 1 \\ X \oplus \mathcal{F}(X), & LSB(X) = 0, \end{cases}$$

and using Floyd's cycle finding algorithm to find a collision of $\mathcal{H}$. In the case when D and T are both close to $2^{n/2}$, with a constant probability such collision yields a pair (X_1, X_2) such that $X_1 \oplus E(X_1) = X_2 \oplus \mathcal{F}(X_2)$, concluding the attack. The problem is that if $D \ll 2^{n/2}$, then with overwhelming probability, a collision in $\mathcal{H}$ is of the form $X_1 \oplus \mathcal{F}(X_1) = X_2 \oplus \mathcal{F}(X_2)$, which is not useful to the adversary. Therefore, we state an additional open problem:

Problem 2. Does there exist a memoryless attack on the Single-Key Even-Mansour scheme with D E-oracle queries and $2^n/D$ $\mathcal{F}$-oracle queries, where $D \ll 2^{n/2}$?

If such memoryless attack can be found only for Single-Key EM and not for the ordinary EM, this will show that at least in some respect, the use of an additional key in EM does make the scheme stronger.

References

1. Biham, E., Dunkelman, O., Keller, N.: Improved Slide Attacks. In: Biryukov, A. (ed.) FSE 2007. LNCS, vol. 4593, pp. 153–166. Springer, Heidelberg (2007)
2. Biham, E., Shamir, A.: Differential Cryptanalysis of the Data Encryption Standard. Springer (1993)
3. Biryukov, A., Wagner, D.: Slide Attacks. In: Knudsen, L.R. (ed.) FSE 1999. LNCS, vol. 1636, pp. 245–259. Springer, Heidelberg (1999)
4. Biryukov, A., Wagner, D.: Advanced Slide Attacks. In: Preneel, B. (ed.) EUROCRYPT 2000. LNCS, vol. 1807, pp. 589–606. Springer, Heidelberg (2000)
5. Daemen, J.: Limitations of the Even-Mansour Construction. In: [11], pp. 495–498
6. Dinur, I., Dunkelman, O., Shamir, A.: Improved Attacks on GOST. Technical report, to appear (2011)
7. Dunkelman, O., Keller, N., Shamir, A.: Minimalism in Cryptography: The Even-Mansour Scheme Revisited. Cryptology ePrint Archive, Report 2011/541 (2011), http://eprint.iacr.org/
8. Even, S., Mansour, Y.: A Construction of a Cipher From a Single Pseudorandom Permutation. In: [11], pp. 210–224
9. Even, S., Mansour, Y.: A Construction of a Cipher from a Single Pseudorandom Permutation. J. Cryptology 10(3), 151–162 (1997)
10. Floyd, R.W.: Nondeterministic Algorithms. J. ACM 14(4), 636–644 (1967)
11. Imai, H., Rivest, R.L., Matsumoto, T. (eds.): ASIACRYPT 1991. LNCS, vol. 739. Springer, Heidelberg (1993)
12. Kilian, J., Rogaway, P.: How to Protect DES Against Exhaustive Key Search (an Analysis of DESX). J. Cryptology 14(1), 17–35 (2001)
13. Kurosawa, K.: Power of a Public Random Permutation and Its Application to Authenticated Encryption. IEEE Transactions on Information Theory 56(10), 5366–5374 (2010)
14. Nivasch, G.: Cycle Detection Using a Stack. Inf. Process. Lett. 90(3), 135–140 (2004)
15. Rivest, R.L.: DESX. Never published (1984)
16. Russian National Bureau of Standards: Federal Information Processing Standard-Cryptographic Protection - Cryptographic Algorithm. GOST 28147-89 (1989)

A   The Mirror Slide Attack

In this section we present the general framework of the *mirror slide* attack, that was presented in Section 5.2 in the special case of the Even-Mansour scheme. We show that the mirror slide attack generalizes the *slide with a twist* attack [4]. We apply the new technique to a 20-round variant of the block cipher GOST [16], other variants of the attack are considered in the extended version of the paper [7].

A.1   The General Framework

The mirror slide attack applies to block ciphers that can be decomposed as a cascade of three sub-ciphers: $E = E_2 \circ E_1 \circ E_0$, where the middle layer E_1 is an involution, i.e., $E_1 = (E_1)^{-1}$.[8]

Let E be such a cipher, and assume that for two plaintext/ciphertext pairs $(P, C), (P^*, C^*)$, we have

$$E_0(P) = E_2^{-1}(C^*). \tag{4}$$

In such case, since E_1 is an involution,

$$E_1(E_0(P)) = E_1^{-1}(E_2^{-1}(C^*)).$$

By the construction, this implies:

$$E_2^{-1}(C) = E_1(E_0(P)) = E_1^{-1}(E_2^{-1}(C^*)) = E_0(P^*). \tag{5}$$

If Equation (4) holds (and thus, Equation (5) also holds, the pair (P, P^*) is called a *mirror slid pair*.

The way to exploit mirror slid pairs in a cryptanalytic attack is similar to standard slide-type attacks [3,4]: The adversary asks for the encryption of $2^{(n+1)/2}$ known plaintexts $P_1, P_2, \ldots$ (where n is the block size of E) and denotes the corresponding ciphertexts by $C_1, C_2, \ldots$. For each pair (P_i, P_j), the adversary assumes that it is a mirror slid pair and tries to solve the system of equations:

$$\begin{cases} C_j = E_2(E_0(P_i)), \\ C_i = E_2(E_0(P_j)) \end{cases}$$

(which is equivalent to Equations (4) and (5)). If E_0 and E_2 are "simple enough", the adversary can solve the system efficiently and recover the key material used in E_0 and E_2.

If the amount of subkey material used in E_0 and E_2 is at most n bits (in total), it is expected that at most a few of the systems of equations generated by the 2^n plaintext pairs are consistent (since the equation system is a $2n$-bit condition). One of them is the system generated by the mirror slid pair, which is expected to exist in the data with a constant probability since the probability of a random pair to be a mirror slid pair is 2^{-n}. Hence, the adversary obtains only a few suggestions for the key, which contain the right key with a constant probability. If the amount of key material used in E_0 and E_2 is bigger than n bits, the adversary can still find the right key, by enlarging the data set by a small factor and using key ranking techniques (exploiting the fact that the right key is suggested by all mirror slid pairs, while the other pairs suggest "random" keys).

The data complexity of the attack is $O(2^{n/2})$ known plaintexts, and its time complexity is $O(2^n)$ (assuming that the system of equations can be solved within constant time).

---

[8] We note that the attack can be applied also if E_1 has some other symmetry properties, as shown in the extended version of the paper.

We note that the attack can be applied even when E_0 and E_2 are not "simple" ciphers using a meet-in-the-middle attack. If both E_0 and E_2 use $\kappa \leq n$ key bits at most, one can try and find the solutions to the above set of equations in time $\min\{O(2^{n+\kappa}), O(2^{n/2+2\kappa})\}$.[9]

A.2   The Slide with a Twist Attack and an Application to 20-Round GOST

The first special case of the mirror slide framework we consider is where in the subdivision of E, we have $E_2 = Identity$. In such case, the system of equations presented above is simplified to:

$$\begin{cases} C_j = E_0(P_i), \\ C_i = E_0(P_j). \end{cases} \qquad (6)$$

It turns out that in this case, the attack is reduced exactly to the slide with a twist attack presented in [4]! (Though, in [4] the attack is described in a different way).

A concrete example of this case is a reduced-round variant of the block cipher GOST [16], that consists of the last 20 of its 32 rounds. It is well-known that the last 16 rounds of GOST compose an involution, and hence, this variant can be represented as $E = E_1 \circ E_0$, where E_0 is 4-round GOST, and E_1 (which is the last 16 rounds of GOST) is an involution.[10] As shown in [6], a 4-round variant of GOST can be broken with two plaintext/ciphertext pairs and time complexity of 2^{12} encryptions. Therefore, the mirror slide attack can break this 20-round variant of GOST with data complexity of 2^{33} known plaintexts (since the block size of GOST is 64 bits), and time complexity of $2^{65} \cdot 2^{12} = 2^{77}$ encryptions.

We note that a similar attack was described in [4] using the slide with a twist technique, but only on a 20-round version of a modified variant of GOST called GOST$\oplus$ in which the key addition is replaced by XOR.

---

[9] One can either take all plaintext/ciphertext pairs and partially encrypt the plaintext under all 2^κ keys for E_0 and partially decrypt the ciphertext under all 2^κ keys for E_2 to find the mirror pairs. Another option is to try for each pair of plaintexts (P_i, P_j) to solve the system

$$\begin{cases} E_2^{-1}(C_j) = E_0(P_i), \\ E_2^{-1}(C_i) = E_0(P_j) \end{cases}$$

which can be easily done in a meet-in-the-middle approach in time 2^κ for each (P_i, P_j).

[10] We note that due to the Feistel structure of GOST, we do not have $E_1 \circ E_1 = Id$, but rather $E_1 \circ swap \circ E_1 = Id$. This can be handled easily by inserting swap to the left hand side of Equation (6). The same correction can be performed in the other Feistel constructions discussed in the sequel.

Message Authentication, Revisited

Yevgeniy Dodis[1], Eike Kiltz[2,*], Krzysztof Pietrzak[2,**], and Daniel Wichs[4]

[1] New York University
[2] Ruhr-Universität Bochum
[3] IST Austria
[4] IBM T.J. Watson Research Center

Abstract. Traditionally, symmetric-key message authentication codes (MACs) are easily built from pseudorandom functions (PRFs). In this work we propose a wide variety of other approaches to building efficient MACs, without going through a PRF first. In particular, unlike deterministic PRF-based MACs, where each message has a unique valid tag, we give a number of *probabilistic* MAC constructions from various other primitives/assumptions. Our main results are summarized as follows:

- We show several new probabilistic MAC constructions from a variety of general assumptions, including CCA-secure encryption, Hash Proof Systems and key-homomorphic weak PRFs. By instantiating these frameworks under concrete number theoretic assumptions, we get several schemes which are more efficient than just using a state-of-the-art PRF instantiation under the corresponding assumption.
- For probabilistic MACs, unlike deterministic ones, unforgeability against a chosen message attack (uf-cma) alone does not imply security if the adversary can additionally make verification queries (uf-cmva). We give an *efficient* generic transformation from any uf-cma secure MAC which is "message-hiding" into a uf-cmva secure MAC. This resolves the main open problem of Kiltz et al. from Eurocrypt'11; By using our transformation on their constructions, we get the first efficient MACs from the LPN assumption.
- While all our new MAC constructions immediately give efficient actively secure, two-round symmetric-key identification schemes, we also show a very simple, three-round actively secure identification protocol from *any weak PRF*. In particular, the resulting protocol is much more efficient than the trivial approach of building a regular PRF from a weak PRF.

1 Introduction

Message Authentication Codes (MACs) are one of the most fundamental primitives in cryptography. Historically, a vast majority of MAC constructions are

---

* Supported by a Sofja Kovalevskaja Award of the Alexander von Humboldt Foundation, funded by the German Federal Ministry for Education and Research.
** Supported by the European Research Council/ERC Starting Grant 259668-PSPC.

D. Pointcheval and T. Johansson (Eds.): EUROCRYPT 2012, LNCS 7237, pp. 355–374, 2012.

based on pseudorandom functions (PRFs).[1] In particular, since a PRF with large output domain is also a MAC, most research on symmetric-key authentication concentrated on designing and improving various PRF constructions. This is done either using very fast heuristic constructions, such as block-cipher based PRFs (e.g., CBC-MAC [6,8] or HMAC [5,4]), or using elegant, but slower number-theoretic constructions, such as the Naor-Reingold (NR) PRF [33]. The former have the speed advantage, but cannot be reduced to simple number-theoretic hardness assumptions (such as the DDH assumption for NR-PRF), and are not friendly to efficient zero-knowledge proofs about authenticated messages and/or their tags, which are needed in some important applications, such as compact e-cash [12]. On the other hand, the latter are comparably inefficient, due to their reliance on number theory. Somewhat surprisingly, the inefficiency of existing number-theoretic PRFs goes beyond what one would expect by the mere fact that "symmetric-key" operations are replaced by the more expensive "public-key" operations. For example, when building a PRF based on discrete-log-type of assumptions, such as DDH, one would naturally expect that the secret key would contain a constant number of group elements/exponents, and the PRF evaluation should cost at most a constant number of exponentiations. In contrast, state-of-the art discrete-log-type PRFs either require a key of quadratic size in the security parameter (e.g. the NR PRF [33]), or a number of exponentiations linear in the security parameter (e.g., tree-type PRFs based on the GGM transform [20] applied to some discrete-log-type pseudorandom generator), or are based on exotic and relatively untested assumptions (e.g., Dodis-Yampolskiy PRF [17] based on the so called "q-DDHI" assumption). In particular, to the best of our knowledge, prior to this work it was unknown how to build a MAC (let alone a PRF) based on the classical DDH assumption, where the secret key consists of a constant number of group elements / exponents and the MAC evaluation only require a constant number of exponentiations.

Of course, one way to improve such deficiencies of existing "algebraic MACs" would be to improve the corresponding "algebraic PRF" constructions. However, as the starting point of our work, we observe that there might exist alternative approaches to building efficient MACs, *without going through a PRF first*. For example, MACs only need to be unpredictable, so we might be able to build efficient MACs from *computational assumptions* (e.g., CDH rather than DDH), without expensive transformations from unpredictability-to-pseudorandomness [34]. Alternatively, even when relying on decisional assumptions (e.g. DDH), MAC constructions are allowed to be *probabilistic*. In contrast, building a PRF effectively forces one to design a MAC where there is only one valid tag for each message, which turns out to be a serious limitation for algebraic constructions.[2]

---

[1] Or block ciphers, which, for the purposes of analysis, are anyway treated as length-preserving PRFs.

[2] The observation that probabilistic MAC might have advantages over the folklore "PRF-is-a-MAC" paradigm is not new, and goes back to at least Wegman and Carter [40], and several other follow-up works (e.g., [30,25,16]). However, most prior probabilistic MACs were still explicitly based on a PRF or a block cipher.

For example, it is instructive to look at the corresponding "public-key domain" of digital signatures, where forcing the scheme to have a unique valid signature appears to be very hard [32,11] and, yet, not necessary for most applications of digital signatures. In particular, prominent digital signature schemes in the standard model[3] [11,39] are all probabilistic. In fact, such signature schemes trivially give MACs. Of course, such MACs are not necessarily as efficient as they could be, since they "unnecessarily" support public verification.[4] However, the point is that such trivial signature-based constructions already give a way to build relatively efficient "algebraic MACs" *without building an "algebraic PRF"* first.

Yet another motivation to building probabilistic MAC comes from the desire of building efficient MACs (and, more generally, symmetric-key authentication protocols) from the *Learning Parity with Noise* [24,26,28,29] (LPN) assumption. This very simple assumption states that one cannot recover a random vector x from any polynomial number of noisy parities $(a, \langle a, x \rangle + e)$, where a is a random vector and e is small *random* noise, and typically leads to very simple and efficient schemes [19,2,38,24,26,28,29]. However, the critical dependence on random errors makes it very hard to design deterministic primitives, such as PRFs, from the LPN assumption. Interestingly, this ambitious challenge was very recently overcome for a more complicated *Learning With Errors* (LWE) assumption by [3], who build a PRF based on a new (but natural) variant of the LWE assumption. However, the resulting PRF has the same deficiencies (e.g., large secret key) as the NR-PRF, and is *much* less efficient than the direct probabilistic MAC constructions from LPN/LWE assumptions recently obtained by [29].

1.1   Our Results

Motivated by the above considerations, in this work we initiate a systematic study of different methods for building efficient probabilistic MACs from a variety assumptions, both general and specific, without going through the PRF route. Our results can be summarized as follows:

Dealing with Verification Queries and other Transformations. The desired notion of security for probabilistic MACs is called "unforgeability against chosen message and verification attack" uf-cmva, where an attacker can arbitrarily interleave tagging queries (also called signing queries) and verification queries. For deterministic MACs, where every message corresponds to exactly one possible tag, this notion is equivalent to just considering a weaker notion called uf-cma (unforgeability under chosen message attack) where the attacker can only make

---

[3] In fact, even in the random oracle model there are noticeable advantages. E.g., full domain hash (FDH) signatures [9] have worse exact security than *probabilistic* FDH signatures, while Fiat-Shamir signatures [18] are inherently probabilistic.

[4] Indeed, one of our results, described shortly, will be about "optimizing" such signature-based constructions.

tagging queries but *no* verification queries. This is because, in the deterministic case, the answers to verification queries are completely predictable to an attacker: for any message for which a tagging query was already made the attacker knows the unique tag on which the verification oracle will answer affirmatively, and for any new message finding such a tag would be equivalent to breaking security without the help of the verification oracle. Unfortunately, as discussed by [7], the situation is more complicated for the case of probabilistic MACs where the attacker might potentially get additional information by modifying a valid tag of some message and seeing if this modified tag is still valid for the same message. In fact, some important MAC constructions, such as the already mentioned "basic" LPN-based construction of [29], suffer from such attacks and are only uf-cma, but not uf-cmva secure.

In Section 3 we give several general transformations for probabilistic MACs. The most important one, illustrated in Figure 1, *efficiently* turns a uf-cma secure (i.e. unforgeable without verification queries) MAC which is "message hiding" (a property we call ind-cma) into a uf-cmva secure (i.e. unforgeable with verification queries) MAC. This transformation is very efficient, requiring just a small amount of extra randomness and one invocation of a pairwise independent hash function with fairly short output.

This transformation solves the main open problem left in Kiltz et al. [29], who construct uf-cmva MACs from the learning parity with noise (LPN) problem. We remark that [29] already implicitly give an uf-cma to uf-cmva transformation, but it is quite inefficient, requiring the evaluation of a pairwise-independent *permutation* over the entire tag of a uf-cma secure MAC. We list the two constructions of uf-cma and suf-cma LPN based MACs from [29] in Section 4.5. Using our transformations, we get uf-cmva secure MACs with basically the same efficiency as these constructions.

Our second transformation extends the domain of an ind-cma secure MAC. A well known technique to extend the domain of PRFs is the "hash then encrypt" approach where one applies an almost universal hash function to the (long) input before applying the PRF. This approach fails for MACs, but we show that it works if the MAC is ind-cma secure. A similar observation has been already made by Bellare [4] for "privacy preserving" MACs.

The last transformation, which actually does nothing except possibly restricting the message domain, states that a MAC which is only selectively secure is also fully secure, albeit with quite a large loss in security. Such a transformation was already proposed in the context of identity based encryption [10], and used implicitly in the construction of LPN based MACs in [29].

New Constructions of Probabilistic MACs. In Section 4, we present a wide variety of new MAC constructions.

First, we show how to build an efficient MAC from any chosen ciphertext attack (CCA) secure (symmetric- or public-key) encryption. At first glance, using CCA-secure encryption seems like a complete "overkill" for building MACs. In fact, in the symmetric-key setting most CCA-secure encryption schemes are actually built *from MACs*; e.g., via the encrypt-then-MAC paradigm. However, if we are interested in obtaining number-theoretic/algebraic MACs using this approach, we would start with *public-key* CCA-secure encryption, such as Cramer-Shoup encryption [15] or many of the subsequent schemes (e.g. [31,22,23,37,21]). Quite remarkably, CCA-secure encryption has received so much attention lately, and the state-of-the-art constructions are so optimized by now, that the MACs resulting from our simple transformation appear to be *better*, at least in certain criteria, than the existing PRF constructions from the same assumptions. For example, by using any state-of-the-art DDH-based scheme, such as those by [15,31,22], we immediately obtain a probabilistic DDH-based MAC where both the secret key and the tag are of constant size, and the tagging/verification each take a constant number of exponentiations. As we mentioned, no such DDH-based MAC was known prior to our work. In fact, several recent constructions built efficient CCA-secure encryption schemes from *computational assumptions*, such as CDH and factoring [13,23,21]. Although those schemes are less efficient than the corresponding schemes based on decisional assumptions, they appear to be more efficient than (or at least comparable with) the best known PRF constructions from the same assumption. For example, the best factoring-based PRF of [35] has a quadratic-size secret key, while our construction based on the Hofheinz-Kiltz [23] CCA-encryption from factoring would have a linear-size (constant number of group elements) secret key.

Second, we give an efficient MAC construction from any *Hash Proof Systems* (HPS) [15]. Hash Proof Systems were originally defined [15] for the purpose of building CCA-secure public-key encryption schemes, but have found many other applications since. Here we continue this trend and give a direct MAC construction from HPS, which is more optimized than building a CCA-secure encryption from HPS, and then applying our prior transformation above.

Third, we give a simple construction of probabilistic MACs from any *key-homomorphic weak PRF* (hwPRF). Recall, a weak PRF [33] is a weakening of a regular PRF, where the attacker can only see the PRF value at random points. This weakening might result in much more efficient instantiations for a variety of number-theoretic assumptions. For example, under the DDH assumption, the basic modulo exponentiation $f_k(m) = m^k$ is already a weak PRF, while the regular NR-PRF from DDH is much less efficient. We say that such a weak PRF $f_k(m)$ is *key homomorphic* (over appropriate algebraic domain and range) if $f_{ak_1+bk_2}(m) = a \cdot f_{k_1}(m) + b \cdot f_{k_2}(m)$. (For example, the DDH-based weak PRF above clearly has this property.) We actually give two probabilistic MACs from any hwPRF. Our basic MAC is very simple and efficient, but only achieves so called *selective* security, meaning that the attacker has to commit to the message to be forged before the start of the attack. It is somewhat reminiscent (in terms of its design and proof technique, but not in any formal way) to the Boneh-Boyen selectively-secure signature scheme [11]. In contrast, our second construction borrows the ideas from (fully secure) Waters signature scheme [39], and builds a less efficient standard MAC from any hwPRF. Interestingly, both constructions are only uf-cma secure, but do not appear to be uf-cmva-secure. Luckily, our MACs are easily seen to be "message-hiding" (i.e., ind-cma-secure), so we can apply our efficient generic transformation to argue full uf-cmva security for both resulting constructions.

Our final MAC constructions are from signature schemes. Recall, any signature scheme trivially gives a MAC which "unnecessarily" supports public verification. This suggests that such constructions might be subject to significant optimizations when "downgraded" into a MAC, both in terms of efficiency and the underlying security assumption. Indeed, we show that this is true for the (selectively-secure) Boneh-Boyen [11] signature scheme, and the (fully-secure) Waters [39] signature schemes. For example, as signatures, both schemes require a bilinear group with a pairing, and are based on the CDH assumption in such a group. We make a simple observation that when public verification is no longer required, no pairing computations are needed, and standard (non-bilinear) groups can be used. However, in doing so we can only prove (selective or full) security under the *gap-Diffie-Hellman assumption*, which states that CDH is still hard even given the DDH oracle. Luckily, we show how to apply the "twinning" technique of Cash et al. [13] to get efficient MAC variants of both schemes which can be proven secure under the standard CDH assumption.

Symmetric-Key Authentication Protocols. While all our new MAC constructions immediately give efficient actively secure, two-round symmetric-key identification schemes, in Section 4.6 we also show a very simple, three-round actively secure identification protocol from *any* weak PRF (wPRF). In particular, the resulting protocol is much more efficient than the trivial approach of building a regular PRF from a weak PRF [33], and then doing the standard PRF-based authentication. Given that all our prior MAC constructions required some algebraic structure (which was indeed one of our motivations), we find a general (and very efficient) construction of actively secure authentication protocols from any wPRF to be very interesting.

Our protocol could be viewed as an abstraction of the LPN-based actively secure authentication protocol of Katz and Shin [27], which in turn consists of a parallel repetition of the HB$^+$ protocol of Juels and Weiss [26]. Although the LPN based setting introduces some complications due to handling of the errors, the high level of our protocol and the security proof abstracts away the corresponding proofs from [27,26]. In fact, we could relax the notion of wPRF slightly to allow for probabilistic computation with approximate correctness, so that the protocol of [27] will become a special case of our wPRF-based protocol.

2   Definitions

2.1   Notation

We denote the set of integers modulo an integer $q \geq 1$ by $\mathbb{Z}_q$. For a positive integer k, $[k]$ denotes the set $\{1, \ldots, k\}$; $[0]$ is the empty set. For a set $\mathcal{X}$, $x \leftarrow_R \mathcal{X}$ denotes sampling x from $\mathcal{X}$ according to the uniform distribution.

2.2   Message Authentication Codes

A message authentication code MAC = {KG, TAG, VRFY} is a triple of algorithms with associated key space $\mathcal{K}$, message space $\mathcal{M}$, and tag space $\mathcal{T}$.

- Key Generation. The probabilistic key-generation algorithm $k \leftarrow \mathsf{KG}(1^\lambda)$ takes as input a security parameter $\lambda \in \mathbb{N}$ (in unary) and outputs a secret key $k \in \mathcal{K}$.
- Tagging. The probabilistic authentication algorithm $\sigma \leftarrow \mathsf{TAG}_k(m)$ takes as input a secret key $k \in \mathcal{K}$ and a message $m \in \mathcal{M}$ and outputs an authentication tag $\sigma \in \mathcal{T}$.
- Verification. The deterministic verification algorithm $\mathsf{VRFY}_k(m, \sigma)$ takes as input a secret key $k \in \mathcal{K}$, a message $m \in \mathcal{M}$ and a tag $\sigma \in \mathcal{T}$ and outputs a decision: {accept, reject}.

If the TAG algorithm is deterministic one does not have to explicitly define VRFY, since it is already defined by the TAG algorithm as $\mathsf{VRFY}_k(m, \sigma)$ = accept iff $\mathsf{TAG}_k(m) = \sigma$. We say that MAC has completeness error α if for all $m \in \mathcal{M}$ and $\lambda \in \mathbb{N}$,

$$\Pr[\mathsf{VRFY}_k(m, \sigma) = \mathsf{reject} \; ; k \leftarrow \mathsf{KG}(1^\lambda) \; , \; \sigma \leftarrow \mathsf{TAG}_k(m)] \leq \alpha.$$

SECURITY. The standard security notion for a *randomized* MAC is unforgeability under chosen message and chosen verification queries attack (uf-cmva). We denote by $\mathsf{Adv}_{\mathsf{MAC}}^{\mathsf{uf\text{-}cmva}}(\mathsf{A}, \lambda, Q_T, Q_V)$, the advantage of the adversary A in forging a message for a random key $k \leftarrow \mathsf{KG}(1^\lambda)$, where A can make Q_T queries to $\mathsf{TAG}_k(\cdot)$ and Q_V queries to $\mathsf{VRFY}_k(\cdot, \cdot)$. Formally this is the probability that the following experiment outputs 1.

Experiment $\mathbf{Exp}_{\mathsf{MAC}}^{\mathsf{uf\text{-}cmva}}(\mathsf{A}, \lambda, Q_T, Q_V)$
  $k \leftarrow \mathsf{KG}(1^\lambda)$
  Invoke $\mathsf{A}^{\mathsf{TAG}_k(\cdot), \mathsf{VRFY}_k(\cdot, \cdot)}$ who can make up to Q_T queries to $\mathsf{TAG}_k(\cdot)$
      and Q_V queries to $\mathsf{VRFY}_k(\cdot, \cdot)$.
  Output 1 if A made a query (m^*, σ^*) to $\mathsf{VRFY}_k(\cdot, \cdot)$ where
      1. $\mathsf{VRFY}_k(m^*, \sigma)$ = accept
      2. A did not already make the query m^* to $\mathsf{TAG}_k(\cdot)$
  Output 0 otherwise.

We also define a weaker notion of *selective security*, captured by the experiment $\mathbf{Exp}_{\mathsf{MAC}}^{\mathsf{suf\text{-}cmva}}$, which is defined in the same way as above with the only difference that A has to specify to the target message m^* (that causes the experiment to output 1) ahead of time, before making any queries to its oracles.

Definition 1 ((Selective) unforgeability under chosen message (& verification) attack.). *A MAC is $(t, Q_T, Q_V, \varepsilon)$-uf-cmva secure if for any A running in time t we have $\Pr[\mathbf{Exp}_{\mathsf{MAC}}^{\mathsf{uf\text{-}cmva}}(\mathsf{A}, \lambda, Q_T, Q_V) = 1] \leq \varepsilon$. It is (t, Q_T, ε)-uf-cma secure if it is $(t, Q_T, 1, \varepsilon)$-uf-cmva-secure. That is, uf-cma security does not allow the adversary to make any verification queries except for the one forgery attempt. We also define the selective security notions suf-cma and suf-cmva security analogously by considering the experiment $\mathbf{Exp}^{\mathsf{suf\text{-}cmva}}(\mathsf{MAC})$.*

In the next section we show a simple generic transformation which turns any uf-cma-secure MAC into a uf-cmva-secure $\overline{\mathsf{MAC}}$. For this transformation to work, we need one extra non-standard property for MAC to hold, namely that tags computationally "hide" the message. A similar notion called "privacy preserving MACs" was considered by Bellare [4]. His notion is for deterministic MACs, whereas our notion can only be achieved for probabilistic MACs.

Definition 2 (ind-cma: indistinguishability under chosen message attack). *A* MAC *is* (t, Q_T, ε)-ind-cma *secure if no adversary* A *running in time* t *can distinguish tags for chosen messages from tags for a fixed message, say* 0, *i.e.*

$$\left| \Pr_{k \leftarrow \mathsf{KG}(1^\lambda)}[\mathsf{A}^{\mathsf{TAG}_k(\cdot)}(1^\lambda) = 1] - \Pr_{k \leftarrow \mathsf{KG}(1^\lambda)}[\mathsf{A}^{\mathsf{TAG}_k(0)}(1^\lambda) = 1] \right| \leq \varepsilon .$$

Here $\mathsf{TAG}_k(0)$ is an oracle which ignores its input, and outputs a tag for some fixed message 0 using key K. Note that a MAC that is secure against ind-cma adversaries must be probabilistic, otherwise A can trivially distinguish by queries on two different messages $m \neq m'$, and checking if the tags she receives are identical, which will be the case iff the oracle implements $\mathsf{TAG}_k(0)$.

3   Transformations for MACs

In this section we give some general transformations for MACs as discussed in the introduction.

3.1   From One to Multiple Verification Queries: uf-cma + ind-cma ⇒ uf-cmva

Let $\mu = \mu(\lambda)$ denote a statistical security parameter and let $\mathcal{H}$ be a family of pairwise independent hash functions $h : \mathcal{T} \to \{0,1\}^\mu$. From MAC = {KG, TAG, VRFY} with key space $\mathcal{K}$, message space $\mathcal{M} \times \{0,1\}^\mu$, and tag space $\mathcal{T}$ we construct $\overline{\mathsf{MAC}} = \{\overline{\mathsf{KG}}, \overline{\mathsf{TAG}}, \overline{\mathsf{VRFY}}\}$ with key space $\mathcal{K} \times \mathcal{H}$, message space $\mathcal{M}$, and tag space $\mathcal{T} \times \{0,1\}^\mu$ as follows.

- Key Generation. Algorithm $\overline{\mathsf{KG}}(1^\lambda)$ runs $k \leftarrow \mathsf{KG}(1^\lambda)$ and samples a pairwise independent hash function $h \leftarrow \mathcal{H}$ with $h : \mathcal{T} \to \{0,1\}^\mu$. It outputs (k,h) as the secret key.

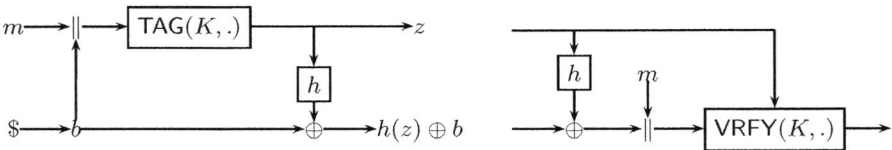

Fig. 1. $\overline{\mathsf{TAG}}$ and $\overline{\mathsf{VRFY}}$ with key (k,h), message m and randomness b

- Tagging. The tagging algorithm $\overline{\mathsf{TAG}}_{(k,h)}(m)$ samples $b \leftarrow_R \{0,1\}^\mu$ and runs $z \leftarrow \mathsf{TAG}_k(m\|b)$. It returns $(z, h(z) \oplus b)$ as the tag.
- Verification. The verification algorithm $\overline{\mathsf{VRFY}}_{(k,h)}(m, (z, y))$ computes $b = y \oplus h(z)$ and outputs $\mathsf{VRFY}_k(m\|b, z)$.

Theorem 1 (uf-cma + ind-cma $\Rightarrow$ uf-cmva). *For any $t, Q_T, Q_V \in \mathbb{N}$, $\varepsilon > 0$, if MAC is*

- (t, Q_T, ε)-uf-cma *secure (unforgeable with no verification queries)*
- (t, Q_T, ε)-ind-cma *secure (indistinguishable)*

then $\overline{\mathsf{MAC}}$ *is* $(t', Q_T, Q_V, \varepsilon')$-uf-cmva *secure (unforgeable with verification queries) where*

$$t' \approx t \qquad \varepsilon' = 2Q_V \varepsilon + 2Q_V Q_T / 2^\mu.$$

The proof of Theorem 1 can be found in the full version of this paper.

3.2   Domain Extension for ind-cma MACs

A simple way to extend the domain of a pseudorandom function from n to $m > n$ bits is the "hash then encrypt" paradigm, where one first hashes the m bit input down to n bits using an ϵ-universal function, before applying the PRF. Unfortunately this simple trick does not work for (deterministic or probabilistic) MACs. Informally, the reason is that the output of a MAC does not "hide" its input, and thus an adversary can potentially learn the key of the hash function used (once she knows the key, she can find collisions for g which allows to break the MAC.) Below we show that, not surprisingly, for MACs where we explicitly require that they hide their input, as captured by the ind-cma notion, extending the domain using a universal hash function is safe.

Proposition 1 (Domain Extension for ind-cma Secure MACs). *Consider* MAC = $\{\mathsf{KG}, \mathsf{TAG}, \mathsf{VRFY}\}$ *with (small) message space* $\mathcal{M} = \{0,1\}^n$, *and let* MAC$'$ = $\{\mathsf{KG}', \mathsf{TAG}', \mathsf{VRFY}'\}$ *for large message space* $\{0,1\}^m$ *be derived from* MAC *by first hashing the message using an β-universal hash function $g : \{0,1\}^\ell \times \{0,1\}^m \to \{0,1\}^n$. (Using existing constructions we can set $\beta = 2^{-n+1}$, $\ell = 4(n + \log m)$, see the full version of the paper for details.) If* MAC *is*

$$(t, Q, \varepsilon) - \mathsf{uf\text{-}cma} \quad secure \ and \ (t, Q, \varepsilon) - \mathsf{ind\text{-}cma} \quad secure$$

then, for any $Q' \leq Q$, MAC$'$ *is*

(1) $(t', Q', 2\varepsilon + Q'\beta) - \mathsf{uf\text{-}cma}$ secure and (2) $(t', Q', \varepsilon) - \mathsf{ind\text{-}cma}$ secure

where $t' \approx t$ can be derived from the proof.

The proof of Proposition 1 can be found in the full version of this paper.

3.3   From Selective to Full Security: suf-cma $\Rightarrow$ uf-cma

In this section we make the simple observation, that every *selectively* chosen-message secure MAC is also a chosen-message secure MAC, as we can simply guess the forgery. This guessing will loose a factor 2^μ in security if the domain is $\{0,1\}^\mu$.

Proposition 2 (From selective to full security). *Consider a MAC* MAC = {KG, TAG, VRFY} *with domain* $\{0,1\}^\mu$. *If* MAC *is* $(t, Q, \varepsilon) -$ suf-cma *secure, then it is* $(t, Q, \varepsilon 2^\mu) -$ uf-cma *secure.*

The proof of Proposition 2 can be found in the full version of this paper.

Remark 1 (Security Loss and Domain Extension). The security loss from the above transformation is 2^μ for MACs with message space $\{0,1\}^\mu$. In order to keep the security loss small, we are better off if we start with a MAC that has a small domain, or if we artificially restrict its domain to the first μ bits. Once we get a fully secure MAC on a small domain, we can always apply the domain-extension trick from Section 3.2 (using $\beta = 2^{-\mu+1}$) to expand this domain back up. Using both transformations together, we can turn any MAC that is (t, Q, ε)-suf-cma and ind-cma secure into a (t', Q', ε')-uf-cma and (t', Q', ε)-ind-cma secure MAC with the same-size (or arbitrarily larger) domain and where $t' \approx t$, and ε' depends on our arbitrary choice of μ as $\varepsilon' = \varepsilon 2^{\mu+1} + Q'/2^{\mu-1}$. In particular, if for some super-polynomial t, Q we assume a known corresponding negligible value ε such that the original MAC is (t, Q, ε)-suf-cma, we can set $\mu = \log(1/\varepsilon)/2$ and the resulting MAC will be secure in the standard asymptotic sense - i.e. (t', Q', ϵ')-uf-cma for all polynomial $t', Q', 1/\epsilon'$.

4   Constructions of Authentication Protocols

In this section we provide a number of MACs from a variety of underlying primitives such as CCA-secure encryption, hash proof systems [15], homomorphic weak PRFs, and digital signatures. For concreteness, the constructions obtained from Diffie-Hellman type assumptions are summarized in Table 1; the constructions we obtain from the LPN assumption are summarized in Table 2. The constructions which are only uf-cma or suf-cma secure can be boosted to full cmva-security using the transformations from Section 3.

Table 1. Overview of MAC constructions over prime-order groups. In all protocols, $\mathsf{TAG}_k(m)$ first generates $U \leftarrow_R \mathbb{G}$ and derives the rest of σ deterministically from U and k.

MAC construction	Secret Key k	Tag σ on m	Security	Assumption
MAC$_{\mathsf{CS}}$ (§4.1)	$(\omega, x, x', y, k_2) \in \mathbb{Z}_p^4 \times \mathbb{G}$	$(U, U^\omega, U^{xH(U,V_1,m)+x'}, U^z \cdot k_2) \in \mathbb{G}^4$	uf-cmva	DDH
MAC$_{\mathsf{HPS}}$ (§4.2)	$(\omega, x, x') \in \mathbb{Z}_p^3$	$(U, U^\omega, U^{xH(U,V_1,m)+x'}) \in \mathbb{G}^3$	uf-cmva	DDH
MAC$_{\mathsf{hwPRF}}$ (§4.3)	$(x, x') \in \mathbb{Z}_p^2$	$(U, U^{xm+x'}) \in \mathbb{G}^2$	suf-cma	DDH
MAC$_{\mathsf{WhwPRF}}$ (§4.3)	$(x, x'_0, \ldots, x'_\lambda) \in \mathbb{Z}_p^{\lambda+2}$	$(U, U^{x+\sum x'_i m_i}) \in \mathbb{G}^2$	uf-cma	DDH
MAC$_{\mathsf{BB}}$ (§4.4)	$(x, x', y) \in \mathbb{Z}_p^3$	$(U, g^{xy} \cdot U^{xm+x'}) \in \mathbb{G}^2$	suf-cmva	gap-CDH
MAC$_{\mathsf{TBB}}$ (§4.4)	$(x_1, x_2, x'_1, x'_2, y) \in \mathbb{Z}_p^5$	$(U, g^{x_1 y} U^{x_1 m + x'_1}, g^{x_2 y} U^{x_2 m + x'_2}) \in \mathbb{G}^3$	suf-cmva	CDH
MAC$_{\mathsf{Waters}}$ (§4.4)	$(x, y, x'_1, \ldots, x'_\lambda) \in \mathbb{Z}_p^{\lambda+2}$	$(U, g^{xy} \cdot U^{x+\sum x'_i m_i}) \in \mathbb{G}^2$	uf-cmva	gap-CDH

Table 2. Overview of MAC constructions from the LPN problem from [29]

MAC construction	Key size	Tag size	Security	Assumption
$\mathsf{MAC}_{\mathsf{LPN}}$ (§4.5)	$\mathbb{Z}_2^{2\ell}$	$\mathbb{Z}_2^{(\ell+1)\times n}$	suf-cma & ind-cma	LPN
$\mathsf{MAC}_{\mathsf{BilinLPN}}$ (§4.5)	$\mathbb{Z}_2^{\ell\times\lambda}$	$\mathbb{Z}_2^{(\ell+1)\times n}$	uf-cma & ind-cma	LPN

4.1 Constructions from CCA-Secure Encryption

Let $\mathsf{E} = (\mathsf{KG_E}, \mathsf{ENC}, \mathsf{DEC})$ be a (t, Q_E, Q_D, ϵ)-CCA secure labeled encryption scheme (see the full version of the paper for a formal definition.) Define $\mathsf{MAC} = (\mathsf{KG_{MAC}}, \mathsf{TAG}, \mathsf{VRFY})$ as follows.

- <u>Key Generation.</u> $k = (k_1, k_2) \leftarrow \mathsf{KG_{MAC}}(1^\lambda)$ samples $k_1 \leftarrow \mathsf{KG_E}(1^\lambda)$ and $k_2 \leftarrow_R \{0,1\}^\lambda$.
- <u>Tagging.</u> $\mathsf{TAG}_{(k_1,k_2)}(m)$ samples $\sigma \leftarrow \mathsf{ENC}_{k_1}(k_2, m)$, i.e., it encrypts the plaintext k_2 using m as a label.
- <u>Verification.</u> $\mathsf{VRFY}_{(k_1,k_2)}(m, \sigma)$ output accept iff $\mathsf{DEC}_{k_1}(c, m) \overset{?}{=} k_2$.

Theorem 2. *Assume that* E *is a* (t, Q_E, Q_D, ϵ)-*CCA secure labeled encryption scheme. Then the construction* MAC *above is* $(t', Q_T, Q_V, \epsilon')$-uf-cmva *secure with* $t' \approx t$, $Q_T = Q_E$, $Q_V = Q_D$ *and* $\epsilon' = Q_T \cdot \epsilon + 2^{-\lambda}$.

The proof of Theorem 2 can be found in the full version of this paper.

Examples. There exists CCA-secure (public-key) encryption schemes from a variety of assumptions such as DDH [14,31,22], Paillier [15], lattices [37], and factoring [23]. In Table 1 we describe $\mathsf{MAC}_{\mathsf{CS}}$, which is $\mathsf{MAC}_{\mathsf{ENC}}$ instantiated with Cramer-Shoup encryption.

4.2 Constructions from Hash Proof Systems

We now give a more direct construction of a MAC from any hash proof system. We recall the notion of (labeled) hash proof systems as introduced by Cramer and Shoup [15]. Let $\mathcal{C}, \mathcal{K}$ be sets and $\mathcal{V} \subset \mathcal{C}$ a language. In the context of public-key encryption (and viewing a hash proof system as a labeled key encapsulation mechanism (KEM) with "special algebraic properties") one may think of $\mathcal{C}$ as the set of all *ciphertexts*, $\mathcal{V} \subset \mathcal{C}$ as the set of all *valid (consistent) ciphertexts*, and $\mathcal{K}$ as the set of all *symmetric keys*. Let $\Lambda_k^\ell : \mathcal{C} \times \mathcal{L} \to \mathcal{K}$ be a labeled hash function indexed with $k \in \mathcal{SK}$ and label $\ell \in \mathcal{L}$, where $\mathcal{SK}$ and $\mathcal{L}$ are sets. A hash function Λ_k is *projective* if there exists a projection $\mu : \mathcal{SK} \to \mathcal{PK}$ such that $\mu(k) \in \mathcal{PK}$ defines the action of Λ_k^ℓ over the subset $\mathcal{V}$. That is, for every $C \in \mathcal{V}$, the value $K = \Lambda_k^\ell(C)$ is uniquely determined by $\mu(k)$, C. In contrast, nothing is guaranteed for $C \in \mathcal{C} \setminus \mathcal{V}$, and it may not be possible to compute $\Lambda_k(C)$ from $\mu(k)$ and C. A projective hash function is *universal2* if for all $C, C^* \in \mathcal{C} \setminus \mathcal{V}$, $\ell, \ell^* \in \mathcal{L}$ with $\ell \neq \ell^*$,

$$(pk, \Lambda_k^{\ell^*}(C^*), \Lambda_k^\ell(C)) = (pk, K, \Lambda_k^\ell(C)) \tag{1}$$

(as joint distributions) where in the above $pk = \mu(k)$ for $k \leftarrow_R \mathcal{SK}$ and $K \leftarrow_R \mathcal{K}$. It is extracting if for all $C \in \mathcal{C}$ (including valid ones) and $\ell \in \mathcal{L}$,

$$\Lambda_k^\ell(C) = K \tag{2}$$

where in the above $k \leftarrow_R \mathcal{SK}$ and $K \leftarrow_R \mathcal{K}$.

A labeled hash proof system $\mathsf{HPS} = (\mathsf{Param}, \mathsf{Pub}, \mathsf{Priv})$ consists of three algorithms. The randomized algorithm $\mathsf{Param}(1^k)$ generates parametrized instances of $params = (group, \mathcal{K}, \mathcal{C}, \mathcal{V}, \mathcal{PK}, \mathcal{SK}, \Lambda_{(\cdot)} : \mathcal{C} \to \mathcal{K}, \mu : \mathcal{SK} \to \mathcal{PK})$, where $group$ may contain some additional structural parameters. The deterministic public evaluation algorithm Pub inputs the projection key $pk = \mu(k)$, $C \in \mathcal{V}$, a witness r of the fact that $C \in \mathcal{V}$, and a label $\ell \in \mathcal{L}$, and returns $K = \Lambda_k^\ell(C)$. The deterministic private evaluation algorithm Priv inputs $k \in \mathcal{SK}$ and returns $\Lambda_k^\ell(C)$, without knowing a witness. We further assume that μ is efficiently computable and that there are efficient algorithms given for sampling $k \in \mathcal{SK}$, sampling $C \in \mathcal{V}$ uniformly (or negligibly close to) together with a witness r, sampling $C \in \mathcal{C}$ uniformly, and for checking membership in $\mathcal{C}$.

As computational problem we require that the *subset membership problem* is (ϵ, t)-hard in HPS which means that for all adversaries B that run in time $\leq t$,

$$\big| \Pr[\mathsf{B}(\mathcal{C}, \mathcal{V}, C_1) = 1] - \Pr[\mathsf{B}(\mathcal{C}, \mathcal{V}, C_0) = 1] \big| \leq \epsilon$$

where $\mathcal{C}$ is taken from the output of $\mathsf{Param}(1^k)$, $C_1 \leftarrow_R \mathcal{C}$ and $C_0 \leftarrow_R \mathcal{C} \setminus \mathcal{V}$.

Construction. We define a MAC $\mathsf{MAC}_{\mathsf{HPS}} = \{\mathsf{KG}, \mathsf{TAG}, \mathsf{VRFY}\}$ with associated key space $\mathcal{K} = \mathcal{SK}$, message space $\mathcal{M} = \mathcal{L}$, and tag space $\mathcal{T} = \mathcal{C} \times \mathcal{K}$ as follows.

- **Key Generation.** The key-generation algorithm KG samples $k \leftarrow_R \mathcal{SK}$ and outputs k.
- **Tagging.** The probabilistic authentication algorithm $\mathsf{TAG}_k(m)$ picks $C \leftarrow_R \mathcal{V}$. It computes $K = \Lambda_k^m(C) \in \mathcal{K}$ and outputs $\sigma = (C, K)$.
- **Verification.** The verification algorithm $\mathsf{VRFY}_k(m, \sigma)$ parses $\sigma = (C, K)$ and outputs accept iff $K = \Lambda_k^m(C)$.

Note that the construction does not use the public evaluation algorithm Pub of HPS. Both tagging and verification only use the private evaluation algorithm Priv.

Theorem 3. *Let HPS be universal$_2$ and extracting. If the subset membership problem is (t, ε)-hard, then $\mathsf{MAC}_{\mathsf{HPS}}$ is $(t', \varepsilon', Q_T, Q_V)$-uf-cmva secure with $\varepsilon' = Q_T \varepsilon + O(Q_T Q_V)/|\mathcal{K}|$ and $t' \approx t$.*

The proof of Theorem 3 can be found in the full version of this paper.

Example. We recall a universal$_2$ HPS by Cramer and Shoup [15], whose hard subset membership problem is based on the DDH assumption. Let $\mathbb{G}$ be a group of prime-order p and let g_1, g_2 be two independent generators of $\mathbb{G}$. Define $\mathcal{L} = \mathbb{Z}_p$, $\mathcal{C} = \mathbb{G}^2$ and $\mathcal{V} = \{(g_1^r, g_2^r) \subset \mathbb{G}^2 \ : \ r \in \mathbb{Z}_p\}$. The value $r \in \mathbb{Z}_p$ is a witness of $C \in \mathcal{V}$. Let $\mathcal{SK} = \mathbb{Z}_p^4$, $\mathcal{PK} = \mathbb{G}^2$, and $\mathcal{K} = \mathbb{G}$. For $k = (x_1, x_2, y_1, y_2) \in \mathbb{Z}_p^4$, define $\mu(k) = (g_1^{x_1} g_2^{x_2}, g_1^{y_1} g_2^{y_2})$. This defines the output of $\mathsf{Param}(1^k)$. For $C = (c_1, c_2) \in \mathcal{C}$ and $\ell \in \mathcal{L}$, define

$$\Lambda_k^\ell(C) := c_1^{x_1 \ell + y_1} c_2^{x_2 \ell + y_2} . \tag{3}$$

This defines $\mathsf{Priv}(k, C)$. Given $pk = \mu(k) = (X_1, X_2)$, $C \in \mathcal{V}$ and a witness $r \in \mathbb{Z}_p$ such that $C = (g_1^r, g_2^r)$ public evaluation $\mathsf{Pub}(pk, C, r)$ computes $K = \Lambda_k(C)$ as $K = (X_1^\ell X_2)^r$. Correctness follows by (3) and the definition of μ. This completes the description of HPS. Clearly, under the DDH assumption, the subset membership problem is hard in HPS. Moreover, this HPS is known to be universal$_2$ [15] and can be verified to be extracting.

Applying our construction from Theorem 3 we get the following MAC which we give in its equivalent (but more efficient) "explicit rejection" variant. Let $\mathbb{G}$ be a group of prime order p and g be a random generator of $\mathbb{G}$. Let $H : \mathbb{G}^2 \times \mathcal{M} \to \mathbb{Z}_p$ be a (target) collision resistant hash function. We define a message authentication code $\mathsf{MAC_{HPS}} = \{\mathsf{KG}, \mathsf{TAG}, \mathsf{VRFY}\}$ with associated key space $\mathcal{K} = \mathbb{Z}_p^3$, message space $\mathcal{M}$, and tag space $\mathcal{T} = \mathbb{G}^3$ as follows.

- Key Generation. The key-generation algorithm KG outputs a secret key $k = (\omega, x, x') \leftarrow_R \mathbb{Z}_p^3$.
- Tagging. The probabilistic authentication algorithm $\mathsf{TAG}_k(m)$ samples $U \leftarrow_R \mathbb{G}$ and outputs an authentication tag $\sigma = (U, V_1, V_2) = (U, U^\omega, U^{x\ell + x'}) \in \mathbb{G}^3$, where $\ell = H(U, V_1, m)$.
- Verification. The verification algorithm $\mathsf{VRFY}_k(m, \sigma)$ parses $\sigma = (U, V_1, V_2) \in \mathbb{G}^3$ and outputs accept iff $A^\omega = V_1$ and $U^{x\ell + x'} = V_2$, where $\ell = H(U, V_1, m)$.

4.3   Construction from Key-Homomorphic Weak-PRFs

Definition 3. *Let $\mathcal{K} = \mathcal{K}(\lambda), \mathcal{X} = \mathcal{X}(\lambda), \mathcal{Y} = \mathcal{Y}(\lambda)$ and $\{f_k : \mathcal{X} \mapsto \mathcal{Y}\}_{k \in \mathcal{K}}$ be a weak PRF. We say that $\{f_k\}$ is key-homomorphic weak PRF if $\mathcal{K}, \mathcal{Y}$ are groups with an efficient group operation (written additively) of prime order $q = q(\lambda)$ and if for any fixed $x \in \mathcal{X}$ the function $f_k(x)$ is a group homomorphism of $\mathcal{K} \mapsto \mathcal{Y}$. In particular, for any $k_1, k_2 \in \mathcal{K}$ and $a, b \in \mathbb{Z}_q$, we have $f_{a \cdot k_1 + b \cdot k_2}(x) = a \cdot f_{k_1}(x) + b \cdot f_{k_2}(x)$.*

Construction. Let $\{f_k : \mathcal{X} \mapsto \mathcal{Y}\}_{k \in \mathcal{K}}$ be a key-homomorphic weak PRF where $\mathcal{K}, \mathcal{Y}$ are of prime order $q = q(\lambda)$. Define $\mathsf{MAC} = (\mathsf{KG}, \mathsf{TAG}, \mathsf{VRFY})$ with keyspace $\mathcal{K} \times \mathcal{K}$ and message-space $\mathbb{Z}_q$ via:

- Key Generation. $\mathsf{KG}(1^\lambda)$ chooses $k_1, k_2 \leftarrow_R \mathcal{K}$ uniformly at random and outputs $k = (k_1, k_2)$.
- Tagging. $\mathsf{TAG}_{(k_1, k_2)}(m)$ chooses $x \leftarrow \mathcal{X}$ uniformly at random and sets $y = f_{m \cdot k_1 + k_2}(x)$. Output $\sigma = (x, y)$.
- Verification. $\mathsf{VRFY}_{(k_1, k_2)}(m, \sigma)$ parses $\sigma = (x, y)$ and outputs accept iff $f_{m \cdot k_1 + k_2}(x) \overset{?}{=} y$.

Theorem 4. *If $\{f_k\}$ is a (t, Q, ϵ)-weak PRF which is key-homomorphic over groups $\mathcal{K}, \mathcal{Y}$ of prime order $q = q(\lambda)$. Then the above construction is a (t', Q, ϵ')-suf-cma-MAC (selective unforgeability, no verification queries) with $t' \approx t$ and $\epsilon' = \epsilon + 1/q$. It is also (t', Q, ϵ)-ind-cma.*

The proof of Theorem 4 can be found in the full version of this paper.

DDH example. To instantiate the above MAC, we can take some DDH group $\mathbb{G}$ of prime order q. Let $\mathcal{K} = \mathbb{Z}_q$, $\mathcal{X} = \mathbb{G}$, $\mathcal{Y} = \mathbb{G}$ (which we now write multiplicatively) and define $f_k(x) = x^k$. This is a weak PRF by the DDH assumption. Furthermore, it is key-homomorphic with $f_{a \cdot k_1 + b \cdot k_2}(x) = (f_{k_1}(x))^a (f_{k_2}(x))^b$. Therefore, the above construction gives us the suf-cma MAC $\mathsf{MAC}_{\mathsf{hwPRF}}$ for messages $m \in \mathbb{Z}_q$, defined by $\mathsf{TAG}_{k_1, k_2}(m) := (g, h)$ with $g \leftarrow \mathbb{G}$ and $h := g^{k_1 \cdot m + k_2}$. See Table 1.

LWE example. To obtain another instantiation from the learning with erros problem, we use a recent construction of a weak PRF implicitly given in [3]. For integers $p < q$ and $x \in \mathbb{Z}_q$, define $\lceil x \rfloor_p = \lceil (p/q) \cdot x \rfloor \bmod p$. For a vector $\mathbf{x} \in \mathbb{Z}_q^m$ we extend this notion component wise to $\lceil \mathbf{x} \rfloor_p \in \mathbb{Z}_p^m$.

We let $\mathcal{K} = \mathbb{Z}_q^{m \times n}$, $\mathcal{X} = \mathbb{Z}_q^n$, $\mathcal{Y} = \mathbb{Z}_p^m$ (written additively) and define $f_{\mathbf{K}}(\mathbf{x}) = \lfloor \mathbf{K} \cdot \mathbf{x} \rfloor_p$. This is a weak PRF under the Learning with Rounding (LWR) assumption of [3]. If p, q are integers such that q/p and the inverse LWE error rate $1/\alpha$ are super-polynomial in n, then the LWE_α assumption implies the LWR assumption [3]. Furthermore, it is key-homomorphic with $f_{a \cdot \mathbf{K}_1 + b \cdot \mathbf{K}_2}(\mathbf{x}) = a f_{\mathbf{K}_1}(\mathbf{x}) + b f_{\mathbf{K}_2}(\mathbf{x})$ for almost all inputs $\mathbf{x} \in \mathcal{X}$. (This is sufficient for our generic construction.) Therefore, the above construction gives us the suf-cma and ind-cma secure MAC for messages $m \in \mathbb{Z}_q$, defined by $\mathsf{TAG}_{\mathbf{K}_1, \mathbf{K}_2}(m) = (\mathbf{x}, \mathbf{y})$ with $\mathbf{x} \leftarrow \mathbb{Z}_q^n$ and $\mathbf{y} = \lfloor (m\mathbf{K}_1 + \mathbf{K}_2)\mathbf{x} \rfloor_p$. (The message space can be extended to $\mathbb{Z}_q^n$ by encoding $\mathbf{m} \in \mathbb{Z}_q^n$ into a matrix $\mathbf{M} \in \mathbb{Z}_q^{n \times n}$ using a full-rank-difference encoding [1,29].)

Full security. As an alternative to the transformation from Section 3.3, we sketch how to use Waters' argument [39] to obtain a (full) uf-cma-secure MAC from a homomorphic weak PRF. Let $\{f_k : \mathcal{X} \mapsto \mathcal{Y}\}_{k \in \mathcal{K}}$ be a key-homomorphic weak PRF where $\mathcal{K}, \mathcal{Y}$ are of prime order $q = q(\lambda)$. Now define $\mathsf{MAC}_{\mathsf{WhwPRF}} = (\mathsf{KG}, \mathsf{TAG}, \mathsf{VRFY})$ with key-space $\mathcal{K}^{\lambda+1}$ and message-space $\{0, 1\}^\lambda$ via:

- $\mathsf{KG}(1^\lambda)$: Choose $k_0 \ldots k_\lambda \leftarrow_R \mathcal{K}$ at random, output $k = (k_0, \ldots, k_\lambda)$.
- $\mathsf{TAG}_k(m)$: Choose $x \leftarrow_R \mathcal{X}$ uniformly at random and set $y = f_{k_0 + \sum k_i m_i}(x)$. Output $\sigma = (x, y)$.
- $\mathsf{VRFY}_k(m, \sigma)$: Parse $\sigma = (x, y)$ and outpt accept iff $f_{k_0 + \sum k_i m_i}(x) \overset{?}{=} y$.

The resulting $\mathsf{MAC}_{\mathsf{WhwPRF}}$ can be proved to be uf-cma and ind-cma-secure. A DDH-based example instantiation is contained in Table 1.

4.4   Constructions from Signatures

Clearly, an uf-cma-secure digital signature scheme directly implies an uf-cmva-secure MAC. In certain cases we can obtain improved efficiency, as we demonstrate with a MAC derived from Boneh-Boyen signatures [11]. Concretely, we can instantiate the MAC in any prime-order groups, no bilinear maps are needed. We define a message authentication code $\mathsf{MAC_{BB}} = \{\mathsf{KG}, \mathsf{TAG}, \mathsf{VRFY}\}$ with associated key space $\mathcal{K} = \mathbb{G} \times \mathbb{Z}_p^2$, message space $\mathcal{M} = \mathbb{Z}_p$, and tag space $\mathcal{T} = \mathbb{G}^2$ as follows.

- Key Generation. The key-generation algorithm KG outputs a secret key $k = (x, x', y) \leftarrow_R \mathbb{Z}_p^3$.
- Tagging. The probabilistic authentication algorithm $\mathsf{TAG}_k(m)$ samples $U \leftarrow_R \mathbb{G}$ and outputs an authentication tag $\sigma = (U, g^{xy} \cdot U^{xm+x'}) \in \mathbb{G}^2$.
- Verification. The verification algorithm $\mathsf{VRFY}_k(m, \sigma)$ parses $\sigma = (U, V) \in \mathbb{G}^2$ and outputs accept iff $g^{xy} \cdot U^{xm+x'} = V$.

Theorem 5. *If the gap-CDH assumption is $(t, Q_T + Q_V, \varepsilon)$-hard, then $\mathsf{MAC_{BB}}$ is $(t', \varepsilon', Q_T, Q_V)$ suf-cmva secure with $\varepsilon' = \varepsilon$ and $t' \approx t$.*

The proof of Theorem 5 can be found in the full version of this paper. The above construction is only secure under the gap-CDH assumption. We now show how to apply the twinning technique [13] to obtain a MAC secure under the standard CDH assumption. We define a message authentication code $\mathsf{MAC_{TBB}} = \{\mathsf{KG}, \mathsf{TAG}, \mathsf{VRFY}\}$ with associated key space $\mathcal{K} = \mathbb{Z}_p^5$, message space $\mathcal{M} = \mathbb{Z}_p$, and tag space $\mathcal{T} = \mathbb{G}^3$ as follows.

- Key Generation. The key-generation algorithm KG outputs a secret key $k = (x_1, x_1', x_2, x_2', y) \leftarrow_R \mathbb{Z}_p^5$.
- Tagging. The probabilistic authentication algorithm $\mathsf{TAG}_k(m)$ picks $U \leftarrow_R \mathbb{G}$ and outputs an authentication tag $\sigma = (U, V_1 = g^{x_1 y} U^{x_1 m + x_1'}, V_2 = g^{x_2 y} U^{x_2 m + x_2'}) \in \mathbb{G}^3$.
- Verification. The verification algorithm $\mathsf{VRFY}_k(m, \sigma)$ parses $\sigma = (U, V_1, V_2)$ and outputs accept iff $g^{x_1 y} U^{x_1 m + x_1'} = V_1$ and $g^{x_2 y} U^{x_2 m + x_2'} = V_2$.

Theorem 6. *If the CDH problem is (t, ε)-hard, then MAC is $(t', \varepsilon', Q_T, Q_V)$ suf-cmva secure with $\varepsilon' = \varepsilon + O((Q_T + Q_V)/p)$ and $t' \approx t$.*

The proof of Theorem 6 can be found in the full version of this paper. We remark that $\mathsf{MAC_{BB}}$ and $\mathsf{MAC_{TBB}}$ are only selectively secure (suf-cmva) MACs. Even though this is sufficient for obtaining man-in-the-middle secure authentication protocols, to obtain a fully secure MAC $\mathsf{MAC_{Waters}}$, one can update the constructions using Waters' hash function [39]. The drawback is that the secret key then contains λ many elements in $\mathbb{Z}_p$ and that the security reduction is not tight anymore. We remark that it is also possible to build slightly more efficient suf-cmva-secure MACs from the (Gap) q-Diffie-Hellman inversion problems.

4.5   Constructions from the LPN Assumption

In this section we review the suf-cma and uf-cma-secure MACs constructions implicitly given in [29, Section 4]. To both constructions can apply the transformations from Section 3 to obtain efficient uf-cmva-secure MACs.

FIRST CONSTRUCTION (suf-cma). Let n denote the number of repetitions, τ the parameter of the Bernoulli distribution, and $\tau' := 1/4 + \tau/2$ controls the correctness error.

We define a message authentication code $\mathsf{MAC_{LPN}} = \{\mathsf{KG}, \mathsf{TAG}, \mathsf{VRFY}\}$ with associated key space $\mathcal{K} = \mathbb{Z}_2^{2\ell}$, message space $\mathcal{M} = \{\mathbf{m} \in \mathbb{Z}_2^{2\ell} : \mathrm{hw}(\mathbf{m}) = \ell\}$, and tag space $\mathcal{T} = \mathbb{Z}_2^{(\ell+1)n}$ as follows.

- Key Generation. The key-generation algorithm KG outputs a secret key a vector $\mathbf{x} \leftarrow_R \mathbb{Z}_2^{2\ell}$.
- Tagging. The probabilistic authentication algorithm $\mathsf{TAG_x}(\mathbf{m})$ samples $\mathbf{R} \leftarrow_R \mathbb{Z}_2^{\ell \times n}$ and outputs an authentication tag $\sigma = (\mathbf{R}, \mathbf{R}^T \cdot \mathbf{x}_{\downarrow \mathbf{m}} + \mathbf{e})$, where $\mathbf{e} \in \mathbb{Z}_2^n$ is sampled according the Bernoulli distribution with parameter τ and $\mathbf{x}_{\downarrow \mathbf{m}} \in \mathbb{Z}_2^{\ell}$ is the vector obtained from $\mathbf{x}$ by deleting all entries where $\mathbf{m}_i = 0$.
- Verification. The verification algorithm $\mathsf{VRFY_x}(\mathbf{m}, \sigma)$ parses $\sigma = (\mathbf{R}, \mathbf{z}) \in \mathbb{Z}_2^{\ell \times n} \times \mathbb{Z}_2^n$ and outputs accept iff $|\mathbf{R}^T \cdot \mathbf{x}_{\downarrow \mathbf{m}} - \mathbf{z}| \le \tau' n$.

Concretely, [29, Th. 4] shows (implicitly)[5] that $\mathsf{MAC_{LPN}}$ has $2^{-O(n)}$ completeness error and is suf-cma and ind-cma-secure under the $\mathsf{LPN}_{\ell,\tau}$ assumption in dimension $\approx \ell$ and Bernoulli parameter τ.

SECOND CONSTRUCTION (uf-cma). We define a message authentication code $\mathsf{MAC_{BilinLPN}} = \{\mathsf{KG}, \mathsf{TAG}, \mathsf{VRFY}\}$ with associated key space $\mathcal{K} = \mathbb{Z}_2^{\ell \times \lambda}$, message space $\mathcal{M} = \mathbb{Z}_2^\lambda$, and tag space $\mathcal{T} = \mathbb{Z}_2^{(\ell+1)n}$ as follows.

- Key Generation. The key-generation algorithm KG outputs a secret key a matrix $\mathbf{X} \leftarrow_R \mathbb{Z}_2^{\ell \times \mu}$.
- Tagging. The probabilistic authentication algorithm $\mathsf{TAG_X}(\mathbf{m})$ samples $\mathbf{R} \leftarrow_R \mathbb{Z}_2^{\ell \times n}$ and outputs an authentication tag $\sigma = (\mathbf{R}, \mathbf{R}^T \cdot \mathbf{X} \cdot \mathbf{m} + \mathbf{e})$, where $\mathbf{e} \in \mathbb{Z}_2^n$ is sampled according the Bernoulli distribution with parameter τ.
- Verification. The verification algorithm $\mathsf{VRFY_X}(\mathbf{m}, \sigma)$ parses $\sigma = (\mathbf{R}, \mathbf{z}) \in \mathbb{Z}_2^{\ell \times n} \times \mathbb{Z}_2^n$ and outputs accept iff $|\mathbf{R}^T \cdot \mathbf{X} \cdot \mathbf{m} - \mathbf{z}| \le \tau' n$.

[29, Th. 5] shows that $\mathsf{MAC_{BilinLPN}}$ is uf-cma and ind-cma-secure under the $\mathsf{LPN}_{\ell,\tau}$ assumption. We remark that $\mathsf{MAC_{BilinLPN}}$ can also be viewed as an instantiation of $\mathsf{MAC_{WhwPRF}}$ of Section 4.3 when generalizing the construction to *randomized* weak PRFs and using $f_\mathbf{x}(\mathbf{R}) = \mathbf{R}^T \mathbf{x} + \mathbf{e}$ which is a randomized weak PRF under LPN.

4.6 Three-Round Authentication from Any Weak PRF

We now state our authentication protocol Π using any wPRF family $\mathcal{F} = \{f_{k_1} : \mathcal{X}_1 \mapsto \mathcal{Y}\}_{k_1 \in \mathcal{K}_1}$ and any weak Almost XOR-Universal (wAXU) family $\mathcal{H} = \{h_{k_2} : \mathcal{X}_2 \mapsto \mathcal{Y}\}_{k_2 \in \mathcal{K}_2}$ (see the full version of the paper for more details on how $\mathcal{H}$ can be instantiated.)

---

[5] [29] give a direct construction of a MAC that is suf-cmva secure. $\mathsf{MAC_{LPN}}$ is the underlying MAC that can be proved only suf-cma secure.

The key generation algorithm $\mathsf{KG}(1^\lambda)$ selects random $k_1 \leftarrow \mathcal{K}_1, k_2 \leftarrow \mathcal{K}_2$ and outputs $k = (k_1, k_2)$. Following this, the three round protocol between a Tag $\mathcal{T}(k)$ and a reader $\mathcal{R}(k)$ is defined below:

- $\mathcal{T} \to \mathcal{R}$: choose random $r \in \mathcal{X}_1$ and send r to $\mathcal{R}$.
- $\mathcal{R} \to \mathcal{T}$: choose random $x \in \mathcal{X}_2$ and send x to $\mathcal{T}$.
- $\mathcal{T} \to \mathcal{R}$: compute $z = f_{k_1}(r) + h_{k_2}(x)$ and send z to $\mathcal{R}$.
- $\mathcal{R}$: accept if and only if $z \overset{?}{=} f_{k_1}(r) + h_{k_2}(x)$.

Theorem 7. *Assuming that $\mathcal{F} = \{f_{k_1}\}$ is a (t, Q, ε)-wPRF and $\mathcal{H} = \{h_{k_2}\}$ is (t, ρ)-wAXU. Then the above authentication protocol is (t', Q, ε')-secure against active adversaries, with $t' = t/2$ and $\varepsilon' = \sqrt{\varepsilon + \rho}$.*

In particular, setting $\mathcal{F} = \mathcal{H}$ and $\mathcal{X}_1 = \mathcal{X}_2 = \mathcal{X}$, we get $\varepsilon' = \sqrt{2\varepsilon + \frac{1}{|\mathcal{X}|} + \frac{1}{|\mathcal{Y}|}}$.

The proof of Theroem 7 can be found in the full version of this paper.

Example. To instantiate the above authentication protocol, we can take some DDH group $\mathbb{G}$ of prime order q. Let $\mathcal{K} = \mathcal{K}_1 = \mathcal{K}_2 = \mathbb{Z}_q$, $\mathcal{X} = \mathcal{X}_1 = \mathcal{X}_2 = \mathbb{G}$, $\mathcal{Y} = \mathbb{G}$ (which we now write multiplicatively). For notational convenience, let us denote $k_1 = a$, $k_2 = b$, $r = g$, and define $f_a(g) := g^a$, $h_b(x) := x^b$ so that $\mathcal{F}$ is a wPRF by DDH, and $\mathcal{H} = \mathcal{F}$ is wAXU by DDH as well. We get the following very simple DDH-based protocol with secret key $k = (a, b)$.

- $\mathcal{T} \to \mathcal{R}$: choose random $g \in \mathbb{G}$ and send g to $\mathcal{R}$.
- $\mathcal{R} \to \mathcal{T}$: choose random $x \in \mathbb{G}$ and send x to $\mathcal{T}$.
- $\mathcal{T} \to \mathcal{R}$: compute $z = g^a x^b \in G$ and send z to $\mathcal{R}$.
- $\mathcal{R}$: accept if and only if $z \overset{?}{=} g^a x^b$.

It is interesting to compare the above actively secure authentication protocol with Okamoto's public-key authentication protocol based on the discrete log assumption [36]. On the one hand, Okamoto's scheme is based on a weaker assumption and works in the public-key setting. On the other hand, our DDH-based protocol is more efficient. Our verifier only has to perform two exponentiations, while Okamoto's verifier needs to do three exponentiations. Also, our last flow z contains one group element, while Okamoto's protocol contains two exponents, which is likely going to be longer.

References

1. Agrawal, S., Boneh, D., Boyen, X.: Efficient Lattice (H)IBE in the Standard Model. In: Gilbert, H. (ed.) EUROCRYPT 2010. LNCS, vol. 6110, pp. 553–572. Springer, Heidelberg (2010)
2. Applebaum, B., Cash, D., Peikert, C., Sahai, A.: Fast Cryptographic Primitives and Circular-Secure Encryption Based on Hard Learning Problems. In: Halevi, S. (ed.) CRYPTO 2009. LNCS, vol. 5677, pp. 595–618. Springer, Heidelberg (2009)
3. Banerjee, A., Peikert, C., Rosen, A.: Pseudorandom functions and lattices. Cryptology ePrint Archive, Report 2011/401 (2011), http://eprint.iacr.org/

4. Bellare, M.: New Proofs for NMAC and HMAC: Security Without Collision-Resistance. In: Dwork, C. (ed.) CRYPTO 2006. LNCS, vol. 4117, pp. 602–619. Springer, Heidelberg (2006)
5. Bellare, M., Canetti, R., Krawczyk, H.: Keying Hash Functions for Message Authentication. In: Koblitz, N. (ed.) CRYPTO 1996. LNCS, vol. 1109, pp. 1–15. Springer, Heidelberg (1996)
6. Bellare, M., Canetti, R., Krawczyk, H.: Pseudorandom functions revisited: The cascade construction and its concrete security. In: 37th Annual Symposium on Foundations of Computer Science, pp. 514–523. IEEE Computer Society Press (October 1996)
7. Bellare, M., Goldreich, O., Mityagin, A.: The power of verification queries in message authentication and authenticated encryption. Cryptology ePrint Archive, Report 2004/309 (2004), http://eprint.iacr.org/
8. Bellare, M., Pietrzak, K., Rogaway, P.: Improved Security Analyses for CBC MACs. In: Shoup, V. (ed.) CRYPTO 2005. LNCS, vol. 3621, pp. 527–545. Springer, Heidelberg (2005)
9. Bellare, M., Rogaway, P.: The Exact Security of Digital Signatures - How to Sign with RSA and Rabin. In: Maurer, U.M. (ed.) EUROCRYPT 1996. LNCS, vol. 1070, pp. 399–416. Springer, Heidelberg (1996)
10. Boneh, D., Boyen, X.: Efficient Selective-ID Secure Identity-Based Encryption Without Random Oracles. In: Cachin, C., Camenisch, J.L. (eds.) EUROCRYPT 2004. LNCS, vol. 3027, pp. 223–238. Springer, Heidelberg (2004)
11. Boneh, D., Boyen, X.: Short Signatures Without Random Oracles. In: Cachin, C., Camenisch, J.L. (eds.) EUROCRYPT 2004. LNCS, vol. 3027, pp. 56–73. Springer, Heidelberg (2004)
12. Camenisch, J.L., Hohenberger, S., Lysyanskaya, A.: Compact E-Cash. In: Cramer, R. (ed.) EUROCRYPT 2005. LNCS, vol. 3494, pp. 302–321. Springer, Heidelberg (2005)
13. Cash, D.M., Kiltz, E., Shoup, V.: The Twin Diffie-Hellman Problem and Applications. In: Smart, N.P. (ed.) EUROCRYPT 2008. LNCS, vol. 4965, pp. 127–145. Springer, Heidelberg (2008)
14. Cramer, R., Shoup, V.: A Practical Public Key Cryptosystem Provably Secure against Adaptive Chosen Ciphertext Attack. In: Krawczyk, H. (ed.) CRYPTO 1998. LNCS, vol. 1462, pp. 13–25. Springer, Heidelberg (1998)
15. Cramer, R., Shoup, V.: Universal Hash Proofs and a Paradigm for Adaptive Chosen Ciphertext Secure Public-Key Encryption. In: Knudsen, L.R. (ed.) EUROCRYPT 2002. LNCS, vol. 2332, pp. 45–64. Springer, Heidelberg (2002)
16. Dodis, Y., Pietrzak, K.: Improving the Security of MACs Via Randomized Message Preprocessing. In: Biryukov, A. (ed.) FSE 2007. LNCS, vol. 4593, pp. 414–433. Springer, Heidelberg (2007)
17. Dodis, Y., Yampolskiy, A.: A Verifiable Random Function with Short Proofs and Keys. In: Vaudenay, S. (ed.) PKC 2005. LNCS, vol. 3386, pp. 416–431. Springer, Heidelberg (2005)
18. Fiat, A., Shamir, A.: How to Prove Yourself: Practical Solutions to Identification and Signature Problems. In: Odlyzko, A.M. (ed.) CRYPTO 1986. LNCS, vol. 263, pp. 186–194. Springer, Heidelberg (1987)
19. Gilbert, H., Robshaw, M., Seurin, Y.: How to Encrypt with the LPN Problem. In: Aceto, L., Damgård, I., Goldberg, L.A., Halldórsson, M.M., Ingólfsdóttir, A., Walukiewicz, I. (eds.) ICALP 2008, Part II. LNCS, vol. 5126, pp. 679–690. Springer, Heidelberg (2008)

20. Goldreich, O., Goldwasser, S., Micali, S.: How to construct random functions. J. ACM 33(4), 792–807 (1986)
21. Haralambiev, K., Jager, T., Kiltz, E., Shoup, V.: Simple and Efficient Public-Key Encryption from Computational Diffie-Hellman in the Standard Model. In: Nguyen, P.Q., Pointcheval, D. (eds.) PKC 2010. LNCS, vol. 6056, pp. 1–18. Springer, Heidelberg (2010)
22. Hofheinz, D., Kiltz, E.: Secure Hybrid Encryption from Weakened Key Encapsulation. In: Menezes, A. (ed.) CRYPTO 2007. LNCS, vol. 4622, pp. 553–571. Springer, Heidelberg (2007)
23. Hofheinz, D., Kiltz, E.: Practical Chosen Ciphertext Secure Encryption from Factoring. In: Joux, A. (ed.) EUROCRYPT 2009. LNCS, vol. 5479, pp. 313–332. Springer, Heidelberg (2009)
24. Hopper, N.J., Blum, M.: Secure Human Identification Protocols. In: Boyd, C. (ed.) ASIACRYPT 2001. LNCS, vol. 2248, pp. 52–66. Springer, Heidelberg (2001)
25. Jaulmes, É., Joux, A., Valette, F.: On the Security of Randomized CBC-MAC Beyond the Birthday Paradox Limit: A New Construction. In: Daemen, J., Rijmen, V. (eds.) FSE 2002. LNCS, vol. 2365, pp. 237–251. Springer, Heidelberg (2002)
26. Juels, A., Weis, S.A.: Authenticating Pervasive Devices with Human Protocols. In: Shoup, V. (ed.) CRYPTO 2005. LNCS, vol. 3621, pp. 293–308. Springer, Heidelberg (2005)
27. Katz, J., Shin, J.S.: Parallel and Concurrent Security of the HB and HB$^+$ Protocols. In: Vaudenay, S. (ed.) EUROCRYPT 2006. LNCS, vol. 4004, pp. 73–87. Springer, Heidelberg (2006)
28. Katz, J., Shin, J.S., Smith, A.: Parallel and concurrent security of the HB and HB+ protocols. Journal of Cryptology 23(3), 402–421 (2010)
29. Kiltz, E., Pietrzak, K., Cash, D., Jain, A., Venturi, D.: Efficient Authentication from Hard Learning Problems. In: Paterson, K.G. (ed.) EUROCRYPT 2011. LNCS, vol. 6632, pp. 7–26. Springer, Heidelberg (2011)
30. Krawczyk, H.: New Hash Functions for Message Authentication. In: Guillou, L.C., Quisquater, J.-J. (eds.) EUROCRYPT 1995. LNCS, vol. 921, pp. 301–310. Springer, Heidelberg (1995)
31. Kurosawa, K., Desmedt, Y.: A New Paradigm of Hybrid Encryption Scheme. In: Franklin, M. (ed.) CRYPTO 2004. LNCS, vol. 3152, pp. 426–442. Springer, Heidelberg (2004)
32. Micali, S., Rabin, M.O., Vadhan, S.P.: Verifiable random functions. In: 40th Annual Symposium on Foundations of Computer Science, pp. 120–130. IEEE Computer Society Press (October 1999)
33. Naor, M., Reingold, O.: Number-theoretic constructions of efficient pseudo-random functions. In: 38th Annual Symposium on Foundations of Computer Science, pp. 458–467. IEEE Computer Society Press (October 1997)
34. Naor, M., Reingold, O.: From Unpredictability to Indistinguishability: A Simple Construction of Pseudo-Random Functions from MACs (Extended Abstract). In: Krawczyk, H. (ed.) CRYPTO 1998. LNCS, vol. 1462, pp. 267–282. Springer, Heidelberg (1998)
35. Naor, M., Reingold, O., Rosen, A.: Pseudo-random functions and factoring (extended abstract). In: 32nd Annual ACM Symposium on Theory of Computing, pp. 11–20. ACM Press (May 2000)
36. Okamoto, T.: Provably Secure and Practical Identification Schemes and Corresponding Signature Schemes. In: Brickell, E.F. (ed.) CRYPTO 1992. LNCS, vol. 740, pp. 31–53. Springer, Heidelberg (1993)

37. Peikert, C.: Public-key cryptosystems from the worst-case shortest vector problem: extended abstract. In: Mitzenmacher, M. (ed.) 41st Annual ACM Symposium on Theory of Computing, pp. 333–342. ACM Press (May/June 2009)
38. Stern, J.: A New Identification Scheme Based on Syndrome Decoding. In: Stinson, D.R. (ed.) CRYPTO 1993. LNCS, vol. 773, pp. 13–21. Springer, Heidelberg (1994)
39. Waters, B.: Efficient Identity-Based Encryption Without Random Oracles. In: Cramer, R. (ed.) EUROCRYPT 2005. LNCS, vol. 3494, pp. 114–127. Springer, Heidelberg (2005)
40. Wegman, M.N., Carter, L.: New hash functions and their use in authentication and set equality. Journal of Computer and System Sciences 22, 265–279 (1981)

Property Preserving Symmetric Encryption

Omkant Pandey[1] and Yannis Rouselakis[2]

[1] Microsoft, Redmond, USA and Microsoft Research, Bangalore, India
omkantp@microsoft.com
[2] The University of Texas at Austin
jrous@cs.utexas.edu

Abstract. Processing on encrypted data is a subject of rich investigation. Several new and exotic encryption schemes, supporting a diverse set of features, have been developed for this purpose. We consider encryption schemes that are suitable for applications such as data clustering on encrypted data. In such applications, the processing algorithm needs to learn certain properties about the encrypted data to make decisions. Often these decisions depend upon multiple data items, which might have been encrypted individually and independently. Current encryption schemes do not capture this setting where computation must be done on multiple ciphertexts to make a decision.

In this work, we seek encryption schemes which allow *public* computation of a pre-specified property P about the encrypted messages. That is, such schemes have an associated property P of fixed arity k, and a publicly computable algorithm Test, such that $\mathsf{Test}(ct_1, \ldots, ct_k) = P(m_1, \ldots, m_k)$, where ct_i is an encryption of m_i for $i = 1, \ldots, k$. Further, this requirement holds even if the ciphertexts $ct_1, \ldots, ct_k$ were generated individually and independently. We call such schemes *property preserving encryption schemes*. Property preserving encryption (PPEnc) makes most sense in the symmetric setting due to the requirement that Test is publicly computable.

In this work, we present a thorough investigation of property preserving symmetric encryption. We start by formalizing several meaningful notions of security for PPEnc. Somewhat surprisingly, we show that there exists a hierarchy of security notions for PPEnc, indexed by integers $\eta \in \mathbb{N}$, which does not collapse. We also present a symmetric PPEnc scheme for encrypting vectors in $\mathbb{Z}_N$ of polynomial length. This construction supports the orthogonality property: for every two vectors $(\vec{x}, \vec{y})$ it is possible to *publicly* learn whether $\vec{x} \cdot \vec{y} = 0 \mod p$. Our scheme is based on bilinear groups of composite order.

1   Introduction

This paper introduces the notion of *property preserving* encryption schemes. The idea is that it should be possible to *publicly* learn the properties of a massive data set, by only looking at the *encrypted* data elements. For simplicity, we model properties as boolean functions P defined over the space $\mathcal{M}^k$ for a fixed natural

D. Pointcheval and T. Johansson (Eds.): EUROCRYPT 2012, LNCS 7237, pp. 375–391, 2012.

number $k \in \mathbb{N}$. The simplest way to capture this idea is by requiring a public algorithm, Test, such that $\forall (m_1, \ldots, m_k) \in \mathcal{M}^k$:

$$P(m_1, \ldots, m_k) = \mathsf{Test}(ct_1, \ldots, ct_k)$$

where ct_i is the encryption of m_i for every $i \in [k]$. An important observation is that the idea makes most sense only for symmetric encryption schemes, which will be the main focus of this work.[1]

Property preserving encryption represents great promise, particularly for developing *private* algorithms for data classification. Of particular interest are the applications that deal with *streaming* data. For example, consider the recipient of a data stream, who receives data-elements arriving one at a time: $m_1, m_2, \ldots$ and so on. The recipient would like to encrypt each of these elements, as they arrive, and store[2] the resulting ciphertexts on an untrusted computing facility, e.g., a public *cloud* [21,33]. The recipient can then instruct the cloud to classify and organize this data—e.g., using data clustering techniques [30,28], for the target application. Current encryption schemes fall short of dealing with this situation. This holds true even for the exotic class of schemes such as predicate encryption [31], functional encryption [15], and fully homomorphic encryption [39,24].

Order Preserving Symmetric Encryption. Property preserving encryption is directly inspired by the recent work of Boldyreva, Chenette, Lee, and O'Neill on *order preserving (symmetric) encryption* [10]. Informally speaking, an encryption scheme is order preserving if the ciphertexts preserve the order of the plaintexts; that is, if m_1, m_2 are two plaintexts integers and $m_1 \geq m_2$, then $ct_1 \geq ct_2$, where ct_1, ct_2 are encryptions of m_1, m_2 respectively. Boldyreva et al. show that order-preserving schemes cannot satisfy the usual "indistinguishability" based notions. In fact, as noted in [10,11], formulating a reasonable notion of security for order preserving encryption is a subtle and involved task. The starting point of our work was to understand the source of this difficulty, and how it affects other properties.

For this purpose, we start by generalizing the idea of preserving the order as follows. First, we do not restrict ourselves only to the ordering relation, and consider arbitrary properties. Second, we do not necessarily require the *same* relation on plaintexts and ciphertexts—e.g., the *greater than or equal to* operation. Instead, we only require a public algorithm to test this relation: $\mathsf{Test}(ct_1, ct_2) = 1$ *if and only if* $m_1 \geq m_2$. With these generalizations, it turns out that there exist nontrivial properties for which we can satisfy indistinguishability-based security notions. This results in very strong and robust security guarantees.

---

[1] For asymmetric (or public-key) encryption, the encrypted message might be recoverable for most properties of interest, simply by using Test and the encryption algorithm. See also section 1.2 for further discussion.

[2] We note that this model is similar to the model considered by Gennaro and Rohatgi [23] for digital signatures. In particular, it is different from the "streaming algorithms" model where the stream cannot be stored, and the computations must be done in a single pass over a small sample of the stream[2,29].

1.1   Our Contribution

It quickly becomes apparent that property preserving encryption is a new notion that requires a thorough investigation. This is the focus of the current work. We present a summary of our results here.

Notions of Security. We start by defining three indistinguishability based notions of security: *(1) find-then-guess* (FTG), *(2) left-or-right* (LoR), and *(3) selective real-versus-random* (sRvR). These notions are directly based upon the work of Bellare, Desai, Jokipii, and Rogaway [5] for defining security of symmetric encryption.

In FTG-security the adversary first participates in a "find" stage in which he receives encryptions of many (adaptively) chosen messages. The adversary then selects two challenges (m_0^*, m_1^*), and receives an encryption of one of them. The adversary is supposed to "guess" which message was encrypted. In LoR-security, the adversary adaptively chooses many pairs of messages $(m_1^0, m_1^1), (m_2^0, m_2^1), \ldots$, and receives encryptions of messages $m_1^b, m_2^b, \ldots$, for a fixed bit b. The adversary is supposed to guess b. In property preserving encryption, the adversary is allowed to learn the value of the property P on various subsets of messages. Therefore we enforce the following "equality pattern" condition (assume P to be binary): in FTG game, we require that for every message m_i that was encrypted, $P(m_0^*, m_i) = P(m_1^*, m_i)$; likewise, in LoR game, we require that for every two indices (i, j): $P(m_i^0, m_j^0) = P(m_i^1, m_j^1)$.

For standard symmetric encryption, these two notions are proven equivalent using a simple hybrid experiment [5]. Quite surprisingly, we show that in case of property preserving encryption, the FTG-security is much weaker than LoR-security. There exist natural properties for which FTG can leak much more about the encrypted messages than LoR. This proof also highlights that in fact FTG is a rather subtle notion: there is a hierarchy of FTG definitions indexed by a natural number $\eta \in \mathbb{N}$, denoted FTG$^\eta$, which lie between FTG and LoR. Roughly speaking, the FTG$^\eta$ notion is like the FTG notion except that the adversary submits at most η pairs of challenges instead of just one: $(m_{0,1}^*, m_{1,1}^*), \ldots, (m_{0,\eta}^*, m_{1,\eta}^*)$. We go on to show that FTG$^\eta$ is weaker than FTG$^{\eta+1}$.

Our final indistinguishability based notion, is an adaptation of the "real-or-random" security presented in [5]. Informally, in this game the attacker submits adaptively chosen messages that form the *real* sequence of messages to an encryption oracle. The oracle either only encrypts the real message sequence or a *random* message sequence. As usual, we want that the adversary should not know which is the case. Adopting this notion to the setting of property-preserving encryption is slightly tricky, due to the equality pattern condition. When returning encryptions of a random sequence, it should be ensured that the random sequence will have the equality pattern of the real sequence. Since the real sequence is chosen adaptively based on the ciphertexts seen so far, the equality pattern of the real sequence "evolves" during the entire experiment. One way to deal with this situation is to require the adversary to select its equality pattern ξ (a binary vector) at the beginning of the game. This choice is motivated by the work on selective security for identity based encryption [18,19]. We require that the encryptions of real sequence with equality pattern ξ, look indistinguishable from a random sequence with the same pattern ξ. The resulting notion is called the *selective real-versus-random* security denoted by sRvR, and is proven equivalent to the selective version of LoR-security, denoted sLoR. The summary of relationships between these security notions is presented in figure 1.

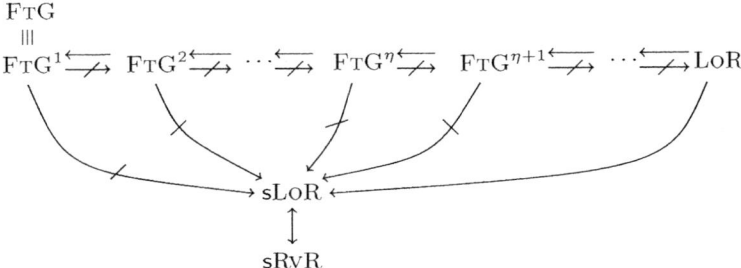

Fig. 1. Relations between all security notions. Solid arrows denote implications for all properties. Cut arrows denote that there exist some properties for which the implication is false.

Our Constructions. We seek interesting properties for which provably secure constructions satisfying our security notions can be obtained. We present constructions that preserve, according to our notion, the orthogonality of encrypted vectors. More formally, let p be a prime number; we construct a property preserving scheme for $P : \mathbb{Z}_p^n \times \mathbb{Z}_p^n \to \{0,1\}$ such that: $P(\vec{u}, \vec{v}) = 0$ if $\vec{u} \cdot \vec{v} = 0$ mod p and 1 otherwise.

First we observe a general approach for constructing property preserving encryption from symmetric predicate-encryption that satisfy two essential properties: (1) predicate privacy in the multi-challenge model, and (2) security in the standard model (as opposed to the *selective* models as defined in [18,19]). Shen, Shi, and Waters [41] formulated the notion of predicate privacy in symmetric encryption, and presented a construction for orthogonality testing. However, their construction is secure only in the selective-security model. At present, there are no known constructions satisfying the two requirements.

We present a new, direct construction, for preserving orthogonality. Our construction is based on composite order groups with bilinear pairings. We prove that our construction satisfies the LoR-security in the generic group model [44]; a provably secure construction in the standard model is left as an important open problem.

1.2   Related Work

Other than the works of Boldyreva et al. [10,11], the work of Bellare, Ristenpart, Rogaway, and Stegers [8] on format preserving encryption is also a related concept which ensures that the ciphertext has the same *format* as the plaintext.

Encryption schemes supporting keyword search on encrypted data are very relevant to our work. They were considered by Song, Wagner, and Perrig in the symmetric setting [45], and by Boneh, Di Crescenzo, Ostrovsky, and Persiano [14] in the public-key setting. We can view these works as testing for the equality property for a fixed keyword(s). Equality tests in symmetric setting are related to oblivious RAM techniques [37]; in the public-key setting they are related to anonymous Identity Based Encryption (IBE) [14,1,17]. Subsequent works developed schemes for complex queries such as conjunctive and range queries [25,16,42], and more efficient constructions [22].

Bellare, Boldyreva, and O'Neill [4] investigated the notion of deterministic encryption to allow search in sub-linear time. These schemes provide meaningful security guarantee only when messages are drawn from high min-entropy distributions. Subsequent works further refined this notion and provided new constructions [7,12,36].

Another notion, closely related to our work, is predicate encryption, introduced by Katz, Sahai and Waters [31], and further generalized to functional encryption [15]. In predicate encryption, messages are encrypted using a set of attributes S, and secret keys can be derived for predicates f, say K_f. A message m encrypted using S can be decrypted using K_f if and only if $f(S) = 1$. The principal difference between our notion and predicate encryption is that the latter only tests *unary* property, i.e., f works only on a single ciphertext. In contrast, property-preserving encryption is required to deal with multiple ciphertexts each generated individually and independently. Predicate encryption is a generalization of previous works on attribute-based encryption [40], further developed in [27,9,20,38,26]. Subsequent works provided improved constructions under a variety of cryptographic assumptions [31,43,41,34].

Our study of relationships between security notions of encryption schemes is inspired by initial works of Bellare, Desai, Jokipii, and Rogaway [5], and Bellare, Desai, Pointcheval, and Rogaway [6]; it has been pursued in many subsequent works since then such as [3,32], as well as previously mentioned works on deterministic encryption.

Somewhat tangentially related to our work is the notion of fully homomorphic encryption (FHE) [39], first realized by Gentry [24]. While FHE allows processing arbitrary computations on any number of ciphertexts, the resulting output is encrypted, and therefore not useful for evaluating properties.

2 Property Preserving Encryption

Standard Notation. We write $s \overset{\$}{\leftarrow} S$ to mean that s is picked uniformly at random from the set S. When multiple elements $x, y, z, \ldots$ are picked uniformly at random from S, we write $x, y, z, \ldots \overset{\$}{\leftarrow} S$. Symbols $\neg$, $\wedge$, and $\oplus$ denote the standard boolean operations: NOT, AND, and XOR, respectively. The set of natural numbers is denoted by $\mathbb{N}$; for $n \in \mathbb{N}$, we write by $[n]$ the set $\{1, 2, \ldots, n\}$. We will often refer to a vector directly by writing its components in order as either

$(a_1, a_2, \ldots, a_n)$ or $\{a_i\}_{i=1}^{n}$. The security parameter is denoted by $\lambda \in \mathbb{N}$, and a function negligible in λ is denoted by $\mathsf{negl}(\lambda)$. All algorithms are assumed to have λ as an implicit input, and run in time polynomial in λ.

Property Preserving Encryption. A property-preserving symmetric encryption scheme, is just like a normal symmetric encryption scheme except that it has an associated property P and a test algorithm, Test. Algorithm Test is a *publicly computable* polynomial time algorithm which operates on ciphertexts. The goal of Test algorithm is to test if the property P is satisfied on the underlying messages of the input ciphertexts. The formal definition of *symmetric property-preserving encryption* is given below; we allow some public-parameters in the system so that Test algorithm can properly operate on the ciphertexts.

Definition 2.1. *A* symmetric property-preserving encryption *scheme, with plaintext - space* $\mathcal{M}$, *consists of four probabilistic polynomial-time algorithms* $\Pi = (\mathsf{Setup}, \mathsf{Enc}, \mathsf{Dec}, \mathsf{Test})$ *and an associated property* $P : \mathcal{M}^k \to \{0, 1\}$, *such that:*

$\mathsf{Setup}(1^\lambda) \to (pp, sk)$:
 This is a randomized algorithm, which on input a security parameter $\lambda \in \mathbb{N}$, *outputs a secret-key* sk, *and public-parameters* pp.
$\mathsf{Enc}(pp, sk, m) \to ct$:
 The (possibly randomized) encryption algorithm takes as input pp, sk, *and the plaintext* m; *it outputs a ciphertext* ct.
$\mathsf{Dec}(pp, sk, ct) \to m$:
 The decryption algorithm takes as input pp, sk, *and the ciphertext* ct; *it outputs the plaintext message* m.
$\mathsf{Test}(pp, ct_1, \ldots, ct_k) \to \{0, 1\}$:
 The testing algorithm takes as input the public parameters pp, *and* k *ciphertexts* $ct_1, \ldots, ct_k$; *it outputs a bit.*

We require that for all possible outputs (pp, sk) *of algorithm* Setup, *and every* $m \in \mathcal{M}$, *it holds that* $\mathsf{Dec}(pp, sk, \mathsf{Enc}(pp, sk, m)) = m$. *Further, we also require that there exist a negligible function* $\mathsf{negl}(\cdot)$ *such that* $\forall (m_1, \ldots, m_k) \in \mathcal{M}^k$:

$$\Pr\left[\begin{matrix} \mathsf{Test}(pp, ct_1, \ldots, ct_k) \\ = P(m_1, m_2, \ldots, m_k) \end{matrix} \;\middle|\; \begin{matrix} (pp, sk) \leftarrow \mathsf{Setup}(1^\lambda) \\ \forall i \in [k] : ct_i \leftarrow \mathsf{Enc}(pp, sk, m_i) \end{matrix} \right] \geq 1 - \mathsf{negl}(\lambda)$$

where the probability is taken over the randomness of all algorithms.

3 Security Notions

We follow the approach of Bellare, Desai, Jokipii, and Rogaway [5], and present three different definitions. We will start by considering the two simplest variants, each of which is obtained by modifying definitions in [5] to accommodate the equality pattern. To do this, we introduce some notation.

Notation. Let $\Pi = (\mathsf{Setup}, \mathsf{Enc}, \mathsf{Dec}, \mathsf{Test})$ be a symmetric property-preserving encryption scheme with plaintext space $\mathcal{M}$. Let P be a k-ary property defined over $\mathcal{M}$ for some fixed positive integer $k \in \mathbb{N}$: $P : \mathcal{M}^k \rightarrow \{0, 1\}$. For a bit b, let the "Left-Right Oracle" be defined as the following function: $\mathsf{LR}(m_0, m_1, b) = m_b$. Let $X = (x_1, \ldots, x_n) \in \mathcal{M}^n$ and $Y = (y_1, \ldots, y_n) \in \mathcal{M}^n$ be two message sequences of polynomial length $n = n(\lambda)$. We say that X and Y have the *same* equality pattern for property P, if and only if: $\forall (i_1, \ldots, i_k) \in [n]^k$, $P(x_{i_1}, \ldots, x_{i_k}) = P(y_{i_1}, \ldots, y_{i_k})$.

It will be convenient to formally define the equality pattern of a sequence X. For integers n, k, let $I_1, \ldots, I_{n^k}$ be *all* sequences of indices $(i_1, \ldots, i_k) \in [n]^k$ in the *lexicographic* order.[3] The equality pattern of a sequence $X \in \mathcal{M}^n$ w.r.t. property $P : \mathcal{M}^k \rightarrow \{0, 1\}$ is a binary vector of length n^k, denoted by $\mathsf{Eqp}(X) := (b_1, \ldots, b_{n^k})$, such that $b_j = P(X_{I_j})$. Here X_{I_j} denotes the projection of X on j^{th}-sequence I_j, for $j \in [n^k]$.

Find-then-Guess Security. The simplest indistinguishability based definition is the "find-then-guess" security. Informally, adversary $\mathcal{A}$ participates in a game, in which first it is given access to an encryption oracle. $\mathcal{A}$ can ask polynomially many encryption queries by adaptively choosing and sending plaintexts $m \in \mathcal{M}$. This is called the "find" stage; at some point, $\mathcal{A}$ produces two equal-length messages (m_0^*, m_1^*). At this point, $\mathcal{A}$ is given a challenge ciphertext ct, which is an encryption of m_b for a random bit b. $\mathcal{A}$ can make more queries to the encryption oracle after receiving ct. At some point, $\mathcal{A}$ outputs a bit b' (as its guess of b), and the game ends. The output of the game is b'.

For convenience, we split $\mathcal{A}$, into two algorithms denoted $\mathcal{A} := (\mathcal{A}_1, \mathcal{A}_2)$. Algorithm $\mathcal{A}_1$ participates in the "find" stage and outputs (m_0^*, m_1^*) and some state information st (which includes public-parameters). Algorithm $\mathcal{A}_2$ represents the actions of $\mathcal{A}$ after the find stage—$\mathcal{A}_2$ receives the challenge ciphertext ct, and the state information st, and outputs the bit b'. Formally, this game is captured by a random process, denoted $\mathsf{Game}_{\Pi, \mathcal{A}, \lambda}^{\mathrm{FTG}}(b)$, which appears in table 1. For succinctness, we adopt the convention that sk includes the public-parameters pp, and we write $\mathsf{Enc}_{sk}(m)$ to mean $\mathsf{Enc}(pp, sk, m)$.

Let the queries of $\mathcal{A}_1$ to the encryption oracle be $(m_1, \ldots, m_\ell)$, and the queries of $\mathcal{A}_2$ be $(m_{\ell+1}, \ldots, m_n)$. We say that $\mathcal{A}$ is a *valid* FTG-adversary if sequences X_0 and X_1 have the same equality pattern, where $X_0 = (m_1, \ldots, m_\ell, m_0^*, m_{\ell+1}, \ldots, m_n)$ and $X_1 = (m_1, \ldots, m_\ell, m_1^*, m_{\ell+1}, \ldots, m_n)$; that is $\mathsf{Eqp}(X_0) = \mathsf{Eqp}(X_1)$. Define the advantage of a valid FTG-adversary $\mathcal{A} = (\mathcal{A}_1, \mathcal{A}_2)$ as follows:

$$\mathsf{Adv}_{\Pi, \mathcal{A}, \lambda}^{\mathrm{FTG}} = \left| \Pr\left[\mathsf{Game}_{\Pi, \mathcal{A}, \lambda}^{\mathrm{FTG}}(1) = 1 \right] - \Pr\left[\mathsf{Game}_{\Pi, \mathcal{A}, \lambda}^{\mathrm{FTG}}(0) = 1 \right] \right|$$

---

[3] Equivalently, every sequence is an ordered multi-set of $[n]^k$. Note that multi-set is important since the property is defined for sequences of the form $P(m, \ldots, m)$. Likewise, order is important since changing the message-order may change the value of P.

Definition 3.1 (FtG Security). *Let Π = (Setup, Enc, Dec, Test) be a symmetric property-preserving encryption scheme with plaintext space $\mathcal{M}$ and associated property $P : \mathcal{M}^k \to \{0,1\}$ for a fixed positive integer $k \in \mathbb{N}$. We say that Π is FtG-secure, if there exists a negligible function $\mathsf{negl}(\cdot)$ such that for all probabilistic polynomial time valid FtG-adversaries $\mathcal{A} = (\mathcal{A}_1, \mathcal{A}_2)$, and for all sufficiently large $\lambda \in \mathbb{N}$, the advantage of $\mathcal{A}$ in game $\mathsf{Game}_{\Pi,\mathcal{A},\lambda}^{\mathrm{FtG}}(b)$ is at most $\mathsf{negl}(\lambda)$. That is, $\mathsf{Adv}_{\Pi,\mathcal{A},\lambda}^{\mathrm{FtG}} \leq \mathsf{negl}(\lambda)$.*

Table 1. Security games for defining the three notions—FtG, LoR, and sRvR

$\mathsf{Game}_{\Pi,\mathcal{A},\lambda}^{\mathrm{FtG}}(b)$	$\mathsf{Game}_{\Pi,\mathcal{A},\lambda}^{\mathrm{LoR}}(b)$	$\mathsf{Game}_{\Pi,\mathcal{A},\lambda}^{\mathrm{sRvR}}(b)$
$(pp, sk) \leftarrow \mathsf{Setup}(1^\lambda)$		$(pp, sk) \leftarrow \mathsf{Setup}(1^\lambda)$
$(m_0^*, m_1^*, st) \leftarrow \mathcal{A}_1^{\mathsf{Enc}_{sk}(\cdot)}(pp)$	$(pp, sk) \leftarrow \mathsf{Setup}(1^\lambda)$	$(\xi, st) \leftarrow \mathcal{A}_1(pp)$
$ct^* \leftarrow \mathsf{Enc}_{sk}(m_b^*)$	$b' \leftarrow \mathcal{A}^{\mathsf{Enc}_{sk}(\mathrm{LR}(\cdot,\cdot,b))}(pp)$	$Z \xleftarrow{\$} \mathcal{S}(\xi)$
$b' \leftarrow \mathcal{A}_2^{\mathsf{Enc}_{sk}(\cdot)}(st, ct^*)$	**return** b'	$b' \leftarrow \mathcal{A}_2^{\mathsf{Enc}_{sk}(\mathrm{LR}(\cdot,Z,b))}(pp)$
return b'		**return** b'

Left-or-Right Security. Define left-or-right *encryption* oracle, denoted by $\mathsf{Enc}(pp, sk, \mathrm{LR}(\cdot, \cdot, b))$, which behaves as follows. On input a pair of equal-length messages $(m_0, m_1) \in \mathcal{M}^2$, the oracle obtains message $\mathrm{LR}(m_0, m_1, b) = m_b$, and then outputs a ciphertext by computing $\mathsf{Enc}(pp, sk, m_b)$. Once again, we drop pp from the notation for succinctness, and denote this oracle by $\mathsf{Enc}_{sk}(\mathrm{LR}(\cdot, \cdot, b))$.

In this security definition, $\mathcal{A}$ participates in a game in which he gets access to $\mathsf{Enc}_{sk}(\mathrm{LR}(\cdot, \cdot, b))$ for a random b. Throughout the execution of the game, $\mathcal{A}$ adaptively submits the queries of the form (m_i^0, m_i^1) to the encryption oracle and receives $ct_i = \mathsf{Enc}_{sk}(m_i^b)$ for $i = 1, \ldots, n$ where $n = n(\lambda)$ is an arbitrary polynomial. At some point, $\mathcal{A}$ outputs a bit b' (as its guess of b), and the game ends. The output of the game is b'. Formally, this game is captured by a random process, denoted $\mathsf{Game}_{\Pi,\mathcal{A},\lambda}^{\mathrm{LoR}}(b)$, which appears in table 1. Let the queries of $\mathcal{A}$ to the oracle be $\left\{ (m_i^0, m_i^1) \right\}_{i=1}^n$, and let $X_0 = (m_1^0, \ldots, m_n^0)$ and $X_1 = (m_1^1, \ldots, m_n^1)$. We say that $\mathcal{A}$ is a *valid* LoR-adversary if sequences X_0 and X_1 have the same equality pattern; that is $\mathsf{Eqp}(X_0) = \mathsf{Eqp}(X_1)$. The advantage of a valid LoR-adversary $\mathcal{A}$ is defined as before:

$$\mathsf{Adv}_{\Pi,\mathcal{A},\lambda}^{\mathrm{LoR}} = \left| \Pr\left[\mathsf{Game}_{\Pi,\mathcal{A},\lambda}^{\mathrm{LoR}}(1) = 1 \right] - \Pr\left[\mathsf{Game}_{\Pi,\mathcal{A},\lambda}^{\mathrm{LoR}}(0) = 1 \right] \right|$$

Definition 3.2 (LoR Security). *Let Π = (Setup, Enc, Dec, Test) be a symmetric property-preserving encryption scheme with plaintext space $\mathcal{M}$ and associated property $P : \mathcal{M}^k \to \{0,1\}$ for a fixed positive integer $k \in \mathbb{N}$. We say that Π is LoR-secure, if there exists a negligible function $\mathsf{negl}(\cdot)$ such that for all probabilistic polynomial time valid LoR-adversaries $\mathcal{A}$, and for all sufficiently large $\lambda \in \mathbb{N}$, the advantage of $\mathcal{A}$ in game $\mathsf{Game}_{\Pi,\mathcal{A},\lambda}^{\mathrm{LoR}}(b)$ is at most $\mathsf{negl}(\lambda)$. That is, $\mathsf{Adv}_{\Pi,\mathcal{A},\lambda}^{\mathrm{LoR}} \leq \mathsf{negl}(\lambda)$.*

We note that in their work on symmetric-key predicate encryption, Shen, Shi, and Waters [41] called the $\mathrm{F\scriptstyle TG}$-security as the "single-challenge" security, and the $\mathrm{L\scriptstyle OR}$-security as the "full-security."

Real-versus-Random Security. Another interesting notion considered in [5] is that of "real-or-random" security, where the attacker instead of giving two sequences gives only one, called the *real*, sequence). In return, it either receives the encryption of the messages from real sequence, or the encryption of *random* messages (which form the *random* sequence). As discussed earlier, adopting this notion to the setting of property-preserving encryption is slightly tricky.

Recalling briefly, the real sequence allows the adversary $\mathcal{A}$ to learn its equality pattern; and therefore indistinguishability makes sense only if a random sequence with the same equality pattern is selected. However, if the real sequence is selected adaptively, its equality pattern also evolves adaptively; but since $\mathcal{A}$ must receive encryptions "on-the-fly," providing encryptions of random messages that "in-the-end" would have the same equality pattern as the real sequence may not always be possible. It is for this reason that defining a meaningful "simulation-based" definition is difficult in this setting.

Nevertheless, a meaningful definition can still be achieved if we do not allow the adversary to *adaptively* evolve the security pattern of the real sequence. That is, we consider a *static* or *selective* setting, where the $\mathcal{A}$ "announces" the equality pattern that the real sequence will have at the beginning of the game (on input the public-parameters). This is much like the the selective-ID model of [18,19].[4]

The *selective* real-versus-random security denoted by sRvR, considers a game that is identical to the game in $\mathrm{L\scriptstyle OR}$-security except for the following difference. The adversary is a pair of algorithms $\mathcal{A} = (\mathcal{A}_1, \mathcal{A}_2)$ such that $\mathcal{A}_1$ on input the public-parameters, outputs a binary vector ξ of length polynomial in λ, and a state information st (which includes public-parameters). Vector ξ represents an equality-pattern and fixes an integer $n \in \mathbb{N}$. A random sequence $Z = (z_1, \ldots, z_n) \in \mathcal{M}^n$ is chosen such that $\mathsf{Eqp}(Z) = \xi$. $\mathcal{A}$ is given access to an encryption oracle which accepts queries of the form $m \in \mathcal{M}$; upon i^{th}-query m_i, the oracle returns the value of $\mathsf{Enc}_{sk}(\mathsf{LR}(m_i, z_i, b))$. We slightly abuse the notation, and denote this special oracle by $\mathsf{Enc}_{sk}(\mathsf{LR}(\cdot, Z, b))$. This game is formally captured by a random process, denoted $\mathsf{Game}^{\mathrm{sRvR}}_{\Pi, \mathcal{A}, \lambda}(b)$, which appears in table 1. Denote by $\mathcal{S}(\xi)$ the set of all message-sequences whose equality pattern is ξ.

We say that $\mathcal{A}$ is a valid sRvR-adversary if the sequence of messages queried by $\mathcal{A}$, denoted $M \in \mathcal{M}^n$ is such that $\mathsf{Eqp}(M) = \xi$. Define the advantage $\mathsf{Adv}^{\mathrm{sRvR}}_{\Pi, \mathcal{A}, \lambda}$ and the sRvR-security of Π for a valid sRvR-adversary $\mathcal{A}$, analogous to $\mathsf{Adv}^{\mathrm{LoR}}_{\Pi, \mathcal{A}, \lambda}$ and $\mathrm{L\scriptstyle OR}$-security by replacing the word $\mathrm{L\scriptstyle OR}$ with sRvR.

Remarks on the Hierarchy. As noted earlier, we show that there is a hierarchy of security notions that does not collapse. The security notion $\mathrm{F\scriptstyle TG}^{\eta}$ is

---

[4] The fact that in our model, the public-parameters pp are given *before* $\mathcal{A}$ decides the equality pattern does not make our model necessarily better. Indeed, pp are irrelevant since we are dealing with symmetric encryption; in particular, pp can be included simply as part of the ciphertext.

identical to FTG except that the adversary has multiple find stages, and sends exactly η pairs of challenges. Likewise, the sRvR notion reduces to the selective variant of the LoR notion, denoted sLoR: the only difference is that in sLoR definition, $\mathcal{A}$ announces the security pattern ξ of the two sequences before seeing any encryptions. Due to space constraints, the formal definitions of FTG$^\eta$, sLoR are given in the full version.

4 Relations among Security Notions

In this section, we will establish relationships between various notions security for symmetric property-preserving encryption (PPEnc). The main result of this section is that FTG$^\eta$ does not imply FTG$^{\eta+1}$. We will start with the simpler case that FTG-security does not imply LoR-security—not even the selective variants sLoR and sRvR. All other implications are rather trivial.

Informally, for a symmetric PPEnc Π for a property P, we say that LoR-security implies FTG-security, denoted LoR $\rightarrow$ FTG, to mean the following statement: "If Π satisfies LoR-security (i.e., definition 3.2) then it also satisfies FTG-security (i.e., definition 3.1)." In [5], it was shown that, for an *ordinary* symmetric encryption scheme, FTG-security and LoR-security, are in fact equivalent (up to a polynomial degradation in security). Which means that FTG implies LoR, and vice-versa. The same proof shows that FTG$^{\eta+1}$ $\rightarrow$ FTG$^\eta$ for every $\eta \in \mathbb{N}$.

4.1 LoR vs. FtG

First off, it is trivial to see that LoR implies FTG. In case of an ordinary[5] scheme, to simulate the FTG-game for an attacker, a simulator participates in an LoR game. To answer encryption queries of $\mathcal{A}$ (in "find" stage and after the challenge ciphertext) which consist of a single message $m \in \mathcal{M}$, the simulator can simply send a query of the form $(m, m) \in \mathcal{M}^2$ to its left-or-right-encryption oracle, and give the answer to $\mathcal{A}$. The challenge-query (m_0^*, m_1^*) can be used directly. This strategy also applies to our setting of symmetric PPEnc, with *no* change. The key observation is that the sequences sent by the simulator to the outside oracle have the same equality pattern, simply because $\mathcal{A}$ is a valid FTG-adversary. This proof is omitted, and we conclude that LoR $\rightarrow$ FTG for all P.

To prove the other direction, i.e., FTG $\rightarrow$ LoR, a simple hybrid experiment is used in [5] in which the left sequence is converted into the right sequence by changing one message at a time. While this works for an ordinary encryption scheme, this approach breaks down in case of PPEnc. In particular, in the i-th hybrid, as we change the encryption of i-th "left" message to the corresponding right message, the equality pattern may change. It might even be true that the right-sequence is not "reachable" from the left-sequence for every property P by changing one message at a time. In this case we say that the two sequences belong in different equivalence classes.

---

[5] That is, it is not necessarily a property-preserving encryption scheme.

Proving the Separation. To separate FTG from LOR, our goal is to think of a property P (preferably, a natural property) and an encryption scheme Π such that: P divides its message space in only a small number of equivalence classes, and Π leaks the "identity" of the equivalence class at the end of the security game. This will not break FTG-security, but by choosing two sequences with same equality pattern but different equivalence classes, LOR-security can be broken.

We will use quadratic residuosity to construct a property. For a prime number p, define by $\mathcal{QR}_p$ and $\mathcal{QNR}_p$ the set of quadratic residues and quadratic non-residues respectively in $\mathbb{Z}_p^*$. It will be convenient to define the following "sign" function $\mathcal{J}$, which outputs whether a message $m \in \mathbb{Z}_p^*$ is a quadratic residue or not:[6] if $m \in \mathcal{QR}_p$ then $\mathcal{J}(m) = 0$, otherwise (i.e., $m \in \mathcal{QNR}_p$), $\mathcal{J}(m) = 1$. For any two messages $(x, y) \in \mathbb{Z}_p^* \times \mathbb{Z}_p^*$, we define the following binary property:

$$P_{\mathrm{qr}}(x, y) = \begin{cases} 1 \text{ if } x \cdot y \in \mathcal{QR}_p \\ 0 \text{ if } x \cdot y \in \mathcal{QNR}_p \end{cases}$$

We now prove the following theorem.

Theorem 4.1 (FtG $\nrightarrow$ LoR). *Suppose there exists a FTG-secure property-preserving symmetric encryption scheme Π for property P_{qr} and plaintext-space $\mathcal{M} = \mathbb{Z}_p^*$. Then there exists another property-preserving symmetric encryption scheme Π^* for property P_{qr} and plaintext space $\mathcal{M}$ such that Π^* is FTG-secure, but it is not LOR-secure.*

Proof. The key-idea in our proof is that the property P_{qr} puts a nice structure on the equality pattern of adversary's queries. We will use a one-time pad to hide crucial information about this structure in the ciphertext, which can be recovered in the LOR-game but not in the FTG-game.

Let $\Pi = (\mathsf{Setup}, \mathsf{Enc}, \mathsf{Dec}, \mathsf{Test})$. We construct a new scheme $\Pi^* = (\mathsf{Setup}^*, \mathsf{Enc}^*, \mathsf{Dec}^*, \mathsf{Test}^*)$, whose algorithms are defined as follows.

1. The Setup^* algorithm calls $\mathsf{Setup} \to (pp, sk)$, it then picks a uniformly random bit $t \xleftarrow{\$} \{0, 1\}$. It outputs pp as the public-parameters and the secret-key is set to the pair $sk^* = (sk, t)$. The bit t will be used as a one-time pad.
2. Algorithm Enc^* encrypts an input $m \in \mathbb{Z}_p^*$ as follows. It calls $\mathsf{Enc}(pp, sk, m) \to ct$. Then it selects a uniformly random bit $b \xleftarrow{\$} \{0, 1\}$. If $b = 0$ the output ciphertext is $ct^* = (ct, b, t)$; otherwise, $b = 1$ and the ciphertext is $ct^* = (ct, b, t \oplus \mathcal{J}(m))$. Namely if $b = 0$ the ciphertext reveals the one-time pad, otherwise the XOR of the pad with the residuosity sign. Compactly, the ciphertext is $ct^* = (\mathsf{Enc}(pp, sk, m), b, t \oplus (b \wedge \mathcal{J}(m)))$.
3. The decryption algorithm, on input (ct, b, c) outputs $\mathsf{Dec}(pp, sk, ct)$. The test algorithm on input (ct_1, b_1, c_1) and (ct_2, b_2, c_2) outputs $\mathsf{Test}(pp, ct_1, ct_2)$.

It is easy to see to see that Π^* satisfies all the correctness properties if Π does. We have to show that Π^* is FTG-secure but not LOR-secure. This follows from lemmas 4.2 and 4.3. This completes the proof.

---

[6] This is essentially the Legendre symbol with -1 replaced by 0.

Lemma 4.2. *For every valid* FTG *adversary* $\mathcal{A}$ *for* Π^*, *there exists a valid* FTG *adversary* $\mathcal{B}$ *for* Π *such that for every* $\lambda \in \mathbb{N}$, $\mathsf{Adv}^{\mathrm{FTG}}_{\Pi^*,\mathcal{A},\lambda} = \mathsf{Adv}^{\mathrm{FTG}}_{\Pi,\mathcal{B},\lambda}$

Proof. We construct adversary (a.k.a. simulator) $\mathcal{B}$, using $\mathcal{A}$. However, before doing so, we first analyze the possible attack sequences for $\mathcal{A}$. Remember that $\mathcal{A}$ participates in an FTG-game against Π^*, and is denoted by $\mathcal{A} = (\mathcal{A}_1, \mathcal{A}_2)$. Further, it must satisfy the equality-pattern condition.

According to the definition of the FTG game, $\mathcal{A}_1$ will query for the messages $m_1, m_2, \ldots, m_\ell$ (in the "find" phase), and output a *challenge* pair (m_0^*, m_1^*) along with some state information. Then $\mathcal{A}_2$, on input a ciphertext and the state, will query for the messages $m_{\ell+1}, m_{\ell+2}, \ldots, m_n$ (in phase 2) and output a guess. There are only two possible cases regarding the challenge pair:

Case 1: $\mathcal{J}(m_0^*) = \mathcal{J}(m_1^*)$. That is, either both messages are quadratic residues, or both are non-residues.

Case 2: $\mathcal{J}(m_0^*) \neq \mathcal{J}(m_1^*)$. That is, one message is a quadratic residue, and the other is a non-residue. Notice that in this case it holds that neither $\mathcal{A}_1$ nor $\mathcal{A}_2$ makes any queries to the encryption oracle. That is, no queries are made either in phase-1 or phase-2. Indeed, suppose that either $\mathcal{A}_1$ or $\mathcal{A}_2$ queries m and receives $ct = \mathsf{Enc}_{sk}(m)$. Then, by the properties of quadratic residues, we have that $P_{\mathrm{qr}}(m, m_0^*) \neq P_{\mathrm{qr}}(m, m_1^*)$. This violates the equality pattern condition since P_{qr} can be learned from ct and ct^* (which $\mathcal{A}_2$ receives).

Now, the adversary $\mathcal{B} = (\mathcal{B}_1, \mathcal{B}_2)$ when participating in the FTG-game for Π, internally simulates the FTG-game for $\mathcal{A}$ (with scheme Π^*) as follows. $\mathcal{B}_1$ on input the public parameters of Π, forwards them to $\mathcal{A}_1$. $\mathcal{A}$ must follow one of the two cases above. Suppose that $\mathcal{A}$ follows Case-1. In this case, if $\mathcal{A}_1$ makes a single-message encryption query, $\mathcal{B}_1$ forwards this query to the outside encryption oracle, and gives $\mathcal{A}_1$ whatever the answer is. At some point, $\mathcal{A}_1$ outputs (m_0^*, m_1^*, st); then $\mathcal{B}_1$ also outputs this triplet and halts.

Algorithm $\mathcal{B}_2$ picks a uniformly random one-time pad $t \xleftarrow{\$} \{0, 1\}$ and stores it. $\mathcal{B}_2$ receives a ciphertext ct' (and state st) as input. Note that ct' is a ciphertext of Π. To construct a ciphertext of Π^*, $\mathcal{B}_2$ picks a random bit b, and sets $ct^* = (ct', b, t)$ if $b = 0$; otherwise it sets $ct^* = (ct', b, (t \oplus \mathcal{J}(m_0^*))$. This is a correctly distributed ciphertext since $\mathcal{J}(m_0^*) = \mathcal{J}(m_1^*)$. $\mathcal{B}_2$ internally provides (ct^*, st) to $\mathcal{A}_2$. Encryption queries of $\mathcal{A}_2$ are answered by $\mathcal{B}_1$ using its encryption oracle. It is clear that the simulation is perfect.

If on the other hand $\mathcal{A}_1$ gives out at the beginning of the game a challenge pair that consists of a residue and a non residue, we are in case-2. This means that no encryption queries are made by $\mathcal{A}_1$, and none will be made by $\mathcal{A}_2$. So $\mathcal{B}_1$ also simply outputs this pair and the state information to outside experiment. Upon receiving a challenge ciphertext and state, it gives the following ciphertext to $\mathcal{A}_2$: (ct, b, c) where both b and c are uniformly random bits. The state information is also given to $\mathcal{A}_2$. In this case also the simulation is perfect, since irrespective of the value of b, c is distributed correctly as in a proper ciphertext (every value of c defines an implicit value for the one-time pad, which is information theoretically hidden since there are no other encryption queries made). This completes the proof.

Lemma 4.3. *There exists a valid polynomial-time* LoR *attacker on* Π^* *with advantage* $1 - 2^{-n+1}$*, where* n *is the number of queries it makes.*

Proof. The attacker proceeds as follows in the LoR-game. It sends queries such that the the left-sequence contains only quadratic-residues, while the right-sequence contains only quadratic-non-residues. Notice that this a valid pair of sequences since the equality patterns are the same with respect to property P_{qr}: the output of the property is always 1 for any pair of messages in each sequence. However if the length of each sequence is n, then with probability $q = 1 - 2 \cdot \left(\frac{1}{2}\right)^n = 1 - 2^{-n+1}$, there will be two ciphertexts (ct_1, b_1, c_1) and (ct_2, b_2, c_2) for which $b_1 \neq b_2$. In this case, the value $c_1 \oplus c_2$ reveals the residuosity-sign of one of the two streams. Since this sign is known to the attacker and it is different for the two streams, it compromises LoR-security. In the unlikely case when $b_1 = b_2$ for all ciphertexts, the attacker fails, say by outputting 0, giving us the required advantage.

Our next goal is to separate $\mathrm{FTG}^{\eta+1}$ from FTG^{η}. The following theorem will be proven in the full version using the same property P_{qr}.

Theorem 4.4 ($\mathbf{FtG}^{\eta} \nrightarrow \mathbf{FtG}^{\eta+1}$). *Let* $\eta \in \mathbb{N}$ *be a fixed positive integer. Suppose there exists a* FTG^{η}*-secure property-preserving symmetric encryption scheme* Π *for property* P_{qr} *and plaintext-space* $\mathcal{M} = \mathbb{Z}_p^*$*. Then there exists another property-preserving symmetric encryption scheme* Π^* *for property* P_{qr} *and plaintext space* $\mathcal{M}$ *such that* Π^* *is* FTG^{η}*-secure, but it is not* $\mathrm{FTG}^{\eta+1}$*-secure.*

5   Constructions of Property-Preserving Encryption

In this section, we present constructions of property preserving encryption (PPEnc) encryption scheme. Instead of constructing the full-fledged scheme, it suffices to construct a slightly weaker variant, called *property-preserving tag scheme* (PPTag). A PPTag scheme allows us to test the property Test, without having a decryption algorithm. We can get correct decryption by utilizing appropriately any IND − CPA secure *symmetric* encryption scheme. We refer the reader to [31,41] for this somewhat standard approach.

To start with, we note that for unary properties P, one can simply include the value of $P(m)$ in the ciphertext, to get a construction. Therefore, we focus on properties of higher arity. In the full version of the paper, we present a generic construction of PPTag for any binary property from adaptively fully secure predicate encryption[41]. The main idea of this construction is that the new encryption algorithm calls the encryption algorithm of the original predicate encryption scheme and the token generation algorithm, both with input the message m. The resulting ciphertext consists of a ciphertext part and a token part. A selectively fully secure scheme is given in [41], which is not sufficient for our LoR security definition. Therefore, we present an explicit PPEnc construction in the following section.

5.1   An Explicit Construction for Testing Orthogonality

This is a construction for testing orthogonality of two vectors. The plaintext space of our scheme is $\mathcal{M} = (\mathbb{Z}_N^* \cup \{0\})^n$ where $N = pq$ for two λ-bit primes p and q, $\mathbb{Z}_N^*$ is the set of invertible elements of $\mathbb{Z}_N$, and $n :\in \mathbb{N} \to \mathbb{N}$ polynomial in λ. [7] The associated property $P : \mathcal{M} \times \mathcal{M} \to \{0, 1\}$ is such that: $P(\vec{u}, \vec{v}) = 0$ if $\vec{u} \cdot \vec{v} = 0 \mod p$ and 1 otherwise. The algorithms of our scheme are the following:

- Setup$(1^\lambda, n) \to (pp, sk)$: Pick two different prime numbers p, q uniformly in the range $[2^{\lambda-1}, 2^\lambda)$, where $\lambda \geq 3$. Pick a group $\mathbb{G}$ of order $N = pq$ with a bilinear map $e : \mathbb{G} \times \mathbb{G} \to \mathbb{G}_T$. Select two random generators g_0, g_1 for subgroups of order p and q respectively.
  Let $\mathcal{S}_n \stackrel{\text{def.}}{=} \{(x_1, \ldots, x_n) \in \mathbb{Z}_q^n \mid \sum_{i=1}^n x_i^2 \in \mathcal{QR}_q\}$ be a set of vectors with n components. Select a vector $\gamma = (\gamma_1, \ldots, \gamma_n)$ uniformly from the set $\mathcal{S}_n$. Finally, let $\delta \in \mathbb{Z}_q$ be such that $\delta^2 = \sum_{i=1}^n \gamma_i^2$ (pick one of the two at random), and $\mathbb{1}_{\mathbb{G}}$ be the identity element of $\mathbb{G}$. The parameters output by the algorithm are:

$$pp = (\lambda, n, N, \mathbb{G}, \mathbb{G}_T, e, \mathbb{1}_{\mathbb{G}}) \qquad sk = (g_0, g_1, \{\gamma_i\}_{i=1}^n, \delta)$$

- Enc$(pp, sk, M) \to ct$: On input a message $M = (m_1, m_2, \ldots, m_n)$ the algorithm picks two random elements of $\mathbb{Z}_N$: $\phi, \psi \stackrel{\$}{\leftarrow} \mathbb{Z}_N$. It outputs the following ciphertext:

$$ct = (ct_0, \{ct_i\}_{i=1}^n) = \left(g_1^{\psi\delta}, \left\{ g_0^{\phi m_i} \cdot g_1^{\psi\gamma_i} \right\}_{i=1}^n \right)$$

- Test$(pp, ct^{(1)}, ct^{(2)}) \to \{0, 1\}$: On input the two ciphertexts $ct^{(1)} = \left(ct_0^{(1)}, \{ct_i^{(1)}\}_{i=1}^n \right)$ and $ct^{(2)} = \left(ct_0^{(2)}, \{ct_i^{(2)}\}_{i=1}^n \right)$, the algorithm outputs 0 if and only if

$$\prod_{i=1}^n e\left(ct_i^{(1)}, ct_i^{(2)} \right) = e\left(ct_0^{(1)}, ct_0^{(2)} \right)$$

Correctness. Correctness is satisfied, except with negligible probability, due to the following:

$$\prod_{i=1}^n e\left(ct_i^{(1)}, ct_i^{(2)} \right) = \prod_{i=1}^n e\left(g_0^{\phi^{(1)} m_i^{(1)}} \cdot g_1^{\psi^{(1)}\gamma_i}, g_0^{\phi^{(2)} m_i^{(2)}} \cdot g_1^{\psi^{(2)}\gamma_i} \right)$$

$$= \prod_{i=1}^n e(g_0, g_0)^{\phi^{(1)}\phi^{(2)} m_i^{(1)} m_i^{(2)}} e(g_1, g_1)^{\psi^{(1)}\psi^{(2)}\gamma_i^2}$$

$$= e(g_0, g_0)^{\phi^{(1)}\phi^{(2)} \vec{m}^{(1)} \cdot \vec{m}^{(2)}} e(g_1, g_1)^{\psi^{(1)}\psi^{(2)} \sum_i \gamma_i^2}$$

---

[7] Since the factorization of N is not public, the plaintext space is not public. However if we assume that factoring is hard, any user that generates messages in $\mathbb{Z}_N$ will, except with negligible probability, generate a message in the correct plaintext space $\mathbb{Z}_N^* \cup \{0\}$.

$$e\left(ct_0^{(1)}, ct_0^{(2)}\right) = e\left(g_1^{\psi^{(1)}\delta}, g_1^{\psi^{(2)}\delta}\right)$$
$$= e\left(g_1, g_1\right)^{\psi^{(1)}\psi^{(2)}\delta^2}$$
$$= e\left(g_1, g_1\right)^{\psi^{(1)}\psi^{(2)}\sum_i \gamma_i^2}$$

In the full version, we prove that our construction satisfies LoR-security in the generic group model. We follow the terminology and proof ideas of [13] and [9]. We assume that the group elements of groups $\mathbb{G}$ and $\mathbb{G}_T$ are encoded by two random encodings $\psi, \psi_T : \mathbb{F}_N \to \{0,1\}^m$. These are injective functions that define the groups $\mathbb{G} = \{\psi(i) | i \in \mathbb{F}_N\}$ and $\mathbb{G}_T = \{\psi_T(i) | i \in \mathbb{F}_N\}$. We are also given functions to compute the group operations on $\mathbb{G}$ and $\mathbb{G}_T$ and a function that computes the non degenerate bilinear mapping e. Then, we prove the following theorem.

Theorem 5.1. *Let $\psi, \psi_T, \mathbb{G}, \mathbb{G}_T$ be as above, and let $\mathcal{A}$ be a generic algorithm, representing a valid LoR-adversary against the scheme described above. Further, suppose that $\mathcal{A}$ makes at most Q encryption queries, and at most W group operations and pairings counted together. Then the advantage of $\mathcal{A}$ in the LoR-game is at most $O\left((nQ + W)^2 \cdot 2^{-\lambda}\right)$.*

Acknowledgments. We are thankful to the Math Overflow online community, especially to the users Noam D. Elkies, GH, and Gerry Myerson, for their swift responses regarding sums of squares modulo a prime number [35], to Brent Waters for useful discussions about predicate encryption and to the anonymous reviewers for their insightful comments.

References

1. Abdalla, M., Bellare, M., Catalano, D., Kiltz, E., Kohno, T., Lange, T., Malone-Lee, J., Neven, G., Paillier, P., Shi, H.: Searchable encryption revisited: Consistency properties, relation to anonymous IBE, and extensions. In: Shoup, V. (ed.) CRYPTO 2005. LNCS, vol. 3621, pp. 205–222. Springer, Heidelberg (2005)
2. Alon, N., Matias, Y., Szegedy, M.: The space complexity of approximating the frequency moments. In: STOC, pp. 20–29 (1996)
3. Bellare, M., Boldyreva, A., Micali, S.: Public-Key Encryption in a Multi-user Setting: Security Proofs and Improvements. In: Preneel, B. (ed.) EUROCRYPT 2000. LNCS, vol. 1807, pp. 259–274. Springer, Heidelberg (2000)
4. Bellare, M., Boldyreva, A., O'Neill, A.: Deterministic and Efficiently Searchable Encryption. In: Menezes, A. (ed.) CRYPTO 2007. LNCS, vol. 4622, pp. 535–552. Springer, Heidelberg (2007)
5. Bellare, M., Desai, A., Jokipii, E., Rogaway, P.: A concrete security treatment of symmetric encryption. In: FOCS, pp. 394–403 (1997)
6. Bellare, M., Desai, A., Pointcheval, D., Rogaway, P.: Relations among Notions of Security for Public-Key Encryption Schemes. In: Krawczyk, H. (ed.) CRYPTO 1998. LNCS, vol. 1462, pp. 26–45. Springer, Heidelberg (1998)

7. Bellare, M., Fischlin, M., O'Neill, A., Ristenpart, T.: Deterministic Encryption: Definitional Equivalences and Constructions without Random Oracles. In: Wagner, D. (ed.) CRYPTO 2008. LNCS, vol. 5157, pp. 360–378. Springer, Heidelberg (2008)
8. Bellare, M., Ristenpart, T., Rogaway, P., Stegers, T.: Format-Preserving Encryption. In: Jacobson Jr., M.J., Rijmen, V., Safavi-Naini, R. (eds.) SAC 2009. LNCS, vol. 5867, pp. 295–312. Springer, Heidelberg (2009)
9. Bethencourt, J., Sahai, A., Waters, B.: Ciphertext-policy attribute-based encryption. In: IEEE Symposium on Security and Privacy, pp. 321–334 (2007)
10. Boldyreva, A., Chenette, N., Lee, Y., O'Neill, A.: Order-Preserving Symmetric Encryption. In: Joux, A. (ed.) EUROCRYPT 2009. LNCS, vol. 5479, pp. 224–241. Springer, Heidelberg (2009)
11. Boldyreva, A., Chenette, N., O'Neill, A.: Order-Preserving Encryption Revisited: Improved Security Analysis and Alternative Solutions. In: Rogaway, P. (ed.) CRYPTO 2011. LNCS, vol. 6841, pp. 578–595. Springer, Heidelberg (2011)
12. Boldyreva, A., Fehr, S., O'Neill, A.: On Notions of Security for Deterministic Encryption, and Efficient Constructions without Random Oracles. In: Wagner, D. (ed.) CRYPTO 2008. LNCS, vol. 5157, pp. 335–359. Springer, Heidelberg (2008)
13. Boneh, D., Boyen, X., Goh, E.-J.: Hierarchical Identity Based Encryption with Constant Size Ciphertext. In: Cramer, R. (ed.) EUROCRYPT 2005. LNCS, vol. 3494, pp. 440–456. Springer, Heidelberg (2005)
14. Boneh, D., Di Crescenzo, G., Ostrovsky, R., Persiano, G.: Public Key Encryption with Keyword Search. In: Cachin, C., Camenisch, J.L. (eds.) EUROCRYPT 2004. LNCS, vol. 3027, pp. 506–522. Springer, Heidelberg (2004)
15. Boneh, D., Sahai, A., Waters, B.: Functional Encryption: Definitions and Challenges. In: Ishai, Y. (ed.) TCC 2011. LNCS, vol. 6597, pp. 253–273. Springer, Heidelberg (2011)
16. Boneh, D., Waters, B.: Conjunctive, Subset, and Range Queries on Encrypted Data. In: Vadhan, S.P. (ed.) TCC 2007. LNCS, vol. 4392, pp. 535–554. Springer, Heidelberg (2007)
17. Boyen, X., Waters, B.: Anonymous Hierarchical Identity-Based Encryption (Without Random Oracles). In: Dwork, C. (ed.) CRYPTO 2006. LNCS, vol. 4117, pp. 290–307. Springer, Heidelberg (2006)
18. Canetti, R., Halevi, S., Katz, J.: A Forward-Secure Public-Key Encryption Scheme. In: Biham, E. (ed.) EUROCRYPT 2003. LNCS, vol. 2656, pp. 255–271. Springer, Heidelberg (2003)
19. Canetti, R., Halevi, S., Katz, J.: Chosen-Ciphertext Security from Identity-Based Encryption. In: Cachin, C., Camenisch, J.L. (eds.) EUROCRYPT 2004. LNCS, vol. 3027, pp. 207–222. Springer, Heidelberg (2004)
20. Chase, M.: Multi-authority Attribute Based Encryption. In: Vadhan, S.P. (ed.) TCC 2007. LNCS, vol. 4392, pp. 515–534. Springer, Heidelberg (2007)
21. Creeger, M.: Cloud computing: An overview. Queue 7, 2:3–2:4 (2009)
22. Curtmola, R., Garay, J.A., Kamara, S., Ostrovsky, R.: Searchable symmetric encryption: improved definitions and efficient constructions. In: ACM Conference on Computer and Communications Security, pp. 79–88 (2006)
23. Gennaro, R., Rohatgi, P.: How to Sign Digital Streams. In: Kaliski Jr., B.S. (ed.) CRYPTO 1997. LNCS, vol. 1294, pp. 180–197. Springer, Heidelberg (1997)
24. Gentry, C.: Fully homomorphic encryption using ideal lattices. In: STOC, pp. 169–178 (2009)
25. Golle, P., Staddon, J., Waters, B.: Secure Conjunctive Keyword Search over Encrypted Data. In: Jakobsson, M., Yung, M., Zhou, J. (eds.) ACNS 2004. LNCS, vol. 3089, pp. 31–45. Springer, Heidelberg (2004)

26. Goyal, V., Jain, A., Pandey, O., Sahai, A.: Bounded Ciphertext Policy Attribute Based Encryption. In: Aceto, L., Damgård, I., Goldberg, L.A., Halldórsson, M.M., Ingólfsdóttir, A., Walukiewicz, I. (eds.) ICALP 2008, Part II. LNCS, vol. 5126, pp. 579–591. Springer, Heidelberg (2008)
27. Goyal, V., Pandey, O., Sahai, A., Waters, B.: Attribute-based encryption for fine-grained access control of encrypted data. In: ACM Conference on Computer and Communications Security, pp. 89–98 (2006)
28. Guha, S., Meyerson, A., Mishra, N., Motwani, R., O'Callaghan, L.: Clustering data streams: Theory and practice. IEEE Trans. Knowl. Data Eng. 15(3), 515–528 (2003)
29. Henzinger, M., Raghavan, P., Rajagopalan, S.: Computing on data streams. Technical report, SRC Palo Alto, CA (1998)
30. Jain, A., Dubes, R.: Algorithms for Clustering Data. Prentice-Hall (1988)
31. Katz, J., Sahai, A., Waters, B.: Predicate Encryption Supporting Disjunctions, Polynomial Equations, and Inner Products. In: Smart, N.P. (ed.) EUROCRYPT 2008. LNCS, vol. 4965, pp. 146–162. Springer, Heidelberg (2008)
32. Katz, J., Yung, M.: Complete characterization of security notions for probabilistic private-key encryption. In: STOC, pp. 245–254 (2000)
33. Klien, M.: Six Benefits of Cloud Computing. Internet Article (2010), http://resource.onlinetech.com/the-six-benefits-of-cloud-computing/
34. Lewko, A., Okamoto, T., Sahai, A., Takashima, K., Waters, B.: Fully Secure Functional Encryption: Attribute-Based Encryption and (Hierarchical) Inner Product Encryption. In: Gilbert, H. (ed.) EUROCRYPT 2010. LNCS, vol. 6110, pp. 62–91. Springer, Heidelberg (2010)
35. Math Overflow. Sum of squares modulo a prime (2011), http://mathoverflow.net/questions/69576/sum-of-squares-modulo-a-prime
36. O'Neill, A.: Deterministic public-key encryption revisited. Cryptology ePrint Archive, Report 2010/533 (2010)
37. Ostrovsky, R.: Efficient computation on oblivious rams. In: STOC, pp. 514–523 (1990)
38. Ostrovsky, R., Sahai, A., Waters, B.: Attribute-based encryption with non-monotonic access structures. In: ACM Conference on Computer and Communications Security, pp. 195–203 (2007)
39. Rivest, R., Adleman, L., Dertouzos, M.: On data banks and privacy homomorphisms. In: Foundations of Secure Computation, pp. 169–177 (1978)
40. Sahai, A., Waters, B.: Fuzzy Identity-Based Encryption. In: Cramer, R. (ed.) EUROCRYPT 2005. LNCS, vol. 3494, pp. 457–473. Springer, Heidelberg (2005)
41. Shen, E., Shi, E., Waters, B.: Predicate Privacy in Encryption Systems. In: Reingold, O. (ed.) TCC 2009. LNCS, vol. 5444, pp. 457–473. Springer, Heidelberg (2009)
42. Shi, E., Bethencourt, J., Chan, H.T.-H., Song, D.X., Perrig, A.: Multi-dimensional range query over encrypted data. In: IEEE Symposium on Security and Privacy, pp. 350–364 (2007)
43. Shi, E., Waters, B.: Delegating Capabilities in Predicate Encryption Systems. In: Aceto, L., Damgård, I., Goldberg, L.A., Halldórsson, M.M., Ingólfsdóttir, A., Walukiewicz, I. (eds.) ICALP 2008, Part II. LNCS, vol. 5126, pp. 560–578. Springer, Heidelberg (2008)
44. Shoup, V.: Lower Bounds for Discrete Logarithms and Related Problems. In: Fumy, W. (ed.) EUROCRYPT 1997. LNCS, vol. 1233, pp. 256–266. Springer, Heidelberg (1997)
45. Song, D.X., Wagner, D., Perrig, A.: Practical techniques for searches on encrypted data. In: IEEE Symposium on Security and Privacy, pp. 44–55 (2000)

Narrow-Bicliques:
Cryptanalysis of Full IDEA

Dmitry Khovratovich[1], Gaëtan Leurent[2], and Christian Rechberger[3]

[1] Microsoft Research, USA
[2] University of Luxembourg, Luxembourg
[3] DTU, Denmark

Abstract. We apply and extend the recently introduced biclique framework to IDEA and for the first time describe an approach to noticeably speed-up key-recovery for the full 8.5 round IDEA.

We also show that the biclique approach to block cipher cryptanalysis not only obtains results on more rounds, but also improves time and data complexities over existing attacks. We consider the first 7.5 rounds of IDEA and demonstrate a variant of the approach that works with practical data complexity.

The conceptual contribution is the narrow-bicliques technique: the recently introduced independent-biclique approach extended with ways to allow for a significantly reduced data complexity with everything else being equal. For this we use available degrees of freedom as known from hash cryptanalysis to *narrow* the relevant differential trails. Our cryptanalysis is of high computational complexity, and does not threaten the practical use of IDEA in any way, yet the techniques are practically verified to a large extent.

Keywords: block ciphers, bicliques, meet-in-the-middle, IDEA, key recovery.

1 Introduction

Since Rijndael has been chosen as a new cipher standard in 2001, block cipher cryptanalysis has been less attractive for the cryptologic community. It may be partly attributed to the eStream and SHA-3 competition, which essentially diverted the attention of cryptanalysts to the design and analysis of new primitives. The most efficient methods — differential and linear cryptanalysis, square attacks, boomerang, meet-in-the-middle and impossible differential attacks — have all been designed in the 90s or earlier, and undergone only a series of evolutionary improvements. Occasional applications of hash function-specific methods like rebound attacks operate mainly in weaker models of security.

The situation seems to change with the recent introduction of biclique attacks on AES [6]. Even considered an extension to meet-in-the-middle attacks, the biclique attack brings new techniques and tools to the world of block ciphers, which were known mainly in the cryptanalysis of hash functions. In contrast

D. Pointcheval and T. Johansson (Eds.): EUROCRYPT 2012, LNCS 7237, pp. 392–410, 2012.

to earlier attempts to cryptanalyze AES[5], the new approach does not use any related keys. To understand the reasons behind the new results and to motivate our work, we proceed with a more detailed story of meet-in-the-middle attacks and their evolution.

Meet-in-the-Middle Attacks on Block Ciphers. The basic idea of meet-in-the-middle attacks is to split an invertible transformation into two parts and separate parameters that are involved in only one part. Then these parameters can be searched independently with a match in the middle as a certificate of a right combination. One of the first applications is the cryptanalysis of DoubleDES $E_{K_2}(E_{K_1}(\cdot))$, which demonstrated that the total security level is not the sum of key lengths [12]. The reason is that given a plaintext/ciphertext pair, an adversary is able to compute the internal middle state of a cipher trying all possible values of K_1 and K_2 independently.

The same principle applies at the round level as well. If there is a sequence of rounds in a block cipher that does not depend on a particular key bit, the meet-in-the-middle attack might work. However, its application has been limited by the design of block ciphers, the majority of which use the full key in the very first rounds of a cipher. As a result, even as little as a half of a cipher is rarely attacked, with four attacked rounds in AES [8] and seven in DES [9,13]. Compared to 7-round attacks on AES [20], and full 16-round attacks on DES [21], the meet-in-the-middle attacks were clearly inferior to other methods in spite of their impressively low data complexity. The widespread use of meet-in-the-middle attacks against the preimage resistance of hash functions follows this argument, as the message schedule of, e.g., SHA-1, admits as many as 15 rounds being independent of some message bits. The block ciphers KTANTAN [7] and GOST [15], recently attacked within the meet-in-the-middle framework, also do not use the full key for large number of rounds.

In this context the recent meet-in-the-middle attacks on the full AES [6] might look as a counterexample. Nevertheless, they have not disclosed any new key schedule properties. However, they are able to cover as many as 6 rounds with a new construction — a biclique, inherited from hash function cryptanalysis [17]. In addition to the aforementioned length of the biclique, its dimension is another important property, and significantly contributes to the computational advantage compared to brute-force approaches. A biclique does not impose constraints on the key schedule decomposition and can be long enough to add a significant number of rounds to a meet-in-the-middle attack. The latter property is, however, difficult to achieve when aiming for a significant advantage over brute-force. When dealing with the full number of rounds, only AES-192 has faced an improvement in a factor of four or larger. We additionally stress that the non-ideal diffusion of a single AES round is an important factor for these results. If AES had a full MDS matrix as the diffusion layer, like SHARK [22], it would be much more difficult to attack. The data complexity would increase greatly.

All these properties are significant issues when one wants to deal with a cipher that achieves full diffusion in a single round. Hence IDEA, which has this property, is a natural challenge for a biclique attack. As explained further, we have to leap over three well-diffusion rounds to successfully cryptanalyze the full version of the cipher.

1.1   Cryptanalytic Attacks on IDEA and Our Contribution

The "International Data Encryption Standard" (IDEA) is one of the longest standing and most analyzed ciphers known. It was designed by Lai and Massey in 1991 [18, 19]. Except for negligibly small classes of weak keys only reduced round variants up to 5 out of its 8.5 rounds have been cryptanalyzed in the most relevant single-key setting. If the cryptanalyst were to choose arbitrary sequences of middle-rounds results go up to 6 rounds, and in the less relevant related-key model up to 7.5 rounds [2].

We consider the starting rounds only as a more difficult and more natural challenge. Attacks on middle-rounds weaken the cipher considerably as there would be no equivalent of a whitening key. Moreover, the full key would be used only after two rounds of the cipher, which is evidently not a property of the actual design.

Our main technical results are a first method for key recovery of full IDEA noticeably faster than brute force search, and improved attacks on round-reduced variants. An overview of our new results as well as a comparison with earlier work is given in Table 1. We list here the conceptually new approaches that eventually led to this result:

- The independent-biclique strategy allows for higher dimensions which in turn can lead to faster key recovery. A straightforward application to IDEA would lead to the full codebook requirement even for a one-round biclique due to the diffusion properties. It drastically differs from AES, where 3-round bicliques may still yield reasonable data complexity. In this paper we extend the independent-biclique framework to allow for lower data complexity requirements. We achieve this by using available degrees of freedom for limiting the diffusion in spite of high dimension. Hence we introduce the prefix *"narrow"*.
- In earlier work on AES the independent-biclique was always combined with a key testing phase that loops over all keys. This combination is however not necessary and many of our attacks do not require testing all keys.
- In previous meet-in-the-middle style attacks the Biryukov-Demirci relation was used in a differential way to cancel out key dependencies, it was termed "keyless Biryukov-Demirci relation". Used in our framework, the BD relation can be used directly, and hence avoids overhead computations.

To illustrate the flexibility of the narrow-biclique approach to IDEA, we also consider round-reduced versions. As an example, consider IDEA reduced to the first 5 rounds, which is the highest number of rounds that allowed results before. For this we simultaneously improve time and memory complexity over other

Table 1. New key recovery for IDEA. By brute-force, the computations/success rate ratio is 2^{128}. For simplicity, only round-reduced variants starting from the actual beginning of the cipher are considered.

Rounds	Data	Comp./succ.rate	Memory	Ref.	Biclique rounds	Matching
first 5	2^{17}	$2^{125.5}$	n.a.	[24]	-	key-dep. BD
first 5	10	2^{119}	2^{24}	[2]	-	differential BD
first 5	2^{64}	$2^{115.3}$	n.a.	[24]	-	key-dep. BD
first 5	2^{25}	2^{110}	2^{16}	Sec. 6	1	direct
first 5	2^{25}	$2^{101.5} + 2^{112}$MA	2^{110}	Sec. 6	1	direct
first 6	2^{41}	$2^{118.9}$	2^{12}	Sec. 7	1	direct BD
first 7.5	2^{18}	$2^{126.5}$	2^{3}	Sec. 8	1.5	direct BD
first 7.5	2^{52}	$2^{123.9}$	2^{7}	Sec. 8	1.5	direct BD
8.5 (full)	2^{52}	$2^{126.06}$	2^{3}	Sec. 5	1.5	direct BD
8.5 (full)	2^{59}	$2^{125.97}$	2^{3}	Sec. 5	1.5	direct BD

pervious attacks, while at the same time achieving a practical data complexity of only 2^{25} chosen plaintexts. We also describe the first attack on the 6 initial rounds of IDEA, with data complexity 2^{41} and time complexity 2^{119}, which is a significant improvement over brute-force.

Independently and concurrently, Biham et al. [3] use similar techniques and also arrive at improvements over previous work. However, for the same variant of IDEA considered, the data complexities we obtain compare favourably to theirs for the reasons outlined above. Note that in [3] middle rounds are considered, which we exclude for reasons outlined above. Hence for the same number of rounds we attack a stronger cipher. For full IDEA, whereas we focus on minimizing time complexity Biham et al. consider a setting with little available data and for this have a different approach that is closer to brute force time complexity.

In [4, 10, 14] classes of weak-keys were found for full IDEA. Even though the class is for all practical purposes negligibly small (only up to a fraction of 2^{-64} of all keys are affected), a small change for IDEA was proposed to get rid of these weak-key properties [10]: a constant addition in the key schedule. Our approach to key recovery even works for those strengthened variants of IDEA, in exactly the same way, because it is independent of such constants.

2   Description of IDEA

In here we give a brief description of IDEA and discuss implementation cost consideration that lead to a cost model in which we evaluate our subsequent cryptanalytic results.

IDEA is a 8.5-round block cipher with a 64-bit state and a 128-bit key. Internal state and subkeys are treated as 16-bit words. Each round is an invertible transformation and follows the key addition (KA) layer (two multiplications,

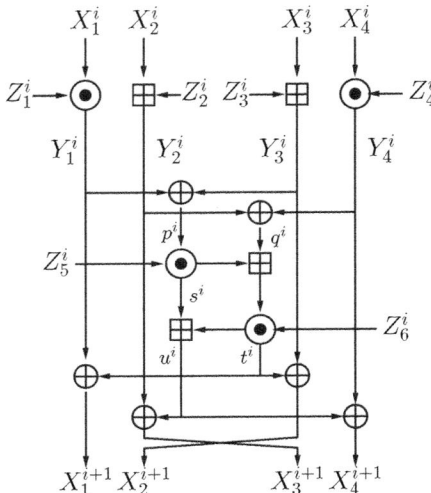

Fig. 1. One round of IDEA

two modular additions) with the multiplication-addition (MA) function (again, two multiplications, two modular additions). Addition is performed modulo 2^{16}. Multiplication is performed modulo $2^{16} + 1$, where 0 is replaced with 2^{16}.

We denote input variables to round i by X^i, the subkeys of round i by Z^i. Additional input variables are depicted in the outline of a single round in Figure 1. Key bits in subkeys are listed in Table 2, where the leftmost bit is the most significant bit in the 16-bit word.

Compared to all other operations of IDEA, the multiplication modulo $2^{16} + 1$ is the most expensive. It can either be realized as a 17-bit multiplier, using a table of size 2^{16}, or with the help of two or three lookup tables of size 2^8. This naturally motivates the model we use to estimate time complexities in this paper: counting multiplications and/or table lookups, and relating them to the number of multiplications needed for IDEA. Each round of IDEA needs four of these multiplications, and the additional key addition layer at the end (often counted as 0.5 round) needs another two. Hence in total 34 multiplications are needed for a single computation. Some of our attacks require the computation of a single output bit of the multiplication which we model with a cost of 0.5 multiplications (see references to complexity estimates in various models in [25]).

3   Biclique Attack

Biclique attacks were introduced for hash function cryptanalysis [17] as an extension to the initial structure technique [23], and later applied to block ciphers [6]. In the biclique attack on block ciphers the full key space is partitioned into groups of keys, so that keys in a group can be efficiently tested in the meet-in-the-middle framework.

The keyspace partition can be described in various ways. For permutation-based key schedules as in IDEA we simply introduce three sets of key bits: K^b, K^f, and K^g. In a key group the value K^g is fixed (and hence enumerates the groups), and K^b and K^f take all possible values.

Biclique. A *biclique* is a set of internal states, which are constructed either in the first or in the last rounds of a cipher and mapped to each other by specifically chosen keys. We consider the former option only in the paper. Let f be the mapping describing the first cipher rounds, then a biclique for a group K^g is a set of states $\{P_i\}, \{S_j\}$ such that

$$P_i \xrightarrow[f]{K^b=i \,||\, K^f=j} S_j.$$

Keys in a group are tested as follows. A cryptanalyst asks for the encryption of plaintexts P_i and gets ciphertexts C_i. Then he checks if

$$\exists\, i, j : \; S_j \xrightarrow[g]{K^b=i \,||\, K^f=j} C_i, \tag{1}$$

where g maps states S_j to ciphertexts. A biclique is said to have dimension d, if both K^b and K^f have d bits.

Key Testing. Each key group is tested separately. There are two approaches to test keys within a group. In the first approach a cryptanalyst uses an intermediate variable v that can be computed in both directions:

$$S_j \xrightarrow[g_1]{K^f=j} v \overset{?}{=} v \xleftarrow[g_2]{K^b=i} C_i.$$

The functions g_1 and g_2 are called *chunks*. This approach is illustrated in Figure 2. The computational complexity of testing a single group is

$$C_{biclique} + 2^{|K^f|}C_{g_1} + 2^{|K^b|}C_{g_2} + C_{recheck},$$

where C_{g_1} and C_{g_2} are the costs of computing v, $C_{biclique}$ is the biclique construction cost, and $C_{recheck}$ is the cost of rechecking key candidates on other state bits or another plaintext/ciphertext pair. The full complexity is derived by the multiplication on the total number of groups.

In the second case a cryptanalyst is unable to find a variable with these properties. Then he tests each key individually

$$S_j \xrightarrow[g_1]{K^b=i \,||\, K^f=j} v \overset{?}{=} v \xleftarrow[g_2]{K^b=i \,||\, K^f=j} C_i,$$

but reuses the computations of chunks, which are defined as parts of g_1 and g_2 that are independent of K^b and K^f, respectively.

This technique was called an *independent-biclique* approach [6] due to use of independent differential trails in the biclique construction.

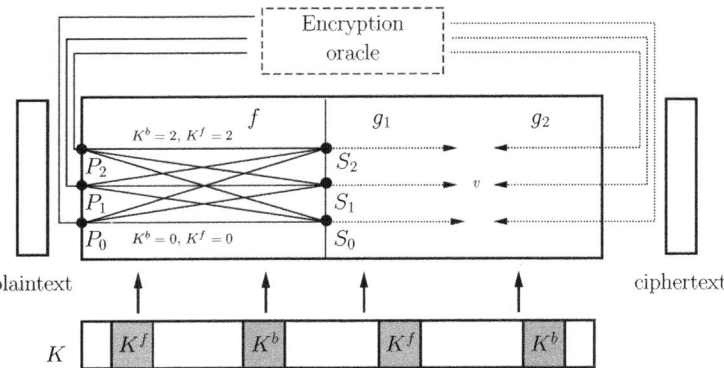

Fig. 2. Key testing with a biclique of three plaintexts and three internal states

Bicliques Based on Independent Differentials. An easy way to construct biclique is to use related-key differentials that do not share active nonlinear components. Let (P, S, K) be a tuple of a plaintext, an internal state, and a key. Let also K^f and K^b be tuples of key bits.

Proposition 1 ([6]). *Suppose that the tuple (P_0, S_0, K_0) conforms to the two sets of related-key differential trails:*

$$0 \xmapsto[f]{\Delta K^f = \Delta_j^K} \Delta_j; \qquad\qquad \nabla_i \xmapsto[f]{\Delta K^b = \nabla_i^K} 0,$$

that share no active non-linear transformations. Then the following states

$$P_i = P_0 \oplus \nabla_i, \qquad\qquad S_j = S_0 \oplus \Delta_j.$$

form a biclique for a group of keys defined by K_0.

Narrow-Bicliques. A straightforward application of the independent differentials technique limits the length of a biclique to the number of rounds needed for the full diffusion. In AES one may use truncated differentials with probability 1, and they would still allow for bicliques over the last three rounds. This is virtually impossible for IDEA, as any one-round truncated differential with probability 1 covers the full state and necessitate the full codebook. Therefore, the biclique differentials must be sparse and hence probabilistic.

We propose to amplify the biclique differentials with guess-and-determine and message modification-like techniques, so that even high-dimensional bicliques would not require the full codebook. Similar techniques have been used for the hash function SHA-2 [17], but aimed only for the independency of trails, but not sparsity. In contrast, for block ciphers the sparsity of a trail is a crucial parameter for the data complexity, as uncontrolled difference results into an uncontrolled plaintexts. It gets much worse in a cipher, whose key is much larger than a plaintext, since the number of bicliques to be constructed greatly exceed the codebook size. As a result, with high probability every possible plaintext becomes involved in a biclique, and this situation we want to avoid.

In our attacks on IDEA we employ various techniques and tools to control the biclique differentials and reduce the data complexity. As a validity certificate, we have implemented a large portion of our biclique construction algorithms on a PC. We experimentally verified the amount of freedom we have in the algorithms and our ability to spend that freedom on setting specific plaintext bits to predefined constants.

A Large-Memory Variant. One of the appealing properties of biclique cryptanalysis is the fact that memory requirements are naturally low, only exponential in the dimension of the biclique, which is usually a small constant. In here we show a rather generic way to speed-up biclique key recovery if a very large, but only sequentially accessible memory is available.

The basic idea is that all those computations that do not depend on replies of the plaintext or ciphertext oracle can be stored and reused for multiple key recoveries. When doing this, the computational complexity to recover the first key remains the same, for subsequent key recoveries however, less computations are needed. In Section 6, for 5-round IDEA we give an example where this gives a noticable speed-up.

4 Biryukov-Demirci Relation

The Biryukov-Demirci relation was introduced in [16] as a combination of two observations by Biryukov (unpublished) and Demirci [11]. Consider two consecutive rounds of IDEA and two lines of computations:

$$X_2^i \to X_3^{i+1} \to X_2^{i+2} \text{ and } X_3^i \to X_2^{i+1} \to X_3^{i+2}$$

For these lines we have:

$$\left((X_2^i \boxplus Z_2^i) \oplus (s^i \boxplus t^i)\right) \boxplus Z_3^{i+1} = X_2^{i+2} \oplus t^{i+1}; \tag{A}$$

$$\left((X_3^i \boxplus Z_3^i) \oplus t^i\right) \boxplus Z_2^{i+1} = X_3^{i+2} \oplus (s^{i+1} \boxplus t^{i+1}). \tag{B}$$

For the least significant bit the modular addition resolves into XOR:

$$LSB(X_2^i \oplus Z_2^i \oplus s^i \oplus t^i \oplus Z_3^{i+1} \oplus t^{i+1}) = LSB(X_2^{i+2});$$
$$LSB(X_3^i \oplus Z_3^i \oplus t^i \oplus Z_2^{i+1} \oplus s^{i+1} \oplus t^{i+1}) = LSB(X_3^{i+2}).$$

Let us sum the equations and redistribute the summands:

$$LSB(X_2^i \oplus X_3^i \oplus Z_2^i \oplus Z_3^i \oplus s^i) = LSB(X_2^{i+2} \oplus X_3^{i+2} \oplus s^{i+1} \oplus Z_2^{i+1} \oplus Z_3^{i+1}). \tag{2}$$

Therefore, for the matching in the MITM attack it is enough to compute X_2^i, X_3^i, s^i in the forward direction, and $X_2^{i+2}, X_3^{i+2}, s^{i+1}$ in the backward direction. To compute s^{i+1} it is enough to compute X_1^{i+2}, X_2^{i+2} and know an appropriate subkey Z_5^{i+1}. If some bits of subkeys Z_2^{i+1} and Z_3^{i+1} belong to K^b or K^f, they are distributed to corresponding sides of the equation (this technique named indirect partial matching [1] was applied to hash functions).

The BD-relation essentially excludes six multiplication operations, or about 1.5 rounds, from the matching part.

Improved Filtering. We can improve the filtering provided by the BD relation by considering more than one bit in equations (A) and (B). More precisely, we consider X_2^i, X_3^i, s^i, X_2^{i+2}, X_3^{i+2}, s^{i+1}, Z_2^i, Z_3^i, Z_2^{i+1} and Z_3^{i+1} as known parameters, and we denote the left hand side and the right hand of (A) and (B) as $L_A(t^i)$, $R_A(t^{i+1})$, $L_B(t^i)$ and $R_B(t^{i+1})$, respectively.

If we know k bits of t^i, we can compute k bits of $L_A(t^i)$ and $L_B(t^i)$, and $k+1$ bits of $L_A(t^i) \oplus L_B(t^i)$. Similarly, if we know k bits of t^{i+1}, we can compute k bits of $R_A(t^{i+1})$ and $R_B(t^{i+1})$, and $k+1$ bits of $R_A(t^{i+1}) \oplus R_B(t^{i+1})$. As seen in the previous section, some values of the parameters are incompatible with any choice of t^i or t^{i+1}. In order to improve the filtering, we will guess some bits of t^i and t^{i+1} and exclude more parameter choices.

For instance, if we guess one bit of t^i and t^{i+1}, we can compute 1 bit of $L_{A,B}$ and 2 bits of $L_A \oplus L_B$ in the forward direction, and 1 bit of $R_{A,B}$ and 2 bits of $R_A \oplus R_B$ in the backward direction. We put those values a hash table for *every* value of t^i and t^{i+1}, and we look for a match between the forward values and the backward values for *some* value of t^i and t^{i+1}. We can show that there is a match with probability 3/8 which means that we have a filtering of 1.41 bits.

More precisely, for given value of X_2^i, X_3^i, s^i, Z_2^i, Z_3^i, Z_2^{i+1}, Z_3^{i+1} in the forward direction, and X_2^{i+2}, X_3^{i+2}, s^{i+1} in the backward direction, there exists a choice for the first bit of t^i, t^{i+1} that result in a match iff:

$$(L_A \oplus L_B)^{[0]} = (R_A \oplus R_B)^{[0]} \text{ and } \begin{cases} (L_A(0) \oplus L_B(0))^{[1]} = (R_A(\alpha) \oplus R_B(\alpha))^{[1]} \\ \textbf{or} \\ (L_A(1) \oplus L_B(1))^{[1]} = (R_A(\bar{\alpha}) \oplus R_B(\bar{\alpha}))^{[1]} \\ \text{where } \alpha = (L_A \oplus R_A)^{[0]} \end{cases}$$

This shows that the parameters will be compatible with probability $1/2 \times 3/4 = 3/8$, and this has been verified experimentally. We can show in the same way that we have a filtering of roughly 2 bits when guessing 3 bits of t^i and t^{i+1}; in this case we have to evaluate $L_{A,B}$ and $R_{A,B}$ for 8 values of t^i and t^{i+1}, which still costs less than one evaluation of the block cipher (we can evaluate four 4-bit values of $L_{A,B}$ or $R_{A,B}$ in parallel using 16-bit operations). We can have 5 bits of filtering using the full 16 bits of t, but we don't see how to use that efficiently.

5   Key Recovery for the Full IDEA

Our approach to cryptanalyze the full IDEA is to construct a short biclique of high dimension, and cover the remaining rounds with the independent-biclique approach. To find an optimal configuration of K^f and K^b bits, and also of the matching position, we have run a short search program. First, we figured out that the longest biclique that is still efficient covers 1.5 rounds. Then we computed the maximum chunk length and hence the minimum matching cost. Then we selected for K^f the bits that form long chunks after round 1, and for K^b the bits that form long chunks ending with the ciphertext.

According to the search results, we have chosen the following key partitioning, which results in a biclique of dimension 3:

- K^g (guess): bits $K_{0...40,\,42...47,\,50...124}$.
- K^f (forward): bits $K_{125...127}$.
- K^b (backward): bits $K_{41,48,49}$.

We have also chosen the partition of the full IDEA into a biclique, chunks, and the matching part according to Table 2. It is also illustrated in Figure 3. By the attack algorithm, each chunk is computed 2^3 times per key group, and the operations in the matching part are computed for each key. The Biryukov-Demirci relation (2) serves as internal variable for the matching in rounds 4–6:

$$\underbrace{LSB(X_2^4 \oplus X_3^4 \oplus Z_2^4 \oplus Z_3^4 \oplus s^4)}_{\text{computed forwards}} \overset{?}{=} \underbrace{LSB(X_2^6 \oplus X_3^6 \oplus s^5) \oplus LSB(Z_2^5 \oplus Z_3^5)}_{\text{computed backwards}}.$$

Biclique. A straightforward way to construct a biclique with our key partition would be as follows. Fix K^g and choose arbitrarily a plaintext P_0. For $K^b = 0$ and each value of K^f compute internal states $S_0, S_1, \ldots, S_7$ that are tuples of variables $(Y_1^2, Y_2^2, Y_3^2, Y_4^2)$. Consider S_0 and for $K^f = 0$ and each value of K^b compute plaintexts $P_0, P_1, \ldots, P_7$. Since differentials resulting from the key differences in K^b and K^f do not interleave, these plaintexts and states form a biclique:

$$P_i \xrightarrow[f]{K^b=i \,||\, K^f=j} S_j.$$

However, we do not control the plaintexts $P_1, \ldots, P_7$. Since we construct 2^{122} bicliques, we are likely to cover the full codebook. To reduce the data complexity, we implement a more complicated biclique construction algorithm, which enforces particular plaintext bits to zero in every biclique.

The improved algorithm works as follows:

1. Fix K^g, $K^f = K^b = 000$.
2. Choose arbitrarily Y_3^2, Y_4^2, p^1.
   - For each K^b (eight options) compute the output of the MA function;
   - For each K^b (eight options) compute X_2^1 — second word of the plaintext.
   - Check if 5 least significant bits of each X_2^1 are zero (many other bit sets would work as well). If not, choose other Y_3^2, Y_4^2, p^1.
   - Note that this implies that the 5 least significant bits of s^1 and t^1 are the same for all K^b. Therefore the 5 least significant bits of X_3^1 will also be the same for all K^b.

Table 2. Round partition for 8.5-round attack. The cipher is splitted into four parts, whose subkey bits are listed. The parts are a biclique, two chunks (where either K^b or K^f) are not used, and matching (where both K^b and K^f are used). The latter part dominates the complexity.

Round	Z_1 $\odot$	Z_2 $\boxplus$	Z_3 $\boxplus$	Z_4 $\odot$	Z_5 $\odot$	Z_6 $\odot$
			Biclique			
1	0–15	16–31	32–47	48–63	64–79	80–95
2	96–111	112–127↓	25–40	41–56↑		
		Chunk 1 (K^b not used)				
2					57–72	73–88
3	89–104	105–120	121–8↓	9–24	50–65	66–81
4	82–97	98–113	114–1	2–17	18–33	
			Matching			
4						34–49↑
5	75–90	91–106	107–122	123–10	11–26	27–42
6	43–58	59–74	100–115	116–3	4–19	20–35
7	36–51	52–67	68–83	84–99	125–12↓	13–28
		Chunk 2 (K^f not used)				
8	29–44	45–60	61–76	77–92	93–108	109–124
9	22–37	38–53↑	54–69	70–85		

3. Choose a value t with the 5 least significant bits set to zero
   - Use t as X_3^1 with $K^b = 000$, and compute X_1^2 and X_2^2.
   - For each K^b (eight options) compute X_1^1 — first word of the plaintext, from X_1^2 and X_2^2.
   - Check if the least significant bit of each X_1^1 — first word of the plaintext — is zero. If not, choose another t. If all the t's have been tried, choose another Y_3^2, Y_4^2, p^1.
4. Compute other plaintext words for each K^b. Derive plaintexts $P_0, P_1, \ldots, P_7$.
5. Vary K^f and derive internal states $S_0, S_1, \ldots, S_7$.

Therefore, a single biclique can be constructed in 2^{40} time, and 11 plaintext bits (15, 27–31 and 43–47) are set to zero. We notice that key bits 16–25, 32–40, 57–63, 96–124 are neutral for the biclique and can be flipped without violating its plaintext property. Therefore, we can reuse the biclique for 2^{55} key groups and hence make the amortized cost negligible.

The first part of the construction require that there exist a choice of Y_3^2, Y_4^2, p^1 so that the 5 least significant bits of each X_2^1 are zero; this is expected to be the case for a proportion $1 - e^{-8} > 99.9$ of the keys. For the full construction, we have a 58-bit condition, which will be satisfied by a proportion $1 - e^{-11} \approx 99.99\%$ of the keys. We can also control one more bit of X_1^1, but this leads to a 61-bit conditions, which is satisfied by 95% of the key. Alternatively, we can look for a six bit match in X_2^1 and X_3^1, and a one bit match in X_1^1. This gives control over

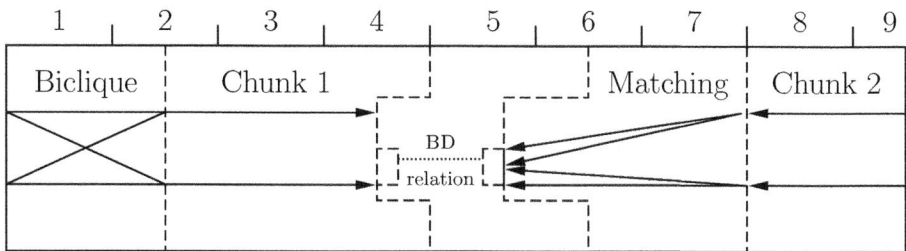

Fig. 3. Attack on the full IDEA (rounds 1–9). Having constructed a biclique, we partially encrypt output states (chunk 1), ask for ciphertexts, partially decrypt them (chunk 2), and match with the help of the BD-relation. The relation allows to ignore about 1.5 rounds of computation.

13 bits of the plaintext, but it only works for less than $(1 - 1/e) \approx 63\%$ of the keys. Finally, we can achieve various tradeoffs by changing the dimension of the biclique; we can control:

- 23 bits of a dim.-2 biclique with succ. rate $(1 - e^{-4})(1 - e^{-5}) > 97.5\%$
- 24 bits of a dim.-2 biclique with succ. rate $(1 - e^{-4})(1 - e^{-2}) > 84\%$
- 11 bits of a dim.-3 biclique with succ. rate $(1 - e^{-8})(1 - e^{-11}) > 99.9\%$
- 12 bits of a dim.-3 biclique with succ. rate $(1 - e^{-8})(1 - e^{-3}) > 94\%$
- 5 bits of a dim.-4 biclique with succ. rate $(1 - e^{-16})(1 - e^{-14}) > 99.9999\%$

Complexity. Each biclique tests 2^6 keys. The first chunk employs 9 multiplications, the second chunk — 6 multiplications, the matching part — 13 multiplications (and hence 7 when we use the relation, of which two compute only a single output bit and are hence counted as a half multiplication is discussed in Section 2). We recheck on Y_1^5, for which we need 2 multiplications: on Z_6^4 and Z_1^5. A negligible proportion of keys is rechecked on the full state and on another plaintext/ciphertext pair. Therefore, 2^6 keys are tested with

$$9 \cdot 8 + 6 \cdot 8 + (5 + \frac{1}{2} + \frac{1}{2}) \cdot 64 + 2 \cdot 32 = 632 \text{ multiplications } = 2^{4.06}$$

calls of IDEA. The total time complexity is hence $2^{126.06}$.

We can expand K^b to 4 bits for the cost of increased data complexity. As we would be able to spend only 4 degrees of freedom per plaintext, the data complexity becomes 2^{59} and the number of multiplications for 2^7 keys is 1064 which yields a total complexity of about $2^{125.97}$ calls of IDEA, as an easily be computed:

$$9 \cdot 8 + 6 \cdot 16 + (5 + \frac{1}{2} + \frac{1}{2}) \cdot 128 + 2 \cdot 64 = 1064 \text{ multiplications.}$$

6   New 5 Round Attack

The only 5-round attacks that start at the beginning of the cipher are the following. Biham et al. mention the possibility of an attack with memory complexity 2^{24} and time complexity of 2^{119}. The fastest attack so far needs $2^{115.5}$ time, but also the full codebook. In here we provide the fastest attack which additionally requires only 2^{25} chosen plaintexts.

- K^g (guess): bits $K_{0...8,\,25...49,\,75...127}$.
- K^f (forward): bits $K_{50...74}$.
- K^b (backward): bits $K_{9...24}$.

We can easily construct a biclique of size $2^{25} \times 2^{16}$ in round 1, since the paths are clearly independent: the backward key only affects X_1^1 and X_2^1, while the forward key affects Y_3^1, p^1, q^1 and everything after the MA function. The matching point is p^3 (16 bits), which can be computed in both chunks.

We can control 39 bits of the plaintext in the following way:

- We start with $X_3^1 = 0$, $X_4^1 = 0$, some arbitrary value for Y_1^1, and $Y_2^1 = 2^9 K_{25...31} + \text{0x1ff}$
- We can then compute X_1^1 and X_2^1 for each choice of K^b. Note that the high 7 bits of $X_2^1 = Y_2^1 \boxminus K_{16...31}$ will be zeros because there will be no carry in the subtraction.
- Finally we compute $X_1^2, X_2^2, X_3^2, X_4^2$ for each choice of K^f.

The biclique construction has negligible cost, as we can use most of the key bits as neutral.

Alternatively, we can see this attack as a basic MITM if we start with state $Y_1^1, Y_2^1, X_3^1, X_4^1$. In the forward part, we can compute p^3 independently of K^b. For the backward part we compute X_1^1 and X_2^1, then we query the oracle on that state, and continue the computation from the ciphertext $X_1^1, X_2^1, X_3^1, X_4^1$, up to p^3.

If we flip key bits 112–127, only 2 multiplications in round 2 are affected. Therefore, in the first chunk we need only 3 multiplications to recompute in total. In the second chunk we recompute 7 multiplications. The matching comes for free, but the 2^{25} key candidates must be rechecked on 1 multiplication. Hence the total complexity of testing $2^{128-25-16} = 2^{87}$ key groups is computed as follows:

$$C = 2^{87} \left(\frac{4}{20} 2^{25} + \frac{7}{20} 2^{16} \right) = 2^{110}.$$

When trying to recover multiple independent keys, and following the large-memory variant outlined in Section 3, the results of 2^{25} forward computations for all 2^{87} bicliques could be precomputed and stored in the form of a table of $2^{25} \cdot 16$ bits (for p^3) for each keygroup together with the plaintext. Hence at the cost of a memory of size equivalent to 2^{110} blocks (of 64 bits each) that needs to be accessed $2^{25} \cdot 2^{87} = 2^{112}$ times in a sequential way for every individual key recovery, the computational cost would drop to about an equivalent of $2^{87} \left(\frac{7}{20} 2^{16} \right) = 2^{101.5}$ IDEA calls, as only the backwards computations need to be performed for every keygroup with the respective oracle responses.

Table 3. Round partition for the 5-round attack

Round	Z_1	Z_2	Z_3	Z_4	Z_5	Z_6
			Biclique			
1	0–15	16–31	32–47	48–63	64–79	80–95
			Chunk 1			
2	96–111	112–127	25–40	41–56	57–72	73–88
3	89–104		121-8			
			Chunk 2			
4	82–97	98–113			18–33	34–49
5	75–90	91–106	107–122	123–10	11–26	27–42

7   New 6 Round Attack

The best currently known 6-round attack, the very recent and still unpublished MITM approach of [2] works for 6-rounds only with a single starting position that does not coincide with the actual start of the cipher. In here we give the first 6-round attack.

The key partition is as follows:

- K^g (guess): bits $K_{0...65,\,75...111}$.
- K^f (forward): bits $K_{66...74}$.
- K^b (backward): bits $K_{112...127}$.

We use a one-round biclique at the end of the cipher, in round 6. Then we compute s^2 in the forward direction and s^3 in the backward direction, and use the BD relation.

For the construction of the biclique, we start with $Y_1^6 = 0$, $X_2^6 = -K_{59...74}$, $Y_3^6 = 0$ and $Y_4^6 = 0$. For each K_f, we compute the final state at the end of round 6. We note that we have $X_1^7 = X_2^7$ and the 7 most significant digits of X_3^7 and X_4^7 are also equal. Thus, we control 23 bits of the ciphertext, and the data complexity is 2^{41}.

In order to filter out enough bad guesses, we will use 9 bicliques for each K_g. The most expensive part of the attack is the backward computation of s^3. For each K_g guess, this costs:

$$9 \cdot \left(2^5 + 2^{16} + 2^{16} + \frac{1}{2} \cdot 2^{16}\right) = 2^{20.5} \text{ multiplications} = 2^{15.9} \text{ IDEA calls}$$

The total complexity is therefore $2^{103} \cdot 2^{15.9} = 2^{118.9}$.

8   New 7.5 Round Attack

In the 7.5-round attack we construct a biclique in the first 1.5 rounds. The key partitioning is defined as follows:

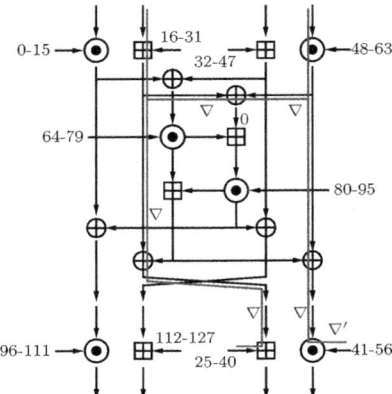

Fig. 4. Biclique ∇-differential in 7.5-round attack

- K^g (guess): bits $K_{0...24,\,25...40,\,42...99,\,125...127}$.
- K^f (forward): bits $K_{100...124}$.
- K^b (backward): bits $K_{25,41}$.

The differentials based on K^f and K^b do not interleave in the first 1.5 rounds. Therefore, we can construct a biclique in a straightforward way similar to the full-round attack. However, the differential generated by K^b affects the full plaintext. To reduce the data complexity we construct two bicliques for a key group so that the differential generated by K^b vanishes at the input of the MA-function.

We proceed as follows. For the first biclique we fix $K_{25} \oplus K_{41} = 0$, and for the second one $K_{25} \oplus K_{41} = 1$. As a result, a difference in K^b generates simultaneous differences in Z_3^2 and Z_4^2. Denote the difference in Z_3^2 by ∇ (generated by bit K_{25}), and in Z_4^2 by ∇' (generated by K_{41}). We want the difference in X_4^2 to be equal to ∇ so that the MA-structure have zero input difference (Figure 4). We fulfill this condition by random trials. Bicliques are hence constructed as follows:

1. Fix $X_1^1 = X_2^1 = X_3^1 = 0$;
2. Choose arbitrarily values for X_4^1:
    - Generate internal states for the biclique;
    - Check whether the MA function has zero input difference. If not, try another value of X_4^1.

Computational and Data Complexity. A single pair of biclique is generated in less than 2^{16} calls of IDEA. This value is amortized since we can derive 2^{19} more bicliques by changing key bits 96–104, (and recomputing Y_1^2), 125–127 (recompute Y_2^2) and bits 57–63 of Z_4^1 (and recomputing the plaintext). Therefore, an amortized cost to construct a biclique is negligible.

Since the ∇-differential affects the most significant bit of X_2^1 only, the plaintexts generated in bicliques have 47 bits fixed to zero. Therefore, the data complexity does not exceed 2^{17}. The full computational complexity of the attack is computed as follows:

$$C = 2^{107} \left(C_{bicl} + 2^{13} C_{chunk1} + 2 C_{chunk2} + 2^{13} C_{recheck} \right),$$

where we test 2^{106} key groups with two bicliques each. The amortized biclique construction cost is negligible. We note that the multiplication by Z_5^2 in round 2 is also amortized as the change in key bits 57–63 does not affect it. Therefore, the total number of multiplications in the first chunk is $1 + 4 + 2 = 7$ multiplications, in the second chunk — $1 + 3 + 4 + 2 = 10$ multiplications, to recheck — 3 multiplications (to compute the full p^5 in both directions). The total complexity is hence $2^{127} \frac{10}{30} = 2^{126.5}$.

We can decrease the time complexity for the cost of the increase in the data complexity. Let us assign one more bit to K^b so that there are 8 values of K^b. We spend 64 bits of freedom in the internal state to fix 13 bits of each biclique plaintext, as shown for the attack on the full IDEA. Then the complexity is estimated as follows:

$$C = 2^{100} \left(2^{25} \frac{8}{30} + 2^3 \frac{10}{30} + 2^{28} \frac{1}{2} \cdot \frac{2}{30} \right) = 2^{124.1}.$$

We can further reduce the complexity using the improved BD filtering on two bits described in Section 4, which filters out $5/8$ of the candidates. First we consider bits 112-113, 105-106, 121-122, 114-115 as part of K^g instead of K^f, so that all the keys involved in the BD relation are part of K^g. We also use some precomputations. For each K^b, we compute s^5, and we evaluate $R_A(t^5)$ and $R_B(t^5)$ for 2 guesses of t^5. Then, we consider the potential candidates from the forward chunk: for each possible 1-bit value of X_2^4, X_3^4 and s^4, plus the second bit of $X_2^4 \oplus X_3^4 \oplus s^4$ we guess 1 bit of t^4 in order to compute $L_A(t^4)$ and $L_B(t^4)$; then we can filter the corresponding candidates for K^b (we expect 3 candidates on average). Then for each K^f, we just use this table to recover the candidates. For the complexity evaluation, we assume that finding a match in the hash table costs the same as one multiplication. This yields a complexity of:

$$C = 2^{108} \left(2^{17} \frac{8}{30} + 2^3 \frac{10 + 2 + 2^5}{30} + 2^{20} \frac{3}{8} \cdot \frac{2}{30} \right) = 2^{123.9}.$$

9   On Practical Verification

Especially for the type of cryptanalysis described in this paper where carrying out an attack in full is computationally infeasible, practical verification of attack details and steps is important in order to get confidence in it. To address this, we explicitly state the following:

- We have implemented the dimension-3 biclique construction of Section 5, which works as expected, and takes a few hours on a desktop PC. An example is given in Appendix A.
- We have implemented the improved matching procedure as described in Section 4. We have verified that we have the expected number of remaining candidates.

10   Concluding Discussion

We showed that a number of extensions to the recently introduced biclique framework and specific properties of IDEA eventually lead to the cryptanalysis of the full version of the cipher. Though IDEA withstood all cryptanalysis attempts in the last 20 years, it is now vulnerable to key recovery methods that are about 4 times faster than brute force search. We also show attacks on the first 7.5 rounds where the attack algorithm does not have to consider each key separately, resulting in a larger complexity advantage. For smaller number of rounds we surpass the best attacks so far, hence refuting the view that biclique attacks lead only to a small advantage over the brute-force.

We emphasize the use of several techniques from hash function cryptanalysis, which are usually associated with the start-in-the-middle framework. Following the recent work on AES, we demonstrate that these techniques are important also in secret-key cryptanalysis. We foresee widespread application of tools aimed for data complexity reduction, which could be based on our concept of narrow-bicliques.

As a natural application of a new concept we would again name AES. Being able to construct high-dimensional bicliques rather deep in the cipher with reasonable data complexity, an adversary might be able to get a significant advantage over brute-force for the full number of rounds, and possibly even present the best attacks on already broken number of rounds. As meet-in-the-middle attacks might potentially work in $2^{n/2}$ time, it would be extremely interesting to figure out the number of AES rounds that could be broken with this almost practical complexity.

This line of work opens up more questions that we feel are important:

1. Bounds for biclique attacks. With a biclique attack in this paper that is 2^{18} times faster than brute force (or $2^{26.5}$ times faster if large sequentially accessible memory is available) an earlier intuition that this class of attacks only allows for rather small speed-ups over brute force search is dismissed. Nevertheless it may be possible to give meaningful bounds on classes of biclique attacks.
2. How to best defend against biclique cryptanalysis? In this paper we see that even a more conservative key schedule design for IDEA is almost as vulnerable. It seems as if only very expensive key schedule designs, i.e. those where the key can not easily be deduced from subkey material, would provide resistance. This remains a topic of future work, though.

Acknowledgements. We thank Orr Dunkelman and Adi Shamir for bringing to our attention their new attack on 6 middle rounds of IDEA during the MSR Symmetric Cryptanalysis Workshop 2011. This was the starting point for our investigations and new results. We also thank Florian Mendel and Andrey Bogdanov for the discussions on possible research directions. Finally, we thank reviewers of Eurocrypt 2012 for their helpful comments.

Part of this work was done while Christian Rechberger was with ENS Ulm and Foundation Chaire France Telecom, and visiting MSR Redmond. Gaëtan Leurent is supported by the AFR grant PDR-10-022 of the FNR Luxembourg. This work was supported in part by the European Commission under contract ICT-2007-216646 ECRYPT NoE phase II.

References

1. Aoki, K., Guo, J., Matusiewicz, K., Sasaki, Y., Wang, L.: Preimages for Step-Reduced SHA-2. In: Matsui, M. (ed.) ASIACRYPT 2009. LNCS, vol. 5912, pp. 578–597. Springer, Heidelberg (2009)
2. Biham, E., Dunkelman, O., Keller, N., Shamir, A.: New data-efficient attacks on 6-round IDEA. Cryptology ePrint Archive, Report 2011/417 (2011), http://eprint.iacr.org/
3. Biham, E., Dunkelman, O., Keller, N., Shamir, A.: New data-efficient attacks on reduced-round idea. Cryptology ePrint Archive, Report 2011/417 (2011), http://eprint.iacr.org/
4. Biryukov, A., Nakahara Jr, J., Preneel, B., Vandewalle, J.: New Weak-Key Classes of IDEA. In: Deng, R.H., Qing, S., Bao, F., Zhou, J. (eds.) ICICS 2002. LNCS, vol. 2513, pp. 315–326. Springer, Heidelberg (2002)
5. Biryukov, A., Khovratovich, D.: Related-Key Cryptanalysis of the Full AES-192 and AES-256. In: Matsui, M. (ed.) ASIACRYPT 2009. LNCS, vol. 5912, pp. 1–18. Springer, Heidelberg (2009)
6. Bogdanov, A., Khovratovich, D., Rechberger, C.: Biclique Cryptanalysis of the Full AES. In: Lee, D.H., Wang, X. (eds.) ASIACRYPT 2011. LNCS, vol. 7073, pp. 344–371. Springer, Heidelberg (2011)
7. Bogdanov, A., Rechberger, C.: A 3-Subset Meet-in-the-Middle Attack: Cryptanalysis of the Lightweight Block Cipher KTANTAN. In: Biryukov, A., Gong, G., Stinson, D.R. (eds.) SAC 2010. LNCS, vol. 6544, pp. 229–240. Springer, Heidelberg (2011)
8. Bouillaguet, C., Derbez, P., Fouque, P.-A.: Automatic Search of Attacks on Round-Reduced AES and Applications. In: Rogaway, P. (ed.) CRYPTO 2011. LNCS, vol. 6841, pp. 169–187. Springer, Heidelberg (2011)
9. Chaum, D., Evertse, J.-H.: Cryptanalysis of DES with a Reduced Number of Rounds. In: Williams, H.C. (ed.) CRYPTO 1985. LNCS, vol. 218, pp. 192–211. Springer, Heidelberg (1986)
10. Daemen, J., Govaerts, R., Vandewalle, J.: Weak Keys for IDEA. In: Stinson, D.R. (ed.) CRYPTO 1993. LNCS, vol. 773, pp. 224–231. Springer, Heidelberg (1994)
11. Demirci, H.: Square-like Attacks on Reduced Rounds of IDEA. In: Nyberg, K., Heys, H.M. (eds.) SAC 2002. LNCS, vol. 2595, pp. 147–159. Springer, Heidelberg (2003)
12. Diffie, W., Hellman, M.: Special feature exhaustive cryptanalysis of the NBS Data Encryption Standard. Computer 10, 74–84 (1977)
13. Dunkelman, O., Sekar, G., Preneel, B.: Improved Meet-in-the-Middle Attacks on Reduced-Round DES. In: Srinathan, K., Rangan, C.P., Yung, M. (eds.) INDOCRYPT 2007. LNCS, vol. 4859, pp. 86–100. Springer, Heidelberg (2007)
14. Hawkes, P.: Differential-Linear Weak Key Classes of IDEA. In: Nyberg, K. (ed.) EUROCRYPT 1998. LNCS, vol. 1403, pp. 112–126. Springer, Heidelberg (1998)

15. Isobe, T.: A Single-Key Attack on the Full GOST Block Cipher. In: Joux, A. (ed.) FSE 2011. LNCS, vol. 6733, pp. 290–305. Springer, Heidelberg (2011)
16. Nakahara Jr., J., Preneel, B., Vandewalle, J.: The Biryukov-Demirci Attack on Reduced-Round Versions of IDEA and MESH Ciphers. In: Wang, H., Pieprzyk, J., Varadharajan, V. (eds.) ACISP 2004. LNCS, vol. 3108, pp. 98–109. Springer, Heidelberg (2004)
17. Khovratovich, D., Rechberger, C., Savelieva, A.: Bicliques for preimages: Attacks on Skein-512 and the SHA-2 family (2011),http://eprint.iacr.org/2011/286.pdf
18. Lai, X., Massey, J.L.: Markov Ciphers and Differential Cryptanalysis. In: Davies, D.W. (ed.) EUROCRYPT 1991. LNCS, vol. 547, pp. 17–38. Springer, Heidelberg (1991)
19. Lai, X., Massey, J.L.: Hash Functions Based on Block Ciphers. In: Rueppel, R.A. (ed.) EUROCRYPT 1992. LNCS, vol. 658, pp. 55–70. Springer, Heidelberg (1993)
20. Mala, H., Dakhilalian, M., Rijmen, V., Modarres-Hashemi, M.: Improved Impossible Differential Cryptanalysis of 7-Round AES-128. In: Gong, G., Gupta, K.C. (eds.) INDOCRYPT 2010. LNCS, vol. 6498, pp. 282–291. Springer, Heidelberg (2010)
21. Matsui, M.: Linear Cryptanalysis Method for DES Cipher. In: Helleseth, T. (ed.) EUROCRYPT 1993. LNCS, vol. 765, pp. 386–397. Springer, Heidelberg (1994)
22. Rijmen, V., Daemen, J., Preneel, B., Bosselaers, A., De Win, E.: The Cipher SHARK. In: Gollmann, D. (ed.) FSE 1996. LNCS, vol. 1039, pp. 99–111. Springer, Heidelberg (1996)
23. Sasaki, Y., Aoki, K.: Finding Preimages in Full MD5 Faster Than Exhaustive Search. In: Joux, A. (ed.) EUROCRYPT 2009. LNCS, vol. 5479, pp. 134–152. Springer, Heidelberg (2009)
24. Sun, X., Lai, X.: The Key-Dependent Attack on Block Ciphers. In: Matsui, M. (ed.) ASIACRYPT 2009. LNCS, vol. 5912, pp. 19–36. Springer, Heidelberg (2009)
25. Wegener, I., Woelfel, P.: New results on the complexity of the middle bit of multiplication. Computational Complexity 16(3), 298–323 (2007)

A   Biclique Example

We give an example of a dimension-3 biclique for the first 1.5 round of IDEA. It is built with the bits of K^g set to the key 0x0102030405060708090a0b0c0d0e0f10, and the bits used for K^f and K^b are $41, 48, 49$ and $116, 117, 118$, respectively. Each plaintext has 11 bits set to zero as explained in Section 5.

$$
\begin{array}{ll}
P_0 \text{ 1754 5580 00c0 d05b} & S_0 \text{ 7092 7352 f5b1 7272} \\
P_1 \text{ 0f10 ca00 a440 aa79} & S_1 \text{ 7092 7152 f5b1 7272} \\
P_2 \text{ bda8 f580 a0a0 c6b7} & S_2 \text{ 7092 6f52 f5b1 7272} \\
P_3 \text{ 17c4 86a0 6f00 6c69} & S_3 \text{ 7092 6d52 f5b1 7272} \\
P_4 \text{ e9fe 6500 5100 143a} & S_4 \text{ 7092 6b52 f5b1 7272} \\
P_5 \text{ 9252 0200 ec00 230c} & S_5 \text{ 7092 6952 f5b1 7272} \\
P_6 \text{ aa8e b5a0 5fc0 16ef} & S_6 \text{ 7092 6752 f5b1 7272} \\
P_7 \text{ 4a9e c520 b040 ecc0} & S_7 \text{ 7092 6552 f5b1 7272}
\end{array}
$$

Cryptanalyses on a Merkle-Damgård Based MAC — Almost Universal Forgery and Distinguishing-H Attacks

Yu Sasaki

NTT Information Sharing Platform Laboratories, NTT Corporation
3-9-11 Midori-cho, Musashino-shi, Tokyo 180-8585 Japan
sasaki.yu@lab.ntt.co.jp

Abstract. This paper presents two types of cryptanalysis on a Merkle-Damgård hash based MAC, which computes a MAC value of a message M by Hash($K\|\ell\|M$) with a shared key K and the message length ℓ. This construction is often called LPMAC. Firstly, we present a distinguishing-H attack against LPMAC instantiating any narrow-pipe Merkle-Damgård hash function with $O(2^{n/2})$ queries, which indicates the incorrectness of the widely believed assumption that LPMAC instantiating a secure hash function should resist the distinguishing-H attack up to 2^n queries. In fact, all of the previous distinguishing-H attacks considered dedicated attacks depending on the underlying hash algorithm, and most of the cases, reduced rounds were attacked with a complexity between $2^{n/2}$ and 2^n. Because it works in generic, our attack updates these results, namely full rounds are attacked with $O(2^{n/2})$ complexity. Secondly, we show that an even stronger attack, which is a powerful form of an almost universal forgery attack, can be performed on LPMAC. In this setting, attackers can modify the first several message-blocks of a given message and aim to recover an internal state and forge the MAC value. For any narrow-pipe Merkle-Damgård hash function, our attack can be performed with $O(2^{n/2})$ queries. These results show that the length prepending scheme is not enough to achieve a secure MAC.

Keywords: LPMAC, distinguishing-H attack, almost universal forgery attack, multi-collision, diamond structure, prefix freeness.

1 Introduction

Message Authentication Code (MAC) is a cryptographic technique which produces the integrity of the data and the authenticity of the communication player. MACs take a message and a key as input and compute a MAC value which is often called tag. Suppose that a sender and a receiver share a secret key K in advance. When the sender sends a message M to the receiver, he computes a tag σ and sends a pair of (M, σ). The receiver computes a tag by using the shared key K and the received M. If the result matches with the received σ, he knows that he received the correct message and it was surely sent by the sender.

D. Pointcheval and T. Johansson (Eds.): EUROCRYPT 2012, LNCS 7237, pp. 411–427, 2012.

MACs are often constructed by using block-ciphers or hash functions. There are three basic MAC constructions based on hash functions which were analyzed by Tsudik [1]. Let $\mathcal{H}$ be a hash function. A *secret-prefix* method computes a tag of a message M by $\mathcal{H}(K\|M)$. A *secret-suffix* method computes a tag by $\mathcal{H}(M\|K)$. A *hybrid* method computes a tag by $\mathcal{H}(K\|M\|K)$. Among the above three, the secret-prefix method is known to be vulnerable when $\mathcal{H}$ processes M block by block by iteratively applying a compression function h. This attack is called *padding attack* in [1] and *length-extension attack* in the recent SHA-3 competition [2]. Assume that the attacker obtains the tag σ for a message M. He then, without knowing the value of K and M, can compute a tag σ' for a message $M\|z$ for any z by computing $\sigma' \leftarrow h(\sigma, z)$.

LPMAC was suggested to avoid the vulnerability of the secret-prefix method [1][1]. In LPMAC, the length of the message to be hashed is prepended before the message is hashed, that is to say, $\sigma \leftarrow \mathcal{H}(K\|\ell\|M)$, where ℓ is the length of M. $K\|\ell$ is often padded to be a multiple of the block-length so that the computation of M can start from a new block. The length prepending scheme can be regarded as a concrete construction of the prefix freeness introduced by Bellare *et al.* [4]. Therefore, by [4], LPMAC was proven to be a secure pseudo-random function (PRF) up to $O(2^{n/2})$ queries.

The security of MAC is usually discussed with respect to the resistance against the following forgery attacks. An *existential forgery attack* creates a pair of valid (M, σ) for a message M which is not queried yet. A *selective forgery attack* creates a pair of valid (M, σ) where M is chosen by the attacker prior to the attack. A *universal forgery attack* creates a pair of valid (M, σ) where M can be any message chosen prior to the attack. Variants of these forgery attacks can also be considered. For example, Dunkelman *et al.* introduced an *almost universal forgery attack* [5] against the ALRED construction, which the attacker can modify one message block in M. In addition, the security against distinguishing attacks are also evaluated on MAC constructions. Kim *et al.* introduced two distinguishing attacks called distinguishing-R and distinguishing-H [6]. Let $\mathcal{R}$ and r be random functions which have the same domain as $\mathcal{H}$ and h, respectively. Moreover, we denote the hash function $\mathcal{H}$ instantiating a compression function f by $\mathcal{H}^f$. In the distinguishing-R attack, the attacker distinguishes a MAC $\mathcal{H}(K, M)$ from $\mathcal{R}(K, M)$. On the other hand, in the distinguishing-H attack, the attacker distinguishes $\mathcal{H}^h(K, M)$ from $\mathcal{H}^r(K, M)$.

Regarding the distinguishing-R attack, Preneel and van Oorschot [7] presented a generic attack against MACs with a Merkle-Damgård like iterative structure, which requires $O(2^{n/2})$ queries. In [7], it was explicitly mentioned that the same attack could be applied to LPMAC. Chang and Nandi later discussed its precise complexity when a long message is used [8]. The distinguishing-R attack is immediately converted to the existential forgery attack with the same complexity. On the other hand, no generic attack is known for the distinguishing-H attack, and it is widely believed that the complexity of the distinguishing-H attack

---

[1] The name "LPMAC" was given by Wang *et al.* [3].

Table 1. Comparison of distinguishing-H attacks against LPMAC

Attack Target	Size(n)	#Rounds	#Queries	Reference
SHA-1	160	43/80	$2^{124.5}$	[3]
SHA-1	160	61/80	$2^{154.5}$	[3]
SHA-1	160	65/80	$2^{80.9}$	[18]
SHA-256	256	39/64	$2^{184.5}$	[20]
RIPEMD	128	48/48 (full)	2^{66}	[19]
RIPEMD-256	256	58/64	$2^{163.5}$	[19]
RIPEMD-320	320	48/80	$2^{208.5}$	[19]
Generic narrow-pipe MD	n	full	$3 \cdot 2^{\frac{n}{2}}$	Ours

Our attack also requires $2^{n/2}$ offline computations and a memory to store $2^{n/2}$ tags. The attack can be memoryless with $6 \cdot 2^{n/2}$ queries and $2^{(n/2)+1}$ offline computations.

against a MAC instantiating a securely designed hash algorithm $\mathcal{H}^h$ should cost 2^n complexity.

There are several cryptanalytic results on MAC constructions. Although several results are known for block-cipher based MACs e.g. [5, 9, 10], in this paper, we focus our attention on hash function based MACs. A notable work in this field is the one proposed by Contini and Yin, which presented distinguishing and key recovery attacks on HMAC/NMAC with several underlying hash functions [11]. After that, several improved results were published [12–16]. Another important work is the one proposed by Wang *et al.* [17], which presented the first distinguishing-H attack on HMAC-MD5 in the single-key setting. In this attack framework, the number of queries principally cannot be below $2^{n/2}$ because the birthday attack is used. With the techniques of [17], a series of distinguishing-H attacks on LPMAC were presented against SHA-1, SHA-256, and the RIPEMD-family [18, 19, 3, 20]. The attack results are summarized in Table 1. Note that the notion of the almost universal forgery attack was firstly mentioned by [5], while some of previous attacks e.g. [10] can directly be applied for this scenario.

Our Contributions

In this paper, we propose generic attacks on LPMAC instantiating any narrow-pipe Merkle-Damgård hash function. We firstly propose a distinguishing-H attack with $3 \cdot 2^{n/2}$ queries, $2^{n/2}$ offline computations, and a memory to store $2^{n/2}$ tags. Our attack updates the previous results, namely full rounds are attacked with $O(2^{n/2})$ complexity. Moreover, the attack can be memoryless by using the technique in [21, Remark9.93]. The complexity of our attack is listed in Table 1.

Our attack is based on a new technique which makes an internal collision starting from two different length-prepend values, and recovers an internal state value with queries of different lengths. This approach is completely different from previous attacks on LPMAC, which utilize the existence of a high-probability

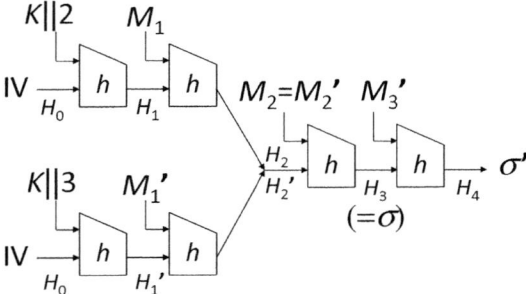

Fig. 1. Sketch of Distinguish-H Attack

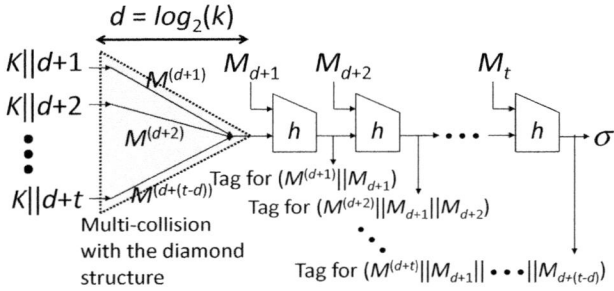

Fig. 2. Sketch of Almost Universal Forgery Attack

differential path of an underlying compression function. The idea of our technique is depicted in Fig. 1. The outline is as follows. We start from two length-prepend values; one is for 2-block messages and the other is for 3-block messages. With a very high probability, the internal states H_1 and H_1' will be different values. Assume that we can easily obtain paired messages (M_1, M_1') such that $h(H_1, M_1)$ and $h(H_1', M_1')$ form an internal collision. Then, we can obtain the value of $H_4(= \sigma')$, which is the output of the last compression function, by querying $M_1'\|M_2'\|M_3'$. In addition, we can obtain the value of $H_3(= \sigma)$, which is the input of the last compression function by querying $M_1\|M_2$. Because we obtain all input and output information for the last compression function, we can judge whether h is the target algorithm or not by comparing $h(\sigma, M_3')$ and σ'.

As mentioned before, the length prepending scheme is known to be prefix-free. However, an internal collision with different length-prepend values have the same effect as the prefix. In fact, in Fig. 1, $M_1\|M_2$ can be regarded as a prefix of $M_1'\|M_2'\|M_3'$. This shows that the core of the security of LPMAC, which is the pre-fix freeness, can be totally broken with $O(2^{n/2})$ queries.

We then further extend the technique to mount an even stronger attack against LPMAC, which is called an almost universal forgery attack. In this attack, for a given t-block message $M_1 \| M_2 \| \cdots \| M_t$, we assume the attacker's ability to modify the first d message blocks where $d = \lceil \log t \rceil$ into the value of his choice $M_1' \| M_2' \| \cdots \| M_d'$. Then, the attacker forges the tag for $M_1' \| \cdots M_d' \| M_{d+1} \| \cdots \| M_t$. Moreover, in our attack, the attacker can reveal the internal state value. The main idea is constructing a multi-collision starting from various different length-prepend values, which is depicted in Fig. 2. This enables the attacker to deal with various undetermined length-prepend values $(1, \ldots, t)$ in advance. To construct the multi-collision, we use the *diamond structure* proposed by Kelsey and Kohno for the herding attack [22].

Paper Outline

In Sect. 2, we introduce LPMAC and briefly summarize related work. In Sect. 3, we explain our generic distinguishing-H attack. In Sect. 4, we explain our generic almost universal forgery attack. In Sect. 5, we conclude this paper.

2   Related Work

2.1   LPMAC with Narrow-Pipe Merkle-Damgård Hash Functions

LPMAC is a hash function based MAC construction observed by Tsudik [1] to prevent the so called length-extension attack on the secret-prefix MAC. In LPMAC, the length of the message to be hashed is prepended before the message is hashed, that is to say, $\sigma \leftarrow \mathcal{H}(K \| \ell \| M)$, where ℓ is the length of M.

Most of widely used hash functions adopt the Merkle-Damgård domain extension with the narrow-pipe structure and the MD-strengthening. In this scheme, the input message M is first padded to be a multiple of the block-length by the padding procedure. Roughly speaking, a single bit '1' is appended to M, and then a necessary number of '0's are appended. Finally, the length of M is appended. Note that this padding scheme is not only the one, and replacing it with another padding scheme e.g. split padding [23] is possible. However, we only use it in this paper because it is the most common. The padded message is divided into message blocks $M_0, M_1, \ldots, M_{t-1}$ with the block size of b bits. Then, the hash value of size n is computed by iteratively updating the chaining variable of size n bits with the compression function h defined as $\{0, 1\}^n \times \{0, 1\}^b \to \{0, 1\}^n$;

$$H_0 \leftarrow \text{IV}, \qquad H_{i+1} \leftarrow h(H_i, M_i) \text{ for } i = 0, 1, \ldots, t-1, \qquad (1)$$

where IV is an n-bit pre-specified value. Finally, H_t is the hash value of M.

In this paper, for simplicity, we assume that $K \| \ell$ is always padded to be 1-block long. Note that all of our attacks can work without this assumption by fixing remaining message bits in the first block to some constant value, say 0.

2.2   Summary of Previous Analyses on MAC Algorithms

To distinguish the target compression function h from a random function r, previous distinguishing-H attacks exploited a high probability differential path on h.

Assume that a good near-collision path exists, namely there exists (Δ, Δ') such that $h(c, m) \oplus h(c, m \oplus \Delta) = \Delta'$ holds with probability p where $p > 2^{-n}$ for a randomly chosen c and m. Then, h is distinguished by querying $1/p$ paired messages with difference Δ and checking whether the output difference is Δ'. To construct a high probability differential path, we usually need the difference of the input chaining variable (pseudo-near collision). However, because the MAC computation starts from the secret information, it is hard to generate a specific difference on intermediate chaining variables, and is also hard to detect it only from the tag values. Wang *et al.* [17] solved these problems by using the birthday attack to generate a specific difference of an intermediate chaining variables and efficiently detect it only by changing the next message block. Previous distinguishing-H attacks on LPMAC [18, 19, 3, 20] used the similar idea as [17]. As long as the birthday attack is used to generate an intermediate difference, the attack complexity is between $2^{n/2}$ and 2^n.

2.3   Multi-collision Attack

The naive method to construct a multi-collision is too expensive. For narrow-pipe Merkle-Damgård hash functions, several generic attacks to construct a multi-collision are known; collisions of sequential blocks [24], collisions with a fixed point [25], *an expandable message* [26], *a diamond structure* [22], and *multi-pipe diamonds* [27]. These are the methods to attack hash functions, which no secret exists in the computation. Because our attack is targeting MAC with secret information, we should choose the most suitable construction carefully.

3   Generic Distinguishing-H Attack on LPMAC

In this section, we present a generic distinguishing-H attack on a narrow-pipe Merkle-Damgård hash function. The attack complexity is $3 \cdot 2^{\frac{n}{2}}$ queries, $2^{\frac{n}{2}}$ offline computations, and a memory to store $2^{\frac{n}{2}}(n + b)$-bit information. Moreover, the attack becomes memoryless with $6 \cdot 2^{\frac{n}{2}}$ queries and $2^{\frac{n}{2}+1}$ offline computations. The results indicate that the hardness of the distinguishing-H attack on LPMAC is almost the same level as the distinguishing-R attack when a narrow-pipe Merkle-Damgård hash function is used.

3.1   Main Idea

Our attack is based on a new technique which makes an internal collision starting from two different length-prepend values t_1 and t_2 where $t_1 < t_2$. The idea is illustrated in Fig. 3. First the attacker generates $2^{n/2}$ t_2-block messages which start from the length-prepend value t_2 and the last $t_2 - t_1$ blocks are fixed to some value. Then he generates $2^{n/2}$ t_1-block messages starting from the length-prepend value t_1 and further computes $t_2 - t_1$ blocks at offline with the fixed value. By checking their match, the inner collision at H_{t_1+1} can be detected with a good probability. The attacker can know the values of H_{t_1+1} and H_{t_2+1}

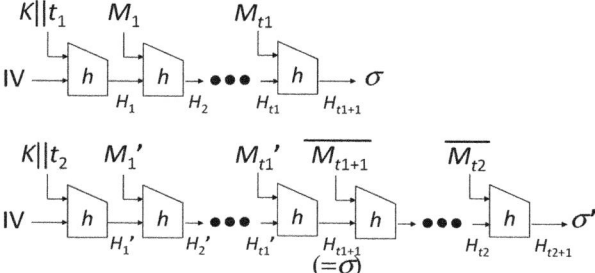

Fig. 3. Internal Collision with Different Length-Prepend Values

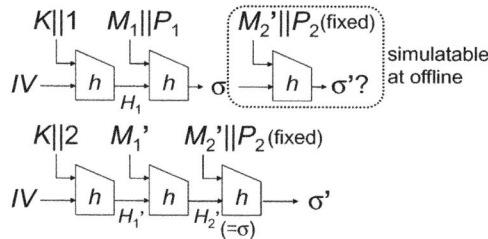

Fig. 4. Procedure of Distinguishing-H Attack

by querying the colliding t_1-block message and t_2-block message, respectively. Therefore, by simulating the last $t_2 - t_1$ blocks at offline, which of h or r is instantiated can be detected.

3.2 Attack Procedure

We assume that the MD-strengthening is used as the padding procedure of the underlying hash function, because it is adopted by most of widely used hash functions. The attack procedure is described in Alg. 1, which is also depicted in Fig. 4. Note that the length-prepend value can be chosen by the distinguisher. The attack procedure returns a bit 1 if the underlying compression function is h and a bit 0 if the underlying compression function is r.

Attack Evaluation. Alg. 1 correctly returns a bit 1 only if the compression function is h.

Let us evaluate the probability that Alg. 1 returns a bit 1 when the compression function is h. In the above procedure, we generate $2^{n/2}$ values for $H_2 (= \sigma)$ and $2^{n/2}$ values for H_2', and thus we have 2^n pairs of (H_2, H_2'). Hence, we have a collision $(H_2 = H_2')$ with probability $1 - e^{-1}$. If $H_2 = H_2'$, the simulated value $temp$ and σ' always become a collision due to the identical message in the last block. Note that we happen to have a collision at Step 9 even if $H_2 \neq H_2'$

Algorithm 1. Distinguishing-H Attack

Input: a compression function algorithm h to be distinguished
Output: a determining bit 0 or 1
1: Randomly choose the value of M_2' so that the padding string P_2 is included in the same block.
2: **for** $2^{n/2}$ different values of M_1' **do**
3:     Query $M_1' \| M_2'$ to obtain the corresponding tag σ'
4:     Store the pair of (M_1', σ') in a table T.
5: **end for**
6: **for** $2^{n/2}$ different values but the same length of M_1 whose padding string P_1 is included in the same block **do**
7:     Query M_1 and obtain the corresponding tag σ.
8:     Compute $temp \leftarrow h(\sigma, M_2' \| P_2)$ at offline.
9:     Check if the same value as $temp$ exists in T.
10:    **if** the same value exists **then**
11:        Replace M_2' with $\overline{M_2'}$ such that $M_2' \neq \overline{M_2'}$ and $|M_2'| = |\overline{M_2'}|$.
12:        Query $M_1' \| \overline{M_2'}$ to obtain the corresponding tag $\overline{\sigma'}$.
13:        Compute $\overline{temp} \leftarrow h(\sigma, \overline{M_2'} \| P_2)$ at offline.
14:        **if** $\overline{\sigma'} = \overline{temp}$ **then**
15:            Return a bit 1.
16:        **end if**
17:    **end if**
18: **end for**
19: Return a bit 0.

with probability $1 - e^{-1}$. This collision is noise in order to detect the internal collision. However, with the additional check at Step 14, the internal collision pair $(H_2 = H_2')$ also generates a collision for $\overline{M_2'}$ with probability 1, whereas the noise collision only produces another collision with probability 2^{-n}. Overall, Alg. 1 returns a bit 1 with a probability of $1 - e^{-1}$ and this is a right pair making an internal collision with probability $1 - 2^{-n} \approx 1$.

Let us evaluate the probability that Alg. 1 returns a bit 1 when the compression function is r. An internal collision $H_2 = H_2'$ occurs with probability $1 - e^{-1}$. However, the collision between H_2 and H_2' is not preserved between $temp$ and σ' at Step 9. This is because σ' is obtained by a query and thus $\sigma' = r(H_2, M_2' \| P_2)$, whereas, $temp$ is computed based on the algorithm of h at offline and thus $temp = h(H_2, M_2' \| P_2)$. The probability that $r(H_2, M_2' \| P_2)$ collides with $h(H_2, M_2' \| P_2)$ is 2^{-n}. As a result, we expect to obtain only one collision at Step 9, and the probability that this pair generates another collision at Step 14 is 2^{-n}. Overall, Alg. 1 returns 1 only with probability 2^{-n}. This is significantly smaller than the probability for the case of h.

Complexity Evaluation. The iteration at Step 2 requires to query $2^{n/2}$ 2-block messages and to store $2^{n/2}$ tag values and corresponding messages. Hence, the query complexity is $2^{(n/2)+1}$ message blocks and the memory complexity is $2^{n/2}(n + b)$ bits, where b is the block size. The iteration at Step 6 requires to query $2^{n/2}$ 1-block messages and to compute h for each of them. Hence, the query complexity is $2^{n/2}$ message blocks and the time complexity is $2^{n/2}$ compression function computations. Overall, the total attack cost is $3 \cdot 2^{n/2}$ message blocks in query, $2^{n/2}$ h computations in time, and $2^{n/2}(n + b)$ bits in memory.

Memoryless Attack. The core of the attack is a collision finding problem on an internal state, thus it can be memoryless with the well-known technique using the cycle structure. Because it needs a collision starting from two different length-prepend values, the problem is a memoryless meet-in-the-middle attack [21, Remark9.93] rather than a memoryless collision attack. To make the cycle, we define the computation from an internal state value s_i to s_{i+1} as follows.

- If the LSB of s_i is 0, apply procedure $\mathcal{F}$. If the LSB of s_i is 1, apply $\mathcal{G}$.
$\mathcal{F}$: Convert an n-bit string s_i into a b-bit string $s_i^{\mathcal{F}}$ e.g. by appending 0s. Query $s_i^{\mathcal{F}} \| M_2'$ to obtain σ' and set $s_{i+1} \leftarrow \sigma'$.
$\mathcal{G}$: Convert an n-bit string s_i into a string $s_i^{\mathcal{G}}$ by some appropriate method so that the padding string P_1 is included in the same block. Query $s_i^{\mathcal{G}}$ to obtain σ and compute $s_{i+1} \leftarrow h(\sigma, M_2'\|P_2)$ at offline.

Finally, the attack can be memoryless with double of the query and time complexities, which are queries of $6 \cdot 2^{n/2}$ message blocks and $2^{(n/2)+1}$ computations of h.

Impact of the Attack. The goal of the attack presented in this section is the distinguishing-H attack. However, the attack not only distinguishes h from r, but also achieves a much stronger result, which is the recovery of the internal state. The knowledge of the internal state leads to much stronger attacks.

The first application is the length-extension attack. We use Fig. 3 to explain the attack. As explained in Sect. 3.1, our attack first recovers the internal state value h_{t_1+1}. That is, we know that any t_2-block message whose first t_1 blocks are fixed to $M_1'\|M_2'\|\cdots\|M_{t_1}'$ will result in the known collision value at h_{t_1+1}. Therefore, for any $(t_2 - t_1)$-block message $\overline{M_{t_1+1}}\|\cdots\|\overline{M_{t_2}}$, the attacker can compute the tag for $M_1'\|\cdots\|M_{t_1}'\|\overline{M_{t_1+1}}\|\cdots\|\overline{M_{t_2}}$ only with one offline computation (of $t_2 - t_1$ blocks) without knowing the value of K. In other words, the length-extension attack can be performed once an internal collision with different length-prepend values is detected.

One may suspect that this length extension attack is the same as the existential forgery because it requires $O(2^{n/2})$ queries. However, obtaining the knowledge of the internal state is actually stronger than the distinguish-R attack [7] because it forges the tag of a message of t_2 blocks long without any more query.

It may be interesting to see our attack from a different viewpoint. The security proof of LPMAC by Bellare et al. assumed the prefix-freeness of the construction, where the prefix-freeness means that any message is not the prefix of other messages. Due to the length-prepend values, LPMAC satisfies the prefix-freeness. However, an internal collision starting from different length-prepend values has the same effect as the prefix. In fact in the above explanation, $M_1\|\cdots\|M_{t_1}$ can be regarded as a prefix of $M_1'\|\cdots\|M_{t_1}'\|\overline{M_{t_1+1}}\|\cdots\|\overline{M_{t_2}}$, and thus the length-extension attack is applied. Note that our attack requires $3\cdot 2^{n/2}$ queries and thus does not contradict with the security proof by Bellare $et\ al.$ where LPMAC is a secure PRF up to $2^{n/2}$ queries.

One limitation of this length extension attack is that it only can forge the tag for messages of t_2 blocks long. We remove this limitation in the next section.

4   Generic almost Universal Forgery Attack on LPMAC

The almost universal forgery attack was mentioned by Dunkelman $et\ al.$ [5]. The original explanation by [5] is as follows.

> we can find in linear time and space the tag of essentially any desired message m chosen in advance, after performing a onetime precomputation in which we query the MAC on $2^{n/2}$ messages which are completely unrelated to m. The only sense in which this is not a universal forgery attack is that we need the ability to modify one message block in an easy to compute way.

In the attack by [5], the first message block of the given message is modified by applying the XOR with two precomputed values.

In our attack, the attacker first determines parameter t, which is the limitation of the block-length of a message to be forged. For parameter t, the block length of the target message must be longer than d blocks and shorter than or equal to $d+t$ blocks, where $d = \lceil \log_2 t \rceil$. We perform a onetime precomputation; query tags for $(t-1)\cdot 2^{n/2}$ messages of at most $d+t$ blocks long (in total $O(2t^2\cdot 2^{n/2})$ message blocks) which are completely unrelated to a target message M, and perform $t\cdot 2^{n/2}$ offline computations of h. At the online stage, we replace the first d blocks of $M(M_1\|M_2\|\cdots\|M_d)$ with the precomputed values $M_1'\|M_2'\|\cdots\|M_d'$. Then, the attacker generates the tag for the modified message only with one offline computation (without any query).

4.1   Easiness and Hardness of almost Universal Forgery Attack

Easiness of almost Universal Forgery Attack on Various MACs. First of all, we point out that the almost universal forgery attack is trivial for various MACs with an iterated structure such as HMAC if the precomputation that queries $O(2^{n/2})$ messages is allowed. To achieve this, we can simply run the distinguishing-R attack by [7]. In details, we generate a pair of one-block messages (X_1, X_1') which forms an internal collision. Then, for any given target message $M_1\|M_2\|\cdots\|M_t$, we replace M_1 with X_1 and query $X_1\|M_2\|\cdots\|M_t$. The corresponding tag is also the valid tag for $X_1'\|M_2\|\cdots\|M_t$.

Hardness of almost Universal Forgery Attack on Prefix-Free MACs. The above attack cannot be applied for prefix-free MACs such as LPMAC in an easy manner. Regarding LPMAC, if the length (not value) of the target message is given to the attacker, the almost universal forgery attack can be performed with the attack presented in Sect. 3.2. However, if the length is not given, performing the precomputation on LPMAC is hard. This is because, in LPMAC, the length-prepend value changes depending on the length of the message to be processed. Hence, without the knowledge of the length of the target message, performing the precomputation seems hard.

A simple method to deal with various length-prepend values in advance is performing the internal state recovery attack in Sect. 3.2 many times. Suppose that you apply the almost universal forgery attack for a message of ℓ blocks where $1 \le \ell \le t$, but the exact value of ℓ is unknown during the precomputation stage. You first assume $\ell = 1$ and apply the internal state recovery attack with the $O(2^{n/2})$ complexity. Then, the value of ℓ is changed into $2, 3, \ldots, t$ and the internal state recovery attack is performed for each of ℓ. Finally, the almost universal forgery attack can be performed for any ℓ-block message where $1 \le \ell \le t$ with t times $O(2^{n/2})$ queries where each query consists of at most t-blocks.

In the following part, we show another approach with the same complexity as the simple method at the order level, but seems to have more applications.

4.2 Overall Strategy

Our idea is constructing an internal multi-collision starting from various length-prepend values as shown in Fig. 5. Assume that a one-block message X makes a multi-collision for $\ell = 2, 3, \ldots, 9$, that is, $h(h(\mathrm{IV}, K\|2), X) = h(h(\mathrm{IV}, K\|3), X) = \cdots = h(h(\mathrm{IV}, K\|9), X)$. Also assume that the attacker knows the collision value denoted by Y. Then, for any message with the block length 2 to 9, we know that replacing the first block with X results in the chaining variable Y, and thus the computation for the remaining message blocks can be simulated at offline.

Finding a multi-collision within one block is very inefficient. For the Merkle-Damgård structure, several constructions of the multi-collision are known as explained in Sect. 2.3. Considering that our multi-collision needs to start from different values due to different length-prepend values, the diamond structure [22] and multi-pipe diamonds [27] are suitable. The diamond structure was proposed for the herding attack and the multi-pipe diamonds were proposed for the herding attack on more complicated structures such as the cascaded construction or zipper hash. So far, we have not discovered the way to utilize the multi-pipe diamonds. We thus construct the multi-collision based on the diamond structure. The construction is described in Fig. 6.

4.3 Multi-collision with Diamond Structure

We explain how to construct a multi-collision with the diamond structure. Because our attack target is a MAC with secret information and the length prepending scheme, we need to detect an internal collision only with queries and tag values. In Alg. 2, we show the procedure to generate a collision starting from $\ell = t$ and $\ell = t + 1$, which is also described in Fig. 7. It is easy to see that if a collision can be generated, the entire diamond can be constructed by iteratively generating collisions.

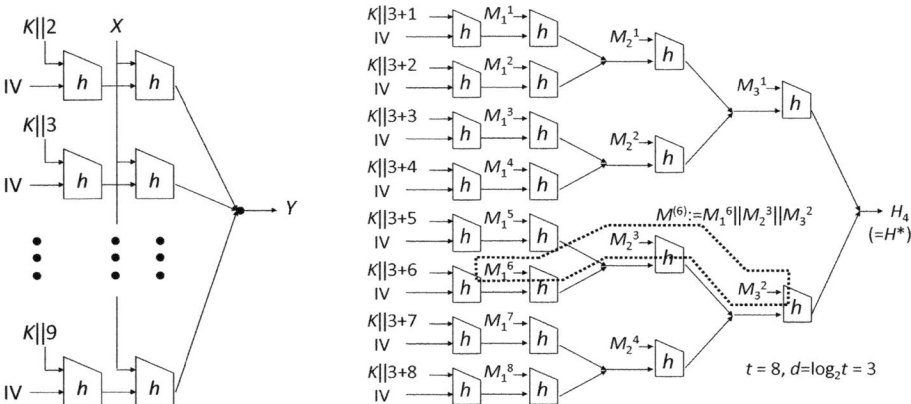

Fig. 5. Length Adjustment   **Fig. 6.** Multi-Collision with Diamond Structure for $t = 8$
with Multi-collision

In this attack we fix all message words but M_1 and M_1' to identical value. Therefore, the collision generated by M_1 and M_1' can be observed as a collision of the tag. Because we try $2^{n/2}$ different M_1 at Step 8 and $2^{n/2}$ different M_1' at Step 4, we will obtain a collision $H_2 = H_2'$ with probability $1 - e^{-1}$. Note that even if $H_2 \neq H_2'$, we have other opportunities of obtaining collisions $H_3 = H_3'$, $H_4 = H_4'$, and so on. These are noise to obtain a collision at H_2, and thus need to be filtered out. For this purpose, for all tag collisions, we replace M_2 and M_2' with another fixed value and check if another collision is generated (Step 17).

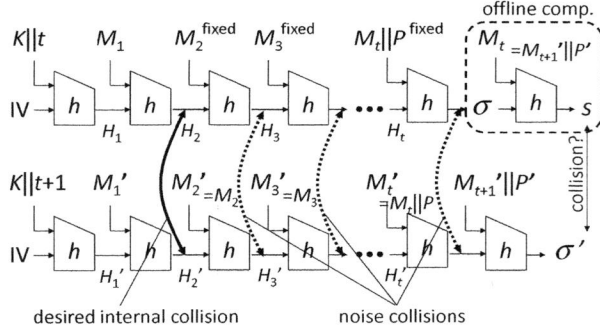

Fig. 7. Construction of an Internal Collision for $\ell = t$ and $t + 1$

The complexity of the attack is as follows. At Step 4, it queries $(t + 1) \cdot 2^{n/2}$ message blocks and requires a memory to store $2^{n/2}$ tags. At Step 8, it queries $t \cdot 2^{n/2}$ message blocks and computes h at offline $2^{n/2}$ times. We expect to obtain

Algorithm 2. Construction of an Internal Collision for $\ell = t$ and $t + 1$

Input: $\ell = t$ (and $t + 1$)

Output: a pair of message M_1^t and M_1^{t+1} such that $h(h(\text{IV}, K\|t), M_1^t) = h(h(\text{IV}, K\|t+1), M_1^{t+1})$

1: Fix the values of $M_2, M_3, \ldots, M_t$ so that the padding string P is included in the same block as M_t.

2: Set $M_i' \leftarrow M_i$ for $i = 2, 3, \ldots, t-1$. Set $M_t' \leftarrow M_t\|P$.

3: Fix the value of M_{t+1}' so that the padding string P' is included in the same block as M_{t+1}'.

4: **for** $2^{n/2}$ different values of M_1' **do**

5:     Query $M_1'\|M_2'\|\cdots\|M_{t+1}'$ to obtain the corresponding tag σ'.

6:     Store the pair of (M_1', σ') in a table T.

7: **end for**

8: **for** $2^{n/2}$ different values of M_1 **do**

9:     Query each $M_1\|M_2\|\cdots\|M_t$ and obtain the corresponding tag σ.

10:     Compute $temp \leftarrow h(\sigma, M_{t+1}'\|P')$ at offline.

11:     Check if the same value as $temp$ exists in T.

12:     **if** the same value exists **then**

13:         Choose a value of $\overline{M_2}$ such that $M_2 \neq \overline{M_2}$.

14:         Set $\overline{M_2'} \leftarrow \overline{M_2}$.

15:         Query $M_1'\|\overline{M_2'}\|M_3'\|\cdots\|M_{t+1}'$ to obtain the corresponding tag $\overline{\sigma'}$.

16:         Query $M_1\|\overline{M_2}\|M_3\|\cdots\|M_t$ to obtain the corresponding tag $\overline{\sigma}$, and compute $\overline{temp} \leftarrow h(\overline{\sigma}, M_{t+1}'\|P')$.

17:         **if** $\overline{\sigma'} = \overline{temp}$ **then**

18:             Return M_1 and M_1'.

19:         **end if**

20:     **end if**

21: **end for**

one desired internal collision and t noise collisions. Therefore Steps 15 and 16 are computed $t + 1$ times and it requires to query $2(t + 1)^2$ message blocks and $2(t + 1)^2$ offline computations. Overall, the cost of Alg. 2 is approximately $(2t + 1) \cdot 2^{n/2}$ queries and $2^{n/2}$ offline computations and a memory for $2^{n/2}$ tags. Note that the attack can be memoryless as discussed in Sect. 3.

Hereafter we denote the value of the multi-collision at H_{d+1} by H^*. We also denote the d-block message starting from $\ell = i$ and reaching H^* by $M^{(i)}$. For example, in Fig. 6, $M^{(1)} := M_1^1\|M_2^1\|M_3^1$ and $M^{(6)} := M_1^6\|M_2^3\|M_3^2$.

Complexity for Entire Diamond Structure. By using Alg. 2, we evaluate the complexity for constructing the entire structure. Assume that we generate a diamond structure which can be used to forge the message whose length is longer than d blocks and shorter than or equal to $d + t$ blocks where $d = \log_2 t$.

The complexity of generating one collision starting from two different ℓ is determined by the bigger value of ℓ. In this example, the biggest value of ℓ is $d + t$ and thus the complexity for generating collisions with other ℓ is smaller than this case. According to Alg. 2, the complexity for the case $\ell = d + t$ is $(2(d + t) + 1) \cdot 2^{n/2}$ queries and $2^{n/2}$ offline computations. The number of the

Algorithm 3. Forging Procedure
Output: a pair of message in which the first d blocks are modified and valid tag

Offline phase
1: Construct the diamond structure which can be used to forge the message whose length is longer than d blocks and shorter than or equal to $d+t$ blocks with $2t^2 \cdot 2^{n/2}$ queries and $t \cdot 2^{n/2}$ offline computations.

Online phase
2: Receive a target message M^* whose length is ℓ blocks where $d < \ell \leq d+t$, that is, $M^* = M_1^* \| M_2^* \| \cdots \| M_\ell^*$.
3: Replace the first d blocks of M^* with $M^{(\ell)}$.
4: Compute the tag value by using $H_{d+1}(= H^*)$ and $M_{d+1}^* \| M_{d+2}^* \| \cdots \| M_\ell^*$, which is denoted by σ^*.
5: **return** a pair of message and valid tag $(M^{(\ell)} \| M_{d+1}^* \| M_{d+2}^* \| \cdots \| M_\ell^*, \sigma^*)$.

leaf in the diamond structure is t, and thus we need to generate a collision $t-1$ times. Hence, the total complexity is less than $(t-1)(2t+2d+1) \cdot 2^{n/2}$ queries and $(t-1) \cdot 2^{n/2}$ offline computations. If only the head term is considered, the complexity is $2t^2 \cdot 2^{n/2}$ queries and $t \cdot 2^{n/2}$ offline computations.

4.4   Forging Procedure

Finally, we show how to produce a forged tag in Alg. 3. The construction of the diamond structure is completely independent of the online phase. As long as the block length ℓ of the given message M^* is in the valid range, by replacing the first $d(= \log_2 t)$ blocks of M^* with $M^{(\ell)}$, the tag for $M^{(\ell)} \| M_{d+1}^* \| M_{d+2}^* \| \cdots \| M_\ell^*$ is computed only with one t-block offline computation.

4.5   Comparison between Simple Method and Diamond Structure

We compare the simple method in Sect. 4.1 (applying the internal state recovery attack t times) and the diamond structure. As long as only the almost universal forgery attack is considered, the simple method is better than the diamond structure. The simple method does not require the offline computation in the precomputation phase, and the number of message blocks we need to replace is only 1 which is shorter than $d = \log_2 t$.

On the other hand, the diamond structure has a unique property; it achieves a common internal state value for various length-prepend values. So far, good applications of this property have not been discovered. However, we show an example that gives some intuition to use the property. Let us consider the connection problem; the goal is finding a message block for the compression function h, which the input chaining variable is fixed to a given value and there are several target output values with different message lengths. The long-message second preimage attack and herding attack in Fig. 8 are such problems, though these problems are for the key-less situations. Finding suitable applications is an open problem.

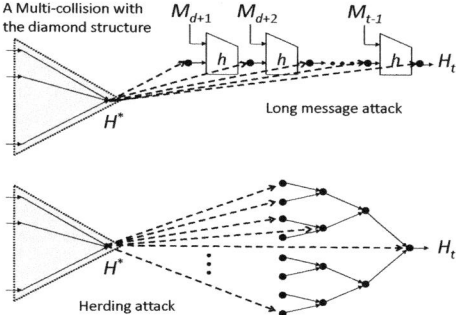

Fig. 8. Examples of potential applications of the diamond structure. The application is finding a message which connects the multi-collision H^* to one of chaining variables. The length information cannot be adjusted in advance.

5   Concluding Remarks

In this paper, we presented two cryptanalyses on LPMAC. Our first result was a generic distinguishing-H attack on LPMAC instantiating a narrow-pipe Merkle-Damgård hash function with a complexity of $O(2^{n/2})$. This showed that the widely believed assumption that a secure hash function should have the n-bit security against the distinguishing-H attack was not correct. Note that although previous results were updated by our generic attack with respect to the distinguish-H attack, the approach was very different. Finding a new problem which the previous differential distinguishers can work faster than a generic attack is an open problem. Our attacks are based on the new technique which generates an internal collision starting from different length-prepend values. One of such colliding messages can be regarded as the prefix of the other colliding message, and thus the core of the security of LPMAC, which is the prefix-freeness, is completely broken.

Our second result was an almost universal forgery attack on LPMAC. With the precomputation of $2t^2 \cdot 2^{n/2}$ queries and $t \cdot 2^{n/2}$ offline computations, we constructed the diamond structure which could realize an identical intermediate value from t different length-prepend values. Hence, by modifying the first $\log_2 t$ message blocks into the value of the attacker's choice, the internal state was recovered and the valid tag of the modified message was forged with only 1 offline computation. Our results show that the security of the length-prepending structure can be totally broken with $O(2^{n/2})$ queries, and thus it is not enough to achieve a secure MAC.

Acknowledgements. The author would like to thank Lei Wang for his comments about the simple method of the almost universal forgery attack on LPMAC. The author also would like to thank the participants of Dagstuhl-Seminar (Jan. 2012), especially Orr Dunkelman and Adi Shamir for their comments on the memoryless meet-in-the-middle attack. Finally, the author would like to thank anonymous referees of Eurocrypt2012 for their helpful comments and suggestions.

References

1. Tsudik, G.: Message authentication with one-way hash functions. ACM SIGCOMM Computer Communication Review 22(5), 29–38 (1992)
2. U.S. Department of Commerce, National Institute of Standards and Technology: Federal Register 72(212) (November 2, 2007), http://csrc.nist.gov/groups/ST/hash/documents/FR_Notice_Nov07.pdf
3. Wang, X., Wang, W., Jia, K., Wang, M.: New Distinguishing Attack on MAC Using Secret-Prefix Method. In: Dunkelman, O. (ed.) FSE 2009. LNCS, vol. 5665, pp. 363–374. Springer, Heidelberg (2009)
4. Bellare, M., Canetti, R., Krawczyk, H.: Pseudorandom functions revisited: The cascade construction and its concrete security. In: FOCS, pp. 514–523 (1996)
5. Dunkelman, O., Keller, N., Shamir, A.: ALRED blues: New attacks on AES-based MACs. Cryptology ePrint Archive, Report 2011/095 (2011), http://eprint.iacr.org/2011/095
6. Kim, J., Biryukov, A., Preneel, B., Hong, S.: On the Security of HMAC and NMAC Based on HAVAL, MD4, MD5, SHA-0 and SHA-1 (Extended Abstract). In: De Prisco, R., Yung, M. (eds.) SCN 2006. LNCS, vol. 4116, pp. 242–256. Springer, Heidelberg (2006)
7. Preneel, B., van Oorschot, P.C.: MDx-MAC and Building Fast MACs from Hash Functions. In: Coppersmith, D. (ed.) CRYPTO 1995. LNCS, vol. 963, pp. 1–14. Springer, Heidelberg (1995)
8. Chang, D., Nandi, M.: General distinguishing attacks on NMAC and HMAC with birthday attack complexity. Cryptology ePrint Archive, Report 2006/441 (2006), http://eprint.iacr.org/2006/441
9. Jia, K., Wang, X., Yuan, Z., Xu, G.: Distinguishing and Second-Preimage Attacks on CBC-Like MACs. In: Garay, J.A., Miyaji, A., Otsuka, A. (eds.) CANS 2009. LNCS, vol. 5888, pp. 349–361. Springer, Heidelberg (2009)
10. Yuan, Z., Wang, W., Jia, K., Xu, G., Wang, X.: New Birthday Attacks on Some MACs Based on Block Ciphers. In: Halevi, S. (ed.) CRYPTO 2009. LNCS, vol. 5677, pp. 209–230. Springer, Heidelberg (2009)
11. Contini, S., Yin, Y.L.: Forgery and Partial Key-Recovery Attacks on HMAC and NMAC Using Hash Collisions. In: Lai, X., Chen, K. (eds.) ASIACRYPT 2006. LNCS, vol. 4284, pp. 37–53. Springer, Heidelberg (2006)
12. Fouque, P.-A., Leurent, G., Nguyen, P.Q.: Full Key-Recovery Attacks on HMAC/NMAC-MD4 and NMAC-MD5. In: Menezes, A. (ed.) CRYPTO 2007. LNCS, vol. 4622, pp. 13–30. Springer, Heidelberg (2007)
13. Lee, E., Chang, D., Kim, J., Sung, J., Hong, S.: Second Preimage Attack on 3-Pass HAVAL and Partial Key-Recovery Attacks on HMAC/NMAC-3-Pass HAVAL. In: Nyberg, K. (ed.) FSE 2008. LNCS, vol. 5086, pp. 189–206. Springer, Heidelberg (2008)
14. Rechberger, C., Rijmen, V.: On Authentication with HMAC and Non-random Properties. In: Dietrich, S., Dhamija, R. (eds.) FC 2007 and USEC 2007. LNCS, vol. 4886, pp. 119–133. Springer, Heidelberg (2007)
15. Rechberger, C., Rijmen, V.: New results on NMAC/HMAC when instantiated with popular hash functions. Journal of Universal Computer Science 14(3), 347–376 (2008)
16. Wang, L., Ohta, K., Kunihiro, N.: New Key-Recovery Attacks on HMAC/NMAC-MD4 and NMAC-MD5. In: Smart, N.P. (ed.) EUROCRYPT 2008. LNCS, vol. 4965, pp. 237–253. Springer, Heidelberg (2008)

17. Wang, X., Yu, H., Wang, W., Zhang, H., Zhan, T.: Cryptanalysis on HMAC/NMAC-MD5 and MD5-MAC. In: Joux, A. (ed.) EUROCRYPT 2009. LNCS, vol. 5479, pp. 121–133. Springer, Heidelberg (2009)
18. Qiao, S., Wang, W., Jia, K.: Distinguishing Attack on Secret Prefix MAC Instantiated with Reduced SHA-1. In: Lee, D., Hong, S. (eds.) ICISC 2009. LNCS, vol. 5984, pp. 349–361. Springer, Heidelberg (2010)
19. Wang, G.: Distinguishing Attacks on LPMAC Based on the Full RIPEMD and Reduced-Step RIPEMD-{256,320}. In: Lai, X., Yung, M., Lin, D. (eds.) Inscrypt 2010. LNCS, vol. 6584, pp. 199–217. Springer, Heidelberg (2011)
20. Yu, H., Wang, X.: Distinguishing Attack on the Secret-Prefix MAC Based on the 39-Step SHA-256. In: Boyd, C., González Nieto, J. (eds.) ACISP 2009. LNCS, vol. 5594, pp. 185–201. Springer, Heidelberg (2009)
21. Menezes, A.J., van Oorschot, P.C., Vanstone, S.A.: Handbook of applied cryptography. CRC Press (1997)
22. Kelsey, J., Kohno, T.: Herding Hash Functions and the Nostradamus Attack. In: Vaudenay, S. (ed.) EUROCRYPT 2006. LNCS, vol. 4004, pp. 183–200. Springer, Heidelberg (2006)
23. Yasuda, K.: How to Fill Up Merkle-Damgård Hash Functions. In: Pieprzyk, J. (ed.) ASIACRYPT 2008. LNCS, vol. 5350, pp. 272–289. Springer, Heidelberg (2008)
24. Joux, A.: Multicollisions in Iterated Hash Functions. Application to Cascaded Constructions. In: Franklin, M. (ed.) CRYPTO 2004. LNCS, vol. 3152, pp. 306–316. Springer, Heidelberg (2004)
25. Dean, R.D.: Formal aspects of mobile code security. Ph.D Dissertation, Princeton University (1999)
26. Kelsey, J., Schneier, B.: Second Preimages on n-Bit Hash Functions for Much Less than 2^n Work. In: Cramer, R. (ed.) EUROCRYPT 2005. LNCS, vol. 3494, pp. 474–490. Springer, Heidelberg (2005)
27. Andreeva, E., Bouillaguet, C., Dunkelman, O., Kelsey, J.: Herding, Second Preimage and Trojan Message Attacks beyond Merkle-Damgård. In: Jacobson Jr., M.J., Rijmen, V., Safavi-Naini, R. (eds.) SAC 2009. LNCS, vol. 5867, pp. 393–414. Springer, Heidelberg (2009)

Statistical Tools Flavor Side-Channel Collision Attacks

Amir Moradi

Horst Görtz Institute for IT Security, Ruhr University Bochum, Germany
moradi@crypto.rub.de

Abstract. By examining the similarity of side-channel leakages, collision attacks evade the indispensable hypothetical leakage models of multi-query based side-channel distinguishers like correlation power analysis and mutual information analysis attacks. Most of the side-channel collision attacks compare two selective observations, what makes them similar to simple power analysis attacks. A multi-query collision attack detecting several collisions at the same time by means of comparing the leakage averages was presented at CHES 2010. To be successful this attack requires the means of the side-channel leakages to be related to the processed intermediate values. It therefore fails in case the mean values and processed data are independent, even though the leakages and the processed values follow a clear relationship. The contribution of this article is to extend the scope of this attack by employing additional statistics to detect the colliding situations. Instead of restricting the analyses to evaluation of means, we propose to employ higher-order statistical moments and probability density functions as the figure of merit to detect collisions. Thus, our new techniques remove the shortcomings of the existing correlation collision attacks using first-order moments. In addition to the theoretical discussion of our approach, practical evidence of its suitability for side-channel evaluation is provided. We provide four case studies, including three FPGA-based masked hardware implementations and a software implementation using boolean masking on a microcontroller, to support our theoretical groundwork.

1 Introduction

Integration of embedded computers into our daily life, e.g., in automotive applications and smartcard applications for financial purposes, led to a widespread deployment of security-sensitive devices. On the downside also adversaries benefit from the resulting easy physical accessibility, as it provides control over the devices and thus simplifies analyses. Consequently, today most sensitive embedded systems need to be considered as operating in a hostile environment. For this reason physical attacks, most notably side-channel analyses, are considered major threats. For instance, power analysis and the closely related electro-magnetic (EM) analysis can easily overcome the security features of unprotected designs by monitoring the power consumption of the executing device. In order to distinguish the correct key hypothesis amongst the others *differential power analysis*

D. Pointcheval and T. Johansson (Eds.): EUROCRYPT 2012, LNCS 7237, pp. 428–445, 2012.

(DPA) [13] and its successor form, the *correlation power analysis* (CPA) [7], use statistical tools: the *difference of means* and the *Pearson correlation coefficient*, respectively. The distinguisher is applied to side-channel observations classified into subsets defined by a boolean partitioning function in the case of a DPA or by means of a hypothetical power model in case of a CPA. The later introduced *mutual information analysis* (MIA) [11] provides a generic distinguisher that lifts the need of sophisticated power models at the cost of an increased number of required side-channel observations. Generally speaking, MIA is able to recover secret information when the CPA fails due to the lack of a suitable power model. However, the efficiency of MIA also relies on the availability of a good hypothetical model that reflects the dependencies of the actual data-dependent leakage provided by the side-channel observations. The loss of efficiency becomes most visible and even critical when the underlying function of the target device is not a many-to-one mapping (see the detailed discussion provided in [26]).

In order to develop an attack method that does not require a device dependent model, a new type of side-channel attacks has been introduced: the side-channel based collision attacks [2, 3, 5, 24, 25]. These methods adopt collision attacks to side-channel analyses and allow efficiently extracting secrets from side-channel measurements using only a small number of observations, especially when the design architecture of the target implementation is known to the adversary (see e.g., [4] where collision attacks are combined with DPA). Collision attacks, however, get infeasible when facing very noisy observations or in presence of both, time-domain and data-domain randomizing countermeasures. Recent works propose a couple of techniques, e.g., in [3] to deal with false-positive collision detections and [10] which reports a successful attack on a mask-protected software implementation which exploits reused masks. Another recent attack method [16] named correlation collision attack exploits conditions that lead to a multitude of collisions whenever a key-dependent relation between the processed input values is fulfilled. More precisely, it compares the sets of leakages (averaged with respect to a fixed relevant input) of one e.g., S-box instance when it processes two distinct input sets, each of which associated with a different part of the secret key. The relation between the inputs of the two sets, that causes all averages to collide, exposes information on the secret key. During the last years the independence of side-channel collision attacks from hypothetical models and the effects of process variations which harden side-channel attacks in nanoscale devices [23] increasingly attracts the attention of the community to the new attack methods leading to new applications and variations as in [17] and [10].

A correlation collision attack [16] applies a statistical tool, i.e., the Pearson correlation coefficient, on the means of side-channel observations that were classified with respect to known input data. This method is successful when the means of the classified side-channel observations are different when they are estimated using a large (but feasible) amount of observations. If the mean values do not show the required dependency to processed data, the attack will fail, even in case there is a clear relation between the processed values and the observed side-channel leakages. To give an example, we refer to *threshold implementations* [19, 20], which claim that the averages of the side-channel leakages are independent of the processed values. In this case a MIA might still be able to exploit the leakage to recover a secret key [21]. Similarly a correlation collision attack is not able to recover the desired secret in this case (as also stated in [18]), as it relies on mean values. Indeed, this was one of the motivations for this work to apply other statistics in side-channel collision attacks in order to enable detection of colliding situations also in cases where the mean values do not provide any exploitable information.

In this article we discuss how to extend the scope of correlation collision attacks to exploit dependencies in different central moments from probability theory and statistics. Furthermore, we elaborate on preprocessing schemes which can be performed to improve correlation collision attacks. We show that in certain situations applying a preprocessing step prior to a correlation collision attack on mean values is equivalent to the same attack targeting a high-order statistical moment. In order to generalize the scheme we moreover propose to compare probability density functions (pdf) instead of any specific moments. Although accurately estimating the pdfs is an expensive task that requires a high number of observations, this generalized approach does not require any assumptions about the type and shape of the leakage distributions and may thus be worth the additional efforts.

In order to practically investigate our proposed schemes on different implementations we have considered both, an FPGA-based platform as well as a microcontroller. Three different masked hardware implementations were mapped to our target FPGA device. These include *i*) an AES encryption engine using the masked S-box presented in [8], *ii*) an implementation of PRESENT [6] using the threshold implementation countermeasure as presented in [22], and *iii*) a threshold implementation of the AES as reported in [18]. Since the masked values and the masks are processed simultaneously in all the aforementioned implementations, a univariate attack method is an applicable choice. We show how to use different statistical moments in a collision attack to recover the desired secret and we discuss their efficiencies. As a fourth case study a software implementation of first-order boolean masking on a microcontroller is analyzed. Here, since the masks and the masked values are processed sequentially, a multivariate attack needs to be applied. We use this case study to illustrate possible solutions including multivariate collision attacks and univariate ones which employ a combining function.

2   Preliminaries

In the following we introduce the notation used in this paper and explain the adopted side-channel model. Afterwards, Section 2.2 provides a short review of linear collision attacks followed by a formal specification of correlation collision attacks.

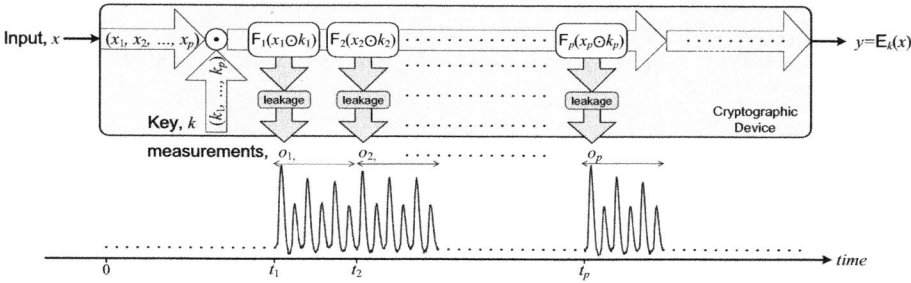

Fig. 1. Side-channel model

2.1   Notations and Side-Channel Model

We consider a cryptographic device that performs the cryptographic operation E on the given input x. E depends on the secret key k and outputs the value $y = \mathsf{E}_k(x)$ (see Fig. 1). The algorithmic computations depending on x and k cause internal state transitions (e.g., bit flips). The internal state transitions affect the side-channel observations o, which are noisy measurements of the leakages.

In Fig. 1 we suppose that the considered cryptographic operation is an iterated symmetric block cipher that starts with a *key whitening* step represented by the general conjunction $\odot$. To denote the small parts of input data and the key used in a *divide-and-conquer* key recovery scheme on the cryptographic operation E, we use the subscripts i, i.e., x_i and k_i, where $i \in \{1, \dots, p\}$ and p is the number of different parts used. Furthermore, we introduce the functions F_i (usually nonlinear), that independently process the key-whitened inputs $x_i \odot k_i$. Although for simplicity we have supposed that each function F_i is performed at a different time t_i[1], sequential or parallel execution of the functions F_i depends on the actual implementations platform and architecture.

Performing q queries to the target device an adversary acquires the side-channel measurements $\boldsymbol{o}^1, \dots, \boldsymbol{o}^q$ corresponding to the device's processing of the supplied inputs $x^1, \dots, x^q$. The j-th measurement $\boldsymbol{o}^j$ consists of p parts $\boldsymbol{o}^j_1, \dots, \boldsymbol{o}^j_p$ corresponding to the computations of the functions F_i at times t_i. Note that each side-channel measurement itself still consists of multiple samples. That is, the i-th part of the j-th measurement, i.e., the vector $\boldsymbol{o}^j_i$, denotes s subsequently measured samples $o^j_{i,1}, \dots, o^j_{i,s}$.[2]

For example in a CPA attack, for a specific portion i the adversary determines $\boldsymbol{w}_{\mathsf{i}}$ as a vector of estimated internal state transitions w^j_{i}, $\forall\, 1 \le j \le q$ using the input portion x^j_{i} and a hypothesis for the key portion k_{i}. Then, he evaluates his guess by comparing the leakage modeled by $\widehat{L}(\boldsymbol{w}_{\mathsf{i}})$ to the actual measurements $o^j_{\mathsf{i},\mathsf{s}}$.

---

[1] Times are measured relative to the the start of each processing of E.

[2] Note that in each measurement j the measurement parts $\boldsymbol{o}^j_{i=1,\dots,p}$ may overlap in some sample points. This is helpful when the exact time instances t_i are uncertain but their distances, e.g., the number of clock cycles between the consecutive t_i, are known.

Hereby the leakages of the sample points $s \in \{1, \ldots, s\}$ are considered independently. The most appealing advantage of the collision attacks, which are restated in the following, is to avoid requiring the hypothetical leakage model $\widehat{L}(\cdot)$.

2.2   Correlation Collision Attack

In the case that two functions F_{i_1} and F_{i_2} ($i_1 \neq i_2 \in \{1, \ldots, p\}$) are identical (see Fig. 1), a collision attack might be possible. Analyzing the measurements o_{i_1} and o_{i_2} a collision attack aims at detecting situations where both functions process the same value. In this case injective functions $\mathsf{F}_{i_1} = \mathsf{F}_{i_2}$ (e.g., the AES S-box) allow concluding

$$\mathsf{F}_{i_1}(x_{i_1} \odot k_{i_1}) = \mathsf{F}_{i_2}(x_{i_2} \odot k_{i_2})$$
$$\Leftrightarrow \qquad x_{i_1} \odot k_{i_1} = x_{i_2} \odot k_{i_2}$$
$$\Leftrightarrow \qquad (x_{i_1})^{-1} \odot x_{i_1} \odot k_{i_1} \odot (k_{i_2})^{-1} = (x_{i_1})^{-1} \odot x_{i_2} \odot k_{i_2} \odot (k_{i_2})^{-1}$$
$$\Leftrightarrow \qquad \Delta k_{i_1,i_2} = k_{i_1} \odot (k_{i_2})^{-1} = (x_{i_1})^{-1} \odot x_{i_2},$$

where $(k_{i_2})^{-1}$ and $(x_{i_1})^{-1}$ are respectively a right inverse of k_{i_2} and a left inverse of x_{i_1}, i.e., $k_{i_2} \odot (k_{i_2})^{-1} = e_r$ and $(x_{i_1})^{-1} \odot x_{i_1} = e_l$, where e_r and e_l are respectively a right and a left identity element of operation $\odot$. Since x_{i_1} and x_{i_2} are supposed to be known to the adversary, $\Delta k_{i_1,i_2}$ gets revealed detecting such a collision. If additional instances of the function F_i are processed within the analyzed algorithm E, all available instances can be pairwise evaluated to reveal terms $\Delta k_{\cdot,\cdot}$ as described above. Depending on the target algorithm this allows an adversary to either determine all parts of the key or to significantly shrink the key space, what allows for feasible exhaustive key searches.

When the target device implements the AES, the functions F_i are AES S-boxes and the conjunction $\odot$ is the first call to the AddRoundKey operation (i.e., $x_i \oplus k_i$, $\oplus$ denoting bitwise XOR) prior to the first round of the encryption. Then, detecting a collision (called linear collision on AES [3]) $\Delta k_{i_1,i_2} = k_{i_1} \oplus k_{i_2} = x_{i_1} \oplus x_{i_2}$ is recovered. In this case, the adversary can recover a maximum of 15 linearly independent relations between the key portions allowing the key search space to be restricted to 2^8.

In the first generation of side-channel collision attacks, e.g., [2–5, 24, 25], the collision detection process is implemented by pairwise comparing measurement parts $(o_{i_1}^{j_1}, o_{i_2}^{j_2})$ where $j_1, j_2 \in \{1, \ldots, q\}$. Also, different methods were used to perform the comparison (e.g., the Euclidean distance in [25]). Although one needs to deal with false-positive comparison results, this attack is feasible when the target device and architecture sequentially processes the algorithm, e.g., a microcontroller. Also, the more clock cycles the observations $(o_{i_1}^{j_1}, o_{i_2}^{j_2})$ include, the more robust the detection gets, leading to a more feasible attack.

However, when attacking a hardware implementation which simultaneously performs multiple operations or when randomizing countermeasures or noise addition schemes are embedded into the target device, examining the similarity of a pair of measurement parts will probably fail to detect the collisions. Also, in these cases each measurement part o_i usually covers only a single clock cycle. The attack introduced in [16] (the so-called *correlation collision attack*) uses a different scheme to overcome such problems. As the instances of the functions F_{i_1} and F_{i_2} always collide whenever the condition $x_{i_2} = x_{i_1} \odot \Delta k_{i_1,i_2}$ holds, $\Delta k_{i_1,i_2}$ can be recovered by means of a hypothesis test. In order to do so, two sets of mean vectors, denoted by μ_{i_1} and μ_{i_2}, are computed. Each set μ_i consists of 2^n mean vectors $\left\{ \boldsymbol{m}_i^0, \ldots, \boldsymbol{m}_i^{2^n-1} \right\}$, where n is the bit-length of a plaintext (or key) portion. Each mean vector $\boldsymbol{m}_i^x$, $x \in \mathbb{F}_{2^n}$ again consists of s mean samples $(m_{i,1}^x, \ldots, m_{i,s}^x)$ which are defined as

$$m_{i,s}^x = \frac{1}{q_i^x} \sum_{j=1, x_i^j = x}^{q} o_{i,s}^j, \qquad s \in \{1, \ldots, s\}, \qquad x \in \mathbb{F}_{2^n},$$

where q_i^x denotes the cardinality of the set $\left\{ j : 1 \leq j \leq q \mid x_i^j = x \right\}$. Now based on $\widehat{\Delta k}$, i.e., a hypothesis for $\Delta k_{i_1,i_2}$, two vectors $\boldsymbol{m}'_{i_1,s}$ and $^{\widehat{\Delta k}}\boldsymbol{m}'_{i_2,s}$ are extracted from the two sets μ_{i_1} and μ_{i_2} defined above:

$$\boldsymbol{m}'_{i_1,s} = (m'^0_{i_1,s}, \ldots, m'^{2^n-1}_{i_1,s}), \qquad m'^x_{i_1,s} = m^x_{i_1,s}, \qquad x \in \mathbb{F}_{2^n}$$

$$^{\widehat{\Delta k}}\boldsymbol{m}'_{i_2,s} = (^{\widehat{\Delta k}}m'^0_{i_2,s}, \ldots, ^{\widehat{\Delta k}}m'^{2^n-1}_{i_2,s}), \qquad ^{\widehat{\Delta k}}m'^x_{i_2,s} = m^{x \odot \widehat{\Delta k}}_{i_2,s}, \qquad x \in \mathbb{F}_{2^n}.$$

Now Pearson's correlation coefficient can be used to measure the similarity of the pair of vectors $\boldsymbol{m}'_{i_1,s}$ and $^{\widehat{\Delta k}}\boldsymbol{m}'_{i_2,s}$. The most similar vectors at the analyzed sample point s indicate the most probable $\widehat{\Delta k}$. This procedure – similar to most of the non-profiled attacks – is repeated for each sample point s independently. The time instances t_{i_1} and t_{i_2} (see Fig. 1) are initially not known to an adversary without detailed information on the implemented architecture of the target device, but [16] proposes a method to reveal this information. The suggested method is to analyze the variance of $\{m_{i,s}^x : \forall\, x \in \mathbb{F}_{2^n}\}$: If the means of the measurements at sample point s depend on the inputs x_i, the variance at sample point s is significantly increased compared to other sample points.

3   Shortcomings and Our Solutions

Since only the mean values contribute to the comparison metric of the original correlation collision attack, it cannot detect collisions whenever the means of the leakages do not depend on processed data, even if the distributions of the leakages show a strong data dependence. As an example consider the distributions in Fig. 2. Since the shown distributions have the same mean, the attack will fail, although the distributions can clearly be distinguished by their shape.

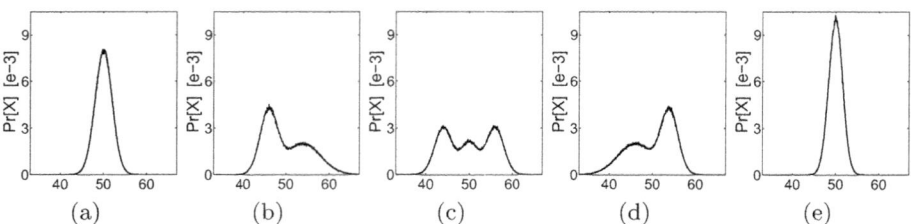

Fig. 2. Examples of probability distributions with the same mean

3.1   Higher-Order Moments

While the mean of all the probability distributions shown in Fig. 2 is the same, their higher-order moments are different. For instance, Figures 2(b), (c), and (d) can be discriminated by their skewnesses. Also, Fig. 2(a) can be distinguished from Figures 2(b), (c), and (d) by the variance, and from Fig. 2(e) by the kurtosis. Therefore, in order to extend the scheme, one can exploit the differences in the higher statistical moments similarly to the analysis of the mean values performed before. In other words, extending the notations given in Section 2.2 we can calculate the sets of the d-th central moments $(d > 1)$ $_d\mu_{i_1}$ and $_d\mu_{i_2}$ of the i_1-th and i_2-th measurement parts. As before, each set $_d\mu_i$ consists of 2^n vectors $\left\{ _d\mu_i^0, \ldots, _d\mu_i^{2^n-1} \right\}$, and each vector $_d\mu_i^x$ includes s elements $(_d\mu_{i,1}^x, \ldots, _d\mu_{i,s}^x)$ which are the d-th central moment values for the different sample points. The d-th central moment for a sample point is calculated by

$$_d\mu_{i,s}^x = \frac{1}{q_i^x} \sum_{j=1,x_i^j=x}^{q} \left(o_{i,s}^j - m_{i,s}^x \right)^d, \qquad s \in \{1,\ldots,s\}, \qquad x \in \mathbb{F}_{2^n}.$$

Note that $_2\mu_i$ indicates the variances, and for $d > 2$ it is recommended to use the standardized central moments defined as $\dfrac{_d\mu_{i,s}^x}{\left(\sqrt{_2\mu_{i,s}^x}\right)^d}$. The remaining task is to create vectors of the sets defined again in order to compare them. Using the same rules as before, a hypothesis $\widehat{\Delta k}$ is used to construct the two vectors

$$_d\mu_{i_1,s}' = (_d\mu_{i_1,s}'^0, \ldots, _d\mu_{i_1,s}'^{2^n-1}), \qquad _d\mu_{i_1,s}'^x = _d\mu_{i_1,s}^x, \qquad x \in \mathbb{F}_{2^n}$$

$$_d^{\widehat{\Delta k}}\mu_{i_2,s}' = (_d^{\widehat{\Delta k}}\mu_{i_2,s}'^0, \ldots, _d^{\widehat{\Delta k}}\mu_{i_2,s}'^{2^n-1}), \qquad _d^{\widehat{\Delta k}}\mu_{i_2,s}'^x = _d\mu_{i_2,s}^{x \odot \widehat{\Delta k}}, \qquad x \in \mathbb{F}_{2^n}.$$

Using the same comparison technique as in the original correlation collision attack, one can compare the aforementioned vectors using the Pearson correlation coefficient at each sample point and for each $\widehat{\Delta k}$ independently.

In fact, the use of high-order central statistical moments is equivalent to perform a preprocessing step on the side-channel observations before running the original correlation collision attack. For instance, for $d = 2$ the use of $_2\mu_i$ (variance) is identical to squaring the mean-free traces and then computing the mean sets (μ_i). $d = 3$ and $d = 4$ (skewness and kurtosis if standardized) are the same as cubing and getting the fourth power of the mean-free traces. As shown later in Section 4 the use of high-order moments leads to efficient attack methods to analyze masked implementations that process the masks and the masked data simultaneously. We should highlight that the higher the moment, the harder it is to estimate. That is, a large number of observations q are required to obtain a reasonably precise estimation. Thus, the use of higher-oder moments ($d > 4$) is very limited in practice. Nevertheless, there might be architectures where the attacks can still benefit from going to these higher-order moments.

3.2   Collision Detection Using Probability Density Functions

In order to generalize the scheme we also evaluated collision detection by comparing pdfs instead of focusing on a particular moment. To do so, we define $\mathfrak{P}_i$ as a family of 2^n sets $\left\{ \mathbb{P}_i^0, \ldots, \mathbb{P}_i^{2^n-1} \right\}$. Each set $\mathbb{P}_i^x$ consists of s probability density functions $\left\{ f_{i,1}^x (O), \ldots, f_{i,s}^x (O) \right\}$ defined as follows.

$$f_{i,s}^x (O = o) = \Pr\left[H(O_{i,s}) = o | X_i = x \right], \qquad \mathsf{s} \in \{1, \ldots, s\}, \qquad \mathsf{x} \in \mathbb{F}_{2^n}$$

Here we introduced the random variables $O_{i,s}$ and X_i describing the distribution of the observed values $o_{i,s}$ and the input portions x_i respectively. Furthermore, we introduced a new random variable O, which is used to estimate the pdf of $O_{i,s}$. We denote the sample space of O as $\mathcal{O}$ and samples as o. We further introduced a function $H(O_{i,s})$ (e.g., bins of a histogram), which maps samples of $O_{i,s}$ to elements of $\mathcal{O}$, i.e., the sample space $\mathcal{O}$ used to estimate the pdf may differ from the sample space of the observed values.

We continue as before and extract the sets $\mathbb{P}'_{i_1,s}$ and $^{\Delta\hat{k}}\mathbb{P}'_{i_2,s}$ from the families $\mathfrak{P}_{i_1}$ and $\mathfrak{P}_{i_2}$, each of which includes 2^n pdfs

$$\mathbb{P}_{i_1,s} = \left\{ f'^0_{i_1,s} (O), \ldots, f'^{2^n-1}_{i_1,s} (O) \right\}, \qquad f'^x_{i_1,s} (O) = f_{i_1,s}^x (O), \qquad \mathsf{x} \in \mathbb{F}_{2^n}$$

$$^{\Delta\hat{k}}\mathbb{P}'_{i_2,s} = \left\{ {}^{\Delta\hat{k}}f'^0_{i_2,s} (O), \ldots, {}^{\Delta\hat{k}}f'^{2^n-1}_{i_2,s} (O) \right\}, \qquad {}^{\Delta\hat{k}}f'^x_{i_2,s} (O) = f_{i_2,s}^{x \odot \Delta\hat{k}} (O), \quad \mathsf{x} \in \mathbb{F}_{2^n}.$$

In contrast to the central moments discussed before, we now need to compare vectors of distributions instead of scalar vectors in order to find a similarity metric that allows distinguishing collisions. Fortunately, comparing pdfs is a well-studied task used in many different research fields, e.g., pattern recognition. The well-known methods include the *Squared Euclidean, Kullback-Leibler, Jeffreys, f-divergence*, and several others (for a comprehensive list see [9]). In the following we summarize the Kullback-Leibler (KL) divergence [14], which is the basis of several other schemes and including the metric we used in our experiments.

Kullback-Leibler Divergence is a non-negative measure of the difference between two probability distributions $p(O)$ and $q(O)$. For the discrete case it is defined as

$$D_{\mathrm{KL}}(p(O)\|q(O)) = \sum_{o \in \mathcal{O}} p(o) \log \frac{p(o)}{q(o)}.$$

In fact, KL divergence is not a true distance metric as it is not symmetric, i.e., $D_{KL}(p(O)\|q(O)) \neq D_{KL}(q(O)\|p(O))$. Therefore, other schemes have been introduced to develop a symmetric metric with similar properties. For instance,

$$D_J(p(O)\|q(O)) = D_{KL}(p(O)\|q(O)) + D_{KL}(q(O)\|p(O)) = \sum_{o \in O} (p(o) - q(o)) \log \frac{p(o)}{q(o)},$$

the symmetric form of the KL divergence is constructed using the addition method. This metric is also known as the *Jeffreys* divergence [12] and is used to perform our experiments. While we use a discrete sample space O for the remainder of this paper, there is an extension of our approach to continuous distributions, which replaces the discrete KL divergence with its continuous equivalent.

Practical Considerations: In this section we want to highlight a few aspects to help practitioners to adopt our approach:

- Methods like this, that rely on estimating pdfs (e.g., MIA) allow for a variety of estimation methods to be used, such as histograms or parameter estimation. In Section 4 we show results derived from histograms.
- As the Jeffrys divergence measures a distance, the smallest value indicates the most similar distributions.
- Any scheme similar to the Jeffreys divergence compares only two pdfs, while our method requires to compare two sets of pdfs. To compensate this, we employ the metric of a weighted mean of the Jeffreys divergence values:

$$D_{i_1,i_2,s}^{\widehat{\Delta k}} = \sum_{x=0}^{2^n-1} \left(D_J \left(f'^x_{i_1,s}(O) \, \| \, {}^{\widehat{\Delta k}} f'^x_{i_2,s}(O) \right) \cdot \Pr \left[X_{i_1} = x \Big| X_{i_2} = X_{i_1} \odot \widehat{\Delta k} \right] \right).$$

While we introduced our approaches for univariate moments and distributions, an extension to multivariate analyses is straightforward. We provide an example of a multivariate analysis in Section 4.4, where we demonstrate an attack on an *all-or-nothing* secret sharing scheme.

4   Practical Experiments

We used two different platforms to perform our practical analyses: the Xilinx Virtex-II Pro FPGA embedded in a SASEBO [1] board and a multi-purpose smartcard based on an Atmel ATMega163 microcontroller. Four implementations, all employing different masking schemes, were used to evaluate our new approach. Three of these implementations ran on the hardware platform (FPGA), the remaining one was a software solution executed on the smartcard. A LeCroy WP715Zi 1.5GHz oscilloscope equipped with a differential probe was used to collect power consumption traces in the VDD path of both platforms. In the following, we first present our results analyzing the hardware implementations. The case study of the protected software implementation is detailed in Section 4.4, which provides a glance at multivariate collision attacks.

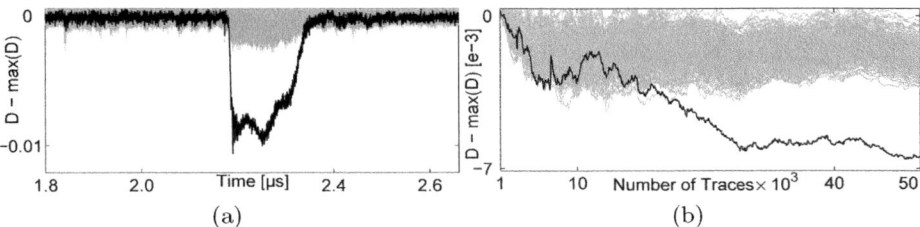

Fig. 3. Result of the collision attack using pdfs on the masked AES implementation based on [8] (a) using 200 000 traces and (b) at point 2.19μs over the number of traces

4.1 Canright-Batina's Masked AES S-Box

In [16] a serialized masked AES encryption is analyzed, where a single masked S-box instance using the design from [8] is used to subsequently process all SubBytes transformations. The interested reader can find an abstract schematic of this architecture in Fig. 7 in the Appendix (architecture is detailed in [16]). The existence of first-order leakage of masked S-boxes implemented in hardware is well-known to the side-channel community [15]. Therefore, a correlation collision attack employing first-order moments (means) can exploit this first-order leakage caused by glitches using around 20 000 measurements. At this, all random masks followed a uniform distribution, and no masks were reused (see [16]).

Since the first-order moments have already shown a dependency on processed data, an analysis of the higher-order moments is not required to perform an attack. Nevertheless, in order to evaluate the feasibility and efficiency of our attack, we implemented the most general form of the attack, the one using pdfs to detect the collisions, on a set of 200 000 measurements. Using histograms with 8 bins we estimated the families of pdfs $\mathfrak{P}_{i_1}$ and $\mathfrak{P}_{i_2}$ for two processed portions (here bytes) i_1 and i_2. The result of computing $D_{i_1,i_2,s}^{\widehat{\Delta k}} \ \forall \ \widehat{\Delta k} \in \{0, \ldots, 255\}$ for each sample point s, is shown in Fig. 3(a).[3] In addition to the increased complexity of the computations, we find that the attack using the pdfs also requires a slightly higher amount of measurements (cf. Fig. 3(b)). One reason, amongst others, of this is the low accuracy of the pdf estimation by means of histograms.

4.2 Threshold Implementation of PRESENT

Threshold implementations were proposed in [19] and later extended in [20] and [21] to overcome the first-order leakage caused by glitches when masks and masked data are processed by combinational hardware circuits. This scheme is a countermeasure at the algorithm level, and a couple of implementations of the PRESENT cipher based on that have been presented in [22]. We selected *profile 2* of [22], where only the data state is shared using 3 shares and only one instance

---

[3] For reasons of visualization we actually show the difference of the Jeffreys divergence to the largest observed value, i.e., $D_{i_1,i_2,s}^{\widehat{\Delta k}} - \max(D_{i_1,i_2,s}^{\Delta K})$.

of the shared S-box is used by the design. Fig. 8 in the Appendix sketches the architecture and shows exemplary measurements. So far only CPA attacks using the straight forward power models, i.e., HW and HD, have been presented [22]. Our analysis provides the first collision attack on this architecture.

We collected 100 million traces of this implementation using uniformly distributed plaintexts and masks. Two plaintext/key portions (here nibbles), that are consecutively processed, are selected. In addition to the general approach using pdfs, the collision attacks using the first three moments (mean, variance, and skewness) have been performed. The corresponding results are shown in Fig. 4. According to Fig. 4(a) the first-order moments do not show any dependency to the processed values, what confirms the claim given in [21]. However, higher-order moments (see Fig. 4(b) and Fig. 4(d)) are strongly dependent on the unmasked values. As expected, also the attack using pdfs (Fig. 4(f)) allows recovering the secret. Since all attacks need roughly the same number of measurements to succeed, i.e., around 5 million (see Fig. 4(c), Fig. 4(e)), and Fig. 4(g)), analyzing statistical moments is to be preferred over the slower pdf approach. Note that using e.g., second-order moments is equivalent to having a preprocessing step squaring the mean-free traces. Successful attacks using high-order moments thus do not contradict the statement given in [21] that threshold implementations prevent first-order leakage.

4.3   Threshold Implementation of AES

The same countermeasure, i.e., threshold implementation, has been applied to AES in [18]. Although this design does not fulfill all the requirements of a threshold implementation, re-masked registers were employed to provide the missing property of *uniformity* (see [21] for the requirements and their meaning). It has been shown that the final design of [18], which applies several internal PRNGs to provide the required fresh masks, is resistant to correlation collision attacks based on means, even when as much as 400 million measurements are used. However, the authors reported that a MIA attack can exploit the leakage using 80 million measurements. Therefore this is a suitable target to evaluate our new methods using higher-order moments and/or pdfs. Similar to the design targeted in Section 4.2 again only one instance of the shared S-box is used in the analyzed architecture. Moreover, the S-box design is based on a four-stage pipeline, thus leakage may appear in several clock cycles. Again, a schematic illustrating the architecture can be found in the Appendix (Fig. 9).

We have collected 100 million measurements of this design implemented on our FPGA platform. We have selected two portions (here bytes) whose corresponding key-whitened plaintext bytes are processed consecutively. Collision attacks using the pdfs and the second- and third-order moments targeting the linear difference between the two selected key bytes have been performed.[4] The results, which are shown in Fig. 5, reveal a dependency on chosen processed data in the second-order moment, but not in the third-order moment. This might be due to the

---

[4] We ignored the first-order moment due to the results reported by in [18].

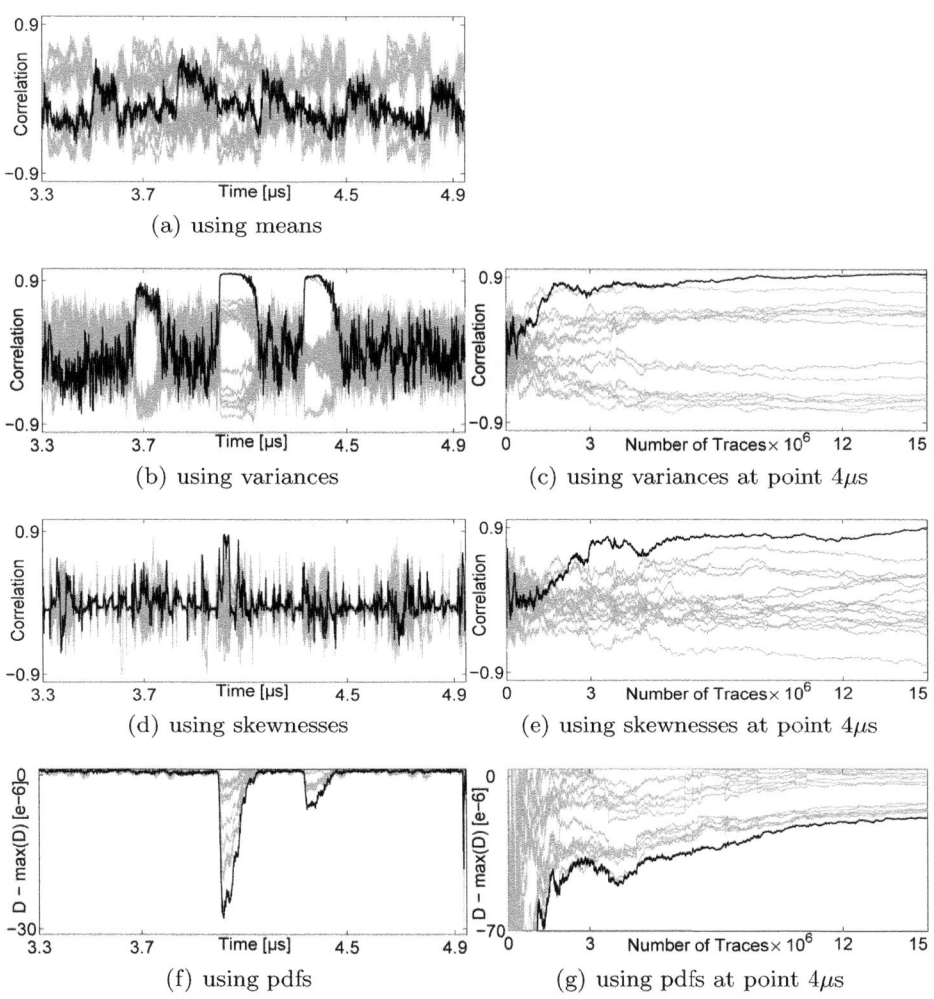

Fig. 4. Result of the collision attacks on a threshold implementation of PRESENT (left) using 100 million traces, (right) over the number of measurements

re-masked registers not present in the design investigated in Section 4.2. The number of required measurements is also interesting. Compared to that shown in [18] our attack needs around 20 million using variances and 50 million using pdfs (see Fig. 5(b) and Fig. 5(e)). This provides another example that employing statistical moments instead of pdfs is not only faster but also is more efficient with respect to the number of required measurements.

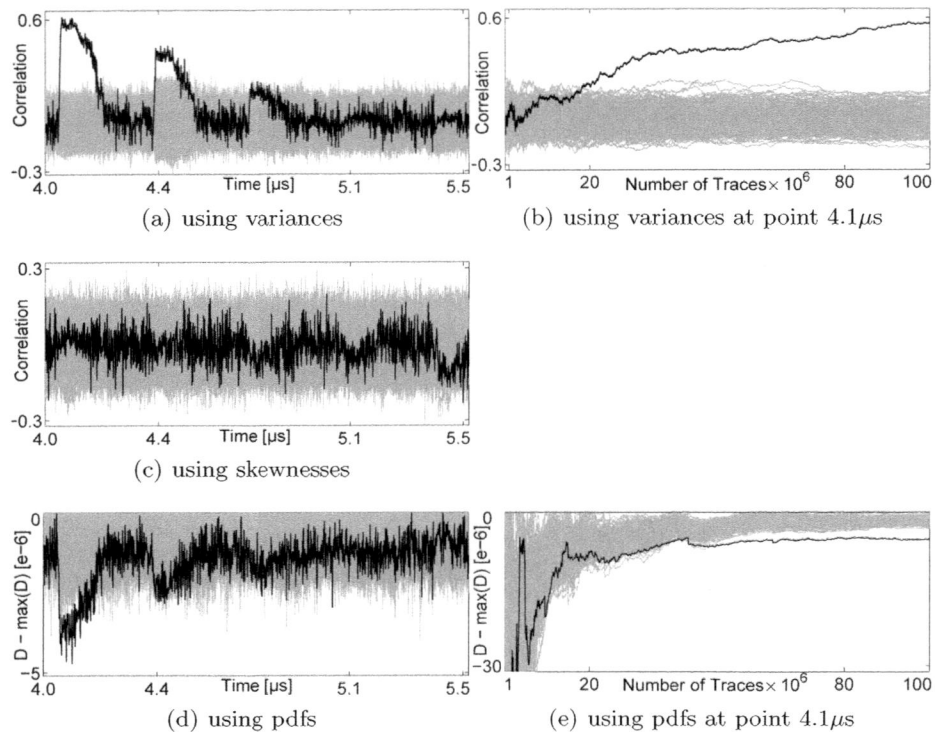

(a) using variances

(b) using variances at point $4.1\mu s$

(c) using skewnesses

(d) using pdfs

(e) using pdfs at point $4.1\mu s$

Fig. 5. Results the collision attacks on a threshold implementation of AES (left) using 100 million traces and (right) over the number of traces

4.4   Boolean Masking in Software

The last case study is a software implementation of the AES based on boolean masking. Two random mask bytes (input mask and output mask) are considered for each plaintext byte (in sum 256 mask bits) at the start of each encryption run. After masking the plaintext bytes using the input masks, the AddRoundkey operation is performed. Afterwards, for each state byte a masked S-box table is constructed in memory, which satisfies the state byte's input and output masks. See Fig. 10 in the appendix for a schematic of the design.

Since every intermediate result is masked by a random value, no univariate attacks can recover a secret. In order to perform a bivariate collision attack using pdfs, we (at the moment) suppose that the two interesting sample points (s_1, s_2) in the measurement parts, that denote the time of processing the masked value and the corresponding mask, are known. Then, a set $\mathbb{P}_i$ consists of joint probability density functions $f^x_{i,s_1,s_2}(O_1, O_2)$. The attack then works analogue to the univariate one, except for the comparison step. Here Jeffreys divergence is extended to measure the distance between two joint pdfs as

$$\mathrm{D_J}(p(O_1,O_2)\|q(O_1,O_2)) = \sum_{o_1 \in \mathcal{O}_1} \sum_{o_2 \in \mathcal{O}_2} (p(o_1,o_2) - q(o_1,o_2)) \log \frac{p(o_1,o_2)}{q(o_1,o_2)}.$$

To use the joint statistical moments, the analysis employs the $(d_1 > 0, d_2 > 0)$

$$_{d_1,d_2}\mu^{\mathrm{x}}_{\mathrm{i,s_1,s_2}} = \frac{1}{q^{\mathrm{x}}_{\mathrm{i}}} \sum_{j=1, x_i^j = \mathrm{x}}^{q} \left(o^j_{\mathrm{i,s_1}} - m^{\mathrm{x}}_{\mathrm{i,s_1}} \right)^{d_1} \left(o^j_{\mathrm{i,s_2}} - m^{\mathrm{x}}_{\mathrm{i,s_2}} \right)^{d_2}.$$

In fact, the attack analyzing $_{1,1}\mu^{\mathrm{x}}_{\mathrm{i,s_1,s_2}}$ is equivalent to combining the corresponding sample points by means of a "multiplication" prior to the averaging step in a univariate collision attack. The dependencies on higher-moments are familiar from traditional higher-order attacks, which exploit them when applying combining functions.

Since finding the interesting sample points $(\mathrm{s_1, s_2})$ in multivariate attacks is always a challenging task, we tried to make use of the moments to mitigate this problem. We collected 250 000 traces from our target implementation using uniformly selected plaintext and mask bytes. Since the construction of the masked S-box tables is time consuming, the measured traces are much longer than the ones of the previously shown case studies. Each trace covers 10 000 clock cycles and was compressed to a vector of 10 000 peaks corresponding to the peaks of the clock cycles. Since the masked value and the mask are processed with a time distance of – most likely – a small number of clock cycles, we defined a window of around 30 clock cycles to sum up adjacent peaks (sliding average). First, we assumed that each measurement part o^j_i covers all summed peak points. Computing the second-order central moments $_2\mu_i$ for two portions i_1 and i_2 and getting the variance of each set at each summed peak point separately led to the two variance curves shown in Fig. 6(a). The graphics clearly exposes the (time) distance between the same process performed on each portion. With this knowledge the measurement parts can be accordingly selected and thus it allows executing a collision attack. The result from a collision attack on second-order moments depicted in Fig. 6(b) confirms our theoretical reasoning and provides evidence of the strength of the attack.

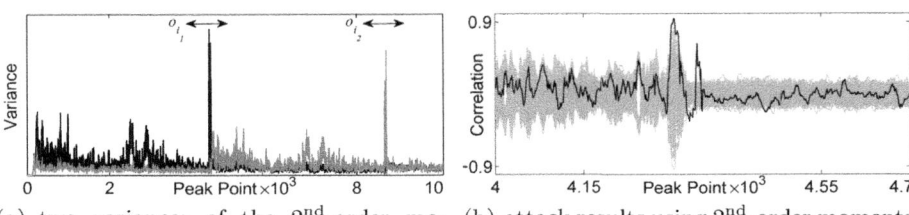

(a) two variances of the 2nd-order moments

(b) attack results using 2nd-order moments

Fig. 6. Result of the attacks on a software implementation of the AES (boolean masking) after two preprocessing steps: 1) peak extraction, 2) sum over a 30 peak point window

5   Conclusions

The attack presented in this work is fundamentally similar to the correlation collision attack presented in [16]. We extended the scheme to employ higher-order moments and introduced a general form of the attack, which makes use of the distribution of side-channel leakages. As supported by the experimental results, the presented methods allow improving univariate collision attacks. We showed that by slightly increasing the computation complexity (e.g., variance vs. mean) the collision attacks can defeat the security provided by one of the most prominent proposed masking schemes for hardware, i.e., threshold implementations. Additionally, we discussed the possible options to perform multivariate collision attacks using either high-order moments or joint probability distributions. We concluded our case studies analyzing a masked software implementation, and presented a scheme to localize the interesting points for a collision attack employing high-order moments.

The majority of the – usually unprotected – devices have a straightforward and known leakage behavior. Thus, in most cases traditional approaches, e.g., CPA using HW model, can be applied. However, in case that masking countermeasures are applied and the leakage points must be combined the leakage model may not be appropriately guessed and the issue addressed in [26] may become critical. In summary, the collision attacks are an essential tool for security evaluations in situations where the leakage model of the target device is not known and cannot be obtained by profiling.

Acknowledgment. The author would like to thank the anonymous reviewers of CHES 2011 for their helpful comments, Kerstin Lemke-Rust for fruitful discussions, Akashi Satoh and RCIS of Japan for the prompt and kind help in obtaining SASEBOs, and especially Markus Kasper for his great help improving the quality of the paper.

References

1. Side-channel Attack Standard Evaluation Board (SASEBO). Further information are available via, http://www.rcis.aist.go.jp/special/SASEBO/index-en.html
2. Bogdanov, A.: Improved Side-Channel Collision Attacks on AES. In: Adams, C., Miri, A., Wiener, M. (eds.) SAC 2007. LNCS, vol. 4876, pp. 84–95. Springer, Heidelberg (2007)
3. Bogdanov, A.: Multiple-Differential Side-Channel Collision Attacks on AES. In: Oswald, E., Rohatgi, P. (eds.) CHES 2008. LNCS, vol. 5154, pp. 30–44. Springer, Heidelberg (2008)
4. Bogdanov, A., Kizhvatov, I.: Beyond the Limits of DPA: Combined Side-Channel Collision Attacks. IEEE Transactions on Computers (2011), (to appear), A draft version at, http://eprint.iacr.org/2010/590
5. Bogdanov, A., Kizhvatov, I., Pyshkin, A.: Algebraic Methods in Side-Channel Collision Attacks and Practical Collision Detection. In: Chowdhury, D.R., Rijmen, V., Das, A. (eds.) INDOCRYPT 2008. LNCS, vol. 5365, pp. 251–265. Springer, Heidelberg (2008)

6. Bogdanov, A., Knudsen, L.R., Leander, G., Paar, C., Poschmann, A., Robshaw, M.J.B., Seurin, Y., Vikkelsoe, C.: PRESENT: An Ultra-Lightweight Block Cipher. In: Paillier, P., Verbauwhede, I. (eds.) CHES 2007. LNCS, vol. 4727, pp. 450–466. Springer, Heidelberg (2007)

7. Brier, E., Clavier, C., Olivier, F.: Correlation Power Analysis with a Leakage Model. In: Joye, M., Quisquater, J.-J. (eds.) CHES 2004. LNCS, vol. 3156, pp. 16–29. Springer, Heidelberg (2004)

8. Canright, D., Batina, L.: A Very Compact "Perfectly Masked" S-Box for AES. In: Bellovin, S.M., Gennaro, R., Keromytis, A.D., Yung, M. (eds.) ACNS 2008. LNCS, vol. 5037, pp. 446–459. Springer, Heidelberg (2008); the corrected version at http://eprint.iacr.org/2009/011

9. Cha, S.-H.: Comprehensive Survey on Distance/Similarity Measures between Probability Density Functions. Journal of Mathematical Models and Methods in Applied Sciences 1, 300–307 (2007)

10. Clavier, C., Feix, B., Gagnerot, G., Roussellet, M., Verneuil, V.: Improved Collision-Correlation Power Analysis on First Order Protected AES. In: Preneel, B., Takagi, T. (eds.) CHES 2011. LNCS, vol. 6917, pp. 49–62. Springer, Heidelberg (2011)

11. Gierlichs, B., Batina, L., Tuyls, P., Preneel, B.: Mutual Information Analysis. In: Oswald, E., Rohatgi, P. (eds.) CHES 2008. LNCS, vol. 5154, pp. 426–442. Springer, Heidelberg (2008)

12. Jeffreys, H.: An Invariant Form for the Prior Probability in Estimation Problems. Royal Society of London Proceedings Series A 186, 453–461 (1946)

13. Kocher, P.C., Jaffe, J., Jun, B.: Differential Power Analysis. In: Wiener, M. (ed.) CRYPTO 1999. LNCS, vol. 1666, pp. 388–397. Springer, Heidelberg (1999)

14. Kullback, S., Leibler, R.A.: On Information and Sufficiency. The Annals of Mathematical Statistics 22(1), 79–86 (1951)

15. Mangard, S., Pramstaller, N., Oswald, E.: Successfully Attacking Masked AES Hardware Implementations. In: Rao, J.R., Sunar, B. (eds.) CHES 2005. LNCS, vol. 3659, pp. 157–171. Springer, Heidelberg (2005)

16. Moradi, A., Mischke, O., Eisenbarth, T.: Correlation-Enhanced Power Analysis Collision Attack. In: Mangard, S., Standaert, F.-X. (eds.) CHES 2010. LNCS, vol. 6225, pp. 125–139. Springer, Heidelberg (2010), The extended version at http://eprint.iacr.org/2010/297

17. Moradi, A., Mischke, O., Paar, C., Li, Y., Ohta, K., Sakiyama, K.: On the Power of Fault Sensitivity Analysis and Collision Side-Channel Attacks in a Combined Setting. In: Preneel, B., Takagi, T. (eds.) CHES 2011. LNCS, vol. 6917, pp. 292–311. Springer, Heidelberg (2011)

18. Moradi, A., Poschmann, A., Ling, S., Paar, C., Wang, H.: Pushing the Limits: A Very Compact and a Threshold Implementation of AES. In: Paterson, K.G. (ed.) EUROCRYPT 2011. LNCS, vol. 6632, pp. 69–88. Springer, Heidelberg (2011)

19. Nikova, S., Rechberger, C., Rijmen, V.: Threshold Implementations Against Side-Channel Attacks and Glitches. In: Ning, P., Qing, S., Li, N. (eds.) ICICS 2006. LNCS, vol. 4307, pp. 529–545. Springer, Heidelberg (2006)

20. Nikova, S., Rijmen, V., Schläffer, M.: Secure Hardware Implementation of Nonlinear Functions in the Presence of Glitches. In: Lee, P.J., Cheon, J.H. (eds.) ICISC 2008. LNCS, vol. 5461, pp. 218–234. Springer, Heidelberg (2009)

21. Nikova, S., Rijmen, V., Schläffer, M.: Secure Hardware Implementation of Nonlinear Functions in the Presence of Glitches. J. Cryptology 24, 292–321 (2011)

22. Poschmann, A., Moradi, A., Khoo, K., Lim, C.-W., Wang, H., Ling, S.: Side-Channel Resistant Crypto for less than 2,300 GE. J. Cryptology 24, 322–345 (2011)

23. Renauld, M., Standaert, F.-X., Veyrat-Charvillon, N., Kamel, D., Flandre, D.: A Formal Study of Power Variability Issues and Side-Channel Attacks for Nanoscale Devices. In: Paterson, K.G. (ed.) EUROCRYPT 2011. LNCS, vol. 6632, pp. 109–128. Springer, Heidelberg (2011)
24. Schramm, K., Leander, G., Felke, P., Paar, C.: A Collision-Attack on AES. In: Joye, M., Quisquater, J.-J. (eds.) CHES 2004. LNCS, vol. 3156, pp. 163–175. Springer, Heidelberg (2004)
25. Schramm, K., Wollinger, T., Paar, C.: A New Class of Collision Attacks and Its Application to DES. In: Johansson, T. (ed.) FSE 2003. LNCS, vol. 2887, pp. 206–222. Springer, Heidelberg (2003)
26. Veyrat-Charvillon, N., Standaert, F.-X.: Generic Side-Channel Distinguishers: Improvements and Limitations. In: Rogaway, P. (ed.) CRYPTO 2011. LNCS, vol. 6841, pp. 354–372. Springer, Heidelberg (2011)

Appendix

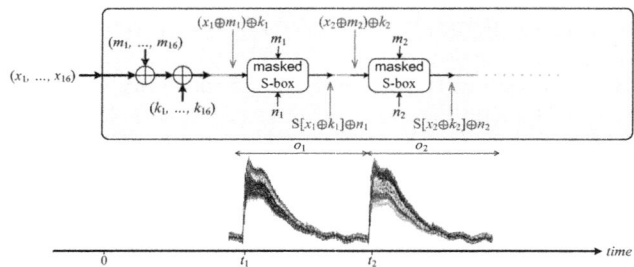

Fig. 7. Schematic of the first case study (a masked AES encryption module using [8])

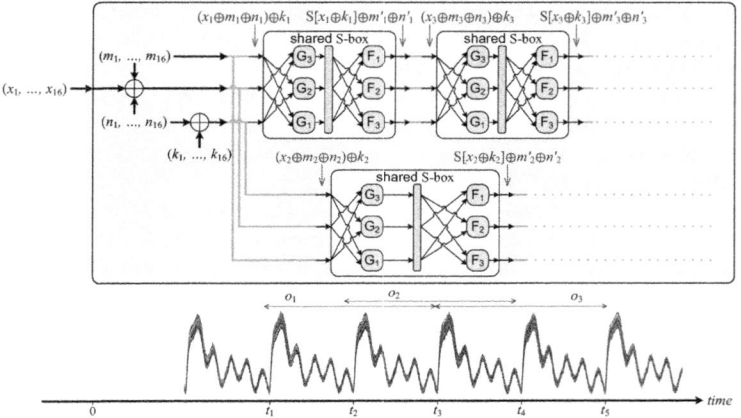

Fig. 8. Schematic of the second case study (a threshold implementation of PRESENT taken from [22])

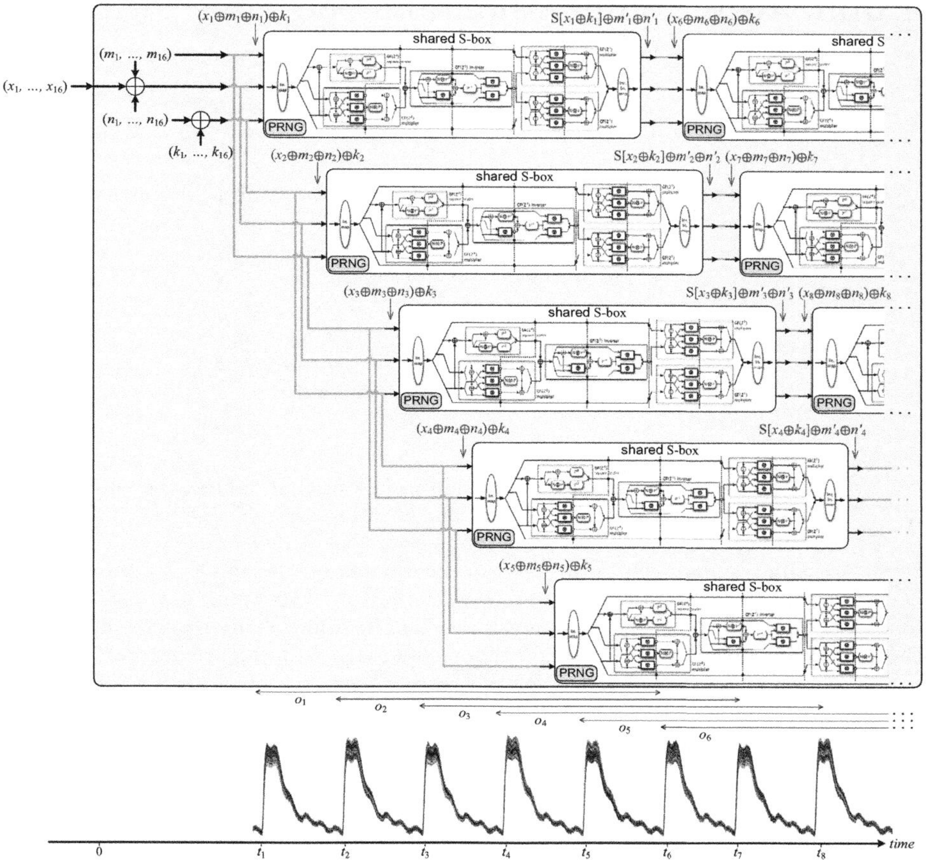

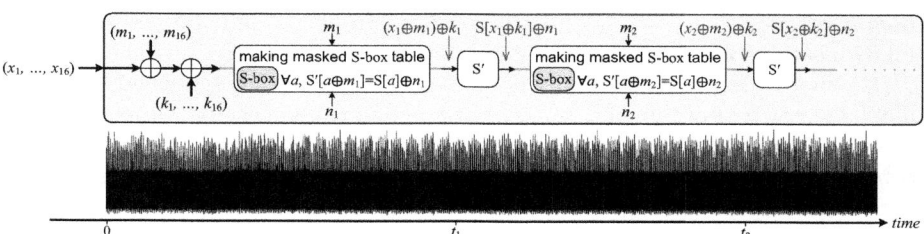

Fig. 9. Schematic of the third case study (a threshold implementation of AES taken from [18])

Fig. 10. Schematic of the forth case study (a (boolean) masked software implementation of AES)

Public Key Compression and Modulus Switching for Fully Homomorphic Encryption over the Integers

Jean-Sébastien Coron[1], David Naccache[2], and Mehdi Tibouchi[3]

[1] Université du Luxembourg
jean-sebastien.coron@uni.lu
[2] École normale supérieure
david.naccache@ens.fr
[3] NTT Information Sharing Platform Laboratories
tibouchi.mehdi@lab.ntt.co.jp

Abstract. We describe a compression technique that reduces the public key size of van Dijk, Gentry, Halevi and Vaikuntanathan's (DGHV) fully homomorphic scheme over the integers from $\tilde{\mathcal{O}}(\lambda^7)$ to $\tilde{\mathcal{O}}(\lambda^5)$. Our variant remains semantically secure, but in the random oracle model. We obtain an implementation of the full scheme with a 10.1 MB public key instead of 802 MB using similar parameters as in [7]. Additionally we show how to extend the quadratic encryption technique of [7] to higher degrees, to obtain a shorter public-key for the basic scheme.

This paper also describes a new modulus switching technique for the DGHV scheme that enables to use the new FHE framework without bootstrapping from Brakerski, Gentry and Vaikuntanathan with the DGHV scheme. Finally we describe an improved attack against the Approximate GCD Problem on which the DGHV scheme is based, with complexity $\tilde{\mathcal{O}}(2^\rho)$ instead of $\tilde{\mathcal{O}}(2^{3\rho/2})$.

1 Introduction

Fully Homomorphic Encryption. An encryption scheme is said to be fully homomorphic when it is possible to perform implicit plaintext additions and multiplications while manipulating only ciphertexts.

The first construction of a fully homomorphic scheme was described by Gentry in [9]. Gentry first obtained a "somewhat homomorphic" scheme, supporting only a limited number of ciphertext multiplications due to the fact that ciphertext contain a certain amount of "noise" which increases with every multiplication, and that decryption fails when noise size passes a certain bound. As a result, in the somewhat homomorphic scheme, the functions that can be homomorphically evaluated on ciphertexts are polynomials of small, bounded degree. The second step in Gentry's framework consists in "squashing" the decryption procedure so that it can be expressed as a low degree polynomial in the bits of the ciphertext and the secret key. Then, Gentry's key idea, called "bootstrapping", is to evaluate this decryption polynomial not on the ciphertext bits and the secret-key

D. Pointcheval and T. Johansson (Eds.): EUROCRYPT 2012, LNCS 7237, pp. 446–464, 2012.

bits (which would yield the plaintext), but homomorphically on the encryption of those bits, which gives another ciphertext of the same plaintext. If the degree of the decryption polynomial is small enough, the noise in the new ciphertext can become smaller than it was the original ciphertext, so that this new ciphertext can be used again in a subsequent homomorphic operation (either addition or multiplication). Using this "ciphertext refresh" procedure the number of permissible homomorphic operations becomes unlimited and one obtains a fully homomorphic encryption scheme. To date, three different fully homomorphic schemes are known:

1. Gentry's original scheme [9], based on ideal lattices. Gentry and Halevi described in [10] the first implementation of Gentry's scheme, using many clever optimizations, including some suggested in a previous work by Smart and Vercauteren [14]. For their most secure setting (claiming 72 bit security) the authors report a public key size of 2.3 GB and a ciphertext refresh procedure taking 30 minutes on a high-end workstation.

2. van Dijk, Gentry, Halevi and Vaikuntanathan's (DGHV) scheme over the integers [8]. This scheme is conceptually simpler than Gentry's scheme, because it operates on integers instead of ideal lattices. Recently it was shown [7] how to reduce the public key size by storing only a small subset of the original public key and generating the full public key on the fly by combining the elements in the small subset multiplicatively. Using some of the optimizations from [10], the authors of [7] report similar performances: a 802 MB public key and a ciphertext refresh in 14 minutes.

3. Brakerski and Vaikuntanathan's scheme based on the Learning with Errors (LWE) and Ring Learning with Errors (RLWE) problems [2,3]. The authors introduce a new dimension reduction technique and a new modulus switching technique to shorten the ciphertext and reduce the decryption complexity. A partial implementation is described in [11], without the fully homomorphic capability.

Recently Brakerski, Gentry and Vaikuntanathan introduced a remarkable new FHE framework, in which the noise ceiling increases only linearly with the multiplicative level instead of exponentially [4]; this implies that bootstrapping is no longer necessary to achieve fully homomorphic encryption. This new framework has the potential to significantly improve the practical FHE performance. The new framework is based on Brakerski and Vaikuntanathan's scheme [2,3], and more specifically on their new modulus switching technique, which efficiently transforms a ciphertext encrypted under a certain modulus p into a ciphertext under a different modulus p' but with reduced noise.

Public Key Compression. The first of our contributions is a technique to reduce the public key size of DGHV-like schemes [8] by several orders of magnitude. In the DGHV scheme the public key is a set of integers of the form:

$$x_i = q_i \cdot p + r_i$$

where p is the secret-key of η bits, q_i is a large random integer of $\gamma - \eta$ bits, and r_i is a small random integer of ρ bits. The scheme's semantic security is based on the Approximate GCD Problem: given a polynomial number of x_i's, recover the secret p. To avoid lattice attacks, the bit-size γ of the x_i's must be very large: [7] takes $\gamma \simeq 2 \cdot 10^7$ for $\eta = 2652$ and $\rho = 39$, and the full public key claims a 802 MB storage.

Our technique proceeds as follows. First generate the secret-key p. Then, use a pseudo-random number generator f with public random seed se to generate a set of γ-bit integers χ_i (i.e. the χ_i's are of the same bit-size as the x_i's). Finally, compute small corrections δ_i to the χ_i's such that $x_i = \chi_i - \delta_i$ is small modulo p, and store only the small corrections δ_i in the public key, instead of the full x_i's. Knowing the PRNG seed se and the δ_i's is sufficient to recover the x_i's.

Therefore instead of storing a set of large γ-bit integers we only have to store a set of much smaller η-bit integers, where η is the bit size of p. The new technique is fully compatible with the DGHV variant described in [7]; with the previous set of parameters from [7] one obtains a public key size of 4.6 MB for the full implementation, instead of the 802 MB required in [7]! The technique can be seen as generating the $\gamma - \eta$ most significant bits of the x_i's with a pseudo-random number generator, and then using the secret key p to fix the η remaining bits so that $x_i \bmod p$ is small. While different, this is somewhat reminiscent of Lenstra's technique [12] for generating an RSA modulus with a predetermined portion.

Under our variant, the encryption scheme can still be proved semantically secure under the Approximate GCD assumption, albeit in the random oracle model. This holds for both the original DGHV scheme form [8] and the variant described in [7] in which the public key elements are first combined multiplicatively to generate the full public key. Unlike [7,8], we need the random oracle model in order to apply the leftover hash lemma in our variant, because the seed of the PRNG is known to the attacker (as part of the public key).

We report the result of an implementation of the new variant with the fully homomorphic capability. As in [7] we use the variant with noise-free $x_0 = q_0 \cdot p$. We also update the parameters from [7] to take into the account the improved attack from Chen and Nguyen against the Approximate GCD problem [5]. We obtain a level of efficiency very similar to [7] but with a 10.1 MB public key instead of a 802 MB one. The source code of this implementation is publicly available [17].

Extension to Higher Degrees. Various techniques have been proposed in [7] to reduce the public key size and increase the efficiency of the DGHV scheme, the most important of which is to use a *quadratic form* instead of a linear form for masking the message when computing a ciphertext. The authors show that the scheme remains semantically secure; the key ingredient is to prove that a certain family of quadratic hash functions is close enough to being pairwise independent, so that the leftover hash lemma can still be applied. The main benefit is a significant reduction in public key size, from $\tau = \tilde{\mathcal{O}}(\lambda^3)$ elements x_i down to $2\beta = \tilde{\mathcal{O}}(\lambda^{1.5})$ elements $x_{i,b}$. In this paper we prove that the natural extension of this quadratic encryption technique to to cubic forms, and more generally forms of arbitrary fixed degree d, remains secure, making it possible to further reduce the public key size.

Modulus Switching and Leveled DGHV Scheme. As a third contribution, we show how to adapt Brakerski, Gentry and Vaikuntanathan's (BGV) new FHE framework [4] to the DGHV scheme over the integers. Under the BGV framework the noise ceiling increases only linearly with multiplicative depth, instead of exponentially. This enables to get a FHE scheme without the costly bootstrapping procedure.

More precisely the new BGV framework is described in [4] with Brakerski and Vaikuntanathan's scheme [2], and the key technical tool is the modulus-switching technique of [2] that transforms a ciphertext c modulo p into a ciphertext c' modulo p' simply by scaling by p'/p and rounding appropriately. This allows to reduce the ciphertext noise by a factor close to p'/p without knowing the secret-key and without bootstrapping. However the modulus switching technique cannot directly apply to DGHV since in DGHV the moduli p and p' are secret. In this paper we explain how this modulus-switching technique can be adapted to DGHV, so as to apply the new BGV framework. We show that the resulting FHE scheme remains semantically secure, albeit under a stronger assumption. We also describe an implementation, showing that the new BGV framework can be applied in practice.

Improved Attack against the Approximate-GCD problem. Finally we consider the security of the Approximate GCD Problem *without* noise-free $x_0 = q_0 \cdot p$. In our leveled DGHV variant under the BGV framework the size of the secret p can become much smaller than in the original Gentry framework ($\eta \simeq 180$ bits for the lowest p in the ladder, instead of $\eta = 2652$ bits in [7]). This implies that the noise-free variant $x_0 = q_0 \cdot p$ cannot be used, since otherwise the prime factor p could easily be extracted using the Elliptic Curve Method for integer factorization [13]. Therefore one must consider the security of the Approximate GCD Problem *without* noise-free x_0. The recent attack by Chen and Nguyen [5] against the Approximate GCD Problem *with* noise-free x_0 has complexity $\tilde{\mathcal{O}}(2^{\rho/2})$, instead of the $\tilde{\mathcal{O}}(2^\rho)$ naive attack; as noted by the authors, this immediately yields an $\tilde{\mathcal{O}}(2^{3\rho/2})$ attack against the Approximate GCD Problem without noise-free x_0, instead of $\tilde{\mathcal{O}}(2^{2\rho})$ for the naive attack. In this paper we exhibit an improved attack with complexity $\tilde{\mathcal{O}}(2^\rho)$. We also describe an implementation showing that this new attack is indeed an improvement in practice.

2   The DGHV Scheme over the Integers

We first recall the somewhat homomorphic encryption scheme described by van Dijk, Gentry, Halevi and Vaikuntanathan (DGHV) in [8]. For a real number x, we denote by $\lceil x \rceil$, $\lfloor x \rfloor$ and $\lceil x \rfloor$ the rounding of x up, down, or to the nearest integer. For integers z, p we denote the reduction of z modulo p by $[z]_p$ with $-p/2 < [z]_p \leq p/2$, and by $\langle z \rangle_p$ with $0 \leq \langle z \rangle_p < p$. Given the security parameter λ, the following parameters are used:

- γ is the bit-length of the x_i's,
- η is the bit-length of the secret key p,
- ρ is the bit-length of the noise r_i,
- τ is the number of x_i's in the public key,
- ρ' is a secondary noise parameter used for encryption.

For a specific η-bit odd integer p, we use the following distribution over γ-bit integers:

$$\mathcal{D}_{\gamma,\rho}(p) = \Big\{ \ Choose \ q \leftarrow \mathbb{Z} \cap [0, 2^\gamma/p), \ r \leftarrow \mathbb{Z} \cap (-2^\rho, 2^\rho) : \ Output \ x = q \cdot p + r \Big\}$$

DGHV. KeyGen(1^λ). Generate a random prime integer p of size η bits. For $0 \leq i \leq \tau$ sample $x_i \leftarrow \mathcal{D}_{\gamma,\rho}(p)$. Relabel the x_i's so that x_0 is the largest. Restart unless x_0 is odd and $[x_0]_p$ is even. Let $pk = (x_0, x_1, \dots x_\tau)$ and $sk = p$.

DGHV. Encrypt($pk, m \in \{0,1\}$). Choose a random subset $S \subseteq \{1, 2, \dots, \tau\}$ and a random integer r in $(-2^{\rho'}, 2^{\rho'})$, and output the ciphertext:

$$c = \left[m + 2r + 2 \sum_{i \in S} x_i \right]_{x_0} \tag{1}$$

DGHV.Evaluate($pk, C, c_1, \dots, c_t$): given the circuit C with t input bits, and t ciphertexts c_i, apply the addition and multiplication gates of C to the ciphertexts, performing all the additions and multiplications over the integers, and return the resulting integer.

DGHV. Decrypt(sk, c). Output $m \leftarrow [c]_p \bmod 2$.

This completes the description of the scheme. As shown in [8] this scheme is somewhat homomorphic, i.e. a limited number of homomorphic operations can be performed on ciphertexts. More precisely given two ciphertexts $c = q \cdot p + 2r + m$ and $c' = q' \cdot p + 2r' + m'$ where r and r' are ρ'-bit integers, the ciphertext $c + c'$ is an encryption of $m + m' \bmod 2$ with $(\rho' + 1)$-bit noise and the ciphertext $c \cdot c'$ is an encryption of $m \cdot m'$ with noise $\simeq 2\rho'$. Since the ciphertext noise must remain smaller than p for correct decryption, the scheme allows roughly η/ρ' multiplications on ciphertexts. As shown in [8] the scheme is semantically secure under the Approximate GCD assumption.

Definition 1 (Approximate GCD). *The (ρ, η, γ)-Approximate GCD Problem is: For a random η-bit odd integer p, given polynomially many samples from $\mathcal{D}_{\gamma,\rho}(p)$, output p.*

3   The New DGHV Public Key Compression Technique

We describe our technique using the variant with noise free $x_0 = q_0 \cdot p$, as suggested in [8] and implemented in [7]. We only describe the basic scheme; we refer to the full version of this paper [6] for a complete description of the fully homomorphic scheme.

3.1   Description

KeyGen(1^λ). Generate a random prime integer p of size η bits. Pick a random odd integer $q_0 \in [0, 2^\gamma/p)$ and let $x_0 = q_0 \cdot p$. Initialize a pseudo-random number generator f with a random seed se. Use $f(\text{se})$ to generate a set of integers $\chi_i \in [0, 2^\gamma)$ for $1 \leq i \leq \tau$. For all $1 \leq i \leq \tau$ compute:

$$\delta_i = \langle \chi_i \rangle_p + \xi_i \cdot p - r_i$$

where $r_i \leftarrow \mathbb{Z} \cap (-2^\rho, 2^\rho)$ and $\xi_i \leftarrow \mathbb{Z} \cap [0, 2^{\lambda+\eta}/p)$. For all $1 \leq i \leq \tau$ compute:

$$x_i = \chi_i - \delta_i \tag{2}$$

Let $pk = (\text{se}, x_0, \delta_1, \dots, \delta_\tau)$ and $sk = p$.

Encrypt($pk, m \in \{0, 1\}$): use $f(\text{se})$ to recover the integers χ_i and let $x_i = \chi_i - \delta_i$ for all $1 \leq i \leq \tau$. Choose a random integer vector $\boldsymbol{b} = (b_i)_{1 \leq i \leq \tau} \in [0, 2^\alpha)^\tau$ and a random integer r in $(-2^{\rho'}, 2^{\rho'})$. Output the ciphertext:

$$c = m + 2r + 2 \sum_{i=1}^{\tau} b_i \cdot x_i \mod x_0$$

Evaluate($pk, C, c_1, \dots, c_t$) and Decrypt(sk, c): same as in the original DGHV scheme, except that ciphertexts are reduced modulo x_0.

This completes the description of our variant. We have the following constraints on the scheme parameters:

- $\rho = \omega(\log \lambda)$ to avoid brute force attack on the noise,
- $\eta \geq \rho \cdot \Theta(\lambda \log^2 \lambda)$ in order to support homomorphic operations for evaluating the "squashed decryption circuit" (see [8]),
- $\gamma = \omega(\eta^2 \cdot \log \lambda)$ in order to thwart lattice-based attacks against the Approximate GCD problem (see [7,8]),
- $\alpha \cdot \tau \geq \gamma + \omega(\log \lambda)$ in order to apply the left-over hash lemma (see [7,8]).
- $\eta \geq \rho + \alpha + 2 + \log_2 \tau$ for correct decryption of a ciphertext,
- $\rho' = \alpha + \rho + \omega(\log \lambda)$ for the secondary noise parameter.

To satisfy the above constraints one can take $\rho = \lambda$, $\eta = \tilde{\mathcal{O}}(\lambda^2)$, $\gamma = \tilde{\mathcal{O}}(\lambda^5)$, $\alpha = \tilde{\mathcal{O}}(\lambda^2)$, $\tau = \tilde{\mathcal{O}}(\lambda^3)$ and $\rho' = \tilde{\mathcal{O}}(\lambda^2)$. The main difference with the original DGHV scheme is that instead of storing the large x_i's in the public key we only store the much smaller δ_i's. The new public key for the somewhat homomorphic scheme has size $\gamma + \tau \cdot (\eta + \lambda) = \tilde{\mathcal{O}}(\lambda^5)$ instead of $(\tau + 1) \cdot \gamma = \tilde{\mathcal{O}}(\lambda^8)$.

Remark 1. We can also compress x_0 by letting $x_0 = \chi_0 - \delta_0$ and storing only $\delta_0 = \langle \chi_0 \rangle_p + \xi_0 \cdot p$ in the public-key.

Remark 2. In the description above we add a random multiple of p to $\langle \chi_i \rangle_p$ in the δ_i's. This is done to obtain a proof of semantic security in the random oracle model (see below). However the scheme seems heuristically secure without adding the random multiple.

Remark 3. For encryption the integers x_i need not be stored in memory as they can be generated on the fly when computing the subset sum.

3.2 Semantic Security

Theorem 1. *The previous encryption scheme is semantically secure under the Approximate GCD assumption with noise-free $x_0 = q_0 \cdot p$, in the random oracle model.*

The proof is almost the same as in [8]. Given a random oracle $H : \{0,1\}^* \to \mathbb{Z} \cap [0, 2^\gamma)$, we assume that the pseudo-random number generation of the χ_i's is defined as $\chi_i = H(\text{se} \| i)$ for all $1 \leq i \leq \tau$ and we show that the integers x_i's generated in (2) have a distribution statistically close to their distribution in the original DGHV scheme. We refer to the full version of the paper [6] for the proof.

4 Extension of DGHV Encryption to Higher Degrees

Various techniques have recently been proposed in [7] to reduce the public key size and increase the efficiency of the DGHV scheme, the most important of which is to use a *quadratic form* instead of a linear form for masking the message when computing a ciphertext. More precisely, one computes:

$$c^* = m + 2r + 2 \sum_{1 \leq i,j \leq \beta} b_{ij} \cdot x_{i,0} \cdot x_{j,1} \mod x_0$$

which is quadratic in the public key elements $x_{i,b}$ instead of linear as in equation (1); here the variant with noise-free $x_0 = q_0 \cdot p$ is used. The main benefit is a significant decrease in the public key size, from $\tau = \tilde{\mathcal{O}}(\lambda^3)$ elements x_i down to $2\beta = \tilde{\mathcal{O}}(\lambda^{1.5})$ elements $x_{i,b}$. Namely the constraint to apply the left-over hash lemma becomes $\alpha \cdot \beta^2 \geq \gamma + \omega(\log \lambda)$, so by taking $\alpha = \tilde{\mathcal{O}}(\lambda^2)$ one can take $\beta = \tilde{\mathcal{O}}(\lambda^{1.5})$. Combined with our compression technique the public-key size of the somewhat homomorphic scheme becomes $(2\beta + 1) \cdot (\eta + \lambda) = \tilde{\mathcal{O}}(\lambda^{3.5})$.

To prove that the scheme remains secure under this modified encryption procedure, the key point in [7] was to prove that the following family of functions $h \colon \{0, \ldots, 2^{\alpha-1}\}^{\beta^2} \to \mathbb{Z}_{q_0}$:

$$h(\boldsymbol{b}) = \sum_{1 \leq i_1,i_2 \leq \beta} b_{i_1 i_2} q_{i_1}^{(1)} q_{i_2}^{(2)} \mod q_0 \qquad (q_i^{(j)} \in \mathbb{Z}_{q_0})$$

is close enough to being a pairwise independent (*i.e.* universal) hash function family (under suitable parameter choices), which in turn makes it possible to apply a variant of the leftover hash lemma.

In this section we show that it is possible to obtain further efficiency gains by using *cubic forms* instead, or more generally forms of higher degree d, if we can prove an analogue of the previous result for the family $\mathcal{H}_d$ of hash functions $h : \{0, \ldots, 2^{\alpha-1}\}^{\beta^d} \to \mathbb{Z}_q$ of the form:

$$h(\boldsymbol{b}) = \sum_{1 \leq i_1, \ldots, i_d \leq \beta} b_{i_1, \ldots, i_d} q_{i_1}^{(1)} \cdots q_{i_d}^{(d)} \mod q \qquad \left(q_i^{(j)} \in \mathbb{Z}_q \right)$$

Such a result also leads to the construction of extractors with relatively short seeds, which is an interesting fact in its own right.

We show that this hash function family is indeed close to being pairwise independent for suitable parameters. As in [7], we can prove this in the simpler case when $q = q_0$ is prime; the result then follows for all q_0 without small prime factors. The main result is as follows (we refer to [7] for the definition of ε-pairwise independence). We provide the proof in the full version of this paper [6].

Theorem 2. *For an odd prime q, the hash function family $\mathcal{H}_d$ is ε-pairwise independent, with:*

$$\varepsilon = \frac{(d-1)(d-2)}{\sqrt{q}} + \frac{5d^{13/3}}{q} + \frac{(d-1) \cdot (2\beta)^d}{2^{\alpha\beta^{d-1}(\beta-2-2/\alpha))}}$$

Using the variant of the leftover hash lemma from [7], this proves the semantic security of the scheme for any encryption degree $d \geq 2$, with the condition $\alpha \cdot \beta^d \geq \gamma + \omega(\log \lambda)$. The constraint for correct decryption becomes $\eta \geq \rho \cdot d + \alpha + 2 + d \cdot \log_2 \beta$, and $\rho' = \rho \cdot d + \alpha + \omega(\log \lambda)$ for the secondary noise parameter. The public-key size for the somewhat homomorphic scheme becomes $(d \cdot \beta + 1) \cdot (\eta + \lambda)$. In particular by taking $\beta = 3$ and $d = \mathcal{O}(\log \lambda)$, we get a public-key size in $\tilde{\mathcal{O}}(\lambda^2)$ for the somewhat homomorphic scheme.

5   Adaptation of the BGV Framework to the DGHV Scheme

5.1   The BGV Framework for Leveled FHE

In this section we first recall the new framework from Brakerski, Gentry and Vaikuntanathan (BGV) [4] for leveled fully homomorphic encryption. Under the BGV framework the noise ceiling increases only linearly with the multiplicative depth, instead of increasing exponentially. This implies that bootstrapping is no longer necessary to achieve fully homomorphic encryption. The new framework is based on the Brakerski and Vaikuntanathan RLWE scheme [2,3]. The key technical tool is the modulus-switching technique from [2] that transforms a ciphertext $\boldsymbol{c}$ modulo p into a ciphertext $\boldsymbol{c}'$ modulo p' simply by scaling by p'/p and rounding appropriately; the noise is also reduced by a factor p'/p.

In the original Gentry framework [9], the multiplication of two mod-p ciphertexts with noise size ρ gives a ciphertext with noise size $\simeq 2\rho$; after a second multiplication level the noise becomes $\simeq 4\rho$, then $\simeq 8\rho$ and so on; the noise size grows exponentially with the number of multiplication levels. The modulus p is a ceiling for correct decryption; therefore if the bit-size of p is $k \cdot \rho$, the noise ceiling is reached after only $\log_2 k$ levels of multiplication. Fully homomorphic encryption is achieved via bootstrapping, *i.e.* homomorphically evaluating the decryption polynomial to obtain a refreshed ciphertext.

The breakthrough idea in the BGV framework [4] is to apply the modulus-switching technique *after every multiplication level*, using a ladder of gradually decreasing moduli p_i. Start with two mod-p_1 ciphertexts with noise ρ; as previously after multiplication one gets a mod-p_1 ciphertext with noise 2ρ. Now switch to a new modulus p_2 such that $p_2/p_1 \simeq 2^{-\rho}$; after the switching one gets a mod-p_2 ciphertext with noise back to $2\rho - \rho = \rho$ again; one can continue by multiplying two mod-p_2 ciphertexts, obtain a 2ρ-noise mod-p_2 ciphertext and switch back to a ρ-noise mod-p_3 ciphertext, and so on. With a ladder of k moduli p_i of decreasing size $(k+1) \cdot \rho, \ldots, 3\rho, 2\rho$ one can therefore perform k levels of multiplication instead of just $\log_2 k$. In other words the (largest) modulus size $(k+1) \cdot \rho$ grows only *linearly* with the multiplicative depth; this is an exponential improvement.

As explained in [4], bootstrapping is no longer strictly necessary to achieve fully homomorphic encryption: namely one can always assume a polynomial upper-bound on the number L of multiplicative levels of the circuit to be evaluated homomorphically. However, bootstrapping is still an interesting operation as a bootstrapped scheme can perform homomorphic evaluations indefinitely without needing to specify at setup time a bound on the multiplicative depth. As shown in [4] bootstrapping becomes also more efficient asymptotically in the BGV framework.

5.2   Modulus-Switching for DGHV

The modulus-switching technique recalled in the previous section is a very lightweight procedure to reduce the ciphertext noise by a factor roughly p/p' without knowing the secret-key and without bootstrapping. However we cannot apply this technique directly to DGHV since in DGHV the moduli p and p' must remain secret.

We now describe a technique for switching moduli in DGHV. We proceed in two steps. Given as input a DGHV ciphertext $c = q \cdot p + r$, we first show in Lemma 1 how to obtain a "virtual" ciphertext of the form $c' = 2^k \cdot q' + r'$ with $[q'] = [q]_2$, given the bits s_i in the following subset-sum sharing of $2^k/p$:

$$\frac{2^k}{p} = \sum_{i=1}^{\Theta} s_i \cdot y_i + \varepsilon \quad \mathrm{mod}\ 2^{k+1}$$

where the y_i's have κ bits of precision after the binary point, with $|\varepsilon| \leq 2^{-\kappa}$. This is done by first "expanding" the initial ciphertext c using the y_i's, as in the "squashed decryption" procedure in [9], and then "collapsing" the expanded ciphertext into c', using the secret-key vector $\boldsymbol{s} = (s_i)$. However we cannot reveal

s in clear, so instead we provide a DGHV encryption under p' of the secret-key bits s_i, as in the bootstrapped procedure. Then as showed in Lemma 2 the expanded ciphertext can be collapsed into a new ciphertext c'' under p' instead of p, for the same underlying plaintext; moreover as in the RLWE scheme the noise is reduced by a factor $\simeq p'/p$.

Lemma 1. *Let p be an odd integer. Let $c = q \cdot p + r$ be a ciphertext. Let k be an integer. Let $\kappa \in \mathbb{Z}$ be such that $|c| < 2^\kappa$. Let $\boldsymbol{y}$ be a vector of Θ numbers with κ bits of precision after the binary point, and let $\boldsymbol{s}$ be a vector of Θ bits such that $2^k/p = \langle \boldsymbol{s}, \boldsymbol{y} \rangle + \varepsilon \bmod 2^{k+1}$, where $|\varepsilon| \leq 2^{-\kappa}$. Let $\boldsymbol{c} = (\lfloor c \cdot y_i \rceil \bmod 2^{k+1})_{1 \leq i \leq \Theta}$. Let $c' = \langle \boldsymbol{s}, \boldsymbol{c} \rangle$. Then $c' = q' \cdot 2^k + r'$ with $[q']_2 = [q]_2$ and $r' = \lfloor r \cdot 2^k/p \rceil + \delta$ where $\delta \in \mathbb{Z}$ with $|\delta| \leq \Theta/2 + 2$.*

Proof. We have:

$$c' = \sum_{i=1}^{\Theta} s_i \lfloor c \cdot y_i \rceil + \Delta \cdot 2^{k+1} = \sum_{i=1}^{\Theta} s_i \cdot c \cdot y_i + \delta_1 + \Delta \cdot 2^{k+1}$$

for some $\Delta \in \mathbb{Z}$ and $|\delta_1| \leq \Theta/2$. Using $\langle \boldsymbol{s}, \boldsymbol{y} \rangle = 2^k/p - \varepsilon - \mu \cdot 2^{k+1}$ for some $\mu \in \mathbb{Z}$ this gives:

$$c' - \delta_1 - \Delta 2^{k+1} = c \cdot \left(\frac{2^k}{p} - \varepsilon - \mu \cdot 2^{k+1} \right) = q \cdot 2^k + r \cdot \frac{2^k}{p} - c \cdot \varepsilon - c \cdot \mu \cdot 2^{k+1}$$

Therefore we can write:

$$c' = q' \cdot 2^k + r'$$

where $[q']_2 = [q]_2$ and $r' = \lfloor r \cdot 2^k/p \rceil + \delta$ for some $\delta \in \mathbb{Z}$ with $|\delta| \leq \Theta/2 + 2$. □

As in [4], given a vector $\boldsymbol{x} \in [0, 2^{k+1}[^\Theta$ we write $\boldsymbol{x} = \sum_{i=0}^{k} 2^j \cdot \boldsymbol{u}_j$ where all the elements in vectors $\boldsymbol{u}_j$ are bits, and we define $\mathsf{BitDecomp}(\boldsymbol{x}, k) := (\boldsymbol{u}_0, \dots, \boldsymbol{u}_k)$. Similarly given a vector $\boldsymbol{z} \in \mathbb{R}^\Theta$ we define $\mathsf{Powersof2}(\boldsymbol{z}, k) := (\boldsymbol{z}, 2 \cdot \boldsymbol{z}, \dots, 2^k \cdot \boldsymbol{z})$. We have for any vectors $\boldsymbol{x}$ and $\boldsymbol{z}$:

$$\langle \mathsf{BitDecomp}(\boldsymbol{x}, k), \mathsf{Powersof2}(\boldsymbol{z}, k) \rangle = \langle \boldsymbol{x}, \boldsymbol{z} \rangle$$

The following lemma shows that given a ciphertext c encrypted under p and with noise r we can compute a new ciphertext c'' under p' with noise $r'' \simeq r \cdot p'/p$, by using an encryption $\boldsymbol{\sigma}$ under p' of the secret-key $\boldsymbol{s}$ corresponding to p.

Lemma 2. *Let p and p' be two odd integers. Let k be an integer such that $p' < 2^k$. Let $c = q \cdot p + r$ be a ciphertext. Let $\kappa \in \mathbb{Z}$ be such that $|c| < 2^\kappa$. Let $\boldsymbol{y}$ be a vector of Θ numbers with κ bits of precision after the binary point, and let $\boldsymbol{s}$ be a vector of Θ bits such that $2^k/p = \langle \boldsymbol{s}, \boldsymbol{y} \rangle + \varepsilon \bmod 2^{k+1}$, where $|\varepsilon| \leq 2^{-\kappa}$. Let $\boldsymbol{\sigma} = p' \cdot \boldsymbol{q} + \boldsymbol{r} + \lfloor \boldsymbol{s}' \cdot p'/2^{k+1} \rceil$ be an encryption of the secret-key $\boldsymbol{s}' = \mathsf{Powersof2}(\boldsymbol{s}, k)$, where $\boldsymbol{q} \leftarrow (\mathbb{Z} \cap [0, 2^\gamma/p'))^{(k+1) \cdot \Theta}$ and $\boldsymbol{r} \leftarrow (\mathbb{Z} \cap (-2^\rho, 2^\rho))^{(k+1) \cdot \Theta}$. Let $\boldsymbol{c} = (\lfloor c \cdot y_i \rceil \bmod 2^{k+1})_{1 \leq i \leq \Theta}$ and let $\boldsymbol{c}' = \mathsf{BitDecomp}(\boldsymbol{c}, k)$ be the expanded ciphertext. Let $c'' = 2 \langle \boldsymbol{\sigma}, \boldsymbol{c}' \rangle + [c]_2$. Then $c'' = q'' \cdot p' + r''$ where $r'' = \lfloor r \cdot p'/p \rceil + \delta'$ for some $\delta' \in \mathbb{Z}$ with $|\delta'| \leq 2^{\rho+2} \cdot \Theta \cdot (k+1)$, and $[r]_2 = [r'']_2$.*

Proof. We have, from $\sigma = p' \cdot q + r + \lfloor s' \cdot p'/2^{k+1} \rceil$:

$$c'' = 2\langle \sigma, c' \rangle + [c]_2 = 2p' \cdot \langle q, c' \rangle + 2\langle r, c' \rangle + 2\left\langle \left\lfloor s' \cdot \frac{p'}{2^{k+1}} \right\rceil, c' \right\rangle + [c]_2 \quad (3)$$

Since the components of c' are bits, we have using $2\lfloor x/2 \rfloor = x + \nu$ with $|\nu| \le 1$:

$$2\left\langle \left\lfloor \frac{p'}{2^{k+1}} \cdot s' \right\rceil, c' \right\rangle = \left\langle \frac{p'}{2^k} \cdot s', c' \right\rangle + \nu_2 = \frac{p'}{2^k} \cdot \langle s', c' \rangle + \nu_2$$

where $|\nu_2| \le \Theta \cdot (k+1)$. Using $\langle s', c' \rangle = \langle s, c \rangle$ and since from Lemma 1 we have $\langle s, c \rangle = q' \cdot 2^k + r'$ with $[q']_2 = [q]_2$ and $r' = \lfloor r \cdot 2^k/p \rfloor + \delta$ where $\delta \in \mathbb{Z}$ with $|\delta| \le \Theta/2 + 2$, we get:

$$2\left\langle \left\lfloor \frac{p'}{2^{k+1}} \cdot s' \right\rceil, c' \right\rangle = \frac{p'}{2^k} \cdot (q' \cdot 2^k + r') + \nu_2 = q' \cdot p' + \frac{p'}{2^k} \cdot r' + \nu_2 = q' \cdot p' + r \cdot \frac{p'}{p} + \nu_3$$

where $|\nu_3| \le |\nu_2| + \Theta/2 + 3 \le 2\Theta \cdot (k+1)$. Therefore we obtain from equation (3):

$$c'' = 2p' \cdot \langle q, c' \rangle + 2\langle r, c' \rangle + q' \cdot p' + r \cdot \frac{p'}{p} + \nu_3 + [c]_2 = q'' \cdot p' + r''$$

where $q'' := q' + 2\langle q, c' \rangle$ and $r'' = \lfloor r \cdot p'/p \rfloor + \delta'$ for some $\delta' \in \mathbb{Z}$ with:

$$|\delta'| \le |2\langle r, c' \rangle| + 1 + |\nu_3| + 1 \le 2^{\rho+1} \cdot \Theta \cdot (k+1) + 2\Theta \cdot (k+1) + 2 \le 2^{\rho+2} \cdot \Theta \cdot (k+1)$$

Eventually from $[c'']_2 = [c]_2$, $[c]_2 = [q]_2 \oplus [r]_2$, $[c'']_2 = [q'']_2 \oplus [r'']_2$ and $[q'']_2 = [q']_2 = [q]_2$, we obtain $[r]_2 = [r'']_2$ as required. $\square$

5.3 The Modulus-Switching Algorithm for DGHV

From Lemma 2 we can now specify the modulus-switching algorithm for DGHV.

SwitchKeyGen(pk, sk, pk', sk'):

1. Take as input two DGHV secret-keys p and p' of size η and η'. Let $\kappa = 2\gamma + \eta$ where γ is the size of the public key integers x_i under p.
2. Generate a vector y of Θ random numbers modulo $2^{\eta'+1}$ with κ bits of precision after the binary point, and a random vector s of Θ bits such that $2^{\eta'}/p = \langle s, y \rangle + \varepsilon \bmod 2^{\eta'+1}$ where $|\varepsilon| \le 2^{-\kappa}$. Generate the expanded secret-key $s' = \text{Powersof2}(s, \eta')$.
3. Compute a vector encryption σ of s' under sk', defined as follows:

$$\sigma = p' \cdot q + r + \left\lfloor s' \cdot \frac{p'}{2^{\eta'+1}} \right\rceil \quad (4)$$

where $q \leftarrow (\mathbb{Z} \cap [0, q_0'))^{(\eta'+1)\cdot\Theta}$ and $r \leftarrow (\mathbb{Z} \cap (-2^\rho, 2^\rho))^{(\eta'+1)\cdot\Theta}$, where q_0' is from $x_0' = q_0' \cdot p' + r'$ in pk'.
4. Output $\tau_{pk \to pk'} = (y, \sigma)$.

SwitchKey($\tau_{pk \to pk'}, c$):

1. Let $\boldsymbol{y}, \boldsymbol{\sigma} \leftarrow \tau_{pk \to pk'}$
2. Compute the expanded ciphertext $\boldsymbol{c} = (\lfloor c \cdot y_i \rceil \bmod 2^{\eta'+1})_{1 \le i \le \Theta}$ and let $\boldsymbol{c'} = \mathsf{BitDecomp}(\boldsymbol{c}, \eta')$.
3. Output $c'' = 2\langle \boldsymbol{\sigma}, \boldsymbol{c'} \rangle + [c]_2$.

5.4   The DGHV Scheme without Bootstrapping

We are now ready to describe our DGHV variant in the BGV framework, that is without bootstrapping. As in [4] we construct a *leveled* fully homomorphic scheme, *i.e.* an encryption scheme whose parameters depend polynomially on the depth of the circuits that the scheme can evaluate.

FHE. KeyGen($1^\lambda, 1^L$). Take as input the security parameter λ and the number of levels L. Let μ be a parameter specified later. Generate a ladder of L decreasing moduli of size $\eta_i = (i+1)\mu$ from $\eta_L = (L+1)\mu$ down to $\eta_1 = 2\mu$. For each η_i run DGHV.KeyGen(1^λ) from Section 2 to generate a random odd integer p_i of size η_i; we take the same parameter γ for all i. Let pk_i be the corresponding public key and $sk_i = p_i$ be the corresponding secret-key. For $j = L$ down to 2 run $\tau_{pk_j \to pk_{j-1}} \leftarrow \mathsf{SwitchKeyGen}(pk_j, sk_j, pk_{j-1}, sk_{j-1})$. The full public key is $pk = (pk_L, \tau_{pk_L \to pk_{L-1}}, \ldots, \tau_{pk_2 \to pk_1})$ and the secret-key is $sk = (p_1, \ldots, p_L)$.

FHE. Encrypt($pk, m \in \{0, 1\}$). Run DGHV. Encrypt(pk_L, m).

FHE. Decrypt(sk, c). Suppose that the ciphertext is under modulus p_j. Output $m \leftarrow [c]_{p_j} \bmod 2$.

FHE. Add(pk, c_1, c_2). Suppose that the two ciphertexts c_1 and c_2 are encrypted under the same pk_j; if they are not, use FHE.Refresh below to make it so. First compute $c_3 \leftarrow c_1 + c_2$. Then output $c_4 \leftarrow$ FHE.Refresh($\tau_{pk_j \to pk_{j-1}}, c_3$), unless both ciphertexts are encrypted under pk_1; in this case, simply output c_3.

FHE. Mult(pk, c_1, c_2). Suppose that the two ciphertexts c_1 and c_2 are encrypted under the same pk_j; if they are not, use FHE.Refresh below to make it so. First compute $c_3 \leftarrow c_1 \cdot c_2$. Then output $c_4 \leftarrow$ FHE.Refresh($\tau_{pk_j \to pk_{j-1}}, c_3$), unless both ciphertexts are encrypted under pk_1; in this case, simply output c_3.

FHE.Refresh($\tau_{pk_{j+1} \to pk_j}, c$). Output $c' \leftarrow$ SwitchKey($\tau_{pk_{j+1} \to pk_j}, c$).

5.5   Correctness and Security

We show in the full version of this paper [6] how to fix the parameter μ so that the ciphertext noise for every modulus in the ladder remains roughly the same, and we prove that FHE is a correct leveled FHE scheme.

Theorem 3. *For some* $\mu = \mathcal{O}(\lambda + \log L)$, FHE *is a correct L-leveled FHE scheme; specifically it correctly evaluates circuits of depth L with* Add *and* Mult *gates over* $GF(2)$.

We show in the full version of this paper [6] that the resulting FHE is semantically secure under the following new assumption.

Definition 2 (Decisional Approximate GCD). *The (ρ, η, γ)-Decisional Approximate GCD Problem is: For a random η-bit odd integer p, given polynomially many samples from $\mathcal{D}_{\gamma,\rho}(p)$, and given an integer $z = x + b \cdot \lfloor 2^j \cdot p/2^{\eta+1} \rceil$ for a given random integer $j \in [0, \eta]$, where $x \leftarrow \mathcal{D}_{\gamma,\rho}(p)$ and $b \leftarrow \{0,1\}$, find b.*

The Decisional Approximate GCD assumption is defined in the usual way. It is clearly stronger than the standard Approximate GCD assumption. We were not able to base the security of the leveled DGHV scheme on the standard Approximate GCD assumption; this is due to equation (4) which requires a non-standard encryption of the secret-key bits.

Theorem 4. FHE *is semantically secure under the Decisional Approximate GCD assumption and under the hardness of subset sum assumption.*

6   Improved Attack against the Approximate GCD Algorithm

Recently, Chen and Nguyen [5] described an improved exponential algorithm for solving the approximate common divisor problem: they obtain a complexity of $\tilde{\mathcal{O}}(2^{\rho/2})$ for the partial version (with an exact multiple $x_0 = q_0 \cdot p$) and $\tilde{\mathcal{O}}(2^{3\rho/2})$ for the general version (with near-multiples only).[1]

In this section, we show that the latter complexity can be heuristically improved to $\tilde{\mathcal{O}}(2^\rho)$ provided that sufficiently many near-multiples are available, which is the case in the DGHV scheme. Our algorithm has memory complexity $\tilde{\mathcal{O}}(2^\rho)$, instead of only $\tilde{\mathcal{O}}(2^{\rho/2})$ for the Chen and Nguyen attack.

Indeed, assume that we have s large near-multiples $x_1, \ldots, x_s$ of a given prime p_0, of the hidden form $x_j = p_0 q_j + r_j$, where $q_j \in [0, 2^\gamma/p_0)$ (for γ polynomial in ρ) and $r_j \in [0, 2^\rho)$ are chosen uniformly and independently at random. We claim that p_0 can then be recovered with overwhelming probability in time $\tilde{\mathcal{O}}(2^{\frac{s+1}{s-1}\rho})$ (and with significant probability in time $\tilde{\mathcal{O}}(2^{\frac{s}{s-1}\rho})$).

The algorithm is as follows. For $j = 1, \ldots, s$, let:

$$y_j = \prod_{i=0}^{2^\rho-1} (x_j - i)$$

Clearly, p_0 divides the GCD $g = \gcd(y_1, \ldots, y_s)$. Each y_i can be computed in time quasilinear in 2^ρ using a product tree, and the GCD can be evaluated as $\gcd(\cdots \gcd(\gcd(y_1, y_2), y_3), \ldots, y_s)$ using $s - 1$ quasilinear GCD computations on numbers of size $\mathcal{O}(2^\rho \cdot \gamma) = \tilde{\mathcal{O}}(2^\rho)$. Hence, the whole computation of g takes time $\tilde{\mathcal{O}}(s \cdot 2^\rho)$.

---

[1] Namely to solve the general version using the partial version algorithm it suffices to do exhaustive search on the ρ bits of noise in $x_0 = q_0 \cdot p + r_0$.

Now, we argue that with high probability on the choice of the (q_j, r_j), all the prime factors of g except p_0 are smaller than a bound B that is not much larger than 2^ρ. Then, p_0 can be recovered as g/g', where g' is the B-smooth part of g, which can in turn be computed in time quasilinear in $\max(B, |g|)$, e.g. using Bernstein's algorithm [1]. Overall, the full time complexity of the attack is thus $\tilde{\mathcal{O}}(\max(B, s \cdot 2^\rho))$, or simply $\tilde{\mathcal{O}}(B)$ assuming that $s = O(\rho)$, and without loss of generality that $B > 2^\rho$. All we need to find is how to choose B to obtain a sufficient success probability.

The probability that all the prime factors of g except p_0 are smaller than B is the probability that, for every prime $p \geq B$ other than p_0, not all the x_j's are congruent to one of $0, 1, \ldots, 2^\rho - 1 \bmod p$. This happens with probability very close to $1 - (2^\rho/p)^s$. Hence, the probability that all the prime factors of g except p_0 are smaller than B is essentially given by the following Euler product:

$$P_{s,\rho}(B) = \prod_{\substack{p \geq B \\ p \neq p_0}} \left(1 - \frac{2^{s\rho}}{p^s}\right)$$

(which clearly converges to some positive value smaller than 1 since $s \geq 2$ and $B > 2^\rho$). We prove in the full version of this paper [6] the following estimate on this Euler product.

Lemma 3. *For any $B > 2^{\rho+1/s}$, we have:*

$$1 - P_{s,\rho}(B) < \frac{2s}{s-1} \cdot \frac{2^{s\rho}}{B^{s-1} \log B}$$

In particular, if we pick $B = 2^{\frac{s}{s-1}\rho}$, we obtain $P_{s,\rho}(B) > 1 - 2/(\rho \log 2)$: thus, the problem can be solved in time $\tilde{\mathcal{O}}(2^{\frac{s}{s-1}\rho})$ with significant success probability. And if we pick $B = 2^{\frac{s+1}{s-1}\rho}$, we get $P_{s,\rho}(B) > 1 - 2^{-\rho}$: hence, the problem can be solved in time $\tilde{\mathcal{O}}(2^{\frac{s+1}{s-1}\rho})$ with an overwhelming success probability.

We see in both cases that for any given $\varepsilon > 0$, the complexity becomes $\mathcal{O}(2^{(1+\varepsilon)\rho})$ if s is large enough. Better yet, if $s = \omega(1)$ (for example $\Theta(\rho)$) near-multiples are available, the problem can be solved in time $\tilde{\mathcal{O}}(2^\rho)$ with overwhelming probability.

As in [5] we can perform a time-memory trade-off. First split the product y_1 into d sub-products z_k's, and guess which of these sub-products $z = z_k$ contains p_0. Let $g = \gcd(z, y_2, \ldots, y_s)$. The first GCD computation $\gcd(z, y_2)$ can be performed in time $\tilde{\mathcal{O}}(2^\rho)$ and memory $\tilde{\mathcal{O}}(2^\rho/d)$ by first computing $y_2 \bmod z$ using a product tree; the remaining gcd's can be computed with the same complexity; the same holds for recovering the B-smooth part of g. Hence p_0 can be recovered in time $\tilde{\mathcal{O}}(d \cdot 2^\rho)$ and memory $\tilde{\mathcal{O}}(2^\rho/d)$.

6.1   Experimental Results

We have implemented the previous attack; see the full version of this paper [6] for the source code. Table 1 shows that our attack performs well in practice; it is roughly 200 times faster than the corresponding attack of Chen and Nguyen for the smallest set of parameters considered in [5].

Table 1. Running time of the attack, on a single core of an Amazon EC2 Cluster Compute Eight Extra Large Instance instance (featuring an Intel Xeon E5 processor at 2.5 GHz and 60.5 GB of memory), with parameter $s = \rho$. For the third instance, the running time of the Chen-Nguyen attack [5] was estimated by multiplying the running time from [5] (1.6 min) by 2^ρ.

Instance	ρ	γ	$\log_2$ mem.	running time	running time [5]
Micro	12	10^4	26.3	40 s	
Toy (Section 8)	13	$61 \cdot 10^3$	29.9	13 min 22 s	
Toy' ([5] without x_0)	17	$1.6 \cdot 10^5$	35.3	17 h 50 min	*3495 hours*

7 Implementation of DGHV with Compressed Public Key

In this section we describe an implementation of the DGHV scheme with the compression technique of Section 3; we use the variant with $x_0 = q_0 \cdot p$. We refer to the full version of this paper [6] for a full description of the resulting scheme, and we provide the source code of our implementation in [17].

Asymptotic Key Size. To prevent lattice attacks against the sparse subset-sum problem, one must have $\Theta^2 = \gamma \cdot \omega(\log \lambda)$; see [7,16] for more details. One can then take $\rho = \lambda$, $\eta = \tilde{\mathcal{O}}(\lambda^2)$, $\gamma = \tilde{\mathcal{O}}(\lambda^5)$, $\alpha = \tilde{\mathcal{O}}(\lambda^2)$, $\tau = \tilde{\mathcal{O}}(\lambda^3)$ and $\Theta = \tilde{\mathcal{O}}(\lambda^3)$. Using our compression technique the public key size is roughly $2\gamma + (\tau + \Theta) \cdot (\eta + \lambda) = \tilde{\mathcal{O}}(\lambda^5)$ bits.

Concrete Key Size and Execution Speed. We have updated the parameters from [7] to take into account the improved approximate-GCD attack from [5]; see Table 2. The attack from [5] is memory bounded; however we took a conservative approach and considered a memory unbounded adversary. As in [7] we take $n = 4$ and $\theta = 15$ for all security levels. We can see in Table 2 that compression reduces the public key size considerably. Table 3 shows no significant performance degradation with respect to [7].

Table 2. The concrete parameters of various test instances and their respective public key sizes, for DGHV with compressed public-key

Instance	λ	ρ	η	$\gamma \times 10^{-6}$	α	τ	Θ	pk size
Toy	42	27	1026	0.15	936	158	144	77 KB
Small	52	41	1558	0.83	1476	572	533	437 KB
Medium	62	56	2128	4.20	2016	2110	1972	2207 KB
Large	72	71	2698	19.35	2556	7659	7897	10.3 MB

Table 3. Timings of our Sage 4.7.2 [15] code (single core of a desktop computer with an Intel Core2 Duo E8400 at 3 GHz), for DGHV with compressed public-key.

Instance	KeyGen	Encrypt	Decrypt	Expand	Recrypt
Toy	0.06 s	0.05 s	0.00 s	0.01 s	0.41 s
Small	1.3 s	1.0 s	0.00 s	0.15 s	4.5 s
Medium	28 s	21 s	0.01 s	2.7 s	51 s
Large	10 min	7 min 15 s	0.05 s	51 s	11 min 34 s

8 Implementation of Leveled DGHV

In this section we describe an implementation of the leveled DGHV scheme described in Section 5 in the BGV framework. We implement the modulus-switching procedure as described in Section 5.3, with an optimization of the ciphertext expansion procedure (see below). We also implement the bootstrapping operation; although not strictly necessary, this enables to get a FHE that can perform homomorphic evaluations indefinitely without needing to specify at setup time a bound on the multiplicative level.

8.1 Faster Ciphertext Expansion

We consider the modulus-switching procedure of Section 5.3. The initial modulus p has size η and the new modulus p' has size $\eta' < \eta$. The first modulus p is shared among the y_i elements as

$$\frac{2^{\eta'}}{p} = \sum_{i=1}^{\Theta} s_i \cdot y_i + \varepsilon \quad \mod 2^{\eta'+1} \tag{5}$$

where the s_i's are bits, the y_i's have κ bits of precision after the binary point, and $|\varepsilon| \leq 2^{-\kappa}$. In practice one can generate the y_i's pseudo-randomly (except y_1), as suggested in [7]. However the ciphertext expansion from Step 2 of SwitchKey algorithm (Section 5.3) is a time-consuming procedure.

Therefore instead of using pseudo-random y_i's we use the following (admittedly aggressive) optimization. Let δ be a parameter specified later. We generate a random y with $\kappa + \delta \cdot \Theta \cdot \eta$ bits of precision after the binary point, and we define the y_i's for $2 \leq i \leq \Theta$ as:

$$y_i = \left[y \cdot 2^{i \cdot \delta \cdot \eta} \right]_{2^{\eta'+1}}$$

keeping only κ bits of precision after the binary point for each y_i as previously. We fix y_1 so that equality (5) holds, assuming $s_1 = 1$. Then the ciphertext expansion from Step 2 of the SwitchKey algorithm (Section 5.3) can be computed as follows, for all $2 \leq i \leq \Theta$:

$$z_i = \lfloor c \cdot y_i \rceil \mod 2^{\eta'+1} = \lfloor c \cdot y \cdot 2^{i \cdot \delta \cdot \eta} \rceil \mod 2^{\eta'+1}$$

Therefore computing all the z_i's (except z_1) is now essentially a single multiplication $c \cdot y$. In the full version of this paper [6] we describe a lattice attack against this optimization; we show that the attack is thwarted by selecting δ such that $\delta \cdot \Theta \cdot \eta \geq 3\gamma$.

Finally we use the following straightforward optimization: instead of using BitDecomp and Powersof2 with bits, we use words of size ω bits instead. This decreases the running time of SwitchKey by a factor of about ω, at the cost of increasing the resulting noise by roughly ω bits. We took $\omega = 32$ in our implementation.

8.2   Bootstrapping: The Decryption Circuit

Recall that the decryption function in the DGHV scheme is:

$$m \leftarrow \left[c - \left\lfloor \sum_{i=1}^{\Theta} s_i \cdot z_i \right\rceil \right]_2 \tag{6}$$

where $z_i = [c \cdot y_i]_2$ for $1 \leq i \leq \Theta$ is the expanded ciphertext, keeping only $n = \lceil \log_2(\theta + 1) \rceil$ bits of precision after the binary point for each z_i. The s_i's form a sparse Θ-dimensional vector of Hamming weight θ, such that $1/p = \sum_{i=1}^{\Theta} s_i \cdot y_i + \varepsilon$ where the y_i's have κ bits of precision after the binary point, and $|\varepsilon| \leq 2^{-\kappa}$. Note that for bootstrapping the decryption circuit is only used for the smallest modulus p in the ladder. The following lemma shows that the message m can be computed using a circuit of multiplicative depth exactly n.

Lemma 4. *Let* $a = [a_0, \ldots, a_n]$ *and* $b = [b_0, \ldots, b_n]$ *be two integers of size* $n + 1$ *bits, where every bit* a_i *and* b_i *has multiplicative depth at most* i. *Then every bit* c_i *of the sum* $c = (a + b) \bmod 2^{n+1} = [c_0, \ldots, c_n]$ *has multiplicative depth at most* i.

Proof. Let δ_i be the i-th carry bit, with $\delta_0 = 0$. We have $c_i = a_i \oplus b_i \oplus \delta_i$ for $0 \leq i \leq n$, where $\delta_i = a_{i-1} \cdot b_{i-1} + a_{i-1} \cdot \delta_{i-1} + b_{i-1} \cdot \delta_{i-1}$ for $1 \leq i \leq n$. Therefore by recursion δ_i has multiplicative depth at most i; this implies that c_i has multiplicative depth at most i. □

Therefore using a simple loop the sum of the Θ numbers $s_i \cdot z_i$ in equation (6) can be computed with a circuit of multiplicative depth n. Since a subsequent homomorphic operation (either addition or multiplication) must be possible between refreshed ciphertexts, the full bootstrapping procedure requires a leveled FHE scheme with multiplicative depth $L = n + 1$. Note that for bootstrapping an encryption of the secret-key bits s_i (corresponding to the last modulus p_1 in the ladder) must be provided under p_L, the first modulus in the ladder, so that the homomorphic evaluation of m in equation (6) can start under the public key pk_L.

8.3   Implementation Results

In this section we describe an implementation of the leveled DGHV scheme, including the bootstrapping operation. As mentioned previously we cannot use the variant with noise-free $x_0 = q_0 \cdot p$ since otherwise p could be recovered using the ECM; namely the smallest modulus in the ladder has size only $2\mu = 164$ bits for the "Large" instance.

We summarize in Tables 4 and 5 the performance of our implementation of the leveled DGHV scheme. We denote by η the size of the largest modulus in the ladder. The running time of the Recrypt operation is disappointing compared to the non-leveled implementation from Section 7; however we think that there is room for improvement.

Table 4. The concrete parameters of various test instances and their respective public-key sizes for leveled DGHV

Instance	λ	ρ	η	μ	$\gamma \times 10^{-6}$	Θ	pk size
Toy	42	14	336	56	0.061	195	354 KB
Small	52	20	390	65	0.27	735	1690 KB
Medium	62	26	438	73	1.02	2925	7.9 MB
Large	72	34	492	82	2.20	5700	18 MB

Table 5. Timings of our Sage 4.7.2 [15] code (single core of a desktop computer with an Intel Core2 Duo E8400 at 3 GHz)

Instance	KeyGen	Encrypt	Decrypt	Mult & Scale	Recrypt
Toy	0.36 s	0.01 s	0.00 s	0.04 s	8.8 s
Small	5.4 s	0.07 s	0.00 s	0.59 s	101 s
Medium	1 min 12 s	0.85 s	0.00 s	9.1 s	32 min 38 s
Large	6 min 18 s	3.4 s	0.00 s	41 s	2 h 27 min

Acknowledgments. We would like to thank Tancrède Lepoint, Phong Nguyen and the EUROCRYPT referees for their helpful comments.

References

1. Bernstein, D.J.: How to Find Smooth Parts of Integers (2004),
   http://cr.yp.to/papers.html#smoothparts
2. Brakerski, Z., Vaikuntanathan, V.: Efficient Fully Homomorphic Encryption from (Standard) LWE. In: Proceedings of FOCS 2011 (2011); Full version available at IACR eprint
3. Brakerski, Z., Vaikuntanathan, V.: Fully Homomorphic Encryption from Ring-LWE and Security for Key Dependent Messages. In: Rogaway, P. (ed.) CRYPTO 2011. LNCS, vol. 6841, pp. 505–524. Springer, Heidelberg (2011)
4. Brakerski, Z., Gentry, C., Vaikuntanathan, V.: Fully Homomorphic Encryption without Bootstrapping. Cryptology ePrint Archive, Report 2011/277

5. Chen, Y., Nguyen, P.Q.: Faster Algorithms for Approximate Common Divisors: Breaking Fully-Homomorphic-Encryption Challenges over the Integers. Cryptology ePrint Archive, Report 2011/436
6. Coron, J.S., Naccache, D., Tibouchi, M.: Public-key Compression and Modulus Switching for Fully Homomorphic Encryption over the Integers. Full version of this paper. Cryptology ePrint Archive, Report 2011/440
7. Coron, J.-S., Mandal, A., Naccache, D., Tibouchi, M.: Fully Homomorphic Encryption over the Integers with Shorter Public Keys. In: Rogaway, P. (ed.) CRYPTO 2011. LNCS, vol. 6841, pp. 487–504. Springer, Heidelberg (2011); Full version available at IACR eprint
8. van Dijk, M., Gentry, C., Halevi, S., Vaikuntanathan, V.: Fully Homomorphic Encryption over the Integers. In: Gilbert, H. (ed.) EUROCRYPT 2010. LNCS, vol. 6110, pp. 24–43. Springer, Heidelberg (2010)
9. Gentry, C.: A fully homomorphic encryption scheme. Ph.D. thesis, Stanford University (2009), http://crypto.stanford.edu/craig
10. Gentry, C., Halevi, S.: Implementing Gentry's Fully-Homomorphic Encryption Scheme. In: Paterson, K.G. (ed.) EUROCRYPT 2011. LNCS, vol. 6632, pp. 129–148. Springer, Heidelberg (2011)
11. Lauter, K., Naehrig, M., Vaikuntanathan, V.: Can Homomorphic Encryption be Practical? Cryptology ePrint Archive, Report 2011/405
12. Lenstra, A.K.: Generating RSA Moduli with a Predetermined Portion. In: Ohta, K., Pei, D. (eds.) ASIACRYPT 1998. LNCS, vol. 1514, pp. 1–10. Springer, Heidelberg (1998)
13. Lenstra, H.W.: Factoring integers with elliptic curves. Annals of Mathematics 126(3), 649–673 (1987)
14. Smart, N.P., Vercauteren, F.: Fully Homomorphic Encryption with Relatively Small Key and Ciphertext Sizes. In: Nguyen, P.Q., Pointcheval, D. (eds.) PKC 2010. LNCS, vol. 6056, pp. 420–443. Springer, Heidelberg (2010)
15. Stein, W.A., et al.: Sage Mathematics Software (Version 4.7.2), The Sage Development Team (2011), http://www.sagemath.org
16. Stehlé, D., Steinfeld, R.: Faster Fully Homomorphic Encryption. In: Abe, M. (ed.) ASIACRYPT 2010. LNCS, vol. 6477, pp. 377–394. Springer, Heidelberg (2010)
17. https://github.com/coron/fhe

Fully Homomorphic Encryption
with Polylog Overhead

Craig Gentry[1], Shai Halevi[1], and Nigel P. Smart[2]

[1] IBM T.J. Watson Research Center,
Yorktown Heights, New York, U.S.A.
[2] Dept. Computer Science, University of Bristol,
Bristol, United Kingdom

Abstract. We show that homomorphic evaluation of (wide enough) arithmetic circuits can be accomplished with only polylogarithmic overhead. Namely, we present a construction of fully homomorphic encryption (FHE) schemes that for security parameter λ can evaluate any width-$\Omega(\lambda)$ circuit with t gates in time $t \cdot \mathrm{polylog}(\lambda)$.

To get low overhead, we use the recent batch homomorphic evaluation techniques of Smart-Vercauteren and Brakerski-Gentry-Vaikuntanathan, who showed that homomorphic operations can be applied to "packed" ciphertexts that encrypt vectors of plaintext elements. In this work, we introduce permuting/routing techniques to move plaintext elements across these vectors efficiently. Hence, we are able to implement general arithmetic circuit in a batched fashion without ever needing to "unpack" the plaintext vectors.

We also introduce some other optimizations that can speed up homomorphic evaluation in certain cases. For example, we show how to use the Frobenius map to raise plaintext elements to powers of p at the "cost" of a linear operation.

1 Introduction

Fully homomorphic encryption (FHE) [1–3] allows a worker to perform arbitrarily-complex dynamically-chosen computations on encrypted data, despite not having the secret decryption key. Processing encrypted data homomorphically requires more computation than processing the data unencrypted. But how much more? What is the *overhead*, the ratio of encrypted computation complexity to unencrypted computation complexity (using a circuit model of computation)? Here, under the ring-LWE assumption, we show that the overhead can be made as low as *polylogarithmic* in the security parameter.

We accomplish this by *packing* many plaintexts into each ciphertext; each ciphertext has $\tilde{\Omega}(\lambda)$ "plaintext slots". Then, we describe a complete set of operations – Add, Mult and Permute – that allows us to evaluate arbitrary circuits *while keeping the ciphertexts packed*. Batch Add and Mult have been done before [4], and follow easily from the Chinese Remainder Theorem within our underlying polynomial ring. Here we introduce the operation Permute, that allows us to

D. Pointcheval and T. Johansson (Eds.): EUROCRYPT 2012, LNCS 7237, pp. 465–482, 2012.

homomorphically move data between the plaintext slots, show how to realize it from our underlying algebra, and how to use it to evaluate arbitrary circuits.

Our approach begins with the observation [4, 5] that we can use an automorphism group $\mathcal{H}$ associated to our underlying ring to "rotate" or "re-align" the contents of the plaintext slots. (These automorphisms were used in a somewhat similar manner by Lyubashevsky et al. [6] in their proof of the pseudorandomness of RLWE.) While $\mathcal{H}$ alone enables only a few permutations (e.g., "rotations"), we show that any permutation can be constructed as a log-depth permutation network, where each level consists of a constant number of "rotations", batch-additions and batch-multiplications. Our method works when the underlying ring has an associated automorphism group $\mathcal{H}$ which is abelian and sharply transitive, a condition that we prove always holds for our scheme's parameters.

Ultimately, the Add, Mult and Permute operations can all be accomplished with $\tilde{O}(\lambda)$ computation by building on the recent Brakerski-Gentry-Vaikuntanathan (BGV) "FHE without bootstrapping" scheme [5], which builds on prior work by Brakerski and Vaikuntanathan and others [7–9]. Thus, we obtain an FHE scheme that can evaluate any circuit that has $\Omega(\lambda)$ average width with only polylog(λ) overhead. For comparison, the smallest overhead for FHE was $\tilde{O}(\lambda^{3.5})$ [10] until BGV recently reduced it to $\tilde{O}(\lambda)$ [5].[1]

In addition to their essential role in letting us move data across plaintext slots, ring automorphisms turn out to have interesting secondary consequences: they also enable more nimble manipulation of data *within* plaintext slots. Specifically, in some cases we can use them to raise the packed plaintext elements to a high power with hardly any increase in the noise magnitude of the ciphertext! In practice, this could permit evaluation of high-degree circuits without resorting to bootstrapping, in applications such as computing AES. See the full version of this paper [12].

1.1   Packing Plaintexts and Batched Homomorphic Computation

Smart and Vercauteren [4, 13] were the first to observe that, by an application the Chinese Remainder Theorem to number fields, the plaintext space of some previous FHE schemes can be partitioned into a vector of "plaintext slots", and that a single homomorphic Add or Mult of a pair of ciphertexts implicitly adds or multiplies (component-wise) the entire plaintext vectors. Each plaintext slot is defined to hold an element in some finite field $\mathbb{K}_n = \mathbb{F}_{p^n}$, and, abstractly, if one has two ciphertexts that hold (encrypt) messages $m_0, \ldots, m_{\ell-1} \in \mathbb{K}_n^\ell$ and $m'_0, \ldots, m'_{\ell-1} \in \mathbb{K}_n^\ell$ respectively in plaintext slots $0, \ldots, \ell - 1$, applying ℓ-Add to the two ciphertexts gives a new ciphertext that holds $m_0 + m'_0, \ldots, m_{\ell-1} + m'_{\ell-1}$ and applying ℓ-Mult gives a new ciphertext that holds $m_0 \cdot m'_0, \ldots, m_{\ell-1} \cdot m'_{\ell-1}$. Smart and Vercauteren used this observation for *batch* (or SIMD [14]) homomorphic operations. That is, they show how to evaluate a function f

---

[1]   However, the polylog factors in our new scheme are rather large. It remains to be seen how much of an improvement this approach yields in practice, as compared to the $\tilde{O}(\lambda^{3.5})$ approach implemented in [10, 11].

homomorphically ℓ times in parallel on ℓ different inputs, with approximately the same cost that it takes to evaluate the function once without batching.

Here is a taste of how these separate plaintext slots are constructed algebraically. As an example, for the ring-LWE-based scheme, suppose we use the polynomial ring $\mathbb{A} = \mathbb{Z}[x]/(x^\ell + 1)$ where ℓ is a power of 2. Ciphertexts are elements of $\mathbb{A}_q^2$ where (as in in [5]) q has only polylog(λ) bits. The "aggregate" plaintext space is $\mathbb{A}_p$ (that is, ring elements taken modulo p) for some small prime $p = 1 \mod 2\ell$. Any prime $p = 1 \mod 2\ell$ *splits* over the field associated to this ring – that is, in $\mathbb{A}$, the ideal generated by p is the product of ℓ ideals $\{\mathfrak{p}_i\}$ each of norm p – and therefore $\mathbb{A}_p \equiv \mathbb{A}_{\mathfrak{p}_0} \times \cdots \times \mathbb{A}_{\mathfrak{p}_{\ell-1}}$. Consequently, using the Chinese remainder theorem, we can encode ℓ independent mod-p plaintexts $m_0, \ldots, m_{\ell-1} \in \{0, \ldots, p-1\}$ as the unique element in $\mathbb{A}_p$ that is in all of the cosets $m_i + \mathfrak{p}_i$. Thus, in a single ciphertext, we may have ℓ independent plaintext "slots".

In this work, we often use ℓ-Add and ℓ-Mult to efficiently implement a Select operation: Given an index set I we can construct a vector $\mathbf{v}_I$ of "select bits" $(v_0, \ldots, v_{\ell-1})$, such that $v_i = 1$ if $i \in I$ and $v_i = 0$ otherwise. Then element-wise multiplication of a packed ciphertext $\mathbf{c}$ with the select vector $\mathbf{v}$ results in a new ciphertext that contains only the plaintext element in the slots corresponding to I, and zero elsewhere. Moreover, by generating two complementing select vectors $\mathbf{v}_I$ and $\mathbf{v}_{\bar{I}}$ we can mix-and-match the slots from two packed ciphertexts $\mathbf{c}_1$ and $\mathbf{c}_2$: Setting $\mathbf{c} = (\mathbf{v}_I \times \mathbf{c}_1) + (\mathbf{v}_{\bar{I}} \times \mathbf{c}_2)$, we pack into $\mathbf{c}$ the slots from $\mathbf{c}_1$ at indexes from I and the slots from $\mathbf{c}_2$ elsewhere.

While batching is useful in many setting, it does not, by itself, yield low-overhead homomorphic computation in general, as it does not help us to reduce the overhead of computing a complicated function just once. Just as in normal program execution of SIMD instructions (e.g., the SSE instructions on x86), one needs a method of moving data between slots in each SIMD word.

1.2   Permuting Plaintexts within the Plaintext Slots

To reduce the overhead of homomorphic computation *in general*, we need a *complete* set of operations over *packed vectors of plaintexts*. The approach above allows us to add or multiply messages that are in the same plaintext slot, but what if we want to add the content of the i-th slot in one ciphertext to the content of the j-th slot of another ciphertext, for $i \neq j$? We can "unpack" the slots into separate ciphertexts (say, using homomorphic decryption[2] [2, 3]), but there is little hope that this approach could yield very efficient FHE. Instead, we complement ℓ-Add and ℓ-Mult with an operation ℓ-Permute to move data efficiently across slots within a a given ciphertext, and efficient procedures to clone slots from a packed ciphertext and move them around to other packed ciphertexts.

Brakerski, Gentry, and Vaikuntanathan [5] observed that for certain parameter settings, one can use *automorphisms* associated with the algebraic ring $\mathbb{A}$

---

[2] This is the approach suggested in [4] for Gentry's original FHE scheme.

to "rotate" all of plaintext spaces simultaneously, sort of like turning a dial on a safe. That is, one can transform a ciphertext that holds $m_0, m_1, \ldots, m_{\ell-1}$ in its ℓ slots into another ciphertext that holds $m_i, m_{i+1}, \ldots, m_{i+\ell-1}$ (for an arbitrary given i, index arithmetic mod ℓ), and this rotation operation takes time quasi-linear in the ciphertext size, which is quasi-linear in the security parameter. They used this tool to construct Pack and Unpack algorithms whereby separate ciphertexts could be aggregated (packed) into a single ciphertext with packed plaintexts before applying bootstrapping (and then the refreshed ciphertext would be unpacked), thereby lowering the amortized cost of bootstrapping.

We exploit these automorphisms more fully, using the basic rotations that the automorphisms give us to construct *permutation networks* that can permute data in the plaintext slots arbitrarily. We also extend the application of the automorphisms to more general underlying rings, beyond the specific parameter settings considered in prior work [5, 7, 8]. This lets us devise low-overhead homomorphic schemes for arithmetic circuits over essentially any small finite field $\mathbb{F}_{p^n}$.

Our efficient implementation of Permute, described in Section 3, uses the Beneš/Waksman permutation network [15, 16]. This network consists of two back-to-back butterfly network of width 2^k, where each level in the network has 2^{k-1} "switch gates" and each switch gate swaps (or not) its two inputs, depending on a control bit. It is possible to realize any permutation of $\ell = 2^k$ items by appropriately setting the control bits of all the switch gates. Viewing this network as acting on k-bit addresses, the i-th level of the network partitions the 2^k addresses into 2^{k-1} pairs, where each pair of addresses differs only in the $|i - k|$-th bit, and then it swaps (or not) those pairs. The fact that the pairs in the i-th level always consist of addresses that differ by exactly $2^{|i-k|}$, makes it easy to implement each level using rotations: All we need is one rotation by $2^{|i-k|}$ and another by $-2^{|i-k|}$, followed by two batched Select operations.

For general rings $\mathbb{A}$, the automorphisms do not always exactly "rotate" the plaintext slots. Instead, they act on the slots in a way that depends on a quotient group $\mathcal{H}$ of the appropriate Galois group. Nonetheless, we use basic theorems from Galois theory, in conjunction with appropriate generalizations of the Beneš/Waksman procedure, to construct a permutation network of depth $O(\log \ell)$ that can realize any permutation over the ℓ plaintext slots, where each level of the network consists of a constant number of permutations from $\mathcal{H}$ and Select operations. As with the rotations considered in [5], applying permutations from $\mathcal{H}$ can be done in time quasi-linear in ciphertext size, which is only quasi-linear in the security parameter. Overall, we find that permutation networks and Galois theory are a surprisingly fruitful combination.

We note that Damgård, Ishai and Krøigaard [17] used permutation networks in a somewhat analogous fashion to perform secure multiparty computation with *packed secret shares*. In their setting, which permits interaction between the parties, the permutations can be evaluated using much simpler mathematical machinery.

1.3   FHE with Polylog Overhead

In our discussion above, we glossed over the fact that ciphertext sizes in a BGV-like cryptosystem [5] depend polynomially on the depth of the circuit being evaluated, because the modulus size must grow with the depth of the circuit (unless bootstrapping [2, 3] is used). So, without bootstrapping, the "polylog overhead" result only applies to circuits of polylog depth. However, decryption itself can be accomplished in log-depth [5], and moreover the parameters can be set so that a ciphertext with $\tilde{\Omega}(\lambda)$ slots can be decrypted using a circuit of size $\tilde{O}(\lambda)$. Therefore, "recryption" can be accomplished with polylog overhead, and we obtain FHE with polylog overhead for arbitrary (wide enough) circuits.

2   Computing on (Encrypted) Arrays

As we explained above, our main tool for low-overhead homomorphic computation is to compute on "packed ciphertexts", namely make each ciphertext hold a vector of plaintext values rather than a single value. Throughout this section we let ℓ be a parameter specifying the number of plaintext values that are packed inside each ciphertext, namely we always work with ℓ-vectors of plaintext values. Let $\mathbb{K}_n = \mathbb{F}_{p^n}$ denote the plaintext space (e.g., $\mathbb{K}_n = \mathbb{F}_2$ if we are dealing with binary circuits directly). It was shown in [4, 5] how to homomorphically evaluate batch addition and multiplication operations on ℓ-vectors:

$$\ell\text{-Add}\big(\langle u_0,\ldots,u_{\ell-1}\rangle, \langle v_0,\ldots,v_{\ell-1}\rangle\big) \stackrel{\text{def}}{=} \langle u_0 + v_0,\ldots,u_{\ell-1} + v_{\ell-1}\rangle$$

$$\ell\text{-Mult}\big(\langle u_0,\ldots,u_{\ell-1}\rangle, \langle v_0,\ldots,v_{\ell-1}\rangle\big) \stackrel{\text{def}}{=} \langle u_0 \times v_0,\ldots,u_{\ell-1} \times v_{\ell-1}\rangle$$

on packed ciphertexts in time $\tilde{O}((\ell + \lambda)(\log|\mathbb{K}_n|))$ where λ is the security parameter (with addition and multiplication in $\mathbb{K}_n$).[3] Specifically, if the size of our plaintext space is polynomially bounded and we set $\ell = \Theta(\lambda)$, then we can evaluate the above operations homomorphically in time $\tilde{O}(\lambda)$.

Unfortunately, component-wise ℓ-Add and ℓ-Mult are not sufficient to perform arbitrary computations on encrypted arrays, since data at different indexes within the arrays can never interact. To get a *complete set of operations for arrays*, we introduce the ℓ-Permute operation that can arbitrarily permute the data within the ℓ-element arrays. Namely, for any permutation π over the indexes $I_\ell = \{0, 1, \ldots, \ell - 1\}$, we want to homomorphically evaluate the function

$$\ell\text{-Permute}_\pi\big(\langle u_0,\ldots,u_{\ell-1}\rangle\big) = \langle u_{\pi(0)},\ldots,u_{\pi(\ell-1)}\rangle.$$

on a packed ciphertext, with complexity similar to the above. We will show how to implement ℓ-Permute homomorphically in Sections 3 and 4 below. For now, we just assume that such an implementation is available and show how to use it to obtain low-overhead implementation of general circuits.

---

[3] To compute L levels of such operations, the complexity expression becomes $\tilde{O}((\ell + \lambda)(L + \log|\mathbb{K}_n|))$.

2.1   Computing with ℓ-Fold Gates

We are interested in computing arbitrary functions using "ℓ-fold gates" that operate on ℓ-element arrays as above. We assume that the function $f(\cdot)$ to be computed is specified using a fan-in-2 arithmetic circuit with t "normal" arithmetic gates (that operate on singletons). Our goal is to implement f using as few ℓ-fold gates as possible, hopefully not much more than t/ℓ of them.

We assume that the input to f is presented in a packed form, namely when computing an r-variate function $f(x_1, \ldots, x_r)$ we get as input $\lceil r/\ell \rceil$ arrays (indexed $A_0, \ldots, A_{\lceil r/\ell \rceil}$) with the j'th array containing the input elements $x_{j\ell}$ through $x_{j\ell+\ell-1}$. The last array may contain less than ℓ elements, and the unused entries contain "don't care" elements. In fact, throughout the computation we allow all of the arrays to contain "don't care" entries. We say that an array is *sparse* if it contains $\ell/2$ or more "don't care" entries. We maintain the invariant that our collection of arrays is always at least half full, i.e., we hold r values using at most $\lceil 2r/\ell \rceil$ ℓ-element arrays.

The gates that we use in the computation are the ℓ-Add, ℓ-Mult, and ℓ-Permute gates from above. The rest of this section is devoted to establishing the following theorem:

Theorem 1. *Let ℓ, t, w and W be parameters. Then any t-gate fan-in-2 arithmetic circuit C with average width w and maximum width W, can be evaluated using a network of $O\big(\lceil t/\ell \rceil \cdot \lceil \ell/w \rceil \cdot \log W \cdot \mathrm{polylog}(\ell)\big)$ ℓ-fold gates of types ℓ-Add, ℓ-Mult, and ℓ-Permute. The depth of this network of ℓ-fold gates is at most $O(\log W)$ times that of the original circuit C, and the description of the network can be computed in time $\tilde{O}(t)$ given the description of C.*

Before turning to proving Theorem 1, we point out that Theorem 1 implies that if the original circuit C has size $t = \mathrm{poly}(\lambda)$, depth L, and average width $w = \Omega(\lambda)$, and if we set the packing parameter as $\ell = \Theta(\lambda)$, then we get an $O(L \cdot \log \lambda)$-depth implementation of C using $O(t/\lambda \cdot \mathrm{polylog}(\lambda))$ ℓ-fold gates. If implementing each ℓ-fold gate takes $\tilde{O}(L\lambda)$ time, then the total time to evaluate C is no more than

$$O\big(\frac{t}{\lambda}\mathrm{polylog}(\lambda) \cdot L \cdot \lambda \cdot \mathrm{polylog}(\lambda)\big) = O(t \cdot L \cdot \mathrm{polylog}(\lambda)).$$

Therefore, with this choice of parameter (and for "wide enough" circuits of average width $\Omega(\lambda)$), our overhead for evaluating depth-L circuits is only $O(L \cdot \mathrm{polylog}(\lambda))$. And if L is also polylogarithmic, as in BGV with bootstrapping [5], then the total overhead is polylogarithmic in the security parameter.

The high-level idea of the proof of Theorem 1 is what one would expect. Consider an arbitrary fan-in two arithmetic circuit C. Suppose that we have $\approx w$ output wire values of level $i-1$ packed into roughly w/ℓ arrays. We need to route these output values to their correct input positions at level i. It should be obvious that the ℓ-Permute gates facilitate this routing, except for two complications:

1. The mapping from outputs of level $i-1$ to inputs of level i is not a permutation. Specifically, level-$(i-1)$ gates may have high fan-out, and so some of the output values may need to be *cloned*.

2. Once the output values are cloned sufficiently (for a total of, say, w' values), routing to level i apparently calls for a *big permutation* over w' elements, not just a small permutation within arrays of ℓ elements.

Below we show that these complications can be handled efficiently.

2.2 Permutations over Hyper-rectangles

First, consider the second complication from above – namely, that we need to perform a permutation over some w elements (possibly $w \gg \ell$) using ℓ-Add, ℓ-Mult, and ℓ-Permute operations that only work on ℓ-element arrays. We use the following basic fact (cf. [18]).

Lemma 1. *Let $S = \{0, \ldots, a-1\} \times \{0, \ldots, b-1\}$ be a set of ab positions, arranged as a matrix of a rows and b columns. For any permutation π over S, there are permutations π_1, π_2, π_3 such that $\pi = \pi_3 \circ \pi_2 \circ \pi_1$ (that is, π is the composition of the three permutations) and such that π_1 and π_3 only permute positions within each column (these permutations only change the row, not the column, of each element) and π_2 only permutes positions within each row. Moreover, there is a polynomial-time algorithm that given π outputs the decomposition permutations π_1, π_2, π_3.*

In our context, Lemma 1 says that if we have w elements packed into $k = \lceil w/\ell \rceil$ ℓ-element arrays, we can express any permutation π of these elements as $\pi = \pi_3 \circ \pi_2 \circ \pi_1$ where π_2 invokes ℓ-Permute (k times in parallel) to permute data within the respective arrays, and π_1, π_3 only permute (ℓ times in parallel) elements that share the same index within their respective arrays. In Section 2.3, we describe how to implement π_1, π_3 using ℓ-Add and ℓ-Mult, and analyze the overall efficiency of implementing π. The following generalization of Lemma 1 to higher dimensions will be used later in this work. It is proved by invoking Lemma 1 recursively.

Lemma 2. *Let $S = I_{n_1} \times \cdots \times I_{n_k}$ where $I_{n_i} = \{0, \ldots, n_i - 1\}$. (Each element in S has k coordinates.) For any permutation π over S, there are permutations $\pi_1, \ldots, \pi_{2k-1}$ such that $\pi = \pi_{2k-1} \circ \cdots \circ \pi_1$ and such that π_i affects only the i-th coordinate for $i \leq k$ and only the $(2k-i)$-th coordinate for $i \geq k$.*

2.3 Batch Selections, Swaps, and Permutation Networks

We now describe how to use ℓ-Add and ℓ-Mult to realize the outer permutations π_1, π_3, which permute (ℓ times in parallel) elements that share the same index within their respective arrays. To perform these permutations, we can apply a *permutation network* à la Beneš/Waksman [15, 16]. Recall that a r-dimensional Beneš network consists of two back-to-back butterfly networks. Namely it is a $(2r-1)$-level network with 2^r nodes in each level, where for $i = 1, 2, \ldots, 2r-1$, we have an edge connecting node j in level $i-1$ to node j' in level i if the indexes j, j' are either equal (a "straight edge") or they differ in only in the $|r-i|$'th bit (a "cross edge"). The following lemma is an easy corollary of Lemma 2.

Lemma 3. [19, Thm 3.11] *Given any one-to-one mapping π of 2^r inputs to 2^r outputs in an r-dimensional Beneš network (one input per level-0 node and one output per level-$(2r-1)$ node), there is a set of node-disjoint paths from the inputs to the outputs connecting input i to output $\pi(i)$ for all i.*

In our setting, to implement our π_1 and π_3 from Lemma 1 we need to evaluate ℓ of these permutation networks in parallel, one for each index in our ℓ-fold arrays. Assume for simplicity that the number of ℓ-fold arrays is a power of two, say 2^r, and denote these arrays by $A_0, \ldots, A_{2^r-1}$, we would have a $(2r-1)$-level network, where the i'th level in the network consists of operating on pairs of arrays $(A_j, A_{j'})$, such that the indexes j, j' differ only in the $|r-i|$'th bit.

The operation applied to two such arrays $A_j, A_{j'}$ works separately on the different indexes of these arrays. For each $k = 0, 1, \ldots, \ell - 1$ the operation will either swap $A_j[k] \leftrightarrow A_{j'}[k]$ or will leave these two entries unchanged, depending on whether the paths in the k'th permutation network uses the cross edges or the straight edges between nodes j and j' in levels $i-1, i$ of the permutation network.

Thus, evaluating ℓ such permutation networks in parallel reduces to the following Select function: Given two arrays $A = [m_0, \ldots, m_{\ell-1}]$ and $A' = [m'_0, \ldots, m'_{\ell-1}]$ and a string $S = s_0 \cdots s_{\ell-1} \in \{0,1\}^\ell$, the operation $\mathsf{Select}_S(A, A')$ outputs an array $A'' = [m''_0, \ldots, m''_{\ell-1}]$ where, for each k, $m''_k = m_k$ if $s_k = 1$ and $m''_k = m'_k$ otherwise. It is easy to implement $\mathsf{Select}_S(A, A')$ using just the ℓ-Add and ℓ-Mult operations – in particular

$$\mathsf{Select}_S(A, A') = \ell\text{-Add}\left(\ \ell\text{-Mult}(A, S),\ \ell\text{-Mult}(A', \bar{S})\ \right)$$

where $\bar{S}$ is the bitwise complement of S. Note that $\mathsf{Select}_{\bar{S}}(A, A')$ outputs precisely the elements that are discarded by $\mathsf{Select}_S(A, A')$. So, $\mathsf{Select}_S(A, A')$ and $\mathsf{Select}_{\bar{S}}(A, A')$ are exactly like the arrays A' and A', except that some pairs of elements with identical indexes have been *swapped* – namely, those pairs at index k where $S_k = 0$. Hence we obtain the following lemma, whose proof is in the full version [12].

Lemma 4. *Evaluating ℓ permutation networks in parallel, each permuting k items, can be accomplished using $O(k \cdot \log k)$ gates of ℓ-Add and ℓ-Mult, and depth $O(\log k)$. Also, evaluating a permutation π over $k \cdot \ell$ elements that are packed into k ℓ-element arrays, can be accomplished using k ℓ-Permute gates and $O(k \log k)$ gates of ℓ-Add and ℓ-Mult, in depth $O(\log k)$. Moreover, there is an efficient algorithm that given π computes the circuit of ℓ-Permute, ℓ-Add, and ℓ-Mult gates that evaluates it, specifically we can do it in time $O(k \cdot \ell \cdot \log(k \cdot \ell))$.*

2.4   Cloning: Handling High Fan-Out in the Circuit

We have described how to efficiently realize a permutation over $w > \ell$ items using ℓ-Add, ℓ-Mult and ℓ-Permute gates that operate on ℓ-element arrays. However, the wiring between adjacent levels of a fan-in-two circuit are typically not permutations, since we typically have gates with high fan-out. We therefore need to clone the output values of these high-fan-out gates before performing a permutation that maps them to their input positions at the next level. We describe an efficient procedure for this "cloning" step.

A Cloning Procedure. The input to the cloning procedure consists of a collection of k arrays, each with ℓ slots, where each slot is either "full" (i.e., contains a value that we want to use) or "empty" (i.e., contains a don't-care value). We assume that initially more than $k \cdot \ell/2$ of the available slots are full, and will maintain a similar invariant throughout the procedure. Denote the number of full slots in the input arrays by w (with $k \cdot \ell/2 < w \leq k \cdot \ell$), and denote the i'th input value by v_i. The ordering of input values is arbitrary – e.g., we concatenate all the arrays and order input values by their index in the concatenated multi-array.

We are also given a set of positive integers $m_1, \ldots, m_w \geq 1$, such that v_1 should be duplicated m_1 times, v_2 should be duplicated m_2 times, etc. We say that m_i is the *intended multiplicity* of v_i. The total number of full slots in the output arrays will therefore be $w' \overset{\text{def}}{=} m_1 + m_2 + \cdots + m_w \geq w$. In more detail, the output of the cloning procedure must consist of some number k' of ℓ-slot arrays, where $k'\ell/2 < w' \leq k'\ell$, such that v_1 appears in at least m_1 of the output slots, v_2 appears in at least m_2 of the output slots, etc.

Denote the largest intended multiplicity of any value by $M = \max_i\{m_i\}$. The cloning procedure works in $\lceil \log M \rceil$ phases, such that after the j'th phase each value v_i is duplicated $\min(m_i, 2^j)$ times. Each phase consists of making a copy of all the arrays, then for values that occur too many times marking the excess slots as empty (i.e., marking the extra occurrences as don't-care values), and finally merging arrays that are "sparse" until the remaining arrays are at least half full. A simple way to merge two sparse arrays is to permute them so that the full slots appear in the left half in one array and the right half in the other, and then apply Select in the obvious way. A pseudo-code description of this procedure is given in Figure 1, whilst the proof of the following lemma is in the full version [12].

Lemma 5. *(i) The cloning procedure from Figure 1 is correct.*

(ii) Assuming that at least half the slots in the input arrays are full, this procedure can be implemented by a network of $O(w'/\ell \cdot \log(w'))$ ℓ-fold gates of type ℓ-Add, ℓ-Mult and ℓ-Permute, where w' is the total number of full slots in the output, $w' = \sum m_i$. The depth of the network is bounded by $O(\log w')$.

(iii) This network can be constructed in time $\tilde{O}(w')$, given the input arrays and the m_i's.

We also describe some more optimizations in the full version, including a different cloning procedure that improves on the complexity bound in Lemma 5. Putting all the above together we can efficiently evaluate a circuit using ℓ-Permute, ℓ-Add and ℓ-Mult, yielding a proof of Theorem 1, see the full version for details [12].

Input: k ℓ-slot arrays, $A_1, \ldots, A_k$, each of the $k \cdot \ell$ slots containing either a value
 or the special symbol '$\perp$', w positive integers $m_1, \ldots, m_w \geq 1$, where w is
 the number of full slots in the input arrays.
Output: k' ℓ-slot arrays, $A'_1, \ldots, A'_{k'}$, with each slot containing either a value or
 the special symbol '$\perp$', where $k'/2 \leq (\sum_i m_i)/\ell \leq k'$ and each input value
 v_i is replicated m_i times in the output arrays

0. Set $M \leftarrow \max_i \lceil m_i \rceil$
1. For $j = 1$ to $\lceil \log M \rceil$ // The j'th phase
2.   Make another copy of all the arrays   // Duplicate everything
3.   While there are values v_i with multiplicity more than m_i:
4.   Replace the excess occurrences of v_i by $\perp$ // Remove redundant entries
5.   While there exist pairs of arrays that have between them ℓ or more slots with $\perp$:
6.   Pick one such pair and merge the two arrays //Merge sparse arrays
7. Output the remaining arrays

Fig. 1. The cloning procedure

3 Permutation Networks from Abelian Group Actions

As we will show in Section 4, the algebra underlying our FHE scheme makes it possible to perform inexpensive operations on packed ciphertexts, that have the effect of permuting the ℓ plaintext slots inside this packed ciphertext. However, not every permutation can be realized this way; the algebra only gives us a small set of "simple" permutations. For example, in some cases, the given automorphisms "rotate" the plaintext slots, transforming a ciphertext that encrypts the vector $\langle v_0, \ldots, v_{\ell-1} \rangle$ into one that encrypts $\langle v_k, \ldots, v_{\ell-1}, v_0, \ldots, v_{k-1} \rangle$, for any value of k of our choosing. (See Section 3.2 for the general case.)

Our goal in this section is therefore to efficiently implement an ℓ-Permute$_\pi$ operation for an arbitrary permutation π using only the simple permutations that the algebra gives us (and also the ℓ-Add and ℓ-Mult operations that we have available). We begin in Section 3.1 by showing how to efficiently realize arbitrary permutations when the small set of "simple permutations" is the set of rotations. In Section 3.2 we generalize this construction to a more general set of simple permutations.

3.1 Permutation Networks from Cyclic Rotations and Swaps

Consider the Beneš permutation network discussed in Lemma 3. It has the interesting property that when the 2^r items being permuted are labeled with r-bit strings, then the i-th level only swaps (or not) pairs whose index differs in the $|r - i|$-th bit. In other words, the i-th level swaps only disjoint pairs that have offset $2^{|r-i|}$ from each other. We call this operation an "offset-swap", since all pairs of elements that might be swapped have the same mutual offset.

Definition 1 (Offset Swap). *Let $I_\ell = \{0, \ldots, \ell - 1\}$. We say that a permutation π over I_ℓ is an i-offset swap if it consists only of 1-cycles and 2-cycles (i.e., $\pi = \pi^{-1}$), and moreover all the 2-cycles in π are of the form $(k, k + i \bmod \ell)$ for different values $k \in I_\ell$.*

Offset swaps modulo ℓ are easy to implement by combining two rotations with the Select operation defined in Section 2.3. Specifically, for an i-offset swap, we need rotations by i and $-i$ mod ℓ and two Select operations. By Lemma 3, a Beneš network can realize any permutation over 2^r elements using $2r - 1$ levels where the i-th level is a $2^{|k-i|}$-offset swap modulo 2^r. An i-offset modulo 2^r, $\ell < 2^r < 2\ell$ can be cobbled together using a constant number of offset swaps modulo ℓ and Select operations, with offsets i and $2\ell - i$. Therefore, given a cyclic group of "simple" permutations $\mathcal{H}$ and Select operations, we can implement any permutation using a Beneš network with low overhead. Specifically, we prove the following lemma in the full version of this paper.

Lemma 6. *Fix an integer ℓ and let $k = \lceil \log \ell \rceil$. Any permutation π over $I_\ell = \{0, \ldots, \ell - 1\}$ can be implemented by a $(2k - 1)$-level network, with each level consisting of a constant number of rotations and Select operations on ℓ-arrays.*

Moreover, regardless of the permutation π, the rotations that are used in level i $(i = 1, \ldots, 2k - 1)$ are always exactly $2^{|k-i|}$ and $\ell - 2^{|k-i|}$ positions, and the network depends on π only via the bits that control the Select operations. Finally, this network can be constructed in time $\tilde{O}(\ell)$ given the description of π.

3.2   Generalizing to Sharply-Transitive Abelian Groups

Below, we extend our techniques above to deal with a more general set of "simple permutations" that we get from our ring automorphisms. (See Section 4)

Definition 2 (Sharply Transitive Permutation Groups). *Denote the ℓ-element symmetric group by $\mathcal{S}_\ell$ (i.e., the group of all permutations over $I_\ell = \{0, \ldots, \ell - 1\}$), and let $\mathcal{H}$ be a subgroup of $\mathcal{S}_\ell$. The subgroup $\mathcal{H}$ is sharply transitive if for every two indexes $i, j \in I_\ell$ there exists a unique permutation $h \in \mathcal{H}$ such that $h(i) = j$.*

Of course, the group of rotations is an example of an abelian and sharply transitive permutation group. It is abelian: rotating by k_1 positions and then by k_2 positions is the same as rotating by k_2 positions and then by k_1 positions. It is also sharply transitive: for all i, j there is a single rotation amount that maps index i to index j, namely rotation by $j - i$. However, it is certainly not the only example. We now explain how to efficiently realize arbitrary permutations using as building blocks the permutations from any sharply-transitive abelian group.

Recall that any abelian group is isomorphic to a direct product of cyclic groups, hence $\mathcal{H} \cong C_{\ell_1} \times \cdots \times C_{\ell_k}$ (where C_{ℓ_i} is a cyclic group with ℓ_i elements for some integers $\ell_i \geq 2$ where ℓ_i divides ℓ_{i+1} for all i). As any cyclic group with ℓ_i elements is isomorphic to $I_{\ell_i} = \{0, 1, \ldots, \ell_i - 1\}$ with the operation of addition mod ℓ_i, we will identify elements in $\mathcal{H}$ with vectors in the box $\mathcal{B} = I_{\ell_1} \times \cdots \times I_{\ell_k}$, where composing two group elements corresponds to adding their associated vectors (modulo the box). The group $\mathcal{H}$ is generated by the k unit vectors $\{e_r\}_{r=1}^k$ (where $e_r = \langle 0, \ldots, 0, 1, 0, \ldots, 0 \rangle$ with 1 in the r-th position). We stress that our group $\mathcal{H}$ has polynomial size, so we can efficiently compute the representation of elements in $\mathcal{H}$ as vectors in $\mathcal{B}$.

Since $\mathcal{H}$ is a sharply transitive group of permutations over the indexes $I_\ell = \{0, \ldots, \ell - 1\}$, we can similarly label the indexes in I_ℓ by vectors in $\mathcal{B}$: Pick an arbitrary index $i_0 \in I_\ell$, then for all $h \in \mathcal{H}$ label the index $h(i_0) \in I_\ell$ with the vector associated with h. This procedure labels every element in I_ℓ with exactly one vector from $\mathcal{B}$, since for every $i \in I_\ell$ there is a unique $h \in \mathcal{H}$ such that $h(i_0) = i$. Also, since $\mathcal{H} \cong \mathcal{B}$, we use all the vectors in $\mathcal{B}$ for this labeling ($|\mathcal{H}| = |\mathcal{B}| = \ell$). Note that with this labeling, applying the generator e_r to an index labeled with vector $\boldsymbol{v} \in \mathcal{B}$, yields an index labeled with $\boldsymbol{v}' = \boldsymbol{v} + e_r \bmod \mathcal{B}$. Namely we increment by one the r'th entry in $\boldsymbol{v}$ (mod ℓ_r), leaving the other entries unchanged.

In other words, rather than a one-dimensional array, we view I_ℓ as a k-dimensional matrix (by identifying it with $\mathcal{B}$). The action of the generator e_r on this matrix is to rotate it by one along the r-th dimension, and similarly applying the permutation $e_r^k \in \mathcal{H}$ to this matrix rotates it by k positions along the r-th dimension. For example, when $k = 2$, we view I_ℓ as an $\ell_1 \times \ell_2$ matrix, and the group $\mathcal{H}$ includes permutations of the form e_1^k that rotate all the columns of this matrix by k positions and also permutations of the form e_2^k that rotate all the rows of this matrix by k positions.

Using Lemma 6, we can now implement arbitrary permutations along the r'th dimension using a permutation network built from offset-swaps along the r'th dimension. Moreover, since the offset amounts used in the network do not depend on the specific permutation that we want to implement, we can use just one such network to implement in parallel different arbitrary permutations on different r'th-dimension sub-matrices. For example, in the 2-dimensional case, we can effect a different permutation on every column, yet realize all these different permutations using just one network of rotations and Selects, by using the same offset amounts but different Select bits for the different columns. More generally we can realize arbitrary (different) ℓ/ℓ_r permutations along all the different "generalized columns" in dimension-r, using a network of depth $O(\log \ell_r)$ consisting of permutations $h \in \mathcal{H}$ and ℓ-fold Select operations (and we can construct that network in time $\ell/\ell_r \cdot \tilde{O}(\ell_r) = \tilde{O}(\ell)$).

Once we are able to realize different arbitrary permutations along the different "generalized columns" in all the dimensions, we can apply Lemma 2. That lemma allows us to decompose any permutation π on I_ℓ into $2k - 1$ permutations $\pi = \pi_i \circ \cdots \circ \pi_{2k-1}$ where each π_i consists only of permuting the generalized columns in dimension $r = |k - i|$. Hence we can realize an arbitrary permutation on I_ℓ as a network of permutations $h \in \mathcal{H}$ and ℓ-fold Select operations, of total depth bounded by $2 \sum_{i=0}^{k-1} O(\log \ell_i) = O(\log \ell)$ (the last bound follows since $\ell = \prod_{i=0}^{k-1} \ell_i$). Also we can construct that network in time bounded by $2 \sum_{i=0}^{k-1} \tilde{O}(\ell_i) = \tilde{O}(\ell)$ (the bound follows since $k \leq \log \ell$). Concluding this discussion, we have:

Lemma 7. *Fix any integer ℓ and any abelian sharply-transitive group of permutations over I_ℓ, $\mathcal{H} \subset \mathcal{S}_\ell$. Then for every permutation $\pi \in \mathcal{S}_\ell$, there is a permutation network of depth $O(\log \ell)$ that realizes π, where each level of the network consists of a constant number of permutations from $\mathcal{H}$ and Select operations on ℓ-arrays.*

Moreover, the permutations used in each level do not depend on the particular permutation π, the network depends on π only via the bits that control the Select operations. Finally, this network can be constructed in time $\tilde{O}(\ell)$ given the description of π and the labeling of elements in $\mathcal{H}, I_\ell$ as vectors in $\mathcal{B}$.   □

Lemma 7 tells us that we can implement an arbitrary ℓ-Permute operation using a log-depth network of permutations $h \in \mathcal{H}$ (in conjunction with ℓ-Add and ℓ-Mult). Plugging this into Theorem 1 we therefore obtain:

Theorem 2. *Let ℓ, t, w and W be parameters, and let $\mathcal{H}$ be an abelian, sharply-transitive group of permutations over I_ℓ.*

Then any t-gate fan-in-2 arithmetic circuit C with average width w and maximum width W, can be evaluated using a network of $O\big(\lceil t/\ell \rceil \cdot \lceil \ell/w \rceil \cdot \log W \cdot \text{polylog}(\ell)\big)$ ℓ-fold gates of types ℓ-Add, ℓ-Mult, and $h \in \mathcal{H}$. The depth of this network of ℓ-fold gates is at most $O(\log W \cdot \log \ell)$ times that of the original circuit C, and the description of the network can be computed in time $\tilde{O}(t \cdot \log \ell)$ given the description of C.   □

4   FHE with Polylog Overhead

Theorem 2 implies that if we could efficiently realize ℓ-Add, ℓ-Mult, and $\mathcal{H}$-actions on packed ciphertexts (where $\mathcal{H}$ is a sharply transitive abelian group of permutations on ℓ-slot arrays), then we can evaluate arbitrary (wide enough) circuits with low overhead. Specifically, if we could set $\ell = \Theta(\lambda)$ and realize ℓ-Add, ℓ-Mult, and $\mathcal{H}$-actions in time $\tilde{O}(\lambda)$, then we can realize any circuit of average width $\Omega(\lambda)$ with just polylog(λ) overhead. It remains only to describe an FHE system that has the required complexity for these basic homomorphic operations.

4.1   The Basic Setting of FHE Schemes Based on Ideal Lattices and Ring LWE

Many of the known FHE schemes work over a polynomial ring $\mathbb{A} = \mathbb{Z}[X]/F(X)$, where $F(X)$ is irreducible monic polynomial, typically a cyclotomic polynomial. Ciphertexts are typically vectors (consisting of one or two elements) over $\mathbb{A}_q = \mathbb{A}/q\mathbb{A}$ where q is an integer modulus, and the plaintext space of the scheme is $\mathbb{A}_p = \mathbb{A}/p\mathbb{A}$ for some integer modulus $p \ll q$ with $\gcd(p, q) = 1$, for example $p = 2$. (Namely, the plaintext is represented as an integer polynomial with coefficients mod p.) Secret keys are also vectors over $\mathbb{A}_q$, and decryption works by taking the inner product $b \leftarrow \langle \mathbf{c}, \mathbf{s} \rangle$ in $\mathbb{A}_q$ (so b is an integer polynomial with coefficients in $(-q/2, q/2]$) then recovering the message as $b \bmod p$. Namely, the decryption formula is $[[\langle \mathbf{c}, \mathbf{s} \rangle \bmod F(X)]_q]_p$ where $[\cdot]_q$ denotes modular reduction into the range $(-q/2, q/2]$. Below we consider ciphertext vectors and secret-key vectors with two entries, since this is indeed the case for the variant of the BGV scheme [5] that we use.

Smart and Vercauteren [4] observed that the underlying ring structure of these schemes makes it possible to realize homomorphic (batch) Add and Mult operations, i.e. our ℓ-Add and ℓ-Mult. Specifically, though $F(X)$ is typically irreducible over $\mathbb{Q}$, it may nonetheless factor modulo p; $F(X) = \prod_{i=0}^{\ell-1} F_i(X) \bmod p$. In this case, the plaintext space of the scheme also factors: $\mathbb{A}_p = \otimes_{j=0}^{\ell-1} \mathbb{A}_{\mathfrak{p}_j}$ where $\mathfrak{p}_i$ is the ideal in $\mathbb{A}$ generated by p and $F_i(X)$. In particular, the Chinese Remainder Theorem applies, and the plaintext space is partitioned into ℓ independent non-interacting "plaintext slots", which is precisely what we need for component-wise ℓ-Add and ℓ-Mult. The decryption formula recovers the "aggregate plaintext" $a \leftarrow [[\langle \mathbf{c}, \mathbf{s} \rangle \bmod F(X)]_q]_p$, and this aggregate plaintext is decoded to get the individual plaintext elements, roughly via $z_j \leftarrow a \bmod (F_i(x), p) \in \mathbb{A}_{\mathfrak{p}_j}$.

4.2   Implementing Group Actions on FHE Plaintext Slots

While component-wise Add and Mult are straightforward, getting different plaintext slots to interact is more challenging. For ease of exposition, suppose at first that $F(X)$ is the degree-$(m-1)$ polynomial $\Phi_m(X) = (X^m - 1)/(X - 1)$ for m prime, and that $p \equiv 1 \pmod{m}$. Thus our ring $\mathbb{A}$ above is the mth cyclotomic number field. In this case $F(X)$ factors to linear terms modulo p, $F(X) = \prod_{i=0}^{\ell-1} (X - \rho_i) \pmod{p}$ with $\rho_i \in \mathbb{F}_p$. Hence we obtain $\ell = m - 1$ plaintext slots, each slot holding an element of the finite field $\mathbb{F}_p$ (i.e. in this case $\mathbb{A}_{\mathfrak{p}_i}$ above is equal to $\mathbb{F}_p$).

To get Φ_m to factor modulo p into linear terms we must have $p \equiv 1 \pmod{m}$, so $p > m$. Also we need $m = \Omega(\lambda)$ to get security (since m is roughly the dimension of the underlying lattice). This means that to get Φ_m to factor into linear terms we must use plaintext spaces that are somewhat large (in particular we cannot directly use $\mathbb{F}_2$). Later in this section we sketch the more elaborate algebra needed to handle the general (and practical) case of non-prime m and $p \ll m$, where Φ_m may not factor into linear terms. This is covered in more detail in the full version of this paper. For now, however, we concentrate on the simple case where Φ_m factors into linear terms modulo p.

Recall that ciphertexts are vectors over $\mathbb{Z}_q[X]/\Phi_m(X)$, so each entry in these vectors corresponds to an integer polynomial. Consider now what happens if we simply replace X with X^i inside all these polynomials, for some exponent $i \in \mathbb{Z}_m^*, i > 1$. Namely, for each polynomial $f(X)$, we consider $f^{(i)}(X) = f(X^i) \bmod \Phi_m(X)$. Notice that if we were using polynomial arithmetic modulo $X^m - 1$ (rather then modulo $\Phi_m(X)$) then this transformation would just permutes the coefficients of the polynomials. Namely $f^{(i)}$ has the same coefficients as f but in a different order, which means that if the coefficient vector of f has small norm then the same holds for the coefficient vector of $f^{(i)}$. In the full version we show that using a different notion of "size" of a polynomial (namely, the norm of the canonical embedding of a polynomial rather than the norm of its coefficient vector), we can conclude the same also for mod-Φ_m polynomial arithmetic. Namely, the mapping $f(X) \mapsto f(X^i) \bmod \Phi_m(X)$ does not change the "size" of the polynomial. To simplify presentation, below we describe everything in terms of coefficient vectors and arithmetic modulo $X^m - 1$. The actual mod-Φ_m implementation that we use is described in the full version of this paper [12].

Let us now consider the effect of the transformation $X \mapsto X^i$ on decryption. Let $\mathbf{c} = (c_0(X), c_1(X))$ and $\mathbf{s} = (s_0(X), s_1(X))$ be ciphertext and secret-key vectors, and let $b = \langle \mathbf{c}, \mathbf{s} \rangle \bmod (X^m - 1, q)$ and $a = b \bmod p$. Denote $\mathbf{c}^{(i)} = (c_0(X^i), c_1(X^i)) \bmod (X^m - 1)$, and define $\mathbf{s}^{(i)}$, $b^{(i)}$ and $a^{(i)}$ similarly. Since $\langle \mathbf{c}, \mathbf{s} \rangle = b \pmod{X^m - 1, q}$, we have that

$$c_0(X)s_0(X) + c_1(X)s_1(X) = b(X) + q \cdot r(X) + (X^m - 1)s(X) \quad \text{(over } \mathbb{Z}[X])$$

for some integer polynomials $r(X), s(X)$, and therefore also

$$c_0(X^i)s_0(X^i) + c_1(X^i)s_1(X^i) = b(X^i) + q \cdot r(X^i) + (X^{mi} - 1)s(X^i) \quad \text{(over } \mathbb{Z}[X]).$$

Since $X^m - 1$ divides $X^{mi} - 1$, then we also have

$$\left\langle \mathbf{c}^{(i)}, \mathbf{s}^{(i)} \right\rangle = b^{(i)} + q \cdot r(X^i) + (X^m - 1)S(X) \quad \text{(over } \mathbb{Z}[X])$$

for some $r(X), S(X)$. That is, $b^{(i)} = \langle \mathbf{c}^{(i)}, \mathbf{s}^{(i)} \rangle \bmod (X^m - 1, q)$. Clearly, we also have $a^{(i)} = b^{(i)} \pmod{p}$. This means that if $\mathbf{c}$ decrypts to the aggregate plaintext a under $\mathbf{s}$, then $\mathbf{c}^{(i)}$ decrypts to $a^{(i)}$ under $\mathbf{s}^{(i)}$! Then using key-switching we can get an encryption of $a^{(i)}$ back under $\mathbf{s}$ (or any other key). See the full version for more details [12].

But how does this new aggregate plaintext $a^{(i)}$ relate to the original a? Here we apply to Galois theory, which tells us that decoding the aggregate $a^{(i)}$ (which we do roughly by setting $z_j \leftarrow a^{(i)} \bmod (F_j, p)$), the set of z_j's that we get is exactly the same as when decoding the original aggregate a, albeit in different order. Roughly, this is because each of our plaintext slots corresponds to a root of the polynomial $F(X)$, and the transformations $X \mapsto X^i$, which are precisely the elements of the Galois group, permute these roots. In other words by transforming $\mathbf{c} \to \mathbf{c}^{(i)}$ (followed by key switching), we can permute the plaintext slots inside the packed ciphertext. Moreover, in our simplified case, the permutations have a single cycle – i.e., they are rotations of the slots. Arranging the slots appropriately we can get that the transformation $\mathbf{c} \to \mathbf{c}^{(i)}$ rotates the slots by exactly i positions, thus we get the group of rotations that we were using in Section 3.1. In general the situation is a little more complicated, but the above intuition still can be made to hold; for more details see the full version [12].

The General Case. In the general case, when m is not a prime, the polynomial $\Phi_m(X)$ has degree $\phi(m)$ (where $\phi(\cdot)$ is Euler's totient function), and it factors mod p into a number of same-degree irreducible factors. Specifically, the degree of the factors is the smallest integer d such that $p^d = 1 \pmod{m}$, and the number of factors is $\ell = \phi(m)/d$ (which is of course an integer), $\Phi_m(X) = \prod_{j=0}^{\ell-1} F_j(X)$. For us, it means that we have ℓ plaintext slots, each isomorphic to the finite field $\mathbb{F}_{p^d}$, and an aggregate plaintext is a degree-$(\phi(m) - 1)$ polynomial over $\mathbb{F}_p$.

Suppose that we want to evaluate homomorphically a circuit over some underlying field $\mathbb{K}_n = \mathbb{F}_{p^n}$, then we need to find an integer m such that $\Phi_m(X)$

factors mod p into degree-d factors, where d is divisible by n. This way we could directly embed elements of the underlying plaintext space $\mathbb{K}_n$ inside our plaintext slots that hold elements of $\mathbb{F}_{p^d}$, and addition and multiplication of plaintext slots will directly correspond to additions and multiplications of elements in $\mathbb{K}_n$. (This follows since $\mathbb{K}_n = \mathbb{F}_{p^n}$ is a subfield of $\mathbb{F}_{p^d}$ when n divides d.)

Note that each plaintext slot will only have $n \log p$ bits of relevant information, i.e., the underlying element of $\mathbb{F}_{p^n}$, but it takes $d \log p$ bits to specify. We thus get an "embedding overhead" factor of d/n even before we encrypt anything. We therefore need to choose our parameter m so as to keep this overhead to a minimum.

Even for a non-prime m, the Galois group $\mathcal{G}\text{al}(\mathbb{Q}[X]/\varPhi_m(X))$ consists of all the transformations $X \mapsto X^i$ for $i \in \mathbb{Z}_m^*$, hence there are exactly $\phi(m)$ of them. As in the simplified case above, if we have a ciphertext $\mathbf{c}$ that decrypts to an aggregate plaintext a under $\mathbf{s}$, then $\mathbf{c}^{(i)}$ decrypts to $a^{(i)}$ under $\mathbf{s}^{(i)}$. Differently from the simple case, however, not all members of the Galois group induce permutations on the plaintext slots, i.e., decoding the aggregate plaintext $a^{(i)}$ does not necessarily give us the same set of (permuted) plaintext elements as decoding the original a. Instead $\mathcal{G}\text{al}(\mathbb{Q}[X]/\varPhi_m(X))$ contains a subgroup $\mathcal{G} = \{(X \mapsto X^{p^j}) : j = 0, 1, \ldots, d-1\}$ corresponding to the Frobenius automorphisms[4] modulo p. This subgroup does not permute the slots at all, but the quotient group $\mathcal{H} = \mathcal{G}\text{al}/\mathcal{G}$ does. Clearly, $\mathcal{G}$ has order d and $\mathcal{H}$ has order $\phi(m)/d = \ell$. In the full version we show that the quotient group $\mathcal{H}$ acts as a transitive permutation group on our ℓ plaintext slots, and since it has order ℓ then it must be sharply transitive. In the general case we therefore use this group $\mathcal{H}$ as our permutation group for the purpose of Lemma 7. Another complication is that the automorphism that we can compute are elements of $\mathcal{G}\text{al}$ and not elements in the quotient group $\mathcal{H}$. In the full version we also show how to emulate the permutations in $\mathcal{H}$, via use of coset representatives in $\mathcal{G}\text{al}$.

4.3 Low-Overhead FHE

Given the background from above (and the modification of the BGV cryptosystem [7] described in the full version), we explain in the full version how to set the parameters for our variant of the BGV scheme so as to get low-overhead FHE scheme. This gives us:

Theorem 3. *For security parameter λ, any t-gate, depth-L arithmetic circuit of average width $\Omega(\lambda)$ over underlying plaintext space $\mathbb{F}_{p^n}$ (with $p^n \leq poly(\lambda)$) can be evaluated homomorphically in time $t \cdot \tilde{O}(L) \cdot polylog(\lambda)$.*

Theorem 3 implies that we can implement shallow arithmetic circuit with low overhead, but when the circuit gets deeper the dependence of the overhead on L causes the overhead to increase. Recall that the reason for this dependence on the depth is that in the BGV cryptosystem [5], the moduli get smaller as we go

---

[4] The group G is called the *decomposition group* at p in the literature.

up the circuit, which means that for the first layers of the circuit we must choose moduli of bitsize $\Omega(L)$.

As explained in [5], the dependence on the depth can be circumvented by using bootstrapping. Namely, we can start with a modulus which is not too large, then reduce it as we go up the circuit, and once the modulus become too small to do further computation we can bootstrap back into the larger-modulus ciphertexts, then continue with the computation.

For our purposes, we need to ensure that we bootstrap often enough to keep the moduli small, and yet that the time we spend on bootstrapping does not significantly impact the overhead. Here we apply to the analysis from [5], that shows that a packed ciphertext with $\tilde{\Omega}(\lambda)$ slots can be decrypted using a circuit of size $\tilde{O}(\lambda)$ and depth polylog(λ). Hence we can even bootstrap after every layer of the circuit and still keep the overhead polylogarithmic, and the moduli never grow beyond polylogarithmic bitsize. We thus get:

Theorem 4. *For security parameter λ, any t-gate arithmetic circuit of average width $\Omega(\lambda)$ over underlying plaintext space $\mathbb{F}_{p^n}$ (with $p^n \leq poly(\lambda)$) can be evaluated homomorphically in time $t \cdot polylog(\lambda)$.*

Acknowledgments. The first and second authors are sponsored by DARPA and ONR under agreement number N00014-11C-0390. The U.S. Government is authorized to reproduce and distribute reprints for Governmental purposes notwithstanding any copyright notation thereon. The views and conclusions contained herein are those of the authors and should not be interpreted as necessarily representing the official policies or endorsements, either expressed or implied, of DARPA, or the U.S. Government. Distribution Statement "A" (Approved for Public Release, Distribution Unlimited).

The third author is sponsored by DARPA and AFRL under agreement number FA8750-11-2-0079. The same disclaimers as above apply. He is also supported by the European Commission through the ICT Programme under Contract ICT-2007-216676 ECRYPT II and via an ERC Advanced Grant ERC-2010-AdG-267188-CRIPTO, by EPSRC via grant COED–EP/I03126X, and by a Royal Society Wolfson Merit Award. The views and conclusions contained herein are those of the authors and should not be interpreted as necessarily representing the official policies or endorsements, either expressed or implied, of the European Commission or EPSRC.

References

1. Rivest, R., Adleman, L., Dertouzos, M.L.: On data banks and privacy homomorphisms. In: Foundations of Secure Computation, pp. 169–180 (1978)
2. Gentry, C.: Fully homomorphic encryption using ideal lattices. In: Mitzenmacher, M. (ed.) STOC, pp. 169–178. ACM (2009)
3. Gentry, C.: A fully homomorphic encryption scheme. PhD thesis, Stanford University (2009), http://crypto.stanford.edu/craig

4. Smart, N.P., Vercauteren, F.: Fully homomorphic SIMD operations (2011) (manuscript), http://eprint.iacr.org/2011/133
5. Brakerski, Z., Gentry, C., Vaikuntanathan, V.: Fully homomorphic encryption without bootstrapping. In: The 3rd Innovations in Theoretical Computer Science Conference, ITCS (2012), Full version at, http://eprint.iacr.org/2011/277
6. Lyubashevsky, V., Peikert, C., Regev, O.: On Ideal Lattices and Learning with Errors over Rings. In: Gilbert, H. (ed.) EUROCRYPT 2010. LNCS, vol. 6110, pp. 1–23. Springer, Heidelberg (2010)
7. Brakerski, Z., Vaikuntanathan, V.: Fully Homomorphic Encryption from Ring-LWE and Security for Key Dependent Messages. In: Rogaway, P. (ed.) CRYPTO 2011. LNCS, vol. 6841, pp. 505–524. Springer, Heidelberg (2011)
8. Brakerski, Z., Vaikuntanathan, V.: Efficient fully homomorphic encryption from (standard) LWE. In: FOCS. IEEE Computer Society (2011)
9. Lauter, K., Naehrig, M., Vaikuntanathan, V.: Can homomorphic encryption be practical? In: ACM Workshop on Cloud Computing Security, pp. 113–124 (2011)
10. Stehlé, D., Steinfeld, R.: Faster Fully Homomorphic Encryption. In: Abe, M. (ed.) ASIACRYPT 2010. LNCS, vol. 6477, pp. 377–394. Springer, Heidelberg (2010)
11. Gentry, C., Halevi, S.: Implementing Gentry's Fully-Homomorphic Encryption Scheme. In: Paterson, K.G. (ed.) EUROCRYPT 2011. LNCS, vol. 6632, pp. 129–148. Springer, Heidelberg (2011)
12. Gentry, C., Halevi, S., Smart, N.P.: Fully homomorphic encryption with polylog overhead (2011), Full version at http://eprint.iacr.org/2011/566
13. Smart, N.P., Vercauteren, F.: Fully Homomorphic Encryption with Relatively Small Key and Ciphertext Sizes. In: Nguyen, P.Q., Pointcheval, D. (eds.) PKC 2010. LNCS, vol. 6056, pp. 420–443. Springer, Heidelberg (2010)
14. Hennessy, J.L., Patterson, D.A.: Computer Architecture: A Quantitative Approach, 4th edn. Morgan Kaufmann (2006)
15. Beneš, V.E.: Optimal rearrangeable multistage connecting networks. Bell System Technical Journal 43, 1641–1656 (1964)
16. Waksman, A.: A permutation network. J. ACM 15, 159–163 (1968)
17. Damgård, I., Ishai, Y., Krøigaard, M.: Perfectly Secure Multiparty Computation and the Computational Overhead of Cryptography. In: Gilbert, H. (ed.) EUROCRYPT 2010. LNCS, vol. 6110, pp. 445–465. Springer, Heidelberg (2010)
18. Lev, G., Pippenger, N., Valiant, L.: A fast parallel algorithm for routing in permutation networks. IEEE Transactions on Computers C-30, 93–100 (1981)
19. Leighton, F.T.: Introduction to parallel algorithms and architectures: arrays, trees, hypercubes, 2nd edn. M. Kaufmann Publishers (1992)

Multiparty Computation
with Low Communication, Computation
and Interaction via Threshold FHE*

Gilad Asharov[1,**], Abhishek Jain[2], Adriana López-Alt[3], Eran Tromer[4,***],
Vinod Vaikuntanathan[5,†], and Daniel Wichs[6]

[1] Bar-Ilan University
[2] University of California Los Angeles (UCLA)
[3] New York University (NYU)
[4] Tel Aviv University
[5] University of Toronto
[6] IBM Research, T.J. Watson

Abstract. Fully homomorphic encryption (FHE) enables secure computation over the encrypted data of a single party. We explore how to extend this to multiple parties, using *threshold* fully homomorphic encryption (TFHE). In such scheme, the parties jointly generate a common FHE public key along with a secret key that is shared among them; they can later cooperatively decrypt ciphertexts without learning anything but the plaintext. We show how to instantiate this approach efficiently, by extending the recent FHE schemes of Brakerski, Gentry and Vaikuntanathan (CRYPTO '11, FOCS '11, ITCS '12) based on the *(ring) learning with errors* assumption. Our main tool is to exploit the property that such schemes are additively homomorphic over their *keys*.

Using TFHE, we construct simple multiparty computation protocols secure against fully malicious attackers, tolerating any number of corruptions, and providing security in the universal composability framework. Our protocols have the following properties: **Low interaction**: 3 rounds of interaction given a common random string, or 2 rounds with a public-key infrastructure. **Low communication**: independent of the function being computed (proportional to just input and output sizes). **Cloud-assisted computation**: the bulk of the computation can be efficiently outsourced to an external entity (e.g. a cloud service) so that the computation of all other parties is independent of the complexity of the evaluated function.

---

* This paper is a merger of two independent but largely overlapping works [3,25]. The respective full versions are posted on ePrint.
** Supported by the European Research Council as part of the ERC project LAST.
*** This work was partially supported by the Check Point Institute for Information Security and by the Israeli Centers of Research Excellence (I-CORE) program (center No. 4/11).
† This work was partially supported by an NSERC Discovery Grant and by DARPA under Agreement number FA8750-11-2-0225.

D. Pointcheval and T. Johansson (Eds.): EUROCRYPT 2012, LNCS 7237, pp. 483–501, 2012.

1   Introduction

Multiparty Computation. Secure multiparty computation (MPC) allows multiple participants to evaluate a common function over their inputs *privately*, without revealing the inputs to each other. This problem was initially studied by Yao [33,34] who gave a protocol for the case of *two semi-honest* that follow the protocol specification honestly but wish to learn as much information as possible, and Goldreich, Micali and Wigderson [19] extended this to *many fully malicious* parties that may arbitrarily deviate from the protocol specification. Since then, the problem of MPC has become a fundamental question in cryptography. Interestingly, on a very high level, most prior results for general MPC can be seen as relying in some way on the original techniques of [34,19].

Fully Homomorphic Encryption. A very different approach to secure computation relies on *fully homomorphic encryption (FHE)*. An FHE scheme allows us to perform arbitrary computations on encrypted data without decrypting it. Although the idea of FHE goes back to Rivest et al. [31], the first implementation is due to the recent breakthrough of Gentry [17], and has now been followed with much exciting activity, most recently with quite simple and efficient schemes [10,9,8]. Using FHE, we immediately get an alternative approach to MPC in the case of two semi-honest parties (Alice and Bob): Alice encrypts her input under her own key and sends the ciphertext to Bob, who then evaluates the desired function homomorphically on Alice's ciphertext and his own input, sending (only) the final encrypted result back to Alice for decryption. This approach has several benefits over prior ones. Perhaps most importantly, the *communication complexity* of the protocol and Alice's *computation* are small and only proportional to Alice's input/output sizes, independent of the complexity of the function being evaluated. Moreover, the protocol consists of only *two rounds of interaction*, which is optimal (matching [34]).[1]

MPC via Threshold FHE. Since FHE solves the secure computation problem for two semi-honest parties, it is natural to ask whether we can extend the above template to the general case of *many fully malicious* parties. Indeed, there is a simple positive answer to this question (as pointed out in e.g. [17]) by using a *threshold fully homomorphic encryption* (TFHE). This consists of a *key generation protocol* where the parties collaboratively agree on a common public key of an FHE scheme and each party also receives a share of the secret key. The parties can then encrypt their individual inputs under the common public key, evaluate the desired function homomorphically on the ciphertexts, and collaboratively execute a *decryption protocol* on the result to learn the output of the computation. Moreover, it is possible to convert any FHE scheme

---

[1] Indeed, Yao's garbled circuits can be thought of as instantiating an FHE with long ciphertexts (see e.g. [18]).

into TFHE by implementing the above key-generation and decryption protocols using general MPC compilers (e.g. [19]). Although this approach already gives the communication/computation savings of FHE, it suffers from two main problems: (1) It does not preserve *round complexity* since generic implementations of the key-generation and decryption protocols will each require many rounds of interaction. (2) It uses the "heavy machinery" of generic MPC compilers and zero-knowledge proofs on top of FHE and is unlikely to yield practical solutions.

1.1   Our Results

In this work, we present an efficient threshold FHE (TFHE) scheme under the learning with errors (LWE) assumption, based on the FHE constructions of Brakerski, Gentry and Vaikuntanathan [9,8]. Our starting observation is that basic LWE-based encryption ([30]) is *key homomorphic*, where summing up several public/secret key pairs (pk_i, sk_i) results in a new valid public/secret key pair $(pk^*, sk^*) = \sum_i (pk_i, sk_i)$. Therefore, if each party broadcasts its own public-key pk_i, and we define the common public key as the sum $pk^* = \sum_i pk_i$, then each party already holds a share sk_i of the common secret key sk^*. Moreover, if each party decrypts a ciphertext c under pk^* with its individual share sk_i, then these partial decryptions can be summed up to recover the message. This gives us simple key-generation and decryption protocols, consisting of one round each. Unfortunately, the above discussion is oversimplified and its implementation raises several challenges, which we are forced to overcome.

Main Challenges of TFHE. The first challenge in instantiating the above idea is that summing-up key pairs as above does not result in a correctly distributed fresh key pair, and summing up decryption shares may reveal more than just the plaintext. Nevertheless, we show the security of this basic approach when augmented with a technique we call *smudging*, in which parties add large noise during important operations so as to "smudge out" small differences in distributions. Perhaps our main challenge is that, in LWE-based FHE schemes, the public key must also contain additional information in the form of an *evaluation key*, which is needed to perform homomorphic operations on ciphertexts. Although the above key-homomorphic properties hold for the public *encryption keys* of the FHE, the evaluation keys have a more complex structure making it harder to combine them. Nevertheless, we show that it is possible to generate the evaluation keys in a threshold manner by having each party carefully release some extra information about its individual secret-key and then cleverly combining this information. Although this forces us to add an extra round to the key-generation protocol in order to generate the evaluation key, the parties can already encrypt their inputs after the first round. Therefore, we get MPC protocol consisting of only 3 broadcast rounds: (Round I) generate encryption key, (Round II) generate evaluation key & encrypt inputs, (Round III) perform homomorphic evaluation locally and decrypt the resulting ciphertext.

Using TFHE for MPC. Our basic TFHE protocol allows us to achieve MPC in the semi-honest model. To transform it to the fully malicious setting, we could use generic techniques consisting of: (1) coin-flipping for the random coins of each party, and (2) having each party prove at each step that it is following the protocol honestly (using the random coins determined by the coin-flip) by a zero-knowledge (ZK) proof of knowledge. Unfortunately, even if we were to use non-interactive zero knowledge (NIZK) in the common-random string (CRS) model for the proofs, the use of coin-flipping would add two extra rounds. Interestingly, we show that coin-flipping is *not* necessary. We do so by showing that our basic MPC protocol is already secure against a stronger class of attackers that we call *semi-malicious*: such attackers follow the protocol honestly but with adaptively and adversarially chosen random coins in each round. We can now generically convert our MPC in the semi-malicious setting to a fully secure one using (UC) NIZKs [32] while *preserving the round complexity*. This gives the first 3 round protocol for general MPC in the CRS model (while achieving UC security for free). Instantiating the above approach with general UC NIZKs proofs might already achieve asymptotic efficiency, but it has little hope of yielding practical protocols. Therefore, we also build efficient Σ-protocols for the necessary relations, which we can then compile into efficient UC NIZKs in the *random-oracle (RO)* model. Therefore, we can get a reasonably *efficient* and very *simple* 3-round protocol for general MPC in the RO model.

Cloud-Assisted MPC. We notice that our protocol can also be easily adapted to the setting of "cloud-assisted computation", where an (untrusted) external entity (e.g. "the cloud") is tasked with performing the homomorphic evaluation over the publicly broadcast ciphertexts. This results in a protocol where the computation of all other parties is small and independent of the size of the evaluated function! This approach only incurs one additional round in which the server broadcasts the ciphertext. To get security against a fully malicious server, we also require the existence of *succinct non-interactive argument systems*.

Public-Key Infrastructure. Our approach also yields 2-round MPC in the *public-key infrastructure* (PKI) setting, by thinking of each party's original (Round I) message as its public key and the randomness used to generate it as the secret key. This gives the first *two*-round MPC construction in the PKI setting. We note that the PKI can be *reused* for many MPC executions of arbitrary functions and arbitrary inputs.

We summarize the above discussion with the following informal theorem:

Main Theorem. (informal) *Under the LWE assumption and the existence of UC NIZKs, for any function f there exists a protocol realizing f that is UC-secure in the presence of a (static) malicious adversary corrupting any number of parties. The protocol consists of **3 rounds** of broadcast in the CRS model, or **2 rounds** in a PKI model. Under an additional "circular security" assumption, its **communication** complexity is independent of the size of the evaluated circuit. In the "cloud-assisted setting" the total **computation** of each party (other than the cloud) is independent of the complexity of f.*

1.2   Related Work

In the context of general MPC, starting from the original proposal of Yao [34], there has been a rich line of work studying the round-complexity of secure multi-party computation protocols. In the semi-honest case, Beaver, Micali and Rogaway [4] gave the first constant-round protocol, which is asymptotically optimal. An alternative approach using randomized polynomials was also given by [22,2]. Although the concrete constants were not explicitly stated, they seem to require at least 4 rounds.In the fully malicious case there is a lower bound of 5 rounds in the plain model (dishonest majority) [24], but it does not seem to extend to the CRS model or other setup models. In the CRS model, we can generically compile semi-honest secure protocols into the fully malicious model using coin-tossing and (UC) NIZKs [32], at the cost of adding *two extra rounds* for the coin-toss. Therefore, the best prior works seem to require at least 6 rounds in the CRS model, although the exact constants were never carefully analyzed.

Recently, Choi et al [11] obtained a UC secure protocol in a *"pre-processing"* model with a 2-round online stage. However, the pre-processing requires "expensive" computation of garbled circuits and can later be only used once for a single online computation (it is *not* reusable). In contrast, our results give a 2-round UC-protocol in the PKI model, which we can think of as "pre-processing" that is only performed once and may be reused for arbitrarily many computations.

The works of [12,15,6,13] use *additively* and *somewhat* homomorphic encryption to get some of the most practically efficient MPC implementations. However, since the schemes are not fully homomorphic, the round and communication complexity in these works is large and linear in the depth of the circuit.

The work of Bendlin and Damgård [5] builds a threshold version of [30] encryption based on LWE and ZK protocols for plaintext knowledge. Indeed, the main ideas behind our *decryption* protocol, such as the idea of using extra noise for "smudging", come from that work. We seem to avoid some of the main difficulties of [5] by analyzing the security of our threshold scheme directly within the application of MPC rather than attempting to realize ideal key-generation and decryption functionalities. However, we face a very different set of challenges in setting up the complicated evaluation key needed for FHE.

In a concurrent and independent work, Myers, Sergi and shelat [28] instantiate a threshold FHE scheme based on the "approximate-integer GCD" problem, and use it to build an explicit MPC protocol whose communication complexity is independent of the circuit size. Perhaps due to the amazing versatility and simplicity of LWE, our scheme enjoys several benefits over that of [28], which only works in the setting of an honest majority and suffers from a large (constant) round-complexity. Most importantly, we believe that our protocol is significantly simpler to describe and understand.

The idea of using a cloud to alleviate the computational efforts of parties was recently explored in the work on "server-aided MPC" by Kamara, Mohassel and Raykova [23]. Their protocols, however, require some of the parties to do a large amount of computation, essentially proportional to the size of the function f being computed. Halevi, Lindell and Pinkas [21] recently considered the model of "secure computation on the web" which gets rid of all interaction between the actual parties, and instead only allows each party to "log in" once to interact with the server. Unfortunately, this necessitates a *weaker* notion of security which is only meaningful for a small class of functions. In contrast, we focus here on *standard* MPC security for arbitrary functions, at the cost of additional interaction. In particular, we achieve full security in the model where the computation occurs in 2 stages (optimal) and each party "logs in" once per stage to post a message to the server. As an additional benefit, the server does not do any processing on messages until the end of each stage. Thus, the parties may, in fact, "log in" *concurrently* in each stage, unlike [21] where the parties must "log in" *sequentially*.

1.3   Organization

In the proceedings version of this work, we follow the exposition of [3], and all omitted proofs can be found there. See [25] for an alternative exposition, using somewhat different abstractions and a variant of the scheme presented here under the *ring LWE* assumption.

In Section 3 we start with a basic LWE-based encryption scheme, highlight its homomorphic properties, and describe how to use it to get the FHE schemes of [9,8]. In Section 4 we then describe our threshold FHE scheme, and in Section 5 we use it to build an MPC protocol. We then discuss several variants of this protocol in Section 6.

2   Preliminaries

Throughout, we let κ denote the *security parameter* and $\mathsf{negl}(\kappa)$ denote a negligible function. For integers n, q, we define $[n]_q$ to be the unique integer $v \in (-q/2, q/2]$ s.t. $n \equiv v \pmod{q}$. Let $\mathbf{x} = (x_1, \ldots, x_n) \in \mathbb{Z}^n$ be a vector. We use the notation $\mathbf{x}[i] \stackrel{\text{def}}{=} x_i$ to denote the ith component scalar. To simplify the descriptions of our schemes, we also abuse notation and define $\mathbf{x}[0] \stackrel{\text{def}}{=} 1$. The ℓ_1-norm of $\mathbf{x}$ is defined as $\ell_1(\mathbf{x}) \stackrel{\text{def}}{=} \sum_{i=1}^n |x_i|$. For a distribution ensemble $\chi = \chi(\kappa)$ over the integers, and integers bounds $B = B(\kappa)$, we say that χ is B-*bounded* if $\Pr_{x \leftarrow \chi(\kappa)}[|x| > B(\kappa)] \leq \mathsf{negl}(\kappa)$. We rely on the following lemma, which says that adding large noise "smudges out" any small values.

Lemma 1 (Smudging). *Let $B_1 = B_1(\kappa)$, and $B_2 = B_2(\kappa)$ be positive integers and let $e_1 \in [-B_1, B_1]$ be a fixed integer. Let $e_2 \stackrel{\$}{\leftarrow} [-B_2, B_2]$ be chosen uniformly at random. Then the distribution of e_2 is statistically indistinguishable from that of $e_2 + e_1$ as long as $B_1/B_2 = \mathsf{negl}(\kappa)$.*

Learning With Errors. The *decisional learning with errors (LWE)* problem, introduced by Regev [30], is defined as follows.

Definition 2 (LWE [30]). *Let κ be the security parameter, $n = n(\kappa), q = q(\kappa)$ be integers and let $\chi = \chi(\kappa), \varphi = \varphi(\kappa)$ be distributions over $\mathbb{Z}$. The* $\mathrm{LWE}_{n,q,\varphi,\chi}$

assumption says that no poly-time distinguisher can distinguish between the following two distributions on tuples $(\mathbf{a}_i, b_i)$, *given polynomially many samples:* **Distribution I.** *Each* $(\mathbf{a}_i, b_i) \xleftarrow{\$} \mathbb{Z}_q^{n+1}$ *is chosen independently, uniformly at random.* **Distribution II.** *Choose* $\mathbf{s} \leftarrow \varphi^n$. *Each sample* $(\mathbf{a}_i, b_i)$ *is chosen as:* $\mathbf{a}_i \xleftarrow{\$} \mathbb{Z}_q^n$, $e_i \leftarrow \chi$, $b_i := \langle \mathbf{a}_i, \mathbf{s} \rangle + e_i$.

The works of [30,29] show that the LWE problem is as hard as approximating short vector problems in lattices (for appropriate parameters) when χ is a Gaussian with "small" standard deviation and $\varphi = U(\mathbb{Z}_q)$ is the uniform distribution over $\mathbb{Z}_q$. The work of [1] shows that, when q is a prime power, then $\mathrm{LWE}_{n,q,\chi,\chi}$ is as hard as $\mathrm{LWE}_{n,q,U(\mathbb{Z}_q),\chi}$. Therefore, we can assume that the secret $\mathbf{s}$ of the LWE problem also comes from a "small" Gaussian distribution. It is also easy to see that, if q is odd, then $\mathrm{LWE}_{n,q,\varphi,(2\chi)}$ is as hard as $\mathrm{LWE}_{n,q,\varphi,\chi}$, where the distribution 2χ samples $e \leftarrow \chi$ and outputs $2e$.

3   Homomorphic Encryption from LWE

In this section, we give a brief description of the FHE schemes of [9,8].

Basic LWE-based Encryption. We start by describing a basic symmetric/public encryption scheme E, which is a variant of [30] encryption scheme based on the LWE problem. This scheme serves as a building block for the more complex FHE schemes of [9,8] and of our threshold FHE scheme.

- **params** $= (1^\kappa, q, m, n, \varphi, \chi)$: The parameters of the scheme are an implicit input to all other algorithms, with: 1^κ is the security parameter, $q = q(k)$ is an odd modulus, $m = m(\kappa), n = n(\kappa)$ are the dimensions, and $\varphi = \varphi(\kappa), \chi = \chi(\kappa)$ are distributions over $\mathbb{Z}_q$.
- **E.SymKeygen(params):** Choose a secret key $\mathbf{s} \leftarrow \varphi^n$.
- **E.PubKeygen(s):** Choose $A \leftarrow \mathbb{Z}_q^{m \times n}$, $\mathbf{e} \leftarrow \chi^m$ and set $\mathbf{p} := A \cdot \mathbf{s} + 2 \cdot \mathbf{e}$. Output the public key $pk := (A, \mathbf{p})$ for the secret key $\mathbf{s}$.
- **E.SymEnc$_{\mathbf{s}}(\mu)$:** To encrypt a message $\mu \in \{0, 1\}$, choose $\mathbf{a} \leftarrow \mathbb{Z}_q^n$, $e \leftarrow \chi$, and set $b \stackrel{\mathrm{def}}{=} \langle \mathbf{a}, \mathbf{s} \rangle + 2 \cdot e + \mu$. Output the ciphertext $c = (\mathbf{a}, b)$.
- **E.PubEnc$_{pk}(\mu)$:** To encrypt a message $\mu \in \{0, 1\}$ under $pk = (A, \mathbf{p})$, choose $\mathbf{r} \leftarrow \{0, 1\}^m$ and set $\mathbf{a} \stackrel{\mathrm{def}}{=} \mathbf{r}^T \cdot A$, $b \stackrel{\mathrm{def}}{=} \langle \mathbf{r}, \mathbf{p} \rangle + \mu$. Output $c = (\mathbf{a}, b)$.
- **E.Dec$_{\mathbf{s}}(c)$ − (decryption):** Parse $c = (\mathbf{a}, b)$, output $[b - \langle \mathbf{a}, \mathbf{s} \rangle]_q \bmod 2$.

Under appropriate parameters and LWE assumption, the above scheme is semantically secure *with pseudorandom ciphertexts*, meaning that, given pk, a ciphertext of a chosen message is indistinguishable from a uniformly random ciphertext over the appropriate domain $\mathbb{Z}_q^{m+1}$.

Theorem 3 ([30]). *Assuming* n, q, $m \geq (n+1)\log(q) + \omega(\log(\kappa))$ *are integers with* q *odd, and that the* $\mathrm{LWE}_{n,q,\varphi,\chi}$ *assumption holds, the above public key encryption scheme* (E.PubKeygen, E.PubEnc, E.Dec) *is semantically secure with pseudorandom ciphertexts.*

Approximate encryption. Although we defined symmetric/public key encryption for the message space $\mu \in \{0, 1\}$, we can (syntactically) extend the same algorithms to any $\mu \in \mathbb{Z}_q$. Unfortunately, if μ is larger than a single bit, it will not be possible to decrypt μ correctly from the corresponding ciphertext. However, we can still think of this as an *approximate encryption* of μ, from which it is possible to recover the value $b - \langle \mathbf{a}, \mathbf{s} \rangle$ which is "close" to μ over $\mathbb{Z}_q$.

Fixing the coefficients. We use $\mathsf{E.PubKeygen}(\mathbf{s}; A)$, $\mathsf{E.PubKeygen}(\mathbf{s}; A; \mathbf{e})$ to denote the execution of the key generation algorithm with fixed coefficients A and (respectively) with fixed A, e. We use $\mathsf{E.SymEnc_s}(\mu; \mathbf{a})$, $\mathsf{E.SymEnc_s}(\mu; \mathbf{a}; e)$ analogously.

Key-Homomorphic Properties of Basic Scheme. It is easy to see that the scheme E is additively homomorphic so that the sum of ciphertexts encrypts the sum of the plaintexts (at least as long as the noise is small enough and does not overflow). We now notice it also satisfies several useful *key-homomorphic* properties, which make it particularly easy to convert into a threshold scheme. In particular, let $\mathbf{s}_1, \mathbf{s}_2$ be two secrets keys, $\mathbf{a}$ be some coefficient vector $(\mathbf{a}, b_1) = \mathsf{E.SymEnc_{s_1}}(\mu_1; \mathbf{a})$, $(\mathbf{a}, b_2) = \mathsf{E.SymEnc_{s_2}}(\mu_2; \mathbf{a})$ be two ciphertexts encrypting the bits μ_1, μ_2 under the keys $\mathbf{s}_1, \mathbf{s}_2$ respectively but using the same randomness $\mathbf{a}$. Then we can write $b_1 = \langle \mathbf{a}, \mathbf{s}_1 \rangle + 2 \cdot e_1 + \mu_1$, $b_2 = \langle \mathbf{a}, \mathbf{s}_2 \rangle + 2 \cdot e_2 + \mu_2$ and

$$b^* := b_1 + b_2 = \langle \mathbf{a}, (\mathbf{s}_1 + \mathbf{s}_2) \rangle + 2(e_1 + e_2) + (\mu_1 + \mu_2).$$

So $(\mathbf{a}, b^*) = \mathsf{E.SymEnc_{s_1+s_2}}(\mu_1 + \mu_2; \mathbf{a})$ is an encryption of $\mu_1 + \mu_2$ under the sum of the keys $(\mathbf{s}_1 + \mathbf{s}_2)$ with a noise level which is just the sum of the noises. Also, if we keep the matrix A fixed, then the sum of two key pairs gives a new valid key pair. That is, if $\mathbf{p}_1 = A\mathbf{s}_1 + 2\mathbf{e}_1$, $\mathbf{p}_2 = A\mathbf{s}_2 + 2\mathbf{e}_2$ are public key with corresponding secret keys $\mathbf{s}_1, \mathbf{s}_2$, then

$$\mathbf{p}^* := \mathbf{p}_1 + \mathbf{p}_2 = A(\mathbf{s}_2 + \mathbf{s}_2) + 2(\mathbf{e}_1 + \mathbf{e}_2)$$

is a public key for the corresponding secret key $\mathbf{s}^* = \mathbf{s}_1 + \mathbf{s}_2$.

Security of Joint Keys. We show a useful security property of combining public keys. Assume that a public key $\mathbf{p} = A\mathbf{s} + 2\mathbf{e}$ is chosen honestly and an attacker can then *adaptively* choose some value $\mathbf{p}' = A\mathbf{s}' + 2\mathbf{e}'$ for which it must *know* the corresponding $\mathbf{s}'$ and a "short" $\mathbf{e}'$. Then the attacker cannot distinguish public-key encryptions under the combined key $\mathbf{p}^* = \mathbf{p} + \mathbf{p}'$ from uniformly random ones.[2] Note that the combined key $\mathbf{p}^*$ may not be at all distributed like a correct public key, and the attacker has a large degree of control over it. Indeed, we can only show that the above holds if the ciphertext under the combined key is "smudged" with additional large noise. We define the above property formally via the following experiment $\mathsf{JoinKeys}_{\mathcal{A}}(\mathsf{params}, B_1, B_2)$:

(-) Challenger chooses $\mathbf{s} \leftarrow \mathsf{E.SymKeygen}(\mathsf{params})$, $(A, \mathbf{p}) \leftarrow \mathsf{E.PubKeygen}(\mathbf{s})$.

(-) $\mathcal{A}$ gets $(A, \mathbf{p})$ and adaptively chooses $\mathbf{p}', \mathbf{s}', \mathbf{e}'$ satisfying $\mathbf{p}' = A\mathbf{s}' + 2\mathbf{e}'$ and

---

[2] A similar idea was used in [16] in the context of threshold ElGamal encryption.

$\ell_1(\mathbf{e}') \le B_1$. It also chooses $\mu \in \{0, 1\}$.

(-) Challenger sets $pk^* := (A, \mathbf{p}^* = \mathbf{p} + \mathbf{p}')$. It chooses a random bit $\beta \xleftarrow{\$} \{0, 1\}$. If $\beta = 0$, it chooses $\mathbf{a}^* \xleftarrow{\$} \mathbb{Z}_q^n$, $b^* \xleftarrow{\$} \mathbb{Z}_q$ uniformly at random. Else it chooses $(\mathbf{a}^*, b) \leftarrow \mathsf{E.PubEnc}_{pk^*}(\mu)$, $e^* \xleftarrow{\$} [-B_2, B_2]$ and sets $b^* = b + 2e^*$.

(-) $\mathcal{A}$ gets $(\mathbf{a}^*, b^*)$ and outputs a bit $\tilde{\beta}$.

The output of the experiment is 1 if $\tilde{\beta} = \beta$, and 0 otherwise.

Lemma 4. *Let* q, m, n, φ, χ *be set as in Theorem 3 and assume that* $\mathrm{LWE}_{n,q,\varphi,\chi}$ *assumption holds. Let* $B_1 = B_1(\kappa)$, $B_2 = B_2(\kappa)$ *be integers s.t.* $B_1/B_2 = \mathsf{negl}(\kappa)$. *Then, for any* PPT $\mathcal{A}$: $|\Pr[\mathsf{JoinKeys}_{\mathcal{A}}(\mathsf{params}, B_1, B_2) = 1] - \frac{1}{2}| = \mathsf{negl}(\kappa)$.

3.1   Fully Homomorphic Encryption from LWE

In this section we present the construction of [9,8]. We start with the syntax of fully homomorphic encryption.

Definition. A *fully homomorphic (public–key) encryption* (FHE) scheme is a quadruple of PPT algorithms $\mathsf{FHE} = (\mathsf{FHE.Keygen}, \mathsf{FHE.Enc}, \mathsf{FHE.Dec}, \mathsf{FHE.Eval})$ defined as follows.

- **FHE.Keygen(1^κ) $\rightarrow$ (pk, evk, sk)**: Outputs a public encryption key pk, a public evaluation key evk and a secret decryption key sk.
- **FHE.Enc$_{pk}(\mu)$,FHE.Dec$_{sk}(c)$**: Have the usual syntax of public-key encryption/decryption.
- **FHE.Eval$_{evk}(f, c_1, \ldots, c_\ell) = c_f$**: The *homomorphic evaluation algorithm* is a *deterministic* poly-time algorithm that takes the evaluation key evk, a boolean circuit $f : \{0, 1\}^\ell \rightarrow \{0, 1\}$, and a set of ℓ ciphertexts $c_1, \ldots, c_\ell$. It outputs the result ciphertext c_f.

We say that an FHE scheme is secure if it satisfies the standard notion of *semantic security* for public-key encryption, where we consider the evaluation key evk as a part of the public key. We say that it is *fully homomorphic* if for any boolean circuit $f : \{0, 1\}^\ell \rightarrow \{0, 1\}$ and respective inputs $\mu_1, \ldots, \mu_\ell \in \{0, 1\}$, keys $(pk, evk, sk) \leftarrow \mathsf{FHE.Keygen}(1^\kappa)$ and ciphertexts $c_i \leftarrow \mathsf{FHE.Enc}_{pk}(\mu_i)$ it holds that: $\mathsf{FHE.Dec}(\mathsf{FHE.Eval}_{evk}(f, c_1, \ldots, c_\ell)) = f(\mu_1, \ldots, \mu_\ell)$. We say that the scheme is a *leveled fully homomorphic* if the FHE.Keygen algorithm gets an additional (arbitrary) input 1^D and the above only holds for circuits f consisting of at most D multiplicative levels.

Construction. We give an overview of the FHE construction of [9,8]. The construction begins with the basic encryption scheme E which is already additively homomorphic. We associate ciphertexts $c = (\mathbf{a}, b)$ under E with symbolic polynomials $\phi_c(\mathbf{x}) \stackrel{\mathsf{def}}{=} b - \langle \mathbf{a}, \mathbf{x} \rangle$: an n-variable degree-1 polynomial over $\mathbf{x}$. so that $\mathsf{Dec}_s(c) = [\phi_c(\mathbf{s})]_q \bmod 2$. If c_1, c_2 encrypt bits μ_1, μ_2 under a secret key $\mathbf{s}$, we can define the polynomial $\phi_{mult}(\mathbf{x}) \stackrel{\mathsf{def}}{=} \phi_{c_1}(\mathbf{x}) \cdot \phi_{c_2}(\mathbf{x})$. This already "encrypts" $\mu_1 \cdot \mu_2$ in the sense that $[\phi_{mult}(\mathbf{s})]_q = \mu_1 \cdot \mu_2 + 2e^*$ where e^* is "small".

Unfortunately, ϕ_{mult} is a degree-2 polynomial and hence its description is much larger than that of the original ciphertexts c_1, c_2.

The main challenge is to *re-linearize* the polynomial ϕ_{mult} to convert it into a degree-1 polynomial ϕ'_{mult} which still encrypts $\mu_1 \cdot \mu_2$. Such re-linearization is possible with two caveats: (1) The polynomial ϕ'_{mult} encrypts $\mu_1 \cdot \mu_2$ under a *new* key $\mathbf{t}$. (2) We need to know additional ciphertexts $\psi_{i,j,\tau}$ that (approximately) encrypt information about the key $\mathbf{s}$ under a new key $\mathbf{t}$ as follows (recall, we define $\mathbf{s}[0] \stackrel{\text{def}}{=} 1$):

$$\{\psi_{i,j,\tau} \leftarrow \mathsf{E.SymEnc_t}(\ 2^\tau \cdot \mathbf{s}[i] \cdot \mathbf{s}[j]\ ) : i,j \in [n] \cup \{0\}, \tau \in \lfloor\{0,\ldots,\log(q)\rfloor\}\}.$$

See [9] for the details of this re-linearization procedure. The above ideas give us *leveled homomorphic encryption scheme* for circuits with D multiplicative levels simply by choosing $D + 1$ secret keys $\mathbf{s}_0, \ldots, \mathbf{s}_D$ and publishing the ciphertexts $\{\psi_{d,i,j,\tau}\}$ which encrypt the required information about the level-d secret $\mathbf{s}_d$ under level-$(d + 1)$ secret $\mathbf{s}_{d+1}$. The public key of the scheme is $pk \leftarrow \mathsf{E.PubKeygen}(\mathbf{s}_0)$, corresponding to the level-0 secret key $\mathbf{s}_0$. The ciphertexts will have an associated level number, which is initially 0. Each time we multiply two ciphertexts with a common level d, we need to perform re-linearization which increases the level to $d + 1$. Using the secret key $\mathbf{s}_D$, we can then decrypt at the top level.

In the above discussion, we left out the crucial question of *noise*, which grows exponentially with the number of multiplications. Indeed, the above template only allows us to evaluate some logarithmic number of levels before the noise gets too large. The work of [8] gives a beautifully simple noise-reduction technique called "modulus reduction". This technique uses progressively smaller moduli q_d for each level d and simply "rescales" the ciphertext to the smaller modulus to reduce its noise level. As an end result, we get a *leveled* FHE scheme, allowing us to evaluate circuits containing at most D multiplicative levels, where D is an arbitrary polynomial, used as a parameter for FHE.Keygen. To get an FHE scheme where key generation does not depend on the number of levels, we can apply the bootstrapping technique of [17], at the expense of having to make an additional "circular security assumption".

4   Threshold Fully Homomorphic Encryption

Syntax. A threshold fully homomorphic encryption scheme (TFHE) is basically a homomorphic encryption scheme, with the difference that the Keygen and Dec are now N-party *protocols* instead of algorithms. We will consider protocols defined in terms of some common setup.

- **TFHE.Keygen(setup) – (key generation protocol):** Initially each party holds setup. At the conclusion of the protocol, each party P_k, for $k \in [N]$ outputs a common public-key pk, a common public evaluation key evk, and a private *share* sk_k of the implicitly defined secret key sk.

- **TFHE.Dec$_{sk_1,\ldots,sk_n}(c)$** – **(decryption protocol)**: Initially, each party P_k holds a common ciphertext c and an individual private share sk_k of the secret key. At the end of the protocol each party receives the decrypted plaintext μ.
- **TFHE.Enc$_{pk}(\mu)$, TFHE.Eval$_{pk}(f, c_1, \ldots, c_\ell)$**: Encryption and evaluation are *non-interactive algorithms* with the same syntax as in FHE.

We do *not* define the security of TFHE on its own. Indeed, requiring that the above protocols securely realize some ideal key-generation and decryption functionalities is unnecessarily restrictive. Instead, we will show that our TFHE scheme is secure in the context of our general MPC protocol in section 5.

4.1 Construction of TFHE

We now give our construction of TFHE, which can be thought of as a threshold version of the [8] FHE scheme. The main difficulty is to generate the evaluation key in a threshold manner, by having each party carefully release some extra information about its key-shares. Another important component of our construction is to require parties to add some additional *smudging* noise during sensitive operations, which will be crucial when analyzing security.

Common Setup. All parties share a common setup consisting of:
1. params $= \left(\{\mathsf{params}_d\}_{0 \leq d \leq D} \, , \, B_\varphi, B_\chi, B_{smdg}^{eval}, B_{smdg}^{enc}, B_{smdg}^{dec} \right)$, where

   - $\mathsf{params}_d = (1^\kappa, q_d, m, n, \varphi, \chi)$ are parameters for the encryption scheme E with differing moduli q_d.
   - $B_\varphi, B_\chi \in \mathbb{Z}$ are bounds s.t. φ is B_φ-bounded and χ is B_χ-bounded.
   - $B_{smdg}^{eval}, B_{smdg}^{enc}, B_{smdg}^{dec} \in \mathbb{Z}$ are bounds for extra "smudging" noise.

2. Randomly chosen common values (i.e. a *common random string* or CRS):

$$\left\{ A_d \xleftarrow{\$} \mathbb{Z}_{q_d}^{m \times n} \right\}_{d \in \{0,\ldots,D\}} \, , \, \left\{ \mathbf{a}_{d,i,\tau}^k \xleftarrow{\$} \mathbb{Z}_{q_d}^n \quad : \quad \begin{array}{l} k \in [N], i \in [n] \\ d \in [D], \tau \in \{0, \ldots, \lfloor \log(q_d) \rfloor\} \end{array} \right\}.$$

Convention. Whenever the protocol specifies that a party is to sample $x \leftarrow \varphi$ (resp. $x \leftarrow \chi$), we assume that it checks that $|x| \leq B_\varphi$ (resp. $|x| \leq B_\chi$) and re-samples if this is not the case (which happens with negligible probability).

TFHE.Keygen(setup). This is a two-round protocol between N parties.
Round 1:

1. Each party P_k invokes the key generation algorithm of the basic scheme E for each level $d \in \{0, \ldots, D\}$ to get $\mathbf{s}_d^k \leftarrow \mathsf{E.SymKeygen}(\mathsf{params}_d)$ and

$$(A_d, \mathbf{p}_d^k) \leftarrow \mathsf{E.PubKeygen}(\mathbf{s}_d^k \; ; \; A_d)$$

so that $\mathbf{p}_d^k = A_d \cdot \mathbf{s}_d^k + 2 \cdot \mathbf{e}_d^k$ for some noise $\mathbf{e}_d^k$. We can think of the values $\mathbf{s}_d^k$ as *individual secret keys* and $\mathbf{p}_d^k$ as *individual encryption keys* of party P_k.

2. For every $d \in [D]$, $i \in [n]$, $\tau \in \{0, \dots, \lfloor \log q \rfloor\}$, the party P_k computes:

$$\left(\mathbf{a}_{d,i,\tau}^{k}, b_{d,i,\tau}^{k,k} \right) \leftarrow \mathsf{E.SymEnc}_{\mathbf{s}_d^k} \left(2^\tau \cdot \mathbf{s}_{d-1}^k[i] \; ; \; \mathbf{a}_{d,i,\tau}^{k} \right)$$

so that $b_{d,i,\tau}^{k,k} = \langle \mathbf{a}_{d,i,\tau}^{k}, \mathbf{s}_d^k \rangle + 2e_{d,i,\tau}^{k,k} + 2^\tau \cdot \mathbf{s}_{d-1}^k[i]$ for some small noise $e_{d,i,\tau}^{k,k}$. In addition, for every d, i, τ as above and $\ell \in [N] \setminus \{k\}$, the party P_k computes "encryptions of 0":

$$\left(\mathbf{a}_{d,i,\tau}^{\ell}, b_{d,i,\tau}^{\ell,k} \right) \leftarrow \mathsf{E.SymEnc}_{\mathbf{s}_d^k} (0 \; ; \; \mathbf{a}_{d,i,\tau}^{\ell})$$

so that $b_{d,i,\tau}^{\ell,k} = \langle \mathbf{a}_{d,i,\tau}^{\ell}, \mathbf{s}_d^k \rangle + 2e_{d,i,\tau}^{\ell,k}$ for some noise $e_{d,i,\tau}^{\ell,k}$. The values $\{b_{d,i,\tau}^{\ell,k}\}$ will be used to create the evaluation key.

3. Each party P_k broadcasts the values $\{\mathbf{p}_d^k\}_d$, $\left\{ b_{d,i,\tau}^{\ell,k} \right\}_{\ell,d,i,\tau}$.

End of Round 1: At the end of round 1, we can define the following values.

1. For every $d \in \{0, \dots, D\}$, define: $\mathbf{p}_d^* := \sum_{\ell=1}^{N} \mathbf{p}_d^\ell$. Let $pk := (A_0, \mathbf{p}_0^*)$ be the common public encryption key of the TFHE scheme.

   Notice that, if all parties act honestly then $(A_d, \mathbf{p}_d^*) = \mathsf{E.PubKeygen}(\mathbf{s}_d^*; A_d; \mathbf{e}_d^*)$. where $\mathbf{s}_d^* := \sum_{\ell=1}^{N} \mathbf{s}_d^\ell$, $\mathbf{e}_d^* := \sum_{\ell=1}^{N} \mathbf{e}_d^\ell$. We can think of these values as the "combined public keys" for each level d.

2. For every $\ell \in [N], d \in [D]$, and all i, τ define $\beta_{d,i,\tau}^{\ell} := \sum_{k=1}^{N} b_{d,i,\tau}^{\ell,k}$.

   Notice that, if all parties follow the protocol then:

$$(\mathbf{a}_{d,i,\tau}^{\ell}, \beta_{d,i,\tau}^{\ell}) = \mathsf{E.SymEnc}_{\mathbf{s}_d^*} \left(2^\tau \cdot \mathbf{s}_{d-1}^\ell[i] \; ; \; \mathbf{a}_{d,i,\tau}^{\ell} \; ; \; e \right) \quad \text{where } e = \sum_{k=1}^{N} e_{d,i,\tau}^{\ell,k}$$

These "approximate encryptions" are already encrypted under the correct combined secret key $\mathbf{s}_d^*$ of level d. However, the "plaintexts" still only correspond to the *individual secret keys* $\mathbf{s}_{d-1}^\ell$ at level $d-1$, instead of the desired combined key $\mathbf{s}_{d-1}^*$. We fix this in the next round.

Round 2:

1. Each party P_k does the following. For all $\ell \in [N], d \in [D]$, $i, j \in [n]$, $\tau \in \{0, \dots, \lfloor \log q \rfloor\}$: sample $(\mathbf{v}_{d,i,j,\tau}^{\ell,k}, w_{d,i,j,\tau}^{\ell,k}) \leftarrow \mathsf{E.PubEnc}_{p_d^*}(0)$ and $e \xleftarrow{\$} [-B_{smdg}^{eval}, B_{smdg}^{eval}]$. Set:

$$(\boldsymbol{\alpha}_{d,i,j,\tau}^{\ell,k}, \beta_{d,i,j,\tau}^{\ell,k}) := \mathbf{s}_{d-1}^k[j] \cdot (\mathbf{a}_{d,i,\tau}^{\ell}, \beta_{d,i,\tau}^{\ell}) + (\mathbf{v}_{d,i,j,\tau}^{\ell,k}, w_{d,i,j,\tau}^{\ell,k} + 2e)$$

Note that, if all parties follow the protocol, then the original tuple $(\mathbf{a}_{d,i,\tau}^{\ell}, \beta_{d,i,\tau}^{\ell})$ approximately encrypts the value $2^\tau \mathbf{s}_{d-1}^\ell[i]$. The above operation has party P_k "multiply in" its component $\mathbf{s}_{d-1}^k[j]$ (and re-randomizing via a public encryption of 0) so that the final tuple $(\boldsymbol{\alpha}_{d,i,j,\tau}^{\ell,k}, \beta_{d,i,j,\tau}^{\ell,k})$ approximately encrypts $2^\tau \cdot \mathbf{s}_{d-1}^\ell[i] \cdot \mathbf{s}_{d-1}^k[j]$.

2. Each party P_k broadcasts the ciphertexts $\left\{(\alpha_{d,i,j,\tau}^{\ell,k}, \beta_{d,i,j,\tau}^{\ell,k})\right\}_{d,i,j,\tau,\ell}$.

End of Round 2: At the end of round 2, we can define the following values.

1. We define the *combined evaluation key* components for all $d \in [D]$ and all $i \in [n], j \in [n] \cup \{0\}, \tau$ as:

$$\psi_{d,i,j,\tau} := \begin{cases} \sum_{\ell=1}^{N} \sum_{k=1}^{N} (\alpha_{d,i,j,\tau}^{\ell,k}, \beta_{d,i,j,\tau}^{\ell,k}) & j \neq 0 \\ \sum_{\ell=1}^{N} (\mathbf{a}_{d,i,\tau}^{\ell}, \beta_{d,i,\tau}^{\ell}) & j = 0 \end{cases}$$

Note that, if all parties follow the protocol, then

$$\psi_{d,i,j,\tau} = \mathsf{E.SymEnc}_{\mathbf{s}_d^*}(2^\tau \cdot \mathbf{s}_{d-1}^*[i] \cdot \mathbf{s}_{d-1}^*[j])$$

where $\mathbf{s}_d^* := \sum_{\ell=1}^{N} \mathbf{s}_d^\ell$ is the combined secret key and all "errors" are "sufficiently small".

Outputs:

1. *Public Evaluation key:* Output $evk = \{\psi_{d,i,j,\tau}\}_{d,i,j,\tau}$.
2. *Public Encryption key:* Output $pk = (A_0, \mathbf{p}_0^*)$ as the public key.
3. *Share of secret key:* Each party P_k has a secret-key share $\mathbf{s}_D^k$.

TFHE.Enc$_{pk}(\mu)$: Once the *first round* of the key-generation protocol is concluded, the public key $pk = (A, \mathbf{p}_0^*)$ is well defined. At this point anybody can encrypt as follows. Choose $(\mathbf{v}, w) \leftarrow \mathsf{E.Enc}_{pk}(\mu)$ using the basic scheme E with the parameters $\mathsf{params}_0 = (1^\kappa, q_0, m, n, \varphi, \chi)$. Choose additional "smudging noise" $e \overset{\$}{\leftarrow} [-B_{smdg}^{enc}, B_{smdg}^{enc}]$ and output the ciphertext $c = ((\mathbf{v}, w + 2e), 0)$ with associated "level" 0.

TFHE.Eval$_{evk}(f, c_1, \ldots, c_t)$: Once the *second round* of the key-generation protocol is concluded, the evaluation key evk is defined. The evaluation algorithm is exactly the same as that of the underlying scheme FHE of [8].

The Decryption Protocol: **TFHE.Dec(c).** This is a one-round protocol between N parties. Initially all parties hold a common ciphertext $c = (\mathbf{v}, w, D)$ with associated "level" D. Moreover, each party P_k holds its share $\mathbf{s}_D^k$ for the joint secret key $\mathbf{s}_D^* = \sum_{k=1}^{N} \mathbf{s}_D^k$. At the end all parties get the decrypted bit μ.

- Each party P_k broadcasts $w^k = \langle \mathbf{v}, \mathbf{s}_D^k \rangle + 2 \cdot e_k$, where $e_k \overset{\$}{\leftarrow} [-B_{smdg}^{dec}, B_{smdg}^{dec}]$.
- Given $w^1, \ldots, w^N$, compute the output bit: $\mu = [w - \sum_{i=1}^{N} w^i]_{q_D} \mod 2$.

5   Secure MPC via TFHE

We now present a protocol for general MPC, using the threshold fully homomorphic scheme TFHE from the previous section. Let $f : (\{0,1\}^{\ell_{in}})^N \rightarrow \{0,1\}^{\ell_{out}}$ be a deterministic function computed by a circuit of multiplicative depth D. Let (TFHE.Keygen, TFHE.Enc, TFHE.Eval, TFHE.Dec) be our TFHE scheme from the previous section, initiated for D levels, and with parameters setup. Our basic MPC protocol π_f for evaluating the function f proceeds as follows.

Initialization: Each party P_k has input $\mathbf{x}_k \in \{0,1\}^{\ell_{in}}$. The parties share the common parameters setup for our D-level TFHE scheme.

Round I. The parties execute the *first* round of the TFHE.Keygen protocol. At the end of this round, each party P_k holds the common public key pk and a secret-key share sk_k.

Round II. The parties execute the *second* round of the TFHE.Keygen protocol. Concurrently, each party P_k also encrypts its input $\mathbf{x}_k$ bit-by-bit under the common public key pk and broadcasts the corresponding ciphertexts $\{\ c_{k,i} \leftarrow$ TFHE.Enc$_{pk}(\mathbf{x}_k[i])\ \}_{i \in \{1,\dots,\ell_{in}\}}$. At the end of this round, each party locally computes the evaluation key evk, and homomorphically evaluate the function f to get the output ciphertexts $\{\ c_j^* := \mathsf{Eval}_{evk}(f_j; \{c_{k,i}\})\ \}_{j \in \{1,\dots,\ell_{out}\}}$ where f_j is the boolean function for the jth bit of f.

Round III. The parties execute the decryption protocol TFHE.Dec on each of the output ciphertexts $\{c_j^*\}$ concurrently. At the end of this invocation, each party learns each of the bits of the underlying plaintext $y = f(x_1,\dots,x_N)$, which it sets as its output.

Security for Semi-Malicious Attackers. We show that the above protocol is secure against a semi-honest attacker corrupting any number of parties. Actually, we show security against a stronger class of attackers which we call *semi-malicious*. A semi-malicious attacker follows the honest protocol specification but with some adversarially chosen random coins (of which it has knowledge). It can choose its malicious random coins adaptively in each round after seeing the protocol messages of all honest parties during that round. We state our main theorem without concrete parameters. We defer the proof to the full version, where we also discuss the settings of the parameters for our protocol and the corresponding LWE assumption required for security.

Theorem 5. *Let f be any deterministic poly-time function with N inputs and single output (same output for all parties). Then there is a setting of parameters* params *such that, under the corresponding LWE assumption, the protocol π_f securely UC-realizes f in the presence of a static* semi-malicious *adversary corrupting any $t \leq N$ parties.*

Proof Intuition. We now give a high-level description of how the proof of security works, and relegate the proof to the full version. The simulator essentially runs rounds I and II honestly on behalf of the honest parties, but encrypts 0s instead of their real inputs. Then, in round III, it tries to force the real-world protocol output to match the idea-world output μ^*, by giving an incorrect decryption share on behalf of some honest party P_h. That is, assume that the combined ciphertext at the end of round II is $c = (\mathbf{v}, w, D)$. The simulator can get the secret keys $\mathbf{s}_D^k$ of all semi-malicious parties P_k at the end of round I (recall that semi-malicious parties follow the protocol honestly up to choosing bad random coins which are available to the simulator). It can therefore approximately compute the decryption shares $w^k \approx \langle \mathbf{v}, \mathbf{s}_D^k \rangle$ of the semi-malicious parties before (round III) starts. It then chooses the decryption share w^h of the honest party P_h by solving the equation

$w - \sum_\ell \langle \mathbf{v}, w^\ell \rangle = 2e + \mu^*$ where $e \xleftarrow{\$} B^{dec}_{smdg}$ is added noise. The decrypted value is therefore μ^*. We claim that the simulation is "good" since:

- The way that the simulator computes the decryption share of party P_h is actually statistically close to the way that the decryption share is given in the real world, when the noise B^{dec}_{smdg} is large enough. This follows by the "smudging" lemma.

- The attacker cannot distinguish encryptions of 0 from the real inputs by the "security of joint keys" (Lemma 4). In particular, the combined public encryption-key pk is derived as the sum of an honestly generated public-key $\mathbf{p}^h_0$ (for party P_h) and several other honestly and semi-maliciously generated keys for which the attacker must "know" a corresponding secret key. Moreover, the secret key $\mathbf{s}^h_0$ of party P_h is now never used during the simulated decryption protocol. Therefore, by the "security of joint keys", encryptions under pk maintain semantic security. There is an added complication here that extra information about the secret key $\mathbf{s}^h_0$ is released during rounds I and II of the protocol to create the evaluation key. However, this extra information essentially consists of ciphertexts under the higher level secret keys $\mathbf{s}^h_d$ for $d = 1, \ldots, D$. Therefore, the full proof consists of several hybrid games where we replace this extra information with random values starting with the top level and going down.

Security for Fully Malicious Attackers. Our basic MPC protocol is only secure in the semi-malicious setting. In the full version, we give a general round-preserving compiler from semi-malicious to fully malicious security using UC NIZKs [32] in the CRS model. In particular, in each round, the attacker must prove (in zero-knowledge) that it is following the protocol consistently with *some* setting of the random coins. Combining this with Theorem 5, we get a 3 round MPC protocol in the CRS model for a fully malicious attacker corrupting any number of parties.

In the full version (see [3]), we also address the question of instantiating such NIZKs *efficiently*. We first present simple, efficient, and statistical Σ-protocols for basic LWE-languages. These Σ-protocols crucially rely on the idea of "smudging" and have an interesting caveat that there is a *gap* between the noise-levels for which zero-knowledge is shown to hold and the ones for which soundness holds. We then use the Σ-protocols for these basic LWE-languages along with a series of AND and OR proofs to convert them into Σ-protocols for the more complicated language showing that a party is behaving "honestly". We can then compile them into UC-NIZKs and obtain general 3-round MPC protocols, in the random oracle model.

6   Variants and Optimizations

We consider several variants and optimizations of our basic MPC protocol.

Two Round MPC under a PKI. An alternative way to present our protocol is as a 2-round protocol with a *public-key infrastructure* (PKI). In particular, we

can think of the (round I) message ($\{\mathbf{p}_d^k\}$, $\{b_{d,i,\tau}^{\ell,k}\}$) sent by party P_k as its
public key and the value $\{\mathbf{s}_D^k\}$ as its secret key (in the fully malicious setting,
the public key would also contain the corresponding NIZKs). The entire MPC
execution then only consists of the remaining two rounds. Note that this PKI
is very simple and does not need a trusted party to set up everything; we just
need a trusted party to choose a CRS and then each party can choose its own
public key individually (possibly maliciously). Moreover, the PKI can be *reused*
for many MPC executions of arbitrary functions f with arbitrary inputs. The
main drawback is that the size of each party's public key is proportional to
the total number of parties, and it would be interesting to remove this. The
security analysis is exactly the same as that of our original three round protocol
in the CRS model, just by noting that the first round there consists of broadcast
message, which does not depend on the inputs of the parties (and hence we can
think of it as a public key).

Cloud-Assisted Computation. Our protocol can be made extremely efficient by
outsourcing large public computations. In particular, the only intensive computation in our protocol, that depends on the circuit size of the evaluated function,
is the homomorphic evaluation at the end of round II. In our basic description
of the protocol, we assumed that each party performs this computation *individually*, but we notice that this computation is the same for all parties and does
not require any secret inputs. Therefore, it is possible to designate one special
party P^* (or even an external entity e.g. a powerful server, or the "cloud") that
does this computation on everyone's behalf and broadcasts the resulting output
ciphertexts to everyone else. Moreover, if P^* is one of the parties, it does not
need to broadcast its input ciphertexts to everyone else in round II, since it
alone needs to know them when performing the evaluation. That is, the communication complexity is only proportional to the output size and the inputs of all
parties *other* than P^*. This may be useful if the MPC computation involves one
powerful party with a huge input and many weaker parties with small inputs.
Broadcasting the output ciphertexts requires an extra round, raising the round
complexity of this variant to 4 rounds in the CRS model, 3 rounds in PKI model.

The above simple idea already achieves security in the semi-honest model,
where we can trust P^* to perform the computation honestly and return the
correct result. However, in the fully malicious setting, we would also require P^*
to prove that the resulting ciphertext is the correct one, using a computationally-
sound proof system with a fixed polynomial (in the security parameter)
verification complexity. Such non-interactive proofs are known to exist in the
random-oracle model or under strong assumptions [27,7,20,14].

Ring LWE. In the full version (see [25]), we show a variant of the protocol using
ring LWE [26]. This variant provides significant practical efficiency savings over
just using standard LWE and the resulting scheme may be even conceptually
simpler than using standard LWE.

Bootstrapping. In our basic MPC protocol, the communication complexity is
proportional to the maximal number of multiplicative-levels in the circuit of

the evaluated function. This is because we start with a leveled TFHE scheme. To make the communication complexity completely independent of circuit size, we can rely on the *bootstrapping* technique of [17]. To apply the bootstrapping technique, each party P_k only needs to encrypt its secret-key share $sk_k = \mathbf{s}_D^k$ (bit-by-bit) under the combined public-key pk in round II of the protocol, and we add these values to the *evaluation key*. With this modification, we can instantiate our TFHE scheme with some *fixed* polynomial D depending on the decryption circuit and maintain the ability to homomorphically evaluate arbitrarily large function f. Therefore, the communication/computation complexity of the key-generation and decryption protocols is completely independent of the circuit size of the function f. For security, however, we must now rely on a non-standard *circular-security assumption* for the basic LWE-based encryption scheme E.

Randomized Functionalities and Individual Outputs. Our basic MPC protocol only considers deterministic functionalities where all the parties receive the same output. However, we can use standard efficient and round-preserving transformations to get a protocol for probabilistic functionalities and where different parties can receive different outputs.

Fairness. Our basic MPC protocol achieves security with abort for any number of corrupted parties. We can also achieve *fairness* for $t < N/2$ corruptions. The main idea is that, in Round I, each party also (threshold) secret-shares its individual secret sk_d^k so that any $\lfloor N/2 \rfloor + 1$ parties can recover it, but any fewer will not get any extra information. If a party P_k aborts in Rounds II or III, the rest of the parties will reconstruct sk_d^k (at the cost of one extra round) and use it to continue the protocol execution on P_k's behalf. Although an honest execution of our fair MPC protocol still uses 3 rounds of interaction, the protocol may now take up to 5 rounds in the worst case when some parties abort, where the extra rounds are needed to reconstruct the keys of the aborted parties.

Acknowledgements. This work was partially supported by an NSERC Discovery Grant and by DARPA under Agreement number FA8750-11-2-0225. The U.S. Government is authorized to reproduce and distribute reprints for Governmental purposes notwithstanding any copyright notation thereon. The views and conclusions contained herein are those of the author and should not be interpreted as necessarily representing the official policies or endorsements, either expressed or implied, of DARPA or the U.S. Government.

References

1. Applebaum, B., Cash, D., Peikert, C., Sahai, A.: Fast cryptographic primitives and circular-secure encryption based on hard learning problems. In: Halevi, S. (ed.) CRYPTO 2009. LNCS, vol. 5677, pp. 595–618. Springer, Heidelberg (2009)
2. Applebaum, B., Ishai, Y., Kushilevitz, E.: Computationally private randomizing polynomials and their applications. In: IEEE Conference on Computational Complexity, pp. 260–274 (2005)

3. Asharov, G., Jain, A., Wichs, D.: Multiparty computation with low communication, computation and interaction via threshold fhe. Cryptology ePrint Archive, Report 2011/613 (2011), http://eprint.iacr.org/

4. Beaver, D., Micali, S., Rogaway, P.: The round complexity of secure protocols (extended abstract). In: STOC, pp. 503–513 (1990)

5. Bendlin, R., Damgård, I.: Threshold Decryption and Zero-Knowledge Proofs for Lattice-Based Cryptosystems. In: Micciancio, D. (ed.) TCC 2010. LNCS, vol. 5978, pp. 201–218. Springer, Heidelberg (2010)

6. Bendlin, R., Damgård, I., Orlandi, C., Zakarias, S.: Semi-homomorphic Encryption and Multiparty Computation. In: Paterson, K.G. (ed.) EUROCRYPT 2011. LNCS, vol. 6632, pp. 169–188. Springer, Heidelberg (2011)

7. Bitansky, N., Canetti, R., Chiesa, A., Tromer, E.: From extractable collision resistance to succinct non-interactive arguments of knowledge, and back again. In: ITCS (2012)

8. Brakerski, Z., Gentry, C., Vaikuntanathan, V.: Fully homomorphic encryption without bootstrapping. In: ITCS (2012)

9. Brakerski, Z., Vaikuntanathan, V.: Efficient fully homomorphic encryption from (standard) lwe. In: FOCS (2011)

10. Brakerski, Z., Vaikuntanathan, V.: Fully Homomorphic Encryption from Ring-LWE and Security for Key Dependent Messages. In: Rogaway, P. (ed.) CRYPTO 2011. LNCS, vol. 6841, pp. 505–524. Springer, Heidelberg (2011)

11. Choi, S.G., Elbaz, A., Malkin, T., Yung, M.: Secure Multi-party Computation Minimizing Online Rounds. In: Matsui, M. (ed.) ASIACRYPT 2009. LNCS, vol. 5912, pp. 268–286. Springer, Heidelberg (2009)

12. Cramer, R., Damgård, I., Nielsen, J.B.: Multiparty Computation from Threshold Homomorphic Encryption. In: Pfitzmann, B. (ed.) EUROCRYPT 2001. LNCS, vol. 2045, pp. 280–300. Springer, Heidelberg (2001)

13. Damgård, I., Pastro, V., Smart, N., Zakarias, S.: Multiparty computation from somewhat homomorphic encryption. Cryptology ePrint Archive, Report 2011/535 (2011), http://eprint.iacr.org/

14. Damgård, I., Faust, S., Hazay, C.: Secure two-party computation with low communication. In: TCC (2012)

15. Damgård, I., Nielsen, J.B.: Universally Composable Efficient Multiparty Computation from Threshold Homomorphic Encryption. In: Boneh, D. (ed.) CRYPTO 2003. LNCS, vol. 2729, pp. 247–264. Springer, Heidelberg (2003)

16. Gennaro, R., Jarecki, S., Krawczyk, H., Rabin, T.: Secure Applications of Pedersen's Distributed Key Generation Protocol. In: Joye, M. (ed.) CT-RSA 2003. LNCS, vol. 2612, pp. 373–390. Springer, Heidelberg (2003)

17. Gentry, C.: Fully homomorphic encryption using ideal lattices. In: STOC, pp. 169–178 (2009)

18. Gentry, C., Halevi, S., Vaikuntanathan, V.: i-Hop Homomorphic Encryption and Rerandomizable Yao Circuits. In: Rabin, T. (ed.) CRYPTO 2010. LNCS, vol. 6223, pp. 155–172. Springer, Heidelberg (2010)

19. Goldreich, O., Micali, S., Wigderson, A.: How to play any mental game or a completeness theorem for protocols with honest majority. In: STOC, pp. 218–229 (1987)

20. Goldwasser, S., Lin, H., Rubinstein, A.: Delegation of computation without rejection problem from designated verifier cs-proofs. Cryptology ePrint Archive, Report 2011/456 (2011), http://eprint.iacr.org/

21. Halevi, S., Lindell, Y., Pinkas, B.: Secure Computation on the Web: Computing without Simultaneous Interaction. In: Rogaway, P. (ed.) CRYPTO 2011. LNCS, vol. 6841, pp. 132–150. Springer, Heidelberg (2011)

22. Ishai, Y., Kushilevitz, E.: Randomizing polynomials: A new representation with applications to round-efficient secure computation. In: FOCS, pp. 294–304 (2000)
23. Kamara, S., Mohassel, P., Raykova, M.: Outsourcing multi-party computation. Cryptology ePrint Archive, Report 2011/272 (2011), http://eprint.iacr.org/
24. Katz, J., Ostrovsky, R.: Round-Optimal Secure Two-Party Computation. In: Franklin, M. (ed.) CRYPTO 2004. LNCS, vol. 3152, pp. 335–354. Springer, Heidelberg (2004)
25. López-Alt, A., Tromer, E., Vaikuntanathan, V.: Cloud-assisted multiparty computation from fully homomorphic encryption. Cryptology ePrint Archive, Report 2011/663 (2011), http://eprint.iacr.org/
26. Lyubashevsky, V., Peikert, C., Regev, O.: On Ideal Lattices and Learning with Errors over Rings. In: Gilbert, H. (ed.) EUROCRYPT 2010. LNCS, vol. 6110, pp. 1–23. Springer, Heidelberg (2010)
27. Micali, S.: Computationally sound proofs. SIAM J. Comput. 30(4), 1253–1298 (2000)
28. Myers, S., Sergi, M., Shelat, A.: Threshold fully homomorphic encryption and secure computation. eprint 2011/454 (2011)
29. Peikert, C.: Public-key cryptosystems from the worst-case shortest vector problem: extended abstract. In: STOC, pp. 333–342 (2009)
30. Regev, O.: On lattices, learning with errors, random linear codes, and cryptography. In: STOC, pp. 84–93 (2005)
31. Rivest, R., Adleman, L., Dertouzos, M.: On data banks and privacy homomorphisms. In: Foundations on Secure Computation, pp. 169–179. Academia Press (1978)
32. De Santis, A., Di Crescenzo, G., Ostrovsky, R., Persiano, G., Sahai, A.: Robust Non-interactive Zero Knowledge. In: Kilian, J. (ed.) CRYPTO 2001. LNCS, vol. 2139, pp. 566–598. Springer, Heidelberg (2001)
33. Yao, A.C.C.: Protocols for secure computations (extended abstract). In: FOCS, pp. 160–164 (1982)
34. Yao, A.C.C.: How to generate and exchange secrets (extended abstract). In: FOCS, pp. 162–167 (1986)

Faster Algorithms for Approximate Common Divisors: Breaking Fully-Homomorphic-Encryption Challenges over the Integers

Yuanmi Chen and Phong Q. Nguyen

[1] ENS, Dept. Informatique, 45 rue d'Ulm, 75005 Paris, France
http://www.eleves.ens.fr/home/ychen/
[2] INRIA, France and Tsinghua University, Institute for Advanced Study, China
http://www.di.ens.fr/~pnguyen/

Abstract. At EUROCRYPT '10, van Dijk *et al.* presented simple fully-homomorphic encryption (FHE) schemes based on the hardness of approximate integer common divisors problems, which were introduced in 2001 by Howgrave-Graham. There are two versions for these problems: the partial version (PACD) and the general version (GACD). The seemingly easier problem PACD was recently used by Coron *et al.* at CRYPTO '11 to build a more efficient variant of the FHE scheme by van Dijk *et al..* We present a new PACD algorithm whose running time is essentially the "square root" of that of exhaustive search, which was the best attack in practice. This allows us to experimentally break the FHE challenges proposed by Coron *et al.* Our PACD algorithm directly gives rise to a new GACD algorithm, which is exponentially faster than exhaustive search. Interestingly, our main technique can also be applied to other settings, such as noisy factoring and attacking low-exponent RSA.

1 Introduction

Following Gentry's breakthrough work [11], there is currently great interest on fully-homomorphic encryption (FHE), which allows to compute arbitrary functions on encrypted data. Among the few FHE schemes known [11,29,9,3,13], the simplest one is arguably the one of van Dijk, Gentry, Halevi and Vaikuntanathan [29] (vDGHV), published at EUROCRYPT '10. The security of the vDGHV scheme is based on the hardness of *approximate integer common divisors problems* introduced in 2001 by Howgrave-Graham [17]. In the general version of this problem (GACD), the goal is to recover a secret number p (typically a large prime number), given polynomially many near-multiples $x_0, \ldots, x_m$ of p, that is, each integer x_i is of the hidden form $x_i = pq_i + r_i$ where each q_i is a very large integer and each r_i is a very small integer. In the partial version of this problem (PACD), the setting is exactly the same, except that x_0 is chosen as an exact multiple of p, namely $x_0 = pq_0$ where q_0 is a very large integer chosen such that no non-trivial factor of x_0 can be found efficiently: for instance, [9] selects q_0 as a rough number, *i.e.* without any small prime factor.

By definition, PACD cannot be harder than GACD, and intuitively, it seems that it should be easier than GACD. However, van Dijk *et al.* [29] mention that there is

D. Pointcheval and T. Johansson (Eds.): EUROCRYPT 2012, LNCS 7237, pp. 502–519, 2012.

currently no PACD algorithm that does not work for GACD. And the usefulness of PACD is demonstrated by the recent construction [9], where Coron, Mandal, Naccache and Tibouchi built a much more efficient variant of the FHE scheme by van Dijk *et al.* [29], whose security relies on PACD rather than GACD. Thus, it is very important to know if PACD is actually easier than GACD.

The hardness of PACD and GACD depends on how the q_i's and the r_i's are exactly generated. For the generation of [29] and [9], the noise r_i is extremely small, and the best attack known is simply gcd exhaustive search: for GACD, this means trying every noise (r_0, r_1) and check whether $\gcd(x_0 - r_0, x_1 - r_1)$ is sufficiently large and allows to recover the secret key; for PACD, this means trying every noise r_1 and check whether $\gcd(x_0, x_1 - r_1)$ is sufficiently large and allows to recover the secret key. In other words, if ρ is the bit-size of the noise r_i, then breaking GACD (resp. PACD) requires $2^{2\rho}$ (resp. 2^ρ) polynomial-time operations, for the parameters of [29,9].

OUR RESULTS. We present new algorithms to solve PACD and GACD, which are exponentially faster in theory and practice than the best algorithms considered in [29,9]. More precisely, the running time of our new PACD algorithm is $2^{\rho/2}$ polynomial-time operations, which is essentially the "square root" of that of gcd exhaustive search. This directly leads to a new GACD algorithm running in $2^{3\rho/2}$ polynomial-time operations, which is essentially the $3/4$-th root of that of gcd exhaustive search. Our PACD algorithm relies on classical algorithms to evaluate univariate polynomials at many points, whose space requirements are not negligible. We therefore present additional tricks, some of which reduce the space requirements, while still providing substantial speedups. This allows us to experimentally break the FHE challenges proposed by Coron *et al.* in [9], which were assumed to have comparable security to the FHE challenges proposed by Gentry and Halevi in [12]: the latter GH-FHE-challenges are based on hard problems with ideal lattices; according to Chen and Nguyen [4], their security level are respectively 52-bit (Toy), 61-bit (Small), 72-bit (Medium) and 100-bit (Large). Table 1 gives benchmarks for our attack on the FHE challenges, and deduces speedups compared to gcd exhaustive search. We can conclude that the FHE challenges of [9] have a much lower security level than those of Gentry and Halevi [14].

Table 1. Time required to break the FHE challenges by Coron *et al.* [9]. Size in bits, running time in seconds for a single 2.27GHz-core with 72 Gb. Timings are extrapolated for RAM > 72 Gb.

Name	Toy	Small	Medium		Large	
Size(public key)	0.95Mb	9.6Mb	89Mb		802Mb	
Size(modulus)	$1.6*10^5$	$0.86*10^6$	$4.2*10^6$		$19*10^6$	
Size(noise)	17	25	33		40	
Expected security level	≥ 42	≥ 52	≥ 62		≥ 72	
Running time of gcd-search	2420	$8.3*10^6$	$1.96*10^{10}$		$1.8*10^{13}$	
	40 mins	96 days	623 years		569193 years	
Concrete security level	≈ 42	≈ 54	≈ 65		≈ 75	
Running time of the new attack	99	25665	$1.64*10^7$	$6.6*10^6$	$6.79*10^{10}$	$2.9*10^8$
	1.6 min	7.1 hours	190 days	76 days	2153 years	9 years
Parameters	$d = 2^8$	$d = 2^{12}$	$d = 2^{13}$	$d = 2^{15}$	$d = 2^{10}$	$d = 2^{19}$
Memory	≤ 130 Mb	≤ 15 Gb	≤ 72 Gb	≈ 240 Gb	≤ 72 Gb	≈ 25 Tb
Speedup	24	324	1202	2977	264	62543
New security level	≤ 37.7	≤ 45.7	≤ 55	≤ 54	≤ 67	≤ 59

Interestingly, we can also apply our technique to different settings, such as noisy factoring, fault attacks on RSA-CRT, and attacking low-exponent RSA encryption. A typical example of noisy factoring is the following: assume that p is a divisor of a public modulus N, and that one is given a noisy version p' of p differing from p by at most k bits at unknown positions, can one recover p from (p', N) faster than exhaustive search? This may have applications in side-channel attacks. Like in the PACD setting, we obtain a square-root attack: for a 1024-bit modulus, the speedup can be as high as 1200 in practice. Similarly, we speed up several exhaustive search attacks on low-exponent RSA encryption.

RELATED WORK. Multipoint evaluation of univariate polynomials has been used in public-key cryptanalysis before. For instance, it is used in factoring (*e.g.* the Pollard-Strassen factorization algorithm [23,28] or in ECM speedup [20]), in the folklore square-root attack on RSA with small CRT exponents (mentioned by [1] and described in [24,21]), as well as in the recent square-root attack [8] by Coron, Joux, Mandal, Naccache and Tibouchi on Groth's RSA Subgroup Assumption [15]. But this does not imply that our attack is trivial, especially since the authors of [9] form a subset of the authors of [8]. In fact, in most cryptanalytic applications (including [8]) of multipoint evaluation, one is interested in the following problem: given two lists $\{a_i\}_i$ and $\{b_j\}_j$ of numbers modulo N, find a pair (a_i, b_j) such that $\gcd(a_i - b_j, N)$ is non-trivial. Instead, we use multipoint evaluation differently, as a way to compute certain products of m elements modulo N in $\tilde{O}(\sqrt{m})$ polynomial-time operations, where $\tilde{O}()$ is the usual notation hiding poly-logarithmic terms. More precisely, it applies to products $\prod_{i=1}^{m} x_i \bmod N$ which can be rewritten under the form $\prod_{j=1}^{m_1} \prod_{k=1}^{m_2} (y_j + z_k) \bmod N$ where both m_1 and m_2 are $O(\sqrt{m})$. The Pollard-Strassen factorization algorithm [23,28] can be viewed as a special case of this technique: it computes $m! \bmod N$ to factor N.

Very recently, Cohn and Heninger [6] announced a new attack on PACD and GACD, based on Coppersmith's small root technique. This attack is interesting from a theoretical point of view, but from a practical point of view, we provide evidence in the full version [5] that for the FHE challenges of [9], it is expected to be slower than gcd exhaustive search, and therefore much slower than our attack.

ROADMAP. In Sect. 2, we describe our square-root algorithm for PACD, and apply it to GACD. In Sect. 3, we discuss implementation issues, present several tricks to speed up the PACD algorithm in practice, and we discuss the impact of our algorithm on the fully-homomorphic challenges of Coron *et al.* [9]. Finally, we apply our main technique to different settings: noisy factoring (Sect. 4) and attacking low-exponent RSA (Sect. 5). More information can be found in the full version [5].

2   A Square-Root Algorithm for Partial Approximate Common Divisors

In this section, we describe our new square-root algorithm for the PACD problem, which is based on evaluating univariate polynomials at many points. In the last subsection, we apply it to GACD.

2.1 Overview

Consider an instance of PACD: $x_0 = pq_0$ and $x_i = pq_i + r_i$ where $0 \leq r_i < 2^\rho, 1 \leq i \leq m$. We start with the following basic observation due to Nguyen (as reported in [9, Sect 6.1]):

$$p = \gcd\left(x_0, \prod_{i=0}^{2^\rho-1}(x_1 - i) \pmod{x_0}\right) \quad (1)$$

which holds with overwhelming probability for the parameters of [9]. At first sight, this observation only allows to replace 2^ρ gcd computations (with numbers of size $\approx \gamma$ bits) with essentially 2^ρ modular multiplications (where the modulus has $\approx \gamma$ bits): the benchmarks of [9] report a speedup of ≈ 5 for the FHE challenges, which is insufficient to impact security estimates.

However, we observe that (1) can be exploited in a much more powerful way as follows. We define the polynomial $f_j(x)$ of degree j, with coefficients modulo x_0:

$$f_j(x) = \prod_{i=0}^{j-1}(x_1 - (x + i)) \pmod{x_0} \quad (2)$$

Letting $\rho' = \lfloor \rho/2 \rfloor$, we notice that:

$$\prod_{i=0}^{2^\rho-1}(x_1 - i) \equiv \prod_{k=0}^{2^{\rho'+(\rho \bmod 2)}-1} f_{2^{\rho'}}(2^{\rho'} k) \pmod{x_0}.$$

We can thus rewrite (1) as:

$$p = \gcd\left(x_0, \prod_{k=0}^{2^{\rho'+(\rho \bmod 2)}-1} f_{2^{\rho'}}(2^{\rho'} k) \pmod{x_0}\right) \quad (3)$$

Clearly, (3) allows to solve PACD using one gcd, $2^{\rho'+(\rho \bmod 2)} - 1$ modular multiplications, and the multi-evaluation of a polynomial (with coefficients modulo x_0) of degree $2^{\rho'}$ at $2^{\rho'+(\rho \bmod 2)}$ points, where $\rho' + (\rho \bmod 2) = \rho - \rho'$. We claim that this costs at most $\tilde{O}(2^{\rho'}) = \tilde{O}(\sqrt{2^\rho})$ operations modulo x_0, which is essentially the square root of gcd exhaustive search. This is obvious for the single gcd and the modular multiplications. For the multi-evaluation part, it suffices to use classical algorithms (see [30,18]) which evaluate a polynomial of degree d at d points, using at most $\tilde{O}(d)$ operations in the coefficient ring. Here, we also need to compute the polynomial $f_{2^{\rho'}}(x)$ explicitly, which can fortunately also be done using $\tilde{O}(\sqrt{2^\rho})$ operations modulo x_0. We give a detailed description of the algorithms in the next subsection.

2.2 Description

We first recall our algorithm to solve PACD, given as Alg. 1, and which was implicitly presented in the overview.

Algorithm 1. Solving PACD by multipoint evaluation of univariate polynomials

Input: An instance (x_0, x_1) of the PACD problem with noise size ρ.
Output: The secret number p such that $x_0 = pq_0$ and $x_1 = pq_1 + r_1$ with appropriate sizes.
1: Set $\rho' \leftarrow \lfloor \rho/2 \rfloor$.
2: Compute the polynomial $f_{2\rho'}(x)$ defined by (2), using Alg. 2.
3: Compute the evaluation of $f_{2\rho'}(x)$ at the $2^{\rho' + (\rho \bmod 2)}$ points $0, 2^{\rho'}, \ldots, 2^{\rho'}(2^{\rho' + (\rho \bmod 2)} - 1)$, using $2^{\rho \bmod 2}$ times Alg. 3 with $2^{\rho'}$ points. Each application of Alg. 3 requires the computation of a product tree, using Alg. 2.

Alg. 1 relies on two classical subroutines (see [30,18]):

- a subroutine to (efficiently) compute a polynomial given as a product of n terms, where n is a power of two: Alg. 2 does this in $\tilde{O}(n)$ ring operations, provided that quasi-linear multiplication of polynomials is available, which can be achieved in our case using Fast Fourier techniques. This subroutine is used in Step 2. The efficiency of Alg. 2 comes from the fact that when the algorithm requires a multiplication, it only multiplies polynomials of similar degree.
- a subroutine to (efficiently) evaluate a univariate degree-n polynomial at n points, where n is a power of two: Alg. 3 does this in $\tilde{O}(n)$ ring operations, provided that quasi-linear polynomial remainder is available, which can be achieved in our case using Fast Fourier techniques. This subroutine is used in Step 3, and requires the computation of a tree product, which is achieved by Alg. 2. Alg. 3 is based on the well-known fact that the evaluation of a univariate polynomial at a point α is the same as its remainder modulo $X - \alpha$, which allows to factor computations using a tree.

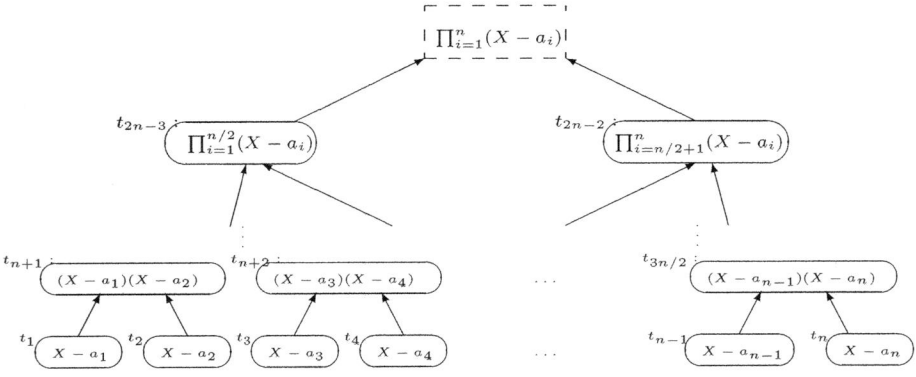

Fig. 1. Polynomial product tree $T = \{t_1, \ldots, t_{2n}\}$ for $\{a_1, \ldots, a_n\}$

Algorithm 2. $[T, D] \leftarrow \texttt{TreeProduct}(A)$

Input: A set of $n = 2^l$ numbers $\{a_1, \ldots, a_n\}$.
Output: The polynomial product tree $T = \{t_1, \ldots, t_{2n-1}\}$, corresponding to the evaluation of
   points $A = \{a_1, \ldots, a_n\}$ as shown in Figure 1.
   $D = [d_1, \ldots, d_{2n-1}]$ descendant indices for non-leaf nodes or 0 for leaf node.
1: **for** $i = 1 \ldots n$ **do**
2:    $t_i \leftarrow X - a_i$; $d_j \leftarrow 0$ {Initializing leaf nodes}
3: **end for**
4: $i \leftarrow 1$; $j \leftarrow n + 1$ {Index of lower and upper levels}
5: **while** $j \leqslant 2n - 1$ **do**
6:    $t_j \leftarrow t_i \cdot t_{i+1}$; $d_j \leftarrow i$; $i \leftarrow i + 2$; $j \leftarrow j + 1$
7: **end while**

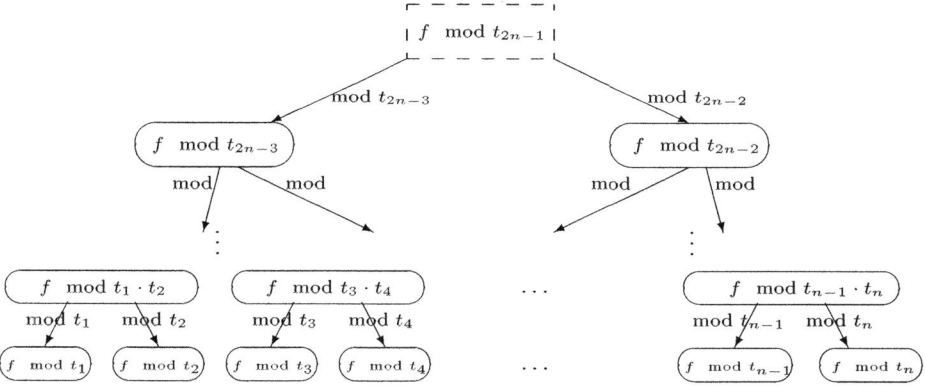

Fig. 2. Evaluation on the polynomial tree $T = \{t_1, \ldots, t_{2n-1}\}$ for $\{a_1, \ldots, a_n\}$

Algorithm 3. $V \leftarrow \texttt{RecursiveEvaluation}(f, t_i, D)$

Input: A polynomial f of degree n.
   A polynomial product tree rooted at t_i, and whose leaves are $\{X - a_k, \ldots, X - a_m\}$
   An array $D = [d_1, \ldots, d_{2n-1}]$ descendant indices for non-leaf nodes or 0 for leaf node.
Output: $V = \{f(a_k), \ldots, f(a_m)\}$
1: **if** $d_i = 0$ **then**
2:    return $\{f(a_i)\}$ {When t_i is a leaf, we apply an evaluation directly.}
3: **else**
4:    $g_1 \leftarrow f \mod t_{d_i}$ {left subtree}
5:    $V_1 \leftarrow \texttt{RecursiveEvaluation}(g_1, t_{d_i}, D)$
6:    $g_2 \leftarrow f \mod t_{d_i+1}$ {right subtree}
7:    $V_2 \leftarrow \texttt{RecursiveEvaluation}(g_2, t_{d_i+1}, D)$
8:    return $V_1 \cup V_2$
9: **end if**

It follows that the running time of Alg. 1 is $\tilde{O}(2^{\rho'}) = \tilde{O}(\sqrt{2^{\rho}})$ operations modulo x_0, which is essentially the "square root" of gcd exhaustive search. But the space requirement is $\tilde{O}(2^{\rho'}) = \tilde{O}(\sqrt{2^{\rho}})$ polynomially many bits: thus, Alg. 1 can be viewed as a time/memory trade-off, compared to gcd exhaustive search.

2.3   Logarithmic Speedup

In the previous analysis, the time complexity $\tilde{O}(n)$ actually stands for $O(n \log^2(n))$ ring multiplications. Interestingly, Bostan, Gaudry and Schost showed in [2] that when the structure of the factors are very regular, there is an algorithm which speeds up the theoretical complexity by a logarithmic term $\log(n)$. This BGS algorithm is tailored for the case where we want to estimate a function f on a set of points with what we call a hypercubic structure. An important subprocedure is ShiftPoly which, given as input a polynomial f of degree at most 2^d, and the evaluations of f on a set of 2^d points with hypercubic structure, outputs the evaluation of f on a shifted set of 2^d points, using $O(2^d)$ ring operations. More precisely:

Theorem 1. *(see Th. 5 of [2]) Let α, β be in ring $\mathbb{P}$ and d be in $\mathbb{N}$ such that $\mathbf{d}(\alpha, \beta, d)$ is invertible, with $\mathbf{d}(\alpha, \beta, d) = \beta \cdot 2 \ldots d \cdot (\alpha - d\beta) \ldots (\alpha + d\beta)$. And suppose also that the inverse of $\mathbf{d}(\alpha, \beta, d)$ is known. Let $F(\cdot) \in \mathbb{P}[X]$ of degree at most d and $x_0 \in \mathbb{P}$. There exists an algorithm ShiftPoly which, given as input $F(x_0)$, $F(x_0 + \beta)$, $\ldots$, $F(x_0 + d\beta)$, outputs $F(x_0 + \alpha)$, $F(x_0 + \alpha + \beta)$, $\ldots$, $F(x_0 + \alpha + d\beta)$ in time $2M(d) + O(d)$ time and space $O(d)$. Here, $M(d)$ is the time of multiplying two polynomial of degree at most d.*

We note $E(k_1, \ldots, k_j)$ for $\left\{ \sum_{i=1}^{j} p_{k_i} 2^{k_i} \right\}$ with each p_{k_i} ranging over $\{0, 1\}$. This is the set enumerating all possibilities of bits $\{k_1, \ldots, k_j\}$. Given a set A and an element and p, $A + p$ is defined as $\{a + p, a \in A\}$. Then we have

$$E(k_1, \ldots, k_{j+1}) = E(k_1, \ldots, k_j) \cup \left(E(k_1, \ldots, k_j) + 2^{k_{j+1}} \right).$$

This is what we call a set with hypercubic structure.

Given a linear polynomial $f(x)$ and a set with hypercubic structure of 2^ρ points, the proposed algorithm iteratively calls Alg.4 which uses ShiftPoly, and calculates the evaluation of $F_i(X) = \prod_{Y \in E(k_1, \ldots, k_i)} f(X + Y)$ on $E(b_{k-i}, \ldots, k_\rho)$ until $i = \lfloor n/2 \rfloor$. The i-th iteration costs $O(2^i)$ ring operations, thus the total complexity amounts to $O(2^{\rho/2})$ ring operations.

Algorithm 4. i-th iteration of the evaluation of $F_i(X)$

Input: For $i = 1, \ldots, \lfloor \rho/2 \rfloor$, the evaluation of $F_i(X)$ on points $X \in E(k_{\rho-i+1}, \ldots, k_\rho)$
Output: the evaluation of $F_{i+1}(X)$ on points $X \in E(k_{\rho-i}, \ldots, k_\rho)$

1: $F_i(X)$ for $X \in E(k_{\rho-i+1}, \ldots, k_\rho) + 2^{k_{\rho-i}} \leftarrow$ ShiftPoly($F_i(X), X \in E(k_{\rho-i+1}, \ldots, k_\rho)$)
2: $F_i(X)$ for $X \in E(k_{\rho-i}, \ldots, k_\rho) + 2^{k_{i+1}} \leftarrow$ ShiftPoly($F_i(X), X \in E(k_{\rho-i}, \ldots, k_\rho)$)

3: $F_{i+1}(X) = F_i(X) \cdot F_i(X + 2^{k_{i+1}})$, for all $X \in E(k_{\rho-i}, \ldots, k_\rho)$

2.4   Application to GACD

Any PACD algorithm can be used to solve GACD, using the trivial reduction from GACD to PACD based on exhaustive search over the noise r_0. More precisely, for an arbitrary instance of GACD:

$$x_i = pq_i + r_i \text{ where } 0 \le r_i < 2^\rho, 0 \le i \le m$$

we apply our PACD algorithm to all $(x_0 - r_0, x_1)$ where r_0 ranges over $\{0, \dots, 2^\rho - 1\}$.

It follows that GACD can be solved in $\tilde{O}(2^{3\rho/2})$ operations modulo x_0, using $\tilde{O}(2^{\rho/2})$ polynomially many bits. This is exponentially faster than the best attack of [29], namely gcd exhaustive search, which required $2^{2\rho}$ gcd operations. Note that in [29], another hybrid attack was described, where one performs exhaustive search over r_0 and factors the resulting number using ECM, but because of the large size of the prime factors (namely, a bit-length $\ge \rho^2$), this attack is not faster: it also requires at least $2^{2\rho}$ operations.

Following our work, it was noted with [10] that one can heuristically beat the GACD bound $\tilde{O}(2^{3\rho/2})$ using more samples x_i, by removing the "smooth part" of $\gcd(y_1, \dots, y_s)$ where $y_i = \prod_{j=0}^{2^\rho - 1}(x_i - j)$ and s is large enough. The choice of s actually gives different time/memory trade-offs. For instance, if $s = \Theta(\rho)$, the running time is heuristically $\tilde{O}(2^\rho)$ poly-time operations and similar memory. From a practical point of view however, our attack is arguably more useful, due to lower memory requirements and better $\tilde{O}()$ constants.

3   Implementation of the Square-Root PACD Algorithm

We implemented both Alg. 1 and the logarithmic speedup using the NTL library [26]. In this section, we describe various tricks that we used to implement efficiently Alg 1. The implementation was not straightforward due to the size of the FHE challenges.

3.1   Obstructions

The main obstruction when implementing Alg. 1 is memory. Consider the Large FHE-challenge from [9]: there, $\rho = 40$, so the optimal parameter is $\rho' = 20$, which implies that $f_{2^{\rho'}}$ is a polynomial of degree 2^{20} with coefficients of size 19×10^6 bits. In other words, simply storing $f_{2^{\rho'}}$ already requires $2^{20} \times 19 \times 10^6$ bits, which is more than 2Tb. This means that in practice, we will have to settle for suboptimal parameters.

More precisely, assume that we select an additional parameter d, which is a power of two less than $2^{\rho'}$. We rewrite (3) as:

$$p = \gcd\left(x_0, \prod_{k=0}^{2^\rho/d-1} f_d(dk) \,(\bmod\ x_0)\right) \tag{4}$$

This gives rise to a constrained version of Alg. 1, called Alg. 5.

Algorithm 5. Solving PACD by multipoint evaluation of univariate polynomials, using fixed memory

Input: An instance (x_0, x_1) of the PACD problem with noise size ρ, and a polynomial degree d (which must be a power of two).

Output: The secret number p such that $x_0 = pq_0$ and $x_1 = pq_1 + r_1$ with appropriate sizes.

1: Compute the polynomial $f_d(x)$ defined by (2), using Alg. 2.
2: Compute the evaluation of $f_d(x)$ at the $2^\rho/d$ points $0, d, 2d, \ldots, d(2^\rho/d - 1)$, using $2^\rho/d^2$ times Alg. 3 with d points. Each application of Alg. 3 requires the computation of a product tree, using Alg. 2.

The running time of Alg. 5 is $\frac{2^\rho \tilde{O}(d)}{d^2}$ elementary operations modulo x_0, and the space requirement is $\tilde{O}(d)$ polynomially many bits. Note that each of the $2^\rho/d^2$ times applications of Alg. 3 can be done in parallel.

3.2   Tricks

The use of Alg. 5 allows several tricks, which we now present.

Minimizing the Product Tree. Each application of Alg. 3 requires the computation of a product tree, using Alg. 2. But this product tree requires to store $2n - 1$ polynomials. Fortunately, these polynomials have coefficients which are in some sense much smaller than the modulus x_0: this is because we evaluate the polynomial $f_d(x)$ at points in $\{0, \ldots, 2^\rho - 1\}$, which is very small compared to the modulus x_0. However, a naive implementation would not exploit this. For instance, consider the polynomial $(X - a_1)(X - a_2) = X^2 - (a_1 + a_2)X + a_1 a_2$, which belongs to the product tree. In a typical library for polynomial computations, the polynomial coefficients would be represented as positive residues modulo x_0. But if $a_1 + a_2$ is small, then $-(a_1 + a_2) + x_0$ is actually big. This means that many coefficients of the product tree polynomials will actually be as big as x_0, if they are represented as positive residues modulo x_0, which drastically reduces the choice of the degree d.

To avoid this problem, we instead slightly modify the polynomial $f_d(X)$, in order to evaluate at small negative numbers inside $\{0, \ldots, 1 - 2^\rho\}$, so that each polynomial of the product tree has "small" positive coefficients. This drastically reduces the storage of the product tree. More precisely, we rewrite (4) as:

$$p = \gcd\left(x_0, \prod_{k=0}^{2^\rho/d^2-1}\prod_{\ell=0}^{d-1} f'_{d,k}(-\ell d) \pmod{x_0}\right) \tag{5}$$

where

$$f'_{d,k}(x) = \prod_{i=0}^{d-1}(x_1 - 2^\rho - x + dk - i) \pmod{x_0} \tag{6}$$

Each product $\prod_{\ell=0}^{d-1} f'_{d,k}(-\ell) \pmod{x_0}$ is computed by applying Alg. 3 once, using the d points $0, -d, -2d, \ldots, -d(d-1)$.

Powers of Two. We need to compute the polynomial $f'_{d,k}(x)$ defined by (6) before each application of Alg. 3, using a simplified version of Alg. 2, which only computes the root rather than the whole product tree. However, notice that the degree of each polynomial of the product tree is exactly a power of two, which is the worst case for the polynomial multiplication implemented in the NTL library [26]. For instance, in NTL, multiplying two 512-degree polynomials with Medium-FHE coefficients takes 50% more time than multiplying two 511-degree polynomials with Medium-FHE coefficients.

To circumvent threshold phenomenons, we notice that each polynomial of the product tree is a monic polynomial, except the leaves (for which the leading coefficient is -1). But the product of two monic polynomials whose degree is a power of two can be derived efficiently from the product of two polynomials with degree strictly less than the power of two, using:

$$(X^n + P(X)) \times (X^n + Q(X)) = X^{2n} + X^n(P(X) + Q(X)) + P(X)Q(X).$$

We apply this trick to speed up the computation of the polynomial $f'_{d,k}(x)$.

Precomputations. Now that we use (5), we change several times the polynomial $f'_{d,k}(x)$, but we keep the same evaluation points $0, -d, -2d, \ldots, -d(d-1)$, and therefore the same product tree. This allows to perform precomputations to speed up Alg. 3. Indeed, the main operation of Alg. 3 is computing the remainder of a polynomial with one of the product tree polynomials, and it is well-known that this can be sped up using precomputations depending on the modulus polynomial. One classical way to do this is to use Newton's method for remainder (Alg. 6). This algorithm requires the following notation: for any polynomial f of degree n and for any integer $m \geqslant n$, we define the m-degree polynomial $\mathrm{rev}(f, m)$ as $\mathrm{rev}(f, m) = f(1/X) \cdot X^m$. In Alg. 6, Line 1 is

Algorithm 6. Remainder using Newton's method (see [18, Sect 7.2])
Input: Polynomials $f \in \mathbb{R}[X]$ of degree $2n - 1$, $g \in \mathbb{R}[X]$ of degree n.
Output: The polynomial $h = f \mod g$
  1: $\bar{g} \leftarrow \mathrm{Inverse}(\mathrm{rev}(g, n)) \mod X^n$
  2: $s \leftarrow \mathrm{rev}(f, 2n - 1) \cdot \bar{g} \mod X^n$
  3: $h \leftarrow f - \mathrm{rev}(s, n - 1) \cdot g$

independent of f. Therefore, whenever one needs to compute many remainders with respect to the same modulus g, it is more efficient to precompute and store h, so that Line 1 does not need to be reexecuted. Hence, in an offline phase, we precompute and store (on a hard disk) the polynomial $\bar{g}$ of Line 1 for each product tree polynomial. And for each remainder required by Alg. 3, we execute the last two lines of Alg. 6.

It follows that each remainder operation of Alg. 3 is reduced to two polynomial multiplications.

The NTL library also contains routines for doing remainders with precomputations, but Alg. 6 turns out to be more efficient for our setting. This is because many factors impact the performance of polynomial arithmetic, such as the size of the modulus and the degree.

3.3 Logarithmic Speedup and Further Tricks

We also implemented the BGS algorithm described in Sect. 2.3, which offers an asymptotical logarithmic speedup, but our implementation was not optimized: a better implementation would require the so-called middle product [2], which we instantiated by a normal product. On the FHE challenges, our implementation turned out to be twice as slow as Alg. 1 for Medium and Large, and marginally slower (resp. faster) for Toy (resp. Small).

Since memory is the main obstruction for choosing d, it is very important to minimize RAM requirements. Since Alg. 3 can be reduced to multiplications using precomputations, one may consider the use of special multiplication algorithms which require less memory than standard algorithms, such as in-place algorithms. We note that there has been recent work [25,16] in this direction, but we did not implement these algorithms. This suggests that our implementation is unlikely to be optimal, and that there is room for improvement.

3.4 New Security Estimates for the FHE Challenges

Table 1 reports benchmarks for our implementation on the fully-homomorphic-encryption challenges of Coron *et al.* [9], which come in four flavours: Toy, Small, Medium and Large. The security level ℓ is defined in [9] is defined as follows: the best attack should require at least 2^{ℓ} clock cycles on a standard single core. The row "Expected security level" is extracted from [9].

Our timings refer to a single 2.27GHz-core with 72Gb of RAM. First, we assessed the cost of gcd exhaustive search, by measuring the running time of the (quasi-linear) gcd routine of the widespread gmp library, which is used in NTL [26]: timings were measured for each modulus size of the four FHE-challenges. This gives the "concrete security level" row, which is slightly higher than the expected security level of [9].

We also report timings for our implementation of our square-root PACD algorithm: these timings are below the expected security level, which breaks all four FHE-challenges of [9]. For the Toy and Small challenges, the parameter d was optimal, and we did not require much memory: the speedup is respectively 24 and 324, compared to gcd exhaustive search. For the Medium and Large challenges, we used a suboptimal parameter d, due to RAM constraints: we used $d = 2^{13}$ (resp. $d = 2^{10}$) for Medium (resp. Large), instead of the optimal $d = 2^{16}$ (resp. $d = 2^{20}$). But the speedups are already significant: 1202 for Medium, and 264 for Large. The timings are obtained by suitably multiplying the running time of a single execution of Alg. 3 and Alg. 2: for instance, in the Large case, this online phase took between 64727s to 65139.4s, for 5 executions, and the precomputation storage was 21Gb.

Table 1 also provides extrapolated figures if the RAM was ≥ 72 Gb, which allows larger values of d: today, one can already buy servers with 4-Tb RAM. For the Large challenge, the potential speedup is over 60,000. Using a more optimized implementation, we believe it is possible to obtain larger speedups, so the "New security level" row should only be interpreted as an upper bound. But our implementation is already sufficient to show that the FHE-challenges of [9] fall short of the expected security level.

Hence, one needs to increase the parameters of the FHE scheme of [9], which makes it less competitive with the FHE implementation of [14]. It can be noted that the new security levels of the challenges of [9] are much lower than those given by [4] on the challenges of Gentry and Halevi [14], namely 52-bit (Toy), 61-bit (Small), 72-bit (Medium) and 100-bit (Large).

4 Applications to Noisy Factoring

Consider a typical "balanced" RSA modulus $N = pq$ where $p, q \leq 2\sqrt{N}$. A celebrated lattice-based cryptanalysis result of Coppersmith [7] states that if one is given half of the bits of p, either in the most significant positions, or the least significant positions, then one can recover p and q in polynomial time. Although this attack has been extended in several works (see [19] for a survey), all these lattice-based results require that the unknown bits are consecutive, or spread across extremely few blocks. This decreases its potential applications to side-channel attacks where errors are likely to be spread unevenly.

This suggests the following setting, which we call noisy factoring. Assume that one is given a noisy version p' of the prime factor p, which differs from p by at most k bits, not necessarily consecutive, under either of the following two cases:

- If the k positions of the noisy bits are known, we can recover p (and therefore q) by exhaustive search using at most 2^k polynomial-time operations: we stress that in this case, we assume that we do not know if each of the k bits has been flipped, otherwise no search would be necessary.
- If instead, none of the positions is known, but we know that exactly k bits have been modified, we can recover p by exhaustive search using at most $\binom{n}{k}$ polynomial-time operations, where n is the bit-length of p. If we only know an upper bound on the number of modified bits, we can simply repeat the attack with decreasing values of k.

These running times do not require that p and q are balanced.

In this section, we show that our previous technique for PACD can be adapted to noisy factoring, yielding new attacks whose running time is essentially the "square root" of exhaustive search, that is, $\tilde{O}(2^{k/2})$ or $\tilde{O}(\sqrt{\binom{n}{k}})$ polynomial-time operations, depending on the case.

4.1 Known Positions

We assume that the prime number p has n bits, so that: $p = \sum_{i=0}^{n-1} p_i 2^i$, where $p_i \in \{0, 1\}$ for $0 \leqslant i \leqslant n - 1$.

In this subsection, we assume that all the bits p_i are known, except possibly at k positions $b_1, \ldots, b_k$, which we sort, so that: $0 \leq b_1 \leqslant \cdots \leqslant b_k < n$. Denote by $p^{(1)}, \ldots, p^{(2^k)}$ the 2^k possibilities for p, when $(p_{b_1}, \ldots, p_{b_k})$ ranges over $\{0, 1\}^k$. With high probability, all the $p^{(i)}$'s are coprime with N, except one, which would imply that:

$$p = \gcd\left(N, \prod_{i=1}^{2^k} p^{(i)} (\text{mod } N)\right) \qquad (7)$$

A naive evaluation of (7) costs 2^k modular multiplications, and one single gcd. We now show that this evaluation can be performed more efficiently using $\tilde{O}(2^{k/2})$ arithmetic operations with numbers with the same size as N.

The unknown bits $p_{b_1}, \ldots, p_{b_k}$ can be regrouped into two sets $\{p_{b_1}, \ldots, p_{b_\ell}\}$, and $\{p_{b_{\ell+1}}, \ldots, p_k\}$ of roughly the same size $\ell = \lfloor k/2 \rfloor$:

- For $1 \leqslant i \leqslant 2^\ell$, let $y^{(i)} = \sum_{j=0}^{n-1} y_j^{(i)} 2^j$, where $y_j^{(i)} = \begin{cases} 0 & \text{if } j > b_\ell \\ t\text{-th bit of } i \text{ if } \exists t \leqslant \ell, j = b_t \\ p_j & \text{otherwise} \end{cases}$

- For $1 \leqslant i \leqslant 2^{k-\ell}$, let $x^{(i)} = \sum_{j=0}^{n-1} x_j 2^j$, where $x_j^{(i)} = \begin{cases} 0 & \text{if } j \leqslant b_l \\ t\text{-th bit of } i \text{ if } \exists t > l, j = b_t \\ p_j & \text{otherwise} \end{cases}$

Hence, by definition of $x^{(i)}$ and $y^{(i)}$, we have:

$$\prod_{i=1}^{2^k} p^{(i)} \equiv \prod_{i=1}^{2^\ell} \prod_{j=1}^{2^{k-\ell}} (x^{(j)} + y^{(i)}) \ (\text{mod } N) \qquad (8)$$

which gives rise to a square-root algorithm (Alg. 7) to solve the noisy factorization problem with known positions.

Algorithm 7. Noisy Factorization With Known Positions

Input: An RSA modulus $N = pq$ and the bits $p_0, \ldots, p_{n-1}$ of p, except the k bits $p_{b_1}, \ldots, p_{b_k}$, where the bit positions $b_1 \leq b_2 \leq \cdots \leq b_k$ are known.

Output: The secret factor $p = \sum_{i=0}^{n-1} p_i 2^i$ of N.

1: Compute the polynomial $f(X) = \prod_{i=1}^{2^\ell} \left(X + y^{(i)}\right) \bmod N$ of degree 2^ℓ, with coefficients modulo N, using Alg. 2.
2: Compute the evaluation of $f(X)$ at the points $\{x^{(1)}, \ldots, x^{(2^{k-\ell})}\}$, using $1 + (k \bmod 2)$ times Alg. 3 with 2^ℓ points.
3: return $p \leftarrow \gcd\left(N, \prod_{i=1}^{2^{k-\ell}} \left(f(x^{(i)})\right) \bmod N\right)$

Similary to Section 2, the cost of Alg. 7 is $\tilde{O}(2^{k/2})$ polynomial-time operations. This is an exponential improvement over naive exhaustive search, but Alg. 7 requires exponential space. In practice, the improvement is substantial. Using our previous implementation, Alg. 7 gives a speedup of about 1200 over exhaustive division to factor a 1024-bit modulus, given a 512-bit noisy factor with 46 unknown bits at known positions.

Furthermore, in this setting, the points to be enumerated happen to satisfy the hypercubic property, thus we may apply the logarithmic speedup described in Sect. 2.3.

11. Gentry, C.: Fully homomorphic encryption using ideal lattices. In: Proc. STOC 2009, pp. 169–178. ACM (2009)

12. Gentry, C., Halevi, S.: Public challenges for fully-homomorphic encryption. The implementation is described in [12] (2010), https://researcher.ibm.com/researcher/view_project.php?id=1548

13. Gentry, C., Halevi, S.: Fully homomorphic encryption without squashing using depth-3 arithmetic circuits. Cryptology ePrint Archive, Report 2011/279 (2011), http://eprint.iacr.org/

14. Gentry, C., Halevi, S.: Implementing Gentry's Fully-Homomorphic Encryption Scheme. In: Paterson, K.G. (ed.) EUROCRYPT 2011. LNCS, vol. 6632, pp. 129–148. Springer, Heidelberg (2011)

15. Groth, J.: Cryptography in Subgroups of $\mathbb{Z}_n^*$. In: Kilian, J. (ed.) TCC 2005. LNCS, vol. 3378, pp. 50–65. Springer, Heidelberg (2005)

16. Harvey, D., Roche, D.S.: An in-place truncated fourier transform and applications to polynomial multiplication. In: Proc. ISSAC 2010, pp. 325–329. ACM (2010)

17. Howgrave-Graham, N.: Approximate Integer Common Divisors. In: Silverman, J.H. (ed.) CaLC 2001. LNCS, vol. 2146, pp. 51–66. Springer, Heidelberg (2001)

18. Mateer, T.: Fast Fourier Transform Algorithms with Applications. PhD thesis, Clemson University (2008)

19. May, A.: Using LLL-reduction for solving RSA and factorization problems: A survey. In: [21] (2010)

20. Montgomery, P.L.: An FFT Extension of the Elliptic Curve Method of Factorization. PhD thesis, University of California Los Angeles (1992)

21. Nguyen, P.Q.: Public-key cryptanalysis. In: Luengo, I. (ed.) Recent Trends in Cryptography. Contemporary Mathematics, vol. 477. AMS–RSME (2009)

22. Nguyen, P.Q., Vallée, B. (eds.): The LLL Algorithm: Survey and Applications. Information Security and Cryptography. Springer, Heidelberg (2010)

23. Pollard, J.M.: Theorems on factorization and primality testing. Proc. Cambridge Philos. Soc. 76, 521–528 (1974)

24. Qiao, G., Lam, K.-Y.: RSA Signature Algorithm for Microcontroller Implementation. In: Schneier, B., Quisquater, J.-J. (eds.) CARDIS 1998. LNCS, vol. 1820, pp. 353–356. Springer, Heidelberg (2000)

25. Roche, D.S.: Space- and time-efficient polynomial multiplication. In: Proc. ISSAC 2009, pp. 295–302. ACM (2009)

26. Shoup, V.: Number Theory C++ Library (NTL) version 5.4.1, http://www.shoup.net/ntl/

27. Stinson, D.R.: Some baby-step giant-step algorithms for the low hamming weight discrete logarithm problem. Math. Comput. 71(237), 379–391 (2002)

28. Strassen, V.: Einige Resultate über Berechnungskomplexität. Jber. Deutsch. Math.-Verein. 78(1), 1–8 (1976/1977)

29. van Dijk, M., Gentry, C., Halevi, S., Vaikuntanathan, V.: Fully Homomorphic Encryption over the Integers. In: Gilbert, H. (ed.) EUROCRYPT 2010. LNCS, vol. 6110, pp. 24–43. Springer, Heidelberg (2010)

30. von Zur Gathen, J., Gerhard, J.: Modern computer algebra, 2nd edn. Cambridge University Press (2003)

Decoding Random Binary Linear Codes in $2^{n/20}$: How $1 + 1 = 0$ Improves Information Set Decoding

Anja Becker[1], Antoine Joux[1,2], Alexander May[3,*], and Alexander Meurer[3,**]

[1] Université de Versailles Saint-Quentin, Laboratoire PRISM
[2] DGA
[3] Ruhr-University Bochum, Horst Görtz Institute for IT-Security
`anja.becker@prism.uvsq.fr`, `antoine.joux@m4x.org`
`{alex.may,alexander.meurer}@rub.de`

Abstract. Decoding random linear codes is a well studied problem with many applications in complexity theory and cryptography. The security of almost all coding and LPN/LWE-based schemes relies on the assumption that it is hard to decode random linear codes. Recently, there has been progress in improving the running time of the best decoding algorithms for binary random codes. The ball collision technique of Bernstein, Lange and Peters lowered the complexity of Stern's information set decoding algorithm to $2^{0.0556n}$. Using *representations* this bound was improved to $2^{0.0537n}$ by May, Meurer and Thomae. We show how to further increase the number of representations and propose a new information set decoding algorithm with running time $2^{0.0494n}$.

Keywords: Information Set Decoding, Representation Technique.

1 Introduction

The NP-hard problem of decoding a random linear code is one of the most promising problems for the design of cryptosystems that are secure even in the presence of quantum computers. Almost all code-based cryptosystems, e.g. McEliece, rely on the fact that random linear codes are hard to decode. In order to embed a trapdoor in coding-based cryptography one usually starts with a well-structured secret code C and linearly transforms it into a code C' that is supposed to be indistinguishable from a random code.

An attacker has two options. Either he tries to distinguish the scrambled version C' of C from a random code by revealing the underlying structure, see [11,28]. Or he directly tries to run a generic decoding algorithm on the scrambled code C'.

Also closely related to random linear codes is the learning parity with noise (LPN) problem that is frequently used in cryptography [1,14,17]. In LPN, one directly starts with a random linear code C and the LPN search problem is a decoding problem in C. It was shown in [27] that the popular LPN decision variant, a very useful tool for many cryptographic constructions, is equivalent to the LPN search problem, and thus

---

* Supported by DFG project MA 2536/7-1 and by ICT-2007-216676 ECRYPT II.
** Ruhr-University Research School, Germany Excellence Initiative [DFG GSC 98/1].

D. Pointcheval and T. Johansson (Eds.): EUROCRYPT 2012, LNCS 7237, pp. 520–536, 2012.

equivalent to decoding a random linear code. The LWE problem of Regev [27] is a generalization of LPN to codes over a larger field. Our decoding algorithm could be adjusted to work for these larger fields (similar to what was done in [9,26]). Since the decoding problem lies at the the heart of coding-based and LPN/LWE-based cryptography it is necessary to study its complexity in order to define proper security parameters for cryptographic constructions.

Let us start by providing some useful notation. A binary linear code C is a k-dimensional subspace of $\mathbb{F}_2^n$ where n is called the length of the code and $R := \frac{k}{n}$ is called its rate. A random k-dimensional linear code C of length n can be defined as the kernel of a random full-rank matrix $\mathbf{H} \in_R \mathbb{F}_2^{(n-k) \times n}$, i.e. $C = \{\mathbf{c} \in \mathbb{F}_2^n \mid \mathbf{Hc}^t = \mathbf{0}^t\}$. The matrix $\mathbf{H}$ is called a parity check matrix of C. For ease of presentation, we use the convention that all vectors are column vectors which allows as to omit all transpositions of vectors.

The distance d of a linear code is defined by the minimal Hamming distance between two codewords. Hence every vector $\mathbf{x}$ whose distance to the closest codeword $\mathbf{c} \in C$ is at most the error correction capacity $\omega = \lfloor \frac{d-1}{2} \rfloor$ can be uniquely decoded to $\mathbf{c}$.

For any point $\mathbf{x} = \mathbf{c} + \mathbf{e} \in \mathbb{F}_2^n$ that differs from a codeword $\mathbf{c} \in C$ by an error vector $\mathbf{e}$, we define its *syndrome* as $s(\mathbf{x}) := \mathbf{Hx} = \mathbf{H}(\mathbf{c} + \mathbf{e}) = \mathbf{He}$. Hence, the syndrome only depends on the error vector $\mathbf{e}$ and not on the codeword $\mathbf{c}$. The *syndrome decoding problem* is to recover $\mathbf{e}$ from $s(\mathbf{x})$. This is equivalent to decoding in C, since the knowledge of $\mathbf{e}$ suffices to recover $\mathbf{c}$ from $\mathbf{x}$.

Usually in cryptographic settings the Hamming weight of $\mathbf{e}$ is smaller than the error correction capability, i.e. $\mathrm{wt}(\mathbf{e}) \leq \omega = \lfloor \frac{d-1}{2} \rfloor$, which ensures unique decoding. This setting is also known as *half/bounded distance decoding*. All known half distance decoding algorithms achieve their worst case behavior for the choice $\mathrm{wt}(\mathbf{e}) = \omega$. As a consequence we assume $\mathrm{wt}(\mathbf{e}) = \omega$ throughout this work. In complexity theory, one also studies the so-called *full decoding* where one has to compute a closest codeword to a given *arbitrary* vector $\mathbf{x} \in \mathbb{F}_2^n$. We also give the complexity of our algorithm for full decoding, but in the following we will focus on half-distance decoding.

The running time of decoding algorithms for linear codes is a function of the three code parameters $[n, k, d]$. However, with overwhelming probability random binary linear codes attain a rate $R := \frac{k}{n}$ which is close to the Gilbert Varshamov bound $1 - H(\frac{d}{n})$ [10]. Therefore, we can express the running time $T(n, R)$ as a function in n, R only. One usually measures the complexity of decoding algorithms asymptotically in the code length n. Since all generic decoding algorithms run in exponential time, a reasonable metric is the complexity coefficient $F(R)$ as defined in [9], i.e. $F(R) = \lim_{n \to \infty} \frac{1}{n} \log T(n, R)$ which suppresses polynomial factors since $\lim \frac{1}{n} \log p(n) = 0$ for any polynomial $p(n)$. Thus, we have $T(n, R) = 2^{nF(R)+o(n)} \leq 2^{n\lceil F(R) \rceil_\rho}$ for large enough n. We obtain the worst-case complexity by taking $\max_{0 < R < 1} \lceil F(R) \rceil_\rho$. Here, $\lceil x \rceil_\rho := \lceil x \cdot 10^\rho \rceil \cdot 10^{-\rho}$ denotes rounding up $x \in \mathbb{R}$ to a certain number of $\rho \in \mathbb{N}$ decimal places.

Related Work. In syndrome decoding one has to compute $\mathbf{e}$ from $s(\mathbf{x})$, which means that one has to find a weight-ω linear combination of the columns of $\mathbf{H}$ that sums to the syndrome $s(\mathbf{x})$ over $\mathbb{F}_2^{n-k}$. Thus, a brute-force algorithm would require to compute $\binom{n}{\omega}$ column sums. Inspired by the work of Prange [25], it was already mentioned in the original work of McEliece [22] and later more carefully studied by Lee and Brickell [19] that the following approach, called *information set decoding*, yields better complexity.

Information set decoding basically proceeds in two steps, an initial transformation step and a search step. Both steps are iterated in a loop until the algorithm succeeds. The initial transformation step starts by randomly permuting the columns of $\mathbf{H}$. In particular, this permutes the ω columns of $\mathbf{H}$ that sum to $s(\mathbf{x})$, and thus permutes the coordinates of e. Then we apply Gaussian elimination on the rows of $\mathbf{H}$ in order to obtain a systematic form $(\mathbf{Q} \mid \mathbf{I}_{n-k})$, where $\mathbf{Q} \in \mathbb{F}_2^{(n-k)\times k}$ and $\mathbf{I}_{n-k}$ is the $(n-k)$-dimensional identity matrix. The Gaussian elimination operations are also applied to $s(\mathbf{x})$ which results in $\tilde{s}(\mathbf{x})$.

Let us fix an integer $p < \omega$. In the search step, we compute for every linear combination of p columns from $\mathbf{Q}$ its Hamming distance to $\tilde{s}(\mathbf{x})$. If the distance is exactly $\omega - p$ then can we add to our p columns those $\omega - p$ unit vectors from $\mathbf{I}_{n-k}$ that exactly yield $\tilde{s}(\mathbf{x})$. Undoing the Gauss elimination recovers the desired error vector e. Obviously, information set decoding can only succeed if the initial column permutation results in a permuted e that has exactly p ones in its first k coordinates and $\omega - p$ ones in its last $n - k$ coordinates. Optimization of p leads to a running time of $2^{0.05752n}$.

Leon[20] and Stern[30] observed in 1989 that one can improve on the running time when replacing in the search step the brute-force search for weight-p linear combinations by a Meet-in-the-middle approach. Let us fix an integer $\ell < n - k$ and let us project $(\mathbf{Q} \mid \mathbf{I}_{n-k})$ to its first ℓ rows. We split the projection of $\mathbf{Q}$ into two matrices $\mathbf{Q}_1$, $\mathbf{Q}_2$ each having $\frac{k}{2}$ columns. Then we create two lists $\mathcal{L}_1, \mathcal{L}_2$ that contain all weight-$\frac{p}{2}$ linear combinations of columns from $\mathbf{Q}_1$ and $\mathbf{Q}_2$, respectively. Moreover, we add the projection of $\tilde{s}(\mathbf{x})$ to every element in $\mathcal{L}_2$ and sort the resulting list.

Then we search for matching elements from $\mathcal{L}_1$ and $\mathcal{L}_2$. These elements define weight-p sums of vectors from $\mathbf{Q}$ that exactly match $\tilde{s}(\mathbf{x})$ in its first ℓ coordinates. As before, if the remaining coordinates differ from $\tilde{s}(\mathbf{x})$ by a weight-$(\omega - p)$ vector, then we can correct these positions by suitable unit vectors from $\mathbf{I}_{n-k}$. The running time of this algorithm is $2^{0.05564n}$.

The ball collision technique of Bernstein, Lange and Peters [5] lowers this complexity to $2^{0.05559n}$ by allowing a non-exact matching of the elements of $\mathcal{L}_1$ and $\mathcal{L}_2$. The same asymptotic complexity can be achieved by transforming $\mathbf{H}$ into $(\mathbf{Q} \mid \begin{smallmatrix} \mathbf{0} \\ \mathbf{I}_{n-k-\ell} \end{smallmatrix})$ with $\mathbf{Q} \in \mathbb{F}_2^{(n-k)\times(k+\ell)}$, as proposed by Finiasz and Sendrier [12]. The lists $\mathcal{L}_1, \mathcal{L}_2$ then each contain all weight-$\frac{p}{2}$ sums out of $\frac{k+\ell}{2}$ columns. The asymptotic analysis of this variant can be found in [23].

Notice that finding a weight-p sum of columns of $\mathbf{Q}$ that exactly matches $\tilde{s}(\mathbf{x})$ in ℓ coordinates is a vectorial version of the subset sum problem in $\mathbb{F}_2^\ell$. This vectorial version was called the *column match problem* by May, Meurer and Thomae (MMT) [23], who adapted the subset sum representation technique from Howgrave-Graham and Joux [15] to the column match problem.

Let $\mathbf{Q} \in \mathbb{F}_2^{(n-k)\times(k+\ell)}$ be as before, where $\mathbf{q}_1,\dots,\mathbf{q}_{k+\ell}$ denote the columns of $\mathbf{Q}$. A Meet-in-the-Middle approach matches the first ℓ coordinates via the identity

$$\sum_{i\in I_1} \mathbf{q}_i = \sum_{i\in I_2} \mathbf{q}_i + \tilde{s}(\mathbf{x}) \ , \tag{1}$$

where $I_1 \subset \left[1, \frac{k+\ell}{2}\right]$, $I_2 \subset \left[\frac{k+\ell}{2} + 1, k+\ell\right]$ and $|I_1| = |I_2| = \frac{p}{2}$.

Using the representation technique, one chooses I_1 and I_2 no longer from half-sized intervals but they both are chosen from the whole interval $[1, k+\ell]$ such that $I_1 \cap I_2 = \emptyset$. Thus, every solution I admits $\binom{p}{p/2}$ representations $I = I_1 \cup I_2$. Notice that increasing the range of I_1, I_2 also increases the size of the lists $\mathcal{L}_1$ and $\mathcal{L}_2$ from $\binom{(k+\ell)/2}{p/2}$ to $\binom{k+\ell}{p/2}$. But constructing only a $\binom{p}{p/2}^{-1}$-fraction of each list suffices to let a single representation of the solution survive on expectation. This approach leads to an algorithm which runs in time $2^{0.05364n}$.

Our Contribution. We propose to choose $|I_1| = |I_2| = \frac{p}{2} + \varepsilon$ for some $\varepsilon > 0$ such that $|I_1 \cap I_2| = \varepsilon$. So we allow for ε columns $\mathbf{q}_i$ that appear on both sides of identity (1). Thus every solution I is written as the symmetric difference $I = I_1 \Delta I_2 := I_1 \cup I_2 \setminus (I_1 \cap I_2)$, where we cancel out all ε elements in the intersection of I_1 and I_2.

Let us compare our approach with the realization of the search step in the algorithms of Stern [30] and MMT [23]. In Stern's algorithm both index sets I_1, I_2 are chosen in a disjoint fashion. Thus every solution I only has a unique representation as the union of I_1 and I_2. MMT choose fully intersecting sets I_1, I_2, but they only consider a union of *disjoint* sets I_1, I_2. Basically, this allows that every of the p elements in $I = I_1 \cup I_2$ can appear either as an element of I_1 or as an element of I_2, so it can appear on both sides of identity (1).

In contrast, we choose fully intersecting sets I_1, I_2 and additionally allow for a union of *intersecting* sets. Thus, we additionally allow that even those $k + \ell - p$ elements that are *outside of* $I = I_1 \cup I_2$ may appear in I_1, I_2 as long as they appear in both sets, and thus cancel out. This drastically increases the number of representations, since for random code instances the number of zeros in an error vector $\mathbf{e}$ is much larger than the number of ones. Whereas MMT only allow to split each 1-entry of $\mathbf{e}$ into two parts, either $1 = 0 + 1$ or $1 = 1 + 0$, we also allow to split each 0-entry of $\mathbf{e}$ into two parts, either $0 = 0 + 0$ or $0 = 1 + 1$. Hence our benefit comes from using the equation $1 + 1 = 0$ in $\mathbb{F}_2$. Notice that our approach therefore increases the number of representation per solution I to $\binom{p}{p/2} \cdot \binom{k+\ell-p}{\varepsilon}$.

Our main algorithmic task that we describe in this work is the construction of two lists $\mathcal{L}_1, \mathcal{L}_2$ such that a single representation of each solution survives. This is realized by a three-level divide-and-conquer algorithm that is similar to Wagner's generalized birthday algorithm [31].

Our enhanced representation technique allows us to significantly lower the asymptotic running time to $2^{0.04934n}$. The following figure shows the curve of the complexity coefficient for the two most recent algorithms [5,23] compared to our new algorithm.

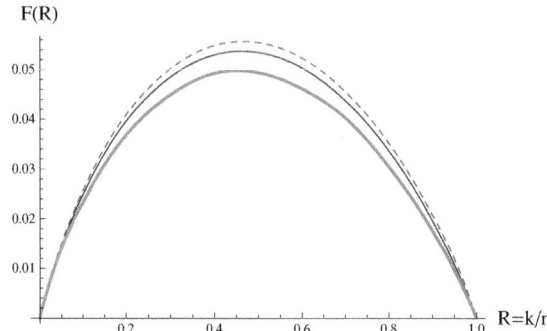

Fig. 1. Comparison of $F(R)$ for code rates $0 < R < 1$ for bounded distance decoding. Our algorithm is represented by the thick curve, MMT is the thin curve and Ball-collision is the dashed curve.

2   Generalized Information Set Decoding

We now give a detailed description of a generalized information set decoding (ISD) framework as described by Finiasz and Sendrier [12] in 2009. Recall that the input to an ISD algorithm is a tuple $(\mathbf{H}, \mathbf{s})$ where $\mathbf{H} \in \mathbb{F}_2^{(n-k) \times n}$ is a parity check matrix of a random linear $[n, k, d]$-code and $\mathbf{s} = \mathbf{He}$ is the syndrome of the unknown error vector $\mathbf{e}$ of weight $\omega := \mathrm{wt}(\mathbf{e}) = \lfloor \frac{d-1}{2} \rfloor$.

ISD is a randomized Las Vegas type algorithm that iterates two steps until the solution $\mathbf{e}$ is found. The first step is an initial linear transformation of the parity check matrix $\mathbf{H}$, followed by a search phase as the second step.

In the initial transformation, we permute the columns of $\mathbf{H}$ by multiplying with a random permutation matrix $\mathbf{P} \in \mathbb{F}_2^{n \times n}$. Then we perform Gaussian elimination on the rows of $\mathbf{HP}$ by multiplying with an invertible matrix $\mathbf{T} \in \mathbb{F}_2^{(n-k) \times (n-k)}$. This yields a parity check matrix $\tilde{\mathbf{H}} = \mathbf{THP}$ in quasi-systematic form containing a 0-submatrix in the right upper corner as illustrated in Fig. 2. Here we denote by $\mathbf{Q}^I$ the projection of $\mathbf{Q}$ to the rows defined by the index set $I \subset \{1, \ldots, n-k\}$. Analogously, we denote by $\mathbf{Q}_I$ the projection of $\mathbf{Q}$ to its columns. In particular we define $[\ell] := \{1, \ldots, \ell\}$ and $[\ell, n-k] = \{\ell, \ldots, n-k\}$. We denote the initial transformation $\mathsf{Init}(\mathbf{H}) := \mathbf{THP}$.

We set $\tilde{\mathbf{s}} := \mathbf{Ts}$ and look for an ISD-solution $\tilde{\mathbf{e}}$ of $(\tilde{\mathbf{H}}, \tilde{\mathbf{s}})$, i.e. we look for an $\tilde{\mathbf{e}}$ satisfying $\tilde{\mathbf{H}}\tilde{\mathbf{e}} = \tilde{\mathbf{s}}$ and $\mathrm{wt}(\tilde{\mathbf{e}}) = \omega$. This yields a solution $\mathbf{e} = \mathbf{P}\tilde{\mathbf{e}}$ for the original problem. Notice that applying the permutation matrix to $\tilde{\mathbf{e}}$ leaves the weight unchanged, i.e. $\mathrm{wt}(\mathbf{e}) = \omega$, and $\mathbf{THe} = \tilde{\mathbf{H}}\mathbf{e} = \tilde{\mathbf{s}} = \mathbf{Ts}$ implies $\mathbf{He} = \mathbf{s}$ as desired. In the search phase, we try to find all error vectors $\tilde{\mathbf{e}}$ that have a specific weight distribution, i.e. we search for vectors that can be decomposed into $\tilde{\mathbf{e}} = (\tilde{\mathbf{e}}_1, \tilde{\mathbf{e}}_2) \in \mathbb{F}_2^{k+\ell} \times \mathbb{F}_2^{n-k-\ell}$

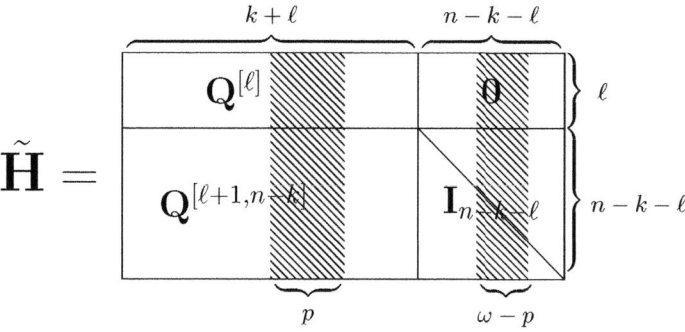

Fig. 2. Parity check matrix $\tilde{\mathbf{H}}$ in quasi-systematic form

where $\mathrm{wt}(\tilde{\mathbf{e}}_1) = p$ and $\mathrm{wt}(\tilde{\mathbf{e}}_2) = \omega - p$. Since $\mathbf{P}$ shuffles $\mathbf{e}$'s coordinates into random positions, $\tilde{\mathbf{e}}$ has the above weight distribution with probability

$$\mathcal{P} = \frac{\binom{k+l}{p}\binom{n-k-l}{\omega-p}}{\binom{n}{\omega}} . \tag{2}$$

The inverse probability $\mathcal{P}^{-1}$ is the expected number of repetitions until $\tilde{\mathbf{e}}$ has the desired distribution. Then it suffices to find the truncated vector $\tilde{\mathbf{e}}_1 \in \mathbb{F}_2^{k+\ell}$ that represents the position of the first p ones. To recover the full error vector $\tilde{\mathbf{e}} = (\tilde{\mathbf{e}}_1, \tilde{\mathbf{e}}_2)$, the missing coordinates $\tilde{\mathbf{e}}_2$ are obtained as the last $n - k - \ell$ coordinates of $\mathbf{Q}\tilde{\mathbf{e}}_1 + \tilde{\mathbf{s}}$. Hence, the goal in the ISD search phase is to compute the truncated error vector $\tilde{\mathbf{e}}_1$ efficiently. For the computation of $\tilde{\mathbf{e}}_1$ we focus on the submatrix $\mathbf{Q}^{[\ell]} \in \mathbb{F}_2^{\ell \times (k+\ell)}$. Since we fixed the $\mathbf{0}$-submatrix in the right-hand part of $\tilde{\mathbf{H}}$, we ensure that $\mathbf{Q}\tilde{\mathbf{e}}_1$ exactly matches the syndrome $\tilde{\mathbf{s}}$ on its first ℓ coordinates. Finding an $\tilde{\mathbf{e}}_1$ with such a property was called the *submatrix matching problem* in [23].

Definition 1 (Submatrix Matching Problem). *Given a random matrix $\mathbf{Q} \in_R \mathbb{F}_2^{\ell \times (k+\ell)}$ and a target vector $\mathbf{s} \in \mathbb{F}_2^\ell$, the* submatrix matching problem (SMP) *consists in finding a set I of size p such that the corresponding columns of $\mathbf{Q}$ sum up to $\mathbf{s}$, i.e. to find $I \subseteq [1, k+\ell], |I| = p$ such that*

$$\sigma(\mathbf{Q}_I) := \sum_{i \in I} \mathbf{q}_i = \mathbf{s}, \text{ where } \mathbf{q}_i \text{ is the } i\text{-th column of } \mathbf{Q}.$$

Note that the SMP itself can be seen as just another syndrome decoding instance with parity check matrix $\mathbf{Q}$, syndrome $\mathbf{s} \in \mathbb{F}_2^\ell$ and parameters $[k + \ell, \ell, p]$.

Our improvement stems from a new algorithm COLUMNMATCH allowing to solve the SMP more efficiently by using more representations of a solution I. In Alg. 1 we describe the resulting ISD algorithm. Here we denote for a vector $\mathbf{x} \in \mathbb{F}_2^n$ and an index set $I \subset [n]$ by $\mathbf{x}_I \in \mathbb{F}_2^{|I|}$ the restriction of $\mathbf{x}$ to the coordinates of I.

Let $T := T(n, R; p, \ell)$ denote the running time of COLUMNMATCH. Then the running time of GENERALIZEDISD is $\mathcal{P}^{-1} \cdot T$.

Algorithm 1. GENERALIZEDISD

Input: Parity check matrix $\mathbf{H} \in \mathbb{F}_2^{(n-k) \times n}$, syndrome $\mathbf{s} = \mathbf{H}\mathbf{e}$ with $\mathrm{wt}(\mathbf{e}) = \omega$.
Output: Error $\mathbf{e} \in \mathbb{F}_2^n$
Parameters: p, ℓ

Repeat

    Compute $\tilde{\mathbf{H}} \leftarrow \mathsf{Init}(\mathbf{H})$ and $\tilde{\mathbf{s}} \leftarrow \mathbf{T}\mathbf{s}$ where $\tilde{\mathbf{H}} = \mathbf{T}\mathbf{H}\mathbf{P}$, $\mathbf{P}$ random permutation.

    Compute $\mathcal{L} = \text{COLUMNMATCH}(\mathbf{Q}^{[\ell]}, \tilde{\mathbf{s}}_{[\ell]}, p)$.

    For all solutions $\tilde{\mathbf{e}}_1 \in \mathcal{L}$ **do**

        If $\mathrm{wt}(\mathbf{Q}\tilde{\mathbf{e}}_1 + \tilde{\mathbf{s}}) = \omega - p$ **then**

            Compute $\widetilde{\mathbf{e}} \leftarrow (\tilde{\mathbf{e}}_1, \tilde{\mathbf{e}}_2) \in \mathbb{F}_2^n$ where $\tilde{\mathbf{e}}_2 \leftarrow (\mathbf{Q}\tilde{\mathbf{e}}_1 + \tilde{\mathbf{s}})_{[\ell+1, n-k]}$

            Output $\mathbf{e} = \widetilde{\mathbf{e}}\mathbf{P}$.

3   The Merge-Join Building Block

In order to realize our improved SMP algorithm, we first introduce an essential building block that realizes the following task. Given a matrix $\mathbf{Q} \in \mathbb{F}_2^{\ell \times (k+\ell)}$ and two lists $\mathcal{L}_1$ and $\mathcal{L}_2$ containing binary vectors $\mathbf{x}_1, \ldots, \mathbf{x}_{|\mathcal{L}_1|}$ and $\mathbf{y}_1, \ldots, \mathbf{y}_{|\mathcal{L}_2|}$ of length $k+\ell$, we aim to join those elements $\mathbf{x}_i$ and $\mathbf{y}_j$ into a new list $\mathcal{L} = \mathcal{L}_1 \bowtie \mathcal{L}_2$ whose sum has weight p, i.e. $\mathrm{wt}(\mathbf{x}_i + \mathbf{y}_j) = p$. Furthermore, we require that the corresponding column-sum of $\mathbf{Q}$ already matches a given target $\mathbf{t} \in \mathbb{F}_2^r$ on its right-most $r \leq \ell$ coordinates, i.e. $(\mathbf{Q}(\mathbf{x}_i + \mathbf{y}_j))_{[r]} = \mathbf{t}$.

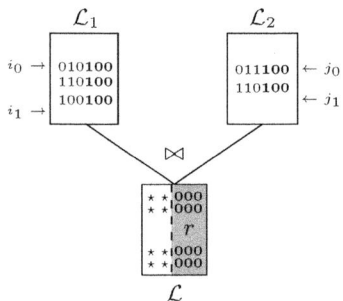

Fig. 3. Illustration of the MERGE-JOIN algorithm to obtain $\mathcal{L} = \mathcal{L}_1 \bowtie \mathcal{L}_2$

Searching for matching vectors $(\mathbf{Q}\mathbf{y}_j)_{[r]} + \mathbf{t}$ and $(\mathbf{Q}\mathbf{x}_i)_{[r]}$ accomplishes this task. We call all matching vectors with weight different from p *inconsistent solutions*. Notice that we might also obtain the same vector sum from two different pairs of vectors from $\mathcal{L}_1, \mathcal{L}_2$. In this case we obtain a matched vector that we already have, which we call a *duplicate*. During our matching process we filter out all inconsistent solutions and duplicates.

The matching process is illustrated in Fig. 3. The complete algorithm is given as Alg. 2 and is based on a classical algorithm from Knuth [18] which realizes the collision search as follows. Sort the first list lexicographically according to the r-bit labels

$L_1(\mathbf{x}_i) := (\mathbf{Q}\mathbf{x}_i)_{[r]}$ and the second list according to the labels $L_2(\mathbf{y}_j) := (\mathbf{Q}\mathbf{y}_j)_{[r]} + \mathbf{t}$. We add $\mathbf{t}$ to the labels of the second list to guarantee $(\mathbf{Q}(\mathbf{x}_i + \mathbf{y}_j))_{[r]} = \mathbf{t}$.

Algorithm 2. MERGE-JOIN

Input: $\mathcal{L}_1, \mathcal{L}_2, r, p$ and $\mathbf{t} \in \mathbb{F}_2^r$
Output: $\mathcal{L} = \mathcal{L}_1 \bowtie \mathcal{L}_2$

Lexicographically sort $\mathcal{L}_1$ and $\mathcal{L}_2$ according to the labels $L_1(\mathbf{x}_i) := (\mathbf{Q}\mathbf{x}_i)_{[r]}$ and $L_2(\mathbf{y}_j) := (\mathbf{Q}\mathbf{y}_j)_{[r]} + \mathbf{t}$.
Set collision counter $C \leftarrow 0$. Let $i \leftarrow 0$ and $j \leftarrow (|\mathcal{L}_2| - 1)$
While $i < |\mathcal{L}_1|$ and $j < |\mathcal{L}_2|$ **do**
    If $L_1(\mathbf{x}_i) <_{lex} L_2(\mathbf{y}_j)$ **then** $i + +$
    If $L_1(\mathbf{x}_i) >_{lex} L_2(\mathbf{y}_j)$ **then** $j + +$
    If $L_1(\mathbf{x}_i) = L_2(\mathbf{y}_j)$ **then**
        Let $i_0, i_1 \leftarrow i$ and $j_0, j_1 \leftarrow j$
        While $i_1 < |\mathcal{L}_1|$ and $L_1(\mathbf{x}_{i_1}) = L_1(\mathbf{x}_{i_0})$ **do** $i_1 + +$
        While $j_1 < |\mathcal{L}_2|$ and $L_2(\mathbf{y}_{j_1}) = L_2(\mathbf{y}_{j_0})$ **do** $j_1 + +$
        For $i \leftarrow i_0$ to $i_1 - 1$ **do**
            For $j \leftarrow j_0$ to $j_1 - 1$ **do**
                $C = C + 1$
                Insert collision $\mathbf{x}_i + \mathbf{y}_j$ into list $\mathcal{L}$ (unless filtered out)
        Let $i \leftarrow i_1$, $j \leftarrow j_1$
Output $\mathcal{L}, C$.

To detect all collisions, one now initializes two counters i and j starting at the beginning of the lists $\mathcal{L}_1$ and $\mathcal{L}_2$ and pointing at elements $\mathbf{x}_i$ and $\mathbf{y}_j$. As long as those elements do not yield a collision, either i or j is increased depending on the relative order of the labels $L_1(\mathbf{x}_i)$ and $L_2(\mathbf{y}_j)$. Once a collision $L_1(\mathbf{x}_i) = L_2(\mathbf{y}_j)$ occurs, four auxiliary counters i_0, i_1 and j_0, j_1 are initialized with i and j, respectively. Then i_1 and j_1 can further be incremented as long as the list elements retain the same labels, while i_0 and j_0 mark the first collision (i, j) between labels $L_1(\mathbf{x}_i)$ and $L_2(\mathbf{y}_j)$. Obviously, this procedure defines two sets $C_1 = \{\mathbf{x}_{i_0}, \ldots, \mathbf{x}_{i_1}\}$ and $C_2 = \{\mathbf{y}_{j_0}, \ldots, \mathbf{y}_{j_1}\}$ such that all possible combinations yield a collision, i.e. the set $C_1 \times C_2$ can be added to the output list $\mathcal{L}$.

    This procedure is then continued with $i \leftarrow i_1$ and $j \leftarrow j_1$ until one of the counters i, j arrives at the end of a list. As mentioned before, we remove on the fly inconsistent solutions with incorrect weight $\mathrm{wt}(\mathbf{x}_i + \mathbf{y}_j) \neq p$ and duplicate elements $\mathbf{x}_i + \mathbf{y}_j = \mathbf{x}_k + \mathbf{y}_\ell$.

    Note that we introduced a collision counter C which allows us to take into account the time that is spent for removing inconsistent solutions and duplicates. The total running time of MERGE-JOIN is given by

$$T = \tilde{\mathcal{O}} \left(\max \{ |\mathcal{L}_1|, |\mathcal{L}_2|, C \} \right) .$$

Assuming uniformly distributed labels $L_1(\mathbf{x}_j)$ and $L_2(\mathbf{y}_j)$ it holds that $\mathbb{E}[C] = |\mathcal{L}_1| \cdot |\mathcal{L}_2| \cdot 2^{-r}$.

4 Our New Algorithm for Solving the Submatrix Matching Problem

As explained in Section 2, improving the submatrix matching problem (SMP) automatically improves information set decoding (ISD).

Our new SMP algorithm is inspired by using *extended representations* similar to Becker, Coron and Joux [2] for the subset sum problem.

In the MMT algorithm [23] a weight-p error vector $\mathbf{e} \in \mathbb{F}_2^{k+\ell}$ is written as the sum $\mathbf{e}_1 + \mathbf{e}_2$. However, MMT only allow that every 1-entry splits to either a 1-entry in $\mathbf{x}_1$ and a 0-entry in $\mathbf{x}_2$, or vice versa. If $\mathrm{wt}(\mathbf{e}_1) = \mathrm{wt}(\mathbf{e}_2) = \frac{p}{2}$ this allows for $\binom{p}{p/2}$ different representations as a sum of two vectors.

Our key observation is that we can also split the 0-entries of $\mathbf{e}$ into either $(0,0)$ or $(1,1)$. Hence if we choose $\mathrm{wt}(\mathbf{e}_1) = \mathrm{wt}(\mathbf{e}_2) = \frac{p}{2} + \varepsilon$ then we gain a factor of $\binom{k+\ell-p}{\varepsilon}$, namely the number of positions where we split as $(1,1)$. Notice that in all coding-based scenarios $\mathrm{wt}(\mathbf{e})$ is relatively small compared to k and n. Thus $\mathbf{e}$ contains many more zeros than ones, from which our new representation heavily profits.

To solve the SMP, we proceed as follows. Let $I \subset [k+\ell]$ be the index set of cardinality p with $\sigma(\mathbf{Q}_I) = \mathbf{s}$ that we want to find.

We represent I by two index sets I_1 and I_2 of cardinality $\frac{p}{2} + \varepsilon$ contained in the whole interval $[k+l]$ and require I_1 and I_2 to intersect in a fixed number of ε coordinates as illustrated in Fig. 4.

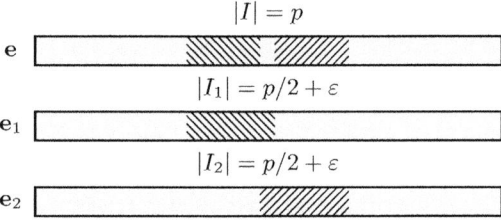

Fig. 4. Decomposition of an index set I into two overlapping index sets

The resulting index set I is then represented as the symmetric difference $I_1 \Delta I_2 := (I_1 \cup I_2) \setminus (I_1 \cap I_2)$ which yields an index set I of cardinality p as long as I_1 and I_2 intersect in exactly ε positions.

It turns out that the optimal running time can be obtained by applying the representation technique twice, i.e. we introduce further representations of the index sets I_1 and I_2 on a second computation layer.

4.1 Our COLUMNMATCH Algorithm

Our algorithm can be described as a computation tree of depth three, see Fig. 5 for an illustration. We enumerate the layers from bottom to top, i.e. the third layer identifies the initial computation of disjoint base lists $\mathcal{B}_1$ and $\mathcal{B}_2$ and the zero layer identifies the final output list $\mathcal{L}$.

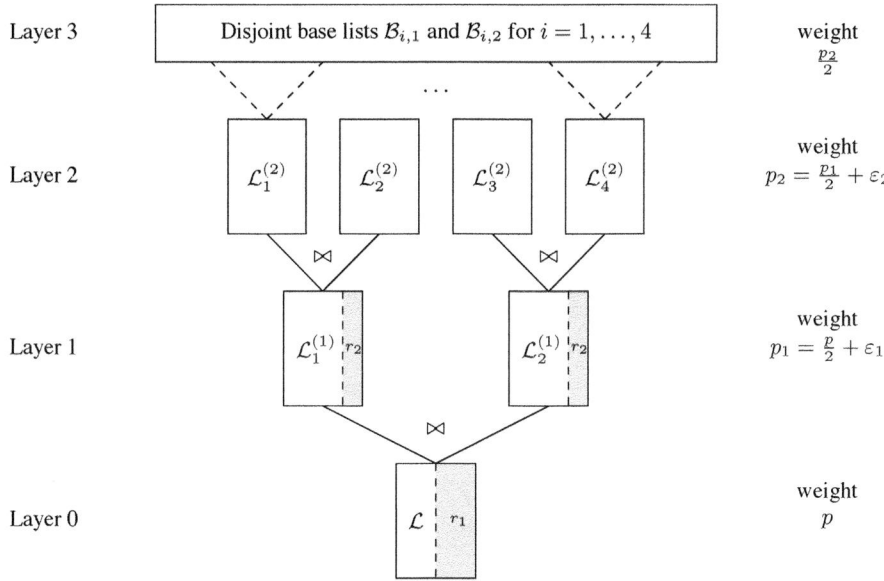

Fig. 5. Illustration of the COLUMNMATCH algorithm

Recall that we aim to find an index set I of size p with $\sum_{i \in I} \mathbf{q}_i = \mathbf{s}$. We introduce parameters ε_1 and ε_2 representing the number of additional 1's we allow on the first and second layer, respectively. In the following description, we equip every object with an upper index that indicates its computation layer, e.g. a list $\mathcal{L}_j^{(2)}$ is contained in the second layer.

On the first layer, we search for index sets $I_1^{(1)}$ and $I_2^{(1)}$ in $[k+\ell]$ of size $p_1 := \frac{p}{2} + \varepsilon_1$ which intersect in exactly ε_1 coordinates such that $I = I_1^{(1)} \Delta I_2^{(1)}$. In other words, we create lists of binary vectors $\mathbf{e}_1^{(1)}$ and $\mathbf{e}_2^{(1)}$ of weight p_1 and search for tuples $(\mathbf{e}_1^{(1)}, \mathbf{e}_2^{(1)})$ such that $\mathrm{wt}(\mathbf{e}_1^{(1)} + \mathbf{e}_2^{(1)}) = p$ and $\mathbf{Q}(\mathbf{e}_1^{(1)} + \mathbf{e}_2^{(1)}) = \mathbf{s}$.

Note that the number of tuples $(\mathbf{e}_1^{(1)}, \mathbf{e}_2^{(1)})$ that represent a single solution vector $\mathbf{e}$ is

$$R_1(p, \ell; \varepsilon_1) := \binom{p}{\frac{p}{2}}\binom{k+l-p}{\varepsilon_1} . \tag{3}$$

To optimize the running time, we impose a constraint on $r_1 \approx \log_2 R_1$ coordinates of the corresponding vectors $\mathbf{Q}\mathbf{e}_i^{(1)}$ such that we can still expect to find one representation of the desired solution $\mathbf{e}$.

More precisely, the algorithm proceeds as follows. We first fix a random vector $\mathbf{t}_1^{(1)} \in_R \mathbb{F}_2^{r_1}$, set $\mathbf{t}_2^{(1)} := \mathbf{s}_{[r_1]} + \mathbf{t}_2^{(1)}$ and compute two lists

$$\mathcal{L}_i^{(1)} = \{\mathbf{e_i}^{(1)} \in \mathbb{F}_2^{k+\ell} \mid \mathrm{wt}(\mathbf{e_i}) = p_1 \text{ and } (\mathbf{Q}\mathbf{e}_i^{(1)})_{[r_1]} = \mathbf{t}_i^{(1)}\} \text{ for } i = 1, 2.$$

Observe that any two elements $e_i^{(1)} \in \mathcal{L}_i^{(1)}$, $i = 1, 2$, already fulfill by construction the equation $(\mathbf{Q}(e_1^{(1)} + e_2^{(1)}))_{[r_1]} = s_{[r_1]}$, i.e. they already match the syndrome s on r_1 coordinates. In order to solve the SMP, we are interested in a solution $\mathbf{e} = e_1^{(1)} + e_2^{(1)}$ that matches the syndrome s on *all* ℓ positions and has weight *exactly* p. Once $\mathcal{L}_1^{(1)}$ and $\mathcal{L}_2^{(1)}$ have been created, this can be accomplished by calling the MERGE-JOIN algorithm from Sect. 3 on input $\mathcal{L}_1^{(1)}, \mathcal{L}_2^{(1)}$ with target s, weight p and parameter ℓ.

It remains to show how to construct $\mathcal{L}_1^{(1)}, \mathcal{L}_2^{(1)}$.

We represent $e_i^{(1)}$ as a sum of two overlapping vectors $e_{2i-1}^{(2)}, e_{2i}^{(2)}$ both of weight $p_2 := \frac{p_1}{2} + \varepsilon_2$, i.e. we require the two vectors to intersect in exactly ε_2 coordinates. Altogether, the solution e is now decomposed as

$$\mathbf{e} = e_1^{(1)} + e_2^{(1)} = e_1^{(2)} + e_2^{(2)} + e_3^{(2)} + e_4^{(2)} \ .$$

Clearly, there are

$$R_2(p, \ell; \varepsilon_1, \varepsilon_2) = \binom{p_1}{p_1/2} \cdot \binom{k + \ell - p_1}{\varepsilon_2}$$

many representations for $e_j^{(1)}$ where $p_1 = \frac{p}{2} + \varepsilon_1$. Similarly to the first layer, this allows us to fix $r_2 \approx \log R_2$ coordinates of the partial sums $\mathbf{Q}e_i^{(2)}$ to some target values $t_i^{(2)}$. More precisely, we draw two target vectors $t_1^{(2)}, t_3^{(2)} \in \mathbb{F}_2^{r_2}$, set $t_{2j}^{(2)} = (t_j^{(1)})_{[r_2]} + t_{2j-1}^{(2)}$ for $j = 1, 2$, and compute four lists

$$\mathcal{L}_i^{(2)} = \{e_i^{(2)} \in \mathbb{F}_2^{k+l} \mid \text{wt}(e_i^{(2)}) = p_2 \text{ and } (\mathbf{Q}e_i^{(2)})_{[r_2]} = t_i^{(2)}\} \text{ for } i = 1, \dots, 4.$$

Notice that by construction all combinations of two elements from either $\mathcal{L}_1^{(2)}, \mathcal{L}_2^{(2)}$ or $\mathcal{L}_3^{(2)}, \mathcal{L}_4^{(2)}$ match their respective target vector $t_j^{(1)}$ on r_2 coordinates.

Creating the Lists $\mathcal{L}_1^{(2)}, \dots, \mathcal{L}_4^{(2)}$. We exemplary explain how to create $\mathcal{L}_1^{(2)}$. The remaining lists can be constructed analogously. We apply a classical Meet-in-the-middle collision search, i.e. we decompose $e_1^{(2)}$ as $e_1^{(2)} = \mathbf{y} + \mathbf{z}$ by two non-overlapping vectors y and z of length $k + \ell$. To be more precise, we first choose a random partition of $[k + \ell]$ into two equal sized sets P_1 and P_2, i.e. $[k + \ell] = P_1 \cup P_2$ with $|P_1| = |P_2| = \frac{k+\ell}{2}$, and force y to have its $\frac{p_2}{2}$ 1-entries in P_1 and z to have its $\frac{p_2}{2}$ 1-entries in P_2. That is we construct two base lists

$$\mathcal{B}_1 := \{\mathbf{y} \in \mathbb{F}_2^{k+\ell} \mid \text{wt}(\mathbf{y}) = \frac{p_2}{2} \text{ and } y_i = 0 \forall i \in P_2\}$$

and

$$\mathcal{B}_2 := \{\mathbf{z} \in \mathbb{F}_2^{k+\ell} \mid \text{wt}(\mathbf{z}) = \frac{p_2}{2} \text{ and } z_i = 0 \forall i \in P_1\}.$$

We invoke MERGE-JOIN to compute $\mathcal{L}_1^{(2)} = \text{MERGE-JOIN}(\mathcal{B}_1, \mathcal{B}_2, r_2, p_2, t_1^{(2)})$. Let $S_3 = |\mathcal{B}_1| = |\mathcal{B}_2|$ denote the size of the base lists and let C_3 be the total number of matched vectors that occur in MERGEJOIN (since the splitting is disjoint, neither duplicates nor inconsistencies can arise). Then MERGEJOIN needs time

$$T_3(p, \ell; \varepsilon_1, \varepsilon_2) = \mathcal{O}\left(\max\{S_3, C_3\}\right).$$

Clearly, we have

$$S_3 := S_3(p, \ell; \varepsilon_1, \varepsilon_2) = \binom{(k+\ell)/2}{p_2/2}.$$

Assuming uniformly distributed partial sums we obtain

$$\mathbb{E}\left[C_3\right] = \frac{S_3^2}{2^{r_2}}.$$

We would like to stress that decomposing $\mathbf{e}_1^{(2)}$ into $\mathbf{x}$ and $\mathbf{y}$ from disjoint sets P_1 and P_2 introduces a probability of loosing the vector $\mathbf{e}_1^{(2)}$ and hence the solution $\mathbf{e} = \mathbf{e}_1^{(2)} + \mathbf{e}_2^{(2)} + \mathbf{e}_3^{(2)} + \mathbf{e}_4^{(2)}$. For a randomly chosen partition P_1, P_2, the probability that $\mathbf{e}_1^{(2)}$ equally distributes its 1-entries over P_1 and P_2 is given by

$$\mathcal{P}_{\text{split}} = \frac{\binom{(k+\ell)/2}{p_2/2}^2}{\binom{k+\ell}{p_2}}$$

which is asymptotically inverse-polynomial in n. Choosing independent partitions $P_{i,1}, P_{i,2}$ and appropriate base lists $\mathcal{B}_{i,1}, \mathcal{B}_{i,2}$ for all four lists $\mathcal{L}_i^{(2)}$, we can guarantee *independent* splitting conditions for all the $\mathbf{e}_i^{(2)}$ yielding a total splitting probability of $\mathcal{P}_{\text{Split}} = (\mathcal{P}_{\text{split}})^4$ which is still inverse-polynomial in n.

After having created the lists $\mathcal{L}_i^{(2)}$, $i = 1, \ldots, 4$ on the second layer, two more applications of the MERGEJOIN algorithm suffice to compute the lists $\mathcal{L}_j^{(1)}$ on the first layer. Eventually, a last application of MERGEJOIN yields $\mathcal{L}$, whose entries are solutions to the SMP. See Alg. 3 for a complete pseudocode description.

Algorithm 3. COLUMNMATCH

Input: $\mathbf{Q} \in \mathbb{F}_2^{\ell \times k+\ell}, \mathbf{s} \in \mathbb{F}_2^\ell, p \leq k + \ell$
Output: List $\mathcal{L}$ of vectors in $\mathbf{e} \in \mathbb{F}_2^{k+\ell}$ with $\text{wt}(\mathbf{e}) = p$ and $\mathbf{Q}\mathbf{e} = \mathbf{s}$
Parameters: Choose optimal $\varepsilon_1, \varepsilon_2$ and set $p_1 = p/2 + \varepsilon_1$ and $p_2 = p_1/2 + \varepsilon_2$.

Choose random partitions $P_{i,1}, P_{i,2}$ of $[k + \ell]$ and create the base lists $\mathcal{B}_{i,1}$ and $\mathcal{B}_{i,2}$.
Choose $\mathbf{t}_1^{(1)} \in_R \mathbb{F}_2^{r_1}$ and set $\mathbf{t}_2^{(1)} = \mathbf{s}_{[r_1]} + \mathbf{t}_1^{(1)}$.
  Choose $\mathbf{t}_1^{(2)}, \mathbf{t}_3^{(2)} \in_R \mathbb{F}_2^{r_2}$. Set $\mathbf{t}_2^{(2)} = (\mathbf{t}_1^{(1)})_{[r_2]} + \mathbf{t}_1^{(2)}$ and $\mathbf{t}_4^{(2)} = (\mathbf{t}_2^{(1)})_{[r_2]} + \mathbf{t}_3^{(2)}$.
  Compute $\mathcal{L}_i^{(2)} = \text{MERGE-JOIN}(\mathcal{B}_{i,1}, \mathcal{B}_{i,2}, r_2, p_2, \mathbf{t}_i^{(2)})$ for $i = 1, \ldots, 4$.
Compute $\mathcal{L}_i^{(1)} = \text{MERGE-JOIN}(\mathcal{L}_{2i-1}^{(2)}, \mathcal{L}_{2i}^{(2)}, r_1, p_1, \mathbf{t}_i^{(1)})$ for $i = 1, 2$.
Compute $\mathcal{L} = \text{MERGE-JOIN}(\mathcal{L}_1^{(1)}, \mathcal{L}_2^{(1)}, \ell, p, \mathbf{s})$.
Output $\mathcal{L}$.

It remains to estimate the complexity of COLUMNMATCH as a function of the parameters $(p, \ell; \varepsilon_1, \varepsilon_2)$, where $(\varepsilon_1, \varepsilon_2)$ are optimization parameters. Notice that the values r_i and p_i are fully determined by $(p, \ell; \varepsilon_1, \varepsilon_2)$. The base lists $\mathcal{B}_1$ and $\mathcal{B}_2$ are of size $S_3(p, \ell; \varepsilon_1, \varepsilon_2)$ as defined above.

The three consecutive calls to the MERGE-JOIN routine create lists $\mathcal{L}_j^{(2)}$ of size $S_2(p, \ell; \varepsilon_1, \varepsilon_2)$, lists $\mathcal{L}_i^{(1)}$ of size $S_1(p, \ell; \varepsilon_1, \varepsilon_2)$ and the final list $\mathcal{L}$ (which has not to be stored). More precisely, we obtain

$$S_i(p, \ell; \varepsilon_1, \varepsilon_2) = \mathbb{E}\left[|\mathcal{L}_j^{(i)}|\right] = \binom{k+\ell}{p_i} \cdot 2^{-r_i} \text{ for } i = 1, 2.$$

Here we assume uniformly distributed partial sums $\mathbf{Q}\mathbf{e}_i^{(j)}$.

Let C_i for $i = 1, 2, 3$ denote the number of all matching vectors (including possible inconsistencies or duplicates) that occur in the three MERGE-JOIN steps. If we set $r_3 = 0$ and $r_0 = \ell$, then

$$\mathbb{E}\left[C_i\right] = S_i^2 \cdot 2^{r_i - r_{i-1}}.$$

Following the analysis of MERGE-JOIN in Sect. 3, the time complexities T_i of the three MERGE-JOIN steps is given by

$$T_i(p, \ell; \varepsilon_1, \varepsilon_2) = \max\left\{S_i, C_i\right\}.$$

The overall time and space complexity is thus given by

$$T(p, \ell; \varepsilon_1, \varepsilon_2) = \max\left\{T_3, T_2, T_1\right\} \tag{4}$$

and

$$S(p, \ell; \varepsilon_1, \varepsilon_2) = \max\left\{S_3, S_2, S_1\right\} .$$

For optimizing $T(p, \ell; \varepsilon_1, \varepsilon_2)$ one has to compute the C_i. Heuristically, we can assume that the C_i achieve their expected values up to a constant factor. Since our heuristic analysis also relies on the fact that projected partial sums of the form $(\mathbf{Q}\mathbf{e})_{[r]}$ yield uniformly distributed vectors in $\mathbb{F}_2^r$, a proper theoretical analysis needs to take care of a certain class of malformed input parity check matrices $\mathbf{H}$. We show how to obtain a provable variant of our algorithm that works for all but a negligible amount of input matrices $\mathbf{H}$ in the full version of the paper [3]. The provable variant simply aborts computation if the C_i differ too much from their expectation.

5   Comparison of Asymptotic Complexity

We now show that we improve information set decoding by an exponential factor in comparison to the latest results [5,23]. To compute the complexity coefficient $F(R)$ for our algorithm for a fixed code rate R, we need to optimize the parameters p, ℓ, ε_1 and ε_2 such that the expression

$$T(p, \ell; \varepsilon_1, \varepsilon_2) \cdot \mathcal{P}(p, \ell)^{-1} \tag{5}$$

is minimized under the natural constraints

$$0 < \ell < \min\{n - k, n - k - \omega - p\}$$
$$0 < p < \min\{\omega, k + \ell\}$$
$$0 < \varepsilon_1 < k + \ell - p$$
$$0 < \varepsilon_2 < k + \ell - p_1$$
$$0 < R_2(p, \ell; \varepsilon_1, \varepsilon_2) < R_1(p, \ell; \varepsilon_1, \varepsilon_2) < \ell \ .$$

The time per iteration T is given by Eq. (4) and the number of iterations $\mathcal{P}^{-1}$ equals $\left(\binom{k+\ell}{p}\binom{n-k-\ell}{\omega-p}/\binom{n}{\omega}\right)^{-1}$ as given in Eq. (2).

For random linear codes, we can relate $R = k/n$ and $D = d/n$ via the Gilbert-Varshamov bound. Thus asymptotically we obtain $D = H^{-1}(1 - R) + o(1)$, where H is the binary entropy function. For *bounded distance decoding*, we set $W := \omega/n = D/2$. We numerically determined the optimal parameters for several equidistant rates R and interpolated $F(R)$. To calculate $F(R)$ we make use of the well known approximation $\binom{\alpha n}{\beta n} = 2^{\alpha H(\beta/\alpha)n + o(n)}$. The results are shown in Fig. 1.

For *full decoding*, in the worst-case we need to decode a highest weight coset leader of the code C, its weight ω corresponds to the *covering radius* of C which is defined as the smallest radius r such that C can be covered by discrete balls of radius r. The Goblick bound [13] ensures that $r \geq nH^{-1}(1 - R) + o(n)$ for *all* linear codes. Independently, Blinovskii [7] and Levitin [21] further proved that this bound is tight for *almost all* linear codes, i.e. $r = nH^{-1}(1 - R) + o(n)$. This justifies our choice $W = H^{-1}(1 - R)$ for the full decoding scenario.

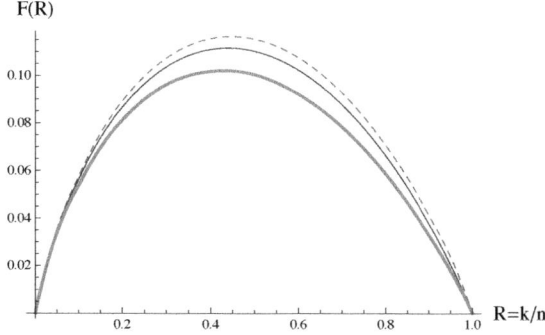

Fig. 6. $F(R)$ for full decoding. Our algorithm is represented by the thick curve, MMT is the thin curve and Ball-collision is the dashed curve.

We conclude by taking a closer look at the *worst-case* complexities of decoding algorithms for random linear codes and a typical McEliece setting with relative distance $D = 0.04$ and rate $R = 0.7577$. Notice that three out of the four parameter sets for security levels between 80 and 256 bit from [4] closely match these code parameters.

Table 1. Comparison of worst-case complexity coefficients, e.g. the time columns represent the maximal complexity coefficient $F(R)$ for $0 < R < 1$

	half-dist.		full dec.		McEliece	
	time	space	time	space	time	space
Lee-Brickell	0.05752	-	0.1208	-	0.0857	-
Stern	0.05564	0.0135	0.1167	0.0318	0.0809	0.0327
Ball-collision	0.05559	0.0148	0.1164	0.0374	0.0807	0.0348
MMT	0.05364	0.0216	0.1116	0.0541	0.0760	0.0482
Our algorithm	0.04934	0.0286	0.1019	0.0769	0.0672	0.0586

All algorithms were optimized for speed, not for memory. For a comparison of full decoding with fixed memory, we can easily restrict Ball-collision, MMT and our new algorithm to the space complexity coefficient 0.0317 of Stern's algorithm which holds for $k \approx 0.446784$. In this case, we obtain time complexities $F_{\text{ball}}(R) = 0.1163$, $F_{\text{MMT}}(R) = 0.1129$ and $F_{\text{our}}(R) = 0.1110$, which shows that our improvement is not a pure time memory tradeoff.

For a better verifiability of our optimization and the resulting complexities, we make all data including the Mathematica code publicly available at `http://cits.rub.de/personen/meurer.html`. If needed, this code may also be used to compute optimal parameters for arbitrary code parameters.

Acknowledgment. We would like to thank Dan Bernstein for several excellent comments, in particular he proposed to use random partitions for generating the base lists in the COLUMNMATCH algorithm.

References

1. Alekhnovich, M.: More on Average Case vs Approximation Complexity. In: 44th Symposium on Foundations of Computer Science (FOCS), pp. 298–307 (2003)
2. Becker, A., Coron, J.-S., Joux, A.: Improved Generic Algorithms for Hard Knapsacks. In: Paterson, K.G. (ed.) EUROCRYPT 2011. LNCS, vol. 6632, pp. 364–385. Springer, Heidelberg (2011)
3. Becker, A., Joux, A., May, A., Meurer, A.: Decoding Random Binary Linear Codes in $2^{n/20}$: How $1 + 1 = 0$ Improves Information Set Decoding. Full Version, `http://eprint.iacr.org`
4. Bernstein, D.J., Lange, T., Peters, C.: Attacking and Defending the McEliece Cryptosystem. In: Buchmann, J., Ding, J. (eds.) PQCrypto 2008. LNCS, vol. 5299, pp. 31–46. Springer, Heidelberg (2008)
5. Bernstein, D.J., Lange, T., Peters, C.: Smaller Decoding Exponents: Ball-Collision Decoding. In: Rogaway, P. (ed.) CRYPTO 2011. LNCS, vol. 6841, pp. 743–760. Springer, Heidelberg (2011)
6. Elwyn, R.J.M., Berlekamp, R., van Tilborg, H.C.: On the inherent intractability of certain coding problems. IEEE Transactions on Information Theory 24, 384–386 (1978)
7. Blinovskii, V.M.: Lower asymptotic bound on the number of linear code words in a sphere of given radius in $\mathbb{F}_q^n$. Probl. Peredach. Inform. 23, 50–53 (1987)

8. Canteaut, A., Chabaud, F.: A new algorithm for finding minimum-weight words in a linear code: Application to mceliece's cryptosystem and to narrow-sense bch codes of length 511. IEEE Transactions on Information Theory 44(1), 367–378 (1998)

9. Coffey, J.T., Goodman, R.M.: The complexity of information set decoding. IEEE Transactions on Information Theory 36, 1031–1037 (1990)

10. Coffey, J.T., Goodman, R.M.: Any code of which we cannot think is good. IEEE Transactions on Information Theory 36 (1990)

11. Faugère, J.-C., Otmani, A., Perret, L., Tillich, J.-P.: A Distinguisher for High Rate McEliece Cryptosystems. In: YACC 2010, full version available as eprint Report 2010/331 (2010)

12. Finiasz, M., Sendrier, N.: Security Bounds for the Design of Code-Based Cryptosystems. In: Matsui, M. (ed.) ASIACRYPT 2009. LNCS, vol. 5912, pp. 88–105. Springer, Heidelberg (2009)

13. Goblick Jr., T.J.: Coding for a discrete information source with a distortion measure. Ph.D. dissertation, Dept. of Elect. Eng. M.I.T., Cambridge, MA (1962)

14. Hopper, N.J., Blum, M.: Secure Human Identification Protocols. In: Boyd, C. (ed.) ASIACRYPT 2001. LNCS, vol. 2248, pp. 52–66. Springer, Heidelberg (2001)

15. Howgrave-Graham, N., Joux, A.: New Generic Algorithms for Hard Knapsacks. In: Gilbert, H. (ed.) EUROCRYPT 2010. LNCS, vol. 6110, pp. 235–256. Springer, Heidelberg (2010)

16. Jordan, J.P.: A variant of a public key cryptosystem based on goppa codes. SIGACT News 15, 61–66 (1983)

17. Kiltz, E., Pietrzak, K., Cash, D., Jain, A., Venturi, D.: Efficient Authentication from Hard Learning Problems. In: Paterson, K.G. (ed.) EUROCRYPT 2011. LNCS, vol. 6632, pp. 7–26. Springer, Heidelberg (2011)

18. Knuth, D.: Art of Computer Programming: Sorting and Searching, 2nd edn., vol. 3. Addison-Wesley Professional (1998)

19. Lee, P.J., Brickell, E.F.: An Observation on the Security of McEliece's Public-Key Cryptosystem. In: Günther, C.G. (ed.) EUROCRYPT 1988. LNCS, vol. 330, pp. 275–280. Springer, Heidelberg (1988)

20. Leon, J.S.: A probabilistic algorithm for computing minimum weights of large error-correcting codes. IEEE Transactions on Information Theory 34(5), 1354–1359 (1988)

21. Levitin, L.B.: Covering radius of almost all linear codes satisfies the Goblick bound. In: IEEE Internat. Symp. on Information Theory, Kobe, Japan (1988)

22. McEliece, R.J.: A public-key cryptosystem based on algebraic coding theory. In: Jet Propulsion Laboratory DSN Progress Report 42–44, pp. 114–116 (1978)

23. May, A., Meurer, A., Thomae, E.: Decoding Random Linear Codes in $\tilde{\mathcal{O}}(2^{0.054n})$. In: Lee, D.H., Wang, X. (eds.) ASIACRYPT 2011. LNCS, vol. 7073, pp. 107–124. Springer, Heidelberg (2011)

24. Nguyen, P.Q., Shparlinski, I.E., Stern, J.: Distribution of modular sums and the security of the server aided exponentiation. In: Progress in Computer Science and Applied Logic. Final Proceedings of Cryptography and Computational Number Theory Workshop, Singapore 1999, vol. 20, pp. 331–224 (2001)

25. Prange, E.: The Use of Information Sets in Decoding Cyclic Codes. IRE Transaction on Information Theory 8(5), 5–9 (1962)

26. Peters, C.: Information-Set Decoding for Linear Codes over $\mathbb{F}_q$. In: Sendrier, N. (ed.) PQCrypto 2010. LNCS, vol. 6061, pp. 81–94. Springer, Heidelberg (2010)

27. Regev, O.: On lattices, learning with errors, random linear codes, and cryptography. In: Proceedings of the 37th Annual ACM Symposium on Theory of Computing (STOC), pp. 84–93 (2005)

28. Sendrier, N.: Finding the permutation between equivalent linear codes: The support splitting algorithm. IEEE Transactions on Information Theory 46, 1193–1203 (2000)

29. Sendrier, N.: On the security of the McEliece public-key cryptosystem. In: Blaum, M., Farrell, P., van Tilborg, H. (eds.) Information, Coding and Mathematics, pp. 141–163. Kluwer (2002); Proceedings of Workshop honoring Prof. Bob McEliece on his 60th birthday
30. Stern, J.: A Method for Finding Codewords of Small Weight. In: Wolfmann, J., Cohen, G. (eds.) Coding Theory 1988. LNCS, vol. 388, pp. 106–113. Springer, Heidelberg (1989)
31. Wagner, D.: A Generalized Birthday Problem. In: Yung, M. (ed.) CRYPTO 2002. LNCS, vol. 2442, pp. 288–303. Springer, Heidelberg (2002)

Optimal Security Proofs
for Full Domain Hash, Revisited

Saqib A. Kakvi and Eike Kiltz

Faculty of Mathematics, Horst Görtz Institute for IT Security
Ruhr University Bochum
{saqib.kakvi,eike.kiltz}@rub.de

Abstract. RSA Full Domain Hash (RSA-FDH) is a digital signature scheme, secure again chosen message attacks in the random oracle model. The best known security reduction from the RSA assumption is non-tight, i.e., it loses a factor of q_s, where q_s is the number of signature queries made by the adversary. It was furthermore proved by Coron (EUROCRYPT 2002) that a security loss of q_s is optimal and cannot possibly be improved. In this work we uncover a subtle flaw in Coron's impossibility result. Concretely, we show that it only holds if the underlying trapdoor permutation is *certified*. Since it is well known that the RSA trapdoor permutation is (for all practical parameters) not certified, this renders Coron's impossibility result moot for RSA-FDH. Motivated by this, we revisit the question whether there is a tight security proof for RSA-FDH. Concretely, we give a new tight security reduction from a stronger assumption, the Phi-Hiding assumption introduced by Cachin et al (EUROCRYPT 1999). This justifies the choice of smaller parameters in RSA-FDH, as it is commonly used in practice. All of our results (positive and negative) extend to the probabilistic signature scheme PSS.

1 Introduction

Among all digital signatures schemes based on the RSA problem, arguably among the most important ones is RSA Full Domain Hash (RSA-FDH) by Bellare and Rogaway [3]. It is extensively used in a wide variety of applications, and serves as the basis of several existing standards such as PKCS #1 [26]. It has been demonstrated by means of a security reduction that, in the random oracle model [2], breaking the security of RSA-FDH (in the sense of existential unforgeability against chosen message attacks) is asymptotically at least as hard as inverting the RSA function.

The seminal work by Bellare and Rogaway introduced the concept of concrete security [3] and highlights the importance of considering the tightness of a security reduction. A security reduction is *tight* if an adversary breaking the scheme yields another adversary breaking the underlying hardness assumption with roughly the same success probability and running time. The current state of RSA-FDH is as follows. Coron's reduction [11] (which improves on earlier results by Bellare and Rogaway [3]) bounds the probability ε of breaking RSA-FDH in time t by $\varepsilon' \cdot q_s$, where ε' is the probability of inverting RSA in time

D. Pointcheval and T. Johansson (Eds.): EUROCRYPT 2012, LNCS 7237, pp. 537–553, 2012.

$t' \approx t$ and q_s is the number of signature queries by the forger. In other words, the security reduction for RSA-FDH is loose (it loses a factor of q_s), which can have great negative impact on the practical parameter choices of the scheme. As a numerical example, for 80 bits of security and assuming that an adversary can make up to $q_s = 2^{30}$ signature queries [3], one should use a large enough RSA modulus N such that inverting the RSA function cannot be done in fewer than $2^{110} = 2^{30} \cdot 2^{80}$ operations. Concretely, using the recommended key sizes from [28], this leads to a modulus N of about 2432 bits, compared to 1248 bits if RSA-FDH had a tight reduction. We further refer to [9] for a recent discussion on the practical impact of non-tight security reductions in cryptography.

It is an interesting question of great practical impact whether or not there is a tight security reduction for general FDH signatures (based on any trapdoor permutation TDP) and, in particular, for RSA-FDH. Unfortunately, this question was already answered to the negative exactly 10 years ago by Coron [12,13] who showed that the above non-tight security reduction is essentially *optimal*. That is, every security reduction from inverting the TDP (i.e., RSA in the case of RSA-FDH) to breaking FDH signatures will inevitably lose a q_s factor. Consequently, for RSA-FDH a large RSA modulus seems unavoidable to obtain a meaningful security proof.

1.1   An Overview of Our Results

REVISITING CORON'S IMPOSSIBILITY RESULT. We uncover a gap in Coron's result about the impossibility of a tight security reduction for FDH signatures [13]. As acknowledged by the author of [13], his impossibility result only holds if the underlying trapdoor permutation (i.e., RSA in the case of RSA-FDH) is a *certified trapdoor permutation*. A trapdoor permutation is certified [5,22] if one can publicly verify that it actually defines a permutation. Unfortunately, the RSA trapdoor permutation is not known to be certified (unless the public exponent e is prime and larger than the modulus N) and therefore the impossibility result does not apply any longer to the case of RSA-FDH.

A TIGHT SECURITY REDUCTION FOR FDH SIGNATURES. In light of the above, we revisit the question whether there exists a tight security reduction for FDH signatures. Unfortunately, we are not able to give such a tight security reduction from the assumption that the TDP is one-way, but from a stronger (yet still non-interactive) assumption, namely that the TDP is lossy (in the sense of Peikert and Waters [25]). Our main result (Theorem 8) shows that there is a *tight* security reduction from the lossiness of the TDP to breaking security of FDH, in the random oracle model.

APPLICATIONS TO RSA-FDH. Recently, Kiltz et al. [20] showed that the RSA trapdoor permutation is lossy under the under the Φ-Hiding Assumption. The Φ-Hiding Assumption was introduced by Cachin, Micali, and Stadler in 1999 [8] and it states that, roughly, (N, e) with $\gcd(\varphi(N), e) = 1$ and $e < N^{1/4}$ is computationally indistinguishable from (N', e') with $e' \mid \varphi(N')$. (Here $\varphi(N)$ is Euler's totient function.) This give a tight security reduction for RSA-FDH from the Φ-Hiding Assumption. We remark that the Φ-Hiding Assumption (or, more generally, the assumption that RSA is lossy) is a stronger assumption than the assumption that RSA is one-way. However, it dates back to 1999 [8] and has ever since been used in a number of cryptographic applications (e.g., [20,7,14,16,23,18]). It has been cryptalanyzed (e.g. [8,7,27]) and for the parameters of interest there is no known algorithm that breaks it without first factoring the modulus $N = pq$. The common interpretation is that the Φ-Hiding Assumption can in practice be viewed as *as hard as factoring* and hence gives a theoretical justification as to why RSA-FDH with a small modulus N is secure in practice.

1.2   Full Domain Hash and Coron's Impossibility Result

Recall that FDH signatures on a message m is $\sigma = f^{-1}(H(m))$, where f is the public description of the TDP and H is a hash function modelled as a random oracle. A reduction $\mathcal{R}$ that reduces inverting the TDP to breaking FDH inputs a challenge instance $(f, y = f(x))$ of the TDP and generates a public-key for FDH that is passed to a forger $\mathcal{F}$ attacking FDH signatures. Next, $\mathcal{F}$ makes a number of signature queries (which are answered by $\mathcal{R}$) and finally outputs a forgery. Finally, $\mathcal{R}$ uses the gathered information to invert the TDP, i.e., to compute $x = f^{-1}(y)$. Reduction $\mathcal{R}$ is *tight* if the success probability of $\mathcal{R}$ is roughly the same as the one of $\mathcal{F}$.

Coron's impossibility result shows that any reduction $\mathcal{R}$ from inverting the TDP f to breaking FDH which is tight (i.e., does not lose more than a factor q_s) can be turned into an efficient inverting algorithm $\mathcal{I}$ for the TDP f (that works without forger $\mathcal{F}$). In a nutshell, the argument is as follows. Given an instance of the TDP, the inverter $\mathcal{I}$ runs reduction $\mathcal{R}$ providing it with a simulated forger $\mathcal{F}$ by making a number of hash queries and then signature queries to $\mathcal{R}$. Next, $\mathcal{I}$ rewinds reduction $\mathcal{R}$ to an earlier state (after the hash queries) and uses one of the signed messages/signature pairs (say (m^*, σ^*)) obtained before the rewind as its forgery. To $\mathcal{R}$, this counts as a valid forgery since after the rewind, $\mathcal{I}$ did not make a signing query on m^*. The central argument is as follows: consider a real forger that is provided with the view as the simulated forger who outputs a forgery σ' on the same message m^*. FDH has *unique signatures*[1] and hence we can argue that σ^* (provided by $\mathcal{R}$ before the rewind) equals σ' (provided by a real forger). Hence $\mathcal{R}$ is convinced it interacts with a real forger and outputs a solution to the TDP instance. Consequently, from $\mathcal{R}$ we were able to construct an algorithm $\mathcal{I}$ that inverts the TDP without using any forger. It is shown by a combinatorial argument that the success probability of $\mathcal{I}$ is non-negative as long as the reduction $\mathcal{R}$ does not loose more than a factor of q_s, the number of signature queries.

THE GAP IN THE PROOF. During the proof of [12, Th. 5] it is silently assumed that the public-key pk generated by reduction $\mathcal{R}$ is a *real public-key*, honestly

---

[1] A signature scheme has unique signatures if for each message there exists exactly one signature that verifies w.r.t. a given (honestly generated) public-key.

generated by the key-generation algorithm of FDH, i.e., it contains f which described a permutation.[2] However, that does not necessarily hold, the public-key generated by $\mathcal{R}$ could be anything. In fact, it is possible that the public-key generated by the reduction $\mathcal{R}$ is *fake* in the sense that the FDH signatures are no longer unique relative to this fake pk. Once signatures are no longer unique (with respect to the fake pk), it is possible that a *real forger* outputs a forgery σ' on m^* which is different from σ^*, the one provided by reduction $\mathcal{R}$ before the rewind. In fact, it could be possible that $\sigma^* \neq \sigma'$ is no longer useful for $\mathcal{R}$ in order to solve the RSA instance after the rewind and hence the impossibility result breaks down. In Section 3 we restate (and prove) a corrected version of Coron's impossibility result. Fortunately, it turns out that Coron's argument can be salvaged by requiring the trapdoor permutation in FDH to be certified. Note that in case of a certified trapdoor permutation it is not longer possible for the reduction $\mathcal{R}$ to generate a fake public-key and hence signatures are guaranteed to be unique.

1.3 A Tight Security Reduction for FDH Signatures

It is precisely the non-uniqueness of FDH signatures with respect to a fake public-key that will allow us to prove a tight security from the lossiness from the lossiness of the TDP (i.e., the Φ-Hiding Assumption in the case of RSA-FDH). Our proof is surprisingly simple and is sketched as follows. In a first step we substitute the trapdoor permutation in public key with a lossy one. We use the programmability of the random oracle to show that this remains unnoticed by the adversary assuming lossiness of the TDP. Note that once the TDP is lossy, FDH signatures (i.e., σ with $f(\sigma) = H(m)$) are not longer unique since, by the definition of lossiness, each $H(m)$ has many pre-images under a lossy f. In the second step we show that any successful forger will be able to find a collision in the TDP, i.e., two values $x \neq \hat{x}$ with $f(x) = f(\hat{x})$, which is again hard assuming lossiness. The full proof is given in Section 3.

For the important case of RSA-FDH this gives a tight security reduction from the Φ-Hiding Assumption, in the random oracle model. The Φ-Hiding Assumption is believed to be true for sufficiently small public RSA exponents $e < N^{1/4-\varepsilon}$ [8]. This in particular includes the important low-exponent cases of $e = 3$ and $e = 2^{16} + 1$ since they allow efficient verification of RSA-FDH signatures.[3]

It is interesting to remark, that, at a conceptual level FDH is the first signature scheme with *unique signatures* and a *tight* security reduction (from a non-interactive assumption).[4] Previously, only tight security reductions for *randomized* signatures were known (e.g., [3,17,6,15]).

---

[2] Such restricted reductions were called *key-preserving reductions* in [24].

[3] We stress that our tight proof technically does not give a counter-example to Coron's impossibility result since our reduction is from the Φ-Hiding Assumption, not the RSA Assumption. However, as corollary the impossibility result would exclude any (even non-tight) equivalence between the Φ-Hiding and the RSA assumption.

[4] Here we do not count tight security proofs from "tautological assumptions" which are essentially assuming that the signature scheme is secure.

1.4   Extensions

Our observations can also be applied to the probabilistic signature scheme (PSS) [3] which is contained in IEEE P1363a [19], ISO/IEC 9796-2, and PKCS#1 v2.1 [26]. Coron proved that, if $\log_2(q_s)$ bits of random salt is used in PSS, then there is a tight security reduction from the one-wayness of the TDP [12,13]. Furthermore, Coron also proved that $\log_2(q_s)$ bits of random salt are essentially optimal for a tight security reduction. Our results for PSS are similar to the ones for FDH. We note that Coron's impossibility proof for PSS contains the same gap as the one in FDH, i.e., it is only correct if the underlying trapdoor permutation is certified. However, since PSS (with random salt of arbitrary length) is at least as secure as FDH, we obtain as a corollary from Section 3 a tight security proof from lossiness to the security of PSS, with random salt of arbitrary (possibly zero) length.

1.5   Related Work

There is a lot of work on FDH and tightly secure signature schemes, we try to summarize part of it relevant to this work.

TIGHT SECURITY REDUCTION FOR RSA-FDH FROM AN INTERACTIVE ASSUMPTION. Kobiltz and Menezes [21, Sec. 3] show a tight reduction from an *interactive assumption* they call the *RSA1 assumption* (which is related to the one-more-RSA assumption RSA-CTI [1]): Given N, e, and a set of $q_s + q_h$ values y_i chosen uniformly from $\mathbb{Z}_N$, the adversary is permitted adaptively to select up to q_s of those y_i for which he is given solutions x_i to $x_i^e = y_i \bmod N$. The adversary wins if he produces a solution $x_i^e = y_i \bmod N$ for one of the remaining y_i. Even though the RSA1 assumption looks plausible, it is an interactive assumption and almost a tautology for expressing that RSA-FDH signatures are secure in the random oracle model. In fact, our tight security proof for RSA-FDH also serves to show a tight reduction from Φ-Hiding to RSA1.

NON-UNIQUE SIGNATURES WITH TIGHT REDUCTIONS. There exists several previous works that build digital signature schemes with a tight security reduction. We stress that all of them have, in contrast to FDH, a randomized signing algorithm, i.e., signatures are not unique. Goh et al. [17] show that adding one single bit of random salt to the hash function of FDH allows to prove a tight security reduction from the RSA assumption. Bernstein [6] shows a tight security reduction for (a certain randomized variant of) Rabin-Williams signature scheme from the factoring assumption. More generally, Gentry et al. [15] introduce the concept of preimage samplable trapdoor functions which are non-injective trapdoor functions with an efficient pre-image sampling algorithm. They further propose a probabilistic variant of FDH and prove it tightly secure. In fact, their proof technique is reminiscent to the second step in our proof of FDH from the lossiness but FDH can not be viewed as an instance of their probabilistic FDH variant.

RSA-OAEP. Recently, [20] used the Φ-Hiding Assumption to show that the RSA function is lossy and used this fact to prove positive instantiability results of RSA-OAEP in the standard model.

1.6   Open Problems

On the one hand the Φ-Hiding Assumption is believed to be true for public exponents $e \leq N^{1/4-\varepsilon}$ and hence for these values we get a tight security reduction for RSA-FDH. On the other hand, Coron's impossibility results holds for prime e with $e > N$. This leaves the interesting open problem whether for public exponents $N^{1/4} \leq e \leq N$ there exists a tight security reduction for RSA-FDH (under a reasonable assumption).

2   Definitions

2.1   Notations and Conventions

We denote our security parameter as k. For all $n \in \mathbb{N}$, we denote by 1^n the n-bit string of all ones. For any element x in a set S, we use $x \in_R S$ to indicate that we choose x uniformly random in S. All algorithms may be randomized. For any algorithm A, we define $x \leftarrow_\$ A(a_1, \ldots, a_n)$ as the execution of A with inputs $a_1, \ldots, a_n$ and fresh randomness and then assigning the output to x. We denote the set of prime numbers by $\mathbb{P}$ and we denote the subset of k-bit primes as $\mathbb{P}_k$. Similarly, we have the integers denoted by $\mathbb{Z}$ and $\mathbb{Z}_k$. We denote by $\mathbb{Z}_N^*$ the multiplicative group modulo $N \in \mathbb{Z}$.

2.2   Games

A game (such as in Figure 2) is defined as a collection of procedures, as per the model of [4]. There is an **Initialize** procedure and a **Finalize** procedure, as well a procedure for each separate oracle. Executing a game G with and adversary $\mathcal{A}$ means running the adversary and using the procedures to answer any oracle queries. The adversary must first make one query to **Initialize**. Then it may query the oracles as many times as allowed by the definition of the game. After this, the adversary must then make 1 query to **Finalize**, which is the final procedure call of the game. The output of **Finalize** is denoted by $G^{\mathcal{A}}$. Where the Finalize procedure simply returns the output of the adversary, we omit the Finalize procedure. We use a strongly typed pseudo-code with implicit initialization. Which means all variables maintain their type throughout the execution of the games and they are all implicitly declared and initialized. Boolean flags are initialized to false, numerical types are initialized to 0, sets are initialized to $\emptyset$. We use the notation $y \leftarrow_\$ \mathcal{A}(a_1, \ldots, a_n)$ to denote invoking the probabilistic algorithm $\mathcal{A}$ with inputs $a_1, \ldots, a_n$ and fresh randomness and assigning the output to y.

2.3   Signature Schemes

A digital signature is a message-dependant bit string σ, which can only be generated by the signer, using a secret signing key sk and is transmitted with the message. The signature can then be verified by the receiver using a public verification key pk. A digital signature scheme is defined as a triple of probabilistic algorithms $\mathsf{SIG} = (\mathsf{KeyGen}, \mathsf{Sign}, \mathsf{Verify})$, which we describe below:

1. KeyGen takes as an input the unary representation of our security parameter (1^k) and outputs a signing key sk and verification key pk.
2. Sign takes as input a signing key sk, message m and outputs a signature σ.
3. Verify is a deterministic algorithm, which on input of a public key and a message-signature pair (m, σ) outputs 1 (accept) or 0 (reject).

We say that SIG is correct if for all public key and secret key pairs generated by KeyGen, we have:

$$\Pr[\mathsf{Verify}(pk, m, \mathsf{Sign}(sk, m)) = 1] = 1.$$

We now define UF-CMA (unforgeability under chosen message attacks) assuming the signature scheme SIG contains a hash function $h : \{0,1\}^* \to \mathsf{Dom}$ which is modeled as a random oracle.

---

procedure **Initialize**
$(pk, sk) \leftarrow_\$ \mathsf{KeyGen}(1^k)$
return pk

procedure **Hash**(m)
if $(m, \cdot) \in \mathcal{H}$ then fetch $(m, y) \in \mathcal{H}$; return y
else $y \in_R \mathsf{Dom}$; $\mathcal{H} \leftarrow \mathcal{H} \cup (m, y)$; return y

procedure **Sign**(m) Game UF-CMA
$\mathcal{M} \leftarrow \mathcal{M} \cup (m)$
return $\sigma \leftarrow_\$ \mathsf{Sign}(sk, y)$

procedure **Finalize**(m^*, σ^*)
if $\mathsf{Verify}(pk, m^*, \sigma^*) = 1 \wedge m^* \notin \mathcal{M}$
then return 1
else return 0

Fig. 1. Game defining UF-CMA security in the random oracle model

---

We say a signature scheme SIG is $(t, \varepsilon, q_h, q_s)$-UF-CMA secure in the random oracle model, if for all adversaries $\mathcal{A}$ running in time upto t, making at most q_h hashing and q_s signing oracle queries, they have an advantage of at most ϵ, where the advantage of $\mathcal{A}$ is defined as:

$$\mathbf{Adv}_{\mathsf{SIG}}^{\mathsf{UF\text{-}CMA}}(\mathcal{A}) = \Pr\left[\mathsf{UF\text{-}CMA}^{\mathcal{A}} \Rightarrow 1\right].$$

2.4 Trapdoor Permutations

We recall the definition of trapdoor permutation families.

Definition 1. A family of trapdoor permutations $\mathsf{TDP} = (\mathsf{Gen}, \mathsf{Eval}, \mathsf{Invert})$ consists of following three polynomial-time algorithms.

1. The probabilistic algorithm Gen, which on input 1^k outputs a public description pub (which includes an efficiently sampleable domain Dom_{pub}) and a trapdoor td.
2. The deterministic algorithm Eval, which on input pub and $x \in \mathsf{Dom}_{pub}$, outputs $y \in \mathsf{Dom}_{pub}$. We write $f(x) = \mathsf{Eval}(pub, x)$.
3. The deterministic algorithm Invert, which on input td and $y \in \mathsf{Dom}_{pub}$, outputs $x \in \mathsf{Dom}_{pub}$. We write $f^{-1}(y) = \mathsf{Invert}(pub, y)$.

We require that for all $k \in \mathbb{N}$ and all (pub, td) output by $\mathsf{Gen}(1^k)$, $f(\cdot) = \mathsf{Eval}(pub, \cdot)$ defines a permutation over Dom_{pub} and that for all $x \in \mathsf{Dom}_{pub}$, $\mathsf{Invert}(td, \mathsf{Eval}(pub, x)) = x$.

We want to point out that $f_{pub}(\cdot) = \mathsf{Eval}(pub, \cdot)$ is only required to be a permutation for correctly generated pub but not every bit-string pub necessarily yields a permutation. A family of trapdoor permutations TDP is said to be *certified* [5] if the fact that it is a permutation can be verified in polynomial time given pub.

Definition 2. A family of trapdoor permutations TDP is called certified if there exists a deterministic polynomial-time algorithm Certify that, on input of 1^k and an arbitrary (polynomially bounded) bit-string pub (potentially not generated by Gen), returns 1 iff $f(\cdot) = \mathsf{Eval}(pub, \cdot)$ defines a permutation over Dom_{pub}.

We now recall security notion for trapdoor permutations. A trapdoor permutation TDP is hard to invert (one-way) if given pub and $f_{pub}(x)$ for uniform $x \in \mathsf{Dom}_{pub}$, it is hard to compute x. More formally, it is (t, ε)-hard to invert if for all adversaries running in time t, $\Pr[\mathcal{A}(pub, \mathsf{Eval}(pub, x)) = x] \leq \varepsilon$, where the probability is taken over $(pub, td) \leftarrow \mathsf{Gen}(1^k)$, $x \in_R \mathsf{Dom}_{pub}$ and the random coin tosses of $\mathcal{A}$. The following security notion, lossiness [25], is a stronger requirement than one-wayness.

Definition 3. Let $l \geq 2$. A trapdoor permutation TDP is a (l, t, ε) lossy trapdoor permutation if the following two conditions hold.[5]

1. There exists a probabilistic polynomial-time algorithm LossyGen, which on input 1^k outputs pub' such that the range of $f_{pub'}(\cdot) := \mathsf{Eval}(pub', \cdot)$ under $\mathsf{Dom}_{pub'}$ is at least a factor of l smaller than the domain $\mathsf{Dom}_{pub'}$: $|\mathsf{Dom}_{pub'}|/|f_{pub'}(\mathsf{Dom}_{pub'})| \geq l$. (Note that we measure the lossiness in its absolute value l, i.e., the function has $\lceil \log_2 l \rceil$ *bits* of lossiness.)
2. All distinguishers $\mathcal{D}$ running in time at most t have an advantage $\mathbf{Adv}^{\mathsf{L}}_{\mathsf{TDP}}(\mathcal{D})$ of at most ε, where

$$\mathbf{Adv}^{\mathsf{L}}_{\mathsf{TDP}}(\mathcal{D}) = \Pr[\mathsf{L}_1^{\mathcal{D}} \Rightarrow 1] - \Pr[\mathsf{L}_0^{\mathcal{D}} \Rightarrow 1].$$

procedure **Initialize** Game L_0	procedure **Initialize** Game L_1
$(pub, td) \leftarrow_{\$} \mathsf{Gen}(1^k)$	$(pub', \bot) \leftarrow_{\$} \mathsf{LossyGen}(1^k)$
return pub	return pub'

Fig. 2. The Lossy Trapdoor Permutation Games

We say TDP is *regular* (l, t, ε) *lossy* if TDP is (l, t, ε) lossy and all functions $f_{pub'}(\cdot) = \mathsf{Eval}(pub', \cdot)$ generated by LossyGen are l-to-1 on $\mathsf{Dom}_{pub'}$.

---

[5] We deviate in two ways from the original definition of lossy trapdoor functions Peikert and Waters [25]. First, we define the permutation over arbitrary domains Dom, rather than $\{0, 1\}^k$; second, we measure the absolute lossiness l, rather than the bits of lossiness $\ell = \log_2(l)$.

2.5   The RSA Trapdoor Permutation

We define the RSA trapdoor permutation $\mathsf{RSA} = (\mathsf{RSAGen}, \mathsf{RSAEval}, \mathsf{RSAInv})$ as follows. The RSA instance generator $\mathsf{RSAGen}(1^k)$ outputs $pub = (N, e)$ and $td = d$, where $N = pq$ is the product of two $k/2$-bit primes, $\gcd(e, \varphi(N)) = 1$, and $d = e^{-1} \bmod \varphi(N)$. The domain is $\mathsf{Dom}_{pub} = \mathbb{Z}_N^*$. The evaluation algorithm $\mathsf{RSAEval}(pub, x)$ returns $f_{pub}(x) = x^e \bmod N$, the inversion algorithm $\mathsf{RSAInv}(td, y)$ returns $f_{pub}^{-1}(y) = y^d \bmod N$. The standard assumption is that RSA is hard to invert. We will review the (regular) lossiness of RSA in Section 4.

3   Full Domain Hash Signatures

3.1   The Scheme

For a familiy of trapdoor permutations $\mathsf{TDP} = (\mathsf{Gen}, \mathsf{Eval}, \mathsf{Invert})$ we define the Full Domain Hash (TDP-FDH) signature scheme [3] in Figure 3.

---

procedure **KeyGen**   TDP-FDH

$(pub, td) \leftarrow_\$ \mathsf{Gen}(1^k)$

Pick a hash function $h : \{0,1\}^* \to \mathsf{Dom}_{pub}$

return $(pk = (h, pub), sk = td)$

procedure **Sign**(sk, m)

return $\sigma = \mathsf{Invert}(td, h(m))$ $// \; \sigma = f_{pub}^{-1}(h(m))$

procedure **Verify**(pk, m, σ)

if $\mathsf{Eval}(pub, \sigma) = h(m)$ then return 1 $// \; f_{pub}(\sigma) \stackrel{?}{=} h(m)$

else return 0

Fig. 3. The Full Domain Hash Signature Scheme TDP-FDH

3.2   Classical Security Results of TDP-FDH

The original reduction by Bellare and Rogaway from one-wayness of TDP loses a factor of $(q_h + q_s)$ [3], which was later improved by Coron to a factor of q_s [11] for the case of the RSA trapdoor permutation.

Theorem 4. *Assume the trapdoor permutation* RSA *is* (t', ε')-*hard to invert. Then for any* (q_h, q_s), *RSA-FDH is* $(t, \varepsilon, q_h, q_s)$-*UF-CMA secure in the Random Oracle Model, where*

$$\varepsilon' = \frac{\varepsilon}{q_s} \cdot \left(1 - \frac{1}{q_s + 1}\right)^{q_s + 1} \approx \frac{\varepsilon}{q_s} \cdot \exp(-1)$$

$$t' = t + (q_h + q_s + 1) \cdot \mathcal{O}(k^3).$$

3.3   A Corrected Version of Coron's Optimality Result

Coron showed that a security loss of a factor q_s (times some constant) is essentially optimal for TDP-FDH [12,13]. To state a corrected version of Coron's impossibility result, we first recall the following definitions [12].

Definition 5. We say a reduction $\mathcal{R}$ $(t_{\mathcal{F}}, t_{\mathcal{R}}, q_h, q_s, \varepsilon_{\mathcal{F}}, \varepsilon_{\mathcal{R}})$-reduces solving a hard problem to breaking SIG = (KeyGen, Sign, Verify) if after running a forger $\mathcal{F}$ that $(t_{\mathcal{F}}, q_h, q_s, \varepsilon_{\mathcal{F}})$-breaks SIG, the reduction outputs a solution of the problem with probability at least $\varepsilon_{\mathcal{R}}$, with running time at most $t_{\mathcal{R}}$.

Definition 6. A signature scheme SIG = (KeyGen, Sign, Verify) is said to be a unique signature scheme if for every public key pk output by KeyGen, for every message m there exists exactly one bit-string $\sigma \in \{0, 1\}^*$ such that Verify$(pk, m, \sigma) = 1$.

We now state the corrected version of Coron's impossibility result which we prove in the full version of this paper.

Theorem 7. *Suppose* TDP *is a certified trapdoor permutation. Let* $\mathcal{R}$ *be a reduction that* $(t_{\mathcal{F}}, t_{\mathcal{R}}, q_h, q_s, \varepsilon_{\mathcal{F}}, \varepsilon_{\mathcal{R}})$-*reduces breaking one-wayness of* TDP *to breaking* UF-CMA *security of* TDP-FDH. *If* $\mathcal{R}$ *runs the forger only once, then we can build an inverter* $\mathcal{I}$ *which* $(t_{\mathcal{I}}, \varepsilon_{\mathcal{I}})$-*breaks one-wayness of* TDP *with:*

$$t_{\mathcal{I}} \leq 2 \cdot t_{\mathcal{R}}$$

$$\varepsilon_{\mathcal{I}} \geq \varepsilon_{\mathcal{R}} - \varepsilon_{\mathcal{F}} \cdot \frac{exp(-1)}{q_s} \cdot \left(1 - \frac{q_s}{q_h}\right)^{-1}.$$

Hence, from a security reduction from one-wayness to the security of TDP-FDH which loses less than a factor of q_s, one obtains an efficient inverter $\mathcal{I}$ for TDP.

3.4   A Tight Security Proof for TDP-FDH

The impossibility result of Theorem 7 only holds for TDP-FDH if TDP is certified trapdoor permutation. However if TDP is not certified, this leaves room for a tight proof for TDP-FDH. We now state our main result, namely that TDP-FDH is tightly secure assuming TDP is regular lossy.

Theorem 8. *Assume* TDP = (Gen, Eval, Invert) *is a regular* (l, t', ε')-*lossy trapdoor permutation for* $l \geq 2$. *Then, for any* (q_h, q_s), TDP-FDH *is* $(t, \varepsilon, q_h, q_s)$-UF-CMA *secure in the Random Oracle Model, where*

$$\varepsilon = \left(\frac{2l - 1}{l - 1}\right) \cdot \varepsilon'$$

$$t = t' - q_h \cdot T_{\mathsf{TDP}}$$

and T_{TDP} *is the time to evaluate* TDP.

Table 1. Games for the proof of Theorem 8

procedure **Initialize** $\qquad$ Game G_0	procedure **Initialize** $\qquad$ Games G_1-G_4
$(pub, td) \leftarrow_\$ \mathsf{Gen}(1^k)$ $\qquad$ $= (\mathsf{UF\text{-}CMA})$	$(pub, td) \leftarrow_\$ \mathsf{Gen}(1^k)$ $\qquad\qquad$ $//G_1, G_4$
Return $pk = pub$	$(pub, \perp) \leftarrow_\$ \mathsf{LossyGen}(1^k)$ $\quad$ $//G_2, G_3$
	Return $pk = pub$

procedure **Hash**(m)	procedure **Hash**(m)
if $(m, \cdot) \in \mathcal{H}$ then fetch (m, y_m); return y_m	if $m \in \mathcal{H}$ lookup $(m, y_m, \sigma_m) \in \mathcal{H}$
else	$\qquad$ return y_m
$\quad y_m \in_R \mathsf{Dom}_{pub}$	else
$\quad \mathcal{H} \leftarrow \mathcal{H} \cup (m, y_m)$; return y_m	$\qquad \sigma_m \in_R \mathsf{Dom}_{pub}$
	$\qquad y_m = \mathsf{Eval}(pub, \sigma_m)$
	$\qquad \mathcal{H} \leftarrow \mathcal{H} \cup (m, y_m, \sigma_m)$; return y_m

procedure **Sign**(m)	procedure **Sign**(m)
$\mathcal{M} \leftarrow \mathcal{M} \cup (m)$	$\mathcal{M} \leftarrow \mathcal{M} \cup (m)$
return $\sigma_m = \mathsf{Invert}(td, h(m))$	lookup $(m, y_m, \sigma_m) \in \mathcal{H}$, return σ_m

Procedure **Finalize**(m^*, σ^*)	Procedure **Finalize**(m^*, σ^*)
if $(\mathsf{Verify}(pub, m^*, \sigma^*) = 1) \wedge (m^* \notin \mathcal{M})$	lookup $(m^*, y_{m^*}, \sigma_{m^*}) \in \mathcal{H}$ $\qquad$ $//G_3, G_4$
return 1	if $\sigma_{m^*} = \sigma^*$ then BAD = true
else return 0	$\qquad$ return 0 $\qquad\qquad\qquad\qquad$ $//G_3, G_4$
	if $\mathsf{Verify}(pub, m^*, \sigma^*) = 1 \wedge (m^* \notin \mathcal{M})$
	return 1
	else return 0

Proof. Let $\mathcal{A}$ be an adversary that runs in time t against TDP-FDH executed in the UF-CMA experiment described in G_0 in Figure 1 with $\varepsilon = \Pr[G_0^{\mathcal{A}} \Rightarrow 1]$. Here we assume wlog that $\mathcal{A}$ always makes a query to $\mathsf{Hash}(m)$ before calling $\mathsf{Sign}(m)$ or $\mathsf{Finalize}(m, \cdot)$.

Lemma 9. $\Pr[G_0^{\mathcal{A}} \Rightarrow 1] = \Pr[G_1^{\mathcal{A}} \Rightarrow 1].$

Proof. In G_0, we modelled the hash function as a random oracle. In G_1 we modify the random oracle and the signing queries. On any m the random oracle now works by evaluating the permutation on a random element $\sigma_m \in \mathsf{Dom}_{pub}$. We then modify the signing oracle to return this element σ_m. Note that signing no longer requires the trapdoor td. It can be seen that all our signatures will verify due to the fact that $\mathsf{Eval}(pub, \sigma_m) = y_m$ for all m. Thus our simulation of the signatures is correct. Since TDP is a permutation, the distribution of our hash queries in G_1 is identical to the distribution in G_0. Thus we have $\Pr[G_0^{\mathcal{A}} \Rightarrow 1] = \Pr[G_1^{\mathcal{A}} \Rightarrow 1].$

Lemma 10. *There exists a distinguisher $\mathcal{D}_1$ against the lossines of TDP, which runs in time $t = t_{\mathcal{A}} + q_h \cdot T_{\mathsf{TDP}}$ and that $\Pr[G_1^{\mathcal{A}} \Rightarrow 1] - \Pr[G_2^{\mathcal{A}} \Rightarrow 1] = \boldsymbol{Adv}_{\mathsf{TDP}}^{\mathsf{L}}(\mathcal{D}_1).$*

Proof. From G_1 to G_2, we change the key generation from a normal permutation to a lossy permutation, however the oracles are identical in both games. We now build a distinguisher $\mathcal{D}_1$ against the lossiness of TDP, using these games. The distinguisher will run $\mathcal{A}$ and simulates the oracles $\mathsf{Sign}(\cdot), \mathsf{Hash}(\cdot)$ as described in games $\mathsf{G}_1\&\mathsf{G}_2$, for which it requires time $q_h \cdot T_{\mathsf{TDP}}$. Note that $\mathcal{D}_1$ does not require the trapdoor td to simulate the oracles. After $\mathcal{A}$ calls $\mathsf{Finalize}$, $\mathcal{D}_1$ returns the inverse of $\mathsf{Finalize}$. Thus we can see that $\Pr[\mathsf{L}_0^{\mathcal{D}_1} \Rightarrow 1] = 1 - \Pr[\mathsf{G}_1^{\mathcal{A}} \Rightarrow 1]$. Similarly, we have $\Pr[\mathsf{L}_1^{\mathcal{D}_1} \Rightarrow 1] = 1 - \Pr[\mathsf{G}_2^{\mathcal{A}} \Rightarrow 1]$. Hence we have $\Pr[\mathsf{G}_1^{\mathcal{A}} \Rightarrow 1] - \Pr[\mathsf{G}_2^{\mathcal{A}} \Rightarrow 1] = (1 - \Pr[\mathsf{L}_0^{\mathcal{D}_1} \Rightarrow 1]) - (1 - \Pr[\mathsf{L}_1^{\mathcal{D}_1} \Rightarrow 1]) = \Pr[\mathsf{L}_1^{\mathcal{D}_1} \Rightarrow 1] - \Pr[\mathsf{L}_0^{\mathcal{D}_1} \Rightarrow 1] = \mathbf{Adv}_{\mathsf{TDP}}^{\mathsf{L}}(\mathcal{D}_1)$.

Lemma 11. $\Pr[\mathsf{G}_3^{\mathcal{A}} \Rightarrow 1] = \left(\frac{l-1}{l}\right) \Pr[\mathsf{G}_2^{\mathcal{A}} \Rightarrow 1]$.

Proof. In G_3, we introduce a new rule, which sets BAD to true if the forgery σ^* provided by $\mathcal{A}$ is the same as the simulated signature σ_{m^*} for the target message m^*. If this is the case, the adversary loses the game, i.e., G_3 outputs 0. σ_{m^*} is independent of $\mathcal{A}$'s view and is uniformly distributed in set of pre-images of y_{m^*}. Due to the l regular lossiness of TDP, the probability of a collision is equal to exactly $1/l$. Thus we see that the BAD rule reduces the probability of the adversary winning the game by $1/l$, hence $\Pr[\mathsf{G}_3^{\mathcal{A}} \Rightarrow 1] = (1 - \frac{1}{l}) \Pr[\mathsf{G}_2^{\mathcal{A}} \Rightarrow 1] = \left(\frac{l-1}{l}\right) \Pr[\mathsf{G}_2^{\mathcal{A}} \Rightarrow 1]$.

Lemma 12. *There exists a distinguisher $\mathcal{D}_2$ against the lossiness of* TDP, *which runs in time* $t = t_{\mathcal{A}} + q_h \cdot T_{\mathsf{TDP}}$ *and that* $\Pr[\mathsf{G}_3^{\mathcal{A}} \Rightarrow 1] - \Pr[\mathsf{G}_4^{\mathcal{A}} \Rightarrow 1] = \mathbf{Adv}_{\mathsf{TDP}}^{\mathsf{L}}(\mathcal{D}_2)$.

Proof. From G_3 to G_4, we change the key generation from a lossy permutation to a normal permutation, however the oracles are identical in both games. We now build a distinguisher $\mathcal{D}_2$ against the lossiness of TDP, using these games. The distinguisher will act as the challenger to $\mathcal{A}$. It will simulate the oracles as described in games $\mathsf{G}_3\&\mathsf{G}_4$, for which it requires time $q_h \cdot T_{\mathsf{TDP}}$. After $\mathcal{A}$ calls $\mathsf{Finalize}$, $\mathcal{D}_2$ returns the output of $\mathsf{Finalize}$. We can see that $\Pr[\mathsf{G}_4^{\mathcal{A}} \Rightarrow 1] = \Pr[\mathsf{L}_0^{\mathcal{D}_2} \Rightarrow 1]$. Similarly, we have $\Pr[\mathsf{G}_3^{\mathcal{A}} \Rightarrow 1] = \Pr[\mathsf{L}_1^{\mathcal{D}_2} \Rightarrow 1]$. Hence we have $\Pr[\mathsf{G}_3^{\mathcal{A}} \Rightarrow 1] - \Pr[\mathsf{G}_4^{\mathcal{A}} \Rightarrow 1] = \Pr[\mathsf{L}_1^{\mathcal{D}_2} \Rightarrow 1] - \Pr[\mathsf{L}_0^{\mathcal{D}_2} \Rightarrow 1] = \mathbf{Adv}_{\mathsf{TDP}}^{\mathsf{L}}(\mathcal{D}_2)$.

Lemma 13. $\Pr[\mathsf{G}_4^{\mathcal{A}} \Rightarrow 1] = 0$.

Proof. In G_4 we again use the original KeyGen such that $\mathsf{Eval}(pub, \cdot)$ defines a permutation. This means that our signing function is now a permutation, thus any forgery implies a collision. Therefore whenever the adversary is able to make a forgery, the game outputs 0 due to the BAD rule. Whenever they are unable to make a forgery, the game outputs 0. Thus we can see that in all cases, the game will output 0, hence $\Pr[\mathsf{G}_4^{\mathcal{A}} \Rightarrow 1] = 0$.

We combine Lemmas 9 to 13 to get:

$$\Pr[\mathsf{G}_0^{\mathcal{A}} \Rightarrow 1] = \mathbf{Adv}_{\mathsf{TDP}}^{\mathsf{L}}(\mathcal{D}_1) + \left(\frac{l}{l-1}\right)\mathbf{Adv}_{\mathsf{TDP}}^{\mathsf{L}}(\mathcal{D}_2).$$

where l is the lossiness of TDP. Because the distinguishers run in the same time, we know that both distinguishers can have at most an advantage of ε', giving us:

$$\varepsilon \le \frac{2l-1}{l-1} \cdot \varepsilon'.$$

This completes the proof.

4   Lossiness of RSA from the Φ-Hiding Assumption

4.1   Lossiness of RSA

The lossiness of RSA for a number of specific instance generators RSAGen was first considered in [20]. We now recall (and extend) some of the results from [20].

First, we recall some definitions from [20]. We denote by $\mathcal{RSA}_k := \{(N, p, q) \mid N = pq, p, q \in \mathbb{P}_{k/2}\}$ the set of all the tuples (N, p, q) such that $N = pq$ is the product of two distinct $k/2$-bit primes. Such an N is called an RSA modulus. By $(N, p, q) \in_R \mathcal{RSA}_k$ we mean the (N, p, q) is sampled according to the uniform distribution on $\mathcal{RSA}_k$. Let R be some relation on p and q. By $\mathcal{RSA}_k[R]$, we denote the subset of $\mathcal{RSA}_k$ such that the relation R holds on p and q. For example, let e be a prime. Then $\mathcal{RSA}_k[p = 1 \bmod e]$ is the set of all (N, p, q), where where $N = pq$ is the product of two distinct $k/2$-bit primes p, q and $p = 1 \bmod e$. That is, the relation $R(p, q)$ is true if $p = 1 \bmod e$ and q is arbitrary. By $(N, p, q) \in_R \mathcal{RSA}_k[R]$ we mean that (N, p, q) is sampled according to the uniform distribution on $\mathcal{RSA}_k[R]$.

α-Φ-HIDING ASSUMPTION. We recall a variant of the Φ-Hiding Assumption introduced by Cachin, Micali and Stadler [8], where we build on a formalization by Kiltz, O'Neil and Smith [20]. The main statement of the assumption is that given an k-bit RSA modulus $N = pq$ and a random $\alpha \cdot k$-bit prime e (where $0 < \alpha < \frac{1}{4}$ is a public constant), it is difficult to decide if $e \mid \varphi(N)$ or if $gcd(e, \varphi(N)) = 1$. We note that if $e \mid \varphi(N)$ with $e \ge N^{1/4}$, then N can be factored using Coppersmith's attacks [10], see [8] for details. Hence for the Φ-Hiding Assumption to hold, the bit-length of e must not exceed one-fourth of the bit length of N.

Consider a distinguisher $\mathcal{D}$ which plays one of the games P_0 or P_1 defined in Table 2, The advantage of $\mathcal{D}$ is defined as:

$$\mathbf{Adv}^{\Phi\mathsf{H}}(\mathcal{D}) = \Pr[\mathsf{P}_1^{\mathcal{D}} \Rightarrow 1] - \Pr[\mathsf{P}_0^{\mathcal{D}} \Rightarrow 1].$$

We say that the α-Φ-Hiding Problem is (t, ϵ)-hard if for all distinguishers $\mathcal{D}$ running in time at most t have and advantage of at most ϵ.

Define an RSA instance generator RSAGen as an algorithm that returns (N, e, p, q) sampled as $e \in_R \mathbb{P}_{\alpha k}$ and $(N, p, q) \in_R \mathcal{RSA}_k[gcd(e, \varphi(N)) = 1]$. (See [20] for details on the sampling algorithm.)

Lemma 14. *If the α-Φ-Hiding Problem is (t, ϵ)-hard, then the* RSA = (RSAGen, RSAEval, RSAInv) *defines a regular $(2^\alpha, t, \epsilon)$-lossy trapdoor permutation.*

Table 2. The α-Φ-Hiding Assumption Games

procedure Initialize Game P_0	**procedure Initialize** Game P_1
$e \in_R \mathbb{P}_{\alpha k}$	$e \in_R \mathbb{P}_{\alpha k}$
$(N,p,q) \in_R \mathcal{RSA}_k[gcd(e,\varphi(N))=1]$	$(N,p,q) \in_R \mathcal{RSA}_k[p=1 \mod e]$
return (N,e)	return (N,e)

Proof. If (N,e) is sampled using RSAGen, then $gcd(e,\varphi(N))=1$ and (N,e) defines a permutation $\mathsf{RSA}(x)=x^e \mod N$ over $\mathbb{Z}_N^*$. We define LossyGen to be an algorithm that returns (N,e) sampled as $e \in_R \mathbb{P}_{\alpha k}$ and $(N,p,q) \in_R \mathcal{RSA}_k[p=1 \mod e]$. If (N,e) is sampled using LossyGen then $e \mid \varphi(N)$ and hence the RSA function is e-to-1 on the domain $\mathsf{Dom}_{pub} = \mathbb{Z}_N^*$. By definition, the outputs of RSAGen and LossyGen are indistinguishable if the α-Φ-Hiding Problem is hard.

FIXED-PRIME Φ-HIDING ASSUMPTION. In practice, e is chosen to be small and is generally fixed to some specific numbers, such as $e=3$ or $e=2^{16}+1$, which allows for fast exponentiation. We now show a minor variant of the α-Φ-Hiding Assumption for *fixed primes* e, where our formalization relies on discussions from [8] and [20, Footnote 9].

First, we discuss the special case of $e=3$. We define our RSA instance RSAGen$_3$ generator as an algorithm that samples (N,p,q) uniformly from $\mathcal{RSA}_k[p=2 \mod 3, q=2 \mod 3]$, which is equivalent to $\mathcal{RSA}_k[gcd(3,\varphi(N))=1]$. We note that $N \mod 3$ is always 1. This means that for the lossy case, we must also ensure the $N \mod 3=1$, otherwise there would be a simple distinguisher. To ensure this is to have 3 divide both $p-1$ and $q-1$. Thus, our lossy keys are sampled from the $\mathcal{RSA}_k[p=1 \mod 3, q=1 \mod 3]$.

Table 3. The Fixed-Prime Φ-Hiding Assumption Games

procedure Initialize Game $3\mathsf{F}_0$	**procedure Initialize** Game $3\mathsf{F}_1$
$(N,p,q) \in_R \mathcal{RSA}_k[gcd(3,\varphi(N))=1]$	$(N,p,q) \in_R \mathcal{RSA}_k[p=1 \mod 3, q=1 \mod 3]$
return $(N,e=3)$	return $(N,e=3)$

Consider a distinguisher $\mathcal{D}$ which plays one of the games in Table 3. The advantage of $\mathcal{D}$ is defined as

$$\mathbf{Adv}^{\mathrm{F}\Phi\mathrm{H}}(\mathcal{D}) = \Pr[3\mathsf{F}_1^{\mathcal{D}} \Rightarrow 1] - \Pr[3\mathsf{F}_0^{\mathcal{D}} \Rightarrow 1].$$

We say that the Fixed-Prime Φ-Hiding Problem, with $e=3$, is (t,ϵ)-hard if all distinguishers running in time at most t have an advantage of at most ϵ.

Lemma 15. *If the Fixed-Prime Φ-Hiding Problem, with $e=3$, is (t,ϵ)-hard, then the* RSA$_3$ = (RSAGen$_3$, RSAEval, RSAInv) *defines a regular $(9,t,\epsilon)$-lossy trapdoor permutation.*

Proof. If $(N, p, q) \in \mathcal{RSA}_k[gcd(3, \varphi(N)) = 1]$ then $(N, 3)$ clearly makes the RSA function a permutation. If $(N, p, q) \in \mathcal{RSA}_k[p = 1 \mod 3, q = 1 \mod 3]$ then $3 \mid \varphi(N)$ and hence the RSA function is 9-to-1 on the domain $\mathsf{Dom}_{pub} = \mathbb{Z}_N^*$.

We now consider the general case of fixed $e > 3$. For this case, we define our RSA instance generator RSAGen_e as an algorithm that samples (N, p, q) from $\mathcal{RSA}_k[gcd(e, \varphi(N)) = 1]$. We note that $N \mod e$ will be some value between 1 and $e - 1$. This means that for the lossy case, we require e to divide $p - 1$ and not $q - 1$, otherwise we would have a simple distinguisher. Our lossy keys are sampled from $\mathcal{RSA}_k[p = 1 \mod e, q \neq 1 \mod e]$. Consider a distinguisher $\mathcal{D}$

Table 4. The Fixed-Prime Φ-Hiding Assumption Games

procedure Initialize Game F_0	procedure Initialize Game F_1
$(N, p, q) \in_R \mathcal{RSA}_k[gcd(e, \varphi(N)) = 1]$	$(N, p, q) \in_R \mathcal{RSA}_k[p = 1 \mod e, q \neq 1 \mod e]$
return (N, e)	return (N, e)

which plays one of the games in Table 4. The advantage of $\mathcal{D}$ is defined as

$$\mathbf{Adv}^{\mathrm{F}\Phi\mathrm{H}}(\mathcal{D}) = \Pr[\mathsf{F}_1^{\mathcal{D}} \Rightarrow 1] - \Pr[\mathsf{F}_0^{\mathcal{D}} \Rightarrow 1].$$

We say that the Fixed-Prime Φ-Hiding Problem, with $e > 3$, is (t, ϵ)-hard if for all distinguishers running in time at most t have an advantage of at most ϵ.

Lemma 16. *If the Fixed-Prime Φ-Hiding Problem, with $e > 3$, is (t, ϵ)-hard, then* $\mathsf{RSA}_e = (\mathsf{RSAGen}_e, \mathsf{RSAEval}, \mathsf{RSAInv})$ *defines a regular (e, t, ϵ)-lossy trapdoor permutation.*

Proof. If $(N, p, q) \in \mathcal{RSA}_k[gcd(e, \varphi(N)) = 1]$ then (N, e) clearly defines a permutation. If $(N, p, q) \in \mathcal{RSA}_k[p = 1 \mod e, q \neq 1 \mod e]$ then $e \mid \varphi(N)$ and hence the RSA function is e-to-1 on the domain $\mathsf{Dom}_{pub} = \mathbb{Z}_N^*$.

Acknowledgements. We thank Mihir Bellare and Dennis Hofheinz for valuable comments on an earlier draft.

References

1. Bellare, M., Namprempre, C., Pointcheval, D., Semanko, M.: The one-more-RSA-inversion problems and the security of Chaum's blind signature scheme. Journal of Cryptology 16(3), 185–215 (2003)
2. Bellare, M., Rogaway, P.: Random oracles are practical: A paradigm for designing efficient protocols. In: Ashby, V. (ed.) ACM CCS 1993: 1st Conference on Computer and Communications Security, pp. 62–73. ACM Press (November 1993)
3. Bellare, M., Rogaway, P.: The Exact Security of Digital Signatures - How to Sign with RSA and Rabin. In: Maurer, U.M. (ed.) EUROCRYPT 1996. LNCS, vol. 1070, pp. 399–416. Springer, Heidelberg (1996)

4. Bellare, M., Rogaway, P.: The Security of Triple Encryption and a Framework for Code-Based Game-Playing Proofs. In: Vaudenay, S. (ed.) EUROCRYPT 2006. LNCS, vol. 4004, pp. 409–426. Springer, Heidelberg (2006)
5. Bellare, M., Yung, M.: Certifying permutations: Noninteractive zero-knowledge based on any trapdoor permutation. Journal of Cryptology 9(3), 149–166 (1996)
6. Bernstein, D.J.: Proving Tight Security for Rabin-Williams Signatures. In: Smart, N.P. (ed.) EUROCRYPT 2008. LNCS, vol. 4965, pp. 70–87. Springer, Heidelberg (2008)
7. Cachin, C.: Efficient private bidding and auctions with an oblivious third party. In: ACM CCS 1999: 6th Conference on Computer and Communications Security, pp. 120–127. ACM Press (November 1999)
8. Cachin, C., Micali, S., Stadler, M.A.: Computationally Private Information Retrieval with Polylogarithmic Communication. In: Stern, J. (ed.) EUROCRYPT 1999. LNCS, vol. 1592, pp. 402–414. Springer, Heidelberg (1999)
9. Chatterjee, S., Menezes, A., Sarkar, P.: Another look at tightness. Cryptology ePrint Archive, Report 2011/442 (2011), http://eprint.iacr.org/
10. Coppersmith, D.: Finding a Small Root of a Univariate Modular Equation. In: Maurer, U.M. (ed.) EUROCRYPT 1996. LNCS, vol. 1070, pp. 155–165. Springer, Heidelberg (1996)
11. Coron, J.-S.: On the Exact Security of Full Domain Hash. In: Bellare, M. (ed.) CRYPTO 2000. LNCS, vol. 1880, pp. 229–235. Springer, Heidelberg (2000)
12. Coron, J.-S.: Optimal security proofs for pss and other signature schemes. Cryptology ePrint Archive, Report 2001/062 (2001), http://eprint.iacr.org/
13. Coron, J.-S.: Optimal Security Proofs for PSS and Other Signature Schemes. In: Knudsen, L.R. (ed.) EUROCRYPT 2002. LNCS, vol. 2332, pp. 272–287. Springer, Heidelberg (2002)
14. Gentry, C., Mackenzie, P.D., Ramzan, Z.: Password authenticated key exchange using hidden smooth subgroups. In: Atluri, V., Meadows, C., Juels, A. (eds.) ACM CCS 2005: 12th Conference on Computer and Communications Security, pp. 299–309. ACM Press (November 2005)
15. Gentry, C., Peikert, C., Vaikuntanathan, V.: Trapdoors for hard lattices and new cryptographic constructions. In: Ladner, R.E., Dwork, C. (eds.) 40th Annual ACM Symposium on Theory of Computing, pp. 197–206. ACM Press (May 2008)
16. Gentry, C., Ramzan, Z.: Single-Database Private Information Retrieval with Constant Communication Rate. In: Caires, L., Italiano, G.F., Monteiro, L., Palamidessi, C., Yung, M. (eds.) ICALP 2005. LNCS, vol. 3580, pp. 803–815. Springer, Heidelberg (2005)
17. Goh, E.-J., Jarecki, S., Katz, J., Wang, N.: Efficient signature schemes with tight reductions to the Diffie-Hellman problems. Journal of Cryptology 20(4), 493–514 (2007)
18. Hemenway, B., Ostrovsky, R.: Public-Key Locally-Decodable Codes. In: Wagner, D. (ed.) CRYPTO 2008. LNCS, vol. 5157, pp. 126–143. Springer, Heidelberg (2008)
19. IEEE P1363a Committee. IEEE P1363a / D9 — standard specifications for public key cryptography: Additional techniques, Draft Version 9 (June 2001), http://grouper.ieee.org/groups/1363/index.html/
20. Kiltz, E., O'Neill, A., Smith, A.: Instantiability of RSA-OAEP under Chosen-Plaintext Attack. In: Rabin, T. (ed.) CRYPTO 2010. LNCS, vol. 6223, pp. 295–313. Springer, Heidelberg (2010)
21. Koblitz, N., Menezes, A.J.: Another look at "provable security". Journal of Cryptology 20(1), 3–37 (2007)

22. Lysyanskaya, A., Micali, S., Reyzin, L., Shacham, H.: Sequential Aggregate Signatures from Trapdoor Permutations. In: Cachin, C., Camenisch, J.L. (eds.) EUROCRYPT 2004. LNCS, vol. 3027, pp. 74–90. Springer, Heidelberg (2004)
23. Micali, S.: Computationally sound proofs. SIAM J. Comput. 30, 1253–1298 (2000)
24. Paillier, P., Villar, J.L.: Trading One-Wayness Against Chosen-Ciphertext Security in Factoring-Based Encryption. In: Lai, X., Chen, K. (eds.) ASIACRYPT 2006. LNCS, vol. 4284, pp. 252–266. Springer, Heidelberg (2006)
25. Peikert, C., Waters, B.: Lossy trapdoor functions and their applications. In: Ladner, R.E., Dwork, C. (eds.) 40th Annual ACM Symposium on Theory of Computing, pp. 187–196. ACM Press (May 2008)
26. PKCS #1: RSA cryptography standard. RSA Data Security, Inc., Version 2.0 (September 1998)
27. Schridde, C., Freisleben, B.: On the Validity of the Φ-Hiding Assumption in Cryptographic Protocols. In: Pieprzyk, J. (ed.) ASIACRYPT 2008. LNCS, vol. 5350, pp. 344–354. Springer, Heidelberg (2008)
28. Smart, N.: Ecrypt ii yearly report on algorithms and keysizes (2009-2010). Framework, p. 116 (March 2010)

On the Exact Security of Schnorr-Type Signatures in the Random Oracle Model

Yannick Seurin

ANSSI, Paris, France
yannick.seurin@m4x.org

Abstract. The Schnorr signature scheme has been known to be provably secure in the Random Oracle Model under the Discrete Logarithm (DL) assumption since the work of Pointcheval and Stern (EUROCRYPT '96), at the price of a very loose reduction though: if there is a forger making at most q_h random oracle queries, and forging signatures with probability ε_F, then the Forking Lemma tells that one can compute discrete logarithms with constant probability by rewinding the forger $\mathcal{O}(q_h/\varepsilon_F)$ times. In other words, the security reduction loses a factor $\mathcal{O}(q_h)$ in its time-to-success ratio. This is rather unsatisfactory since q_h may be quite large. Yet Paillier and Vergnaud (ASIACRYPT 2005) later showed that under the One More Discrete Logarithm (OMDL) assumption, any *algebraic* reduction must lose a factor at least $q_h^{1/2}$ in its time-to-success ratio. This was later improved by Garg *et al.* (CRYPTO 2008) to a factor $q_h^{2/3}$. Up to now, the gap between $q_h^{2/3}$ and q_h remained open. In this paper, we show that the security proof using the Forking Lemma is essentially the best possible. Namely, under the OMDL assumption, any algebraic reduction must lose a factor $f(\varepsilon_F)q_h$ in its time-to-success ratio, where $f \leq 1$ is a function that remains close to 1 as long as ε_F is noticeably smaller than 1. Using a formulation in terms of expected-time and queries algorithms, we obtain an optimal loss factor $\Omega(q_h)$, independently of ε_F. These results apply to other signature schemes based on one-way group homomorphisms, such as the Guillou-Quisquater signature scheme.

Keywords: Schnorr signatures, discrete logarithm, Forking Lemma, Random Oracle Model, meta-reduction, one-way group homomorphism.

1 Introduction

Schnorr Signatures. The Schnorr signature scheme [25,26], derived from the Schnorr identification scheme (an honest-verifier zero-knowledge proof of knowledge of a discrete logarithm) through the Fiat-Shamir transform [12], is one of the earliest discrete log-based signature schemes proposed in the literature. Its simplicity and efficiency (short signature length and the possibility of precomputing exponentiations for very quick on-line signature generation) has attracted considerable attention. Its security has been analyzed in the Random Oracle Model (ROM) [2] under the Discrete Logarithm (DL) assumption by

D. Pointcheval and T. Johansson (Eds.): EUROCRYPT 2012, LNCS 7237, pp. 554–571, 2012.

Pointcheval and Stern [23,24]. The main idea of the proof is to have the forger output two distinct forgeries corresponding to the same random oracle query, but for two distinct answers of the random oracle. The so-called Forking Lemma shows that by rewinding the forger $\mathcal{O}(q_h/\varepsilon_F)$ times, where q_h is the maximal number of random oracle queries of the forger and ε_F its success probability, then one finds two such forgeries with constant probability, which enables to compute the discrete logarithm of the public key. Said otherwise, the reduction loses a factor $\mathcal{O}(q_h)$ in its time-to-success ratio. This results in a very loose security assurance since q_h may be quite large (*e.g.* 2^{60}), which implies to increase the problem parameters length in order to achieve an appropriate provable security level.

Previous Negative Results. Whether the loss of this factor q_h is unavoidable remained obscure until Paillier and Vergnaud [22] showed that under the One More Discrete Logarithm (OMDL) assumption[1], any *algebraic*[2] reduction from the DL problem to forging Schnorr signatures in the ROM must lose a factor $\Omega(q_h^{1/2})$ in its time-to-success ratio. Starting from a reduction from the DL problem to forging Schnorr signatures in the ROM, [22] builds a *meta-reduction* that solves the OMDL problem without using any forger (it simulates the forger using the discrete log oracle it can access to solve the OMDL problem). This result was later improved by Garg *et al.* [14] to a factor $\Omega(q_h^{2/3})$, using the same meta-reduction (only the *analysis* of its success probability was improved). Interestingly, [14] also showed that under a simple assumption on the forger (namely that the distribution of the random oracle query index ℓ corresponding to the forged signature is uniformly random in $[1..q_h]$), the factor lost in the time-to-success ratio of the reduction of [24] can be reduced from $\mathcal{O}(q_h)$ to $\mathcal{O}(q_h^{2/3})$. Since the meta-reduction used in [22,14] simulates a forger that obeys this assumption, one cannot hope to improve the analysis of this particular meta-reduction to show that a factor $\Omega(q_h)$ must be lost by any algebraic reduction.

Contributions of this Work. Up to now, the gap between the security reduction of [24] loosing a factor $\mathcal{O}(q_h)$ and the lower bound $\Omega(q_h^{2/3})$ of [14] remained open. Basically two possible directions were conceivable in order to narrow it: either improve the security reduction of [24] for a *general* forger, or find a better meta-reduction enabling to overcome the $q_h^{2/3}$ bound. We essentially close this gap in the second direction by showing that under the OMDL assumption, any algebraic reduction from the DL problem to forging Schnorr signatures in the ROM must lose a factor $f(\varepsilon_F)q_h$ in its time-to-success ratio, where f is a function that remains close to 1 as long as the success probability ε_F of the forger is noticeably smaller than 1. Our meta-reduction is different from the one used in [22,14] (this is unavoidable by the previous considerations). In particular, the random oracle query index ℓ corresponding to the forged signature is

[1] The OMDL problem consists in solving $n+1$ discrete logarithms by making at most n calls to a discrete log oracle (cf. Section 2).
[2] An algebraic reduction is limited to perform group operations when it manipulates group elements (cf. Section 4).

not uniformly distributed in $[1..q_h]$ (it has a truncated geometric distribution), nor is it independent for two distinct executions of the forger (as we argue later, a uniformly distributed forgery index ℓ is in fact quite unnatural). Though the description of our new meta-reduction is slightly more complicated, its analysis is arguably simpler (the analysis of [14] uses advanced results on the statistics of random permutations). Curiously, our bound vanishes when ε_F is negligibly close to 1. We argue however that this shortcoming is due to the formulation in terms of strictly bounded adversaries. By considering definitions using expected-time (and queries) algorithms, we are able to show that any algebraic reduction must lose a factor $\Omega(q_h)$, independently of ε_F, in its expected-time-to-success ratio.

Interpretation of Our Results. Interpreting our results is quite delicate (as is often the case for results in the ROM). The conservative point of view would be to consider that breaking Schnorr signatures in the ROM is strictly easier than solving the DL problem (which our results do not prove), and to increase security parameters adequately. Yet taking into account that no one has been able to find a better forgery attack than by solving the DL problem, another possible interpretation is that they point out the limitations of black-box reduction techniques. For example, consider the (t, q_h, ε)-forger $\mathcal{F}$ obtained as follows: starting from any algorithm that (t, ε)-solves the DL problem, $\mathcal{F}$ first recovers the secret key, and then forges a signature corresponding to one of its $q_h > 1$ random oracle queries (*e.g.* uniformly chosen at random). This adversary is arguably artificial since it could forge a signature for any message *with a single random oracle query*. Yet any black-box reduction will lose a huge factor when using such a forger, whereas a non-black-box one, accessing the DL-subroutine of the forger, would yield back an algorithm solving the DL problem with the same time-to-success ratio as the forger.

Related Work. Techniques similar to the ones of [22,14] and this paper were used to separate one-more computational problems independently by Brown [7] (who termed such results *irreductions*) and Bresson *et al.* [6].

Coron [10] gave a result close in spirit to ours for the RSA with Full Domain Hash (FDH) signature scheme [3]: he showed that the security of RSA-FDH in the ROM cannot be proved tightly equivalent to the hardness of inverting RSA. This was generalized by Dodis and Reyzin [11] to FDH used with any trapdoor one-way permutation induced by a family of claw-free permutations. There are however two main differences between these results and ours. First, the result of [10,11] is specific to chosen-message attacks (FDH is tightly secure for no-message attacks), whereas in our case the result holds even for no-message attacks. Second, the factor necessarily lost by any reduction for FDH is $\Omega(q_s)$, where q_s is the maximal number of *signature* queries asked by the forger. A security proof matching this $\Omega(q_s)$ bound had been previously given by Coron [9].

The security of the Schnorr signature scheme in the standard model remains elusive (beyond the obvious fact that key-recovery is as hard as the DL problem under no-message attacks).[3] Paillier and Vergnaud [22] showed that under the OMDL assumption, it is immune to key-recovery under chosen-message attacks (whatever the hash function used), but that it cannot be proved universally unforgeable under no-message attacks with respect to an algebraic reduction (again under the OMDL assumption). Neven *et al.* [21] gave necessary conditions on the hash function for the Schnorr signature scheme to be existentially unforgeable under chosen-message attacks, and also showed that these conditions are sufficient in the generic group model. To the best of our knowledge, these are the only results up to now. All practical[4] discrete log-based signature schemes provably secure in the standard model rely on bilinear groups [4,28].

Faced with the apparent impossibility to obtain tight security reductions in the ROM for discrete log-based schemes, two main research options emerged. The first was to rely on weaker assumptions, with proposals such as the EDL scheme [15] and subsequent improvements [8] relying on the Computational Diffie-Hellman assumption, and the proposal by Katz and Wang [18] relying on the Decisional Diffie-Hellman assumption (see also [16]). The second option was to find alternatives to the Fiat-Shamir transform with tighter security reductions, as explored by Micali and Reyzin [20] (but their technique is inapplicable to discrete log-based schemes) and Fischlin [13] (but the resulting scheme is relatively inefficient).

Open Problems. We leave the problem of eliminating the dependency in ε_F for strictly bounded adversaries as an intriguing (though minor) open question. This paper more or less settles the case of algebraic reductions; a natural question is what can be said for arbitrary reductions. More generally, an interesting research subject is to build an efficient signature scheme with a tight reduction in the ROM under the DL assumption (and not under weaker related ones), or to prove a general impossibility result. Another important challenge is to say anything meaningful about the security of Schnorr signatures in the standard model, or to propose a practical scheme based on DL-like assumptions provably secure in the standard model and not relying on bilinear groups.

Organization. In Section 2, we give the necessary background on Schnorr signatures and the DL and OMDL problems. In Section 3, we recall the security proof of [24] for Schnorr signatures through the Forking Lemma. In Section 4, we describe our new meta-reduction and show in Section 5 that it implies a necessary loss of a factor $f(\varepsilon_F)q_h$ for any algebraic reduction. In the full version of the paper [27], we put our results in a more general framework based on one-way group homomorphisms, and extend them to other related signature schemes (such as Modified ElGamal). We also treat the expected-time and queries scenario in the full version.

---

[3] We note that the Fiat-Shamir transform is known to be intrinsically problematic in the standard model [17].

[4] General constructions of signature schemes from any one-way function are known, but are quite impractical.

2   Preliminaries

$[i..j]$ will denote the set of integers k such that $i \leq k \leq j$. When $\mathcal{X}$ is a non-empty finite set, we write $x \leftarrow_\$ \mathcal{X}$ to mean that a value is sampled uniformly at random from $\mathcal{X}$ and assigned to x. We denote Ber_μ the Bernoulli distribution of parameter $\mu \in [0,1]$ (*i.e.* $\delta \leftarrow \text{Ber}_\mu$ is such that $\Pr[\delta = 1] = \mu$ and $\Pr[\delta = 0] = 1 - \mu$), and for $\mu \in [0,1]$ and a non-zero positive integer q, we denote $\text{Bin}_{\mu,q}$ the binomial distribution of parameters μ and q (*i.e.* $X \leftarrow \text{Bin}_{\mu,q}$ is such that $\Pr[X = k] = \binom{q}{k}\mu^k(1-\mu)^{q-k}$). The security parameter will be denoted κ. We will write $f = \text{poly}(\cdot)$ to denote a polynomially bounded function and $f = \text{negl}(\cdot)$ to denote a negligible function. We assume the existence of an adequate group generation algorithm, which on input 1^κ returns a cyclic group $\mathbb{G}$ of prime order $q \in [2^{\kappa-1}, 2^\kappa[$ and a generator g of $\mathbb{G}$. We will assume that all algorithms are given $(\mathbb{G}, q, g)$ as input and will sometimes not mention it explicitly.

The Schnorr signature scheme is obtained by applying the Fiat-Shamir transform [12] to the Schnorr identification scheme [25,26].

Definition 1 (Schnorr Signature Scheme). *Let $\mathbb{G}$ be a cyclic group of prime order q and g be a generator of $\mathbb{G}$. Let $H : \{0,1\}^* \times \mathbb{G} \to \mathbb{Z}_q$ be a hash function. The Schnorr signature scheme is defined as follows:*

- *Key generation: Let $x \leftarrow_\$ \mathbb{Z}_q \setminus \{0\}$, and $y = g^x$. The private key is x and the public key is y.*
- *Signature: To sign a message $m \in \{0,1\}^*$, draw $a \leftarrow_\$ \mathbb{Z}_q$, compute $r = g^a$, $c = H(m,r)$, and $s = a + cx \mod q$. The signature is (s,c).*
- *Verification: Given a message $m \in \{0,1\}^*$, and a claimed signature (s,c), compute $r = g^s y^{-c}$ and check that $c = H(m,r)$.*

From a practical point of view, the Schnorr signature scheme is more usually defined with a hash function mapping its inputs to $\{0,1\}^k$ (interpreted as integers in $[0..(2^k-1)]$) rather than $\mathbb{Z}_q$. There is no difficulty in extending our results to this case (q must simply be replaced by 2^k in Theorem 2). When we talk of the Schnorr signature scheme in the Random Oracle Model (ROM), we mean the scheme obtained when H is replaced by a random oracle.

In this work we focus on security against universal forgery under no-message attacks (UF-NM-security) in the ROM. This a weak security notion, but this makes our negative result of Section 4 stronger than considering a more constraining notion such as security against existential forgery under chosen-message attacks.

Definition 2 (UF-NM Forger). *A forger $\mathcal{F}$ is said to be $(t_F, q_h, \varepsilon_F)$-UF-NM-break Schnorr signatures in the ROM if on input any message $m \in \{0,1\}^*$ and a public key $y \leftarrow_\$ \mathbb{G}$, $\mathcal{F}$ runs in time at most t_F, makes at most q_h queries to the random oracle, and returns a valid forgery (s,c) for m with probability at least ε_F (where the probability is taken over the random choice of y, the random tape of $\mathcal{F}$, and the answers of the random oracle).*
Moreover, we will say that the forgery (s,c) corresponds to the random oracle query index $\ell \in [1..q_h]$ if the ℓ-th query/answer of $\mathcal{F}$ to the random oracle was $H(m, g^s y^{-c}) = c$.

In all the following, we will assume *wlog* the following: when $\mathcal{F}$ returns a forgery (s, c), and made the query $(m, g^s y^{-c})$ to the random oracle, the corresponding answer was c (in other words, the forger never returns a forgery that it knows to be invalid: we assume it returns $\perp$ in this case). For clarity, when the forger returns a forgery corresponding to the random oracle query index ℓ, we will assume it outputs the triplet (ℓ, s, c). Note that the forger may return a random forgery that does not correspond to any of its random oracle queries, in which case it is valid with probability $1/q$. We will denote $(\emptyset, s, c)$ the output of the forger in that case. In all the following, when we say that the forger returns a forgery (ℓ, s, c), we mean $\ell \neq \emptyset$ unless otherwise stated.

As we will see in Section 3, the security of Schnorr signatures in the ROM can be proved under the assumption that the Discrete Logarithm (DL) problem, that we formalize below, is hard.

Definition 3 (DL Problem). *Let $\mathbb{G}$ be a cyclic group of order q and g be a generator of $\mathbb{G}$. An algorithm $\mathcal{A}$ is said to (t, ε)-solve the DL problem if on input (G, q, g) and $r \leftarrow_\$ \mathbb{G}$, it runs in time at most t and returns the discrete logarithm of r in base g with probability at least ε (where the probability is taken over the random choice of r and the random tape of $\mathcal{A}$).*

The One-More Discrete Logarithm (OMDL) problem, introduced under the name Known-Target DL problem in [1], is defined as follows. Note that Koblitz and Menezes [19] argue that the ODML problem might be easier than the DL problem for some groups.

Definition 4 (OMDL Problem). *Let $\mathbb{G}$ be a cyclic group of order q and g be a generator of $\mathbb{G}$. Let Θ be an oracle taking no input and returning a random element of $\mathbb{G}$ (named the* challenge *oracle). Let $\mathrm{DLog}_g(\cdot)$ be the oracle returning the discrete logarithm in base g of its input. An algorithm $\mathcal{A}$ is said to (t, n, ε)-solve the OMDL problem if on input $(\mathbb{G}, q, g)$, it runs in time at most t, makes $m \leq n + 1$ queries $r_1, \ldots, r_m \leftarrow \Theta$, and returns the discrete logarithm of all r_i's in base g while making* strictly less *than m queries to $\mathrm{DLog}_g(\cdot)$, with probability at least ε (where the probability is taken over the random challenges of Θ and the random tape of $\mathcal{A}$).*

3   Security Proof with the Forking Lemma

In this section, we recall the analysis of the security of the Schnorr signature scheme using the Forking Lemma [23,24]. We focus on UF-NM-security, but there is no difficulty in extending the result to existential forgery and to chosen-message attacks using the honest-verifier zero-knowledge property of the Schnorr identification scheme [24].

The main idea is to obtain from the forger two valid forgeries (ℓ, s, c) and (ℓ, s', c') corresponding to the same random oracle query (m, r), but for distinct answers of the random oracle $c \neq c'$. Indeed this implies $r = g^s y^{-c} = g^{s'} y^{-c'}$, which yields the discrete logarithm of the public key $\mathrm{DLog}_g(y) = (s - s')/(c - c')$ mod q. For this, the reduction runs the forger with input some message m, public key y (the target element of the reduction), and some uniformly chosen random tape ω, answering the random oracle queries of the forger uniformly at random, until it returns a forgery corresponding to some random oracle query index $\ell \in [1..q_h]$. Then, it replays the forger, using the same input (m, y), the same random tape ω and the same answers to random oracle queries up to the $(\ell - 1)$-th one as for the successful execution. Consequently, the ℓ-th random oracle query of the forger is the same as in the successful execution. Starting from the ℓ-th random oracle query, the reduction draws the answers uniformly at random again (using the terminology of Section 4, we will say that such an execution *forks* from the successful one at point ℓ). It repeats this until the forger returns another forgery corresponding to the same random oracle query index $\ell \in [1..q_h]$. The Forking Lemma gives a lower bound on the probability that this strategy succeeds.

The security result for Schnorr signatures can be concretely stated as the following theorem, from which it can easily be seen that the security reduction loses a factor $\mathcal{O}(q_h)$ in its time-to-success ratio t_R/ε_R compared with the one of the forger t_F/ε_F.

Theorem 1 ([24]). *Assume there is a forger which $(t_F, q_h, \varepsilon_F)$-UF-NM-breaks Schnorr signatures in the ROM for some group parameters $(\mathbb{G}, q, g)$. Assume moreover that $\varepsilon_F \geq \max(2/(q + 1), 16q_h/q)$. Then there is a reduction $\mathcal{R}$ which (t_R, ε_R)-solves the DL problem (for the same group parameters), where $t_R \simeq (16q_h + 2)t_F/\varepsilon_F$ and $\varepsilon_R > 0.099$.*

Proof. We give a slightly adapted proof in the full version of the paper [27]. □

4   Description of the New Meta-reduction

In the next section we will prove the following result, that we state informally for now.

Theorem (Informal). *Under the OMDL assumption, any algebraic reduction from the DL problem to UF-NM-breaking Schnorr signatures in the ROM must lose a factor $f(\varepsilon_F)q_h$ in its time-to-success ratio, where q_h is the maximal number of random oracle queries of the forger, ε_F its success probability, and $f(\varepsilon_F) = \varepsilon_F/\ln\big((1 - \varepsilon_F)^{-1}\big)$.*

In order to prove this result, we will start from an algebraic reduction $\mathcal{R}$ (the meaning of algebraic will be explained shortly) that turns a UF-NM-forger for Schnorr signatures in the ROM into a solver for the DL problem, and describe a meta-reduction $\mathcal{M}$ that uses the reduction $\mathcal{R}$ to solve the OMDL problem without using any forger (the meta-reduction will actually simulate the forger to the reduction thanks to its discrete log oracle). In order to formalize this, we need a precise definition of a reduction.

Definition 5. *A reduction* $\mathcal{R}$ *is said to* $(t_R, n, \varepsilon_R, q_h, \varepsilon_F)$*-reduce the DL problem to UF-NM-breaking Schnorr signatures in the ROM if upon input* $r_0 \leftarrow_\$ \mathbb{G}$ *and after running at most* n *times any forger which* $(t_F, q_h, \varepsilon_F)$*-UF-NM-breaks Schnorr signatures,* $\mathcal{R}$ *outputs* $\mathtt{DLog}_g(r_0)$ *with probability greater than* ε_R, *within an additional running time* t_R *(meaning that the total running time of* $\mathcal{R}$ *is at most* $t_R + n t_F$*).*

The probability ε_R is taken as in Definition 3 over the random choice of r_0 and the random tape of $\mathcal{R}$ (the random tape of $\mathcal{F}$ is assumed under control of $\mathcal{R}$). The reduction described in the proof of Theorem 1 is a $(\mathcal{O}(1), (16 q_h + 2)/\varepsilon_F, 0.099, q_h, \varepsilon_F)$-reduction.

Similarly to previous work [22,14], we will only consider *algebraic* reductions (originally introduced in [5]). An algorithm $\mathcal{R}$ is algebraic with respect to some group $\mathbb{G}$ if the only operations it can perform on group elements are group operations (see [22] for details). We characterize such reductions by the existence of a procedure $\mathtt{Extract}$ which, given the group elements $(g_1, \ldots, g_k)$ input to $\mathcal{R}$, other inputs σ to $\mathcal{R}$, $\mathcal{R}$'s code, and any group element y produced by $\mathcal{R}$ during its computation in at most t steps, outputs $\alpha_1, \ldots, \alpha_k \in \mathbb{Z}_q$ such that $y = g_1^{\alpha_1} \ldots g_k^{\alpha_k}$. We require that $\mathtt{Extract}$ runs in time $\mathtt{poly}(t, |\mathcal{R}|, \lfloor \log_2 q \rfloor)$, where $|\mathcal{R}|$ is the code size of $\mathcal{R}$. As will appear clearly later, the need to restrict the reduction to be algebraic arises from the fact that $\mathcal{R}$ can run the forger on arbitrary public keys, and the meta-reduction will need to extract the discrete logarithm of these public keys (assuming $\mathcal{R}$ returns the discrete logarithm of its input r_0). This can also be interpreted as saying that $\mathcal{R}$ runs $\mathcal{F}$ on public keys that are derived from its input r_0 through group operations, which does not seem an overly restrictive assumption. Note in particular that the reduction of [24] using the Forking Lemma is algebraic: it repeatedly runs the forger on the same public key $y = r_0$ (or, in the variant described in the full version of the paper [27], on public keys $y = (r_0)^\alpha$ for α's randomly chosen during the first phase of the reduction).

We now describe the new meta-reduction $\mathcal{M}$. It has access to an OMDL challenge oracle Θ returning random elements from $\mathbb{G}$, and to an oracle $\mathtt{DLog}_g(\cdot)$ returning the discrete logarithm in base g of its input. It also has access[5] to a $(t_R, n, \varepsilon_R, q_h, \varepsilon_F)$-algebraic reduction $\mathcal{R}$, which expects access to a forger $\mathcal{F}$, and offers a random oracle interface that we denote $\mathcal{R}.H$. We assume $t_R, n, q_h = \mathtt{poly}(\kappa)$ and $\varepsilon_R, \varepsilon_F = 1/\mathtt{poly}(\kappa)$. Recall that the goal of $\mathcal{M}$ is to return the discrete logarithm of all challenge elements it queries to Θ, by making strictly less queries to $\mathtt{DLog}_g(\cdot)$. In all the following we assume $0 < \varepsilon_F < 1$, we fix $\alpha \in]0, (1 - \varepsilon_F)^{1/q_h}[$ and we define the quantities μ_0 and $\mu \in]0, 1[$ (whose meaning will appear clearer in view of Lemmata 2 and 3) as:

$$\mu_0 = 1 - (1 - \varepsilon_F)^{1/q_h} \quad \text{and} \quad \mu = \frac{\mu_0}{1 - \alpha} = \frac{1}{1 - \alpha} \left(1 - (1 - \varepsilon_F)^{1/q_h} \right) .$$

---

[5] By access we essentially mean black-box access, but $\mathcal{M}$ also needs the code of $\mathcal{R}$ to run procedure $\mathtt{Extract}$.

$\mathcal{M}$ first queries the OMDL challenge oracle Θ, receiving a random element $r_0 \in \mathbb{G}$, and runs $\mathcal{R}$ on input r_0 and some uniformly chosen random tape. Then it simulates (at most) n sequential executions of the forger that we denote $\mathcal{F}_i(m_i, y_i, \omega_i)$, $1 \le i \le n$, where m_i is the input message, y_i the input public key, and ω_i the random tape of the forger received from the reduction.[6] Depending on how $\mathcal{R}$ chooses (m_i, y_i, ω_i) and the answers to queries of $\mathcal{M}$ to $\mathcal{R}.H$, these successive executions may be identical up to some point, that we will call a *forking point*.

Definition 6 (Forking Point). *Consider two distinct simulated executions of the forger $\mathcal{F}_i(m_i, y_i, \omega_i)$ and $\mathcal{F}_j(m_j, y_j, \omega_j)$, $1 \le j < i \le n$. We say that execution $\mathcal{F}_i$ forks from execution $\mathcal{F}_j$ at point $t_{i/j} = 0$ if $(m_i, y_i, \omega_i) \ne (m_j, y_j, \omega_j)$, or at point $t_{i/j} \in [1..q_h]$ if all the following holds:*

- $(m_i, y_i, \omega_i) = (m_j, y_j, \omega_j)$;
- *for $k \in [1..(t_{i/j} - 1)]$, the k-th query and answer to $\mathcal{R}.H$ are the same in both executions;*
- *the $t_{i/j}$-th query to $\mathcal{R}.H$ is the same in both executions, but the answers are distinct.*

We also define the point where execution $\mathcal{F}_i$ forks from all previous executions as $t_i = \max\{t_{i/j}, 1 \le j < i\}$.

We assume *wlog* that all simulated executions are distinct, *i.e.* they fork at some point.

The simulation of the forger works as follows. The meta-reduction will dynamically construct two (initially empty) disjoint sets $\Gamma_{\text{good}}, \Gamma_{\text{bad}} \subset \mathbb{G}$. Γ_{good} will be the set of elements $z \in \mathbb{G}$ whose discrete logarithm is known from $\mathcal{M}$ because it has made the corresponding query to its discrete log oracle (we assume the discrete logarithm of elements in Γ_{good} are adequately stored by $\mathcal{M}$), while Γ_{bad} will be the set of elements $z \in \mathbb{G}$ such that $\mathcal{M}$ will never make the corresponding query to its discrete log oracle. The main idea of the simulation of the forger on input (m, y, ω) is that $\mathcal{M}$ will return a forgery corresponding to the *first* query $\mathcal{R}.H(m, r)$ such that the answer c satisfies $ry^c \in \Gamma_{\text{good}}$. Whether an element $z \in \mathbb{G}$ will be in Γ_{good} or Γ_{bad} will be determined by drawing a random coin $\delta_z \leftarrow \text{Ber}_\mu$ during the simulation. If $\delta_z = 1$ (resp. $\delta_z = 0$), z will be added to Γ_{good} (resp. Γ_{bad}).

We now describe in details the i-th execution of the forger $\mathcal{F}_i(m_i, y_i, \omega_i)$ (see also Figure 1). Before the simulation begins, $\mathcal{M}$ queries a challenge r_i from Θ and initializes a flag $\texttt{forge} = \texttt{false}$. Let t_i denote the point where execution $\mathcal{F}_i$ forks from all previous executions. Assume first that $t_i = 0$, meaning that (m_i, y_i, ω_i) is distinct from the input to all previous executions. Then $\mathcal{M}$ proceeds as follows. For $k = 1, \ldots, q_h$, and while $\texttt{forge} = \texttt{false}$, it makes queries $(m_i, r_i^{\beta_{ik}})$ to $\mathcal{R}.H$ using arbitrary[7] randomization exponents $\beta_{ik} \in \mathbb{Z}_q \setminus \{0\}$. Denoting c_{ik} the answer received from $\mathcal{R}.H$, $\mathcal{M}$ computes $z_{ik} = r_i^{\beta_{ik}} y_i^{c_{ik}}$. Three distinct cases may occur:

---

[6] We stress that $\mathcal{F}_i$, $i = 1, \ldots, n$, denote *distinct executions* of the *same* forger $\mathcal{F}$.

[7] The only constraint is that the β_{ik}'s be distinct in order to avoid making twice the same query.

i) If $z_{ik} \in \Gamma_{\text{bad}}$, then $\mathcal{M}$ simply continues with the next query to $\mathcal{R}.H$.
ii) If $z_{ik} \in \Gamma_{\text{good}}$, then by definition $\mathcal{M}$ already requested $\text{DLog}_g(z_{ik})$ to its discrete log oracle. In that case, it sets $\ell_i = k$, $s_i = \text{DLog}_g(z_{ik})$, $c_i = c_{ik}$, and sets the flag $\texttt{forge}$ to $\texttt{true}$.
iii) If $z_{ik} \notin \Gamma_{\text{good}} \cup \Gamma_{\text{bad}}$, then $\mathcal{M}$ draws a random coin $\delta_{z_{ik}} \leftarrow \text{Ber}_\mu$. If $\delta_{z_{ik}} = 0$, z_{ik} is added to Γ_{bad} and $\mathcal{M}$ continues with the next query to $\mathcal{R}.H$. If $\delta_{z_{ik}} = 1$, then $\mathcal{M}$ queries $\text{DLog}_g(z_{ik})$ and adds z_{ik} to Γ_{good}. It then proceeds exactly as in case ii), and moreover stores the value of β_{ik} as β_i.

Once the flag $\texttt{forge}$ has been set to $\texttt{true}$, $\mathcal{M}$ completes the sequence of queries to $\mathcal{R}.H$ arbitrarily.[8] When the q_h queries to $\mathcal{R}.H$ have been issued, if $\texttt{forge} = \texttt{false}$, then $\mathcal{M}$ returns $\bot$ to $\mathcal{R}$, meaning that execution $\mathcal{F}_i$ fails to forge. Else, $\texttt{forge} = \texttt{true}$ and $\mathcal{M}$ returns (ℓ_i, s_i, c_i) as set at step ii) as forgery for m_i to $\mathcal{R}$. Moreover, if $\mathcal{M}$ did not query its discrete log oracle during the simulation (either because no forgery was returned or because z_{ik} was already in Γ_{good}), then $\mathcal{M}$ directly queries $\text{DLog}_g(r_i)$ (a more economic strategy could be used, but this simplifies notations).

The simulation for the case $t_i \geq 1$ is quite similar to the case $t_i = 0$, with one important difference though. By definition of the forking point, the t_i first queries to $\mathcal{R}.H$ are determined by previous executions, and $\mathcal{M}$ must simulate the forger accordingly. In particular, it cannot embed the current challenge r_i before the $(t_i + 1)$-th query. If there is some query $\mathcal{R}.H(m_i, r)$ of index $k \in [1..(t_i - 1)]$ such that the answer c satisfies $z = ry_i^c \in \Gamma_{\text{good}}$, then $\mathcal{M}$ sets the flag $\texttt{forge}$ to $\texttt{true}$ and will return a forgery corresponding to the first such query (without having to query its discrete log oracle since z is already in Γ_{good}). Note that this same forgery was necessarily already returned in at least one previous execution. At the end of the simulation, $\mathcal{M}$ directly queries $\text{DLog}_g(r_i)$.

Assume now that the flag $\texttt{forge}$ is still set to $\texttt{false}$ when arrived at the t_i-th query. By definition of the forking point, this query was first issued during a previous execution $j < i$, so that $\mathcal{M}$ cannot choose it freshly. The answer of $\mathcal{R}.H$, however, differs from the one received in all previous executions from which $\mathcal{F}_i$ forks exactly at point t_i. Denote $(m_i, \hat{r})$ this t_i-th query to $\mathcal{R}.H$ ($\hat{r} = r_j^{\beta_{jt_i}}$, where r_j was the challenge used during the j-th execution), $\hat{c}$ the corresponding new answer, and $\hat{z} = \hat{r} y_i^{\hat{c}}$. If $\hat{z} \in \Gamma_{\text{bad}}$, then $\mathcal{M}$ can resume the simulation as described for $t_i = 0$, starting from the $(t_i + 1)$-th query to $\mathcal{R}.H$. If $\hat{z} \in \Gamma_{\text{good}}$, then $\mathcal{M}$ can forge a signature for this query without calling its discrete log oracle (and hence will be able to query directly $\text{DLog}_g(r_i)$ at the end of the simulation). If $\hat{z} \notin \Gamma_{\text{good}} \cup \Gamma_{\text{bad}}$, then $\mathcal{M}$ draws a fresh coin $\delta_{\hat{z}} \leftarrow \text{Ber}_\mu$. If $\delta_{\hat{z}} = 0$, then $\mathcal{M}$ can also resume the simulation as described for $t_i = 0$, starting from the $(t_i + 1)$-th query to $\mathcal{R}.H$. The problematic case arises if $\delta_{\hat{z}} = 1$, since $\mathcal{M}$ must return a forgery for the t_i-th query but does not know the discrete logarithm of $\hat{z}$ yet. Hence, $\mathcal{M}$ queries $\hat{s} = \text{DLog}_g(\hat{z})$, completes the sequence of queries to $\mathcal{R}.H$ arbitrarily for $k = t_i + 1$ to q_h, and outputs $(\ell_i = t_i, \hat{s}, \hat{c})$ as forgery for message m_i. After the simulation of $\mathcal{F}_i$, $\mathcal{M}$ makes the additional query $\text{DLog}_g(r_i)$. For

---

[8] Alternatively, we could let $\mathcal{M}$ stop its queries here since queries after the forgery point are irrelevant.

the sake of the discussion in Section 5, we will say that event Bad happens if this last case occurs during one of the n simulations. As we will see shortly, event Bad makes $\mathcal{M}$ fail since in total $\mathcal{M}$ makes two calls to $\mathrm{DLog}_g(\cdot)$ related to the same challenge r_j.[9]

Once the n calls to the forger have been simulated, the reduction $\mathcal{R}$ returns either $\perp$ (in which case $\mathcal{M}$ returns $\perp$ as well), or the discrete logarithm a_0 of r_0. In the latter case, $\mathcal{M}$ uses the procedure Extract to retrieve[10] $x_i = \mathrm{DLog}_g(y_i)$ for $i = 1$ to n. For each challenge r_i received from Θ, either $\mathcal{M}$ queried directly $a_i = \mathrm{DLog}_g(r_i)$, or during the simulation of $\mathcal{F}_i$, $\mathcal{M}$ returned (ℓ_i, s_i, c_i) as forgery, with $s_i = \mathrm{DLog}_g(r_i^{\beta_i} y_i^{c_i})$. Hence $\mathcal{M}$ can compute the discrete logarithm of r_i as $a_i = (s_i - c_i x_i)/\beta_i \mod q$. Finally, $\mathcal{M}$ returns a_0 and $(a_i)_{i=1..n}$. This concludes the description of the meta-reduction.

Differences with the Previous Meta-reduction. In [22,14], the distribution of the indexes ℓ_i returned by the meta-reduction was uniform in $[1..q_h]$ and independent for each execution. On the contrary, for our meta-reduction, it is not difficult to see that for an execution such that all $z_{ik} = r_i^{\beta_{ik}} y_i^{c_{ik}}$ are fresh, ℓ_i is distributed according to a *truncated geometric distribution*:

$$\Pr[\ell_i = k] = \mu(1-\mu)^{k-1} \text{ for } k \in [1..q_h] \text{ and } \Pr[\ell_i = \perp] = 1 - \sum_{k=1}^{q_h} \mu(1-\mu)^{1-k} .$$

Moreover, when an execution forks from previous ones at $t_i > 0$, the distribution of ℓ_i is obviously not independent from the previous forgery indexes ℓ_j. In fact, returning a forgery for independently and uniformly chosen ℓ_i's leads to counter-intuitive behaviors. Consider two distinct executions of a forger $\mathcal{F}$. Assume that some execution $\mathcal{F}_1$ returns a forgery corresponding to some random oracle query index ℓ_1. Then, if another execution $\mathcal{F}_2$ forks from the first one at $t_{2/1} > \ell_1$, it seems more natural for $\mathcal{F}_2$ to return the same forgery as $\mathcal{F}_1$ rather than a new one since the forger "knows" the corresponding signature. Such events cannot happen with our meta-reduction because it simulates a forger that has a natural interpretation: when run on input (m, y), it returns a forgery for the first query $H(m, r)$ such that the answer c satisfies $ry^c \in \Gamma_{\mathrm{good}}$, where Γ_{good} is a set of size $\sim \mu q$ such that the forger can compute the discrete logarithm of elements of Γ_{good} efficiently.

5   Proof of the Main Theorem

We will now prove a sequence of lemmata from which our main result will easily follow. The following lemma will be useful. It results from a simple function analysis and is stated without proof.

---

[9] We could simply let $\mathcal{M}$ abort in that case, but for simplicity of the analysis we prefer to let it make an additional call to $\mathrm{DLog}_g(\cdot)$.

[10] More precisely, for each $i \in [1..n]$, Extract returns γ_i and γ_i' such that $y_i = g^{\gamma_i} r_0^{\gamma_i'} = g^{\gamma_i + a_0 \gamma_i'}$.

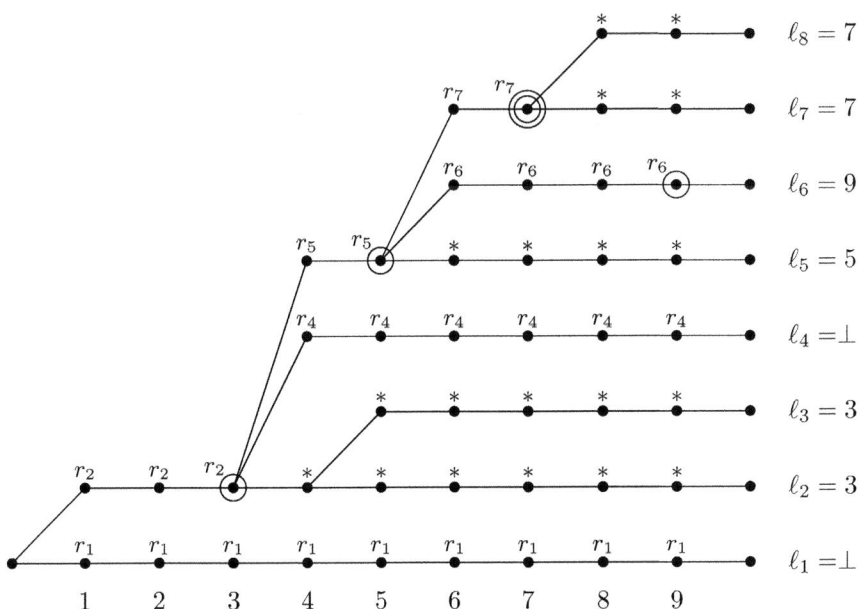

Fig. 1. A possible execution tree of the simulated forger for $q_h = 9$ and $n = 8$. Execution paths go from the root to the leaves. The root symbolizes the beginning of each simulation of the forger. Vertices originating from the root symbolizes the input (m, y, ω) received from $\mathcal{R}$: execution paths sharing the same vertex correspond to the same input. Then, each internal node symbolizes a query to the random oracle $\mathcal{R}.H$, and the vertex originating from this node symbolizes the corresponding answer. Again, execution paths sharing a node, resp. a vertex, share the same query, resp. answer. The label above each query node represents the challenge r_i from Θ used by $\mathcal{M}$ to construct the query (we do not indicate the randomization exponent β_{ik}). Stars indicate that the query is arbitrary since it comes after the forgery point for the execution. Finally, leaves symbolize the output of the forger (a forgery or $\perp$). Here, we simply label leaves with the index ℓ_i of the random oracle query corresponding to the forgery (with the convention that $\ell_i = \perp$ in case the simulated forger returns $\perp$) and we circle the corresponding random oracle query in the execution path. The first execution is run on some input (m_1, y_1, ω_1) and returns no forgery. All subsequent executions are run on the same input $(m_2, y_2, \omega_2) \neq (m_1, y_1, \omega_1)$. The second execution returns some forgery for $\ell_2 = 3$. The third execution forks from the second one at $t_3 = 4 > \ell_2$ so that it returns the same forgery as the second execution. The fourth and fifth executions both fork from previous ones at $t_4 = t_5 = 3$. The fourth one returns no forgery while the fifth one returns a forgery for $l_5 = 5$. The sixth and seventh executions both fork from previous ones at $t_6 = t_7 = 5$, both returning a forgery for resp. $l_6 = 9$ and $l_7 = 7$. Finally, execution 8 forks from previous ones at $t_8 = 7$, and returns a forgery for $l_8 = 7$: since two forgeries related to the same challenge r_7 are returned, event **Bad** happens (assuming $\mathcal{M}$ has to make two queries to its discrete log oracle to forge the signatures).

Lemma 1. *Let $\varepsilon_F \in]0,1[$, and $\mu_0 = 1 - (1 - \varepsilon_F)^{1/q_h}$. Then for any $q_h \geq 1$, one has:*

$$\varepsilon_F \leq q_h \mu_0 \leq \ln\left((1 - \varepsilon_F)^{-1}\right) \ .$$

5.1   Successful Simulation of the Forger

The first thing to do is to lower bound the probability that $\mathcal{R}$ succeeds in returning $\mathtt{DLog}_g(r_0)$. For this, we will show that with sufficiently high probability, $\mathcal{M}$ simulates a "good" forger, *i.e.* a forger that would succeed with probability greater than ε_F when interacting with a real random oracle (rather than $\mathcal{R}.H$).

Definition 7 (Good Forger). *We say that a forger $\mathcal{F}$ making q_h random oracle queries is μ_0-good if for any input (m, y, ω), the distribution over uniform sequences of random oracle answers $(c_1, \ldots, c_{q_h})$ of the forgery index ℓ follows a truncated geometric law of parameter $\tilde{\mu} \geq \mu_0$, i.e. $\Pr[\ell = k] = \tilde{\mu}(1 - \tilde{\mu})^{k-1}$ for $k \in [1..q_h]$.*

Lemma 2. *Let $\mu_0 = 1 - (1 - \varepsilon_F)^{1/q_h}$. Then a μ_0-good forger making q_h random oracle queries $(t_F, q_h, \varepsilon_F)$-UF-NM-breaks Schnorr signatures in the ROM (for some t_F).*

Proof. Fix any message m. Then for any (y, ω), the probability over the answers $(c_1, \ldots, c_{q_h})$ of the random oracle that $\mathcal{F}$ returns a valid forgery is

$$\sum_{k=1}^{q_h} \tilde{\mu}(1 - \tilde{\mu})^{k-1} = 1 - (1 - \tilde{\mu})^{q_h} \geq 1 - (1 - \mu_0)^{q_h} = \varepsilon_F \ .$$

This remains true for the probability over (y, ω) and the answers of the random oracle. □

The success probability of the forger simulated by $\mathcal{M}$ when interacting with a real random oracle depends on the random tape of $\mathcal{M}$ through the draws of the coins δ_z. We will now show that with overwhelming probability, $\mathcal{M}$ simulates a μ_0-good forger. Note that the oracle answers c of $\mathcal{R}.H$ may be determined by the random tape of $\mathcal{R}$, which is set uniformly at random by $\mathcal{M}$. Hence elements $z = ry^c$ may range over all $\mathbb{G}$, and $\mathcal{M}$ must be able to draw δ_z independently for *any* $z \in \mathbb{G}$. In order to avoid using an exponential amount of randomness, $\mathcal{M}$ should derive the coins δ_z from a secure pseudorandom number generator. In all the following, we will assume that the coins δ_z are truly random. By a standard hybrid argument, this assumption cannot affect the success probability of $\mathcal{M}$ by more than a negligible quantity (since otherwise $\mathcal{M}$ would constitute a distinguisher for the pseudorandom number generator).

Lemma 3. *Set $\alpha = q^{-1/4}$. Then there is a negligible function ν such that for any challenges $(r_1, \ldots, r_n)$ received from Θ and any randomization exponents β_{ik}, $\mathcal{M}$ simulates a μ_0-good forger with probability greater that $(1 - \nu)$ over its random tape.*

Proof. Assume that all coins δ_z for $z \in \mathbb{G}$ are drawn before the simulation starts rather than by lazy sampling (this does not change the success probability of the simulated forger). By definition, $\Gamma_{\text{good}} = \{z \in \mathbb{G} : \delta_z = 1\}$. Clearly, the size of Γ_{good} is distributed according to the binomial distribution $\text{Bin}_{\mu,q}$. A Chernoff bound hence gives:

$$\nu \overset{\text{def}}{=} \Pr_{\delta_z} \left[|\Gamma_{\text{good}}| \leq (1 - \alpha)\mu q \right] \leq e^{-\mu q \alpha^2 / 2} \ .$$

Fix an arbitrary input (m, y, ω). For any $r \in \mathbb{G}$, the probability over $c \leftarrow_\$ \mathbb{Z}_q$ that $ry^c \in \Gamma_{\text{good}}$ is equal to $\tilde{\mu} = |\Gamma_{\text{good}}|/q$. Recall that the simulated forger returns a forgery corresponding to the first random oracle query $H(m, r)$ such that the answer c satisfies $ry^c \in \Gamma_{\text{good}}$. Hence, independently of the sequence of queries of the simulated forger, the distribution over uniform sequences of random oracle answers $(c_1, \ldots, c_{q_h})$ of the forgery index ℓ follows a truncated geometric law of parameter $\tilde{\mu}$. When $|\Gamma_{\text{good}}| > (1 - \alpha)\mu q = \mu_0 q$, then $\tilde{\mu} > \mu_0$. This holds for any input (m, y, ω) and any sequence of queries of the simulated forger, so that for any challenges $(r_1, \ldots, r_n)$ received from Θ and any randomization exponents β_{ik}, with probability greater than $(1 - \nu)$ over the draws of the coins δ_z, $\mathcal{M}$ simulates a μ_0-good forger. Moreover, we have:

$$e^{-\mu q \alpha^2 / 2} = e^{-\frac{q_h \mu_0 q \alpha^2}{2 q_h (1 - \alpha)}} \leq e^{-\frac{q_h \mu_0 q \alpha^2}{2 q_h}} \leq e^{-\frac{\varepsilon_F \sqrt{q}}{2 q_h}} \ ,$$

where for the last inequality we used Lemma 1 and $\alpha = q^{-1/4}$. Since by assumption $q_h = \text{poly}(\kappa)$ and $\varepsilon_F = 1/\text{poly}(\kappa)$, we see that ν is negligible, hence the result. □

5.2 Success of the Meta-reduction

The next step is to analyze the probability that $\mathcal{M}$ succeeds given that $\mathcal{R}$ does. It is straightforward to verify that the computation of the discrete logarithm of all challenges $(r_1, \ldots, r_n)$ received from Θ by $\mathcal{M}$ is correct. Consequently, given that $\mathcal{R}$ returns the discrete logarithm of r_0, $\mathcal{M}$ may only fail because it did not make strictly less queries to $\text{DLog}_g(\cdot)$ than to Θ. However, it is not hard to see from the description of $\mathcal{M}$ that if event Bad does not happen, then $\mathcal{M}$ makes *exactly* one query to its discrete log oracle per simulation of the forger, and hence returns the discrete logarithm of $n + 1$ challenges while making n queries to $\text{DLog}_g(\cdot)$. Hence, given that $\mathcal{R}$ returns $a_0 = \text{DLog}_g(r_0)$, and that event Bad does not happen, then $\mathcal{M}$ is successful.

The last step towards proving our main theorem is to bound the probability of event Bad.

Lemma 4. *Event Bad happens with probability less than*

$$n\mu \leq \frac{n \ln \left((1 - \varepsilon_F)^{-1} \right)}{(1 - \alpha) q_h} \ .$$

Proof. Consider the i-th simulation of the forger by $\mathcal{M}$. Let t_i be the point where this execution forks from all previous executions. By construction of $\mathcal{M}$, Bad can only happen if $t_i \geq 1$, and the output of the fresh coin $\delta_{\hat{2}}$ (we refer to notations of Section 4) drawn to decide whether a signature must be forged for the t_i-th query is 1, which happens with probability μ. An union bound on the n simulated executions and Lemma 1 give the result. $\qquad\square$

5.3   Main Theorem and Discussion

We are now ready to state and prove the main theorem of this paper.

Theorem 2. *Assume there is an algebraic reduction $\mathcal{R}$ that $(t_R, n, \varepsilon_R, q_h, \varepsilon_F)$-reduces the DL problem to UF-NM-breaking Schnorr signatures in the ROM, with $\varepsilon_F < 1$. Set $\alpha = q^{-1/4}$. Then there is a negligible function ν such that the meta-reduction $\mathcal{M}$ (t_M, n, ε_M)-solves the OMDL problem, where:*

$$\varepsilon_M \geq \varepsilon_R \left(1 - \nu - \frac{n \ln\left((1 - \varepsilon_F)^{-1}\right)}{(1 - \alpha)q_h} \right)$$

$$t_M \leq \mathtt{poly}(t_R, |\mathcal{R}|, n, q_h, \lfloor \log_2(q) \rfloor) \ .$$

Proof. Denote Sim the event that $\mathcal{M}$ simulates a μ_0-good forger. By Lemma 2 and by definition of a $(t_R, n, \varepsilon_R, q_h, \varepsilon_F)$-reduction, when Sim happens, $\mathcal{R}$ returns $\mathtt{DLog}_g(r_0)$ with probability greater than ε_R (over r_0 and its own random tape). Provided that $\mathcal{R}$ returns the discrete logarithm of r_0 and that Bad does not happen, the meta-reduction is successful. Hence, one has $\varepsilon_M \geq \varepsilon_R(1 - \Pr[\overline{\mathtt{Sim}}] - \Pr[\mathtt{Bad}])$. Combining Lemmata 3 and 4 yields the lower bound on ε_M. Taking into account the fact that $\mathcal{M}$ uses a secure pseudorandom number generator rather than truly random coins cannot modify ε_M by more than a negligible amount (otherwise $\mathcal{M}$ would constitute a distinguisher), that we can incorporate in ν. The running time of $\mathcal{M}$ is upper bounded by the sum of the time needed to simulate the n executions of the forger which is $\mathtt{poly}(n, q_h, \lfloor \log_2 q \rfloor)$, the additional running time t_R of $\mathcal{R}$, and the time to run Extract which is $\mathtt{poly}(t_R, |\mathcal{R}|, \lfloor \log_2 q \rfloor)$, hence the result. $\qquad\square$

Remark 1. As already noted by [22] for their meta-reduction, the above proof can be straightforwardly extended to reductions of the OMDL problem to forging Schnorr signatures in the ROM. Hence the security of Schnorr signatures cannot be proved tightly equivalent to the OMDL problem either (under the OMDL assumption).

Interpretation. Recall that the total running time of the reduction is at most $t_R + nt_F$. Denote $\rho_F = t_F/\varepsilon_F$ and $\rho_R = (t_R + nt_F)/\varepsilon_R \geq nt_F/\varepsilon_R$ the time-to-success ratio of resp. the forger and the reduction. Then some computation gives:

$$\frac{n \ln\left((1 - \varepsilon_F)^{-1}\right)}{(1 - \alpha)q_h} \leq \frac{\varepsilon_R \rho_R}{(1 - \alpha)f(\varepsilon_F)q_h \rho_F} \leq \frac{\rho_R}{(1 - \alpha)f(\varepsilon_F)q_h \rho_F} \ ,$$

where $f(\varepsilon_F) = \varepsilon_F / \ln\left((1 - \varepsilon_F)^{-1}\right)$. Hence one has:

$$\varepsilon_M \geq \varepsilon_R \left(1 - \nu - \frac{\rho_R}{(1 - \alpha) f(\varepsilon_F) q_h \rho_F}\right) .$$

Since $t_R, |\mathcal{R}|, n, q_h, \lfloor \log_2(q) \rfloor = \texttt{poly}(\kappa), t_M = \texttt{poly}(\kappa)$, so that under the OMDL assumption, one must have ε_M negligible. Then the inequality above yields (using $\varepsilon_R = 1/\texttt{poly}(\kappa)$ and $\nu, \alpha = \texttt{negl}(\kappa)$):

$$\rho_R \geq f(\varepsilon_F) q_h \rho_F - \texttt{negl}(\kappa) .$$

Hence one must have that ρ_R is negligibly close to $f(\varepsilon_F) q_h \rho_F$: the reduction essentially loses a factor $f(\varepsilon_F) q_h$ in its time-to-success ratio.

The function $f(\varepsilon_F)$ is depicted below. For small ε_F, one has $f(\varepsilon_F) \simeq 1 - \varepsilon_F/2$ (which is a good approximation up to $\varepsilon_F \simeq 0.5$). For ε_F close to 1, writing $\varepsilon_F = 1 - u$, one has $f(\varepsilon_F) \simeq -1/\ln(u)$. In particular, for $\varepsilon_F = 1 - 1/\texttt{poly}(\kappa)$, $f(\varepsilon_F) \simeq C/\ln(\kappa)$ for some constant C, which shows that f approaches 0 very slowly. For $f(\varepsilon_F) \leq q_h^{-1/3}$, our bound becomes worse than the one by Garg *et al.* [14]. However, for large q_h (which is the case of interest), this implies that ε_F is very close to 1 (*e.g.* for $q_h = 2^{60}$, a rough estimation shows that our bound is not worse than $q_h^{2/3}$ before $\varepsilon_F > 1 - e^{-2^{19}}$).

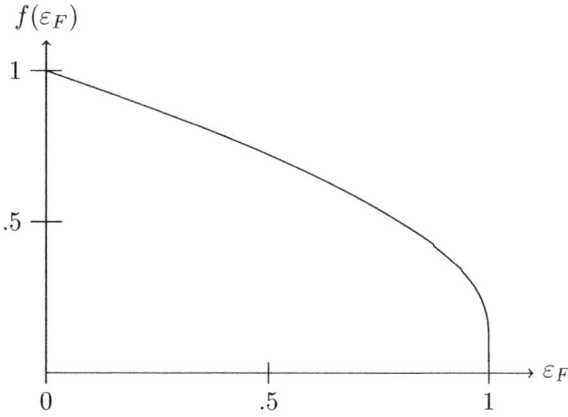

It is interesting to consider what happens when $\varepsilon_F = 1$ since our bound vanishes in that case, while both the security reduction of [24] and the necessary loss $\Omega(q_h^{2/3})$ of [14] hold. In that case one has by definition $\mu = 1$, which means that the meta-reduction simulates an adversary which always returns a forgery corresponding to its *first* random oracle query (in which case there is a reduction which succeeds by running the forger only twice). However, this singularity seems to be an artifact due to definitions in terms of strictly bounded-time and queries algorithms and we can escape it by considering expected-time and queries algorithms. This is developed in the full version of the paper [27]. The main idea is that when simulating a forger making an expected number of random oracle queries q_h, one can choose the distribution of the forgery index ℓ to be a geometric distribution of parameter $\mu \simeq 1/q_h$. This is not possible when the number of oracle queries must be strictly less than q_h, in which case we had to appeal to a truncated geometric distribution. It remains nevertheless that in the special case of a forger making strictly less than q_h random oracle queries and forging with probability $\varepsilon_F = 1$, we do not know of any better simulation strategy than choosing the forgery index uniformly at random in $[1..q_h]$ as was done in the meta-reduction of [22,14], in which case one gets a loss factor $\Omega(q_h^{2/3})$ at best.

References

1. Bellare, M., Namprempre, C., Pointcheval, D., Semanko, M.: The One-More-RSA-Inversion Problems and the Security of Chaum's Blind Signature Scheme. Journal of Cryptology 16(3), 185–215 (2003)
2. Bellare, M., Rogaway, P.: Random Oracles are Practical: A Paradigm for Designing Efficient Protocols. In: ACM Conference on Computer and Communications Security, pp. 62–73 (1993)
3. Bellare, M., Rogaway, P.: The Exact Security of Digital Signatures - How to Sign with RSA and Rabin. In: Maurer, U.M. (ed.) EUROCRYPT 1996. LNCS, vol. 1070, pp. 399–416. Springer, Heidelberg (1996)
4. Boneh, D., Boyen, X.: Short Signatures Without Random Oracles. In: Cachin, C., Camenisch, J.L. (eds.) EUROCRYPT 2004. LNCS, vol. 3027, pp. 56–73. Springer, Heidelberg (2004)
5. Boneh, D., Venkatesan, R.: Breaking RSA May Not Be Equivalent to Factoring. In: Nyberg, K. (ed.) EUROCRYPT 1998. LNCS, vol. 1403, pp. 59–71. Springer, Heidelberg (1998)
6. Bresson, E., Monnerat, J., Vergnaud, D.: Separation Results on the "One-More" Computational Problems. In: Malkin, T. (ed.) CT-RSA 2008. LNCS, vol. 4964, pp. 71–87. Springer, Heidelberg (2008)
7. Brown, D.R.L.: Irreducibility to the One-More Evaluation Problems: More May Be Less. ePrint Archive Report 2007/435 (2007),
   http://eprint.iacr.org/2007/435.pdf
8. Chevallier-Mames, B.: An Efficient CDH-Based Signature Scheme with a Tight Security Reduction. In: Shoup, V. (ed.) CRYPTO 2005. LNCS, vol. 3621, pp. 511–526. Springer, Heidelberg (2005)
9. Coron, J.-S.: On the Exact Security of Full Domain Hash. In: Bellare, M. (ed.) CRYPTO 2000. LNCS, vol. 1880, pp. 229–235. Springer, Heidelberg (2000)
10. Coron, J.-S.: Optimal Security Proofs for PSS and Other Signature Schemes. In: Knudsen, L.R. (ed.) EUROCRYPT 2002. LNCS, vol. 2332, pp. 272–287. Springer, Heidelberg (2002)
11. Dodis, Y., Reyzin, L.: On the Power of Claw-Free Permutations. In: Cimato, S., Galdi, C., Persiano, G. (eds.) SCN 2002. LNCS, vol. 2576, pp. 55–73. Springer, Heidelberg (2003)
12. Fiat, A., Shamir, A.: How to Prove Yourself: Practical Solutions to Identification and Signature Problems. In: Odlyzko, A.M. (ed.) CRYPTO 1986. LNCS, vol. 263, pp. 186–194. Springer, Heidelberg (1987)
13. Fischlin, M.: Communication-Efficient Non-interactive Proofs of Knowledge with Online Extractors. In: Shoup, V. (ed.) CRYPTO 2005. LNCS, vol. 3621, pp. 152–168. Springer, Heidelberg (2005)

14. Garg, S., Bhaskar, R., Lokam, S.V.: Improved Bounds on Security Reductions for Discrete Log Based Signatures. In: Wagner, D. (ed.) CRYPTO 2008. LNCS, vol. 5157, pp. 93–107. Springer, Heidelberg (2008)
15. Goh, E.-J., Jarecki, S.: A Signature Scheme as Secure as the Diffie-Hellman Problem. In: Biham, E. (ed.) EUROCRYPT 2003. LNCS, vol. 2656, pp. 401–415. Springer, Heidelberg (2003)
16. Goh, E.-J., Jarecki, S., Katz, J., Wang, N.: Efficient Signature Schemes with Tight Reductions to the Diffie-Hellman Problems. Journal of Cryptology 20(4), 493–514 (2007)
17. Goldwasser, S., Kalai, Y.T.: On the (In)security of the Fiat-Shamir Paradigm. In: Symposium on Foundations of Computer Science, FOCS 2003, pp. 102–115. IEEE Computer Society (2003)
18. Katz, J., Wang, N.: Efficiency improvements for signature schemes with tight security reductions. In: Jajodia, S., Atluri, V., Jaeger, T. (eds.) ACM Conference on Computer and Communications Security, pp. 155–164. ACM (2003)
19. Koblitz, N., Menezes, A.: Another Look at Non-Standard Discrete Log and Diffie-Hellman Problems. ePrint Archive Report 2007/442 (2007), http://eprint.iacr.org/2007/442.pdf
20. Micali, S., Reyzin, L.: Improving the Exact Security of Digital Signature Schemes. Journal of Cryptology 15(1), 1–18 (2002)
21. Neven, G., Smart, N.P., Warinschi, B.: Hash Function Requirements for Schnorr Signatures. J. Math. Crypt. 3(1), 69–87 (2009)
22. Paillier, P., Vergnaud, D.: Discrete-Log-Based Signatures May Not Be Equivalent to Discrete Log. In: Roy, B. (ed.) ASIACRYPT 2005. LNCS, vol. 3788, pp. 1–20. Springer, Heidelberg (2005)
23. Pointcheval, D., Stern, J.: Security Proofs for Signature Schemes. In: Maurer, U.M. (ed.) EUROCRYPT 1996. LNCS, vol. 1070, pp. 387–398. Springer, Heidelberg (1996)
24. Pointcheval, D., Stern, J.: Security Arguments for Digital Signatures and Blind Signatures. Journal of Cryptology 13(3), 361–396 (2000)
25. Schnorr, C.-P.: Efficient Identification and Signatures for Smart Cards. In: Brassard, G. (ed.) CRYPTO 1989. LNCS, vol. 435, pp. 239–252. Springer, Heidelberg (1990)
26. Schnorr, C.-P.: Efficient Signature Generation by Smart Cards. Journal of Cryptology 4(3), 161–174 (1991)
27. Seurin, Y.: On the Exact Security of Schnorr-Type Signatures in the Random Oracle Model. Full version of this paper, http://eprint.iacr.org
28. Waters, B.: Efficient Identity-Based Encryption Without Random Oracles. In: Cramer, R. (ed.) EUROCRYPT 2005. LNCS, vol. 3494, pp. 114–127. Springer, Heidelberg (2005)

Tightly-Secure Signatures
from Lossy Identification Schemes

Michel Abdalla[1], Pierre-Alain Fouque[1],
Vadim Lyubashevsky[1], and Mehdi Tibouchi[2]

[1] École Normale Supérieure
{michel.abdalla,pierre-alain.fouque,vadim.lyubashevsky}@ens.fr
[2] NTT Information Sharing Platform Laboratories
tibouchi.mehdi@lab.ntt.co.jp

Abstract. In this paper we present three digital signature schemes with tight security reductions. Our first signature scheme is a particularly efficient version of the short exponent discrete log based scheme of Girault et al. (J. of Cryptology 2006). Our scheme has a tight reduction to the *decisional* Short Discrete Logarithm problem, while still maintaining the non-tight reduction to the *computational* version of the problem upon which the original scheme of Girault et al. is based. The second signature scheme we construct is a modification of the scheme of Lyubashevsky (Asiacrypt 2009) that is based on the worst-case hardness of the shortest vector problem in ideal lattices. And the third scheme is a very simple signature scheme that is based directly on the hardness of the Subset Sum problem. We also present a general transformation that converts, what we term *lossy* identification schemes, into signature schemes with tight security reductions. We believe that this greatly simplifies the task of constructing and proving the security of such signature schemes.

Keywords: Signature schemes, tight reductions, Fiat-Shamir.

1 Introduction

Due to the widespread use of digital signature schemes in practical applications, their construction and security analysis comprises an important area of modern cryptography. While there exist many digital signatures that are secure in the *standard* model (e.g. [16,9,25,6,5]), they are usually less efficient than those that are proved secure in the *random oracle* model, and so are not as suitable for practical applications. Signature schemes secure in the random oracle model generally fall into one of two categories. In the first category are schemes constructed using the *Full Domain Hash* (FDH) approach [4], and in the second are schemes based on the *Fiat-Shamir* technique [12]. Our current work focuses on the latter type.

Proving the security of schemes that are designed using the Fiat-Shamir heuristic (e.g. [24,48,20]) generally involves an invocation of the *forking lemma* [43]. Reductions with this feature entail getting one forgery from the adversary, then rewinding him back to a particular point, and then re-running the adversary

D. Pointcheval and T. Johansson (Eds.): EUROCRYPT 2012, LNCS 7237, pp. 572–590, 2012.
© International Association for Cryptologic Research 2012

from that point with the hope of getting another forgery. Using these two related forgeries, the reduction can extract an answer to some underlying hard problem such as discrete log or factorization. Due to the fact that two related forgeries are required and one also needs to guess on which of the q_h of his random oracle query the adversary will forge on, a reduction using an adversary that succeeds with probability ε in forging a signature will have probability ε^2/q_h of breaking the hardness assumption. Asymptotically, this does not cause a problem, since the reduction only incurs a polynomial loss in the success probability. The reduction does not, however, provide us with useful guidance for setting concrete parameters because it is unclear whether the efficiency loss is just an artifact of the proof or whether it represents an actual weakness of the scheme. It is therefore preferable to construct protocols that have a tight proof of security by avoiding the use of the forking lemma.

1.1   Related Work and Contributions

Constructing number-theoretic signature schemes with tight security reductions has received some attention in the past. The first work in this direction is due to Bellare and Rogaway [4], who proposed an RSA-based signature scheme known as PSS whose security is tightly related to the security of the RSA function. Later, in the context of signature schemes based on the Fiat-Shamir heuristic, Micali and Reyzin [36] showed that it is sometimes possible to modify the Fiat-Shamir transform in order to achieve tighter reductions. In more recent work, Goh and Jarecki [21] and Katz and Wang [27,22] constructed digital signatures with tight security reductions based on the Computational and Decisional Diffie-Hellman problems. These latter two schemes are versions of the Schnorr signature scheme, and thus inherit most of its characteristics. In particular, the scheme based on the DDH problem has a very simple construction and a rather short signature size. There are other signature schemes, though, that possess other desirable features, but do not yet have a tight security reduction. A notable example of such a scheme is the one of Girault, Poupard, and Stern [20] which is extremely efficient when the signer is allowed to perform pre-processing before receiving the signature. One of the contributions of this paper is a construction of a scheme that possesses all the advantages of the scheme in [20] in addition to having a tight security reduction.

As far as we are aware, there has not been any previous work that specifically considered tight reductions for lattice-based signatures. Similar to number-theoretic constructions, lattice-based signatures secure in the random oracle model are built using either the Full Domain Hash [18,50,39] or the Fiat-Shamir [40,31,28,32,33] methodologies. While FDH-based lattice signatures have tight reductions, the currently most efficient lattice-based schemes (in terms of both the signature size and the running time) are those based on the Fiat-Shamir framework [32,33]. And so it is an interesting problem whether it's possible to construct an efficient Fiat-Shamir based scheme that has tight reductions. The construction of such a scheme is another contribution of this work, though it is unfortunately a little less efficient than the ones in [32,33].

574   M. Abdalla et al.

The third scheme that we construct in our work is based on the hardness of the low-density subset sum problem. Due to a known reduction from subset sum to lattice problems [30,13], all signature schemes based on lattices are already based on subset sum. The aforementioned reduction, however, incurs a loss, and so the lattice-based schemes are not based on as hard a version of subset sum as we achieve in this paper by building a scheme directly on subset sum. Additionally, our scheme is surprisingly simple (to describe and to prove) and we believe that it could be of theoretical interest.

Proving schemes secure using the Fiat-Shamir heuristic is usually done by first building a 3-move identification scheme secure against passive adversaries, and then applying the Fiat-Shamir transformation, which was proven in [1] to yield provably secure signatures. The advantage of building schemes using this modular approach is that one does not have to deal with any (usually messy) issues pertaining to random oracles when building the identification scheme – all mention of random oracles is delegated to the black-box transformation. For signature schemes with tight security reductions, however, this construction method does not work. The reason is that the transformation of [1] inherently loses a factor of q_h in the success probability of the impersonator to the ID scheme in relation to the forger of the signature scheme, which results in a non-tight security reduction.

In this paper, we give a black-box transformation analogous to that of [1] that converts what we call, *lossy identification schemes* into signature schemes with tight security reductions. Roughly speaking, a lossy identification scheme is a three move commit-challenge-response identification scheme that satisfies the following four simple properties:

1. Completeness: the verification algorithm must accept a valid interaction with non-negligible probability.
2. Simulatability: there is a simulator, who does not have access to the secret key, who is able to produce valid interaction transcripts that are statistically indistinguishable from real ones.
3. Key indistinguishability: there is an algorithm that produces *lossy* keys that are computationally indistinguishable from the real keys.
4. Lossiness: when the keys are lossy, it is statistically impossible to provide a valid response to a random challenge after making a commitment.

Properties 1 and 2 are generally true of all identification schemes, whereas properties 3 and 4 are particular to the lossy case and are crucially required for obtaining a tight black-box transformation. Our transformation converts a lossy identification scheme into a signature scheme and proves that a successful forger can be converted into a successful impersonator to the identification scheme. Since the only non-statistical property in the definition above is property 3, it means that the successful impersonator breaks this property, which is where we will plant the instance of the hard problem that we are trying to solve. We demonstrate the usefulness and generality of this approach by building our signature schemes in this way.

1.2   Overview of Our Signature Schemes

Construction Based on the (decisional) Short Discrete Logarithm Problem. The (computational) c-Discrete Logarithm with Short Exponent (c-DLSE) problem in a cyclic group $\mathbb{G}$ with generator g is the well-studied problem of recovering the discrete logarithm x of a given group element g^x when x is a c-bit long integer, c being typically much smaller than the bit-size of $\mathbb{G}$. Pollard's lambda algorithm [44] solves this problem in time $O(2^{c/2})$, but when $\mathbb{G}$ is a subgroup of prime order in $\mathbb{Z}_p^*$ and c is at least twice the security parameter ($c = 160$ for the 80-bit security level, say), the c-DLSE problem is believed to be as hard as the full-length discrete logarithm problem [51,41]. A number of cryptographic schemes are based on the hardness of the c-DLSE problem, including pseudo-random bit generators [41,14,15], key agreement protocols [17] and signature schemes including Girault-Poupard-Stern (GPS) signatures [45,20].

Like other discrete log-based schemes [48,27,7], GPS is an online/offline scheme in the sense of Even, Goldreich and Micali [10,11]: when preprocessing can be done prior to receiving the message to be signed, signature generation becomes very efficient. The main advantage of GPS signatures, however, is that this online signature generation step doesn't even require a modular reduction, which according to the work of [47], can save as much as 60% of the signing time, which makes the scheme extremely well-suited for situations where processing time is at a premium.

Our scheme, described in Section 4, is very similar to the scheme of [20], but with some tweaks making it possible to choose smaller parameters. Moreover, while the security proof for GPS is a very loose reduction to the computational c-DLSE problem, our security proof provides a *tight* reduction, which is however to the *decisional* short discrete log problem (c-DSDL). Informally, the c-DSDL problem asks to distinguish between a pair (g, g^x) where x is c-bit long and a pair (g, h) where h is uniformly random. No better algorithm is known for solving this problem than actually computing the discrete logarithm and checking whether it is small—in fact, a search-to-decision reduction was established by Koshiba and Kurosawa [29].

Given the pair (g, g^x), we set it as the public key, which by our assumption is computationally indistinguishable from (g, g^x) where x is random (i.e. not small). We then build an identification scheme that satisfies our simulatability requirement, and furthermore show that it is information-theoretically impossible to respond to a random challenge if x is not small. Using our transformation to signatures, this implies that if a forger can produce a valid forgery, then he can respond to a random challenge, which would mean that x is small.

In the end, we obtain a tightly-secure scheme which is quite efficient in terms of size (signatures are around 320-bits long at the 80-bit security level) and speed, especially when used with coupons (in which case signature generation only requires a single multiplication between integers of 80 and 160 bits respectively).

Construction Based on the Shortest Vector Problem in Ideal Lattices. In Section 5, we give a construction of a signature scheme based on the hardness of the approximate worst-case shortest vector problem in ideal lattices. Our scheme is a modification of the scheme in [32] that eliminates the need to use the forking lemma. The scheme in [32] was shown to be secure based on the hardness of the RING-SIS problem, which was previously shown to be as hard as worst-case ideal lattice problems [34,42]. In this work, we construct a similar scheme, but instead have it based on the hardness of the RING-LWE problem, which was recently shown to also be as hard as the worst-case shortest vector problem under quantum reductions [35].

The secret key in our scheme consists of two vectors s_1, s_2 with small coefficients in the ring $\mathcal{R} = \mathbb{Z}_q[\mathbf{x}]/(\mathbf{x}^n + 1)$, and the public key consists of a random element $\mathbf{a} \in \mathcal{R}$ and $\mathbf{t} = \mathbf{a}s_1 + s_2$. The RING-LWE reduction states that distinguishing $(\mathbf{a}, \mathbf{t})$ from a uniformly random pair in $\mathcal{R} \times \mathcal{R}$ is as hard as solving worst-case lattice problems. In our identification scheme, the commitment is the polynomial $\mathbf{a}\mathbf{y}_1 + \mathbf{y}_2$ where $\mathbf{y}_1, \mathbf{y}_2$ are elements in $\mathcal{R}$ chosen with a particular distribution. The challenge is an element $\mathbf{c} \in \mathcal{R}$ with small coefficients, and the response is $(\mathbf{z}_1, \mathbf{z}_2)$ where $\mathbf{z}_1 = \mathbf{y}_1 + s_1\mathbf{c}$ and $\mathbf{z}_2 = \mathbf{y}_2 + s_2\mathbf{c}$. As in [32], the procedure sometimes aborts in order to make sure that the distribution of $(\mathbf{z}_1, \mathbf{z}_2)$ is independent of the secret keys. The verification procedure checks that $\mathbf{z}_1, \mathbf{z}_2$ have "small" coefficients, and that $\mathbf{a}\mathbf{z}_1 + \mathbf{z}_2 - \mathbf{c}\mathbf{t} = \mathbf{a}\mathbf{y}_1 + \mathbf{y}_2$.

The crux of the security proof lies in showing that whenever $(\mathbf{a}, \mathbf{t})$ is truly random, it is information-theoretically impossible to produce a valid response to a random challenge. Proving this part in our security reduction requires analyzing the ideal structure of the ring $\mathcal{R}$ using techniques similar to the ones in [37]. This analysis is somewhat loose, however, so that the resulting signature scheme is not as efficient as the one in [32]. We believe that improving the analysis (possibly using some recent techniques in [49]) and obtaining a more efficient signature scheme is an interesting research direction.

Construction Based on Subset Sum. In Section 6, we present a very simple scheme based on the hardness of the subset sum problem. The secret key consists of an $n \times k$ 0/1 matrix $\mathbf{X}$, and the public key consists of a random vector $\mathbf{a} \in \mathbb{Z}_M^n$, as well as a k-dimensional vector of subset sums $\mathbf{t} = \mathbf{a}^T\mathbf{X} \bmod M$ that use $\mathbf{a}$ as weights. The main idea for constructing the lossy identification scheme is to achieve the property that if the vector $\mathbf{t}$ is uniformly random, rather than being a vector of valid subset sums, then it should be impossible (except with a small probability) to produce a valid response to a random challenge. And so an adversary who is able to break the resulting signature scheme can be used to distinguish vectors $\mathbf{t}$ that are valid subset sums of the elements in $\mathbf{a}$ from those that are just uniformly random. We defer further details to Section 6.

2   Preliminaries

2.1   The Decisional Short Discrete Logarithm Problem

Let $\mathbb{G}$ be a finite, cyclic group of prime order q whose group operation is noted multiplicatively, and g a fixed generator of $\mathbb{G}$. Let further c be a size parameter.

The *c-decisional* discrete logarithm (*c*-DSDL) problem may be informally described as the problem of distinguishing between tuples of the form (g, h) for a uniformly random $h \in \mathbb{G}$ and tuples of the form (g, g^x) with x uniformly random in $\{0, \ldots, 2^c - 1\}$. More precisely:

Definition 1. *A distinguishing algorithm $\mathscr{D}$ is said to (t, ε)-solve the c-DSDL problem in group $\mathbb{G}$ if $\mathscr{D}$ runs in time at most t and satisfies:*

$$\left| \Pr[x \xleftarrow{\$} \mathbb{Z}_q : \mathscr{D}(g, g^x) = 1] - \Pr[x \xleftarrow{\$} \{0, \ldots, 2^c - 1\} : \mathscr{D}(g, g^x) = 1] \right| \geq \varepsilon$$

We say that $\mathbb{G}$ is a (t, ε)-c-DSDL group if no algorithm (t, ε)-solves the c-DSDL problem in $\mathbb{G}$.

This problem is related to the well-known (computational) *c*-discrete logarithm with short exponent (*c*-DLSE) problem. In fact, for the groups where that problem is usually considered, namely prime order subgroups of $\mathbb{Z}_p^*$ where p is a safe prime, a search-to-decision reduction is known for all c [29]: if the *c*-DLSE problem is hard, then so is the *c*-DSDL problem. The reduction is not tight, however, so while the signature scheme presented in the next section admits a tight reduction to the decisional problem, there is a polynomial loss in the reduction to the search problem.

2.2   The Ring-LWE Problem and Lattices

For any positive integer n and any positive real σ, the distribution $D_{\mathbb{Z}^n, \sigma}$ assigns the probability proportional to $e^{-\pi \|\mathbf{y}\|^2 / \sigma^2}$ to every $\mathbf{y} \in \mathbb{Z}^n$ and 0 everywhere else. For any odd prime p, the ring $\mathcal{R} = \mathbb{Z}_p[\mathbf{x}]/(\mathbf{x}^n + 1)$ is represented by polynomials of degree at most $n - 1$ with coefficients in the range $\left[-\frac{p-1}{2}, \frac{p-1}{2} \right]$. As an additive group, $\mathcal{R}$ is isomorphic to $\mathbb{Z}_p^n$, and we use the notation $\mathbf{y} \xleftarrow{\$} D_{\mathcal{R}, \sigma}$ to mean that a vector $\mathbf{y}$ is chosen from the distribution $D_{\mathbb{Z}^n, \sigma}$ and then mapped to a polynomial in $\mathcal{R}$ in the natural way (i.e. position i of the vector corresponds to the coefficient of the $\mathbf{x}^i$ term of the polynomial). The (decisional) Ring Learning With Errors Problem (RING-LWE) over the ring $\mathcal{R}$ with standard deviation σ is to distinguish between the following two oracles: $\mathcal{O}_0$ outputs random elements in $\mathcal{R} \times \mathcal{R}$, while the oracle $\mathcal{O}_1$ has a secret $\mathbf{s} \in \mathcal{R}$ where $\mathbf{s} \xleftarrow{\$} D_{\mathcal{R}, \sigma}$, and on every query it chooses a uniformly random element $\mathbf{a} \xleftarrow{\$} \mathcal{R}$, $\mathbf{e} \xleftarrow{\$} D_{\mathcal{R}, \sigma}$, and outputs $(\mathbf{a}, \mathbf{as} + \mathbf{e})$. The RING-LWE problem is a natural generalization of the LWE problem [46] to rings and it was recently shown in [35] that if $p = poly(n)$ is a prime congruent to 1 mod $2n$, then solving the RING-LWE problem over the ring $\mathcal{R}$ with standard deviation[1] σ is as hard as finding an approximate shortest vector in all ideal lattices in the ring $\mathbb{Z}[\mathbf{x}]/(\mathbf{x}^n + 1)$. Intuitively, the smaller the ratio between p and σ is, the smaller the vector the reduction is able to find, and thus it is preferable to keep this ratio low.

---

[1] In the actual reduction of [35], the standard deviation is itself chosen from a somewhat complicated probability distribution, but if the number of times the RING-LWE oracle is queried is bounded (in this paper it only needs to provide one output), then the standard deviation can be fixed.

2.3   The Subset Sum Problem

In the search version of the random subset sum problem, $SS(n, M)$, one is given
n elements a_i generated uniformly at random in $\mathbb{Z}_M$ (in this paper, we will
only deal with *low-density* instances of the problem, where $M > 2^n$) and an
element $t = \sum a_i s_i \bmod M$, where the s_i are randomly chosen from $\{0, 1\}$, and
is asked to find the s_i (with high probability, there is only one possible set
of s_i). The decision version of the problem, which was shown to be as hard
as the search version [26,38], is to distinguish an instance $(a_1, \ldots, a_n, t)$ where
$t = a_1 x_1 + \ldots + a_n s_n \bmod M$ from the instance $(a_1, \ldots, a_n, t)$ where t is uniformly
random in $\mathbb{Z}_M$. The low-density $SS(n, M)$ problem is hardest when $M \approx 2^n$, in
which case the best algorithm runs in time $2^{\Omega(n)}$ (see for example [3]), but the
best known algorithms for the problem when $M = n^{O(n)}$, still require time
$2^{\Omega(n)}$. As M increases, however, the problem becomes easier, until it is solvable
in polynomial-time when $M = 2^{\Omega(n^2)}$ [30,13].

2.4   Signature Schemes

Definition 2. *A signature scheme* Sig *is composed of three algorithms* (GenKey,
Sign, Verify) *such that:*

- *The key generation algorithm* GenKey *takes as input the security parameter
  in unary notation and outputs a pair* (pk, sk) *containing the public verification key and the secret signing key.*
- *The signing algorithm* Sign *takes as input a message* m *and the signing key
  sk and outputs a signature* σ. *This algorithm can be probabilistic so that
  many signatures can be computed for the same message.*
- *The verification algorithm* Verify *takes as input a message* m, *a signature* σ
  and the public key pk and outputs 1 *if the signature is correct and* 0 *otherwise.*

The standard security notion for signature scheme is *strong existential unforgeability against adaptive chosen-message attacks* [23] which informally means
that, after obtaining signatures on polynomially many arbitrary messages of his
choice, an adversary cannot produce a new valid signature, even for a message
m for which he already knows a correct signature.

Definition 3. *Let* Sig $=$ (GenKey, Sign, Verify) *be a signature scheme and let
H be a random oracle. We say that* Sig *is* $(t, q_h, q_s, \varepsilon)$-*strongly existentially unforgeable against adaptive chosen-message attacks, if there is no algorithm* $\mathcal{D}$
that runs in time at most t, *while making at most* q_h *hash queries and at most*
q_s *signing queries, such that*

$$\Pr[(pk, sk) \leftarrow \mathsf{GenKey}(1^k); (m, \sigma) \leftarrow \mathcal{D}^{\mathsf{Sign}(sk, \cdot), H(\cdot)}(pk) :$$
$$(m, \sigma) \notin \mathcal{S} \wedge \mathsf{Verify}(m, \sigma, pk) = 1] \geq \varepsilon,$$

where $\mathcal{S}$ *is the set of message-signature pairs returned by the signing oracle.*

3 Lossy Identification Schemes

In order to unify the security proofs of our signature schemes without sacrificing the tightness of the reduction, we introduce in this section a new class of identification schemes, called lossy identification schemes. In these schemes, the public key associated with the prover can take one of two indistinguishable forms, called *normal* and *lossy*. When the public key is normal, the scheme behaves as a standard identification scheme with similar security guarantees against impersonation attacks. However, in the lossy case, the public key may not have a corresponding secret key and no prover (even computationally unbounded ones) should be able to make the verifier accept with non-negligible probability.

As with other identification schemes used to build signature schemes via the Fiat-Shamir transform, the identification schemes that we consider in this paper consist of a canonical three-move protocol, as defined in [1]. In these protocols, the verifier's move consists in choosing a random string from the challenge space and sending it to the prover. Moreover, its final decision is a *deterministic* function of the conversation transcript and the public key. Since our results can be seen as a generalization of the results of Abdalla *et al.* [1] to the lossy setting, we use their definitions as the basis for ours below.

Definition 4. *A lossy identification scheme* ID *is defined by a tuple* (KeyGen, LosKeyGen, Prove, c, Verify) *such that:*

- KeyGen *is the normal key generation algorithm which takes as input the security parameter in unary notation and outputs a pair* (pk, sk) *containing the publicly available verification key and the prover's secret key.*
- LosKeyGen *is the lossy key generation algorithm which takes as input the security parameter in unary notation and outputs a lossy verification key* pk.
- Prove *is the prover algorithm which takes as input the current conversation transcript and outputs the next message to be sent to the verifier.*
- $c(k)$ *is a function of the security parameter which determines the length of the challenge sent by the verifier.*
- Verify *is a deterministic algorithm which takes the conversation transcript as input and outputs 1 to indicate acceptance or 0 otherwise.*

Following [1], we associate to ID, k, and (pk, sk) a randomized *transcript generation oracle* $\mathsf{Tr}^{\mathsf{ID}}_{pk,sk,k}$ which takes no inputs and returns a random transcript of an "honest" execution. However, to adapt it to specific setting of our schemes, we modify to the original definition to take into account the possibility that the prover may fail and output $\perp$ as response during the execution of the identification protocol. Moreover, when this happens, instead of outputting $(cmt, ch, \perp)$, our transcript generation oracle will simply return a triplet $(\perp, \perp, \perp)$ to simulate the scenario in which the verifier simply *forgets* failed identification attempts. Interestingly, as we show later in this section, this weaker requirement is sufficient for building secure signature schemes as failed impersonation attempts will be kept hidden from the adversary since the tasks of generating the commitment and challenge are performed by the signer. More precisely, the transcript generation oracle $\mathsf{Tr}^{\mathsf{ID}}_{pk,sk,k}$ is defined as follows:

$\mathsf{Tr}^{\mathsf{ID}}_{pk,sk,k}()$:

1: $cmt \xleftarrow{\$} \mathsf{Prove}(sk)$
2: $ch \xleftarrow{\$} \{0,1\}^{c(k)}$
3: $rsp \xleftarrow{\$} \mathsf{Prove}(sk, cmt, ch)$
4: **if** $rsp = \perp$ **then** $(cmt, ch) \leftarrow (\perp, \perp)$
5: **return** (cmt, ch, rsp)

Definition 5. *An identification scheme is said to be lossy if it has the following properties:*

1. **Completeness of Normal Keys.** *We say that ID is ρ-complete, where ρ is a non-negligible function of k, if for every security parameter k and all honestly generated keys $(pk, sk) \xleftarrow{\$} \mathsf{KeyGen}(1^k)$, $\mathsf{Verify}(pk, cmt, ch, rsp) = 1$ holds with probability ρ when $(cmt, ch, rsp) \xleftarrow{\$} \mathsf{Tr}^{\mathsf{ID}}_{pk,sk,k}()$.*
2. **Simulatability of Transcripts.** *Let (pk, sk) be the output of $\mathsf{KeyGen}(1^k)$ for a security parameter k. Then, we say that ID is ε-simulatable if there exists a PPT algorithm $\widetilde{\mathsf{Tr}}^{\mathsf{ID}}_{pk,k}$ with no access to the secret key sk which can generate transcripts $\{(cmt, ch, rsp)\}$ whose distribution is statistically indistinguishable from the transcripts output by $\mathsf{Tr}^{\mathsf{ID}}_{pk,sk,k}$, where ε is an upper-bound for the statistical distance. When $\varepsilon = 0$, then ID is said to simulatable.*
3. **Indistinguishability of Keys.** *Consider the experiments $\mathbf{Exp}^{\text{ind-keys-real}}_{\mathsf{ID},\mathscr{D}}(k)$ and $\mathbf{Exp}^{\text{ind-keys-lossy}}_{\mathsf{ID},\mathscr{D}}(k)$ in which we generate pk via $\mathsf{KeyGen}(1^k)$, respectively $\mathsf{LosKeyGen}(1^k)$, and provide it as input to the distinguishing algorithm $\mathscr{D}$. We say that $\mathscr{D}$ can (t, ε)-solve the key-indistinguishability problem if $\mathscr{D}$ runs in time t and*

$$\left| \Pr[\mathbf{Exp}^{\text{ind-keys-real}}_{\mathsf{ID},\mathscr{D}}(k) = 1] - \Pr[\mathbf{Exp}^{\text{ind-keys-lossy}}_{\mathsf{ID},\mathscr{D}}(k) = 1] \right| \geq \varepsilon.$$

Furthermore, we say that ID is (t, ε)-key-indistinguishable if no algorithm (t, ε)-solves the key-indistinguishability problem.
4. **Lossiness.** *Let $\mathscr{I}$ be an impersonator, st be its state, and k be a security parameter. Let $\mathbf{Exp}^{\text{los-imp-pa}}_{\mathsf{ID},\mathscr{I}}(k)$ be the following experiment played between $\mathscr{I}$ and a hypothetical challenger:*

$\mathbf{Exp}^{\text{los-imp-pa}}_{\mathsf{ID},\mathscr{I}}(k)$:

1: $pk \xleftarrow{\$} \mathsf{LosKeyGen}(1^k)$
2: $(st, cmt) \xleftarrow{\$} \mathscr{I}^{\widetilde{\mathsf{Tr}}^{\mathsf{ID}}_{pk,k}}(pk)$; $ch \xleftarrow{\$} \{0,1\}^{c(k)}$; $rsp \xleftarrow{\$} \mathscr{I}(st, ch)$
3: **return** $\mathsf{Verify}(pk, cmt, ch, rsp)$

Sign(sk, m):	Verify(pk, m, σ):
1: $ctr \leftarrow 0$	1: parse σ as (cmt, rsp)
2: **while** $ctr \leq \ell$ and $rsp = \perp$ **do**	2: $ch \leftarrow H(cmt, m)$
3: $ctr \leftarrow ctr + 1$	3: $d \leftarrow$ Verify(pk, cmt, ch, rsp)
4: $cmt \leftarrow$ Prove(sk)	4: **return** d
5: $ch \leftarrow H(cmt, m)$	
6: $rsp \leftarrow$ Prove(sk, cmt, ch)	
7: **end while**	
8: **if** $rsp = \perp$ **then** $cmt \leftarrow \perp$	
9: $\sigma \leftarrow (cmt, rsp)$	
10: **return** σ	

Fig. 1. Description of our signature scheme Sig[ID, ℓ] = (GenKey, Sign, Verify), where ID = (KeyGen, LosKeyGen, Prove, c, Verify) is a lossy identification scheme, H is a random oracle, and ℓ is a bound on the number of signing attempts

We say $\mathscr{I}$ ε-solves the impersonation problem with respect to lossy keys if

$$\Pr[\,\mathbf{Exp}_{\mathsf{ID},\mathscr{I}}^{\mathrm{los\text{-}imp\text{-}pa}}(k) = 1\,] \geq \varepsilon.$$

Furthermore, we say that ID is ε-lossy if no (computationally unrestricted) algorithm ε-solves the impersonation problem with respect to lossy keys.

As in [1], we need to use the concept of min-entropy [8] to measure the maximum likelihood that a commitment generated by the prover collides with a fixed value. The precise definition of min-entropy can be found in Definition 3.2 in [1].

Transform. The signature schemes that we consider in this paper are built from lossy identification schemes via the Fiat-Shamir transform [12], in which the challenge becomes the hash of the message together with the commitment. However, since we do not assume perfect completeness of normal keys for the underlying lossy identification scheme, the signing algorithm will differ slightly from those considered in [1] in order to decrease the probability of abort during signing. More precisely, let ID = (KeyGen, LosKeyGen, Prove, c, Verify) be a lossy identification scheme and let H be a random oracle. Let ℓ be a parameter defining the maximum number of signing attempts. We can construct a signature scheme Sig[ID, ℓ] = (GenKey, Sign, Verify), where GenKey simply calls KeyGen from the ID scheme, and Sign, Verify are depicted in Figure 1.

We remark that the signature length of the scheme in Figure 1 can sometimes be optimized by setting $\sigma = (ch, rsp)$. However, this is only possible when the commitment value cmt is uniquely defined by (ch, rsp), which is the case for all the schemes considered in this paper.

Theorem 1. *Let* ID = (KeyGen, LosKeyGen, Prove, c, Verify) *be a lossy identification scheme whose commitment space has min-entropy $\beta(k)$, let H be a random oracle, and let* Sig[ID] = (GenKey, Sign, Verify) *be the signature scheme obtained* *via the transform in Figure 1. If* ID *is* ε_s*-simulatable,* ρ*-complete,* (t', ε_k)*-key-indistinguishable, and* ε_ℓ*-lossy, then* Sig[ID] *is* $(t, q_h, q_s, \varepsilon)$*-strongly existentially unforgeable against adaptive chosen-message attacks in the random oracle model for:*

$$\varepsilon = \varepsilon_k + q_s \varepsilon_s + (q_h + 1)\varepsilon_\ell + \ell(q_s + q_h + 1)q_s/2^\beta$$
$$t \approx t' - O(q_s \cdot t_{\mathsf{Sign}})$$

where t_{Sign} *denotes the average signing time. Furthermore, the probability that* Sig[ID] *outputs a valid signature is* $1 - (1 - \rho)^\ell$.

Proof Overview. In order to prove the security of the signature scheme based on the security properties of the underlying lossy identification scheme, the main idea is to use honest transcripts generated by the identification scheme to answer signature queries made the adversary by appropriately programming the random oracle. More precisely, let (cmt, ch, rsp) be a valid transcript (i.e., Verify$(pk, cmt, ch, rsp) = 1$). To answer a query m to the signing oracle, we need to program the random oracle to set $H(cmt, m) = ch$ so that (cmt, rsp) is a valid signature for m. Unfortunately, this programming may conflict with previous values outputted by the hash oracle. To address this problem, the first step of the proof is to show that such collisions happen with with probability at most $\ell(q_s + q_h + 1)q_s/2^\beta$.

Next, we make a sequence of small changes to the security experiment to be able to bound the success probability of the forger. The first significant modification is to change the simulation of the signing oracle so that it no longer uses the secret key. This is done by replacing the transcript generation oracle $\mathsf{Tr}^{\mathsf{ID}}_{pk,sk,k}$ with its simulated version $\widetilde{\mathsf{Tr}}^{\mathsf{ID}}_{pk,k}$. Since we make at most q_s calls to $\widetilde{\mathsf{Tr}}^{\mathsf{ID}}_{pk,k}$, the difference in the success probability of the forger changes by at most $q_s \varepsilon_s$ due to the simulatability of ID.

The second important modification is to replace the key generation algorithm with its lossy version. Since the secret key is no longer needed in the simulation of the signing oracle, the difference in the success probability of the forger changes by at most ε_k due to the key-indistinguishability of ID.

Finally, we can bound the success probability of the forger in this final experiment by relating this probability with that of solving the impersonation problem with respect to lossy keys. Since we need to guess the hash query which will be used in the forgery to be able to break the underlying impersonation problem, we lose a factor $q_h + 1$ in the reduction, resulting in the term $(q_h + 1)\varepsilon_\ell$ in the theorem. For more details, please refer to the full version of this paper [2].

4   A Signature Scheme Based on the DSDL Problem

In this section we describe our short discrete log based signature scheme. While it looks similar to the prime-order version of the Girault-Poupard-Stern identification scheme [19,45,20], the proof strategy is in fact closer to the one used by

Katz and Wang for their DDH-based signature scheme [27,22]. We first present a lossy identification scheme and then use the generic transformation from the previous section to obtain the signature scheme.

The public parameters of the identification scheme are a cyclic group $\mathbb{G}$ of prime order q (typically chosen as the subgroup of order q in $\mathbb{Z}_p^*$ where p is prime), a generator g of $\mathbb{G}$, and size parameters c, k, k'. The secret key is a small (relative to q) integer x and the public key consists of a single group element $h = g^x \bmod p$. The prover's first move is to generate a small (but larger than x) random integer y and send $u = g^y$ as a commitment to the verifier. Next, the (honest) verifier picks a value e uniformly in $\{0, \ldots, 2^k - 1\}$ and sends it to the prover. After receiving e from the verifier, the prover computes $z = ex + y$ (without any modular reduction), and checks whether z is in the range $\{2^{k+c}, \ldots, 2^{k+k'+c}-1\}$. If z is in the "correct" range, then the prover sends z to the verifier, who can check the verifying equation $u = g^z/h^e$ to authenticate the prover. If z is outside the correct range, the prover sends $\bot$ to indicate failure—as in [31,32], this check is important to ensure that the distribution of the value z is independent of the secret key x. In the full version of this paper [2], we prove:

Theorem 2. *If $\mathbb{G}$ is a (t, ε)-c-DSDL group, then the identification scheme described above is perfectly simulatable, ρ-complete, (t, ε)-key-indistinguishable, and ε_ℓ-lossy, for $\rho = 1 - 2^{-k'}$ and $\varepsilon_\ell \le 2^{2k+k'+c+2}/q + 1/2^k$.*

In order to obtain our signature scheme based on the DSDL problem, we apply the transform provided in the previous section to the identification scheme described above. The full description of the resulting scheme is provided in Figure 2. In addition to those of the underlying identification scheme, the public parameters of the signature scheme also include the maximum number of signing attempts ℓ and a random oracle $H: \{0,1\}^* \to \{0, \ldots, 2^k - 1\}$. The key pair is as before. To sign a message m, we generate a small (but larger than x) random integer y and compute $e \gets H(g^y \bmod p, m)$. Finally, we set $z = ex+y$ and check whether z is in the correct range. If it's not, we restart the signature process. In case of ℓ failures, the signing algorithm simply outputs $(\bot, \bot)$ to indicate failure. Otherwise, the signature will consist of the pair $\sigma = (z, e)$. Since the probability that z is not in the correct range is smaller than $1/2^{k'}$, the signing algorithm will fail with probability at most $(1 - 1/2^{k'})^\ell$. Moreover, the average number of iterations is $1/(1 - 1/2^{k'})$. As a direct consequence of Theorems 1 and 2, we get:

Theorem 3. *If $\mathbb{G}$ is a (t', ε')-c-DSDL group, then this signature scheme is $(t, q_h, q_s, \varepsilon)$-strongly existentially unforgeable against adaptive chosen-message attacks in the random oracle model for:*

$$\varepsilon = \varepsilon' + (q_h + 1) \cdot \frac{2^{2k+k'+c+2}}{q} + \ell(q_s + q_h + 1) \cdot \frac{q_s}{2^k}$$

$$t \approx t' - O(q_s \cdot t_1)$$

(where t_1 is the cost of an exponentiation in $\mathbb{G}$), and it outputs a valid signature with probability $1 - 2^{k'\ell}$.

KeyGen(): Pick $x \xleftarrow{\$} \{0, \ldots, 2^c - 1\}$ as the private key, and $X \leftarrow g^x \bmod p$ as the public key.

Sign(m, x):
1: $ctr \leftarrow 0$
2: $y \xleftarrow{\$} \{0, \ldots, 2^{k+k'+c} - 1\}$
3: $e \leftarrow H(g^y \bmod p, m)$
4: $z \leftarrow ex + y$
5: if $z \notin \{2^{k+c}, \ldots, 2^{k+k'+c} - 1\}$ and $ctr < \ell$ then
6:     $ctr \leftarrow ctr + 1$
7:     goto Step 2
8: if $z \notin \{2^{k+c}, \ldots, 2^{k+k'+c} - 1\}$ then $(z, e) \leftarrow (\bot, \bot)$
9: return $\sigma = (z, e)$

Verify($m, X, \sigma = (z, e)$): accept if and only if $z \in \{2^{k+c}, \ldots, 2^{k+k'+c} - 1\}$ and $e = H(g^z \cdot X^{-e} \bmod p)$.

Fig. 2. DSDL-Based Signature Scheme

Remarks

1. The scheme in Figure 2 uses (z, e) instead of (z, g^y) as the signature since (z, e) can be used to recover g^y, but the length of e is shorter than that of g^y.
2. This is an online/offline signature scheme: it can be used with coupons by pre-computing $(y, g^y \bmod p)$ independently of the message. In the rare case when z is not in the right interval (which can be checked without even computing a multiplication), it suffices to use another coupon.
3. The reduction is not *completely* tight: there is a small loss of $\ell \cdot q_s$. As in [22], this loss can be avoided by ensuring that the masking parameter y is always the same for a given message, either by making the scheme stateful (keeping track of the randomness on signed messages) or by generating y as a deterministic, pseudorandom function of the signed message and the private key(but the resulting scheme is no longer online/offline).

Suggested Parameters. We propose the following parameters for an instantiation of our scheme with an 80-bit security level. The group $\mathbb{G}$ is a subgroup of order q in $\mathbb{Z}_p^*$, where p is a 1024-bit prime and q a prime factor of $p - 1$ of length ≥ 490 bits. Moreover, we set $(c, k, k') = (160, 80, 8)$. The size of the public key $g^x \bmod p$ is then 1024 bits and the size of the signature (z, e) is $k+k'+c+k = 328$ bits.

A full signature requires a single exponentiation of 248 bits in $\mathbb{Z}_p^*$ with fixed base, which is about as efficient as comparable schemes (faster than the two 160-bit exponentiations in the Katz-Wang DDH scheme, for example). In our scheme, there is a $1/2^{k'} = 1/256$ chance that the signing algorithm will have to be repeated, but this has little effect on the expected running time.

Parameter	Definition
n	integer that is a power of 2
σ	standard deviation of the secret key coefficients
p	"small" prime equal to 1 mod $2n$
$\mathcal{R}$	ring $\mathbb{Z}_p[\mathbf{x}]/\langle \mathbf{x}^n + 1\rangle$
$\mathcal{C}$	$\{\mathbf{g} \in \mathcal{R} : \|\mathbf{g}\|_\infty \leq \log n\}$
$\mathcal{M}$	$\{\mathbf{g} \in \mathcal{R} : \|\mathbf{g}\|_\infty \leq n^{3/2}\sigma\log^3 n\}$
$\mathcal{G}$	$\{\mathbf{g} \in \mathcal{R} : \|\mathbf{g}\|_\infty \leq (n-1)\sqrt{n}\sigma\log^3 n\}$

Fig. 3. Parameter Definitions

When used with coupons, the scheme is possibly the fastest option available, with an online cost of one single integer multiplication between a 80-bit number and a 160-bit number, and no modular reduction.

5   A Signature Scheme Based on Lattices

In this section, we present a signature scheme whose security is based on the hardness of the RING-LWE problem. Towards this goal, we first describe a lossy identification scheme based on the RING-LWE problem and then use our generic transformation in Section 3 to obtain the signature scheme.

Our identification scheme depends on some public parameters defined in Figure 3. The secret key consists of two polynomials $\mathbf{s}_1, \mathbf{s}_2$ with "small" coefficients chosen from the distribution $D_{\mathcal{R},\sigma}$, and the public key consists of a randomly-chosen element $\mathbf{a} \in \mathcal{R}$ and of the value $\mathbf{t} = \mathbf{as}_1 + \mathbf{s}_2$. Under the RING-LWE assumption in the ring $\mathcal{R}$, the public key is thus indistinguishable from a uniformly random element of $\mathcal{R}^2$.

In our protocol, the prover's first move is to create two "small" polynomials $\mathbf{y}_1, \mathbf{y}_2$ (larger than $\mathbf{s}_1, \mathbf{s}_2$ by a factor $\approx n$) from the set $\mathcal{M}$, and then send the value $\mathbf{u} = \mathbf{ay}_1 + \mathbf{y}_2$ to the verifier. Upon receipt of $\mathbf{u}$, the (honest) verifier chooses a value $\mathbf{c}$ uniformly at random in the set $\mathcal{C}$ and sends it to the prover. After receiving $\mathbf{c}$ from the verifier, the prover sets $\mathbf{z}_1 \leftarrow \mathbf{s}_1\mathbf{c} + \mathbf{y}_1$ and $\mathbf{z}_2 \leftarrow \mathbf{s}_2\mathbf{c} + \mathbf{y}_2$ and checks whether the $\mathbf{z}_i$'s are both in $\mathcal{G}$. If they are, the prover then sends the response $(\mathbf{z}_1, \mathbf{z}_2)$ to the verifier. If one (or both) of the $\mathbf{z}_i$ are outside of $\mathcal{G}$ (which happens with probability approximately $1 - 1/e^2$), then the prover simply sends $(\perp, \perp)$. Finally, the verifier simply checks whether the $\mathbf{z}_i$'s are in $\mathcal{G}$ and that $\mathbf{az}_1 + \mathbf{z}_2 = \mathbf{tc} + \mathbf{u}$.

At this point, we would like to point out that using the recent techniques in [33], it is possible to lower the bitsize of the response $(\mathbf{z}_1, \mathbf{z}_2)$ by choosing the polynomials $\mathbf{y}_1, \mathbf{y}_2$ from a normal distribution and then doing a somewhat more involved rejection sampling when deciding whether to send $(\mathbf{z}_1, \mathbf{z}_2)$ or $(\perp, \perp)$ to the verifier.

In the full version of this paper [2], we prove:

Theorem 4. *If $p \gg \sigma^{2/\alpha} \cdot n^{3/\alpha+\eta}$ for some $\eta > 0$, and the RING-LWE problem over $\mathcal{R}$ with standard deviation σ is (ε, t)-hard, then the identification scheme*

KeyGen(): Pick $\mathbf{s}_1, \mathbf{s}_2 \xleftarrow{\$} D_{\mathcal{R},\sigma}$ and set $(\mathbf{s}_1, \mathbf{s}_2)$ as the private key. Select $\mathbf{a} \xleftarrow{\$} \mathcal{R}$ and let the public key be $(\mathbf{a}, \mathbf{t})$, where $\mathbf{t} \leftarrow \mathbf{a}\mathbf{s}_1 + \mathbf{s}_2$. Let H be a random oracle mapping to the range $\mathcal{C}$.

Sign($m, \mathbf{a}, \mathbf{s}_1, \mathbf{s}_2$):
1: $ctr \leftarrow 0$
2: $\mathbf{y}_1, \mathbf{y}_2 \xleftarrow{\$} \mathcal{M}$
3: $\mathbf{c} \leftarrow H(\mathbf{a}\mathbf{y}_1 + \mathbf{y}_2, m)$
4: $\mathbf{z}_1 \leftarrow \mathbf{s}_1\mathbf{c} + \mathbf{y}_1, \mathbf{z}_2 \leftarrow \mathbf{s}_2\mathbf{c} + \mathbf{y}_2$
5: if $\mathbf{z}_1$ or $\mathbf{z}_2 \notin \mathcal{G}$ and $ctr < \ell$ then
6:     $ctr \leftarrow ctr + 1$
7:     goto Step 2
8: if $\mathbf{z}_1$ or $\mathbf{z}_2 \notin \mathcal{G}$ then $(\mathbf{z}_1, \mathbf{z}_1, \mathbf{c}) \leftarrow (\bot, \bot, \bot)$
9: return $(\mathbf{z}_1, \mathbf{z}_2, \mathbf{c})$

Verify($m, \mathbf{z}_1, \mathbf{z}_2, \mathbf{c}, \mathbf{a}, \mathbf{t}$): accept if and only if $\mathbf{z}_1, \mathbf{z}_2 \in \mathcal{G}$ and $\mathbf{c} = H(\mathbf{a}\mathbf{z}_1 + \mathbf{z}_2 - \mathbf{t}\mathbf{c}, m)$.

Fig. 4. Lattice-Based Signature Scheme

described above is ε_s-simulatable, ρ-complete, (t, ε)-key-indistinguishable and ε_ℓ-lossy, for $\rho \geq 1/e^2 - 2/(en)$ and $\varepsilon_s, \varepsilon_\ell \leq \mathrm{negl}(n)$.

In order to obtain our signature scheme based on lattices, we apply our generic transform to the identification scheme described above. The full description of the resulting scheme is provided in Figure 4.

6   A Signature Scheme Based on Subset Sum

In this section, we construct a lossy identification scheme based on the hardness of the random $SS(n, M)$ problem for $M > (2kn + 1)^n \cdot 3^{2k}$, where k is a security parameter. The secret key is a random matrix $\mathbf{X} \xleftarrow{\$} \{0, 1\}^{n \times k}$, and the public key consists of a vector $\mathbf{a} \xleftarrow{\$} \mathbb{Z}_M^n$, and a vector $\mathbf{t} = \mathbf{a}^T \mathbf{X} \bmod M$. In the first step of the protocol, the prover selects a vector $\mathbf{y} \xleftarrow{\$} \{-kn, \dots, kn\}^n$ and sends an integer commitment $u = \langle \mathbf{a}, \mathbf{y} \rangle \bmod M$ to the verifier. The verifier selects a random challenge vector $\mathbf{c} \xleftarrow{\$} \{0, 1\}^k$, and sends it to the prover, who checks that $\mathbf{c}$ is indeed a valid challenge vector. The prover then computes a possible response $\mathbf{z} = \mathbf{X}\mathbf{c} + \mathbf{y}$ (note that there is no modular reduction here), and sends it to the verifier if it is in the range $\{-kn + k, \dots, kn - k\}^n$. If $\mathbf{z}$ is not in this range, then the prover sends $\bot$. Upon receiving a $\mathbf{z}$, the verifier accepts the interaction if $\mathbf{z} \in \{-kn + k, \dots, kn - k\}^n$ and $\langle \mathbf{a}, \mathbf{z} \rangle - \langle \mathbf{t}, \mathbf{c} \rangle \bmod M = u$.

It is easy to see that in the case that the prover does not send $\bot$, he will be accepted by the verifier since

$$\langle \mathbf{a}, \mathbf{z} \rangle - \langle \mathbf{t}, \mathbf{c} \rangle \bmod M = \mathbf{a}^T \mathbf{X}\mathbf{c} + \langle \mathbf{a}, \mathbf{y} \rangle - \mathbf{a}^T \mathbf{X}\mathbf{c} \bmod M = u.$$

Then, we observe that the probability that for any element $\bar{\mathbf{z}} \in \{-kn + k, \ldots,$ $kn - k\}^n$, the probability that the response will be $\mathbf{z} = \bar{\mathbf{z}}$ is

$$Pr[\mathbf{z} = \bar{\mathbf{z}}] = Pr[\mathbf{y} = \bar{\mathbf{z}} - \mathbf{X}\mathbf{c}] = 1/\left|\{-kn, \ldots, kn\}^n\right|,$$

since all the coefficients of the vector $\mathbf{X}\mathbf{c}$ have absolute value at most k. Therefore every element $\mathbf{z}$ in the set $\{-kn + k, \ldots, kn - k\}^n$ has an equal probability of being outputted and the probability that $\mathbf{z} \neq \perp$ is

$$\rho = \left|\{-kn + k, \ldots, kn - k\}^n\right| / \left|\{-kn, \ldots, kn\}^n\right| \approx (1 - 1/n)^n \approx 1/e.$$

And thus the simulatability property of the scheme is satisfied since one can create a valid transcript by generating $(\perp, \perp, \perp)$ with probability $1 - \rho$, and otherwise pick a random $\mathbf{z} \in \{-kn + k, \ldots, kn - k\}^n$, a random $\mathbf{c} \in \{0, 1\}^k$, and output $(\langle \mathbf{a}, \mathbf{z}\rangle - \langle \mathbf{t}, \mathbf{c}\rangle \bmod M, \mathbf{c}, \mathbf{z})$.

The lossy public keys are just two uniformly random vectors $\mathbf{a}$ and $\mathbf{t}$, and so the indistinguishability of these keys from the real keys is directly based on the hardness of the $SS(n, M)$ problem using a standard hybrid argument.

To show lossiness, we observe that if $\mathbf{t}$ is uniformly random in $\mathbb{Z}_M^k$, then it can be shown that with high probability, for any choice of $u \in \mathbb{Z}_M$, there is at most one value $\mathbf{c}$ such that u can be written as $\langle \mathbf{a}, \mathbf{z}\rangle - \langle \mathbf{t}, \mathbf{c}\rangle \bmod M$. Indeed, if there exist two pairs $(\mathbf{z}, \mathbf{c})$, $(\mathbf{z}', \mathbf{c}')$, such that

$$\langle \mathbf{a}, \mathbf{z}\rangle - \langle \mathbf{t}, \mathbf{c}\rangle = \langle \mathbf{a}, \mathbf{z}'\rangle - \langle \mathbf{t}, \mathbf{c}'\rangle \bmod M,$$

then we have

$$\langle \mathbf{a}, \mathbf{z} - \mathbf{z}'\rangle - \langle \mathbf{t}, \mathbf{c} - \mathbf{c}'\rangle \bmod M = 0. \tag{1}$$

The set of valid pairs $(\mathbf{z} - \mathbf{z}', \mathbf{c} - \mathbf{c}')$ consists of $(2kn + 1)^n \cdot 3^k$ elements. If $(\mathbf{a}, \mathbf{t})$ is chosen completely at random, then for each of those valid pairs, the probability that Equation (1) is satisfied is $1/M$ (this assumes that either $\mathbf{a}$ or $\mathbf{t}$ has at least one element that is invertible modulo M, which is the case with extremely high probability), and so the probability over the randomness of $\mathbf{a}$ and $\mathbf{t}$ that Equation (1) is satisfied for any of the valid pairs is at most $(2kn + 1)^n \cdot 3^k/M$, which by our choice of M, is at most 3^{-k}.

To convert this lossy identification scheme to a signature scheme, one would simply perform the transformation described in Figure 1, as we did for the other schemes in this paper. And as for the lattice-based scheme in Section 5, we point out that the technique in [33] can be used to reduce the coefficients of the signature by about a factor of $\sqrt{n}$ to make them fall in the range $\{-O(k\sqrt{n}), \ldots,$ $O(k\sqrt{n})\}^n$ by sampling the vector $\mathbf{y}$ from a normal distribution and performing a somewhat more involved rejection sampling procedure when deciding whether or not to send the response $\mathbf{z}$. This would also allow us to reduce the modulus M to approximately $M = O(k\sqrt{n})^n \cdot 3^{2k}$, which makes the $SS(n, M)$ problem more difficult. Another possible optimization could include making k larger, but making the vector $\mathbf{c}$ sparser (while still making sure that it comes from a large enough set), which would result in a shorter vector $\mathbf{X}\mathbf{c}$.

Acknowledgments. This work was supported in part by the European Research Council and by the European Commission through the ICT Program under Contract ICT-2007-216676 ECRYPT II.

We thank Mihir Bellare and Eike Kiltz for helpful comments on a preliminary version of this paper.

References

1. Abdalla, M., An, J.H., Bellare, M., Namprempre, C.: From Identification to Signatures via the Fiat-Shamir Transform: Minimizing Assumptions for Security and Forward-Security. In: Knudsen, L.R. (ed.) EUROCRYPT 2002. LNCS, vol. 2332, pp. 418–433. Springer, Heidelberg (2002)
2. Abdalla, M., Fouque, P.-A., Lyubashevsky, V., Tibouchi, M.: Tightly-secure signatures from lossy identification schemes (2012); Full version of this paper.
3. Becker, A., Coron, J.-S., Joux, A.: Improved Generic Algorithms for Hard Knapsacks. In: Paterson, K.G. (ed.) EUROCRYPT 2011. LNCS, vol. 6632, pp. 364–385. Springer, Heidelberg (2011)
4. Bellare, M., Rogaway, P.: The Exact Security of Digital Signatures - How to Sign with RSA and Rabin. In: Maurer, U.M. (ed.) EUROCRYPT 1996. LNCS, vol. 1070, pp. 399–416. Springer, Heidelberg (1996)
5. Boyen, X.: Lattice Mixing and Vanishing Trapdoors: A Framework for Fully Secure Short Signatures and More. In: Nguyen, P.Q., Pointcheval, D. (eds.) PKC 2010. LNCS, vol. 6056, pp. 499–517. Springer, Heidelberg (2010)
6. Cash, D., Hofheinz, D., Kiltz, E., Peikert, C.: Bonsai Trees, or How to Delegate a Lattice Basis. In: Gilbert, H. (ed.) EUROCRYPT 2010. LNCS, vol. 6110, pp. 523–552. Springer, Heidelberg (2010)
7. Chevallier-Mames, B.: An Efficient CDH-Based Signature Scheme with a Tight Security Reduction. In: Shoup, V. (ed.) CRYPTO 2005. LNCS, vol. 3621, pp. 511–526. Springer, Heidelberg (2005)
8. Chor, B., Goldreich, O.: Unbiased bits from sources of weak randomness and probabilistic communication complexity. In: 26th FOCS, pp. 429–442. IEEE Computer Society Press (October 1985)
9. Cramer, R., Shoup, V.: Signature schemes based on the strong RSA assumption. ACM Trans. Inf. Syst. Secur. 3(3), 161–185 (2000)
10. Even, S., Goldreich, O., Micali, S.: On-Line/Off-Line Digital Signatures. In: Brassard, G. (ed.) CRYPTO 1989. LNCS, vol. 435, pp. 263–275. Springer, Heidelberg (1990)
11. Even, S., Goldreich, O., Micali, S.: On-line/off-line digital signatures. Journal of Cryptology 9(1), 35–67 (1996)
12. Fiat, A., Shamir, A.: How to Prove Yourself: Practical Solutions to Identification and Signature Problems. In: Odlyzko, A.M. (ed.) CRYPTO 1986. LNCS, vol. 263, pp. 186–194. Springer, Heidelberg (1987)
13. Frieze, A.M.: On the lagarias-odlyzko algorithm for the subset sum problem. SIAM J. Comput. 15(2), 536–539 (1986)
14. Gennaro, R.: An Improved Pseudo-random Generator Based on Discrete Log. In: Bellare, M. (ed.) CRYPTO 2000. LNCS, vol. 1880, pp. 469–481. Springer, Heidelberg (2000)
15. Gennaro, R.: An improved pseudo-random generator based on the discrete logarithm problem. Journal of Cryptology 18(2), 91–110 (2005)

16. Gennaro, R., Halevi, S., Rabin, T.: Secure Hash-and-Sign Signatures without the Random Oracle. In: Stern, J. (ed.) EUROCRYPT 1999. LNCS, vol. 1592, pp. 123–139. Springer, Heidelberg (1999)

17. Gennaro, R., Krawczyk, H., Rabin, T.: Secure Hashed Diffie-Hellman over Non-DDH Groups. In: Cachin, C., Camenisch, J.L. (eds.) EUROCRYPT 2004. LNCS, vol. 3027, pp. 361–381. Springer, Heidelberg (2004)

18. Gentry, C., Peikert, C., Vaikuntanathan, V.: Trapdoors for hard lattices and new cryptographic constructions. In: Ladner, R.E., Dwork, C. (eds.) 40th ACM STOC, pp. 197–206. ACM Press (May 2008)

19. Girault, M.: An Identity-Based Identification Scheme Based on Discrete Logarithms Modulo a Composite Number. In: Damgård, I.B. (ed.) EUROCRYPT 1990. LNCS, vol. 473, pp. 481–486. Springer, Heidelberg (1991)

20. Girault, M., Poupard, G., Stern, J.: On the fly authentication and signature schemes based on groups of unknown order. Journal of Cryptology 19(4), 463–487 (2006)

21. Goh, E.-J., Jarecki, S.: A Signature Scheme as Secure as the Diffie-Hellman Problem. In: Biham, E. (ed.) EUROCRYPT 2003. LNCS, vol. 2656, pp. 401–415. Springer, Heidelberg (2003)

22. Goh, E.-J., Jarecki, S., Katz, J., Wang, N.: Efficient signature schemes with tight reductions to the Diffie-Hellman problems. Journal of Cryptology 20(4), 493–514 (2007)

23. Goldwasser, S., Micali, S., Rivest, R.L.: A digital signature scheme secure against adaptive chosen-message attacks. SIAM Journal on Computing 17(2), 281–308 (1988)

24. Guillou, L.C., Quisquater, J.-J.: A "Paradoxical" Identity-Based Signature Scheme Resulting from Zero-Knowledge. In: Goldwasser, S. (ed.) CRYPTO 1988. LNCS, vol. 403, pp. 216–231. Springer, Heidelberg (1990)

25. Hohenberger, S., Waters, B.: Short and Stateless Signatures from the RSA Assumption. In: Halevi, S. (ed.) CRYPTO 2009. LNCS, vol. 5677, pp. 654–670. Springer, Heidelberg (2009)

26. Impagliazzo, R., Naor, M.: Efficient cryptographic schemes provably as secure as subset sum. Journal of Cryptology 9(4), 199–216 (1996)

27. Katz, J., Wang, N.: Efficiency improvements for signature schemes with tight security reductions. In: Jajodia, S., Atluri, V., Jaeger, T. (eds.) ACM CCS 2003, pp. 155–164. ACM Press (October 2003)

28. Kawachi, A., Tanaka, K., Xagawa, K.: Concurrently Secure Identification Schemes Based on the Worst-Case Hardness of Lattice Problems. In: Pieprzyk, J. (ed.) ASIACRYPT 2008. LNCS, vol. 5350, pp. 372–389. Springer, Heidelberg (2008)

29. Koshiba, T., Kurosawa, K.: Short Exponent Diffie-Hellman Problems. In: Bao, F., Deng, R., Zhou, J. (eds.) PKC 2004. LNCS, vol. 2947, pp. 173–186. Springer, Heidelberg (2004)

30. Lagarias, J.C., Odlyzko, A.M.: Solving low-density subset sum problems. In: 24th FOCS, pp. 1–10 (1983)

31. Lyubashevsky, V.: Lattice-Based Identification Schemes Secure Under Active Attacks. In: Cramer, R. (ed.) PKC 2008. LNCS, vol. 4939, pp. 162–179. Springer, Heidelberg (2008)

32. Lyubashevsky, V.: Fiat-Shamir with Aborts: Applications to Lattice and Factoring-Based Signatures. In: Matsui, M. (ed.) ASIACRYPT 2009. LNCS, vol. 5912, pp. 598–616. Springer, Heidelberg (2009)

33. Lyubashevsky, V.: Lattice Signatures without Trapdoors. In: Pointcheval, D., Johansson, T. (eds.) EUROCRYPT 2012. LNCS, vol. 7237, pp. 738–755. Springer, Heidelberg (2012)

34. Lyubashevsky, V., Micciancio, D.: Generalized Compact Knapsacks Are Collision Resistant. In: Bugliesi, M., Preneel, B., Sassone, V., Wegener, I. (eds.) ICALP 2006. LNCS, vol. 4052, pp. 144–155. Springer, Heidelberg (2006)

35. Lyubashevsky, V., Peikert, C., Regev, O.: On Ideal Lattices and Learning with Errors over Rings. In: Gilbert, H. (ed.) EUROCRYPT 2010. LNCS, vol. 6110, pp. 1–23. Springer, Heidelberg (2010)

36. Micali, S., Reyzin, L.: Improving the exact security of digital signature schemes. Journal of Cryptology 15(1), 1–18 (2002)

37. Micciancio, D.: Generalized compact knapsacks, cyclic lattices, and efficient one-way functions. Computational Complexity 16(4), 365–411 (2007)

38. Micciancio, D., Mol, P.: Pseudorandom Knapsacks and the Sample Complexity of LWE Search-to-Decision Reductions. In: Rogaway, P. (ed.) CRYPTO 2011. LNCS, vol. 6841, pp. 465–484. Springer, Heidelberg (2011)

39. Micciancio, D., Peikert, C.: Trapdoors for Lattices: Simpler, Tighter, Faster, Smaller. In: Pointcheval, D., Johansson, T. (eds.) EUROCRYPT 2012. LNCS, vol. 7237, pp. 700–718. Springer, Heidelberg (2012)

40. Micciancio, D., Vadhan, S.P.: Statistical Zero-Knowledge Proofs with Efficient Provers: Lattice Problems and More. In: Boneh, D. (ed.) CRYPTO 2003. LNCS, vol. 2729, pp. 282–298. Springer, Heidelberg (2003)

41. Patel, S., Sundaram, G.S.: An Efficient Discrete Log Pseudo Random Generator. In: Krawczyk, H. (ed.) CRYPTO 1998. LNCS, vol. 1462, pp. 304–317. Springer, Heidelberg (1998)

42. Peikert, C., Rosen, A.: Efficient Collision-Resistant Hashing from Worst-Case Assumptions on Cyclic Lattices. In: Halevi, S., Rabin, T. (eds.) TCC 2006. LNCS, vol. 3876, pp. 145–166. Springer, Heidelberg (2006)

43. Pointcheval, D., Stern, J.: Security arguments for digital signatures and blind signatures. Journal of Cryptology 13(3), 361–396 (2000)

44. Pollard, J.M.: Kangaroos, monopoly and discrete logarithms. Journal of Cryptology 13(4), 437–447 (2000)

45. Poupard, G., Stern, J.: Security Analysis of a Practical "On the Fly" Authentication and Signature Generation. In: Nyberg, K. (ed.) EUROCRYPT 1998. LNCS, vol. 1403, pp. 422–436. Springer, Heidelberg (1998)

46. Regev, O.: On lattices, learning with errors, random linear codes, and cryptography. J. ACM 56(6) (2009)

47. Santoso, B., Ohta, K., Sakiyama, K., Hanaoka, G.: Improving Efficiency of an 'On the Fly' Identification Scheme by Perfecting Zero-Knowledgeness. In: Pieprzyk, J. (ed.) CT-RSA 2010. LNCS, vol. 5985, pp. 284–301. Springer, Heidelberg (2010)

48. Schnorr, C.-P.: Efficient signature generation by smart cards. Journal of Cryptology 4(3), 161–174 (1991)

49. Stehlé, D., Steinfeld, R.: Making NTRU as Secure as Worst-Case Problems over Ideal Lattices. In: Paterson, K.G. (ed.) EUROCRYPT 2011. LNCS, vol. 6632, pp. 27–47. Springer, Heidelberg (2011)

50. Stehlé, D., Steinfeld, R., Tanaka, K., Xagawa, K.: Efficient Public Key Encryption Based on Ideal Lattices. In: Matsui, M. (ed.) ASIACRYPT 2009. LNCS, vol. 5912, pp. 617–635. Springer, Heidelberg (2009)

51. van Oorschot, P.C., Wiener, M.: On Diffie-Hellman Key Agreement with Short Exponents. In: Maurer, U.M. (ed.) EUROCRYPT 1996. LNCS, vol. 1070, pp. 332–343. Springer, Heidelberg (1996)

Adaptively Attribute-Hiding
(Hierarchical) Inner Product Encryption

Tatsuaki Okamoto[1] and Katsuyuki Takashima[2]

[1] NTT
okamoto.tatsuaki@lab.ntt.co.jp
[2] Mitsubishi Electric
Takashima.Katsuyuki@aj.MitsubishiElectric.co.jp

Abstract. This paper proposes the first inner product encryption (IPE) scheme that is adaptively secure and fully attribute-hiding (attribute-hiding in the sense of the definition by Katz, Sahai and Waters), while the existing IPE schemes are either fully attribute-hiding but selectively secure or adaptively secure but weakly attribute-hiding. The proposed IPE scheme is proven to be adaptively secure and fully attribute-hiding under the decisional linear assumption in the standard model. The IPE scheme is comparably as efficient as the existing attribute-hiding IPE schemes. We also present a variant of the proposed IPE scheme with the same security that achieves shorter public and secret keys. A hierarchical IPE scheme can be constructed that is also adaptively secure and fully attribute-hiding under the same assumption. In this paper, we extend the dual system encryption technique by Waters into a more general manner, in which new forms of ciphertext and secret keys are employed and new types of information theoretical tricks are introduced along with several forms of computational reduction.

1 Introduction

1.1 Background

Functional encryption (FE) is an advanced class of encryption and it covers identity-based encryption (IBE)[3,4,7,11], hidden-vector encryption (HVE) [8], inner-product encryption (IPE) [15], predicate encryption (PE) and attribute-based encryption (ABE) [2,13,23,16,22,24,19]. In FE, there is a relation $R(v, x)$ which determines what a secret key with parameter v can decrypt a ciphertext encrypted under parameter x. The enhanced functionality and flexibility provided by FE systems are very appealing for many practical applications.

For some applications, the parameters for encryption are required to be hidden from ciphertexts. One of such applications is an advanced notion of PKE with keyword search (PEKS) [6], which we call *PKE with functional search* (PEFS) in this paper. In PEFS, a parameter x (not just a keyword) embedded in a ciphertext is searched (checked) whether $R(v, x)$ holds or not by using a secret key with parameter v. Here, keyword search is a special case of functional search

D. Pointcheval and T. Johansson (Eds.): EUROCRYPT 2012, LNCS 7237, pp. 591–608, 2012.

$R(v, x)$ when $R(v, x) \Leftrightarrow [x = v]$. Parameter x of a ciphertext is often private information and should be hidden from ciphertexts in such applications.

To capture the security requirement, Katz, Sahai and Waters [15] introduced *attribute-hiding* (based on the same notion for HVE by Boneh and Waters [8]), a security notion for FE that is stronger than the basic security requirement, *payload-hiding*. Roughly speaking, attribute-hiding requires that a ciphertext conceal the associated parameter as well as the plaintext, while payload-hiding only requires that a ciphertext conceal the plaintext. Attribute-hiding FE is often called predicate encryption (PE).

The widest class of relations of a FE system in the literature is general nonmonotone (span program) relations, which can be expressed using AND, OR, Threshold and NOT gates [19]. FE systems supporting such a wide class of relations, however, have one limitation in that the parameter x of the ciphertext should be revealed to users to decrypt. That is, such FE systems do not satisfy the attribute-hiding security.

To the best of our knowledge, the widest class of relations supported by attribute-hiding FE systems are *inner-product predicates* in [15,16,19], which we call the KSW08, LOS⁺10 and OT10 schemes. Parameters of inner-product predicates are expressed as vectors $\vec{x}$ (for a ciphertext) and $\vec{v}$ (for a secret key), where $R(\vec{v}, \vec{x})$ holds iff $\vec{v} \cdot \vec{x} = 0$. (Here, $\vec{v} \cdot \vec{x}$ denotes the standard inner-product.) In this paper we call FE for inner-product predicates *inner product encryption* (IPE).

Inner-product predicates represent a fairly wide class of relations including equality tests as the simplest case (i.e., anonymous IBE and HVE are very special classes of attribute-hiding IPE), disjunctions or conjunctions of equality tests, and, more generally, CNF or DNF formulas. We note, however, that inner product predicates are less expressive than general (even monotone span program) relations of FE. To use inner product predicates for such general relations, formulas must be written in CNF or DNF form, which can cause a super-polynomial blowup in size for arbitrary formulas.

Among the existing attribute-hiding IPEs, the KSW08 IPE scheme [15] is proven to be only *selectively* secure. Although the LOS⁺10 and OT10 IPE schemes [16,19] are proven to be *adaptively* secure, the achieved attribute-hiding security is limited or weaker than that defined in [15]. Here, we call the attribute-hiding security defined in [15] *fully attribute-hiding* and that achieved in [16,19] *weakly attribute-hiding*. In the fully attribute-hiding security definition [15], adversary $\mathcal{A}$ is allowed to ask a key-query for $\vec{v}$ such that $\vec{v} \cdot \vec{x}^{(0)} = \vec{v} \cdot \vec{x}^{(1)} = 0$ provided that $m^{(0)} = m^{(1)}$ ($\vec{x}^{(b)}$ and $m^{(b)}$ ($b = 0, 1$) are for the challenge ciphertext in the security definition), while in the weakly attribute-hiding security definition [16,19], $\mathcal{A}$ is only allowed to ask a key-query for $\vec{v}$ such that $\vec{v} \cdot \vec{x}^{(b)} \neq 0$ for all $b \in \{0, 1\}$.

Let us explain the difference between the fully and weakly attribute-hiding definitions in a PEFS system. User Alice provides her secret key, $\mathsf{sk}_{\vec{v}}$, to proxy server Bob, who checks whether $\vec{v} \cdot \vec{x} = 0$ or not for an incoming ciphertext, $\mathsf{ct}_{\vec{x}}$, encrypted with parameter $\vec{x}$. In the weakly attribute-hiding security, privacy of

$\vec{x}$ from $\mathsf{ct}_{\vec{x}}$ is ensured only if $\vec{v} \cdot \vec{x} \neq 0$, but cannot be ensured or some privacy on $\vec{x}$ may be revealed if $\vec{v} \cdot \vec{x} = 0$. Here note that there still exists $(n-1)$-dimensional freedom (or room of privacy) of n-dimensional vector $\vec{x}$, even if $\vec{v}$ and the fact that $\vec{v} \cdot \vec{x} = 0$ is revealed. For example, let $\vec{v}$ express formula on an email message attributes, [[Subject $= X$] $\vee$ [Subject $= Y$]] $\wedge$ [[Receiver $=$ Alice] $\vee$ [Receiver $=$ Alice's secretary]], and $\vec{x}$ express ciphertext attribute (Subject $= X$, Receiver $=$ Alice). In this case, $\vec{v} \cdot \vec{x} = 0$, since the ciphertext attribute expressed by $\vec{x}$ satisfies the formula expressed by $\vec{v}$. Although Bob knows $\mathsf{sk}_{\vec{v}}$ and $\vec{v}$, Bob has no idea which attribute $\vec{x}$ is embedded in $\mathsf{ct}_{\vec{x}}$ except that the ciphertext attribute satisfies the formula, i.e., $\vec{v} \cdot \vec{x} = 0$, if the fully attribute-hiding security is achieved. On the other hand, Bob may obtain some additional information on the attribute (e.g., Bob may know that the subject is X, not Y), if only the weakly attribute-hiding security is guaranteed.

The KSW08 IPE scheme is fully attribute-hiding but selectively secure, and the LOS$^+$10 and OT10 IPE schemes are adaptively secure but weakly attribute-hiding. Therefore, there is no IPE scheme that is adaptively secure and fully attribute-hiding simultaneously. As for a more limited class of schemes, HVE (as mentioned above, HVE is a very special class of attribute-hiding IPE), an adaptively secure and fully attribute-hiding HVE scheme has been proposed [10]. For hierarchical IPE (HIPE), the LOS$^+$10 and OT10 HIPE schemes [16,19] are adaptively secure but weakly attribute-hiding, i.e., there is no HIPE scheme that is adaptively secure and fully attribute-hiding simultaneously.

It is a technically challenging task to achieve an adaptively secure and fully attribute-hiding (H)IPE scheme. Even if we use the powerful dual system encryption technique by Waters, the main difficulty resides in how to change a (normal) secret key queried with $\vec{v}$ to a semi-functional secret key, without knowing $\vec{x}^{(b)}$ $(b = 0, 1)$ for the challenge ciphertext, i.e., without knowing whether $\vec{v} \cdot \vec{x}^{(b)} = 0$ or not, since an adversary may issue key queries with $\vec{v}$ before issuing the challenge ciphertext query with $\vec{x}^{(b)}$ $(b = 0, 1)$ and two possible cases, $\vec{v} \cdot \vec{x}^{(b)} = 0$ (for all $b \in \{0, 1\}$) and $\vec{v} \cdot \vec{x}^{(b)} \neq 0$ (for all $b \in \{0, 1\}$), are allowed in *fully* attribute-hiding IPE. Note that in *weakly* attribute-hiding IPE, it is always required that $\vec{v} \cdot \vec{x}^{(b)} \neq 0$. At a first glance, it looks hard to achieve it, since the form of semi-functional secret key may be different (e.g., canceled or randomized) depending on whether $\vec{v} \cdot \vec{x}^{(b)} = 0$ or not. Another technically challenging target in this paper is to prove the security under the decisional linear (DLIN) assumption (on prime order pairing groups) in the standard model.

1.2   Our Results

This paper proposes the first IPE scheme that is adaptively secure and fully attribute-hiding simultaneously. The proposed IPE scheme is proven to be adaptively secure and fully attribute-hiding under the DLIN assumption in the standard model (Section 4). We also present a variant of the proposed IPE scheme with the same security that achieves shorter master public keys and shorter secret keys (Section 5). A hierarchical IPE (HIPE) scheme can be realized that is also adaptively secure and fully attribute-hiding under the same assumption (see the full version of this paper [21] for the HIPE scheme). Table 2 in Section 6 compares the proposed IPE schemes with several existing attribute-hiding IPE schemes.

1.3 Key Techniques

To overcome the above-mentioned difficulty, we extend the dual system encryption technique into a more general manner, in which various forms of ciphertext and secret keys are introduced ('normal', 'temporal 0', 'temporal 1', 'temporal 2' and 'unbiased' forms for a ciphertext, and 'normal', 'temporal 1' and 'temporal 2' forms for a secret key), and new types (Types 1, 2, 3) of information theoretical tricks are employed with several forms of computational reduction (the security of Problems 1, 2 and 3 to DLIN). See Table 1 and Figure 1 in Section 4.2 for the outline.

In our approach, all forms ('normal', 'temporal 1' and 'temporal 2') of a secret key do not depend on whether $\vec{v} \cdot \vec{x}^{(b)} = 0$ or not. Although the aim of a 'semi-functional' secret key in the original dual system encryption method is to randomize the semi-functional part, the aim of these forms of a secret-key in our approach is just to encode $\vec{v}$ in a (hidden) subspace for a secret-key.

Another key point in our approach is that we transform a challenge ciphertext to an 'unbiased' ciphertext whose advantage is 0 in the final game, and $\vec{x}^{(b)}$ is randomized to a random vector in a two-dimensional subspace, $\mathsf{span}\langle\vec{x}^{(0)}, \vec{x}^{(1)}\rangle$. In contrast, $\vec{x}^{(b)}$ is randomized to a random vector in the n-dimensional whole space, $\mathbb{F}_q^n$, in [16,19] for weakly attribute-hiding IPE based on the original dual system encryption technique.

Therefore, in our approach, only $\vec{v}$ is encoded in a (hidden) subspace of the temporal forms of a secret-key, and a random vector in $\mathsf{span}\langle\vec{x}^{(0)}, \vec{x}^{(1)}\rangle$ is encoded in the corresponding (hidden) subspace for the temporal and unbiased forms of a ciphertext.

To realize this approach, our construction is based on the dual pairing vector spaces (DPVS) (Section 2) [16,19]. A nice property of DPVS is that we can set a hidden linear subspace by concealing the basis of a subspace from the public key. Typically, a pair of dual (or orthonormal) bases, $\mathbb{B}$ and $\mathbb{B}^*$, are randomly generated using random linear transformation, and a part of $\mathbb{B}$ (say $\hat{\mathbb{B}}$) is used as a public key and the corresponding part of $\mathbb{B}^*$ (say $\hat{\mathbb{B}}^*$) is used as a secret key or trapdoor. Therefore, the basis, $\mathbb{B} - \hat{\mathbb{B}}$, is information theoretically concealed against an adversary, i.e., even an infinite power adversary has no idea on which basis is selected as $\mathbb{B} - \hat{\mathbb{B}}$ when $\hat{\mathbb{B}}$ is published. It provides a framework for information theoretical tricks in the public-key setting.

In the proposed (basic) IPE scheme, $\mathsf{span}\langle\mathbb{B}\rangle$ and $\mathsf{span}\langle\mathbb{B}^*\rangle$, are $(4n + 2)$-dimensional (where the dimension of inner-product vectors is n), and, as for public parameter $\hat{\mathbb{B}}$, $\mathsf{span}\langle\hat{\mathbb{B}}\rangle$ is $(2n + 2)$-dimensional, i.e., the basis for the remaining $2n$-dimensional space is information theoretically concealed (ambiguous). We use the $2n$-dimensional hidden subspace to realize the various forms of ciphertext and secret keys and make elaborate game transformations over these forms towards the final goal, the 'unbiased' ciphertext.

The game transformations are alternating over computational and conceptual (information theoretical), and the combinations of three types of information theoretical tricks and three computational tricks (Problems 1, 2 and 3) play a central role in our approach, as shown in Figure 1. Type 1 is a (conceptual) linear transformation inside a (hidden) subspace for a ciphertext, Type 2 is a (conceptual) linear transformation inside a (hidden) subspace for a ciphertext with preserving the corresponding secret key value, and Type 3 is a (conceptual) linear transformation across (hidden and partially public) subspaces. The security of Problems 1, 2 and 3 is reduced to the DLIN assumption.

See Section 4.2 for the details of our techniques, in which the game transformations as well as the form changes of ciphertext and secret keys are summarized in Table 1 and Figure 1.

1.4   Notations

When A is a random variable or distribution, $y \xleftarrow{\mathsf{R}} A$ denotes that y is randomly selected from A according to its distribution. When A is a set, $y \xleftarrow{\mathsf{U}} A$ denotes that y is uniformly selected from A. $y := z$ denotes that y is set, defined or substituted by z. When a is a fixed value, $A(x) \to a$ (e.g., $A(x) \to 1$) denotes the event that machine (algorithm) A outputs a on input x. A function $f : \mathbb{N} \to \mathbb{R}$ is *negligible* in λ, if for every constant $c > 0$, there exists an integer n such that $f(\lambda) < \lambda^{-c}$ for all $\lambda > n$.

We denote the finite field of order q by $\mathbb{F}_q$, and $\mathbb{F}_q \setminus \{0\}$ by $\mathbb{F}_q^\times$. A vector symbol denotes a vector representation over $\mathbb{F}_q$, e.g., $\vec{x}$ denotes $(x_1, \ldots, x_n) \in \mathbb{F}_q^n$. For two vectors $\vec{v} = (v_1, \ldots, v_n)$ and $\vec{x} = (x_1, \ldots, x_n)$, $\vec{v} \cdot \vec{x}$ denotes the inner-product $\sum_{i=1}^n x_i v_i$. The vector $\vec{0}$ is abused as the zero vector in $\mathbb{F}_q^n$ for any n. X^{T} denotes the transpose of matrix X. I_ℓ denotes the $\ell \times \ell$ identity matrix. A bold face letter denotes an element of vector space $\mathbb{V}$, e.g., $\boldsymbol{x} \in \mathbb{V}$. When $\boldsymbol{b}_i \in \mathbb{V}$ $(i = 1, \ldots, n)$, $\mathsf{span}\langle \boldsymbol{b}_1, \ldots, \boldsymbol{b}_n \rangle \subseteq \mathbb{V}$ (resp. $\mathsf{span}\langle \vec{x}_1, \ldots, \vec{x}_n \rangle$) denotes the subspace generated by $\boldsymbol{b}_1, \ldots, \boldsymbol{b}_n$ (resp. $\vec{x}_1, \ldots, \vec{x}_n$). For bases $\mathbb{B} := (\boldsymbol{b}_1, \ldots, \boldsymbol{b}_N)$ and $\mathbb{B}^* := (\boldsymbol{b}_1^*, \ldots, \boldsymbol{b}_N^*)$, $(x_1, \ldots, x_N)_{\mathbb{B}} := \sum_{i=1}^N x_i \boldsymbol{b}_i$ and $(v_1, \ldots, v_N)_{\mathbb{B}^*} := \sum_{i=1}^N v_i \boldsymbol{b}_i^*$. $GL(n, \mathbb{F}_q)$ denotes the general linear group of degree n over $\mathbb{F}_q$.

2   Dual Pairing Vector Spaces (DPVS) and the Decisional Linear (DLIN) Assumption

Definition 1. *"Symmetric bilinear pairing groups"* $(q, \mathbb{G}, \mathbb{G}_T, G, e)$ *are a tuple of a prime q, cyclic additive group $\mathbb{G}$ and multiplicative group $\mathbb{G}_T$ of order q, $G \neq 0 \in \mathbb{G}$, and a polynomial-time computable nondegenerate bilinear pairing $e : \mathbb{G} \times \mathbb{G} \to \mathbb{G}_T$ i.e., $e(sG, tG) = e(G, G)^{st}$ and $e(G, G) \neq 1$. Let $\mathcal{G}_{\mathsf{bpg}}$ be an algorithm that takes input 1^λ and outputs a description of bilinear pairing groups $(q, \mathbb{G}, \mathbb{G}_T, G, e)$ with security parameter λ.*

In this paper, we concentrate on the symmetric version of dual pairing vector spaces [17,18]. constructed by using symmetric bilinear pairing groups given in

Definition 1. For the asymmetric version of DPVS, $(q, \mathbb{V}, \mathbb{V}^*, \mathbb{G}_T, \mathbb{A}, \mathbb{A}^*, e)$, see the full version of this paper. The following symmetric version is obtained by identifying $\mathbb{V} = \mathbb{V}^*$ and $\mathbb{A} = \mathbb{A}^*$ in the asymmetric version.

Definition 2. *"Dual pairing vector spaces (DPVS)"* $(q, \mathbb{V}, \mathbb{G}_T, \mathbb{A}, e)$ *by a direct product of symmetric pairing groups* $(q, \mathbb{G}, \mathbb{G}_T, G, e)$ *are a tuple of prime* q, N-*dimensional vector space* $\mathbb{V} := \overbrace{\mathbb{G} \times \cdots \times \mathbb{G}}^{N}$ *over* $\mathbb{F}_q$, *cyclic group* $\mathbb{G}_T$ *of order* q, *canonical basis* $\mathbb{A} := (\boldsymbol{a}_1, \ldots, \boldsymbol{a}_N)$ *of* $\mathbb{V}$, *where* $\boldsymbol{a}_i := (\overbrace{0, \ldots, 0}^{i-1}, G, \overbrace{0, \ldots, 0}^{N-i})$, *and pairing* $e : \mathbb{V} \times \mathbb{V} \to \mathbb{G}_T$. *The pairing is defined by* $e(\boldsymbol{x}, \boldsymbol{y}) := \prod_{i=1}^{N} e(G_i, H_i) \in \mathbb{G}_T$ *where* $\boldsymbol{x} := (G_1, \ldots, G_N) \in \mathbb{V}$ *and* $\boldsymbol{y} := (H_1, \ldots, H_N) \in \mathbb{V}$. *This is nondegenerate bilinear i.e.,* $e(s\boldsymbol{x}, t\boldsymbol{y}) = e(\boldsymbol{x}, \boldsymbol{y})^{st}$ *and if* $e(\boldsymbol{x}, \boldsymbol{y}) = 1$ *for all* $\boldsymbol{y} \in \mathbb{V}$, *then* $\boldsymbol{x} = \boldsymbol{0}$. *For all* i *and* j, $e(\boldsymbol{a}_i, \boldsymbol{a}_j) = e(G, G)^{\delta_{i,j}}$ *where* $\delta_{i,j} = 1$ *if* $i = j$, *and* 0 *otherwise, and* $e(G, G) \neq 1 \in \mathbb{G}_T$.

DPVS also has linear transformations $\phi_{i,j}$ *on* $\mathbb{V}$ *s.t.* $\phi_{i,j}(\boldsymbol{a}_j) = \boldsymbol{a}_i$ *and* $\phi_{i,j}(\boldsymbol{a}_k) = \boldsymbol{0}$ *if* $k \neq j$, *which can be easily achieved by* $\phi_{i,j}(\boldsymbol{x}) := (\overbrace{0, \ldots, 0}^{i-1}, G_j, \overbrace{0, \ldots, 0}^{N-i})$ *where* $\boldsymbol{x} := (G_1, \ldots, G_N)$. *We call* $\phi_{i,j}$ *"canonical maps". DPVS generation algorithm* $\mathcal{G}_{\mathsf{dpvs}}$ *takes input* 1^{λ} $(\lambda \in \mathbb{N})$ *and* $N \in \mathbb{N}$, *and outputs a description of* $\mathsf{param}'_{\mathbb{V}} := (q, \mathbb{V}, \mathbb{G}_T, \mathbb{A}, e)$ *with security parameter* λ *and* N-*dimensional* $\mathbb{V}$. *It can be constructed by using* $\mathcal{G}_{\mathsf{bpg}}$.

We describe random dual orthonormal basis generator $\mathcal{G}_{\mathsf{ob}}$ below, which is used as a subroutine in the proposed (H)IPE scheme.

$$\mathcal{G}_{\mathsf{ob}}(1^{\lambda}, N) : \mathsf{param}'_{\mathbb{V}} := (q, \mathbb{V}, \mathbb{G}_T, \mathbb{A}, e) \xleftarrow{\mathsf{R}} \mathcal{G}_{\mathsf{dpvs}}(1^{\lambda}, N), \psi \xleftarrow{\mathsf{U}} \mathbb{F}_q^{\times}, g_T := e(G, G)^{\psi},$$

$$X := (\chi_{i,j}) \xleftarrow{\mathsf{U}} GL(N, \mathbb{F}_q), (\vartheta_{i,j}) := \psi \cdot (X^{\mathsf{T}})^{-1}, \mathsf{param}_{\mathbb{V}} := (\mathsf{param}'_{\mathbb{V}}, g_T),$$

$$\boldsymbol{b}_i := \sum_{j=1}^{N} \chi_{i,j} \boldsymbol{a}_j, \mathbb{B} := (\boldsymbol{b}_1, \ldots, \boldsymbol{b}_N), \boldsymbol{b}_i^* := \sum_{j=1}^{N} \vartheta_{i,j} \boldsymbol{a}_j, \mathbb{B}^* := (\boldsymbol{b}_1^*, \ldots, \boldsymbol{b}_N^*),$$

$$\text{return } (\mathsf{param}_{\mathbb{V}}, \mathbb{B}, \mathbb{B}^*).$$

Definition 3 (DLIN: Decisional Linear Assumption [5]). *The DLIN problem is to guess* $\beta \in \{0, 1\}$, *given* $(\mathsf{param}_{\mathbb{G}}, G, \xi G, \kappa G, \delta\xi G, \sigma\kappa G, Y_{\beta}) \xleftarrow{\mathsf{R}} \mathcal{G}_{\beta}^{\mathsf{DLIN}}(1^{\lambda})$, *where* $\mathcal{G}_{\beta}^{\mathsf{DLIN}}(1^{\lambda})$: $\mathsf{param}_{\mathbb{G}} := (q, \mathbb{G}, \mathbb{G}_T, G, e) \xleftarrow{\mathsf{R}} \mathcal{G}_{\mathsf{bpg}}(1^{\lambda}), \kappa, \delta, \xi, \sigma \xleftarrow{\mathsf{U}} \mathbb{F}_q, Y_0 := (\delta + \sigma)G, Y_1 \xleftarrow{\mathsf{U}} \mathbb{G}$, *return* $(\mathsf{param}_{\mathbb{G}}, G, \xi G, \kappa G, \delta\xi G, \sigma\kappa G, Y_{\beta})$, *for* $\beta \xleftarrow{\mathsf{U}} \{0, 1\}$. *For a probabilistic machine* $\mathcal{E}$, *we define the advantage of* $\mathcal{E}$ *for the DLIN problem as:* $\mathsf{Adv}_{\mathcal{E}}^{\mathsf{DLIN}}(\lambda) := \left| \Pr\left[\mathcal{E}(1^{\lambda}, \varrho) \to 1 \,\middle|\, \varrho \xleftarrow{\mathsf{R}} \mathcal{G}_0^{\mathsf{DLIN}}(1^{\lambda})\right] - \Pr\left[\mathcal{E}(1^{\lambda}, \varrho) \to 1 \,\middle|\, \varrho \xleftarrow{\mathsf{R}} \mathcal{G}_1^{\mathsf{DLIN}}(1^{\lambda})\right] \right|$. *The DLIN assumption is: For any probabilistic polynomial-time adversary* $\mathcal{E}$, *the advantage* $\mathsf{Adv}_{\mathcal{E}}^{\mathsf{DLIN}}(\lambda)$ *is negligible in* λ.

3   Definition of Inner Product Encryption (IPE)

This section defines predicate encryption (PE) for the class of inner-product predicates, i.e., inner product encryption (IPE) and its security.

An attribute of inner-product predicates is expressed as a vector $\vec{x} \in \mathbb{F}_q^n \setminus \{\vec{0}\}$ and a predicate $f_{\vec{v}}$ is associated with a vector $\vec{v}$, where $f_{\vec{v}}(\vec{x}) = 1$ iff $\vec{v} \cdot \vec{x} = 0$. Let $\Sigma := \mathbb{F}_q^n \setminus \{\vec{0}\}$, i.e., the set of the attributes, and $\mathcal{F} := \{f_{\vec{v}} | \vec{v} \in \mathbb{F}_q^n \setminus \{\vec{0}\}\}$ i.e., the set of the predicates.

Definition 4. *An inner product encryption scheme (for predicates $\mathcal{F}$ and attributes Σ) consists of probabilistic polynomial-time algorithms* Setup, KeyGen, Enc *and* Dec. *They are given as follows:*

- Setup *takes as input security parameter 1^λ outputs (master) public key* pk *and (master) secret key* sk.
- KeyGen *takes as input the master public key* pk, *secret key* sk, *and predicate vector $\vec{v}$. It outputs a corresponding secret key* sk$_{\vec{v}}$.
- Enc *takes as input the master public key* pk, *plaintext m in some associated plaintext space,* msg, *and attribute vector $\vec{x}$. It returns ciphertext* ct$_{\vec{x}}$.
- Dec *takes as input the master public key* pk, *secret key* sk$_{\vec{v}}$ *and ciphertext* ct$_{\vec{x}}$. *It outputs either plaintext m or the distinguished symbol $\bot$.*

An IPE scheme should have the following correctness property: for all (pk, sk) $\overset{\text{R}}{\leftarrow}$ Setup($1^\lambda, n$), all $f_{\vec{v}} \in \mathcal{F}$ and $\vec{x} \in \Sigma$, all sk$_{\vec{v}}$ $\overset{\text{R}}{\leftarrow}$ KeyGen(pk, sk, $\vec{v}$), all messages m, all ciphertext ct$_{\vec{x}}$ $\overset{\text{R}}{\leftarrow}$ Enc(pk, $m, \vec{x}$), it holds that $m = $ Dec(pk, sk$_{\vec{v}}$, ct$_{\vec{x}}$) if $f_{\vec{v}}(\vec{x}) = 1$. Otherwise, it holds with negligible probability.

We then define the security notion of IPE, that was called "*adaptively* secure and *fully attribute-hiding*" in Abstract and Section 1. Since we will deal with only this security notion hereafter, we shortly call it "*adaptively attribute-hiding.*"

Definition 5. *The model for defining the adaptively attribute-hiding security of IPE against adversary $\mathcal{A}$ (under chosen plaintext attacks) is given as follows:*

1. Setup *is run to generate keys* pk *and* sk, *and* pk *is given to $\mathcal{A}$.*
2. *$\mathcal{A}$ may adaptively make a polynomial number of key queries for predicate vectors, $\vec{v}$. In response, $\mathcal{A}$ is given the corresponding key* sk$_{\vec{v}}$ $\overset{\text{R}}{\leftarrow}$ KeyGen(pk, sk, $\vec{v}$).
3. *$\mathcal{A}$ outputs challenge attribute vector $(\vec{x}^{(0)}, \vec{x}^{(1)})$ and challenge plaintexts $(m^{(0)}, m^{(1)})$, subject to the following restrictions:*
   - *$\vec{v} \cdot \vec{x}^{(0)} \neq 0$ and $\vec{v} \cdot \vec{x}^{(1)} \neq 0$ for all the key queried predicate vectors, $\vec{v}$.*
   - *Two challenge plaintexts are equal, i.e., $m^{(0)} = m^{(1)}$, and any key query $\vec{v}$ satisfies $f_{\vec{v}}(\vec{x}^{(0)}) = f_{\vec{v}}(\vec{x}^{(1)})$, i.e., one of the following conditions.*
     - *$\vec{v} \cdot \vec{x}^{(0)} = 0$ and $\vec{v} \cdot \vec{x}^{(1)} = 0$,*
     - *$\vec{v} \cdot \vec{x}^{(0)} \neq 0$ and $\vec{v} \cdot \vec{x}^{(1)} \neq 0$,*
4. *A random bit b is chosen. $\mathcal{A}$ is given* ct$_{\vec{x}^{(b)}}$ $\overset{\text{R}}{\leftarrow}$ Enc(pk, $m^{(b)}, \vec{x}^{(b)}$).
5. *The adversary may continue to issue key queries for additional predicate vectors, $\vec{v}$, subject to the restriction given in step 3. $\mathcal{A}$ is given the corresponding key* sk$_{\vec{v}}$ $\overset{\text{R}}{\leftarrow}$ KeyGen(pk, sk, $\vec{v}$).
6. *$\mathcal{A}$ outputs a bit b', and wins if $b' = b$.*

The advantage of $\mathcal{A}$ in the above game is defined as $\mathsf{Adv}_{\mathcal{A}}^{\mathsf{IPE,AH}}(\lambda) := \Pr[\mathcal{A} \text{ wins }] - 1/2$ for any security parameter λ. An IPE scheme is adaptively attribute-hiding

(AH) against chosen plaintext attacks *if all probabilistic polynomial-time adversaries $\mathcal{A}$ have at most negligible advantage in the above game.*

For each run of the game, the variable s is defined as $s := 0$ if $m^{(0)} \neq m^{(1)}$ for challenge plaintexts $m^{(0)}$ and $m^{(1)}$, and $s := 1$ otherwise.

4   Proposed (Basic) IPE Scheme

4.1   Construction

In the description of the scheme, we assume that the first coordinate, x_1, of input vector, $\vec{x} := (x_1, \ldots, x_n)$, is nonzero. Random dual basis generator $\mathcal{G}_{ob}(1^\lambda, N)$ is defined at the end of Section 2. We refer to Section 1.4 for notations on DPVS.

$\mathsf{Setup}(1^\lambda,\ n)$:

$\quad (\mathsf{param}_{\mathbb{V}}, \mathbb{B} := (\boldsymbol{b}_0, \ldots, \boldsymbol{b}_{4n+1}), \mathbb{B}^* := (\boldsymbol{b}_0^*, \ldots, \boldsymbol{b}_{4n+1}^*)) \xleftarrow{\mathsf{R}} \mathcal{G}_{ob}(1^\lambda, 4n+2),$

$\quad \widehat{\mathbb{B}} := (\boldsymbol{b}_0, \ldots, \boldsymbol{b}_n, \boldsymbol{b}_{4n+1}),\ \ \widehat{\mathbb{B}}^* := (\boldsymbol{b}_0^*, \ldots, \boldsymbol{b}_n^*, \boldsymbol{b}_{3n+1}^*, \ldots, \boldsymbol{b}_{4n}^*),$

$\quad \text{return } \mathsf{pk} := (1^\lambda, \mathsf{param}_{\mathbb{V}}, \widehat{\mathbb{B}}),\ \ \mathsf{sk} := \widehat{\mathbb{B}}^*.$

$\mathsf{KeyGen}(\mathsf{pk}, \mathsf{sk}, \vec{v} \in \mathbb{F}_q^n \setminus \{\vec{0}\})$: $\quad \sigma \xleftarrow{\mathsf{U}} \mathbb{F}_q,\ \vec{\eta} \xleftarrow{\mathsf{U}} \mathbb{F}_q^n,$

$$\boldsymbol{k}^* := (\ \overbrace{1,}^{1} \quad \overbrace{\sigma\vec{v},}^{n} \quad \overbrace{0^{2n},}^{2n} \quad \overbrace{\vec{\eta},}^{n} \quad \overbrace{0}^{1}\ )_{\mathbb{B}^*},$$

$\quad \text{return } \mathsf{sk}_{\vec{v}} := \boldsymbol{k}^*.$

$\mathsf{Enc}(\mathsf{pk},\ m, \vec{x} \in \mathbb{F}_q^n \setminus \{\vec{0}\})$: $\quad \omega, \varphi, \zeta \xleftarrow{\mathsf{U}} \mathbb{F}_q,$

$$\boldsymbol{c}_1 := (\ \overbrace{\zeta,}^{1} \quad \overbrace{\omega\vec{x},}^{n} \quad \overbrace{0^{2n},}^{2n} \quad \overbrace{0^n,}^{n} \quad \overbrace{\varphi}^{1}\ )_{\mathbb{B}}, \quad \boldsymbol{c}_2 := g_T^\zeta m,$$

$\quad \text{return } \mathsf{ct}_{\vec{x}} := (\boldsymbol{c}_1, \boldsymbol{c}_2).$

$\mathsf{Dec}(\mathsf{pk},\ \mathsf{sk}_{\vec{v}} := \boldsymbol{k}^*,\ \mathsf{ct}_{\vec{x}} := (\boldsymbol{c}_1, \boldsymbol{c}_2))$: $\quad m' := \boldsymbol{c}_2 / e(\boldsymbol{c}_1, \boldsymbol{k}^*),\quad \text{return } m'.$

[Correctness] If $\vec{v} \cdot \vec{x} = 0$, then $e(\boldsymbol{c}_1, \boldsymbol{k}^*) = g_T^{\zeta + \omega\sigma\vec{v}\cdot\vec{x}} = g_T^\zeta.$

4.2   Security

Main Theorem (Theorem 1) and Main Lemma (Lemma 1)

Theorem 1. *The proposed IPE scheme is adaptively attribute-hiding against chosen plaintext attacks under the DLIN assumption.*

For any adversary $\mathcal{A}$, there exist probabilistic machines $\mathcal{E}_{0\text{-}1}, \mathcal{E}_{0\text{-}2}, \mathcal{E}_{1\text{-}1}, \mathcal{E}_{1\text{-}2\text{-}1}$ and $\mathcal{E}_{1\text{-}2\text{-}2}$, whose running times are essentially the same as that of $\mathcal{A}$, such that for any security parameter λ,

$$\mathsf{Adv}_{\mathcal{A}}^{\mathsf{IPE},\mathsf{AH}}(\lambda) \leq \mathsf{Adv}_{\mathcal{E}_{0\text{-}1}}^{\mathsf{DLIN}}(\lambda) + \mathsf{Adv}_{\mathcal{E}_{1\text{-}1}}^{\mathsf{DLIN}}(\lambda)$$

$$+ \sum_{h=1}^{\nu} \left(\mathsf{Adv}_{\mathcal{E}_{0\text{-}2\text{-}h}}^{\mathsf{DLIN}}(\lambda) + \mathsf{Adv}_{\mathcal{E}_{1\text{-}2\text{-}h\text{-}1}}^{\mathsf{DLIN}}(\lambda) + \mathsf{Adv}_{\mathcal{E}_{1\text{-}2\text{-}h\text{-}2}}^{\mathsf{DLIN}}(\lambda) \right) + \epsilon,$$

where $\mathcal{E}_{0\text{-}2\text{-}h}(\cdot) := \mathcal{E}_{0\text{-}2}(h,\cdot)$, $\mathcal{E}_{1\text{-}2\text{-}h\text{-}1}(\cdot) := \mathcal{E}_{1\text{-}2\text{-}1}(h,\cdot)$, $\mathcal{E}_{1\text{-}2\text{-}h\text{-}2}(\cdot) := \mathcal{E}_{1\text{-}2\text{-}2}(h,\cdot)$, ν is the maximum number of $\mathcal{A}$'s key queries and $\epsilon := (18\nu + 17)/q$.

Proof. First, we execute a preliminary game transformation from Game 0 (original security game in Definition 5) to Game 0', which is the same as Game 0 except that flip a coin $t \xleftarrow{\mathsf{U}} \{0, 1\}$ before setup, and the game is aborted in step 3 if $t \neq s$. We define that $\mathcal{A}$ wins with probability $1/2$ when the game is aborted (and the advantage in Game 0' is $\Pr[\mathcal{A} \text{ wins }] - 1/2$ as well). Since t is independent from s, the game is aborted with probability $1/2$. Hence, the advantage in Game 0' is a half of that in Game 0, i.e., $\mathsf{Adv}_{\mathcal{A}}^{\mathsf{IPE},\mathsf{AH},0'}(\lambda) = 1/2 \cdot \mathsf{Adv}_{\mathcal{A}}^{\mathsf{IPE},\mathsf{AH}}(\lambda)$. Moreover, $\Pr[\mathcal{A} \text{ wins}] = 1/2 \cdot (\Pr[\mathcal{A} \text{ wins} \mid t = 0] + \Pr[\mathcal{A} \text{ wins} \mid t = 1])$ in Game 0' since t is uniformly and independently generated.

As for the conditional probability with $t = 0$, it holds that, for any adversary $\mathcal{A}$, there exist probabilistic machines $\mathcal{E}_1$ and $\mathcal{E}_2$, whose running times are essentially the same as that of $\mathcal{A}$, such that for any security parameter λ, in Game 0',

$$\Pr[\mathcal{A} \text{ wins} \mid t = 0] - 1/2 \leq \mathsf{Adv}_{\mathcal{E}_1}^{\mathsf{DLIN}}(\lambda) + \sum_{h=1}^{\nu} \mathsf{Adv}_{\mathcal{E}_{2\text{-}h}}^{\mathsf{DLIN}}(\lambda) + \epsilon,$$

where $\mathcal{E}_{2\text{-}h}(\cdot) := \mathcal{E}_2(h, \cdot)$ and ν is the maximum number of $\mathcal{A}$'s key queries and $\epsilon := (6\nu + 5)/q$. This is obtained in the same manner as the weakly attribute-hiding security of the OT10 IPE in the full version of [19]: Since the difference between our IPE and the OT10 IPE is only the dimension of the hidden subspaces, i.e., the former has $2n$ and the latter has n, the weakly attribute-hiding security of the OT10 IPE implies the security with $t = 0$ of our IPE.

As for the conditional probability with $t = 1$, i.e., $\Pr[\mathcal{A} \text{ wins} \mid t = 1]$, Lemma 1 (Eq. (1)) holds. Therefore,

$$\begin{aligned}
\mathsf{Adv}_{\mathcal{A}}^{\mathsf{IPE},\mathsf{AH}}(\lambda) &= 2 \cdot \mathsf{Adv}_{\mathcal{A}}^{\mathsf{IPE},\mathsf{AH},0'}(\lambda) = \Pr[\mathcal{A} \text{ wins} \mid t = 0] + \Pr[\mathcal{A} \text{ wins} \mid t = 1] - 1 \\
&= (\Pr[\mathcal{A} \text{ wins} \mid t = 0] - 1/2) + (\Pr[\mathcal{A} \text{ wins} \mid t = 1] - 1/2) \\
&\leq \mathsf{Adv}_{\mathcal{E}_{0\text{-}1}}^{\mathsf{DLIN}}(\lambda) + \sum_{h=1}^{\nu} \mathsf{Adv}_{\mathcal{E}_{0\text{-}2\text{-}h}}^{\mathsf{DLIN}}(\lambda) + \mathsf{Adv}_{\mathcal{E}_{1\text{-}1}}^{\mathsf{DLIN}}(\lambda) \\
&\quad + \sum_{h=1}^{\nu} \left(\mathsf{Adv}_{\mathcal{E}_{1\text{-}2\text{-}h\text{-}1}}^{\mathsf{DLIN}}(\lambda) + \mathsf{Adv}_{\mathcal{E}_{1\text{-}2\text{-}h\text{-}2}}^{\mathsf{DLIN}}(\lambda) \right) + \epsilon, \text{ where } \epsilon := (18\nu + 17)/q. \quad \square
\end{aligned}$$

Lemma 1 (Main Lemma). *For any adversary $\mathcal{A}$, there exist probabilistic machines $\mathcal{E}_1, \mathcal{E}_{2\text{-}1}$ and $\mathcal{E}_{2\text{-}2}$, whose running times are essentially the same as that of $\mathcal{A}$, such that for any security parameter λ, in Game 0' (described in the proof of Theorem 1),*

$$\begin{aligned}
&\Pr[\mathcal{A} \text{ wins} \mid t = 1] - 1/2 \\
&\quad \leq \mathsf{Adv}_{\mathcal{E}_1}^{\mathsf{DLIN}}(\lambda) + \sum_{h=1}^{\nu} \left(\mathsf{Adv}_{\mathcal{E}_{2\text{-}h\text{-}1}}^{\mathsf{DLIN}}(\lambda) + \mathsf{Adv}_{\mathcal{E}_{2\text{-}h\text{-}2}}^{\mathsf{DLIN}}(\lambda) \right) + \epsilon, \quad (1)
\end{aligned}$$

where $\mathcal{E}_{2\text{-}h\text{-}1}(\cdot) := \mathcal{E}_{2\text{-}1}(h, \cdot)$, $\mathcal{E}_{2\text{-}h\text{-}2}(\cdot) := \mathcal{E}_{2\text{-}2}(h, \cdot)$, ν is the maximum number of $\mathcal{A}$'s key queries and $\epsilon := (12\nu + 12)/q$.

Proof Outline of Lemma 1 At the top level strategy of the security proof, an extended form of the dual system encryption by Waters [25] is employed, where ciphertexts and secret keys have three forms, *normal, temporal 1* and *temporal 2*. The real system uses only normal ciphertexts and normal secret keys, and temporal 1 and 2 ciphertexts and keys are used only in a sequence of security games for the security proof. (Additionally, ciphertexts have temporal 0 and unbiased forms. See below.)

To prove this lemma, we only consider the $t = 1$ case. We employ Game 0' (described in the proof of Theorem 1) through Game 3. In Game 1, the challenge ciphertext is changed to temporal 0 form. When at most ν secret key queries are issued by an adversary, there are 4ν game changes from Game 1 (Game 2-0-4), Game 2-1-1, Game 2-1-2, Game 2-1-3, Game 2-1-4 through Game 2-ν-1, Game 2-ν-2, Game 2-ν-3, Game 2-ν-4.

In Game 2-h-1, the challenge ciphertext is changed to temporal 1 form, and the first $h - 1$ keys are temporal 2 form, while the remaining keys are normal. In Game 2-h-2, the h-th key is changed to temporal 1 form while the remaining keys and the challenge ciphertext is the same as in Game 2-h-1. In Game 2-h-3, the challenge ciphertext is changed to temporal 2 form while all the queried keys are the same as in Game 2-h-2. In Game 2-h-4, the h-th key is changed to temporal 2 form while the remaining keys and the challenge ciphertext is the same as in Game 2-h-3. At the end of the Game 2 sequence, in Game 2-ν-4, all the queried keys are temporal 2 forms (and the challenge ciphertext is temporal 2 form), which allows the next conceptual change to Game 3. In Game 3, the challenge ciphertext is changed to *unbiased* form (while all the queried keys are temporal 2 form). In the final game, advantage of the adversary is zero.

We summarize these changes in Table 1, where shaded parts indicate the challenge ciphertext or queried key(s) which were changed in a game from the previous game

As usual, we prove that the advantage gaps between neighboring games are negligible.

For $\mathsf{ct}_{\vec{x}} := (c_1, c_2)$, we focus on c_1, and ignore the other part of $\mathsf{ct}_{\vec{x}}$, i.e., c_2, (and call c_1 ciphertext) in this proof outline. In addition, we ignore a negligible factor in the (informal) descriptions of this proof outline. For example, we say "A is bounded by B" when $A \leq B + \epsilon(\lambda)$ where $\epsilon(\lambda)$ is negligible in security parameter λ.

A normal secret key, $\boldsymbol{k}^{*\,\mathsf{norm}}$ (with vector $\vec{v}$), is the correct form of the secret key of the proposed IPE scheme, and is expressed by Eq. (2). Similarly, a normal ciphertext (with vector $\vec{x}$), $\boldsymbol{c}_1^{\mathsf{norm}}$, is expressed by Eq. (3). A temporal 0 ciphertext is expressed by Eq. (4). A temporal 1 ciphertext, $\boldsymbol{c}_1^{\mathsf{temp1}}$, is expressed by Eq. (5) and a temporal 1 secret key, $\boldsymbol{k}^{*\,\mathsf{temp1}}$, is expressed by Eq. (6). A temporal 2 ciphertext, $\boldsymbol{c}_1^{\mathsf{temp2}}$, is expressed by Eq. (7) and a temporal 2 secret key, $\boldsymbol{k}^{*\,\mathsf{temp2}}$, is expressed by Eq. (8). An unbiased ciphertext, $\boldsymbol{c}_1^{\mathsf{unbias}}$, is expressed by Eq. (9).

To prove that the advantage gap between Games 0' and 1 is bounded by the advantage of Problem 1 (to guess $\beta \in \{0, 1\}$), we construct a simulator of the challenger of Game 0' (or 1) (against an adversary $\mathcal{A}$) by using an instance with

Table 1. Outline of Game Descriptions

Game	Challenge ciphertext	1	⋯	h − 1	h	h + 1	⋯	ν
0'	normal				normal			
1	temporal 0				normal			
2-1-1	temporal 1				normal			
2-1-2	temporal 1	temporal 1			normal			
2-1-3	temporal 2	temporal 1			normal			
2-1-4	temporal 2	temporal 2			normal			
			⋮					
2-h-1	temporal 1			temporal 2		normal		
2-h-2	temporal 1			temporal 2	temporal 1	normal		
2-h-3	temporal 2			temporal 2	temporal 1	normal		
2-h-4	temporal 2			temporal 2	temporal 2	normal		
			⋮					
2-ν-4	temporal 2				temporal 2			temporal 2
3	unbiased				temporal 2			

$\beta \xleftarrow{\mathsf{U}} \{0,1\}$ of Problem 1. We then show that the distribution of the secret keys and challenge ciphertext replied by the simulator is equivalent to those of Game 0' when $\beta = 0$ and those of Game 1 when $\beta = 1$. That is, the advantage of Problem 1 is equivalent to the advantage gap between Games 0' and 1 (Lemma 6). The advantage of Problem 1 is proven to be equivalent to that of the DLIN assumption (Lemma 2).

We then show that Game $2\text{-}(h-1)\text{-}4$ can be conceptually changed to Game $2\text{-}h\text{-}1$ (Lemma 7), by using the fact that parts of bases, $(\boldsymbol{b}_{n+1}, \ldots, \boldsymbol{b}_{2n})$ and $(\boldsymbol{b}^*_{n+1}, \ldots, \boldsymbol{b}^*_{2n})$, are unknown to the adversary. In particular, when $h = 1$, it means that Game 1 can be conceptually changed to Game 2-1-1. When $h \geq 2$, we notice that temporal 2 key and temporal 1 challenge ciphertext, $(\boldsymbol{k}^{*\,\mathrm{temp2}}, \boldsymbol{c}_1^{\mathrm{temp1}})$, are equivalent to temporal 2 key and temporal 2 challenge ciphertext, $(\boldsymbol{k}^{*\,\mathrm{temp2}}, \boldsymbol{c}_1^{\mathrm{temp2}})$, except that $\vec{x}^{(b)}$ is used in $\boldsymbol{c}_1^{\mathrm{temp1}}$ instead of $\omega_0' \vec{x}^{(0)} + \omega_1' \vec{x}^{(1)}$ (with $\omega_0', \omega_1' \xleftarrow{\mathsf{U}} \mathbb{F}_q$) for some coefficient vector in $\boldsymbol{c}_1^{\mathrm{temp2}}$. This change of coefficient vectors can be done conceptually since zero vector 0^n is used for the corresponding part in $\boldsymbol{k}^{*\,\mathrm{temp2}}$.

The advantage gap between Games $2\text{-}h\text{-}1$ and $2\text{-}h\text{-}2$ is shown to be bounded by the advantage of Problem 2, i.e., advantage of the DLIN assumption (Lemmas 8 and 3).

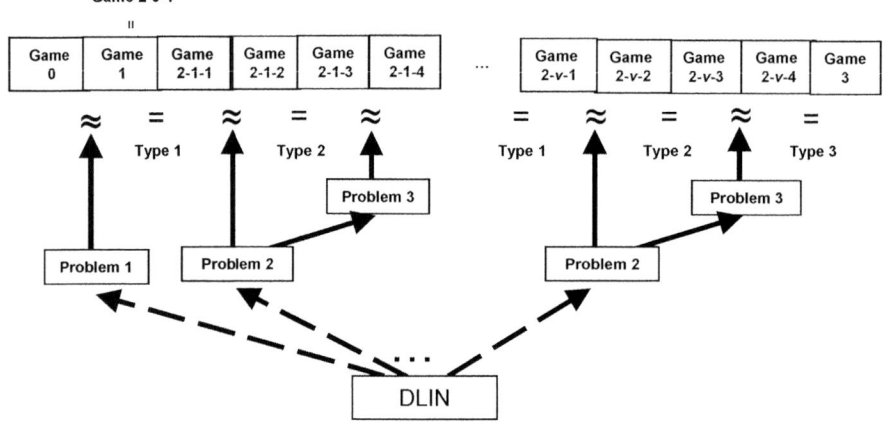

Fig. 1. Structure of Reductions

We then show that Game 2-*h*-2 can be conceptually changed to Game 2-*h*-3 (Lemma 9), again by using the fact that parts of bases, $(\boldsymbol{b}_{n+1}, \ldots, \boldsymbol{b}_{2n})$ and $(\boldsymbol{b}_{n+1}^*, \ldots, \boldsymbol{b}_{2n}^*)$, are unknown to the adversary. In this conceptual change, we use the fact that all key queries $\vec{v}$ satisfy $\vec{v} \cdot \vec{x}^{(0)} = \vec{v} \cdot \vec{x}^{(1)} = 0$ or $\vec{v} \cdot \vec{x}^{(0)} \neq 0$ and $\vec{v} \cdot \vec{x}^{(1)} \neq 0$. Here, we notice that temporal 1 key and temporal 1 challenge ciphertext, $(\boldsymbol{k}^{*\,\text{temp1}}, \boldsymbol{c}_1^{\text{temp1}})$, are equivalent to temporal 1 key and temporal 2 challenge ciphertext, $(\boldsymbol{k}^{*\,\text{temp1}}, \boldsymbol{c}_1^{\text{temp2}})$, except that random linear combination $\omega_0' \vec{x}^{(0)} + \omega_1' \vec{x}^{(1)}$ (with $\omega_0', \omega_1' \xleftarrow{\mathsf{U}} \mathbb{F}_q$) is used in $\boldsymbol{c}_1^{\text{temp2}}$ instead of $\vec{x}^{(b)}$ for some coefficient vector in $\boldsymbol{c}_1^{\text{temp1}}$. This conceptual change is proved by using Lemma 5.

The advantage gap between Games 2-*h*-3 and 2-*h*-4 is similarly shown to be bounded by the advantage of Problem 3, i.e., advantage of the DLIN assumption (Lemmas 10 and 4).

We then show that Game 2-*v*-4 can be conceptually changed to Game 3 (Lemma 11) by using the fact that parts of bases, $(\boldsymbol{b}_{n+1}, \ldots, \boldsymbol{b}_{3n})$ and $(\boldsymbol{b}_1^*, \ldots, \boldsymbol{b}_{2n}^*)$, are unknown to the adversary.

Figure 1 shows the structure of the security reduction, where the security of the scheme is hierarchically reduced to the intractability of the DLIN problem. The reduction steps indicated by dotted arrows can be shown in the same manner as that in (the full version of) [19].

Proof of Lemma 1. To prove Lemma 1, we consider the following $4\nu + 3$ games when $t = 1$. In Game 0', a part framed by a box indicates coefficients to be changed in a subsequent game. In the other games, a part framed by a box indicates coefficients which were changed in a game from the previous game.

Game 0' : Same as Game 0 except that flip a coin $t \xleftarrow{\mathsf{U}} \{0,1\}$ before setup, and the game is aborted in step 3 if $t \neq s$. In order to prove Lemma 1, we consider the case with $t = 1$. The reply to a key query for $\vec{v}$ is:

$$\boldsymbol{k}^* := (\ 1, \ \sigma\vec{v}, \ \boxed{0^n}, \ \boxed{0^n}, \ \vec{\eta}, \ 0\ )_{\mathbb{B}^*}, \tag{2}$$

where $\sigma \xleftarrow{\mathsf{U}} \mathbb{F}_q$ and $\vec{\eta} \xleftarrow{\mathsf{U}} \mathbb{F}_q^n$. The challenge ciphertext for challenge plaintext $m := m^{(0)} = m^{(1)}$ and vectors $(\vec{x}^{(0)}, \vec{x}^{(1)})$ is:

$$\boldsymbol{c}_1 := (\ \zeta, \ \boxed{\omega\vec{x}^{(b)}}, \ \boxed{0^n}, \ \boxed{0^n}, \ 0^n, \ \varphi\ )_{\mathbb{B}}, \quad \boldsymbol{c}_2 := g_T^\zeta m, \tag{3}$$

where $b \xleftarrow{\mathsf{U}} \{0,1\}$ and $\zeta, \omega, \varphi \xleftarrow{\mathsf{U}} \mathbb{F}_q$. Here, we note that $\boldsymbol{c}_2$ is independent from bit b.

Game 1 : Game 1 is the same as Game 0' except that $\boldsymbol{c}_1$ of the challenge ciphertext for (challenge plaintext $m := m^{(0)} = m^{(1)}$ and) vectors $(\vec{x}^{(0)}, \vec{x}^{(1)})$ is:

$$\boldsymbol{c}_1 := (\ \zeta, \ \omega\vec{x}^{(b)}, \ \boxed{zx_1^{(b)}, 0^{n-1}}, \ 0^n, \ 0^n, \ \varphi\ )_{\mathbb{B}}, \tag{4}$$

where $x_1^{(b)} \neq 0$ is the first coordinate of $\vec{x}^{(b)}$, $z \xleftarrow{\mathsf{U}} \mathbb{F}_q$ and all the other variables are generated as in Game 0'.

Game 2-h-1 ($h = 1, \ldots, \nu$) : Game 2-0-4 is Game 1. Game 2-h-1 is the same as Game 2-$(h-1)$-4 except that $\boldsymbol{c}_1$ of the challenge ciphertext for (challenge plaintext $m := m^{(0)} = m^{(1)}$ and) vectors $(\vec{x}^{(0)}, \vec{x}^{(1)})$ is:

$$\boldsymbol{c}_1 := (\ \zeta, \ \omega\vec{x}^{(b)}, \ \boxed{\omega'\vec{x}^{(b)}}, \ \boxed{\omega_0''\vec{x}^{(0)} + \omega_1''\vec{x}^{(1)}}, \ 0^n, \ \varphi\ )_{\mathbb{B}}, \tag{5}$$

where $\omega', \omega_0'', \omega_1'' \xleftarrow{\mathsf{U}} \mathbb{F}_q$ and all the other variables are generated as in Game 2-$(h-1)$-4.

Game 2-h-2 ($h = 1, \ldots, \nu$) : Game 2-h-2 is the same as Game 2-h-1 except that the reply to the h-th key query for $\vec{v}$ is:

$$\boldsymbol{k}^* := (\ 1, \ \sigma\vec{v}, \ \boxed{\sigma'\vec{v}}, \ 0^n, \ \vec{\eta}, \ 0\ )_{\mathbb{B}^*}, \tag{6}$$

where $\sigma' \xleftarrow{\mathsf{U}} \mathbb{F}_q$ and all the other variables are generated as in Game 2-h-1.

Game 2-h-3 ($h = 1, \ldots, \nu$) : Game 2-h-3 is the same as Game 2-h-2 except that $\boldsymbol{c}_1$ of the challenge ciphertext for (challenge plaintexts $m := m^{(0)} = m^{(1)}$ and) vectors $(\vec{x}^{(0)}, \vec{x}^{(1)})$ is:

$$\boldsymbol{c}_1 := (\ \zeta, \ \omega\vec{x}^{(b)}, \ \boxed{\omega_0'\vec{x}^{(0)} + \omega_1'\vec{x}^{(1)}}, \ \omega_0''\vec{x}^{(0)} + \omega_1''\vec{x}^{(1)}, \ 0^n, \ \varphi\ )_{\mathbb{B}}, \tag{7}$$

where $\omega_0', \omega_1' \xleftarrow{\mathsf{U}} \mathbb{F}_q$ and all the other variables are generated as in Game 2-h-2.

Game 2-h-4 ($h = 1, \ldots, \nu$) : Game 2-h-4 is the same as Game 2-h-3 except that the reply to the h-th key query for $\vec{v}$ is:

$$\boldsymbol{k}^* := (\ 1, \ \sigma\vec{v}, \ \boxed{0^n}, \ \boxed{\sigma''\vec{v}}, \ \vec{\eta}, \ 0\ )_{\mathbb{B}^*}, \tag{8}$$

where $\sigma'' \xleftarrow{\mathsf{U}} \mathbb{F}_q$ and all the other variables are generated as in Game 2-h-3.

Game 3 : Game 3 is the same as Game 2-ν-4 except that c_1 of the challenge ciphertext for (challenge plaintexts $m := m^{(0)} = m^{(1)}$ and) vectors $(\vec{x}^{(0)}, \vec{x}^{(1)})$ is:

$$c_1 := (\, \zeta, \boxed{\omega_0 \vec{x}^{(0)} + \omega_1 \vec{x}^{(1)}}, \; \omega_0' \vec{x}^{(0)} + \omega_1' \vec{x}^{(1)}, \; \omega_0'' \vec{x}^{(0)} + \omega_1'' \vec{x}^{(1)}, \; 0^n, \; \varphi \,)_{\mathbb{B}}, \quad (9)$$

where $\omega_0, \omega_1 \xleftarrow{\mathsf{U}} \mathbb{F}_q$ and all the other variables are generated as in Game 2-ν-4. Here, we note that c_1 is independent from bit $b \xleftarrow{\mathsf{U}} \{0, 1\}$.

Let $\mathsf{Adv}_{\mathcal{A}}^{(0')}(\lambda), \mathsf{Adv}_{\mathcal{A}}^{(1)}(\lambda), \mathsf{Adv}_{\mathcal{A}}^{(2\text{-}h\text{-}1)}(\lambda), \ldots, \mathsf{Adv}_{\mathcal{A}}^{(2\text{-}h\text{-}4)}(\lambda)$ and $\mathsf{Adv}_{\mathcal{A}}^{(3)}(\lambda)$ be the advantage of $\mathcal{A}$ in Game $0', 1, 2\text{-}h\text{-}1, \ldots, 2\text{-}h\text{-}4$ and 3 when $t = 1$, respectively. $\mathsf{Adv}_{\mathcal{A}}^{(0')}(\lambda)$ is equivalent to the left-hand side of Eq. (1). We will show six lemmas (Lemmas 6–11) that evaluate the gaps between pairs of neighboring games. From these lemmas and Lemmas 2–4, we obtain $\mathsf{Adv}_{\mathcal{A}}^{(0')}(\lambda) \leq \left| \mathsf{Adv}_{\mathcal{A}}^{(0')}(\lambda) - \mathsf{Adv}_{\mathcal{A}}^{(1)}(\lambda) \right| +$
$\sum_{h=1}^{\nu} \sum_{\iota=1}^{4} \left| \mathsf{Adv}_{\mathcal{A}}^{(2\text{-}h\text{-}(\iota-1))}(\lambda) - \mathsf{Adv}_{\mathcal{A}}^{(2\text{-}h\text{-}\iota)}(\lambda) \right| + \left| \mathsf{Adv}_{\mathcal{A}}^{(2\text{-}\nu\text{-}4)}(\lambda) - \mathsf{Adv}_{\mathcal{A}}^{(3)}(\lambda) \right| +$
$\mathsf{Adv}_{\mathcal{A}}^{(3)}(\lambda) \leq \mathsf{Adv}_{\mathcal{B}_1}^{\mathsf{P1}}(\lambda) + \sum_{h=1}^{\nu} \left(\mathsf{Adv}_{\mathcal{B}_{2\text{-}h\text{-}1}}^{\mathsf{P2}}(\lambda) + \mathsf{Adv}_{\mathcal{B}_{2\text{-}h\text{-}2}}^{\mathsf{P3}}(\lambda) \right) + (2\nu + 1)/q \leq$
$\mathsf{Adv}_{\mathcal{E}_1}^{\mathsf{DLIN}}(\lambda) + \sum_{h=1}^{\nu} \left(\mathsf{Adv}_{\mathcal{E}_{2\text{-}h\text{-}1}}^{\mathsf{DLIN}}(\lambda) + \mathsf{Adv}_{\mathcal{E}_{2\text{-}h\text{-}2}}^{\mathsf{DLIN}}(\lambda) \right) + (12\nu + 12)/q.$ $\qquad \square$

The definitions of Problems 1–3 and the advantages $(\mathsf{Adv}_{\mathcal{B}}^{\mathsf{P1}}(\lambda), \mathsf{Adv}_{\mathcal{B}}^{\mathsf{P2}}(\lambda), \mathsf{Adv}_{\mathcal{B}}^{\mathsf{P3}}(\lambda))$, and the proofs of Lemmas 2–12 are given in the full version [21].

Lemma 2 (resp. 3, 4). *For any adversary $\mathcal{B}$, there is a probabilistic machine $\mathcal{E}$, whose running time is essentially the same as that of $\mathcal{B}$, such that for any security parameter λ, $\mathsf{Adv}_{\mathcal{B}}^{\mathsf{P1}}(\lambda) \leq \mathsf{Adv}_{\mathcal{E}}^{\mathsf{DLIN}}(\lambda) + 6/q$, (resp. $\mathsf{Adv}_{\mathcal{B}}^{\mathsf{P2}}(\lambda) \leq \mathsf{Adv}_{\mathcal{E}}^{\mathsf{DLIN}}(\lambda) + 5/q$, $\mathsf{Adv}_{\mathcal{B}}^{\mathsf{P3}}(\lambda) \leq \mathsf{Adv}_{\mathcal{E}}^{\mathsf{DLIN}}(\lambda) + 5/q$).*

Lemma 5 is the same as Lemma 3 in [19].

Lemma 6. *For any adversary $\mathcal{A}$, there exists a probabilistic machine $\mathcal{B}_1$, whose running time is essentially the same as that of $\mathcal{A}$, such that for any security parameter λ, $|\mathsf{Adv}_{\mathcal{A}}^{(0')}(\lambda) - \mathsf{Adv}_{\mathcal{A}}^{(1)}(\lambda)| \leq \mathsf{Adv}_{\mathcal{B}_1}^{\mathsf{P1}}(\lambda)$.*

Lemma 7. *For any adversary $\mathcal{A}$, $|\mathsf{Adv}_{\mathcal{A}}^{(2\text{-}(h-1)\text{-}4)}(\lambda) - \mathsf{Adv}_{\mathcal{A}}^{(2\text{-}h\text{-}1)}(\lambda)| \leq 2/q$.*

Lemma 8. *For any adversary $\mathcal{A}$, there exists a probabilistic machine $\mathcal{B}_{2\text{-}1}$, whose running time is essentially the same as that of $\mathcal{A}$, such that for any security parameter λ, $|\mathsf{Adv}_{\mathcal{A}}^{(2\text{-}h\text{-}1)}(\lambda) - \mathsf{Adv}_{\mathcal{A}}^{(2\text{-}h\text{-}2)}(\lambda)| \leq \mathsf{Adv}_{\mathcal{B}_{2\text{-}h\text{-}1}}^{\mathsf{P2}}(\lambda)$, where $\mathcal{B}_{2\text{-}h\text{-}1}(\cdot) := \mathcal{B}_{2\text{-}1}(h, \cdot)$.*

Lemma 9. *For any adversary $\mathcal{A}$, $\mathsf{Adv}_{\mathcal{A}}^{(2\text{-}h\text{-}2)}(\lambda) = \mathsf{Adv}_{\mathcal{A}}^{(2\text{-}h\text{-}3)}(\lambda)$.*

Lemma 10. *For any adversary $\mathcal{A}$, there exists a probabilistic machine $\mathcal{B}_{2\text{-}2}$, whose running time is essentially the same as that of $\mathcal{A}$, such that for any security parameter λ, $|\mathsf{Adv}_{\mathcal{A}}^{(2\text{-}h\text{-}3)}(\lambda) - \mathsf{Adv}_{\mathcal{A}}^{(2\text{-}h\text{-}4)}(\lambda)| \leq \mathsf{Adv}_{\mathcal{B}_{2\text{-}h\text{-}2}}^{\mathsf{P3}}(\lambda)$, where $\mathcal{B}_{2\text{-}h\text{-}2}(\cdot) := \mathcal{B}_{2\text{-}2}(h, \cdot)$.*

Lemma 11. *For any adversary $\mathcal{A}$, $|\mathsf{Adv}_{\mathcal{A}}^{(2\text{-}\nu\text{-}4)}(\lambda) - \mathsf{Adv}_{\mathcal{A}}^{(3)}(\lambda)| \leq 1/q$.*

Lemma 12. *For any adversary $\mathcal{A}$, $\mathsf{Adv}_{\mathcal{A}}^{(3)}(\lambda) = 0$.*

5   A Variant for Achieving Shorter Public and Secret Keys

A variant of the proposed (basic) IPE scheme with the same security, that achieves a shorter ($O(n)$-size) master public key and shorter ($O(1)$-size) secret keys (excluding the description of $\vec{v}$), can be constructed by combining with the techniques in [20], where n is the dimension of vectors of the IPE scheme. This variant also enjoys more efficient decryption. Here, we show this variant. See the key idea, performance and the security proof of this scheme in the full versions of this paper [21] and [20]. Let $N := 5n + 1$ and

$$
\mathcal{H}(n, \mathbb{F}_q) := \left\{ \begin{pmatrix} \mu & & & \mu_1' \\ & \ddots & & \vdots \\ & & \mu & \mu_{n-1}' \\ & & & \mu_n' \end{pmatrix} \middle| \begin{array}{l} \mu, \mu_l' \in \mathbb{F}_q \text{ for } l = 1, \ldots, n, \\ \text{a blank element in the matrix} \\ \text{denotes } 0 \in \mathbb{F}_q \end{array} \right\}, \quad (10)
$$

$$
\mathcal{L}^+(5, n, \mathbb{F}_q) :=
$$
$$
\left\{ X := \begin{pmatrix} \chi_{0,0} & \chi_{0,1}\vec{e}_n & \cdots & \chi_{0,5}\vec{e}_n \\ \vec{\chi}_{1,0}^{\mathrm{T}} & X_{1,1} & \cdots & X_{1,5} \\ \vdots & \vdots & & \vdots \\ \vec{\chi}_{5,0}^{\mathrm{T}} & X_{5,1} & \cdots & X_{5,5} \end{pmatrix} \middle| \begin{array}{l} X_{i,j} \in \mathcal{H}(n, \mathbb{F}_q), \\ \vec{\chi}_{i,0} := (\chi_{i,0,l})_{l=1,\ldots,n} \in \mathbb{F}_q^n, \\ \chi_{0,0}, \chi_{0,j} \in \mathbb{F}_q \\ \text{for } i, j = 1, \ldots, 5 \end{array} \right\}
$$
$$
\bigcap GL(N, \mathbb{F}_q). \quad (11)
$$

We note that $\mathcal{L}^+(5, n, \mathbb{F}_q)$ is a subgroup of $GL(N, \mathbb{F}_q)$. Random dual orthonormal basis generator $\mathcal{G}_{\mathsf{ob}}^{\mathsf{ZIPE,SK}}$ below is used as a subroutine in the proposed IPE.

$\mathcal{G}_{\mathsf{ob}}^{\mathsf{ZIPE,SK}}(1^\lambda, 5, n)$: $\mathsf{param}_{\mathbb{G}} := (q, \mathbb{G}, \mathbb{G}_T, G, e) \xleftarrow{\mathsf{R}} \mathcal{G}_{\mathsf{bpg}}(1^\lambda)$, $N := 5n + 1$,

$\psi \xleftarrow{\mathsf{U}} \mathbb{F}_q^\times$, $g_T := e(G, G)^\psi$, $\mathsf{param}_{\mathbb{V}} := (q, \mathbb{V}, \mathbb{G}_T, \mathbb{A}, e) := \mathcal{G}_{\mathsf{dpvs}}(1^\lambda, N, \mathsf{param}_{\mathbb{G}})$,

$\mathsf{param}_n := (\mathsf{param}_{\mathbb{V}}, g_T)$, $X \xleftarrow{\mathsf{U}} \mathcal{L}^+(5, n, \mathbb{F}_q)$, $(\vartheta_{i,j})_{i,j=0,\ldots,5n} := \psi \cdot (X^{\mathrm{T}})^{-1}$,

hereafter, $\{\chi_{0,0}, \chi_{0,j}, \chi_{i,0,l}, \mu_{i,j}, \mu_{i,j,l}'\}_{i,j=1,\ldots5;l=1,\ldots,n}$ denotes non-zero entries of X, where $\{\mu_{i,j}, \mu_{i,j,l}'\}$ are non-zero entries of submatrices $X_{i,j}$ of X as given in Eqs. (11) and (10),

$\boldsymbol{b}_i := (\vartheta_{i,0}, \ldots, \vartheta_{i,5n})_{\mathbb{A}} = \sum_{j=0}^{5n} \vartheta_{i,j} \boldsymbol{a}_j$ for $i = 0, \ldots, 5n$, $\mathbb{B} := (\boldsymbol{b}_0, \ldots, \boldsymbol{b}_{5n})$,

$B_{0,0}^* := \chi_{0,0}G$, $B_{0,j}^* := \chi_{0,j}G$, $B_{i,0,l}^* := \chi_{i,0,l}G$, $B_{i,j}^* := \mu_{i,j}G$, $B_{i,j,l}'^* := \mu_{i,j,l}'G$

for $i, j = 1, \ldots, 5; l = 1, \ldots, n$,

return $(\mathsf{param}_n, \mathbb{B}, \{B_{0,0}^*, B_{0,j}^*, B_{i,0,l}^*, B_{i,j}^*, B_{i,j,l}'^*\}_{i,j=1,\ldots,5;l=1,\ldots,n})$.

Remark 1. Let $\boldsymbol{b}_0^* := (\, B_{0,0}^*, \, 0^{n-1}, B_{0,1}^*, \ldots, 0^{n-1}, B_{0,5}^* \,)$,

$$
\begin{pmatrix} \boldsymbol{b}_{(i-1)n+1}^* \\ \vdots \\ \boldsymbol{b}_{in}^* \end{pmatrix} := \begin{pmatrix} B_{i,0,1}^* & B_{i,1}^* & & B_{i,1,1}'^* & & B_{i,5}^* & & B_{i,5,1}'^* \\ \vdots & & \ddots & \vdots & \cdots & & \ddots & \vdots \\ B_{i,0,n-1}^* & B_{i,1}^* & B_{i,1,n-1}'^* & & & B_{i,5}^* & B_{i,5,n-1}'^* \\ B_{i,0,n}^* & & B_{i,1,n}'^* & & & & B_{i,5,n}'^* \end{pmatrix}
$$

for $i = 1, \ldots, 5$, and $\mathbb{B}^* := (\boldsymbol{b}_0^*, \ldots, \boldsymbol{b}_{5n}^*)$, where a blank element in the matrix denotes $0 \in \mathbb{G}$. $\mathbb{B}^*$ is the dual orthonormal basis of $\mathbb{B}$, i.e., $e(\boldsymbol{b}_i, \boldsymbol{b}_i^*) = g_T$ and $e(\boldsymbol{b}_i, \boldsymbol{b}_j^*) = 1$ for $0 \leq i \neq j \leq 5n$.

Here, we assume that input vector, $\vec{v} := (v_1, \ldots, v_n)$, has an index l ($1 \leq l \leq n - 1$) with $v_l \neq 0$, and that input vector, $\vec{x} := (x_1, \ldots, x_n)$, satisfies $x_n \neq 0$.

Setup(1^λ, n) :

$(\text{param}_n, \mathbb{B}, \{B_{0,0}^*, B_{0,j}^*, B_{i,0,l}^*, B_{i,j}^*, B_{i,j,l}'^*\}_{i,j=1,\ldots,5;l=1,\ldots,n}) \xleftarrow{\mathsf{R}} \mathcal{G}_{\mathrm{ob}}^{\mathsf{ZIPE,SK}}(1^\lambda, 5, n),$

$\widehat{\mathbb{B}} := (\boldsymbol{b}_0, \ldots, \boldsymbol{b}_n, \boldsymbol{b}_{4n+1}, \ldots, \boldsymbol{b}_{5n}),$

return $\mathsf{pk} := (1^\lambda, \text{param}_n, \widehat{\mathbb{B}})$, $\mathsf{sk} := \{B_{0,0}^*, B_{0,j}^*, B_{i,0,l}^*, B_{i,j}^*, B_{i,j,l}'^*\}_{i=1,4;j=1,\ldots,5;l=1,\ldots,n}.$

KeyGen(pk, sk, $\vec{v}$) : $\sigma, \eta \xleftarrow{\mathsf{U}} \mathbb{F}_q$, $K_0^* := B_{0,0}^* + \sum_{l=1}^n v_l(\sigma B_{1,0,l}^* + \eta B_{4,0,l}^*),$

$K_{1,j}^* := \sigma B_{1,j}^* + \eta B_{4,j}^*$, $K_{2,j}^* := B_{0,j}^* + \sum_{l=1}^n v_l(\sigma B_{1,j,l}'^* + \eta B_{4,j,l}'^*)$ for $j = 1, \ldots, 5$,

return $\mathsf{sk}_{\vec{v}} := (\vec{v}, K_0^*, \{K_{1,j}^*, K_{2,j}^*\}_{j=1,\ldots,5}).$

Enc(pk, m, $\vec{x}$) : $\omega, \zeta \xleftarrow{\mathsf{U}} \mathbb{F}_q$, $\vec{\varphi} \xleftarrow{\mathsf{U}} \mathbb{F}_q^n$, $\boldsymbol{c}_1 := (\ \zeta, \ \overbrace{\omega\vec{x}}^{n}, \ \overbrace{0^{2n}}^{2n}, \ \overbrace{0^n}^{n}, \ \overbrace{\vec{\varphi}}^{n}\ )_{\mathbb{B}},$

$c_2 := g_T^\zeta m$,   return $\mathsf{ct}_{\vec{x}} := (\boldsymbol{c}_1, c_2).$

Dec(pk, $\mathsf{sk}_{\vec{v}} := (\vec{v}, K_0^*, \{K_{1,j}^*, K_{2,j}^*\}_{j=1,\ldots,5})$, $\mathsf{ct}_{\vec{x}} := (\boldsymbol{c}_1, c_2))$:

Parse $\boldsymbol{c}_1$ as a $(5n + 1)$-tuple $(C_0, \ldots, C_{5n}) \in \mathbb{G}^{5n+1}$,

$D_j := \sum_{l=1}^{n-1} v_l C_{(j-1)n+l}$ for $j = 1, \ldots, 5$,

$F := e(C_0, K_0^*) \cdot \prod_{j=1}^5 \left(e(D_j, K_{1,j}^*) \cdot e(C_{jn}, K_{2,j}^*) \right)$,   return $m' := c_2/F.$

Remark 2. A part of output of Setup($1^\lambda, n$), $\{B_{0,0}^*, B_{0,j}^*, B_{i,0,l}^*, B_{i,j}^*, B_{i,j,l}'^*\}_{i=1,4;j=1,\ldots,5;l=1,\ldots,n}$, can be identified with $\widehat{\mathbb{B}}^* := (\boldsymbol{b}_0^*, \ldots, \boldsymbol{b}_n^*, \boldsymbol{b}_{3n+1}^*, \ldots, \boldsymbol{b}_{4n}^*)$, while $\mathbb{B}^* := (\boldsymbol{b}_0^*, \ldots, \boldsymbol{b}_{5n}^*)$ is identified with $\{B_{0,0}^*, B_{0,j}^*, B_{i,0,l}^*, B_{i,j}^*, B_{i,j,l}'^*\}_{i=1,\ldots,5;j=1,\ldots,5;l=1,\ldots,n}$ in Remark 1. Decryption Dec can be alternatively described as:

Dec'(pk, $\mathsf{sk}_{\vec{v}} := (\vec{v}, K_0^*, \{K_{1,j}^*, K_{2,j}^*\}_{j=1,\ldots,5})$, $\mathsf{ct}_{\vec{x}} := (\boldsymbol{c}_1, c_2))$:

$$\boldsymbol{k}^* := (\ \overbrace{K_0^*, v_1 K_{1,1}^*, \ldots, v_{n-1} K_{1,1}^*, K_{2,1}^*}^{n}, \ \ldots, \ \overbrace{v_1 K_{1,5}^*, \ldots, v_{n-1} K_{1,5}^*, K_{2,5}^*}^{n}\ ),$$

that is, $\boldsymbol{k}^* = (1, \overbrace{\sigma\vec{v},}^{n} \overbrace{0^{2n},}^{2n} \overbrace{\eta\vec{v}}^{n} \overbrace{0^n}^{n})_{\mathbb{B}^*}$, $F := e(\boldsymbol{c}_1, \boldsymbol{k}^*),$

return $m' := c_2/F.$

Theorem 2. *The proposed IPE scheme is adaptively attribute-hiding against chosen plaintext attacks under the DLIN assumption.*

6   Comparison

Table 2 compares the proposed IPE schemes in Sections 4 and 5 with existing attribute-hiding IPE schemes in [15,18,16,19].

Table 2. Comparison with IPE schemes in [15,18,16,19], where $|\mathbb{G}|$ and $|\mathbb{G}_T|$ represent size of an element of $\mathbb{G}$ and that of $\mathbb{G}_T$, respectively. AH, PK, SK, CT, GSD, DSP and eDDH stand for attribute-hiding, master public key, secret key, ciphertext, general subgroup decision [1], decisional subspace problem [18], and extended decisional Diffie-Hellman [16], respectively.

	KSW08 [15]	OT09 [18]	LOS$^+$10 [16]	OT10 [19]	Proposed (basic)	Proposed (variant)
Security	selective & fully-AH	selective & weakly-AH	adaptive & weakly-AH	adaptive & weakly-AH	adaptive & fully-AH	adaptive & fully-AH
Order of $\mathbb{G}$	composite	prime	prime	prime	prime	prime
Assump.	2 variants of GSD	2 variants of DSP	n-eDDH	DLIN	DLIN	DLIN
PK size	$O(n)|\mathbb{G}|$	$O(n^2)|\mathbb{G}|$	$O(n^2)|\mathbb{G}|$	$O(n^2)|\mathbb{G}|$	$O(n^2)|\mathbb{G}|$	$O(n)|\mathbb{G}|$
SK size	$(2n+1)|\mathbb{G}|$	$(n+3)|\mathbb{G}|$	$(2n+3)|\mathbb{G}|$	$(3n+2)|\mathbb{G}|$	$(4n+2)|\mathbb{G}|$	$11|\mathbb{G}|$
CT size	$(2n+1)|\mathbb{G}|$ $+|\mathbb{G}_T|$	$(n+3)|\mathbb{G}|$ $+|\mathbb{G}_T|$	$(2n+3)|\mathbb{G}|$ $+|\mathbb{G}_T|$	$(3n+2)|\mathbb{G}|$ $+|\mathbb{G}_T|$	$(4n+2)|\mathbb{G}|$ $+|\mathbb{G}_T|$	$(5n+1)|\mathbb{G}|$ $+|\mathbb{G}_T|$

References

1. Bellare, M., Waters, B., Yilek, S.: Identity-Based Encryption Secure against Selective Opening Attack. In: Ishai, Y. (ed.) TCC 2011. LNCS, vol. 6597, pp. 235–252. Springer, Heidelberg (2011)
2. Bethencourt, J., Sahai, A., Waters, B.: Ciphertext-policy attribute-based encryption. In: IEEE Symposium on Security and Privacy, pp. 321–334. IEEE Computer Society (2007)
3. Boneh, D., Boyen, X.: Efficient selective-ID secure identity-based encryption without random oracles. In: Cachin, Camenisch (eds.) [9], pp. 223–238
4. Boneh, D., Boyen, X.: Secure identity based encryption without random oracles. In: Franklin (ed.) [12], pp. 443–459
5. Boneh, D., Boyen, X., Shacham, H.: Short group signatures. In: Franklin (ed.) [12], pp. 41–55
6. Boneh, D., Crescenzo, G.D., Ostrovsky, R., Persiano, G.: Public key encryption with keyword search. In: Cachin, Camenisch (eds.) [9], pp. 506–522
7. Boneh, D., Franklin, M.: Identity-Based Encryption from the Weil Pairing. In: Kilian, J. (ed.) CRYPTO 2001. LNCS, vol. 2139, pp. 213–229. Springer, Heidelberg (2001)
8. Boneh, D., Waters, B.: Conjunctive, Subset, and Range Queries on Encrypted Data. In: Vadhan, S.P. (ed.) TCC 2007. LNCS, vol. 4392, pp. 535–554. Springer, Heidelberg (2007)
9. Cachin, C., Camenisch, J.L. (eds.): EUROCRYPT 2004. LNCS, vol. 3027. Springer, Heidelberg (2004)
10. Caro, A.D., Iovino, V., Persiano, G.: Hidden vector encryption fully secure against unrestricted queries. IACR Cryptology ePrint Archive 2011, 546 (2011)
11. Cocks, C.: An Identity Based Encryption Scheme Based on Quadratic Residues. In: Honary, B. (ed.) Cryptography and Coding 2001. LNCS, vol. 2260, pp. 360–363. Springer, Heidelberg (2001)

12. Franklin, M. (ed.): CRYPTO 2004. LNCS, vol. 3152. Springer, Heidelberg (2004)
13. Goyal, V., Pandey, O., Sahai, A., Waters, B.: Attribute-based encryption for fine-grained access control of encrypted data. In: Juels, et al. (eds.) [14], pp. 89–98
14. Juels, A., Wright, R.N., di Vimercati, S.D.C. (eds.): Proceedings of the 13th ACM Conference on Computer and Communications Security, CCS 2006, Alexandria, VA, USA, October 30 - November 3. ACM (2006)
15. Katz, J., Sahai, A., Waters, B.: Predicate Encryption Supporting Disjunctions, Polynomial Equations, and Inner Products. In: Smart, N.P. (ed.) EUROCRYPT 2008. LNCS, vol. 4965, pp. 146–162. Springer, Heidelberg (2008)
16. Lewko, A.B., Okamoto, T., Sahai, A., Takashima, K., Waters, B.: Fully Secure Functional Encryption: Attribute-Based Encryption and (Hierarchical) Inner Product Encryption. In: Gilbert, H. (ed.) EUROCRYPT 2010. LNCS, vol. 6110, pp. 62–91. Springer, Heidelberg (2010), http://eprint.iacr.org/2010/110
17. Okamoto, T., Takashima, K.: Homomorphic Encryption and Signatures from Vector Decomposition. In: Galbraith, S.D., Paterson, K.G. (eds.) Pairing 2008. LNCS, vol. 5209, pp. 57–74. Springer, Heidelberg (2008)
18. Okamoto, T., Takashima, K.: Hierarchical Predicate Encryption for Inner-Products. In: Matsui, M. (ed.) ASIACRYPT 2009. LNCS, vol. 5912, pp. 214–231. Springer, Heidelberg (2009)
19. Okamoto, T., Takashima, K.: Fully secure functional encryption with general relations from the decisional linear assumption. In: Rabin, T. (ed.) CRYPTO 2010. LNCS, vol. 6223, pp. 191–208. Springer, Heidelberg (2010), http://eprint.iacr.org/2010/563
20. Okamoto, T., Takashima, K.: Achieving Short Ciphertexts or Short Secret-Keys for Adaptively Secure General Inner-Product Encryption. In: Lin, D., Tsudik, G., Wang, X. (eds.) CANS 2011. LNCS, vol. 7092, pp. 138–159. Springer, Heidelberg (2011), http://eprint.iacr.org/2011/648
21. Okamoto, T., Takashima, K.: Adaptively attribute-hiding (hierarchical) inner product encryption. IACR Cryptology ePrint Archive 2011, 543 (2011), the full version of this paper, http://eprint.iacr.org/2011/543
22. Ostrovsky, R., Sahai, A., Waters, B.: Attribute-based encryption with non-monotonic access structures. In: Ning, P., di Vimercati, S.D.C., Syverson, P.F. (eds.) ACM Conference on Computer and Communications Security, pp. 195–203. ACM (2007)
23. Pirretti, M., Traynor, P., McDaniel, P., Waters, B.: Secure attribute-based systems. In: Juels, V., et al. (eds.) [14], pp. 99–112
24. Sahai, A., Waters, B.: Fuzzy Identity-Based Encryption. In: Cramer, R. (ed.) EUROCRYPT 2005. LNCS, vol. 3494, pp. 457–473. Springer, Heidelberg (2005)
25. Waters, B.: Dual System Encryption: Realizing Fully Secure IBE and HIBE under Simple Assumptions. In: Halevi, S. (ed.) CRYPTO 2009. LNCS, vol. 5677, pp. 619–636. Springer, Heidelberg (2009)

Scalable Group Signatures with Revocation

Benoît Libert[1,*], Thomas Peters[1,**], and Moti Yung[2]

[1] Université Catholique de Louvain, ICTEAM Institute, Belgium
[2] Google Inc. and Columbia University, USA

Abstract. Group signatures are a central cryptographic primitive, simultaneously supporting accountability and anonymity. They allow users
to anonymously sign messages on behalf of a group they are members of.
The recent years saw the appearance of several constructions with security proofs in the standard model (*i.e.*, without appealing to the random
oracle heuristic). For a digital signature scheme to be adopted, an efficient
revocation scheme (as in regular PKI) is absolutely necessary. Despite
over a decade of extensive research, membership revocation remains a
non-trivial problem in group signatures: all existing solutions are not
truly scalable due to either high overhead (e.g., large group public key
size), or limiting operational requirement (the need for all users to follow
the system's entire history). In the standard model, the situation is even
worse as many existing solutions are not readily adaptable. To fill this gap
and tackle this challenge, we describe a new revocation approach based,
perhaps somewhat unexpectedly, on the Naor-Naor-Lotspiech framework
which was introduced for a different problem (namely, that of broadcast
encryption). Our mechanism yields efficient and scalable revocable group
signatures in the standard model. In particular, the size of signatures and
the verification cost are independent of the number of revocations and the
maximal cardinality N of the group while other complexities are at most
polylogarithmic in N. Moreover, the schemes are history-independent:
unrevoked group members do not have to update their keys when a revocation occurs.

Keywords: Group signatures, revocation, standard model, efficiency.

1 Introduction

As suggested by Chaum and van Heyst in 1991 [31], *group signatures* allow
members of a group to anonymously sign messages on behalf of a population
group members managed by a group authority. Using some trapdoor information,
a tracing authority must be able to "open" signatures and identify the signer.
A complex problem in group signatures is the revocation of members whose
signing capability should be disabled (either because they misbehaved or they
intentionally leave the group).

---

* This author was supported by the Belgian Fund for Scientific Research (F.R.S.-
  F.N.R.S.) via a "Chargé de recherches" fellowship.
** Supported by the IUAP B-Crypt Project and the Walloon Region Camus Project.

D. Pointcheval and T. Johansson (Eds.): EUROCRYPT 2012, LNCS 7237, pp. 609–627, 2012.
© International Association for Cryptologic Research 2012

1.1   Related Work

GROUP SIGNATURES WITHOUT REVOCATION. The first efficient and provably coalition-resistant group signature was described by Ateniese, Camenisch, Joye and Tsudik in 2000 [7]. At that time, the security of group signatures was not totally understood and proper security definitions were given later on by Bellare, Micciancio and Warinschi [9] (BMW) whose model captures all the requirements of group signatures in three properties. In (a relaxation of) this model, Boneh, Boyen and Shacham [16] obtained a construction in the random oracle model [10] with signatures shorter than 200 bytes [13].

In the BMW model, the population of users is frozen after the setup phase beyond which no new member can be added. Dynamic group signatures were independently formalized by Kiayias and Yung [42] and Bellare-Shi-Zhang [11]. In these models, pairing-based schemes with relatively short signatures were put forth in [50,32]. Ateniese *et al.* [6] also gave a construction without random oracles using interactive assumptions. In the BMW model [9], Boyen and Waters independently came up with a different standard model proposal [19] using more classical assumptions and they subsequently refined their scheme [20] to obtain constant-size signatures. In the dynamic model [11], Groth [37] described a system with constant-size signatures without random oracles but this scheme was rather a feasibility result than an efficient construction. Later on, Groth gave [38] a fairly efficient realization – with signatures consisting of about 50 group elements – in the standard model with the strongest anonymity level.

REVOCATION. In group signatures, membership revocation has received much attention in the last decade [21,8,28,18] since revocation is central to digital signature schemes. One simple solution is to generate a new group public key and deliver a new signing key to each unrevoked member. However, in large groups, it may be inconvenient to change the public key and send a new secret to signers after they joined the group. An alternative approach taken by Bresson and Stern [21] is to have the signer prove that his membership certificate does not appear in a public list or revoked certificates. Unfortunately, the signer's workload and the size of signatures grow with the number of expelled users.

Song [51] presented an approach handling revocation in forward-secure group signatures. However, verification takes linear time in the number of revocations.

Using accumulators[1] [12], Camenisch and Lysyanskaya [28] proposed a method (followed by [55,26]) to revoke users in the ACJT group signature [7] while keeping $O(1)$ costs for signing and verifying. While elegant, this approach is history-dependent and requires users to keep track of all changes in the population of the group: at each modification of the accumulator value, unrevoked users need to update their membership certificates before signing new messages, which may require up to $O(r)$ exponentiations if r is the number of revoked users.

Brickell [22] suggested the notion of *verifier-local revocation* group signatures, which was formalized by Boneh and Shacham [18] and further studied

---

[1] An accumulator allows hashing a set of values into a short string of constant size while allowing to efficiently prove that a specific value was accumulated.

in [47,56,45]. In their systems, revocation messages are only sent to verifiers (making the signing algorithm independent of the number of revocations). The group manager maintains a revocation list (RL) which is used by verifiers to make sure that signatures were not generated by a revoked member. The RL contains a token for each revoked user and the verification algorithm has to verify signatures w.r.t. each token (a similar revocation mechanism is used in [23]). As a result, the verification cost is inevitably linear in the number of expelled users.

More recently, Nakanishi, Fuji, Hira and Funabiki [46] described a construction with constant complexities for signing/verifying and where group members never have to update their credentials. On the other hand, their proposal has the disadvantage of linear-size group public keys (in the maximal number N of users), although a tweak allows reducing the size to $O(N^{1/2})$.

In the context of anonymous credentials, Tsang et $al.$ [53,54] showed how to blacklist users without compromising their anonymity or involving a trusted third party. Their protocols either have linear proving complexity in the number of revocations or rely on accumulators (which may be problematic for our purposes). Camenisch, Kohlweiss and Soriente [27] handle revocations by periodically updating users credentials in which a specific attribute indicates a validity period. While useful in certain applications of anonymous credentials, in group signatures, their technique would place quite a burden on the group manager who would have to generate updates for each unrevoked individual credential.

1.2   Our Contribution

For the time being and despite years of research efforts, group signatures in the standard model have no revocation mechanism allowing for scalable (*i.e.*, constant or polylogarithmic) verification time without dramatically degrading the efficiency in other metrics and without being history-dependent. In pairing-based group signatures, accumulator-based approaches are unlikely to result in solutions supporting very large groups. The reason is that, in known pairing-based accumulators [49,26], public keys have linear size in the maximal number of accumulated values (unless one sacrifices the constant size of proofs of non-membership as in [5]), which would result in linear-size group public keys in straightforward implementations. Recently [34], Fan et $al.$ suggested a different way to use the accumulator of [26] and announced constant-size group public keys but their scheme still requires the group manager to publicize $O(N)$ values at each revocation. In a revocation mechanism along the lines of [28], Boneh, Boyen and Shacham [16] managed to avoid linear dependencies. However, their technique seems hard to combine[2] with Groth-Sahai proofs [39] so as to work

---

[2] In [16], signing keys consist of pairs $(g^{1/(\omega+s)}, s) \in \mathbb{G} \times \mathbb{Z}_p$, where $\omega \in \mathbb{Z}_p$ is the private key of the group manager, and the revocation mechanism relies on the availability of the exponent $s \in \mathbb{Z}_p$. In the standard model, the Groth-Sahai techniques would require to turn the membership certificates into triples $(g^{1/(\omega+s)}, g^s, u^s)$, for some $u \in \mathbb{G}$ (as in [20]), which is no longer compatible with the revocation technique.

in the standard model. In addition, we would like to save unrevoked users from having to update their keys after each revocation. To this end, it seems possible to adapt the approach of [46] in the standard model. However, merely replacing sigma-protocols by Groth-Sahai proofs in the scheme of [46] would result in group public keys of size $O(N^{1/2})$ in the best case.

In this paper, we describe a novel and scalable revocation technique that interacts nicely with Groth-Sahai proofs and gives constructions in the standard model with $O(1)$ verification cost and at most polylogarithmic complexity in other metrics. Our approach bears similarities with the one of Nakanishi *et al.* [46] in that it does not require users to update their membership certificates at any time but, unlike [46], our group public key size is either $O(\log N)$ or constant. Like the scheme of [46], our main system uses revocation lists (RLs) of size $O(r)$ – which is in line with RLs of standard PKIs – and we emphasize that these are *not* part of the group public key: verifiers only need to know the number of the latest revocation epoch and they do not have to read RLs entirely.

To obtain our constructions, we turn to the area of broadcast encryption and build on the Subset Cover framework of Naor, Naor and Lotspiech [48] (NNL). In a nutshell, the idea is to use the NNL ciphertext as a revocation list and have non-revoked signers prove their ability to decrypt in order to convince verifiers that they are not revoked. In its public-key variant, due to Dodis and Fazio [33], the Subset Cover framework relies on hierarchical identity-based encryption (HIBE) [41,36] and each NNL ciphertext consists of several HIBE encryptions. To anonymously sign a message, we let group members commit to the specific HIBE ciphertext that they can decrypt (which gives constant-size signatures since only one ciphertext is committed to), and provide a non-interactive proof that: (i) they hold a private key which decrypts the committed HIBE ciphertext. (ii) The latter belongs to the revocation list.

By applying this approach to the Subset Difference (SD) method [48], we obtain a scheme with $O(1)$-size signatures, $O(\log N)$-size group public keys, membership certificates of size $O(\log^3 N)$ and revocation lists of size $O(r)$. The Layered Subset Difference method [40] can be used in the same way to obtain membership certificates of size $O(\log^{2.5} N)$. Using the Complete Subtree method, we obtain a tradeoff with $O(r \cdot \log N)$ revocation lists, log-size membership certificates and constant-size group public keys.

A natural question is whether our SD-based revocable group signatures can generically use any HIBE scheme. The answer is negative as the Boneh-Boyen-Goh (BBG) construction [15] is currently the only suitable candidate. Indeed, for anonymity reasons, ciphertexts should be of constant size and our security proof requires the HIBE system to satisfy a new and non-standard security property which is met by [15]. As we will see, the proof can hardly rely on the standard security notion for HIBE schemes [36].

We note that the new revocation mechanism can find applications in contexts other than group signatures. For example, it seems that it can be used in the oblivious transfer with access control protocol of [25], which also uses the technique of Nakanishi *et al.* [46] to revoke credentials.

2 Background

2.1 Bilinear Maps and Complexity Assumptions

We use bilinear maps $e : \mathbb{G} \times \mathbb{G} \to \mathbb{G}_T$ over groups of prime order p where $e(g, h) \neq 1_{\mathbb{G}_T}$ whenever $g, h \neq 1_{\mathbb{G}}$. We assume the hardness of several problems.

Definition 1 ([16]). *The **Decision Linear Problem** (DLIN) in $\mathbb{G}$, is to distinguish the distributions $(g^a, g^b, g^{ac}, g^{bd}, g^{c+d})$ and $(g^a, g^b, g^{ac}, g^{bd}, g^z)$, with $a, b, c, d \xleftarrow{R} \mathbb{Z}_p^*$, $z \xleftarrow{R} \mathbb{Z}_p^*$.*

Definition 2 ([13]). *The q-**Strong Diffie-Hellman Problem** (q-SDH) in $\mathbb{G}$ is, given $(g, g^a, \ldots, g^{(a^q)})$, for some $g \xleftarrow{R} \mathbb{G}$ and $a \xleftarrow{R} \mathbb{Z}_p$, to find a pair $(g^{1/(a+s)}, s) \in \mathbb{G} \times \mathbb{Z}_p$.*

We appeal to yet another "q-type" assumption introduced by Abe *et al.* [2].

Definition 3 ([2]). *In a group $\mathbb{G}$, the q-**Simultaneous Flexible Pairing Problem** (q-SFP) is, given $\left(g_z,\ h_z,\ g_r,\ h_r,\ a,\ \tilde{a},\ b,\ \tilde{b} \in \mathbb{G} \right)$ and $q \in \mathsf{poly}(\lambda)$ tuples $(z_j, r_j, s_j, t_j, u_j, v_j, w_j) \in \mathbb{G}^7$ such that*

$$e(a, \tilde{a}) = e(g_z, z_j) \cdot e(g_r, r_j) \cdot e(s_j, t_j), \qquad e(b, \tilde{b}) = e(h_z, z_j) \cdot e(h_r, u_j) \cdot e(v_j, w_j),$$

to find a new tuple $(z^\star, r^\star, s^\star, t^\star, u^\star, v^\star, w^\star) \in \mathbb{G}^7$ satisfying the above relation and such that $z^\star \neq 1_{\mathbb{G}}$ and $z^\star \neq z_j$ for $j \in \{1, \ldots, q\}$.

2.2 Groth-Sahai Proof Systems

In the following notations, for equal-dimension vectors or matrices A and B containing group elements, $A \odot B$ stands for their entry-wise product.

In their instantiations based on the DLIN assumption, the Groth-Sahai (GS) techniques [39] make use of prime order groups and a common reference string comprising vectors $\vec{f_1}, \vec{f_2}, \vec{f_3} \in \mathbb{G}^3$, where $\vec{f_1} = (f_1, 1, g)$, $\vec{f_2} = (1, f_2, g)$ for some $f_1, f_2 \in \mathbb{G}$. To commit to an element $X \in \mathbb{G}$, one sets $\vec{C} = (1, 1, X) \odot \vec{f_1}^{\,r} \odot \vec{f_2}^{\,s} \odot \vec{f_3}^{\,t}$ with $r, s, t \xleftarrow{R} \mathbb{Z}_p^*$. When the CRS is configured to give perfectly sound proofs, we have $\vec{f_3} = \vec{f_1}^{\,\xi_1} \odot \vec{f_2}^{\,\xi_2}$ where $\xi_1, \xi_2 \in \mathbb{Z}_p^*$. Commitments to group elements $\vec{C} = (f_1^{r+\xi_1 t}, f_2^{s+\xi_2 t}, X \cdot g^{r+s+t(\xi_1+\xi_2)})$ are then Boneh-Boyen-Shacham (BBS) ciphertexts [16] that can be decrypted using $\beta_1 = \log_g(f_1)$, $\beta_2 = \log_g(f_2)$. In the witness indistinguishability (WI) setting, vectors $\vec{f_1}, \vec{f_2}, \vec{f_3}$ are linearly independent and $\vec{C}$ is a perfectly hiding commitment. Under the DLIN assumption, the two kinds of CRS are computationally indistinguishable.

To commit to a scalar $x \in \mathbb{Z}_p$, one computes $\vec{C} = \vec{\varphi}^{\,x} \odot \vec{f_1}^{\,r} \odot \vec{f_2}^{\,s}$, where $r, s \xleftarrow{R} \mathbb{Z}_p^*$, using a CRS comprising vectors $\vec{\varphi}, \vec{f_1}, \vec{f_2}$. In the soundness setting, $\vec{\varphi}, \vec{f_1}, \vec{f_2}$ are linearly independent (typically $\vec{\varphi} = \vec{f_3} \odot (1, 1, g)$ where $\vec{f_3} = \vec{f_1}^{\,\xi_1} \odot \vec{f_2}^{\,\xi_2}$) whereas, in the WI setting, choosing $\vec{\varphi} = \vec{f_1}^{\,\xi_1} \odot \vec{f_2}^{\,\xi_2}$ gives a perfectly hiding commitment since $\vec{C}$ is always a BBS encryption of $1_{\mathbb{G}}$, no matter which exponent x is committed to.

To prove that committed variables satisfy a set of relations, the prover computes one commitment per variable and one proof element (made of a constant number of group elements) per relation.

Such proofs are available for pairing-product equations, which are of the type

$$\prod_{i=1}^{n} e(\mathcal{A}_i, \mathcal{X}_i) \cdot \prod_{i=1}^{n} \cdot \prod_{j=1}^{n} e(\mathcal{X}_i, \mathcal{X}_j)^{a_{ij}} = t_T, \tag{1}$$

for variables $\mathcal{X}_1, \ldots, \mathcal{X}_n \in \mathbb{G}$ and constants $t_T \in \mathbb{G}_T$, $\mathcal{A}_1, \ldots, \mathcal{A}_n \in \mathbb{G}$, $a_{ij} \in \mathbb{Z}_p$, for $i, j \in \{1, \ldots, n\}$. Efficient proofs also exist for multi-exponentiation equations

$$\prod_{i=1}^{m} \mathcal{A}_i^{y_i} \cdot \prod_{j=1}^{n} \mathcal{X}_j^{b_j} \cdot \prod_{i=1}^{m} \cdot \prod_{j=1}^{n} \mathcal{X}_j^{y_i \gamma_{ij}} = T, \tag{2}$$

for variables $\mathcal{X}_1, \ldots, \mathcal{X}_n \in \mathbb{G}$, $y_1, \ldots, y_m \in \mathbb{Z}_p$ and constants $T, \mathcal{A}_1, \ldots, \mathcal{A}_m \in \mathbb{G}$, $b_1, \ldots, b_n \in \mathbb{Z}_p$ and $\gamma_{ij} \in \mathbb{G}$, for $i \in \{1, \ldots, m\}, j \in \{1, \ldots, n\}$.

In pairing-product equations, proofs for quadratic equations require 9 group elements whereas linear equations (*i.e.*, where $a_{ij} = 0$ for all i, j in equation (1)) only take 3 group elements each. Linear multi-exponentiation equations of the type $\prod_{i=1}^{m} \mathcal{A}_i^{y_i} = T$ demand 2 group elements.

Multi-exponentiation equations admit zero-knowledge (NIZK) proofs at no additional cost. On a simulated CRS (prepared for the WI setting), a trapdoor makes it is possible to simulate proofs without knowing witnesses and simulated proofs have the same distribution as real proofs.

2.3   Structure-Preserving Signatures

Several applications (see [2,3,35,30,4] for examples) require to sign groups elements while preserving the feasibility of efficiently proving that a committed signature is valid for a committed group element.

In [2,3], Abe, Haralambiev and Ohkubo showed how to conveniently sign n group elements at once using signatures consisting of $O(1)$ group elements. Their scheme (which is referred to as the AHO signature in the paper) makes use of bilinear groups of prime order. In the context of symmetric pairings, the description below assumes public parameters $\mathsf{pp} = \big((\mathbb{G}, \mathbb{G}_T),\ g\big)$ consisting of groups $(\mathbb{G}, \mathbb{G}_T)$ of order $p > 2^\lambda$, where $\lambda \in \mathbb{N}$ is a security parameter, with a bilinear map $e : \mathbb{G} \times \mathbb{G} \to \mathbb{G}_T$ and a generator $g \in \mathbb{G}$.

Keygen(pp, n): given an upper bound $n \in \mathbb{N}$ on the number of group elements that can be signed altogether, choose generators $G_r, H_r \xleftarrow{R} \mathbb{G}$. Pick $\gamma_z, \delta_z \xleftarrow{R} \mathbb{Z}_p$ and $\gamma_i, \delta_i \xleftarrow{R} \mathbb{Z}_p$, for $i = 1$ to n. Then, compute $G_z = G_r^{\gamma_z}$, $H_z = H_r^{\delta_z}$ and $G_i = G_r^{\gamma_i}$, $H_i = H_r^{\delta_i}$ for each $i \in \{1, \ldots, n\}$. Finally, choose $\alpha_a, \alpha_b \xleftarrow{R} \mathbb{Z}_p$ and define $A = e(G_r, g^{\alpha_a})$ and $B = e(H_r, g^{\alpha_b})$. The public key is

$$pk = \big(G_r,\ H_r,\ G_z,\ H_z,\ \{G_i, H_i\}_{i=1}^{n},\ A,\ B\big) \in \mathbb{G}^{2n+4} \times \mathbb{G}_T^2$$

while the private key consists of $sk = \big(\alpha_a, \alpha_b, \gamma_z, \delta_z, \{\gamma_i, \delta_i\}_{i=1}^{n}\big)$.

Sign$(sk, (M_1, \ldots, M_n))$: to sign a vector $(M_1, \ldots, M_n) \in \mathbb{G}^n$ using the private key $sk = (\alpha_a, \alpha_b, \gamma_z, \delta_z, \{\gamma_i, \delta_i\}_{i=1}^n)$, choose $\zeta, \rho, \tau, \nu, \omega \xleftarrow{R} \mathbb{Z}_p$ and compute $\theta_1 = g^\zeta$ as well as

$$\theta_2 = g^{\rho - \gamma_z \zeta} \cdot \prod_{i=1}^n M_i^{-\gamma_i}, \qquad \theta_3 = G_r^\tau, \qquad \theta_4 = g^{(\alpha_a - \rho)/\tau},$$

$$\theta_5 = g^{\nu - \delta_z \zeta} \cdot \prod_{i=1}^n M_i^{-\delta_i}, \qquad \theta_6 = H_r^\omega, \qquad \theta_7 = g^{(\alpha_b - \nu)/\omega},$$

The signature consists of $\sigma = (\theta_1, \theta_2, \theta_3, \theta_4, \theta_5, \theta_6, \theta_7)$.

Verify$(pk, \sigma, (M_1, \ldots, M_n))$: parse σ as $(\theta_1, \theta_2, \theta_3, \theta_4, \theta_5, \theta_6, \theta_7) \in \mathbb{G}^7$ and return 1 iff these equalities hold:

$$A = e(G_z, \theta_1) \cdot e(G_r, \theta_2) \cdot e(\theta_3, \theta_4) \cdot \prod_{i=1}^n e(G_i, M_i), \qquad (3)$$

$$B = e(H_z, \theta_1) \cdot e(H_r, \theta_5) \cdot e(\theta_6, \theta_7) \cdot \prod_{i=1}^n e(H_i, M_i). \qquad (4)$$

The scheme was proved [2,3] existentially unforgeable under chosen-message attacks under the q-SFP assumption, where q is the number of signing queries.

Abe *et al.* [2,3] also showed that signatures can be publicly randomized to obtain a different signature $\{\theta_i'\}_{i=1}^7 \leftarrow \mathsf{ReRand}(pk, \sigma)$ on $(M_1, \ldots, M_n)$. After randomization, we have $\theta_1' = \theta_1$ while $\{\theta_i'\}_{i=2}^7$ are uniformly distributed among the values satisfying the equalities $e(G_r, \theta_2') \cdot e(\theta_3', \theta_4') = e(G_r, \theta_2) \cdot e(\theta_3, \theta_4)$ and $e(H_r, \theta_5') \cdot e(\theta_6', \theta_7') = e(H_r, \theta_5) \cdot e(\theta_6, \theta_7)$. Moreover, $\{\theta_i'\}_{i \in \{3,4,6,7\}}$ are statistically independent of $(M_1, \ldots, M_n)$ and the rest of the signature. This implies that, in anonymity-related protocols, re-randomized $\{\theta_i'\}_{i \in \{3,4,6,7\}}$ can be safely revealed as long as $(M_1, \ldots, M_n)$ and $\{\theta_i'\}_{i \in \{1,2,5\}}$ are given in committed form.

In [4], Abe, Groth, Haralambiev and Ohkubo described a more efficient structure-preserving signature based on interactive assumptions. Here, we use the scheme of [2,3] so as to rely on non-interactive assumptions.

2.4   The NNL Framework for Broadcast Encryption

The Subset Cover framework [48] considers secret-key broadcast encryption schemes with $N = 2^\ell$ registered receivers. Each one of them is associated with a leaf of a complete binary tree T of height ℓ and each tree node is assigned a secret key. If $\mathcal{N}$ denotes the universe of users and $\mathcal{R} \subset \mathcal{N}$ is the set of revoked receivers, the idea of the framework is to partition the set of non-revoked users into m disjoint subsets $S_1, \ldots, S_m$ such that $\mathcal{N} \backslash \mathcal{R} = S_1 \cup \ldots \cup S_m$. Depending on the way to partition $\mathcal{N} \backslash \mathcal{R}$ and the distribution of keys to users, different instantiations and tradeoffs are possible.

THE COMPLETE SUBTREE METHOD. In this technique, each subset S_i consists of the leaves of a complete subtree rooted at some node x_i of T. Upon registration, each user obtains secret keys for all nodes on the path connecting his leaf to the root of T (and thus $O(\ell)$ keys overall). By doing so, users in $\mathcal{N} \backslash \mathcal{R}$ can decrypt the content if the latter is enciphered using symmetric keys $K_1, \ldots, K_m$ corresponding to the roots of subtrees $S_1, \ldots, S_m$. As showed in [48], the CS partitioning method entails at most $m \leq r \cdot \log(N/r)$ subsets, where $r = |\mathcal{R}|$. Each transmission requires to send $O(r \cdot \log N)$ symmetric encryptions while, at each user, the storage complexity is $O(\log N)$.

As noted in [48,33], a single-level identity-based encryption scheme allows implementing a public-key variant of the CS method. The master public key of the IBE scheme forms the public key of the broadcast encryption system, which allows for public keys of size $O(1)$ (instead of $O(N)$ in instantiations using ordinary public-key encryption). When users join the system, they obtain $O(\ell)$ IBE private keys (in place of symmetric keys) associated with the "identities" of nodes on the path between their leaf and the root.

THE SUBSET DIFFERENCE METHOD. The SD method reduces the transmission cost to $O(r)$ at the expense of increased storage requirements. For each node $x_j \in$ T, we call T_{x_j} the subtree rooted at x_j. The set $\mathcal{N} \backslash \mathcal{R}$ is now divided into disjoint subsets $S_{k_1,u_1}, \ldots, S_{k_m,u_m}$. For each $i \in \{1, \ldots, m\}$, the subset S_{k_i,u_i} is determined by a node x_{k_i} and one of its descendants x_{u_i} – which are called *primary* and *secondary* roots of S_{k_i,u_i}, respectively – and it consists of the leaves of $\mathsf{T}_{x_{k_i}}$ that are not in $\mathsf{T}_{x_{u_i}}$. Each user thus belongs to much more generic subsets than in the CS method and this allows reducing the maximal number of subsets to $m = 2r - 1$ (see [48] for a proof of this bound).

A more complex key distribution is necessary here. Each subset S_{k_i,u_i} is assigned a "proto-key" $P_{x_{k_i},x_{u_i}}$ that allows deriving the actual symmetric encryption key K_{k_i,u_i} for S_{k_i,u_i} and as well as proto-keys $P_{x_{k_i},x_{u_l}}$ for any descendant x_{u_l} of x_{u_i}. At the same time, $P_{x_{k_i},x_{u_l}}$ should be hard to compute without a proto-key $P_{x_{k_i},x_{u_i}}$ for an ancestor x_{u_i} of x_{u_l}. The key distribution phase then proceeds as follows. Let user i be assigned a leaf v_i and let $\epsilon = x_0, x_1, \ldots, x_\ell = v_i$ denote the path from the root ϵ to v_i. For each subtree T_{x_j} (with $j \in \{1, \ldots, \ell\}$), if copath$_{x_j}$ denotes the set of all siblings of nodes on the path from x_j to v_i, user i must obtain proto-keys $P_{x_j,w}$ for each node $w \in$ copath$_{x_j}$ because he belongs to the generic subset whose primary root is x_j and whose secondary root is w. By storing $O(\ell^2)$ proto-keys (i.e., $O(\ell)$ for each subtree T_{x_j}), users will be able to derive keys for all generic subsets they belong to.

In [33], Dodis and Fazio extended the SD method to the public-key setting using hierarchical identity-based encryption. In the tree, each node w at depth $\leq \ell$ has a label $\langle w \rangle$ which is defined by assigning the label ε to the root (at depth 0). The left and right children of w are then labeled with $\langle w \rangle \| 0$ and $\langle w \rangle \| 1$, respectively. For each subset S_{k_i,u_i} of $\mathcal{N} \backslash \mathcal{R}$, the sender considers the primary and secondary roots x_{k_i}, x_{u_i} and parses the label $\langle x_{u_i} \rangle$ as $\langle x_{k_i} \rangle \| u_{i,\ell_{i,1}} \ldots u_{i,\ell_{i,2}}$, with $u_{i,j} \in \{0,1\}$ for each $j \in \{\ell_{i,1}, \ldots, \ell_{i,2}\}$. Then, he computes a HIBE ciphertext for the hierarchical identity $(\langle x_{k_i} \rangle, u_{i,\ell_{i,1}}, \ldots, u_{i,\ell_{i,2}})$ at level $\ell_{i,2} - \ell_{i,1} + 2$. Upon registration, if $\epsilon = x_0, \ldots, x_\ell = v_i$ denotes the path from the root to his leaf v_i, for each subtree T_{x_j}, user i receives exactly one HIBE private key for each $w \in \mathsf{copath}_{x_j}$: namely, for each $w \in \mathsf{copath}_{x_j}$, there exist $\ell_1, \ell_2 \in \{1, \ldots, \ell\}$ such that $\langle w \rangle = \langle x_j \rangle || w_{\ell_1} \ldots w_{\ell_2}$ with $w_j \in \{0, 1\}$ for all $j \in \{\ell_1, \ldots, \ell_2\}$ and user i obtains a HIBE private key for the hierarchical identity $(\langle x_j \rangle, w_{\ell_1}, \ldots, w_{\ell_2})$. By construction, this key will allow user i to decrypt any HIBE ciphertext encrypted for a subset whose primary root is x_j and whose secondary root is a descendant of w. Overall, each user thus has to store $O(\log^2 N)$ HIBE private keys.

2.5   Revocable Group Signatures

We consider schemes that have their lifetime divided into revocation epochs at the beginning of which group managers update their revocation lists.

The syntax and the security model are similar to [46] but they build on those defined by Kiayias and Yung [42]. Like the Bellare-Shi-Zhang model [11], the latter assumes an interactive join protocol between the group manager and the user. This protocol provides the user with a membership certificate and a membership secret. Such protocols may consist of several rounds of interaction.

SYNTAX. We denote by $N \in \mathsf{poly}(\lambda)$ the maximal number of group members. At the beginning of each revocation epoch t, the group manager publicizes an up-to-date revocation list RL_t and we denote by $\mathcal{R}_t \subset \{1, \ldots, N\}$ the corresponding set of revoked users (we assume that $\mathcal{R}_t$ is part of RL_t). A revocable group signature (R-GS) scheme consists of the following algorithms or protocols.

Setup(λ, N): given a security parameter $\lambda \in \mathbb{N}$ and a maximal number of members $N \in \mathbb{N}$, this algorithm (which is run by a trusted party) generates a group public key $\mathcal{Y}$, the group manager's private key $\mathcal{S}_{\mathsf{GM}}$ and the opening authority's private key $\mathcal{S}_{\mathsf{OA}}$. $\mathcal{S}_{\mathsf{GM}}$ and $\mathcal{S}_{\mathsf{OA}}$ are given to the appropriate authority while $\mathcal{Y}$ is publicized. The algorithm initializes a public state St containing a set data structure $St_{users} = \emptyset$ and a string structure $St_{trans} = \epsilon$.

Join: is an interactive protocol between the group manager GM and a user $\mathcal{U}_i$ where the latter becomes a group member. The protocol involves two interactive Turing machines $\mathsf{J}_{\mathsf{user}}$ and J_{GM} that both take as input $\mathcal{Y}$. The execution, denoted as $[\mathsf{J}_{\mathsf{user}}(\lambda, \mathcal{Y}), \mathsf{J}_{\mathsf{GM}}(\lambda, St, \mathcal{Y}, \mathcal{S}_{\mathsf{GM}})]$, terminates with user $\mathcal{U}_i$ obtaining a membership secret sec_i, that no one else knows, and a membership certificate cert_i. If the protocol successfully terminates, the group manager updates the public state St by setting $St_{users} := St_{users} \cup \{i\}$ as well as $St_{trans} := St_{trans} || \langle i, \mathsf{transcript}_i \rangle$.

Revoke: is a (possibly randomized) algorithm allowing the GM to generate an updated revocation list RL_t for the new revocation epoch t. It takes as input a public key $\mathcal{Y}$ and a set $\mathcal{R}_t \subset St_{users}$ that identifies the users to be revoked. It outputs an updated revocation list RL_t for epoch t.

Sign: given a revocation epoch t with its revocation list RL_t, a membership certificate cert_i, a membership secret sec_i and a message M, this algorithm outputs $\perp$ if $i \in \mathcal{R}_t$ and a signature σ otherwise.

Verify: given a signature σ, a revocation epoch t, the corresponding revocation list RL_t, a message M and a group public key $\mathcal{Y}$, this deterministic algorithm returns either 0 or 1.

Open: takes as input a message M, a valid signature σ w.r.t. $\mathcal{Y}$ for the indicated revocation epoch t, the opening authority's private key $\mathcal{S}_{\mathsf{OA}}$ and the public state St. It outputs $i \in St_{users} \cup \{\perp\}$, which is the identity of a group member or a symbol indicating an opening failure.

Each membership certificate contains a unique tag that identifies the user.

A R-GS scheme must satisfy three security notions, that are formally defined in the full version of the paper. The first one is called *security against misidentification attacks*. It requires that, even if the adversary can introduce and revoke users at will, it cannot produce a signature that traces outside the set of unrevoked adversarially-controlled users.

As in ordinary (*i.e.*, non-revocable) group signatures, the notion of *security against framing attacks* mandates that, even if the whole system colludes against a user, that user will not bear responsibility for messages that he did not sign. Finally, the notion of *anonymity* is also defined (in the presence of a signature opening oracle) as in the models of [11,42].

3   A Revocable Group Signature Based on the Subset Difference Method

The idea is to turn the NNL global ciphertext into a revocation list in the group signature. Each member is assigned to a leaf of a binary tree of height ℓ and the outcome of the join protocol is the user obtaining a membership certificate that contains the same key material as in the public-key variant of the SD method (*i.e.*, $O(\ell^2)$ HIBE private keys). To ensure traceability and non-frameability, these NNL private keys are linked to a group element X, that only the user knows the discrete logarithm of, by means of structure-preserving signatures.

At each revocation epoch t, the group manager generates an up-to-date revocation list RL_t consisting of $O(r)$ HIBE ciphertexts, each of which is signed using a structure-preserving signature. When it comes to sign a message, the user $\mathcal{U}_i$ proves that he is not revoked by providing evidence that he is capable of decrypting one of the HIBE ciphertexts in RL_t. To this end, $\mathcal{U}_i$ commits to that HIBE ciphertext C_l and proves that he holds a key that decrypts C_l. To convince the verifier that C_l belongs to RL_t, he proves knowledge of a signature on the committed HIBE ciphertext C_l (this technique is borrowed from the set membership proofs of [52,24]). Of course, to preserve the anonymity of signers, we need a HIBE scheme with constant-size ciphertexts (otherwise, the length of the committed ciphertext could betray the signer's location in the tree), which is why the Boneh-Boyen-Goh construction [15] is the ideal candidate.

The scheme is made anonymous and non-frameable using the same techniques as Groth [38] in steps 4-6 of the signing algorithm. As for the security against misidentification attacks, we cannot prove it by relying on the standard collusion-resistance (captured by the definition of [36]) of the HIBE scheme. In the proof of security against misidentification attacks, the problem appears in the treatment of forgeries that open to a revoked user: while this user cannot have obtained a private key that decrypts the committed HIBE ciphertext of the forgery (because he is revoked), unrevoked adversarially-controlled users can. To solve this problem, we need to rest on a non-standard security property (formally defined in the full version of the paper) called "key-robustness". This notion asks that, given a private key generated for some hierarchical identity using specific random coins, it be infeasible to compute the private key of a different identity for the *same random coins* and even *knowing* the *master secret key* of the HIBE scheme. While unusual, this property can be proved (as shown in the full version of the paper) under the Diffie-Hellman assumption for the BBG construction.

Perhaps surprisingly, even though we rely on the BBG HIBE, we do not need its underlying q-type assumption [15]. The reason is that the master secret key of the scheme is unnecessary here as its role is taken over by the private key of a structure-preserving signature. In the ordinary BBG system, private keys contain components of the form $(g_2^{\alpha} \cdot F(\mathsf{ID})^r, g^r)$, for some $r \in \mathbb{Z}_p$, where g_2^{α} is the master secret key and $F(\mathsf{ID})$ is a function of the hierarchical identity. In the join protocol, the master key g_2^{α} disappears: the user obtains a private key of the form $(F(\mathsf{ID})^r, g^r)$ and an AHO signature is used to bind the user's membership public key X to g^r. The latter can be thought of as a public key for a one-time variant of the Boneh-Lynn-Shacham signature [17]. The underlying one-time private key $r \in \mathbb{Z}_p$ is used to compute $F(\mathsf{ID})^r$ as well as a number of delegation components allowing to derive signatures for messages that ID is a prefix of (somewhat in the fashion of wildcard signatures [1][Section 6]).

3.1   Construction

As in Section 2.4, $\langle x \rangle$ denotes the label of node $x \in \mathsf{T}$ and, for any sub-tree T_{x_j} rooted at x_j and any leaf v_i of T_{x_j}, copath_{x_j} denotes the set of all siblings of nodes on the path from x_j to v_i, not counting x_j itself.

As is standard in group signatures, the description below assumes that, before joining the group, user $\mathcal{U}_i$ chooses a long term key pair $(\mathsf{usk}[i], \mathsf{upk}[i])$ and registers it in some PKI.

Setup(λ, N): given $\lambda \in \mathbb{N}$ and the permitted number of users $N = 2^{\ell}$,

1. Choose bilinear groups $(\mathbb{G}, \mathbb{G}_T)$ of prime order $p > 2^{\lambda}$, with $g \xleftarrow{R} \mathbb{G}$.
2. Generate two key pairs $(sk_{\mathsf{AHO}}^{(0)}, pk_{\mathsf{AHO}}^{(0)})$ and $(sk_{\mathsf{AHO}}^{(1)}, pk_{\mathsf{AHO}}^{(1)})$ for the AHO signature to sign messages of two group elements. These pairs consist of

$$pk_{\mathsf{AHO}}^{(d)} = \left(G_r^{(d)}, \ H_r^{(d)}, \ G_z^{(d)} = G_r^{\gamma_z^{(d)}}, \ H_z^{(d)} = H_r^{\delta_z^{(d)}}, \right.$$

$$\left. \{G_i^{(d)} = G_r^{\gamma_i^{(d)}}, H_i^{(d)} = H_r^{\delta_i^{(d)}}\}_{i=1}^{2}, \ A^{(d)}, \ B^{(d)} \right)$$

and $sk_{\mathsf{AHO}}^{(d)} = \left(\alpha_a^{(d)}, \alpha_b^{(d)}, \gamma_z^{(d)}, \delta_z^{(d)}, \{\gamma_i^{(d)}, \delta_i^{(d)}\}_{i=1}^{2} \right)$, where $d \in \{0, 1\}$.

3. As a CRS for the NIWI proof system, select vectors $\mathbf{f} = (\vec{f}_1, \vec{f}_2, \vec{f}_3)$ s.t.
   $\vec{f}_1 = (f_1, 1, g) \in \mathbb{G}^3$, $\vec{f}_2 = (1, f_2, g) \in \mathbb{G}^3$, and $\vec{f}_3 = \vec{f}_1^{\,\xi_1} \cdot \vec{f}_2^{\,\xi_2}$, with
   $f_1 = g^{\beta_1}, f_2 = g^{\beta_2} \xleftarrow{R} \mathbb{G}$ and $\beta_1, \beta_2, \xi_1, \xi_2 \xleftarrow{R} \mathbb{Z}_p^*$.
4. Choose $(U, V) \xleftarrow{R} \mathbb{G}^2$ that, together with f_1, f_2, g, will form a public
   encryption key.
5. Generate a master public key mpk_{BBG} for the Boneh-Boyen-Goh HIBE.
   Such a public key consists[3] of $mpk_{\mathsf{BBG}} = \left(\{h_i\}_{i=0}^{\ell} \right)$, where $\ell = \log_2(N)$,
   and no master secret key is needed.
6. Select an injective encoding[4] function $\mathcal{H} : \{0, 1\}^{\leq \ell} \to \mathbb{Z}_p^*$ and a strongly
   unforgeable one-time signature $\Sigma = (\mathcal{G}, \mathcal{S}, \mathcal{V})$.
7. Set $\mathcal{S}_{\mathsf{GM}} := \left(sk_{\mathsf{AHO}}^{(0)}, sk_{\mathsf{AHO}}^{(1)} \right)$, $\mathcal{S}_{\mathsf{OA}} := (\beta_1, \beta_2)$ as authorities' private keys
   and the group public key is

$$\mathcal{Y} := \left(g, \; pk_{\mathsf{AHO}}^{(0)}, \; pk_{\mathsf{AHO}}^{(1)}, \; mpk_{\mathsf{BBG}}, \; \mathbf{f}, \; (U, V), \; \mathcal{H}, \; \Sigma \right).$$

Join$^{(\mathsf{GM}, \mathcal{U}_i)}$: the GM and the prospective user $\mathcal{U}_i$ run the following protocol
$[\mathsf{J}_{\mathsf{user}}(\lambda, \mathcal{Y}), \mathsf{J}_{\mathsf{GM}}(\lambda, St, \mathcal{Y}, \mathcal{S}_{\mathsf{GM}})]$:

1. $\mathsf{J}_{\mathsf{user}}(\lambda, \mathcal{Y})$ computes $X = g^x$, for a randomly chosen $x \xleftarrow{R} \mathbb{Z}_p$, and sends
   it to $\mathsf{J}_{\mathsf{GM}}(\lambda, St, \mathcal{Y}, \mathcal{S}_{\mathsf{GM}})$. If the value X already appears in some entry
   $\mathsf{transcript}_j$ of the database St_{trans}, J_{GM} aborts and returns $\perp$ to $\mathsf{J}_{\mathsf{user}}$.
2. J_{GM} assigns to $\mathcal{U}_i$ an available leaf v_i of label $\langle v_i \rangle = v_{i,1} \ldots v_{i,\ell} \in \{0, 1\}^{\ell}$
   in the tree T. Let $x_0 = \epsilon, x_1, \ldots, x_{\ell-1}, x_{\ell} = v_i$ be the path from v_i to
   the root ϵ of T. For $j = 0$ to ℓ, J_{GM} does the following.

   a. Consider the sub-tree T_{x_j} rooted at node x_j. Let copath_{x_j} be the
      co-path from x_j to v_i.
   b. For each node $w \in \mathsf{copath}_{x_j}$, since x_j is an ancestor of w, $\langle x_j \rangle$ is a
      prefix of $\langle w \rangle$ and we denote by $w_{\ell_1} \ldots w_{\ell_2} \in \{0, 1\}^{\ell_2 - \ell_1 + 1}$, for some
      $\ell_1 \leq \ell_2 \leq \ell$, the suffix of $\langle w \rangle$ coming right after $\langle x_j \rangle$.
   b.1 Choose a random $r \xleftarrow{R} \mathbb{Z}_p$ and compute a HIBE private key

$$d_w = (D_{w,1}, D_{w,2}, K_{w,\ell_2-\ell_1+3}, \ldots, K_{w,\ell})$$
$$= \left(\left(h_0 \cdot h_1^{\mathcal{H}(\langle x_j \rangle)} \cdot h_2^{\mathcal{H}(w_{\ell_1})} \cdots h_{\ell_2-\ell_1+2}^{\mathcal{H}(w_{\ell_2})} \right)^r, g^r, h_{\ell_2-\ell_1+3}^r, \ldots, h_{\ell}^r \right)$$

   for the identity $(\mathcal{H}(\langle x_j \rangle), \mathcal{H}(w_{\ell_1}), \ldots, \mathcal{H}(w_{\ell_2})) \in (\mathbb{Z}_p^*)^{\ell_2-\ell_1+2}$.
   b.2 Using $sk_{\mathsf{AHO}}^{(0)}$, generate an AHO signature $\sigma_w = (\theta_{w,1}, \ldots, \theta_{w,7})$
       on $(X, D_{w,2}) \in \mathbb{G}^2$ so as to bind the HIBE private key d_w to the
       value X that identifies $\mathcal{U}_i$.

---

[3] In comparison with the original HIBE scheme where mpk_{BBG} includes $(g_1 = g^{\alpha}, g_2)$
and $msk_{\mathsf{BBG}} = g_2^{\alpha}$, the public elements g_1 and g_2 have disappeared.

[4] This encoding allows making sure that "identities" will be non-zero at each level.
Since the set $\{0, 1\}^{\leq \ell}$ is of cardinality $\sum_{i=0}^{\ell} 2^i = 2^{\ell+1} - 1 < p - 1$, such a function
can be efficiently constructed without any intractability assumption.

3. J_{GM} sends $\langle v_i \rangle \in \{0,1\}^{\ell}$, and the HIBE private keys $\{\{d_w\}_{w \in \mathsf{copath}_{x_j}}\}_{j=0}^{\ell}$ to J_{user} that verifies their validity. If these keys are all well-formed, J_{user} acknowledges them by generating an ordinary digital signature $sig_i = \mathsf{Sign}_{\mathsf{usk}[i]}(X\|\{\{d_w\}_{w \in \mathsf{copath}_{x_j}}\}_{j=0}^{\ell})$ and sends it back to J_{GM}.

4. J_{GM} checks that $\mathsf{Verify}_{\mathsf{upk}[i]}(X\|\{\{d_w\}_{w \in \mathsf{copath}_{x_j}}\}_{j=0}^{\ell}, sig_i) = 1$. If not, then J_{GM} aborts. Otherwise, J_{GM} returns the set of AHO signatures $\{\{\sigma_w\}_{w \in \mathsf{copath}_{x_j}}\}_{j=0}^{\ell}$ to J_{user} and stores the entire conversation transcript $\mathsf{transcript}_i = (X, \{\{d_w, \sigma_w\}_{w \in \mathsf{copath}_{x_j}}\}_{j=0}^{\ell}, sig_i)$ in the database St_{trans}.

5. J_{user} defines user $\mathcal{U}_i$'s membership certificate cert_i to be the tuple $\mathsf{cert}_i = (\langle v_i \rangle, \{\{d_w, \sigma_w\}_{w \in \mathsf{copath}_{x_j}}\}_{j=0}^{\ell}, X)$, where X will serve as the tag that identifies $\mathcal{U}_i$. The membership secret sec_i is defined to be $\mathsf{sec}_i = x$.

Revoke$(\mathcal{Y}, \mathcal{S}_{\mathsf{GM}}, t, \mathcal{R}_t)$: Parse $\mathcal{S}_{\mathsf{GM}}$ as $\mathcal{S}_{\mathsf{GM}} := (sk_{\mathsf{AHO}}^{(0)}, sk_{\mathsf{AHO}}^{(1)})$. Using the SD covering algorithm, find a cover of the unrevoked user set $\{1, \ldots, N\} \backslash \mathcal{R}_t$ as the union of disjoint subsets $S_{k_1, u_1}, \ldots, S_{k_m, u_m}$, with $m \leq 2 \cdot |\mathcal{R}_t| - 1$. Then, for $i = 1$ to m, do the following.

a. Consider S_{k_i, u_i} as the difference between sub-trees rooted at an internal node x_{k_i} and one of its descendants x_{u_i}. The label of x_{u_i} can be written $\langle x_{u_i} \rangle = \langle x_{k_i} \rangle \| u_{i, \ell_{i,1}} \ldots u_{i, \ell_{i,2}}$ for some $\ell_{i,1} < \ell_{i,2} \leq \ell$ and where $u_{i,\kappa} \in \{0, 1\}$ for each $\kappa \in \{\ell_{i,1}, \ldots, \ell_{i,2}\}$. Then, compute an encoding of S_{k_i, u_i} as a group element

$$C_i = h_0 \cdot h_1^{\mathcal{H}(\langle x_{k_i} \rangle)} \cdot h_2^{\mathcal{H}(u_{i, \ell_{i,1}})} \cdots h_{\ell_{i,2} - \ell_{i,1} + 2}^{\mathcal{H}(u_{i, \ell_{i,2}})},$$

which can be seen as a de-randomized HIBE ciphertext for the hierarchical identity $(\mathcal{H}(\langle x_{k_i} \rangle), \mathcal{H}(u_{i, \ell_{i,1}}), \ldots, \mathcal{H}(u_{i, \ell_{i,2}})) \in (\mathbb{Z}_p^*)^{\ell_{i,2} - \ell_{i,1} + 2}$.

b. To authenticate the HIBE ciphertext C_i and bind it to the revocation epoch t, use $sk_{\mathsf{AHO}}^{(1)}$ to generate an AHO signature $\Theta_i = (\Theta_{i,1}, \ldots, \Theta_{i,7}) \in \mathbb{G}^7$ on the pair $(C_i, g^t) \in \mathbb{G}^2$, where the epoch number t is interpreted as an element of $\mathbb{Z}_p$.

Return the revocation data RL_t which is defined to be

$$RL_t = \left(t, \ \mathcal{R}_t, \ \{\langle x_{k_i} \rangle, \ \langle x_{u_i} \rangle, \ (C_i, \Theta_i = (\Theta_{i,1}, \ldots, \Theta_{i,7}))\}_{i=1}^{m} \right) \qquad (5)$$

Sign$(\mathcal{Y}, t, RL_t, \mathsf{cert}_i, \mathsf{sec}_i, M)$: return $\perp$ if $i \in \mathcal{R}_t$. Otherwise, to sign $M \in \{0,1\}^*$, generate a one-time signature key pair $(\mathsf{SK}, \mathsf{VK}) \leftarrow \mathcal{G}(\lambda)$. Parse cert_i as $(\langle v_i \rangle, \{\{(d_w, \sigma_w)\}_{w \in \mathsf{copath}_{x_j}}\}_{j=0}^{\ell}, X)$ and sec_i as $x \in \mathbb{Z}_p$.

1. Using RL_t, determine the set S_{k_l, u_l}, with $l \in \{1, \ldots, m\}$, that contains the leaf v_i (this subset must exist since $i \notin \mathcal{R}_t$) and let x_{k_l} and x_{u_l} denote the primary and secondary roots of S_{k_l, u_l}. Since x_{k_l} is an ancestor of x_{u_l}, we can write $\langle x_{u_l} \rangle = \langle x_{k_l} \rangle \| u_{l, \ell_1} \ldots u_{l, \ell_2}$, for some $\ell_1 < \ell_2 \leq \ell$ and with $u_{l,\kappa} \in \{0, 1\}$ for each $\kappa \in \{\ell_1, \ldots, \ell_2\}$. The signer $\mathcal{U}_i$ computes a HIBE decryption key of the form

$$(D_{l,1}, D_{l,2}) = \left(\left(h_0 \cdot h_1^{\mathcal{H}(\langle x_{k_l} \rangle)} \cdot h_2^{\mathcal{H}(u_{l, \ell_1})} \cdots h_{\ell_2 - \ell_1 + 2}^{\mathcal{H}(u_{l, \ell_2})} \right)^r, \ g^r \right). \qquad (6)$$

This is possible since, if we denote by $\langle x_{k,l}\rangle \| u_{l,\ell_1}\ldots u_{l,\ell_1'}$ the shortest prefix of $\langle x_{u_l}\rangle$ that is not a prefix of $\langle v_i\rangle$, the key material $\{d_w\}_{w\in\mathrm{copath}_{x_{k_l}}}$ corresponding to the sub-tree rooted at x_{k_l} contains a HIBE private key $d_w = (D_{w,1}, D_{w,2}, K_{w,\ell_1'-\ell_1+3},\ldots,K_{w,\ell})$ such that

$$d_w = \left(\left(h_0\cdot h_1^{\mathcal{H}(\langle x_{k_l}\rangle)}\cdot h_2^{\mathcal{H}(u_{l,\ell_1})}\cdots h_{\ell_1'-\ell_1+2}^{\mathcal{H}(u_{l,\ell_1'})}\right)^r,\ g^r,\ h_{\ell_1'-\ell_1+3}^r,\ldots,h_\ell^r\right),$$

which allows deriving a key of the form (6) such that $D_{l,2} = D_{w,2}$.

2. To prove his ability to "decrypt" C_l, user $\mathcal{U}_i$ first re-randomizes Θ_l as $\{\Theta_{l,i}'\}_{i=1}^7 \leftarrow \mathsf{ReRand}(pk_{\mathsf{AHO}}^{(1)}, \Theta_l)$. Then, he computes a Groth-Sahai commitment com_{C_l} to C_l as well as commitments $\{com_{\Theta_{l,i}'}\}_{i\in\{1,2,5\}}$ to $\{\Theta_{l,i}'\}_{i\in\{1,2,5\}}$. He generates a proof π_{C_l} that C_l is a certified HIBE ciphertext for epoch t: $i.e.$, π_{C_l} provides evidence that

$$A^{(1)}\cdot e(\Theta_{l,3}',\Theta_{l,4}')^{-1}\ \cdot\ e(G_2^{(1)},g^t)^{-1} \tag{7}$$
$$= e(G_z^{(1)},\Theta_{l,1}')\cdot e(G_r^{(1)},\Theta_{l,2}')\cdot e(G_1^{(1)},C_l),$$
$$B^{(1)}\cdot e(\Theta_{l,6}',\Theta_{l,7}')^{-1}\ \cdot\ e(H_2^{(1)},g^t)^{-1}$$
$$= e(H_z^{(1)},\Theta_{l,1}')\cdot e(H_r^{(1)},\Theta_{l,5}')\cdot e(H_1^{(1)},C_l). \tag{8}$$

Then, $\mathcal{U}_i$ generates commitments $\{com_{D_{l,i}}\}_{i=1}^2$ to the HIBE key components $\{D_{l,i}\}_{i=1}^2$ derived at step 1 and computes a proof π_{D_l} that $e(D_{l,1},g) = e(C_l, D_{l,2})$. The latter is quadratic and requires 9 group elements. Since $\{\Theta_{l,i}'\}_{i\in\{3,4,6,7\}}$ are constants, equations (7) are linear and require 3 elements each. So, π_{C_l} and π_{D_l} take 15 elements altogether.

3. Let $\sigma_l = (\theta_{l,1},\ldots,\theta_{l,7})$ be the AHO signature on $(X, D_{l,2})$. Compute $\{\theta_{l,i}'\}_{i=1}^7 \leftarrow \mathsf{ReRand}(pk_{\mathsf{AHO}}^{(0)}, \sigma_l)$ as well as commitments $\{com_{\theta_{l,i}'}\}_{i\in\{1,2,5\}}$ to $\{\theta_{l,i}'\}_{i\in\{1,2,5\}}$ and a commitment com_X to X. Then, generate a proof π_{σ_l} that committed variables satisfy the verification equations

$$A^{(0)}\cdot e(\theta_{l,3}',\theta_{l,4}')^{-1} = e(G_z^{(0)},\theta_{l,1}')\cdot e(G_r^{(0)},\theta_{l,2}')\cdot e(G_1^{(0)},X)\cdot e(G_2^{(0)},D_{l,2}),$$
$$B^{(0)}\cdot e(\theta_{l,6}',\theta_{l,7}')^{-1} = e(H_z^{(0)},\theta_{l,1})\cdot e(H_r^{(0)},\theta_{l,5}')\cdot e(H_1^{(0)},X)\cdot e(H_2^{(0)},D_{l,2}).$$

Since these equations are linear, π_{σ_l} requires 6 group elements.

4. Using VK as a tag (we assume that it is first hashed onto $\mathbb{Z}_p$ in such a way that it can be interpreted as a $\mathbb{Z}_p$ element), compute a tag-based encryption [44] of X by drawing $z_1, z_2 \xleftarrow{R} \mathbb{Z}_p$ and setting

$$(\Psi_1,\Psi_2,\Psi_3,\Psi_4,\Psi_5) = \left(f_1^{z_1}, f_2^{z_2}, X\cdot g^{z_1+z_2}, (g^{\mathsf{VK}}\cdot U)^{z_1}, (g^{\mathsf{VK}}\cdot V)^{z_2}\right).$$

5. Generate a NIZK proof that $com_X = (1,1,X)\cdot \vec{f_1}^{\phi_{X,1}}\cdot \vec{f_2}^{\phi_{X,2}}\cdot \vec{f_3}^{\phi_{X,3}}$ and (Ψ_1,Ψ_2,Ψ_3) are BBS encryptions of the same value X. If we write

$\vec{f_3} = (f_{3,1}, f_{3,2}, f_{3,3})$, the Groth-Sahai commitment com_X can be written as $(f_1^{\phi_{X,1}} \cdot f_{3,1}^{\phi_{X,3}}, f_2^{\phi_{X,2}} \cdot f_{3,2}^{\phi_{X,3}}, X \cdot g^{\phi_{X,1}+\phi_{X,2}} \cdot f_{3,3}^{\phi_{X,3}})$, so that we have

$$com_X \odot (\Psi_1, \Psi_2, \Psi_3)^{-1} = \left(f_1^{\tau_1} \cdot f_{3,1}^{\tau_3}, \ f_2^{\tau_2} \cdot f_{3,2}^{\tau_3}, \ g^{\tau_1+\tau_2} \cdot f_{3,3}^{\tau_3} \right) \qquad (9)$$

with $\tau_1 = \phi_{X,1} - z_1$, $\tau_2 = \phi_{X,2} - z_2$, $\tau_3 = \phi_{X,3}$. The signer $\mathcal{U}_i$ commits to $\tau_1, \tau_2, \tau_3 \in \mathbb{Z}_p$ (by computing $com_{\tau_j} = \vec{\varphi}^{\tau_j} \cdot \vec{f_1}^{\phi_{\tau_j,1}} \cdot \vec{f_2}^{\phi_{\tau_j,2}}$, for $j \in \{1,2,3\}$, using the vector $\vec{\varphi} = \vec{f_3} \cdot (1,1,g)$ and random $\{\phi_{\tau_j,1}, \phi_{\tau_j,2}\}_{j=1}^3$), and generates proofs $\{\pi_{eq\text{-}com,j}\}_{j=1}^3$ that τ_1, τ_2, τ_3 satisfy the three relations (9). Since these are linear equations, proofs $\{\pi_{eq\text{-}com,j}\}_{j=1}^3$ cost 2 elements each.

6. Compute $\sigma_{\mathsf{VK}} = g^{1/(x+\mathsf{VK})}$ and generate a commitment $com_{\sigma_{\mathsf{VK}}}$ to σ_{VK}. Then, generate a NIWI proof that committed variables σ_{VK} and X satisfy $e(\sigma_{\mathsf{VK}}, X \cdot g^{\mathsf{VK}}) = e(g,g)$. This relation is quadratic and costs 9 group elements to prove. We denote this proof by $\pi_{\sigma_{\mathsf{VK}}} = (\vec{\pi}_{\sigma_{\mathsf{VK}},1}, \vec{\pi}_{\sigma_{\mathsf{VK}},2}, \vec{\pi}_{\sigma_{\mathsf{VK}},3})$.

7. Compute $\sigma_{ots} = \mathcal{S}(\mathsf{SK}, (M, RL_t, \Psi_1, \Psi_2, \Psi_3, \Psi_4, \Psi_5, \Omega, \mathbf{com}, \mathbf{\Pi}))$, where we define $\Omega = \{\Theta'_{l,i}, \theta'_{l,i}\}_{i \in \{3,4,6,7\}}$, and

$$\mathbf{com} = \left(com_{C_l}, \{com_{D_{l,i}}\}_{i=1}^2, com_X, \{com_{\Theta'_{l,i}}\}_{i \in \{1,2,5\}}, \right.$$
$$\left. \{com_{\theta'_{l,i}}\}_{i \in \{1,2,5\}}, \{com_{\tau_i}\}_{i=1}^3, com_{\sigma_{\mathsf{VK}}} \right)$$
$$\mathbf{\Pi} = \left(\pi_{C_l}, \pi_{D_l}, \pi_{\sigma_l}, \pi_{eq\text{-}com,1}, \pi_{eq\text{-}com,2}, \pi_{eq\text{-}com,3}, \pi_{\sigma_{\mathsf{VK}}} \right)$$

Return the signature $\sigma = \left(\mathsf{VK}, \Psi_1, \Psi_2, \Psi_3, \Psi_4, \Psi_5, \Omega, \mathbf{com}, \mathbf{\Pi}, \sigma_{ots} \right)$.

Verify$(\sigma, M, t, RL_t, \mathcal{Y})$: parse σ as above and do the following.

1. If $\mathcal{V}(\mathsf{VK}, (\Psi_1, \Psi_2, \Psi_3, \Psi_4, \Psi_5, \Omega, \mathbf{com}, \mathbf{\Pi}), \sigma_{ots}) = 0$, return 0.
2. Return 0 if $e(\Psi_1, g^{\mathsf{VK}} \cdot U) \neq e(f_1, \Psi_4)$ or $e(\Psi_2, g^{\mathsf{VK}} \cdot V) \neq e(f_2, \Psi_5)$.
3. Return 1 if all proofs properly verify. Otherwise, return 0.

Open$(M, t, RL_t, \sigma, \mathcal{S}_{\mathsf{OA}}, \mathcal{Y}, St)$: given $\mathcal{S}_{\mathsf{OA}} = (\beta_1, \beta_2)$, parse the signature σ as above and return $\perp$ if $\mathsf{Verify}(\sigma, M, t, RL_t, \mathcal{Y}) = 0$. Otherwise, compute the value $\tilde{X} = \Psi_3 \cdot \Psi_1^{-1/\beta_1} \cdot \Psi_2^{-1/\beta_2}$. In the database of transcripts St_{trans}, find a record $\langle i, \mathsf{transcript}_i = (X, \{\{d_w, \sigma_w\}_{w \in \mathsf{copath}_{x_j}}\}_{j=0}^{\ell}, sig_i) \rangle$ such that $X = \tilde{X}$. If no such record exists in St_{trans}, return $\perp$. Otherwise, return i.

From an efficiency point of view, for each $i \in \{1 \ldots, m\}$, RL_t comprises 8 group elements plus the labels of nodes that identify S_{k_i, u_i}. If $\lambda_{\mathbb{G}}$ denotes the bitlength of a group element, the number of bits of RL_t is thus bounded by $2 \cdot |\mathcal{R}_t| \cdot (8 \cdot \lambda_{\mathbb{G}} + 2 \log N) < 2 \cdot |\mathcal{R}_t| \cdot (9\lambda_{\mathbb{G}})$ bits (as $\log N < \lambda_{\mathbb{G}}/2$ since $\lambda \leq \lambda_{\mathbb{G}}$ and N is polynomial). The size of revocation lists thus amounts to that of at most $18 \cdot |\mathcal{R}_t|$ group elements.

Users need $O(\log^3 N)$ group elements to store their membership certificate. As far as the size of signatures goes, $\mathbf{com}$ and $\mathbf{\Pi}$ require 42 and 36 group elements, respectively. If the one-time signature of [37] is used, σ consists of 96 group elements, which is less than twice the size of Groth's signatures [38]. At the 128-bit security level, a signature takes 6 kB.

Verifying signatures takes constant time. The cost of each signature generation is dominated by at most $\ell = \log N$ exponentiations to derive a HIBE private key at step 1. However, this step only has to be executed once per revocation epoch, at the first signature of that epoch.

The scheme is proved secure against misidentification attacks assuming the hardness of the q-SFP problem, where q is a polynomial function of $\ell = \log_2 N$, the number of adversarially-controlled users and the number of revocations. The security against framing attacks is proved under the SDH assumption and assuming that the one-time signature is strongly unforgeable. As for the anonymity property, we prove it under the DLIN assumption and assuming the strong unforgeability of the one-time signature. Due to space limitation, all security proofs are deferred to the full version of the paper.

References

1. Abdalla, M., Kiltz, E., Neven, G.: Generalized Key Delegation for Hierarchical Identity-Based Encryption. In: Biskup, J., López, J. (eds.) ESORICS 2007. LNCS, vol. 4734, pp. 139–154. Springer, Heidelberg (2007)
2. Abe, M., Haralambiev, K., Ohkubo, M.: Signing on Elements in Bilinear Groups for Modular Protocol Design. Cryptology ePrint Archive: Report 2010/133 (2010)
3. Abe, M., Fuchsbauer, G., Groth, J., Haralambiev, K., Ohkubo, M.: Structure-Preserving Signatures and Commitments to Group Elements. In: Rabin, T. (ed.) CRYPTO 2010. LNCS, vol. 6223, pp. 209–236. Springer, Heidelberg (2010)
4. Abe, M., Groth, J., Haralambiev, K., Ohkubo, M.: Optimal Structure-Preserving Signatures in Asymmetric Bilinear Groups. In: Rogaway, P. (ed.) CRYPTO 2011. LNCS, vol. 6841, pp. 649–666. Springer, Heidelberg (2011)
5. Acar, T., Nguyen, L.: Revocation for Delegatable Anonymous Credentials. In: Catalano, D., Fazio, N., Gennaro, R., Nicolosi, A. (eds.) PKC 2011. LNCS, vol. 6571, pp. 423–440. Springer, Heidelberg (2011)
6. Ateniese, G., Camenisch, J., Hohenberger, S., de Medeiros, B.: Practical group signatures without random oracles. Cryptology ePrint Archive: Report 2005/385 (2005)
7. Ateniese, G., Camenisch, J., Joye, M., Tsudik, G.: A Practical and Provably Secure Coalition-Resistant Group Signature Scheme. In: Bellare, M. (ed.) CRYPTO 2000. LNCS, vol. 1880, pp. 255–270. Springer, Heidelberg (2000)
8. Ateniese, G., Song, D., Tsudik, G.: Quasi-Efficient Revocation in Group Signatures. In: Blaze, M. (ed.) FC 2002. LNCS, vol. 2357, pp. 183–197. Springer, Heidelberg (2003)
9. Bellare, M., Micciancio, D., Warinschi, B.: Foundations of Group Signatures: Formal Definitions, Simplified Requirements, and a Construction Based on General Assumptions. In: Biham, E. (ed.) EUROCRYPT 2003. LNCS, vol. 2656, pp. 614–629. Springer, Heidelberg (2003)
10. Bellare, M., Rogaway, P.: Random Oracles are Practical: A Paradigm for Designing Efficient Protocols. In: 1st ACM Conference on Computer and Communications Security, pp. 62–73. ACM Press (1993)
11. Bellare, M., Shi, H., Zhang, C.: Foundations of Group Signatures: The Case of Dynamic Groups. In: Menezes, A. (ed.) CT-RSA 2005. LNCS, vol. 3376, pp. 136–153. Springer, Heidelberg (2005)

12. Benaloh, J.C., de Mare, M.: One-Way Accumulators: A Decentralized Alternative to Digital Signatures. In: Helleseth, T. (ed.) EUROCRYPT 1993. LNCS, vol. 765, pp. 274–285. Springer, Heidelberg (1994)
13. Boneh, D., Boyen, X.: Short Signatures Without Random Oracles. In: Cachin, C., Camenisch, J.L. (eds.) EUROCRYPT 2004. LNCS, vol. 3027, pp. 56–73. Springer, Heidelberg (2004)
14. Boneh, D., Boyen, X.: Efficient Selective-ID Secure Identity-Based Encryption Without Random Oracles. In: Cachin, C., Camenisch, J.L. (eds.) EUROCRYPT 2004. LNCS, vol. 3027, pp. 223–238. Springer, Heidelberg (2004)
15. Boneh, D., Boyen, X., Goh, E.-J.: Hierarchical Identity Based Encryption with Constant Size Ciphertext. In: Cramer, R. (ed.) EUROCRYPT 2005. LNCS, vol. 3494, pp. 440–456. Springer, Heidelberg (2005)
16. Boneh, D., Boyen, X., Shacham, H.: Short Group Signatures. In: Franklin, M. (ed.) CRYPTO 2004. LNCS, vol. 3152, pp. 41–55. Springer, Heidelberg (2004)
17. Boneh, D., Lynn, B., Shacham, H.: Short Signatures from the Weil Pairing. In: Boyd, C. (ed.) ASIACRYPT 2001. LNCS, vol. 2248, pp. 514–532. Springer, Heidelberg (2001)
18. Boneh, D., Shacham, H.: Group signatures with verifier-local revocation. In: ACM-CCS 2004, pp. 168–177. ACM Press (2004)
19. Boyen, X., Waters, B.: Compact Group Signatures Without Random Oracles. In: Vaudenay, S. (ed.) EUROCRYPT 2006. LNCS, vol. 4004, pp. 427–444. Springer, Heidelberg (2006)
20. Boyen, X., Waters, B.: Full-Domain Subgroup Hiding and Constant-Size Group Signatures. In: Okamoto, T., Wang, X. (eds.) PKC 2007. LNCS, vol. 4450, pp. 1–15. Springer, Heidelberg (2007)
21. Bresson, E., Stern, J.: Efficient Revocation in Group Signatures. In: Kim, K.-c. (ed.) PKC 2001. LNCS, vol. 1992, pp. 190–206. Springer, Heidelberg (2001)
22. Brickell, E.: An efficient protocol for anonymously providing assurance of the container of the private key. Submission to the Trusted Computing Group (April 2003)
23. Brickell, E., Camenisch, J., Chen, L.: Direct Anonymous Attestation. In: ACM-CCS 2004, pp. 132–145 (2004)
24. Camenisch, J.L., Chaabouni, R., Shelat, A.: Efficient Protocols for Set Membership and Range Proofs. In: Pieprzyk, J. (ed.) ASIACRYPT 2008. LNCS, vol. 5350, pp. 234–252. Springer, Heidelberg (2008)
25. Camenisch, J., Dubovitskaya, M., Neven, G., Zaverucha, G.: Oblivious Transfer with Hidden Access Control Policies. In: Catalano, D., Fazio, N., Gennaro, R., Nicolosi, A. (eds.) PKC 2011. LNCS, vol. 6571, pp. 192–209. Springer, Heidelberg (2011)
26. Camenisch, J., Kohlweiss, M., Soriente, C.: An Accumulator Based on Bilinear Maps and Efficient Revocation for Anonymous Credentials. In: Jarecki, S., Tsudik, G. (eds.) PKC 2009. LNCS, vol. 5443, pp. 481–500. Springer, Heidelberg (2009)
27. Camenisch, J., Kohlweiss, M., Soriente, C.: Solving Revocation with Efficient Update of Anonymous Credentials. In: Garay, J.A., De Prisco, R. (eds.) SCN 2010. LNCS, vol. 6280, pp. 454–471. Springer, Heidelberg (2010)
28. Camenisch, J., Lysyanskaya, A.: Dynamic Accumulators and Application to Efficient Revocation of Anonymous Credentials. In: Yung, M. (ed.) CRYPTO 2002. LNCS, vol. 2442, pp. 61–76. Springer, Heidelberg (2002)
29. Canetti, R., Halevi, S., Katz, J.: A Forward-Secure Public-Key Encryption Scheme. In: Biham, E. (ed.) EUROCRYPT 2003. LNCS, vol. 2656, pp. 254–271. Springer, Heidelberg (2003)

30. Cathalo, J., Libert, B., Yung, M.: Group Encryption: Non-Interactive Realization in the Standard Model. In: Matsui, M. (ed.) ASIACRYPT 2009. LNCS, vol. 5912, pp. 179–196. Springer, Heidelberg (2009)
31. Chaum, D., van Heyst, E.: Group Signatures. In: Davies, D.W. (ed.) EUROCRYPT 1991. LNCS, vol. 547, pp. 257–265. Springer, Heidelberg (1991)
32. Delerablée, C., Pointcheval, D.: Dynamic Fully Anonymous Short Group Signatures. In: Nguyên, P.Q. (ed.) VIETCRYPT 2006. LNCS, vol. 4341, pp. 193–210. Springer, Heidelberg (2006)
33. Dodis, Y., Fazio, N.: Public Key Broadcast Encryption for Stateless Receivers. In: Feigenbaum, J. (ed.) DRM 2002. LNCS, vol. 2696, pp. 61–80. Springer, Heidelberg (2003)
34. Fan, C.-I., Hsu, R.-H., Manulis, M.: Group Signature with Constant Revocation Costs for Signers and Verifiers. In: Lin, D., Tsudik, G., Wang, X. (eds.) CANS 2011. LNCS, vol. 7092, pp. 214–233. Springer, Heidelberg (2011)
35. Fuchsbauer, G.: Automorphic Signatures in Bilinear Groups and an Application to Round-Optimal Blind Signatures. Cryptology ePrint Archive: Report 2009/320 (2009)
36. Gentry, C., Silverberg, A.: Hierarchical ID-Based Cryptography. In: Zheng, Y. (ed.) ASIACRYPT 2002. LNCS, vol. 2501, pp. 548–566. Springer, Heidelberg (2002)
37. Groth, J.: Simulation-Sound NIZK Proofs for a Practical Language and Constant Size Group Signatures. In: Lai, X., Chen, K. (eds.) ASIACRYPT 2006. LNCS, vol. 4284, pp. 444–459. Springer, Heidelberg (2006)
38. Groth, J.: Fully Anonymous Group Signatures Without Random Oracles. In: Kurosawa, K. (ed.) ASIACRYPT 2007. LNCS, vol. 4833, pp. 164–180. Springer, Heidelberg (2007)
39. Groth, J., Sahai, A.: Efficient Non-interactive Proof Systems for Bilinear Groups. In: Smart, N.P. (ed.) EUROCRYPT 2008. LNCS, vol. 4965, pp. 415–432. Springer, Heidelberg (2008)
40. Halevy, D., Shamir, A.: The LSD Broadcast Encryption Scheme. In: Yung, M. (ed.) CRYPTO 2002. LNCS, vol. 2442, pp. 47–60. Springer, Heidelberg (2002)
41. Horwitz, J., Lynn, B.: Toward Hierarchical Identity-Based Encryption. In: Knudsen, L.R. (ed.) EUROCRYPT 2002. LNCS, vol. 2332, pp. 466–481. Springer, Heidelberg (2002)
42. Kiayias, A., Yung, M.: Secure scalable group signature with dynamic joins and separable authorities. International Journal of Security and Networks (IJSN) 1(1/2), 24–45 (2006)
43. Kiayias, A., Yung, M.: Group Signatures with Efficient Concurrent Join. In: Cramer, R. (ed.) EUROCRYPT 2005. LNCS, vol. 3494, pp. 198–214. Springer, Heidelberg (2005)
44. Kiltz, E.: Chosen-Ciphertext Security from Tag-Based Encryption. In: Halevi, S., Rabin, T. (eds.) TCC 2006. LNCS, vol. 3876, pp. 581–600. Springer, Heidelberg (2006)
45. Libert, B., Vergnaud, D.: Group Signatures with Verifier-Local Revocation and Backward Unlinkability in the Standard Model. In: Garay, J.A., Miyaji, A., Otsuka, A. (eds.) CANS 2009. LNCS, vol. 5888, pp. 498–517. Springer, Heidelberg (2009)
46. Nakanishi, T., Fujii, H., Hira, Y., Funabiki, N.: Revocable Group Signature Schemes with Constant Costs for Signing and Verifying. In: Jarecki, S., Tsudik, G. (eds.) PKC 2009. LNCS, vol. 5443, pp. 463–480. Springer, Heidelberg (2009)
47. Nakanishi, T., Funabiki, N.: Verifier-Local Revocation Group Signature Schemes with Backward Unlinkability from Bilinear Maps. In: Roy, B. (ed.) ASIACRYPT 2005. LNCS, vol. 3788, pp. 533–548. Springer, Heidelberg (2005)

48. Naor, D., Naor, M., Lotspiech, J.: Revocation and Tracing Schemes for Stateless Receivers. In: Kilian, J. (ed.) CRYPTO 2001. LNCS, vol. 2139, pp. 41–62. Springer, Heidelberg (2001)
49. Nguyen, L.: Accumulators from Bilinear Pairings and Applications. In: Menezes, A. (ed.) CT-RSA 2005. LNCS, vol. 3376, pp. 275–292. Springer, Heidelberg (2005)
50. Nguyen, L., Safavi-Naini, R.: Efficient and Provably Secure Trapdoor-Free Group Signature Schemes from Bilinear Pairings. In: Lee, P.J. (ed.) ASIACRYPT 2004. LNCS, vol. 3329, pp. 372–386. Springer, Heidelberg (2004)
51. Song, D.: Practical forward secure group signature schemes. In: ACM-CCS 2001, pp. 225–234 (2001)
52. Teranishi, I., Sako, K.: k-Times Anonymous Authentication with a Constant Proving Cost. In: Yung, M., Dodis, Y., Kiayias, A., Malkin, T. (eds.) PKC 2006. LNCS, vol. 3958, pp. 525–542. Springer, Heidelberg (2006)
53. Tsang, P., Au, M.-H., Kapadia, A., Smith, S.: Blacklistable anonymous credentials: blocking misbehaving users without TTPs. In: ACM-CCS 2007, pp. 72–81 (2007)
54. Tsang, P., Au, M.-H., Kapadia, A., Smith, S.: PEREA: towards practical TTP-free revocation in anonymous authentication. In: ACM-CCS 2008, pp. 333–344 (2008)
55. Tsudik, G., Xu, S.: Accumulating Composites and Improved Group Signing. In: Laih, C.-S. (ed.) ASIACRYPT 2003. LNCS, vol. 2894, pp. 269–286. Springer, Heidelberg (2003)
56. Zhou, S., Lin, D.: Shorter Verifier-Local Revocation Group Signatures from Bilinear Maps. In: Pointcheval, D., Mu, Y., Chen, K. (eds.) CANS 2006. LNCS, vol. 4301, pp. 126–143. Springer, Heidelberg (2006)

Incremental Deterministic
Public-Key Encryption

Ilya Mironov[1], Omkant Pandey[2], Omer Reingold[1], and Gil Segev[1]

[1] Microsoft Research Silicon Valley, Mountain View, CA 94043, USA
{mironov,gil.segev,omer.reingold}@microsoft.com
[2] Microsoft, Redmond, USA and Microsoft Research, Bangalore, India
omkantp@microsoft.com

Abstract. Motivated by applications in large storage systems, we initiate the study of incremental deterministic public-key encryption. Deterministic public-key encryption, introduced by Bellare, Boldyreva, and O'Neill (CRYPTO '07), provides a realistic alternative to randomized public-key encryption in various scenarios where the latter exhibits inherent drawbacks. A deterministic encryption algorithm, however, cannot satisfy any meaningful notion of security for low-entropy plaintexts distributions, and Bellare et al. demonstrated that a strong notion of security can in fact be realized for relatively high-entropy plaintext distributions.

In order to achieve a meaningful level of security, a deterministic encryption algorithm should be typically used for encrypting rather long plaintexts for ensuring a sufficient amount of entropy. This requirement may be at odds with efficiency constraints, such as communication complexity and computation complexity in the presence of small updates. Thus, a highly desirable property of deterministic encryption algorithms is incrementality: small changes in the plaintext translate into small changes in the corresponding ciphertext.

We present a framework for modeling the incrementality of deterministic public-key encryption. Within our framework we propose two schemes, which we prove to enjoy an optimal tradeoff between their security and incrementality up to small polylogarithmic factors. Our first scheme is a generic method which can be based on any deterministic public-key encryption scheme, and in particular, can be instantiated with any semantically-secure (randomized) public-key encryption scheme in the random oracle model. Our second scheme is based on the Decisional Diffie-Hellman assumption in the standard model.

The approach underpinning our schemes is inspired by the fundamental "sample-then-extract" technique due to Nisan and Zuckerman (JCSS '96) and refined by Vadhan (J. Cryptology '04), and by the closely related notion of "locally-computable extractors" due to Vadhan. Most notably, whereas Vadhan used such extractors to construct *private-key* encryption schemes in the bounded-storage model, we show that techniques along these lines can also be used to construct incremental *public-key* encryption schemes.

D. Pointcheval and T. Johansson (Eds.): EUROCRYPT 2012, LNCS 7237, pp. 628–644, 2012.

1   Introduction

The fundamental notion of *semantic security* for public-key encryption schemes was introduced by Goldwasser and Micali [19]. While semantic security provides strong privacy guarantees, it inherently requires a *randomized* encryption algorithm. Unfortunately, randomized encryption breaks several assumptions of large storage systems that are crucial in efficient implementation of search (and, more generally, of indexing) and de-duplication [9,23]. Further, randomized encryption necessarily expands the length of the plaintext, which may be undesirable in some applications, such as legacy code or in-place encryption.

Deterministic Encryption. To deal with these and other drawbacks, Bellare, Boldyreva, and O'Neill [2] initiated the study of deterministic public-key encryption schemes. These are public-key encryption schemes where the encryption algorithm is deterministic. Bellare et al. formulate meaningful, and essentially "best possible", security requirements for such schemes which are inspired by and very close to semantic security. Clearly, in this setting, no meaningful notion of security can be achieved if the space of plaintexts is small. Therefore, Bellare et al. [2] required security to hold only when the plaintexts are drawn from a high min-entropy distribution.

Deterministic encryption already alleviates many of the above mentioned problems when dealing with large data volumes. For example, since the encryption algorithm is deterministic, we can now do indexing and perform fast search on encrypted data. Further, schemes that have length-preserving ciphertexts are possible as well [2]. Also, unlike randomized encryption, there is no fundamental reason that precludes noticeable savings in storage by using de-duplication techniques (which can be as large as 97% [27]); although one may not get the same amount of savings as with usual plaintext.

We emphasize that security of deterministic encryption is contingent on a very strong assumption about the underlying data distribution, namely that the plaintext has high min-entropy from the adversary's point of view. One possibility for improving security margin is to encrypt longer plaintexts whenever possible, for example, by not cutting files into smaller pieces or using larger blocks for in-place encryption. If, however, changing the plaintext requires re-computation of the ciphertext, doing that for any update may quickly negate all efficiency gains from using deterministic encryption. For a remedy we turn to *incremental cryptography*, explained below.

Incremental Cryptography. Given that we are dealing with large plaintexts, computing the ciphertext from scratch for the modified plaintext can be quite an expensive operation. One such example is maintaining an (encrypted) daily backup of your hard-disk on an untrusted server. The disk may contain gigabytes of data, most of which is likely to remain unchanged between two successive backups. The problem is further intensified in various client-server settings where *all* of previous plaintext might not be available when the modification request is made. In such settings where plaintext is really large, downloading old data can be a serious problem. This issue is clearly not specific to (deterministic) encryption, and is of very general interest.

To address this issue, Bellare, Goldreich and Goldwasser [5] introduced and developed the notion of *incremental* cryptography, first in application to digital signatures. The idea is that, once we have signed a document M, signing new versions of M should be rather quick. For example, if we only flip a single bit of M, we should be able to update the signature in time polynomial in $\log |M|$ (instead of $|M|$) and the security parameter λ. Clearly, incrementality is an attractive feature to have for any cryptographic primitive such as encryption, signatures, hash functions, and so on [6,20,15,7,11].

It is clear from our discussion that when dealing with deterministic encryption over large databases, where we are forced to encrypt rather long plaintexts for ensuring their min-entropy, what we really need is an *incremental* encryption scheme. That is, the scheme should allow quickly updating the ciphertexts to reflect small changes. In light of the observation that deterministic encryption is most desirable when dealing with large data volumes, perhaps it is not exaggerating to suggest that incrementality should be an important *design goal* for deterministic encryption rather than merely a "nice to have" feature.

1.1   Our Contributions

In this work we formalize the notion of *incremental* deterministic public-key encryption. We view incrementality and security as two orthogonal objectives, which together have a great potential in improving the deployment of deterministic encryption schemes with provable security properties in real-world applications.

Modeling Incremental Updates. Intuitively, a deterministic public-key encryption scheme is *incremental* if any small modification of a plaintext m resulting in a plaintext m' can be efficiently carried over for updating the encryption $c = \mathsf{Enc}_{pk}(m)$ of m to the encryption $c' = \mathsf{Enc}_{pk}(m')$ of m'. For capturing the efficiency of such an update operation we consider two natural complexity measures: (1) *input locality* (i.e., the number of ciphertexts bits that are affected when flipping a single plaintext bit), and (2) *query complexity* (i.e., the number of public-key, plaintext, and ciphertext bits that have to be read in order to update the ciphertext).

We note that modeling updates for deterministic public-key encryption is slightly different than for other primitives. For example, suppose that we allow "replacements" as considered by [5]. These are queries of the form (j, b) that replace the j-th bit of a given plaintext m by $b \in \{0, 1\}$. Then, if there exists a public algorithm Update for updating the ciphertext, then one can recover the entire plaintext from the ciphertext[1]. Therefore, we focus on the *bit flipping* operation instead. This operation is specified by an index j, and sets the current value of $m[j]$ to $\neg m[j]$.

---

[1] The encryption algorithm is deterministic, and hence the ciphertext for every message is unique. The operation $\mathsf{Update}(j, 0)$ changes the ciphertext if and only if the jth bit of m is 1.

For capturing the above measures of efficiency we model the update operation as a probabilistic polynomial-time algorithm Update that receives as input the index i^* of a plaintext bit to be flipped, and has oracle access to the individual bits of the public key pk, the plaintext m to be modified, and to its encryption $c = \mathsf{Enc}_{pk}(m)$. That is, the algorithm Update can submit queries of the form (pk, i), (m, i) or (c, i), which are answered with the ith bit of pk, m, or c, respectively. We refer the reader to Section 3 for the formal description of our model, which considers also update in a "private" fashion in which the update algorithm can access the secret key but not the plaintext.

Locality Lower Bound. An important insight is that deterministic encryption cannot have very small incrementality. Deterministic encryption schemes require high min-entropy messages to provide any meaningful guarantee, and we show that any scheme with low incrementality can be secure only for messages with much higher entropy. Specifically, we show that for every deterministic public-key encryption scheme that satisfies the minimal notion of PRIV1-IND security for plaintext distributions of min-entropy k, plaintext length n, and ciphertext length t, the incrementality Δ of the scheme must satisfy: $\Delta \geq \frac{n-3}{k \log t}$.

Ignoring the lower-order $\log t$ factor, our proof shows in particular that the input locality of the encryption algorithm must be roughly n/k. This should be compared with the case of randomized encryption, where flipping a single plaintext bit may require to flip only a single ciphertext bit. Indeed, consider encrypting a plaintext m as the pair $(\mathsf{Enc}_{pk}(r), r \oplus m)$ for a randomly chosen mask r. Flipping a single bit of m requires flipping only a single bit of the ciphertext.

Constructions with Optimal Incrementality. We construct two deterministic public-key encryption schemes with optimal incrementality (up to lower-order polylogarithmic factors). Our first construction is a general transformation from any deterministic encryption scheme to an incremental one. Following the terminology developed in [2,4,8], the resulting scheme from this approach is PRIV1-IND secure if the underlying scheme is PRIV-IND secure. As a result, using the construction of Bellare et al. [2] in the random oracle model, we can instantiate our approach in the random oracle model based on any semantically-secure (randomized) public-key encryption scheme, and obtain a deterministic scheme with optimal incrementality.

Our second, more direct construction, avoids the random oracle model. It is based on the Decisional Diffie-Hellman assumption in the standard model, and enjoys optimal incrementality. The scheme relies on the notion of *smooth trapdoor functions* that we introduce (and was implicitly used by Boldyreva et al. [8]), and realize it in an incremental manner based on the Decisional Diffie-Hellman assumption. Both of our constructions guarantee PRIV1-IND security when encrypting n-bit plaintexts with min-entropy $k \geq n^\epsilon$, where $\epsilon > 0$ is any pre-specified constant.

1.2   Related Work

The problem of composing public-key encryption and de-duplication was addressed by Doucer et al. [14] via the concept of *convergent encryption*, in which files are encrypted using their own hash values as keys. Security of the scheme is argued in the random-oracle model and under implicit assumption of the plaintext's high min-entropy. The formal goal of leveraging entropy of the source to achieve information-theoretic security with a short symmetric key was articulated by Russell and Wang [24], followed by Dodis and Smith [13].

The notion of public-key deterministic encryption was introduced by Bellare, Boldyreva, and O'Neill [2], and then further studied by Bellare, Fischlin, O'Neill, and Ristenpart [4], Boldyreva, Fehr, and O'Neill [8], Brakerski and Segev [10], Wee [26], and Fuller, O'Neill and Reyzin [18]. Bellare et al. [2] proved their constructions in the random oracle model; subsequent papers demonstrated schemes secure in the standard model based on trapdoor permutations [4] and lossy trapdoor functions [8]. Brakerski and Segev [10] and Wee [26] address the question of security of public-key deterministic encryption in the presence of auxiliary input. Fuller et al. [18] presented a construction based on any trapdoor function that admits a large number of simultaneous hardcore bits, and a construction that is secure for a bounded number of possibly related plaintexts.

Constructions of deterministic public-key encryption found an intriguing application in "hedged" public-key encryptions [3]. These schemes remain secure even if the randomness used during the encryption process is not perfect (controlled by or leaked to the adversary) as long as the joint distribution of plaintext-randomness has sufficient min-entropy.

The concept of incremental cryptography started with the work of Bellare, Goldreich, and Goldwasser [5], who considered the case of hashing and signing. They also provided discrete-logarithm based constructions for incremental collision-resistant hash and signatures, that support block *replacement* operation. Constructions supporting block *insertion* and *deletion* were first developed in [6], with further refinements and new issues concerning incrementality such as tamper-proof updates, privacy of updates, and incrementality in symmetric encryption. In subsequent work, Fischlin presented an incremental signature schemes supporting insertion/deletion of blocks, and tamper-proof updates [15], and proved a $\Omega(\sqrt{n})$ lower bound on the signature size of schemes that support substitution and replacement operations (the bound can be improved to $\Omega(n)$ in certain special cases) [16]. Bellare and Micciancio [7] revisited the case of hashing, and provided new constructions for the same based on discrete logarithms and lattices. Buonanno, Katz, and Yung [11] considered the issue of incrementality in symmetric unforgeable encryption and suggested three modes of operations for AES achieving this notion.

The goal of incremental cryptography, i.e., *input locality*, can be contrasted with the dual question of placing cryptography in the NC^0 complexity class, i.e., identifying cryptographic primitives with constant *output locality*. This problem has essentially been resolved for public-key encryption in the positive by Applebaum, Ishai, and Kushilevitz [1], who construct schemes based on standard number-theoretic assumptions and lattice problems where each bit of the encryption operation depends on at most four bits of the input. Applebaum et al. also argue impossibility of semantically-secure public-key encryption scheme with constant input locality [1, Section C.1].

1.3   Overview of Our Approach

In this section we present a high-level overview of our two constructions. First, we describe the well-known "sample-then-extract" approach [21,25] that serves as our inspiration for constructing incremental schemes. Then, we describe the main ideas underlying our schemes, each of which is based on a different realization of the "sample-then-extract" approach.

"Sample-then-Extract". A fundamental fact in the theory of pseudorandomness is that a random sample of bits from a string of high min-entropy essentially preserves the min-entropy rate. This was initially proved by Nisan and Zuckerman [21] and then refined by Vadhan [25] that captured the optimal parameters. Intuitively, the "sample-then-extract" lemma states that if $\mathcal{X} \in \{0,1\}^n$ has min-entropy rate δ, and $\mathcal{X}_S \in \{0,1\}^t$ is the projection of $\mathcal{X}$ onto a random set $S \subseteq [n]$ of t positions, then $\mathcal{X}_S$ is statistically-close to a source with min-entropy rate $\delta' = \Omega(\delta)$.

This lemma serves as a fundamental tool in the design of randomness extractors. Moreover, in the cryptographic setting, it was used by Vadhan [25] to construct *locally-computable extractors*, which allow to compute their output by examining a small number of input bits. Such extractors were used by Vadhan to design private-key encryption schemes in the bounded-storage model. In this work we demonstrate for the first time that the "sample-then-extract" approach can be leveraged to design not only *private-key* encryption schemes, but also *public-key* encryption schemes.

A Generic Construction via Random Partitioning. In the setting of randomized encryption, a promising approach for ensuring incrementality is to divide each plaintext m into consecutive and rather small blocks $m = m_1||\cdots||m_\ell$, and to separately encrypt each block m_i. Thus, changing a single bit of m affects only a single block of the ciphertext. Moreover, the notion of semantic security is sufficiently powerful to even allow each block m_i to be as small as a single bit. In the setting of deterministic encryption, however, security can hold only when each encrypted block has a sufficient amount of min-entropy. At this point we note that even if a plaintext $m = m_1||\cdots||m_\ell$ has high min-entropy, it may clearly be the case that some of its small blocks have very low min-entropy (or even fixed). Thus, this approach seems to fail for deterministic encryption.

As an alternative, however, we propose the following approach: instead of dividing the plaintext m into fixed blocks, we project it onto a *uniformly chosen partition* $S_1, \ldots, S_\ell$ of the plaintext positions to sets of equal sizes, and then separately encrypt each of the projections $m_{S_1}, \ldots, m_{S_\ell}$ using an underlying (possibly non-incremental) deterministic encryption scheme[2]. By the fact that we use a partition of the plaintext positions we ensure on the one hand that the plaintext m can be fully recovered, and on the other that each plaintext position appears in only one set (and thus the scheme is incremental). In terms of security, since we use a uniformly chosen partition, the distribution of each individual set S_i is uniform, and therefore by carefully choosing the size of the sets the "sample-and-extract" lemma guarantees that with overwhelming probability each projection m_{S_i} preserves the min-entropy rate of m. Therefore, the scheme is secure as long as the underlying scheme guarantees PRIV-IND security (see Section 2.2 for the notions of security for deterministic encryption).

By instantiating this approach with the constructions of Bellare et al. [2] in the random oracle model, we obtain as a corollary a deterministic public-key encryption scheme with optimal incrementality based either on any semantically-secure (randomized) public-key encryption scheme, or on RSA-OAEP which yields a *length-preserving* incremental scheme.

A Construction Based on Smooth Trapdoor Functions. Although our first construction is a rather generic one, constructions of PRIV-IND-secure schemes are known only in the random oracle model. In the standard model, Boldyreva et al. [8] introduced the slightly weaker notion of PRIV1-IND security, which considers plaintexts that have high min-entropy even when conditioned on other plaintexts, and showed that it can be realized by composing any lossy trapdoor function with a pairwise independent permutation. This approach, however, does not seem useful for constructing incremental schemes, since pairwise independence is inherently non-incremental. A simple observation, however, shows that the approach of Boldyreva et al. [8] requires in fact trapdoor functions with weaker properties, that we refer to as *smooth trapdoor functions* (this is implicit in [8]).

Informally, a collection of smooth trapdoor functions consists of two families of functions. Functions in one family are injective and can be efficiently inverted using a trapdoor. Functions in the other family are "smooth" in the sense that their output distribution on any source of input with high min-entropy is statistically close to their output distribution on a uniformly sampled input. The only security requirement is that a description of a randomly chosen function from the family of injective functions is computationally indistinguishable from a description of a randomly chosen function from the family of smooth functions. We show that any collection of smooth trapdoor functions is a PRIV1-IND-secure deterministic encryption scheme (again, this is implicit in [8]).

Next, we construct a collection of *incremental* smooth trapdoor functions based on the Decisional Diffie-Hellman (DDH) assumption, by significantly refining the DDH-based lossy trapdoor functions of Freeman et al. [17] (which in turned generalized those of Peikert and Waters [22]). Our collection is parameterized by a group G of prime order p that is generated by an element $g \in G$. A

---

[2] A minor technical detail is that we would also like to ensure that we always encrypt distinct values, and therefore we concatenate the block number i to each projection m_{S_i}.

public key is of the form g^A, where $A \in \mathbb{Z}^{n \times n}$ is sampled from one distribution for injective keys, and from a different distribution for smooth keys[3]. Evaluating a function on an input $x \in \{0,1\}^n$ is done by computing $g^{Ax} \in G^n$ and inversion for injective keys is done using the secret key A^{-1}.

The key point in our scheme is the distribution of the matrix A for injective and smooth keys. For smooth keys the matrix A is generated to satisfy two properties. The first is that each of its first ℓ rows has t randomly chosen entries with values that are chosen uniformly from $\mathbb{Z}_p$, and all other $n-t$ entries are zeros (where ℓ and t are carefully chosen depending on the min-entropy rate). Looking ahead, when computing the inner product of such a sparse row with a source of min-entropy larger than $\log p$, the "sample-then-extract" lemma guarantees that the output is statistically close to uniform. In a sense, this is a realization of a locally-computable extractor that is embedded in our functions. The second property, is that each of its last $n - \ell$ rows are linear combinations of the first ℓ rows, and therefore the image of its corresponding linear map is determined by the first ℓ rows. This way, we can argue that smooth keys hide essentially all information on the underlying input distribution.

For injective keys, we sample a matrix A from the distribution of smooth keys, and then re-sample all its non-zero entries with independently and uniformly distributed elements of $\mathbb{Z}_p$. A subtle complication arises since such a matrix is not necessarily invertible, as required for injective keys, but this is easily resolved (without hurting the smooth keys – see Section 5 for more details). Observing that for injective keys each column of A contains roughly t non-zero entries, this yields a PRIV1-IND-secure scheme with optimal incrementality.

Paper Organization. In Section 2 we introduce the notation and tools that are used in this paper. In Section 3 we present a framework for modeling the incrementality of deterministic public-key encryption schemes. In Section 4 we present our generic construction, and in Section 5 we present our DDH-based construction. Due to space limitations we refer the reader to the full version for the proof of the lower bound.

2   Preliminaries

2.1   Probability Distributions

For a distribution $\mathcal{X}$ we denote by $x \leftarrow \mathcal{X}$ the process of sampling a value x according to $\mathcal{X}$. Similarly, for a set Ω we denote by $\omega \leftarrow \Omega$ the process of sampling a value ω from the uniform distribution over Ω. If $\mathcal{X}$ is a distribution and f is a function defined over its support, then $f(\mathcal{X})$ denotes the outcome of the experiment where $f(x)$ is evaluated on x sampled from $\mathcal{X}$. For any $n \in \mathbb{N}$ we denote by $\mathcal{U}_n$ the uniform distribution over the set $\{0,1\}^n$.

The *min-entropy* of a distribution $\mathcal{X}$ that is defined over a set Ω is defined as $H_\infty(\mathcal{X}) = \min_{\omega \in \Omega} \log \left(1 / \Pr[\mathcal{X} = \omega] \right)$. A k-*source* is distribution $\mathcal{X}$ with

---

[3] For any matrix $A = \{a_{ij}\}_{i \in [n], j \in [n]} \in \mathbb{Z}_p^{n \times n}$ we denote by $g^A \in G^{n \times n}$ the matrix $\{g^{a_{ij}}\}_{i \in [n], j \in [n]}$.

$H_\infty(\mathcal{X}) \geq k$, and the *min-entropy rate* of a k-source over the set $\{0,1\}^n$ is k/n. The *statistical distance* between two distributions $\mathcal{X}$ and $\mathcal{Y}$ over a set Ω is defined as $\mathrm{SD}(\mathcal{X}, \mathcal{Y}) = \max_{S \subseteq \Omega} |\mathrm{Pr}[\mathcal{X} \in S] - \mathrm{Pr}[\mathcal{Y} \in S]|$. A distribution $\mathcal{X}$ is ϵ-*close* to a k-source if there exists a k-source $\mathcal{Y}$ such that $\mathrm{SD}(\mathcal{X}, \mathcal{Y}) \leq \epsilon$. The following standard lemma (see, for example, [12]) essentially states that revealing r bits of information on a random variable may reduce its min-entropy by roughly r.

Lemma 2.1. *Let $\mathcal{Z}$ be a distribution over at most 2^r values, then for any distribution $\mathcal{X}$ and for any $\epsilon > 0$ it holds that*

$$\mathrm{Pr}_{z \leftarrow \mathcal{Z}} \left[H_\infty(\mathcal{X}|\mathcal{Z} = z) \geq H_\infty(\mathcal{X}) - r - \log(1/\epsilon) \right] \geq 1 - \epsilon .$$

We say that two families of distributions $\mathcal{X} = \{\mathcal{X}_\lambda\}_{\lambda \in \mathbb{N}}$ and $\mathcal{Y} = \{\mathcal{Y}_\lambda\}_{\lambda \in \mathbb{N}}$ are *statistically close*, denoted by $\mathcal{X} \approx \mathcal{Y}$, if there exists a negligible function $\nu(\lambda)$ such that $\mathrm{SD}(\mathcal{X}, \mathcal{Y}) \leq \nu(\lambda)$ for all sufficiently large $\lambda \in \mathbb{N}$. Two families of distributions $\mathcal{X} = \{\mathcal{X}_\lambda\}_{\lambda \in \mathbb{N}}$ and $\mathcal{Y} = \{\mathcal{Y}_\lambda\}_{\lambda \in \mathbb{N}}$ are *computationally indistinguishable*, denoted by $\mathcal{X} \overset{c}{\approx} \mathcal{Y}$, if for any probabilistic polynomial-time algorithm A there exists a negligible function $\nu(\lambda)$ such that

$$\left| \mathrm{Pr}_{x \leftarrow \mathcal{X}_\lambda} \left[A(1^\lambda, x) = 1 \right] - \mathrm{Pr}_{y \leftarrow \mathcal{Y}_\lambda} \left[A(1^\lambda, y) = 1 \right] \right| \leq \nu(\lambda)$$

for all sufficiently large $\lambda \in \mathbb{N}$.

The "Sample-then-Extract" Lemma. The following lemma due to Vadhan [25] plays a major role in our constructions. This is a refinement of the fundamental "sample-then-extract" lemma that was originally proved by Nisan and Zuckerman [21], stating that a random of sample of bits from a string essentially preserves its min-entropy rate. Vadhan's refinement shows that the min-entropy rate is in fact preserved up to an arbitrarily small additive loss, whereas the original lemma loses a logarithmic factor. Intuitively, the lemma states that if $\mathcal{X} \in \{0,1\}^n$ is a δn-source, and $\mathcal{X}_S \in \{0,1\}^t$ is the projection of $\mathcal{X}$ onto a random set $S \subseteq [n]$ of t positions, then, with high probability, $\mathcal{X}_S$ is statistically-close to a $\delta' t$-source, where $\delta' = \Omega(\delta)$. Whereas Nisan and Zuckerman [21] and Vadhan [25] were concerned with the amount of randomness that is required for sampling the t positions, in our case we can allow ourselves to sample the set S uniformly at random, and this leads to the following simplified form of the lemma:

Lemma 2.2 ([25] – simplified). *Let $\mathcal{X}$ be a δn-source over $\{0,1\}^n$, let $t \in [n]$, and let $\mathcal{S}$ denote the uniform distribution over sets $S \subseteq [n]$ of size t. Then, there exists a distribution $\mathcal{W}$ over $\{0,1\}^t$, jointly distributed with $\mathcal{S}$, such that the following hold:*

1. *$(\mathcal{S}, \mathcal{X}_\mathcal{S})$ is $2^{-\Omega(\delta t / \log(1/\delta))}$-close to $(\mathcal{S}, \mathcal{W})$.*
2. *For any set $S \subseteq [n]$ of size t it holds that $\mathcal{W}|_{\mathcal{S}=S}$ is a $\delta' t$-source for $\delta' = \delta/4$.*

2.2   Deterministic Public-Key Encryption

A deterministic public-key encryption scheme is almost identical to a (randomized) public-key encryption scheme, where the only difference is that the encryption algorithm is deterministic. More specifically, a deterministic public-key encryption scheme is a triple of polynomial-time algorithms $\Pi = (\mathsf{KG}, \mathsf{Enc}, \mathsf{Dec})$. The key-generation algorithm KG is a randomized algorithm which takes as input the security parameter 1^λ, where $\lambda \in \mathbb{N}$, and outputs a pair (pk, sk) of a public key pk and a secret key sk. The encryption algorithm Enc takes as input the security parameter 1^λ, a public key pk, and a plaintext $m \in \{0,1\}^{n(\lambda)}$, and outputs a ciphertext $c \in \{0,1\}^{t(\lambda)}$. The (possibly deterministic) decryption algorithm Dec takes as input the security parameter 1^λ, a secret key sk, and a ciphertext $c \in \{0,1\}^{t(\lambda)}$, and outputs either a plaintext $m \in \{0,1\}^{n(\lambda)}$ or the special symbol $\bot$. For succinctness, we will always assume 1^λ as an implicit input to all algorithms and refrain from explicitly specifying it.

In terms of security, in this paper we follow the standard approach for formalizing the security of deterministic public-key encryption schemes introduced by Bellare, Boldyreva and O'Neill [2] and further studied by Bellare, Fischlin, O'Neill and Ristenpart [4] and by Boldyreva, Fehr and O'Neill [8]. Specifically, we consider the PRIV-IND notion of security asking that any efficient algorithm has only a negligible advantage in distinguishing between encryptions of different sequences of plaintexts as long as each plaintext is sampled from high-entropy sources. We also consider the PRIV1-IND notion of security that focuses on a single plaintext, and asks that any efficient algorithm has only a negligible advantage in distinguishing between encryptions of different plaintexts that are sampled from high-entropy sources. This notion of security was shown by Boldyreva, Fehr and O'Neill [8] to guarantee security for block-sources of messages (that is, for sequences of messages where each message has high-entropy even when conditioned on the previous messages).

For defining these notions of security we rely on the following notation. We denote by $\mathbf{m} = (m_1, \ldots, m_\ell)$ a sequence of plaintexts, and by $\mathbf{c} = \mathsf{Enc}_{pk}(\mathbf{m})$ the sequence of their encryptions $(\mathsf{Enc}_{pk}(m_1), \ldots, \mathsf{Enc}_{pk}(m_\ell))$ under a public key pk.

Definition 2.3 (k-source ℓ-message adversary). *Let $A = (A_1, A_2)$ be a probabilistic polynomial-time algorithm, and let $k = k(\lambda)$ and $\ell = \ell(\lambda)$ be functions of the security parameter $\lambda \in \mathbb{N}$. For any $\lambda \in \mathbb{N}$ denote by $(\mathcal{M}_\lambda^{(0)}, \mathcal{M}_\lambda^{(1)}, \mathcal{STATE}_\lambda)$ the distribution corresponding to the output of $A_1(1^\lambda)$. Then, A is a k-source ℓ-message adversary if the following properties hold:*

1. *$\mathcal{M}_\lambda^{(b)} = \left(\mathcal{M}_{1,\lambda}^{(b)}, \ldots, \mathcal{M}_{\ell,\lambda}^{(b)} \right)$ is a distribution over sequences of ℓ plaintexts for each $b \in \{0,1\}$.*
2. *For any $\lambda \in \mathbb{N}$, $i, j \in [\ell]$, and $\left(\left(m_1^{(0)}, \ldots, m_\ell^{(0)} \right), \left(m_1^{(1)}, \ldots, m_\ell^{(1)} \right), \mathsf{state} \right)$ that is produced by $A_1(1^\lambda)$ it holds that $m_i^{(0)} = m_j^{(0)}$ if and only if $m_i^{(1)} = m_j^{(1)}$.*
3. *For any $\lambda \in \mathbb{N}$, $b \in \{0,1\}$, $i \in [\ell]$, and $\mathsf{state} \in \{0,1\}^*$ it holds that $\mathcal{M}_{i,\lambda}^{(b)} | \mathcal{STATE}_\lambda = \mathsf{state}$ is a $k(\lambda)$-source.*

Definition 2.4 (PRIV-IND). *A deterministic public-key encryption scheme $\Pi = (\mathsf{KG}, \mathsf{Enc}, \mathsf{Dec})$ is PRIV-IND-secure for $k(\lambda)$-source $\ell(\lambda)$-message* adversaries *if for any probabilistic polynomial-time $k(\lambda)$-source $\ell(\lambda)$-message adversary $A = (A_1, A_2)$ there exists a negligible function $\nu(\lambda)$ such that*

$$\mathsf{Adv}_{\Pi,A,\lambda}^{\mathsf{PRIV-IND}} \stackrel{\mathrm{def}}{=} \left| \Pr\left[\mathsf{Expt}_{\Pi,A,\lambda}^{\mathsf{PRIV-IND}}(0) = 1 \right] - \Pr\left[\mathsf{Expt}_{\Pi,A,\lambda}^{\mathsf{PRIV-IND}}(1) = 1 \right] \right| \leq \nu(\lambda)$$

for all sufficiently large $\lambda \in \mathbb{N}$, where $\mathsf{Expt}_{\Pi,A,\lambda}^{\mathsf{PRIV-IND}}(b)$ is defined as follows:

1. $(pk, sk) \leftarrow \mathsf{KG}(1^\lambda)$.
2. $(\mathbf{m}_0, \mathbf{m}_1, \mathsf{state}) \leftarrow A_1(1^\lambda)$.
3. $\mathbf{c} \leftarrow \mathsf{Enc}_{pk}(\mathbf{m}_b)$.
4. *Output* $A_2(1^\lambda, pk, \mathbf{c}, \mathsf{state})$.

Definition 2.5 (PRIV1-IND). *A deterministic public-key encryption scheme $\Pi = (\mathsf{KG}, \mathsf{Enc}, \mathsf{Dec})$ is PRIV1-IND-secure for $k(\lambda)$-source adversaries if for any probabilistic polynomial-time $k(\lambda)$-source 1-message adversary $A = (A_1, A_2)$ there exists a negligible function $\nu(\lambda)$ such that*

$$\mathsf{Adv}_{\Pi,A,\lambda}^{\mathsf{PRIV1-IND}} \stackrel{\mathrm{def}}{=} \left| \Pr\left[\mathsf{Expt}_{\Pi,A,\lambda}^{\mathsf{PRIV1-IND}}(0) = 1 \right] - \Pr\left[\mathsf{Expt}_{\Pi,A,\lambda}^{\mathsf{PRIV1-IND}}(1) = 1 \right] \right| \leq \nu(\lambda)$$

for all sufficiently large $\lambda \in \mathbb{N}$, where $\mathsf{Expt}_{\Pi,A,\lambda}^{\mathsf{PRIV1-IND}}(b)$ is defined as follows:

1. $(pk, sk) \leftarrow \mathsf{KG}(1^\lambda)$.
2. $(m_0, m_1, \mathsf{state}) \leftarrow A_1(1^\lambda)$.
3. $c \leftarrow \mathsf{Enc}_{pk}(m_b)$.
4. *Output* $A_2(1^\lambda, pk, c, \mathsf{state})$.

3   Modeling Incremental Deterministic Public-Key Encryption

In this section we present a framework for modeling the incrementality of deterministic public-key encryption schemes. Intuitively, a deterministic public-key encryption scheme is *incremental* if any small modification of a plaintext m resulting in a plaintext m' can be efficiently carried over for updating the encryption $c = \mathsf{Enc}_{pk}(m)$ of m to the encryption $c' = \mathsf{Enc}_{pk}(m')$ of m'. For capturing the efficiency of such an update operation we consider two natural complexity measures[4]:

- Input locality: The number of ciphertexts bits that are affected when flipping a single plaintext bit.
- Query complexity: The number of public-key, plaintext, and ciphertext bits that have to be read in order to update the ciphertext when flipping a single plaintext bit.

---

[4] For simplicity we focus on the case where both plaintexts and ciphertexts are represented as bit strings. We note, however, that our approach easily generalizes to arbitrary message and ciphertext spaces.

For capturing the above measures of efficiency we model the update operation as a probabilistic polynomial-time algorithm Update that receives as input the index i^* of a plaintext bit to be flipped, and has oracle access to the individual bits of the public key pk, the plaintext m to be modified, and to its encryption $c = \mathsf{Enc}_{pk}(m)$. That is, the algorithm Update can submit queries of the form (pk, i), (m, i) or (c, i), which are answered with the ith bit of pk, m, or c, respectively.

More formally, let $\Pi = (\mathsf{KG}, \mathsf{Enc}, \mathsf{Dec})$ be a deterministic public-key encryption scheme with message space $\{0,1\}^n$ and ciphertext space $\{0,1\}^t$ (where $n = n(\lambda)$ and $t = t(\lambda)$ are functions of the security parameter $\lambda \in \mathbb{N}$), and let Update be its corresponding update algorithm. We denote by $S \leftarrow \mathsf{Update}^{pk,m,c}(1^\lambda, i^*)$ the process in which the update algorithm with input $i^* \in [n]$ and oracle access to the individual bits of the public key pk, the plaintext m to be modified, and to its encryption $c = \mathsf{Enc}_{pk}(m)$, outputs a set $S \subseteq [t]$ of positions indicating which bits of the ciphertext c have to be flipped.

Definition 3.1 (Incremental deterministic PKE). *Let $\Pi = (\mathsf{KG}, \mathsf{Enc}, \mathsf{Dec})$ be a deterministic public-key encryption scheme with message space $\{0,1\}^n$ and ciphertext space $\{0,1\}^t$, where $n = n(\lambda)$ and $t = t(\lambda)$ are functions of the security parameter $\lambda \in \mathbb{N}$. The scheme Π is $\Delta(\lambda)$-incremental is there exists a probabilistic polynomial-time algorithm Update satisfying the following requirements:*

1. *Correctness: There exists a negligible function $\nu(\lambda)$ such that for all sufficiently large $\lambda \in \mathbb{N}$, for any plaintext $m \in \{0,1\}^n$ and for any index $i^* \in [n]$ it holds that*

$$\Pr\left[c' = \mathsf{Enc}_{pk}(m') \middle| \begin{array}{c} c = \mathsf{Enc}_{pk}(m),\ S \leftarrow \mathsf{Update}^{pk,m,c}(1^\lambda, i^*) \\ m'[i^*] = \neg m[i^*] \\ m'[i] = m[i] \text{ for all } i \in [n] \setminus \{i^*\} \\ c'[j] = \neg c[j] \text{ for all } j \in S \\ c'[j] = c[j] \text{ for all } j \in [t] \setminus S \end{array} \right] \geq 1 - \nu(\lambda),$$

   where the probability is taken over the internal coin tosses of KG and Update.
2. *Efficiency: For all sufficiently large $\lambda \in \mathbb{N}$ the algorithm $\mathsf{Update}^{(\cdot)}(1^\lambda, \cdot)$ issues at most $\Delta(\lambda)$ oracle queries and outputs sets of size at most $\Delta(\lambda)$.*

Access to the Plaintext. When providing the update algorithm with oracle access to the bits of the plaintext $m \in \{0,1\}^n$ we can assume without loss of generality that the only update operations are to flip the ith bit of m for $i \in [n]$. That is, one can also consider the operation of setting the ith bit of m to 0 or 1, but this can be handled by first querying the ith bit of m and then flipping it if it is different than the required value. We note, however, that for supporting only flipping operations it is not clear that access to the plaintext must be provided.

An important observation is that when access to the plaintext is not provided (i.e., when the update algorithm can query only the public key and the ciphertext), it is impossible to support the operation of setting a bit to 0 and 1 while providing PRIV1-IND security. That is, any such update algorithm can be used to attack the PRIV1-IND security of the scheme by distinguishing between encryptions of high-entropy messages (and this holds for any level of incrementality)[5].

Privately-Incremental Schemes. In various scenarios it may be natural to provide the update algorithm with access not to the plaintext m but rather to the secret key sk (and thus indirect access to the plaintext which may be less efficient in terms of query complexity). Consider for example, a scenario in which a client stores an encrypted version $\bar{F}$ of a file F on a remote and untrusted server. In this the client does not have direct access to the file F, but only indirect access by using its secret key to recover parts of the file. In such a scenario it is required to capture the efficiency of the client by considering its query complexity to the secret key (and ciphertext) and not to the plaintext. This leads to a natural variant of Definition 3.1 in which the update algorithm is given oracle access to the public key pk, the secret key sk, and the ciphertext c (but no direct access to the plaintext).

4 A Generic Construction via Random Partitioning

In this section we present a generic construction of an incremental PRIV1-IND-secure deterministic public-key encryption scheme from any PRIV-IND-secure deterministic public-key encryption scheme. As discussed in Section 1.3 our approach is a "randomized" alternative to the commonly-used approach of dividing the plaintext into small blocks and encrypting each block. Instead of dividing an n-bit plaintext m into fixed blocks, we project it onto a uniformly chosen partition $S_1, \ldots, S_{n/t}$ of the plaintext positions $\{1, \ldots, n\}$ to sets of size t each, and then separately encrypt each of the projections $m_{S_1}, \ldots, m_{S_{n/t}}$ using the underlying encryption scheme. Thus, when flipping a single bit of m we only need to update the encryption of the projection m_{S_i} for which the corresponding position belongs to the set S_i. Therefore, the resulting scheme enjoys the same incrementality that the underlying scheme has for small blocks. A more formal description follows.

The Scheme. Let $\Pi' = (\mathsf{KG}', \mathsf{Enc}', \mathsf{Dec}')$ be a deterministic public-key encryption scheme for n'-bit plaintexts that is IND-PRIV-secure for k'-source ℓ'-message adversaries, where $n' = n'(\lambda)$, $k' = k'(\lambda)$ and $\ell' = \ell'(\lambda)$ are functions of the security parameter $\lambda \in \mathbb{N}$. We construct a deterministic public-key encryption scheme $\Pi = (\mathsf{KG}, \mathsf{Enc}, \mathsf{Dec})$ for n-bit plaintexts that is PRIV1-IND-secure

---

[5] Consider the adversary $A = (A_1, A_2)$ that is defined as follows. The algorithm A_1 outputs $(m_0, m_1, \mathsf{state})$ where $m_0 \leftarrow \mathcal{U}_k \| 0^{n-k}$ and $m_1 \leftarrow \mathcal{U}_n$ are sampled independently at random, and $\mathsf{state} = \perp$. That is, m_0 is a distributed uniformly conditioned on ending with 0^{n-k}, and m_1 is distributed uniformly. The algorithm A_2 on input $c = \mathsf{Enc}_{pk}(m_b)$ invokes the update algorithm to set the leftmost k bits of the plaintext corresponding to c to 0, and then compares the resulting ciphertext to $\mathsf{Enc}_{pk}(0^n)$. Note that if $b = 0$ then the two ciphertexts are always equal, and if $b = 1$ then they are equal only with probability $2^{-(n-k)}$.

for k-source adversaries, where $n = n(\lambda)$ and $k = k(\lambda)$ are functions of the security parameter $\lambda \in \mathbb{N}$ as follows:

- The algorithm KG on input the security parameter 1^λ samples $(pk', sk') \leftarrow$ KG$'(1^\lambda)$ together with a uniformly chosen partition $S_1, \ldots, S_{n/t}$ of $[n]$, where each set in the partition is of size $t = \Theta(\frac{n}{k} \cdot k')$. It then outputs $pk = (pk', S_1, \ldots, S_{n/t})$ and $sk = sk'$.[6]
- The algorithm $\mathsf{Enc}_{pk}(\cdot)$ on input a plaintext $m \in \{0,1\}^n$ outputs the ciphertext $(\mathsf{Enc}'_{pk'}(1||m_{S_1}), \ldots, \mathsf{Enc}'_{pk'}(n/t||m_{S_{n/t}}))$.
- The algorithm $\mathsf{Dec}_{sk}(\cdot)$ on input a ciphertext $(c_1, \ldots, c_{n/t})$ computes $m_{S_i} = \mathsf{Dec}'_{sk'}(c_i)$ for every $i \in [n/t]$, and outputs the plaintext m defined by the projections $m_{S_1}, \ldots, m_{S_{n/t}}$.

We establish the security and incrementality of this scheme by proving the following theorem (due to space limitations the proof appears in the full version):

Theorem 4.1. *Assuming that Π' encrypts n'-bit plaintexts, for $n' = t + \log(n/t)$, and is IND-PRIV-secure for k'-source ℓ'-message adversaries, for some $k' = \omega(\log^2 n)$ and for $\ell' = n/t$, the scheme Π is PRIV1-IND-secure for k-sources.*

5 A Construction Based on the Decisional Diffie-Hellman Assumption

In this section we construct a deterministic public-key encryption scheme that enjoys essentially optimal incrementality, and guarantees PRIV1-IND security based on the Decisional Diffie-Hellman (DDH) assumption. We begin by introducing rather standard notation and then describe the scheme.

Notation. Let G be a group of prime order p that is generated by $g \in G$. For any matrix $A = \{a_{ij}\}_{i \in [n], j \in [n]} \in \mathbb{Z}_p^{n \times n}$ we denote by $g^A \in G^{n \times n}$ the matrix $\{g^{a_{ij}}\}_{i \in [n], j \in [n]}$. In addition, for a column vector $m = (m_1, \ldots, m_n)^\top \in \mathbb{Z}_p^n$ and a matrix $A = \{a_{ij}\}_{i \in [n], j \in [n]} \in \mathbb{Z}_p^{n \times n}$ we define

$$A \odot g^m \stackrel{\text{def}}{=} g^A \odot m \stackrel{\text{def}}{=} g^{Am} = (g^{\sum_i a_{1,i} m_i}, \ldots, g^{\sum_i a_{n,i} m_i})^\top \in G^n \ .$$

The Scheme. Let GroupGen be a probabilistic polynomial-time algorithm that takes as input the security parameter 1^λ, and outputs a triplet (G, p, g) where G is a group of prime order p that is generated by $g \in G$, and p is a λ-bit prime number. The scheme is parameterized by the security parameter λ, the message length $n = n(\lambda)$, and the min-entropy $k = k(\lambda)$ for which the scheme is secure. Both n and k are polynomials in the security parameter. The scheme $\Pi = (\mathsf{KG}, \mathsf{Enc}, \mathsf{Dec})$ is defined as follows:

---

[6] Without loss of generality we can assume that t divides n, as otherwise we can pad plaintexts with at most t zeros, and for our choice of parameters this would only have a minor effect on the min-entropy rate.

- **Key generation.** The algorithm KG on input the security parameter 1^λ samples $(G, p, g) \leftarrow \mathsf{GroupGen}(1^\lambda)$, and a matrix $A \leftarrow \mathcal{A}_{n,k,p}$, where $\mathcal{A}_{n,k,p}$ is a distribution over $\mathbb{Z}_p^{n \times n}$ which is defined below. It then outputs $pk = (G, p, g, g^A)$ and $sk = A^{-1}$.
- **Encryption.** The algorithm $\mathsf{Enc}_{pk}(\cdot)$ on input a plaintext $m \in \{0,1\}^n$ outputs the ciphertext $g^A \odot m = g^{Am} \in G^n$.
- **Decryption.** The algorithm $\mathsf{Dec}_{sk}(\cdot)$ on input a ciphertext $g^c = (g^{c_1}, \ldots, g^{c_n}) \in G^n$ first computes $w = A^{-1} \odot g^c = g^{A^{-1}c} \in G^n$, and lets $w = (g^{m_1}, \ldots, g^{m_n})$. If $m = (m_1, \ldots, m_n) \in \{0,1\}^n$ (note that this test can be computed efficiently) then it outputs m, and otherwise it outputs $\bot$.

The Distribution $\mathcal{A}_{n,k,p}$. For completing the description of our scheme it remains to specify the distribution $\mathcal{A}_{n,k,p}$ that is defined over $\mathbb{Z}_p^{n \times n}$. Looking ahead this distribution will be used to define the distribution of injective keys in our collection of smooth trapdoor functions. In fact, we find it convenient to first specify the distribution $\widetilde{\mathcal{A}}_{n,k,p}$ that will be used to define the distribution of smooth keys. These two distributions rely on the following distributions as building block:

- $\mathcal{R}_{n,k,p}$: **sparse random $\ell \times n$ matrices.** The distribution $\mathcal{R}_{n,k,p}$ is defined as a random sample from $\mathbb{Z}_p^{\ell \times n}$ matrices that have exactly $t = \Theta(\frac{n}{k} \cdot \log^3 n)$ non-zero entries in each row, where $\ell = \Theta(k/\log p)$.
- $\mathcal{D}_{n,k,p}$: **diagonally-striped $\ell \times n$ matrices.** The distribution $\mathcal{D}_{n,k,p}$ is defined as a random sample from $\mathbb{Z}_p^{\ell \times n}$ matrices whose elements d_{ij} are non-zero if and only if $i \equiv j \pmod{\ell}$ (for simplicity we assume that n is divisible by ℓ).

The distribution $\widetilde{\mathcal{A}}_{n,k,p}$ over $\mathbb{Z}_p^{n \times n}$ is defined as matrices $\widetilde{A}$ obtained by independently sampling $R \leftarrow \mathcal{R}_{n,k,p}$, $D_1 \leftarrow \mathcal{D}_{n,k,p}$, and $D_2 \leftarrow \mathcal{D}_{n,k,p}$, and letting $\widetilde{A} \overset{\text{def}}{=} D_2^{\mathsf{T}} \times (R + D_1)$. Then, the distribution $\mathcal{A}_{n,k,p}$ is defined as matrices A obtained by sampling a matrix $\widetilde{A} \leftarrow \widetilde{\mathcal{A}}_{n,k,p}$ and then *re-sampling* all its non-zero entries from $\mathbb{Z}_p$ independently and uniformly at random. In other words, the resulting matrix A preserves zeroes of the matrix $\widetilde{A}$, while randomizing all other elements (and thus linear dependencies between rows) of the original matrix. See Figure 1 for an illustration of the distributions $\mathcal{R}_{n,k,p}$, $\mathcal{D}_{n,k,p}$ and $\widetilde{\mathcal{A}}_{n,k,p}$.

Intuitively, the matrix D_1 is only meant to ensure that such the resulting matrix A is invertible. Indeed, the matrix D_1 guarantees that with an overwhelming probability all the elements on the main diagonal of A are non-zeros. Now, ignoring the matrix D_1, the matrix $\widetilde{A}$ is generated to satisfy two properties. The first is that each of its first ℓ rows has t randomly chosen entries with values that are chosen uniformly from $\mathbb{Z}_p$, and all other $n - t$ entries are zeros. Looking ahead, when computing the inner product of such a row with a source of min-entropy larger than $\log p$, the "sample-then-extract" lemma (see Lemma 2.2) guarantees that the output is statistically close to uniform. The second property, is that each of its last $n - \ell$ rows are linear combinations of the first ℓ rows, and therefore the image of its corresponding linear map is determined by the first ℓ rows.

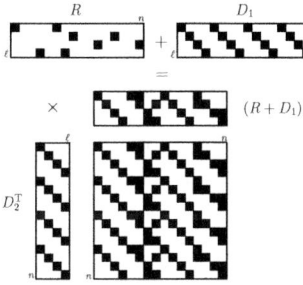

Fig. 1. The distributions $\mathcal{R}_{n,k,p}$, $\mathcal{D}_{n,k,p}$ and $\widetilde{\mathcal{A}}_{n,k,p}$

The following theorem establishes the security of the scheme (due to space limitations the proof appears in the full version):

Theorem 5.1. *Under the Decisional Diffie-Hellman assumption the scheme Π is PRIV1-IND-secure for k-sources.*

Acknowledgements. We thank Salil Vadhan for useful discussions regarding Lemma 2.2.

References

1. Applebaum, B., Ishai, Y., Kushilevitz, E.: Cryptography in NC^0. SIAM Journal on Computing 36(4), 845–888 (2006)
2. Bellare, M., Boldyreva, A., O'Neill, A.: Deterministic and Efficiently Searchable Encryption. In: Menezes, A. (ed.) CRYPTO 2007. LNCS, vol. 4622, pp. 535–552. Springer, Heidelberg (2007)
3. Bellare, M., Brakerski, Z., Naor, M., Ristenpart, T., Segev, G., Shacham, H., Yilek, S.: Hedged Public-Key Encryption: How to Protect against Bad Randomness. In: Matsui, M. (ed.) ASIACRYPT 2009. LNCS, vol. 5912, pp. 232–249. Springer, Heidelberg (2009)
4. Bellare, M., Fischlin, M., O'Neill, A., Ristenpart, T.: Deterministic Encryption: Definitional Equivalences and Constructions without Random Oracles. In: Wagner, D. (ed.) CRYPTO 2008. LNCS, vol. 5157, pp. 360–378. Springer, Heidelberg (2008)
5. Bellare, M., Goldreich, O., Goldwasser, S.: Incremental Cryptography: The Case of Hashing and Signing. In: Desmedt, Y.G. (ed.) CRYPTO 1994. LNCS, vol. 839, pp. 216–233. Springer, Heidelberg (1994)
6. Bellare, M., Goldreich, O., Goldwasser, S.: Incremental cryptography and application to virus protection. In: Proceedings of the 27th Annual ACM Symposium on Theory of Computing, pp. 45–56 (1995)
7. Bellare, M., Micciancio, D.: A New Paradigm for Collision-Free Hashing: Incrementality at Reduced Cost. In: Fumy, W. (ed.) EUROCRYPT 1997. LNCS, vol. 1233, pp. 163–192. Springer, Heidelberg (1997)
8. Boldyreva, A., Fehr, S., O'Neill, A.: On Notions of Security for Deterministic Encryption, and Efficient Constructions without Random Oracles. In: Wagner, D. (ed.) CRYPTO 2008. LNCS, vol. 5157, pp. 335–359. Springer, Heidelberg (2008)

9. Bolosky, W.J., Corbin, S., Goebel, D., Douceur, J.R.: Single instance storage in Windows 2000. In: Proceedings of the 4th USENIX Windows Systems Symposium, pp. 13–24 (2000)
10. Brakerski, Z., Segev, G.: Better Security for Deterministic Public-Key Encryption: The Auxiliary-Input Setting. In: Rogaway, P. (ed.) CRYPTO 2011. LNCS, vol. 6841, pp. 543–560. Springer, Heidelberg (2011)
11. Buonanno, E., Katz, J., Yung, M.: Incremental Unforgeable Encryption. In: Matsui, M. (ed.) FSE 2001. LNCS, vol. 2355, pp. 109–124. Springer, Heidelberg (2002)
12. Dodis, Y., Ostrovsky, R., Reyzin, L., Smith, A.: Fuzzy extractors: How to generate strong keys from biometrics and other noisy data. SIAM Journal on Computing 38(1), 97–139 (2008)
13. Dodis, Y., Smith, A.: Entropic security and the encryption of high entropy messages. In: Proceedings of the 2nd Theory of Cryptography Conference, pp. 556–577 (2005)
14. Douceur, J.R., Adya, A., Bolosky, W.J., Simon, D., Theimer, M.: Reclaiming space from duplicate files in a serverless distributed file system. In: Proceedings of the 22nd International Conference on Distributed Computing Systems, pp. 617–624 (2002)
15. Fischlin, M.: Incremental Cryptography and Memory Checkers. In: Fumy, W. (ed.) EUROCRYPT 1997. LNCS, vol. 1233, pp. 393–408. Springer, Heidelberg (1997)
16. Fischlin, M.: Lower bounds for the signature size of incremental schemes. In: Proceedings of the 38th Annual IEEE Symposium on Foundations of Computer Science, pp. 438–447 (1997)
17. Freeman, D.M., Goldreich, O., Kiltz, E., Rosen, A., Segev, G.: More Constructions of Lossy and Correlation-Secure Trapdoor Functions. In: Nguyen, P.Q., Pointcheval, D. (eds.) PKC 2010. LNCS, vol. 6056, pp. 279–295. Springer, Heidelberg (2010)
18. Fuller, B., O'Neill, A., Reyzin, L.: A unified approach to deterministic encryption: New constructions and a connection to computational entropy. Cryptology ePrint Archive, Report 2012/005 (2012)
19. Goldwasser, S., Micali, S.: Probabilistic encryption. Journal of Computer and System Sciences 28(2), 270–299 (1984)
20. Micciancio, D.: Oblivious data structures: Applications to cryptography. In: Proceedings of the 29th Annual ACM Symposium on the Theory of Computing, pp. 456–464 (1997)
21. Nisan, N., Zuckerman, D.: Randomness is linear in space. Journal of Computer and System Sciences 52(1), 43–52 (1996)
22. Peikert, C., Waters, B.: Lossy trapdoor functions and their applications. In: Proceedings of the 40th Annual ACM Symposium on Theory of Computing, pp. 187–196 (2008)
23. Quinlan, S., Dorward, S.: Venti: A new approach to archival storage. In: Long, D.D.E. (ed.) FAST, pp. 89–101. USENIX (2002)
24. Russell, A., Wang, H.: How to Fool an Unbounded Adversary with a Short Key. In: Knudsen, L.R. (ed.) EUROCRYPT 2002. LNCS, vol. 2332, pp. 133–148. Springer, Heidelberg (2002)
25. Vadhan, S.P.: Constructing locally computable extractors and cryptosystems in the bounded-storage model. Jounal of Cryptology 17(1), 43–77 (2004)
26. Wee, H.: Dual Projective Hashing and Its Applications - Lossy Trapdoor Functions and More. In: Pointcheval, D., Johansson, T. (eds.) EUROCRYPT 2012, LNCS, vol. 7237, pp. 246–262. Springer, Heidelberg (2012)
27. Zhu, B., Li, K., Patterson, R.H.: Avoiding the disk bottleneck in the data domain deduplication file system. In: Proceedings of the 6th USENIX Conference on File and Storage Technologies, pp. 269–282 (2008)

Standard Security Does Not Imply Security against Selective-Opening

Mihir Bellare[1], Rafael Dowsley[1], Brent Waters[2], and Scott Yilek[3]

[1] Department of Computer Science & Engineering, University of California San Diego
http://cseweb.ucsd.edu/~mihir, http://cseweb.ucsd.edu/~rdowsley
[2] Department of Computer Science, University of Texas at Austin
http://www.cs.utexas.edu/~bwaters
[3] Department of Computer and Information Sciences, University of St. Thomas
http://personal.stthomas.edu/yile5901

Abstract. We show that no commitment scheme that is hiding and binding according to the standard definition is semantically-secure under selective opening attack (SOA), resolving a long-standing and fundamental open question about the power of SOAs. We also obtain the first examples of IND-CPA encryption schemes that are not secure under SOA, both for sender corruptions where encryption coins are revealed and receiver corruptions where decryption keys are revealed. These results assume only the existence of collision-resistant hash functions.

1 Introduction

A *commitment scheme* $\mathcal{E}$ can be applied to a message m and coins r to (deterministically) produce a commitment $c \leftarrow \mathcal{E}(m; r)$ that is sent to a receiver. The sender can later "open" the commitment by providing m, r and the receiver checks that $\mathcal{E}(m; r) = c$. The first security requirement, often called hiding, is formalized as IND-CPA, namely an adversary knowing m_0, m_1 and $\mathcal{E}(m_b; r)$ for random b, r has negligible advantage in computing challenge bit b. The second requirement, binding, asks that it be hard for an adversary to produce r_0, r_1 and *distinct* m_0, m_1 such that $\mathcal{E}(m_0; r_0) = \mathcal{E}(m_1; r_1) \neq \bot$. Let us refer to a commitment scheme as HB-secure (Hiding and Binding) if it satisfies both these properties. HB-security is the standard requirement and HB-secure commitment schemes are a fundamental tool in cryptography in general and in protocol design in particular. HB-secure commitment implies PRGs [31], PRFs [21] and ZK proofs for **NP** [24].

Suppose there are n committers, the i-th computing its commitment $\mathbf{c}[i] \leftarrow \mathcal{E}(\mathbf{m}[i]; \mathbf{r}[i])$ to its message $\mathbf{m}[i]$ using coins $\mathbf{r}[i]$, the coins of different committers being of course not only random but also independent of each other. An adversary computes, as a function of the vector $\mathbf{c}$ of commitments, a subset $I \subseteq \{1, \ldots, n\}$ of the senders, and obtains the corresponding openings, namely $\langle \mathbf{m}[i] : i \in I \rangle$ and $\langle \mathbf{r}[i] : i \in I \rangle$. This is called a selective opening attack (SOA). We say that $\mathcal{E}$ is SOA-secure if privacy of the un-opened messages is preserved, meaning the adversary, after its SOA, cannot learn anything about $\langle \mathbf{m}[i] : i \notin I \rangle$

D. Pointcheval and T. Johansson (Eds.): EUROCRYPT 2012, LNCS 7237, pp. 645–662, 2012.

other than it would from possession of $\langle \mathbf{m}[i] : i \in I \rangle$. (That is, the coins are unhelpful.) SOAs arise quite naturally in multi-party cryptographic protocols and SOA-security is desirable in many such settings.

A fundamental question that was posed in this area is whether (standard) HB-security implies SOA-security, meaning, is a HB-secure commitment scheme also SOA-secure? So far, the question has received neither a positive nor a negative answer. Intuitively, the answer would appear to be "yes," for how could the coins accompanying the opened messages help, beyond the opened messages themselves, in revealing something about the un-opened messages? Yet attempts to prove SOA-security of a commitment scheme based on its HB-security have failed. But attempts to find a counter-example have also failed. We do not have a single example, even artificial, of a HB-secure commitment scheme that is demonstrably not SOA-secure. This situation has vexed and intrigued cryptographers for many years and been the subject or inspiration for much work [12,19,13,35,3,20,29,7,28].

This paper answers this long-standing open question. We show that the answer is negative. We give an example of a HB-secure commitment scheme which we prove is not SOA-secure. In fact our result is much stronger. It shows that *no* HB-secure commitment scheme is SOA-secure. Given any HB-secure commitment scheme, we present an attack showing it is not SOA-secure. Before going on to our results on encryption let us expand on this result on commitment including its implications and its relation to previous work.

SOA-SECURE COMMITMENT. Dwork, Naor, Reingold and Stockmeyer (DNRS) [19] gave a definition of SOA-secure commitments, henceforth referred to as SS-SOA, that captures semantic security for relations via a simulation-based formalization. Suitable for applications and widely accepted as the right definition, SS-SOA is what we use in our results. We show that no HB-secure commitment scheme is SS-SOA-secure by presenting, for any given HB-secure commitment scheme $\mathcal{E}$, an adversary for which we prove that there is no successful simulator. We do *not* assume the simulation is blackbox. The only assumption made is the existence of a collision-resistant (CR) hash function.

This general result rules out SS-SOA security for particular schemes. For example, a widely employed way to commit to $m \in \mathbb{Z}_p$ is by picking $r \in \mathbb{Z}_p$ at random and returning $\mathcal{E}(m; r) = g^m h^r \in \mathbb{G}$ where g, h are generators of a group $\mathbb{G}$ of prime order p [36]. This scheme is binding if the DL problem is hard in $\mathbb{G}$ and it is unconditionally hiding. Our results imply that it is not SS-SOA secure. They yield a specific attack, in the form of an adversary for which there is no simulator. Since CR hash functions exist if DL is hard, one does not even need extra assumptions. We stress that this is just an example; our result rules out SS-SOA security for *all* HB-secure schemes.

IMPLICATIONS FOR IND-SOA-CRS. An indistinguishability-based definition of SOA-secure commitment is given in [3,29]. It only applies when the message vector $\mathbf{m}$ is drawn from what's called a "conditionally re-samplable (CRS) distribution," and accordingly we denote it IND-SOA-CRS. This definition is of limited use in applications because message distributions there are often not CRS, but for CRS distributions the definition is intuitively compelling and sound.

Letting SS-SOA-CRS denote the restriction of SS-SOA to CRS distributions, [3,29] had noted that SS-SOA-CRS implies IND-SOA-CRS and asked whether the converse was true. We settle this question in the negative, showing that SS-SOA-CRS is strictly stronger. We arrive at this separation by combining two facts. First, the message distribution underlying our negative result is CRS, meaning we say that there does not exist a HB-secure commitment scheme that is SS-SOA-CRS, not just SS-SOA. Second, it is known that there does exist a HB-secure commitment scheme that is IND-SOA-CRS [3,29].

Hofheinz [3,29] shows that any commitment scheme that is *statistically* hiding and binding is IND-SOA-CRS. This positive result does not contradict our result, because, as we have just seen (indeed, invoking this positive result to do so), IND-SOA-CRS is a strictly weaker requirement than SS-SOA or SS-SOA-CRS. A question that still remains open is whether HB-security implies IND-SOA-CRS security.

MESSAGE DISTRIBUTION. It has been suggested that the difficulty in showing that HB-security implies SS-SOA is that the messages in the vector **m** may be related to each other. Our results imply that although showing HB-security implies SS-SOA-security is not just hard but impossible, it is not for this reason. We have already noted that our negative result holds for a message distribution that is CRS. In fact, the message distribution is uniform, meaning the messages in the vector are uniformly and independently distributed strings. Even for this uniform distribution, no HB-secure commitment scheme is SS-SOA secure. This may at first glance appear to contradict known results, for DNRS [19] showed that HB-security implied SOA-security for independently distributed messages. The difference is that they only showed this for what they called semantic security for functions, a notion implied by, but not known to imply their main notion of semantic security for relations that we call SS-SOA. Thus, not only is there no contradiction, but our results settle an open question from [19]. Namely we show that their result does not extend to SS-SOA and also that SS-SOA is strictly stronger than semantic security for functions.

RANDOM ORACLES. Our result holds in the standard model and in the non-programmable random oracle (RO) model [32]. (In the latter the simulator is given oracle access to the RO and cannot define it.) In the standard (programmable) RO model [5], where the simulator can define the RO, our result is not true: there *do* exist HB-secure schemes that are SS-SOA secure. As an example, commitment scheme $\mathcal{E}^H(m;r) = H(m;r)$, where H is the RO, is HB-secure in the non-programmable RO. Our results show it is not SS-SOA in this model. However, it can be easily shown SS-SOA in the programmable RO model. Consequently, our results yield another separation between the programmable and non-programmable RO models complementing that of [32].

PREVIOUS NEGATIVE RESULTS. Hofheinz [3,29] shows that no HB-secure scheme can be proven SS-SOA secure via blackbox reduction to "standard" assumptions.

(A "standard" assumption as defined in [17,3,29] is one specified by a certain type of game.) However, it might still be possible to prove that a particular HB-secure scheme was SS-SOA in some ad hoc and non-blackbox way. The blackbox separation does not yield a single example of an HB-secure scheme that is not SS-SOA secure, let alone show, as we do, that all HB-secure schemes fail to be SS-SOA secure.

INTERACTION. Our result applies to non-interactive commitment schemes. When commitment involves an interactive protocol between sender and receiver the corresponding claim is not true. There *does* exist an interactive HB and SS-SOA secure commitment scheme. Specifically, Hofheinz [3,29] presents a particular construction of such a scheme based on one-way permutations. Further results on interactive SOA-secure commitment are [39,34].

SOA-SECURE ENCRYPTION FOR SENDER CORRUPTIONS. Turning now to encryption, consider a setting with n senders and one receiver, the latter having public encryption key ek. Sender i picks random coins $\mathbf{r}[i]$, encrypts its message $\mathbf{m}[i]$ via $\mathbf{c}[i] \leftarrow \mathcal{E}(ek, \mathbf{m}[i]; \mathbf{r}[i])$, and sends ciphertext $\mathbf{c}[i]$ to the receiver. The adversary selects, as a function of $\mathbf{c}$, a set $I \subseteq \{1, \ldots, n\}$ of the senders and corrupts them, obtaining their messages $\langle \mathbf{m}[i] : i \in I \rangle$ and coins $\langle \mathbf{r}[i] : i \in I \rangle$. As before, we say that $\mathcal{E}$ is SOA-secure if privacy of the un-opened messages is preserved. An SS-SOA definition analogous to the one for commitment was given in [3,8].

The standard and accepted security condition for encryption since [26] is of course IND-CPA. SOA-security was identified upon realizing that it is necessary to implement the assumed-secure channels in multi-party secure computation protocols like those of [9,14]. The central open question was whether or not IND-CPA implies SS-SOA. Neither a proof showing the implication is true, nor a counter-example showing it is false, had been given. We show that IND-CPA does not imply SS-SOA by exhibiting a large class of IND-CPA encryption schemes that we prove are not SS-SOA. The class includes many natural and existing schemes.

DNRS [19] had pointed out that the obstacle to proving that IND-CPA implies SS-SOA is that most encryption schemes are "committing." Our results provide formal support for this intuition. We formalize a notion of binding-security for encryption. Our result is that no binding encryption scheme is SS-SOA secure. As with commitment, it holds when the distribution on messages is uniform.

The existence of a decryption algorithm corresponding to the encryption algorithm means that for any ek created by honest key-generation, there do not exist r_0, r_1 and distinct m_0, m_1 such that $\mathcal{E}(ek, m_0; r_0) = \mathcal{E}(ek, m_1; r_1)$. Binding strengthens this condition to also hold when ek is adversarially chosen, while also relaxing it from unconditional to computational. It is thus a quite natural condition and is met by many schemes.

Inability to show that IND-CPA implies SS-SOA led to the search for specific SS-SOA secure encryption schemes. Non-commiting encryption [12] yields a solution when the number of bits encrypted is bounded by the length of the public key. The first full solution was based on lossy encryption [3,8]. Deniable encryption [11] was used to obtain further solutions [20,7]. More lossy-encryption based solutions appear in [28]. In all these solutions, the encryption scheme is *not* binding. Our results show that this is necessary to achieve SS-SOA security.

SOA-security has so far been viewed as a theoretical rather than practical issue because even if there was no proof that IND-CPA implies SS-SOA, there were no attacks on standard, practical schemes such as ElGamal. Our results change this situation for they show that ElGamal and other practical schemes are not SS-SOA secure. Thus, the above-mentioned schemes that achieve SS-SOA in more involved ways are necessary if we want SS-SOA security.

IND-CCA doesn't help: The Cramer-Shoup scheme [15] meets our definition of binding and is thus not SS-SOA secure. As with commitment, our results imply that IND-SOA-CRS security is *strictly* weaker than SS-SOA-CRS security, answering an open question from [3,8]. Subsequent to our work, the relations between different notions of SOA-security under sender corruptions were further clarified in [10] but whether there exist schemes that are IND-CPA but not IND-SOA-CRS secure remains open.

SOA-SECURE ENCRYPTION FOR RECEIVER CORRUPTIONS. In a dual of the above setting, there are n receivers and one sender, receiver i having public encryption key $\mathbf{ek}[i]$ and secret decryption key $\mathbf{dk}[i]$. For each i the sender picks random coins $\mathbf{r}[i]$, encrypts message $\mathbf{m}[i]$ via $\mathbf{c}[i] \leftarrow \mathcal{E}(\mathbf{ek}[i], \mathbf{m}[i]; \mathbf{r}[i])$, and sends ciphertext $\mathbf{c}[i]$ to receiver i. The adversary selects, as a function of $\mathbf{c}$, a set $I \subseteq \{1, \ldots, n\}$ of the receivers and corrupts them, obtaining not only the messages $\langle \mathbf{m}[i] : i \in I \rangle$ but also the decryption keys $\langle \mathbf{dk}[i] : i \in I \rangle$. As usual, we say that $\mathcal{E}$ is SOA-secure if privacy of the un-opened messages is preserved. An SS-SOA definition analogous to the ones for commitment and sender-corruptions in encryption is given in Section 5.

The status and issues are analogous to what we have seen above, namely that it has been open whether IND-CPA security implies SS-SOA for receiver corruptions, neither a proof nor a counter-example ever being given. We settle this with the first counter-examples. We define a notion of decryption verifiability for encryption that can be seen as a weak form of robustness [1]. It asks that there is an algorithm $\mathcal{W}$ such that it is hard to find ek, dk_0, dk_1 and distinct m_0, m_1 such that $\mathcal{W}(ek, dk_0, m_0)$ and $\mathcal{W}(ek, dk_1, m_1)$ both accept. We show that no IND-CPA and decryption-verifiable encryption scheme is SS-SOA secure. Standard encryption schemes like ElGamal are decryption verifiable (even though they are not robust) so our result continues to rule out SS-SOA security for many natural schemes.

Non-committing encryption [12] yields an SS-SOA scheme secure for receiver corruptions when the number of bits encrypted is bounded by the length of the secret key. Nielsen [32] showed that any non-committing encryption scheme has keys larger than the total number of message bits it can securely encrypt. This result is not known to extend to SS-SOA, meaning the existence of an SS-SOA scheme for receiver corruptions without this restriction is open. Our results do not rule out such a full solution but indicate that the scheme must not be decryption-verifiable.

2   Technical Approach

We provide a high-level description of our approach, focusing for simplicity on commitment schemes and the claim that no HB-secure commitment scheme is SS-SOA secure. We then discuss extensions and variants of our results.

THE DEFINITION. Let $\mathcal{E}$ be a commitment scheme. To compact notation, we extend it to vector inputs by letting $\mathcal{E}(\mathbf{m}; \mathbf{r})$ be the vector whose i-th component is $\mathcal{E}(\mathbf{m}[i]; \mathbf{r}[i])$. Let $\mathcal{M}$ be a message sampler that outputs a vector $\mathbf{m}$ of messages and let R be a relation. Adversary A, given ciphertext vector $\mathbf{c} = \mathcal{E}(\mathbf{m}; \mathbf{r})$ will corrupt a subset I of the senders, get their messages and coins, and output a value w. It is said to win if $R(\mathbf{m}, I, w)$ is true. The simulator, given no ciphertexts, can also corrupt a subset I of senders but gets back only the corresponding messages, and outputs a value w. It too is said to win if $R(\mathbf{m}, I, w)$ is true. Security requires that for every $\mathcal{M}, R$ and adversary A there is a simulator S such that S wins with about the same probability as A. DNRS [19, Sec 7.1] require this to be true even for any auxiliary input a given initially to A and also to S. See Section 4 for a formal definition.

THE ATTACK. Let $\mathcal{E}$ be any, given HB-secure commitment scheme. We construct $\mathcal{M}, R, A$ for which we prove there is no simulator. We let $\mathcal{M}$ output $n = 2h$ randomly and independently distributed messages, each of length ℓ. Our adversary A applies to the vector $\mathbf{c} = \mathcal{E}(\mathbf{m}; \mathbf{r})$ of commitments a hash function H to get back an h-bit string $b[1] \ldots b[h]$ and then corrupts the set of indices $I = \{2j - 1 + b[j] : 1 \leq j \leq h\}$ to get back $\langle \mathbf{m}[i] : i \in I \rangle$ and $\langle \mathbf{r}[i] : i \in I \rangle$. Its output w consists of $\mathbf{c}$ and $\langle \mathbf{r}[i] : i \in I \rangle$. We define R, on inputs $\mathbf{m}, I$ and w, to check two constraints. The *opening constraint* is that $\mathcal{E}(\mathbf{m}[i]; \mathbf{r}[i]) = \mathbf{c}[i]$ for all $i \in I$. The *hash constraint* is that $I = \{2j - 1 + b[j] : 1 \leq j \leq h\}$ for $b[1] \ldots b[h] = H(\mathbf{c})$. A detailed description of A and R is in Fig. 3.

The simulator gets no ciphertexts. It must corrupt some set I of indices to get back $\langle \mathbf{m}[i] : i \in I \rangle$. Now it must create a ciphertext vector $\mathbf{c}$ and a list $\langle \mathbf{r}[i] : i \in I \rangle$ of coins to output as w to R, and to satisfy the latter it must satisfy both constraints. Intuitively, the simulator faces a Catch-22. It is helpful for the intuition to think of H as a random oracle. The simulator could first pick I in some way, get $\langle \mathbf{m}[i] : i \in I \rangle$ from its oracle, and compute $\mathbf{c}$ and $\langle \mathbf{r}[i] : i \in I \rangle$ to satisfy the opening constraint. But it is unlikely, given only poly$(\cdot)$ queries to H, to satisfy the hash constraint. On the other hand it could pick some $\mathbf{c}$, define I to satisfy the hash constraint, and get $\langle \mathbf{m}[i] : i \in I \rangle$ from its oracle. But now it would have a hard time satisfying the opening constraint *because the commitment scheme is binding.*

This intuition that the simulator's task is hard is, however, not a proof that a simulator does not exist. Furthermore, the intuition relies on the hash function being a random oracle and we only want to assume collision-resistance. Our proof takes an arbitrary simulator and proves that the probability that it makes the relation true is small unless it finds a hash collision or violates binding. The proof involves backing up the simulator, feeding it different, random responses to its corruption query, and applying a Reset Lemma analogous to that of [4]. We do not assume the simulation is blackbox. See Theorem 2.

RELATED WORK. The strategy of specifying challenges by a hash of commitments arose first in showing failure of parallel-repetition to preserve zero-knowledge [22,23]. The model, goals and techniques are however quite different. Also in [23] the simulator is assumed to make only blackbox calls to the adversary (verifier) and we make no such assumption, and they use a pairwise independent hash rather than a CR one. We point out that although the seed of our technique can be traced back 20 years it was not noted until now that it could be of use in settling the long-standing open question of whether HB-secure commitments are SS-SOA-secure.

ADAPTIVE SECURITY. Our definition of SS-SOA, following [19,3,7] is one-shot, meaning the adversary gets all the ciphertexts at once and performs all its corruptions in parallel. A definition where the adversary can make adaptive ciphertext-creation and corruption requests is more suitable for applications. But our result is negative so using a restricted adversary only makes it stronger. (We are saying there is an attack with a one-shot adversary so certainly there is an attack with an adaptive adversary.)

The flip side is that if the adversary is allowed to be adaptive, so is the simulator. Our theorems only consider (and rule out) one-shot simulators for simplicity, but the proofs can be extended to also rule out adaptive simulators. We discuss briefly how to do this following the proof of Theorem 2.

AUXILIARY INPUTS. As indicated above, the definition of DNRS [19] that we use allows both the adversary and simulator to get an auxiliary input, denoted "z" in [19, Sec 7.1]. The simplest and most basic form of our result exploits the auxiliary input to store the key describing the CR hash function. (If the simulator can pick this key the function will not be CR.)

Auxiliary inputs model history. They were introduced in the context of zero-knowledge by Goldreich and Oren [25] who showed that in their presence ZK had natural and desirable composability properties absent under the original definition of [27]. They have since become standard in zero-knowledge and also in simulation-based definitions in other contexts [18,19] to provide composability. Their inclusion in the SS-SOA definition of commitment by DNRS [19] was thus correct and justified and we put them to good use.

Later definitions [3,29] however appear to have dropped the auxiliary inputs. Although this appears to be only for notational simplicity (modern works on ZK also often drop auxiliary inputs since it is well understood how to extend the definition to include them) it does raise an interesting technical question, namely what negative results can we prove without auxiliary inputs?

A simple solution is to use one of the messages as a key. The adversary would corrupt the corresponding party to get this key, thereby defining the hash function, and then proceed as above. This however makes the adversary adaptive, and while this is still a significant result, we ask whether anything can be shown for one-shot adversaries without using auxiliary inputs.

This turns out to be technically challenging. The difficulty is that the simulator can control the hash key. In [2] we present a construction relying on a new primitive we call an encrypted hash scheme (EHS). The idea is that there is an underlying core hash function whose keys are messages and an encrypted hash function whose keys are ciphertexts. We show how to build an EHS based on DDH.

We remark that from a practical perspective these distinctions are moot since hash functions like SHA-256 are keyless. Also, it is possible to work theoretically with keyless hash functions [38]. But in classical asymptotic theoretical cryptography, hash functions are keyed and we were interested in results in this setting.

3 Preliminaries

NOTATION AND CONVENTIONS. If $n \in \mathbb{N}$ then let 1^n denote the string of n ones and $[n]$ the set $\{1, \ldots, n\}$. The empty string is denoted by ε. By $a \parallel b$ we denote the concatenation of strings a, b. If a is tuple then $(a_1, \ldots, a_n) \leftarrow a$ means we parse a into its constituents. We use boldface letters for vectors. If $\mathbf{x}$ is a vector then we let $|\mathbf{x}|$ denote the number of components of $\mathbf{x}$ and for $1 \leq i \leq |\mathbf{x}|$ we let $\mathbf{x}[i]$ denote its i-th component. For a set $I \subseteq [|\mathbf{x}|]$ we let $\mathbf{x}[I]$ be the $|\mathbf{x}|$-vector whose i-th component is $\mathbf{x}[i]$ if $i \in I$ and $\perp$ otherwise. We let $\perp_n$ denote the n-vector all of whose components are $\perp$. We define the Embedding subroutine Emb to take 1^n, $I \subseteq [n]$, a $|I|$-vector $\mathbf{x}^*$ and a n-vector $\overline{\mathbf{x}}$ and return the n-vector that consists of $\overline{\mathbf{x}}$ with $\mathbf{x}^*$ embedded in the positions indexed by I. More precisely,

$$\underline{\text{Subroutine Emb}(1^n, I, \mathbf{x}^*, \overline{\mathbf{x}})}$$
$$j \leftarrow 0 \, ; \text{ For } i = 1, \ldots, n \text{ do If } i \in I \text{ then } j \leftarrow j + 1 \, ; \, \overline{\mathbf{x}}[i] \leftarrow \mathbf{x}^*[j]$$
$$\text{Return } \overline{\mathbf{x}}.$$

All algorithms are randomized, unless otherwise specified as being deterministic. We use the abbreviation PT for polynomial-time. If A is an algorithm then $y \leftarrow A(x_1, \ldots, x_n; r)$ represents the act of running the algorithm A with inputs $x_1, \ldots, x_n$ and coins r to get an output y and $y \leftarrow_\$ A(x_1, \ldots, x_n)$ represents the act of picking r at random and letting $y \leftarrow A(x_1, \ldots, x_n; r)$. By $[A(x_1, \ldots, x_n)]$ we denote the set of all y for which there exists r such that $y = A(x_1, \ldots, x_n; r)$.

GAMES. We use the language of code-based game-playing [6]. A game (see Fig. 1 for examples) has an INITIALIZE procedure, procedures to respond to adversary oracle queries, and a FINALIZE procedure. A game G is executed with an adversary A and security parameter λ as follows. A is given input 1^λ and can then call game procedures. Its first oracle query must be INITIALIZE(1^λ) and its last oracle query must be to FINALIZE, and it must make exactly one query to each of these oracles. In between it can query the other procedures as oracles as it wishes. The output of FINALIZE, denoted $\mathrm{G}^A(\lambda)$, is called the output of the game, and we let "$\mathrm{G}^A(\lambda)$" denote the event that this game output takes value true.

$$\begin{array}{l|l}
\underline{\text{INITIALIZE}(1^\lambda)} & \\
b \leftarrow\!\!{\scriptstyle\$}\, \{0,1\}\ ;\ \pi \leftarrow\!\!{\scriptstyle\$}\, \mathcal{P}(1^\lambda) & \underline{\text{INITIALIZE}(1^\lambda)} \\
(ek, dk) \leftarrow\!\!{\scriptstyle\$}\, \mathcal{K}(\pi) & \pi \leftarrow\!\!{\scriptstyle\$}\, \mathcal{P}(1^\lambda) \\
\text{Return } (\pi, ek) & \text{Return } \pi \\[4pt]
\underline{\text{LR}(m_0, m_1)} & \underline{\text{FINALIZE}(ek, c, m_0, m_1, r_0, r_1)} \\
c \leftarrow\!\!{\scriptstyle\$}\, \mathcal{E}(1^\lambda, \pi, ek, m_b) & d_0 \leftarrow \mathcal{V}(1^\lambda, \pi, ek, c, m_0, r_0) \\
\text{Return } c & d_1 \leftarrow \mathcal{V}(1^\lambda, \pi, ek, c, m_1, r_1) \\[4pt]
\underline{\text{FINALIZE}(b')} & \text{Return } (d_0 \wedge d_1 \wedge (m_0 \neq m_1)) \\
\text{Return } (b' = b) &
\end{array}$$

Fig. 1. Game IND_Π (left) and game BIND_Π (right) defining, respectively, IND-CPA privacy and binding security of CE scheme $\Pi = (\mathcal{P}, \mathcal{K}, \mathcal{E}, \mathcal{V})$

CE SCHEMES. We introduce CE (Committing Encryption) schemes as a way to unify commitment and encryption schemes under a single syntax and avoid duplicating similar definitions and results for the two cases. A *CE scheme* $\Pi = (\mathcal{P}, \mathcal{K}, \mathcal{E}, \mathcal{V})$ is specified by four PT algorithms. Via $\pi \leftarrow\!\!{\scriptstyle\$}\, \mathcal{P}(1^\lambda)$ the parameter-generation algorithm $\mathcal{P}$ generates system parameters such as a description of a group. Via $(ek, dk) \leftarrow\!\!{\scriptstyle\$}\, \mathcal{K}(\pi)$ the key-generation algorithm $\mathcal{K}$ generates an encryption key ek and decryption key dk. Via $c \leftarrow \mathcal{E}(1^\lambda, \pi, ek, m; r)$ the encryption algorithm deterministically maps a message m and coins $r \in \{0,1\}^{\rho(\lambda)}$ to a ciphertext $c \in \{0,1\}^* \cup \{\perp\}$ where $\rho\colon \mathbb{N} \to \mathbb{N}$ is the *randomness length* associated to Π and $c \neq \perp$ iff $|m| = \ell(\lambda)$ where $\ell\colon \mathbb{N} \to \mathbb{N}$ is the *message length* associated to Π. Via $d \leftarrow \mathcal{V}(1^\lambda, \pi, ek, c, m, r)$, deterministic verification algorithm $\mathcal{V}$ returns true or false. We require that $\mathcal{V}(1^\lambda, \pi, ek, \mathcal{E}(1^\lambda, \pi, ek, m; r), m, r) = \text{true}$ for all $\lambda \in \mathbb{N}$, all $\pi \in [\mathcal{P}(1^\lambda)]$, all $(ek, dk) \in [\mathcal{K}(\pi)]$, all $r \in \{0,1\}^{\rho(\lambda)}$ and all $m \in \{0,1\}^*$ such that $\mathcal{E}(1^\lambda, \pi, ek, m; r) \neq \perp$. We say that the verification algorithm $\mathcal{V}$ is *canonical* if $\mathcal{V}(1^\lambda, \pi, ek, c, m, r)$ returns the boolean $(\mathcal{E}(1^\lambda, \pi, ek, m; r) = c \neq \perp)$.

Game IND_Π of Fig. 1 captures the standard notion of indistinguishability under chosen-plaintext attack (IND-CPA) [26] and serves to define privacy for CE schemes. The adversary is allowed only one LR query and the messages m_0, m_1 involved must be of the same length. Game BIND_Π captures binding security. For adversaries A, B we let

$$\mathbf{Adv}_{\Pi,A}^{\text{indcpa}}(\lambda) = 2\Pr[\text{IND}_\Pi^A(\lambda)] - 1 \quad \text{and} \quad \mathbf{Adv}_{\Pi,B}^{\text{bind}}(\lambda) = \Pr[\text{BIND}_\Pi^B(\lambda)] .$$

We say that Π is IND-CPA secure if $\mathbf{Adv}_{\Pi,A}^{\text{indcpa}}(\cdot)$ is negligible for all PT A, *binding* if $\mathbf{Adv}_{\Pi,B}^{\text{bind}}(\cdot)$ is negligible for all PT B and *perfectly binding* if $\mathbf{Adv}_{\Pi,B}^{\text{bind}}(\cdot) = 0$ for all (not necessarily PT) B.

DISCUSSION. Commitment and encryption schemes can be recovered as special cases of CE schemes as follows. We say that Π is a *commitment scheme* if $\mathcal{K}$ always returns $(\varepsilon, \varepsilon)$. We see that our two security requirements capture the standard hiding and binding properties. In Section 1 we had simplified by assuming the verification algorithm is canonical and there were no parameters but here we are more general. We say that $\mathcal{D}$ is a *decryption algorithm* for CE scheme Π if

$\mathcal{D}(1^\lambda, \pi, dk, \mathcal{E}(1^\lambda, \pi, ek, m; r)) = m$ for all $\lambda \in \mathbb{N}$, all $\pi \in [\mathcal{P}(1^\lambda)]$, all $(ek, dk) \in [\mathcal{K}(\pi)]$, all $r \in \{0, 1\}^{\rho(\lambda)}$ and all $m \in \{0, 1\}^*$ such that $\mathcal{E}(1^\lambda, \pi, ek, m; r) \neq \perp$. We say that Π *admits decryption* if it has a PT decryption algorithm and in that case we say Π is an *encryption scheme*. IND-CPA is then, of course, the standard privacy goal.

Typical encryption schemes are perfectly binding under canonical verification with some added checks. For example, the ElGamal encryption scheme over a order-p group $\mathbb{G}$ with generator g (these quantities in the parameters) is binding under a verification algorithm that performs the re-encryption check and then also checks that quantities that should be in $\mathbb{G}$ or $\mathbb{Z}_p$ really are. RSA-based schemes can be made binding by requiring the encryption exponent to be a prime larger than the modulus.

Lossy encryption schemes [3,30,37] are not binding because the adversary could provide a lossy encryption key and, under this, be able to generate encryption collisions. Non-committing [12,16] and deniable [11,33] encryption schemes are intentionally not binding. These types of encryption schemes have been shown to have SOA security. Our results show that the lack of binding was necessary for their success at this task.

HASH FUNCTIONS. A hash function $\Gamma = (\mathcal{A}, \mathcal{H})$ with associated output length $h \colon \mathbb{N} \to \mathbb{N}$ is a tuple of PT algorithms. Via $a \leftarrow_\$ \mathcal{A}(1^\lambda)$ the key-generation algorithm $\mathcal{A}$ produces a key a. Via $y \leftarrow \mathcal{H}(a, x)$ the deterministic hashing algorithm $\mathcal{H}$ produces the $h(\lambda)$-bit hash of a string x under key a. Collision-resistance is defined via game CR_Γ whose INITIALIZE(1^λ) procedure returns $a \leftarrow_\$ \mathcal{A}(1^\lambda)$ and whose FINALIZE procedure on input (x, x') returns $(x \neq x') \land (\mathcal{H}(a, x) = \mathcal{H}(a, x'))$. There are no other procedures. The advantage of an adversary C is defined by $\mathbf{Adv}_{\Gamma, C}^{\mathrm{cr}}(\lambda) = \Pr\left[\mathrm{CR}_\Gamma^C(\lambda)\right]$. We say that Γ is collision-resistant (CR) if $\mathbf{Adv}_{\Gamma, C}^{\mathrm{cr}}(\cdot)$ is negligible for every PT C. The following says that CR hash functions must have super-logarithmic output length and will be useful later:

Proposition 1. Let $\Gamma = (\mathcal{A}, \mathcal{H})$ be a hash function with associated output length $h \colon \mathbb{N} \to \mathbb{N}$. If Γ is collision-resistant then the function $2^{-h(\cdot)}$ is negligible.

4   SOA-C Insecurity of CE Schemes

Here we show that no CE-scheme that is binding is SOA-C secure. This implies that no HB-secure commitment scheme is SOA-secure and that no binding IND-CPA encryption scheme is SOA-secure under sender corruptions. In [2] we establish similar results for SOA-K to show that no robust IND-CPA encryption scheme is SOA-secure for receiver corruptions.

SOA-C SECURITY. A *relation* is a PT algorithm with boolean output. A *message sampler* is a PT algorithm $\mathcal{M}$ taking input 1^λ and a string α and returning a vector over $\{0, 1\}^*$. There must exist a function $n \colon \mathbb{N} \to \mathbb{N}$ (called the number of messages) and a function $\ell \colon \mathbb{N} \times \{0, 1\}^* \times \mathbb{N} \to \mathbb{N}$ (called the message length) such that $|\mathbf{m}| = n(\lambda)$ and $|\mathbf{m}[i]| = \ell(\lambda, \alpha, i)$ for all $\mathbf{m} \in [\mathcal{M}(1^\lambda, \alpha)]$ and all $i \in [n]$. An *auxiliary-input generator* is a PT algorithm.

$\text{INITIALIZE}(1^\lambda)$

$\pi \leftarrow\!\!{}^{\$} \mathcal{P}(1^\lambda) \,;\, a \leftarrow \mathcal{A}(1^\lambda) \,;\, (ek, dk) \leftarrow\!\!{}^{\$} \mathcal{K}(\pi)$

Return (a, π, ek)

$\text{ENC}(\alpha)$

$\mathbf{m} \leftarrow\!\!{}^{\$} \mathcal{M}(1^\lambda, \alpha)$

For $i = 1, \ldots, n(\lambda)$ do

  $\mathbf{r}[i] \leftarrow\!\!{}^{\$} \{0, 1\}^{\rho(\lambda)} \,;\, \mathbf{c}[i] \leftarrow \mathcal{E}(1^\lambda, \pi, ek, \mathbf{m}[i]; \mathbf{r}[i])$

Return $\mathbf{c}$

$\text{CORRUPT}(I)$

Return $\mathbf{m}[I], \mathbf{r}[I]$

$\text{FINALIZE}(w)$

Return $\mathsf{R}(1^\lambda, a, \pi, \mathbf{m}, \alpha, I, w)$

$\text{INITIALIZE}(1^\lambda)$

$\pi \leftarrow\!\!{}^{\$} \mathcal{P}(1^\lambda) \,;\, a \leftarrow \mathcal{A}(1^\lambda)$

Return (a, π)

$\text{MSG}(\alpha)$

$\mathbf{m} \leftarrow\!\!{}^{\$} \mathcal{M}(1^\lambda, \alpha)$

$\text{CORRUPT}(I)$

Return $\mathbf{m}[I]$

$\text{FINALIZE}(w)$

Return $\mathsf{R}(1^\lambda, a, \pi, \mathbf{m}, \alpha, I, w)$

Fig. 2. Game $\text{RSOAC}_{\Pi, \mathcal{M}, \mathsf{R}, \mathcal{A}}$ capturing the real-world SOA-C attack to be mounted by an adversary (left) and game $\text{SSOAC}_{\Pi, \mathcal{M}, \mathsf{R}, \mathcal{A}}$ capturing the simulated-world SOA-C attack to be mounted by a simulator (right)

Let $\Pi = (\mathcal{P}, \mathcal{K}, \mathcal{E}, \mathcal{V})$ be a CE-scheme, R a relation, $\mathcal{M}$ a message sampler and $\mathcal{A}$ an auxiliary-input generator. We define SOA-C security via the games of Fig. 2. "Real" game $\text{RSOAC}_{\Pi, \mathcal{M}, \mathsf{R}, \mathcal{A}}$ will be executed with an adversary A. An soa-c adversary's (mandatory, starting) $\text{INITIALIZE}(1^\lambda)$ call results in its being returned an auxiliary input, parameters, and an encryption key, the latter corresponding to the single receiver modeled here. The adversary is then required to make exactly one $\text{ENC}(\alpha)$ call. This results in production of a message vector whose encryption is provided to the adversary. Now the adversary is required to make exactly one $\text{CORRUPT}(I)$ call to get back the messages *and coins* corresponding to the senders named in the set $I \subseteq [n(\lambda)]$. It then calls FINALIZE with some value w of its choice and wins if the relation returns true on the inputs shown. A soa-c simulator S runs with the simulator game $\text{SSOAC}_{\Pi, \mathcal{M}, \mathsf{R}, \mathcal{A}}$ and gets back only auxiliary input and parameters from its $\text{INITIALIZE}(1^\lambda)$ call, there being no encryption key in its world. It is then required to make exactly one $\text{MSG}(\alpha)$ call resulting in creation of a message vector but the simulator is returned nothing related to it. It must then make its $\text{CORRUPT}(I)$ and $\text{FINALIZE}(w)$ calls like the adversary and wins under the same conditions. The soa-c-advantage of an soa-c-adversary A with respect to CE-scheme Π, message sampler $\mathcal{M}$, relation R, auxiliary input generator $\mathcal{A}$ and soa-c simulator S is defined by

$$\mathbf{Adv}^{\text{soa-c}}_{\Pi, \mathcal{M}, \mathsf{R}, \mathcal{A}, A, S}(\lambda) = \Pr\left[\text{RSOAC}^A_{\Pi, \mathcal{M}, \mathsf{R}, \mathcal{A}}(\lambda) \right] - \Pr\left[\text{SSOAC}^S_{\Pi, \mathcal{M}, \mathsf{R}, \mathcal{A}}(\lambda) \right].$$

We say that Π is $(\mathcal{M}, \mathcal{A})$-SOA-C-secure if for every PT R and every PT soa-c adversary A there exists a PT soa-c simulator S such that $\mathbf{Adv}^{\text{soa-c}}_{\Pi, \mathcal{M}, \mathsf{R}, \mathcal{A}, A, S}(\cdot)$ is negligible. We say that Π is SOA-C-secure if it is $(\mathcal{M}, \mathcal{A})$-SOA-C-secure for every PT $\mathcal{M}, \mathcal{A}$.

RESULT. The following implies that any binding CE-scheme is not SOA-C-secure.

Theorem 2. *Let $\Pi = (\mathcal{P}, \mathcal{K}, \mathcal{E}, \mathcal{V})$ be a binding CE-scheme with message length ℓ: $\mathbb{N} \to \mathbb{N}$. Let $\Gamma = (\mathcal{A}, \mathcal{H})$ be a collision-resistant hash function with associated output length h: $\mathbb{N} \to \mathbb{N}$. Let $n(\cdot) = 2h(\cdot)$ and let $\mathcal{M}$ be the message sampler that on input $1^\lambda, \alpha$ (ignores α and) returns a $n(\lambda)$-vector whose components are uniformly and independently distributed over $\{0,1\}^{\ell(\lambda)}$. Then there exists a PT soa-c adversary A and a PT relation R such that for all PT simulators S there is a negligible function ν such that $\mathbf{Adv}^{\text{soa-c}}_{\Pi, \mathcal{M}, \mathsf{R}, \mathcal{A}, A, S}(\lambda) \geq 1 - \nu(\lambda)$ for all $\lambda \in \mathbb{N}$.* ∎

Thus, Π is not $(\mathcal{M}, \mathcal{A})$-SOA-C-secure and hence cannot be SOA-C-secure. Moreover, this is true when the distribution on messages is uniform. These claims would only require $\mathbf{Adv}^{\text{soa-c}}_{\Pi, \mathcal{M}, \mathsf{R}, \mathcal{A}, A, S}(\cdot)$ in the theorem to be non-negligible, but we show more, namely that it is almost one. Note that ℓ is arbitrary and could even be $\ell(\cdot) = 1$, meaning we rule out SOA-C-security even for bit-commitment and encryption of 1-bit messages. The proof will make use of the following variant of the Reset Lemma of [4].

Lemma 3. *Let $V = \{V_\lambda\}_{\lambda \in \mathbb{N}}$ be a collection of non-empty sets. Let P_1, P_2 be algorithms, the second with boolean output. The single-execution acceptance probability $\mathbf{AP}_1(P_1, P_2, V, \lambda)$ is defined as the probability that $d = \mathsf{true}$ in the single execution experiment $\overline{St} \leftarrow\!\!\$\, P_1(1^\lambda)$; $\mathbf{m}^* \leftarrow\!\!\$\, V_\lambda$; $d \leftarrow\!\!\$\, P_2(\overline{St}, \mathbf{m}^*)$. The double-execution acceptance probability $\mathbf{AP}_2(P_1, P_2, V, \lambda)$ is defined as the probability that $d_1 = d_2 = \mathsf{true}$ and $\mathbf{m}_0^* \neq \mathbf{m}_1^*$ in the double execution experiment $\overline{St} \leftarrow\!\!\$\, P_1(1^\lambda); \mathbf{m}_0^*, \mathbf{m}_1^* \leftarrow\!\!\$\, V_\lambda; d_0 \leftarrow\!\!\$\, P_2(\overline{St}, \mathbf{m}_0^*); d_1 \leftarrow\!\!\$\, P_2(\overline{St}, \mathbf{m}_1^*)$. Then $\mathbf{AP}_1(P_1, P_2, V, \lambda) \leq 1/|V_\lambda| + \sqrt{\mathbf{AP}_2(P_1, P_2, V, \lambda)}$ for all $\lambda \in \mathbb{N}$.* ∎

The two executions in the double-execution experiment are not independent because $\overline{St}$ is the same for both, which is why the lemma is not trivial.

Proof (Lemma 3). Let $\delta = 1/|V_\lambda|$. Let $\mathsf{X}(\omega) = \Pr[d = \mathsf{true}]$ in the experiment $\overline{St} \leftarrow P_1(1^\lambda; \omega)$; $\mathbf{m}^* \leftarrow\!\!\$\, V_\lambda$; $d \leftarrow\!\!\$\, P_2(\overline{St}, \mathbf{m}^*)$. So $\mathbf{E}[\mathsf{X}] = \mathbf{AP}_1(P_1, P_2, V, \lambda)$ where the expectation is over the coins ω of P_1. Let $a_1 = \mathbf{AP}_1(P_1, P_2, V, \lambda)$ and $a_2 = \mathbf{AP}_2(P_1, P_2, V, \lambda)$. Then

$$a_2 \geq \mathbf{E}[\mathsf{X}(\mathsf{X} - \delta)] = \mathbf{E}[\mathsf{X}^2] - \delta \cdot \mathbf{E}[\mathsf{X}] \geq \mathbf{E}[\mathsf{X}]^2 - \delta \cdot \mathbf{E}[\mathsf{X}] = a_1^2 - \delta \cdot a_1$$

where the third step above is by Jensen's inequality. Now $a_1^2 - \delta \cdot a_1 = (a_1 - \delta/2)^2 - \delta^2/4$ so

$$a_1 \leq \delta/2 + \sqrt{a_2 + \delta^2/4} \leq \delta/2 + \sqrt{a_2} + \sqrt{\delta^2/4} = \delta + \sqrt{a_2}$$

which yields the lemma. ∎

Proof (Theorem 2). The adversary A and relation R are depicted in Fig. 3. Let S be any PT soa-c simulator. In the real game the adversary always makes the relation return true hence

$$\mathbf{Adv}^{\text{soa-c}}_{\Pi, \mathcal{M}, \mathsf{R}, \mathcal{A}, A, S}(\lambda) = 1 - \Pr\left[\text{SSOAC}^S_{\Pi, \mathcal{M}, \mathsf{R}, \mathcal{A}}(\lambda)\right] .$$

Adversary $A(1^\lambda)$	Relation $\mathrm{R}(1^\lambda, a, \pi, \mathbf{m}, \alpha, I, w)$
$(a, \pi, ek) \leftarrow \text{INITIALIZE}(1^\lambda)$	If $\alpha \neq \varepsilon$ then return false
$\mathbf{c} \leftarrow \text{ENC}(\varepsilon)$	$(ek, \mathbf{c}, \bar{\mathbf{r}}) \leftarrow w$; $b[1] \ldots b[h(\lambda)] \leftarrow \mathcal{H}(a, ek \parallel \mathbf{c})$
$b[1] \ldots b[h(\lambda)] \leftarrow \mathcal{H}(a, ek \parallel \mathbf{c})$	If $(I \neq \{2j - 1 + b[j] : 1 \leq j \leq h(\lambda)\})$ then
$I \leftarrow \{2j - 1 + b[j] : 1 \leq j \leq h(\lambda)\}$	return false
$(\overline{\mathbf{m}}, \bar{\mathbf{r}}) \leftarrow \text{CORRUPT}(I)$	If $|\mathbf{c}| \neq n(\lambda)$ or $|\bar{\mathbf{r}}| \neq n(\lambda)$ then return false
$w \leftarrow (ek, \mathbf{c}, \bar{\mathbf{r}})$	For all $i \in I$ do
$\text{FINALIZE}(w)$	If $\mathcal{V}(1^\lambda, \pi, ek, \mathbf{c}[i], \mathbf{m}[i], \bar{\mathbf{r}}[i]) = \text{false}$ then
	return false
	Return true

Fig. 3. Adversary A and relation R for the proof of Theorem 2

We will construct a binding-adversary B and cr-adversary C such that

$$\Pr\left[\text{SSOAC}^S_{\Pi, \mathcal{M}, \mathsf{R}, \mathcal{A}}(\lambda)\right] \leq 2^{-h(\lambda)\ell(\lambda)} + \sqrt{\mathbf{Adv}^{cr}_{\Gamma, C}(\lambda) + \mathbf{Adv}^{bind}_{\Pi, B}(\lambda)}. \quad (1)$$

The assumptions that Γ is collision-resistant, Π is binding, together with Proposition 1, imply that the RHS of Eq. (1) is negligible, which proves the theorem. It remains to construct B and C. Given S we can define sub-algorithms S_1, S_2 such that S can be written in terms of S_1, S_2 as follows:

Simulator $S(1^\lambda)$
$(a, \pi) \leftarrow \text{INITIALIZE}(1^\lambda)$; $\text{MSG}(1^\lambda, \varepsilon)$; $(St, I) \leftarrow_\$ S_1(a, \pi)$
$\overline{\mathbf{m}} \leftarrow \text{CORRUPT}(I)$; $w \leftarrow S_2(St, \overline{\mathbf{m}})$; $\text{FINALIZE}(w)$

We clarify that we are not defining S; the latter is given and arbitrary. Rather, *any* S has the form above for *some* S_1, S_2 that can be determined given S. Specifically, S_1 runs S until S makes its $\text{CORRUPT}(I)$ query, returning I along with the current state St of S. Then S_2, given the response $\overline{\mathbf{m}}$ to the query, feeds it back to S and continues executing S from St. By having S_1 put all S's coins in St we can assume S_2 is deterministic. We may assume wlog that $|I|$ is always $h(\lambda)$ and that the argument α in S's MSG call is ε since otherwise R rejects. We now define adversary B. The embedding subroutine Emb it calls and the notation $\perp_{n(\lambda)}$ were defined in Section 3:

Adversary $B(1^\lambda)$
$\pi \leftarrow \text{INITIALIZE}(1^\lambda)$; $a \leftarrow_\$ \mathcal{A}(1^\lambda)$; $(St, I) \leftarrow_\$ S_1(a, \pi)$
$\mathbf{m}_0^*, \mathbf{m}_1^* \leftarrow_\$ (\{0, 1\}^{\ell(\lambda)})^{h(\lambda)}$
$\overline{\mathbf{m}}_0 \leftarrow \text{Emb}(1^{n(\lambda)}, I, \mathbf{m}_0^*, \perp_{n(\lambda)})$; $\overline{\mathbf{m}}_1 \leftarrow \text{Emb}(1^{n(\lambda)}, I, \mathbf{m}_1^*, \perp_{n(\lambda)})$
$w_0 \leftarrow S_2(St, \overline{\mathbf{m}}_0)$; $(ek_0, \mathbf{c}_0, \bar{\mathbf{r}}_0) \leftarrow w_0$
$w_1 \leftarrow S_2(St, \overline{\mathbf{m}}_1)$; $(ek_1, \mathbf{c}_1, \bar{\mathbf{r}}_1) \leftarrow w_1$; $t \leftarrow_\$ I$
For all $i \in I$ do If $\overline{\mathbf{m}}_0[i] \neq \overline{\mathbf{m}}_1[i]$ then $t \leftarrow i$
$\text{FINALIZE}(ek_0, \mathbf{c}_0[t], \overline{\mathbf{m}}_0[t], \overline{\mathbf{m}}_1[t], \bar{\mathbf{r}}_0[t], \bar{\mathbf{r}}_1[t])$

Adversary B is running S to get its CORRUPT query I and then, by backing it up, providing two different responses. Adversary C has a similar strategy, only deviating in how the final values are used:

Adversary $C(1^\lambda)$
$a \leftarrow \text{INITIALIZE}(1^\lambda)$; $\pi \leftarrow\!\!{}_{\$} \mathcal{P}(1^\lambda)$; $(St, I) \leftarrow\!\!{}_{\$} S_1(a, \pi)$
$\mathbf{m}_0^*, \mathbf{m}_1^* \leftarrow\!\!{}_{\$} (\{0, 1\}^{\ell(\lambda)})^{h(\lambda)}$
$\overline{\mathbf{m}}_0 \leftarrow \text{Emb}(1^{n(\lambda)}, I, \mathbf{m}_0^*, \perp_{n(\lambda)})$; $\overline{\mathbf{m}}_1 \leftarrow \text{Emb}(1^{n(\lambda)}, I, \mathbf{m}_1^*, \perp_{n(\lambda)})$
$w_0 \leftarrow S_2(St, \overline{\mathbf{m}}_0)$; $(ek_0, \mathbf{c}_0, \overline{\mathbf{r}}_0) \leftarrow w_0$
$w_1 \leftarrow S_2(St, \overline{\mathbf{m}}_1)$; $(ek_1, \mathbf{c}_1, \overline{\mathbf{r}}_1) \leftarrow w_1$
$\text{FINALIZE}((ek_0 \parallel \mathbf{c}_0, ek_1 \parallel \mathbf{c}_1))$.

The analysis will use Lemma 3. Let $V_\lambda = (\{0, 1\}^{\ell(\lambda)})^{h(\lambda)}$ and $V = \{V_\lambda\}_{\lambda \in \mathbb{N}}$. Define P_1, P_2 via:

Algorithm $P_1(1^\lambda)$
$\pi \leftarrow \mathcal{P}(1^\lambda)$; $a \leftarrow\!\!{}_{\$} \mathcal{A}(1^\lambda)$; $(St, I) \leftarrow\!\!{}_{\$} S_1(a, \pi)$
$\mathbf{m} \leftarrow\!\!{}_{\$} (\{0, 1\}^{\ell(\lambda)})^{n(\lambda)}$; $\overline{St} \leftarrow (1^\lambda, a, \pi, \mathbf{m}, I, St)$
Return $\overline{St}$

Algorithm $P_2(\overline{St}, \mathbf{m}^*)$
$(1^\lambda, a, \pi, \mathbf{m}, I, St) \leftarrow \overline{St}$
$\overline{\mathbf{m}} \leftarrow \text{Emb}(1^{n(\lambda)}, I, \mathbf{m}^*, \perp_{n(\lambda)})$
$w \leftarrow S_2(St, \overline{\mathbf{m}})$
$\mathbf{m} \leftarrow \text{Emb}(1^{n(\lambda)}, I, \mathbf{m}^*, \mathbf{m})$
Return $\mathsf{R}(1^\lambda, a, \pi, \mathbf{m}, \varepsilon, I, w)$

Above the argument $\mathbf{m}^*$ to P_2 is drawn from V_λ. Now

$$\Pr\left[\text{SSOAC}_{\Pi, \mathcal{M}, \mathsf{R}, \mathcal{A}}^S(\lambda)\right] = \mathbf{AP}_1(P_1, P_2, V, \lambda)$$

$$\leq 2^{-h(\lambda)\ell(\lambda)} + \sqrt{\mathbf{AP}_2(P_1, P_2, V, \lambda)} \qquad (2)$$

Above the equality is from the definitions and the inequality is by Lemma 3. Finally we claim that

$$\mathbf{AP}_2(P_1, P_2, V, \lambda) \leq \mathbf{Adv}_{\Gamma, C}^{\text{cr}}(\lambda) + \mathbf{Adv}_{\Pi, B}^{\text{bind}}(\lambda) . \qquad (3)$$

Eqs. (2) and (3) imply Eq. (1) and conclude the proof. We now justify Eq. (3). To do so it is helpful to write down the double-execution experiment underlying $\mathbf{AP}_2(P_1, P_2, V, \lambda)$:

$\pi \leftarrow \mathcal{P}(1^\lambda)$; $a \leftarrow\!\!{}_{\$} \mathcal{A}(1^\lambda)$; $(St, I) \leftarrow\!\!{}_{\$} S_1(a, \pi)$; $\mathbf{m} \leftarrow\!\!{}_{\$} (\{0, 1\}^{\ell(\lambda)})^{n(\lambda)}$
$\mathbf{m}_0^*, \mathbf{m}_1^* \leftarrow\!\!{}_{\$} (\{0, 1\}^{\ell(\lambda)})^{h(\lambda)}$
$\overline{\mathbf{m}}_0 \leftarrow \text{Emb}(1^{n(\lambda)}, I, \mathbf{m}_0^*, \perp_{n(\lambda)})$; $\overline{\mathbf{m}}_1 \leftarrow \text{Emb}(1^{n(\lambda)}, I, \mathbf{m}_1^*, \perp_{n(\lambda)})$
$w_0 \leftarrow S_2(St, \overline{\mathbf{m}}_0)$; $w_1 \leftarrow S_2(St, \overline{\mathbf{m}}_1)$; $(ek_0, \mathbf{c}_0, \overline{\mathbf{r}}_0) \leftarrow w_0$; $(ek_1, \mathbf{c}_1, \overline{\mathbf{r}}_1) \leftarrow w_1$
$\mathbf{m}_0 \leftarrow \text{Emb}(1^{n(\lambda)}, I, \mathbf{m}_0^*, \mathbf{m})$; $\mathbf{m}_1 \leftarrow \text{Emb}(1^{n(\lambda)}, I, \mathbf{m}_1^*, \mathbf{m})$
Return $\mathsf{R}(1^\lambda, a, \pi, \mathbf{m}_0, \varepsilon, I, w_0) \wedge \mathsf{R}(1^\lambda, a, \pi, \mathbf{m}_1, \varepsilon, I, w_1) \wedge (\mathbf{m}_0^* \neq \mathbf{m}_1^*)$.

Assume this experiment returns true. By definition of R it must be that $I = \{2j - 1 + b_0[j] : 1 \leq j \leq h(\lambda)\}$ where $b_0[1] \ldots b_0[h(\lambda)] = \mathcal{H}(a, ek_0 \parallel \mathbf{c}_0)$ and also $I = \{2j - 1 + b_1[j] : 1 \leq j \leq h(\lambda)\}$ where $b_1[1] \ldots b_1[h(\lambda)] = \mathcal{H}(a, ek_1 \parallel \mathbf{c}_1)$. However, I is the same in both cases, so we must have $\mathcal{H}(a, ek_0 \parallel \mathbf{c}_0) = \mathcal{H}(a, ek_1 \parallel \mathbf{c}_1)$, meaning we have a hash collision. This means that C succeeds unless $ek_0 \parallel \mathbf{c}_0 = ek_1 \parallel \mathbf{c}_1$. But we now argue that in the latter case, B succeeds. We know $\mathbf{m}_0^* \neq \mathbf{m}_1^*$ so there is some $t \in I$ such that $\overline{\mathbf{m}}_0[t] \neq \overline{\mathbf{m}}_1[t]$. The definition of R implies that $\mathcal{V}(1^\lambda, \pi, ek_0, \mathbf{c}_0[t], \overline{\mathbf{m}}_0[t], \overline{\mathbf{r}}_0[t]) = \text{true}$ and also $\mathcal{V}(1^\lambda, \pi, ek_1, \mathbf{c}_1[t], \overline{\mathbf{m}}_1[t], \overline{\mathbf{r}}_1[t]) = \text{true}$. But since $ek_0 \parallel \mathbf{c}_0 = ek_1 \parallel \mathbf{c}_1$ we have $\mathcal{V}(1^\lambda, \pi, ek_0, \mathbf{c}_0[t], \overline{\mathbf{m}}_0[t], \overline{\mathbf{r}}_0[t]) = \text{true}$ and also $\mathcal{V}(1^\lambda, \pi, ek_0, \mathbf{c}_0[t], \overline{\mathbf{m}}_1[t], \overline{\mathbf{r}}_1[t]) = \text{true}$ with $\overline{\mathbf{m}}_0[t] \neq \overline{\mathbf{m}}_1[t]$ so B wins. ∎

EXTENSIONS, APPLICATIONS AND REMARKS. The SOA-C definition could be weakened by allowing the simulator's corruptions to be adaptive, meaning S is allowed multiple queries to procedure CORRUPT that now would take input $i \in [n(\lambda)]$ and return $\mathbf{m}[i]$. The proof strategy of Theorem 2 no longer works but can be extended to also rule out adaptive simulators. We would back S up to its last CORRUPT query and give a new response only to this query. We would now require $\ell(\cdot)$ to be super-logarithmic so that collisions are rare on single messages. We omit the details.

Theorem 2 applies to all commitment schemes since they are binding by definition. Not all encryption schemes are binding, but many popular ones are. For example, the ElGamal scheme is binding. The Cramer-Shoup scheme [15] is also binding, showing that IND-CCA is not a panacea against SOAs.

Our model allows a scheme to have system parameters π that effectively function as auxiliary input. This means the simulator cannot modify them. This is not necessary but merely makes the results more general. If one wishes to view commitment, as in DNRS [19], as having no parameters, just restrict attention to schemes where π is always 1^λ. Our result applies to these as a special case.

5   SOA-K Insecurity of Encryption Schemes

Here we show that no decryption-verifiable IND-CPA encryption scheme is SOA-secure for receiver corruptions.

SOA-K SECURITY. This is the dual of SOA-C where there are multiple receivers and a single sender rather than a single receiver and multiple senders, and corruptions reveal decryption keys rather than coins. The definition uses games $\mathrm{RSOAK}_{\Pi,\mathcal{M},\mathsf{R},\mathcal{A}}$ and $\mathrm{SSOAK}_{\Pi,\mathcal{M},\mathsf{R},\mathcal{A}}$ of Fig. 4. The soa-k-advantage of an soa-k-adversary A with respect to the encryption scheme Π, message sampler $\mathcal{M}$, relation R, auxiliary input generator $\mathcal{A}$ and soa-k simulator S is defined by

$$\mathbf{Adv}^{\mathrm{soa\text{-}k}}_{\Pi,\mathcal{M},\mathsf{R},\mathcal{A},A,S}(\lambda) \;=\; \Pr\left[\,\mathrm{RSOAK}^A_{\Pi,\mathcal{M},\mathsf{R},\mathcal{A}}\,\right] - \Pr\left[\,\mathrm{SSOAK}^S_{\Pi,\mathcal{M},\mathsf{R},\mathcal{A}}\,\right].$$

We say that Π is $(\mathcal{M},\mathcal{A})$-SOA-K-secure if for every PT R and every PT soa-k adversary A there exists a PT soa-k simulator S such that $\mathbf{Adv}^{\mathrm{soa\text{-}k}}_{\Pi,\mathcal{M},\mathsf{R},\mathcal{A},A,S}(\cdot)$ is negligible. We say that Π is SOA-K-secure if it is $(\mathcal{M},\mathcal{A})$-SOA-K-secure for every PT $\mathcal{M},\mathcal{A}$.

RESULT. The following implies that any decryption-verifiable encryption scheme is not SOA-K-secure. Decryption-verifiable encryption schemes are defined in [2] and include many common schemes. The proof is in [2].

Theorem 4. *Let $\Pi = (\mathcal{P},\mathcal{K},\mathcal{E},\mathcal{V})$ be a decryption-verifiable encryption scheme with decryption verifier $\mathcal{W}$ and message length $\ell\colon \mathbb{N} \to \mathbb{N}$. Let $\Gamma = (\mathcal{A},\mathcal{H})$ be a collision-resistant hash function with associated output length $h\colon \mathbb{N} \to \mathbb{N}$. Let $n(\cdot) = 2h(\cdot)$ and let $\mathcal{M}$ be the message sampler that on input $1^\lambda, \alpha$ (ignores α and) returns a $n(\lambda)$-vector whose components are uniformly and independently*

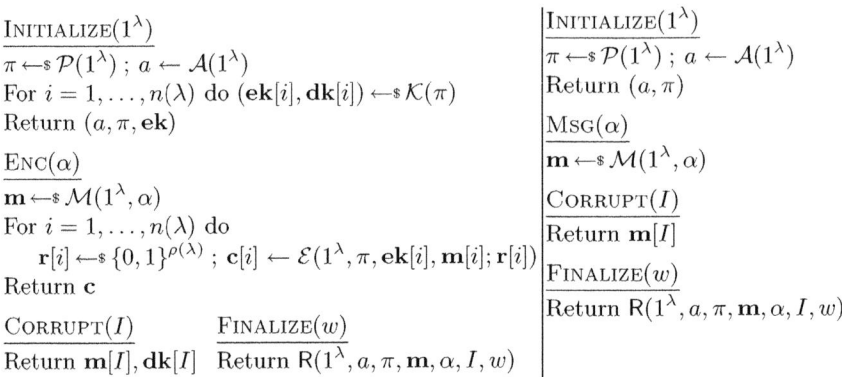

Fig. 4. Game RSOAK$_{\Pi,\mathcal{M},\mathsf{R},\mathcal{A}}$ capturing the real-world SOA-K attack to be mounted by an adversary (left) and game SSOAK$_{\Pi,\mathcal{M},\mathsf{R},\mathcal{A}}$ capturing the simulated-world SOA-K attack to be mounted by a simulator (right)

distributed over $\{0,1\}^{\ell(\lambda)}$. Then there exists a PT soa-k adversary A and a PT relation R such that for all PT soa-k simulators S there is a negligible function ν such that $\mathbf{Adv}^{\mathrm{soa\text{-}k}}_{\Pi,\mathcal{M},\mathsf{R},A,S}(\lambda) \geq 1 - \nu(\lambda)$ for all $\lambda \in \mathbb{N}$. ∎

Acknowledgments. The first two authors were supported in part by NSF grants CNS-0627779 and CCF-0915675. The third author was supported in part by NSF grants CNS-0915361 and CNS-0952692, AFOSR grant FA9550-08-1-0352, DARPA PROCEED, DARPA N11AP20006, a Google Faculty Research award, the Alfred P. Sloan Fellowship, a Microsoft Faculty Fellowship, and a Packard Foundation Fellowship.

References

1. Abdalla, M., Bellare, M., Neven, G.: Robust Encryption. In: Micciancio, D. (ed.) TCC 2010. LNCS, vol. 5978, pp. 480–497. Springer, Heidelberg (2010)
2. Bellare, M., Dowsley, R., Waters, B., Yilek, S.: Standard security does not imply security against selective-opening. Cryptology ePrint Archive, Report 2011/581 (2011), Full version of this abstract, http://eprint.iacr.org/
3. Bellare, M., Hofheinz, D., Yilek, S.: Possibility and Impossibility Results for Encryption and Commitment Secure under Selective Opening. In: Joux, A. (ed.) EUROCRYPT 2009. LNCS, vol. 5479, pp. 1–35. Springer, Heidelberg (2009)
4. Bellare, M., Palacio, A.: GQ and Schnorr Identification Schemes: Proofs of Security against Impersonation under Active and Concurrent Attacks. In: Yung, M. (ed.) CRYPTO 2002. LNCS, vol. 2442, pp. 162–177. Springer, Heidelberg (2002)
5. Bellare, M., Rogaway, P.: Random oracles are practical: A paradigm for designing efficient protocols. In: Ashby, V. (ed.) ACM CCS 1993, pp. 62–73. ACM Press (November 1993)
6. Bellare, M., Rogaway, P.: The Security of Triple Encryption and a Framework for Code-Based Game-Playing Proofs. In: Vaudenay, S. (ed.) EUROCRYPT 2006. LNCS, vol. 4004, pp. 409–426. Springer, Heidelberg (2006)

7. Bellare, M., Waters, B., Yilek, S.: Identity-Based Encryption Secure against Selective Opening Attack. In: Ishai, Y. (ed.) TCC 2011. LNCS, vol. 6597, pp. 235–252. Springer, Heidelberg (2011)
8. Bellare, M., Yilek, S.: Encryption schemes secure under selective opening attack. Cryptology ePrint Archive, Report 2009/101 (2009), http://eprint.iacr.org/
9. Ben-Or, M., Goldwasser, S., Wigderson, A.: Completeness theorems for noncryptographic fault-tolerant distributed computations. In: 20th ACM STOC, pp. 1–10. ACM Press (May 1988)
10. Böhl, F., Hofheinz, D., Kraschewski, D.: On definitions of selective opening security. Cryptology ePrint Archive, Report 2011/678 (2011), http://eprint.iacr.org/
11. Canetti, R., Dwork, C., Naor, M., Ostrovsky, R.: Deniable Encryption. In: Kaliski Jr., B.S. (ed.) CRYPTO 1997. LNCS, vol. 1294, pp. 90–104. Springer, Heidelberg (1997)
12. Canetti, R., Feige, U., Goldreich, O., Naor, M.: Adaptively secure multi-party computation. In: 28th ACM STOC, pp. 639–648. ACM Press (May 1996)
13. Canetti, R., Halevi, S., Katz, J.: Adaptively-Secure, Non-interactive Public-Key Encryption. In: Kilian, J. (ed.) TCC 2005. LNCS, vol. 3378, pp. 150–168. Springer, Heidelberg (2005)
14. Chaum, D., Crépeau, C., Damgård, I.: Multiparty unconditionally secure protocols. In: 20th ACM STOC, pp. 11–19. ACM Press (May 1988)
15. Cramer, R., Shoup, V.: Design and analysis of practical public-key encryption schemes secure against adaptive chosen ciphertext attack. SIAM Journal on Computing 33(1), 167–226 (2003)
16. Damgård, I.B., Nielsen, J.B.: Improved Non-committing Encryption Schemes Based on a General Complexity Assumption. In: Bellare, M. (ed.) CRYPTO 2000. LNCS, vol. 1880, pp. 432–450. Springer, Heidelberg (2000)
17. Dodis, Y., Oliveira, R., Pietrzak, K.: On the Generic Insecurity of the Full Domain Hash. In: Shoup, V. (ed.) CRYPTO 2005. LNCS, vol. 3621, pp. 449–466. Springer, Heidelberg (2005)
18. Dolev, D., Dwork, C., Naor, M.: Nonmalleable cryptography. SIAM Journal on Computing 30(2), 391–437 (2000)
19. Dwork, C., Naor, M., Reingold, O., Stockmeyer, L.J.: Magic functions. Journal of the ACM 50(6), 852–921 (2003)
20. Fehr, S., Hofheinz, D., Kiltz, E., Wee, H.: Encryption Schemes Secure against Chosen-Ciphertext Selective Opening Attacks. In: Gilbert, H. (ed.) EUROCRYPT 2010. LNCS, vol. 6110, pp. 381–402. Springer, Heidelberg (2010)
21. Goldreich, O., Goldwasser, S., Micali, S.: How to construct random functions. Journal of the ACM 33, 792–807 (1986)
22. Goldreich, O., Krawczyk, H.: On the Composition of Zero-Knowledge Proof Systems. In: Paterson, M. (ed.) ICALP 1990. LNCS, vol. 443, pp. 268–282. Springer, Heidelberg (1990)
23. Goldreich, O., Krawczyk, H.: On the composition of zero-knowledge proof systems. SIAM Journal on Computing 25(1), 169–192 (1996)
24. Goldreich, O., Micali, S., Wigderson, A.: Proofs that yield nothing but their validity or all languages in NP have zero-knowledge proof systems. Journal of the ACM 38(3), 691–729 (1991)
25. Goldreich, O., Oren, Y.: Definitions and properties of zero-knowledge proof systems. Journal of Cryptology 7(1), 1–32 (1994)
26. Goldwasser, S., Micali, S.: Probabilistic encryption. Journal of Computer and System Sciences 28(2), 270–299 (1984)

27. Goldwasser, S., Micali, S., Rackoff, C.: The knowledge complexity of interactive proof systems. SIAM Journal on Computing 18(1), 186–208 (1989)
28. Hemenway, B., Libert, B., Ostrovsky, R., Vergnaud, D.: Lossy Encryption: Constructions from General Assumptions and Efficient Selective Opening Chosen Ciphertext Security. In: Lee, D.H., Wang, X. (eds.) ASIACRYPT 2011. LNCS, vol. 7073, pp. 70–88. Springer, Heidelberg (2011)
29. Hofheinz, D.: Possibility and impossibility results for selective decommitments. Journal of Cryptology 24(3), 470–516 (2011)
30. Kol, G., Naor, M.: Cryptography and Game Theory: Designing Protocols for Exchanging Information. In: Canetti, R. (ed.) TCC 2008. LNCS, vol. 4948, pp. 320–339. Springer, Heidelberg (2008)
31. Naor, M.: Bit commitment using pseudorandomness. Journal of Cryptology 4(2), 151–158 (1991)
32. Nielsen, J.B.: Separating Random Oracle Proofs from Complexity Theoretic Proofs: The Non-committing Encryption Case. In: Yung, M. (ed.) CRYPTO 2002. LNCS, vol. 2442, pp. 111–126. Springer, Heidelberg (2002)
33. O'Neill, A., Peikert, C., Waters, B.: Bi-Deniable Public-Key Encryption. In: Rogaway, P. (ed.) CRYPTO 2011. LNCS, vol. 6841, pp. 525–542. Springer, Heidelberg (2011)
34. Ostrovsky, R., Rao, V., Scafuro, A., Visconti, I.: Revisiting lower and upper bounds for selective decommitments. Cryptology ePrint Archive, Report 2011/536 (2011), http://eprint.iacr.org/
35. Panjwani, S.: Tackling Adaptive Corruptions in Multicast Encryption Protocols. In: Vadhan, S.P. (ed.) TCC 2007. LNCS, vol. 4392, pp. 21–40. Springer, Heidelberg (2007)
36. Pedersen, T.P.: Non-interactive and Information-Theoretic Secure Verifiable Secret Sharing. In: Feigenbaum, J. (ed.) CRYPTO 1991. LNCS, vol. 576, pp. 129–140. Springer, Heidelberg (1992)
37. Peikert, C., Vaikuntanathan, V., Waters, B.: A Framework for Efficient and Composable Oblivious Transfer. In: Wagner, D. (ed.) CRYPTO 2008. LNCS, vol. 5157, pp. 554–571. Springer, Heidelberg (2008)
38. Rogaway, P.: Formalizing Human Ignorance. In: Nguyên, P.Q. (ed.) VIETCRYPT 2006. LNCS, vol. 4341, pp. 211–228. Springer, Heidelberg (2006)
39. Xiao, D.: (Nearly) Round-Optimal Black-Box Constructions of Commitments Secure against Selective Opening Attacks. In: Ishai, Y. (ed.) TCC 2011. LNCS, vol. 6597, pp. 541–558. Springer, Heidelberg (2011)

Detecting Dangerous Queries: A New Approach for Chosen Ciphertext Security

Susan Hohenberger[1,*], Allison Lewko[2,**], and Brent Waters[3,***]

[1] Johns Hopkins University
susan@cs.jhu.edu
[2] University of Texas at Austin
{alewko,bwaters}@cs.utexas.edu

Abstract. We present a new approach for creating chosen ciphertext secure encryption. The focal point of our work is a new abstraction that we call *Detectable Chosen Ciphertext Security* (DCCA). Intuitively, this notion is meant to capture systems that are not necessarily chosen ciphertext attack (CCA) secure, but where we can detect whether a certain query CT can be useful for decrypting (or distinguishing) a challenge ciphertext CT^*.

We show how to build chosen ciphertext secure systems from DCCA security. We motivate our techniques by describing multiple examples of DCCA systems including creating them from 1-bit CCA secure encryption — capturing the recent Myers-shelat result (FOCS 2009). Our work identifies DCCA as a new target for building CCA secure systems.

1 Introduction

A central goal of public key cryptography is to design encryption systems that are secure against chosen ciphertext attacks. Public key encryption systems that are chosen ciphertext attack (CCA) secure are robust against powerful adversaries that are able to leverage interaction with a decryptor. Such an attacker is modeled by allowing him to query for the decryption of any ciphertext except a challenge ciphertext for which he is trying to break. This includes ciphertexts derived from the challenge ciphertext[1]. Due to its robustness against powerful

---

* Sponsored by the Defense Advanced Research Projects Agency (DARPA) and the Air Force Research Laboratory (AFRL) under contract FA8750-11-2-0211, the Office of Naval Research under contract N00014-11-1-0470, a Microsoft Faculty Fellowship and a Google Faculty Research Award. Applying to all authors, the views expressed are those of the authors and do not reflect the official policy or position of the Department of Defense or the U.S. Government.

** Sponsored by a Microsoft Research Ph.D. Fellowship.

*** Supported by NSF CNS-0915361 and CNS-0952692, AFOSR Grant No: FA9550-08-1-0352, DARPA PROCEED, DARPA N11AP20006, Google Faculty Research award, the Alfred P. Sloan Fellowship, Microsoft Faculty Fellowship, and Packard Foundation Fellowship.

[1] We use "CCA" and "CCA2" interchangeably in this paper.

D. Pointcheval and T. Johansson (Eds.): EUROCRYPT 2012, LNCS 7237, pp. 663–681, 2012.
© International Association for Cryptologic Research 2012

attackers, chosen ciphertext security has become the accepted goal for building secure encryption. For this reason, building chosen ciphertext secure systems has been a central pursuit of cryptographers for over twenty years and we have seen many distinct approaches to achieving CCA security.

Early pioneering work in chosen ciphertext security [23,14,26] introduced the technique of leveraging Non-Interactive Zero Knowledge Proofs (NIZKs) [5] to build CCA-secure encryption systems from chosen plaintext secure encryption systems. Roughly, a NIZK is used to prove that a ciphertext is "well-formed" or legal. Later Cramer and Shoup [12,13] introduced the first practical CCA-secure systems that were built on specific number theoretic assumptions such as Decisional Diffie Hellman. These techniques implicitly embed a certain form of designated verifier Non-Interactive Zero Knowledge proofs in them. More recently, different methods for building chosen ciphertext security from Identity-Based Encryption [7] and Lossy Trapdoor Functions [25] have emerged. In addition, Myers and shelat [22] described general methods for amplifying CCA encryption of 1 bit to many bits.

In this work, we introduce a new approach to obtaining chosen ciphertext secure systems. The focal point of our work is a new abstraction that we call *Detectable Chosen Ciphertext Security* (DCCA). Intuitively, this notion is meant to capture systems that are not necessarily CCA secure, but where we can detect whether a certain query CT can be useful for decrypting (or distinguishing) a challenge ciphertext CT*.

A system that is DCCA secure will be associated with a boolean function F that takes in three inputs: a public key pk, a challenge ciphertext CT* and a query ciphertext CT. The function will output 1 if the query CT is "dangerous" for the an attacker wishing to distinguish CT*. A DCCA secure system must have the following two properties stated here informally:

- **Unpredictability.** Without seeing CT* it should be hard to find a ciphertext CT such that $F(PK, CT^*, CT) = 1$. In other words, an attacker must first see a challenge ciphertext in order to discover a dangerous query for it.
- **Indistinguishability.** The system will be secure under a detectable chosen ciphertext attack *if* the attacker is limited to decryption queries of ciphertexts CT where $F(pk, CT^*, CT) = 0$ for challenge ciphertext CT*. I.e. the system is CCA secure if the attacker does not make dangerous queries.

The goal of our work will be to construct fully chosen ciphertext secure systems from detectable CCA-secure systems. We first motivate this goal by observing multiple DCCA systems that naturally occur:

- **Many Bit Encryption from 1-bit CCA.** Suppose we have a 1-bit CCA-secure system and we wish to encrypt multiple bits by concatenating multiple 1-bit encryptions together. The resulting system is no longer chosen ciphertext secure, but is DCCA secure. The detecting function F is 1 iff any of the 1-bit ciphertext components between CT* and CT are equal. This scenario is akin to the problem of showing that bit encryption is complete considered by Myers and shelat [22], where they worried about such "quoting" attacks.

- **Tag-Based Encryption Systems.** MacKenzie, Reiter and Yang [21] and Kiltz [19] define a tag-based encryption scheme as an encryption scheme that takes in an additional "tag" parameter on encryption and decryption. The security game allows an attacker to make decryption queries with any tag parameter t, except for the tag t^* that the challenge ciphertext is encrypted under. Several examples of tag-based schemes exist. Kiltz [19] gave a direct construction from the linear assumption. The CCA1-secure encryption variant of the Canetti, Halevi and Katz [7] construction where the tag is an IBE identity is an additional example. One can also view the CCA1-secure variant of Peikert and Waters [25] as a tag-based scheme, where the tag is the "branch" in an all-but-one encryption scheme.

  Most of the above examples of tag-based encryption can be proven selectively secure, where an attacker must commit to the tag of the challenge ciphertext before seeing the public key. However, if we are willing to utilize complexity leveraging arguments, we can argue that these are adaptively secure. In addition, the CHK-lite transformation will be an adaptively secure tag-based scheme if used with an adaptively secure Identity-Based Encryption system. We observe that adaptively-secure tag-based encryption immediately gives rise to DCCA-secure encryption. A ciphertext of the DCCA-secure system consists of a random tag t plus a tag-based encryption of the message under the tag t. Decryption follows analogously and the function F simply tests if two ciphertexts have the same tag. Unpredictability follows from having a large tag space. Although it is already possible to transform tag-based encryption into CCA-secure encryption using a strongly unforgeable signature [19], these examples demonstrate natural DCCA systems.

- **"Sloppy" CCA Encryption.** One can envision that in practice an encryption system is CCA secure, but an implementation of it is not due to certain nuances. For instance, suppose a number theoretic library had a slack bit in its representation of group elements (e.g. a bit that was supposed to be 0, but if set to 1 does not affect any computations.) A CCA attacker could exploit this weakness in an implementation, however, it is possible that the system would still be DCCA secure. One might use our techniques as a hedge against such problems. This is somewhat analogous to recent work [2] on applying deterministic encryption as a hedge against faulty random bit generation.

In addition to the examples listed above, we believe that it is useful to identify DCCA security as a new "target" for achieving chosen ciphertext security.

Overview of Our Techniques. We now give an overview of our construction and proof. Our construction will build a chosen ciphertext secure system from three components: a chosen plaintext secure system, 1-bounded CCA-secure system[2], and a detectable CCA-secure system. Since DCCA security (trivially) implies CPA, and we can build 1-bounded CCA from CPA encryption [24,11,10], it follows that all components are realizable from DCCA as a building block.

---

[2] A 1-bounded CCA-secure encryption system is secure against one chosen ciphertext query.

A public key from our system consists of three components. An "inner" public key PK_{in} which is a DCCA public key and two "outer" keys PK_A, PK_B respectively from 1-bounded CCA and CPA secure systems. To encrypt a message M, one first chooses the randomness r_A, r_B to be used for the outer encryptions and then encrypts the tuple (r_A, r_B, M) under the inner (detectable) key to compute an inner ciphertext CT_{in}. Next, the encryption algorithm encrypts CT_{in} under the outer public key PK_A using randomness r_A to get CT_A. It then analogously creates CT_B as the encryption of CT_{in} under key PK_B and randomness r_B. The output ciphertext is $CT = (CT_A, CT_B)$.

The structure of our ciphertexts is that the two outer ciphertexts both encrypt the same message — the inner ciphertext. This ciphertext itself encrypts the message and the randomness used to create the outer ciphertexts. Thus, the outer ciphertexts indirectly encrypt their own randomness.[3] The decryption algorithm will receive $CT = (CT_A, CT_B)$ and first decrypt CT_A to get CT'_{in} and decrypt this to get (r_A', r_b', M') using the appropriate secret keys. Finally, it will check that the ciphertext is well formed by itself encrypting CT'_{in} under PK_A, PK_B and the respective randomness r_A', r_B' and validating that the output matches CT_A and CT_B before accepting M' as the message. Our encryption system has elements both of the Naor-Yung [23] two key method for our two outer keys and the Myers-shelat [22] method of embedding outer randomness in inner ciphertexts.

Security of our system depends on the premise that no attacker is able to learn the message encrypted in the inner ciphertext. This will follow from the Detectable CCA security *if* we are able to guarantee that an attacker is unable to make any ciphertext queries CT_A, CT_B where the decryption of CT_A, denoted CT_{in}, is related to the inner component of our challenge ciphertext CT^*_{in} according to to the DCCA function F. Intuitively, we hope to achieve this from the combination of two features of our system. First, the 1-bounded CCA security of PK_A will (hopefully) make it difficult to create an encryption under PK_A related to CT^*_{in}. Second, the embedded randomness will allow us to check that ciphertexts are well formed and thus answer multiple ciphertext queries under the Naor-Yung two key type manner.

The trickiness in proving security lies with the embedded randomness which is a two-edge sword. On one hand, forcing the attacker queries to embed randomness allows a reduction algorithm to decrypt if it knows either one of the two outer keys. On the other hand, it is not clear how such a reduction can

---

[3] This construction implicitly assumes that the length of the random string needed for encryption is dependent only on the security parameter and is independent (or at least smaller than) the message size of the outer ciphertexts. We can justify this assumption with the common technique of using a seed to a (variable length) Pseudo Random Generator (PRG) as the input to each encryption algorithm. The PRG can then extend the randomness to whatever length is required by the underlying encryption system. By using this justified assumption in our definitions, we are able to simplify the presentation of our construction and proofs. In contrast, Myers and shelat [22] explicitly carry the PRG technique through their exposition. This choice gives our exposition and proof an advantage in simplicity.

create valid ciphertexts while playing the 1-bounded CCA game, since a reduction algorithm will not know the randomness r_A to embed. Thus, this circularity creates a fundamental barrier similar to difficulties encountered in attempts to create trapdoor functions from encryption [15].

We deal with this by arguing security in an indirect way that steps around this barrier. We first define a security game specific to our construction called nested indistinguishability. In this game, an attacker will receive a public key and is allowed to make decryption queries. The attacker at some point submits a single message M. The challenger will flip a coin z. If $z = 0$, the challenger creates a valid encryption of M; otherwise, if $z = 1$ the challenger creates a encryption where the innermost message is all 0's — it neither includes the message nor the embedded randomness. The attacker continues to make decryption queries (other than the challenge ciphertext) and wins if it is successfully able to guess z. It follows that if no attacker is successful in this game, then our system is chosen ciphertext secure.

To prove security of this nested indistinguishability game, we begin by defining a "bad event". The bad event is defined to be when the attacker submits a query $(\text{CT}_A, \text{CT}_B)$ such that $\text{CT}_A \neq \text{CT}_A^*$ where CT_A^* is from the challenge ciphertext and the decryption of CT_A gives a ciphertext that is related to the inner challenge ciphertext according to F. If we can argue that such bad events only occur with negligible probability, then security of the nested indistinguishability game follows straightforwardly from DCCA security.

The crux of our proof is how we eliminate the possibility of a bad event. We do so in an indirect manner. We begin by arguing this event cannot happen in the case where $z = 1$, which is where all 0's are encrypted and the randomness is not embedded. In this case, we get the best of both worlds. We are able to require that the attacker's queries have the randomness embedded in them, so that we can check ciphertext well-formedness, however, *the challenge ciphertext is not required to embed the outer randomness*. We argue that the bad event does not happen by applying a set of hybrid experiments. First, we change CT_B^* to be an encryption of all 1's. Next, we change the decryption algorithm to decrypt using the secret key for PK_B. Finally, we change CT_A^* to be an encryption of all 1's. In each experiment we argue that the chance of a bad event must be very close to that of the prior experiment. For the last step we leverage the 1-bounded CCA property of the first component. Finally, we note that in the last experiment the probability of a bad event is negligible since the inner challenge ciphertext CT_{in}^* is replaced by all 1's and is not even present.

One interesting question is why is 1-bounded CCA security needed for the PK_A since at the last step in the proof we can use the secret key SK_B to execute decryption. While this is true, it is actually possible for the bad event to occur on a malformed ciphertext that will not decrypt. We need the 1-bounded CCA property to detect the occurrence of the bad event in this case during the security reduction.

We are not able to argue the lack of a bad event in a similar manner for the $z = 0$ (embedded randomness) case due to the aforementioned circularity problems. Instead, we can infer this from the lack of event in the $z = 1$ case along with

DCCA security. To prove this, we can create an algorithm that plays the DCCA indistinguishability game while simulating the nested indistinguishability game to the attacker. The simulator will choose the outer keys and outer randomness for the challenge ciphertext itself. It submits the message and outer randomness as one inner message and the 0's string as another. Then it will be able to decrypt all ciphertext queries until a bad event happens using its keys in addition to the DCCA decryption oracle. Once a bad event query is made though, it is stuck. However, it need not go any further! The fact that the attacker was able to create a bad event at all must mean that the message and randomness were embedded. It can then break the DCCA distinguishing game. Thus, we can infer that the bad event happens with negligible probability in either case. The remainder of the proof follows straightforwardly.

Comparison to Myers-shelat. Myers and shelat [22] showed how to achieve many-bit chosen ciphertext security from 1-bit chosen ciphertext security and motivated us to explore the notion of detectability. They created a system using an inner/outer structure where the inner ciphertext encrypted the outer random coins. Their inner scheme, built from 1-bit CCA, is what they call "unquoteable" secure. Their concept is roughly analogous to a specific instance of a DCCA scheme. Encryptions of many-bit messages are concatenations of 1-bit encryptions; the system is chosen ciphertext secure as long as queries do not copy a 1-bit ciphertext component of the underlying scheme. For the outer scheme, they use a notion of security that is an amalgam of unquoteability and non-malleability. Their outer construction follows a specific adaptation of the Choi et. al. [10] methods applied to the 1-bit primitive. (No two key structure is used.) Their proof relies on defining quoting attacks on *both* the inner and outer layers and then establishing a certain order that outer quoting attacks must happen before inner quoting attacks.

We believe our methods offer benefits in terms of generality, simplicity, and efficiency. First, our general notion of Detectable Chosen Ciphertext Security can be realized by multiple systems. These include the 1-bit to many-bit examples, the tag-based encryption class and future systems that can leverage this as a new target path for creating CCA secure encryption.

Another key difference is that the outer layer of our scheme is built from simple 1-bounded CCA and CPA-secure parts. We argue these provide simpler concepts and are easier to work with. In addition, one can instantiate them from *any* 1-bounded encryption system. For instance, we can apply any candidate 1-bounded CCA-secure system and do not need to work through the Choi et. al. [10] construction. Instead we can apply the 1-bounded CCA system of Cramer et. al. [11], which is signfcantly more efficient and simpler than the non-malleable systems of either PSV [24] or Choi et. al. [10]. We also regard avoiding a combination security definition between 1-bounded CCA (or non-malleability) and detection as a benefit for simplicity. This simplification will also improve efficiency in the case where there is a candidate CPA primitive that is more efficient than the candidate DCCA primtive, since we can build the 1-bounded scheme out of the CPA primitive.

Our choice of abstractions and structure allow us to have a simple proof. We can eliminate the possibility of a bad event using a basic Naor-Yung two key argument. Then once we are able to eliminate this, the rest of the proof follows in a straightforward manner.

Why not CCA1? One intriguing possibility is to try to leverage our techniques to build full chosen ciphertext security from CCA1 security. A natural direction would be to use a CCA1 system for the inner component in place of the detectable encryption scheme. The intuitive rationale would be if the outer keys are 1-bounded CCA or non-malleable then the queries produced by the attacker should not be related to the inner challenge ciphertext and thus CCA1 might suffice. Unfortunately, we were able to create an attack oracle which breaks full CCA security in our scheme, yet does not perturb the 1-bounded CCA or CCA1 primitives, giving evidence that this approach may not work. However, the oracle we use is quite strong and "exotic". This suggests that there might be primitives that lie somewhere in between DCCA and CCA1. One interesting example is the CCA-1 secure "Cramer-Shoup lite" [12] cryptosystem. There exists a malleability attack on a challenge CT^* that produces a query ciphertext which has the same distribution as a fresh encryption of a random message. Hence the CS-lite system is not CCA secure. However, it would be very interesting and surprising if there existed attack algorithms that matched the above oracle. We expand on this in the full version of this paper.

1.1   Related Work

Relaxations of CCA. Multiple relaxations of chosen ciphertext security have been proposed in the literature.

One interesting class of relaxations is the notion of Replayable Chosen Ciphertext Security [8] and other similar works [29,1]. These works aim to capture the concept that some malleability attacks might intuitively be benign. In particular, consider a cryptosystem where an attacker is only able to maul a ciphertext CT encrypting a message M into a different ciphertext C' that encrypts the *same* message M. If an application (or user) makes all decisions based on the decrypted plaintexts as opposed to the representation of the ciphertext such notions might be sufficient.

The primary goal of RCCA is to formally capture a form of "good enough" security under ciphertext attacks. In contrast, Detectable CCA inherently does not have good enough security on its own. In DCCA systems, it may be possible to maul ciphertexts to be encryptions of different messages or even create attack ciphertexts that each target a single bit of a target ciphertext. Thus, our primary focus is to create CCA security from a less secure DCCA building block.

We observe that DCCA does not imply RCCA. In [8], the authors gave an example of an RCCA scheme that could not be publicly detected. Conversely, not all DCCA schemes will be RCCA secure. Our bit encryption instance serves as an example. We also note that [8] discusses a notion of detectability and introduces a definition that combines replayable and detectable properties. This combined definition is a particular instance of DCCA. However, they do not explore the notion of detectability in isolation or how to build CCA security from it. Canetti, Krawcyzk, and Nielsen [8] do show how to create CCA security from RCCA security using the KEM/DEM framework.

Finally, Hofheinz and Kiltz [17] introduce a notion they call Constrained CCA security particular to developing Key Encapsulation Mechanisms. In their definition an attacker must include a predicate p along with each query ciphertext CT. The challenger will only answer the query if the predicate evaluated on the decrypted key of the ciphertext is true and the predicate is false for all but a negligible fraction of possible KEM keys. While this notion is weaker than CCA security, they show that when combined with a (symmetric) authenticated encryption scheme, the resulting system is CCA secure.

Other Related Work. Goldwasser and Micali [16] gave the first formal definition of security for public key encryption systems. Naor and Yung [23] and Rackoff and Simon [26] extended this to include chosen ciphertext attacks.

Naor and Yung [23] initiated the approach of leveraging NIZKs to build chosen ciphertext security by introducing their "two key" method. A NIZK would guarantee the integrity of the ciphertext by giving a proof that the same message was encrypted to two keys. While their system gave security against lunchtime or CCA1 attacks, Dolev, Dwork and Naor [14] showed how to achieve full CCA2 security. In addition, they introduced the fundamental concept of non-malleability. Sahai [28] introduced a concept of simulation sound NIZKs that could be used to achieve CCA security through the NY two key structure. Bellare and Sahai [4] gave relations between non-malleability [14] chosen ciphertext security.

Since then, different approaches to achieving CCA security have been proposed. Cramer and Shoup [12,13] showed techniques for proving ciphertexts were well-structured and abstracted this into projective hash functions. Several other novel cryptosystems make use of specific number-theoretic techniques (e.g. [19,9,18]). Boneh, Canetti, Halevi and Katz [6] showed a generic method of achieving chosen ciphertext security from IBE systems. Peikert and Waters [25] gave a new avenue for achieving CCA security with the introduction of Lossy Trapdoor Functions (TDFs). Notably, this gave the first chosen ciphertext secure systems from lattice-based assumptions. Subsequently, various refinements of weaker conditions on the trapdoor functions were introduced [27,20].

The above techniques are proven secure in the standard model. Bellare and Rogaway [3] show that in the random oracle model chosen ciphertext security can be built from chosen plaintext security.

2   Detectable Chosen Ciphertext Security

In this section, we define *detectable chosen ciphertext security*. An encryption scheme satisfying this definition is called a *detectable* encryption system. Our discussions assume a familiarity with CPA, CCA1 and CCA2 security as well as bounded CCA security and non-malleability. A reader wishing to review these definitions can find them in the full version of this paper.

2.1   Detectable Encryption

We define a *detectable* encryption scheme as having the usual algorithms (KeyGen, Enc, Dec) together with an efficiently-computable boolean function F. Informally, F tests for a "detectable" relationship between two ciphertexts. The security game will mirror that of CCA2 security, except that decryption queries in the second phase will not be answered for ciphertexts detectably-related to the challenge ciphertext. Our formal definition follows below.

Definition 1 (Detectable Encryption System). *A detectable encryption system is a tuple of probabilistic polynomial-time algorithms* (KeyGen, Enc, Dec, F) *such that:*

1. (KeyGen, Enc, Dec) *have the usual input and output, although we sometimes denote* Enc($pk, m; r$) *as a deterministic function of the public key pk, the message m and randomness r, and*
2. $F(pk, c', c) \to \{0, 1\}$: *the detecting function F takes as input a public key pk and two ciphertexts c' and c, and outputs a bit.*

Correctness is the same as a regular encryption system.

A detectable encryption system must have the following two properties.

Unpredictability of the Detecting Function F. Informally, given the description of F and a public key pk, for an unknown ciphertext c, it should be hard to find a second ciphertext c' that is "related" to c; i.e., such that $F(pk, c', c) = 1$. We consider both a basic and a strong formalization.

Basic Unpredictability Experiment. Consider the experiment $\mathsf{Exp}_{\mathcal{A}, \Pi}^{\mathrm{predict.basic}}(\lambda)$ defined for a detectable encryption scheme $\Pi = $ (KeyGen, Enc, Dec, F) and an adversary $\mathcal{A}$:

1. Setup: KeyGen(1^λ) is run to obtain keys (pk, sk).
2. Queries: Adversary $\mathcal{A}$ is given pk and access to a decryption oracle Dec($sk, \cdot$). The adversary outputs a message m in the message space associated with pk and a ciphertext c in the ciphertext space associated with pk.
3. Challenge: A ciphertext $c^* \leftarrow$ Enc(pk, m) is computed.
4. Output: The output of the experiment is defined to be 1 if $F(pk, c^*, c)$, and 0 otherwise.

We also define a stronger variant $\mathsf{Exp}_{\mathcal{A}, \Pi}^{\mathrm{predict.strong}}(\lambda)$ of the unpredictability experiment where the adversary is additionally given sk. We observe that strong unpredictability implies basic unpredictability since the adversary can simulate the decryption oracle using the secret key.

Indistinguishability of Encryptions. Next, we formalize the confidentiality guarantee. Consider the following experiment $\mathsf{Exp}_{\mathcal{A}, \Pi}^{\mathrm{indist}}(\lambda)$ defined for a detectable encryption scheme $\Pi = $ (KeyGen, Enc, Dec, F) and an adversary $\mathcal{A}$:

1. Setup: KeyGen(1^λ) is run to obtain keys (pk, sk).
2. Phase 1: Adversary $\mathcal{A}$ is given pk and access to a decryption oracle Dec($sk, \cdot$). $\mathcal{A}$ outputs a pair of messages m_0, m_1 of the same length in the message space associated with pk.
3. Challenge: A random bit $b \leftarrow \{0,1\}$ is chosen, and then a ciphertext $c^* \leftarrow$ Enc(pk, m_b) is computed and given to $\mathcal{A}$. We call c^* the challenge ciphertext.
4. Phase 2: $\mathcal{A}$ continues to have access to Dec($sk, \cdot$), but may not request a decryption of a ciphertext c such that $F(pk, c^*, c) = 1$. Finally, $\mathcal{A}$ outputs a bit b'.
5. Output: The output of the experiment is defined to be 1 if $b' = b$, and 0 otherwise.

Definition 2 (Detectable Chosen Ciphertext Security). *A detectable encryption scheme $\Pi = (\mathsf{KeyGen}, \mathsf{Enc}, \mathsf{Dec}, F)$ has an* unpredictable detecting function *and* indistinguishable encryptions under a detectable chosen-ciphertext attack *(or is DCCA-secure) if for all probabilistic polynomial-time adversaries $\mathcal{A}$ there exists a negligible function negl such that:*

1. *(F is unpredictable:)* $\Pr[\mathsf{Exp}_{\mathcal{A}, \Pi}^{\text{predict.basic}}(\lambda) = 1] \leq negl(\lambda)$ *and*
2. *(Encryptions are indistinguishable:)* $\Pr[\mathsf{Exp}_{\mathcal{A}, \Pi}^{\text{indist}}(\lambda) = 1] \leq \frac{1}{2} + negl(\lambda)$.

2.2 Facts about DCCA Security

For space reasons, we omit the simple proofs of the first two lemmas. We conjecture that the converse of Lemma 2 is not true. Indeed if the DDH assumption holds, then the CCA-1 secure Cramer-Shoup lite system would separate these two notions as discussed in the introduction.

Lemma 1 (CCA2 $\Longrightarrow$ DCCA). *If $\Pi = (\mathsf{KeyGen}, \mathsf{Enc}, \mathsf{Dec})$ is a CCA2-secure encryption scheme, then $\Pi' = (\mathsf{KeyGen}, \mathsf{Enc}, \mathsf{Dec}, F)$ is a DCCA-secure encryption scheme where F outputs 0 on all inputs except those of the form $(\cdot, c, c)$.*

Lemma 2 (DCCA $\Longrightarrow$ CCA1). *If $\Pi = (\mathsf{KeyGen}, \mathsf{Enc}, \mathsf{Dec}, F)$ is a DCCA-secure encryption scheme, then $\Pi' = (\mathsf{KeyGen}, \mathsf{Enc}, \mathsf{Dec})$ is a CCA1-secure encryption scheme.*

We also claim that one-bit DCCA-secure encryption implies arbitrary-length DCCA-secure encryption. Say $\Pi = (\mathsf{KeyGen}, \mathsf{Enc}, \mathsf{Dec}, F)$ is a detectable encryption system with plaintext space $\{0,1\}$. We can construct a new scheme $\Pi' = (\mathsf{KeyGen}, \mathsf{Enc}', \mathsf{Dec}', F')$ with plaintext space $\{0,1\}^*$ by defining Enc$'$ as:

$$\mathsf{Enc}'(pk, m) = \mathsf{Enc}(pk, m_1), \ldots, \mathsf{Enc}(pk, m_n)$$

where $m = m_1 \ldots m_n$. The decryption algorithm Dec$'$ decrypts each ciphertext piece using Dec. The function F' performs n^2 invocations of F, testing each ciphertext piece of C with each ciphertext piece of C', and outputting 1 if any invocation of F returned 1, and 0 otherwise.

Lemma 3 (1-bit DCCA Encryption Implies Many-bit DCCA Encryption). *Let Π and Π' be as above. If Π is DCCA-secure, then so is Π'.*

We defer the proof of this lemma to the full version of the paper.

3   The Construction: CCA2 Security from DCCA Security

An overview of the techniques used for our construction is provided in Section 1.

The Construction Description We now construct a CCA2-secure public-key encryption scheme $\Pi = (\mathsf{KeyGen}, \mathsf{Enc}, \mathsf{Dec})$ using three building blocks[4]:

1. a DCCA-secure encryption scheme, denoted $\Pi_{\mathrm{dcca}} = (\mathsf{KeyGen}_{\mathrm{dcca}}, \mathsf{Enc}_{\mathrm{dcca}}, \mathsf{Dec}_{\mathrm{dcca}}, F)$.
2. a 1-bounded CCA-secure encryption scheme with perfect correctness, denoted $\Pi_{\mathrm{1b-cca}} = (\mathsf{KeyGen}_{\mathrm{1b-cca}}, \mathsf{Enc}_{\mathrm{1b-cca}}, \mathsf{Dec}_{\mathrm{1b-cca}})$.
3. a CPA-secure encryption scheme with perfect correctness, denoted $\Pi_{\mathrm{cpa}} = (\mathsf{KeyGen}_{\mathrm{cpa}}, \mathsf{Enc}_{\mathrm{cpa}}, \mathsf{Dec}_{\mathrm{cpa}})$.

We assume that the message space of each system is $\{0,1\}^*$ and that messages of the form (x, y, z) can be uniquely and efficiently encoded as strings in $\{0,1\}^*$, where the encoding length is the same for all inputs of the same length. We assume that λ bits will be sufficient randomness for the encryption algorithm of each system, where 1^λ is the security parameter. We assume that $\Pi_{\mathrm{1b-cca}}$ and Π_{cpa} have perfect correctness for decryption. Finally, we assume that for Π_{dcca} the ciphertext length is a deterministic function of the security parameter and the message length. We further justify these assumptions in the full version.

$\mathsf{KeyGen}(1^\lambda)$ Run $\mathsf{KeyGen}_{\mathrm{dcca}}(1^\lambda)$ to produce $(\mathrm{PK}_{\mathrm{in}}, \mathrm{SK}_{\mathrm{in}})$, $\mathsf{KeyGen}_{\mathrm{1b-cca}}(1^\lambda)$ to produce $(\mathrm{PK}_A, \mathrm{SK}_A)$, and $\mathsf{KeyGen}_{\mathrm{cpa}}(1^\lambda)$ to produce $(\mathrm{PK}_B, \mathrm{SK}_B)$. Set the public key as $\mathrm{PK} := (\mathrm{PK}_{\mathrm{in}}, \mathrm{PK}_A, \mathrm{PK}_B)$ and the secret key as $\mathrm{SK} := (\mathrm{SK}_{\mathrm{in}}, \mathrm{SK}_A, \mathrm{SK}_B)$.

$\mathsf{Enc}(\mathrm{PK}, M)$ The encryption algorithm first chooses random strings $r_{\mathrm{in}}, r_A, r_B \in \{0,1\}^\lambda$. Next, it computes the ciphertext $\mathrm{CT}_{\mathrm{in}} := \mathsf{Enc}_{\mathrm{dcca}}(\mathrm{PK}_{\mathrm{in}}, (r_A, r_B, M); r_{\mathrm{in}})$. It treats this ciphertext as the message and computes $\mathrm{CT}_A := \mathsf{Enc}_{\mathrm{1b-cca}}(\mathrm{PK}_A, \mathrm{CT}_{\mathrm{in}}; r_A)$ and $\mathrm{CT}_B := \mathsf{Enc}_{\mathrm{cpa}}(\mathrm{PK}_B, \mathrm{CT}_{\mathrm{in}}; r_B)$. Finally, it outputs $(\mathrm{CT}_A, \mathrm{CT}_B)$.

$\mathsf{Dec}(\mathrm{SK}, \mathrm{CT})$ The decryption algorithm takes a ciphertext $\mathrm{CT} := (\mathrm{CT}_A, \mathrm{CT}_B)$. It decrypts the first ciphertext as $\mathrm{CT}_{\mathrm{in}} := \mathsf{Dec}_{\mathrm{1b-cca}}(\mathrm{SK}_A, \mathrm{CT}_A)$. It then decrypts this output as $(r_A, r_B, M) := \mathsf{Dec}_{\mathrm{dcca}}(\mathrm{SK}_{\mathrm{in}}, \mathrm{CT}_{\mathrm{in}})$. It then checks that

$$\mathrm{CT}_A = \mathsf{Enc}_{\mathrm{1b-cca}}(\mathrm{PK}_A, \mathrm{CT}_{\mathrm{in}}; r_A) \text{ and } \mathrm{CT}_B = \mathsf{Enc}_{\mathrm{cpa}}(\mathrm{PK}_B, \mathrm{CT}_{\mathrm{in}}; r_B).$$

If all checks pass, it outputs M; otherwise, it outputs $\perp$.

---

[4] A 1-bounded CCA-secure encryption system is secure if an attacker makes at most one decryption query. One-bounded CCA security can be constructed from CPA security [24,10]. CPA security is trivially implied by DCCA security. Thus, there is really only one necessary building block: a DCCA-secure system.

4   Proof of Security

We will now argue that the Section 3 construction is CCA2 secure, assuming
the respective security properties of the underlying building blocks. To do so, it
will be easier to consider a slight variant of the CCA2 security game, which we
call *nested indistinguishability*, where the challenger either encrypts one of the
two challenge messages or encrypts a string of zeros. The experiment involves
three encryption schemes and combines them in the same manner as our main
construction.

Nested Indistinguishability. Consider the experiment $\mathsf{Exp}^{\text{nested}}_{\mathcal{A}, \Pi_{\text{dcca}}, \Pi_{\text{1b-cca}}, \Pi_{\text{cpa}}}(\lambda)$
defined for detectable encryption scheme Π_{dcca}, encryption schemes $\Pi_{\text{1b-cca}}, \Pi_{\text{cpa}}$
and an adversary $\mathcal{A}$:

1. Setup: Run $\mathsf{KeyGen}_{\text{dcca}}$, $\mathsf{KeyGen}_{\text{1b-cca}}$ and $\mathsf{KeyGen}_{\text{cpa}}$ to obtain key pairs
   $(\text{PK}_{\text{in}}, \text{SK}_{\text{in}})$, $(\text{PK}_A, \text{SK}_A)$ and $(\text{PK}_B, \text{SK}_B)$ respectively. Set $pk := (\text{PK}_{\text{in}}, \text{PK}_A, \text{PK}_B)$ and $sk := (\text{SK}_{\text{in}}, \text{SK}_A, \text{SK}_B)$.
2. Phase 1: Adversary $\mathcal{A}$ is given pk and access to a decryption oracle $\mathsf{Dec}(sk, \cdot)$,
   which executes the decryption algorithm as defined in Section 3. $\mathcal{A}$ outputs a
   pair of messages m_0, m_1 of the same length in the message space associated
   with pk.
3. Challenge: Randomness $\beta, z \leftarrow \{0,1\}$ and $r_A, r_B \leftarrow \{0,1\}^\lambda$ are chosen. Let
   ℓ denote the length of the encoding of (r_A, r_B, m_β). Then compute:

$$\text{CT}^*_{\text{in}} := \begin{cases} \mathsf{Enc}_{\text{dcca}}(\text{PK}_{\text{in}}, (r_A, r_B, m_\beta)) & \text{if } z = 0; \\ \mathsf{Enc}_{\text{dcca}}(\text{PK}_{\text{in}}, 0^\ell) & \text{if } z = 1. \end{cases} \qquad (1)$$

   Next compute $\text{CT}^*_A := \mathsf{Enc}_{\text{1b-cca}}(\text{PK}_A, \text{CT}^*_{\text{in}}; r_A)$ and $\text{CT}^*_B := \mathsf{Enc}_{\text{cpa}}(\text{PK}_B, \text{CT}^*_{\text{in}}; r_B)$. Return to $\mathcal{A}$ the ciphertext $\text{CT}^* := (\text{CT}^*_A, \text{CT}^*_B)$.
4. Phase 2: $\mathcal{A}$ continues to have access to $\mathsf{Dec}(sk, \cdot)$, but may not request a
   decryption of the challenge ciphertext CT^*. Finally, $\mathcal{A}$ outputs a bit z'.
5. Output: The output of the experiment is defined to be 1 if $z' = z$, and 0
   otherwise.

Definition 3 (Nested Indistinguishability). *A tuple of encryption systems*
$(\Pi_{\text{dcca}}, \Pi_{\text{1b-cca}}, \Pi_{\text{cpa}})$ *has* nested indistinguishable encryptions under a chosen-
ciphertext attack *if for all probabilistic polynomial-time adversaries $\mathcal{A}$ there exists a negligible function negl such that:*

$$\Pr[\mathsf{Exp}^{\text{nested}}_{\mathcal{A}, \Pi_{\text{dcca}}, \Pi_{\text{1b-cca}}, \Pi_{\text{cpa}}}(\lambda) = 1] \leq \frac{1}{2} + negl(\lambda).$$

It is important to observe that the nested indistinguishability experiment com-
bines $\Pi_{\text{dcca}}, \Pi_{\text{1b-cca}}, \Pi_{\text{cpa}}$ in exactly the same manner as the Section 3 construction. When $z = 1$, it encrypts "properly" and when $z = 0$, it encrypts all
zeros.

   With a goal of proving CCA2 security, our main task is to argue that our
Section 3 construction provides nested indistinguishability. To do this, we must

first establish that a certain event does not happen, except with negligible probability. We define this event as follows.

Definition 4 (The Bad Query Event). *Let* Π_{dcca}, $\Pi_{1b-\text{cca}}$, *and* Π_{cpa} *be the schemes parameterizing the experiment* $\text{Exp}^{\text{nested}}$. *Let* PK_{in} *be the public key output by running* $\text{KeyGen}_{\text{det}}$ *during the course of the experiment. We say that a bad query event* has occurred *during an execution of this experiment if in Phase 2, the adversary* $\mathcal{A}$ *makes a decryption query of the form* $\text{CT} := (\text{CT}_A, \text{CT}_B)$ *such that*

- *(Query inner is "related" to challenge inner:)*
  $F(\text{PK}_{in}, \text{CT}^*_{in}, \text{Dec}_{1b-\text{cca}}(\text{SK}_A, \text{CT}_A)) = 1$, *and*
- *(Query ciphertext differs from challenge ciphertext in first half):*
  $\text{CT}^*_A \neq \text{CT}_A$.

where $\text{CT}^* := (\text{CT}^*_A, \text{CT}^*_B)$ *is the challenge ciphertext and* CT^*_A *is an encryption of* CT^*_{in}. *We note that this event is well defined in both the cases where* $z = 0$ *and* $z = 1$.

4.1   Proof That Bad Query Event Does Not Happen

Lemma 4 (No Bad Query Event when $z = 1$ (all zeros encrypted)).
Suppose that Π_{dcca} *is DCCA secure,* $\Pi_{1b-\text{cca}}$ *is 1-bounded CCA secure, and* Π_{cpa} *is CPA secure, all with perfect correctness. Then for all probabilistic polynomial-time adversaries* $\mathcal{A}$, *during a run of experiment* $\text{Exp}^{\text{nested}}_{\mathcal{A},\Pi_{\text{dcca}},\Pi_{1b-\text{cca}},\Pi_{\text{cpa}}}(\lambda)$ *with* $z = 1$, *a bad query event does not take place except with negligible probability in* λ *where the probability is taken over the coins of the adversary and the experiment.*

Proof. We proceed via a series of hybrids. Let BQE denote a bad query event.

Step 1: $\Pr[\text{BQE in Nested}] \sim \Pr[\text{BQE in Right-Erased}]$ *from CPA-security of* Π_{cpa}. We first define a variation of the nested indistinguishability experiment with $z = 1$, which we call the *right-erased* experiment. In this experiment, CT^*_B is formed as $\text{CT}^*_B := \text{Enc}_{\text{cpa}}(\text{PK}_B, 1^k; r_B)$ where k denotes the length of CT^*_{in}. CT^*_A is formed the same as in the nested indistinguishability experiment with $z = 1$. We suppose there exists a PPT adversary $\mathcal{A}$ for the nested indistinguishability experiment which causes the bad query event to occur with non-negligibly different probability in the usual experiment with $z = 1$ compared to the right-erased experiment. We construct a PPT algorithm $\mathcal{B}$ which violates the CPA-security of Π_{cpa}.

$\mathcal{B}$ is given PK_B. $\mathcal{B}$ then runs $\text{KeyGen}_{\text{dcca}}$ and $\text{KeyGen}_{1b-\text{cca}}$ for itself to produce $\text{PK}_{in}, \text{SK}_{in}$ and PK_A, SK_A respectively. It gives $\mathcal{A}$ $pk = (\text{PK}_{in}, \text{PK}_A, \text{PK}_B)$. $\mathcal{B}$ can simulate the decryption oracle $\text{Dec}(sk, \cdot)$ for $\mathcal{A}$ by running the usual decryption algorithm (note that this does not require SK_B).

The adversary $\mathcal{A}$ outputs a pair of messages m_0, m_1 of the same length in the message space associated with pk. $\mathcal{B}$ chooses $r_A \in \{0,1\}^{\lambda}$ and computes $\text{CT}^*_{in} = \text{Enc}_{\text{dcca}}(\text{PK}_{in}, 0^{\ell})$, where ℓ is the length of the encoding of (r_A, r_A, m_0).

It then computes $\mathrm{CT}_A^* = \mathsf{Enc}_{1b-cca}(\mathrm{PK}_A, \mathrm{CT}_{in}^*; r_A)$. It submits CT_{in}^* and 1^k to its challenger as its two messages. It receives CT_B^* as the ciphertext. It gives $\mathrm{CT}^* := (\mathrm{CT}_A^*, \mathrm{CT}_B^*)$ to $\mathcal{A}$.

To respond to remaining decryption queries $\mathcal{A}$ makes, $\mathcal{B}$ runs the usual decryption algorithm (after checking that the query is not equal to the challenge ciphertext). In addition, $\mathcal{B}$ checks for the bad query event by first checking if $\mathrm{CT}_A \neq \mathrm{CT}_A^*$ and then computing $F(\mathrm{PK}_{in}, \mathrm{CT}_{in}^*, \mathsf{Dec}_{1b-cca}(\mathrm{SK}_A, \mathrm{CT}_A))$. We recall that $\mathcal{B}$ generated $\mathrm{SK}_A, \mathrm{PK}_A$ for itself, so it can compute $\mathsf{Dec}_{1b-cca}(\mathrm{SK}_A, \mathrm{CT}_A)$.

If CT_B^* is an encryption of CT_{in}^*, then $\mathcal{B}$ has properly simulated the usual experiment with $z = 1$. If it is instead an encryption of 1^k, then $\mathcal{B}$ has properly simulated the right-erased experiment. We note that the bad query event occurs in the simulation if and only if it is detected by $\mathcal{B}$.

We let ϵ denote the probability that the bad query event occurs in the usual experiment with $z = 1$ and δ denote this probability in the right-erased experiment. We suppose $\epsilon - \delta$ is positive and non-negligible (the opposite case is analogous). Now, if $\mathcal{B}$ detects the bad query event, it guesses that CT_A^* is an encryption of CT_{in}^*. Otherwise, it guesses the opposite. $\mathcal{B}$'s probability of guessing correctly in the CPA security game for Π_{cpa} is then equal to $\frac{\epsilon}{2} + \frac{1}{2}(1 - \delta) = \frac{1}{2} + \frac{1}{2}(\epsilon - \delta)$. The quantity $\epsilon - \delta$ is non-negligible, so $\mathcal{B}$ violates the CPA-security of Π_{cpa}. Hence we may conclude that the probability of the bad query event happening in the usual experiment with $z = 1$ is the same (up to a negligible difference) as the probability of the bad query event happening in the right-erased experiment for any PPT adversary.

Step 2: $\Pr[BQE$ in Full-Erased] is negligible from the unpredictability of the detecting function of Π_{dcca}. We now define an additional variation of the experiment, which we call the *full-erased* experiment. This is like the right-erased experiment, except that CT_A^* is also an encryption of 1^k, instead of an encryption of CT_{in}^*. We claim that in the full-erased experiment, the bad query event can only occur with negligible probability. To see this, we suppose we have a PPT adversary $\mathcal{A}$ which causes the bad query event to occur with non-negligible probability in the full-erased experiment. We will build a PPT adversary $\mathcal{B}$ for the basic unpredictability experiment which violates unpredictability of the detecting function for Π_{dcca}.

$\mathcal{B}$ is given PK_{in} and access to a decryption oracle $\mathsf{Dec}(\mathrm{SK}_{in}, \cdot)$. It runs KeyGen_{1b-cca} and KeyGen_{cpa} for itself to produce $\mathrm{PK}_A, \mathrm{SK}_A$ and $\mathrm{PK}_B, \mathrm{SK}_B$. It gives $(\mathrm{PK}_{in}, \mathrm{PK}_A, \mathrm{PK}_B)$ to $\mathcal{A}$. $\mathcal{B}$ can simulate the decryption oracle for $\mathcal{A}$ using SK_A and its own decryption oracle. $\mathcal{A}$ outputs m_0, m_1. $\mathcal{B}$ then computes $\mathrm{CT}_A^* = \mathsf{Enc}_{1b-cca}(\mathrm{PK}_A, 1^k)$ and $\mathrm{CT}_B^* = \mathsf{Enc}_{cpa}(\mathrm{PK}_B, 1^k)$ and gives $\mathrm{CT}^* = (\mathrm{CT}_A^*, \mathrm{CT}_B^*)$ to $\mathcal{A}$. We let q denote the number of Phase 2 queries made by $\mathcal{A}$. $\mathcal{B}$ can respond to these queries as before. $\mathcal{B}$ chooses a random $i \in \{1, 2, \ldots, q\}$ and a random bit $b \in \{0, 1\}$. It takes the ith Phase 2 query of $\mathcal{A}$, denoted by $(\mathrm{CT}_A^i, \mathrm{CT}_B^i)$, and computes $\mathrm{CT}_{in}^i = \mathsf{Dec}_{1b-cca}(\mathrm{SK}_A, \mathrm{CT}_A^i)$. It submits m_b and CT_{in}^i to its challenger. Then, the distribution of $c^* = \mathsf{Enc}_{dcca}(\mathrm{PK}_{in}, m_b)$ in the basic unpredictability experiment is precisely the distribution of CT_{in}^*. Hence, the bad query event for query i corresponds to an output of 1 for basic unpredictability experiment. Thus, if the bad query event occurs with some non-negligible probability ϵ, $\mathcal{B}$ will cause an output of 1 in the basic unpredictability experiment with probability at least $\frac{\epsilon}{q}$, which is non-negligible.

Step 3: Pr[BQE in Right-Erased] $\sim$ Pr[BQE in Full-Erased] from the 1-bounded CCA security of $\Pi_{\text{1b-cca}}$. We now return to considering a PPT adversary $\mathcal{A}$ in the right-erased experiment. We let q denote the number of Phase 2 queries made by $\mathcal{A}$. We suppose that $\mathcal{A}$ causes the bad query event with non-negligible probability. Then there exists some index $i \in \{1, \ldots, q\}$ such that $\mathcal{A}$ causes the bad query event to occur with non-negligible probability on its ith Phase 2 query. In other words, if there exists a PPT adversary $\mathcal{A}$ for which the bad query event occurs with non-negligible probability in the right-erased experiment, then for each value of the security parameter, there exists an index i such that $\mathcal{A}$ causes the BQE to occur on its ith Phase 2 query with non-negligible probability. We note that for *any* i, the probability that $\mathcal{A}$ causes the BQE to occur on its ith Phase 2 query in the full-erased experiment is negligible, as we proved above.

We fix such an i, and we define a PPT algorithm $\mathcal{B}$ which violates the 1-bounded CCA security of $\Pi_{\text{1b-cca}}$. $\mathcal{B}$ receives PK_A from its challenger. It runs $\text{KeyGen}_{\text{dcca}}$ and $\text{KeyGen}_{\text{cpa}}$ for itself to produce $\text{PK}_{\text{in}}, \text{SK}_{\text{in}}$ and PK_B, SK_B. It gives $(\text{PK}_{\text{in}}, \text{PK}_A, \text{PK}_B)$ to $\mathcal{A}$ as the public key.

$\mathcal{B}$ simulates the decryption oracle for $\mathcal{A}$ as follows. Upon receiving a ciphertext $(\text{CT}_A, \text{CT}_B)$, $\mathcal{B}$ decrypts CT_B using Dec_{cpa} with SK_B, and we let CT_{in} denote the output. It then decrypts CT_{in} using Dec_{dcca} with SK_{in}, and parses the output as r_A, r_B, M. It checks if $\text{CT}_A = \text{Enc}_{\text{1b-cca}}(\text{PK}_A, \text{CT}_{\text{in}}; r_A)$ and if $\text{CT}_B = \text{Enc}_{\text{cpa}}(\text{PK}_B, \text{CT}_{\text{in}}; r_B)$. If both checks pass, it outputs M. Else, it outputs $\perp$.

We claim that this matches the output of the usual decryption algorithm, even though $\mathcal{B}$ is first decrypting CT_B instead of CT_A. To see this, note that the outputs are the same whenever $\text{Dec}_{\text{1b-cca}}(\text{CT}_A, \text{SK}_A) = \text{Dec}_{\text{cpa}}(\text{CT}_B, \text{SK}_B)$. Whenever these are unequal, both decryption methods will output $\perp$. This is because $\text{CT}_A = \text{Enc}_{\text{1b-cca}}(\text{PK}_A, \text{CT}_{\text{in}}; r_A)$ and $\text{CT}_B = \text{Enc}_{\text{cpa}}(\text{PK}_B, \text{CT}_{\text{in}}; r_B)$ imply that $\text{Dec}_{\text{1b-cca}}(\text{CT}_A, \text{SK}_A) = \text{CT}_{\text{in}} = \text{Dec}_{\text{cpa}}(\text{CT}_B, \text{SK}_B)$. (Recall here that we have assumed $\Pi_{\text{1b-cca}}$ and Π_{cpa} have perfect correctness.)

At some point, $\mathcal{A}$ outputs m_0, m_1. $\mathcal{B}$ forms $\text{CT}_{\text{in}}^* = \text{Enc}_{\text{dcca}}(\text{PK}_{\text{in}}, 0^\ell)$ and $\text{CT}_B^* = \text{Enc}_{\text{cpa}}(\text{PK}_B, 1^k)$. It outputs the messages CT_{in}^* and 1^k to its challenger, and receives a ciphertext which it sets as CT_A^*. It gives the ciphertext $(\text{CT}_A^*, \text{CT}_B^*)$ to $\mathcal{A}$. It can then respond to $\mathcal{A}$'s Phase 2 decryption queries in the same way as before. When it receives the ith Phase 2 query of $\mathcal{A}$, denoted by $(\text{CT}_A^i, \text{CT}_B^i)$, $\mathcal{B}$ checks for the bad query event by first checking if $\text{CT}_A^i \neq \text{CT}_A^*$ and if so, submitting CT_A^i as its one decryption query to its decryption oracle for PK_A. It can compute $F(\text{PK}_{\text{in}}, \text{CT}_{\text{in}}^*, \text{Dec}(\text{SK}_A, \text{CT}_A^i))$. This equals 1 if and only if the bad query event has occurred for query i, and in this case $\mathcal{B}$ guesses that CT_A^* is an encryption of CT_{in}^*. Otherwise, $\mathcal{B}$ guesses the opposite.

We observe that when CT_A^* is an encryption of CT_{in}^*, then $\mathcal{B}$ has properly simulated the right-erased experiment, and when CT_A^* is an encryption of 0^k, then $\mathcal{B}$ has properly simulated the full-erased experiment. We let ϵ denote the non-negligible probability that $\mathcal{A}$ causes the bad query event to occur on (Phase 2) query i in the right-erased experiment, and we let δ denote the corresponding probability for the full-erased experiment. We know that δ must be negligible, therefore $\epsilon - \delta$ is positive and non-negligible. The probability that $\mathcal{B}$ guesses correctly is: $\frac{1}{2}(1-\delta) + \frac{1}{2}\epsilon = \frac{1}{2} + \frac{1}{2}(\epsilon-\delta)$, so $\mathcal{B}$ achieves a non-negligible advantage in the 1-bounded CCA security game for Π_{1b-cca}.

Thus, it must be the case that for all PPT algorithms $\mathcal{A}$, the BQE occurs with only negligible probability in the right-erased experiment, and hence also in the nested experiment with $z = 1$.

Lemma 5 (No Bad Query Event when $z=0$ (Real Message Encrypted)).
As a consequence of Lemma 4 and the DCCA security of Π_{dcca}, it holds that for all probabilistic polynomial-time adversaries $\mathcal{A}$, during a run of experiment $\mathsf{Exp}^{nested}_{\mathcal{A},\Pi_{dcca},\Pi_{1b-cca},\Pi_{cpa}}(\lambda)$ with $z = 0$, a bad query event does not take place except with negligible probability in λ where the probability is taken over the coins of the adversary and the experiment.

Proof. In Lemma 4, we established that bad query events happen with at most negligible probability when $z = 1$. We will use this fact to argue that they cannot happen much more frequently when $z = 0$. Suppose to the contrary that there exists a PPT adversary $\mathcal{A}$ that forces bad query events to happen with non-negligible probability ϵ when $z = 0$. We create an PPT adversary $\mathcal{B}$ who interacts with $\mathcal{A}$ in a run of the nested indistinguishability experiment to break the DCCA security of Π_{dcca} with detecting function F with probability negligibly-close to $\frac{1}{2} + \frac{\epsilon}{2}$ as follows:

1. Setup: $\mathcal{B}$ obtains $\mathrm{PK_{in}}$ from the Exp^{indist} challenger. It runs KeyGen_{1b-cca} to obtain $(\mathrm{PK}_A, \mathrm{SK}_A)$ and KeyGen_{cpa} to obtain $(\mathrm{PK}_B, \mathrm{SK}_B)$.
2. Phase 1: $\mathcal{B}$ gives to $\mathcal{A}$ the public key $\mathrm{PK} = (\mathrm{PK_{in}}, \mathrm{PK}_A, \mathrm{PK}_B)$. When $\mathcal{A}$ queries the decryption oracle on CT, $\mathcal{B}$ can simulate the normal decryption algorithm using SK_A and the phase 1 oracle $\mathsf{Dec}(\mathrm{SK_{in}}, \cdot)$. Eventually, $\mathcal{A}$ outputs a pair of messages m_0, m_1.
3. Challenge: Choose random $\beta \in \{0,1\}$ and $r_A, r_B \in \{0,1\}^\lambda$. Send to the Exp^{indist} challenger the messages $M_0 = (r_A, r_B, m_\beta)$ and $M_1 = 0^{|M_0|}$, and obtain from the challenger $\mathrm{CT_{in}^*}$. Compute $\mathrm{CT}_A^* := \mathsf{Enc}_{1b-cca}(\mathrm{PK}_A, \mathrm{CT_{in}^*}; r_A)$ and $\mathrm{CT}_B^* := \mathsf{Enc}_{cpa}(\mathrm{PK}_B, \mathrm{CT_{in}^*}; r_B)$. Return $\mathrm{CT}^* := (\mathrm{CT}_A^*, \mathrm{CT}_B^*)$ to $\mathcal{A}$.
4. Phase 2: When $\mathcal{A}$ queries the decryption oracle on $\mathrm{CT} := (\mathrm{CT}_A, \mathrm{CT}_B)$, compute $\mathrm{CT_{in}} := \mathsf{Dec}_{1b-cca}(\mathrm{SK}_A, \mathrm{CT}_A)$. If
   (a) Case 1 (a bad query event): $\mathrm{CT}_A \neq \mathrm{CT}_A^*$ and yet $F(\mathrm{PK_{in}}, \mathrm{CT_{in}^*}, \mathrm{CT_{in}}) = 1$, then abort and output the bit 0.
   (b) Case 2 (partial match with challenge): $\mathrm{CT}_A = \mathrm{CT}_A^*$, then return $\bot$ to $\mathcal{A}$.
   Otherwise, query the phase 2 oracle, $\mathsf{Dec}(\mathrm{SK_{in}}, \cdot)$, to decrypt $\mathrm{CT_{in}}$, and return its response to $\mathcal{A}$.
5. Output: When $\mathcal{A}$ outputs a bit, $\mathcal{B}$ echos the bit as its output.

Analysis. We begin our analysis by arguing that $\mathcal{B}$ correctly answers all decryption queries except when it aborts. First, we show that a partial match with the challenge, causing the $\perp$ response in Case 2, is correct because that query must be invalid. Since a decryption query on the challenge is forbidden by the experiment, if $\mathrm{CT}_A = \mathrm{CT}_A^*$, then $\mathrm{CT}_B \neq \mathrm{CT}_B^*$. However, we argue that this must be an invalid ciphertext, i.e., one on which the main construction's decryption algorithm would return $\perp$. We see this as follows. Since decryption is deterministic, we have $T := \mathsf{Dec}_{1b-cca}(\mathrm{SK}_A, \mathrm{CT}_A) = \mathsf{Dec}_{1b-cca}(\mathrm{SK}_A, \mathrm{CT}_A^*)$ and $(r_A, r_B, m) := \mathsf{Dec}_{dcca}(\mathrm{SK}_{in}, T)$. By the checks enforced by the main construction's decryption algorithm, there is only one "second half" that matches $\mathrm{CT}_A = \mathrm{CT}_A^*$, that is $\mathsf{Enc}_{cpa}(\mathrm{PK}_B, T; r_B)$. Since the challenge is a valid ciphertext, CT_B^* must be this value and CT_B must cause an error.

When neither Case 1 or Case 2 applies in phase 2, the inner decryption query will succeed since the ciphertext is not detectably related to the challenge. This allows $\mathcal{B}$ to respond correctly.

When a bad query event occurs in Phase 2, $\mathcal{B}$ cannot query Exp^{indist}'s decryption oracle to decrypt the ciphertext. At first glance, one seems stuck. However, we assumed bad query events happen only when $z = 0$ with all but negligible probability. Thus, $\mathcal{B}$ can guess that $\mathcal{A}$ thinks $z = 0$, which corresponds to M_0 being encrypted in our reduction. Thus, $\mathcal{B}$ can abort and guess 0 at this point.

When $\mathcal{B}$ aborts, it causes the Exp^{indist} experiment to output 1 with high probability. When $\mathcal{B}$ does not abort, it causes Exp^{indist} experiment to output 1 with probability $\frac{1}{2}$. Since $\mathcal{B}$ aborts with non-negligible probability ϵ when $z = 0$, then $\mathcal{B}$ causes the experiment's output to be 1 with probability non-negligibly greater than $\frac{1}{2}$.

4.2 Putting the Proof of the Main Theorem together

Theorem 1 (Main Construction is Nested Indistinguishable). *Our main construction in Section 3, comprised of the three building blocks $\Pi_{dcca}, \Pi_{1b-cca}, \Pi_{cpa}$, has nested indistinguishable encryptions under a chosen-ciphertext attack under the assumptions that Π_{dcca} is DCCA secure, Π_{1b-cca} is 1-bounded CCA secure, and Π_{cpa} is CPA secure, all with perfect correctness.*

Proof of Theorem 1 appears in the full version. The crux of the argument is that bad query events do not happen (except with negligible probability). This was already established in Lemmas 4 and 5. Armed with this fact, we can prove the nested indistinguishability of the main construction based on the indistinguishability property of the DCCA-security of Π_{dcca}. The reduction and its analysis are similar to those in the proof of Lemma 5.[5]

The following corollary follows from Theorem 1. Informally, if the adversary cannot distinguish an encryption of a message from an encryption of zeros, then she also cannot distinguish between the encryptions of two different messages.

---

[5] We alternatively could have merged the proofs of Lemma 5 and Theorem 1. However, we chose to keep the bad event analysis separate for pedagogical purposes at the expense of some redundancy in the description of the related reductions.

Corollary 1 (Main Construction is CCA2 Secure). *Our main construction in Section 3, comprised of the three building blocks $\Pi_{\mathrm{dcca}}, \Pi_{\mathrm{1b-cca}}, \Pi_{\mathrm{cpa}}$, is CCA2 secure under the assumptions that Π_{dcca} is DCCA secure, $\Pi_{\mathrm{1b-cca}}$ is 1-bounded CCA secure, and Π_{cpa} is CPA secure, all with perfect correctness.*

Acknowledgments. The authors thank Steven Myers and the anonymous reviewers for helpful comments.

References

1. An, J.H., Dodis, Y., Rabin, T.: On the Security of Joint Signature and Encryption. In: Knudsen, L.R. (ed.) EUROCRYPT 2002. LNCS, vol. 2332, pp. 83–107. Springer, Heidelberg (2002)
2. Bellare, M., Brakerski, Z., Naor, M., Ristenpart, T., Segev, G., Shacham, H., Yilek, S.: Hedged Public-Key Encryption: How to Protect against Bad Randomness. In: Matsui, M. (ed.) ASIACRYPT 2009. LNCS, vol. 5912, pp. 232–249. Springer, Heidelberg (2009)
3. Bellare, M., Rogaway, P.: Random oracles are practical: A paradigm for designing efficient protocols. In: ACM Conference on Computer and Communications Security, pp. 62–73 (1993)
4. Bellare, M., Sahai, A.: Non-malleable Encryption: Equivalence between Two Notions, and an Indistinguishability-Based Characterization. In: Wiener, M. (ed.) CRYPTO 1999. LNCS, vol. 1666, pp. 519–536. Springer, Heidelberg (1999)
5. Blum, M., Feldman, P., Micali, S.: Non-interactive zero-knowledge and its applications (extended abstract). In: STOC, pp. 103–112 (1988)
6. Boneh, D., Canetti, R., Halevi, S., Katz, J.: Chosen-ciphertext security from identity-based encryption. SIAM J. Comput. 36(5), 1301–1328 (2007)
7. Canetti, R., Halevi, S., Katz, J.: Chosen-Ciphertext Security from Identity-Based Encryption. In: Cachin, C., Camenisch, J.L. (eds.) EUROCRYPT 2004. LNCS, vol. 3027, pp. 207–222. Springer, Heidelberg (2004)
8. Canetti, R., Krawczyk, H., Nielsen, J.B.: Relaxing Chosen-Ciphertext Security. In: Boneh, D. (ed.) CRYPTO 2003. LNCS, vol. 2729, pp. 565–582. Springer, Heidelberg (2003)
9. Cash, D.M., Kiltz, E., Shoup, V.: The Twin Diffie-Hellman Problem and Applications. In: Smart, N.P. (ed.) EUROCRYPT 2008. LNCS, vol. 4965, pp. 127–145. Springer, Heidelberg (2008)
10. Choi, S.G., Dachman-Soled, D., Malkin, T., Wee, H.M.: Black-Box Construction of a Non-malleable Encryption Scheme from Any Semantically Secure One. In: Canetti, R. (ed.) TCC 2008. LNCS, vol. 4948, pp. 427–444. Springer, Heidelberg (2008)
11. Cramer, R., Hanaoka, G., Hofheinz, D., Imai, H., Kiltz, E., Pass, R., Shelat, A., Vaikuntanathan, V.: Bounded CCA2-Secure Encryption. In: Kurosawa, K. (ed.) ASIACRYPT 2007. LNCS, vol. 4833, pp. 502–518. Springer, Heidelberg (2007)
12. Cramer, R., Shoup, V.: A Practical Public Key Cryptosystem Provably Secure against Adaptive Chosen Ciphertext Attack. In: Krawczyk, H. (ed.) CRYPTO 1998. LNCS, vol. 1462, pp. 13–25. Springer, Heidelberg (1998)
13. Cramer, R., Shoup, V.: Universal Hash Proofs and a Paradigm for Adaptive Chosen Ciphertext Secure Public-Key Encryption. In: Knudsen, L.R. (ed.) EUROCRYPT 2002. LNCS, vol. 2332, pp. 45–64. Springer, Heidelberg (2002)

14. Dolev, D., Dwork, C., Naor, M.: Non-malleable cryptography (extended abstract). In: STOC, pp. 542–552 (1991)
15. Gertner, Y., Malkin, T., Reingold, O.: On the impossibility of basing trapdoor functions on trapdoor predicates. In: FOCS, pp. 126–135 (2001)
16. Goldwasser, S., Micali, S.: Probabilistic encryption. J. Comput. Syst. Sci. 28(2), 270–299 (1984)
17. Hofheinz, D., Kiltz, E.: Secure Hybrid Encryption from Weakened Key Encapsulation. In: Menezes, A. (ed.) CRYPTO 2007. LNCS, vol. 4622, pp. 553–571. Springer, Heidelberg (2007)
18. Hofheinz, D., Kiltz, E.: Practical Chosen Ciphertext Secure Encryption from Factoring. In: Joux, A. (ed.) EUROCRYPT 2009. LNCS, vol. 5479, pp. 313–332. Springer, Heidelberg (2009)
19. Kiltz, E.: Chosen-Ciphertext Security from Tag-Based Encryption. In: Halevi, S., Rabin, T. (eds.) TCC 2006. LNCS, vol. 3876, pp. 581–600. Springer, Heidelberg (2006)
20. Kiltz, E., Mohassel, P., O'Neill, A.: Adaptive Trapdoor Functions and Chosen-Ciphertext Security. In: Gilbert, H. (ed.) EUROCRYPT 2010. LNCS, vol. 6110, pp. 673–692. Springer, Heidelberg (2010)
21. MacKenzie, P.D., Reiter, M.K., Yang, K.: Alternatives to Non-malleability: Definitions, Constructions, and Applications (Extended Abstract). In: Naor, M. (ed.) TCC 2004. LNCS, vol. 2951, pp. 171–190. Springer, Heidelberg (2004)
22. Myers, S., Shelat, A.: Bit encryption is complete. In: FOCS, pp. 607–616 (2009)
23. Naor, M., Yung, M.: Public-key cryptosystems provably secure against chosen ciphertext attacks. In: STOC, pp. 427–437 (1990)
24. Pass, R., Shelat, A., Vaikuntanathan, V.: Construction of a Non-malleable Encryption Scheme from Any Semantically Secure One. In: Dwork, C. (ed.) CRYPTO 2006. LNCS, vol. 4117, pp. 271–289. Springer, Heidelberg (2006)
25. Peikert, C., Waters, B.: Lossy trapdoor functions and their applications. In: STOC, pp. 187–196 (2008)
26. Rackoff, C., Simon, D.R.: Non-interactive Zero-Knowledge Proof of Knowledge and Chosen Ciphertext Attack. In: Feigenbaum, J. (ed.) CRYPTO 1991. LNCS, vol. 576, pp. 433–444. Springer, Heidelberg (1992)
27. Rosen, A., Segev, G.: Chosen-Ciphertext Security via Correlated Products. In: Reingold, O. (ed.) TCC 2009. LNCS, vol. 5444, pp. 419–436. Springer, Heidelberg (2009)
28. Sahai, A.: Non-malleable non-interactive zero knowledge and adaptive chosen-ciphertext security. In: FOCS, pp. 543–553 (1999)
29. Shoup, V.: A proposal for an ISO standard for public key encryption. Cryptology ePrint Archive, Report 2001/112 (2001), http://eprint.iacr.org/

Security of Symmetric Encryption
in the Presence of Ciphertext Fragmentation[*]

Alexandra Boldyreva[1],[**], Jean Paul Degabriele[2],[***], Kenneth G. Paterson[2],[†],
and Martijn Stam[3]

[1] Georgia Institute of Technology
[2] Royal Holloway, University of London
[3] University of Bristol

Abstract. In recent years, a number of standardized symmetric encryption schemes have fallen foul of attacks exploiting the fact that in some real world scenarios ciphertexts can be delivered in a fragmented fashion. We initiate the first general and formal study of the security of symmetric encryption against such attacks. We extend the SSH-specific work of Paterson and Watson (Eurocrypt 2010) to develop security models for the fragmented setting. We also develop security models to formalize the additional desirable properties of ciphertext boundary hiding and robustness against Denial-of-Service (DoS) attacks for schemes in this setting. We illustrate the utility of each of our models via efficient constructions for schemes using only standard cryptographic components, including constructions that simultaneously achieve confidentiality, ciphertext boundary hiding and DoS robustness.

1 Introduction

Despite the existence of proofs guaranteeing security, deployed schemes do get compromised sometimes. Consider for example SSH, one of the most widely used secure protocols. Bellare et al. [4] have formally analysed variants of SSH's Binary Packet Protocol (BPP) and showed that these variants are secure. Yet a few years later, Albrecht et al. [1] presented plaintext recovery attacks against these provably secure SSH BPP variants. These attacks exploited the fact that encrypted data can be delivered to the receiver in a fragmented, byte-by-byte manner, and that the attacker can observe the receiver's behaviour at each point (in particular how long it takes to reject certain carefully crafted faulty ciphertexts). On the other hand, formal security definitions, including the one used

---

[*] This work has been supported in part by the European Commission through the ICT programme under contract ICT-2007-216676 ECRYPT II.

[**] This author is supported in part by NSF CAREER award 0545659 and NSF Cyber Trust award 0831184.

[***] This author is supported by Vodafone Group Services Limited, a Thomas Holloway Research Studentship, and the Strategic Educational Pathways Scholarship Scheme (Malta), part-financed by the European Union European Social Fund.

[†] This author is supported by EPSRC Leadership Fellowship EP/H005455/1.

D. Pointcheval and T. Johansson (Eds.): EUROCRYPT 2012, LNCS 7237, pp. 682–699, 2012.

to prove SSH secure, traditionally treat plaintexts and ciphertexts as *atomic*, meaning that the entire ciphertext is offered for decryption and a plaintext (or error symbol) is instantly returned.

To bridge this gap between theory and practice and to have schemes with security guarantees that hold not only on paper but also in reality, one has to design security definitions which are integrated better with the environments in which the protocols are deployed. Paterson and Watson [14] recently took a first step in this direction by showing that certain SSH BPP variants meet a newly introduced security notion that takes the aforementioned attacks into account. However, their security definition itself is heavily intertwined with the SSH BPP specification and too complex to be extended easily to apply to different schemes. We provide a more detailed critique of this precursor [14] in the full version [6].

Overview of Contributions. In this work we initiate a general study of security of symmetric encryption schemes against fragmentation attacks. Our study goes beyond just message privacy, and also includes length-hiding (or, more precisely, hiding ciphertext boundaries in a ciphertext stream) and the prevention of fragmentation-enabled Denial-of-Service (DoS) attacks against the receiver. These two properties have not been previously studied, partly because the corresponding threats are not present if encryption is treated as being atomic. The framework we develop can be used to provide *meaningful* provable security analyses of practical schemes when deployed in environments that permit ciphertext fragmentation attacks (including but not limited to the ones from [14]).

We complement our new security definitions with efficient cryptographic constructions based on standard primitives meeting the new goals. While it may be relatively easy to achieve each security goal independently, it transpires that it is not straightforward to achieve two or three of the aforementioned goals simultaneously and one of our schemes is the first to do so.

Let us now describe our focus and results in a little more detail.

Data Fragmentation. Data sent over networks is often fragmented, meaning that it is broken up into smaller pieces, or packets. If the data is encrypted, the receiver first has to determine what constitutes a complete ciphertext in order to decrypt it and obtain the underlying message. Reconstruction of the original ciphertext by the receiver can be accomplished by various methods. For example, SSH uses a length field that tells the receiver how many bytes are needed before the complete ciphertext has arrived; this length field is encrypted, ostensibly to increase the security of the protocol against traffic analysis.

During transmission, the packets can be accidentally delayed or delivered out of order. But there also may be malicious tampering of legitimate fragmentation, such as breaking the encrypted data into adversarially selected packets, or maliciously delaying their delivery. We already mentioned an attack of this kind on SSH [1]. Another example is an attack by Degabriele and Paterson [8] on the IPsec protocol. While fragmentation is used adversarially in both attacks, there are some notable differences. In particular, in SSH for honest users, the ciphertext can be effectively regarded as a bitstring (the result of say CBC-and-MAC) and it is only the adversary who starts fragmenting this string. In IPsec, further protocol layers already forcibly introduce fragmentation; on the one hand this ties the adversary, but on the other hand the interaction of this protocol layer with the cryptographic layer can offer new avenues of attack (as exploited by the attack on IPsec [8]). Our treatment addresses both scenarios.

Syntax for Encryption Supporting Fragmentation. We start our analysis with defining encryption in the presence of fragmentation. A *symmetric encryption scheme supporting fragmentation* is defined similarly to a regular atomic (fragmentation devoid) encryption scheme, except the decryption algorithm is always stateful, mainly to model decryption algorithms that may, for example, combine data coming from multiple ciphertext fragments before outputting any plaintext. In addition to state, decryption takes input fragments, one-by-one. Depending on the scheme, the minimal fragment length can be one bit, a byte or a block (of some fixed length). The correctness requirement is defined more intricately than that for atomic encryption. It requires that regardless of how one fragments the ciphertexts, the original messages are returned, with correct message boundaries indicated.

In this paper we focus on two subclasses of symmetric encryption schemes for fragmented ciphertexts, for which we give separate security definitions. One subclass, to which we will simply refer to as *stateful*, covers most practical encryption schemes, whose encryption and decryption algorithms are both stateful. For example, encryption and decryption can both use a counter that increases depending on the number of messages or ciphertexts processed.

The other subclass we consider is an extension of standard (atomic) encryption schemes that makes handling fragmented ciphertexts possible. Namely, the decryption algorithm is now stateful, but the state just models the buffer the receiver keeps to store the ciphertext fragments before a complete ciphertext is received (thus allowing the decryption oracle to perform operations on the entire ciphertext, even if it arrives in fragments). We call such schemes *stateless beyond buffering (sbb)*. Because of space constraints, we focus here on the stateful case, with details for the sbb case appearing in the full version [6].

Message Privacy in the Presence of Fragmentation. We observe that fragmentation becomes relevant for security only in the case of chosen-ciphertext attacks (CCA). We extend the existing CCA security notions for regular atomic (IND-CCA) and atomic and stateful (IND-sfCCA [4]) schemes to the case of ciphertext fragmentation (denoted CFA).

Recall that for IND-sfCCA there is no restriction on the decryption queries, but if the adversary forwards the challenge ciphertexts returned by the left-or-right encryption oracle to the decryption oracle in order, this is considered in-sync, and the decryption output is suppressed. Otherwise, it is declared out-of-sync, and the decryptions are returned to the adversary. This allows the adversary to advance the state of both encryption and decryption algorithms to potentially favourable values. When dealing with fragments, the challenge is to decide when to enter the out-of-sync state. We found the need to declare part of the fragment in-sync, and part of it out-of-sync and resolve the ambiguity with regards to the exact boundary to use. We provide our IND-sfCFA definition and more discussion in Section 3.2.

Ciphertext Boundary Hiding. It is conventional wisdom in cryptographic security definitions that an encryption scheme is allowed to leak the length of the ciphertext; it is often regarded as inevitable. However, real schemes try to achieve another goal as well: they try to hide the lengths of encrypted messages, with a view to frustrating traffic analysis based on these lengths. This is generally achieved in practice by two distinct mechanisms.

Firstly, an encryption scheme for which the ciphertext length does *not* deterministically depend on the message length may be used (e.g. by using variable-length padding). The SSH Binary Packet Protocol and the TLS Record Protocol both adopt this approach. This mechanism has recently received attention from differing perspectives [16, 18]. Secondly, an encryption scheme may be designed in such a way that it is hard to distinguish where the boundaries between ciphertexts lie in a *stream* of ciphertexts. TLS, with its explicit length field in the header of each TLS Record Protocol message, does not achieve this. But SSH's Binary Packet Protocol (BPP) does attempt to achieve boundary hiding. This necessitated the introduction of an encrypted length field in SSH, which is used by the receiver to determine how many bytes are required before a complete ciphertext has arrived and a MAC on the plaintext can be checked. However, this design decision, coupled with the use of CBC mode encryption, is precisely what enabled recent fragmentation attacks against SSH [1]. Thus having boundary hiding as a security goal can act in opposition to achieving other, more standard security goals.

In Section 4, we formalize the goal of boundary hiding for ciphertext streams. We give definitions for both the passive and the active adversary cases, which we call BH-CPA and BH-sfCFA. The passive case is very common in the traffic analysis literature. Here the adversary merely monitors encrypted traffic and tries to infer information from ciphertext lengths and other information such as network packet timings, but without giving away its presence by actively modifying network traffic. By hiding the ciphertext boundaries, the adversary no longer has access to fine-grained ciphertext lengths (our solution of course does not help to hide the total volume being sent). We also define boundary hiding in the active case and find out that it is much more challenging to achieve.

Denial-of-Service. Next, we focus on the very important goal of preventing fragmentation-related Denial-of-Service (DoS) attacks against the receiver. This is, to the best of our knowledge, the first formal treatment of DoS prevention as a property of encryption. For an example of such an attack, consider the SSH-CTR scheme (see [14] for a description) and the adversary who changes the length field that occupies the first 32 bits of plaintext by bit flipping in the ciphertext. If the length is maliciously increased to a very large value (say, $2^{32} - 1$, the maximum possible value for a 32-bit field), then the receiver will

Table 1. Stateful schemes and their security properties

Scheme	Ref.	IND-sfCFA	BH-CPA	BH-sfCFA	DOS-sfCFA
Prefix free ($\mathcal{EC} \circ \mathcal{E}$)	(Sec. 3.3)	+	×	×	×
Keyed Prefix Free (KPF)	(full version [6])	+	+	×	×
InterMAC	(Sec. 5)	+	+	+	+

continue listening for ciphertext fragments awaiting message completion, until 2^{32} bytes of data have been received. Only then will SSH-CTR's MAC verification be conducted and the message rejected. The application (or user) receiving data from the SSH connection experiences this as an SSH connection hang, a form of Denial-of-Service.

We provide a security definition DOS-sfCFA for DoS attacks in Section 5 that is sufficiently flexible to capture the SSH attack and others like it. Essentially, we measure the attacker's ability to create a sequence of ciphertext fragments for which the decryption algorithm of a scheme does not output any message or failure symbol within a reasonable timeframe, measured in terms of the number of symbols submitted to a decryption oracle by the adversary.

Constructions and Their Security. So far, our emphasis has been on developing security models and notions. However, as we proceed, we demonstrate how each of the security notions we provide can be met in practice by efficient schemes using only standard symmetric components. These constructions are illustrative rather than definitive. Table 1 lists our main schemes for the stateful setting and their properties; for definitions and further discussions, see the referenced sections. We note that the scheme InterMAC is able to simultaneously achieve all three of our active security notions IND-sfCFA, BH-sfCFA, and DOS-sfCFA.

Further Related Work. Our fragmented approach bears more than a passing resemblance to work on on-line encryption [2, 3, 7, 9, 10, 11, 12]. However, whereas the on-line setting concerns a single continuous message and ciphertext, with each block of plaintext leading to a block of ciphertext being output during encryption (and vice-versa during decryption), our setting concerns atomic encryption (reflecting how many secure protocols operate) but allows fragmented decryption of ciphertexts. Moreover, we extensively treat the case of active adversaries, a topic that has not achieved much attention in the on-line literature, and we consider more than just confidentiality security notions. This said, our ultimate construction, InterMAC, can be seen as a kind of online scheme with a large block size.

2   Preliminaries

Since manipulating sequences of symbols (strings) in various ways will be crucial to our later exposition, we begin with some standard and not-so-standard definitions relating to strings.

Let $\mathcal{B}$ be a set and $\mathcal{B}^*$ denote the set of finite strings with symbols from $\mathcal{B}$ (including the empty string ε). Let $\mathcal{B}^+$ denote $\mathcal{B}^* \setminus \{\varepsilon\}$. Denote by $|\cdot|: \mathcal{B}^* \to \mathbb{Z}^+$ the length function which counts the number of symbols in a string. Typically we will have that $\mathcal{B}$ is the set of bitstrings of some fixed length n, that is $\mathcal{B} = \{0,1\}^n$, leading to the usual notation of $\{0,1\}^*$ for the set of all binary strings of finite length. In this case, if X is a string, then $|X|$ denotes its bit-length. If $\mathbf{X}$ is a vector of strings, then $|\mathbf{X}|$ denotes the number of its components. Given two strings $X, Y \in \mathcal{B}^*$, we write $X \parallel Y$ for the concatenation of the two strings. Given a sequence of strings, we define the operator $\parallel$ that simply concatenates all the constituent strings. For example, if $X = (00)$ and $Y = (11)$, then $X \parallel Y = (0011) \in \{0,1\}^*$ and $\parallel ((00),(11)) = (0011) \in \{0,1\}^*$.

For two elements $a, b \in \mathcal{B}^*$ we call a a *prefix* of b if there exists $c \in \mathcal{B}^*$ such that $b = a \parallel c$. For two elements $a, b \in \mathcal{B}^*$ we denote with $a \star b$ the greatest common prefix of a and b and by $a \% b$ the remainder string of a with respect to b (so in particular, $a = (a \star b) \parallel (a \% b)$ and $b = (a \star b) \parallel (b \% a)$). A subset S of $\mathcal{B}^*$ is called prefix-free if for all distinct $a, b \in S$ it holds that a is not a prefix of b.

If A is finite, we can identify it with $\mathbb{Z}_{|A|}$. In the specific case of $A = \{0,1\}^n$ we use the notation $\langle \cdot \rangle_n : \mathbb{Z}_{2^n} \to \{0,1\}^n$ for the corresponding mapping. We extend this notion to a more general map from $\mathbb{N}$ to A^* or A^+.

3 Symmetric Encryption Supporting Fragmentation

3.1 Unified Syntax

Morphology. We extend the standard definition of symmetric encryption for the case of fragmented ciphertexts. For fragmentation to make sense, we will restrict our attention to ciphertexts that are strings, so CphSp $= \mathcal{B}^*$ where e.g. $\mathcal{B} = \{0,1\}$ (bits), $\mathcal{B} = \{0,1\}^8$ (bytes), or $\mathcal{B} = \{0,1\}^{128}$ (blocks). Furthermore, we assume that the message space consists of strings, so MsgSp $= \mathcal{B}^*$. The move to fragmentation results in some complications. For instance, a single ciphertext can be split up in multiple fragments or a single fragment can contain multiple ciphertexts.

Definition 1. *A symmetric encryption scheme supporting fragmentation $\mathcal{SE} = (\mathcal{K}, \mathcal{E}, \mathcal{D})$ with associated message space MsgSp $= \mathcal{B}^*$, ciphertext space CphSp $= \mathcal{B}^*$ and error messages $\mathcal{S}_\perp$ is defined by three algorithms:*

- *The randomized key generation algorithm $\mathcal{K}$ returns a secret key K and initial states σ_0 and τ_0.*
- *The randomized or stateful (or both) encryption algorithm*

$$\mathcal{E} : \mathcal{K} \times \mathrm{MsgSp} \times \Sigma \to \mathrm{CphSp} \times \Sigma$$

takes input the secret key $K \in \mathcal{K}$, a plaintext $m \in \mathrm{MsgSp}$, and optional state $\sigma \in \Sigma$, and returns a ciphertext in CphSp together with an updated state. For $\mathbf{m} = (m_1, \ldots, m_\ell) \in (\mathcal{B}^*)^*$ and $\mathbf{c} = (c_1, c_2, \ldots, c_\ell)$, we write

$(\mathbf{c}, \sigma) \leftarrow \mathcal{E}_K(\mathbf{m}, \sigma_0)$ *as shorthand for* $(c_1, \sigma_1) \leftarrow \mathcal{E}_K(m_1, \sigma_0)$, $(c_2, \sigma_2) \leftarrow \mathcal{E}_K(m_2, \sigma_1), \ldots (c_\ell, \sigma_\ell) \leftarrow \mathcal{E}_K(m_\ell, \sigma_{\ell-1})$ *where* $\sigma = \sigma_\ell$.

- *The deterministic and stateful decryption algorithm*

$$\mathcal{D} : \mathcal{K} \times \mathcal{B}^* \times \Sigma \to (\mathcal{B} \cup \{\P\} \cup \mathcal{S}_\perp)^* \times \Sigma$$

takes the secret key K, *a ciphertext fragment* $f \in \mathcal{B}^*$, *and the current state* τ *to return the corresponding plaintext fragment* m *(which can be the empty string* ε *or an error from error space* $\mathcal{S}_\perp$*) and also the updated state* τ. *For* $\mathbf{f} = (f_1, \ldots, f_\ell) \in (\mathcal{B}^*)^*$, *we write* $(m, \tau) \leftarrow \mathcal{D}_K(\mathbf{f}, \tau_0)$ *as shorthand for* $(m_1, \tau_1) \leftarrow \mathcal{D}_K(f_1, \tau_0)$, $(m_2, \tau_2) \leftarrow \mathcal{D}_K(f_2, \tau_1), \ldots (m_\ell, \tau_\ell) \leftarrow \mathcal{D}_K(f_\ell, \tau_{\ell-1})$, *where* $m = m_1 \| \ldots \| m_\ell$ *and* $\tau = \tau_\ell$.

This definition requires a little unpacking. Firstly, and in contrast to the usual definitions, our decryption algorithm is stateful, mainly to model decryption algorithms that may, for example, combine data coming from multiple ciphertext fragments before outputting any plaintext.

Secondly, note that the decryption algorithm is assumed to be able to handle ciphertexts which decrypt to multiple plaintext messages, or to a mixture of plaintexts and error symbols, or possibly to nothing at all (perhaps because the input ciphertext is insufficient to enable decryption to yet output anything, giving a significant difference from the atomic setting where decryption always outputs something). We use $\P \notin \mathcal{B} \cup \mathcal{S}_\perp$ to denote the end of plaintext messages, enabling an application making use of the decryption algorithm to parse the output uniquely into a sequence of elements of $\mathcal{B}^*$ and errors from $\mathcal{S}_\perp$. Our introduction of an explicit symbol $\P$ to help delineate messages during decryption seems novel. This is not because our solution is in any way innovative, but rather because the problem does not arise in earlier works.

Thirdly, note that, when failing, the decryption algorithm can output one of possibly many error messages from the set $\mathcal{S}_\perp$. This reflects the fact that real schemes may fail in more than one way, with the different failure modes being visible to both legitimate users and adversaries. Our definition is sufficiently flexible to model schemes (such as those used in SSH and TLS) that tear down secure sessions and destroy session keys as soon as an error is detected during decryption, by having the decryption algorithm maintain an extra "abort" status flag, setting the flag once a first error is encountered, and always outputting a failure symbol once the flag is set. The definition can also handle schemes (such as those used in IPsec and DTLS) which are more tolerant of errors.

While we enforce that from a decryption of a sequence of ciphertext fragments, the corresponding message boundaries are easy to distinguish, we make no such requirement for ciphertexts. Indeed, given a sequence of ciphertext fragments, it will not be a priori clear what the constituent ciphertexts are (and in fact, in Section 4, we want to model schemes which hide these boundaries as a security goal). Looking ahead, the absence of clear ciphertext boundaries (in a sequence of fragments) will cause challenging parsing problems for our CCA definitions:

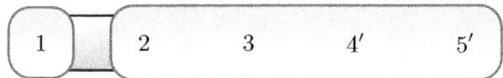

Fig. 1. Two consecutive fragments $f_1 = (1)$ and $f_2 = (234'5')$. The second fragment completes the first ciphertext $c_1 = (12)$, so we expect that to be decrypted at this point, even though ciphertext $c_2 = (345)$ in the second fragment has been modified to produce a possibly invalid ciphertext.

in order to 'forbid' decryption of the challenge ciphertext, a prerequisite is that this challenge ciphertext can be located accurately in the sequence of ciphertext fragments!

Correctness. If a single message is encrypted and the corresponding ciphertext is subsequently decrypted, we expect that the message is returned. When multiple messages are encrypted and the fragments correspond *exactly* to the ciphertext, again we expect to retrieve the original messages. However, we expect something stronger, namely that regardless of how we fragment the ciphertext(s), the original message(s) are returned. Moreover, we require correct decryption, even when an extra string $\mathcal{B}^*$ is added to the original (string of) ciphertexts. This forces correct decryption once a complete valid ciphertext has been received, even if what intuitively might remain in the buffer is invalid. For instance, in the situation depicted in Fig. 1 two ciphertexts $c_1 = (12)$ and $c_2 = (345)$ are produced by the encryption oracle, the adversary subsequently submits fragments $f_1 = (1)$ and $f_2 = (234'5')$ to its decryption oracle, and we still want to see the first ciphertext decrypted properly.

With this intuition in mind, we are almost ready to give our definition of correctness for a symmetric encryption scheme with fragmented ciphertexts. We first define a map $\P : (\mathcal{B}^* \cup \mathcal{S}_\perp)^* \to (\mathcal{B} \cup \{\P\} \cup \mathcal{S}_\perp)^*$ by $\P(m_1, \ldots, m_\ell) = m_1 \parallel \P \parallel \ldots \P \parallel m_\ell \parallel \P$. Note that $\P$ is injective but not surjective.

Definition 2 (Correctness Requirement). *For all (K, σ_0, τ_0) that can be output by $\mathcal{K}$ and for all $\mathbf{m} \in \mathrm{MsgSp}^*$ and $\mathbf{f} \in (\mathcal{B}^*)^*$, it holds (with probability 1) that if $(\mathbf{c}, \sigma) \leftarrow \mathcal{E}_K(\mathbf{m}, \sigma_0)$ and $\|(\mathbf{c})$ prefixes $\|(\mathbf{f})$, then $(m', \tau) \leftarrow \mathcal{D}_K(\mathbf{f}, \tau_0)$ satisfies m' is prefixed by $\P(\mathbf{m})$.*

Stateful Versus Stateless Schemes. As noted in the introduction, we mainly study two subclasses of symmetric encryption schemes supporting fragmentation, *stateful*, which covers most practical encryption schemes, and *stateless beyond buffering*, an extension of standard (atomic), stateless encryption schemes that makes handling fragmented ciphertexts possible. The former case is covered by Definition 1 above. In the latter case, the decryption algorithm is still stateful, but we impose that after receiving any valid ciphertext it returns to the original state (output by key generation). More formally, we have:

$\mathbf{Exp}_{\mathcal{SE}}^{\text{IND-sfCFA-}b}(\mathcal{A})$:
   $C \leftarrow \varepsilon, F \leftarrow \varepsilon, M \leftarrow \varepsilon$
   $\mathsf{C} \leftarrow (), \mathsf{M} \leftarrow (), i \leftarrow 0, j \leftarrow 0$
   active $\leftarrow$ false
   $(K, \sigma, \tau) \xleftarrow{\$} \mathcal{K}$
   $b' \xleftarrow{\$} \mathcal{A}^{\text{LoR}(\cdot,\cdot),\text{Dec}(\cdot)}$
   return b'

$\text{LoR}(m_0, m_1)$:
   if $|m_0| \neq |m_1|$ then return $\lightning$
   $(c, \sigma) \leftarrow \mathcal{E}_K(m_b, \sigma)$
   $i \leftarrow i + 1$
   $\mathsf{C}_i \leftarrow c, \mathsf{M}_i \leftarrow m_b$
   return c

$\text{Dec}(f)$:
   $(m, \tau) \leftarrow \mathcal{D}_K(f, \tau)$
   $F \leftarrow F\|f$ and $M \leftarrow M\|m$
   if $\neg$active then
     while C is a prefix of F and $j < i$
       $j \leftarrow j + 1$
       $C \leftarrow C \| \mathsf{C}_j$
     if F is prefix of C then
       $m \leftarrow \varepsilon$
     else
       active $\leftarrow$ true
       determine $m' \leftarrow \P(\mathsf{M}_1, \ldots, \mathsf{M}_{j-1})$
       extract $m \leftarrow M \% m'$
   return m

Fig. 2. The experiment defining the IND-sfCFA security notion for fragmented decryption of stateful schemes

Definition 3. *A symmetric encryption scheme with fragmented ciphertexts is called* stateless beyond buffering *(or* sbb *for short) if it is correct (Definition 2) and satisfies the additional conditions*

1. *The initial decryption state is empty, that is for all (K, σ_0, τ_0) that can be output by $\mathcal{K}$, $\tau_0 = \varepsilon$; for simplicity's sake, we will often simply write $(K, \sigma) \xleftarrow{\$} \mathcal{K}$ for sbb schemes.*
2. *The decryption state is empty after decryption of each ciphertext obtained from encryption, i.e. for all K that can be output by $\mathcal{K}$, for all $\sigma \in \Sigma$, for all $m \in \text{MsgSp}$, it holds (with probability 1) that if $(c, \sigma) \leftarrow \mathcal{E}_K(m, \sigma)$ then $(m', \tau) \leftarrow \mathcal{D}_K(c, \varepsilon)$ satisfies $\tau = \varepsilon$.*
3. *The scheme satisfies* literal decryption: *for all $K \in \mathcal{K}$ and for all $\mathbf{f} = (f_1, \ldots, f_\ell)$, when $f' = f_1 \| \ldots \| f_\ell$, then $\mathcal{D}_K(\mathbf{f}, \varepsilon) = \mathcal{D}_K(f', \varepsilon)$.*

For schemes with literal decryption we assume, without loss of generality, that the decryption algorithm only keeps a buffer ρ as state, where a buffer is understood to be a suffix of the stream of ciphertext fragments received so far. Moreover, if the scheme is sbb as well, this buffer will be emptied after each valid ciphertext. Essentially, the scheme is stateless beyond the necessary buffering (to keep track of the *current* ciphertext).

3.2 Security for Stateful Schemes

When discussing a security notion a scheme supporting fragmentation, the first thing to note is that this only makes sense in the CCA setting: if there is no decryption oracle, then whether decryption is fragmented or atomic is immaterial to the security of the scheme. In the context of fragmentation, we will replace the usual notion of chosen-ciphertext attacks by chosen-fragment attacks (CFA). Our first notion, IND-sfCFA is tailored for stateful schemes and it is inspired by

Bellare et al.'s notion of IND-sfCCA (for atomic schemes) from [4]. Recall that for IND-sfCCA, an adversary has unlimited access to the decryption oracle; there are no 'prohibited' queries. Instead, to avoid trivial attacks (by the adversary simply relaying its challenge ciphertext for decryption) a syncing mechanism is used. Initially the decryption oracle is in-sync and its output (to the adversary) will be suppressed. Only when the adversary causes the decryption oracle to be out-of-sync (by deviating from the ciphertext stream output by the encryption oracle) will the purported plaintexts (or error messages) be returned.

For atomic schemes, this is relatively straightforward to define, but for schemes supporting fragmentation, some ambiguity arises. Consider again the scenario sketched in Fig. 1. The first fragment is in-sync and its output will be suppressed. In the second fragment a deviation from the challenge ciphertext stream occurs. However, *part* of the fragment is still in-sync and certainly outputting the full decryption would—mindful of the correctness requirement—reveal (part of) the plaintext (12). We will need to formalize this by officially declaring part of the fragment in-sync, and part of it out-of-sync. The ambiguity arises with regards to the boundary we should use: is sync lost already at '3' (being the first symbol of a ciphertext that is not completed properly) or only at '4' (being the first symbol of the fragment that actually deviates)?

In our definition of IND-sfCFA (Definition 4) we opted for the strongest interpretation, namely where synchronization is lost at the ciphertext boundary. Since this results in synchronization potentially being lost *earlier*, the decryption oracle consequently suppresses *less* of its output, making it the stronger option.

Definition 4. *Let* $\mathcal{SE} = (\mathcal{K}, \mathcal{E}, \mathcal{D})$ *be an encryption scheme supporting fragmentation. For an adversary* $\mathcal{A}$ *and a bit* b, *define the experiments* $\mathbf{Exp}_{\mathcal{SE}}^{\mathsf{IND\text{-}sfCFA}\text{-}b}(\mathcal{A})$ *as depicted in Fig. 2.*

In both experiments, first the key K *is generated by* $\mathcal{K}$. *The adversary* $\mathcal{A}$ *is given access to two oracles. The first is the left-or-right encryption oracle* $\mathsf{LoR}(\cdot, \cdot)$ *that it can query on any pair of messages of equal length. The second oracle is the stateful decryption oracle* $\mathsf{Dec}(\cdot)$ *that it can query on any sequence of ciphertext fragments, but for certain sequences the output is artificially suppressed.*

The adversary's goal is to output a bit b', *as its guess of the challenge bit* b, *and the experiment returns* b' *as well. The* IND-sfCFA *advantage of an adversary* $\mathcal{A}$ *is defined as:*

$$\mathbf{Adv}_{\mathcal{SE}}^{\mathsf{IND\text{-}sfCFA}}(\mathcal{A}) = \Pr\left[\mathbf{Exp}_{\mathcal{SE}}^{\mathsf{IND\text{-}sfCFA}\text{-}1}(\mathcal{A}) = 1\right] - \Pr\left[\mathbf{Exp}_{\mathcal{SE}}^{\mathsf{IND\text{-}sfCFA}\text{-}0}(\mathcal{A}) = 1\right].$$

The scheme with fragmentation $\mathcal{SE}$ *is said to be* indistinguishable against chosen-ciphertext-fragments attack *or* IND-sfCFA *secure, if for every adversary* $\mathcal{A}$ *with reasonable resources its* IND-sfCFA *advantage is small.*

Security for Stateless Schemes. In the full version [6], we define security of stateless beyond buffering encryption schemes by appropriately modifying the standard notion of indistinguishability against chosen-ciphertext attacks (IND-CCA). We also provide the details and discuss the subtleties regarding the definition.

Algorithm $\mathcal{K}^f$:	Algorithm $\mathcal{E}^f_K(m,\sigma)$:	Algorithm $\mathcal{D}^f_K(f,(\tau,\rho))$:
$\rho \leftarrow \varepsilon$	$(c,\sigma') \leftarrow \mathcal{E}^a_K(m,\sigma)$	$w \leftarrow f,\ m' \leftarrow \varepsilon$
$(K,\sigma,\tau) \overset{\$}{\leftarrow} \mathcal{K}^a$	$v \leftarrow \mathcal{EC}(c)$	$\rho \leftarrow \rho \parallel f$
$\mathbf{return}\ (K,\sigma,(\tau,\rho))$	$\mathbf{return}\ (v,\sigma')$	$\mathbf{while}\ (w \neq \varepsilon)$
		$\quad (w,\rho) \leftarrow \mathcal{DC}(\rho)$
		$\quad \mathbf{if}\ (w \neq \varepsilon)\ \mathbf{then}$
		$\qquad (m,\tau) \leftarrow \mathcal{D}^a_K(w,\tau)$
		$\qquad m' \leftarrow m' \parallel m \parallel \P$
		$\mathbf{return}\ (m',(\tau,\rho))$

Fig. 3. Construction of encryption schemes supporting fragmentation

3.3 Realizations and Non-realizations

In the full version [6], we simplify an attack by Albrecht et al. [1] to show that schemes meeting traditional notions of security in the atomic setting can fail to be secure in the fragmented setting. This establishes that our whole approach is not vacuous.

Prefix-Free Postprocessing. Next we give a simple transformation that converts any secure atomic scheme $\mathcal{SE}^a$ with MsgSp $= \mathcal{B}^*$ into a secure scheme $\mathcal{SE}^f$ supporting fragmentation. One of the challenges that has to be overcome is ensuring correct decryption of the fragmented scheme. We solve this by encoding ciphertexts (originating from $\mathcal{SE}^a$) using a prefix-free encoding scheme $\mathcal{EC}$. This allows the decrypting algorithm to correctly parse a concatenation of ciphertexts into discrete ciphertexts which it can then decrypt in an atomic fashion.

A prefix-free encoding scheme $\mathcal{EC} : \mathcal{B}^+ \to \mathcal{B}^+$ is a (deterministic) function whose image (viewed as a multiset) is prefix-free and that can be evaluated efficiently. A useful property of a prefix-free encoding scheme is that an arbitrary concatenation of encoded strings can be *uniquely* decoded, moreover this can be done instantaneously and we will assume efficiently. This property, dubbed *instantaneous decodability*, is defined below.

Definition 5. *A prefix-free encoding scheme $\mathcal{EC} : \mathcal{B}^+ \to \mathcal{B}^+$ has* instantaneous decodability *iff there exists an* efficient deterministic *algorithm $\mathcal{DC} : \mathcal{B}^* \to \mathcal{B}^* \times \mathcal{B}^*$ such that:*

1. *For all $w \in \mathcal{B}^+$ and all $s \in \mathcal{B}^*$ it holds that if $v \leftarrow \mathcal{EC}(w)$ then $(w,s) = \mathcal{DC}(v \parallel s)$.*
2. *For all $x \in \mathcal{B}^*$, if no $v \in \mathcal{EC}(\mathcal{B}^+)$ is a prefix of x then $(\varepsilon,x) = \mathcal{DC}(x)$.*

A prefix-free encoding scheme $\mathcal{EC}$ can be combined with an atomic encryption scheme with message space $\mathcal{B}^*$ to yield an encryption scheme supporting fragmentation as in Construction 6.

Construction 6 (Encrypt-then-prefix-free-encode). Let $\mathcal{SE}^a = (\mathcal{K}^a, \mathcal{E}^a, \mathcal{D}^a)$ be an atomic encryption scheme with $\mathcal{E}^a : \mathcal{B}^* \to \mathcal{B}^+$ and let $\mathcal{EC} : \mathcal{B}^+ \to \mathcal{B}^+$ be a prefix-free encoding scheme with associated (instantaneous) decoding algorithm $\mathcal{DC}$. Then Fig. 3 defines encryption scheme supporting fragmentation $\mathcal{SE}^f = (\mathcal{K}^f, \mathcal{E}^f, \mathcal{D}^f)$.

Proposition 7. *Construction 6 provides an encryption scheme supporting fragmentation with message space* $\mathrm{MsgSp} = \mathcal{B}^*$, *ciphertext space* $\mathrm{CphSp} = \mathcal{B}^+$, *the same* $\mathcal{S}_\perp$ *as* $\mathcal{SE}^a$ *and it satisfies the correctness requirement given by Definition 2 (assuming* $\mathcal{SE}^a$ *itself is correct). Furthermore if* $\mathcal{D}^a$ *is stateless, then Construction 6 is stateless beyond buffering.*

Theorem 8. *If* $\mathcal{SE}^a$ *is* IND-sfCCA *secure then* $\mathcal{SE}^f$ *from Construction 6 is* IND-sfCFA *secure. More precisely, for any adversary* $\mathcal{A}_{\mathsf{sfCFA}}$ *there exists an equally efficient adversary* $\mathcal{A}_{\mathsf{sfCCA}}$ *such that*

$$\mathbf{Adv}^{\mathsf{IND\text{-}sfCFA}}_{\mathcal{SE}^f}(\mathcal{A}_{\mathsf{sfCFA}}) \le \mathbf{Adv}^{\mathsf{IND\text{-}sfCCA}}_{\mathcal{SE}^a}(\mathcal{A}_{\mathsf{sfCCA}}).$$

Theorem 9. *Let* $\mathcal{SE}^a$ *have stateless decryption. If* $\mathcal{SE}^a$ *is* IND-CCA *secure then* $\mathcal{SE}^f$ *from Construction 6 is* IND-sbbCFA *secure. More precisely, for any adversary* $\mathcal{A}_{\mathsf{sbbCFA}}$ *there exists an equally efficient adversary* $\mathcal{A}_{\mathsf{CCA}}$ *such that*

$$\mathbf{Adv}^{\mathsf{IND\text{-}sbbCFA}}_{\mathcal{SE}^f}(\mathcal{A}_{\mathsf{sbbCFA}}) \le \mathbf{Adv}^{\mathsf{IND\text{-}CCA}}_{\mathcal{SE}^a}(\mathcal{A}_{\mathsf{CCA}}).$$

The proofs of these results (and the definition of IND-sbbCFA security) can be found in the full version [6].

4 Boundary Hiding

In this section, we focus on formalizing the goal of boundary hiding for ciphertext streams, giving security definitions and constructions achieving these definitions. While the boundaries should be hidden to an adversary, they should of course *not* lead to decryption problems: a stream (i.e. concatenation) of ciphertexts should still lead to the correct sequence of plaintexts. The correctness requirement for an encryption scheme with fragmented decryption already ensures that everything goes well here.

Definition 10. *Let* $\mathcal{SE} = (\mathcal{K}, \mathcal{E}, \mathcal{D})$ *be an encryption scheme supporting fragmentation. For an adversary* $\mathcal{A}$ *and a bit b, define experiments* $\mathbf{Exp}^{\mathsf{BH\text{-}sfCFA\text{-}b}}_{\mathcal{SE}}(\mathcal{A})$ *as depicted in Fig. 4.*

In these experiments, the adversary $\mathcal{A}$ *is given access to a special left-or-right oracle: on input two vectors of messages, either the left or the right result is returned, but with the caveat that the concatenated ciphertext is returned only if in both worlds the same length ciphertext is produced (but note that we do not insist that the two vectors of messages contain the same number of components). The adversary is also given access to a decryption oracle that is identical to the one provided in the* IND-sfCFA *security experiment.*

The adversary's goal is to output a bit b', *as its guess of the challenge bit b, and the experiment returns* b' *as well. The* BH-sfCFA *advantage of an adversary* $\mathcal{A}$ *is defined as:*

$$\mathbf{Adv}^{\mathsf{BH\text{-}sfCFA}}_{\mathcal{SE}}(\mathcal{A}) = \Pr\left[\mathbf{Exp}^{\mathsf{BH\text{-}sfCFA\text{-}1}}_{\mathcal{SE}}(\mathcal{A}) = 1\right] - \Pr\left[\mathbf{Exp}^{\mathsf{BH\text{-}sfCFA\text{-}0}}_{\mathcal{SE}}(\mathcal{A}) = 1\right].$$

Exp$_{\mathcal{SE}}^{\text{BH-sfCFA-}b}(\mathcal{A})$:
  $C \leftarrow \varepsilon, F \leftarrow \varepsilon, M \leftarrow \varepsilon$
  $\mathtt{C} \leftarrow (), \mathtt{M} \leftarrow (), i \leftarrow 0, j \leftarrow 0$
  active $\leftarrow$ **false**
  $(K, \sigma, \tau) \xleftarrow{\$} \mathcal{K}$
  $b' \xleftarrow{\$} \mathcal{A}^{\text{LoR}(\cdot,\cdot),\text{Dec}(\cdot)}$
  return b'

LoR$(\mathbf{m}_0, \mathbf{m}_1)$:
  $\sigma_0 \leftarrow \sigma, \sigma_1 \leftarrow \sigma$
  $(\mathbf{c}_0, \sigma_0) \leftarrow \mathcal{E}_K(\mathbf{m}_0, \sigma_0)$
  $(\mathbf{c}_1, \sigma_1) \leftarrow \mathcal{E}_K(\mathbf{m}_1, \sigma_1)$
  $c_0 \leftarrow ||(\mathbf{c}_0), c_1 \leftarrow ||(\mathbf{c}_1)$
  if $|c_0| \neq |c_1|$ **then return** $\cancel{\ }$
  $\sigma \leftarrow \sigma_b$
  for $\iota = 1$ **to** $\iota = |\mathbf{c}_b|$
    $i \leftarrow i + 1$
    $\mathtt{C}_i \leftarrow \mathbf{c}_b(\iota), \mathtt{M}_i \leftarrow \mathbf{m}_b(\iota)$
  return c_b

Dec(f):
  $(m, \tau) \leftarrow \mathcal{D}_K(f, \tau)$
  $F \leftarrow F||f$ and $M \leftarrow M||m$
  if $\neg$active **then**
    while C is a prefix of F and $j < i$
      $j \leftarrow j + 1$
      $C \leftarrow C \parallel \mathtt{C}_j$
    if F is prefix of C **then**
      $m \leftarrow \varepsilon$
    else
      active $\leftarrow$ **true**
      determine $m' \leftarrow \P(\mathtt{M}_1, \ldots, \mathtt{M}_{j-1})$
      extract $m \leftarrow M \% m'$
  return m

Fig. 4. Experiment **Exp**$_{\mathcal{SE}}^{\text{BH-sfCFA-}b}(\mathcal{A})$ for defining boundary hiding security for stateful schemes and an active adversary

We say that $\mathcal{SE}$ is boundary-hiding against chosen-ciphertext-fragments attack *or* BH-sfCFA *secure, if for every adversary $\mathcal{A}$ with reasonable resources its* BH-sfCFA *advantage is small.*

Boundary-hiding notions for the case of passive adversaries can be obtained simply by removing the adversary's access to the relevant decryption oracle. In this case, we refer to BH-CPA security (this notion is implied by the notion IND$-CPA as introduced by Rogaway [17], see the full version [6]). The notion of boundary-hiding security for sbb schemes for active attacks BH-sbbCFA can also be developed, by replacing the decryption oracle Dec in Fig. 4 by the decryption oracle Dec from the corresponding sbb security game and by appropriately modifying how ciphertexts generated by queries to the encryption oracle are tracked. The details are in the full version [6].

Constructions. In order to achieve BH-CPA security we extend Construction 6 by using *keyed* encoding schemes. The use of a key in the encoding scheme enables the encryption algorithm to disguise the ciphertext boundaries from a passive adversary. Details of this construction, dubbed KPF (Keyed Prefix Free), can be found in the full version [6]. Achieving BH-sfCFA requires more work. We show that this can be done at the same time as achieving DoS security in the next section. In the sbb setting, however, we will only achieve BH-CPA security (we show this and discuss the difficulty of meeting BH-sbbCFA security in the next section as well). It remains an open problem to design a practical BH-sbbCFA secure sbb scheme.

$\mathbf{Exp}_{\mathcal{SE}}^{N\text{-DOS-sfCFA}}(\mathcal{A})$:
   $C \leftarrow \varepsilon, F \leftarrow \varepsilon, M \leftarrow \varepsilon$
   $\mathtt{C} \leftarrow (), \mathtt{M} \leftarrow (), i \leftarrow 0, j \leftarrow 0$
   active $\leftarrow$ **false**
   $(K, \sigma, \tau) \xleftarrow{\$} \mathcal{K}$
   run $\mathcal{A}^{\mathsf{Enc}(\cdot), \mathsf{Dec}(\cdot)}$
   return 1

$\mathsf{Enc}(m)$:
   $(c, \sigma) \leftarrow \mathcal{E}_K(m, \sigma)$
   $i \leftarrow i + 1$
   $\mathtt{C}_i \leftarrow c, \mathtt{M}_i \leftarrow m$
   return c

$\mathsf{Dec}(f)$:
   $(m, \tau) \leftarrow \mathcal{D}_K(f, \tau)$
   $F \leftarrow F \| f$ and $M \leftarrow M \| m$
   if $\neg$active **then**
      while C is a prefix of F and $j < i$
         $j \leftarrow j + 1$
         $C \leftarrow C \| \mathtt{C}_j$
      if F is not a prefix of C **then**
         active $\leftarrow$ **true**
         $m' \leftarrow \P(\mathtt{M}_1, \ldots, \mathtt{M}_{j-1})$
         $m \leftarrow M \% m'$
   if active **then**
      if $m \neq \varepsilon$ **then**
         exit Exp with 0
      else if $|F \% C| \geq N$ **then**
         exit Exp with 1
   return m

Fig. 5. The experiment defining the N-DOS-sfCFA security notion for Denial of Service attack against stateful schemes

5   Denial-of-Service Attacks

In this section we study fragmentation-related Denial-of-Service (DoS) attacks. In the introduction we mentioned an example of a fragmentation-related attack that constitutes DoS. Here we provide formal definitions that are general enough to capture all such attacks. We focus on the stateful setting.

Definition 11. *Let $\mathcal{SE} = (\mathcal{K}, \mathcal{E}, \mathcal{D})$ be a stateful encryption scheme with fragmented ciphertexts. For an adversary $\mathcal{A}$ and $N \in \mathbb{N}$ define the experiment* $\mathbf{Exp}_{\mathcal{SE}}^{N-\text{DOS-sfCFA}}(\mathcal{A})$ *as in Fig. 5. In the experiment, $\mathcal{A}$ is given access to two oracles. The first is a regular encryption oracle $\mathsf{Enc}(\cdot)$ that it can query on any message in the message space. The second oracle is a special stateful decryption oracle $\mathsf{Dec}(\cdot)$ that it can query on any string treated as a ciphertext fragment. The adversary's goal is to submit to the special decryption oracle $\mathsf{Dec}(\cdot)$ a fragment or a sequence of fragments of length at least N, which is not a valid replay of legitimate ciphertexts and such that $\mathsf{Dec}(\cdot)$ does not return a non-empty message m or a failure symbol from $\mathcal{S}_\perp$. In this case the oracle exits the experiment with a value 1. In other out-of-sync cases the oracle exits with 0. When decryption is still in-sync, the oracle just returns the correct decryption m. The N-DOS-sfCFA advantage of $\mathcal{A}$ is defined to be:*

$$\mathbf{Adv}_{\mathcal{SE}}^{N\text{-DOS-sfCFA}}(\mathcal{A}) = \Pr\left[\mathbf{Exp}_{\mathcal{SE}}^{N\text{-DOS-sfCFA}}(\mathcal{A}) = 1\right].$$

The scheme $\mathcal{SE}$ is said to be N-DOS-sfCFA secure if for every legitimate adversary $\mathcal{A}$ with reasonable resources, its N-DOS-sfCFA advantage is small.

Note that, to win the above game, the adversary need not make his attack in the first out-of-sync query to $\mathsf{Dec}(\cdot)$. Also note that in order to win the adversary must submit at least N symbols after the longest common prefix with a valid ciphertext stream obtained from the encryption oracle without provoking any output from the decryption oracle.

The parameter N in the definition above measures the shortest fragment length below which a DoS attack cannot be prevented by a scheme; since all reasonable schemes that meet our other security notions must do some degree of buffering before outputting any message symbols, we cannot hope to make N as small as we please. Our objective then, when designing a scheme, is to make the scheme N-DOS-sfCFA secure for N as small as possible.

We develop a similar definition for the sbb setting in the full version [6].

InterMAC: Construction in the Stateful Case and its Security. Our idea for DoS prevention in the stateful setting is to break the ciphertexts into equal-sized segments and authenticate all of them. We could use this idea to modify an IND-sfCFA scheme, but we propose a more efficient construction that uses an IND-CPA (possibly stateless) scheme and a SUF-CMA MAC. (We defer standard definitions for syntax and security of MACs to the full version [6].) In our construction, the sender and receiver keep a state which contains a message and a segment number. The encryption algorithm MACs this state together with the encryption of the segment, but the state does not have to be transmitted, as the receiver maintains it for himself. Each segment uses a bit flag to indicate the last segment in a message. We now provide the details.

Construction 12 (InterMAC). Let $\mathcal{SE} = (\mathcal{K}_e, \mathcal{E}, \mathcal{D})$ be an encryption scheme with associated message space MsgSp_e and let $\mathcal{MAC} = (\mathcal{K}_t, \mathcal{T}, \mathcal{V})$ be a message authentication code with associated message space MsgSp_t. Let $N \in \mathbb{N}$ be a DoS parameter. We assume that $\mathrm{MsgSp}_t = \{0,1\}^*$ and, for simplicity, that $\mathrm{MsgSp}_e = \{\{0,1\}^{N-el-1-tl}\}^*$ where $\mathcal{T}$ always outputs tags of fixed length tl and $\mathcal{E}$ always produces ciphertexts which are el bits longer than the messages. This restriction on the message space can be relaxed by introducing an appropriate padding scheme (such as abit padding, as analysed in [15]) but we omit this detail for simplicity. Define a new stateful encryption scheme with fragmentation $\mathcal{SE}^f = (\mathcal{K}^f, \mathcal{E}^f, \mathcal{D}^f)$ as in Figure 6.

It is not hard to check that the scheme is correct.

The proofs of the following two theorems are in the full version [6].

Theorem 13. *If $\mathcal{MAC}$ is SUF-CMA, then $\mathcal{SE}^f$ constructed as per Construction 12 is N-DOS-sfCFA secure. More precisely, for any adversary $\mathcal{A}$ there exists an equally efficient adversary $\mathcal{A}'$ so that*

$$\mathbf{Adv}_{\mathcal{MAC}}^{\mathrm{uf\text{-}cma}}(\mathcal{A}') \geq \mathbf{Adv}_{\mathcal{SE}^f}^{N\text{-}\mathrm{DOS\text{-}sfCFA}}(\mathcal{A}).$$

Algorithm $\mathcal{E}^f_{K_e\|K_t,\sigma}(m;\sigma)$:

    $\sigma \leftarrow \sigma + 1$

    parse m as $m_1 \| \ldots \| m_\ell$

        where for each $1 \le i \le \ell$,

        $|m_i| = N - el - 1 - tl$

    for $j = 1$ to ℓ do

        if $j = \ell$ then

            $bm_j \leftarrow 1 \| m_j$

        else $bm_j \leftarrow 0 \| m_j$

        $c_j \leftarrow \mathcal{E}_{K_e}(bm_j)$

        $t_j \leftarrow \mathcal{T}_{K_t}(\sigma \| j \| c_j)$

    $c \leftarrow c_1 \| t_1 \ldots \| c_\ell \| t_\ell$

    return (c, σ)

Algorithm $\mathcal{D}^f_{K_e\|K_t}(f, \tau)$:

    parse τ as (i_m, i_s, bad, m, F)

    if $|F \| f| < N$ then

        $F \leftarrow F \| f$

        return $(\varepsilon, (i_m, i_s, bad, m, F))$

    else parse $F \| f$ as $c_1 \| t_1 \| \ldots \| c_\ell \| t_\ell \| s$

        where $|s| < N$ and for each $1 \le i \le \ell$,

        $|c_i \| t_i| = N$ and $|t_i| = tl$

    for $j = 1$ to ℓ do

        $i_s \leftarrow i_s + 1$

        if $bad = 1$ then

            output $\bot$

        else if $\mathcal{V}_{K_t}(i_m \| i_s \| c_j, t_j) = 0$ then

            $bad \leftarrow 1$; output $\bot$

        else

            $bm_j \leftarrow \mathcal{D}_{K_e}(c_j)$; parse bm_j as $b \| m_j$

            $m \leftarrow m \| m_j$

            if $b = 1$ then

                $i_m \leftarrow i_m + 1; i_s \leftarrow 0$

                if $bad = 0$ then

                    output $(m\P)$; $m \leftarrow \varepsilon$

    return$(\varepsilon, (i_m, i_s, bad, m, s))$

Fig. 6. The stateful construction $\mathcal{SE}^f$. Key generation $\mathcal{K}^f$ picks $K_e \xleftarrow{\$} \mathcal{K}_e, K_t \xleftarrow{\$} \mathcal{K}_t$, sets $\sigma \leftarrow 0$ and $\tau = (i_m, i_s, bad, m, F) \leftarrow (0, 0, 0, \varepsilon, \varepsilon)$, and returns (c, σ, τ).

Theorem 14. *If $\mathcal{SE}$ is* IND\$-CPA[1] *and $\mathcal{MAC}$ is* SUF-CMA *and* PRF *secure, then $\mathcal{SE}^f$ constructed as per Construction 12 is* BH-sfCFA *and* IND-sfCFA *secure.*

We provide the concrete security statements in [6].

Construction in the sbb Case and its Security. In the full version [6] we provide an analogous scheme for the sbb setting. We also use the idea of authenticating the ciphertext segments, however, the solution becomes more complex as we cannot keep message and segment numbers as state and it is not efficient to keep them as part of the segments. To prevent the re-ordering attacks we need to authenticate the previous segments as well. To improve efficiency we only authenticate the tags from the previous segments. The security results we get are similar, but the sbb scheme does not provide BH-sbbCFA security.

The reason for the difficulty in achieving BH-sbbCFA security in the sbb setting is as follows. An adversary can always query a valid ciphertext to the stateful decryption oracle fragment by fragment and flip the last bit at the end. Observing when the decryption algorithm returns $\bot$ gives the adversary information about

---

[1] This notion captures indistinguishability of ciphertexts from random strings and is introduced by Rogaway [17]. We recall the formal definition in [6].

the ciphertext boundary. Prohibiting the decryption algorithm to ever return $\bot$ is not very practical and is subject to DoS attacks. It is an open question to provide a practical scheme with both BH-sbbCFA and DOS-sbbCFA security.

6   Conclusions

In this paper, we have initiated the formal study of fragmentation attacks against symmetric encryption schemes. We also developed security models to formalise the additional desirable properties of ciphertext boundary-hiding and robustness against Denial-of-Service (DoS) attacks for schemes in this setting. We illustrated the utility of each of our models via efficient constructions for schemes using only standard cryptographic components. This work raises many interesting open questions, amongst which we list:

- We have focussed on confidentiality notions here, and suitable integrity notions remain to be developed. Can such notions then be combined to provide more general notions of security as seen in authenticated encryption?
- Some of our constructions build fragmented schemes from atomic schemes. What general relationships are there between schemes in the two settings?
- In the sbb case, the properties of DoS resistance and ciphertext boundary hiding appear to be in opposition to one another. Can this be formally proven? Is there a fundamental reason (beyond the ability to keep state) why this does not seem to arise in the stateful setting?

References

[1] Albrecht, M.R., Paterson, K.G., Watson, G.J.: Plaintext recovery attacks against SSH. In: IEEE Symposium on Security and Privacy, pp. 16–26. IEEE Computer Society (2009)
[2] Bard, G.V.: A challenging but feasible blockwise-adaptive chosen-plaintext attack on SSL. In: Malek, M., Fernandez-Medina, E., Hernando, J. (eds.) SECRYPT, pp. 99–109. INSTICC Press (2006)
[3] Bard, G.V.: Blockwise-Adaptive Chosen-Plaintext Attack and Online Modes of Encryption. In: Galbraith, S.D. (ed.) Cryptography and Coding 2007. LNCS, vol. 4887, pp. 129–151. Springer, Heidelberg (2007)
[4] Bellare, M., Kohno, T., Namprempre, C.: Breaking and provably repairing the SSH authenticated encryption scheme: A case study of the encode-then-encrypt-and-MAC paradigm. ACM Transactions on Information and Systems Security 7(2), 206–241 (2004)
[5] Bellare, M., Namprempre, C.: Authenticated Encryption: Relations among Notions and Analysis of the Generic Composition Paradigm. In: Okamoto, T. (ed.) ASIACRYPT 2000. LNCS, vol. 1976, pp. 531–545. Springer, Heidelberg (2000)
[6] Boldyreva, A., Degabriele, J.P., Paterson, K.G., Stam, M.: Security of symmetric encryption in the presence of ciphertext fragmentation. Full version of this paper. Cryptology ePrint Archive (2012), http://eprint.iacr.org
[7] Boldyreva, A., Taesombut, N.: Online Encryption Schemes: New Security Notions and Constructions. In: Okamoto, T. (ed.) CT-RSA 2004. LNCS, vol. 2964, pp. 1–14. Springer, Heidelberg (2004)

[8] Degabriele, J.P., Paterson, K.G.: Attacking the IPsec standards in encryption-only configurations. In: IEEE Symposium on Security and Privacy, pp. 335–349. IEEE Computer Society Press (2007)

[9] Fouque, P.-A., Joux, A., Martinet, G., Valette, F.: Authenticated On-line Encryption. In: Matsui, M., Zuccherato, R.J. (eds.) SAC 2003. LNCS, vol. 3006, pp. 145–159. Springer, Heidelberg (2004)

[10] Fouque, P.-A., Joux, A., Poupard, G.: Blockwise Adversarial Model for On-line Ciphers and Symmetric Encryption Schemes. In: Handschuh, H., Hasan, M.A. (eds.) SAC 2004. LNCS, vol. 3357, pp. 212–226. Springer, Heidelberg (2004)

[11] Fouque, P.-A., Martinet, G., Poupard, G.: Practical Symmetric On-Line Encryption. In: Johansson, T. (ed.) FSE 2003. LNCS, vol. 2887, pp. 362–375. Springer, Heidelberg (2003)

[12] Joux, A., Martinet, G., Valette, F.: Blockwise-Adaptive Attackers: Revisiting the (In)Security of Some Provably Secure Encryption Models: CBC, GEM, IACBC. In: Yung, M. (ed.) CRYPTO 2002. LNCS, vol. 2442, pp. 17–30. Springer, Heidelberg (2002)

[13] Krawczyk, H.: The Order of Encryption and Authentication for Protecting Communications (or: How Secure Is SSL?). In: Kilian, J. (ed.) CRYPTO 2001. LNCS, vol. 2139, pp. 310–331. Springer, Heidelberg (2001)

[14] Paterson, K.G., Watson, G.J.: Plaintext-Dependent Decryption: A Formal Security Treatment of SSH-CTR. In: Gilbert, H. (ed.) EUROCRYPT 2010. LNCS, vol. 6110, pp. 345–361. Springer, Heidelberg (2010)

[15] Paterson, K.G., Watson, G.J.: Immunising CBC Mode Against Padding Oracle Attacks: A Formal Security Treatment. In: Ostrovsky, R., De Prisco, R., Visconti, I. (eds.) SCN 2008. LNCS, vol. 5229, pp. 340–357. Springer, Heidelberg (2008)

[16] Paterson, K.G., Ristenpart, T., Shrimpton, T.: Tag Size *Does* Matter: Attacks and Proofs for the TLS Record Protocol. In: Lee, D.H., Wang, X. (eds.) ASIACRYPT 2011. LNCS, vol. 7073, pp. 372–389. Springer, Heidelberg (2011)

[17] Rogaway, P.: Nonce-Based Symmetric Encryption. In: Roy, B., Meier, W. (eds.) FSE 2004. LNCS, vol. 3017, pp. 348–359. Springer, Heidelberg (2004)

[18] Tezcan, C., Vaudenay, S.: On Hiding a Plaintext Length by Preencryption. In: Lopez, J., Tsudik, G. (eds.) ACNS 2011. LNCS, vol. 6715, pp. 345–358. Springer, Heidelberg (2011)

Trapdoors for Lattices:
Simpler, Tighter, Faster, Smaller

Daniele Micciancio[1,*] and Chris Peikert[2,**]

[1] University of California, San Diego
[2] Georgia Institute of Technology

Abstract. We give new methods for generating and using "strong trapdoors" in cryptographic lattices, which are simultaneously simple, efficient, easy to implement (even in parallel), and asymptotically optimal with very small hidden constants. Our methods involve a new kind of trapdoor, and include specialized algorithms for inverting LWE, randomly sampling SIS preimages, and securely delegating trapdoors. These tasks were previously the main bottleneck for a wide range of cryptographic schemes, and our techniques substantially improve upon the prior ones, both in terms of practical performance and quality of the produced outputs. Moreover, the simple structure of the new trapdoor and associated algorithms can be exposed in applications, leading to further simplifications and efficiency improvements. We exemplify the applicability of our methods with new digital signature schemes and CCA-secure encryption schemes, which have better efficiency and security than the previously known lattice-based constructions.

1 Introduction

Cryptography based on lattices has several attractive and distinguishing features:

- On the *security* front, the best attacks on the underlying problems require exponential $2^{\Omega(n)}$ time in the main security parameter n, even for quantum adversaries. By contrast, for example, mainstream factoring-based cryptography can be broken in subexponential $2^{\tilde{O}(n^{1/3})}$ time classically, and even in polynomial $n^{O(1)}$ time using quantum algorithms. Moreover, lattice cryptography is supported by strong worst-case/average-case security reductions,

* This material is based on research sponsored by NSF under Award CNS-1117936 and DARPA under agreement number FA8750-11-C-0096. The U.S. Government is authorized to reproduce and distribute reprints for Governmental purposes notwithstanding any copyright notation thereon.

** This material is based upon work supported by the National Science Foundation under Grant CNS-0716786 and CAREER Award CCF-1054495, by the Alfred P. Sloan Foundation, and by the Defense Advanced Research Projects Agency (DARPA) and the Air Force Research Laboratory (AFRL) under Contract No. FA8750-11-C-0098. The views expressed are those of the authors and do not necessarily reflect the official policy or position of the National Science Foundation, the Sloan Foundation, DARPA or the U.S. Government.

D. Pointcheval and T. Johansson (Eds.): EUROCRYPT 2012, LNCS 7237, pp. 700–718, 2012.

which provide solid theoretical evidence that the random instances used in cryptography are indeed asymptotically hard, and do not suffer from any unforeseen "structural" weaknesses.

- On the *efficiency* and *implementation* fronts, lattice cryptography operations can be extremely simple, fast and parallelizable. Typical operations are the selection of uniformly random integer matrices $\mathbf{A}$ modulo some small $q = \text{poly}(n)$, and the evaluation of simple linear functions like

$$f_{\mathbf{A}}(\mathbf{x}) := \mathbf{A}\mathbf{x} \bmod q \quad \text{and} \quad g_{\mathbf{A}}(\mathbf{s}, \mathbf{e}) := \mathbf{s}^t \mathbf{A} + \mathbf{e}^t \bmod q$$

on short integer vectors $\mathbf{x}, \mathbf{e}$.[1] (For commonly used parameters, $f_{\mathbf{A}}$ is surjective while $g_{\mathbf{A}}$ is injective.) Often, the modulus q is small enough that all the basic operations can be directly implemented using machine-level arithmetic. By contrast, the analogous operations in number-theoretic cryptography (e.g., generating huge random primes, and exponentiating modulo such primes) are much more complex, admit only limited parallelism in practice, and require the use of "big number" arithmetic libraries.

In recent years lattice-based cryptography has also been shown to be extremely versatile, leading to a large number of theoretical applications ranging from (hierarchical) identity-based encryption [20, 13, 1, 2], to fully homomorphic encryption schemes [17, 16, 45, 12, 11, 18, 10], and much more (e.g., [29, 40, 26, 38, 39, 35, 6, 42, 9, 19, 22]).

Not all lattice cryptography is as simple as selecting random matrices $\mathbf{A}$ and evaluating linear functions like $f_{\mathbf{A}}(\mathbf{x}) = \mathbf{A}\mathbf{x} \bmod q$, however. In fact, such operations yield only collision-resistant hash functions, public-key encryption schemes that are secure under passive attacks, and little else. Richer and more advanced lattice-based cryptographic schemes, including chosen ciphertext-secure encryption, "hash-and-sign" digital signatures, and identity-based encryption also require generating a matrix $\mathbf{A}$ together with some *"strong" trapdoor*, typically in the form of a nonsingular square matrix (a basis) $\mathbf{S}$ of short integer vectors such that $\mathbf{A}\mathbf{S} = \mathbf{0} \bmod q$. (The matrix $\mathbf{S}$ is usually interpreted as a basis of a lattice defined by using $\mathbf{A}$ as a "parity check" matrix.) Applications of such strong trapdoors also require certain efficient inversion algorithms for the functions $f_{\mathbf{A}}$ and $g_{\mathbf{A}}$, using $\mathbf{S}$. Appropriately inverting $f_{\mathbf{A}}$ can be particularly complex, as it typically requires sampling *random preimages* of $f_{\mathbf{A}}(\mathbf{x})$ according to a Gaussian-like probability distribution (see [20]).

Theoretical solutions for all the above tasks (generating $\mathbf{A}$ with strong trapdoor $\mathbf{S}$ [3, 5], trapdoor inversion of $g_{\mathbf{A}}$ and preimage sampling for $f_{\mathbf{A}}$ [20]) are known, but they are rather complex and not very suitable for practice, in either runtime or the "quality" of their outputs. (The quality of a trapdoor $\mathbf{S}$ roughly corresponds to the Euclidean lengths of its vectors — shorter is better.) The

---

[1] Inverting these functions corresponds to solving the "short integer solution" (SIS) problem [4] for $f_{\mathbf{A}}$, and the "learning with errors" (LWE) problem [41] for $g_{\mathbf{A}}$, both of which are widely used in lattice cryptography and enjoy provable worst-case hardness.

current best method for trapdoor generation [5] is conceptually and algorithmically complex, and involves costly computations of Hermite normal forms and matrix inverses. And while the dimensions and quality of its output are *asymptotically* optimal (or nearly so, depending on the precise notion of quality), the hidden constant factors are rather large. Similarly, the standard methods for inverting $g_{\mathbf{A}}$ and sampling preimages of $f_{\mathbf{A}}$ [7, 24, 20] are inherently sequential and time-consuming, as they are based on an orthogonalization process that uses high-precision real numbers. A more efficient and parallelizable method for preimage sampling (which uses only small-integer arithmetic) has recently been discovered [36], but it is still more complex than is desirable for practice, and the quality of its output can be slightly worse than that of the sequential algorithm when using the same trapdoor $\mathbf{S}$.

More compact and efficient trapdoors appear necessary for bringing advanced lattice-based schemes to practice, not only because of the current unsatisfactory runtimes, but also because the concrete security of lattice cryptography can be quite sensitive to changes in the main parameters, and improvements by even small constant factors can have a significant impact on concrete security. (See, e.g., [15, 34], and the full version for a more detailed discussion.)

1.1   Contributions

The first main contribution of this paper is a new method of trapdoor generation for cryptographic lattices, which is simultaneously simple, efficient, easy to implement (even in parallel), and asymptotically optimal with small hidden constants. The new trapdoor generator strictly subsumes the prior ones of [3, 5], in that it proves the main theorems from those works, but with improved concrete bounds for all the relevant quantities (simultaneously), and via a conceptually simpler and more efficient algorithm. To accompany our trapdoor generator, we also give specialized algorithms for trapdoor inversion (for $g_{\mathbf{A}}$) and preimage sampling (for $f_{\mathbf{A}}$), which are simpler and more efficient in our setting than the prior general solutions [7, 24, 20, 36].

Our methods yield large constant-factor improvements, and in some cases even small asymptotic improvements, in the lattice dimension m, trapdoor quality[2] s, and storage size of the trapdoor. Because trapdoor generation and inversion algorithms are the main operations in many lattice cryptography schemes, our algorithms can be plugged in as 'black boxes' to deliver significant concrete improvements in all such applications. Moreover, it is often possible to expose the special (and very simple) structure of our trapdoor directly in cryptographic schemes, yielding additional improvements and potentially new applications. In the full version we detail several improvements to existing applications. We now give a detailed comparison of our results with the most relevant prior works [3, 5, 20, 36]. The quantitative improvements are summarized in Figure 1.

---

[2] There are several notions quality for lattice trapdoors, of varying strength. For now, the reader can think of the quality as a measure of the norm of the vectors in $\mathbf{S}$, where smaller values are better.

Simpler, Faster Trapdoor Generation and Inversion Algorithms. Our trapdoor generator is exceedingly simple, especially as compared with the prior constructions [3, 5]. It essentially amounts to just one multiplication of two random matrices, whose entries are chosen independently from appropriate probability distributions. Surprisingly, this method is nearly identical to Ajtai's original method [4] of generating a random lattice together with a "weak" trapdoor of one or more short vectors (but *not* a full basis), with one added twist. And while there are no detailed runtime analyses or public implementations of [3, 5], it is clear from inspection that our new method is significantly more efficient, since it does not involve any expensive Hermite normal form or matrix inversion computations. Our specialized, parallel inversion algorithms for $f_{\mathbf{A}}$ and $g_{\mathbf{A}}$ are also simpler and more practically efficient than the general solutions of [7, 24, 20, 36] (though we note that our trapdoor generator is entirely compatible with those general algorithms as well). In particular, we give the first *parallel* algorithm for inverting $g_{\mathbf{A}}$ under asymptotically optimal error rates (previously, handling such large errors required the sequential "nearest-plane" algorithm of [7]), and our preimage sampling algorithm for $f_{\mathbf{A}}$ works with smaller integers and requires much less offline storage than the one from [36].

Tighter Parameters. To generate a matrix $\mathbf{A} \in \mathbb{Z}_q^{n \times m}$ that is within negligible statistical distance of uniform, our new trapdoor construction improves the lattice dimension from $m > 5n \lg q$ [5] down to $m \approx 2n \lg q$. (In both cases, the base of the logarithm is a tunable parameter that appears as a multiplicative factor in the quality of the trapdoor; here we fix upon base 2 for concreteness.) In addition, we give the first known *computationally pseudorandom* construction (under the LWE assumption), where the dimension can be as small as $m = n(1 + \lg q)$, although at the cost of an $\Omega(\sqrt{n})$ factor worse quality s.

Our construction also greatly improves the quality s of the trapdoor. The best prior construction [5] produces a basis whose Gram-Schmidt quality (i.e., the maximum length of its Gram-Schmidt orthogonalized vectors) was loosely bounded by $20\sqrt{n \lg q}$. However, the Gram-Schmidt notion of quality is useful only for less efficient, sequential inversion algorithms [7, 20] that use high-precision real arithmetic. For the more efficient, parallel preimage sampling algorithm of [36] that uses small-integer arithmetic, the parameters guaranteed by [5] are asymptotically worse, at $m > n \lg^2 q$ and $s \geq 16\sqrt{n \lg^2 q}$. By contrast, our (statistically secure) trapdoor construction achieves the "best of both worlds:" asymptotically optimal dimension $m \approx 2n \lg q$ and quality $s \approx 1.6\sqrt{n \lg q}$ or better, with a parallel preimage sampling algorithm that is slightly more efficient than the one of [36].

Altogether, for any n and typical values of $q \geq 2^{16}$, we conservatively estimate that the new trapdoor generator and inversion algorithms collectively provide at least a $7 \lg q \geq 112$-*fold improvement* in the length bound $\beta \approx s\sqrt{m}$ for $f_{\mathbf{A}}$ preimages (generated using an efficient algorithm). We also obtain similar improvements in the size of the error terms that can be handled when efficiently inverting $g_{\mathbf{A}}$.

New, Smaller Trapdoors. As an additional benefit, our construction actually produces a *new kind of trapdoor* — not a basis — that is at least 4 times smaller in storage than a basis of corresponding quality, and is at least as powerful, i.e., a good basis can be efficiently derived from the new trapdoor. We stress that our specialized inversion algorithms using the new trapdoor provide almost exactly the same quality as the inefficient, sequential algorithms using a derived basis, so there is no trade-off between efficiency and quality. (This is in contrast with [36] when using a basis generated according to [5].) Moreover, the storage size of the new trapdoor grows only linearly in the lattice dimension m, rather than quadratically as a basis does. This is most significant for applications like hierarchical ID-based encryption [13, 1] that *delegate* trapdoors for increasing values of m. The new trapdoor also admits a very simple and efficient delegation mechanism, which unlike the prior method [13] does not require any costly operations like linear independence tests, or conversions from a full-rank set of lattice vectors into a basis. In summary, the new type of trapdoor and its associated algorithms are *strictly preferable* to a short basis in terms of algorithmic efficiency, output quality, and storage size (simultaneously).

Ring-Based Constructions. Finally, and most importantly for practice, all of the above-described constructions and algorithms extend immediately to the *ring* setting, where functions analogous to $f_{\mathbf{A}}$ and $g_{\mathbf{A}}$ require only quasi-linear $\tilde{O}(n)$ space and time to specify and evaluate (respectively), which is a factor of $\tilde{\Omega}(n)$ improvement over the matrix-based functions defined above. See the representative works [32, 37, 28, 30, 44, 31] for more details on these functions and their security foundations.

Applications. Our improved trapdoor generator and inversion algorithms can be plugged into any scheme that uses such tools as a "black box," and the resulting scheme will inherit all the efficiency improvements. (Every application we know of admits such a black-box replacement.) Moreover, the special properties of our methods allow for further improvements to the design, efficiency, and security reductions of existing schemes. In the full version we describe new and improved applications, with a focus on signature schemes and chosen ciphertext-secure encryption.

To illustrate the kinds of concrete improvements that our methods provide, in Figure 2 we give representative parameters for the canonical application of GPV signatures [20], comparing the old and new trapdoor constructions for nearly equal levels of concrete security. We stress that these parameters are not highly optimized, and making adjustments to some of the tunable parameters in our constructions may provide better combinations of efficiency and concrete security. We leave this effort for future work.

1.2   Techniques

The main idea behind our new method of trapdoor generation is as follows. Instead of building a random matrix $\mathbf{A}$ through some specialized and complex process, we start from a carefully crafted *public* matrix $\mathbf{G}$ (and its associated lattice), for which

	[3, 5] constructions	This work (fast $f_{\mathbf{A}}^{-1}$)	Impr. Factor
Dimension m	slow $f_{\mathbf{A}}^{-1}$ [24, 20]: $> 5n \lg q$ fast $f_{\mathbf{A}}^{-1}$ [36]: $> n \lg^2 q$	$\approx 2n \lg q$ $(\stackrel{s}{\approx})$ $n(1 + \lg q)$ $(\stackrel{c}{\approx})$	2.5 to $\lg q$
Quality s	slow $f_{\mathbf{A}}^{-1}$: $\approx 20\sqrt{n \lg q}$ fast $f_{\mathbf{A}}^{-1}$: $\approx 16\sqrt{n \lg^2 q}$	$\approx 1.6\sqrt{n \lg q}$ $(\stackrel{s}{\approx})$	12.5 to $10\sqrt{\lg q}$
Length $\beta \approx s\sqrt{m}$	slow $f_{\mathbf{A}}^{-1}$: $> 45n \lg q$ fast $f_{\mathbf{A}}^{-1}$: $> 16n \lg^2 q$	$\approx 2.3\,n \lg q$ $(\stackrel{s}{\approx})$	19 to $7 \lg q$

Fig. 1. Summary of parameters for our constructions versus prior ones. The symbols $\stackrel{s}{\approx}$ and $\stackrel{c}{\approx}$ denote constructions producing public keys $\mathbf{A}$ that are statistically and computationally close to uniform, respectively. All quality terms s and length bounds β omit the same "smoothing" factor for $\mathbb{Z}$, which is about 4–5 in practice.

	[5] with fast $f_{\mathbf{A}}^{-1}$	This work	Improvement Factor
Sec param n	436	284	1.53
Modulus logarithm $\log_2(q)$	32	24	1.33
Dimension m	446,644	13,812	32.3
Quality s	10.7×10^3	418	25.6
Length β	12.9×10^6	91.6×10^3	141
Key size (bits)	6.22×10^9	92.2×10^6	**67.5**
Key size (ring-based)	$\approx 16 \times 10^6$	$\approx 361 \times 10^3$	$\approx$ **44.3**

Fig. 2. Representative parameters for GPV signatures (using fast inversion algorithms) estimated using the methodology from [34] with $\delta \leq 1.007$, which is estimated to require about 2^{46} core-years on a 64-bit 1.86GHz Xeon [15, 14]. We used $\omega_n = 4.5$ for $\mathbb{Z}$, which corresponds to statistical error $< 2^{-90}$ for each randomized-rounding operation during signing. Key sizes for *ring-based* GPV signatures are approximated to be smaller by a factor of about $0.9n$.

the associated functions $f_{\mathbf{G}}$ and $g_{\mathbf{G}}$ admit very efficient (in both sequential and parallel complexity) and high-quality inversion algorithms. In particular, preimage sampling for $f_{\mathbf{G}}$ and inversion for $g_{\mathbf{G}}$ can be performed in essentially $O(n \log n)$ sequential time, and can even be performed by n parallel $O(\log n)$-time operations or table lookups. (This should be compared with the general algorithms for these tasks, which require at least quadratic $\Omega(n^2 \log^2 n)$ time, and are not always parallelizable for optimal noise parameters.) We emphasize that $\mathbf{G}$ is *not* a cryptographic key, but rather a fixed and public matrix that may be used by all parties, so the implementation of all its associated operations can be highly optimized, in both software and hardware. We also mention that the simplest and most practically efficient choices of $\mathbf{G}$ work for a modulus q that is a power of a small prime, such as $q = 2^k$, but no LWE search/decision reduction for such q was known till recently, despite its obvious practical utility. The only such result we are aware of is the recent sample preserving reduction of [33], which applies to arbitrary q (including powers of 2), but requires the error distribution to be polynomially bounded. In the full version we provide a different and very general reduction (generalizing and extending [8, 41, 35, 6],) that also covers the $q = 2^k$ case and others, and is incomparable to [33], as it requires all prime factors of q to be polynomially bounded, but does not impose this restriction on the errors.

To generate a *random* matrix $\mathbf{A}$ with a trapdoor, we take two additional steps: first, we extend $\mathbf{G}$ into a semi-random matrix $\mathbf{A}' = [\bar{\mathbf{A}} \mid \mathbf{G}]$, for uniform $\bar{\mathbf{A}} \in \mathbb{Z}_q^{n \times \bar{m}}$ and sufficiently large $\bar{m}$. (Concretely, $\bar{m} \approx n \lg q$ for the statistically secure construction, and $\bar{m} = 2n$ or n for computational security.) As shown in [13], inversion of $g_{\mathbf{A}'}$ and preimage sampling for $f_{\mathbf{A}'}$ reduce very efficiently to the corresponding tasks for $g_{\mathbf{G}}$ and $f_{\mathbf{G}}$. Finally, we simply apply to $\mathbf{A}'$ a certain random unimodular transformation defined by the matrix $\mathbf{T} = \left[\begin{smallmatrix} \mathbf{I} & -\mathbf{R} \\ \mathbf{0} & \mathbf{I} \end{smallmatrix} \right]$, for a random "short" secret matrix $\mathbf{R}$ that will serve as the trapdoor, to obtain

$$\mathbf{A} = \mathbf{A}' \cdot \mathbf{T} = [\bar{\mathbf{A}} \mid \mathbf{G} - \bar{\mathbf{A}}\mathbf{R}].$$

The transformation given by $\mathbf{T}$ has the following properties:

- It is very easy to compute and invert, requiring essentially just one multiplication by $\mathbf{R}$ in both cases. (Note that $\mathbf{T}^{-1} = \left[\begin{smallmatrix} \mathbf{I} & \mathbf{R} \\ \mathbf{0} & \mathbf{I} \end{smallmatrix} \right]$.)
- It results in a matrix $\mathbf{A}$ that is distributed essentially uniformly at random, as required by the security reductions (and worst-case hardness proofs) for lattice-based cryptographic schemes.
- For the resulting functions $f_{\mathbf{A}}$ and $g_{\mathbf{A}}$, preimage sampling and inversion very simply and efficiently reduce to the corresponding tasks for $f_{\mathbf{G}}, g_{\mathbf{G}}$. The overhead of the reduction is essentially just a single matrix-vector product with the secret matrix $\mathbf{R}$ (which, when inverting $f_{\mathbf{A}}$, can largely be precomputed even before the target value is known).

As a result, the cost of the inversion operations ends up being very close to that of computing $f_{\mathbf{A}}$ and $g_{\mathbf{A}}$ in the forward direction. Moreover, the fact that the running time is dominated by matrix-vector multiplications with the *fixed* trapdoor matrix $\mathbf{R}$ yields theoretical (but asymptotically significant) improvements in the context of batch execution of several operations relative to the same secret key $\mathbf{R}$: instead of evaluating several products $\mathbf{R}\mathbf{z}_1, \mathbf{R}\mathbf{z}_2, \ldots, \mathbf{R}\mathbf{z}_n$ individually at a total cost of $\Omega(n^3)$, one can employ fast matrix multiplication techniques to evaluate $\mathbf{R}[\mathbf{z}_1, \ldots, \mathbf{z}_n]$ as a whole in subcubic time. Batch operations can be exploited in applications like the multi-bit IBE of [20] and its extensions to HIBE [13, 1, 2].

Related Techniques. At the surface, our trapdoor generator appears similar to the original "GGH" approach of [21] for generating a lattice together with a short basis. That technique works by choosing some random short vectors as the secret "good basis" of a lattice, and then transforms them into a public "bad basis" for the *same* lattice, via a unimodular matrix having large entries. (Note, though, that this does not produce a lattice from Ajtai's worst-case-hard family.) A closer look reveals, however, that (worst-case hardness aside) our method is actually *not* an instance of the GGH paradigm: in our case, the initial short basis of the lattice defined by $\mathbf{G}$ (or the semi-random matrix $[\bar{\mathbf{A}}|\mathbf{G}]$) is *fixed* and *public*, while the random unimodular matrix $\mathbf{T} = \left[\begin{smallmatrix} \mathbf{I} & -\mathbf{R} \\ \mathbf{0} & \mathbf{I} \end{smallmatrix}\right]$ actually produces a *new* lattice by applying a (reversible) linear transformation to the original one. In other words, in contrast with GGH, we multiply a (short) unimodular matrix on the "other side" of the original short basis, thus changing the lattice it generates. Moreover, it is crucial in our setting that the transformation matrix $\mathbf{T}$ has small entries, while with GGH the transformation matrix can be arbitrary.

A more appropriate comparison is to Ajtai's original method [4] for generating a random $\mathbf{A}$ together with a "weak" trapdoor of one or more short lattice vectors (but not a full basis). There, one simply chooses a semi-random matrix $\mathbf{A}' = [\bar{\mathbf{A}} \mid \mathbf{0}]$ and outputs $\mathbf{A} = \mathbf{A}' \cdot \mathbf{T} = [\bar{\mathbf{A}} \mid -\bar{\mathbf{A}}\mathbf{R}]$, with short vectors $\left[\begin{smallmatrix} \mathbf{R} \\ \mathbf{I} \end{smallmatrix}\right]$. Perhaps surprisingly, our strong trapdoor generator is just a simple twist on Ajtai's original weak generator, replacing $\mathbf{0}$ with the gadget $\mathbf{G}$. We remark that Ajtai's method to generate strong trapdoors [3] and follow-up work [5] are quite different and much more complex.

Our constructions and inversion algorithms also draw upon several other techniques from throughout the literature. The trapdoor basis generator of [5] and the LWE-based "lossy" injective trapdoor function of [40] both use a fixed "gadget" matrix analogous to $\mathbf{G}$, whose entries grow geometrically in a structured way. In both cases, the gadget is concealed (either statistically or computationally) in the public key by a small combination of uniformly random vectors. Our method for adding tags to the trapdoor is very similar to a technique for doing the same with the lossy TDF of [40], and is identical to the method used in [1] for constructing compact (H)IBE. Finally, in our preimage sampling algorithm for $f_{\mathbf{A}}$, we use the "convolution" technique from [36] to correct for some statistical skew that arises when converting preimages for $f_{\mathbf{G}}$ to preimages for $f_{\mathbf{A}}$, which would otherwise leak information about the trapdoor $\mathbf{R}$.

Other Related Work. Concrete parameter settings for a variety "strong" trapdoor applications are given in [43]. Those parameters are derived using the previous suboptimal generator of [5], and using the methods from this work would yield substantial improvements. The recent work of [25] also gives improved key sizes and concrete security for LWE-based cryptosystems; however, that work deals only with IND-CPA-secure encryption, and not at all with strong trapdoors or the further applications they enable (CCA security, digital signatures, (H)IBE, etc.). In a concurrent and independent work, Lyubashevsky [27] constructs a signature scheme in the random oracle model "without (strong) trapdoors," i.e., without relying on short bases or a gadget matrix $\mathbf{G}$. The form and sizes of his public and secret keys are very similar to ours, but the schemes and their security proofs work entirely differently.

2 Primitive Lattices

At the heart of our new trapdoor generation algorithm (described in Section 3) is the construction of a very special family of lattices which have excellent geometric properties, and admit very fast and parallelizable decoding algorithms.

The lattices are defined by means of what we call a *primitive matrix*. We say that a matrix $\mathbf{G} \in \mathbb{Z}_q^{n \times w}$ is primitive $\mathbf{G} \cdot \mathbb{Z}^w = \mathbb{Z}_q^n$.[3] The main results of this section are summarized in the following theorem.

Theorem 1. *For any integers $q \geq 2$, $n \geq 1$, $k = \lceil \log_2 q \rceil$ and $w = nk$, there is a primitive matrix $\mathbf{G} \in \mathbb{Z}_q^{n \times w}$ such that*

- *The lattice $\Lambda^\perp(\mathbf{G})$ has a known basis $\mathbf{S} \in \mathbb{Z}^{w \times w}$ with $\|\widetilde{\mathbf{S}}\| \leq \sqrt{5}$ and $\|\mathbf{S}\| \leq \max\{\sqrt{5}, \sqrt{k}\}$. Moreover, when $q = 2^k$, we have $\widetilde{\mathbf{S}} = 2\mathbf{I}$ (so $\|\widetilde{\mathbf{S}}\| = 2$) and $\|\mathbf{S}\| = \sqrt{5}$.*
- *Both $\mathbf{G}$ and $\mathbf{S}$ require little storage. In particular, they are sparse (with only $O(w)$ nonzero entries) and highly structured.*
- *Inverting $g_\mathbf{G}(\mathbf{s}, \mathbf{e}) := \mathbf{s}^t \mathbf{G} + \mathbf{e}^t \bmod q$ can be performed in quasilinear $O(n \cdot \log^c n)$ time for any $\mathbf{s} \in \mathbb{Z}_q^n$ and any $\mathbf{e} \in \mathcal{P}_{1/2}(q \cdot \mathbf{B}^{-t})$, where $\mathbf{B}$ can denote either $\mathbf{S}$ or $\widetilde{\mathbf{S}}$. Moreover, the algorithm is perfectly parallelizable, running in polylogarithmic $O(\log^c n)$ time using n processors. When $q = 2^k$, the polylogarithmic term $O(\log^c n)$ is essentially just the cost of k additions and shifts on k-bit integers.*
- *Preimage sampling for $f_\mathbf{G}(\mathbf{x}) = \mathbf{G}\mathbf{x} \bmod q$ with Gaussian parameter $s \geq \|\widetilde{\mathbf{S}}\| \cdot \omega_n$ can be performed in quasilinear $O(n \cdot \log^c n)$ time, or parallel polylogarithmic $O(\log^c n)$ time using n processors. When $q = 2^k$, the polylogarithmic term is essentially just the cost of k additions and shifts on k-bit integers, plus the (offline) generation of about w random integers drawn from $D_{\mathbb{Z},s}$.*

More generally, for any integer $b \geq 2$, all of the above statements hold with $k = \lceil \log_b q \rceil$, $\|\widetilde{\mathbf{S}}\| \leq \sqrt{b^2 + 1}$, and $\|\mathbf{S}\| \leq \max\{\sqrt{b^2 + 1}, (b-1)\sqrt{k}\}$; and when $q = b^k$, we have $\widetilde{\mathbf{S}} = b\mathbf{I}$ and $\|\mathbf{S}\| = \sqrt{b^2 + 1}$.

Let $q \geq 2$ be an integer modulus and $k \geq 1$ be an integer dimension. Our construction starts with a *primitive vector* $\mathbf{g} \in \mathbb{Z}_q^k$, i.e., a vector such that $\gcd(g_1, \ldots, g_k, q) = 1$. The vector $\mathbf{g}$ defines a k-dimensional lattice $\Lambda^\perp(\mathbf{g}^t) \subset \mathbb{Z}^k$ having determinant $|\mathbb{Z}^k / \Lambda^\perp(\mathbf{g}^t)| = q$, because the residue classes of $\mathbb{Z}^k / \Lambda^\perp(\mathbf{g}^t)$ are in bijective correspondence with the possible values of $\langle \mathbf{g}, \mathbf{x} \rangle \bmod q$ for $\mathbf{x} \in \mathbb{Z}^k$, which cover all of $\mathbb{Z}_q$ since $\mathbf{g}$ is primitive. Notice that when $q = \text{poly}(n)$, we have $k = O(\log q) = O(\log n)$ and so $\Lambda^\perp(\mathbf{g}^t)$ is a very low-dimensional lattice. In the full version, we prove that the vector $\mathbf{g} = (1, 2, 4, \ldots, 2^{k-1}) \in \mathbb{Z}_q^k$ for $k = \lceil \lg q \rceil$ admits a short basis for the lattice $\Lambda^\perp(\mathbf{g}^t)$, and we also describe specialized inversion and sampling algorithms that are both very simple and more efficient than generic solutions.

Let $\mathbf{S}_k \in \mathbb{Z}^{k \times k}$ be a basis of $\Lambda^\perp(\mathbf{g}^t)$, that is, $\mathbf{g}^t \cdot \mathbf{S}_k = \mathbf{0} \in \mathbb{Z}_q^{1 \times k}$ and $|\det(\mathbf{S}_k)| = q$. The primitive vector $\mathbf{g}$ and associated basis $\mathbf{S}_k$ are used to define the parity-check matrix $\mathbf{G}$ and basis $\mathbf{S} \in \mathbb{Z}_q$ as $\mathbf{G} := \mathbf{I}_n \otimes \mathbf{g}^t \in \mathbb{Z}_q^{n \times nk}$ and $\mathbf{S} := \mathbf{I}_n \otimes \mathbf{S}_k \in \mathbb{Z}^{nk \times nk}$. Equivalently, $\mathbf{G}$, $\Lambda^\perp(\mathbf{G})$, and $\mathbf{S}$ are the direct sums of n

---

[3] We do not say that $\mathbf{G}$ is "full-rank," because $\mathbb{Z}_q$ is not a field when q is not prime, and the notion of rank for matrices over $\mathbb{Z}_q$ is not well defined.

copies of $\mathbf{g}^t$, $\Lambda^\perp(\mathbf{g}^t)$, and $\mathbf{S}_k$, respectively. It follows that $\mathbf{G}$ is a primitive matrix, the lattice $\Lambda^\perp(\mathbf{G}) \subset \mathbb{Z}^{nk}$ has determinant q^n, and $\mathbf{S}$ is a basis for this lattice. It also follows (and is clear by inspection) that $\|\mathbf{S}\| = \|\mathbf{S}_k\|$ and $\|\widetilde{\mathbf{S}}\| = \|\widetilde{\mathbf{S}_k}\|$.

By this direct sum construction, it is immediate that inverting $g_\mathbf{G}(\mathbf{s}, \mathbf{e})$ and sampling preimages of $f_\mathbf{G}(\mathbf{x})$ can be accomplished by performing the same operations n times in parallel for $g_{\mathbf{g}^t}$ and $f_{\mathbf{g}^t}$ on the corresponding portions of the input, and concatenating the results. For preimage sampling, if each of the $f_{\mathbf{g}^t}$-preimages has Gaussian parameter $\sqrt{\Sigma}$, then by independence, their concatenation has parameter $\mathbf{I}_n \otimes \sqrt{\Sigma}$. Likewise, inverting $g_\mathbf{G}$ will succeed whenever all the n independent $g_{\mathbf{g}^t}$-inversion subproblems are solved correctly. Theorem 1 follows by substituting appropriate primitive vectors $\mathbf{g}$ and bases $\mathbf{S}_k$ into the definitions of $\mathbf{G}$ and $\mathbf{S}$.

3 Trapdoor Generation and Operations

In this section we describe our new trapdoor generation, inversion and sampling algorithms for hard random lattices. Recall that these are lattices $\Lambda^\perp(\mathbf{A})$ defined by an (almost) uniformly random matrix $\mathbf{A} \in \mathbb{Z}_q^{n \times m}$, and that the standard notion of a "strong" trapdoor for these lattices (put forward in [20] and used in a large number of subsequent applications) is a short lattice basis $\mathbf{S} \in \mathbb{Z}^{m \times m}$ for $\Lambda^\perp(\mathbf{A})$. There are several measures of quality for the trapdoor $\mathbf{S}$, the most common ones being (in nondecreasing order): the maximal Gram-Schmidt length $\|\widetilde{\mathbf{S}}\|$; the maximal Euclidean length $\|\mathbf{S}\|$; and the maximal singular value $s_1(\mathbf{S})$. Algorithms for generating random lattices together with high-quality trapdoor bases are given in [3, 5] (and in [44], for the ring setting). In this section we give much simpler, faster and tighter algorithms to generate a hard random lattice with a trapdoor, and to use a trapdoor for performing standard tasks like inverting the LWE function $g_\mathbf{A}$ and sampling preimages for the SIS function $f_\mathbf{A}$. We also give a new, simple algorithm for delegating a trapdoor, i.e., using a trapdoor for $\mathbf{A}$ to obtain one for a matrix $[\mathbf{A} \mid \mathbf{A}']$ that extends $\mathbf{A}$, in a secure and non-reversible way.

The following theorem summarizes the main results of this section. Here we state just one typical instantiation with only asymptotic bounds. More general results and exact bounds are presented throughout the section.

Theorem 2. *There is an efficient randomized algorithm* GenTrap$(1^n, 1^m, q)$ *that, given any integers $n \geq 1$, $q \geq 2$, and sufficiently large $m = O(n \log q)$, outputs a parity-check matrix $\mathbf{A} \in \mathbb{Z}_q^{n \times m}$ and a 'trapdoor' $\mathbf{R}$ such that the distribution of $\mathbf{A}$ is* negl(n)-*far from uniform. Moreover, there are efficient algorithms* Invert *and* SampleD *that with overwhelming probability over all random choices, do the following:*

- *For $\mathbf{b}^t = \mathbf{s}^t \mathbf{A} + \mathbf{e}^t$, where $\mathbf{s} \in \mathbb{Z}_q^n$ is arbitrary and $\|\mathbf{e}\| < q/O(\sqrt{n \log q})$ or $\mathbf{e} \leftarrow D_{\mathbb{Z}^m, \alpha q}$ for $1/\alpha \geq \sqrt{n \log q} \cdot \omega_n$, the deterministic algorithm* Invert$(\mathbf{R}, \mathbf{A}, \mathbf{b})$ *outputs $\mathbf{s}$ and $\mathbf{e}$.*

– For any $\mathbf{u} \in \mathbb{Z}_q^n$ and large enough $s = O(\sqrt{n \log q})$, the randomized algorithm
SampleD($\mathbf{R}, \mathbf{A}, \mathbf{u}, s$) samples from a distribution within negl(n) statistical
distance of $D_{\Lambda_{\mathbf{u}}^{\perp}(\mathbf{A}), s \cdot \omega_n}$.

Throughout this section, we let $\mathbf{G} \in \mathbb{Z}_q^{n \times w}$ denote some fixed primitive matrix
that admits efficient inversion and preimage sampling algorithms, as described
in Theorem 1. (Recall that typically, $w = n \lceil \log q \rceil$ for some appropriate base of
the logarithm.) All our algorithms and efficiency improvements are based on the
primitive matrix $\mathbf{G}$ and associated algorithms described in Section 2, and a new
notion of trapdoor that we define next.

3.1   A New Trapdoor Notion

We begin by defining the new notion of trapdoor, establish some of its most
important properties, and give a simple and efficient algorithm for generating
hard random lattices together with high-quality trapdoors.

Definition 1. *Let* $\mathbf{A} \in \mathbb{Z}_q^{n \times m}$ *and* $\mathbf{G} \in \mathbb{Z}_q^{n \times w}$ *be matrices with* $m \geq w \geq n$. *A*
$\mathbf{G}$*-trapdoor for* $\mathbf{A}$ *is a matrix* $\mathbf{R} \in \mathbb{Z}^{(m-w) \times w}$ *such that* $\mathbf{A}\begin{bmatrix} \mathbf{R} \\ \mathbf{I} \end{bmatrix} = \mathbf{HG}$ *for some*
invertible matrix $\mathbf{H} \in \mathbb{Z}_q^{n \times n}$. *We refer to* $\mathbf{H}$ *as the* tag *or* label *of the trapdoor.*
The quality *of the trapdoor is measured by its largest singular value* $s_1(\mathbf{R})$.

We remark that, by definition of $\mathbf{G}$-trapdoor, if $\mathbf{G}$ is a primitive matrix and $\mathbf{A}$ admits a $\mathbf{G}$-trapdoor, then $\mathbf{A}$ is primitive as well. In particular, $\det(\Lambda^{\perp}(\mathbf{A})) = q^n$.
Since the primitive matrix $\mathbf{G}$ is typically fixed and public, we usually omit refer-
ences to it, and refer to $\mathbf{G}$-trapdoors simply as trapdoors. Since $\mathbf{G}$ is primitive,
the tag $\mathbf{H}$ in the above definition is uniquely determined by (and efficiently
computable from) $\mathbf{A}$ and the trapdoor $\mathbf{R}$.

In the full version we show that a good basis for $\Lambda^{\perp}(\mathbf{A})$ may be obtained from
knowledge of the trapdoor $\mathbf{R}$. This is not used anywhere in the rest of the paper,
but it establishes that our new definition of trapdoor is at least as powerful as
the traditional one of a short basis. Our algorithms for Gaussian sampling and
LWE inversion do not need a full basis, and make direct (and more efficient) use
of the new type of trapdoor.

We also make the following simple but useful observations: (1) The rows of
$\begin{bmatrix} \mathbf{R} \\ \mathbf{I} \end{bmatrix}$ in Definition 1 can appear in any order, since this just induces a permutation
of $\mathbf{A}$'s columns. (2) If $\mathbf{R}$ is a trapdoor for $\mathbf{A}$, then it can be made into an equally
good trapdoor for any extension $[\mathbf{A} \mid \mathbf{B}]$, by padding $\mathbf{R}$ with zero rows; this
leaves $s_1(\mathbf{R})$ unchanged. (3) If $\mathbf{R}$ is a trapdoor for $\mathbf{A}$ with tag $\mathbf{H}$, then $\mathbf{R}$ is
also a trapdoor for $\mathbf{A}' = \mathbf{A} - [\mathbf{0} \mid \mathbf{H'G}]$ with tag $(\mathbf{H} - \mathbf{H'})$ for any $\mathbf{H'} \in \mathbb{Z}_q^{n \times n}$,
as long as $(\mathbf{H} - \mathbf{H'})$ is invertible modulo q. This is the main idea behind the
compact IBE of [1], and can be used to give a family of "tag-based" trapdoor
functions [23]. In the full version we recall explicit families of matrices $\mathbf{H}$ having
suitable properties for applications.

3.2 Trapdoor Generation

We now give an algorithm to generate a (pseudo)random matrix $\mathbf{A}$ together with a $\mathbf{G}$-trapdoor. The algorithm is straightforward, and in fact it can be easily derived from the definition of $\mathbf{G}$-trapdoor itself. A random lattice is built by first extending the primitive matrix $\mathbf{G}$ into a semi-random matrix $\mathbf{A}' = [\bar{\mathbf{A}} \mid \mathbf{HG}]$ (where $\bar{\mathbf{A}} \in \mathbb{Z}_q^{n \times (m-w)}$ is chosen at random, and $\mathbf{H} \in \mathbb{Z}_q^{n \times n}$ is the desired tag), and then applying a random transformation $\mathbf{T} = \begin{bmatrix} \mathbf{I} & \mathbf{R} \\ \mathbf{0} & \mathbf{I} \end{bmatrix} \in \mathbb{Z}^{m \times m}$ to the semi-random lattice $\Lambda^\perp(\mathbf{A}')$. Since $\mathbf{T}$ is unimodular with inverse $\mathbf{T}^{-1} = \begin{bmatrix} \mathbf{I} & -\mathbf{R} \\ \mathbf{0} & \mathbf{I} \end{bmatrix}$, this yields the lattice $\mathbf{T} \cdot \Lambda^\perp(\mathbf{A}') = \Lambda^\perp(\mathbf{A}' \cdot \mathbf{T}^{-1})$ associated with the parity-check matrix $\mathbf{A} = \mathbf{A}' \cdot \mathbf{T}^{-1} = [\bar{\mathbf{A}} \mid \mathbf{HG} - \bar{\mathbf{A}}\mathbf{R}]$. Moreover, the distribution of $\mathbf{A}$ is close to uniform (either statistically, or computationally) as long as the distribution of $[\bar{\mathbf{A}} \mid \mathbf{0}]\mathbf{T}^{-1} = [\bar{\mathbf{A}} \mid -\bar{\mathbf{A}}\mathbf{R}]$ is. For details, see Algorithm 1.

Algorithm 1. Efficient algorithm $\mathsf{GenTrap}^{\mathcal{D}}(\bar{\mathbf{A}}, \mathbf{H})$ for generating a parity-check matrix $\mathbf{A}$ with trapdoor $\mathbf{R}$.

Input: Matrix $\bar{\mathbf{A}} \in \mathbb{Z}_q^{n \times \bar{m}}$ for some $\bar{m} \geq 1$, invertible matrix $\mathbf{H} \in \mathbb{Z}_q^{n \times n}$, and distribution $\mathcal{D}$ over $\mathbb{Z}^{\bar{m} \times w}$.
 (If no particular $\bar{\mathbf{A}}$, $\mathbf{H}$ are given as input, then the algorithm may choose them itself, e.g., picking $\bar{\mathbf{A}} \in \mathbb{Z}_q^{n \times \bar{m}}$ uniformly at random, and setting $\mathbf{H} = \mathbf{I}$.)
Output: A parity-check matrix $\mathbf{A} = [\bar{\mathbf{A}} \mid \mathbf{A}_1] \in \mathbb{Z}_q^{n \times m}$, where $m = \bar{m} + w$, and trapdoor $\mathbf{R}$ with tag $\mathbf{H}$.
 1: Choose a matrix $\mathbf{R} \in \mathbb{Z}^{\bar{m} \times w}$ from distribution $\mathcal{D}$.
 2: Output $\mathbf{A} = [\bar{\mathbf{A}} \mid \mathbf{HG} - \bar{\mathbf{A}}\mathbf{R}] \in \mathbb{Z}_q^{n \times m}$ and trapdoor $\mathbf{R} \in \mathbb{Z}^{\bar{m} \times w}$.

We next describe two types of $\mathsf{GenTrap}$ instantiations. The first type generates a trapdoor $\mathbf{R}$ for a statistically near-uniform output matrix $\mathbf{A}$ using dimension $\bar{m} \approx n \log q$ or less (there is a trade-off between $\bar{m}$ and the trapdoor quality $s_1(\mathbf{R})$). The second type generates a computationally *pseudorandom* $\mathbf{A}$ (under the LWE assumption) using dimension $\bar{m} = 2n$ (this pseudorandom construction is the first of its kind in the literature). Some applications allow for an optimization that additionally decreases $\bar{m}$ by an additive n term; this is most significant in the computationally secure construction because it yields $\bar{m} = n$.

Statistical instantiation. This instantiation works for any parameter $\bar{m}$ and distribution $\mathcal{D}$ over $\mathbb{Z}^{\bar{m} \times w}$ having the following two properties:

1. *Subgaussianity*: $\mathcal{D}$ is subgaussian with some parameter $s > 0$ (or δ-subgaussian for some small δ). This implies that $\mathbf{R} \leftarrow \mathcal{D}$ has $s_1(\mathbf{R}) = s \cdot O(\sqrt{\bar{m}} + \sqrt{w})$, except with probability $2^{-\Omega(\bar{m}+w)}$. (Recall that the constant factor hidden in the $O(\cdot)$ expression is $\approx 1/\sqrt{2\pi}$.)
2. *Regularity*: for $\bar{\mathbf{A}} \leftarrow \mathbb{Z}_q^{n \times \bar{m}}$ and $\mathbf{R} \leftarrow \mathcal{D}$, $\mathbf{A} = [\bar{\mathbf{A}} \mid \bar{\mathbf{A}}\mathbf{R}]$ is δ-uniform for some $\delta = \mathrm{negl}(n)$. In fact, there is no loss in security if $\bar{\mathbf{A}}$ contains an identity

matrix $\mathbf{I}$ as a submatrix and is otherwise uniform, since this corresponds with the Hermite normal form of the SIS and LWE problems. See, e.g., [34, Section 5] for further details.

For example, let $\mathcal{D} = \mathcal{P}^{\bar{m} \times w}$ where $\mathcal{P}$ is the distribution over $\mathbb{Z}$ that outputs 0 with probability $1/2$, and ± 1 each with probability $1/4$. Then $\mathcal{P}$ (and hence $\mathcal{D}$) is 0-subgaussian with parameter $\sqrt{2\pi}$, and satisfies the regularity condition (for any q) for $\delta \leq \frac{w}{2}\sqrt{q^n/2^{\bar{m}}}$, by a version of the leftover hash lemma (see, e.g., [5, Section 2.2.1]). Therefore, we can use any $\bar{m} \geq n \lg q + 2 \lg \frac{w}{2\delta}$. Other statistical instantiations are presented in the full version.

Computational Instantiation. Let $\bar{\mathbf{A}} = [\mathbf{I}_n \mid \hat{\mathbf{A}}] \in \mathbb{Z}_q^{n \times \bar{m}}$ for $\bar{m} = 2n$, and let $\mathcal{D} = D_{\mathbb{Z},s}^{\bar{m} \times w}$ for some $s = \alpha q$, where $\alpha > 0$ is an LWE relative error rate (and typically $\alpha q > \sqrt{n}$). Clearly, $\mathcal{D}$ is 0-subgaussian with parameter αq. Also, $[\bar{\mathbf{A}} \mid \bar{\mathbf{A}}\mathbf{R} = \hat{\mathbf{A}}\mathbf{R}_2 + \mathbf{R}_1]$ for $\mathbf{R} = \begin{bmatrix} \mathbf{R}_1 \\ \mathbf{R}_2 \end{bmatrix} \leftarrow \mathcal{D}$ is exactly an instance of decision-LWE$_{n,q,\alpha}$ in its normal form, and hence is pseudorandom (ignoring the identity submatrix) assuming that the problem is hard.

Further Optimizations. In applications that use only a single tag $\mathbf{H} = \mathbf{I}$ (e.g., GPV signatures [20]), we can save an additive n term in the dimension $\bar{m}$ (and hence in the total dimension m): instead of putting an identity submatrix in $\bar{\mathbf{A}}$, we can instead use the identity submatrix from $\mathbf{G}$ (which exists without loss of generality, since $\mathbf{G}$ is primitive) and conceal the remainder of $\mathbf{G}$ using either of the above methods.

All of the above ideas also translate immediately to the ring setting, using an appropriate regularity lemma (e.g., the ones from [44] or [31]) for a statistical instantiation, and the ring-LWE problem for a computationally secure instantiation.

3.3   LWE Inversion

Algorithm 2 below shows how to use a trapdoor to solve LWE relative to $\mathbf{A}$. Given a trapdoor $\mathbf{R}$ for $\mathbf{A} \in \mathbb{Z}_q^{n \times m}$ and an LWE instance $\mathbf{b}^t = \mathbf{s}^t \mathbf{A} + \mathbf{e}^t \bmod q$ for some short error vector $\mathbf{e} \in \mathbb{Z}^m$, the algorithm recovers $\mathbf{s}$ (and $\mathbf{e}$). This naturally yields an inversion algorithm for the injective trapdoor function $g_{\mathbf{A}}(\mathbf{s}, \mathbf{e}) = \mathbf{s}^t \mathbf{A} + \mathbf{e}^t \bmod q$, which is hard to invert (and whose output is pseudorandom) if LWE is hard.

Theorem 3. *Suppose that oracle $\mathcal{O}$ in Algorithm 2 correctly inverts $g_{\mathbf{G}}(\hat{\mathbf{s}}, \hat{\mathbf{e}})$ for any error vector $\hat{\mathbf{e}} \in \mathcal{P}_{1/2}(q \cdot \mathbf{B}^{-t})$ for some $\mathbf{B}$. Then for any $\mathbf{s}$ and $\mathbf{e}$ of length $\|\mathbf{e}\| < q/(2\|\mathbf{B}\|s)$ where $s = \sqrt{s_1(\mathbf{R})^2 + 1}$, Algorithm 2 correctly inverts $g_{\mathbf{A}}(\mathbf{s}, \mathbf{e})$. Moreover, for any $\mathbf{s}$ and random $\mathbf{e} \leftarrow D_{\mathbb{Z}^m, \alpha q}$ where $1/\alpha \geq 2\|\mathbf{B}\|s \cdot \omega_n$, the algorithm inverts successfully with overwhelming probability over the choice of $\mathbf{e}$.*

Note that using our constructions from Section 2, we can implement $\mathcal{O}$ so that either $\|\mathbf{B}\| = 2$ (for q a power of 2, where $\mathbf{B} = \tilde{\mathbf{S}} = 2\mathbf{I}$) or $\|\mathbf{B}\| = \sqrt{5}$ (for arbitrary q).

Algorithm 2. Efficient algorithm $\mathsf{Invert}^{\mathcal{O}}(\mathbf{R}, \mathbf{A}, \mathbf{b})$ for inverting the function $g_{\mathbf{A}}(\mathbf{s}, \mathbf{e})$.

Input: An oracle $\mathcal{O}$ for inverting the function $g_{\mathbf{G}}(\hat{\mathbf{s}}, \hat{\mathbf{e}})$ when $\hat{\mathbf{e}} \in \mathbb{Z}^w$ is suitably small.
- parity-check matrix $\mathbf{A} \in \mathbb{Z}_q^{n \times m}$;
- $\mathbf{G}$-trapdoor $\mathbf{R} \in \mathbb{Z}^{\bar{m} \times kn}$ for $\mathbf{A}$ with invertible tag $\mathbf{H} \in \mathbb{Z}_q^{n \times n}$;
- vector $\mathbf{b}^t = g_{\mathbf{A}}(\mathbf{s}, \mathbf{e}) = \mathbf{s}^t \mathbf{A} + \mathbf{e}^t$ for any $\mathbf{s} \in \mathbb{Z}_q^n$ and suitably small $\mathbf{e} \in \mathbb{Z}^m$.

Output: The vectors $\mathbf{s}$ and $\mathbf{e}$.
1: Compute $\hat{\mathbf{b}}^t = \mathbf{b}^t \begin{bmatrix} \mathbf{R} \\ \mathbf{I} \end{bmatrix}$.
2: Get $(\hat{\mathbf{s}}, \hat{\mathbf{e}}) \leftarrow \mathcal{O}(\hat{\mathbf{b}})$.
3: **return** $\mathbf{s} = \mathbf{H}^{-t}\hat{\mathbf{s}}$ and $\mathbf{e} = \mathbf{b} - \mathbf{A}^t \mathbf{s}$, interpreted in $\mathbb{Z}^m$ with entries in $[-\frac{q}{2}, \frac{q}{2})$.

Proof. Let $\bar{\mathbf{R}} = [\mathbf{R}^t \mid \mathbf{I}]$, and note that $s = s_1(\bar{\mathbf{R}})$. By the above description, the algorithm works correctly when $\bar{\mathbf{R}} \mathbf{e} \in \mathcal{P}_{1/2}(q \cdot \mathbf{B}^{-t})$; equivalently, when $(\mathbf{b}_i^t \bar{\mathbf{R}})\mathbf{e}/q \in [-\frac{1}{2}, \frac{1}{2})$ for all i. By definition of s, we have $\|\mathbf{b}_i^t \bar{\mathbf{R}}\| \le s\|\mathbf{B}\|$. If $\|\mathbf{e}\| < q/(2\|\mathbf{B}\|s)$, then $|(\mathbf{b}_i^t \bar{\mathbf{R}})\mathbf{e}/q| < 1/2$ by Cauchy-Schwarz. Moreover, if $\mathbf{e}$ is chosen at random from $D_{\mathbb{Z}^m, \alpha q}$, then by the fact that $\mathbf{e}$ is 0-subgaussian with parameter αq, the probability that $|(\mathbf{b}_i^t \bar{\mathbf{R}})\mathbf{e}/q| \ge 1/2$ is negligible, and the second claim follows by the union bound.

3.4 Gaussian Sampling

Here we show how to use a trapdoor for efficient Gaussian preimage sampling for the function $f_{\mathbf{A}}$, i.e., sampling from a discrete Gaussian over a desired coset of $\Lambda^{\perp}(\mathbf{A})$. Our precise goal is, given a $\mathbf{G}$-trapdoor $\mathbf{R}$ (with tag $\mathbf{H}$) for matrix $\mathbf{A}$ and a syndrome $\mathbf{u} \in \mathbb{Z}_q^n$, to sample from the spherical discrete Gaussian $D_{\Lambda_{\mathbf{u}}^{\perp}(\mathbf{A}), s}$ for relatively small parameter s. As we show next, this task can be reduced, via some efficient pre- and post-processing, to sampling from any sufficiently narrow (not necessarily spherical) Gaussian over the primitive lattice $\Lambda^{\perp}(\mathbf{G})$.

The main ideas behind our algorithm, which is described formally in the full version, are as follows. For simplicity, suppose that $\mathbf{R}$ has tag $\mathbf{H} = \mathbf{I}$, so $\mathbf{A}\begin{bmatrix} \mathbf{R} \\ \mathbf{I} \end{bmatrix} = \mathbf{G}$, and suppose we have a subroutine for Gaussian sampling from any desired coset of $\Lambda^{\perp}(\mathbf{G})$ with some small, fixed parameter $\sqrt{\Sigma_{\mathbf{G}}} \ge \eta_{\epsilon}(\Lambda^{\perp}(\mathbf{G}))$. For example, Section 2 describes algorithms for which $\sqrt{\Sigma_{\mathbf{G}}}$ is either 2 or $\sqrt{5}$. (Throughout this summary we omit the small smoothing factor ω_n from all Gaussian parameters.) The algorithm for sampling from a coset $\Lambda_{\mathbf{u}}^{\perp}(\mathbf{A})$ follows from two main observations:

1. If we sample a Gaussian $\mathbf{z}$ with parameter $\sqrt{\Sigma_{\mathbf{G}}}$ from $\Lambda_{\mathbf{u}}^{\perp}(\mathbf{G})$ and produce $\mathbf{y} = \begin{bmatrix} \mathbf{R} \\ \mathbf{I} \end{bmatrix} \mathbf{z}$, then $\mathbf{y}$ is Gaussian over the (non-full-rank) set $\begin{bmatrix} \mathbf{R} \\ \mathbf{I} \end{bmatrix} \Lambda_{\mathbf{u}}^{\perp}(\mathbf{G}) \subsetneq \Lambda_{\mathbf{u}}^{\perp}(\mathbf{A})$ with parameter $\begin{bmatrix} \mathbf{R} \\ \mathbf{I} \end{bmatrix} \sqrt{\Sigma_{\mathbf{G}}}$ (i.e., covariance $\begin{bmatrix} \mathbf{R} \\ \mathbf{I} \end{bmatrix} \Sigma_{\mathbf{G}}[\mathbf{R}^t \mid \mathbf{I}]$). The (strict) inclusion holds because for any $\mathbf{y} = \begin{bmatrix} \mathbf{R} \\ \mathbf{I} \end{bmatrix} \mathbf{z}$ where $\mathbf{z} \in \Lambda_{\mathbf{u}}^{\perp}(\mathbf{G})$, we have

$$\mathbf{A}\mathbf{y} = (\mathbf{A}\begin{bmatrix} \mathbf{R} \\ \mathbf{I} \end{bmatrix})\mathbf{z} = \mathbf{G}\mathbf{z} = \mathbf{u}.$$

Note that $s_1(\left[\begin{smallmatrix} R \\ I \end{smallmatrix}\right] \cdot \sqrt{\Sigma_G}) \leq s_1(\left[\begin{smallmatrix} R \\ I \end{smallmatrix}\right]) \cdot s_1(\sqrt{\Sigma_G}) \leq \sqrt{s_1(R)^2 + 1} \cdot s_1(\sqrt{\Sigma_G})$, so y's distribution is only about an $s_1(R)$ factor wider than that of z over $\Lambda_u^\perp(G)$. However, y lies in a non-full-rank subset of $\Lambda_u^\perp(A)$, and its distribution is 'skewed' (non-spherical). This leaks information about the trapdoor R, so we cannot just output y.

2. To sample from a *spherical* Gaussian over all of $\Lambda_u^\perp(A)$, we use the 'convolution' technique from [36] to correct for the above-described problems with the distribution of y. Specifically, we first choose a Gaussian perturbation $p \in \mathbb{Z}^m$ having covariance $s^2 - \left[\begin{smallmatrix} R \\ I \end{smallmatrix}\right] \Sigma_G [R^t \mid I]$, which is well-defined as long as $s \geq s_1(\left[\begin{smallmatrix} R \\ I \end{smallmatrix}\right] \cdot \sqrt{\Sigma_G})$. We then sample $y = \left[\begin{smallmatrix} R \\ I \end{smallmatrix}\right] z$ as above for an adjusted syndrome $v = u - Ap$, and output $x = p + y$. Now the support of x is all of $\Lambda_u^\perp(A)$, and because the covariances of p and y are additive (subject to some mild hypotheses), the overall distribution of x is spherical with Gaussian parameter s that can be as small as $s \approx s_1(R) \cdot s_1(\sqrt{\Sigma_G})$.

Quality Analysis. Our algorithm can sample from a discrete Gaussian with parameter $s \cdot \omega_n$ where s can be as small as $\sqrt{s_1(R)^2 + 1} \cdot \sqrt{s_1(\Sigma_G) + 2}$. We stress that this is only very slightly larger — a factor of at most $\sqrt{6/4} \leq 1.23$ — than the bound $(s_1(R) + 1) \cdot \|\widetilde{S}\|$ on the largest Gram-Schmidt norm of a lattice basis derived from the trapdoor R. (Recall that our constructions from Section 2 give $s_1(\Sigma_G) = \|\widetilde{S}\|^2 = 4$ or 5.) In the iterative "randomized nearest-plane" sampling algorithm of [24, 20], the Gaussian parameter s is bounded from below by the largest Gram-Schmidt norm of the orthogonalized input basis (times the same ω_n factor used in our algorithm). Therefore, the efficiency and parallelism of our algorithm comes at almost no cost in quality versus slower, iterative algorithms that use high-precision arithmetic. (It seems very likely that the corresponding small loss in security can easily be mitigated with slightly larger parameters, while still yielding a significant net gain in performance.)

Runtime Analysis. We now analyze the computational cost of the sampling algorithm, with a focus on optimizing the online runtime and parallelism (sometimes at the expense of the offline phase, which we do not attempt to optimize).

The offline phase is dominated by sampling from $D_{\mathbb{Z}^m, \sqrt{\Sigma} \cdot \omega_n}$ for some fixed (typically non-spherical) covariance matrix $\Sigma > I$. By [36, Theorem 3.1], this can be accomplished (up to any desired statistical distance) simply by sampling a continuous Gaussian $D_{\sqrt{\Sigma - I} \cdot \omega_n}$ with sufficient precision, then independently randomized-rounding each entry of the sampled vector to $\mathbb{Z}$ using Gaussian parameter $\omega_n \geq \eta_\epsilon(\mathbb{Z})$.

Naively, the online work is dominated by the computation of $H^{-1}(u - \bar{w})$ and Rz (plus the call to $\mathcal{O}(v)$, which as described in Section 2 requires only $O(\log^c n)$ work, or one table lookup, by each of n processors in parallel). In general, the first computation takes $O(n^2)$ scalar multiplications and additions in $\mathbb{Z}_q$, while the latter takes $O(\bar{m} \cdot w)$, which is typically $\Theta(n^2 \log^2 q)$. (Obviously, both computations are perfectly parallelizable.) However, the special form of z, and often of H,

Algorithm 3. Efficient algorithm $\mathsf{DelTrap}^{\mathcal{O}}(\mathbf{A}' = [\mathbf{A} \mid \mathbf{A}_1], \mathbf{H}', s')$ for delegating a trapdoor.

Input: an oracle $\mathcal{O}$ for discrete Gaussian sampling over cosets of $\Lambda = \Lambda^{\perp}(\mathbf{A})$ with
  parameter $s' \geq \eta_{\epsilon}(\Lambda)$.
  – parity-check matrix $\mathbf{A}' = [\mathbf{A} \mid \mathbf{A}_1] \in \mathbb{Z}_q^{n \times m} \times \mathbb{Z}_q^{n \times w}$;
  – invertible matrix $\mathbf{H}' \in \mathbb{Z}_q^{n \times n}$;
Output: a trapdoor $\mathbf{R}' \in \mathbb{Z}^{m \times w}$ for $\mathbf{A}'$ with tag $\mathbf{H}' \in \mathbb{Z}_q^{n \times n}$.
  1: Using $\mathcal{O}$, sample each column of $\mathbf{R}'$ independently from a discrete Gaussian with
     parameter s' over the appropriate coset of $\Lambda^{\perp}(\mathbf{A})$, so that $\mathbf{A}\mathbf{R}' = \mathbf{H}'\mathbf{G} - \mathbf{A}_1$.

allow for some further asymptotic and practical optimizations: since $\mathbf{z}$ is typically produced by concatenating n independent dimension-k subvectors that are sampled offline, we can precompute much of $\mathbf{R}\mathbf{z}$ by pre-multiplying each subvector by each of the n blocks of k columns in $\mathbf{R}$. This reduces the online computation of $\mathbf{R}\mathbf{z}$ to the summation of n dimension-$\bar{m}$ vectors, or $O(n^2 \log q)$ scalar additions (and no multiplications) in $\mathbb{Z}_q$. As for multiplication by $\mathbf{H}^{-1}$, in some applications (like GPV signatures) $\mathbf{H}$ is always the identity $\mathbf{I}$, in which case multiplication is unnecessary; in all other applications we know of, $\mathbf{H}$ actually represents multiplication in a certain extension field/ring of $\mathbb{Z}_q$, which can be computed in $O(n \log n)$ scalar operations and depth $O(\log n)$. In conclusion, the asymptotic cost of the online phase is still dominated by computing $\mathbf{R}\mathbf{z}$, which takes $\tilde{O}(n^2)$ work, but the hidden constants are small and many practical speedups are possible.

3.5   Trapdoor Delegation

Here we describe very simple and efficient mechanism for securely delegating a trapdoor for $\mathbf{A} \in \mathbb{Z}_q^{n \times m}$ to a trapdoor for an extension $\mathbf{A}' \in \mathbb{Z}_q^{n \times m'}$ of $\mathbf{A}$. Our method has several advantages over the previous basis delegation algorithm of [13]: first and most importantly, the size of the delegated trapdoor grows only linearly with the dimension m' of $\Lambda^{\perp}(\mathbf{A}')$, rather than quadratically. Second, the algorithm is much more efficient, because it does not require testing linear independence of Gaussian samples, nor computing the expensive $\mathsf{ToBasis}$ and Hermite normal form operations. Third, the resulting trapdoor $\mathbf{R}$ has a 'nice' Gaussian distribution that is easy to analyze and may be useful in applications. We do note that while the delegation algorithm from [13] works for *any* extension $\mathbf{A}'$ of $\mathbf{A}$ (including $\mathbf{A}$ itself), ours requires $m' \geq m + w$. Fortunately, this is frequently the case in applications such as HIBE and others that use delegation.

   Usually, the oracle $\mathcal{O}$ needed by Algorithm 3 would be implemented (up to $\mathrm{negl}(n)$ statistical distance) by our Gaussian sampling algorithm above, using a trapdoor $\mathbf{R}$ for $\mathbf{A}$ where $s_1(\mathbf{R})$ is sufficiently small relative to s'. The following is immediate from the fact that the columns of $\mathbf{R}'$ are independent and $\mathrm{negl}(n)$-subgaussian.

Lemma 1. *For any valid inputs $\mathbf{A}'$ and $\mathbf{H}'$, Algorithm 3 outputs a trapdoor $\mathbf{R}'$ for $\mathbf{A}'$ with tag $\mathbf{H}'$, whose distribution is the same for any valid implementation of $\mathcal{O}$, and $s_1(\mathbf{R}') \leq s' \cdot O(\sqrt{m} + \sqrt{w})$ except with negligible probability.*

References

[1] Agrawal, S., Boneh, D., Boyen, X.: Efficient Lattice (H)IBE in the Standard Model. In: Gilbert, H. (ed.) EUROCRYPT 2010. LNCS, vol. 6110, pp. 553–572. Springer, Heidelberg (2010)

[2] Agrawal, S., Boneh, D., Boyen, X.: Lattice Basis Delegation in Fixed Dimension and Shorter-Ciphertext Hierarchical IBE. In: Rabin, T. (ed.) CRYPTO 2010. LNCS, vol. 6223, pp. 98–115. Springer, Heidelberg (2010)

[3] Ajtai, M.: Generating Hard Instances of the Short Basis Problem. In: Wiedermann, J., Van Emde Boas, P., Nielsen, M. (eds.) ICALP 1999. LNCS, vol. 1644, pp. 1–9. Springer, Heidelberg (1999)

[4] Ajtai, M.: Generating hard instances of lattice problems. Quaderni di Matematica 13, 1–32 (1996); Preliminary version in STOC 1996

[5] Alwen, J., Peikert, C.: Generating shorter bases for hard random lattices. Theory of Computing Systems 48(3), 535–553 (2011); Preliminary version in STACS 2009

[6] Applebaum, B., Cash, D., Peikert, C., Sahai, A.: Fast Cryptographic Primitives and Circular-Secure Encryption Based on Hard Learning Problems. In: Halevi, S. (ed.) CRYPTO 2009. LNCS, vol. 5677, pp. 595–618. Springer, Heidelberg (2009)

[7] Babai, L.: On Lovász' lattice reduction and the nearest lattice point problem. Combinatorica 6(1), 1–13 (1986); Preliminary version in STACS 1985

[8] Blum, A., Furst, M.L., Kearns, M., Lipton, R.J.: Cryptographic Primitives Based on Hard Learning Problems. In: Stinson, D.R. (ed.) CRYPTO 1993. LNCS, vol. 773, pp. 278–291. Springer, Heidelberg (1994)

[9] Boyen, X.: Lattice Mixing and Vanishing Trapdoors: A Framework for Fully Secure Short Signatures and More. In: Nguyen, P.Q., Pointcheval, D. (eds.) PKC 2010. LNCS, vol. 6056, pp. 499–517. Springer, Heidelberg (2010)

[10] Brakerski, Z., Gentry, C., Vaikuntanathan, V.: Fully homomorphic encryption without bootstrapping. Cryptology ePrint Archive, Report 2011/277 (2011), http://eprint.iacr.org/

[11] Brakerski, Z., Vaikuntanathan, V.: Efficient fully homomorphic encryption from (standard) LWE. In: FOCS (2011)

[12] Brakerski, Z., Vaikuntanathan, V.: Fully Homomorphic Encryption from Ring-LWE and Security for Key Dependent Messages. In: Rogaway, P. (ed.) CRYPTO 2011. LNCS, vol. 6841, pp. 505–524. Springer, Heidelberg (2011)

[13] Cash, D., Hofheinz, D., Kiltz, E., Peikert, C.: Bonsai Trees, or How to Delegate a Lattice Basis. In: Gilbert, H. (ed.) EUROCRYPT 2010. LNCS, vol. 6110, pp. 523–552. Springer, Heidelberg (2010)

[14] Chen, Y., Nguyen, P.Q.: BKZ 2.0: Better Lattice Security Estimates. In: Lee, D.H., Wang, X. (eds.) ASIACRYPT 2011. LNCS, vol. 7073, pp. 1–20. Springer, Heidelberg (2011)

[15] Gama, N., Nguyen, P.Q.: Predicting Lattice Reduction. In: Smart, N.P. (ed.) EUROCRYPT 2008. LNCS, vol. 4965, pp. 31–51. Springer, Heidelberg (2008)

[16] Gentry, C.: A fully homomorphic encryption scheme. PhD thesis, Stanford University (2009), http://crypto.stanford.edu/craig

[17] Gentry, C.: Fully homomorphic encryption using ideal lattices. In: STOC, pp. 169–178 (2009)

[18] Gentry, C., Halevi, S.: Fully homomorphic encryption without squashing using depth-3 arithmetic circuits. In: FOCS (2011)

[19] Gentry, C., Halevi, S., Vaikuntanathan, V.: A Simple BGN-Type Cryptosystem from LWE. In: Gilbert, H. (ed.) EUROCRYPT 2010. LNCS, vol. 6110, pp. 506–522. Springer, Heidelberg (2010)

[20] Gentry, C., Peikert, C., Vaikuntanathan, V.: Trapdoors for hard lattices and new cryptographic constructions. In: STOC, pp. 197–206 (2008)

[21] Goldreich, O., Goldwasser, S., Halevi, S.: Public-Key Cryptosystems from Lattice Reduction Problems. In: Kaliski Jr., B.S. (ed.) CRYPTO 1997. LNCS, vol. 1294, pp. 112–131. Springer, Heidelberg (1997)

[22] Gordon, S.D., Katz, J., Vaikuntanathan, V.: A Group Signature Scheme from Lattice Assumptions. In: Abe, M. (ed.) ASIACRYPT 2010. LNCS, vol. 6477, pp. 395–412. Springer, Heidelberg (2010)

[23] Kiltz, E., Mohassel, P., O'Neill, A.: Adaptive Trapdoor Functions and Chosen-Ciphertext Security. In: Gilbert, H. (ed.) EUROCRYPT 2010. LNCS, vol. 6110, pp. 673–692. Springer, Heidelberg (2010)

[24] Klein, P.N.: Finding the closest lattice vector when it's unusually close. In: SODA, pp. 937–941 (2000)

[25] Lindner, R., Peikert, C.: Better Key Sizes (and Attacks) for LWE-Based Encryption. In: Kiayias, A. (ed.) CT-RSA 2011. LNCS, vol. 6558, pp. 319–339. Springer, Heidelberg (2011)

[26] Lyubashevsky, V.: Lattice-Based Identification Schemes Secure Under Active Attacks. In: Cramer, R. (ed.) PKC 2008. LNCS, vol. 4939, pp. 162–179. Springer, Heidelberg (2008)

[27] Lyubashevsky, V.: Lattice Signatures without Trapdoors. In: Pointcheval, D., Johansson, T. (eds.) EUROCRYPT 2012. LNCS, vol. 7237, pp. 738–755. Springer, Heidelberg (2012)

[28] Lyubashevsky, V., Micciancio, D.: Generalized Compact Knapsacks Are Collision Resistant. In: Bugliesi, M., Preneel, B., Sassone, V., Wegener, I. (eds.) ICALP 2006, Part II. LNCS, vol. 4052, pp. 144–155. Springer, Heidelberg (2006)

[29] Lyubashevsky, V., Micciancio, D.: Asymptotically Efficient Lattice-Based Digital Signatures. In: Canetti, R. (ed.) TCC 2008. LNCS, vol. 4948, pp. 37–54. Springer, Heidelberg (2008)

[30] Lyubashevsky, V., Micciancio, D., Peikert, C., Rosen, A.: SWIFFT: A Modest Proposal for FFT Hashing. In: Nyberg, K. (ed.) FSE 2008. LNCS, vol. 5086, pp. 54–72. Springer, Heidelberg (2008)

[31] Lyubashevsky, V., Peikert, C., Regev, O.: On Ideal Lattices and Learning with Errors over Rings. In: Gilbert, H. (ed.) EUROCRYPT 2010. LNCS, vol. 6110, pp. 1–23. Springer, Heidelberg (2010)

[32] Micciancio, D.: Generalized compact knapsacks, cyclic lattices, and efficient one-way functions. Computational Complexity 16(4), 365–411 (2007); Preliminary version in FOCS 2002

[33] Micciancio, D., Mol, P.: Pseudorandom Knapsacks and the Sample Complexity of LWE Search-to-Decision Reductions. In: Rogaway, P. (ed.) CRYPTO 2011. LNCS, vol. 6841, pp. 465–484. Springer, Heidelberg (2011)

[34] Micciancio, D., Regev, O.: Lattice-based cryptography. In: Post Quantum Cryptography, pp. 147–191. Springer, Heidelberg (2009)

[35] Peikert, C.: Public-key cryptosystems from the worst-case shortest vector problem. In: STOC, pp. 333–342 (2009)

[36] Peikert, C.: An Efficient and Parallel Gaussian Sampler for Lattices. In: Rabin, T. (ed.) CRYPTO 2010. LNCS, vol. 6223, pp. 80–97. Springer, Heidelberg (2010)

[37] Peikert, C., Rosen, A.: Efficient Collision-Resistant Hashing from Worst-Case Assumptions on Cyclic Lattices. In: Halevi, S., Rabin, T. (eds.) TCC 2006. LNCS, vol. 3876, pp. 145–166. Springer, Heidelberg (2006)

[38] Peikert, C., Vaikuntanathan, V.: Noninteractive Statistical Zero-Knowledge Proofs for Lattice Problems. In: Wagner, D. (ed.) CRYPTO 2008. LNCS, vol. 5157, pp. 536–553. Springer, Heidelberg (2008)

[39] Peikert, C., Vaikuntanathan, V., Waters, B.: A Framework for Efficient and Composable Oblivious Transfer. In: Wagner, D. (ed.) CRYPTO 2008. LNCS, vol. 5157, pp. 554–571. Springer, Heidelberg (2008)

[40] Peikert, C., Waters, B.: Lossy trapdoor functions and their applications. In: STOC, pp. 187–196 (2008)

[41] Regev, O.: On lattices, learning with errors, random linear codes, and cryptography. J. ACM 56(6), 1–40 (2005); Preliminary version in STOC 2005

[42] Rückert, M.: Strongly Unforgeable Signatures and Hierarchical Identity-Based Signatures from Lattices without Random Oracles. In: Sendrier, N. (ed.) PQCrypto 2010. LNCS, vol. 6061, pp. 182–200. Springer, Heidelberg (2010)

[43] Rückert, M., Schneider, M.: Selecting secure parameters for lattice-based cryptography. Cryptology ePrint Archive, Report 2010/137 (2010), http://eprint.iacr.org/

[44] Stehlé, D., Steinfeld, R., Tanaka, K., Xagawa, K.: Efficient Public Key Encryption Based on Ideal Lattices. In: Matsui, M. (ed.) ASIACRYPT 2009. LNCS, vol. 5912, pp. 617–635. Springer, Heidelberg (2009)

[45] van Dijk, M., Gentry, C., Halevi, S., Vaikuntanathan, V.: Fully Homomorphic Encryption over the Integers. In: Gilbert, H. (ed.) EUROCRYPT 2010. LNCS, vol. 6110, pp. 24–43. Springer, Heidelberg (2010)

Pseudorandom Functions and Lattices

Abhishek Banerjee[1,*], Chris Peikert[1,**], and Alon Rosen[2,***]

[1] Georgia Institute of Technology
[2] IDC Herzliya

Abstract. We give direct constructions of pseudorandom function (PRF) families based on conjectured hard lattice problems and learning problems. Our constructions are asymptotically efficient and highly parallelizable in a practical sense, i.e., they can be computed by simple, relatively *small* low-depth arithmetic or boolean circuits (e.g., in NC^1 or even TC^0). In addition, they are the first low-depth PRFs that have no known attack by efficient quantum algorithms. Central to our results is a new "derandomization" technique for the learning with errors (LWE) problem which, in effect, generates the error terms deterministically.

1 Introduction and Main Results

The past few years have seen significant progress in constructing public-key, identity-based, and homomorphic cryptographic schemes using lattices, e.g., [35, 33, 15, 14, 13, 1] and many more. Part of their appeal stems from provable worst-case hardness guarantees (starting with the seminal work of Ajtai [3]), good asymptotic efficiency and parallelism, and apparent resistance to quantum attacks (unlike the classical problems of factoring integers or computing discrete logarithms).

Perhaps surprisingly, there has been comparatively less progress in using lattices for *symmetric* cryptography, e.g., message authentication codes, block ciphers, and the like, which are widely used in practice. While in principle most symmetric objects of interest can be obtained generically from any one-way function, and hence from lattices, these generic constructions are usually very inefficient, which puts them at odds with the high performance demands of most applications. In addition, generic constructions often use their underlying primitives (e.g., one-way functions) in an inherently inefficient and *sequential*

---

* Research supported in part by an ARC Fellowship and the second author's grants.
** This material is based upon work supported by the National Science Foundation under Grant CNS-0716786 and CAREER Award CCF-1054495, by the Alfred P. Sloan Foundation, by the Defense Advanced Research Projects Agency (DARPA) and the Air Force Research Laboratory (AFRL) under Contract No. FA8750-11-C-0098, and by BSF grant 2010296. The views expressed are those of the authors and do not necessarily reflect the official policy or position of the National Science Foundation, the Sloan Foundation, DARPA or the U.S. Government, or the BSF.
*** Research supported in part by BSF grant 2010296.

D. Pointcheval and T. Johansson (Eds.): EUROCRYPT 2012, LNCS 7237, pp. 719–737, 2012.

manner. While most lattice-based primitives are relatively efficient and highly parallelizable in a practical sense (i.e., they can be evaluated by small, low-depth circuits), those advantages are completely lost when plugging them into generic sequential constructions. This motivates the search for specialized constructions of symmetric objects that have comparable efficiency and parallelism to their lower-level counterparts.

Our focus in this work is on *pseudorandom function* (PRF) families, a central object in symmetric cryptography first rigorously defined and constructed by Goldreich, Goldwasser, and Micali ("GGM") [16]. Given a PRF family, most central goals of symmetric cryptography (e.g., encryption, authentication, identification) have simple solutions that make efficient use of the PRF. Informally, a family of deterministic functions is pseudorandom if no efficient adversary, given adaptive oracle access to a randomly chosen function from the family, can distinguish it from a uniformly random function. The seminal GGM construction is based generically on any length-doubling pseudorandom generator (and hence on any one-way function), but it requires k *sequential* invocations of the generator when operating on k-bit inputs.

In contrast, by relying on a generic object called a "pseudorandom *synthesizer*," or directly on concrete number-theoretic problems (such as decision Diffie-Hellman, RSA, and factoring), Naor and Reingold [28, 29] and Naor, Reingold, and Rosen [30] (see also [23, 9]) constructed very elegant and more efficient PRFs, which can in principle be computed in parallel by low-depth circuits (e.g., in NC^2 or TC^0). However, achieving such low depth for their number-theoretic constructions requires extensive preprocessing and enormous circuits, so their results serve mainly as a proof of theoretical feasibility rather than practical utility.

In summary, thus far all parallelizable PRFs from commonly accepted cryptographic assumptions rely on exponentiation in large multiplicative groups, and the functions (or at least their underlying hard problems) can be broken by polynomial-time quantum algorithms. While lattices appear to be a natural candidate for avoiding these drawbacks, and there has been some partial progress in the form of *randomized* weak PRFs [4] and randomized MACs [34, 21], constructing an efficient, parallelizable (deterministic) PRF under lattice assumptions has, frustratingly, remained open for some time now.

1.1   Results and Techniques

In this work we give the first direct constructions of PRF families based on lattices, via the *learning with errors* (LWE) [35] and *ring*-LWE [25] problems, and some new variants. Our constructions are highly parallelizable in a *practical* sense, i.e., they can be computed by relatively *small* low-depth circuits, and the runtimes are also potentially practical. (However, their performance and key sizes are still far from those of heuristically designed functions like AES.) In addition, (at least) one of our constructions can be evaluated in the circuit class TC^0 (i.e., constant-depth, poly-sized circuits with unbounded fan-in and threshold gates), which asymptotically matches the shallowest known PRF constructions based on the decision Diffie-Hellman and factoring problems [29, 30].

As a starting point, we recall that in their work introducing *synthesizers* as a foundation for PRFs [28], Naor and Reingold described a synthesizer based on a simple, conjectured hard-to-learn function. At first glance, this route seems very promising for obtaining PRFs from lattices, using LWE as the hard learning problem (which is known to be as hard as worst-case lattice problems [35, 31]). However, a crucial point is that Naor and Reingold's synthesizer uses a *deterministic* hard-to-learn function, whereas LWE's hardness depends essentially on adding *random, independent* errors to every output of a mod-q "parity" function. (Indeed, without any error, parity functions are trivially easy to learn.) Probably the main obstacle so far in constructing efficient lattice/LWE-based PRFs has been in finding a way to introduce (sufficiently independent) error terms into each of the exponentially many function outputs, while still keeping the function deterministic and its key size a fixed polynomial. As evidence, consider that recent constructions of weaker primitives such as symmetric authentication protocols [18, 19, 20], randomized weak PRFs [4], and message-authentication codes [34, 21] from noisy-learning problems are all inherently *randomized* functions, where security relies on introducing fresh noise at every invocation. Unfortunately, this is not an option for deterministic primitives like PRFs.

Derandomizing LWE. To resolve the above-described issues, our first main insight is a way of partially "derandomizing" the LWE problem, i.e., generating the *errors* efficiently and deterministically, while preserving hardness. This technique immediately yields a deterministic synthesizer and hence a simple and parallelizable PRF, though with a few subtleties specific to our technique that we elaborate upon below.

Before we explain the derandomization idea, first recall the learning with errors problem $\mathsf{LWE}_{n,q,\alpha}$ in dimension n (the main security parameter) with modulus q and error rate α. We are given many independent pairs $(\mathbf{a}_i, b_i) \in \mathbb{Z}_q^n \times \mathbb{Z}_q$, where each $\mathbf{a}_i$ is uniformly random, and the b_i are all either "noisy inner products" of the form $b_i = \langle \mathbf{a}_i, \mathbf{s} \rangle + e_i \bmod q$ for a random secret $\mathbf{s} \in \mathbb{Z}_q^n$ and "small" random error terms $e_i \in \mathbb{Z}$ of magnitude $\approx \alpha q$, or are uniformly random and independent of the $\mathbf{a}_i$. The goal of the (decision) LWE problem is to distinguish between these two cases, with any non-negligible advantage. In the *ring*-LWE problem [25], we are instead given noisy ring products $b_i \approx a_i \cdot s$, where s and the a_i are random elements of a certain polynomial ring R_q (the canonical example being $R_q = \mathbb{Z}_q[z]/(z^n + 1)$ for n a power of 2), and the error terms are "small" in a certain basis of the ring; the goal again is to distinguish these from uniformly random pairs. While the dimension n is the main hardness parameter, the error rate α also plays a very important role in both theory and practice: as long as the "absolute" error αq exceeds $\sqrt{n}$ or so, (ring-)LWE is provably as hard as approximating conjectured hard problems on (ideal) lattices to within $\tilde{O}(n/\alpha)$ factors in the worst case [35, 31, 25]. Moreover, known attacks using lattice basis reduction (e.g., [22, 37]) or combinatorial/algebraic methods [8, 5] require time $2^{\tilde{\Omega}(n/\log(1/\alpha))}$, where the $\tilde{\Omega}(\cdot)$ notation hides polylogarithmic factors in n. We emphasize that without the error terms, (ring-)LWE would become trivially easy, and that all prior hardness results for LWE and its many variants (e.g., [35, 31, 17, 25, 34]) require random, independent errors.

Our derandomization technique for LWE is very simple: instead of adding a small random error term to each inner product $\langle \mathbf{a}_i, \mathbf{s} \rangle \in \mathbb{Z}_q$, we just deterministically *round* it to the nearest element of a sufficiently "coarse" public subset of $p \ll q$ well-separated values in $\mathbb{Z}_q$ (e.g., a subgroup). In other words, the "error term" comes solely from deterministically rounding $\langle \mathbf{a}_i, \mathbf{s} \rangle$ to a relatively nearby value. Since there are only p possible rounded outputs in $\mathbb{Z}_q$, it is usually easier to view them as elements of $\mathbb{Z}_p$ and denote the rounded value by $\lfloor \langle \mathbf{a}_i, \mathbf{s} \rangle \rceil_p \in \mathbb{Z}_p$. We call the problem of distinguishing such rounded inner products from uniform samples the *learning with rounding* ($\mathsf{LWR}_{n,q,p}$) problem. Note that the problem can be hard only if $q > p$ (otherwise no error is introduced), that the "absolute" error is roughly q/p, and that the "error rate" relative to q (i.e., the analogue of α in the LWE problem) is on the order of $1/p$.

We show that for appropriate parameters, $\mathsf{LWR}_{n,q,p}$ is at least as hard as $\mathsf{LWE}_{n,q,\alpha}$ for an error rate α proportional to $1/p$, giving us a worst-case hardness guarantee for LWR. In essence, the reduction relies on the fact that with high probability, we have $\lfloor \langle \mathbf{a}, \mathbf{s} \rangle + e \rceil_p = \lfloor \langle \mathbf{a}, \mathbf{s} \rangle \rceil_p$ when e is small relative to q/p, while $\lfloor U(\mathbb{Z}_q) \rceil_p \approx U(\mathbb{Z}_p)$ where U denotes the uniform distribution. Therefore, given samples $(\mathbf{a}_i, b_i)$ of an unknown type (either LWE or uniform), we can simply round the b_i terms to generate samples of a corresponding type (LWR or uniform, respectively). (The formal proof is somewhat more involved, because it has to deal with the rare event that the error term changes the rounded value.) In the ring setting, the derandomization technique and hardness proof based on ring-LWE all go through without difficulty as well. While our proof needs both the ratio q/p and the inverse LWE error rate $1/\alpha$ to be slightly super-polynomial in n, the state of the art in attack algorithms indicates that as long as q/p is an integer (so that $\lfloor U(\mathbb{Z}_q) \rceil_p = U(\mathbb{Z}_p)$) and is at least $\Omega(\sqrt{n})$, LWR may be exponentially hard (even for quantum algorithms) for any $p = \mathrm{poly}(n)$, and superpolynomially hard when $p = 2^{n^\epsilon}$ for any $\epsilon < 1$.

We point out that in LWE-based cryptosystems, rounding to a fixed, coarse subset is a common method of removing noise and recovering the plaintext when decrypting a "noisy" ciphertext; here we instead use it to avoid having to introduce any random noise in the first place. We believe that this technique should be useful in many other settings, especially in symmetric cryptography. For example, the LWR problem immediately yields a simple and practical pseudorandom generator that does not require extracting biased (e.g., Gaussian) random values from its input seed, unlike the standard pseudorandom generators based on the LWE or LPN (learning parity with noise) problems. In addition, the rounding technique and its implications for PRFs are closely related to the "modulus reduction" technique from a concurrent and independent work of Brakerski and Vaikuntanathan [11] on fully homomorphic encryption from LWE, and a very recent follow-up work of Brakerski, Gentry, and Vaikuntanathan [10]; see Section 1.3 below for a discussion and comparison.

LWR-*based synthesizers and PRFs.* Recall from [28] that a pseudorandom *synthesizer* is a two-argument function $S(\cdot, \cdot)$ such that, for random and independent sequences $x_1, \ldots, x_m$ and $y_1, \ldots, y_m$ of inputs (for any $m = \text{poly}(n)$), the matrix of all m^2 values $z_{i,j} = S(x_i, y_j)$ is pseudorandom (i.e., computationally indistinguishable from uniform). A synthesizer can be seen as an (almost) length-*squaring* pseudorandom generator with good locality properties, in that it maps $2m$ random "seed" elements (the x_i and y_j) to m^2 pseudorandom elements, and any component of its output depends on only two components of the input seed.

Using synthesizers in a recursive tree-like construction, Naor and Reingold gave PRFs on k-bit inputs, which can be computed using a total of about k synthesizer evaluations, arranged nicely in only $\lg k$ levels (depth). Essentially, the main idea is that given a synthesizer $S(\cdot, \cdot)$ and two independent PRF instances F_0 and F_1 on t input bits each, one gets a PRF on $2t$ input bits, defined as

$$F(x_1 \cdots x_{2t}) = S\big(F_0(x_1 \cdots x_t), F_1(x_{t+1} \cdots x_{2t}) \big). \tag{1}$$

The base case of a 1-bit PRF can trivially be implemented by returning one of two random strings in the function's secret key. Using particular NC^1 synthesizers based on a variety of both concrete and general assumptions, Naor and Reingold therefore obtain k-bit PRFs in NC^2, i.e., having circuit depth $O(\log^2 k)$.

We give a very simple and computationally efficient $\text{LWR}_{n,q,p}$-based synthesizer $S_{n,q,p} \colon \mathbb{Z}_q^n \times \mathbb{Z}_q^n \to \mathbb{Z}_p$, defined as

$$S_{n,q,p}(\mathbf{a}, \mathbf{s}) = \lfloor \langle \mathbf{a}, \mathbf{s} \rangle \rceil_p. \tag{2}$$

(In this and what follows, products of vectors or matrices over $\mathbb{Z}_q$ are always performed modulo q.) Pseudorandomness of this synthesizer under LWR follows by a standard hybrid argument, using the fact that the $\mathbf{a}_i$ vectors given in the LWR problem are public. (In fact, the synthesizer outputs $S(\mathbf{a}_i, \mathbf{s}_j)$ are pseudorandom even given the $\mathbf{a}_i$.) To obtain a PRF using the tree construction of [28], we need the synthesizer output length to roughly match its input length, so we actually use the synthesizer $T_{n,q,p}(\mathbf{S}_1, \mathbf{S}_2) = \lfloor \mathbf{S}_1 \cdot \mathbf{S}_2 \rceil_p \in \mathbb{Z}_p^{n \times n}$ for $\mathbf{S}_i \in \mathbb{Z}_q^{n \times n}$. Note that the matrix multiplication can be done with a constant-depth, size-$O(n^2)$ arithmetic circuit over $\mathbb{Z}_q$. Or for better space and time complexity, we can instead use the ring-LWR synthesizer $S_{R,q,p}(s_1, s_2) = \lfloor s_1 \cdot s_2 \rceil_p$, since the ring product $s_1 \cdot s_2 \in R_q$ is the same size as $s_1, s_2 \in R_q$. The ring product can also be computed with a constant depth, size-$O(n^2)$ circuit over $\mathbb{Z}_q$, or in $O(\log n)$ depth and only $O(n \log n)$ scalar operations using Fast Fourier Transform-like techniques [24, 25].

Using the recursive input-doubling construction from Equation (1) above, we get the following concrete PRF with input length $k = 2^d$. Let $q_d > q_{d-1} > \cdots > q_0 \geq 2$ be a chain of moduli where each q_j / q_{j-1} is a sufficiently large integer, e.g., $q_j = q^{j+1}$ for some $q \geq \sqrt{n}$. The secret key is a set of $2k$ matrices $\mathbf{S}_{i,b} \in \mathbb{Z}_{q_d}^{n \times n}$ for each $i \in \{1, \ldots, k\}$ and $b \in \{0, 1\}$. Each pair $(\mathbf{S}_{i,0}, \mathbf{S}_{i,1})$ defines a 1-bit PRF $F_i(b) = \mathbf{S}_{i,b}$, and these are combined in a tree-like fashion according to Equation (1) using the appropriate synthesizers $T_{n,q_j,q_{j-1}}$ for $j = d, \ldots, 1$. As a concrete example, when $k = 4$ (so $x = x_1 \cdots x_4$ and $d = 2$), we have

$$F_{\{\mathbf{S}_{i,b}\}}(x) \;=\; \left\lfloor \left\lfloor \mathbf{S}_{1,x_1} \cdot \mathbf{S}_{2,x_2} \right\rceil_{q_1} \cdot \left\lfloor \mathbf{S}_{3,x_3} \cdot \mathbf{S}_{4,x_4} \right\rceil_{q_1} \right\rceil_{q_0} . \tag{3}$$

(In the ring setting, we just use random elements $s_{i,b} \in R_{q_d}$ in place of the matrices $\mathbf{S}_{i,b}$.) Notice that the function involves $d = \lg k$ levels of matrix (or ring) products, each followed by a rounding operation. In the exemplary case where $q_j = q^{j+1}$, the rounding operations essentially drop the "least-significant" base-q digit, so they can be implemented very easily in practice, especially if every q_j is a power of 2. The function is also amenable to all of the nice time/space trade-offs, seed-compression techniques, and incremental computation ideas described in [28].

In the security proof, we rely on the conjectured hardness of $\mathsf{LWR}_{q_j,q_{j-1}}$ for $j = d, \dots, 1$. The strongest of these assumptions appears to be for $j = d$, and this is certainly the case when relying on our reduction from LWE to LWR. For the example parameters $q_j = q^{j+1}$ where $q \approx \sqrt{n}$, the dominating assumption is therefore the hardness of $\mathsf{LWR}_{q^{d+1},q^d}$, which involves a quasi-polynomial inverse error rate of $1/\alpha \approx q^d = n^{O(\lg k)}$. However, because the strongest assumptions are applied to the "innermost" layers of the function, it is unclear whether security actually *requires* such strong assumptions, or even whether the innermost layers need to be rounded at all. We discuss these issues further in Section 1.2 below.

Degree-k synthesizers and shallower PRFs. One moderate drawback of the above function is that it involves $\lg k$ levels of rounding operations, which appears to lower-bound the depth of any circuit computing the function by $\Omega(\lg k)$. Is it possible to do better?

Recall that in later works, Naor and Reingold [29] and Naor, Reingold, and Rosen [30] gave direct, more efficient number-theoretic PRF constructions which, while still requiring exponentiation in large multiplicative groups, can in principle be computed in very shallow circuit classes like NC^1 or even TC^0. Their functions can be interpreted as "degree-k" (or k-argument) synthesizers for arbitrary $k = \mathrm{poly}(n)$, which immediately yield k-bit PRFs without requiring any composition. With this in mind, a natural question is whether there are direct $\mathsf{LWE}/\mathsf{LWR}$-based synthesizers of degree $k > 2$.

We give a positive answer to this question. Much like the functions of [29, 30], ours have a subset-product structure. We have public moduli $q \gg p$, and the secret key is a set of k matrices $\mathbf{S}_i \in \mathbb{Z}_q^{n \times n}$ (whose distributions may not necessarily be uniform; see below) for $i = 1, \dots, k$, along with a uniformly random $\mathbf{a} \in \mathbb{Z}_q^n$.[1] The function $F = F_{\mathbf{a},\{\mathbf{S}_i\}} \colon \{0,1\}^k \to \mathbb{Z}_p^n$ is defined as the "rounded subset-product"

$$F_{\mathbf{a},\{\mathbf{S}_i\}}(x_1 \cdots x_k) = \left\lfloor \mathbf{a}^t \cdot \prod_{i=1}^{k} \mathbf{S}_i^{x_i} \right\rceil_p . \tag{4}$$

---

[1] To obtain longer function outputs, we can replace $\mathbf{a} \in \mathbb{Z}_q^n$ with a uniformly random matrix $\mathbf{A} \in \mathbb{Z}_q^{n \times m}$ for any $m = \mathrm{poly}(n)$.

The ring variant is analogous, replacing $\mathbf{a}$ with uniform $a \in R_q$ and each $\mathbf{S}_i$ by some $s_i \in R_q$ (or R_q^*, the set of invertible elements modulo q). This function is particularly efficient to evaluate using the discrete Fourier transform, as is standard with ring-based primitives (see, e.g., [24, 25]). In addition, similarly to [29, 30], one can optimize the subset-product operation via pre-processing, and evaluate the function in TC^0. We elaborate on these optimizations in the full version of the paper [7].

For the security analysis of construction (4), we have meaningful security proofs under various conditions on the parameters and computational assumptions, including standard LWE. In our LWE-based proof, two important issues are the distribution of the secret key components $\mathbf{S}_i$, and the choice of moduli q and p. For the former, it turns out that our proof needs the $\mathbf{S}_i$ matrices to be *short*, i.e., their entries should be drawn from the LWE error distribution. (LWE is no easier to solve for such short secrets [4].) This appears to be an artifact of our proof technique, which can be viewed as a variant of our LWE-to-LWR reduction, enhanced to handle adversarial queries. Summarizing the approach, define

$$G(x) = G_{\mathbf{a},\{\mathbf{S}_i\}}(x) := \mathbf{a}^t \cdot \prod_i \mathbf{S}_i^{x_i}$$

to be the subset-product function inside the rounding operation of (4). The fact that $F = \lfloor G \rceil_p$ lets us imagine adding *independent error terms* to each distinct output of G, but *only as part of a thought experiment* in the proof. More specifically, we consider a related *randomized* function $\tilde{G} = \tilde{G}_{\mathbf{a},\{\mathbf{S}_i\}} \colon \{0,1\}^k \to \mathbb{Z}_q^n$ that computes the subset-product by multiplying by each $\mathbf{S}_i^{x_i}$ in turn, but then also adds a fresh error term immediately following each multiplication. Using the LWE assumption and induction on k, we can show that the randomized function $\tilde{G}$ is itself pseudorandom (over $\mathbb{Z}_q$), hence so is $\lfloor \tilde{G} \rceil_p$ (over $\mathbb{Z}_p$). Moreover, we show that for every queried input, with high probability $\lfloor \tilde{G} \rceil_p$ coincides with $\lfloor G \rceil_p = F$, because G and $\tilde{G}$ differ only by a cumulative error term that is small relative to q—this is where we need to assume that the entries of $\mathbf{S}_i$ are *small*. Finally, because $\lfloor \tilde{G} \rceil_p$ is a (randomized) pseudorandom function over $\mathbb{Z}_p$ that coincides with the deterministic function F on all queries, we can conclude that F is pseudorandom as well.

In the above-described proof strategy, the gap between G and $\tilde{G}$ grows *exponentially* in k, because we add a separate noise term following each multiplication by an $\mathbf{S}_i$, which gets enlarged when multiplied by all the later $\mathbf{S}_i$. So in order to ensure that $\lfloor \tilde{G} \rceil_p = \lfloor G \rceil_p$ on all queries, our LWE-based proof needs both the modulus q and inverse error rate $1/\alpha$ to exceed $n^{\Omega(k)}$. In terms of efficiency and security, this compares rather unfavorably with the quasipolynomial $n^{O(\lg k)}$ bound in the proof for our tree-based construction, though on the positive side, the direct degree-k construction has better circuit depth. However, just as with construction (3) it is unclear whether such strong assumptions and large parameters are actually *necessary* for security, or whether the matrices $\mathbf{S}_i$ really need to be short.

In particular, it would be nice if the function in (4) were secure if the $\mathbf{S}_i$ matrices were *uniformly random* over $\mathbb{Z}_q^{n \times n}$, because we could then recursively compose the function in a k-ary tree to rapidly extend its input length.[2] It would be even better to have a security proof for a smaller modulus q and inverse error rate $1/\alpha$, ideally both polynomial in n even for large k. While we have been unable to find such a security proof under standard LWE, we do give a very tight proof under a new, interactive "*related samples*" LWE/LWR assumption. Roughly speaking, the assumption says that LWE/LWR remains hard even when the sampled $\mathbf{a}_i$ vectors are related by adversarially chosen subset-products of up to k given random matrices (drawn from some known distribution). This provides some evidence that the function may indeed be secure for appropriately distributed $\mathbf{S}_i$, small modulus q, and large k. For further discussion, see Section 1.2, and for full details see the full version of the paper [7].

PRFs via the GGM construction. The above constructions aim to minimize the depth of the circuit evaluating the PRF. However, if parallel complexity is not a concern, and one wishes to minimize the total amount of work per PRF evaluation (or the seed length), then the original GGM construction with an LWR-based pseudorandom generator may turn out to be even more efficient in practice. We elaborate in the full version [7].

1.2 Discussion and Open Questions

The quasipolynomial $n^{O(\log k)}$ or exponential $n^{O(k)}$ moduli and inverse error rates used in our LWE-based security proofs are comparable to those used in recent fully homomorphic encryption (FHE) schemes (e.g., [14, 38, 12, 11, 10]), hierarchical identity-based encryption (HIBE) schemes (e.g., [13, 1, 2]), and other lattice-based constructions. However, there appears to be a major difference between our use of such strong assumptions, and that of schemes such as FHE/HIBE in the public-key setting. Constructions of the latter systems actually reveal LWE samples having very small error rates (which are needed to ensure correctness of decryption) to the attacker, and the attacker can break the cryptosystems by solving those instances. Therefore, the underlying assumptions and the true security of the schemes are essentially equivalent. In contrast, our PRF uses (small) errors *only as part of a thought experiment* in the security proof, not for any purpose in the operation of the function itself. This leaves open the possibility that our functions (or slight variants) remain secure even for much larger input lengths and smaller moduli than our proofs require. We conjecture that this is the case, even though we have not yet found security proofs (under standard assumptions) for these more efficient parameters. Certainly, determining whether there are effective cryptanalytic attacks is a very interesting and important research direction.

Note that in our construction (4), if we draw the secret key components from the uniform (or error) distribution and allow k to be too large relative to q,

---

[2] Note that we can always compose the degree-k function with our degree-2 synthesizers from above, but this would only yield a tree with 2-ary internal nodes.

then the function can become insecure via a simple attack (and our new "interactive" LWR assumption, which yields a tight security proof, becomes false). This is easiest to see for the ring-based function: representing each $s_i \in R_q$ by its vector of "Fourier coefficients" over $\mathbb{Z}_q^n$, each coefficient is 0 with probability about $1/q$ (depending on the precise distribution of s_i). Therefore, with noticeable probability the product of $k = O(q \log n)$ random s_i will have all-0 Fourier coefficients, i.e., will be $0 \in R_q$. In this case our function will return zero on the all-1s input, in violation of the PRF requirement. (A similar but more complicated analysis can also be applied to the matrix-based function.) Of course, an obvious countermeasure is just to restrict the secret key components to be *invertible*; to our knowledge, this does not appear to have any drawback in terms of security. In fact, it is possible to show that the decision-(ring-)LWE problem remains hard when the secret is restricted to be invertible (and otherwise drawn from the uniform or error distribution), and this fact may be useful in further analysis of the function with more efficient parameters.

In summary, our work raises several interesting concrete questions, including:

- Is $\mathsf{LWR}_{n,q,p}$ really exponentially hard for $p = \mathrm{poly}(n)$ and sufficiently large integer $q/p = \mathrm{poly}(n)$? Are there stronger worst-case hardness guarantees than our current proof based on LWE?
- Is there a security proof for construction (4) (with $k = \omega(1)$) for $\mathrm{poly}(n)$-bounded moduli and inverse error rates, under a non-interactive assumption?
- In construction (4), is there a security proof (under a non-interactive assumption) for uniformly random $\mathbf{S}_i$? Is there any provable security advantage to using *invertible* $\mathbf{S}_i$?
- Is there an efficient, low-depth PRF family based on the conjectured average-case hardness of the *subset-sum* problem?
- Our derandomization technique and LWR problem require working with moduli q greater than 2. Is there an efficient, parallel PRF family based on the learning parity with noise (LPN) problem?

1.3 Other Related Work

In a companion paper [6], we have defined and implemented practically efficient variants of our functions, using rounding over the ring $\mathbb{Z}_N$ of integers modulo large powers-of-2 N. The functions have throughput and security levels that appear comparable with (or even exceed) those of ASE.

Most closely related to the techniques in this work are two very recent results of Brakerski and Vaikuntanathan [11] and a follow-up work of Brakerski, Gentry, and Vaikuntanathan [10] on fully homomorphic encryption from LWE. In particular, the former work includes a "modulus reduction" technique for LWE-based cryptosystems, which maps a large-modulus ciphertext to a small-modulus one; this induces a shallower decryption circuit and allows the system to be "bootstrapped" into a fully homomorphic scheme using the techniques of [14]. The modulus-reduction technique involves a rounding operation much like the one we use to derandomize LWE; while they use it on ciphertexts that are already "noisy," we apply it to noise-free LWE samples. Our discovery of the rounding/derandomization technique in the PRF context was independent of [11]. In fact, the first PRF and security proof we found were for the direct degree-k construction defined in (4), not the synthesizer-based construction in (3). As another point of comparison, the "somewhat homomorphic" cryptosystem from [11] that supports degree-k operations (along with all prior ones, e.g., [14, 38]) involves an inverse error rate of $n^{O(k)}$, much like the LWE-based proof for our degree-k synthesizer.

Building on the modulus reduction technique of [11], Brakerski *et al.* [10] showed that homomorphic cryptosystems can support certain degree-k functions using a much smaller modulus and inverse error rate of $n^{O(\log k)}$. The essential idea is to interleave the homomorphic operations with several "small" modulus-reduction steps in a tree-like fashion, rather than performing all the homomorphic operations followed by one "huge" modulus reduction. This very closely parallels the difference between our direct degree-k synthesizer and the Naor-Reingold-like [28] composed synthesizer defined in (3). Indeed, after we found construction (4), the result of [10] inspired our search for a PRF having similar tree-like structure and quasipolynomial error rates. Given our degree-2 synthesizer, the solution turned out to largely be laid out in the work of [28]. We find it very interesting that the same quantitative phenomena arise in two seemingly disparate settings (PRFs and FHE).

2   Preliminaries

For a probability distribution X over a domain D, let X^n denote its n-fold product distribution over D^n. The uniform distribution over a finite domain D is denoted by $U(D)$. The discrete Gaussian probability distribution over $\mathbb{Z}$ with parameter $r > 0$, denoted $D_{\mathbb{Z},r}$, assigns probability proportional to $\exp(-\pi x^2/r^2)$ to each $x \in \mathbb{Z}$. It is possible to efficiently sample from this distribution (up to negl(n) statistical distance) via rejection [15].

For any integer modulus $q \geq 2$, $\mathbb{Z}_q$ denotes the quotient ring of integers modulo q. We define a 'rounding' function $\lfloor \cdot \rceil_p \colon \mathbb{Z}_q \to \mathbb{Z}_p$, where $q \geq p \geq 2$ will be apparent from the context, as

$$\lfloor x \rceil_p = \lfloor (p/q) \cdot \bar{x} \rceil \bmod p, \tag{5}$$

where $\bar{x} \in \mathbb{Z}$ is any integer congruent to $x \bmod q$. We extend $\lfloor \cdot \rceil_p$ component-wise to vectors and matrices over $\mathbb{Z}_q$, and coefficient-wise (with respect to the "power basis") to the quotient ring R_q defined in the next subsection. Note that we can use any other common rounding method, like the floor $\lfloor \cdot \rfloor$, or ceiling $\lceil \cdot \rceil$ functions, in Equation 5 above, with only minor changes to our proofs. In implementations, it may be advantageous to use the floor function $\lfloor \cdot \rfloor$ when q and p are both powers of some common base b (e.g., 2). In this setting, computing $\lfloor \cdot \rceil_p$ is equivalent to dropping the least-significant digit(s) in base b.

Learning With Errors. We recall the learning with errors (LWE) problem due to Regev [35] and its ring analogue by Lyubashevsky, Peikert, and Regev [25]. For positive integer dimension n (the security parameter) and modulus $q \geq 2$, a probability distribution χ over $\mathbb{Z}$, and a vector $\mathbf{s} \in \mathbb{Z}_q^n$, define the LWE distribution $A_{\mathbf{s},\chi}$ to be the distribution over $\mathbb{Z}_q^n \times \mathbb{Z}_q$ obtained by choosing a vector $\mathbf{a} \leftarrow \mathbb{Z}_q^n$ uniformly at random, an error term $e \leftarrow \chi$, and outputting $(\mathbf{a}, b = \langle \mathbf{a}, \mathbf{s} \rangle + e \bmod q)$. We use the following "normal form" of the *decision-*LWE$_{n,q,\chi}$ problem, which is to distinguish (with advantage non-negligible in n) between any desired number $m = \text{poly}(n)$ of independent samples $(\mathbf{a}_i, b_i) \leftarrow A_{\mathbf{s},\chi}$ where $\mathbf{s} \leftarrow \chi^n \bmod q$ is chosen from the (folded) error distribution, and the same number of samples from the uniform distribution $U(\mathbb{Z}_q^n \times \mathbb{Z}_q)$. This form of the problem is as hard as the one where $\mathbf{s} \in \mathbb{Z}_q^n$ is chosen uniformly at random [4].

We extend the LWE distribution to $w \geq 1$ secrets, defining $A_{\mathbf{S},\chi}$ for $\mathbf{S} \in \mathbb{Z}_q^{n \times w}$ to be the distribution obtained by choosing $\mathbf{a} \leftarrow \mathbb{Z}_q^n$, an error vector $\mathbf{e}^t \leftarrow \chi^w$, and outputting $(\mathbf{a}, \mathbf{b}^t = \mathbf{a}^t \mathbf{S} + \mathbf{e}^t \bmod q)$. By a standard hybrid argument, distinguishing such samples (for $\mathbf{S} \leftarrow \chi^{n \times w}$) from uniformly random is as hard as decision-LWE$_{n,q,\chi}$, for any $w = \text{poly}(n)$. It is often convenient to group many (say, m) sample pairs together in matrices. This allows us to express the LWE problem as: distinguish any desired number of pairs $(\mathbf{A}^t, \mathbf{B}^t = \mathbf{A}^t \mathbf{S} + \mathbf{E} \bmod q) \in \mathbb{Z}_q^{m \times n} \times \mathbb{Z}_q^{m \times w}$, for the same $\mathbf{S}$, from uniformly random.

For certain moduli q and (discrete) Gaussian error distributions χ, the decision-LWE problem is as hard as the search problem, where the goal is to find $\mathbf{s}$ given samples from $A_{\mathbf{s},\chi}$ (see, e.g., [35, 31, 4, 26], and [27] for the mildest known requirements on q, which include the case where q is a power of 2). In turn, for $\chi = D_{\mathbb{Z},r}$ with $r = \alpha q \geq 2\sqrt{n}$, the search problem is as hard as approximating worst-case lattice problems to within $\tilde{O}(n/\alpha)$ factors; see [35, 31] for precise statements.[3]

Ring-LWE. For simplicity of exposition, we use the following special case of the ring-LWE problem. (Our results can be extended to the more general form defined in [25].) Throughout the paper we let R denote the cyclotomic polynomial ring $R = \mathbb{Z}[z]/(z^n + 1)$ for n a power of 2. (Equivalently, R is the ring of integers $\mathbb{Z}[\omega]$ for $\omega = \exp(\pi i/n)$.) For any integer modulus q, define the quotient ring $R_q = R/qR$. An element of R can be represented as a polynomial (in z) of degree less than n having integer coefficients; in other words, the "power basis" $\{1, z, \ldots, z^{n-1}\}$ is a $\mathbb{Z}$-basis for R. Similarly, it is a $\mathbb{Z}_q$-basis for R_q.

For a modulus q, a probability distribution χ over R, and an element $s \in R_q$, the ring-LWE (RLWE) distribution $A_{s,\chi}$ is the distribution over $R_q \times R_q$ obtained by choosing $a \in R_q$ uniformly at random, an error term $x \leftarrow \chi$, and outputting $(a, b = a \cdot s + x \bmod qR)$. The normal form of the decision-RLWE$_{R,q,\chi}$ problem is to distinguish (with non-negligible advantage) between any desired number $m = \text{poly}(n)$ of independent samples $(a_i, b_i) \leftarrow A_{s,\chi}$ where $s \leftarrow \chi \bmod q$, and

---

[3] It is important to note that the original hardness result of [35] for search-LWE is for a *continuous* Gaussian error distribution, which when rounded naïvely to the nearest integer does not produce a true discrete Gaussian $D_{\mathbb{Z},r}$. Fortunately, a suitable randomized rounding method does so [32].

the same number of samples drawn from the uniform distribution $U(R_q \times R_q)$. We will use the error distribution χ over R where each coefficient (with respect to the power basis) is chosen independently from the discrete Gaussian $D_{\mathbb{Z},r}$ for some $r = \alpha q \geq \omega(\sqrt{n \log n})$.

For a prime modulus $q = 1 \bmod 2n$ and the error distribution χ described above, the decision-RLWE problem is as hard as the search problem, via a reduction that runs in time $q \cdot \mathrm{poly}(n)$ [25]. In turn, the search problem is as hard as quantumly approximating worst-case problems on ideal lattices.[4]

3   The Learning with Rounding Problem

We now define the "learning with rounding" (LWR) problem and its ring analogue, which are like "derandomized" versions of the usual (ring)-LWE problems, in that the error terms are chosen deterministically.

Definition 1. *Let $n \geq 1$ be the main security parameter and moduli $q \geq p \geq 2$ be integers.*

- *For a vector $\mathbf{s} \in \mathbb{Z}_q^n$, define the LWR distribution $L_{\mathbf{s}}$ to be the distribution over $\mathbb{Z}_q^n \times \mathbb{Z}_p$ obtained by choosing a vector $\mathbf{a} \leftarrow \mathbb{Z}_q^n$ uniformly at random, and outputting $(\mathbf{a}, b = \lfloor \langle \mathbf{a}, \mathbf{s} \rangle \rceil_p)$.*
- *For $s \in R_q$ (defined in Section 2), define the ring-LWR (RLWR) distribution L_s to be the distribution over $R_q \times R_p$ obtained by choosing $a \leftarrow R_q$ uniformly at random and outputting $(a, b = \lfloor a \cdot s \rceil_p)$.*

For a given distribution over $\mathbf{s} \in \mathbb{Z}_q^n$ (e.g., the uniform distribution), the decision-$\mathsf{LWR}_{n,q,p}$ problem is to distinguish (with advantage non-negligible in n) between any desired number of independent samples $(\mathbf{a}_i, b_i) \leftarrow L_{\mathbf{s}}$, and the same number of samples drawn uniformly and independently from $\mathbb{Z}_q^n \times \mathbb{Z}_p$. The decision-$\mathsf{RLWR}_{R,q,p}$ problem is defined analogously.

Note that we have defined LWR exclusively as a decision problem, as this is the only form of the problem we will need. By a simple (and by now standard) hybrid argument, the (ring-)LWR problem is no easier, up to a $\mathrm{poly}(n)$ factor in advantage, if we reuse each public $\mathbf{a}_i$ across several independent secrets. That is, distinguishing samples $(\mathbf{a}_i, \lfloor \langle \mathbf{a}_i, \mathbf{s}_1 \rangle \rceil_p, \ldots, \lfloor \langle \mathbf{a}_i, \mathbf{s}_\ell \rangle \rceil_p) \in \mathbb{Z}_q^n \times \mathbb{Z}_p^\ell$ from uniform, where each $\mathbf{s}_j \in \mathbb{Z}_q^n$ is chosen independently for any $\ell = \mathrm{poly}(n)$, is at least as hard as decision-LWR for a single secret $\mathbf{s}$. An analogous statement also holds for ring-LWR.

---

[4] More accurately, to prove that the search problem is hard for an a priori *unbounded* number of RLWE samples, the worst-case connection from [25] requires the error distribution's parameters to themselves be chosen at random from a certain distribution. Our constructions are easily modified to account for this subtlety, but for simplicity, we ignore this issue and assume hardness for a fixed, public error distribution.

3.1   Reduction from LWE

We now show that for appropriate parameters, decision-LWR is at least as hard as decision-LWE. We say that a probability distribution χ over $\mathbb{R}$ (more precisely, a family of distributions χ_n indexed by the security parameter n) is *B-bounded* (where $B = B(n)$ is a function of n) if $\Pr_{x \leftarrow \chi}[|x| > B] \leq \mathrm{negl}(n)$. Similarly, a distribution over the ring R is B-bounded if the marginal distribution of every coefficient (with respect to the power basis) of an $x \leftarrow \chi$ is B-bounded.

Theorem 1. *Let χ be any efficiently sampleable B-bounded distribution over $\mathbb{Z}$, and let $q \geq p \cdot B \cdot n^{\omega(1)}$. Then for any distribution over the secret $\mathbf{s} \in \mathbb{Z}_q^n$, solving decision-$\mathsf{LWR}_{n,q,p}$ is at least as hard as solving decision-$\mathsf{LWE}_{n,q,\chi}$ for the same distribution over $\mathbf{s}$. The same holds true for $\mathsf{RLWR}_{R,q,p}$ and $\mathsf{RLWE}_{R,q,\chi}$, for any B-bounded χ over R.*

We note that although our proof uses a super-polynomial $q = n^{\omega(1)}$, as long as $q/p \geq \sqrt{n}$ is an integer, the LWR problem appears to be exponentially hard (in n) for any $p = \mathrm{poly}(n)$, and super-polynomially hard for $p \leq 2^{n^{\epsilon}}$ for any $\epsilon < 1$, given the state of the art in noisy learning algorithms [8, 5] and lattice reduction algorithms [22, 37]. We also note that in our proof, we do not require the error terms drawn from χ in the LWE samples to be independent; we just need them all to have magnitude bounded by B with overwhelming probability.

Proof (Sketch, Theorem 1). We give a rough proof sketch for the LWR case; the one for RLWR proceeds essentially identically. For the full and detailed proof, we refer the reader to the full version of the paper. The main idea behind the reduction is simple: given pairs $(\mathbf{a}_i, b_i) \in \mathbb{Z}_q^n \times \mathbb{Z}_q$ which are distributed either according to an LWE distribution $A_{\mathbf{s},\chi}$ or are uniformly random, we translate them into the pairs $(\mathbf{a}_i, \lfloor b_i \rceil_p) \in \mathbb{Z}_q^n \times \mathbb{Z}_p$, which we show will be distributed according to the LWR distribution $L_{\mathbf{s}}$ (with overwhelming probability) or uniformly random, respectively.

4   Synthesizer-Based PRFs

We now describe the LWR-based synthesizer and our construction of a PRF from it. We first define a *pseudorandom synthesizer*, slightly modified from the definition proposed by Naor and Reingold [28].

Let $S : A \times A \to B$ be a function (where A and B are finite domains, which along with S are implicitly indexed by the security parameter n) and let $X = (x_1, \ldots, x_k) \in A^k$ and $Y = (y_1, \ldots, y_\ell) \in A^\ell$ be two sequences of inputs. Then $\mathbf{C}_S(X, Y) \in B^{k \times \ell}$ is defined to be the matrix with $S(x_i, y_j)$ as its (i, j)th entry. (Here $\mathbf{C}$ stands for combinations.)

Definition 2 (Pseudorandom Synthesizer). *We say that a function $S : A \times A \to B$ is a* pseudorandom synthesizer *if it is polynomial-time computable, and if for every $\mathrm{poly}(n)$-bounded $k = k(n)$, $\ell = \ell(n)$,*

$$\mathbf{C}_S\big(U(A^k), U(A^\ell)\big) \stackrel{c}{\approx} U\big(B^{k \times \ell}\big).$$

That is, the matrix $\mathbf{C}_S(X, Y)$ for uniform and independent $X \leftarrow A^k$, $Y \leftarrow A^\ell$ is computationally indistinguishable from a uniformly random k-by-ℓ matrix over B.

4.1   Synthesizer Constructions

We now describe synthesizers whose security is based on the (ring-)LWR problem.

Definition 3 ((Ring-)LWR Synthesizer). *For moduli $q > p \geq 2$, the* LWR *synthesizer is the function* $S_{n,q,p} \colon \mathbb{Z}_q^n \times \mathbb{Z}_q^n \to \mathbb{Z}_p$ *defined as*

$$S_{n,q,p}(\mathbf{x}, \mathbf{y}) = \lfloor \langle \mathbf{x}, \mathbf{y} \rangle \rceil_p.$$

The RLWR *synthesizer is the function* $S_{R,q,p} \colon R_q \times R_q \to R_p$ *defined as*

$$S_{R,q,p}(x, y) = \lfloor x \cdot y \rceil_p.$$

Theorem 2. *Assuming the hardness of decision-*LWR$_{n,q,p}$ *(respectively, decision-*RLWR$_{R,q,p}$*) for a uniformly random secret, the function $S_{n,q,p}$ (respectively, $S_{R,q,p}$) given in Definition 3 above is a pseudorandom synthesizer.*

It follows generically from this theorem that the function $T_{n,q,p} \colon \mathbb{Z}_q^{n \times n} \times \mathbb{Z}_q^{n \times n} \to \mathbb{Z}_p^{n \times n}$, defined as $T_{n,q,p}(\mathbf{X}, \mathbf{Y}) = \lfloor \mathbf{X} \cdot \mathbf{Y} \rceil_p$, is also a pseudorandom synthesizer, since by the definition of matrix multiplication, we only incur a factor of n increase in the length of the input sequences. This is the synthesizer that we use below in the construction of a PRF.

4.2   The PRF Construction

Definition 4 ((Ring-)LWR PRF). *For parameters $n \in \mathbb{N}$, input length $k = 2^d \geq 1$, and moduli $q_d \geq q_{d-1} \geq \ldots \geq q_0 \geq 2$, the* LWR *family $\mathcal{F}^{(j)}$ for $0 \leq j \leq d$ is defined inductively to consist of functions from $\{0,1\}^{2^j}$ to $\mathbb{Z}_{q_{d-j}}^{n \times n}$. We define $\mathcal{F} = \mathcal{F}^{(d)}$.*

- *For $j = 0$, a function $F \in \mathcal{F}^{(0)}$ is indexed by $\mathbf{S}_b \in \mathbb{Z}_{q_d}^{n \times n}$ for $b \in \{0,1\}$, and is defined simply as $F_{\{\mathbf{S}_b\}}(x) = \mathbf{S}_x$. We endow $\mathcal{F}^{(0)}$ with the distribution where the $\mathbf{S}_b$ are uniform and independent.*
- *For $j \geq 1$, a function $F \in \mathcal{F}^{(j)}$ is indexed by some $F_0, F_1 \in \mathcal{F}^{(j-1)}$, and is defined as*

$$F_{F_0, F_1}(x_0, x_1) = T^{(j)}\big(F_0(x_0), F_1(x_1)\big)$$

where $|x_0| = |x_1| = 2^{j-1}$ and $T^{(j)} = T_{n, q_{d-j+1}, q_{d-j}}$ is the appropriate synthesizer. We endow $\mathcal{F}^{(j)}$ with the distribution where F_0 and F_1 are chosen independently from $\mathcal{F}^{(j-1)}$.

*The ring-*LWR *family $\mathcal{RF}^{(j)}$ is defined similarly to consist of functions from $\{0,1\}^{2^j}$ to $R_{q_{d-j}}$, where in the base case ($j = 0$) we replace each $\mathbf{S}_b$ with a uniformly random $s_b \in R_{q_d}$, and in the inductive case ($j \geq 1$) we use the ring-*LWR *synthesizer $S^{(j)} = S_{R, q_{d-j+1}, q_{d-j}}$.*

We remark that the recursive LWR-based construction above does not have to use *square* matrices; any legal dimensions would be acceptable with no essential change to the security proof. Square matrices appear to give the best combination of seed size, computational efficiency, and input/output lengths.

4.3 Security

The security proof for our PRF hinges on the fact that the functions $T^{(j)} = T_{n,q_{d-j+1},q_{d-j}}$ are synthesizers for appropriate choices of the moduli. In fact, the proof is essentially identical to Naor and Reingold's [28] for their PRF construction from pseudorandom synthesizers; the only reason we cannot use their theorem exactly as stated is because they assume that the synthesizer output is exactly the same size as its two inputs, which is not quite the case with our synthesizer due to the modulus reduction. This is a minor detail that does not change the proof in any material way; it only limits the number of times we may compose the synthesizer, and hence the input length of the PRF. We thus refer the reader to the full version for the proof.

Theorem 3. *Assuming that $T^{(j)} = T_{n,q_{d-j+1},q_{d-j}}$ is a pseudorandom synthesizer for every $j \in [d]$, the LWR family $\mathcal{F}$ from Definition 4 is a pseudorandom function family.*

The same is also true for the ring-LWR family $\mathcal{RF}$, assuming that $S^{(j)} = S_{R,q_{d-j+1},q_{d-j}}$ is a pseudorandom synthesizer for every $j \in [d]$.

5 Direct PRF Constructions

Here we present another, potentially more efficient construction of a pseudorandom function family whose security is based on the intractibility of the LWE problem.

Definition 5 ((Ring-)LWE degree-k PRF). *For parameters $n \in \mathbb{N}$, moduli $q \geq p \geq 2$, positive integer $m = \text{poly}(n)$, and input length $k \geq 1$, the family $\mathcal{F}$ consists of functions from $\{0,1\}^k$ to $\mathbb{Z}_p^{m \times n}$. A function $F \in \mathcal{F}$ is indexed by some $\mathbf{A} \in \mathbb{Z}_q^{n \times m}$ and $\mathbf{S}_i \in \mathbb{Z}^{n \times n}$ for each $i \in [k]$, and is defined as*

$$F(x) = F_{\mathbf{A},\{\mathbf{S}_i\}}(x_1 \cdots x_k) := \left\lfloor \mathbf{A}^t \cdot \prod_{i=1}^{k} \mathbf{S}_i^{x_i} \right\rceil_p . \tag{6}$$

We endow $\mathcal{F}$ with the distribution where $\mathbf{A}$ is chosen uniformly at random, and below we consider a number of natural distributions for the $\mathbf{S}_i$.

The ring-based family $\mathcal{RF}$ is defined similarly to consist of functions from $\{0,1\}^k$ to R_p, where we replace $\mathbf{A}$ with uniformly random $a \in R_q$ and each $\mathbf{S}_i$ with some $s_i \in R$.

5.1 Efficiency

Consider a function $F \in \mathcal{F}$ as in Definition 5. Using ideas from [36], we see that both binary matrix product and rounding can be implemented with simple depth-2 arithmetic circuits, and hence in TC^0, so at worst F can be computed in TC^1 by computing the subset product in a tree-like fashion, followed by a final rounding step.

The ring variant of Construction 6 appears to be more efficient to evaluate, by storing the ring elements in the discrete Fourier transform or "Chinese remainder" representation modulo q (see, e.g., [24, 25]), so that multiplication of two ring elements just corresponds to a coordinate-wise product of their vectors. Then to evaluate the function, one would just compute a subset-product of the appropriate vectors, then interpolate the result to the power-basis representation, using essentially an n-dimensional Fast Fourier Transform over $\mathbb{Z}_q$, in order to perform the rounding operation. In terms of theoretical depth, the multi-product of vectors can be performed in TC^0, as can the Fast Fourier Transform and rounding steps [36]. This implies that the entire function can be computed in TC^0, matching (asymptotically) the shallowest known PRFs based on the DDH and factoring problems [29, 30].

5.2 Security under LWE

Theorem 4. *Let $\chi = D_{\mathbb{Z},r}$ for some $r > 0$, and let $q \geq p \cdot k(Cr\sqrt{n})^k \cdot n^{\omega(1)}$ for a suitable universal constant C. Endow the family $\mathcal{F}$ from Definition 4 with the distribution where each $\mathbf{S}_i$ is drawn independently from $\chi^{n \times n}$. Then assuming the hardness of decision-$\mathsf{LWE}_{n,q,\chi}$, the family $\mathcal{F}$ is pseudorandom.*

An analogous theorem holds for the ring-based family $\mathcal{RF}$, under decision-RLWE.

Theorem 5. *Let χ be the distribution over the ring R where each coefficient (with respect to the power basis) is chosen independently from $D_{\mathbb{Z},r}$ for some $r > 0$, and let $q \geq p \cdot k(r\sqrt{n} \cdot \omega(\sqrt{\log n}))^k \cdot n^{\omega(1)}$. Endow the family $\mathcal{RF}$ from Definition 4 with the distribution where each s_i is drawn independently from χ. Then assuming the hardness of decision-$\mathsf{RLWE}_{n,q,\chi}$, the family $\mathcal{RF}$ is pseudorandom.*

Proof (Sketch, Theorem 4). To aid the proof, it helps to define a family $\mathcal{G}$ of functions $G \colon \{0,1\}^k \to \mathbb{Z}_q^{n \times n}$, which are simply the unrounded counterparts of the functions in $\mathcal{F}$. That is, for $\mathbf{A} \in \mathbb{Z}_q^{n \times m}$ and $\mathbf{S}_i \in \mathbb{Z}^{n \times n}$ for $i \in [k]$, we define $G_{\mathbf{A},\{\mathbf{S}_i\}}(x_1 \cdots x_k) := \mathbf{A}^t \cdot \prod_{i=1}^{k} \mathbf{S}_i^{x_i}$. We endow $\mathcal{G}$ with the same distribution over $\mathbf{A}$ and the $\mathbf{S}_i$ as $\mathcal{F}$ has.

We proceed via a sequence of games, much like in the proof of Theorem 1. First as a "thought experiment" we define a new family $\tilde{\mathcal{G}}$ of functions from $\{0,1\}^k$ to $\mathbb{Z}_q^{m \times n}$. This family is a counterpart to $\mathcal{G}$, but with two important differences: it is a PRF family *without* any rounding (and hence, with rounding as well), but each function in the family has an exponentially large key. Alternatively, one may think of the functions in $\tilde{\mathcal{G}}$ as *randomized* functions with small keys. Then we show that with overwhelming probability, the rounding of $\tilde{G} \leftarrow \tilde{\mathcal{G}}$ agrees with the rounding of the corresponding $G \in \mathcal{G}$ on all the attacker's queries, because the outputs of the two functions are relatively close. It follows that the rounding of $G \leftarrow \mathcal{G}$ (i.e., $F \leftarrow \mathcal{F}$) cannot be distinguished from a uniformly random function, as desired. We again refer the reader to the full version of the paper for the formal proof.

References

[1] Agrawal, S., Boneh, D., Boyen, X.: Efficient Lattice (H)IBE in the Standard Model. In: Gilbert, H. (ed.) EUROCRYPT 2010. LNCS, vol. 6110, pp. 553–572. Springer, Heidelberg (2010)

[2] Agrawal, S., Boneh, D., Boyen, X.: Lattice Basis Delegation in Fixed Dimension and Shorter-Ciphertext Hierarchical IBE. In: Rabin, T. (ed.) CRYPTO 2010. LNCS, vol. 6223, pp. 98–115. Springer, Heidelberg (2010)

[3] Ajtai, M.: Generating hard instances of lattice problems. Quaderni di Matematica 13, 1–32 (2004); Preliminary version in STOC 1996

[4] Applebaum, B., Cash, D., Peikert, C., Sahai, A.: Fast Cryptographic Primitives and Circular-Secure Encryption Based on Hard Learning Problems. In: Halevi, S. (ed.) CRYPTO 2009. LNCS, vol. 5677, pp. 595–618. Springer, Heidelberg (2009)

[5] Arora, S., Ge, R.: New Algorithms for Learning in Presence of Errors. In: Aceto, L., Henzinger, M., Sgall, J. (eds.) ICALP 2011, Part I. LNCS, vol. 6755, pp. 403–415. Springer, Heidelberg (2011)

[6] Banerjee, A., Ben-Zvi, N., Peikert, C., Rosen, A.: SPRINT: Efficient pseudorandomness via rounded integer products (2011) (manuscript)

[7] Banerjee, A., Peikert, C., Rosen, A.: Pseudorandom functions and lattices. Cryptology ePrint Archive, Report 2011/401 (2011), http://eprint.iacr.org/

[8] Blum, A., Kalai, A., Wasserman, H.: Noise-tolerant learning, the parity problem, and the statistical query model. J. ACM 50(4), 506–519 (2003)

[9] Boneh, D., Montgomery, H.W., Raghunathan, A.: Algebraic pseudorandom functions with improved efficiency from the augmented cascade. In: ACM Conference on Computer and Communications Security, pp. 131–140 (2010)

[10] Brakerski, Z., Gentry, C., Vaikuntanathan, V.: Fully homomorphic encryption without bootstrapping. Cryptology ePrint Archive, Report 2011/277 (2011), http://eprint.iacr.org/

[11] Brakerski, Z., Vaikuntanathan, V.: Efficient fully homomorphic encryption from (standard) LWE. In: FOCS, pp. 97–106 (2011)

[12] Brakerski, Z., Vaikuntanathan, V.: Fully Homomorphic Encryption from Ring-LWE and Security for Key Dependent Messages. In: Rogaway, P. (ed.) CRYPTO 2011. LNCS, vol. 6841, pp. 505–524. Springer, Heidelberg (2011)

[13] Cash, D., Hofheinz, D., Kiltz, E., Peikert, C.: Bonsai Trees, or How to Delegate a Lattice Basis. In: Gilbert, H. (ed.) EUROCRYPT 2010. LNCS, vol. 6110, pp. 523–552. Springer, Heidelberg (2010)

[14] Gentry, C.: Fully homomorphic encryption using ideal lattices. In: STOC, pp. 169–178 (2009)

[15] Gentry, C., Peikert, C., Vaikuntanathan, V.: Trapdoors for hard lattices and new cryptographic constructions. In: STOC, pp. 197–206 (2008)

[16] Goldreich, O., Goldwasser, S., Micali, S.: How to construct random functions. J. ACM 33(4), 792–807 (1986); Preliminary version in FOCS 1984

[17] Goldwasser, S., Kalai, Y.T., Peikert, C., Vaikuntanathan, V.: Robustness of the learning with errors assumption. In: ICS, pp. 230–240 (2010)

[18] Hopper, N.J., Blum, M.: Secure Human Identification Protocols. In: Boyd, C. (ed.) ASIACRYPT 2001. LNCS, vol. 2248, pp. 52–66. Springer, Heidelberg (2001)

[19] Juels, A., Weis, S.A.: Authenticating Pervasive Devices with Human Protocols. In: Shoup, V. (ed.) CRYPTO 2005. LNCS, vol. 3621, pp. 293–308. Springer, Heidelberg (2005)

[20] Katz, J., Shin, J.S., Smith, A.: Parallel and concurrent security of the HB and HB^+ protocols. J. Cryptology 23(3), 402–421 (2010); Preliminary version in Eurocrypt 2006

[21] Kiltz, E., Pietrzak, K., Cash, D., Jain, A., Venturi, D.: Efficient Authentication from Hard Learning Problems. In: Paterson, K.G. (ed.) EUROCRYPT 2011. LNCS, vol. 6632, pp. 7–26. Springer, Heidelberg (2011)

[22] Lenstra, A.K., Lenstra Jr., H.W., Lovász, L.: Factoring polynomials with rational coefficients. Mathematische Annalen 261(4), 515–534 (1982)

[23] Lewko, A.B., Waters, B.: Efficient pseudorandom functions from the decisional linear assumption and weaker variants. In: ACM Conference on Computer and Communications Security, pp. 112–120 (2009)

[24] Lyubashevsky, V., Micciancio, D., Peikert, C., Rosen, A.: SWIFFT: A Modest Proposal for FFT Hashing. In: Nyberg, K. (ed.) FSE 2008. LNCS, vol. 5086, pp. 54–72. Springer, Heidelberg (2008)

[25] Lyubashevsky, V., Peikert, C., Regev, O.: On Ideal Lattices and Learning with Errors over Rings. In: Gilbert, H. (ed.) EUROCRYPT 2010. LNCS, vol. 6110, pp. 1–23. Springer, Heidelberg (2010)

[26] Micciancio, D., Mol, P.: Pseudorandom Knapsacks and the Sample Complexity of LWE Search-to-Decision Reductions. In: Rogaway, P. (ed.) CRYPTO 2011. LNCS, vol. 6841, pp. 465–484. Springer, Heidelberg (2011)

[27] Micciancio, D., Peikert, C.: Trapdoors for Lattices: Simpler, Tighter, Faster, Smaller. In: Pointcheval, D., Johansson, T. (eds.) EUROCRYPT 2012. LNCS, vol. 7237, pp. 700–718. Springer, Heidelberg (2012)

[28] Naor, M., Reingold, O.: Synthesizers and their application to the parallel construction of pseudo-random functions. J. Comput. Syst. Sci. 58(2), 336–375 (1999); Preliminary version in FOCS 1995

[29] Naor, M., Reingold, O.: Number-theoretic constructions of efficient pseudorandom functions. J. ACM 51(2), 231–262 (2004); Preliminary version in FOCS 1997

[30] Naor, M., Reingold, O., Rosen, A.: Pseudorandom functions and factoring. SIAM J. Comput. 31(5), 1383–1404 (2002); Preliminary version in STOC 2000

[31] Peikert, C.: Public-key cryptosystems from the worst-case shortest vector problem. In: STOC, pp. 333–342 (2009)

[32] Peikert, C.: An Efficient and Parallel Gaussian Sampler for Lattices. In: Rabin, T. (ed.) CRYPTO 2010. LNCS, vol. 6223, pp. 80–97. Springer, Heidelberg (2010)

[33] Peikert, C., Waters, B.: Lossy trapdoor functions and their applications. In: STOC, pp. 187–196 (2008)

[34] Pietrzak, K.: Subspace LWE (2010) (manuscript), http://homepages.cwi.nl/~pietrzak/publications/SLWE.pdf (Last retrieved from June 28, 2011)

[35] Regev, O.: On lattices, learning with errors, random linear codes, and cryptography. J. ACM 56(6), 1–40 (2009); Preliminary version in STOC 2005
[36] Reif, J.H., Tate, S.R.: On threshold circuits and polynomial computation. SIAM J. Comput. 21(5), 896–908 (1992)
[37] Schnorr, C.-P.: A hierarchy of polynomial time lattice basis reduction algorithms. Theor. Comput. Sci. 53, 201–224 (1987)
[38] van Dijk, M., Gentry, C., Halevi, S., Vaikuntanathan, V.: Fully Homomorphic Encryption over the Integers. In: Gilbert, H. (ed.) EUROCRYPT 2010. LNCS, vol. 6110, pp. 24–43. Springer, Heidelberg (2010)

Lattice Signatures without Trapdoors

Vadim Lyubashevsky[*]

INRIA / École Normale Supérieure

Abstract. We provide an alternative method for constructing lattice-based digital signatures which does not use the "hash-and-sign" methodology of Gentry, Peikert, and Vaikuntanathan (STOC 2008). Our resulting signature scheme is secure, in the random oracle model, based on the worst-case hardness of the $\tilde{O}(n^{1.5})$-SIVP problem in general lattices. The secret key, public key, and the signature size of our scheme are smaller than in all previous instantiations of the hash-and-sign signature, and our signing algorithm is also quite simple, requiring just a few matrix-vector multiplications and rejection samplings. We then also show that by slightly changing the parameters, one can get even more efficient signatures that are based on the hardness of the Learning With Errors problem. Our construction naturally transfers to the ring setting, where the size of the public and secret keys can be significantly shrunk, which results in the most practical to-date provably secure signature scheme based on lattices.

1 Introduction

The versatility of lattice-based cryptography has elevated it to the status of a promising potential alternative to cryptography based on standard security assumptions such as factoring and discrete log. But before lattices can become a viable replacement for number-theoretic schemes, it is crucial to have efficient lattice-based constructions of the most ubiquitous cryptographic primitives in practical applications, which are arguably encryption schemes and digital signatures.

On the encryption front, lattice-based schemes have been making a lot of progress with recent provably-secure schemes [39,28,24,41] being almost as practical as (and actually looking quite similar to) the deployed NTRU [20] encryption scheme, which in turn has many advantages over number theory-based schemes. Lattice-based signatures, on the other hand, have been a different story. An early attempt at lattice-based signatures was the GGH scheme [18] but it was almost immediately shown to be weaker than expected [33], and eventually completely broken [34]. The NTRU signature scheme had an even more more tumultuous history since its introduction in 2001 [21], with attacks [17] being followed by fixes [19], until its basic version was also completely broken by Nguyen and Regev [34].

---

[*] Work supported in part by the European Research Council.

D. Pointcheval and T. Johansson (Eds.): EUROCRYPT 2012, LNCS 7237, pp. 738–755, 2012.

Provably secure lattice-based signature schemes were finally constructed in 2008, when Gentry, Peikert, and Vaikuntanathan [16] constructed a "hash-and-sign" signature scheme based on the hardness of worst-case lattice problems and Lyubashevsky and Micciancio [27] constructed a one-time signature based on the hardness of worst-case ideal lattice problems. The hash-and-sign signatures were rather inefficient (with signatures being megabytes long) and the one-time signature, while being relatively short, still required Merkle trees to become a full-fledged signature. Building on [27], Lyubashevsky proposed a digital signature, using the Fiat-Shamir framework [12] based on the hardness of ideal lattice problems [26]. This latter scheme has signature lengths on the order of 60000 bits for reasonable security parameters, and while closer to being practical, it is still not as small as one would like. Subsequently, lattice-based signature schemes without random oracles were also constructed [11,8], but they are all much less efficient in practice than their random oracle-using counterparts.

1.1 Related Work and Our Results

A common thread running through constructions of digital signatures in the random oracle model, whether using the hash-and-sign or the Fiat-Shamir technique [12], is to force the distribution of the signature to be statistically independent of the secret key. If this property is achieved, then by programming the random oracle, one can hope to produce the valid signatures requested by the potential forger in the security reduction, without knowing the secret key. Then, when the forger produces a signature of a new message, it can be used to solve the underlying hard problem. In the case of lattices, the underlying hard problem is usually the Small Integer Solution (SIS) problem in which one is given a matrix $\mathbf{A}$ and is asked to find a *small* vector $\mathbf{v}$ such that $\mathbf{Av} = 0 \bmod q$. The length of $\mathbf{v}$ is very close to the length of signatures in the scheme, and thus the challenge for improving lattice-based signatures based on SIS is to reduce the norm of the signatures produced by the signing algorithm.

In lattice-based hash-and-sign signatures [16], every signer has a personal uniformly random public matrix $\mathbf{A} \in \mathbb{Z}_q^{n \times m}$ and an associated secret "trapdoor" $\mathbf{S} \in \mathbb{Z}_q^{m \times m}$ with small coefficients such that $\mathbf{AS} = 0 \bmod q$. To sign a message μ, the signer uses his secret key $\mathbf{S}$ to produce a short signature vector $\mathbf{z}$, whose distribution is independent of $\mathbf{S}$, such that $\mathbf{Az} = \mathrm{H}(\mu) \bmod q$, where H is a cryptographic hash function. Since the length of $\mathbf{z}$ roughly depends on the norms of the columns of $\mathbf{S}$, improving the hash-and-sign signature scheme involves coming up with better algorithms for generating the pairs $(\mathbf{A}, \mathbf{S})$ such that $\mathbf{S}$ has smaller dimensions and smaller coefficients. Using the original algorithm due to Ajtai [1], the signature scheme of [16] produced signatures of norm $\tilde{O}(n^{1.5})$. A subsequent improvement of the key-generation algorithm by Alwen and Peikert [3] lowered the signature length to $\tilde{O}(n)$, and the very recent algorithm of Micciancio and Peikert [30] further reduces the constants (and removes some logarithmic factors) from the previous algorithms.

There has been much less progress in the direction of building lattice-based signature schemes using the Fiat-Shamir technique. In fact, the only such scheme[1] is the ring-based one of Lyubashevsky [26], in which the signature vectors are of norm $\tilde{O}(n^{1.5})$. The first contribution of this current work is adapting the ring-SIS based scheme from [26] to one based on the hardness of the regular SIS problem which results in signatures of the same $\tilde{O}(n^{1.5})$ length[2]. Our second contribution is analogous to what the works [3,36,30] did for hash-and-sign signatures – reduce the signature length to $\tilde{O}(n)$ (of course the issues that have to be dealt with are completely different). Our third contribution is showing that the parameters of our scheme can be set so that the resulting scheme produces much shorter signatures, but is now based on the hardness of the Learning With Errors (LWE) problem [39] or on the hardness of a low-density version of the SIS problem. All our results very naturally carry over to the ring setting, where the key bit-size is reduced by a factor of approximately n (some sample parameters are given in Figure 2).

Our signature scheme is also quite simple, requiring no pre-image sampling over arbitrary lattices. All we do is sample the Normal distribution over $\mathbb{Z}^m$, compute a vector-matrix product, do a random oracle query, compute another vector-matrix product (this time the vector is sparse), and rejection sample. In fact, in an online/offline setting where we can do pre-computations before being given the message to sign, the online phase simply consists of doing a few vector additions (since the matrix is being multiplied by a sparse vector) and rejection sampling.

1.2   Techniques

We now briefly sketch our signature scheme and describe the issues involved in lowering the size of the signature. The secret key is a matrix $\mathbf{S} \in \mathbb{Z}_q^{m \times k}$ with small coefficients, and the public key consists of the matrices $\mathbf{A} \in \mathbb{Z}_q^{n \times m}$ and $\mathbf{T} = \mathbf{AS} \bmod q$. The matrix $\mathbf{A}$ can be shared among all users, but the matrix $\mathbf{T}$ is individual. To sign a message, the signer first picks a vector $\mathbf{y} \in \mathbb{Z}_q^m$ according to some distribution D. Then he computes $\mathbf{c} \in \mathbb{Z}_q^k$ where $\mathbf{c} \leftarrow \mathrm{H}(\mathbf{Ay} \bmod q, \mu)$, and computes the potential signature vector $\mathbf{z} = \mathbf{Sc} + \mathbf{y}$ (there is no reduction modulo q in this step). The vector $\mathbf{z}$, along with $\mathbf{c}$, will then be output as the signature based on some criteria with the end goal being that the distribution of $(\mathbf{z}, \mathbf{c})$ should be independent of the secret key matrix $\mathbf{S}$.

---

[1] We mention that the lattice-based identification schemes of Lyubashevsky [25] and Kawachi et al. [23], while may be converted into signature schemes, are inherently inefficient because every round of the ID scheme has soundness error at least $1/2$.

[2] As a side note to this first result, we think that it is interesting to point out that the ring-structure, which seemed so native to [26] (and to [27]), turns out to not actually provide any additional functionality, with its purpose being only to shorten the key-sizes and make operations more efficient. This somewhat resembles the recent developments in constructions of fully-homomorphic encryption schemes, where the additional structure of ideal lattices was crucially used in earlier constructions [14,15,10], but was subsequently shown to be unnecessary [9,2].

Choosing when to output the pair $(\mathbf{z}, \mathbf{c})$ can be seen as a kind of *rejection sampling*. If f and g are probability distributions and $M \in \mathbb{R}$ is such that for all x, $f(x) \leq Mg(x)$, then if one samples elements z from g and outputs them with probability $f(z)/(Mg(z))$, the resulting distribution is exactly f, and the expected amount of time needed to output a sample is M.

Our goal, in the signature scheme above, is to come up with distributions f and D so that for all $\mathbf{x}$, two properties are satisfied: there is a small constant M such that $f(\mathbf{x}) \leq Mg(\mathbf{x})$, where g is the distribution generated by first picking $\mathbf{y}$ from D and adding it to $\mathbf{Sc}$ for some random $\mathbf{c}$; and the expected value of vectors distributed according to f (which is the length of the signature) is as small as possible. The idea in [26], when put into the above framework, was to choose $\mathbf{y}$ uniformly from an m-dimensional sphere[3] β_{r+v} of radius $r + v$, where r is some number and v is the maximum possible length of the vector $\mathbf{Sc}$, and only output $\mathbf{z}$ if it fell into a sphere β_r of radius r. It's not hard to check that if f is the uniform distribution over the sphere β_r, then by setting $M = vol(\beta_{r+v}/\beta_r) \approx (1 + v/r)^m$, the distribution of $\mathbf{z}$ is exactly f. But in order to keep M small, we need $r > mv = \tilde{\Theta}(m^{1.5}) = \tilde{\Theta}(n^{1.5})$, and so the vectors $\mathbf{z}$ have length $\tilde{O}(n^{1.5})$.

In our present work we show that we can do better by choosing f and D to be the m-dimensional Normal distribution with standard deviation $\sigma = \tilde{\Theta}(v) = \tilde{\Theta}(\sqrt{m})$, and only require that $f(\mathbf{x}) \leq Mg(\mathbf{x})$ for the $\mathbf{x}$ that are not too big. We can then show that M can be set to a constant, and the rejection sampling algorithm produces a distribution that is statistically close to the distribution of f. This means that the expected value of the length of the signature of $\mathbf{z}$ is $\sigma\sqrt{m} = \tilde{O}(m) = \tilde{O}(n)$. We prove the technical rejection sampling theorem in Section 3 and then prove the security of the above signature scheme based on the hardness of the SIS problem in Section 4.

Notice that the length of the signature is greatly affected by the parameter m, and lowering m, while leaving everything else the same would produce even shorter signatures. The danger of doing this is that the problem of recovering $\mathbf{S}$ when given $\mathbf{A}$ and $\mathbf{AS} \bmod q$ now becomes easier (and is no longer based on the SIS problem). The intuition is then to set all the parameters so that the hardness of recovering the secret key is equal, in practice, to the hardness of forging a signature. In Section 5 we explain how the parameters can be significantly lowered by making our scheme be based on the LWE problem instead of on SIS.

1.3   A Practical Comparison with Hash-and-Sign Signatures

On the theoretical side, both the scheme constructed in this paper and the hash-and-sign scheme that uses the trapdoor sampling algorithms of [3,36] are based on the hardness of finding a vector of length $\tilde{O}(n)$ in SIS instances, which by the worst-case to average-case reduction of Micciancio and Regev [31] is as hard as solving approximate SIVP with a factor of $\tilde{O}(n^{1.5})$ in all n-dimensional

---

[3] In [26], it was actually a box, but it does not make a difference for the analysis here.

lattices. On the practical side, however, the bit-length of our signature and keys (see Figure 2) are approximately two orders of magnitude smaller for the same security level (see [40] and also Figure 2 in [30]). This is mostly due to the constants that are hidden in the big-Oh notation of the trapdoor generation algorithms of [3] and [36].

As mentioned earlier, in a concurrent and independent work, Micciancio and Peikert greatly improved the constants, and in some cases even removed some logarithmic factors, in the trapdoor sampling algorithms [30]. While the proof techniques are completely different, there are some high-level similarities between the two schemes. The public key in our scheme is $(\mathbf{A}, \mathbf{AS})$ where $\mathbf{A}$ is a random matrix mod q and $\mathbf{S}$ is a secret matrix with small coefficients. In [30], the public key is $(\mathbf{A}, \mathbf{AS} + \mathbf{G})$ where $\mathbf{G}$ is an additional public matrix with a very "simple" form. In our scheme, the signature of a message is an ordered pair $(\mathbf{Sc} + \mathbf{y}, \mathbf{c})$ where $\mathbf{c}$ is a function (that invokes a random oracle) of the message and the vector $\mathbf{y}$ is there to "hide" the shift $\mathbf{Sc}$; while in [30], the signature is $(\mathbf{Sc} + \mathbf{y}_1, \mathbf{c} + \mathbf{y}_2)$ where $\mathbf{c}$ is a (different, random oracle-invoking) function of the message and $\mathbf{y}_i$ also serve the purpose of hiding the shift $\mathbf{Sc}$ (and $\mathbf{c}$ itself). While the schemes may look similar, under the surface they behave rather differently.

The most interesting and significant difference occurs in the way the signatures are generated. In our scheme, the vector $\mathbf{c}$ is a very sparse $-1/0/1$ vector whose entropy is as small as the security parameter, but we *must* output it as part of the signature. In [30], however, the size of the elements in $\mathbf{c}$ depends *inversely* on the number of columns of $\mathbf{S}$, but one only outputs a *perturbed* version of $\mathbf{c}$ as part of the signature. Notice that the size of our signature is therefore dominated by the number of rows of $\mathbf{S}$ multiplied by the number of bits needed to represent elements in the vector $\mathbf{Sc} + \mathbf{y}$, whereas in [30], the number of columns of $\mathbf{S}$ may also play a significant role in the signature length.

The advantage in [30] due to the fact that $\mathbf{c}$ is never output in the clear is that they may tailor the perturbations $\mathbf{y}_1, \mathbf{y}_2$ to the particular $\mathbf{S}$ that they are supposed to hide, which allows these perturbations to be smaller than ours in the case that $\mathbf{S}$ has enough columns to allow $\mathbf{c}$ to be "small enough". When instantiating both signature schemes based on the worst-case hardness of the SIS problem, $\mathbf{S}$ needs to have a large number of rows, and thus the fact that the bit-size of the entries of the signature from [30] is smaller than of those in our scheme, may make the scheme from [30] more compact. On the other hand, if one is to instantiate the more practical version of the schemes based on the hardness of the LWE problem, then the number of rows in $\mathbf{S}$ could be significantly smaller, and thus the fact that the size of our signature does not depend on the number of columns of $\mathbf{S}$ gives it an advantage over the one in [30]. We direct the reader to our sample instantiations in Figure 2 where one can see the signature size rapidly decreasing as the number of rows (denoted by m) shrinks. The trade-off is that as the number of rows shrinks, the worst-case hardness assumption becomes stronger, but it is still believed that the security of the average-case problem remains the same (see Section 2 and the full version of this work).

Additionally, the number of columns in our secret key $\mathbf{S}$ needs to only be large enough to support multiplication by $\mathbf{c}$, which allows the number of columns to be significantly smaller than in the secret key of [30], where, for technical reasons, reducing the number of columns of $\mathbf{S}$ ends up increasing the coefficients of $\mathbf{c}$, and thus possibly increasing the size of the signature. This allows our secret key to be smaller that the one in [30]. Compared to the one concrete instantiation (based on the hardness of the SIS problem) provided in [30], where the key size is approximately $2^{26.5}$ bits and the signature is a 13800 dimensional vector of length 92000, thus requiring at least $13800 \cdot \log(92000/\sqrt{13800}) \approx 130000$ bits to represent, for the same security level, some of our instantiations have the signature bit-length about 25% longer, with the benefit of having the keys be about 10 times smaller (column I of Figure 2). For different instantiations, we can have the signature bit-length be about 45% shorter and have the same key size (column III of Figure 2).

1.4   Notation

Throughout the paper, we will assume that q is a small (i.e. polynomial-size) prime number and elements in $\mathbb{Z}_q$ are represented by integers in the range $\left[-\frac{q-1}{2}, \frac{q-1}{2}\right]$. We will represent vectors by bold-face letters, and matrices by bold-face capital letters. We will assume that all vectors are column vectors, and $\mathbf{v}^T$ will denote the transpose of the vector $\mathbf{v}$. The ℓ_p norm of a vector $\mathbf{v}$ is denoted by $\|\mathbf{v}\|_p$, and we will usually avoid writing the p for the ℓ_2 norm. Whenever dealing with elements that are in $\mathbb{Z}_q$, we always explicitly assume that all operations in which they are involved end with a reduction modulo q. Thus for a matrix $\mathbf{A} \in \mathbb{Z}_q^{n \times n}$ and a vector $\mathbf{s} \in \mathbb{Z}^n$, the product $\mathbf{As}$ is a vector in $\mathbb{Z}_q^n$. For a distribution $\mathcal{D}$, we use the notation $x \xleftarrow{} \mathcal{D}$ to mean that x is chosen according to the distribution $\mathcal{D}$. If S is a set, then $x \xleftarrow{\$} S$ means that x is chosen uniformly at random from S. For an event E, we write $Pr[E; x_1 \xleftarrow{\$} \mathcal{D}_1, \ldots, x_k \xleftarrow{\$} \mathcal{D}_k]$ to mean the probability that E occurs when the x_i are chosen from distributions $\mathcal{D}_i$. All logarithms are base 2.

2   The SIS Problem and Its Variants

In this section, we will define the average-case problems upon whose security our signature schemes will be based. All these problems fall into the category of the Small Integer Solution (SIS) problem, which is essentially the knapsack problem over elements in $\mathbb{Z}_q^n$.

Definition 2.1 (ℓ_2-SIS$_{q,n,m,\beta}$ problem). *Given a random matrix $\mathbf{A} \xleftarrow{\$} \mathbb{Z}_q^{n \times m}$ find a vector $\mathbf{v} \in \mathbb{Z}^m \setminus \{0\}$ such that $\mathbf{Av} = 0$ and $\|\mathbf{v}\| \leq \beta$.*

In order for the above problem to not be vacuously hard, we need to have $\beta \geq \sqrt{m}q^{n/m}$ in order for there to exist a solution $\mathbf{v}$. The signature scheme that we construct in Section 4 is based on the presumed hardness of the above problem. In Section 5, we construct a more efficient signature scheme based on the hardness of SIS variants defined below.

Definition 2.2 ($\mathrm{SIS}_{q,n,m,d}$ **distribution**). *Choose a random matrix* $\mathbf{A} \xleftarrow{\$} \mathbb{Z}_q^{n \times m}$ *and a vector* $\mathbf{s} \xleftarrow{\$} \{-d,\ldots,0,\ldots,d\}^m$ *and output* $(\mathbf{A}, \mathbf{As})$.

Definition 2.3 ($\mathrm{SIS}_{q,n,m,d}$ **search problem**). *Given a pair* $(\mathbf{A}, \mathbf{t})$ *from the* $\mathrm{SIS}_{q,n,m,d}$ *distribution, find a* $\mathbf{s} \in \{-d,\ldots,0,\ldots,d\}^m$ *such that* $\mathbf{As} = \mathbf{t}$.

Definition 2.4 ($\mathrm{SIS}_{q,n,m,d}$ **decision problem**). *Given a pair* $(\mathbf{A}, \mathbf{t})$ *decide, with non-negligible advantage, whether it came from the* $\mathrm{SIS}_{q,n,m,d}$ *distribution or whether it was generated uniformly at random from* $\mathbb{Z}_q^{n \times m} \times \mathbb{Z}_q^n$.

Depending on the relationship between its parameters, the $\mathrm{SIS}_{q,n,m,d}$ search (and decision) problem has somewhat different characteristics. If, for example, we have $d \ll q^{n/m}$, then with very high probability there is only one vector $\mathbf{s}$ whose coefficients have absolute value at most d such that $\mathbf{As} = \mathbf{t}$, and such instances of the $\mathrm{SIS}_{q,n,m,d}$ problem are said to be *low-density* instances (borrowing from terminology used to describe instances of the random subset sum problem). On the other hand, if $d \gg q^{n/m}$ then the $\mathrm{SIS}_{q,n,m,d}$ distribution is actually statistically close to uniform over $\mathbb{Z}_q^{n \times m} \times \mathbb{Z}_q^n$ (by the leftover hash lemma) and there are many possible solutions $\mathbf{s}$ for which $\mathbf{As} = \mathbf{t}$. These instances are traditionally called *high-density* instances. The hardness of the SIS problem is discussed in the full version of this work, but we will just mention that the hardest instances are those in which $d \approx q^{n/m}$.

Notice that if $m \geq 2n$, then the matrix $\mathbf{A} \xleftarrow{\$} \mathbb{Z}_q^{n \times m}$ will, with high probability, contain n columns that are linearly independent over $\mathbb{Z}_q$ (when $m \geq 2n$ and q is a prime of size at least $2m$, this will be true with probability $e^{-\Omega(n)}$). Without loss of generality, assume that the last n columns of $\mathbf{A}$ are linearly independent, and so $\mathbf{A} = [\mathbf{A}_1 \| \mathbf{A}_2]$ where $\mathbf{A}_2$ is an $n \times n$ invertible matrix. If we consider the matrix $\mathbf{A}' = \mathbf{A}_2^{-1}\mathbf{A} = [\mathbf{A}_2^{-1}\mathbf{A}_1 \| \mathbf{I}]$, where $\mathbf{I}$ is an $n \times n$ identity matrix, then we have $\mathbf{Av} = 0$ iff $\mathbf{A}'\mathbf{v} = 0$, and so the $\ell_2\text{-}\mathrm{SIS}_{q,n,m,\beta}$ problem is equally hard if the last n columns of the matrix $\mathbf{A}$ form the identity matrix. Similarly, given an instance $(\mathbf{A}, \mathbf{t})$ of the $\mathrm{SIS}_{q,n,m,d}$ problem, we can change it to $(\mathbf{A}_2^{-1}\mathbf{A}, \mathbf{A}_2^{-1}\mathbf{t})$, and a solution for one will be exactly the same as the solution for the other. Therefore throughout this paper we will assume, without loss of generality, that the matrix $\mathbf{A} \in \mathbb{Z}_q^{n \times m}$ is of the form $\mathbf{A} = [\bar{\mathbf{A}} \| \mathbf{I}]$, where $\bar{\mathbf{A}}$ is uniformly generated in $\mathbb{Z}_q^{n \times (m-n)}$. For reasons related to lattices, when $\mathbf{A}$ is in this form, we will refer to it as being in *Hermite Normal Form* [32].

2.1 Relations between the SIS Variants

We now state some results about the relationship between the SIS variants defined above. The first relationship is an adaptation of a classic theorem of Impagliazzo and Naor [22], who showed that the decisional version of the random subset sum problem is as hard as the search version. This theorem has been recently generalized by Micciancio and Mol [29].

Theorem 2.5. [22,29] *If d is polynomial in n, then there is a polynomial-time reduction from the $\mathrm{SIS}_{q,n,m,d}$ search problem to the $\mathrm{SIS}_{q,n,m,d}$ decision problem.*

The next lemma shows that the *decision* $\mathrm{SIS}_{q,n,m,d}$ problem gets harder when the value of d increases. This is a rather intuitive result since the decision $\mathrm{SIS}_{q,n,m,d}$ problem becomes vacuously hard when $d \gg q^{n/m}$ since the $\mathrm{SIS}_{q,n,m,d}$ distribution will be statistically close to uniform.

Lemma 2.6. *For any non-negative integer α such that $\gcd(2\alpha + 1, q) = 1$, there is a polynomial-time reduction from the $\mathrm{SIS}_{q,n,m,d}$ decision problem to the $\mathrm{SIS}_{q,n,m,(2\alpha+1)d+\alpha}$ decision problem.*

The final lemma that we prove shows that if $m = 2n$ and one can solve the can solve $\ell_2\text{-}\mathrm{SIS}_{q,n,m,\beta}$ problem for a small-enough β, then one can solve the decision $\mathrm{SIS}_{q,n,m,d}$ problem. This result is essentially folklore (see [32]), but we state it here for completeness.

Lemma 2.7. *If $m = 2n$ and $4d\beta \leq q$, then there is a polynomial-time reduction from solving the $\mathrm{SIS}_{q,n,m,d}$ decision problem to the $\ell_2\text{-}\mathrm{SIS}_{q,n,m,\beta}$ problem.*

3 Rejection Sampling and the Normal Distribution

Definition 3.1. *The continuous Normal distribution over $\mathbb{R}^m$ centered at $\mathbf{v}$ with standard deviation σ is defined by the function $\rho_{\mathbf{v},\sigma}^m(\mathbf{x}) = \left(\frac{1}{\sqrt{2\pi\sigma^2}}\right)^m e^{\frac{-\|\mathbf{x}-\mathbf{v}\|^2}{2\sigma^2}}$*

When $\mathbf{v} = 0$, we will just write $\rho_\sigma^m(\mathbf{x})$. We will define the *discrete* Normal distribution over $\mathbb{Z}^m$ as follows:

Definition 3.2. *The discrete Normal distribution over $\mathbb{Z}^m$ centered at some $\mathbf{v} \in \mathbb{Z}^m$ with standard deviation σ is defined as $D_{\mathbf{v},\sigma}^m(\mathbf{x}) = \rho_{\mathbf{v},\sigma}^m(\mathbf{x})/\rho_\sigma^m(\mathbb{Z}^m)$.*

In the above definition, the quantity $\rho_\sigma^m(\mathbb{Z}^m) = \sum\limits_{\mathbf{z} \in \mathbb{Z}^m} \rho_\sigma^m(\mathbf{z})$ is just a scaling quantity needed to make the function into a probability distribution. Also note that for all $\mathbf{v} \in \mathbb{Z}^m$, $\rho_{\mathbf{v},\sigma}^m(\mathbb{Z}^m) = \rho_\sigma^m(\mathbb{Z}^m)$, thus the scaling factor is the same for all $\mathbf{v}$.

The below lemma collects some basic facts about the discrete Normal distribution over $\mathbb{Z}^m$. These results are special cases of more general results about the discrete Normal distribution over arbitrary lattices from [6,31,37].

Lemma 3.3

1. $Pr[|z| > \omega(\sigma\sqrt{\log m}); z \xleftarrow{\$} D_\sigma^1] = 2^{-\omega(\log m)}$, *and more specifically,*
   $Pr[|z| > 12\sigma; z \xleftarrow{\$} D_\sigma^1] < 2^{-100}$.
2. *For any* $\mathbf{z} \in \mathbb{Z}^m$, *and* $\sigma \geq \sqrt{\log 3m}$, $D_\sigma^m(\mathbf{z}) \leq 2^{-m+1}$.
3. $Pr[\|\mathbf{z}\| > 2\sigma\sqrt{m}; \mathbf{z} \xleftarrow{\$} D_\sigma^m] < 2^{-m}$.

We now state the main theorem of this section whose proof is given in the full version of this paper.

Signing Key: $\mathbf{S} \xleftarrow{\$} \{-d, \ldots, 0, \ldots, d\}^{m \times k}$
Verification Key: $\mathbf{A} \xleftarrow{\$} \mathbb{Z}_q^{n \times m}, \mathbf{T} \leftarrow \mathbf{AS}$
Random Oracle: $\mathrm{H} : \{0, 1\}^* \rightarrow \{\mathbf{v} : \mathbf{v} \in \{-1, 0, 1\}^k, \|\mathbf{v}\|_1 \leq \kappa\}$

$\mathrm{Sign}(\mu, \mathbf{A}, \mathbf{S})$
  1: $\mathbf{y} \xleftarrow{\$} D_\sigma^m$
  2: $\mathbf{c} \leftarrow \mathrm{H}(\mathbf{Ay}, \mu)$
  3: $\mathbf{z} \leftarrow \mathbf{Sc} + \mathbf{y}$
  4: output $(\mathbf{z}, \mathbf{c})$ with probability
    $\min \left(\frac{D_\sigma^m(\mathbf{z})}{M D_{\mathbf{Sc}, \sigma}^m(\mathbf{z})}, 1 \right)$

$\mathrm{Verify}(\mu, \mathbf{z}, \mathbf{c}, \mathbf{A}, \mathbf{T})$
  1: Accept iff
    $\|\mathbf{z}\| \leq 2\sigma\sqrt{m}$ and $\mathbf{c} = \mathrm{H}(\mathbf{Az} - \mathbf{Tc}, \mu)$

Fig. 1. Signature Scheme

Theorem 3.4. *Let V be a subset of $\mathbb{Z}^m$ in which all elements have norms less than T, σ be some element in $\mathbb{R}$ such that $\sigma = \omega(T\sqrt{\log m})$, and $h : V \rightarrow \mathbb{R}$ be a probability distribution. Then there exists a constant $M = O(1)$ such that the distribution of the following algorithm $\mathcal{A}$:*

1: $\mathbf{v} \xleftarrow{\$} h$
2: $\mathbf{z} \xleftarrow{\$} D_{\mathbf{v}, \sigma}^m$
3: output $(\mathbf{z}, \mathbf{v})$ with probability $\min \left(\frac{D_\sigma^m(\mathbf{z})}{M D_{\mathbf{v}, \sigma}^m(\mathbf{z})}, 1 \right)$

is within statistical distance $\frac{2^{-\omega(\log m)}}{M}$ of the distribution of the following algorithm $\mathcal{F}$:

1: $\mathbf{v} \xleftarrow{\$} h$
2: $\mathbf{z} \xleftarrow{\$} D_\sigma^m$
3: output $(\mathbf{z}, \mathbf{v})$ with probability $1/M$

Moreover, the probability that $\mathcal{A}$ outputs something is at least $\frac{1 - 2^{-\omega(\log m)}}{M}$.

More concretely, if $\sigma = \alpha T$ for any positive α, then $M = e^{12/\alpha + 1/(2\alpha^2)}$, the output of algorithm $\mathcal{A}$ is within statistical distance $\frac{2^{-100}}{M}$ of the output of $\mathcal{F}$, and the probability that $\mathcal{A}$ outputs something is at least $\frac{1 - 2^{-100}}{M}$.

4   Signature Scheme Based on SIS

In this section we present our main theoretical result – a signature scheme based, in the random oracle model, on the average-case hardness of the ℓ_2-$\mathrm{SIS}_{q,n,m,\beta}$ problem for $\beta = \tilde{O}(n)$. The scheme is presented in Figure 1 and the definition of its parameters and some sample instantiations are in Figure 2. We will now explain the workings of the scheme and sketch the intuition for its security.

The secret key is an $m \times k$ matrix $\mathbf{S}$ of random integers of absolute value at most d, and the public key consists of a random matrix $\mathbf{A} \in \mathbb{Z}_q^{n \times m}$ and another matrix $\mathbf{T} \in \mathbb{Z}_q^{m \times k}$ which is equal to $\mathbf{AS}$. For concreteness, we will consider distributions to be statistically close if they are $\approx 2^{-100}$ apart, and we will also

	I	II	III	IV	V
n	512	512	512	512	512
q	2^{27}	2^{25}	2^{33}	2^{24}	2^{33}
d	1	1	31	1	31
k	80	512	512	512	512
$m \approx 64 + n \cdot \log q / \log(2d+1)$	8786	8139	3253	-	-
$m = 2n$ (used in Section 5)	-	-	-	1024	1024
κ s.t. $2^{\kappa} \cdot \binom{n}{\kappa} \geq 2^{100}$	28	14	14	14	14
$\sigma \approx 12 \cdot d \cdot \kappa \cdot \sqrt{m}$	31495	15157	300926	5376	166656
$M \approx \exp\left(12d\kappa\sqrt{m}/\sigma + (d\kappa\sqrt{m}/2\sigma)^2\right)$	2.72	2.72	2.72	2.72	2.72
signature size (bits) $\approx m \log(12\sigma)$	163000	142300	73000	16500	20500
secret key size (bits) $\approx m \cdot k \cdot \log(2d+1)$	2^{20}	$2^{22.5}$	2^{23}	$2^{19.5}$	$2^{21.5}$
public key size (bits) $\approx n \cdot k \cdot \log q$	2^{20}	$2^{22.5}$	2^{23}	$2^{22.5}$	2^{23}

Fig. 2. Signature Scheme Parameters. The parameters in columns I, II, and III are based on the hardness of the ℓ_2-$\mathrm{SIS}_{q,n,m,\beta}$ problem where for the β in Theorem 4.1. Columns IV and V are based on the hardness of the $\mathrm{SIS}_{q,n,m,d}$ search problem (see Section 5). Furthermore, the parameters in column V are also compatible with the LWE assumption (see Section 5.1. The security level for all the instantiations is for $\delta \approx 1.007$ (see the full version of this paper). For the ring-based instantiations, described in the full version, the key sizes are smaller by a factor of k.

want ≈ 100 bits of security from our cryptographic hash function H, and so we will assume that the output of H is 100 bits.[4]

To sign a message μ, the signer first picks an m-dimensional vector $\mathbf{y}$ from the distribution D_σ^m, for some standard deviation σ, then computes $\mathbf{c} = \mathrm{H}(\mathbf{Ay}, \mu)$, and finally computes $\mathbf{z} = \mathbf{Sc} + \mathbf{y}$ (there is no reduction modulo q in this step!). The potential signature which he outputs is $(\mathbf{z}, \mathbf{c})$, but he only outputs it with probability $\min\left(\frac{D_\sigma^m(\mathbf{z})}{MD_{\mathbf{Sc},\sigma}^m(\mathbf{z})}, 1\right)$. If nothing was output, the signer runs the signing algorithm again until some signature is outputted.

The main idea behind this structure of the signing algorithm is to make the distribution of the signature $(\mathbf{z}, \mathbf{c})$ *independent* of the secret key $\mathbf{S}$. The target distribution for the $\mathbf{z}$'s that we will be aiming for is D_σ^m, but the elements $\mathbf{z}$ in the signature scheme come from the distribution $D_{\mathbf{v},\sigma}^m$, where $\mathbf{v} = \mathbf{Sc}$. This is where we will apply the rejection sampling theorem, Theorem 3.4, from Section 3 to show that for an appropriately-chosen value of M and σ, the signature algorithm will output something with probability approximately $1/M$ and the statistical distance between its output is statistically close to the distribution in which $\mathbf{z}$ is chosen from D_σ^m.

---

[4] It is generally considered folklore that for obtaining signatures with λ bits of security using the Fiat-Shamir transform, one only needs random oracles that output λ bits (i.e. collision-resistance is not a requirement). While finding collisions in the random oracle does allow the *valid* signer to produce two distinct messages that have the same signature, this does not constitute a break.

Once we decoupled the distribution of the signature from the distribution of the secret key, we can use a forger who successfully breaks the signature to solve the ℓ_2-$\mathrm{SIS}_{q,n,m,\beta}$ problem for $\beta \approx \tilde{O}(\|\mathbf{z}\|)$. The idea is that given an $\mathbf{A}$, one can create a secret key $\mathbf{S}$ and publish the public key $(\mathbf{A}, \mathbf{AS})$. Then one can reply to signing queries of the forger by either using the key $\mathbf{S}$, or simply by producing signatures by generating $\mathbf{z}$ from the distribution D_σ^m and programming the random oracle accordingly. In our proof (Lemma 4.4), we choose the latter approach because in Section 5, we will not know a valid secret key, but we would like to be able to still use the the the same lemma there. Once we have a way to reply to signing queries, we use the forking lemma [38,7] to use the forger's valid signatures to recover a short vector $\mathbf{v}$ such that $\mathbf{Av} = 0$. One important caveat is that to prove that $\mathbf{v} \neq 0$, there needs to be a second (unknown to us) valid secret key $\mathbf{S}'$ such that $\mathbf{AS} = \mathbf{AS}'$, and the forger cannot know which secret key we know. To satisfy the existence of another secret key requires a particular relationship between n, m, and q (Lemma 4.2), and the indistinguishability of $\mathbf{S}$ and $\mathbf{S}'$ is clearly satisfied because the distribution of the signature is independent of the secret key.

We now discuss the verification procedure. Since we tailored $\mathbf{z}$ to be distributed according to D_σ^m, by Lemma 3.3, we know that with probability at least $1 - 2^{-m}$, we have $\|\mathbf{z}\| < 2\sigma\sqrt{m}$. And since $\mathbf{Ay} = \mathbf{Az} - \mathbf{Tc}$, the second part of the verification will accept a valid signature.

Theorem 4.1. *If there is a polynomial-time forger, who makes at most s queries to the signing oracle and h queries to the random oracle H, who breaks the signature in Figure 1 (with the relationship between the parameters as in Figure 2) with probability δ, then there is a polynomial-time algorithm who can solve the ℓ_2-$\mathrm{SIS}_{q,n,m,\beta}$ problem for $\beta = (4\sigma + 2d\kappa)\sqrt{m} = \tilde{O}(dn)$ with probability $\approx \frac{\delta^2}{2(h+s)}$. Moreover, the signing algorithm produces a signature with probability $\approx 1/M$ and the verifying algorithm accepts a signature produced by an honest signer with probability at least $1 - 2^{-m}$.*

Proof. The theorem is proved in a sequence of two Lemmas. In Lemma 4.3, we show that our signing algorithm can be replaced by the one in Hybrid 2 of Figure 3, and the statistical distance between the two outputs will be at most $\epsilon = s(h+s) \cdot 2^{-n+1} + s \cdot \frac{2^{-100}}{M}$. Since Hybrid 2 produces an output with probability exactly $1/M$, the signing algorithm produces an output with probability at least $(1 - \epsilon)/M$. Then in Lemma 4.4, we show that if a forger can produce a forgery with probability δ when when the signing algorithm is replaced by one in Hybrid 2, then we can use him to recover a vector $\mathbf{v}$ such that $\|\mathbf{v}\| \leq (4\sigma + 2d\kappa)\sqrt{m}$ and $\mathbf{Av} = 0$ with probability at least $\left(\frac{1}{2} - 2^{-100}\right)\left(\delta - 2^{-100}\right)\left(\frac{\delta - 2^{-100}}{h+s} - 2^{-100}\right) \approx \frac{\delta^2}{2(h+s)}$. $\qquad\square$

Lemma 4.2. *For any $\mathbf{A} \in \mathbb{Z}_q^{n \times m}$ where $m > 64 + n \cdot \log q / \log(2d+1)$, for randomly chosen $\mathbf{s} \overset{\$}{\leftarrow} \{-d, \ldots, 0, \ldots, d\}^m$, with probability $1 - 2^{-100}$, there exists another $\mathbf{s}' \in \{-d, \ldots, 0, \ldots, d\}^m$ such that $\mathbf{As} = \mathbf{As}'$.*

Hybrid 1

Sign(μ, $\mathbf{A}$, $\mathbf{S}$)
1: $\mathbf{y} \xleftarrow{\$} D_\sigma^m$
2: $\mathbf{c} \xleftarrow{\$} \{\mathbf{v} : \mathbf{v} \in \{-1, 0, 1\}^k, \|\mathbf{v}\|_1 \leq \kappa\}$
3: $\mathbf{z} \leftarrow \mathbf{Sc} + \mathbf{y}$
4: with probability $\min \left(\frac{D_\sigma^m(\mathbf{z})}{MD_{\mathbf{Sc},\sigma}^m(\mathbf{z})}, 1 \right)$,
5:     output $(\mathbf{z}, \mathbf{c})$
6:     Program $\mathrm{H}(\mathbf{Az} - \mathbf{Tc}, \mu) = \mathbf{c}$

Hybrid 2

Sign(μ, $\mathbf{A}$, $\mathbf{S}$)
1: $\mathbf{c} \xleftarrow{\$} \{\mathbf{v} : \mathbf{v} \in \{-1, 0, 1\}^k, \|\mathbf{v}\|_1 \leq \kappa\}$
2: $\mathbf{z} \xleftarrow{\$} D_\sigma^m$
3: with probability $1/M$,
4:     output $(\mathbf{z}, \mathbf{c})$
5:     Program $\mathrm{H}(\mathbf{Az} - \mathbf{Tc}, \mu) = \mathbf{c}$

Fig. 3. Signing Hybrids

Lemma 4.3. *Let $\mathcal{D}$ be a distinguisher who can query the random oracle H and either the actual signing algorithm in Figure 1 or Hybrid 2 in Figure 3. If he makes h queries to H and s queries to the signing algorithm that he has access to, then for all but a $e^{-\Omega(n)}$ fraction of all possible matrices $\mathbf{A}$, his advantage of distinguishing the actual signing algorithm from the one in Hybrid 2 is at most $s(h + s) \cdot 2^{-n+1} + s \cdot \frac{2^{-\omega(\log m)}}{M}$, or more concretely, $s(h + s) \cdot 2^{-n+1} + s \cdot \frac{2^{-100}}{M}$.*

Proof. We first show that the distinguisher $\mathcal{D}$ has advantage of at most $s(h + s)2^{-n+1}$ of distinguishing between the real signature scheme and Hybrid 1. The only difference between the actual signing algorithm and the algorithm in Hybrid 1 is that in Hybrid 1, the output of the random oracle H is chosen at random from $\{\mathbf{v} : \mathbf{v} \in \{-1, 0, 1\}^k, \|\mathbf{v}\|_1 \leq \kappa\}$ and then programmed as the answer to $\mathrm{H}(\mathbf{Az} - \mathbf{Tc}, \mu) = \mathrm{H}(\mathbf{Ay}, \mu)$ without checking whether the value for $(\mathbf{Ay}, \mu)$ was already set. Since $\mathcal{D}$ calls H h times, and the signing algorithm s times, at most $s + h$ values of $(\mathbf{Ay}, \mu)$ will ever be set. We now show that each time the Hybrid 1 procedure is called, the probability of generating a $\mathbf{y}$ such that $\mathbf{Ay}$ is equal to one of the previous values that was queried is at most 2^{-n+1}. With probability at least $1 - e^{-\Omega(n)}$, the matrix $\mathbf{A}$ can be written in "Hermite Normal Form" (see Section 2) as $\mathbf{A} = [\bar{\mathbf{A}}\|\mathbf{I}]$. Then, for any $\mathbf{t} \in \mathbb{Z}_q^n$,

$$Pr[\mathbf{Ay} = \mathbf{t}; \mathbf{y} \xleftarrow{\$} D_\sigma^m] = Pr[\mathbf{y}_1 = (\mathbf{t} - \bar{\mathbf{A}}\mathbf{y}_0); \mathbf{y} \xleftarrow{\$} D_\sigma^m]$$
$$\leq \max_{\mathbf{t}' \in \mathbb{Z}_q^n} Pr[\mathbf{y}_1 = \mathbf{t}'; \mathbf{y}_1 \xleftarrow{\$} D_\sigma^n] \leq 2^{-n+1},$$

where the last inequality follows from Lemma 3.3. Thus if Hybrid 1 is accessed s times, and the probability of getting a collision each time is at most $(s+h)2^{-n+1}$, the probability that a collision occurs after s queries is at most $s(s + h)2^{-n+1}$.

We next show that the statistical distance between the outputs of Hybrid 1 and Hybrid 2 is at most $\frac{2^{-\omega(\log m)}}{M}$. The proof of this fact is almost a direct consequence of Theorem 3.4. Notice that if both Hybrids simply outputted $(\mathbf{z}, \mathbf{v} = \mathbf{Sc})$ with probability $\min \left(\frac{D_\sigma^m(\mathbf{z})}{MD_{\mathbf{Sc},\sigma}^m(\mathbf{z})}, 1 \right)$ for Hybrid 1 and probability $1/M$ for Hybrid 2, then Hybrid 1 exactly plays the role of the algorithm $\mathcal{A}$ in Theorem 3.4 and Hybrid 2 corresponds to $\mathcal{F}$ (where the maximum T in Theorem 3.4 corresponds to $d\kappa\sqrt{m}$).

But instead of outputting $\mathbf{v} = \mathbf{Sc}$, the Hybrids output just $\mathbf{c}$. But this does not increase the statistical distance because given $\mathbf{v}$, one can generate $\mathbf{c}$ by picking a random element $\mathbf{c} \in \{\mathbf{w} : \mathbf{w} \in \{-1,0,1\}^k, \|\mathbf{w}\|_1 \le \kappa\}$ such that $\mathbf{Sc} = \mathbf{v}$ (for our choice of parameters in this paper, there will actually be only one possible $\mathbf{c}$, with very high probability), and this will have the exact same distribution as the $\mathbf{c}$ in both Hybrids. And finally, since the signing oracle is called s times, the statistical distance is no more than $s \cdot \frac{2^{-\omega(\log m)}}{M}$, or more concretely, $s \cdot \frac{2^{-100}}{M}$, (since we set $\sigma = 12T$), and we obtain the claim in the lemma.   □

Lemma 4.4. *Suppose there exists a polynomial-time forger $\mathcal{F}$ who makes at most h queries to the signer in Hybrid 2, s queries to the random oracle H, and succeeds in forging with probability δ. Then there exists an algorithm of the same time-complexity as $\mathcal{F}$ that for a given $\mathbf{A} \xleftarrow{\$} \mathbb{Z}_q^{n \times m}$ finds a non-zero $\mathbf{v} \in \mathbb{Z}^m$ such that $\|\mathbf{v}\| \le (4\sigma + 2d\kappa)\sqrt{m}$ and $\mathbf{Av} = 0$ with probability at least*

$$\left(\frac{1}{2} - 2^{-100}\right)\left(\delta - 2^{-100}\right)\left(\frac{\delta - 2^{-100}}{h+s} - 2^{-100}\right).$$

4.1   Setting the Parameters

In Figure 2, we set some sample parameters to demonstrate the influence of their interplay on the sizes of the signature length and the key size. The secret key is an $m \times k$ matrix with coefficients having absolute value at most d, and so it can be represented by $mk \log (2d + 1)$ bits. The public key $\mathbf{A}, \mathbf{T}$ can be spit into two parts – the matrix $\mathbf{A}$ can be shared by all users (and so can be considered as part of the function), whereas the matrix $\mathbf{T}$ is individual. The part of the public key that is individual for each user requires $nk \log q$ bits of storage. The signature size is dominated by the vector $\mathbf{z}$, since $\mathbf{c}$ is just a small bit-string that is the output of the cryptographic hash function H. By design, the vector $\mathbf{z}$ is distributed according to D_σ^m, and by Lemma 3.3, we know that with probability at least $1 - 2^{-100}$, each coefficient of $\mathbf{z}$ is of length at most 12σ. Thus $\mathbf{z}$ can be represented by $m \log (12\sigma)$ bits.

For security, we use the analysis of [13,32] (also discussed in the full version of this paper), where it is shown that the smallest vector $\mathbf{v}$ such that $\mathbf{Av} = 0$ can be produced has length $\min\left(q, 2^{2\sqrt{n \log q \log \delta}}\right)$. We would like this vector $\mathbf{v}$ to have a larger size than the vector that can be extracted from the successful forger, which is given in Lemma 4.4. There are some trade-offs between the sizes of signatures and keys that can be achieved for the same security level. For example, if we change the value of k from 80 in column I to 512 in column II, it has the effect of making the keys larger by a factor of around 6, and at the same time reducing the signature size by a little over 10%. Another interesting trade-off is achieved by raising the value of d as in column III. Notice that what most affects the length of the signature size is the parameter m. By raising the value of d and q, we can lower m, and can reduce the signature size by almost 50% at the expense of slightly increasing the key sizes.

5   Signatures Based on Low-Density SIS and LWE

From the sample instantiations in the previous section, we saw that m is the one parameter that most affects the signature size. In this section we explore the results of breaking the requirement that $m \approx 64 + n \cdot \log q / \log (2d+1)$ (which is required for Lemma 4.2) and show that this still gives us a provably-secure signature scheme (based on the low-density $\text{SIS}_{q,n,m,d}$ problem), but with much smaller signature and key sizes. Let us consider, for example, taking instantiation III in Figure 2 and lowering the value of d from 31 to, say, 1, *without* changing the value of m. The potential advantage of this modification is that the value of σ goes down by a factor of d, which has the effect of making the signature vector $\mathbf{z}$ smaller (by a factor d), which in turn makes it harder for the adversary to produce a forgery, since he now needs to find a vector that is d times smaller than before. This in turn allow us to lower other parameters, such as q and m, which leads to a "virtuous cycle" of reducing the length of the signature.

We now look at what happens to the security proofs if we proceed as described above. The main problem is that Lemma 4.2 is no longer true since for every $\mathbf{T}$, there will now be, with extremely high probability, only one $\mathbf{S}$ for which $\mathbf{AS} = \mathbf{T}$. The fact that there were multiple $\mathbf{S}$'s was crucially used at the end of Lemma 4.4 to argue that a successful forger can be used to extract a small vector $\mathbf{v}$ such that $\mathbf{Av} = 0$. On the other hand, the proof of Lemma 4.3 is not affected by the relationship between d and m, and so the real signature scheme is still indistinguishable from one that uses Hybrid 2 as its signing algorithm. And since Hybrid 2 *does not* use the secret key to produce signatures, for a given $\mathbf{A}$, we can use the secret key $\mathbf{S}$ with small coefficients in the actual signature, but use an $\mathbf{S}'$ with large coefficients (so that there exists an $\mathbf{S}''$ such that $\mathbf{AS}' = \mathbf{AS}''$) in the proof (see Figure 4). If the distribution of the verification key $(\mathbf{A}, \mathbf{AS})$ is computationally indistinguishable from that of $(\mathbf{A}, \mathbf{AS}')$ (and it is, based on the hardness of the low-density $\text{SIS}_{q,n,m,d}$ problem from Definition 2.4), the distinguisher will not be able to tell that he is given an invalid key pair. And since we never use the secret key to provide signatures to the forger in Lemma 4.4, the forger should act in the same way, and we will be able to find a non-zero $\mathbf{v}$ such that $\mathbf{Av} = 0$.

Using the above framework, we can obtain a signature scheme that is based on the hardness of two problems (i.e. both problems need to be hard for our scheme to be secure): the $\text{SIS}_{q,n,m,d}$ decisional (and by Theorem 2.5, also computational) problem and the $\ell_2\text{-SIS}_{q,n,m,\beta}$ problem with $\beta = (4\sigma + 2d'\kappa)\sqrt{m}$. Thus the optimal parameter settings will be where the two problems are equally hard. Furthermore, if we set $m = 2n$ and have $4d\beta \le q$, then the $\text{SIS}_{q,n,m,d}$ problem reduces to the $\ell_2\text{-SIS}_{q,n,m,\beta}$ one (Lemma 2.7), and we end up with just one simple computational hardness assumption – $\text{SIS}_{q,n,m,d}$. We formalize the above intuition in two lemmas analogous to Lemmas 4.3 and 4.4 from Section 4.

Lemma 5.1. *Let $\mathcal{D}$ be a distinguisher who can query the random oracle H and either the actual key-generation/signing algorithms in Figure 1 or those in*

Hybrid 2

Signing Key: $\mathbf{S} \xleftarrow{\$} \{-d, \ldots, 0, \ldots, d\}^{m \times k}$
Verification Key: $\mathbf{A} \xleftarrow{\$} \mathbb{Z}_q^{n \times m}, \mathbf{T} \leftarrow \mathbf{AS}$

$\mathrm{Sign}(\mu, \mathbf{A}, \mathbf{S})$
1: $\mathbf{c} \xleftarrow{\$} \{\mathbf{v} : \mathbf{v} \in \{-1, 0, 1\}^k, \|\mathbf{v}\|_1 \le \kappa\}$
2: $\mathbf{z} \xleftarrow{\$} D_\sigma^m$
3: with probability $1/M$,
4:     output $(\mathbf{z}, \mathbf{c})$
5:     Program $H(\mathbf{Az} - \mathbf{Tc}, \mu) = \mathbf{c}$

Hybrid 3

Signing Key: $\mathbf{S} \xleftarrow{\$} \{-d', \ldots, 0, \ldots, d'\}^{m \times k}$
Verification Key: $\mathbf{A} \xleftarrow{\$} \mathbb{Z}_q^{n \times m}, \mathbf{T} \leftarrow \mathbf{AS}$

$\mathrm{Sign}(\mu, \mathbf{A}, \mathbf{S})$
1: $\mathbf{c} \xleftarrow{\$} \{\mathbf{v} : \mathbf{v} \in \{-1, 0, 1\}^k, \|\mathbf{v}\|_1 \le \kappa\}$
2: $\mathbf{z} \xleftarrow{\$} D_\sigma^m$
3: with probability $1/M$,
4:     output $(\mathbf{z}, \mathbf{c})$
5:     Program $H(\mathbf{Az} - \mathbf{Tc}, \mu) = \mathbf{c}$

Fig. 4. Key-Generation and Signing Hybrids: d' is set so that $d' = (2\alpha + 1)d + \alpha$ for some positive integer α and $m \ge 64 + n \cdot \log q / \log(2d' + 1)$

Hybrid 3 in Figure 4. If he makes h queries to H and s queries to the signing algorithm that he has access to, and can distinguish the real world from Hybrid 3 with advantage δ, then he has advantage $\Omega(\delta/k) - \left(s(h+s) \cdot 2^{-n+1} + s \cdot \frac{2^{-\omega(\log m)}}{M}\right)$ in solving the $\mathrm{SIS}_{q,n,m,d}$ decision problem.

Lemma 5.2. *Suppose there exists a polynomial-time forger $\mathcal{F}$ who is given the verification key and access to the signing algorithm from Hybrid 3, and makes at most h queries to the signing algorithm, s queries to the random oracle H, and succeeds in forging with probability δ. Then there exists an algorithm of the same time-complexity as $\mathcal{F}$ that for a given $\mathbf{A} \xleftarrow{\$} \mathbb{Z}_q^{n \times m}$ finds a $\mathbf{v} \in \mathbb{Z}^m$ such that $\|\mathbf{v}\| \le (4\sigma + 2d'\kappa)\sqrt{m}$ and $\mathbf{Av} = 0$ with probability at least*

$$\left(\frac{1}{2} - 2^{-100}\right) \left(\delta - 2^{-100}\right) \left(\frac{\delta - 2^{-100}}{h+s} - 2^{-100}\right).$$

Proof. The proof is exactly the same as the one of Lemma 4.4, with d' playing the role of d. □

5.1 The LWE Problem

In the Learning With Errors (LWE) problem, one is given an oracle that produces ordered pairs of the form $(\mathbf{a}_i, b_i) \in \mathbb{Z}_q^n \times \mathbb{Z}$ where the $\mathbf{a}_i$ are uniformly random in $\mathbb{Z}_q^n$, and $b_i = \mathbf{a}_i \cdot \mathbf{s} + e_i$ where $\mathbf{s}$ is some secret vector in $\mathbb{Z}_q^n$ and e_i is some "error" of small absolute value. Regev [39] showed that there is a quantum reduction from approximating SIVP in all lattices to solving random instances of LWE when the errors e_i come from the discrete Normal distribution D_ψ, and Peikert later showed a classical reduction to LWE from some different lattice problems [35].

An equivalent version of LWE, as shown in [4], is if the secret key is selected from the distribution D_ψ^n rather than from the uniform distribution. In addition, Regev also showed that the decisional version of the LWE problem, where one is asked to decide whether the ordered pairs $(\mathbf{a}_i, b_i)$ come from the uniform distribution or whether they are generated such that $b_i = \mathbf{a}_i \cdot \mathbf{s} + e_i$, is as hard as the search version.

Using the above definitions, observe that if we have a matrix $\mathbf{A} = [\bar{\mathbf{A}}\|\mathbf{I}] \in \mathbb{Z}_q^{n \times 2n}$, where $\bar{A} \xleftarrow{\$} \mathbb{Z}_q^{n \times n}$, then distinguishing pairs $(\mathbf{A}, \mathbf{As})$, where each $\mathbf{s} \xleftarrow{\$} D_\psi^{2n}$, from uniformly distributed pairs in $\mathbb{Z}_q^{n \times 2n} \times \mathbb{Z}_q^{2n}$ is exactly the decisional LWE problem. By the hybrid argument, distinguishing $(\mathbf{A}, \mathbf{AS})$, where each column of the k columns of $\mathbf{S}$ is distributed according to D_ψ^{2n}, from uniformly distributed pairs in $\mathbb{Z}_q^{n \times 2n} \times \mathbb{Z}_q^{2n \times k}$ is also as hard as LWE. Therefore, except for the distribution of the secret key $\mathbf{S}$, the LWE problem is exactly the low-density $\mathrm{SIS}_{q,n,2n,d}$ problem, and so we can easily change the scheme in the previous section based on the hardness of low-density SIS to be based on LWE instead.

The most important feature of the secret key $\mathbf{S}$ that is used in the proofs is the norm of each of its columns. If the norm of $\mathbf{s} \xleftarrow{\$} D_\psi^m$ is approximately the same as that of a vector $\mathbf{s}' \xleftarrow{\$} \{-d, \ldots, 0, \ldots, d\}^m$, then the security and correctness of the scheme from this section will go through almost entirely unchanged. It can be seen that if $\psi \approx \sqrt{\frac{d \cdot (d+1)}{3}}$, then the length of $\mathbf{s}$ is approximately the same as that of a vector $\mathbf{s}'$ (since $\|\mathbf{s}\|$ is tightly concentrated around $\psi\sqrt{m}$ and $\|\mathbf{s}'\|$ around $\sqrt{d(d+1)m/3}$). So a scheme based on LWE where $\psi \approx 18$ would have approximately the same signature size and key lengths as the scheme in column V of Figure 2 where $d = 31$.

Notice that the LWE-based scheme in column V produces signatures that are slightly longer than those produced by the scheme in column IV that is based on the $\mathrm{SIS}_{q,n,2n,1}$ problem. At this point, we are not aware of any algorithms that specifically attack $\mathrm{SIS}_{q,n,2n,1}$ which would justify making the signature longer just so that it is based on the hardness of the LWE problem. But in view of the recent algorithm of Arora and Ge [5], which uses algebraic attacks to attack the LWE problem with very small errors, there may be reasons to think that the instantiation in column V could be more secure because it uses larger coefficients.

Acknowledgements. I am very grateful to Oded Regev for all his great ideas and suggestions that greatly improved this work. I would also like to thank Daniele Micciancio and Chris Peikert for illuminating discussions about the results of [30]. And finally, I thank the anonymous Eurocrypt reviewers for their valuable comments and suggestions.

References

1. Ajtai, M.: Generating Hard Instances of the Short Basis Problem. In: Wiedermann, J., Van Emde Boas, P., Nielsen, M. (eds.) ICALP 1999. LNCS, vol. 1644, pp. 1–9. Springer, Heidelberg (1999)
2. Albrecht, M.R., Farshim, P., Faugère, J.-C., Perret, L.: Polly Cracker, Revisited. In: Lee, D.H., Wang, X. (eds.) ASIACRYPT 2011. LNCS, vol. 7073, pp. 179–196. Springer, Heidelberg (2011)
3. Alwen, J., Peikert, C.: Generating shorter bases for hard random lattices. Theory Comput. Syst. 48(3), 535–553 (2011)
4. Applebaum, B., Cash, D., Peikert, C., Sahai, A.: Fast Cryptographic Primitives and Circular-Secure Encryption Based on Hard Learning Problems. In: Halevi, S. (ed.) CRYPTO 2009. LNCS, vol. 5677, pp. 595–618. Springer, Heidelberg (2009)

5. Arora, S., Ge, R.: New Algorithms for Learning in Presence of Errors. In: Aceto, L., Henzinger, M., Sgall, J. (eds.) ICALP 2011, Part I. LNCS, vol. 6755, pp. 403–415. Springer, Heidelberg (2011)

6. Banaszczyk, W.: New bounds in some transference theorems in the geometry of numbers. Mathematische Annalen 296, 625–635 (1993)

7. Bellare, M., Neven, G.: Multi-signatures in the plain public-key model and a general forking lemma. In: ACM Conference on Computer and Communications Security, pp. 390–399 (2006)

8. Boyen, X.: Lattice Mixing and Vanishing Trapdoors: A Framework for Fully Secure Short Signatures and More. In: Nguyen, P.Q., Pointcheval, D. (eds.) PKC 2010. LNCS, vol. 6056, pp. 499–517. Springer, Heidelberg (2010)

9. Brakerski, Z., Vaikuntanathan, V.: Efficient fully homomorphic encryption from (standard) LWE. In: FOCS (2011)

10. Brakerski, Z., Vaikuntanathan, V.: Fully Homomorphic Encryption from Ring-LWE and Security for Key Dependent Messages. In: Rogaway, P. (ed.) CRYPTO 2011. LNCS, vol. 6841, pp. 505–524. Springer, Heidelberg (2011)

11. Cash, D., Hofheinz, D., Kiltz, E., Peikert, C.: Bonsai Trees, or How to Delegate a Lattice Basis. In: Gilbert, H. (ed.) EUROCRYPT 2010. LNCS, vol. 6110, pp. 523–552. Springer, Heidelberg (2010)

12. Fiat, A., Shamir, A.: How to Prove Yourself: Practical Solutions to Identification and Signature Problems. In: Odlyzko, A.M. (ed.) CRYPTO 1986. LNCS, vol. 263, pp. 186–194. Springer, Heidelberg (1987)

13. Gama, N., Nguyen, P.Q.: Predicting Lattice Reduction. In: Smart, N.P. (ed.) EUROCRYPT 2008. LNCS, vol. 4965, pp. 31–51. Springer, Heidelberg (2008)

14. Gentry, C.: Fully homomorphic encryption using ideal lattices. In: STOC, pp. 169–178 (2009)

15. Gentry, C.: Toward Basing Fully Homomorphic Encryption on Worst-Case Hardness. In: Rabin, T. (ed.) CRYPTO 2010. LNCS, vol. 6223, pp. 116–137. Springer, Heidelberg (2010)

16. Gentry, C., Peikert, C., Vaikuntanathan, V.: Trapdoors for hard lattices and new cryptographic constructions. In: STOC, pp. 197–206 (2008)

17. Gentry, C., Szydlo, M.: Cryptanalysis of the Revised NTRU Signature Scheme. In: Knudsen, L.R. (ed.) EUROCRYPT 2002. LNCS, vol. 2332, pp. 299–320. Springer, Heidelberg (2002)

18. Goldreich, O., Goldwasser, S., Halevi, S.: Public-Key Cryptosystems from Lattice Reduction Problems. In: Kaliski Jr., B.S. (ed.) CRYPTO 1997. LNCS, vol. 1294, pp. 112–131. Springer, Heidelberg (1997)

19. Hoffstein, J., Howgrave-Graham, N., Pipher, J., Silverman, J.H., Whyte, W.: NTRUSIGN: Digital Signatures Using the NTRU Lattice. In: Joye, M. (ed.) CT-RSA 2003. LNCS, vol. 2612, pp. 122–140. Springer, Heidelberg (2003)

20. Hoffstein, J., Pipher, J., Silverman, J.H.: NTRU: A Ring-Based Public Key Cryptosystem. In: Buhler, J.P. (ed.) ANTS 1998. LNCS, vol. 1423, pp. 267–288. Springer, Heidelberg (1998)

21. Hoffstein, J., Pipher, J., Silverman, J.H.: NSS: An NTRU Lattice-Based Signature Scheme. In: Pfitzmann, B. (ed.) EUROCRYPT 2001. LNCS, vol. 2045, pp. 211–228. Springer, Heidelberg (2001)

22. Impagliazzo, R., Naor, M.: Efficient cryptographic schemes provably as secure as subset sum. J. Cryptology 9(4), 199–216 (1996)

23. Kawachi, A., Tanaka, K., Xagawa, K.: Concurrently Secure Identification Schemes Based on the Worst-Case Hardness of Lattice Problems. In: Pieprzyk, J. (ed.) ASIACRYPT 2008. LNCS, vol. 5350, pp. 372–389. Springer, Heidelberg (2008)

24. Lindner, R., Peikert, C.: Better Key Sizes (and Attacks) for LWE-Based Encryption. In: Kiayias, A. (ed.) CT-RSA 2011. LNCS, vol. 6558, pp. 319–339. Springer, Heidelberg (2011)
25. Lyubashevsky, V.: Lattice-Based Identification Schemes Secure Under Active Attacks. In: Cramer, R. (ed.) PKC 2008. LNCS, vol. 4939, pp. 162–179. Springer, Heidelberg (2008)
26. Lyubashevsky, V.: Fiat-Shamir with Aborts: Applications to Lattice and Factoring-Based Signatures. In: Matsui, M. (ed.) ASIACRYPT 2009. LNCS, vol. 5912, pp. 598–616. Springer, Heidelberg (2009)
27. Lyubashevsky, V., Micciancio, D.: Asymptotically Efficient Lattice-Based Digital Signatures. In: Canetti, R. (ed.) TCC 2008. LNCS, vol. 4948, pp. 37–54. Springer, Heidelberg (2008)
28. Lyubashevsky, V., Peikert, C., Regev, O.: On Ideal Lattices and Learning with Errors over Rings. In: Gilbert, H. (ed.) EUROCRYPT 2010. LNCS, vol. 6110, pp. 1–23. Springer, Heidelberg (2010)
29. Micciancio, D., Mol, P.: Pseudorandom Knapsacks and the Sample Complexity of LWE Search-to-Decision Reductions. In: Rogaway, P. (ed.) CRYPTO 2011. LNCS, vol. 6841, pp. 465–484. Springer, Heidelberg (2011)
30. Micciancio, D., Peikert, C.: Trapdoors for Lattices: Simpler, Tighter, Faster, Smaller. In: Pointcheval, D., Johansson, T. (eds.) EUROCRYPT 2012. LNCS, vol. 7237, pp. 700–718. Springer, Heidelberg (2012), Preliminary version, http://eprint.iacr.org/2011/501
31. Micciancio, D., Regev, O.: Worst-case to average-case reductions based on gaussian measures. SIAM J. Comput. 37(1), 267–302 (2007)
32. Micciancio, D., Regev, O.: Lattice-based cryptography. In: Bernstein, D.J., Buchmann, J., Dahmen, E. (eds.) Chapter in Post-quantum Cryptography, pp. 147–191. Springer, Heidelberg (2008)
33. Nguyên, P.Q.: Cryptanalysis of the Goldreich-Goldwasser-Halevi Cryptosystem from Crypto'97. In: Wiener, M. (ed.) CRYPTO 1999. LNCS, vol. 1666, pp. 288–304. Springer, Heidelberg (1999)
34. Nguyen, P.Q., Regev, O.: Learning a parallelepiped: Cryptanalysis of GGH and NTRU signatures. J. Cryptology 22(2), 139–160 (2009)
35. Peikert, C.: Public-key cryptosystems from the worst-case shortest vector problem: extended abstract. In: STOC, pp. 333–342 (2009)
36. Peikert, C.: An Efficient and Parallel Gaussian Sampler for Lattices. In: Rabin, T. (ed.) CRYPTO 2010. LNCS, vol. 6223, pp. 80–97. Springer, Heidelberg (2010)
37. Peikert, C., Rosen, A.: Efficient Collision-Resistant Hashing from Worst-Case Assumptions on Cyclic Lattices. In: Halevi, S., Rabin, T. (eds.) TCC 2006. LNCS, vol. 3876, pp. 145–166. Springer, Heidelberg (2006)
38. Pointcheval, D., Stern, J.: Security arguments for digital signatures and blind signatures. J. Cryptology 13(3), 361–396 (2000)
39. Regev, O.: On lattices, learning with errors, random linear codes, and cryptography. J. ACM 56(6) (2009)
40. Rückert, M., Schneider, M.: Estimating the security of lattice-based cryptosystems. Cryptology ePrint Archive, Report 2010/137 (2010), http://eprint.iacr.org/
41. Stehlé, D., Steinfeld, R.: Making NTRU as Secure as Worst-Case Problems over Ideal Lattices. In: Paterson, K.G. (ed.) EUROCRYPT 2011. LNCS, vol. 6632, pp. 27–47. Springer, Heidelberg (2011)

Author Index

GPSR Compliance

*The European Union's (EU) General Product Safety Regulation (GPSR)
is a set of rules that requires consumer products to be safe and our
obligations to ensure this.*

*If you have any concerns about our products, you can contact us on
ProductSafety@springernature.com*

In case Publisher is established outside the EU, the EU authorized
representative is:

Springer Nature Customer Service Center GmbH
Europaplatz 3
69115 Heidelberg, Germany

Batch number: 09467180

Printed by Printforce, the Netherlands